Ergebnisse der Mathematik und ihrer Grenzgebiete 94

A Series of Modern Surveys in Mathematics

Herbert Heyer

Probability Measures on Locally Compact Groups

Springer-Verlag
Berlin Heidelberg New York 1977

Herbert Heyer

Mathematisches Institut, Universität Tübingen,
Auf der Morgenstelle 10
D-7400 Tübingen 1

AMS Subject Classification (1970): Primary 60-02, 60B15
Secondary 43-02, 43A05, 43A30, 43A60, 43A65, 47D05, 60F05,
60G50, 60J15, 60J30

ISBN-13: 978-3-642-66708-4 e-ISBN-13: 978-3-642-66706-0
DOI: 10.1007/978-3-642-66706-0

Library of Congress Cataloging in Publication Data. Heyer, Herbert. Probability measures on locally compact groups. (Ergebnisse der Mathematik und ihrer Grenzgebiete; 94). Bibliography: p. Includes index. 1. Probabilities. 2. Measure theory. 3. Locally compact groups. I. Title. II. Series.
QA273.43.H49. 519.2'6. 77-24147.

Softcover reprint of the hardcover 1st edition 1977
2141/3140-543210

... les probabilités sur les structures algébriques, sujet neuf et passionnant.

Pierre Lelong

Preface

Probability measures on algebraic-topological structures such as topological semigroups, groups, and vector spaces have become of increasing importance in recent years for probabilists interested in the structural aspects of the theory as well as for analysts aiming at applications within the scope of probability theory. In order to obtain a natural framework for a first systematic presentation of the most developed part of the work done in the field we restrict ourselves to probability measures on locally compact groups. At the same time we stress the non-Abelian aspect. Thus the book is concerned with a set of problems which can be regarded either from the probabilistic or from the harmonic-analytic point of view. In fact, it seems to be the synthesis of these two viewpoints, the initial inspiration coming from probability and the refined techniques from harmonic analysis which made this newly established subject so fascinating.

The goal of the presentation is to give a fairly complete treatment of the central limit problem for probability measures on a locally compact group. In analogy to the classical theory the discussion is centered around the infinitely divisible probability measures on the group and their relationship to the convergence of infinitesimal triangular systems. In particular we emphasize the problem of embedding infinitely divisible probability measures in continuous convolution semigroups (Chapter III), the canonical representations in the sense of Lévy and Khintchine of continuous convolution semigroups (Chapter IV) and the rôle of the Gauss distribution in connection with the central limit theorems (Chapters V and VI). To make the book nearly self-contained the first two chapters have been devoted to general tools from the harmonic analysis of almost periodic locally compact groups (Chapter I) and from the elementary theory of convergence of convolution sequences of probability measures on the group (Chapter II). The preliminaries on almost periodic locally compact groups are designed to collect notations and basic facts concerning general locally compact groups, Lie groups, and almost periodic locally compact groups with the additional aim of formulating the main structure theorems for locally compact groups as they will be quoted throughout the book.

The exposition covers by no means all of the theory of probability on locally compact groups. Interesting areas had to be excluded. We mention the theory of random walks on groups and its implications for boundary and potential theory (as for example the work of Furstenberg, Azencott and Berg, Forst resp.), and also the general theory of additive processes with values in Lie groups as initiated recently by Stroock, Varadhan, and Feinsilver.

This book supplements the irreplacable monographs of Grenander [182] and Parthasarathy [377] and concentrates on advances of the theory which, on the basis of the research done during the last ten years, proved to be milestones for future development.

The method applied in discussing the general form of the central limit problem is partly representation-theoretic and partly operator-theoretic. Within the framework of finite-dimensional representation theory we are able to achieve results which are valid for all almost periodic locally compact groups. The tools of operator theory enable us to extend part of the theory via the solution of Hilbert's fifth problem to arbitrary locally compact groups.

While the main text of the book is somewhat selective, the references and comments added to each chapter and the bibliography contain as much information as could be assembled for a full description of the problems. Naturally there will be omissions and insufficiencies; the reader is asked to accept the author's fallibility and to communicate any improvement.

The author has benefitted from the work and comments of several colleagues. Egbert Dettweiler, Wilfried Hazod and Eberhard Siebert studied parts of the draft and proposed numerous ameliorations. Walter Maxones checked the final form of the manuscript, supplied significant criticism and took care of the index. Thomas Barth and Egbert Dettweiler helped with the tedious job of proof reading. To all of them go the author's heartfelt thanks.

The invitation to write this book was extended to the author by Reinhold Remmert. His initiative deserves special acknowledgement. Thanks are also due to the Deutsche Forschungsgemeinschaft for financial support granted to Thomas Barth, Wilfried Hazod and Walter Maxones, and to the Kultusministerium des Landes Baden-Württemberg for a research semester which enabled the author to start working on the manuscript on a full-time basis.

Last not least the author is grateful to Springer-Verlag and its staff for an excellent cooperation in the process of producing the book.

Tübingen
September 1977

Herbert Heyer

Table of Contents

Introduction

The idea of studying probability measures on spheres in Euclidean space $\mathbb{R}^p$ rather than on the Euclidean space itself is as old as the beginnings of probability theory and statistics. In 1734 Daniel Bernoulli looked at the orbital planes of the planets known at his time as random points on the surface of a sphere and asserted their uniform distribution. In the first quarter of this century Rayleigh and Karl Pearson started investigations on the resultant length of normal vectors, in connection with approximation problems for large samples, within the framework of random walks on spheres. Until then the distributions appearing in the work of the pioneers were all uniform.

In the early papers of Perrin [387] and von Mises [355] also non-uniform distributions on the sphere in $\mathbb{R}^3$ and on the circle resp. entered the discussion. A first probabilistic study of the Brownian motion process on the circle and on the orthogonal group carried out by Perrin led to the definition of the wrapped normal distribution. A little earlier von Mises had investigated the problem whether the atomic weights were integers subject to errors by introducing a normal distribution on the torus group $\mathbb{T} := \mathbb{R}/\mathbb{Z}$. Von Mises and later R.A. Fisher laid the foundations of a new field of research which nowadays is called the statistics of directional data and has received remarkable acknowledgement in recent years, at least since the publication of the monograph [344] of Mardia. While the work of Perrin, von Mises and Fisher was primarily oriented to particular applied problems arising in physics, earth science, meteorology, and biology, a first theoretical treatment of probability measures on the torus is due to P. Lévy [320]. In his paper Lévy succeeds in extending a number of results from the theory for the real line $\mathbb{R}$ to the torus $\mathbb{T}$ on the basis of the classical reasoning. The methodical breakthrough into a completely new direction of probability theory was achieved by K. Ito and Y. Kawada who, in their paper [287] of 1940, established the fundamentals of a probability theory on general compact groups. It turned out that the methods and results of this new branch of probability theory differed essentially from those available in the classical setup. In order to make the differences more deeply understandable and to provide a common framework for the special cases of the Euclidean group and of a general compact group Bochner, in his basic work [28] and [29], studied for the first time probability measures on locally compact Abelian groups. His results parallel the classical ones as they are collected in the books of Gnedenko and Kolmogorov [172] and Bergström [14] and contain extensions of the case of classical central limit theory. With the work summarized by Grenander in [181] and [182] on probability measures on locally compact groups and of

Hannan [191] on the relationship between the theory of probability measures on groups and the theory of group representations we arrive at the starting point for the material to be treated in the present book.

The classical central limit problem for a sequence $(X_j)_{j\geq 1}$ of (stochastically) independent real-valued random variables on a probability space $(\Omega, \mathfrak{A}, P)$ concerns the limiting behavior (in the sense of convergence in distribution) of the corresponding sequence $(S_n)_{n\geq 1}$ of partial sums $S_n := \sum_{j=1}^n X_j$. More precisely, the aim of studies is to obtain conditions under which the sequences $(T_n)_{n\geq 1}$ of normed sums $T_n := \frac{1}{b_n} S_n - a_n$ (for norming sequences $(b_n)_{n\geq 1}$ and $(a_n)_{n\geq 1}$ in $\mathbb{R}_+^*$ and $\mathbb{R}$ resp.) converge in distribution to the degenerate, Poisson or Gauss random variable. In accordance with the particular limit distribution the corresponding theorems are called the law of large numbers, the Poisson and the normal convergence theorems. It was P. Lévy who stated and solved the central limit problem for sequences of normed sums by finding all possible limit distributions of the sequence $(T_n)_{n\geq 1}$ under the additional hypothesis that the random variables X_j are identically distributed. Later the condition of identical distribution could be weakened considerably. One considered the triangular system $(X_{nj})_{j=1,\ldots,n;n\geq 1}$ of random variables $X_{nj} := \frac{1}{b_n} X_j - \frac{1}{n} a_n$ and posed the condition of infinitesimality (uniform asymptotic negligeability) in the sense that $(X_{nj})_{n\geq 1}$ converges stochastically to the zero variable as n tends to ∞, uniformly in j. Under this assumption the central limit problem consists of the following two partial problems:

(α) To determine the structure of the set $\mathscr{L}$ of all possible limit distributions of sequences $(S_n)_{n\geq 1}$ of sums $S_n := \sum_{j=1}^n X_{nj}$.

(β) To exhibit conditions for the convergence of such sequences $(S_n)_{n\geq 1}$ to random variables with specified distributions of the set $\mathscr{L}$.

In the language of measures the setup of the central limit problem can be phrased as follows: One considers the topological semigroup $\mathscr{M}^1(\mathbb{R})$ of all probability measures on $\mathbb{R}$ (furnished with vague topology and convolution) and studies triangular systems $(\mu_{nj})_{j=1,\ldots,n;\, n\geq 1}$ in $\mathscr{M}^1(\mathbb{R})$ which are infinitesimal in the sense that

$$\lim_{n\to\infty} \max_{1\leq j\leq n} \mu_{nj}(\complement U) = 0$$

holds for all neighbourhoods U of 0, and convergent in the sense that the sequence $(\mu_n)_{n\geq 1}$ of n-fold convolution products $\mu_n := \mu_{n1} * \cdots * \mu_{nn}$ tends to a limit measure $\mu \in \mathscr{M}^1(\mathbb{R})$ as n goes to ∞. It turns out that μ is infinitely divisible, i.e., for each $n \geq 1$ there exists an n-th root $\mu_n \in \mathscr{M}^1(\mathbb{R})$ satisfying $\mu_n^n = \mu$. The full solution of problem (α) is the complete description of the set $\mathscr{I}(\mathbb{R})$ of all infinitely divisible probability measures on $\mathbb{R}$. Clearly (compound) Poisson and Gauss measures belong to $\mathscr{I}(\mathbb{R})$. A classification of the measures in $\mathscr{I}(\mathbb{R})$ has been achieved with the Lévy-Khintchine representation, from which characterizations of the Poisson and Gauss measures can be deduced. An important auxiliary result needed in the proof of the Lévy-Khintchine representation is the embedding theorem (due to P. Lévy) stating that every measure $\mu \in \mathscr{I}(\mathbb{R})$ lies on a continuous one-parameter semigroup $(\mu_t)_{t\in\mathbb{R}_+}$ in $\mathscr{M}^1(\mathbb{R})$ satisfying $\mu_1=\mu$. In view of the embedding theorem the Lévy-Khintchine representation of $\mu\in\mathscr{I}(\mathbb{R})$ with embedding one-parameter semigroup is related to the generating functional

$$A:=\lim_{t\downarrow 0}\frac{1}{t}(\mu_t-\varepsilon_0)$$

of $(\mu_t)_{t\in\mathbb{R}_+}$ defined at least on the characters of $\mathbb{R}$ by the formula $\hat{\mu}_t=\exp(tA)$ for all $t\in\mathbb{R}_+$. Poisson and Gauss measures can now be characterized as embeddable measures by their embedding one-parameter semigroups or alternatively by their generating functionals. These two classes of probability measures play a prominent rôle in the context of problem (β), which yields the Poisson and normal convergence theorems.

So far the measure-theoretic description of the central limit problem was given in terms of the group $\mathbb{R}$ of the real line. In the present book infinitesimal triangular systems $(\mu_{nj})_{j=1,\ldots,n;\,n\geq 1}$ in the semigroup $\mathscr{M}^1(G)$ of probability measures on an arbitrary locally compact group G are considered, and the structure of the set $\mathscr{I}(G)$ of infinitely divisible probability measures on G is analyzed. The program of the discussion within the general framework follows the outline of the classical theory reviewed above. In Chapter II the special cases of not necessarily infinitesimal triangular systems $(\mu_{nj})_{j=1,\ldots,n;\,n\geq 1}$ in $\mathscr{M}^1(G)$ with $\mu_{nj}:=\mu\in\mathscr{M}^1(G)$ or $\mu_{nj}:=\mu_j\in\mathscr{M}^1(G)$ for all $j=1,\ldots,n$; $n\geq 1$ are considered. The problem of their convergence appears to be intimately related to the structure of the underlying group G. In the first case one studies the convergence of sequences $(\mu^n)_{n\geq 1}$ of powers of μ, in the second case the convergence of sequences $(\nu_{0,n})_{n\geq 1}$ of n-fold products $\nu_{0,n}:=\mu_1*\cdots*\mu_n$.

In Chapter III the general embedding problem is posed. It suggests a characterization of all locally compact groups G for which every measure $\mu\in\mathscr{I}(G)$ lies on at least one continuous one-parameter semigroup in $\mathscr{M}^1(G)$. Although substantial contributions have been made, in its full generality the problem is still unsolved. The following sample result is typical of the aim of our presentation: A strongly root compact, locally compact Abelian group G has the embedding property iff the connected component of the neutral element of G is locally arcwise connected. This theorem not only generalizes a classical result from the groups $\mathbb{R}$ or $\mathbb{T}$ to more general locally compact groups, but also determines the domain of validity of the embedding property. Results of this kind seem to be optimal in the sense that probabilistic properties can be used to characterize classes of locally compact groups. Chapter IV includes an extended discussion of the representation of the generating functional A of a continuous one-parameter semigroup $(\mu_t)_{t\in\mathbb{R}_+}$ in $\mathscr{M}^1(G)$ and gives rise to the characterizations of Poisson and Gauss measures and semigroups to be treated in detail in Chapters V and VI. In Chapter VI we then present answers to the questions implied by problems (α) and (β) of the central limit problem for general locally compact groups G.

Finally a word should be said concerning applications of the general theory.

In recent years it became evident that problems of non-commuting random evolutions are favorably tackled within the framework of continuous one-parameter semigroups of probability measures on a non-Abelian locally compact group. Here, the contributions of Hersh, Keller, Papanicolao and Pinsky in connection with the Feynman-Kac theory generated great interest especially among physicists. Similar problems appeared in the theory of motions of bacteria, where the Poisson semigroup on an arbitrary compact group and its generator entered the discussion in a natural way. Work in this context has been done by Baggett and Stroock, Kohler and Papanicolao. See for example [4]. On the other hand, there exists a well-developed theory of non-Hamiltonian systems and their dynamical semigroups induced by Markov processes on a locally compact group with applications to the harmonic oscillator, which started with the work of Ingarden and was carried on by Kossakowski.

In order to give a more precise explanation of how continuous one-parameter semigroups of probability measures on a non-Abelian locally compact group can be used to solve a physical problem, we shall describe Kac's probabilistic approach to the solution of the telegraphist's equation improved recently by Kisyński. Let α be a positive real number and let $(N_t^{(\alpha)})_{t\in\mathbb{R}_+}$ be a homogeneous Poisson process defined by

$$P[N_t^{(\alpha)}=k]:=e^{-\alpha t}\frac{(\alpha t)^k}{k!} \quad \text{for all } k\in\mathbb{Z}_+ \text{ and } t\in\mathbb{R}_+.$$

For any $t\in\mathbb{R}_+$ one considers the random variables

$$Y_t^{(\alpha)}:=(-1)^{N_t^{(\alpha)}} \text{ and } X_t^{(\alpha)}:=\int_0^t Y_\tau^{(\alpha)}\,d\tau,$$

which have a natural interpretation: let a point of $\mathbb{R}$ move with velocity $+1$ or -1 changing at random such that the number of changes within the interval $[0,t]$ equals $N_t^{(\alpha)}$. If at $t=0$ the point is at the origin 0 of $\mathbb{R}$ and has velocity $+1$, then $X_{t_0}^{(\alpha)}$ describes its position and $Y_{t_0}^{(\alpha)}$ its velocity at $t=t_0$. The state space of the random motion described is the topological semidirect product $G:=\mathbb{R}\times_\eta\mathbb{Z}_2$ of the groups $\mathbb{R}$ and $\mathbb{Z}_2=\{-1,+1\}$ with defining homomorphism η: $\mathbb{Z}_2\to\mathrm{Aut}(\mathbb{R})$ given by $\eta(k)(y):=k\,y$ for all $k\in\mathbb{Z}_2$ and $y\in\mathbb{R}$. One now looks at the stochastic process $(X_t)_{t\in\mathbb{R}_+}$ with

$$X_t:=X_t^{(\alpha)}\otimes Y_t^{(\alpha)} \quad \text{for all } t\in\mathbb{R}_+,$$

which takes values in the non-Abelian locally compact group G. For every $t\in\mathbb{R}_+$ the distribution μ_t of X_t is a measure in $\mathscr{M}^1(G)$, has compact support in the set $\{(x,k)\in G: |x|\le t,\ k\in\mathbb{Z}_2\}$ and admits a representation

$$\mu_t=e^{-\alpha t}\varepsilon_{(t,1)}+\nu_t$$

with a non-negative measure ν_t on G satisfying

$$\nu_t(G)=P[N_t^{(\alpha)}\ge 1]=1-e^{-\alpha t}.$$

One shows that $(\mu_t)_{t\in\mathbb{R}_+}$ is in fact a continuous one-parameter semigroup in $\mathscr{M}^1(G)$. Its generating functional is the key tool in solving the telegraphist's equation.

This example, though only indicated, was intended to give an insight into possible applications of the theory to fields neighbouring mathematics. Further applications might come up in the future, and the author would be happy to learn some day that the influence of the present book went beyond the range of abstract probability theory.

Preliminaries

Almost Periodic Locally Compact Groups

Locally Compact Groups. A *topological group* $G:=(G,\mathcal{T})$ is a group G together with a topology $\mathcal{T}$ on G such that the mapping $(x,y)\to x y^{-1}$ from $G\times G$ into G is continuous. G will be called T_2, compact, locally compact, connected etc., if $\mathcal{T}$ is T_2, compact, locally compact, connected etc. resp. For every $a\in G$ the *right translation* (left translation) $R(a)$: $x\to x a$ ($L(a)$: $x\to a x$) *by* $a\in G$ and the *inversion* S: $x\to x^{-1}$ of G are homeomorphisms of G. Let $\mathfrak{B}$ be a basis (of neighborhoods) of the neutral element (identity) e of G. Then the systems $a\mathfrak{B}$ and $\mathfrak{B}a$ are bases of $a\in G$.

Let $\mathfrak{B}(e):=\mathfrak{B}_G(e)$ denote the neighborhood filter of e in G. We shall always assume that G is a T_2-group, i.e., $\bigcap_{U\in\mathfrak{B}(e)} U=\{e\}$.

In this case G is completely regular in the sense that for $a\in G$ and $V\in\mathfrak{B}(a)$ there exists a bounded real-valued continuous function f on G such that $0\le f\le 1$, $f(a)=0$ and $f(\complement V)=1$ hold. On G the right and left *uniform structures* U_r and U_l are defined by

$$U_r:=\{R_U: U\in\mathfrak{B}(e)\} \quad \text{and} \quad U_l:=\{L_U: U\in\mathfrak{B}(e)\} \quad \text{with}$$
$$R_U:=\{(x,y)\in G\times G: y x^{-1}\in U\} \quad \text{and}$$
$$L_U:=\{(x,y)\in G\times G: x^{-1}y\in U\} \qquad (U\in\mathfrak{B}(e)).$$

The completely regular groups are exactly those for which the topologies induced by U_r or U_l coincide with the initial topology of G. These groups are the *uniformizable* groups. A topological group G is *metrizable* iff $\mathfrak{B}(e)$ admits a countable basis. Metrizable groups can be furnished with a left invariant metric. It is a well-known and often applied fact that for open (compact) subsets A, B of G the subset AB is open (compact), that for closed A, B the set AB is not necessarily closed, but that closedness of AB can be achieved if A or B is compact. For every compact subset F of G and every $U\in\mathfrak{B}(e)$ there exists a $V\in\mathfrak{B}(e)$ such that $xVx^{-1}\subset U$ for all $x\in F$. If F is compact and U is an open subset of G with $F\subset U$, then there is a $V\in\mathfrak{B}(e)$ satisfying $(FV)\cup(VF)\subset U$. If, in addition, G is locally compact, then V can be chosen such that $(FV)\cup(VF)$ is relatively compact in G. (Topological) *subgroups* of topological groups G are defined as (abstract) subgroups H of G together with the topology induced in H. For every subgroup (normal subgroup) H of a topological group G the closure $\bar{H}$ is again a subgroup (normal subgroup). Moreover, open subgroups are always closed. Subgroups of particular interest for analytic applications are the *normalizers* $N(H)$ and *centralizers* $Z(H)$ of (closed) subgroups H of G; in particular, the *center* $Z(G)$ of G and the *component* G_0 of G, i.e., the connected component of e, which is always a closed characteristic subgroup of G.

By $K(G):=[G,G]^-$ we denote the closed *commutator* (subgroup) of G. We further define subgroups $\bar{D}_n(G)$ and $\bar{C}_n(G)$ of G by

$$\bar{C}_0(G):=G, \quad \bar{C}_n(G):=[G,\bar{C}_{n-1}(G)]^- \quad \text{and}$$
$$\bar{D}_0(G):=G, \quad \bar{D}_n(G):=[\bar{D}_{n-1}(G),\bar{D}_{n-1}(G)]^- \quad (n\geq 1) \text{ resp.}$$

G is said to be *nilpotent* or *solvable*, if $\bar{C}_n(G)=\{e\}$ or $\bar{D}_n(G)=\{e\}$ resp. for a sufficiently large $n\geq 1$.

Homomorphisms (isomorphisms) ϕ: $G\to G'$ between topological groups are always defined to be continuous (bicontinuous). Moreover, any (continuous) homomorphism ϕ: $G\to G'$ is uniformly continuous for the uniform structures U_r and U_l on G. If G and G' are locally compact groups, G is σ-compact and ϕ: $G\to G'$ surjective, then ϕ is open. A *local isomorphism* between locally compact groups G and G' is a homeomorphism ϕ of a neighborhood $U\in\mathfrak{B}_G(e)$ onto a neighborhood $U'\in\mathfrak{B}_{G'}(e')$ with the properties that if $a,b\in U$ and $ab\in U$, then $\phi(ab)=\phi(a)\phi(b)$ holds and that $a,b\in U$ and $\phi(a)\phi(b)\in U'$ imply $ab\in U$.

Let G be a topological group, H a subgroup of G and p the canonical projection from G onto the quotient space G/H. The latter becomes a topological space if one introduces the final topology of p. Its open sets are just the sets $p(A)$ for open sets A in G. p is open and continuous, but not in general closed. If H is a normal subgroup of G, then G/H becomes a topological group, the *quotient group* of G with respect to H. It is known that G/H is discrete iff H is open in G, and that G/H is T_2 iff H is closed. Since we want to restrict ourselves to T_2-groups, we shall always work with closed (or open) subgroups H of G. Results of particular interest are the following:

1. Given a topological group G and a closed subgroup H of G one has that G is compact (locally compact) iff G/H and H are compact (locally compact).

2. If G is a connected group and H a subgroup of G, then G/H is connected. Conversely, the connectedness of G/H and H implies that of G.

Let $\mathbb{A}$ be any set and $(G_\alpha)_{\alpha\in\mathbb{A}}$ a family of topological groups $G_\alpha\,(\alpha\in\mathbb{A})$. One defines the topological *product* group $\prod_{\alpha\in\mathbb{A}} G_\alpha$ as the abstract product group $\prod_{\alpha\in\mathbb{A}} G_\alpha$ together with the product topology. Similarly one defines the (topological) *weak direct product* $\prod^*_{\alpha\in\mathbb{A}} G_\alpha$ of the family $(G_\alpha)_{\alpha\in\mathbb{A}}$, which is a dense subgroup of $\prod_{\alpha\in\mathbb{A}} G_\alpha$.

Let $\mathbb{A}$ be ordered by the relation $<$. We are given for every $\alpha\in\mathbb{A}$ a topological group G_α and for $\alpha,\beta\in\mathbb{A}$ with $\alpha<\beta$ a homomorphism $p_{\alpha\beta}: G_\beta\to G_\alpha$ such that $p_{\alpha\gamma}=p_{\alpha\beta}\circ p_{\beta\gamma}$ holds for all $\alpha,\beta,\gamma\in\mathbb{A}$ with $\alpha<\beta<\gamma$. Then the triplet $(G_\alpha, p_{\alpha\beta}, \mathbb{A})$ is called a *projective system* (of topological groups). The *projective limit* $\varprojlim_{\alpha\in\mathbb{A}} G_\alpha$ of the projective system $(G_\alpha, p_{\alpha\beta}, \mathbb{A})$ is defined as the closed subgroup

$$\{(x_\alpha)\in\prod_{\alpha\in\mathbb{A}} G_\alpha : x_\alpha=p_{\alpha\beta}(x_\beta) \text{ for all } \alpha<\beta\}$$

of the product group $\prod_{\alpha\in\mathbb{A}} G_\alpha$. Obviously $\varprojlim_{\alpha\in\mathbb{A}} G_\alpha$ is locally compact if G_α is locally compact for all $\alpha\in\mathbb{A}$ and all but a finite number of the groups G_α are compact.

Given a locally compact group G and a projective system $(G_\alpha, p_{\alpha\beta}, \mathbb{A})$ of locally compact groups we call a system $(p_\alpha)_{\alpha\in\mathbb{A}}$ of (continuous) homomorphisms $p_\alpha: G \to G_\alpha$ *consistent* for $(G_\alpha, p_{\alpha\beta}, \mathbb{A})$, if $p_{\alpha\beta} \circ p_\beta = p_\alpha$ for all $\alpha < \beta$, and *separating*, if $\bigcap_{\alpha\in\mathbb{A}} \ker p_\alpha = \{e\}$ holds. One plainly notes that there exists a consistent and separating system $(p_\alpha)_{\alpha\in\mathbb{A}}$ for $(G_\alpha, p_{\alpha\beta}, \mathbb{A})$ iff G can be monomorphically embedded into $\varprojlim_{\alpha\in\mathbb{A}} G_\alpha$.

Let G be a topological group and $A(G) := \mathrm{Aut}(G)$ the set of all (topological) automorphisms of G. For any compact subset K of G and every $U \in \mathfrak{V}(e)$ one introduces the set

$$W(K, U) := \{\tau \in A(G) : \tau(x) \in Ux \text{ and } \tau^{-1}(x) \in Ux \text{ for all } x \in K\}.$$

The system $\mathscr{W} := \{W(K, U) : K \text{ compact} \subset G, U \in \mathfrak{V}(e)\}$ defines a topology $\mathscr{T}_A$ in $A(G)$, which in the case of a locally compact group G makes $A(G)$ a completely regular (but not necessarily locally compact) topological group. Of special interest is the subgroup $I(G) := \mathrm{Int}(G)$ of all inner topological automorphisms of G.

Let now G be a locally compact group, H an arbitrary topological group and η a homomorphism from H into $A(G)$. On the product set $G \times H$ one defines a group structure by introducing for elements (x, h) and (y, k) of $G \times H$ the composition $(x, h)(y, k) := (x(\eta(h)(y)), hk)$. It is shown that $G \times H$ becomes an abstract group $G \times_\eta H$ which together with the product topology of $G \times H$ is a topological group iff η is continuous. In this case $G \times_\eta H$ will be called the (topological) *semidirect product of the groups G and H with defining homomorphism η*. We recall a few properties. Let e denote the neutral element of G, and of H as well, and put $K := \ker \eta$, $G_1 := \{(x, e) : x \in G\}$ and $H_1 := \{(e, h) : h \in H\}$.

1. G_1 is a closed normal subgroup of $G \times_\eta H$ with $G_1 \cong G$ and H_1 is a closed subgroup of $G \times_\eta H$ with $H_1 \cong H$.
2. H_1 is a normal subgroup of $G \times_\eta H$ iff $K = H$. In this case $G \times_\eta H = G \times H$ is a direct product.

We finally mention that a locally compact group G is always complete (for U_r and U_l) and a locally compact subgroup H of a locally compact group G is closed, since it is complete.

For locally compact Abelian groups an extended *structure theory* is available based primarily on the harmonic analysis of such groups.

A topological group G is called *compactly generated*, if there exists a compact subset F of G such that

$$G = \{e\} \cup \bigcup_{n\geq 1} (F \cup F^{-1})^n$$

holds. Clearly, every compact group and every connected locally compact group is compactly generated. Moreover, we have for a locally compact group G and a closed normal subgroup H of G that G is compactly generated if H and G/H are compactly generated.

A Theorem (Van Kampen, Weil). *For any locally compact Abelian group G the following statements are equivalent:*

(i) *G is compactly generated*;

(ii) *there exist* $m, n \geq 0$ *and a (largest) compact (sub)group* H *such that* $G \cong \mathbb{R}^m \times \mathbb{Z}^n \times H$;
(iii) *the character group* $G^{\wedge}$ *of* G *is locally isomorphic to* $\mathbb{R}^p$ *for some* $p \geq 0$;
(iv) *there are* $m, n \geq 0$ *and a discrete group* D *such that* $G^{\wedge} \cong \mathbb{R}^m \times \mathbb{T}^n \times D$.

B Theorem (Dixmier). *Let* G *be a locally compact Abelian group having a countable basis of its topology. Then the following statements are equivalent:*
(i) $G \cong \mathbb{R}^n \times \mathbb{T}^{\mathfrak{a}}$ *with* $n \geq 0$ *and* $\mathfrak{a} \geq 0$ *or* $\mathfrak{a} = \mathbb{N}$;
(ii) G *is connected and locally arcwise connected*;
(iii) G *is arcwise connected*;
(iv) G *is connected and locally connected.*

A locally compact group is said to be *locally Euclidean* of dimension $d \geq 1$ if for every $x \in G$ there is a $U \in \mathfrak{V}(x)$ which is homeomorphic to an open subset of $\mathbb{R}^d$.

The significance of the class of locally Euclidean groups becomes apparent from the fact that every connected, locally Euclidean group is isomorphic to a group carrying an analytic structure in a sense that will be made precise in the context of

Lie Groups. A *Lie group* $G := (G, \mathscr{V})$ (of dimension $d \geq 1$) is a group G together with the structure of an analytic manifold $\mathscr{V}$ (of dimension $d \geq 1$) on G such that the mapping $(x, y) \to x y^{-1}$ from (the product manifold) $G \times G$ into G is analytic.

A Lie group G of dimension $d \geq 1$ is a locally Euclidean group of dimension d and is hence locally compact and locally arcwise connected with respect to the topology induced by the analytic structure of G. If G is connected, then its topology has a countable basis. It is clear that the right and left translations $R(a)$ and $L(a)$ $(a \in G)$ as well as the inversion S of G are analytic.

A *Lie subgroup* of a Lie group G is a Lie group H which is an abstract subgroup and an analytic submanifold of G. A subgroup H of a Lie group G is a Lie subgroup of G if H is an analytic submanifold of G and if H is a topological subgroup of G with respect to the topology induced by the analytic structure of H.

Let G be a Lie group, $T(G)$ the *tangent bundle* of G with fibers $T(G, x)$ for $x \in G$ and projection π and $D(G)$ the set of *analytic vector fields* $X: G \to T(G)$ with $\pi \circ X = \mathrm{Id}$. Every $X \in D(G)$ maps the space $\mathscr{C}^{\infty}(G) := \mathscr{C}^{\infty}_{\mathbb{R}}(G)$ of real analytic functions on G into itself. Hence, for $X, Y \in D(G)$ one defines $[X, Y] := XY - YX$ and observes that the bilinear form $[\cdot, \cdot]$ on the vector space $D(G)$ satisfies the following conditions of a *Lie algebra*

(a) $[X, Y] = -[Y, X]$ and
(b) $[X, [Y, Z]] + [Y, [Z, X]] + [Z, [X, Y]] = 0$, whenever $X, Y, Z \in D(G)$.

Let G, G' be two Lie groups and $\phi: G \to G'$ a *homomorphism of Lie groups*, i.e., an analytic homomorphism. There exist a bundle homomorphism $d\phi: T(G) \to T(G')$ and for each $x \in G$ a vector space homomorphism $d\phi(x): T(G, x) \to T(G', \phi(x))$ induced by ϕ. The linear mapping $d\phi(e)$ is called the *differential* of ϕ. For every $X \in D(G)$ we define

$$d\phi(X): G \to T(G') \text{ by } d\phi(X)(x) := [d\phi(x)]\, X(x) \quad \text{for all } x \in G.$$

Application of this construction to the analytic homomorphism $L(a): G \to G$ for $a \in G$ yields the definition of the mapping $dL(a)(X)$ and enables us to define $X \in D(G)$ to be *(G-)invariant* if $dL(a)(X)(e) = X(a)$ holds for all $x \in G$. The set $\mathfrak{L}(G)$ of all invariant analytic vector fields on G is a Lie algebra with respect to $[\cdot, \cdot]$, and there is a Lie algebra isomorphism between $\mathfrak{L}(G)$ and $T(G, e)$ defined in the following way: For $X \in T(G, e)$ there exists a unique invariant analytic vector field $\tilde{X} \in \mathfrak{L}(G)$ with $\tilde{X}(e) = X$ defined by

$$(\tilde{X} f)(x) := X(f \circ L(x)) \quad \text{for all } f \in \mathscr{C}^\infty(G),\ x \in G.$$

The mapping

$$\psi: \mathfrak{L}(G) \to T(G, e) \quad \text{given by} \quad \psi(\tilde{X}) = X \quad \text{for all } \tilde{X} \in \mathfrak{L}(G)$$

is the desired isomorphism. $\mathfrak{L}(G)$ is said to be the *Lie algebra of the Lie group G*.

In order to investigate the correspondence between a Lie group G and its Lie algebra $\mathfrak{L}(G)$ more closely we introduce the *exponential mapping* $\exp_G$ for G as a mapping from $\mathfrak{L}(G)$ into G as follows: For $X \in T(G, e)$ and corresponding $\tilde{X} \in \mathfrak{L}(G)$ there exists a unique analytic homomorphism $f_{\tilde{X}}: \mathbb{R} \to G$ such that $\dot{f}_{\tilde{X}}(0) = X$ holds. $\exp := \exp_G$: $\mathfrak{L}(G) \to G$ is defined as the analytic homomorphism $\tilde{X} \to f_{\tilde{X}}(1)$. Its differential $d(\exp_G)(0)$ equals the identity of End $\mathfrak{L}(G)$. One shows that there exist a bounded, connected open neighborhood $U_0 \in \mathfrak{V}_{\mathfrak{L}(G)}(0)$ and an open neighborhood $U_e \in \mathfrak{V}_G(e)$ such that $X \to \exp X$ is an analytic diffeomorphism from U_0 onto U_e. Let $\log: U_e \to U_0$ be the analytic inverse of exp and $\eta: \mathfrak{L}(G) \to \mathbb{R}^d$ a vector space isomorphism. Then $D_0 := \eta(U_0)$ is open in $\mathbb{R}^d$. The pair $(U_e, \eta \circ \log)$ or briefly, $(U_e, \log)$, is called a *canonical chart* in e (with canonical neighborhood U_e).

If $\{X_1, \ldots, X_d\}$ is a basis of $\mathfrak{L}(G)$, then

$$D_0 = \{(x_1, \ldots, x_d) \in \mathbb{R}^d : \textstyle\sum_{i=1}^d x_i X_i \in U_0\}$$

and the *coordinate mapping* $\log: U_e \to U_0$ is given by

$$\log(\exp(\textstyle\sum_{i=1}^d x_i X_i)) = (x_1, \ldots, x_d).$$

The system $\{x_1, \ldots, \ldots x_d\}$ of analytic functions on G with the property that for each $x \in U_e$ one has $x = \exp(\sum_{i=1}^d x_i(x) X_i)$ is called *a system of canonical coordinates with respect to the basis* $\{X_1, \ldots, X_d\}$.

By studying (analytic) homomorphisms or isomorphisms of Lie groups $\phi: G \to G'$ one obtains corresponding Lie algebra homomorphisms or isomorphisms $d\phi: \mathfrak{L}(G) \to \mathfrak{L}(G')$ defined by $\phi \circ \exp_G = \exp_{G'} \circ\, d\phi$. It can be shown that given Lie groups G and G', any homomorphism $\phi: G \to G'$ of topological groups is in fact an analytic homomorphism. A vector subspace $\mathfrak{S}$ of an abstract Lie algebra $\mathfrak{L}$ is called a *Lie subalgebra* of $\mathfrak{L}$ if $[X, Y] \in \mathfrak{S}$ for all $X, Y \in \mathfrak{S}$. The correspondence between Lie subgroups of a Lie group G and Lie subalgebras of $\mathfrak{L}(G)$ is established in two steps.

1. Let H be a Lie subgroup of G. Then $\mathfrak{L}(H)$ is a Lie subalgebra of $\mathfrak{L}(G)$.

2. For every Lie subalgebra $\mathfrak{S}$ of $\mathfrak{L}(G)$ there exists a unique connected Lie subgroup H of G satisfying $\mathfrak{L}(H)=\mathfrak{S}$. Here H is the smallest Lie subgroup of G containing $\exp_G \mathfrak{L}(H)$.

Furthermore, the Lie algebras of two Lie groups G and G' are isomorphic (as Lie algebras) iff G and G' are locally isomorphic. On the basis of the results developed one proves the famous result of Elie Cartan that locally compact groups admitting a continuous homomorphism into a Lie group which is injective on a neighborhood of the neutral element are Lie groups. Since every closed subgroup of a Lie group is locally compact, this implies that a closed subgroup H of a Lie group G with Lie algebra $\mathfrak{L}(G)$ is itself a Lie subgroup with Lie algebra

$$\mathfrak{L}(H)=\{X\in\mathfrak{L}(G):\exp_G tX\in H \text{ for all } t\in\mathbb{R}\}.$$

It follows that two Lie groups which are isomorphic as topological groups are also isomorphic as Lie groups.

We note that given a Lie group G and a closed subgroup H of G the homogeneous space G/H can be made naturally into an analytic manifold. If H is a normal subgroup, then G/H becomes in fact a Lie group and $\mathfrak{L}(G/H)$ is isomorphic to the quotient $\mathfrak{L}(G)/\mathfrak{L}(H)$ by the ideal $\mathfrak{L}(H)$ of $\mathfrak{L}(G)$.

Let G be an Abelian Lie group with Lie algebra $\mathfrak{L}(G)$ considered as an (additive) Lie group $\mathbb{R}^d$ for some $d\geq 0$. Then $\exp_G$ is a homomorphism of Lie groups such that $\ker(\exp_G)$ is a discrete subgroup of $\mathfrak{L}(G)$. This implies

C Theorem. *Every connected Abelian Lie group G is of the form $G\cong\mathbb{R}^m\times\mathbb{T}^n$ for $m,n\geq 0$.*

A connected Lie group G with Lie algebra $\mathfrak{L}(G)$ is nilpotent iff $\mathfrak{L}(G)$ is an abstract *nilpotent Lie algebra* $\mathfrak{L}$ in the sense of the following equivalent definitions:

(i) There exists a decreasing sequence $(\mathfrak{L}_i)_{0\leq i\leq m}$ of ideals of $\mathfrak{L}$ with $\mathfrak{L}=\mathfrak{L}_0\supset\mathfrak{L}_1\supset\cdots\supset\mathfrak{L}_{m-1}\supset\mathfrak{L}_m=\{0\}$ such that $[\mathfrak{L},\mathfrak{L}_i]\subset\mathfrak{L}_{i+1}$ for $0\leq i\leq m-1$;

(ii) there exists a decreasing sequence $(\mathfrak{L}'_i)_{0\leq i\leq r}$ of ideals of $\mathfrak{L}$ with $\mathfrak{L}=\mathfrak{L}'_0\supset\mathfrak{L}'_1\supset\cdots\supset\mathfrak{L}'_{r-1}\supset\mathfrak{L}'_r=\{0\}$ such that $[\mathfrak{L},\mathfrak{L}'_i]\subset\mathfrak{L}'_{i+1}$ and $\dim(\mathfrak{L}'_i/\mathfrak{L}'_{i+1})=1$ for all $0\leq i\leq r-1$.

The center $Z(G)$ of a connected nilpotent Lie group G is always connected. If, in addition, G is simply connected, then it is an *exponential Lie group*, i.e., $\exp_G$ is an analytic manifold isomorphism from $\mathfrak{L}(G)$ onto G, and $Z(G)$ is also simply connected.

A connected Lie group G with Lie algebra $\mathfrak{L}(G)$ is solvable iff $\mathfrak{L}(G)$ is an abstract *solvable Lie algebra* $\mathfrak{L}$ in the sense of the following equivalent definitions:

(i) There exists a decreasing sequence $(\mathfrak{L}_k)_{0\leq k\leq m}$ of ideals of $\mathfrak{L}$ such that $\mathfrak{L}=\mathfrak{L}_0\supset\mathfrak{L}_1\supset\cdots\supset\mathfrak{L}_{m-1}\supset\mathfrak{L}_m=\{0\}$ and the algebra $\mathfrak{L}_k/\mathfrak{L}_{k+1}$ is Abelian for all $0\leq k\leq m-1$;

(ii) there exists a decreasing sequence $(\mathfrak{L}'_k)_{0\leq k\leq r}$ of subalgebras of $\mathfrak{L}$ such that $\mathfrak{L}=\mathfrak{L}'_0\supset\mathfrak{L}'_1\supset\cdots\supset\mathfrak{L}'_{r-1}\supset\mathfrak{L}'_r=\{0\}$, $\mathfrak{L}'_{k+1}$ is an ideal in $\mathfrak{L}'_k$ and $\dim(\mathfrak{L}'_k/\mathfrak{L}'_{k+1})=1$ for all $0\leq k\leq r-1$.

D Theorem (Chevalley). *Let G be a simply connected, connected, solvable Lie group of dimension $d \geq 1$ and H a Lie subgroup of G. Then there exist bases $\{X_1, \ldots, X_d\}$ of $\mathfrak{L}(G)$ and $\{X_{i_1}, \ldots, X_{i_m}\} \subset \{X_1, \ldots, X_d\}$ of $\mathfrak{L}(H)$ resp. such that the mapping $\Phi: \mathfrak{L}(G) \to G$ defined by*

$$\Phi\left(\sum_{i=1}^{d} s_i X_i\right) := \prod_{i=1}^{d} \exp_G(s_i X_i)$$

with $s_1, \ldots, s_d \in \mathbb{R}$ is an analytic manifold isomorphism which maps $\mathfrak{L}(H)$ onto H. In particular, H is closed and simply connected.

It follows that any connected solvable Lie group G of dimension $n \geq 1$ is analytically isomorphic to a product $\mathbb{R}^m \times \mathbb{T}^{n-m}$ for some $m \geq 0$.

An abstract Lie algebra $\mathfrak{L}$ is called *semisimple* if the only Abelian ideal of $\mathfrak{L}$ is $\{0\}$. In analogy to the characterizations of nilpotency and solvability we call a connected Lie group G *semisimple* if its Lie algebra $\mathfrak{L}(G)$ is semisimple in the above sense.

E Theorem. *Let G be a connected compact group. Then there exist a compact connected Lie group L, an Abelian compact connected group M and a normal subgroup H of $L \times M$ such that $G \cong (L \times M)/H$ holds. Moreover, for L there are an $m \geq 0$, a semisimple compact connected Lie group N and a finite subgroup D of $Z(\mathbb{T}^m \times N)$ satisfying the following conditions:*

(i) $L \cong (\mathbb{T}^m \times N)/D$;

(ii) $D \cap \mathbb{T}^m = D \cap N = \{e\}$.

Given a connected Lie group G and its Lie algebra $\mathfrak{L}(G)$ one introduces for every $X \in \mathfrak{L}(G)$ the endomorphism $\operatorname{ad} X: Y \to [X, Y]$ of the vector space $\mathfrak{L}(G)$. The mapping $\operatorname{ad}: X \to \operatorname{ad} X$ is a homomorphism of $\mathfrak{L}(G)$ onto the subalgebra $\operatorname{ad} \mathfrak{L}(G)$ of the Lie algebra $\operatorname{End} \mathfrak{L}(G)$ in the sense that for all $X, Y \in \mathfrak{L}(G)$ one has $\operatorname{ad}[X, Y] = [\operatorname{ad} X, \operatorname{ad} Y]$; this mapping is called the *adjoint representation* of $\mathfrak{L}(G)$.

Clearly, there exists a Lie subgroup $\operatorname{Int} \mathfrak{L}(G)$ of $\mathfrak{GL}(n, \mathbb{R})$ with corresponding Lie algebra $\operatorname{ad} \mathfrak{L}(G)$. For every $x \in G$ one denotes by τ_x the inner automorphism of G defined by x. τ_x induces a mapping $\operatorname{Ad} x \in \operatorname{Aut} \mathfrak{L}(G)$, and one has $\exp_G(\operatorname{Ad} x)(X) = x(\exp_G X)x^{-1}$ for all $x \in G$, $X \in \mathfrak{L}(G)$. The mapping $\operatorname{Ad}: x \to \operatorname{Ad} x$ is a homomorphism from G into $\mathfrak{GL}(\mathfrak{L}(G), \mathbb{R})$, which by definition is the *adjoint representation* of G. For connected Lie groups the differential of the adjoint representation of G is just the adjoint representation of $\mathfrak{L}(G)$. The formula $(\operatorname{Ad} \exp_G Y) X = e^{\operatorname{ad} Y}(X)$ holds for all $Y, X \in \mathfrak{L}(G)$. In addition $\operatorname{Ad} G \cong G/Z(G)$ and $\mathfrak{L}(\operatorname{Ad} G) = \operatorname{ad} \mathfrak{L}(G)$.

The Lie analysis given above yields further results on the *structure of locally compact groups.*

A locally compact group G is said to be *without small subgroups* if there exists a $U \in \mathfrak{B}(e)$ containing nonontrivial subgroup of G.

F Theorem (Gleason, Montgomery, Zippin). *For every locally compact group G the following statements are equivalent:*

(i) *G is without small subgroups;*

(ii) *G is locally Euclidean (of dimension* $d \geq 1$*);*
(iii) *G is a Lie group (of dimension* $d \geq 1$*).*

We now discuss a few more details on this theorem, which represents the solution to Hilbert's fifth problem.

A locally compact group G is called *Lie projective* if it is the projective limit $\varprojlim_{\alpha\in\mathbb{A}} G_\alpha$ of Lie groups $G_\alpha := G/K_\alpha$ with a descending family $(K_\alpha)_{\alpha\in\mathbb{A}}$ of compact normal subgroups K_α of G satisfying $\bigcap_{\alpha\in\mathbb{A}} K_\alpha = \{e\}$.

Let G be a locally compact group which admits a compact quotient group G/G_0. Then G is Lie projective. More generally, one has

G Theorem (Gluškov). *In every locally compact group G there exists an open Lie projective subgroup H such that H/G_0 is compact.*

An immediate consequence of this theorem is the fact that a locally compact group G is a Lie group iff every compact subgroup of G is a Lie group.

A locally compact group G with compact quotient G/G_0 is called *finite-dimensional* if there exists a totally disconnected compact normal subgroup H of G such that G/H is a Lie group.

H Theorem (N.W. Rickert). *Let G be a connected, finite-dimensional locally compact group. There exists a totally disconnected, compact Abelian group K, a connected Lie group L and a discrete normal subgroup D of $H := K \times L$ with $G = H/D$. The restriction to L of the canonical mapping from H onto G is a continuous monomorphism from L onto the arc component of the identity of G which is a dense subgroup of G. In addition, G is arcwise connected (locally connected) iff G is a Lie group.*

Since a connected locally compact group is locally connected iff every finite-dimensional factor group is locally connected, one concludes that any arcwise connected locally compact group is locally connected.

Moreover, let G be a locally connected, locally compact group such that $(Z(G_0))_0$ is metrizable. Then G_0 is arcwise connected, and G is locally arcwise connected.

Another fundamental structure theorem based on the solution of Hilbert's fifth problem is the following generalization of Chevalley's result.

K Theorem (Iwasawa, Malcev). *Let G be a connected, locally compact group. There exist maximal compact subgroups, and all maximal compact subgroups are conjugate to each other. Given a maximal compact subgroup H of G there exists an $m \geq 0$ such that $\mathbb{R}^m \times H$ is homeomorphic to G under the mapping $(x, h) \to x\,h$.*

Almost Periodic Locally Compact Groups. Let G be a locally compact group. A function $f \in \mathscr{C}^b_{\mathbb{C}}(G)$ is called *almost periodic* if one of the following equivalent conditions is satisfied:

(i) $\{{}_a f := f \circ L(a) : a \in G\}$ is relatively compact in $\mathscr{C}^b_{\mathbb{C}}(G)$.

(ii) $\{f_a := f \circ R(a) : a \in G\}$ is relatively compact in $\mathscr{C}^b_{\mathbb{C}}(G)$.
(iii) $\{{}_a f_b := f \circ R(b) \circ L(a) : a, b \in G\}$ is relatively compact in $\mathscr{C}^b_{\mathbb{C}}(G)$.

The collection of all almost periodic functions on G will be abbreviated by $\mathfrak{A}(G)$. It is easily proved that $\mathfrak{A}(G)$ is a commutative C^*-algebra (with respect to pointwise operations) with unit containing along with f also $\operatorname{Re} f$, $\operatorname{Im} f$ and the translates ${}_a f$, f_a and ${}_a f_b (a, b \in G)$. In order to characterize the algebra $\mathfrak{A}(G)$ in various ways one efficiently uses a compactification procedure due to H. Bohr. Let $(G, \mathscr{T})$ be a topological group. The *Bohr compactification* $(\tilde{G}, \beta) := ((\tilde{G}, \tilde{\mathscr{T}}), \beta)$ of $(G, \mathscr{T})$ (with *Bohr group* $(\tilde{G}, \tilde{\mathscr{T}})$ and *Bohr homomorphism* β) is defined by the following conditions:

(a) $(\tilde{G}, \tilde{\mathscr{T}})$ is a compact group.

(b) β is a continuous homomorphism from $(G, \mathscr{T})$ onto some dense subgroup of $(\tilde{G}, \tilde{\mathscr{T}})$.

(c) (Universality) For every continuous homomorphism γ from $(G, \mathscr{T})$ into a dense subgroup of some compact group $(H, \mathscr{S})$ there exists a continuous homomorphism $\tilde{\gamma} : \tilde{G} \to H$ such that $\gamma = \tilde{\gamma} \circ \beta$ holds.

One shows that for every topological group $(G, \mathscr{T})$ the Bohr compactification $(\tilde{G}, \beta)$ exists, and that it is unique up to topological isomorphisms.

The Bohr compactification $\tilde{}$ defines a *covariant functor* from the category of topological groups into the category of compact groups (with continuous mappings as corresponding morphisms). In fact, if G_1 and G_2 are two topological groups with Bohr compactifications $(\tilde{G}_1, \beta_1)$ and $(\tilde{G}_2, \beta_2)$ resp., and if $\phi : G_1 \to G_2$ is a homomorphism of topological groups, then there exists a homomorphism of compact groups $\tilde{\phi} : \tilde{G}_1 \to \tilde{G}_2$ such that $\beta_2 \circ \phi = \tilde{\phi} \circ \beta_1$ holds. The functor $\tilde{}$ is *right exact* in the following sense: If in the subsequent diagram the upper sequence is exact, then the lower sequence is exact at the second and third place only, i.e., we have

(a) $\operatorname{im} \tilde{i} = \ker \tilde{p}$ and

(b) $\tilde{p}$ is an epimorphism (or $\tilde{G}_2 \cong \tilde{G} / \ker \tilde{p}$)

$$\begin{array}{ccccccccc} e & \longrightarrow & G_1 & \xrightarrow{\;i\;} & G & \xrightarrow{\;p\;} & G_2 & \longrightarrow & e \\ & & \downarrow{\scriptstyle \beta_1} & & \downarrow{\scriptstyle \beta} & & \downarrow{\scriptstyle \beta_2} & & \\ & & \tilde{G}_1 & \xrightarrow[\tilde{i}]{} & \tilde{G} & \xrightarrow[\tilde{p}]{} & \tilde{G}_2 & \longrightarrow & e. \end{array}$$

If in addition G_1 is a compact normal subgroup of G and β is surjective, then the whole lower sequence is exact. More generally, one has for every closed normal subgroup G_1 of G the isomorphism $(G/G_1)^{\sim} \cong \tilde{G}/\overline{\beta(G_1)}$ and, if G_1 is compact and β injective, then $(G/G_1)^{\sim} \cong \tilde{G}/\tilde{G}_1$, as expected.

Given a locally compact group we consider (continuous, unitary) *representations* of G defined as continuous homomorphisms D from G into the group $\mathfrak{U}(\mathscr{H}(D))$ of all unitary operators of a (complex) Hilbert space $\mathscr{H}(D)$. A representation D of G with representing Hilbert space $\mathscr{H}(D)$ is called *irreducible* if there exists no nontrivial closed D-invariant subspace of $\mathscr{H}(D)$. By the Gelfand-

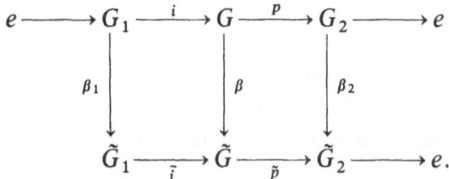

Raikov theorem the set of all irreducible representations of G separates the points of G.

For any representation D of G with representing Hilbert space $\mathscr{H}(D)$ and $\xi, \eta \in \mathscr{H}(D)$ the complex-valued function $d_{\xi,\eta}(D)$ on G defined by

$$d_{\xi,\eta}(D)(x) := \langle D(x)\,\xi, \eta \rangle \quad \text{for all } x \in G$$

is called a *coefficient* (function) of D.

From now on we concentrate on the discussion of finite-dimensional representations of G. The collection $\mathfrak{K}(G)$ of coefficients (of finite-dimensional representations of G) is an involutive algebra over $\mathbb{C}$ and is called the *coefficient algebra* of G.

The *characterization of the algebra* $\mathfrak{A}(G)$ via Bohr compactification is contained in the following statement: For any topological group G with Bohr group $\tilde{G}$ and any $f \in \mathscr{C}_{\mathbb{C}}^b(G)$ the following statements are equivalent:

(i) $f \in \mathfrak{A}(G)$;

(ii) there exists $\tilde{f} \in \mathscr{C}_{\mathbb{C}}^b(\tilde{G})$ such that $f = \tilde{f} \circ \beta$ holds;

(iii) f is the (uniform) limit of complex linear combinations of coefficients of (finite-dimensional) representations of G.

One notes that any function $f \in \mathfrak{A}(G)$ is uniformly continuous for the uniform structures U_r and U_l on G. If G is compact, then clearly (G, Id) is a Bohr compactification of G and $\mathfrak{A}(G)$ coincides with the space $\mathscr{C}_{\mathbb{C}}^b(G)$ which in this case equals the space $\mathscr{C}_{\mathbb{C}}(G)$ of all complex continuous functions on G.

A topological group G is called (maximally) *almost periodic* (an *MAP*-group), if $\mathfrak{A}(G)$ separates the points of G. Let $\mathbf{A}$ denote the class of all almost periodic locally compact groups.

The above characterization of $\mathfrak{A}(G)$ yields the equivalence of the following statements:

(i) $G \in \mathbf{A}$;

(ii) the set of all finite-dimensional representations of G separates the points of G;

(iii) if $(\tilde{G}, \beta)$ is a Bohr compactification of G, then β is a monomorphism;

(iv) G is *injective*, i.e., there exist a compact group H and a continuous monomorphism from G into H.

The class $\mathbf{A}$ enjoys the following fundamental *properties*:

(1) If $G \in \mathbf{A}$ and H is a closed subgroup of G, then $H \in \mathbf{A}$.

(2) Locally compact products and projective limits of families of groups in $\mathbf{A}$ are themselves in $\mathbf{A}$.

(3) Let G be a topological group and H a closed subgroup of G with $[G:H] < \infty$. Then $G \in \mathbf{A}$ iff $H \in \mathbf{A}$.

(4) Let G be a topological group and H a closed normal subgroup of G which is either compact or equals $Z(G)$. Then $G \in \mathbf{A}$ implies that $G/H \in \mathbf{A}$.

Clearly, all compact groups and all locally compact Abelian groups are elements of $\mathbf{A}$.

It should be noted that any compactly generated group $G \in \mathbf{A}$ is Lie projective. Moreover, G is the projective limit of a family of Lie groups which are locally isomorphic to compact Lie groups. These facts make up the proof of the following

L Theorem (Freudenthal, Weil). *For every connected locally compact group G the following statements are equivalent:*
(i) $G \in \mathbf{A}$;
(ii) *there are an* $m \geq 0$ *and a compact connected group K such that* $G \cong \mathbb{R}^m \times K$;
(iii) *G is the projective limit of Lie groups which are locally isomorphic to compact groups;*
(iv) $G/Z(G)$ *is compact;*
(v) *there exists a basis of compact neighborhoods of e each of which is (inner) invariant.*

Locally compact groups G with compact quotient $G/Z(G)$ are said to be *central groups* (Z-groups). Those with a basis of invariant neighborhoods of e are called *groups with small invariant neighborhoods* (SIN-groups). Clearly, all compact groups and all locally compact Abelian groups are Z-groups, and Z-groups are SIN-groups. Furthermore, SIN-groups are Lie projective.

M Theorem (Grosser, Moskowitz). *For every Z-group there are an* $m \geq 0$ *and a locally compact group H containing an open compact subgroup such that* $G \cong \mathbb{R}^m \times H$ *holds.*

We introduce finally a class of locally compact groups whose structure will be basic for various results in the development of the theory.

A locally compact group G is said to be a *Moore group* if all its irreducible representations are finite-dimensional. The class of Moore groups will be abbreviated by $\mathbf{M}$. The Gelfand-Raikov theorem implies immediately that $\mathbf{M} \subset \mathbf{A}$.

N Theorem (C.C. Moore). *For any locally compact group G the following statements are equivalent:*
(i) $G \in \mathbf{M}$;
(ii) $G = \varprojlim_{\alpha \in \mathbb{A}} G_\alpha$ *with Lie groups* $G_\alpha \in \mathbf{M}$ *which are finite extensions of central open subgroups* ($\alpha \in \mathbb{A}$).

The statement of this theorem and its proof imply that $\mathbf{M}$ is closed under the formation of finite direct products, projective limits and finite extensions. In particular, every Moore group is Lie projective.

A group $G \in \mathbf{A}$ is called a *Takahashi group* (T-group), if $K(G)$ is compact. One gets immediately that every Z-group is a T-group. The connection between T-groups and Moore groups is expressed in

O Theorem (L.C. Robertson). *For any locally compact group G the following statements are equivalent:*
(i) $G \in \mathbf{M}$;
(ii) *G is a finite extension of a characteristic T-subgroup.*

In particular, every closed subgroup of a Moore group is itself a Moore group.

Denoting the classes of Z-groups, T-groups and groups with small invariant neighborhoods by **Z**, **T** and **SIN** resp., we obtain in summary the chain $\mathbf{Z}\subset\mathbf{T}\subset\mathbf{M}\subset\mathbf{SIN}$.

References and Comments

References for the fundamental concepts of the theory of topological groups can be taken from the standard encyclopedia by Hewitt and Ross [218], also from Pontryagin [389] and Weil [499]. The structure of compactly generated locally compact Abelian groups is systematically developed in [218]; the formulation of Theorem A is due to Bourbaki [46]. Theorem B found by Dixmier in [117] has not yet entered the textbook literature. The theory of Lie groups will be used often as a basic tool of the theory of probability on groups. Monographs of special actuality are the books by Bourbaki [38], [39], by Helgason [213], by Hochschild [246], the small, but well-written book by Pichon [388], the book by Sagle and Walde [422], and finally the presentation by Varadarajan [489]. This more or less subjective selection has been made on the basis of the particular requirements of various special problems. Nilpotent and solvable locally compact groups have been studied by N.W. Rickert in [401]. Within the framework of Lie groups their structure is treated in [38] and [39]. For solvable Lie groups we refer to Chevalley [72] and Dixmier [116], [119] and also to a selfcontained presentation in the lecture notes [250] of Hofmann and in Varadajan's monograph [489]. The state of the structure theory of solvable Lie groups can be taken from a collection of exposés published as [15]. The formulation of Theorems D and E concerning the structure of solvable and compact groups resp. are taken from Hochschild [246]. For the structure of compact groups see also [251]. Theorem F is a result due to Gleason, Montgomery and Zippin, which in its mature form appears in the book by Montgomery and Zippin [356], and also in Gluškov [171] or Kaplansky [283]. In these references one also finds Theorem G. Theorem H is a result on the structure of finite-dimensional groups due to Rickert [401]. For its consequences also see Rickert [402]. A proof of Theorem K contributed by Iwasawa and Malcev is contained in [356]. In what follows (maximally) almost periodic groups are treated along the lines of Heyer [227]. More about the representation theory of locally compact groups is given in [218] by Hewitt and Ross or [118] by Dixmier. For the general problem of splitting in locally compact groups see the booklet by Hofmann and Mostert [252]. A complete proof of Theorem L based on Theorem E can be found in [246] or [118]. See also [227]. Central groups have been the aim of various extended studies by Grosser and Moskowitz in [183], [184] and [185]. The first mentioned reference contains Theorem M. Moore groups have been of great interest for the development of our topic. They were introduced by Moore in [357] and independently by Štern in [454]. Theorem N gives a characterization of Moore groups in terms of central groups. An extension of Moore's main theorem adapted to our needs is due to Robertson [407]. It appears as Theorem O. A complete proof of it has been indicated in Heyer [231].

Chapter I

Harmonic Analysis of Almost Periodic Locally Compact Groups

We start with a presentation of the fundamental notions and facts from the theory of measures on a locally compact space with emphasis on the study of compact and tight sets of measures.

We choose the Bourbaki concept of measure as the basic notion, but also discuss its relationship to Borel measures and their integrals. As an application we treat invariant measures and convolutions on a locally compact group. Here the main interest will be the discussion of the convolution semigroup of all probability measures on the group and some of its arithmetic. An introduction to the representation theory of almost periodic locally compact groups via finite-dimensional representations prepares us for the notion of the Fourier transform of a bounded measure on the group and its first properties. The following studies are devoted to a thorough analysis of the domain of validity of the well-known theorems of P. Lévy on the bicontinuity of the Fourier transform and of Bochner on the representation of Fourier transforms of measures by positive-definite functions. The aim of the chapter is a detailed treatment of the correspondence between negative-definite forms and continuous one-parameter semigroups of probability measures on the group.

1.1 Measures on a Locally Compact Space

Measures. Let E be a locally compact space, for every compact subset K of E let $\mathscr{K}_{\mathbb{C}}(E, K)$ be the subspace of $\mathscr{C}^b_{\mathbb{C}}(E)$ containing all continuous complex functions f on E with compact support $\operatorname{supp}(f)$ in K and let $\mathscr{K}_{\mathbb{C}}(E)$ be the union of the spaces $\mathscr{K}_{\mathbb{C}}(E, K)$, where K runs through the system of all compact subsets K of E, which is the space of all complex continuous functions on E with compact support. A (complex) *Radon measure* on E is defined as a linear functional μ on $\mathscr{K}_{\mathbb{C}}(E)$ with the following property: For any compact subset K of E there exists a number $M_K \in \mathbb{R}_+$ such that for every $f \in \mathscr{K}_{\mathbb{C}}(E, K)$ one has $|\mu(f)| \leq M_K \|f\|$. For $f \in \mathscr{K}_{\mathbb{C}}(E)$ the symbol $\langle \mu, f \rangle = \int f \, d\mu := \mu(f)$ denotes the (μ-)*measure* of f. The collection of Radon measures on E will be abbreviated by $\mathscr{M}_{\mathbb{C}}(E)$. In analogy to $\mathscr{K}_{\mathbb{C}}(E, K)$, for a compact subset K of E, and $\mathscr{K}_{\mathbb{C}}(E)$ we define the real vector spaces $\mathscr{K}(E, K) := \mathscr{K}_{\mathbb{R}}(E, K)$ of continuous real functions on E with compact support in K and $\mathscr{K}(E) := \mathscr{K}_{\mathbb{R}}(E)$ of continuous real functions on E with compact support. A measure $\mu \in \mathcal{M}_{\mathbb{C}}(E)$ is called *real* if $\mu(f) \in \mathbb{R}$ for all $f \in \mathcal{K}_{\mathbb{R}}(E)$. The set of all real Radon measures on E will be denoted by $\mathcal{M}_{\mathbb{R}}(E)$. Defining for each measure $\mu \in \mathcal{M}_{\mathbb{C}}(E)$ its *conjugate* $\bar{\mu}$ as the linear functional $f \to \overline{\mu(\bar{f})}$ on $\mathcal{K}_{\mathbb{C}}(E)$ one sees that μ is real iff $\mu = \bar{\mu}$.

Moreover, every $\mu \in \mathcal{M}_{\mathbb{C}}(E)$ can be written in the form

$$\mu = \operatorname{Re}\mu + i \operatorname{Im}\mu, \text{ where } \operatorname{Re}\mu := \tfrac{1}{2}(\mu + \bar{\mu}) \text{ and } \operatorname{Im}\mu := \frac{1}{2i}(\mu - \bar{\mu})$$

are measures in $\mathcal{M}_{\mathbb{R}}(E)$. It is known that a linear functional μ on $\mathcal{K}_{\mathbb{C}}(E)$ is a Radon measure on E iff it is continuous with respect to the inductive limit of the topologies of uniform convergence on $\mathcal{K}_{\mathbb{C}}(E, K)$ for compact subsets K of E. Since most of the time we shall be working with real measures, we extend the above notation by putting $\mathcal{M}(E) := \mathcal{M}_{\mathbb{R}}(E)$. Defining a measure $\mu \in \mathcal{M}(E)$ to be *positive* ($\mu \geq 0$) if $\mu(f) \geq 0$ for all $f \in \mathcal{K}_+(E)$, and abbreviating the set of all positive measures on E by $\mathcal{M}_+(E)$ we observe that the set $\mathcal{M}(E)$ inherits the structure of a vector lattice from the vector lattice $\mathcal{K}(E)$. Moreover, every $\mu \in \mathcal{M}(E)$ can be written in the form $\mu = \mu^+ - \mu^-$ with μ^+, $\mu^- \in \mathcal{M}_+(E)$ and $\mu^+ \wedge \mu^- = 0$, and one has the *absolute value* of μ defined by $|\mu| := \mu^+ + \mu^-$.

The measure $|\mu|$ is the smallest measure $\rho \in \mathcal{M}_+(E)$ such that $|\mu(f)| \leq \rho(|f|)$ holds for all $f \in \mathcal{K}(E)$. This definition extends to the notion of absolute value of a measure in $\mathcal{M}_{\mathbb{C}}(E)$. With this modification one obtains

(a) $|\operatorname{Re}\mu| \leq |\mu|$, $|\operatorname{Im}\mu| \leq |\mu|$ and $|\mu| \leq |\operatorname{Re}\mu| + |\operatorname{Im}\mu|$ for $\mu \in \mathcal{M}_{\mathbb{C}}(E)$ as well as

(b) $|\mu + \nu| \leq |\mu| + |\nu|$ for all $\mu, \nu \in \mathcal{M}_{\mathbb{C}}(E)$.

For $\mathcal{M}_+(E)$ one now develops the foundations of integration theory by extending the measures $\mu \in \mathcal{M}_+(E)$ as linear functionals on $\mathcal{K}(E)$ to *upper integrals* μ^* on the space $\mathcal{L}^1_{\mathbb{R}}(E, \mu)$ of *μ-integrable functions* on E. The procedure is outlined in what follows.

Let $\mathcal{I}_+(E)$ denote the set of all nonnegative lower semicontinuous numerical functions on E. Clearly, every $f \in \mathcal{I}_+(E)$ is of the form

$$f = \sup\{g : g \in \mathcal{K}_+(E),\ g \leq f\}.$$

If $\mu \in \mathcal{M}_+(E)$, then for any ascending (filtering increasing) family $(f_\alpha)_{\alpha \in \mathbf{A}}$ in $\mathcal{K}_+(E)$ with

$$f := \sup_{\alpha \in \mathbf{A}} f_\alpha \in \mathcal{K}_+(E) \quad \text{one obtains} \quad \mu(f) = \sup_{\alpha \in \mathbf{A}} \mu(f_\alpha).$$

Thus one can define the (μ-)*upper integral* of $f \in \mathcal{I}_+(E)$ by putting

$$\mu^*(f) := \sup\{\mu(g) : g \in \mathcal{K}_+(E),\ g \leq f\}.$$

Properties. (1) For $f \in \mathcal{I}_+(E)$ and $\alpha \in \mathbb{R}_+$ one has $\mu^*(\alpha f) = \alpha\, \mu^*(f)$.

(2) If $f, g \in \mathcal{I}_+(E)$ and $f \leq g$, then $\mu^*(f) \leq \mu^*(g)$.

(3) If $f, g \in \mathcal{I}_+(E)$, then $\mu^*(f+g) = \mu^*(f) + \mu^*(g)$.

(4) For any ascending family $(f_\alpha)_{\alpha \in \mathbf{A}}$ in $\mathcal{I}_+(E)$ with $f := \sup_{\alpha \in \mathbf{A}} f_\alpha$ one has $\mu^*(f) = \sup_{\alpha \in \mathbf{A}} \mu^*(f_\alpha)$.

Given $f \in \overline{\mathbb{R}}^E_+$ we now put $\mu^*(f) := \inf\{\mu^*(g) : g \in \mathcal{I}_+(E),\ g \geq f\}$ and hence extend the notion of (μ-)upper integral for all functions in $\overline{\mathbb{R}}^E_+$.

Further Properties. (5) For $f \in \overline{\mathbb{R}}_+^E$, $\alpha \in \overline{\mathbb{R}}_+$ one has $\mu^*(\alpha f) = \alpha\, \mu^*(f)$.
(6) If $f, g \in \overline{\mathbb{R}}_+^E$, $f \leq g$, then $\mu^*(f) \leq \mu^*(g)$.
(7) For $f, g \in \overline{\mathbb{R}}_+^E$ one obtains $\mu^*(f+g) \leq \mu^*(f) + \mu^*(g)$.

In the course of the development one now establishes the theorems of Beppo Levi and Fatou, as well as the theory of spaces of μ-integrable functions.

Putting $\mu^*(X) := \mu^*(1_X)$ for any set $X \subset E$ one can define X to be μ-*negligeable* if $\mu^*(X) = 0$. For any $f \in \overline{\mathbb{R}}_+^E$ one has $\mu^*(f) = 0$ iff $f = 0$ μ-a.e. or equivalently, that $[f > 0] := \{x \in E : f(x) > 0\}$ is μ-negligeable. A measure $\mu \in \mathcal{M}_+(E)$ will be *carried* (supported) by a set $X \subset E$ if $\mu^*(\complement X) = 0$ holds. Moreover, one defines the *support* $\operatorname{supp}(\mu)$ of any measure $\mu \in \mathcal{M}(E)$ as the complement of the union of all open subsets U of E on which μ vanishes in the sense that $f \in \mathcal{K}(E)$ with $\operatorname{supp}(f) \subset U$ implies $\mu(f) = 0$.

Let $\mu \in \mathcal{M}_+(E)$. For $1 \leq p < \infty$ one studies with the aid of the seminorm N_p introduced by

$$N_p(f) := (\mu^*(|f|^p))^{\frac{1}{p}} \qquad \text{for all } f \in \overline{\mathbb{R}}^E$$

the spaces

$$\mathcal{F}_{\mathbb{R}}^p(E) := \{f \in \mathbb{R}^E : N_p(f) < \infty\}, \qquad \mathcal{L}_{\mathbb{R}}^p(E, \mu) := \overline{\mathcal{K}_{\mathbb{R}}(E)}$$

(the closure being taken in the sense of the above seminorm within the space $\mathcal{F}_{\mathbb{R}}^p(E)$) and $L_{\mathbb{R}}^p(E, \mu) := \mathcal{L}_{\mathbb{R}}^p(E, \mu)/R$, where the relation R is defined by $f R g :\Leftrightarrow N_p(f-g) = 0$.

Bounded measures. For every $\mu \in \mathcal{M}_{\mathbb{C}}(E)$ we put

$$\|\mu\| := \sup\{|\mu(f)| : f \in \mathcal{K}_{\mathbb{C}}(E), \|f\| \leq 1\}$$

and observe $\|\mu\| = |\mu|^*(1)$. We have $\|\mu\| < \infty$ iff $|\mu|$ is bounded in the sense that

$$|\mu|^*(E) = |\mu|^*(1) < \infty$$

holds. Hence,

$$\|\mu\| = |\mu|^*(1) = |\mu|^*(E)$$

defines the *total mass* of E for $|\mu|$. A measure $\mu \in \mathcal{M}_{\mathbb{C}}(E)$ is called *bounded* if $\|\mu\| < \infty$. Clearly, μ is bounded iff $|\mu|$ is bounded, and $\||\mu|\| = \|\mu\|$. The collection $\mathcal{M}_{\mathbb{C}}^b(E)$ of all bounded measures on E is a vector subspace of $\mathcal{M}_{\mathbb{C}}(E)$, whose elements can be interpreted as linear functionals on $\mathcal{K}_{\mathbb{C}}(E)$ which are continuous with respect to the uniform norm $\|\cdot\|$ on $\mathcal{K}_{\mathbb{C}}(E)$. Then $\mu \to \|\mu\|$ is the usual norm on the dual $\mathcal{M}_{\mathbb{C}}^b(E)$ of the normed space $\mathcal{K}_{\mathbb{C}}(E)$. It should be noted that $\mathcal{M}_{\mathbb{C}}^b(E)$ is complete with respect to this norm and is therefore a Banach space.

If $\mu \in \mathcal{M}_{\mathbb{C}}^b(E)$, then every function $f \in \mathcal{C}_{\mathbb{C}}^b(E)$ is μ-integrable and one obtains $|\mu(f)| \leq \|\mu\|\, \|f\|$. In other words; the mapping $f \to \int f\, d\mu = \mu(f)$ is a continuous linear functional on the Banach space $\mathcal{C}_{\mathbb{C}}^b(E)$. It should be noted, however, that not every continuous linear functional on $\mathcal{C}_{\mathbb{C}}^b(E)$ is of the form $f \to \int f\, d\mu$ for some $\mu \in \mathcal{M}_{\mathbb{C}}(E)$. Finally, we remark that $\mathcal{M}_{\mathbb{C}}^b(E)$ is also the dual of the space $\mathcal{C}_{\mathbb{C}}^0(E)$ of all continuous complex functions on E vanishing at infinity which is the closure of $\mathcal{K}_{\mathbb{C}}(E)$ in $\mathcal{C}_{\mathbb{C}}^b(E)$. In what follows we shall specialize the discussion to the space

$$\mathscr{M}^b_{\mathbb{R}}(E) := \mathscr{M}^b_{\mathbb{C}}(E) \cap \mathscr{M}_{\mathbb{R}}(E)$$

of bounded real measures on E and its cone $\mathscr{M}^b_+(E)$ of positive elements.

Borel Measures. Let $\mathfrak{B}(E)$ denote the Borel σ-algebra of E and $\mathfrak{B}_0(E)$ the σ-ring generated in $\mathfrak{B}(E)$ by the system of all compact subsets of E. Clearly, $\mathfrak{B}_0(E)$ consists exactly of the σ-bounded sets in $\mathfrak{B}(E)$. By definition $\mathfrak{B}_0(E) \subset \mathfrak{B}(E)$. If, in addition, E is σ-compact, then we have $\mathfrak{B}_0(E) = \mathfrak{B}(E)$. In particular, $\mathfrak{B}_0(E)$ is a σ-algebra in E and it is the σ-algebra generated by the system of compact subsets of E.

Let μ be a (positive) premeasure on $(E, \mathfrak{B}_0(E))$ such that $\mu(K) < \infty$ for every compact $K \subset E$. For every $A \in \mathfrak{B}_0(E)$ the following statements are equivalent:
(i) (Outer regularity) $\mu(A) = \inf\{\mu(U): U \supset A,\ U \in \mathfrak{B}_0(E) \text{ open}\}$.
(ii) (Inner regularity) $\mu(A) = \sup\{\mu(K): K \subset A,\ K \text{ compact}\}$.

Every premeasure μ on $(E, \mathfrak{B}_0(E))$ with $\mu(K) < \infty$ for all compact $K \subset E$, and satisfying one of the conditions (i) or (ii) is called a *Borel premeasure* on $(E, \mathfrak{B}_0(E))$.

If $\mu \in \mathscr{M}_+(E)$ with upper integral μ^* and if $\hat{\mu}$ is defined on $\mathfrak{B}(E)$ by

$$\hat{\mu}(A) := \mu^*(1_A) = \mu^*(A) \qquad \text{for all } A \in \mathfrak{B}(E),$$

then $\hat{\mu}$ is an abstract measure on $(E, \mathfrak{B}(E))$ and $\mu^*(f) = \int f \, d\hat{\mu}$ for every non-negative Borel measurable function or every $\hat{\mu}$-integrable function f on E. In particular, we obtain $\mu(f) = \int f \, d\hat{\mu}$ for all $f \in \mathscr{K}(E)$.

We note a few consequences:

(a) For open $U \subset E$ one has

$$\hat{\mu}(U) = \sup\{\hat{\mu}(K): K \subset U,\ K \text{ compact}\}$$
$$= \sup\{\mu(g): g \in \mathscr{K}_+(E),\ g \le 1,\ \operatorname{supp}(g) \subset U\}.$$

(b) The abstract premeasure $\operatorname{Res}_{\mathfrak{B}_0(E)} \hat{\mu}$ is a Borel premeasure on $(E, \mathfrak{B}_0(E))$, and for all compact $K \subset E$ one has

$$\hat{\mu}(K) = \inf\{\mu(f): f \in \mathscr{K}_+(E),\ 1_K \le f\}.$$

(c) For every $A \subset E$ we obtain

$$\mu^*(A) = \inf\{\hat{\mu}(U): A \subset U,\ U \text{ open}\}.$$

These results enable us to associate to any $\mu \in \mathscr{M}_+(E)$ a measure $\hat{\mu}$ on $(E, \mathfrak{B}(E))$ such that $\mu(f) = \int f \, d\hat{\mu}$ holds for all $f \in \mathscr{K}(E)$. The converse will follow from the subsequent discussion.

Let μ be a Borel premeasure on $(E, \mathfrak{B}_0(E))$. By

$$\bar{\mu}(A) := \sup\{\mu(K): K \subset A,\ K \text{ compact}\} \qquad \text{for all } A \in \mathfrak{B}(E)$$

we define an extension $\bar{\mu}$ of μ to a measure on $(E, \mathfrak{B}(E))$ with the additional property that for every extension ν of μ one has $\nu \le \bar{\mu}$. Then

$$L(f) := \int f \, d\bar{\mu} \qquad \text{for all } f \in \mathscr{K}(E)$$

defines a Radon measure L on E satisfying

$$\hat{L}(A) = \mu(A) \qquad \text{for all } A \in \mathfrak{B}_0(E).$$

The measure $\bar{\mu}$ on $(E, \mathfrak{B}(E))$ is called the *essential extension* of μ. The measure $\hat{\mu}$ on $(E, \mathfrak{B}(E))$ defined by

$$\hat{\mu}(A) := \inf\{\bar{\mu}(U): A \subset U,\ U \text{ open}\} \qquad \text{for all } A \in \mathfrak{B}(E)$$

is said to be the *principal extension* of μ. Measures μ on $(E, \mathfrak{B}(E))$ whose restrictions to $\mathfrak{B}_0(E)$ are Borel premeasures on $(E, \mathfrak{B}_0(E))$ are called *Borel measures* on $(E, \mathfrak{B}(E))$. The following special cases illuminate the construction:

(α) If E is σ-compact, then $\hat{\mu} = \bar{\mu} = \mu$.

(β) If $\mu(A) < \infty$ for all $A \in \mathfrak{B}(E)$, then $\hat{\mu} = \bar{\mu}$ and

$$\hat{\mu}(E) = \bar{\mu}(E) = \sup\{\mu(A): A \in \mathfrak{B}_0(E)\}.$$

We are now ready to present the *Riesz representation theorem*. Let E be a locally compact space and μ a Borel premeasure on $(E, \mathfrak{B}_0(E))$ with principal extension $\hat{\mu}$. The definition $L_\mu(f) := \int f \, d\hat{\mu}$ for all $f \in \mathscr{K}(E)$ establishes a one-to-one correspondence $\mu \to L_\mu$ between the set of all Borel premeasures on $(E, \mathfrak{B}_0(E))$ and the set $\mathscr{M}_+(E)$ of all positive Radon measures on E. In terms of (β) this one-to-one correspondence extends to a one-to-one correspondence between the set of all bounded Borel measures on $(E, \mathfrak{B}(E))$ and the set $\mathscr{M}_+^b(E)$ of all bounded positive Radon measures on E. In the sequel we shall not distinguish between the notation of Borel measures μ on $(E, \mathfrak{B}(E))$ and that of their principal (= essential) extensions $\hat{\mu}$.

Analogously, we drop the distinction between Radon measures and their upper integrals.

The Riesz representation theorem makes it possible to introduce operations on measures like taking products of measures with functions (measures with densities), forming images of measures under various kinds of mappings, induced measures on subsets (subspaces) and projective limits of systems of measures as is done in abstract measure theory. For details see [45] or [73]. Product measures will be discussed separately.

The Vague Topology. Since $\mathscr{M}_{\mathbb{C}}(E)$ is a subspace of the space of all mappings from $\mathscr{K}_{\mathbb{C}}(E)$ into $\mathbb{C}$, we can endow it with the topology σ $(\mathscr{M}_{\mathbb{C}}(E), \mathscr{K}_{\mathbb{C}}(E))$ of simple convergence on $\mathscr{K}_{\mathbb{C}}(E)$ which is defined as the initial topology on $\mathscr{M}_{\mathbb{C}}(E)$ with respect to the family of all functions $\mu \to \mu(f)$ $(f \in \mathscr{K}_{\mathbb{C}}(E))$ on $\mathscr{M}_{\mathbb{C}}(E)$ and called the *vague topology* in $\mathscr{M}_{\mathbb{C}}(E)$. In the following discussion we restrict ourselves to studying the vague topology $\mathscr{T}_v$ on $\mathscr{M}(E) := \mathscr{M}_{\mathbb{R}}(E)$. Clearly $\mathscr{M}(E)$ and $\mathscr{M}_+(E)$ admit natural uniform structures derived from the family $\{p_f: f \in \mathscr{K}(E)\}$ of seminorms p_f defined for every $f \in \mathscr{K}(E)$ by $p_f(\mu) := |\mu(f)|$ for all μ in $\mathscr{M}(E)$ and $\mathscr{M}_+(E)$ resp. $\mathscr{M}_+(E)$ is ($\mathscr{T}_v$-)complete and thus closed in $\mathscr{M}(E)$. In general, $\mathscr{M}(E)$ is not complete, but only sequentially complete.

1.1.1 Theorem. *Let E be a locally compact space. For any $\mathscr{N} \subset \mathscr{M}(E)$ the following statements are equivalent:*

(i) *$\bar{\mathscr{N}}$ is compact.*

(ii) *$\mathscr{N}$ is bounded.*

(iii) *For each compact $K \subset E$ there exists an $M_K \in \mathbb{R}_+$ such that for all $\mu \in \mathscr{N}$ one has $|\mu(f)| \leq M_K \|f\|$ whenever $f \in \mathscr{K}(E, K)$.*

Proof. 1. (i)$\Rightarrow$(ii) is clear since the image of $\bar{\mathcal{N}}$ under the mapping $\mu\to\mu(f)$ $(f\in\mathcal{K}(E))$ is compact and hence bounded.

2. (ii)$\Rightarrow$(iii) follows from the Banach-Steinhaus theorem applied to the family $\{\mathrm{Res}_{\mathcal{K}(E,K)}\,\mu:\mu\in\mathcal{N}\}$ in the topological dual $\mathcal{K}(E,K)'$ of $\mathcal{K}(E,K)$ for some compact subset K of E. Then the boundedness of $\{|\mu(f)|:\mu\in\mathcal{N}\}$ for every $f\in\mathcal{K}(E,K)$ implies the uniform boundedness of $\{\|\mu\|:\mu\in\mathcal{N}\}$.

3. (iii)$\Rightarrow$(i). Let $\mathfrak{F}$ be an ultrafilter in $\bar{\mathcal{N}}$. Since $\bar{\mathcal{N}}$ is vaguely bounded, the image of $\mathfrak{F}$ under the mapping $\mu\to\mu(f)$ for any $f\in\mathcal{K}(E)$ is an ultrafilter on a bounded subset of $\mathbb{R}$, which converges to $\mu_0(f)$ say. Clearly, μ_0 is a linear functional on $\mathcal{K}(E)$, and by the inequality in (iii) we obtain $\mu_0\in\mathcal{M}(E)$, i.e., $\mathfrak{F}$ converges to μ_0. □

1.1.2 Corollary. *For any $a\in\mathbb{R}_+^*$ the set*

$$\mathcal{M}^{(a)}(E):=\{\mu\in\mathcal{M}(E):\|\mu\|\le a\}\quad \textit{is compact in } \mathcal{M}(E).$$

Proof. Firstly we note that the set $\mathcal{N}:=\mathcal{M}^{(a)}(E)$ is closed in $\mathcal{M}(E)$. Let $(\mu_\alpha)_{\alpha\in\mathbf{A}}$ be a net in $\mathcal{N}$ with $\lim_{\alpha\in\mathbf{A}}\mu_\alpha=\mu$ and $f\in\mathcal{K}_+(E)$ with $f\le 1$. Then $|\mu_\alpha|(f)\le a$ and hence $|\mu|(f)\le a$. Thus $\|\mu\|=\|\,|\mu|\,\|\le a$ and $\mu\in\mathcal{N}$. But

$$|\mu(f)|\le|\mu|(|f|)\le|\mu|(\|f\|\,1_K)=|\mu|(K)\,\|f\|$$

for every compact $K\subset E$ and each $f\in\mathcal{K}(E,K)$ $(\mu\in\mathcal{N})$ yields (iii) of the theorem, and hence the compactness of $\mathcal{N}$. □

For every $x\in E$ we denote by ε_x the *Dirac measure* in x. The set $\{\varepsilon_x:x\in E\}$ will be abbreviated by $\mathcal{D}(E)$.

1.1.3 Theorem. *Let E be a locally compact space.*

(i) *The mapping $x\to\varepsilon_x$ is a homeomorphism from E into the closed subset $\mathcal{D}(E)\cup\{0\}$ of $\mathcal{M}_+(E)$.*

(ii) *If E is not compact, then $\lim_{x\to\infty}\varepsilon_x=0$ in the sense that for $\varepsilon>0$ and $f\in\mathcal{K}(E)$ there exists a compact $K\subset E$ such that $|\varepsilon_x(f)|<\varepsilon$ holds for all $x\in\complement K$.*

Proof. It remains to show (i). The mapping $x\to\varepsilon_x$ from E onto $\mathcal{D}(E)$ is clearly continuous. Let $(\varepsilon_{x_\alpha})_{\alpha\in\mathbf{A}}$ be a net in $\mathcal{D}(E)$ with $\mathcal{T}_v-\lim_{\alpha\in\mathbf{A}}\varepsilon_{x_\alpha}=\mu\in\mathcal{M}_+(E)$ and suppose that $\mu\neq0$ (hence $\mathrm{supp}(\mu)\neq\emptyset$). For $x_0\in\mathrm{supp}(\mu)$ and any neighborhood $U\in\mathfrak{B}(x_0)$ we choose $f\in\mathcal{K}_+(E)$ with $f\le 1_U$ and $f(x_0)>0$. But then

$$\lim_{\alpha\in\mathbf{A}}\varepsilon_{x_\alpha}(f)=\mu(f)>0$$

implies $f(x_\alpha)>0$ for sufficiently large $\alpha\ge\alpha_0$ (with $\alpha_0\in\mathbf{A}$). Hence, $x_\alpha\in U$ for all $\alpha\ge\alpha_0$, i.e., $(x_\alpha)_{\alpha\in\mathbf{A}}$ converges to $x_0\in E$ say. Plainly, $\mu=\varepsilon_{x_0}$ so that $\mathcal{D}(E)\cup\{0\}$ is a $\mathcal{T}_v$-closed subset of $\mathcal{M}_+(E)$, and E is embedded in $\mathcal{D}(E)$. □

1.1.4 Corollary. *For any $a\in\mathbb{R}_+^*$ the set*

$$\mathcal{M}^a(E):=\{\mu\in\mathcal{M}_+(E):\|\mu\|=a\}$$

is compact iff E is compact.

Proof. 1. Let E be compact. Then $1 \in \mathcal{K}(E)$ and $\mathcal{M}^a(E)$ is closed as the inverse image in $\mathcal{M}_+(E)$ of $\{a\}$ under the continuous mapping $\mu \to \mu(1)$. Moreover, $\mathcal{M}^a(E)$ is compact since the unit ball of $\mathcal{K}(E)' = \mathcal{M}^b(E)$ is compact by the Alaoglu-Bourbaki theorem.

2. If $\mathcal{M}^a(E)$ is compact, then by the theorem $\mathcal{D}(E)$ and hence E is compact. □

1.1.5 Theorem. *Let E be a locally compact space and $\mathcal{A}$ a subspace of $\mathcal{K}(E)$ such that for each compact $K \subset E$ the subset $\mathcal{A} \cap \mathcal{K}_+(E, K)$ is dense in $\mathcal{K}_+(E, K)$. Then*

(i) *any positive linear functional μ on $\mathcal{A}$ is uniquely extendable to a Radon measure on E;*

(ii) *any net $(\mu_\alpha)_{\alpha \in \mathbf{A}}$ in $\mathcal{M}_+(E)$ such that $(\mu_\alpha(\phi))_{\alpha \in \mathbf{A}}$ converges for all $\phi \in \mathcal{A}$, converges in $\mathcal{M}_+(E)$;*

(iii) *if $(\mu_\alpha)_{\alpha \in \mathbf{A}}$ is a net in $\mathcal{M}(E)$ such that $(|\mu_\alpha(f)|)_{\alpha \in \mathbf{A}}$ is bounded for every $f \in \mathcal{K}(E)$ and $(\mu_\alpha(\phi))_{\alpha \in \mathbf{A}}$ converges for every $\phi \in \mathcal{A}$, then $(\mu_\alpha)_{\alpha \in \mathbf{A}}$ converges in $\mathcal{M}(E)$.*

Proof. We content ourselves with the proof of (ii) and (iii).

(ii) For every compact $K \subset E$ there exists an $f \in \mathcal{A}_+$ with $f(x) \geq 1$ for all $x \in K$. Thus there is an $M_K \in \mathbb{R}_+$ with the property $\mu_\alpha(1_K) \leq \mu_\alpha(f) \leq M_K$ for all sufficiently large $\alpha \geq \alpha_0$. For $\phi \in \mathcal{K}(E)$ and $\varepsilon > 0$ there is a $\psi \in \mathcal{A}$ with $\|\phi - \psi\| < \varepsilon$. Therefore, for $\operatorname{supp}(\phi) \cup \operatorname{supp}(\psi) \subset K$ and $\beta > \alpha \geq \alpha_0$ we obtain

$$\begin{aligned} |\mu_\alpha(\phi) - \mu_\beta(\phi)| &\leq |\mu_\alpha(\phi) - \mu_\alpha(\psi)| + |\mu_\alpha(\psi) - \mu_\beta(\psi)| + |\mu_\beta(\psi) - \mu_\beta(\phi)| \\ &\leq 2 M_K \varepsilon + |\mu_\alpha(\psi) - \mu_\beta(\psi)|, \end{aligned}$$

and hence the assertion.

Statement (iii) is proved similarly if one has for every compact $K \subset E$ the existence of a constant $M_K \in \mathbb{R}_+$ such that $\phi \in \mathcal{K}(E, K)$ implies $|\mu_\alpha(\phi)| \leq M_K \|\phi\|$ independently of $\alpha \in \mathbf{A}$. But this follows from the hypothesis with the help of Theorem 1.1.1. □

1.1.6 Theorem. *Let E be a locally compact space. The following statements are equivalent:*

(i) *$\mathcal{M}_+(E)$ is metrizable and separable;*

(ii) *E admits a countable basis of its topology.*

Proof. 1. The implication (i) ⇒ (ii) follows immediately from Theorem 1.1.3. It remains to prove

2. (ii) ⇒ (i). As a locally compact space having a countable basis of its topology E is σ-compact. There exists a sequence $(K_n)_{n \geq 1}$ of compact subsets of E with $K_n \subset \mathring{K}_{n+1}$ for all $n \geq 1$ and $\bigcup_{n \geq 1} K_n = E$. For every $n \geq 1$ let $(f_{n,k})_{k \geq 1}$ be a dense family in $\mathcal{K}_+(K_n)$. The topology $\mathcal{T}_v$ in $\mathcal{M}_+(E)$ is the initial topology with respect to the family of functions $\mu \to \mu(f_{n,k})$ $(n, k \geq 1)$, as follows from Theorem 1.1.5. Hence, $\mathcal{M}_+(E)$ is homeomorphic to a subspace of the space of all mappings $\mathbb{N} \times \mathbb{N} \to \mathbb{R}$ which is metrizable and separable. □

It should be noted that $\mathcal{M}(E)$ is not metrizable iff E contains an infinite compact set, even if the topology of E has a countable basis.

1.1.7 Theorem. *Let E be a locally compact space.*

(i) *The convex cone generated in $\mathcal{M}(E)$ by the set $\mathcal{D}(E)$ of all Dirac measures is dense in $\mathcal{M}_+(E)$;*

(ii) *For every compact subset K of E and $\mu \in \mathcal{M}_+(E)$ with $\operatorname{supp}(\mu) \subset K$ there exists a net $(\mu_\alpha)_{\alpha \in \mathbf{A}}$ of discrete measures $\mu_\alpha \in \mathcal{M}_+(E)$ with finite $\operatorname{supp}(\mu_\alpha) \subset K$ and $\|\mu_\alpha\| = \|\mu\|$ $(\alpha \in \mathbf{A})$ such that $\lim_{\alpha \in \mathbf{A}} \mu_\alpha = \mu$.*

Proof. (i) follows immediately from the bi-polar theorem, which says that the bi-polar B^{00} of a subset B of the topological dual X of a topological vector space coincides with the closed convex hull of $B \cup \{0\}$. In fact; if $X := \mathcal{M}(E)$ and B denotes the cone generated by $\mathcal{D}(E)$, then $B^{00} = \mathcal{M}_+(E)$, whence the assertion follows.

(ii) Since every measure $\mu \in \mathcal{M}_+(E)$ with $\operatorname{supp}(\mu) \subset K$ can be viewed as a measure in $\mathcal{M}_+(K)$, we obtain from (i) the existence of a net $(\mu'_\alpha)_{\alpha \in \mathbf{A}}$ in $\mathcal{M}_+(K)$ with finite $\operatorname{supp}(\mu'_\alpha) \subset K$ for all $\alpha \in \mathbf{A}$ and $\lim_{\alpha \in \mathbf{A}} \mu_\alpha = \mu$. Replacing $\mu_\alpha := \frac{\|\mu\|}{\|\mu'_\alpha\|} \mu'_\alpha$ for $\alpha \in \mathbf{A}$ and observing that $\lim_{\alpha \in \mathbf{A}} \|\mu'_\alpha\| = \|\mu\| < \infty$, since K is compact and μ can be evaluated at the constant function 1, we obtain $\lim_{\alpha \in \mathbf{A}} \mu_\alpha = \mu$. □

The proof of the theorem shows that an analogous result can be established for measures in $\mathcal{M}_{\mathbb{C}}(E)$.

The Weak Topology. For any locally compact space E we also introduce the *weak topology* $\mathcal{T}_w$ in $\mathcal{M}^b(E)$ defined as the topology $\sigma(\mathcal{M}^b(E), \mathcal{C}^b(E))$. Plainly the topology $\mathcal{T}_w$ is finer than the topology $\operatorname{Res}_{\mathcal{M}^b(E)} \mathcal{T}_v$, and $\mathcal{T}_w = \operatorname{Res}_{\mathcal{M}^b(E)} \mathcal{T}_v$ holds iff E is compact. It is also evident that $\mathcal{M}^b_+(E)$ is $\mathcal{T}_w$-closed in $\mathcal{M}^b(E)$. We further note that for every lower semicontinuous function $f: E \to \overline{\mathbb{R}}_+$ the function $\mu \to \mu(f)$ is lower semicontinuous on $\mathcal{M}^b_+(E)$. In fact, $f = \sup\{g \in \mathcal{C}^b(E): g \leq f\}$ so that $\mu \to \mu(f)$ is the supremum of the family $\{\mu \to \mu(g): g \in \mathcal{C}^b(E),\ g \leq f\}$ of continuous functions $\mu \to \mu(g)$ and is hence lower semicontinuous. It follows that for every bounded upper semicontinuous function $f: E \to \overline{\mathbb{R}}_+$ the function $\mu \to \mu(f)$ on $\mathcal{M}^b_+(E)$ is upper semicontinuous. Obviously $\|f\| - f$ is lower semicontinuous so that by the preceding assertion $\mu \to \mu(\|f\| - f)$ is lower semicontinuous on $\mathcal{M}^b_+(E)$ and therefore $\mu \to \mu(f)$ is upper semicontinuous.

1.1.8 Theorem. *Let F be a locally compact space, E a locally compact subspace of F and i the canonical injection from E into F. The mapping $\mu \to i(\mu)$ from $\mathcal{M}^b_+(E)$ into $\mathcal{M}^b_+(F)$ is a homeomorphism from $\mathcal{M}^b_+(E)$ onto the set*

$$\mathcal{W}(F) := \{\nu \in \mathcal{M}^b_+(F) : \nu(\complement_F E) = 0\}.$$

Proof. Evidently $i: \mathcal{M}^b_+(E) \to \mathcal{M}^b_+(F)$ is injective and maps $\mathcal{M}^b_+(E)$ onto $\mathcal{W}(F)$. For every $\nu \in \mathcal{W}(F)$ one has $\nu = i(\nu_E)$ with $\nu_E := \operatorname{Res}_E \nu$. Thus i is a bijection with inverse $j: \nu \to \nu_E$ on $\mathcal{W}(F)$.

Since for every $\mu \in \mathcal{M}^b_+(E)$ and $f \in \mathcal{C}^b(F)$ we have $\langle i(\mu), f \rangle = \langle \mu, f \circ i \rangle$ and $f \circ i \in \mathcal{C}^b(E)$, i is continuous. It remains to be proved that j is continuous, i.e., that for all $\nu \in \mathcal{W}(F)$ and all $f \in \mathcal{C}^b_+(E)$ one obtains

$$\lim_{\substack{\nu\to\mu \\ \nu\in\mathscr{W}(F)}} \nu_E(f)=\mu_E(f).$$

This is a consequence of the fact that every function $f\in\mathscr{C}^b(E)$ can be extended to a bounded lower or upper semicontinuous function f_1 or f_2 on F resp. One defines

$$\tilde{f}(y):=\begin{cases} f(y), & \text{if } y\in E \\ \|f\|, & \text{if } y\in F\setminus E \end{cases}$$

or

$$\bar{f}(y):=\begin{cases} f(y), & \text{if } y\in E \\ -\|f\|, & \text{if } y\in F\setminus E \end{cases}$$

and then puts $f_1(x):=\underline{\lim}_{y\to x}\tilde{f}(y)$ or $f_2(x):=\overline{\lim}_{y\to x}\bar{f}(y)$ resp. for all $x\in F$.

By the preceding discussion the mappings $\mu\to\mu(f_1)$ and $\mu\to\mu(f_2)$ from $\mathscr{M}_+^b(F)$ into $\mathbb{R}$ are lower and upper semicontinuous resp. Therefore we obtain

$$\overline{\lim}_{\substack{\nu\to\mu \\ \nu\in\mathscr{W}(F)}} \nu(f_2)\le\mu(f_2)=\mu(f_1)\le \underline{\lim}_{\substack{\nu\to\mu \\ \nu\in\mathscr{W}(F)}} \nu(f_1) \quad \text{for } \mu\in\mathscr{W}(F).$$

But $\nu(f_2)=\nu(f_1)=\nu_E(f)$ since $\nu\in\mathscr{W}(F)$ and $f_1(x)=f_2(x)=f(x)$ for all $x\in E$. Hence, $\lim_{\substack{\nu\to\mu \\ \nu\in\mathscr{W}(F)}} \nu_E(f)=\mu_E(f)$, which is the desired assertion. □

1.1.9 Theorem. *Let E be a locally compact space, $(\mu_\alpha)_{\alpha\in\mathbf{A}}$ a net in $\mathscr{M}_+^b(E)$ and $\mu\in\mathscr{M}_+^b(E)$. The following statements are equivalent:*

(i) $\mathscr{T}_w-\lim_{\alpha\in\mathbf{A}}\mu_\alpha=\mu$;

(ii) $\mathscr{T}_v-\lim_{\alpha\in\mathbf{A}}\mu_\alpha=\mu$ and $\lim_{\alpha\in\mathbf{A}}\|\mu_\alpha\|=\|\mu\|$.

Proof. It remains to be proved that (ii)$\Rightarrow$(i). Let F denote the one-point compactification $E\cup\{\infty\}$ of E and i the canonical injection from E into F. We have to show that the net $(i(\mu_\alpha))_{\alpha\in\mathbf{A}}$ $\mathscr{T}_w$-converges to $i(\mu)\in\mathscr{M}_+^b(F)$.

Since $\|\mu\|<\infty$ and $\lim_{\alpha\in\mathbf{A}}\|\mu_\alpha\|=\|\mu\|$, there is an $\alpha_0\in\mathbf{A}$ such that $\|\mu_\alpha\|\le M$ holds for all $\alpha\ge\alpha_0$, where M is a constant $\in\mathbb{R}_+$. It suffices to show that

$$\lim_{\alpha\in\mathbf{A}} i(\mu_\alpha)(g)=i(\mu)(g)$$

for all g in a total subset $\mathscr{A}$ of $\mathscr{C}^b(F)$. But the set

$$\{g\in\mathscr{K}(F):\operatorname{supp}(g)\subset E\}\cup\{g\in\mathscr{C}^b(F):g \text{ constant}\}$$

is a total subset $\mathscr{A}$ of $\mathscr{C}^b(F)$ for which by assumption the above limit relation holds. □

1.1.10 Corollary. *For every $a\in\mathbb{R}_+^*$ we have* $\operatorname{Res}_{\mathscr{M}^a(E)}\mathscr{T}_w=\operatorname{Res}_{\mathscr{M}^a(E)}\mathscr{T}_v$.

The *proof* is obvious. □

Let E be a locally compact space. A subset $\mathscr{N}$ of $\mathscr{M}_+^b(E)$ is called *uniformly tight* if $\sup_{\mu\in\mathscr{N}}\|\mu\|<\infty$ holds and if for every $\varepsilon>0$ there exists a compact subset $K:=K_\varepsilon$ of E with the property that $\mu(\complement K)<\varepsilon$ for all $\mu\in\mathscr{N}$.

We observe that if $\mathscr{N} \subset \mathscr{M}_+^b(E)$ is uniformly tight, then $\bar{\mathscr{N}}$ (in the sense of $\mathscr{T}_w$) is also. This follows from the lower semicontinuity of the functions $\mu \to \|\mu\| = \mu(1)$ and $\mu \to \mu(\complement K)$ (for compact $K \subset E$).

1.1.11 Theorem (Prohorov). *Let E be a locally compact space and $\mathscr{N}$ a subset of $\mathscr{M}_+^b(E)$. The following statements are equivalent:*
(i) *$\mathscr{N}$ is uniformly tight;*
(ii) *$\mathscr{N}$ is $\mathscr{T}_w$-relatively compact in $\mathscr{M}_+^b(E)$.*

Proof. 1. (i) ⇒ (ii). Without loss of generality we suppose that E is a locally compact subspace of a compact space F. Let i be the canonical injection from E into F. By the fact preceding the theorem, we may further assume that $\mathscr{N}$ is $\mathscr{T}_w$-closed in $\mathscr{M}_+^b(E)$. It remains to be shown that every universal net $(\mu_\alpha)_{\alpha \in \mathbf{A}}$ in $\mathscr{N}$ converges in $\mathscr{M}_+^b(E)$.

Since the set $\{\|\mu\| : \mu \in \mathscr{N}\}$ is bounded by hypothesis, $(i(\mu_\alpha))_{\alpha \in \mathbf{A}}$ $\mathscr{T}_v$-converges to a measure $\nu \in \mathscr{M}_+(F)$ by Theorem 1.1.1. Using Theorem 1.1.8 we still have to show that $\nu(\complement_F E) = 0$. In fact, let $\varepsilon > 0$ and let $K := K_\varepsilon$ be a compact subset of E with the property $\mu(\complement K) \leq \varepsilon$ for all $\mu \in \mathscr{N}$. Since $\complement_F K$ is open in F, we obtain

$$\nu(\complement_F E) \leq \nu(\complement_F K) \leq \underline{\lim}_{\alpha \in \mathbf{A}}\, i(\mu_\alpha)(\complement_F K) = \underline{\lim}_{\alpha \in \mathbf{A}}\, \mu_\alpha(\complement_E K) \leq \varepsilon$$

and thus the assertion, since $\varepsilon > 0$ is arbitrary.

2. (ii) ⇒ (i). Again we assume without loss of generality that $\mathscr{N}$ is compact. Since $\mu \to \|\mu\|$ is continuous, we conclude that $\sup_{\mu \in \mathscr{N}} \|\mu\| < \infty$.

Now, let $\varepsilon > 0$ be given. Since E is locally compact, for every $\mu \in \mathscr{M}_+^b(E)$ there exists an open set U_μ such that $K_\mu := \bar{U}_\mu$ is compact and $\mu(\complement U_\mu) \leq \varepsilon$.

Since the function $\nu \to \nu(\complement U_\mu)$ on $\mathscr{M}_+^b(E)$ is upper semicontinuous, there exists an open neighborhood V_μ of μ such that for all $\nu \in V_\mu$ we have $\nu(\complement U_\mu) \leq \varepsilon$. In particular, $\nu(\complement K_\mu) \leq \varepsilon$. Since $\mathscr{N}$ is compact it can be written as $\bigcup_{i=1}^n V_{\mu_i}$ for some $n \geq 1$. Putting $K := K_\varepsilon := \bigcup_{i=1}^n K_{\mu_i}$ we obtain $\nu(\complement K) \leq \varepsilon$ for all $\nu \in \mathscr{N}$. □

An application of the preceding theorem to the set

$$\mathscr{M}^1(E) := \{\mu \in \mathscr{M}_+(E) : \|\mu\| = 1\}$$

of all *probability measures* on E yields that in general $\mathscr{M}^1(E)$ does not admit any compactness property.

1.1.12 Theorem. *Let E be a locally compact space. The following statements are equivalent:*
(i) *E is compact;*
(ii) *$\mathscr{M}^1(E)$ is σ-compact;*
(iii) *$\mathscr{M}^1(E)$ is locally compact.*

Proof. 1. (i)⇒(ii) and (i)⇒(iii) follow directly from Corollary 1.1.4.

2. (ii)⇒(i). Let $(\mathscr{N}_n)_{n \geq 1}$ be a sequence of $\mathscr{T}_v$-(or $\mathscr{T}_w$-)compact subsets of $\mathscr{M}^1(E)$ with $\bigcup_{n \geq 1} \mathscr{N}_n = \mathscr{M}^1(E)$. Without loss of generality we assume that $\mathscr{N}_n \subset \mathscr{N}_{n+1}$ for all $n \geq 1$. Theorem 1.1.11 implies that for every $n \geq 1$ there exists a

compact subset K_n of E with $\mu(\complement K_n)\leq\frac{1}{n}$ for all $\mu\in\mathcal{N}_n$. Again without loss of generality one assumes $K_n\subset K_{n+1}$ for all $n\geq 1$. Suppose now that E were not compact. Then we assume $K_n\neq K_{n+1}$ for all $n\geq 1$ and choose for every $n\geq 2$ an $x_n\in K_n\setminus K_{n-1}$. Furthermore, let $(a_n)_{n\geq 1}$ be a sequence in $\mathbb{R}_+$ with

$$\textstyle\sum_{n\geq 1} a_n=1 \quad\text{and}\quad \sum_{n=1}^{k} a_n=1-\frac{2}{k} \qquad \text{for all } k\geq 2.$$

Such a sequence exists; for example, the sequence $(a_n)_{n\geq 1}$ with

$$a_1:=a_2:=0, \qquad a_n:=\frac{2}{n(n-1)} \qquad \text{for all } n\geq 3$$

has these properties. Putting $\mu:=\sum_{n\geq 1} a_n\,\varepsilon_{x_n}$ we observe that $\mu\in\mathcal{M}^1(E)$ and

$$\mu(K_m)=\textstyle\sum_{n=1}^{m} a_n\,\varepsilon_{x_n}(K_m)=\sum_{n=1}^{m} a_n=1-\frac{2}{m}<1-\frac{1}{m}$$

for all $m\geq 1$. Thus $\mu\notin\mathcal{N}_m$ for all $m\geq 1$ which is the desired contradiction.

3. (iii)$\Rightarrow$(i). Let E be noncompact. Then by the Hahn-Banach theorem there exists a normed positive linear functional $L\in\mathscr{C}^b(E)'$ satisfying $L(f)=0$ for all $f\in\mathscr{C}^0(E)$. Plainly L is not a measure in $\mathcal{M}^1(E)$. We now pick an $x\in E$ and consider the $\mathscr{T}_v$-neighborhood

$$V:=\{\mu\in\mathcal{M}^1(E):|\mu(f_j)-f_j(x)|<\varepsilon \text{ for all } j=1,\ldots,n\}$$

(with $f_1,\ldots,f_n\in\mathscr{C}^b(E)$) of $\varepsilon_x\in\mathcal{M}^1(E)$. One can choose $p\in[0,1[$ such that $2(1-p)\,\|f_j\|<\varepsilon$ holds for all $j=1,\ldots,n$, and hence, $p\,\varepsilon_x+(1-p)\,\mu\in V$ for all $\mu\in\mathcal{M}^1(E)$. From the bi-polar theorem we conclude that there exists a net $(\mu_\alpha)_{\alpha\in\mathbb{A}}$ in $\mathcal{M}^1(E)$ which $\sigma(\mathscr{C}^b(E)', \mathscr{C}^b(E))$-converges to L. We therefore have $(1-p)\,\mu_\alpha+p\,\varepsilon_x\in V$ for all $\alpha\in\mathbb{A}$. But

$$\mathscr{T}_v-\lim_{\alpha\in\mathbb{A}}[(1-p)\,\mu_\alpha+p\,\varepsilon_x]=(1-p)\,L+p\,\varepsilon_x\notin\mathcal{M}^1(E).$$

Thus, V is not $\mathscr{T}_v$-relatively compact in $\mathcal{M}^1(E)$ and so $\mathcal{M}^1(E)$ is not locally compact. □

Product Measures. Let E and F be two locally compact spaces, $\mu\in\mathcal{M}_{\mathbb{C}}(E)$ and $\nu\in\mathcal{M}_{\mathbb{C}}(F)$. It is known that there exists exactly one measure $\pi\in\mathcal{M}_{\mathbb{C}}(E\times F)$ such that for all $f\in\mathscr{K}_{\mathbb{C}}(E)$ and $g\in\mathscr{K}_{\mathbb{C}}(F)$ one has

$$\int f(x)\,g(y)\,\pi(d(x,y))=\left(\int f(x)\,\mu(dx)\right)\left(\int g(y)\,\nu(dy)\right).$$

The existence proof for the measure π is based on the fact that if K is compact in E and L is compact in F, and if $h\in\mathscr{K}_{\mathbb{C}}(E\times F, K\times L)$, then the function

$$y\to g(y):=\int h(x,y)\,\mu(dx)$$

is an element of $\mathscr{K}_{\mathbb{C}}(F, L)$. We shall denote the expression

$$\pi(h) = \nu(\int h(x,y)\,\mu(dx)) \quad \text{by} \ \int \nu(dy) \int h(x,y)\,\mu(dx).$$

Since the rôles of E and F can be interchanged, we obtain for all $h \in \mathscr{K}_{\mathbb{C}}(E \times F)$ the formula

$$\int h(x,y)\,\pi(d(x,y)) = \int \mu(dx) \int h(x,y)\,\nu(dy) = \int \nu(dy) \int h(x,y)\,\mu(dx)$$

and write

$$\iint h\,d\mu\,d\nu \quad \text{or} \quad \iint h(x,y)\mu(dx)\,\nu(dy)$$

instead of

$$\int h(x,y)\,\pi(d(x,y)).$$

The measure π so defined is called the *product* (measure) of μ and ν. It will be written $\mu \otimes \nu$. Clearly, the mapping $(\mu, \nu) \to \mu \otimes \nu$ from $\mathscr{M}_{\mathbb{C}}(E) \times \mathscr{M}_{\mathbb{C}}(F)$ into $\mathscr{M}_{\mathbb{C}}(E \times F)$ is bilinear. The extension to $\mu \otimes \nu$-integrable functions on $E \times F$ of the product formula above is embodied in Fubini's theorem. In the course of the theory one shows that for measures $\mu \in \mathscr{M}_{\mathbb{C}}(E)$ and $\nu \in \mathscr{M}_{\mathbb{C}}(F)$ one has

$$|\mu \otimes \nu| = |\mu| \otimes |\nu| \quad \text{and deduces} \quad \|\mu \otimes \nu\| = \|\mu\|\,\|\nu\|$$

(with the obvious conventions). Clearly

$$\mathscr{M}^b_{\mathbb{C}}(E) \otimes \mathscr{M}^b_{\mathbb{C}}(F) \subset \mathscr{M}^b_{\mathbb{C}}(E \times F).$$

As in the preceding parts of this section we finally restrict ourselves to the study of $\mathscr{M}(E \times F) = \mathscr{M}_{\mathbb{R}}(E \times F)$ $(= \overline{\mathscr{M}_{\mathbb{R}}(E) \otimes \mathscr{M}_{\mathbb{R}}(F)})$.

We shall need the following

1.1.13 Theorem. *Let E and F be locally compact spaces and $(\mu, \nu) \in \mathscr{M}(E) \times \mathscr{M}(F)$. Then the formula*

$$\operatorname{supp}(\mu \otimes \nu) = \operatorname{supp}(\mu) \times \operatorname{supp}(\nu) \quad \textit{holds}.$$

Proof. First of all we note that if $\mu \in \mathscr{M}(E)$ is concentrated on a set $A \subset E$ in the sense that $\complement A$ is $|\mu|$-negligeable and if $\nu \in \mathscr{M}(F)$ is concentrated on a set $B \subset F$, then $\mu \otimes \nu$ is concentrated on $A \times B \subset E \times F$. This follows immediately from the fact that the set $(E \times F) \setminus (A \times B)$ can be written as the union of the two $|\mu \otimes \nu|$-negligeable sets $(E \setminus A) \times F$ and $E \times (F \setminus B)$. From this we conclude that

$$\operatorname{supp}(\mu \otimes \nu) \subset \operatorname{supp}(\mu) \times \operatorname{supp}(\nu).$$

It remains to prove the converse inclusion.

Let $x \in \operatorname{supp}(\mu)$ and $y \in \operatorname{supp}(\nu)$. For every compact neighborhood U of x in E and every compact neighborhood V of y in F one has $|\mu|(U) > 0$ and $|\nu|(V) > 0$ resp. Hence,

$$|\mu \otimes \nu|(U \times V) = |\mu|(U)\,|\nu|(V) > 0,$$

which implies the assertion. □

1.1.14 Theorem. *Let E and F be locally compact spaces and denote the mapping* $(\mu, \nu) \to \mu \otimes \nu$ *from* $\mathscr{M}(E) \times \mathscr{M}(F)$ *into* $\mathscr{M}(E \times F)$ *by* Ψ.

(i) *If* $\mathscr{L}$ *and* $\mathscr{N}$ *are bounded subsets of* $\mathscr{M}(E)$ *and* $\mathscr{M}(F)$ *resp., then* $\mathrm{Res}_{\mathscr{L} \times \mathscr{N}} \Psi$ *is (jointly* $\mathscr{T}_v$*-)continuous.*

Moreover,

(ii) *the mapping* $\mathrm{Res}_{\mathscr{M}_+(E) \times \mathscr{M}_+(F)} \Psi$ *is continuous.*

Proof. (i) Suppose that $(\mu_\alpha)_{\alpha \in \mathbb{A}}$ and $(\nu_\alpha)_{\alpha \in \mathbb{A}}$ are nets in $\mathscr{L}$ and $\mathscr{N}$ resp. with

$$\lim_{\alpha \in \mathbb{A}} \mu_\alpha = \mu \quad \text{and} \quad \lim_{\alpha \in \mathbb{A}} \nu_\alpha = \nu.$$

We have to show that

$$\lim_{\alpha \in \mathbb{A}} \mu_\alpha \otimes \nu_\alpha = \mu \otimes \nu.$$

Let $f \in \mathscr{K}(E \times F)$ with $\mathrm{supp}(f) \subset \mathring{K} \times \mathring{L}$ for compact sets $K \subset E$ and $L \subset F$.

Since for compact spaces K and L the algebra $\mathscr{K}(K) \otimes \mathscr{K}(L)$ is dense in $\mathscr{K}(K \times L)$ one can find for $\varepsilon > 0$ functions $f_k \in \mathscr{K}(E)$ with $\mathrm{supp}(f_k) \subset K$ and functions $g_k \in \mathscr{K}(F)$ with $\mathrm{supp}(g_k) \subset L$ $(k = 1, \ldots, n)$ such that

$$\|f - \textstyle\sum_{k=1}^n f_k \otimes g_k\| < \varepsilon \quad \text{holds.}$$

Since $\mathscr{L}$ and $\mathscr{N}$ are bounded, by Theorem 1.1.1 there exists an $\alpha_0 \in \mathbb{A}$ such that for $\alpha \geq \alpha_0$ we have

$$\textstyle\sum_{k=1}^n |\mu_\alpha(f_k)| \, |\nu_\alpha(g_k) - \nu(g_k)| < \varepsilon$$

and

$$\textstyle\sum_{k=1}^n |\nu(g_k)| \, |\mu_\alpha(f_k) - \mu(f_k)| < \varepsilon.$$

Again by the boundedness of $\mathscr{L}$ and $\mathscr{N}$ one obtains via Theorem 1.1.1 the existence of a constant $M_{K,L} \in \mathbb{R}_+$ independent of μ and ν such that for $h \in \mathscr{K}(E \times F)$ the inclusion $\mathrm{supp}(h) \subset K \times L$ implies $|\mu_\alpha \otimes \nu_\alpha(h)| \leq M_{K,L} \|h\|$ whenever $\alpha \geq \alpha_0$ is sufficiently large. The following chain of inequalities, valid for all $\alpha \geq \alpha_0$, yields the assertion:

$$\begin{aligned}
&|\mu_\alpha \otimes \nu_\alpha(f) - \mu \otimes \nu(f)| \leq |\mu_\alpha \otimes \nu_\alpha(f) - \textstyle\sum_{k=1}^n \mu_\alpha \otimes \nu_\alpha(f_k \otimes g_k)| \\
&+ |\textstyle\sum_{k=1}^n \mu_\alpha \otimes \nu_\alpha(f_k \otimes g_k) - \textstyle\sum_{k=1}^n \mu_\alpha \otimes \nu(f_k \otimes g_k)| \\
&+ |\textstyle\sum_{k=1}^n \mu_\alpha \otimes \nu(f_k \otimes g_k) - \textstyle\sum_{k=1}^n \mu \otimes \nu(f_k \otimes g_k)| \\
&+ |\textstyle\sum_{k=1}^n \mu \otimes \nu(f_k \otimes g_k) - \mu \otimes \nu(f)| \leq 2\varepsilon(M_{K,L} + 1).
\end{aligned}$$

(ii) The proof can be reduced to an application of (i) by showing that for any net $(\mu_\alpha)_{\alpha \in \mathbb{A}}$ in $\mathscr{M}_+(E)$ with $\lim_{\alpha \in \mathbb{A}} \mu_\alpha = \mu$ and $f \in \mathscr{K}(E)$ there exists an $\alpha_0 \in \mathbb{A}$ such that the set $\{f \cdot \mu_\alpha \colon \alpha \geq \alpha_0\}$ is bounded in $\mathscr{M}_+(E)$. To see this we may suppose without loss of generality that $f \in \mathscr{K}_+(E)$. Given $\phi \in \mathscr{K}_+(E)$ one has $0 \leq \lim_{\alpha \in \mathbb{A}} \mu_\alpha(f\phi) = \mu(f\phi)$.

But $\mu_\alpha(f\phi) \leq \mu_\alpha(f \|\phi\|) = \|\phi\| \mu_\alpha(f)$ is bounded for sufficiently large $\alpha \geq \alpha_0$ since $\lim_{\alpha \in \mathbb{A}} \mu_\alpha(f) = \mu(f)$ by assumption, and the assertion follows. □

1.2 Convolution of Measures on a Locally Compact Group

Convolution of Measures. Let G be a locally compact group. A pair $(\mu, \nu) \in \mathcal{M}_{\mathbb{C}}(G) \times \mathcal{M}_{\mathbb{C}}(G)$ is said to be *convolvable* if for every $f \in \mathcal{K}_{\mathbb{C}}(G)$ the function $(x, y) \to f(xy)$ on $G \times G$ is $\mu \otimes \nu$-integrable. Clearly the convolvability of $(\mu, \nu) \in \mathcal{M}_{\mathbb{C}}(G) \times \mathcal{M}_{\mathbb{C}}(G)$ is equivalent to the convolvability of $(|\mu|, |\nu|) \in \mathcal{M}_+(G) \times \mathcal{M}_+(G)$. It is evident that the mapping $f \to \iint f(xy)\,|\mu|(dx)\,|\nu|(dy)$ is a positive linear functional on $\mathcal{K}_{\mathbb{R}}(G)$, i.e., a positive measure on G. Moreover, for every $f \in \mathcal{K}_{\mathbb{C}}(G)$ one has the inequality

$$|\iint f(xy)\,\mu(dx)\,\nu(dy)| \leq \iint |f(xy)|\,|\mu|(dx)\,|\nu|(dy).$$

Hence, the mapping $f \to \iint f(xy)\,\mu(dx)\,\nu(dy)$ is a (complex) measure on G, the *convolution* $\mu * \nu$ of μ and ν. Clearly, $|\mu * \nu| \leq |\mu| * |\nu|$. One notes that $(\mu, \nu) \in \mathcal{M}_{\mathbb{C}}(G) \times \mathcal{M}_{\mathbb{C}}(G)$ is convolvable iff for every compact subset K of G the (closed) set $\{(x, y) \in G \times G : xy \in K\}$ is $\mu \otimes \nu$-integrable. If $(\mu, \nu) \in \mathcal{M}_{\mathbb{C}}(G) \times \mathcal{M}_{\mathbb{C}}(G)$ is convolvable, one also says that μ and ν (in this order) are convolvable or μ is left convolvable with ν or ν is right convolvable with μ. The extension of convolvability and convolution to an arbitrary finite number of measures is obvious.

1.2.1 Theorem. *Let G be a locally compact group, and (μ, ν) a convolvable pair in $\mathcal{M}(G) \times \mathcal{M}(G)$. Then we have*

(i) $\mathrm{supp}(\mu * \nu) \subset \overline{\mathrm{supp}(\mu)\,\mathrm{supp}(\nu)}$.

(ii) *If, in addition, $(\mu, \nu) \in \mathcal{M}_+(G) \times \mathcal{M}_+(G)$, one gets*

$$\mathrm{supp}(\mu * \nu) = \overline{\mathrm{supp}(\mu)\,\mathrm{supp}(\nu)}.$$

Proof. (i) Let $z \notin \overline{\mathrm{supp}(\mu)\,\mathrm{supp}(\nu)}$ and let U be an open neighborhood of z with $U \cap (\mathrm{supp}(\mu)\,\mathrm{supp}(\nu)) = \varnothing$. For every $f \in \mathcal{C}(G)$ with $\mathrm{supp}(f) \subset U$ one has by Theorem 1.1.13

$$\iint f(xy)\,\mu(dx)\,\nu(dy) = \int_{\mathrm{supp}(\mu)} \mu(dx) \int_{\mathrm{supp}(\nu)} f(xy)\,\nu(dy).$$

But for $x \in \mathrm{supp}(\mu)$ and $y \in \mathrm{supp}(\nu)$ we have $f(xy) = 0$ by hypothesis, and hence $z \notin \mathrm{supp}(\mu * \nu)$.

(ii) Clearly, $\mu * \nu \in \mathcal{M}_+(G)$, if $(\mu, \nu) \in \mathcal{M}_+(G) \times \mathcal{M}_+(G)$. Let U be an open and $\mu * \nu$-negligeable subset of G and let K be a compact subset of U. There exists a function $f \in \mathcal{K}_+(G)$, $f \leq 1$, $f(K) = 1$ and $f(\complement U) = 0$. By assumption we get

$$\iint f(xy)\,\mu(dx)\,\nu(dy) = 0,$$

and hence the open set

$$A := \{(x, y) \in G \times G : f(xy) > 1/2\}$$

is $\mu \otimes \nu$-negligeable and

$$A \cap (\mathrm{supp}(\mu) \times \mathrm{supp}(\nu)) = \varnothing.$$

Since the multiplication in G is continuous,

$$K\cap\overline{\operatorname{supp}(\mu)\operatorname{supp}(\nu)}=\emptyset \quad\text{and}\quad \overline{\operatorname{supp}(\mu)\operatorname{supp}(\nu)}\subset\operatorname{supp}(\mu*\nu).\quad\square$$

Considering, in particular, bounded measures one sees immediately that any pair $(\mu,\nu)\in\mathcal{M}^b_{\mathbb{C}}(G)\times\mathcal{M}^b_{\mathbb{C}}(G)$ is convolvable and $\mu*\nu\in\mathcal{M}^b_{\mathbb{C}}(G)$. In this case

$$\|\mu*\nu\|\le\|\mu\|\,\|\nu\|,\quad\text{or}\quad \|\mu*\nu\|=\|\mu\|\,\|\nu\|\quad\text{if}\quad (\mu,\nu)\in\mathcal{M}_+(G)\times\mathcal{M}_+(G).$$

We summarize by noting that the set $\mathcal{M}^b_{\mathbb{C}}(G)$ is an algebra over $\mathbb{C}$ with respect to convolution with unit element ε_e. $\mathcal{M}^b_{\mathbb{C}}(G)$ is commutative iff G is Abelian. Moreover, $\mathcal{M}^b_{\mathbb{C}}(G)$ is a normed algebra with respect to the norm $\|\cdot\|$ introduced at an earlier stage and is also complete. Hence it is a Banach algebra.

For every $\mu\in\mathcal{M}^b_{\mathbb{C}}(G)$ one defines the *adjoint* $\mu^\sim$ by $\mu^\sim(f):=\overline{\mu(\overline{f^*})}$ for all $f\in\mathcal{K}_{\mathbb{C}}(G)$, where $f^*:=f\circ S$ $(f\in\mathcal{K}_{\mathbb{C}}(G))$. The mapping $\mu\to\mu^\sim$ from $\mathcal{M}^b_{\mathbb{C}}(G)$ into itself is an involution, and hence $\mathcal{M}^b_{\mathbb{C}}(G)$ becomes an involutive Banach algebra with unit, the *measure algebra* of G.

Clearly, $\mathcal{M}^b_+(G)$ and $\mathcal{M}^1(G)$ are semigroups in $\mathcal{M}^b(G)$. We shall study some of their properties in greater detail.

For measures $\mu,\nu\in\mathcal{M}^b_+(G)$ and Borel sets $B\in\mathfrak{B}(G)$ the useful formulae

$$\mu*\nu(B)=\int\mu(By^{-1})\,\nu(dy)=\int\nu(x^{-1}B)\,\mu(dx)$$

obtain.

1.2.2 Theorem. *Let G be a locally compact group. Then the sets $\mathcal{M}^b_+(G)$ and $\mathcal{M}^1(G)$ are $\mathcal{T}_w$-topological semigroups.*

Proof. We have to show that the mapping $(\mu,\nu)\to\mu*\nu$ from $\mathcal{M}^b_+(G)\times\mathcal{M}^b_+(G)$ into $\mathcal{M}^b_+(G)$ is $\mathcal{T}_w$-continuous. This, however, follows from Theorem 1.1.14 in the following way. By the same theorem we obtain the $\mathcal{T}_v$-continuity of the mapping $(\mu,\nu)\to\mu\otimes\nu$ from $\mathcal{M}^b_+(G)\times\mathcal{M}^b_+(G)$ into $\mathcal{M}^b_+(G\times G)$. The continuity of the multiplication in G implies the $\mathcal{T}_w$-continuity of the mapping $\mu\otimes\nu\to\mu*\nu$ from $\mathcal{M}^b_+(G\times G)$ into $\mathcal{M}^b_+(G)$. The final implication is a consequence of Theorem 1.1.9 and the formula $\|\mu\otimes\nu\|=\|\mu\|\,\|\nu\|$, which is valid for all $\mu,\nu\in\mathcal{M}^b(G)$. $\square$

Invariant Measures. We continue to assume that G is a locally compact group. For every measure $\mu\in\mathcal{M}_+(G)$ and every $x\in G$ the pairs (ε_x,μ) and (μ,ε_x) in $\mathcal{M}_+(G)\times\mathcal{M}_+(G)$ are convolvable. Let H be an arbitrary subset of G. A measure $\mu\in\mathcal{M}_+(G)$ is called *left* (right) *H-invariant* if $\varepsilon_x*\mu=\mu$ $(\mu*\varepsilon_x=\mu)$ for all $x\in H$ and *H-invariant* (without any further reference) if it is left and right H-invariant. Left (right) G-invariant measures in $\mathcal{M}_+(G)$ are said to be simply *left* (right) *invariant*. For the Borel measure corresponding to a left (right) invariant measure $\mu\in\mathcal{M}_+(G)$ we clearly have

$$\mu(x^{-1}B)=\mu(B)\quad(\mu(Bx^{-1})=\mu(B))\qquad\text{for all } B\in\mathfrak{B}(G).$$

It is well-known that on a locally compact group G there always exists a left invariant measure $\mu\neq0$, and all other left invariant measures are of the form $a\,\mu$

for some $a \in \mathbb{R}_+$. In this sense left (right) invariant measures on G are uniquely determined (within multiplication by a nonnegative factor). Left (right) invariant measures $\mu \neq 0$ on G are called left (right) *Haar measures* on G and will be denoted by $\omega_G^l(\omega_G^r)$ or briefly by ω_G, if there is no danger of confusion.

1.2.3 Properties. (1) $\mu \in \mathcal{M}_+(G)$ is left invariant iff $\mu^\sim$ is right invariant.
(2) Every left invariant measure $\mu \in \mathcal{M}_+(G)$ is right convolvable with any measure $\nu \in \mathcal{M}_+^b(G)$, and one has $\nu * \mu = \|\nu\| \mu$.
(3) Left Haar measures ω_G on G are not convolvable with themselves, unless G is compact.
(4) $\operatorname{supp}(\omega_G) = G$.
[This is a consequence of the fact that $\operatorname{supp}(\varepsilon_x * \omega_G) = x \operatorname{supp}(\omega_G)$ for all $x \in G$ (Theorem 1.2.1) and $\operatorname{supp}(\omega_G) \neq \emptyset$.]
(5) If $G = G_1 \times G_2$ is the product of two locally compact groups, then $\omega_{G_1} \otimes \omega_{G_2}$ is a Haar measure of G.
(6) $\omega_G(\{e\}) > 0$ iff G is discrete.
[If G is discrete, then $\{e\}$ is an open neighborhood of e, and $\omega_G(\{e\}) > 0$ since $\operatorname{supp}(\omega_G) = G$. Conversely, let U be compact in $\mathfrak{V}(e)$ and assume that $\omega_G(\{e\}) > 0$. Then by the (left) invariance of ω_G we obtain $\omega_G(\{x\}) = \omega_G(\{e\})$ for all $x \in G$. The number of points in U is therefore finite and no greater than $\dfrac{\omega(U)}{\omega(\{e\})}$. Hence G (being T_2) is discrete.]

The following result and its proof are basic for arithmetic in $\mathcal{M}_+^b(G)$ for a locally compact group G.

For every measure $\mu \in \mathcal{M}_+(G)$ we define the *left* (right) *invariance subgroup*

$$K(\mu) := \{x \in G : \mu = \varepsilon_x * \mu\} \quad (H(\mu) := \{x \in G : \mu = \mu * \varepsilon_x\})$$

of μ. The subgroup $I(\mu) := K(\mu) \cap H(\mu)$ is called the *invariance subgroup* of μ. For $\mu \in \mathcal{M}_+^b(G)$ the subgroups $K(\mu)$, $H(\mu)$ and $I(\mu)$ are closed in G. This follows from the (separate) continuity of the convolution operation in $\mathcal{M}_+^b(G)$. Moreover, we have

1.2.4 Theorem. *Let G be a locally compact group and let μ be in $\mathcal{M}_+^b(G)$. Then the subgroups $K(\mu)$, $H(\mu)$ and $I(\mu)$ are compact.*

Proof. It suffices to prove the compactness of $K(\mu)$: If $K(\mu)$ were not compact, then for every compact subset C of G with $\mu(C) > 0$ there would exist an $x \in K(\mu) \cap \complement C$ and thus a sequence $(x_k)_{k \geq 1}$ in $K(\mu)$ with $x_i C \cap x_j C = \emptyset$ for all $i \neq j$, $i, j \geq 1$. In fact, let $x_1, \ldots, x_{n-1} \in K(\mu)$ be constructed such that

$$x_i C \cap x_j C = \emptyset \quad \text{for } i \neq j,\ i, j = 1, \ldots, n-1.$$

Then there is an

$$x_n \in K(\mu) \cap \complement\left(\textstyle\bigcup_{i=1}^{n-1} x_i C C^{-1}\right)$$

with the property that

$$\left(\textstyle\bigcup_{i=1}^{n-1} x_i C\right) \cap x_n C = \emptyset,$$

which implies that $x_i C \cap x_n C = \varnothing$ for all $i=1,\dots,n-1$. Since the sequence $(x_k C)_{k\geq 1}$ is pairwise disjoint, one deduces the desired contradiction from

$$\|\mu\| = \mu(G) \geq \sum_{k\geq 1} \mu(x_k C) = \sum_{k\geq 1} \mu(C) = \infty. \quad \square$$

1.2.5 Corollary. *$\omega_G \in \mathcal{M}_+^b(G)$ iff G is compact.*

The *proof* is obvious, since $K(\omega_G) = G$. $\square$

If G is compact, ω_G will always be normed in such a way that it becomes an element of $\mathcal{M}^1(G)$.

1.2.6 Definition. A measure $\mu \in \mathcal{M}^1(G)$ is called the *normed Haar measure of a compact subgroup H* of G if it is H-invariant and satisfies $\operatorname{supp}(\mu) = H$.

In order to characterize the class $\mathcal{J}(G)$ of all normed Haar measures ω_H of compact subgroups H of G within the semigroup $\mathcal{M}^1(G)$ we discuss a result connected with the solution of the *convolution equation.*

1.2.7 Theorem. *Let G be a locally compact group and $\mu, \nu \in \mathcal{M}^1(G)$. The following statements are equivalent:*
(i) $\nu * \mu = \mu$ $(\mu * \nu = \mu)$;
(ii) $\operatorname{supp}(\nu) \subset K(\mu)$ $(\operatorname{supp}(\nu) \subset H(\mu))$.

The *proof* of the theorem will require the following

1.2.8 Lemma. *Let G be a locally compact group and $\mu \in \mathcal{M}_+^b(G)$. For any compact $K \subset G$ the functions ϕ_K and ${}_K\phi$ on G defined resp. by*

$$\phi_K(x) := \mu(xK) \quad \textit{and} \quad {}_K\phi(x) := \mu(Kx) \quad \textit{for all } x \in G$$

are upper semicontinuous and vanish at infinity.

Proof. We restrict ourselves to the discussion of the function $\phi := \phi_K$.

1. ϕ is upper semicontinuous. We shall show that for any $a \in \mathbb{R}_+^*$ the set $A_a := \{x \in G : \phi(x) < a\}$ is open in G. Indeed, for $x \in A_a$ one has $\mu(xK) = \phi(x) < a$. By the regularity of μ there is an open subset U of G with $U \supset xK$ satisfying $\mu(U) < a$. But there exists a symmetric open neighborhood $V \in \mathfrak{B}(e)$ such that $VxK \subset U$ holds. For all $y \in V$ we obtain

$$\mu(yxK) \leq \mu(U) < a \quad \text{or} \quad Vx \subset A_a \quad \text{for all } x \in A_a.$$

2. ϕ vanishes at infinity. For every $\varepsilon > 0$ there exists a compact subset $C := C_\varepsilon$ of G with $\mu(\complement C) < \varepsilon$. Let K be an arbitrary compact subset of G and put $K_0 := CK^{-1}$. Then for all $x \in \complement K_0$ one has $xK \cap C = \varnothing$, hence $xK \subset \complement C$, and so

$$\phi(x) = \mu(xK) \leq \mu(\complement C) < \varepsilon. \quad \square$$

Proof of the theorem. It suffices to show the implication (i)$\Rightarrow$(ii). Let $\mu, \nu \in \mathcal{M}^1(G)$ satisfy $\nu * \mu = \mu$. By the regularity of μ it remains to be shown that for any compact subset C of G and all $x \in \operatorname{supp}(\nu)$ one obtains $\mu(x^{-1}C)=\mu(C)$. For any compact subset K of G we consider the function ϕ_K on G defined in Lemma 1.2.8. We know that ϕ_K is upper semicontinuous and vanishes at infinity so that it attains its supremum in some $x_0 \in G$. This implies that $\mu(x_0 K) - \mu(yK) \geq 0$ for all $y \in G$ and in particular, $\mu(x_0 K) - \mu(x^{-1} x_0 K) \geq 0$ for all $x \in G$. The assumption (i) implies

$$\mu(x_0 K) = \int_{\operatorname{supp}(\nu)} \mu(x^{-1} x_0 K)\, \nu(dx)$$

and thus

$$\int (\mu(x_0 K) - \mu(x^{-1} x_0 K))\, \nu(dx) = 0$$

or

$$\mu(x^{-1} x_0 K) = \mu(x_0 K)$$

for ν-a.a. $x \in \operatorname{supp}(\nu)$. By the upper semicontinuity of ϕ_K this implies

$$\mu(x^{-1} x_0 K) = \mu(x_0 K) \quad \text{for all } x \in \operatorname{supp}(\nu).$$

Since the compact subset K of G was arbitrary, we obtain $\mu(x^{-1}C) = \mu(C)$ for all $x \in \operatorname{supp}(\nu)$ and every compact $C \subset G$. □

1.2.9 Corollary. *Under the condition* (i) *of the theorem the closed semigroup* $S := \langle \operatorname{supp}(\nu) \rangle^-$ *generated by* $\operatorname{supp}(\nu)$ *is a compact subgroup of* G.

Proof. By the theorem, $\operatorname{supp}(\nu) \subset K(\mu)$. Theorem 1.2.4 yields that $S := \langle \operatorname{supp}(\nu) \rangle^-$ is a compact semigroup in G. But as a compact semigroup embedded in a group, S turns out to be a compact subgroup in G. Indeed: We have to show that $x \in S$ implies $x^{-1} \in S$. Let $x \in S$. Since S is compact, the sequence $(x^k)_{k \geq 1}$ admits an accumulation point $x_0 \in S$. Let $U, V \in \mathfrak{V}(e)$ with $VV^{-1} \subset U$. By $x^n, x^m \in Vx_0$ for sufficiently large $n, m \geq 1$ with $m \geq n+2$ we conclude $x^{m-n} \in (Vx_0)(Vx_0)^{-1}$ and hence $x^{m-n-1} \in Ux^{-1}$, thus $x^{-1} \in S$, since S is closed. □

1.2.10 Theorem. *Let G be a locally compact group. The elements of* $\mathscr{J}(G)$ *are precisely the idempotents in* $\mathscr{M}_+^b(G) \setminus \{0\}$.

Proof. It suffices to show that the measures of $\mathscr{J}(G)$ are precisely the idempotents of $\mathscr{M}^1(G)$.

1. If $\mu \in \mathscr{M}^1(G)$ is idempotent, i.e., if $\mu^2 = \mu$ holds, then by the support formula of Theorem 1.2.1 we obtain $\operatorname{supp}(\mu)^2 \subset \operatorname{supp}(\mu)$. Since $\operatorname{supp}(\mu) \neq \emptyset$ one gets that $\operatorname{supp}(\mu)$ is a semigroup in G and hence

$$S := \langle \operatorname{supp}(\mu) \rangle^- = \operatorname{supp}(\mu).$$

But Corollary 1.2.9 implies that $\operatorname{supp}(\mu)$ is a compact subgroup H of G. Since by Theorem 1.2.7 $\mu \in \mathscr{M}^1(G)$ is also H-invariant, the result $\mu = \omega_H \in \mathscr{M}^1(G)$ follows.

2. That any normed Haar measure ω_H of a compact subgroup H of G is an idempotent in $\mathscr{M}^1(G)$ follows directly from the definition. □

1.2.11 Remark. Measures $\mu\in\mathcal{M}^1(G)$ satisfying the convolution equation (i) of Theorem 1.2.7 are in general not normed Haar measures ω_H of a compact subgroup H of G. In fact, they are just H-invariant for the compact subgroup H generated by $\operatorname{supp}(\nu)$, but not supported by it. As an example one chooses $G:=\mathbb{T}\times\mathbb{R}^d$ for $d\geq 1$ and considers the translation ω_0 by $x_0\in\mathbb{R}^d$ of $\omega_{\mathbb{T}}$ as a measure on $\mathbb{T}+x_0$. Obviously $\omega_0=\omega_{\mathbb{T}}*\omega_0$ and ω_0 is $\mathbb{T}$-invariant, but $\operatorname{supp}(\omega_0)\neq\mathbb{T}$.

We add to the discussion of the convolution equation a generalization to unbounded measures of Theorem 1.2.7 for the special case of a locally compact Abelian group.

1.2.12 Theorem. *Let G be a locally compact Abelian group, $\mu\in\mathcal{M}_+(G)$ and $\nu\in\mathcal{M}^1(G)$ such that the closed subgroup* $[\operatorname{supp}(\nu)]^-$ *generated by* $\operatorname{supp}(\nu)$ *equals G. The following statements are equivalent:*

(i) (a) $\nu*\mu=\mu$ *and*

(b) $\{\varepsilon_x*\mu : x\in G\}$ *is $\mathcal{T}_v$-relatively compact;*

(ii) $\operatorname{supp}(\nu)\subset K(\mu)=H(\mu)$.

Proof. Again it suffices to verify the implication (i) $\Rightarrow$ (ii). For any $\phi\in\mathcal{K}(G)$ we consider the function $f:=f_\phi$ on G defined by $f(x):=\int\phi(xy)\,\mu(dy)$ for all $x\in G$. Clearly, f is uniformly continuous with respect to the uniform structure of G, and it is bounded by Condition (i)(b). Moreover, f satisfies the integral equation

(2.1) $$f(x)=\int f(xz)\,\nu(dz)\qquad(\text{for all } x\in G).$$

Let $a\in\operatorname{supp}(\nu)$. Then the functions ${}_af$ and $g:={}_af-f$ are uniformly continuous, bounded and satisfy the integral equation (2.1). Let $c\in\mathbb{R}_+$ with $|g|\leq c$ and $\gamma:=\frac{1}{2}\sup_{x\in G}g(x)$. There exists a sequence $(x_n)_{n\geq1}$ in G satisfying

$$\lim_{n\to\infty}g(x_n)=2\gamma.$$

We put $g_n:={}_{x_n}g$ for all $n\geq1$. Then the set $\{g_n : n\in\mathbb{N}\}$ is equicontinuous since g is uniformly continuous. Moreover, $|g_n|\leq c$ for all $n\geq1$. Hence, the Arzela-Ascoli theorem implies the existence of a subnet $(g_{n_\alpha})_{\alpha\in\mathbf{A}}$ of $(g_n)_{n\geq1}$ which converges to a function h on G uniformly on the compact subsets of G. Plainly h is continuous and fulfills $|h|\leq c$. Furthermore, h satisfies the integral equation (2.1). But then

$$2\gamma=\lim_{\alpha\in\mathbf{A}}g(x_{n_\alpha})=\lim_{\alpha\in\mathbf{A}}g_{n_\alpha}(e)=h(e)=\int h(z)\,\nu(dz)=\int h\,d\nu.$$

From $h\leq2\gamma$ one concludes that $h(x)=2\gamma$ for all $x\in\operatorname{supp}(\nu)$. Given $x\in\operatorname{supp}(\nu)$ we further obtain $2\gamma=h(x)=\int h(xz)\,\nu(dz)$ and thus $h(xz)=2\gamma$ for all $z\in\operatorname{supp}(\nu)$. An iteration procedure yields $h(x)=2\gamma$ for all $x\in\langle\operatorname{supp}(\nu)\rangle^-$ and so $h(a^k)=2\gamma$ for all $a\in\operatorname{supp}(\nu)$ and $k\geq1$. For every $l\geq1$ there exists an $\alpha\in\mathbf{A}$ with $g_{n_\alpha}(a^k)>\gamma$ for all $k=1,\dots,l$, if $\gamma>0$.

We now have $g_{n_\alpha}(a^k)=f(a\,x_{n_\alpha}a^k)-f(x_{n_\alpha}a^k)=f(x_{n_\alpha}a^{k+1})-f(x_{n_\alpha}a^k)$ for all $k=1,\dots,l$ and after summation for $k=1,\dots,l$, $f(x_{n_\alpha}a^{l+1})-f(x_{n_\alpha}a)>l\gamma$. Since f is bounded, this yields a contradiction. Hence, $\gamma\leq0$, i.e., $g\leq0$. Replacing g by $-g$ one gets $-g\leq0$, and thus altogether $g=0$, i.e. $f={}_af$ for all $a\in\operatorname{supp}(\nu)$, from which follows

$$\int \phi(a\,y)\,\mu(dy) = \int \phi(y)\,\mu(dy)$$

and finally $\varepsilon_a * \mu = \mu$ for all $a \in \operatorname{supp}(\nu)$. □

Let G be a locally compact group and H a compact subgroup of G. We define the sets

$$\mathcal{M}_H^b(G) := \{\mu \in \mathcal{M}^b(G) : \mu * \omega_H = \omega_H * \mu = \mu\}$$

and

$$\mathcal{M}_H^1(G) = \mathcal{M}_H^b(G) \cap \mathcal{M}^1(G).$$

They are closed subsemigroups of $\mathcal{M}^b(G)$ and $\mathcal{M}^1(G)$ resp.

The following result needed in the future discussion concerns the arithmetic in $\mathcal{M}^1(G)$.

1.2.13 Theorem. *Let G be a locally compact group, H a compact subgroup of G and $\mu \in \mathcal{M}_H^1(G)$, $\nu \in \mathcal{M}^1(G)$ such that $\mu * \nu = \omega_H$ holds. Then there exists an $x \in G$ with $x H x^{-1} \subset H$ such that $\mu = \omega_H * \varepsilon_x$ holds. If, in addition, $\nu \in \mathcal{M}_H^1(G)$, then $\nu = \varepsilon_{x^{-1}} * \omega_H$ and $x \in N(H)$.*

Proof. We restrict ourselves to the proof of the first statement. From the support formula in Theorem 1.2.1 we deduce that $\operatorname{supp}(\mu)\operatorname{supp}(\nu) \subseteq H$. Hence, there exists a $z \in \operatorname{supp}(\nu)$ satisfying $\operatorname{supp}(\mu)\, z \subset H$, and $\operatorname{supp}(\mu)$ turns out to be compact. Consequently, $\operatorname{supp}(\mu)\operatorname{supp}(\nu) = H$, and there is an $x \in \operatorname{supp}(\mu)$ such that $x^{-1} \in \operatorname{supp}(\nu)$ and $\operatorname{supp}(\mu) \subseteq Hx$ hold. Using the fact that $\mu \in \mathcal{M}_H^1(G)$ one concludes $\omega_H = \omega_H * \mu * \varepsilon_{x^{-1}}$ so that $\mu = \omega_H * \varepsilon_x$.

From $\omega_H * \varepsilon_x = \mu = \mu * \omega_H = \omega_H * \varepsilon_x * \omega_H$ follows $Hx = HxH$. For $y, z \in H$ there exists a $w \in H$ such that $y x z = w x$, whence $x z x^{-1} = y^{-1} w \in H$ or $x H x^{-1} \subset H$. □

1.2.14 Remark. Theorem 1.2.13 implies that solutions $\mu, \nu \in \mathcal{M}_H^1(G)$ of the equation $\mu * \nu = \omega_H$ are necessarily *trivial* in the sense that they are translates of ω_H. The following example shows that there exist nontrivial solutions of the equation which do not belong to $\mathcal{M}_H^1(G)$. One just considers the permutation group

$$G := \mathfrak{S}_3 := \{e, x_1, x_2, \ldots, x_5\}$$

of three objects with neutral element e,

$$x_1 = (2,3,1), \quad x_2 = (3,2,1), \quad x_3 = (1,2),$$
$$x_4 = (2,3) \text{ and } x_5 = (1,3).$$

Considering the measure

$$\mu := \tfrac{1}{6}[(\varepsilon_{x_1} + \varepsilon_{x_3}) - (\varepsilon_{x_2} + \varepsilon_{x_4})] \in \mathcal{M}^b(G)$$

one obtains $\mu^2 = 0$. Hence, the measure $\mu_1 := \omega_G + \mu \in \mathcal{M}^1(G)$ satisfies $\mu_1^2 = \omega_G$, but $I(\mu_1) \neq G$ and $\operatorname{supp}(\mu_1) \neq G$.

1.2.15 Theorem. *Let G be a locally compact group, H a closed subgroup of G and p the canonical homomorphism from G onto $\dot{G} := G/H$ which induces a continuous (semigroup) homomorphism from $\mathscr{M}^1(G)$ into $\mathscr{M}^1(\dot{G})$, which we shall denote again by p. Then*

(i) *p is a surjection from $\mathscr{M}^1(G)$ onto $\mathscr{M}^1(\dot{G})$.*

If, in addition, H is a compact subgroup of G, then

(ii) *for any relatively compact subset $\mathscr{N}$ of $\mathscr{M}^1(\dot{G})$ the set $p^{-1}(\mathscr{N})$ is relatively compact in $\mathscr{M}^1(G)$ (with respect to the vague topology in $\mathscr{M}^1(\dot{G})$ and $\mathscr{M}^1(G)$ resp.) and*

(iii) $\mathrm{Res}_{\mathscr{M}^1_H(G)}\, p$ *is a topological isomorphism from $\mathscr{M}^1_H(G)$ onto $\mathscr{M}^1(\dot{G})$.*

Proof. (i)(a) Let $\dot{\mu} \in \mathscr{M}^b_+(\dot{G})$ be a measure with finite support, i.e., of the form

$$\dot{\mu} := \sum_{j=1}^n a_j \varepsilon_{\dot{x}_j} \quad \text{for } \dot{x}_j \in \dot{G} \text{ and } a_j \in \mathbb{R}^*_+ \ (j=1, \dots, n).$$

Choosing $x_j \in G$ with $p(x_j) = \dot{x}_j$ $(j=1, \dots, n)$ and putting $\mu := \sum_{j=1}^n a_j \varepsilon_{x_j}$ one observes that $\mu \in \mathscr{M}^b_+(G)$ and $p(\mu) = \dot{\mu}$.

(b) Let $\dot{\mu} \in \mathscr{M}^b_+(\dot{G})$ be a measure with compact support $\dot{K}$. Then by Theorem 1.1.7 there exists a net $(\dot{\mu}_\alpha)_{\alpha \in \mathbb{A}}$ of measures in $\mathscr{M}^b_+(\dot{G})$ with finite $\mathrm{supp}(\dot{\mu}_\alpha)$ in $\dot{K}$ and $\|\dot{\mu}_\alpha\| = \|\dot{\mu}\|$ $(\alpha \in \mathbb{A})$ satisfying $\mathscr{T}_v\text{-}\lim_{\alpha \in \mathbb{A}} \dot{\mu}_\alpha = \dot{\mu}$.

Let K be a compact subset of G with $p(K) = \dot{K}$. By (a), for every $\alpha \in \mathbb{A}$ there is a $\mu_\alpha \in \mathscr{M}^b_+(G)$ with $p(\mu_\alpha) = \dot{\mu}_\alpha$ and $\mathrm{supp}(\mu_\alpha) \subset K$. Moreover, $\|\mu_\alpha\| = \|\dot{\mu}_\alpha\| = \|\dot{\mu}\|$ holds for all $\alpha \in \mathbb{A}$. Hence, Prohorov's theorem implies that $\{\mu_\alpha : \alpha \in \mathbb{A})$ is $\mathscr{T}_w$-relatively compact so that there exists a subnet $(\mu_{\alpha(\beta)})_{\beta \in \mathbb{B}}$ of $(\mu_\alpha)_{\alpha \in \mathbb{A}}$ with $\mathscr{T}_w - \lim_{\beta \in \mathbb{B}} \mu_{\alpha(\beta)} =: \mu \in \mathscr{M}^b_+(G)$.

Consequently, $p(\mu) = \mathscr{T}_v\text{-}\lim_{\beta \in \mathbb{B}} p(\mu_{\alpha(\beta)}) = \mathscr{T}_v\text{-}\lim_{\beta \in \mathbb{B}} \dot{\mu}_{\alpha(\beta)} = \dot{\mu}$, i.e., $\dot{\mu}$ admits a pre-image under p.

(c) Let $\dot{\mu} \in \mathscr{M}^1(\dot{G})$ be arbitrary. There exists an increasing sequence $(\dot{K}_n)_{n \geq 1}$ of compact subsets $\dot{K}_n$ of $\dot{G}$ satisfying $\dot{\mu}(\dot{K}_n) \geq 1 - \dfrac{1}{n}$ $(n \geq 1)$. Defining $\dot{\mu}_n := 1_{\dot{K}_n} \cdot \dot{\mu}$ for all $n \geq 1$ one observes $\dot{\mu}_n \leq \dot{\mu}_{n+1} \leq \dot{\mu}$ for all $n \geq 1$ and obtains $\mathscr{T}_w\text{-}\lim_{n \to \infty} \dot{\mu}_n = \dot{\mu}$.

Let $\dot{\mu}_0 := 0$ and $\dot{\nu}_n := \dot{\mu}_n - \dot{\mu}_{n-1} = 1_{\dot{K}_n \setminus \dot{K}_{n-1}} \cdot \dot{\mu} \in \mathscr{M}^b_+(\dot{G})$ for $n \geq 1$. Then

$$\sum_{j=1}^n \dot{\nu}_j = \dot{\mu}_n.$$

But $\mathrm{supp}(\dot{\nu}_n)$ is compact since $\dot{\nu}_n(\dot{G} \setminus \dot{K}_n) = 0$ $(n \geq 1)$. Hence, by (b) to any $n \geq 1$ there exists a $\nu_n \in \mathscr{M}^b_+(G)$ with $p(\nu_n) = \dot{\nu}_n$. Defining $\mu_n := \sum_{j=1}^n \nu_j$ one obtains for any given $\varepsilon > 0$ an $n_\varepsilon \geq 1$ such that for all $m, n \geq 1$ with $n_\varepsilon < m < n$ the inequality $\sum_{j=m}^n \dot{\nu}_j(1) \leq \varepsilon$ holds. From $\nu_j(1) = p(\nu_j)(1) = \dot{\nu}_j(1)$ we deduce that

$$\Big\|\sum_{j=m}^n \nu_j\Big\| = \sum_{j=m}^n \|\nu_j\| = \sum_{j=m}^n \nu_j(1) = \sum_{j=m}^n \dot{\nu}_j(1) \leq \varepsilon$$

or $(\mu_n)_{n \geq 1}$ is a Cauchy sequence in $\mathscr{M}^b(G)$. Thus $\mathscr{T}_w\text{-}\lim_{n \to \infty} \mu_n = \mu \in \mathscr{M}^b_+(G)$, and one has

$$\begin{aligned} p(\mu) &= \mathscr{T}_w\text{-}\lim p(\mu_n) = \mathscr{T}_w\text{-}\lim_{n \to \infty} \sum_{j=1}^n p(\nu_j) \\ &= \mathscr{T}_w\text{-}\lim_{n \to \infty} \sum_{j=1}^n \dot{\nu}_j = \mathscr{T}_w\text{-}\lim_{n \to \infty} \dot{\mu}_n = \dot{\mu}. \end{aligned}$$

Since $\mu(1) = \dot{\mu}(1) = 1$, μ is a pre-image of $\dot{\mu}$ under p in $\mathscr{M}^1(G)$.

(ii) Let $\mathcal{N}$ be a relatively compact subset of $\mathcal{M}^1(\dot{G})$. By Prohorov's theorem $\mathcal{N}$ is uniformly tight, and for every $\varepsilon>0$ there exists a compact subset $\dot{K}:=\dot{K}_\varepsilon$ of $\dot{G}$ with $\dot{\mu}(\complement\dot{K})\leq\varepsilon$ for all $\dot{\mu}\in\mathcal{N}$. $C:=p^{-1}(\dot{K})$ is a compact subset of G. Let $\mu\in p^{-1}(\mathcal{N})$ or $p(\mu)\in\mathcal{N}$. Then

$$\mu(\complement C)=\mu(p^{-1}(\complement\dot{K}))=p(\mu)(\complement\dot{K})\leq\varepsilon.$$

Thus $p^{-1}(\mathcal{N})$ is uniformly tight and, again by Prohorov's theorem, it is relatively compact in $\mathcal{M}^1(G)$.

(iii) Let q denote the restriction $\operatorname{Res}_{\mathcal{M}^1_H(G)}p$ of p to the closed subsemigroup $\mathcal{M}^1_H(G)$ of $\mathcal{M}^1(G)$. q is a continuous homomorphism from $\mathcal{M}^1_H(G)$ into $\mathcal{M}^1(\dot{G})$.

Furthermore, q is surjective. Indeed, for $\dot{\mu}\in\mathcal{M}^1(\dot{G})$ there is by (i) a measure $\mu\in\mathcal{M}^1(G)$ with $p(\mu)=\dot{\mu}$. Clearly, $\omega_H*\mu*\omega_H\in\mathcal{M}^1_H(G)$ and $q(\omega_H*\mu*\omega_H)=\dot{\mu}$. For $f\in\mathscr{C}^b(G)$ we define $\bar{f}\in\mathscr{C}^b(\dot{G})$ by

$$\bar{f}(p(x)):=\int f(x\,h)\,\omega_H(dh)\qquad\text{for all } x\in G.$$

We note that for any $\mu\in\mathcal{M}^1_H(G)$ and $f\in\mathscr{C}^b(G)$ one has $\mu(f)=q(\mu)(\bar{f})$, which follows from

$$\begin{aligned}\mu(f)&=\mu*\omega_H(f)=\iint f(x\,h)\,\mu(dx)\,\omega_H(dh)=\int\Big(\int f(x\,h)\,\omega_H(dh)\Big)\,\mu(dx)\\&=\int\bar{f}(p(x))\,\mu(dx)=p(\mu)(\bar{f})=q(\mu)(\bar{f}).\end{aligned}$$

q is also injective. To see this note that for $\mu,\nu\in\mathcal{M}^1_H(G)$ with $q(\mu)=q(\nu)$ the above formula implies

$$\mu(f)=q(\mu)(\bar{f})=q(\nu)(\bar{f})=\nu(f)\qquad\text{for all } f\in\mathscr{C}^b(G),$$

hence $\mu=\nu$. Finally, q^{-1} is continuous. Indeed, let $(\mu_\alpha)_{\alpha\in\mathbf{A}}$ be a net in $\mathcal{M}^1_H(G)$ and $\mu\in\mathcal{M}^1_H(G)$ with $\mathscr{T}_v\text{-}\lim_{\alpha\in\mathbf{A}}q(\mu_\alpha)=q(\mu)$. Then for every $f\in\mathscr{C}^b(G)$ one obtains

$$\mu(f)=q(\mu)(\bar{f})=\lim_{\alpha\in\mathbf{A}}q(\mu_\alpha)(\bar{f})=\lim_{\alpha\in\mathbf{A}}\mu_\alpha(f),$$

hence $\mathscr{T}_v\text{-}\lim_{\alpha\in\mathbf{A}}\mu_\alpha=\mu$. □

Let G be a locally compact group which is the projective limit of a projective system $(G_\alpha,p_{\alpha\beta},\mathbf{A})$ of groups $G_\alpha:=G/K_\alpha$ $(\alpha\in\mathbf{A})$, where $(K_\alpha)_{\alpha\in\mathbf{A}}$ is a descending system of compact normal subgroups of G satisfying $\bigcap_{\alpha\in\mathbf{A}}K_\alpha=\{e\}$.

1.2.16 Definition. Given a family $(\mu_\alpha)_{\alpha\in\mathbf{A}}$ of measures $\mu_\alpha\in\mathcal{M}_+(G_\alpha)$ we call the triplet $(\mu_\alpha,p_{\alpha\beta},\mathbf{A})$ a *projective system of measures* if $p_{\beta\alpha}(\mu_\alpha)=\mu_\beta$ for all $\alpha>\beta$. Any measure $\mu\in\mathcal{M}_+(G)$ satisfying $p_\alpha(\mu)=\mu_\alpha$ for all $\alpha\in\mathbf{A}$ is said to be a *projective limit* of the projective system $(\mu_\alpha,p_{\alpha\beta},\mathbf{A})$.

1.2.17 Theorem. *Let G be a locally compact group which is the projective limit of the projective system $(G_\alpha,p_{\alpha\beta},\mathbf{A})$ above and let $(\mu_\alpha,p_{\alpha\beta},\mathbf{A})$ be a projective system of measures $\mu_\alpha\in\mathcal{M}_+(G_\alpha)$. Then there exists exactly one projective limit $\varprojlim_{\alpha\in\mathbf{A}}\mu_\alpha$ of the projective system $(\mu_\alpha,p_{\alpha\beta},\mathbf{A})$.*

The *proof* will require the following

1.2.18 Lemma. *Let G be the locally compact group* $\varprojlim_{\alpha\in\mathbb{A}} G_\alpha$ *of the theorem,* $f\in\mathscr{K}_+(G)$, *S a compact subset of G with* $\operatorname{supp}(f)\subset S$, *U an open neighborhood of S in G and* $\varepsilon>0$. *Then there exist an* $\alpha\in\mathbb{A}$ *and a function* $g\in\mathscr{K}_+(G)$ *with the following properties:*

(i) $g(\complement U)=0$,
(ii) *g is constant on the (left) cosets of* K_α *and*
(iii) $|f-g|\le\varepsilon$.

Proof. Since there exists an index $\alpha_0\in\mathbb{A}$ with $K_{\alpha_0}S\cap K_{\alpha_0}\complement U=\varnothing$, we may without loss of generality suppose that S and U are unions of cosets of K_{α_0}.

Let $\mathscr{A}$ be the set of all continuous numerical functions h on S admitting the property that there exists an $\alpha\ge\alpha_0$ such that h is constant on the cosets of K_α. Then $\mathscr{A}$ is a subalgebra of $\mathscr{K}(S)$ which contains the constants and separates the points of S. The separation property can be seen as follows: Let $x,y\in S$ with $x\neq y$. Since $\bigcap_{\alpha\in\mathbb{A}}K_\alpha=\{e\}$, there exists an $\alpha\ge\alpha_0$ satisfying $p_\alpha(x)\neq p_\alpha(y)$, and hence a continuous numerical function u on $p_\alpha(S)$ with $u(p_\alpha(x))\neq u(p_\alpha(y))$. By the Stone-Weierstrass theorem there exist an $\alpha\ge\alpha_0$ and a continuous function $h:S\to\overline{\mathbb{R}}_+$, constant on the cosets of K_α, such that $|f-h|\le\frac{\varepsilon}{2}$ holds (on S). For $t\in\mathbb{R}$ we define $\delta(t):=\left(t-\frac{\varepsilon}{2}\right)^+$, and we put $h':=\delta\circ h$. Then $h':S\to\overline{\mathbb{R}}_+$ is continuous and constant on the cosets of K_α, and it satisfies $|h-h'|\le\frac{\varepsilon}{2}$ on S, and hence $|f-h'|\le\varepsilon$ on S. On the other hand, $h'(\partial S)=0$ since $h(\partial S)\le\frac{\varepsilon}{2}$.

Putting $h'(\complement S)=0$ one obtains a function $g\in\mathscr{K}_+(G)$ fulfilling all the requirements of the lemma. □

Proof of Theorem 1.2.17. 1. In order to show the uniqueness of $\varprojlim_{\alpha\in\mathbb{A}}\mu_\alpha$ we show more generally that whenever μ,μ' are measures in $\mathscr{M}_+(G)$ satisfying $p_\alpha(\mu)=p_\alpha(\mu')$ for all $\alpha\in\mathbb{A}$, then $\mu=\mu'$. In fact, let μ,μ' be as assumed and let $v\in\mathscr{K}(G)$ be a function constant on the cosets of K_α for some $\alpha\in\mathbb{A}$. Then there exists a $w\in\mathscr{K}(G_\alpha)$ with $v=w\circ p_\alpha$. But this implies that

$$\mu(v)=(p_\alpha(\mu))(w)=(p_\alpha(\mu'))(w)=\mu'(v).$$

Lemma 1.2.18 then yields $\mu(f)=\mu'(f)$ for all $f\in\mathscr{K}(G)$ or $\mu=\mu'$.

2. We now prove the existence of $\varprojlim_{\alpha\in\mathbb{A}}\mu_\alpha$. Let $\mathscr{V}$ be the linear subspace of $\mathscr{K}(G)$ containing the functions which are constant on cosets of K_α for some $\alpha\in\mathbb{A}$. By Lemma 1.2.18 $\mathscr{V}$ is a positive rich linear subspace of $\mathscr{K}(G)$. Let $f\in\mathscr{V}$. Then there exists an $\alpha\in\mathbb{A}$ such that f is constant on the cosets of K_α. Hence, there is a function $f_\alpha\in\mathscr{K}(G_\alpha)$ with $f=f_\alpha\circ p_\alpha$. The number $\mu(f):=\mu_\alpha(f_\alpha)$ is independent of the choice of α. In fact, let $\beta\in\mathbb{A}$ be such that f is constant on the cosets of K_β, and let $\gamma\in\mathbb{A}$ satisfy $\gamma>\alpha,\beta$. Then f defines functions

$$f_\beta\in\mathscr{K}(G_\beta)\quad\text{and}\quad f_\gamma\in\mathscr{K}(G_\gamma)\ \text{with}\ f=f_\beta\circ p_\beta=f_\gamma\circ p_\gamma.$$

One has $f_\alpha \circ p_{\alpha\gamma} = f_\gamma$ and hence

$$\mu_\gamma(f_\gamma) = (p_{\alpha\gamma}(\mu_\gamma))(f_\alpha) = \mu_\alpha(f_\alpha).$$

Similarly, one obtains $\mu_\gamma(f_\gamma) = \mu_\beta(f_\beta)$ and the independence of $\mu(f)$ is established.

Plainly, μ is a positive linear functional on $\mathscr{V}$ and thus extends to a measure $\mu \in \mathscr{M}_+(G)$. By construction we get $p_\alpha(\mu) = \mu_\alpha$ for every $\alpha \in \mathbb{A}$, so $\mu = \varprojlim_{\alpha \in \mathbb{A}} \mu_\alpha$. □

We continue with a few remarks on unimodular groups. Given a locally compact group G and a left Haar measure $\omega_G \in \mathscr{M}_+(G)$ one observes the equalities

$$\varepsilon_y * (\omega_G * \varepsilon_{x^{-1}}) = (\varepsilon_y * \omega_G) * \varepsilon_{x^{-1}} = \omega_G * \varepsilon_{x^{-1}}$$

which are valid for all $x, y \in G$.

It follows that for every $x \in G$ the measure $\omega_G * \varepsilon_{x^{-1}}$ is a left Haar measure on G. Hence, by the existence and uniqueness of the Haar measure there exists $\Delta(x) \in \mathbb{R}_+^*$ such that $\omega_G * \varepsilon_{x^{-1}} = \Delta(x)\,\omega_G$ holds. Evidently $\Delta(x)$ is independent of the special choice of the Haar measure ω_G. It turns out that the mapping $x \to \Delta_G(x) := \Delta(x)$ is a continuous homomorphism from G onto $\mathbb{R}_+^*$, the *modular function* of G. A locally compact group G is said to be *unimodular* if $\Delta_G = 1$. For unimodular groups left and right Haar measures coincide, and they are simply called *Haar measures* on G. If G admits a compact neighborhood $U \in \mathfrak{V}(e)$ which is (inner) invariant, then G is unimodular. Indeed, for such a group G and a neighborhood $U \in \mathfrak{V}(e)$ we have

$$\omega_G(U) = \omega_G(x^{-1} U x) = \Delta_G(x)\,\omega_G(U) \quad \text{for all } x \in G,$$

which by $0 < \omega_G(U) < \infty$ implies the assertion.

Consequently, all discrete groups, the compact groups and the Abelian locally compact groups are unimodular. More generally, one can show that every almost periodic (locally compact) group is unimodular. Finally, we note that for unimodular groups G and $a \in G$ one has

$$\omega_G({}_a f) = \omega_G(f_a) = \omega_G(f^*) = \omega_G(f) \quad \text{for all } f \in \mathscr{K}(G)$$

or

$$\omega_G(aB) = \omega_G(Ba) = \omega_G(B^{-1}) = \omega_G(B) \quad \text{whenever } B \in \mathfrak{B}(G).$$

Convolution with Functions. Let G be a locally compact group and $\omega := \omega_G$ a left Haar measure on G. Furthermore, assume given a measure $\mu \in \mathscr{M}_{\mathbb{C}}(G)$ and a complex-valued function f on G that is locally ω-integrable. Then the convolvability of the pair $(\mu, f \cdot \omega) \in \mathscr{M}_{\mathbb{C}}(G) \times \mathscr{M}_{\mathbb{C}}(G)$ is equivalent to the properties that there exists an ω-negligeable subset N of G such that the function $y \to {}_{y^{-1}}f(x)$ is ω-integrable for all $x \in \complement N$ and that the function $x \to \int |{}_{y^{-1}}f(x)|\,|\mu|(dy)$ (defined ω-a.e. in G) is locally ω-integrable.

In this case the function $g: x \to \int {}_{y^{-1}}f(x)\,\mu(dy)$ (defined ω-a.e. on G) is ω-integrable, and one has $\mu * (f \cdot \omega) = g$. One says that the measure $\mu \in \mathscr{M}_{\mathbb{C}}(G)$ and the function f on G are *convolvable*. Every function ω-a.e. equal to the function g above is called the *convolution of* μ *with* f and is denoted by $\mu * f$. Clearly one has for ω-a.a. $x \in G$ the formula

$$\mu * f(x) = \int {}_{y^{-1}}f(x)\, \mu(dy).$$

If, in particular, one representative of the ω-equivalence class of g is continuous, then it is the only such function in the class, and is denoted by $\mu * f$. In an analogous way one defines convolvability of the pair $(f \cdot \omega, \mu) \in \mathcal{M}_{\mathbb{C}}(G) \times \mathcal{M}_{\mathbb{C}}(G)$ and introduces the class $f * \mu = (f \cdot \omega) * \mu$ for ω-a.a. $x \in G$ by putting

$$f * \mu(x) = \int f(y^{-1}x)\, \Delta(y^{-1})\, \mu(dy),$$

where $\Delta := \Delta_G$ denotes the modular function of G.

We list the results of some special cases as

1.2.19 Properties. (1) If $\mu \in \mathcal{M}_{\mathbb{C}}(G)$ has compact support and f on G is locally ω-integrable, then μ and f are convolvable.

(2) If, in addition to (1) one has given $f \in \mathcal{C}_{\mathbb{C}}(G)$ or $f \in \mathcal{K}_{\mathbb{C}}(G)$, then $\mu * f$ is defined on all of G and lies in $\mathcal{C}_{\mathbb{C}}(G)$ or $\mathcal{K}_{\mathbb{C}}(G)$ resp.

(3) If $\mu \in \mathcal{M}^b_{\mathbb{C}}(G)$ and $f \in \mathcal{L}^p_{\mathbb{C}}(G, \omega)$ (for $p \in [1, \infty]$), then $\mu * f \in \mathcal{L}^p_{\mathbb{C}}(G, \omega)$, and for the $\mathcal{L}^p$-seminorm N_p we have $N_p(\mu * f) \leq \|\mu\|\, N_p(f)$.

(4) If $\mu \in \mathcal{M}^b_{\mathbb{C}}(G)$ and $f \in \mathcal{C}^b_{\mathbb{C}}(G)$, then $\mu * f$ is defined on all of G and $\mu * f \in \mathcal{C}^b_{\mathbb{C}}(G)$.

Plainly $\mathcal{L}^p_{\mathbb{C}}(G, \omega)$ is a left module on $\mathcal{M}^b_{\mathbb{C}}(G)$, and if G is unimodular, also a right module on $\mathcal{M}^b_{\mathbb{C}}(G)$ $(p \in [1, \infty])$.

Let f and g be locally ω-integrable complex functions on G. For the measures $f \cdot \omega$ and $g \cdot \omega$ to be convolvable, it is necessary and sufficient that there exists an ω-negligeable subset N of G such that the function $y \rightarrow {}_{y^{-1}}g(x)\, f(y)$ is ω-integrable for all $x \in \complement N$ and that the function

$$x \rightarrow \int |{}_{y^{-1}}g(x)\, f(y)|\, \omega(dy)$$

defined ω-a.e. on G is locally ω-integrable. In this case the function

$$x \rightarrow h(x) := \int {}_{y^{-1}}g(x)\, f(y)\, \omega(dy)$$

is defined ω-a.e. on G, it is locally ω-integrable and $(f \cdot \omega) * (g \cdot \omega) = h \cdot \omega$.

One says that f and g are *convolvable*, and every function ω-a.e. equal to h is called the *convolution* $f * g$ *of* f *and* g. Clearly, for ω-a.a. $x \in G$ we obtain

$$(f * g)(x) = \int g(y^{-1}x)\, f(y)\, \omega(dy) = \int f(xy^{-1})\, g(y)\, \Delta(y^{-1})\, \omega(dy).$$

The properties 1.2.19 extend to the present framework in a natural way. In particular, one gets for $p = 1, 2$ or ∞ and $f \in \mathcal{L}^1_{\mathbb{C}}(G, \omega)$, $g \in \mathcal{L}^p_{\mathbb{C}}(G, \omega)$ that

$$f * g \in \mathcal{L}^p_{\mathbb{C}}(G, \omega) \quad \text{and that} \quad N_p(f * g) \leq N_1(f)\, N_p(g)$$

holds.

It can be shown that after identification of ω-a.e. equal functions $\mathcal{L}^1_{\mathbb{C}}(G, \omega)$ becomes a closed two-sided ideal $L^1_{\mathbb{C}}(G, \omega)$ of the Banach algebra $\mathcal{M}^b_{\mathbb{C}}(G)$. The ideal $L^1_{\mathbb{C}}(G, \omega)$ can be identified with the space of all measures in $\mathcal{M}^b_{\mathbb{C}}(G)$ which are absolutely ω-continuous since for every $f \in L^1_{\mathbb{C}}(G, \omega)$ one has $N_1(f) = \|f \cdot \omega\|$.

$L^1_{\mathbb{C}}(G,\omega)$ is called the *group algebra* of G. Clearly, $L^1_{\mathbb{C}}(G,\omega)$ is commutative iff G is Abelian and one shows that $L^1_{\mathbb{C}}(G,\omega)$ coincides with $\mathscr{M}^b_{\mathbb{C}}(G)$ iff G is discrete.

Tight Nets of Measures. Let E be a locally compact space and $(\mu_\alpha)_{\alpha\in\mathbf{A}}$ a net in $\mathscr{M}^b_+(E)$ with $\overline{\lim}_{\alpha\in\mathbf{A}}\|\mu_\alpha\|<\infty$. $(\mu_\alpha)_{\alpha\in\mathbf{A}}$ is said to be a *tight net* if for every $\varepsilon>0$ there exist a compact subset $K:=K_\varepsilon$ of E and an $\alpha_0:=\alpha_0(\varepsilon)\in\mathbf{A}$ such that $\mu_\alpha(\complement K)\leq\varepsilon$ holds for all $\alpha>\alpha_0$. Obviously every uniformly tight net in $\mathscr{M}^b_+(E)$ (in the sense of Section 1) is a tight net. $(\mu_\alpha)_{\alpha\in\mathbf{A}}$ is called a $\mathscr{T}_w$-*compact net* if every subnet of $(\mu_\alpha)_{\alpha\in\mathbf{A}}$ admits a $\mathscr{T}_w$-convergent subnet. Plainly $(\mu_\alpha)_{\alpha\in\mathbf{A}}$ is a tight net iff $\inf\{\overline{\lim}_{\alpha\in\mathbf{A}}\mu_\alpha(\complement K): K \text{ compact}\subset E\}=0$ and it admits a tight subnet iff

$$\inf\{\underline{\lim}_{\alpha\in\mathbf{A}}\mu_\alpha(\complement K): K \text{ compact}\subset E\}=0.$$

1.2.20 Properties. (1) Every tight net $(\mu_\alpha)_{\alpha\in\mathbf{A}}$ in $\mathscr{M}^b_+(E)$ is a $\mathscr{T}_w$-compact net.

In fact, we assume without loss of generality that $(\mu_\alpha)_{\alpha\in\mathbf{A}}$ is universal. For every $n\geq 1$ there exist a compact subset K_n of E and an $\alpha(n)\in\mathbf{A}$ such that $\mu_\alpha(\complement K_n)\leq\frac{1}{n}$ for all $\alpha>\alpha(n)$. We also assume that $K_n\subset K_{n+1}$ and $\alpha(n)<\alpha(n+1)$ for all $n\geq 1$ as well as $\|\mu_\alpha\|\leq M<\infty$ for all $\alpha\in\mathbf{A}$, where $M\in\mathbb{R}^*_+$.

For each $n\geq 1$ we have

$$\lim_{\alpha\in\mathbf{A}}\operatorname{Res}_{K_n}\mu_\alpha=:\mu^{(n)}\in\mathscr{M}^b_+(K_n) \quad\text{and}\quad \mu^{(n+1)}\geq\mu^{(n)}.$$

Let $\mu:=\sup_{n\geq 1}\mu^{(n)}$. Then $\|\mu\|\leq M$ and $\mu(\complement K_m)\leq\frac{1}{m}$ for all $m\geq 1$ and hence $\mu\in\mathscr{M}^b_+(E)$.

Finally one shows that $\mathscr{T}_w\text{-}\lim\mu_\alpha=\mu$.

(2) Every $\mathscr{T}_w$-convergent net $(\mu_\alpha)_{\alpha\in\mathbf{A}}$ in $\mathscr{M}^b_+(E)$ is a tight net.

Indeed, let $\mathscr{T}_w\text{-}\lim_{\alpha\in\mathbf{A}}\mu_\alpha=\mu\in\mathscr{M}^b_+(E)$. Then, of course,

$$\overline{\lim}_{\alpha\in\mathbf{A}}\|\mu_\alpha\|=\|\mu\|<\infty.$$

For every $\varepsilon>0$ there exists an $f\in\mathscr{K}_+(E)$ satisfying $f\leq 1$ and $\|\mu\|-\mu(f)<\varepsilon$. Since $\lim_{\alpha\in\mathbf{A}}\mu_\alpha=\mu$, there exists an $\alpha_0\in\mathbf{A}$ with $\|\mu_\alpha\|-\mu_\alpha(f)<\varepsilon$ for all $\alpha>\alpha_0$. Choosing $K:=\operatorname{supp}(f)$ we obtain $\mu_\alpha(\complement K)=\|\mu_\alpha\|-\mu_\alpha(K)\leq\|\mu_\alpha\|-\mu_\alpha(f)<\varepsilon$ for all $\alpha>\alpha_0$, i.e., $(\mu_\alpha)_{\alpha\in\mathbf{A}}$ is a tight net.

We shall now apply the notions of tight and compact nets to measures on a group.

1.2.21 Theorem. *Let G be a locally compact group and let $(\mu_\alpha)_{\alpha\in\mathbf{A}}$, $(\nu_\alpha)_{\alpha\in\mathbf{A}}$ and $(\lambda_\alpha)_{\alpha\in\mathbf{A}}$ be nets in $\mathscr{M}^1(G)$ such that $\mu_\alpha=\nu_\alpha*\lambda_\alpha$ holds for all $\alpha\in\mathbf{A}$.*

(i) *If $(\nu_\alpha)_{\alpha\in\mathbf{A}}$ and $(\lambda_\alpha)_{\alpha\in\mathbf{A}}$ are tight (uniformly tight), then so is $(\mu_\alpha)_{\alpha\in\mathbf{A}}$.*

(ii) *If $(\mu_\alpha)_{\alpha\in\mathbf{A}}$ and $(\nu_\alpha)_{\alpha\in\mathbf{A}}$ are tight (uniformly tight), then so is $(\lambda_\alpha)_{\alpha\in\mathbf{A}}$.*

(iii) *If $(\mu_\alpha)_{\alpha\in\mathbf{A}}$ is tight (uniformly tight), then for every $\alpha\in\mathbf{A}$ there is an $x_\alpha\in G$ such that $(\nu_\alpha*\varepsilon_{x_\alpha})_{\alpha\in\mathbf{A}}$ and $(\varepsilon_{x_\alpha^{-1}}*\lambda_\alpha)_{\alpha\in\mathbf{A}}$ are tight (uniformly tight).*

Proof. 1. (i) follows from the inequality $\mu_\alpha(BC)\geq\nu_\alpha(B)\,\lambda_\alpha(C)$, which is valid for all $B, C\in\mathfrak{B}(G)$ and each $\alpha\in\mathbf{A}$.

2. (ii) is an immediate consequence of the relations

$$\mu_\alpha(B) = \int \nu_\alpha(Bx^{-1})\,\lambda_\alpha(dx) \leq \nu_\alpha(BC^{-1}) + \lambda_\alpha(\complement C),$$

which are valid for all $B, C \in \mathfrak{B}(G)$ and $\alpha \in \mathbb{A}$.

3. Let $(\varepsilon_n)_{n \geq 1}$ be a sequence in $\mathbb{R}_+^*$ such that $\sum_{n \geq 1} \varepsilon_n < \infty$ holds. By assumption there exist a compact subset K_n of G and an index $\alpha(n) \in \mathbb{A}$ satisfying $\mu_\alpha(\complement K_n) < \varepsilon_n$ for all $\alpha > \alpha(n)$. We put

$$\mathbb{A}_n := \{\alpha \in \mathbb{A} : \alpha > \alpha(n)\}, \quad \mathbb{D}_n := \mathbb{A}_n \setminus \mathbb{A}_{n+1} \quad \text{for every } n \geq 1 \text{ and}$$

$$\mathbb{A}_\infty := \bigcap_{n \geq 1} \mathbb{A}_n.$$

Without loss of generality we may assume $\alpha(n) < \alpha(n+1)$ and also $\mathbb{A}_1 = \mathbb{A}$. But then $\mathbb{A}_{n+1} \subset \mathbb{A}_n$ for all $n \geq 1$ and $\bigcup_{n \geq 1} \mathbb{D}_n = \mathbb{A} \setminus \mathbb{A}_\infty$.

Let $(\delta_n)_{n \geq 1}$ be another sequence in $\mathbb{R}_+^*$ satisfying $\lim_{n \to \infty} \delta_n = 0$ and $\sum_{n \geq 1} \frac{\varepsilon_n}{\delta_n} < \frac{1}{2}$. For every $n \geq 1$ we introduce the sets

$$B_{\alpha n} := \{x \in G : \nu_\alpha(K_n x^{-1}) \geq 1 - \delta_n\},$$

$F_\alpha := \bigcap_{m=1}^n B_{\alpha m}$ for $\alpha \in \mathbb{D}_n$ and $F_\alpha := \bigcap_{n \geq 1} B_{\alpha n}$ for $\alpha \in \mathbb{A}_\infty$.

We have to show that $F_\alpha \neq \varnothing$ for all $\alpha \in \mathbb{A}$. In fact, the chain of inequalities

$$\begin{aligned} 1 - \varepsilon_n &\leq \mu_\alpha(K_n) \\ &= \int_{B_{\alpha n}} \nu_\alpha(K_n x^{-1})\,\lambda_\alpha(dx) + \int_{\complement B_{\alpha n}} \nu_\alpha(K_n x^{-1})\,\lambda_\alpha(dx) \\ &\leq \lambda_\alpha(B_{\alpha n}) + (1 - \delta_n)\,\lambda_\alpha(\complement B_{\alpha n}) \end{aligned}$$

yields $\lambda_\alpha(\complement B_{n\alpha}) \leq \frac{\varepsilon_n}{\delta_n}$ for all $\alpha \geq \alpha(n)$, $n \geq 1$. Hence, $\lambda_\alpha(\complement F_\alpha) \leq \sum_{n \geq 1} \frac{\varepsilon_n}{\delta_n} < \frac{1}{2}$ for all $\alpha \in \mathbb{A}$. But this implies that $F_\alpha \neq \varnothing$. For each $\alpha \in \mathbb{A}$ we now choose an element x_α of F_α.

Then $(\nu_\alpha * \varepsilon_{x_\alpha})_{\alpha \in \mathbb{A}}$ is a tight net in $\mathscr{M}^1(G)$. By (ii) $(\varepsilon_{x_\alpha^{-1}} * \lambda_\alpha)_{\alpha \in \mathbb{A}}$ is also a tight net.

The statement concerning uniformly tight nets is shown analogously. □

1.2.22 Theorem. *Let G be a σ-compact locally compact group. The mapping $(\mu, \nu) \to \mu * \nu$ from $\mathscr{M}^1(G) \times \mathscr{M}^1(G)$ into $\mathscr{M}^1(G)$ is $\mathscr{T}_w$-σ-closed in the sense that any $\mathscr{T}_w$-closed subset of $\mathscr{M}^1(G) \times \mathscr{M}^1(G)$ is mapped onto a $\mathscr{T}_w$-F_σ-subset of $\mathscr{M}^1(G)$.*

Proof. 1. Let K be a compact subset of G, $\varepsilon \in]0, 1[$ and

$$\mathscr{N} := \{\mu \in \mathscr{M}^1(G) : \mu(K) \geq \varepsilon\}.$$

We denote the convolution mapping $(\mu, \nu) \to \mu * \nu$ from $\mathscr{M}^1(G) \times \mathscr{M}^1(G)$ into $\mathscr{M}^1(G)$ by Ψ and show that Ψ maps closed subsets of $\mathscr{N} \times \mathscr{N}$ onto closed subsets of $\mathscr{M}^1(G)$. In fact, let $\mathscr{F}$ be a closed subset of $\mathscr{N} \times \mathscr{N}$ and let $((\mu_\alpha, \nu_\alpha))_{\alpha \in \mathbb{A}}$ be a net in $\mathscr{F}$ with $\lim_{\alpha \in \mathbb{A}} \mu_\alpha * \nu_\alpha =: \lambda \in \mathscr{M}^1(G)$. It is to be shown that $\lambda \in \Psi(\mathscr{F})$. Plainly $(\mu_\alpha * \nu_\alpha)_{\alpha \in \mathbb{A}}$ is tight. By Theorem 1.2.21 there is a net $(x_\alpha)_{\alpha \in \mathbb{A}}$ in G such that $(\mu_\alpha * \varepsilon_{x_\alpha})_{\alpha \in \mathbb{A}}$ and $(\varepsilon_{x_\alpha^{-1}} * \nu_\alpha)_{\alpha \in \mathbb{A}}$ are tight nets. Thus there exist an $\alpha_0 := \alpha_0(\varepsilon) \in \mathbb{A}$ and a compact subset C of G with

$$\mu_\alpha(Cx_\alpha^{-1}) = \mu_\alpha * \varepsilon_{x_\alpha}(C) > 1 - \varepsilon \quad \text{for all } \alpha > \alpha_0.$$

On the other hand, we have by hypothesis $\mu_\alpha(K)\geq\varepsilon$ for all $\alpha\in\mathbf{A}$. Therefore, $K\cap Cx_\alpha^{-1}\neq\emptyset$ and so $x_\alpha\in K^{-1}C$ for all $\alpha>\alpha_0$. By Theorem 1.2.21 $(\mu_\alpha)_{\alpha\in\mathbf{A}}$ and $(\nu_\alpha)_{\alpha\in\mathbf{A}}$ are tight nets.

Let μ and ν be accumulation points of $(\mu_\alpha)_{\alpha\in\mathbf{A}}$ and $(\nu_\alpha)_{\alpha\in\mathbf{A}}$ resp. Then obviously $(\mu,\nu)\in\mathscr{F}$, since $\mathscr{F}$ is closed in $\mathscr{M}^1(G)\times\mathscr{M}^1(G)$. Hence,

$$\lambda:=\mu*\nu=\Psi(\mu,\nu).$$

2. Let $(K_n)_{n\geq 1}$ be an increasing sequence of compact sets in G with $G=\bigcup_{n\geq 1}K_n$. For each $n\geq 1$ we define $\mathscr{N}_n:=\{\mu\in\mathscr{M}^1(G):\mu(K_n)\geq\frac{1}{2}\}$. Then $(\mathscr{N}_n)_{n\geq 1}$ is an increasing sequence of closed subsets of $\mathscr{M}^1(G)$ with the property that

$$\mathscr{M}^1(G)\times\mathscr{M}^1(G)=\bigcup_{n\geq 1}(\mathscr{N}_n\times\mathscr{N}_n).$$

If now $\mathscr{F}$ is a closed subset of $\mathscr{M}^1(G)\times\mathscr{M}^1(G)$, then

$$\Psi(\mathscr{F})=\Psi\Big(\bigcup_{n\geq 1}(\mathscr{F}\cap(\mathscr{N}_n\times\mathscr{N}_n))\Big)=\bigcup_{n\geq 1}\Psi(\mathscr{F}\cap(\mathscr{N}_n\times\mathscr{N}_n))$$

is an F_σ-set since $\Psi(\mathscr{F}\cap(\mathscr{N}_n\times\mathscr{N}_n))$ is closed in $\mathscr{M}^1(G)$ by Part 1. □

We now give an application of Theorem 1.2.22 to measures indecomposable with respect to convolution.

Let G be a locally compact group. A measure $\mu\in\mathscr{M}^1(G)\setminus\mathscr{D}(G)$ is called a *factor* of a measure $\nu\in\mathscr{M}^1(G)$ if there exists a measure $\lambda\in\mathscr{M}^1(G)$ such that either $\nu=\mu*\lambda$ or $\nu=\lambda*\mu$. $\mu\in\mathscr{M}^1(G)$ is said to be a *translate* of $\nu\in\mathscr{M}^1(G)$ if there exists an $x\in G$ such that either $\mu=\varepsilon_x*\nu$ or $\mu=\nu*\varepsilon_x$ is satisfied. Finally, a measure $\mu\in\mathscr{M}^1(G)$ is called *prime* if its only factors are translates of μ. The set of all prime measures in $\mathscr{M}^1(G)$ will be denoted by $\mathscr{M}_p^1(G)$.

It can be shown that for an infinite locally compact group G one has $\mathscr{M}_p^1(G)\neq\emptyset$ ([377], p. 69).

1.2.23 Theorem. *Let G be a locally compact group with a countable basis of its topology. Then $\mathscr{M}_p^1(G)$ is a $(\mathscr{T}_w$-$)G_\delta$-subset of $\mathscr{M}^1(G)$.*

Proof. We first observe that the set

$$\Delta:=\{(\mu,\nu)\in\mathscr{M}^1(G)\times\mathscr{M}^1(G):\text{either } \mu \text{ or } \nu \text{ is in } \mathscr{D}(G)\}$$

is $\mathscr{T}_w$-homeomorphic to $(G\times\mathscr{M}^1(G))\cup(\mathscr{M}^1(G)\times G)$. Since by Theorem 1.1.6 the space $\mathscr{M}^1(G)\times\mathscr{M}^1(G)$ is metrizable, the open subset $\complement\Delta$ of $\mathscr{M}^1(G)\times\mathscr{M}^1(G)$ is an F_σ-set, and since by Theorem 1.2.22 the convolution mapping $\Psi:(\mu,\nu)\to\mu*\nu$ from $\mathscr{M}^1(G)\times\mathscr{M}^1(G)$ into $\mathscr{M}^1(G)$ is σ-closed, $\Psi(\complement\Delta)$ is an F_σ-subset of $\mathscr{M}^1(G)$. Thus,

$$\mathscr{M}_p^1(G)=\complement_{\mathscr{M}^1(G)}\Psi(\complement_{\mathscr{M}^1(G)\times\mathscr{M}^1(G)}\Delta)$$

is a G_δ-subset of $\mathscr{M}^1(G)$. □

1.3 Fourier Transforms of Bounded Measures

Dual and Coefficient Algebra of a Locally Compact Group. Let G be a locally compact group. In the following discussion we shall be concerned mainly with finite-dimensional representations of G. For each $n \geq 1$ we denote by $\mathrm{Rep}_n(G)$ the totality of all (continuous, unitary) representations D of G with n-dimensional representation space $\mathscr{H}(n) := \mathscr{H}(n(D))$, where $n(D)$ is the dimension of D. In $\mathrm{Rep}_n(G)$ the compact open topology $\mathscr{T}_{co}$ is introduced in the following fashion: For each $D \in \mathrm{Rep}_n(G)$, every compact subset K of G and given $\varepsilon > 0$ one defines the set

$$W(D; K, \varepsilon) := \{D' \in \mathrm{Rep}_n(G) : \|D'(x) - D(x)\| < \varepsilon \text{ for all } x \in K\},$$

the collection

$$\mathscr{W}(D) := \{W(D; K, \varepsilon) : K \text{ compact} \subset G, \ \varepsilon > 0\},$$

and takes the family

$$\mathscr{W} := \{\mathscr{W}(D) : D \in \mathrm{Rep}_n(G)\}$$

as a basis for a topology in $\mathrm{Rep}_n(G)$. This topology is inherited by the subset $\mathrm{Irr}_n(G)$ of all irreducible representations in $\mathrm{Rep}_n(G)$ ($n \geq 1$). Now let $\mathrm{Rep}(G)$ and $\mathrm{Irr}(G)$ be the (disjoint) topological sums of the spaces $\mathrm{Rep}_n(G)$ and $\mathrm{Irr}_n(G)$ resp. $\mathrm{Rep}(G)$ is called the (Chu) *dual* of G.

1.3.1 Properties of the dual. (1) $\mathrm{Rep}(G)$ is locally compact and uniformizable with respect to the topology $\mathscr{T}_{co}$.

(2) If G is a discrete group, then $\mathrm{Rep}_n(G)$ is compact for every $n \geq 1$ and hence $\mathrm{Rep}(G)$ is σ-compact.

(3) If the topology of G admits a countable basis, then also $\mathrm{Rep}(G)$ and $\mathrm{Irr}(G)$ do. In this case $\mathrm{Rep}(G)$ and $\mathrm{Irr}(G)$ are in fact Polish spaces.

For $n \geq 1$ let $\mathrm{Rep}_n(G)/\sim$ be the set of (unitary) equivalence classes of elements of $\mathrm{Rep}_n(G)$ furnished with the quotient topology derived from $\mathscr{T}_{co}$ on $\mathrm{Rep}_n(G)$.

(4) If G is compact, then $\mathrm{Rep}_n(G)/\sim$ is discrete for every $n \geq 1$.

For any closed subgroup H of G one defines the orthogonal complement or (first) *annihilator* of H with respect to $\mathrm{Rep}_n(G)$ ($n \geq 1$) or $\mathrm{Rep}(G)$ by

$$H_n^{\perp} := \{D \in \mathrm{Rep}_n(G) : D(z) = E_{n(D)} \text{ for all } z \in H\} \text{ or}$$
$$H^{\perp} := \{D \in \mathrm{Rep}(G) : D(z) = E_{n(D)} \text{ for all } z \in H\} \text{ resp.}$$

(5) $H_n^{\perp}$ and $H^{\perp}$ are closed subsets of $\mathrm{Rep}_n(G)$ (for $n \geq 1$) and $\mathrm{Rep}(G)$ resp.

(6) If H is a normal subgroup of G, then $H_n^{\perp} \cong \mathrm{Rep}_n(G/H)$ (for $n \geq 1$) and $H^{\perp} \cong \mathrm{Rep}(G/H)$.

Let $\Sigma'(G)$ stand for the system $\mathrm{Rep}(G)/\sim$ of all equivalence classes of elements in $\mathrm{Rep}(G)$ and denote by $\Sigma(G)$ the subsystem $\mathrm{Irr}(G)/\sim$ of $\Sigma'(G)$. Then the above definitions become more concrete. Let $\sigma \in \Sigma(G)$ and $D^{(\sigma)}$ be a representative of σ. Putting

$$n(\sigma) := \dim \mathscr{H}(D^{(\sigma)}) \quad \text{and} \quad \mathscr{H}(\sigma) := \mathscr{H}(D^{(\sigma)}) = \mathscr{H}(n(\sigma))$$

and choosing an orthonormal basis $\{\xi_1^{(\sigma)}, \ldots, \xi_{n(\sigma)}^{(\sigma)}\}$ of $\mathscr{H}(\sigma)$ one obtains the coefficients $d_{ij}(D) := d_{\xi_i^{(\sigma)}, \xi_j^{(\sigma)}}(D)$ of $D \in \sigma$ which one also writes as

$$d_{ij}^{(\sigma)} := d(\sigma; i, j) = d_{ij}(D^{(\sigma)}) \qquad (i, j = 1, \ldots, n(\sigma)).$$

In this notation we sometimes interpret $H^\perp$ for any closed subgroup H of G as the subset

$$\{\sigma \in \Sigma(G) : D^{(\sigma)}(x) = E_{n(\sigma)} \text{ for all } x \in H\}$$

of $\Sigma(G)$ and carry over Properties (5) and (6) above in the obvious way.

Let G be a compact group. For any class $\sigma \in \Sigma(G)$ we denote by $\mathfrak{R}^{(\sigma)}(G)$ the linear space generated by the set of all coefficients of σ and for any subset P of $\Sigma(G)$, by $\mathfrak{R}^P(G)$ the linear hull

$$\langle \bigcup_{\sigma \in P} \mathfrak{R}^{(\sigma)}(G) \rangle \text{ of the linear spaces } \mathfrak{R}^{(\sigma)}(G) \text{ for } \sigma \in P.$$

Clearly, $\mathfrak{R}(G) = \langle \bigcup_{\sigma \in \Sigma(G)} \mathfrak{R}^{(\sigma)}(G) \rangle$. For classes $\sigma_1, \sigma_2 \in \Sigma(G)$ with direct sum decomposition $D^{(\sigma_1)} \otimes D^{(\sigma_2)} = m_1 D^{(\tau_1)} \oplus \cdots \oplus m_s D^{(\tau_s)}$ of the tensor product $D^{(\sigma_1)} \otimes D^{(\sigma_2)}$ (with $m_j \in \mathbb{N}$, $\tau_j \in \Sigma(G)$ for $j = 1, \ldots, s$) one introduces the set $\sigma_1 \times \sigma_2 := \{\tau_1, \ldots, \tau_s\}$. Moreover, given subsets P, P_1 and P_2 of $\Sigma(G)$ one defines

$$P_1 \times P_2 := \bigcup \{\sigma_1 \times \sigma_2 : \sigma_1 \in P_1, \sigma_2 \in P_2\}$$

and $\bar{P} := \{\bar{\sigma} : \sigma \in P\}$, where $\bar{\sigma}$ denotes the conjugate of $\sigma \in \Sigma(G)$.

By $[P]$ we denote the smallest subset of $\Sigma(G)$ containing P and being closed under the operations $\times$ and $^-$. Then for any subset P of $\Sigma(G)$ with $P = [P]$ the following statements are equivalent:

(i) $P = \Sigma(G)$;

(ii) $\mathfrak{R}^P(G)$ is a (uniformly) dense subalgebra of $\mathscr{C}_\mathbb{C}(G)$;

(iii) for each $x \in G \setminus \{e\}$ there exists $\sigma \in P$ such that $D^{(\sigma)}(x) \neq E_{n(\sigma)}$ for some $D^{(\sigma)} \in \sigma$.

In fact, implication (i) $\Rightarrow$ (iii) is just the Gelfand-Raikov theorem; (iii) $\Rightarrow$ (i) follows from the Stone-Weierstrass theorem. It is a result due to Peter and Weyl that the system

$$\{\sqrt{n(\sigma)}\, d(\sigma; i, j) : \sigma \in \Sigma(G),\ i, j = 1, \ldots, n(\sigma)\}$$

is an orthonormal basis of $\mathscr{L}_\mathbb{C}^2(G, \omega)$. If G is infinite, then $\Sigma(G)$ is infinite and card $\Sigma(G) = \dim \mathscr{L}_\mathbb{C}^2(G, \omega)$.

A more profound analysis of the structure of the coefficient algebra $\mathfrak{R}(G)$ of a compact group yields its interpretation as a Krein algebra or more generally as a Hopf algebra with unit, co-unit and symmetry. The following result is part of the duality of compact groups G or their coefficient algebra $\mathfrak{R}(G)$: The multiplicative linear functionals N on $\mathfrak{R}(G)$ which are real in the sense that $N(\bar{f}) = \overline{N(f)}$ for all $f \in \mathfrak{R}(G)$ are evaluation functionals E_a for points $a \in G$.

In order to present a further characterization of the class $\mathbf{A}$ of all almost periodic (locally compact) groups we proceed as follows: Clearly two representations of a class $\sigma \in \Sigma(G)$ admit the same kernels. For every $D^{(\sigma)}$ in the class $\sigma \in \Sigma(G)$ we put $K^{(\sigma)} := \ker D^{(\sigma)}$ and define the closed normal subgroup $G^0 := \bigcap_{\sigma \in \Sigma(G)} K^{(\sigma)}$ of G to be the *von Neumann kernel* of G. It becomes evident now that $G \in \mathbf{A}$ iff $G^0 = \{e\}$, which means that the dual Rep(G) of G separates the points of G. The von Neumann kernel G^0 of G turns out to be the intersection of all normal subgroups H of G with $G/H \in \mathbf{A}$.

1.3.2 Further Properties of the dual. Let $G \in \mathbf{A}$.

(1) If $\mathrm{Rep}_n(G)$ is connected for all $n \geq 1$, then G has no nontrivial compact subgroup.

Let $(K_\alpha)_{\alpha \in \mathbb{A}}$ be a descending family of compact normal subgroups of G with $\bigcap_{\alpha \in \mathbb{A}} K_\alpha = \{e\}$. For every $\alpha \in \mathbb{A}$ let p_α be the canonical mapping from G onto $G_\alpha := G/K_\alpha$.

(2) $(K_\alpha^\perp)_{\alpha \in \mathbb{A}}$ is an ascending family of open subsets of $\mathrm{Rep}(G)$ with

$$\bigcup_{\alpha \in \mathbb{A}} K_\alpha^\perp = \mathrm{Rep}(G).$$

[For every $\alpha \in \mathbb{A}$ the set $K_\alpha^\perp$ is open in $\mathrm{Rep}(G)$. In fact, let $D \in K_\alpha^\perp$. Since K_α is compact in G, we have $W := W\left(D; K_\alpha, \frac{1}{n(D)}\right) \in \mathscr{W}(D)$. For $D' \in W$ one obtains

$$\mathrm{Res}_{K_\alpha} D' = \mathrm{Res}_{K_\alpha} D = I_{n(D)}$$

since K_α is compact and hence $D' \in K_\alpha^\perp$, so that $W \subset K_\alpha^\perp$. Plainly $(K_\alpha^\perp)_{\alpha \in \mathbb{A}}$ is ascending since $(K_\alpha)_{\alpha \in \mathbb{A}}$ is descending. Finally, for every $D \in \mathrm{Rep}(G)$ there exist an $\alpha \in \mathbb{A}$ and a $D_\alpha \in \mathrm{Rep}(G_\alpha)$ with $D = D_\alpha \circ p_\alpha$ and thus $D \in K_\alpha^\perp$, which implies the last assertion.]

(3) For every compact subset C of $\mathrm{Rep}(G)$ there exists an $\alpha \in \mathbb{A}$ with $C \subset K_\alpha^\perp$. [This follows from Property (2) since $(K_\alpha^\perp)_{\alpha \in \mathbb{A}}$ is an open covering of C.]

For any closed subgroup H of an arbitrary locally compact group G one defines the *second annihilator* by putting

$$H^{\perp\perp} := \{x \in G : D(x) = E_{n(D)} \text{ for all } D \in H^\perp\}.$$

(4) $G/H \in \mathbf{A}$ iff $H^{\perp\perp} = H$.

Pseudo- and quasirepresentations of $\mathrm{Rep}(G)$. Let G be a locally compact group. A mapping

$$F : \mathrm{Rep}(G) \to \bigcup_{n \geq 1} \mathfrak{M}(n, \mathbb{C})$$

is called an *operation* on $\mathrm{Rep}(G)$ if the following axioms hold:

(O1) $F(D) \in \mathfrak{M}(n(D), \mathbb{C})$ for all $D \in \mathrm{Rep}(G)$.

(O2) $F(D_1 \oplus D_2) = F(D_1) \oplus F(D_2)$ for all $D_1, D_2 \in \mathrm{Rep}(G)$.

(O3) $F(U^{-1}DU) = U^{-1}F(D)\,U$ for all $D \in \mathrm{Rep}(G)$ and $U \in \mathfrak{U}(n(D))$.

F is called a *pseudorepresentation* of $\mathrm{Rep}(G)$ if one has in addition

(O4) $F(D_1 \otimes D_2) = F(D_1) \otimes F(D_2)$ for all $D_1, D_2 \in \mathrm{Rep}(G)$,

and a *quasirepresentation* of $\mathrm{Rep}(G)$ if (O1) is strengthened to

(O5) $F(D) \in \mathfrak{U}(n(D))$ for all $D \in \mathrm{Rep}(G)$.

The collection of all operations on $\mathrm{Rep}(G)$ will be denoted by $\mathfrak{O}(G)$, the collection of pseudorepresentations of $\mathrm{Rep}(G)$ by $\mathfrak{P}(G)$ and the set of quasirepresentations of $\mathrm{Rep}(G)$ by $\mathfrak{Q}(G)$.

Defining for operations F, $F' \in \mathfrak{O}(G)$ the sum $F + F'$ and the product $F \cdot F'$ (pointwise) by

$$(F + F')(D) := F(D) + F'(D) \quad \text{and} \quad (F \cdot F')(D) := F(D)\,F'(D)$$

for all $D \in \operatorname{Rep}(G)$ one makes $\mathfrak{D}(G)$ an (associative) algebra over $\mathbb{C}$. The algebraic structure of $\mathfrak{D}(G)$ can be transferred to $\mathfrak{R}(G)^*$ according to the following result.

1.3.3 Theorem. *Let $G \in \mathbf{A}$. For every $N \in \mathfrak{R}(G)^*$ one defines an element $F_N \in \mathfrak{D}(G)$ by*

$$\langle F_N(D)\, \xi, \eta\rangle = N(d_{\xi,\eta}(D)) \quad \text{for all } D \in \operatorname{Rep}(G),\ \xi, \eta \in \mathscr{H}(D).$$

The mapping $\rho: \mathfrak{R}(G)^* \to \mathfrak{D}(G)$ given by $\rho(N) := F_N$ for all $N \in \mathfrak{R}(G)^*$ is a linear bijection from $\mathfrak{R}(G)^*$ onto $\mathfrak{D}(G)$.

Proof. Since transition to the Bohr group $\tilde{G}$ of G leaves the sets $\operatorname{Rep}(G)$, $\mathfrak{R}(G)$ and $\mathfrak{D}(G)$ isomorphic, we may assume without loss of generality that G is compact.

1. For $N \in \mathfrak{R}(G)^*$ and $D \in \operatorname{Rep}(G)$ the mapping $(\xi, \eta) \to N(d_{\xi,\eta}(D))$ from $\mathscr{H}(D) \times \mathscr{H}(D)$ into $\mathbb{C}$ is obviously bilinear. Thus there exists an operator $F_N(D)$ on $\mathscr{H}(D)$ with the property

$$N(d_{\xi,\eta}(D)) = \langle F_N(D)\, \xi, \eta\rangle \quad \text{for all } \xi, \eta \in \mathscr{H}(D).$$

We restrict ourselves to proving Axiom (O3) for the operator-valued mapping $D \to F_N(D)$ on $\operatorname{Rep}(G)$. Let U be a linear isometry of a Hilbert space $\mathscr{H}'$ onto $\mathscr{H}(D)$ and $W := U^{-1} D U$. For $\xi, \eta \in \mathscr{H}'$ and $x \in G$ one has

$$d_{\xi,\eta}(W)(x) = \langle U^{-1} D(x)\, U\xi, \eta\rangle = d_{U\xi, U\eta}(D)(x)$$

and therefore

$$\begin{aligned} \langle F_N(U^{-1} D U)\, \xi, \eta\rangle &= N(d_{\xi,\eta}(W)) = N(d_{U\xi, U\eta}(D)) \\ &= \langle F_N(D)\, U\xi, U\eta\rangle = \langle (U^{-1} F_N(D)\, U)\, \xi, \eta\rangle, \end{aligned}$$

which proves the assertion.

2. Let $F \in \mathfrak{D}(G)$, $\sigma \in \Sigma(G)$, $D^{(\sigma)}$ a representative of σ and $\{\xi_1^{(\sigma)}, \ldots, \xi_{n(\sigma)}^{(\sigma)}\}$ an orthonormal basis of $\mathscr{H}(\sigma)$. We define

$$N_F(d_{ij}^{(\sigma)}) := \langle F(D^{(\sigma)})\, \xi_i^{(\sigma)}, \xi_j^{(\sigma)}\rangle \quad \text{for } i, j = 1, \ldots, n(\sigma)$$

and obtain exactly one element $N_F \in \mathfrak{R}(G)^*$ satisfying this relationship. Indeed, considering any $D \in \operatorname{Rep}(G)$ with representing Hilbert space $\mathscr{H}(D)$ we have

$$\mathscr{H}(D) = \mathscr{H}_{1,1} \oplus \cdots \oplus \mathscr{H}_{1,m_1} \oplus \cdots \oplus \mathscr{H}_{n,1} \oplus \cdots \oplus \mathscr{H}_{n,m_n},$$

where $\mathscr{H}_{r,s}$ is a D-invariant subspace of $\mathscr{H}(D)$ and D is equivalent to $D^{(\sigma_r)}$ on $\mathscr{H}_{r,s}$ ($r = 1, \ldots, n$; $s = 1, \ldots, m_r$). Plainly, there exists a linear isometry $U_{r,s}$ from $\mathscr{H}(\sigma_r)$ into $\mathscr{H}_{r,s}$ with the property that $(U_{r,s} D^{(\sigma_r)}(x)\, U_{r,s}^{-1})\, \xi_{r,s} = D(x)\, \xi_{r,s}$ for all $x \in G$, $\xi_{r,s} \in \mathscr{H}_{r,s}$. Writing $\xi := \sum_{r=1}^{n} \sum_{s=1}^{m_r} \xi_{r,s}$ and $\eta := \sum_{r=1}^{n} \sum_{s=1}^{m_r} \eta_{r,s}$ with $\xi_{r,s} \in \mathscr{H}_{r,s}$ and $\eta_{r,s} \in \mathscr{H}_{r,s}$ resp., we obtain

$$\langle D(x)\, \xi, \eta\rangle = \sum_{r=1}^{n} \sum_{s=1}^{m_r} \langle D^{(\sigma_r)}(x)\, U_{r,s}^{-1}\, \xi_{r,s}, U_{r,s}^{-1}\, \eta_{r,s}\rangle.$$

Axioms (O2) and (O3) yield

$$\langle F(D)\, \xi, \eta\rangle = \sum_{r=1}^{n} \sum_{s=1}^{m_r} \langle F(D^{(\sigma_r)})\, U_{r,s}^{-1}\, \xi_{r,s}, U_{r,s}^{-1}\, \eta_{r,s}\rangle.$$

Expanding $U_{r,s}^{-1}\xi_{r,s}$ and $U_{r,s}^{-1}\eta_{r,s}$ with respect to the given basis $\{\xi_1^{(\sigma_1)}, \dots, \xi_{n(\sigma_r)}^{(\sigma_r)}\}$ one immediately verifies that $\langle F(D)\,\xi, \eta\rangle = N_F(d_{\xi,\eta}(D))$. Obviously N_F is uniquely determined since this relationship can be written for the coefficients of $D^{(\sigma)}$ in $\sigma\in\Sigma(G)$ with respect to the basis $\{\xi_1^{(\sigma)}, \dots, \xi_{n(\sigma_r)}^{(\sigma_r)}\}$ and hence, N_F can be extended to $\mathfrak{R}^{(\sigma)}(G)$ or $\mathfrak{R}(G)$.

3. The equations $F_{N_F}=F$ and $N_{F_N}=N$ complete the proof of the theorem. □

1.3.4 Theorem. *Let $G\in\mathbf{A}$ and let $\rho: N\to F_N$ be the linear bijection from $\mathfrak{R}(G)^*$ onto $\mathfrak{O}(G)$ defined in the above theorem. The following statements are equivalent:*
(i) *N is multiplicative and $\neq 0$;*
(ii) *$\rho(N)\in\mathfrak{P}(G)$.*

Proof. Again we assume that G is compact.

1. (i)$\Rightarrow$(ii). Let $N\in\mathfrak{R}(G)^*$ be multiplicative. Since $N(1)=1$ and $F_N(I)=E$, we have $F_N\neq 0$. For representations D_1, $D_2\in\mathrm{Rep}(G)$ with representing Hilbert spaces $\mathscr{H}_1$, $\mathscr{H}_2$ resp. the tensor product $W:=D_1\otimes D_2$ with representing Hilbert space $\mathscr{H}_1\otimes\mathscr{H}_2$ and ξ_j, $\eta_j\in\mathscr{H}_j (j=1,2)$ one computes:

$$d_{\xi_1\otimes\xi_2,\,\eta_1\otimes\eta_2}(W)=d_{\xi_1,\eta_1}(D_1)\,d_{\xi_2,\eta_2}(D_2)$$

and using the definition of F_N one obtains

$$\begin{aligned}&\langle F_N(D_1\otimes D_2)(\xi_1\otimes\xi_2), \eta_1\otimes\eta_2\rangle = N(d_{\xi_1\otimes\xi_2,\,\eta_1\otimes\eta_2}(W))\\ &=N(d_{\xi_1,\eta_1}(D_1)\,d_{\xi_2,\eta_2}(D_2))=N(d_{\xi_1,\eta_1}(D_1))\,N(d_{\xi_2,\eta_2}(D_2))\\ &=\langle F_N(D_1)\otimes F_N(D_2)(\xi_1\otimes\xi_2), \eta_1\otimes\eta_2\rangle\end{aligned}$$

since N is multiplicative, i.e., $F_N(D_1\otimes D_2)=F_N(D_1)\otimes F_N(D_2)$.

2. (ii)$\Rightarrow$(i). Let $F\in\mathfrak{O}(G)$ with $F\not\equiv 0$. For any 1-dimensional representation $D\in\mathrm{Rep}(G)$ the operator $F(D)$ is 1-dimensional by (O1) and can therefore be viewed as (multiplication by) a complex number. The identity representation I of G will be identified with the function 1 on G. Since $F\not\equiv 0$, one has $F(1)=1$. Let $\sigma,\tau\in\Sigma(G)$, $i,j\in\{1,\dots,n(\sigma)\}$, $p,q\in\{1,\dots,n(\tau)\}$. We choose orthonormal bases in $\mathscr{H}(\sigma)$ and $\mathscr{H}(\tau)$ as in the proof of the preceding theorem and get

$$N_F(d_{ij}^{(\sigma)})=\langle F(D^{(\sigma)})\,\xi_i^{(\sigma)}, \xi_j^{(\sigma)}\rangle \quad\text{and}\quad N_F(d_{pq}^{(\tau)})=\langle F(D^{(\tau)})\,\xi_p^{(\tau)}, \xi_q^{(\tau)}\rangle$$

resp. For all $x\in G$ one computes

$$\langle (D^{(\sigma)}\otimes D^{(\tau)})(x)(\xi_i^{(\sigma)}\otimes\xi_p^{(\tau)}), \xi_j^{(\sigma)}\otimes\xi_q^{(\tau)}\rangle = d_{ij}^{(\sigma)}(x)\,d_{pq}^{(\tau)}(x).$$

Applying the definition of N_F together with (O4) one obtains

$$\begin{aligned}&N_F(d_{ij}^{(\sigma)}\,d_{pq}^{(\tau)})=\langle F(D^{(\sigma)}\otimes D^{(\tau)})(\xi_i^{(\sigma)}\otimes\xi_p^{(\tau)}), \xi_j^{(\sigma)}\otimes\xi_q^{(\tau)}\rangle\\ &=\langle (F(D^{(\sigma)})\,\xi_i^{(\sigma)})\otimes(F(D^{(\tau)})\,\xi_p^{(\tau)}), \xi_j^{(\sigma)}\otimes\xi_q^{(\tau)}\rangle\\ &=\langle F(D^{(\sigma)})\,\xi_i^{(\sigma)}, \xi_j^{(\sigma)}\rangle\langle F(D^{(\tau)})\,\xi_p^{(\tau)}, \xi_q^{(\tau)}\rangle = N_F(d_{ij}^{(\sigma)})\,N_F(d_{pq}^{(\tau)}).\end{aligned}$$

Thus, N_F is multiplicative on products of basis elements in $\mathfrak{R}(G)$ and hence on all of $\mathfrak{R}(G)$. □

1.3.5 Theorem. *Let $G \in \mathbf{A}$ and let $\rho: N \to F_N$ be the linear bijection from $\mathfrak{R}(G)^*$ onto $\mathfrak{D}(G)$ defined above. The following statements are equivalent:*
(i) *N is multiplicative and real;*
(ii) *$\rho(N) \in \mathfrak{Q}(G)$.*

Proof. Again G is assumed to be compact.

1. (i) $\Rightarrow$ (ii). Let $N \in \mathfrak{R}(G)^*$ be multiplicative and such that $N(\bar{f}) = \overline{N(f)}$ holds for all $f \in \mathfrak{R}(G)$. Then we have noted that $N = E_a$ for some $a \in G$. For $F_a := \rho(E_a)$ we obtain

$$\langle F_a(D)\,\xi, \eta\rangle = E_a(d_{\xi,\eta}(D)) = \langle D(a)\,\xi, \eta\rangle$$

whenever $D \in \operatorname{Rep}(G)$ and $\xi, \eta \in \mathscr{H}(D)$. Since $F_N = \rho(N) = \rho(E_a) = F_a$, we conclude that $F_a(D) = D(a)$ is unitary for all $D \in \operatorname{Rep}(G)$, which implies $\rho(N)(D) \in \mathfrak{U}(n(D))$ for all $D \in \operatorname{Rep}(G)$ and thus $\rho(N) \in \mathfrak{Q}(G)$.

2. (ii) $\Rightarrow$ (i). Since $\rho(N)(D^{(\sigma)})$ is unitary for all $\sigma \in \Sigma(G)$ and $D^{(\sigma)} \in \sigma$ with representing Hilbert space $\mathscr{H}(\sigma)$ admitting an orthonormal basis $\{\xi_1^{(\sigma)}, \ldots, \xi_{n(\sigma)}^{(\sigma)}\}$, we have

$$\rho(N)(D^{(\sigma)})\,\rho(N)(D^{(\sigma)})^\sim\,\xi_i = \xi_i \qquad \text{for all } i = 1, \ldots, n(\sigma),$$

and thus,

$$\begin{aligned}\delta_{ij} &= \langle \xi_i, \xi_j\rangle = \langle \rho(N)(D^{(\sigma)})\,\rho(N)(D^{(\sigma)})^\sim\,\xi_i, \xi_j\rangle \\ &= \textstyle\sum_{r=1}^{n(\sigma)} \langle \rho(N)(D^{(\sigma)})\,\xi_r, \xi_j\rangle \overline{\langle \rho(N)(D^{(\sigma)})\xi_r, \xi_i\rangle} \\ &= \textstyle\sum_{r=1}^{n(\sigma)} N(d_{rj}^{(\sigma)})\,\overline{N(d_{ri}^{(\sigma)})} \qquad \text{for all } i = 1, \ldots, n(\sigma).\end{aligned}$$

This implies that each matrix $(N(d_{ij}^{(\sigma)}))_{i,j=1,\ldots,n(\sigma)}$ is unitary. Clearly,

$$(N(\overline{d_{ij}^{(\sigma)}}))_{i,j=1,\ldots,n(\sigma)}^T = (N(d_{ij}^{(\sigma)}))_{i,j=1,\ldots,n(\sigma)}^{-1}$$

and hence

$$N(\overline{d_{ij}^{(\sigma)}}) = \overline{N(d_{ij}^{(\sigma)})} \qquad \text{for all } \sigma \in \Sigma(G) \quad \text{and} \quad D^{(\sigma)} := (d_{ij}^{(\sigma)})_{i,j=1,\ldots,n(\sigma)} \in \sigma.$$

Since N is linear and $\{d_{ij}^{(\sigma)} : \sigma \in \Sigma(G),\ i, j = 1, \ldots, n(\sigma)\}$ is a basis of $\mathfrak{R}(G)$, N is in fact real on $\mathfrak{R}(G)$. □

Theorem 1.3.3 yields the definition of a multiplication $\odot$ in $\mathfrak{R}(G)^*$ such that $\mathfrak{R}(G)^*$ becomes an algebra. For every $N \in \mathfrak{R}(G)^*$ and $D \in \operatorname{Rep}(G)$ we write

$$N(D) := \rho(N)(D) = F_N(D) = (N(d_{ij}(D)))_{i,j=1,\ldots,n(D)},$$

where

$$D := (d_{ij}(D))_{i,j=1,\ldots,n(D)}$$

is the matrix representation of D with respect to the basis $\{\xi_1^{(D)}, \ldots, \xi_{n(D)}^{(D)}\}$ of $\mathscr{H}(D)$. N can be regarded as a linear form on $\mathfrak{R}(G)$ or as an operation on $\operatorname{Rep}(G)$. $N \in \mathfrak{R}(G)^*$ is called *continuous* if N is continuous (as an element of $\mathfrak{D}(G)$) on $\operatorname{Rep}(G)$. The set of all continuous elements N of $\mathfrak{R}(G)^*$ will be denoted by $\mathscr{S}(G)$. Plainly $\mathscr{S}(G)$ is a subalgebra of $\mathfrak{R}(G)^*$.

1.3.6 Properties of $\mathfrak{K}(G)^*$. For every $D \in \operatorname{Rep}(G)$ we denote by $\mathscr{V}(D)$ the vector space over $\mathbb{C}$ generated by the coefficients of D, $\bar{D}$ and by 1. Then $\mathscr{V}(D)$ is a finite-dimensional subspace of $\mathfrak{K}(G)$ invariant by $R(x)$, $L(x)$ (for $x \in G$) and S. The family $(\mathscr{V}(D))_{D \in \operatorname{Rep}(G)}$ is (filtering) ascending, and one has

$$\bigcup_{D \in \operatorname{Rep}(G)} \mathscr{V}(D) = \mathfrak{K}(G).$$

For every $N \in \mathfrak{K}(G)^*$ we define a linear mapping $R_N : \mathfrak{K}(G) \to \mathfrak{K}(G)$ by

$$(R_N f)(x) := N \circ R(x) f = N({}_x f) \qquad \text{for all } f \in \mathfrak{K}(G), x \in G.$$

Plainly

(1) $R_N \mathscr{V}(D) \subset \mathscr{V}(D)$ for all $D \in \operatorname{Rep}(G)$ and

(2) for $N_1, N_2 \in \mathfrak{K}(G)^*$ and $f \in \mathfrak{K}(G)$ one has

$$(N_1 \odot N_2) f = (R_{N_1} \circ R_{N_2} f)(e) = N_1(R_{N_2} f).$$

If N_1, N_2 are positive, then $N_1 \odot N_2$ is also.

Furthermore, we introduce for every $N \in \mathfrak{K}(G)^*$ its *exponential* $\operatorname{Exp} N \in \mathfrak{D}(G)$ by $(\operatorname{Exp} N)(D) := \exp N(D)$ for all $D \in \operatorname{Rep}(G)$.

(3) Given $N_1, N_2 \in \mathfrak{K}(G)^*$ with $N_1 \odot N_2 = N_2 \odot N_1$ we obtain the formula

$$(\operatorname{Exp} N_1) \odot (\operatorname{Exp} N_2) = \operatorname{Exp}(N_1 + N_2).$$

For $N \in \mathfrak{K}(G)^*$ the linear mapping R_N gives rise to a linear operator

$$\exp R_N := I + \sum_{n \geq 1} \frac{R_N^{\circ n}}{n!} \qquad \text{on } \mathscr{V}(D).$$

(4) Since for every $f \in \mathfrak{K}(G)$ there exist $D \in \operatorname{Rep}(G)$ and $M \in \mathfrak{M}(n(D), \mathbb{C})$ with $f = \operatorname{tr}(MD)$, one establishes the formulae

$$(\exp R_N) f = \operatorname{tr}(MD \exp N(D))$$

and

$$(\operatorname{Exp} N) f = [(\exp R_N) f](e) = \operatorname{tr}(M \exp N(D)).$$

Finally, we define for every $N \in \mathfrak{K}(G)^*$ the functional

$$N^\# := N \circ S \in \mathfrak{K}(G)^*.$$

N is called *symmetric* (*skew symmetric*) if $N^\# = N$ ($N^\# = -N$).

(5) Every $N \in \mathfrak{K}(G)^*$ admits a unique representation $N = N' + N''$ with skew symmetric and symmetric summands $N' = \frac{1}{2}(N - N^\#)$ and $N'' = \frac{1}{2}(N + N^\#)$.

Fourier transform and Fourier topology. Let G be an arbitrary locally compact group.

1.3.7 Definition. For every $\mu \in \mathscr{M}_{\mathbb{C}}^b(G)$ the linear functional $\hat{\mu} := \operatorname{Res}_{\mathfrak{K}(G)} \mu$ on $\mathfrak{K}(G)$ is called the *Fourier transform* of μ.

If G is an Abelian locally compact group, the Fourier transform $\hat{\mu}$ of a measure $\mu \in \mathscr{M}_{\mathbb{C}}^b(G)$ can be viewed as a function on the character group $G^\wedge$ of G

since in this case the algebra $\mathfrak{R}(G)$ of trigonometric polynomials of G is generated by $G^\wedge$.

In the case of a compact (not necessarily Abelian) group G it is convenient to consider the Fourier transform $\hat{\mu}$ of a measure $\mu \in \mathscr{M}_{\mathbb{C}}^b(G)$ as a function on $\Sigma(G)$, and we shall write

$$\hat{\mu}(\sigma) = \hat{\mu}(D^{(\sigma)}) := \left(\int d_{ij}^{(\sigma)}(x)\, \mu(dx)\right)_{i,j=1,\ldots,n(\sigma)}$$

for all $\sigma \in \Sigma(G)$.

The Fourier transform $\hat{\phi}$ of a function $\phi \in \mathscr{L}_{\mathbb{C}}^1(G, \omega_G)$ is naturally defined as the Fourier transform $\widehat{\phi \cdot \omega_G}$ of the measure $\phi \cdot \omega_G \in \mathscr{M}_{\mathbb{C}}^b(G)$.

1.3.8 Theorem. *Let $G \in \mathbf{A}$. The Fourier mapping $\mathscr{F}: \mathscr{M}_{\mathbb{C}}^b(G) \to \mathfrak{R}(G)^*$ defined by $\mathscr{F}(\mu) := \hat{\mu}$ for all $\mu \in \mathscr{M}_{\mathbb{C}}^b(G)$ is an injective algebra homomorphism from $\mathscr{M}_{\mathbb{C}}^b(G)$ into $\mathscr{S}(G)$.*

Proof. 1. $\mathscr{F}$ maps $\mathscr{M}_{\mathbb{C}}^b(G)$ into $\mathscr{S}(G)$. Let $\mu \in \mathscr{M}_{\mathbb{C}}^b(G)$. We have to show the continuity of $\hat{\mu}$ on $\operatorname{Rep}(G)$. Indeed, for every $D \in \operatorname{Rep}(G)$ and $\varepsilon > 0$ there exists a compact subset K of G such that $|\mu|(\complement K) \leq \delta := \dfrac{\varepsilon}{2 + \|\mu\|}$.

Then for $D' \in W(D, K, \delta)$ one obtains $\|\hat{\mu}(D) - \hat{\mu}(D')\|_s \leq \varepsilon$, and hence $\hat{\mu} \in \mathscr{S}(G)$. Here $\|M\|_s$ denotes the spectral norm of the matrix $M \in \mathfrak{M}(n, \mathbb{C})$ defined as $\max_{1 \leq j \leq n} \sqrt{\beta_j}$, where $\beta_1, \ldots, \beta_n$ are the eigenvalues of the matrix $M\overline{M}^T$.

2. $\mathscr{F}$ is an algebra homomorphism from $\mathscr{M}_{\mathbb{C}}^b(G)$ into $\mathscr{S}(G)$ by Theorem 1.3.3.

3. It remains to show the injectivity of $\mathscr{F}$. Without loss of generality we assume that $\mu, \nu \in \mathscr{M}_{+}^b(G)$ are given with $\hat{\mu} = \hat{\nu}$. Let $(\tilde{G}, \beta)$ be a Bohr compactification of G. Since $\mathfrak{R}(G)$ is dense in $\mathfrak{A}(G)$, we immediately conclude $\beta(\mu) = \beta(\nu)$. Hence, for every compact subset K of G we get

$$\mu(K) = \mu(\beta^{-1}(\beta(K))) = \beta(\mu)(\beta(K)) = \beta(\nu)(\beta(K)) = \nu(K),$$

which implies that $\mu = \nu$. □

1.3.9. In addition to the compact open topology $\mathscr{T}_{co}$ the coefficient algebra $\mathfrak{R}(G)$ of a group $G \in \mathbf{A}$ carries the norm topology $\mathscr{T}_n$. One observes that $\mathscr{T}_n < \mathscr{T}_{co}$, but that $\mathscr{T}_n \neq \mathscr{T}_{co}$ in general. Also, linear functionals N on $\mathfrak{R}(G)$ which are continuous with respect to $\mathscr{T}_{co}$ or $\mathscr{T}_n$ are in general not Fourier transforms of measures in $\mathscr{M}_{\mathbb{C}}^b(G)$.

(a) Let $(\tilde{G}, \beta)$ be a Bohr compactification of $G \in \mathbf{A}$ and let $N \in (\mathfrak{R}(G), \mathscr{T}_n)'$ be defined by $N(f) := \omega_{\tilde{G}}(\tilde{f})$ for all $f \in \mathfrak{R}(G)$, where for $f \in \mathfrak{R}(G)$ the extended function $\tilde{f}$ is given by $f = \tilde{f} \circ \beta$. If in this case G is not compact, then there does not exist a measure $\mu \in \mathscr{M}_{\mathbb{C}}^b(G)$ with $\hat{\mu} = N$.

[Otherwise μ would be a finite Haar measure on G in contradiction to the hypothesis.]

(b) Let $\mu \in \mathscr{M}_{\mathbb{C}}^b(G)$ be chosen such that $\hat{\mu} \in (\mathfrak{R}(G), \mathscr{T}_{co})'$. By the equality

$$(\mathfrak{R}(G), \mathscr{T}_{co})^- = (\mathscr{C}_{\mathbb{C}}(G), \mathscr{T}_{co})$$

the linear functional $\hat{\mu}$ can be uniquely extended to an element L of $(\mathscr{C}_{\mathbb{C}}(G), \mathscr{T}_{co})'$. Therefore there exists exactly one $\nu \in \mathscr{M}^b_{\mathbb{C}}(G)$ with $\nu(f) = L(f)$ for all $f \in \mathscr{C}^b_{\mathbb{C}}(G)$. But Theorem 1.3.8 implies that $\nu = \mu$. Since μ is a $\mathscr{T}_{co}$-continuous linear functional on $\mathscr{C}_{\mathbb{C}}(G)$, there exist a compact subset K of G and a constant $c \in \mathbb{R}^*_+$ satisfying

$$|\mu(f)| \leq c\, p_K(f) \quad \text{with } p_K(f) := \sup_{x \in K} |f(x)|$$

for all $f \in \mathscr{C}_{\mathbb{C}}(G)$. Hence, $\operatorname{supp}(\mu) \subset K$ and so $\operatorname{supp}(\mu)$ is compact.

In other words, on a noncompact group $G \in \mathbf{A}$ there exist measures $\mu \in \mathscr{M}^b_{\mathbb{C}}(G)$ with a non-$\mathscr{T}_{co}$-continuous Fourier transform $\hat{\mu}$.

1.3.10 Definition. A net $(f_\alpha)_{\alpha \in \mathbb{A}}$ in $\mathfrak{K}(G)$ is called *F-convergent* to $f \in \mathfrak{K}(G)$ if
(i) $\mathscr{T}_{co}\text{-}\lim_{\alpha \in \mathbb{A}} f_\alpha = f$ and
(ii) $\{\|f_\alpha\| : \alpha \in \mathbb{A}\}$ is bounded.

A subset B of $\mathfrak{K}(G)$ is called *F-closed* if every F-convergent net in B has a limit in B.

The topology in $\mathfrak{K}(G)$ defined by the system of F-closed subsets of $\mathfrak{K}(G)$ is called the *Fourier* (*F-*) *topology* in $\mathfrak{K}(G)$ and will be abbreviated by $\mathscr{T}_F$. One shows without difficulty that $(\mathfrak{K}(G), \mathscr{T}_F)$ is a separated topological vector space and that the topologies in $\mathfrak{K}(G)$ are related via $\mathscr{T}_n < \mathscr{T}_F < \mathscr{T}_{co}$.

Moreover, the F-convergent nets and the $\mathscr{T}_F$-convergent nets in $\mathfrak{K}(G)$ coincide.

1.3.11 Theorem. *Let* $G \in \mathbf{A}$ *and* $N \in (\mathfrak{K}(G), \mathscr{T}_n)'$. *The following statements are equivalent:*
(i) *There exists a* $\mu \in \mathscr{M}^b_{\mathbb{C}}(G)$ $(\mu \in \mathscr{M}^b_+(G))$ *with* $\hat{\mu} = N$;
(ii) N *is* $\mathscr{T}_F$*-continuous (positive and* $\mathscr{T}_F$*-continuous);*
(iii) *writing* $B_0 := \{g \in \mathfrak{K}(G) : \|g\| \leq 1\}$ *we have that* $\operatorname{Res}_{B_0} N$ *is* $\mathscr{T}_{co}$*-continuous (positive and* $\mathscr{T}_{co}$*-continuous).*

Proof. 1. (i)$\Rightarrow$(ii). Let $N = \hat{\mu}$ with $\mu \in \mathscr{M}^b_{\mathbb{C}}(G)$ and let B' be a closed subset of $\mathbb{C}$. It is to be shown that $B := N^{-1}(B')$ is $\mathscr{T}_F$-closed. In fact, let $(f_\alpha)_{\alpha \in \mathbb{A}}$ be a $\mathscr{T}_F$-convergent net in B with $\lim_{\alpha \in \mathbb{A}} f_\alpha =: f \in \mathfrak{K}(G)$. For $\varepsilon > 0$ there exist a compact subset K of G with $|\mu|(\complement K) < \varepsilon$ and an $\alpha_0 \in \mathbb{A}$ with $|f(x) - f_\alpha(x)| \leq \varepsilon$ for all $x \in K$ and $\alpha > \alpha_0$.

Putting $c := \sup_{\alpha \in \mathbb{A}} \|f_\alpha\|$ we further obtain

$$\begin{aligned} |N(f) - N(f_\alpha)| &= |\textstyle\int (f - f_\alpha)\, d\mu| \\ &\leq \textstyle\int_K |f - f_\alpha|\, d|\mu| + \int_{\complement K} |f - f_\alpha|\, d|\mu| \leq \varepsilon\, \|\mu\| + (\|f\| + c)\,\varepsilon \\ &= \varepsilon(\|\mu\| + \|f\| + c) \end{aligned}$$

for all $\alpha > \alpha_0$, and hence $\lim_{\alpha \in \mathbb{A}} N(f_\alpha) = N(f)$. From $N(f_\alpha) \in B'$ for all $\alpha \in \mathbb{A}$ and the closedness of B' we conclude that $N(f) \in B'$ and thus $f \in B$. If $\mu \geq 0$, then clearly $\hat{\mu} = N \geq 0$.

2. (ii)$\Rightarrow$(iii). Let $(f_\alpha)_{\alpha \in \mathbb{A}}$ be a net in B_0 which is $\mathscr{T}_{co}$-convergent to $f \in B_0$. Since $\|f_\alpha\| \leq 1$ for all $\alpha \in \mathbb{A}$, we have $\mathscr{T}_F\text{-}\lim_{\alpha \in \mathbb{A}} f_\alpha = f$. By the $\mathscr{T}_F$-continuity of N it follows that $\lim_{\alpha \in \mathbb{A}} N(f_\alpha) = N(f)$. If N is positive, then so is $\operatorname{Res}_{B_0} N$.

3. (iii)$\Rightarrow$(i). The $\mathcal{T}_{co}$-closure of B_0 coincides with the unit ball of $\mathfrak{A}(G)$, which is dense in $B_1 := \{g \in \mathscr{C}_{\mathbb{C}}^b(G) : \|g\| \leq 1\}$, and $\mathrm{Res}_{B_0} N$ can be extended to a $\mathcal{T}_{co}$-continuous linear functional L on $\mathscr{C}_{\mathbb{C}}^b(G)$. But $\mathrm{Res}_{\mathfrak{R}(G)} L = N$.

Since $\mathrm{Res}_{B_1} L$ is $\mathcal{T}_{co}$-continuous, there exists $\mu \in \mathscr{M}_{\mathbb{C}}^b(G)$ with $\mu = L$, and thus $\hat{\mu} = N$. If $N \geq 0$, then also $L \geq 0$, and so $\mu \in \mathscr{M}_+^b(G)$. □

1.4 The Theorems of Lévy and Bochner

Let $G \in \mathbf{A}$. If $\mathfrak{R}(G)^*$ is furnished with the topology $\mathcal{T}_s$ of simple convergence on $\mathfrak{R}(G)$, then the Fourier mapping $\mathscr{F} : \mathscr{M}_{\mathbb{C}}^b(G) \to \mathscr{S}(G)$ is $\mathcal{T}_w$-$\mathcal{T}_s$-continuous. We shall now discuss the sequential bicontinuity of $\mathscr{F}$, which in the special case $G := \mathbb{R}$ was studied for the first time by P. Lévy.

For the following we introduce the notation

$$\mathscr{C}(\mathrm{Rep}(G)) := \mathscr{C}(\mathrm{Rep}(G), \textstyle\bigcup_{n \geq 1} \mathfrak{M}(n, \mathbb{C})).$$

1.4.1 Definition. A group $G \in \mathbf{A}$ is called a *Lévy group* (L-group) if for every sequence $(\mu_n)_{n \geq 1}$ in $\mathscr{M}_+^b(G)$ and $\phi \in \mathscr{C}(\mathrm{Rep}(G))$ with the property

$$\lim_{n \to \infty} \hat{\mu}_n(D) = \phi(D) \qquad \text{for all } D \in \mathrm{Rep}(G)$$

there exists a measure $\mu \in \mathscr{M}_+^b(G)$ such that one has
(a) $\hat{\mu} = \phi$ and
(b) $\mathcal{T}_w$-$\lim_{n \to \infty} \mu_n = \mu$.

The class of L-groups will be denoted by $\mathbf{L}$.
In order to obtain first examples of L-groups we establish the following theorem whose proof is of considerably greater generality than its statement.

1.4.2 Theorem. *Every locally compact Abelian group G is an L-group.*

Proof. Let G be a locally compact Abelian group with character group $G^\wedge$ and Haar measure $\omega := \omega_{G^\wedge}$ on $G^\wedge$.

1. We show that for a net $(\mu_\alpha)_{\alpha \in \mathbf{A}}$ in $\mathscr{M}_+^b(G)$ and a function $\phi \in \mathscr{C}(G^\wedge)$ such that

$$\mathcal{T}_v\text{-}\lim_{\alpha \in \mathbf{A}} \hat{\mu}_\alpha \cdot \omega = \phi \cdot \omega \quad \text{and} \quad \lim_{\alpha \in \mathbf{A}} \hat{\mu}_\alpha(1) = \phi(1)$$

hold there exists a measure $\mu \in \mathscr{M}_+^b(G)$ satisfying $\hat{\mu} = \phi$ and

$$\mathcal{T}_w\text{-}\lim_{\alpha \in \mathbf{A}} \mu_\alpha = \mu.$$

Indeed, for $\mu \in \mathscr{M}_+^b(G)$ and $f \in \mathscr{K}(G^\wedge)$ we have $\int f \hat{\mu} \, d\omega = \int \hat{f} \, d\mu$. From

$$\lim_{\alpha \in \mathbf{A}} \hat{\mu}_\alpha(1) = \phi(1)$$

we conclude that the set $\{\|\mu_\alpha\| : \alpha \in \mathbb{A}, \alpha > \alpha_0\}$ is bounded for some $\alpha_0 \in \mathbb{A}$. Hence, there exists a subnet $(\mu_{\alpha(\beta)})_{\beta \in \mathbb{B}}$ of $(\mu_\alpha)_{\alpha \in \mathbb{A}}$ which $\mathcal{T}_v$-converges to μ. From the Riemann-Lebesgue lemma for Abelian groups we obtain $\hat{f} \in \mathscr{C}^0_{\mathbb{C}}(G)$, and hence,

$$\int f\phi\, d\omega = \lim_{\beta \in \mathbb{B}} \int f \hat{\mu}_{\alpha(\beta)}\, d\omega = \lim_{\beta \in \mathbb{B}} \int \hat{f}\, d\mu_{\alpha(\beta)}$$
$$= \int \hat{f}\, d\mu = \int f \hat{\mu}\, d\omega.$$

It follows $\int f(\phi - \hat{\mu})\, d\omega = 0$ for all $f \in \mathscr{K}(G^\wedge)$. The continuity of ϕ and $\hat{\mu}$ implies $\phi = \hat{\mu}$. Furthermore, we have

$$\lim_{\beta \in \mathbb{B}} \mu_{\alpha(\beta)}(1) = \mu(1),$$

and the net $(\mu_{\alpha(\beta)})_{\beta \in \mathbb{B}}$ $\mathcal{T}_w$-converges to μ. Since by Theorem 1.3.8 the Fourier mapping is injective, every $\mathcal{T}_w$-convergent subnet $(\mu_{\alpha(\beta)})_{\beta \in \mathbb{B}}$ of $(\mu_\alpha)_{\alpha \in \mathbb{A}}$ converges to μ.

2. Let $(\mu_n)_{n \geq 1}$ be a sequence in $\mathscr{M}^b_+(G)$ and $\phi \in \mathscr{C}(G^\wedge)$ such that

$$\lim_{n \to \infty} \hat{\mu}_n(\chi) = \phi(\chi)$$

for ω-a.a. $\chi \in G^\wedge$ and $\lim_{n \to \infty} \hat{\mu}_n(1) = \phi(1)$ hold. Then there exists a measure $\mu \in \mathscr{M}^b_+(G)$ with the properties (a) and (b) of the above definition. Indeed, for every $f \in \mathscr{K}(G^\wedge)$ the sequence $(f\hat{\mu}_n)_{n \geq 1}$ in $\mathscr{L}^1_{\mathbb{C}}(G^\wedge, \omega)$ is bounded by $a|f|$ with an appropriate $a \in \mathbb{R}^*_+$ and it converges ω-a.e. on $G^\wedge$ to $f\phi$. The dominated convergence theorem yields $\lim_{n \to \infty} \int f \hat{\mu}_n\, d\omega = \int f\phi\, d\omega$ and thus the assumption of Part 1. The assertion follows. □

1.4.3 Theorem. *Let $G \in \mathbf{A}$ and let K be a compact normal subgroup of G such that $G/K \in \mathbf{L}$. Then $G \in \mathbf{L}$.*

Proof. Let $(\mu_n)_{n \geq 1}$ be a sequence in $\mathscr{M}^b_+(G)$ and let $\phi \in \mathscr{C}(\mathrm{Rep}(G))$ be such that

$$\lim_{n \to \infty} \hat{\mu}(D) = \phi(D) \qquad \text{for all } D \in \mathrm{Rep}(G).$$

By p we denote as usual the canonical homomorphism from G onto $\dot{G} := G/K$. For all $D \in \mathrm{Rep}(G)$ we obtain

$$p(\phi)(D) = \phi(D \circ p) \quad \text{and hence} \quad \lim_{n \to \infty} p(\mu_n)^\wedge(D) = p(\phi)(D).$$

Since $\mathrm{Rep}(\dot{G}) \cong K^\perp$, one has $p(\phi) \in \mathscr{C}(\mathrm{Rep}(\dot{G}))$. Thus there exists a measure $\dot{\mu} \in \mathscr{M}^b_+(\dot{G})$ such that $\hat{\dot{\mu}} = p(\phi)$ and $\lim_{n \to \infty} p(\mu_n) = \dot{\mu}$ hold. We conclude that the set $\{p(\mu_n) : n \geq 1\} \cup \{\dot{\mu}\}$ is compact.

Consequently, to every $\varepsilon > 0$ there is a compact subset $\dot{C} := \dot{C}_\varepsilon$ of $\dot{G}$ with $p(\mu_n)(\complement \dot{C}) \leq \varepsilon$ for all $n \geq 1$. Since $C := p^{-1}(\dot{C})$ is compact in G we obtain $\mu_n(\complement C) \leq \varepsilon$ for all $n \geq 1$ and therefore by Prohorov's theorem the relative compactness of $(\mu_n)_{n \geq 1}$. The assumption together with the injectivity of the Fourier transform yield the result. □

1.4.4 Theorem. *Let $G \in \mathbf{A}$ and let H be a closed subgroup of G with $[G:H] < \infty$. If $H \in \mathbf{L}$, then also $G \in \mathbf{L}$.*

Proof. 1. Let $G=a_1H\cup\cdots\cup a_nH$ be a decomposition of G as a disjoint union of H-left cosets such that $a_1=e$. For $f\in\mathscr{C}^b(H)$ we define $f'\in\mathscr{C}^b(G)$ by $f'(x):=f(x)$ if $x\in H$, and $f'(x)=0$ otherwise.

Furthermore, we define for every $g\in\mathscr{C}^b(G)$ and every $i=1,\ldots,n$ the function

$$g_i:=\operatorname{Res}_H g\circ L(a_i).$$

Clearly, $g_i\in\mathscr{C}^b(H)$ for $i=1,\ldots,n$, and one has the representation

$$g=\textstyle\sum_{i=1}^n g_i'\circ L(a_i^{-1}).$$

If, in particular, $f\in\mathfrak{A}(H)$, then obviously $f'\in\mathfrak{A}(G)$, and for $g\in\mathfrak{A}(G)$ the above representation is given with functions $g_i\in\mathfrak{A}(H)$. Let ψ be a positive linear functional on $\mathfrak{A}(G)$. Then for every $i=1,\ldots,n$ there exists a positive linear functional ψ_i on $\mathfrak{A}(H)$ defined by $\psi_i(h):=\psi(h'\circ L(a_i^{-1}))$ for all $h\in\mathfrak{A}(H)$ such that

$$\psi(g)=\textstyle\sum_{i=1}^n\psi_i(g_i)\quad\text{for all } g\in\mathfrak{A}(G). \tag{4.1}$$

2. Let $(\mu_m)_{m\geq 1}$ be a sequence in $\mathscr{M}_+^b(G)$ and $\phi\in\mathscr{C}(\operatorname{Rep}(G))$ such that

$$\lim_{m\to\infty}\hat\mu_m(D)=\phi(D)\qquad\text{holds for all } D\in\operatorname{Rep}(G).$$

Extending ϕ to a positive linear functional ψ on $\mathfrak{A}(G)$ we get the representation (4.1) of Part 1 with positive linear functionals ψ_i on $\mathfrak{A}(H)$ whose restrictions ϕ_i to $\operatorname{Rep}(H)$ are continuous ($i=1,\ldots,n$). In fact, for every $D\in\operatorname{Rep}(H)$ we denote by $\operatorname{Ind}(D)$ the representation induced on G by D. Let $D_0\in\operatorname{Rep}(H)$ and $\varepsilon>0$ be given. Since ϕ is continuous on $\operatorname{Rep}(G)$ there exist a compact subset K of G and a number $\delta\in\mathbb{R}_+^*$ such that for all $D\in W(\operatorname{Ind}(D_0);K,\delta)$ one has $\|\phi(D)-\phi(\operatorname{Ind}(D_0))\|<\varepsilon$.

The subset $C:=\bigcup_{i,j=1,\ldots,n}(a_i^{-1}Ka_j\cap H)$ of H is compact, and for all $D\in W\left(D_0;C,\frac{\delta}{n}\right)$ we obtain $\|\phi(\operatorname{Ind}(D))-\phi(\operatorname{Ind}(D_0))\|<\varepsilon$. Indeed, let $x\in K$. Then for every pair $(i,j)\in\{1,\ldots,n\}^2$ one has either $a_i^{-1}xa_i\in C$ or $a_i^{-1}xa_j\notin H$. The definition of $\operatorname{Ind}(D)$ yields

$$\begin{aligned}&\|\operatorname{Ind}(D)(x)-\operatorname{Ind}(D_0)(x)\|^2\\&=\textstyle\sum_{a_i^{-1}xa_j\in C,\,i,j=1,\ldots,n}\|D(a_i^{-1}xa_j)-D_0(a_i^{-1}xa_j)\|^2\\&\leq\textstyle\sum_{i,j=1,\ldots,n}\left(\frac{\delta}{n}\right)^2=\delta^2,\quad\text{and thus}\quad \operatorname{Ind}(D)\in W(\operatorname{Ind}(D_0);K,\delta).\end{aligned}$$

Now let $D\in W\left(D_0;C,\frac{\delta}{n}\right)$ and put $D':=(d_{ij}(D)')_{i,j=1,\ldots,n(D)}$. Then $D'\circ L(a_i^{-1})$ is a block of $\operatorname{Ind}(D)$ and hence

$$\begin{aligned}&\|\phi_i(D)-\phi_i(D_0)\|^2\\&=\|\phi(D'\circ L(a_i^{-1}))-\phi(D_0'\circ L(a_i^{-1}))\|^2\leq\|\phi(\operatorname{Ind}(D))-\phi(\operatorname{Ind}(D_0))\|^2\leq\varepsilon^2.\end{aligned}$$

3. Let $\psi^{(m)}:=\hat\mu_m$ in the sense that $\hat\mu_m(f)=\int f\,d\mu_m$ for all $f\in\mathfrak{A}(G)$. For every $i=1,\ldots,n$ we put $\mu_m^{(i)}:=\operatorname{Res}_{a_iH}\mu_m$ and obtain

$$\psi_i^{(m)} = [L(a_i^{-1})\mu_m^{(i)}]^\wedge, \quad \text{where } L(a_i^{-1})(\mu_m^{(i)}) \in \mathcal{M}_+^b(H).$$

On the other hand,

$$\psi_i^{(m)}(h) = \int h' \circ L(a_i^{-1})\, d\mu_m \quad \text{for all } h \in \mathfrak{A}(H)$$

and therefore

$$\lim_{m\to\infty} [L(a_i^{-1})(\mu_m^{(i)})]^\wedge(h) = \phi_i(h) \quad \text{for all } h \in \mathfrak{A}(H).$$

Since $H \in \mathbf{L}$, there exists a $\mu_i \in \mathcal{M}_+^b(H)$ with $\hat{\mu}_i = \phi_i$ and

$$\lim_{m\to\infty} L(a_i^{-1})(\mu_m^{(i)}) = \mu_i \quad \text{for } i = 1, \ldots, n.$$

Then $\mu := \sum_{i=1}^n L(a_i)(\mu_i) \in \mathcal{M}_+^b(G)$ with $\hat{\mu} = \phi$. Indeed, for any $g \in \mathfrak{A}(G)$ one obtains

$$\hat{\mu}(g) = \sum_{i=1}^n [L(a_i)\mu_i]^\wedge(g) = \sum_{i=1}^n \hat{\mu}_i(g \circ L(a_i))$$

$$= \sum_{i=1}^n \hat{\mu}_i(\mathrm{Res}_H(g \circ L(a_i))) = \sum_{i=1}^n \phi_i(g_i) = \phi(g).$$

But for $g \in \mathscr{C}^b(G)$ one has

$$\mu(g) = \sum_{i=1}^n \mu_i(g_i) \quad \text{and} \quad \mu_m(g) = \sum_{i=1}^n L(a_i^{-1})\mu_m^{(i)}(g_i) \quad (m \geq 1)$$

and hence

$$\lim_{m\to\infty} \mu_m(g) = \lim_{m\to\infty} \sum_{i=1}^n L(a_i^{-1})\mu_m^{(i)}(g_i) = \sum_{i=1}^n \mu_i(g_i) = \mu(g).$$

Thus, $G \in \mathbf{L}$. □

1.4.5 Theorem. *Every Moore group is an L-group.*

Proof. Every Moore group G is by Theorem O a finite extension of a Takahashi group H. By Theorem 1.4.4 it therefore suffices to prove the assertion for a Takahashi group H.

This, however, follows from Theorem 1.4.2 with the help of Theorem 1.4.3, since $H \in \mathbf{A}$, $K(H)$ is compact and $H/K(H)$ is Abelian. □

For a locally compact Abelian group G Bochner's theorem can be reformulated as follows: A linear functional ϕ on $\mathfrak{K}(G)$ is the Fourier transform of the measure $\mu \in \mathcal{M}_+^b(G)$ iff ϕ is continuous on $\mathrm{Rep}(G)$ and satisfies the condition $\phi(f\bar{f}) \geq 0$ for all $f \in \mathfrak{K}(G)$. This statement motivates the following

1.4.6 Definition. For any group $G \in \mathbf{A}$ a linear functional ϕ on $\mathfrak{K}(G)$ is called a *positive-definite* form on $\mathfrak{K}(G)$ if $\phi(f\bar{f}) \geq 0$ for all $f \in \mathfrak{K}(G)$.

In order to obtain generalizations of Bochner's theorem to larger classes of almost periodic groups we axiomatize the statement of the theorem of Bochner in

1.4.7 Definition. A group $G \in \mathbf{A}$ is said to be a *Bochner group* (B-group) if every continuous positive-definite form on $\mathfrak{K}(G)$ is the Fourier transform of a measure in $\mathcal{M}_+^b(G)$. The class of all B-groups will be denoted by $\mathbf{B}$.

The above mentioned classical result shows that every locally compact Abelian group is a B-group. In order to provide further examples of B-groups we have to prove

1.4.8 Theorem. *Let $G \in \mathbf{A}$. Then a linear functional ϕ on $\mathfrak{R}(G)$ is a positive-definite form on $\mathfrak{R}(G)$ iff ϕ is positive. In either case ϕ is continuous.*

Proof. It remains to prove that given $\phi \in \mathfrak{R}(G)^*$ the condition
(a) $\phi(|f|^2)=\phi(f\bar{f})\geq 0$ for all $f\in\mathfrak{R}(G)$ implies the condition
(b) $\phi(f)\geq 0$ for all $f\in\mathfrak{R}_+(G)$.

Without loss of generality we assume G to be compact.

First of all we observe that for any continuous $\phi\in\mathfrak{R}(G)^*$ the desired implication (a)$\Rightarrow$(b) holds. In fact, for every $g\in\mathfrak{R}_+(G)$ there exists a sequence $(t_n)_{n\geq 1}$ in $\mathfrak{R}(G)$ with the property that

$$\lim_{n\to\infty}\|g-t_n\bar{t}_n\|=0,$$

which by the continuity of ϕ implies the limit relationship

$$\lim_{n\to\infty}(\phi(g)-\phi(t_n\bar{t}_n))=0.$$

Thus by assumption $\phi(g)\geq 0$.

2. Let $\phi\in\mathfrak{R}(G)^*$ be given such that (a) holds. For any $h\in\mathfrak{R}_+(G)$ we define $\phi^{(h)}\in\mathfrak{R}(G)^*$ by $\phi^{(h)}(f):=\int h(\underline{\phi} f)^*\,d\omega$ for all $f\in\mathfrak{R}(G)$, where $\underline{\phi}$ is given by

$$\underline{\phi}(f)(x):=\phi(f_x)\quad\text{for all } f\in\mathfrak{R}(G) \text{ and } x\in G.$$

By hypothesis, one obtains $\phi^{(h)}(|f|^2)\geq 0$ for all $f\in\mathfrak{R}(G)$.

3. Let X be a finite subset of $\Sigma(G)$ such that $h\in\mathfrak{R}^X(G)$. By the orthogonality relations one has $\phi^{(h)}(d_{jk}^{(\sigma)})\neq 0$ only for $\sigma\in X$, j, $k=1,\dots,n(\sigma)$. Every $f\in\mathfrak{R}(G)$ is a finite sum of the form

$$f=\sum_{\sigma\in\Sigma(G)}\sum_{j,k=1}^{n(\sigma)}\Big(n(\sigma)\int_G f\,\overline{d_{jk}^{(\sigma)}}\,d\omega\Big)\,d_{jk}^{(\sigma)}\quad\text{and hence}$$
$$|\phi^{(h)}(f)|\leq c\,\|f\|\max\{\|d_{jk}^{(\sigma)}\|:\sigma\in X, j,k=1,\dots,n(\sigma)\}$$
$$\text{with } c:=c(h)\in\mathbb{R}_+^*.$$

Consequently, $\phi^{(h)}$ is continuous and satisfies (a). By Part 1 it therefore satisfies (b).

4. An arbitrary $g\in\mathfrak{R}(G)$ is of the form $g=\sum_{\sigma\in X}\sum_{j,k=1}^{n(\sigma)}\alpha_{jk}^{(\sigma)}d_{jk}^{(\sigma)}$ with a finite subset X of $\Sigma(G)$ and $\alpha_{jk}^{(\sigma)}\in\mathbb{C}$ for $\sigma\in X$; j, $k=1,\dots,n(\sigma)$. Any $h\in\mathfrak{R}_+(G)$ such that $h(xy)=h(yx)$ for all $x,y\in G$ is of the form $h=\sum_{\tau\in Y}\beta_\tau\chi^{(\tau)}$ with a finite subset Y of $\Sigma(G)$, $\beta_\tau\in\mathbb{C}$ and characters $\chi^{(\tau)}$ of the classes $\tau\in Y$ defined by $\chi^{(\tau)}(x):=\mathrm{tr}(D^{(\tau)}(x))$ for all $x\in G$. This representation is a consequence of the Peter-Weyl theorem ([218], (28.50)). Using again the orthogonality relations one obtains

$$\text{(4.2)}\qquad \phi^{(h)}(g)=\sum_{\sigma\in X\cap Y}\sum_{j,k=1}^{n(\sigma)}\beta_\sigma\frac{1}{n(\sigma)}\alpha_{jk}^{(\sigma)}\phi(d_{jk}^{(\sigma)}).$$

If $g\in\mathfrak{R}_+(G)$, then by Part 2 we have $\phi^{(h)}(g)\geq 0$. Now let $P:=[X]$ and let $(h_n)_{n\geq 1}$ be an approximate unit in $\mathfrak{R}_+^P(G)$ in the sense of [218] (28.57) satisfying

(α) $h_n(xy) = h_n(yx)$ for all $x, y \in G, n \geq 1$,
(β) $\hat{h}_n(\sigma)$ is a nonnegative multiple of $E_{n(\sigma)}$ for all $\sigma \in \Sigma(G)$, $n \geq 1$,
(γ) $\lim_{n\to\infty} \hat{h}_n(\sigma) = E_{n(\sigma)}$ for all $\sigma \in P$.

For every $n \geq 1$ and all $\sigma \in \Sigma(G)$ we therefore obtain the existence of $\beta_\sigma^{(n)} \in \mathbb{R}_+$ with

$$\hat{h}_n(\sigma) = \frac{1}{n(\sigma)} \beta_\sigma^{(n)} E_{n(\sigma)},$$

and for all $\sigma \in P$ we get $\lim_{n\to\infty} \frac{1}{n(\sigma)} \beta_\sigma^{(n)} = 1$. Another application of the Peter-Weyl theorem yields $h_n = \sum_{\sigma \in \Sigma(G)} \beta_\sigma^{(n)} \chi^{(\sigma)}$ for all $n \geq 1$, and (4.2) with h_n instead of h implies that

$$\lim_{n\to\infty} \phi^{(h_n)}(g) = \sum_{\sigma \in X} \sum_{j,k=1}^{n(\sigma)} \alpha_{jk}^{(\sigma)} \phi(d_{jk}^{(\sigma)}) = \phi(g) \geq 0. \quad \square$$

1.4.9 Theorem. *Any compact group is a B-group.*

Proof. Let G be a compact group. By Theorem 1.4.8 any positive-definite form ϕ on $\mathfrak{K}(G)$ is uniquely extendable to a positive linear form on $\mathfrak{A}(G)$ which by $\mathfrak{A}(G) = \mathscr{K}(G)$ for a compact group G is a Radon measure on G. $\square$

1.4.10 Theorem. *Let $G \in \mathbf{A}$ and K a compact normal subgroup of G such that $G/K \in \mathbf{B}$. Then $G \in \mathbf{B}$.*

Proof. Let ϕ be a positive linear functional on $\mathfrak{A}(G)$ which is continuous on $\operatorname{Rep}(G)$. By p we denote the canonical homomorphism from G onto $\dot{G} := G/K$. We define a positive linear functional $p(\phi)$ on $\mathfrak{K}(\dot{G})$ by $p(\phi)(\dot{D}) := \phi(\dot{D} \circ p)$ for all $\dot{D} \in \operatorname{Rep}(\dot{G})$ and observe that $p(\phi)$ is in fact continuous on $\operatorname{Rep}(\dot{G})$. Now let $\dot{G} \in \mathbf{B}$. Then there exists a $\dot{\mu} \in \mathscr{M}_+^b(\dot{G})$ such that $\hat{\dot{\mu}} = p(\phi)$ holds. Let $(\tilde{G}, \beta)$ denote a Bohr compactification of G. For every $D \in \operatorname{Rep}(G)$ we consider the representation $\tilde{D} \in \operatorname{Rep}(\tilde{G})$ defined by $D = \tilde{D} \circ \beta$. Since $\tilde{G} \in \mathbf{B}$ by Theorem 1.4.9, there exists a $\tilde{\mu} \in \mathscr{M}_+^b(\tilde{G})$ with $\hat{\tilde{\mu}}(\tilde{D}) = \phi(D)$ for all $D \in \operatorname{Rep}(G)$. Denoting the canonical homomorphism from $\tilde{G}$ onto $\tilde{G}/\beta(K)$ by $\tilde{p}$ and defining a mapping $\dot{\beta}$ from $\dot{G}$ into $\tilde{G}/\beta(K)$ by $\dot{\beta} \circ p = \tilde{p} \circ \beta$ we obtain a Bohr compactification $(\tilde{G}/\beta(K), \dot{\beta})$ of $\dot{G}$. For all $\dot{D} \in \operatorname{Rep}(\dot{G})$ we then have

$$\begin{aligned} \tilde{p}(\tilde{\mu})^\wedge(\tilde{\dot{D}} \circ \tilde{p}) &= \hat{\tilde{\mu}}(\tilde{\dot{D}} \circ \tilde{p}) = \hat{\tilde{\mu}}((\dot{D} \circ p)^\sim) \\ &= \phi(\dot{D} \circ p) = p(\phi)(\dot{D}) = \hat{\dot{\mu}}(\dot{D}) = \hat{\dot{\mu}}(\tilde{\dot{D}} \circ \dot{\beta}) = \dot{\beta}(\dot{\mu})^\wedge(\tilde{\dot{D}}). \end{aligned}$$

The injectivity of the Fourier transform yields $\tilde{p}(\tilde{\mu}) = \dot{\beta}(\dot{\mu})$. Without loss of generality we assume that $\phi(1) = 1$. For every $\varepsilon > 0$ there exists a compact subset $\dot{C} := \dot{C}_\varepsilon$ of $\dot{G}$ with the property $\dot{\mu}(\dot{C}) > 1 - \varepsilon$. Then $C := p^{-1}(\dot{C})$ is a compact subset of G satisfying the chain of relations

$$\begin{aligned} 1 - \varepsilon &< \dot{\beta}(\dot{\mu})(\dot{\beta}(\dot{C})) = \tilde{p}(\tilde{\mu})(\dot{\beta}(p(C))) = \tilde{p}(\tilde{\mu})(\tilde{p}(\beta(C))) \\ &= \tilde{\mu}(\beta(C)\,\beta(K)) = \tilde{\mu}(\beta(CK)) = \tilde{\mu}(\beta(C)). \end{aligned}$$

Thus there exists a $\mu \in \mathcal{M}_+^b(G)$ such that $\beta(\mu)=\tilde{\mu}$ holds. From the equalities

$$\hat{\mu}(D)=\hat{\mu}(\tilde{D}\circ\beta)=\beta(\mu)^\wedge(\tilde{D})=\hat{\tilde{\mu}}(\tilde{D})=\phi(D)$$

valid for any $D\in\mathrm{Rep}(G)$ one concludes that $\hat{\mu}=\phi$. □

1.4.11 Theorem. *Let $G\in\mathbf{A}$ and let H be a closed subgroup of G with $[G:H]<\infty$. If $H\in\mathbf{B}$, then also $G\in\mathbf{B}$.*

The *proof* is established in complete analogy to the proof of Theorem 1.4.4. First of all one shows that given a coset decomposition $G=a_1H\cup\cdots\cup a_nH$ of G any positive linear functional ϕ on $\mathfrak{A}(G)$ continuous on $\mathrm{Rep}(G)$ can be written in the form $\phi(g)=\sum_{i=1}^n \phi_i(g_i)$ for $g\in\mathfrak{R}(G)$, where

$$g_i := \mathrm{Res}_H\, g\circ L(a_i) \text{ and}$$
$$\phi_i(h) := \phi(h'\circ L(a_i^{-1})) \quad \text{for all } h\in\mathfrak{R}(H)$$

with $h'(x):=h(x)$, if $x\in H$, and $h'(x):=0$ otherwise. For fixed $i=1,\ldots,n$ the function ϕ_i is a positive linear functional on $\mathfrak{A}(H)$ continuous on $\mathrm{Rep}(H)$. Since $H\in\mathbf{B}$, there exists a $\mu_i\in\mathcal{M}_+^b(H)$ with $\hat{\mu}_i=\phi_i$. Considering μ_i as a measure on G one obtains for

$$\mu := \sum_{i=1}^n L(a_i)(\mu_i)\in\mathcal{M}_+^b(G)$$

the desired equality $\hat{\mu}=\phi$. Thus $G\in\mathbf{B}$. □

1.4.12 Theorem. *Any Moore group is a B-group.*

The *proof* follows like that of Theorem 1.4.5 from Theorems 1.4.10 and 1.4.11 by the structure theorem O for Moore groups. □

We add two more

1.4.13 Permanence Properties of the class **B**.

(1) Let G be a locally compact group which is the projective limit $\varprojlim_{\alpha\in\mathbb{A}} G_\alpha$ of a projective system $(G_\alpha, p_{\alpha\beta}, \mathbb{A})$ of groups $G_\alpha := G/K_\alpha\,(\alpha\in\mathbb{A})$, where $(K_\alpha)_{\alpha\in\mathbb{A}}$ is a descending system of compact normal subgroups of G satisfying $\bigcap_{\alpha\in\mathbb{A}} K_\alpha=\{e\}$.

If $G_\alpha\in\mathbf{B}$ for all $\alpha\in\mathbb{A}$, then $G\in\mathbf{B}$.

This statement is an immediate consequence of Theorem 1.4.10.

(2) If $G_1, G_2\in\mathbf{B}$, then $G:=G_1\times G_2\in\mathbf{B}$.

In fact, for $i=1,2$ let the symbol p_i denote the canonical projection from G onto G_i and let $(\tilde{G}_i,\beta_i)$ be a Bohr compactification of G_i. Clearly $(\tilde{G}_1\times\tilde{G}_2, \beta_1\times\beta_2)$ is a Bohr compactification $(\tilde{G},\beta)$ of G with the property $\beta_i\circ p_i=\tilde{p}_i\circ\beta$ for $i=1,2$. Let ϕ be a positive linear functional on $\mathfrak{A}(G)$ continuous on $\mathrm{Rep}(G)$. By assumption there exists for every $i=1,2$ a measure $\mu_i\in\mathcal{M}_+^b(G_i)$ such that $\hat{\mu}_i=p_i(\phi)$. Furthermore, there exists a $\tilde{\mu}\in\mathcal{M}_+^b(\tilde{G})$ satisfying $\hat{\tilde{\mu}}=\beta(\phi)$. Analogous to the proof of Theorem 1.4.10 one obtains $\tilde{p}_i(\tilde{\mu})=\beta_i(\mu_i)$ for $i=1,2$. For every $\varepsilon>0$ and $i=1,2$ there exists a compact subset C_i of G_i with $\mu_i(\complement C_i)<\frac{\varepsilon}{2}$. Therefore one obtains a compact subset $C := C_1 \times C_2$ of $G := G_1 \times G_2$ satisfying

$$\begin{aligned}\tilde{\mu}(\complement\, \beta(C)) &= \tilde{\mu}(\complement\, (\beta_1(C_1) \times \beta_2(C_2)))\\ &\leq \tilde{\mu}(\complement\, (\beta_1(C_1) \times \tilde{G}_2)) + \tilde{\mu}(\complement\, (\tilde{G}_1 \times \beta_2(C_2)))\\ &= \sum_{i=1}^{2} \tilde{\mu}(\tilde{p}_i^{-1}(\complement\, \beta_i(C_i))) = \sum_{i=1}^{2} \tilde{p}_i(\tilde{\mu})(\complement\, \beta_i(C_i))\\ &= \sum_{i=1}^{2} \beta_i(\mu_i)(\complement\, \beta_i(C_i)) = \sum_{i=1}^{2} \mu_i(\complement\, C_i) < \varepsilon.\end{aligned}$$

Thus there is a $\mu \in \mathscr{M}_+^b(G)$ with $\beta(\mu) = \tilde{\mu}$. For any $D \in \mathrm{Rep}(G)$ one has

$$\hat{\mu}(D) = \hat{\mu}(\tilde{D} \circ \beta) = \hat{\tilde{\mu}}(\tilde{D}) = \beta(\phi)(\tilde{D}) = \phi(\tilde{D} \circ \beta) = \phi(D) \text{ or } \phi = \hat{\mu}.$$

1.4.14. Let G be a locally compact group with dual $\mathrm{Rep}(G)$. The set $\mathrm{Rep}(G)^\vee$ of all continuous quasirepresentations of $\mathrm{Rep}(G)$ is made a group by defining for elements Q, Q' of $\mathrm{Rep}(G)^\vee$ the (pointwise) product QQ' and inverse Q^{-1} by $QQ'(D) := Q(D)\,Q'(D)$ and $Q^{-1}(D) := Q(D)^{-1}$ for all $D \in \mathrm{Rep}(G)$ resp. Furnished with the compact open topology $\mathscr{T}_{co}$ the set $\mathrm{Rep}(G)^\vee$ becomes a Hausdorff (topological) group which is called the (Chu) *quasidual* of $\mathrm{Rep}(G)$. One notes that $\mathrm{Rep}(G)^\vee \in \mathbf{A}$. For each $x \in G$ one further defines the evaluation mapping Ω_x on $\mathrm{Rep}(G)$ by $\Omega_x(D) = D(x)$ for all $D \in \mathrm{Rep}(G)$ and the canonical homomorphism $\Omega\colon G \to \mathrm{Rep}(G)^\vee$ by $\Omega(x) := \Omega_x$ for all $x \in G$. Clearly, Ω is a continuous homomorphism which is injective iff $G \in \mathbf{A}$.

Definition. A locally compact group G with dual $\mathrm{Rep}(G)$ and quasidual $\mathrm{Rep}(G)^\vee$ is said to be a *Chu group* if the canonical homomorphism $\Omega\colon G \to \mathrm{Rep}(G)^\vee$ is surjective.

It is known that Abelian locally compact groups and arbitrary compact groups are Chu groups. The duality theorems of Pontryagin and of Tannaka yield the stronger statement that the canonical homomorphism is in fact a topological isomorphism. It can be shown that every Moore group is a Chu group in this stronger sense.

1.4.15 Theorem. *Every B-group is a Chu-group.*

Proof. Let $G \in \mathbf{B}$ and $Q \in \mathrm{Rep}(G)^\vee$. By Theorem 1.3.3 there exists a $Q' \in \mathfrak{R}(G)^*$ such that $\rho(Q') = Q$ (for the bijection $\rho\colon \mathfrak{R}(G)^* \to \mathfrak{O}(G)$ defined in that theorem) which by Theorem 1.3.5 is multiplicative and real. For every $f \in \mathfrak{R}(G)$ we have

$$Q'(f\bar{f}) = Q'(f)\,Q'(\bar{f}) = |Q'(f)|^2 \geq 0.$$

Since Q is assumed to be continuous on $\mathrm{Rep}(G)^\vee$, Q' is a continuous positive-definite form on $\mathfrak{R}(G)$. From $G \in \mathbf{B}$ follows the existence of $\mu \in \mathscr{M}_+^b(G)$ with $\hat{\mu} = Q$. Furthermore, there exists a $\nu \in \mathscr{M}_+^b(G)$ with $\hat{\nu} = Q^{-1}$. Hence, for every $D \in \mathrm{Rep}(G)$ we conclude that

$$(\mu * \nu)^\wedge\,(D) = \hat{\mu}(D)\,\hat{\nu}(D) = Q(D)\,Q^{-1}(D) = E_{n(D)} = \hat{\varepsilon}_e(D)$$

and thus $\mu * \nu = \varepsilon_e$.

Therefore, there exists an $x \in G$ such that $\mu = \varepsilon_x$ holds. Since

$$Q(D) = \hat{\mu}(D) = D(x) = \Omega_x(D) \quad \text{for all } D \in \operatorname{Rep}(G),$$

one finally obtains the assertion $Q = \Omega(x)$ for some $x \in G$. □

For measures on B-groups the Fourier transform admits a continuity property which completes the discussion of Theorem 1.3.11.

1.4.16 Theorem. *Let $G \in \mathbf{B}$ and $\phi \in \mathfrak{R}(G)^*$. The following statements are equivalent:*
(i) *There exists a $\mu \in \mathcal{M}^b(G)$ such that $\phi = \hat{\mu}$ holds;*
(ii) *$\phi \in \mathcal{S}(G)$, and there is a $c \in \mathbb{R}_+^*$ with $|\phi(f)| \leq c \|f\|$ for all $f \in \mathfrak{R}(G)$.*

Proof. It remains to prove (ii) ⇒ (i). By hypothesis, ϕ is extendable to a continuous linear functional on $\mathfrak{A}(G)$. Let $(\tilde{G}, \beta)$ be a Bohr compactification of G. There exists a $\tilde{\mu} \in \mathcal{M}^b(\tilde{G})$ with $\tilde{\mu}(f) = \phi(f)$ for all $f \in \mathfrak{A}(G)$. From the Radon-Nikodym theorem we obtain a measurable function g on $\tilde{G}$ with $|g| = 1$ such that $|\tilde{\mu}| = g \cdot \tilde{\mu}$. And by the Peter-Weyl theorem there is a sequence $(f_n)_{n \geq 1}$ *in* $\mathfrak{R}(G)$ with

$$\lim_{n \to \infty} \int |\tilde{f}_n - g| \, d|\tilde{\mu}| = 0.$$

For every $n \geq 1$ one defines a function ϕ_n on $\mathfrak{A}(G)$ by $\phi_n(f) := \int_{\tilde{G}} \tilde{f} \tilde{f}_n \, d\tilde{\mu}$ for all $f \in \mathfrak{R}(G)$, which is proved to be continuous on $\operatorname{Rep}(G)$. Defining also

$$\phi'(f) := \int_{\tilde{G}} \tilde{f} \, d|\tilde{\mu}| = \int_{\tilde{G}} \tilde{f} g \, d\tilde{\mu}$$

for all $f \in \mathfrak{A}(G)$ one gets

$$|\phi_n(f) - \phi'(f)| \leq \|f\| \int_{\tilde{G}} |\tilde{f}_n - g| \, d|\tilde{\mu}|$$

for all $f \in \mathfrak{A}(G)$, and thus convergence in the topology $\mathcal{T}_{co}$ on $\operatorname{Rep}(G)$ to ϕ'. It follows that ϕ' is continuous on $\operatorname{Rep}(G)$. Since ϕ' is a positive-definite form on $\mathfrak{R}(G)$ and $G \in \mathbf{B}$ by assumption, there is a measure $v \in \mathcal{M}_+^b(G)$ with $\hat{v} = \phi'$. For $f \in \mathfrak{A}(G)$ one obtains

$$\beta(v)(f) = \hat{v}(\tilde{f} \circ \beta) = \hat{v}(f) = \phi'(f) = |\tilde{\mu}|(\tilde{f})$$

and hence $\beta(v) = |\tilde{\mu}|$. We now define $h := g^{-1} \circ \beta$ and $\mu := h \cdot v$. Then

$$\begin{aligned} \hat{\mu}(f) &= \int_G f \, d\mu = \int_G f h \, dv = \int_G (\tilde{f} g^{-1}) \circ \beta \, dv \\ &= \int_{\tilde{G}} \tilde{f} g^{-1} \, d\beta(v) = \int_{\tilde{G}} f g^{-1} \, d|\tilde{\mu}| = \int_{\tilde{G}} \tilde{f} \, d\tilde{\mu} = \phi(f) \end{aligned}$$

for every $f \in \mathfrak{R}(G)$, or $\hat{\mu} = \phi$. □

1.5 Convolution Semigroups and Negative-Definite Forms

We complete the chapter with a discussion of various kinds of continuity arising in the study of convolution semigroups. A useful tool in this context is the notion of a measure operator.

Let E and T be two locally compact spaces. For a given function $f \in \mathscr{C}^b(T \times E)$ and any measure $\mu \in \mathscr{M}^b(E)$ we introduce the function F_μ on T by

$$F_\mu(t) := \int f(t, x)\, \mu(dx) \qquad \text{for all } t \in T.$$

We shall establish properties of the mapping $\mu \to F_\mu$ from $\mathscr{M}^b(E)$ into $\mathbb{R}^T$.

1.5.1. Let B denote the unit ball of the Banach space $\mathscr{C}^b(E)$. By $\mathscr{C}^b_c(E)$ we denote the space $\mathscr{C}^b(E)$ equipped with the compact open topology $\mathscr{T}_{co}$. Given any measure $\mu \in \mathscr{M}^b(E)$ we conclude that $\mathrm{Res}_B\, \mu$ is a continuous linear functional on $B \cap \mathscr{C}^b_c(E)$. In fact, let $\varepsilon > 0$ and K be a compact subset of E with $\mu(\complement K) < \varepsilon$. For a given $f \in B$ we pick a neighborhood $U \in \mathfrak{B}_B(f)$ with respect to the topology $\mathscr{T}_{co}$ on $\mathscr{C}^b(E)$ of the form

$$\{g \in B : \sup_{x \in K} |g(x) - f(x)| < \varepsilon\}.$$

Then for any given $g \in U$ one obtains

$$|\mu(g) - \mu(f)| \le \int_E |g - f|\, d\mu \le \varepsilon\, \mu(K) + 2\mu(\complement K) \le (\|\mu\| + 2)\, \varepsilon$$

since for all $x \in K$ we have

$$|g(x) - f(x)| \le \varepsilon \quad \text{and for all } y \in \complement K \text{ just } \quad |g(y) - f(y)| \le 2.$$

1.5.2. Let H be a bounded subset of the space $\mathscr{C}^b(E)$. Then the mapping $(\mu, f) \to \mu(f)$ from $\mathscr{M}^b_+(E) \times H$ into $\mathbb{R}$ is continuous ($\mathscr{M}^b(E)$ being furnished with the topology $\mathscr{T}_w$). Let $\mu \in \mathscr{M}^b_+(E)$, $f \in H$ and $M \in \mathbb{R}_+$, such that $\|\mu\| < M$ and $|g| \le M$ hold for all $g \in H$. For $\varepsilon > 0$ one chooses a subset K of E with $\mu(\complement K) < \varepsilon$. Moreover, let W be an open relatively compact neighborhood of K. Plainly the set

$$U := \{\lambda \in \mathscr{M}^b_+(E) : \lambda(E) < M, \lambda(\complement W) < \varepsilon \text{ and } |\lambda(f) - \mu(f)| < \varepsilon\}$$

is a neighborhood of μ in $\mathscr{M}^b_+(E)$. Furthermore, let V be the neighborhood of f in H of the form $\{g \in H : \sup_{x \in W} |g(x) - f(x)| < \varepsilon\}$.

Given $\lambda \in U$ and $g \in V$ and the fact that

$$|g(x) - f(x)| \le \varepsilon \qquad \text{for all } x \in W,$$
$$|g(x) - f(x)| \le 2M \qquad \text{for all } x \in \complement W$$

one obtains

$$|\lambda(g) - \lambda(f)| \le \int_E |g - f|\, d\lambda \le \varepsilon\, \lambda(W) + 2M\, \lambda(\complement W) \le 3M\varepsilon.$$

Hence,

$$|\lambda(g) - \mu(f)| \le |\lambda(g) - \lambda(f)| + |\lambda(f) - \mu(f)| \le (3M + 1)\, \varepsilon.$$

This, however, proves the continuity of $(\lambda, g) \to \lambda(g)$ at the point (μ, f) of $\mathscr{M}^b_+(E) \times H$.

1.5.3 Lemma. *Let E and T be locally compact spaces, and let f be in $\mathscr{C}^b(T \times E)$. The mapping $\mu \to F_\mu$ introduced above maps $\mathscr{M}^b(E)$ into $\mathscr{C}^b(T)$. Its restriction to $\mathscr{M}^b_+(E)$ is continuous from $\mathscr{M}^b_+(E)$ into $\mathscr{C}^b_c(T)$.*

Proof. For every $t \in T$ we denote the function $x \to f(t,x)$ by ${}_tf$. Clearly, ${}_tf \in \mathscr{C}^b(E)$ for all $t \in T$. The mapping $t \to {}_tf$ from T into $\mathscr{C}^b(E)$ has a bounded image and is continuous from T into $\mathscr{C}^b_c(E)$. One has $F_\mu(t) = \mu({}_tf)$ for all $t \in T$. Hence, by 1.5.1 $F_\mu \in \mathscr{C}^b(T)$. On the other hand, we know from 1.5.2 that the mapping $(\mu, t) \to F_\mu(t)$ from $\mathscr{M}^b_+(E) \times T$ into $\mathbb{R}$ is continuous, which yields the desired assertion. □

1.5.4. Let G be a locally compact group. For any $\mu \in \mathscr{M}^b_+(G)$ the *convolution operator* T_μ of μ is defined by $T_\mu f(x) := \mu({}_xf) = \int f(xy)\,\mu(dy)$ for all $f \in \mathscr{C}^b(G)$ and $x \in G$.

One notes that $T_\mu \mathscr{C}^b(G) \subset \mathscr{C}^b(G)$ and that T_μ is a bounded linear operator on $\mathscr{C}^b(G)$ with $\|T_\mu\| = \|\mu\|$.

If, in particular, $\mu \in \mathscr{M}^1(G)$, then T_μ is called the *probability operator* of μ. Moreover, in the general framework we have

(i) $T_\mu \mathscr{C}_u(G) \subset \mathscr{C}_u(G)$, $T_\mu \mathscr{C}^0(G) \subset \mathscr{C}^0(G)$ for $\mu \in \mathscr{M}^b_+(G)$ and

(ii) $T_{\mu * \nu} = T_\mu T_\nu$ whenever $\mu, \nu \in \mathscr{M}^b_+(G)$.

1.5.5 Theorem. *Let G be a locally compact group. For any net $(\mu_\alpha)_{\alpha \in \mathbb{A}}$ in $\mathscr{M}^1(G)$ the following statements are equivalent:*

(i) $\lim_{\alpha \in \mathbb{A}} \mu_\alpha = \mu \in \mathscr{M}^1(G)$ *in the sense of* $\mathscr{T}_w$;

(ii) $\lim_{\alpha \in \mathbb{A}} T_{\mu_\alpha} = T \in \mathscr{L}(\mathscr{C}^0(G))$ *with* $\|T\| = 1$ *in the sense of the strong operator topology.*

Proof. 1. (i) ⇒ (ii). Let $T := \mathrm{Res}_{\mathscr{C}^0(G)} T_\mu$. By $G_\infty := G \cup \{\infty\}$ we denote the one-point compactification of G. Putting $\infty x := \infty$ for all $x \in G$ we obtain a continuous mapping $(x, y) \to xy$ from $G_\infty \times G$ into G_∞. Extending every function $g \in \mathscr{C}^0(G)$ to a function $\bar{g} \in \mathscr{C}(G_\infty)$ by $g(\infty) := 0$ we conclude that the mapping $(t, x) \to f(t, x) := \bar{g}(tx)$ from $G_\infty \times G$ into $\mathbb{R}$ is continuous and bounded. By assumption $(\mu_\alpha)_{\alpha \in \mathbb{A}}$ converges. Hence, Lemma 1.5.3 implies that $(F_{\mu_\alpha})_{\alpha \in \mathbb{A}}$ converges to F_μ on $\mathscr{C}(G_\infty)$. From

$$F_{\mu_\alpha}(x) = (T_{\mu_\alpha} g)(x) \qquad \text{for all } x \in G$$

and

$$F_{\mu_\alpha}(\infty) = \int f(\infty, x)\,\mu_\alpha(dx) = \int \bar{g}(\infty)\,\mu_\alpha(dx) = 0$$

for all $\alpha \in \mathbb{A}$ the assertion follows.

2. (ii) ⇒ (i). By assumption T is a positive operator in $\mathscr{L}(\mathscr{C}^0(G))$ with $\|T\| = 1$. Thus there exists a measure $\mu \in \mathscr{M}^1(G)$ with $\mathrm{Res}_{\mathscr{C}^0(G)} T_\mu = T$. For every $f \in \mathscr{C}^0(G)$ we have

$$\lim_{\alpha \in \mathbb{A}} \mu_\alpha(f) = \lim_{\alpha \in \mathbb{A}} T_{\mu_\alpha} f(e) = T_\mu f(e) = \mu(f),$$

and so $\mathscr{T}_w$-$\lim_{\alpha \in \mathbb{A}} \mu_\alpha = \mu$ by Theorem 1.1.9. □

1.5.6. Let $\mathbb{D}$ be a dense subsemigroup of $\mathbb{R}^*_+$ with the property that for all $r, s \in \mathbb{D}$ such that $r < s$ holds one has $s - r \in \mathbb{D}$. $\mathbb{D}_0$ denotes the semigroup $\mathbb{D} \cup \{0\}$. If, in addition, G is a locally compact group and f is a homomorphism from $\mathbb{D}$ into $\mathscr{M}^1(G)$, we define $\mu_r := f(r)$ for all $r \in \mathbb{D}$ and call $S := (\mu_r)_{r \in \mathbb{D}}$ a *convolution semigroup* in $\mathscr{M}^1(G)$.

$S := (\mu_r)_{r \in \mathbb{D}}$ is called *continuous* if the homomorphism f is continuous, and 0-*continuous* if the limit $\mu_0 := \lim_{\substack{r \downarrow 0 \\ r \in \mathbb{D}}} \mu_r$ exists in $\mathscr{M}^1(G)$. It should be noted that the limit measure μ_0 is an idempotent element of $\mathscr{M}^1(G)$ and hence by Theorem 1.2.10 it is of the form $\mu_0 = \omega_H$ for some compact subgroup H of G. In this case we often speak of an *H-continuous* convolution semigroup in $\mathscr{M}^1(G)$. For $H = \{e\}$ we obtain $\{e\}$-continuous convolution semigroups.

We assume given a convolution semigroup $S := (\mu_r)_{r \in \mathbb{D}_0}$ in $\mathscr{M}^1(G)$ with corresponding semigroup $(T_{\mu_r})_{r \in \mathbb{D}_0}$ of convolution operators on $\mathscr{C}^b(G)$.

For each $r \in \mathbb{D}_0$ we define $S_r := \mathrm{Res}_{\mathscr{C}_u(G)} T_{\mu_r}$.

Putting $\phi(r) := S_r$ for each $r \in \mathbb{D}$ (or $\mathbb{D}_0$) one obtains a homomorphism ϕ from $\mathbb{D}$ (or $\mathbb{D}_0$) into $\mathscr{L}(\mathscr{C}_u(G))$.

1.5.7 Theorem. *Let G be a locally compact group, $S := (\mu_r)_{r \in \mathbb{D}_0}$ a 0-continuous convolution semigroup in $\mathscr{M}^1(G)$ and $\mu_0 \in \mathscr{M}^1(G)$ such that $\mu_0 * \mu_r = \mu_r * \mu_0 = \mu_r$ holds for all $r \in \mathbb{D}$. By $(S_r)_{r \in \mathbb{D}_0}$ we denote the corresponding semigroup of operators on $\mathscr{C}_u(G)$. Then*

(i) *the semigroup $(S_r)_{r \in \mathbb{D}_0}$ is uniformly continuous and*

(ii) *the semigroup S is continuous.*

Proof. (ii) follows from (i) since $\mu_r(f) = (S_r f)(e)$ for all $f \in \mathscr{C}_u(G)$ and $r \in \mathbb{D}_0$.

We shall prove (i). Since $\|S_r\| = \|\mu_r\| = 1$ and $S_r S_0 = S_0 S_r = S_r$ for all $r \in \mathbb{D}_0$, we have

$$\|S_{r+h} f - S_r f\| = \|S_r S_h f - S_r S_0 f\| \leq \|S_h f - S_0 f\|$$

and

$$\|S_{r-h} f - S_r f\| = \|S_{r-h} S_0 f - S_{r-h} S_h f\| \leq \|S_0 f - S_h f\|$$

for all $r, h \in \mathbb{D}_0$ and $f \in \mathscr{C}_u(G)$. It therefore suffices to show that ϕ defined in 1.5.6 is continuous in $r = 0$. For μ_0 there exists a compact subgroup H of G with $\mu_0 = \omega_H$ by 1.5.6. S_0 is therefore a projector on $\mathscr{C}_u(G)$.

1. Let $f \in \mathscr{C}_u(G)$ with $S_0 f = f$. For $\varepsilon > 0$ there exists a $U \in \mathfrak{V}(e)$ with

$$|f(x) - f(xy)| \leq \varepsilon \quad \text{for all } x \in G \quad \text{and} \quad y \in U.$$

Without loss of generality we assume $U = UH$, since $f(xz) = f(x)$ for all $z \in H$, $x \in G$. Then there is an $r_0 \in \mathbb{D}$ with $\mu_r(\complement U) \leq \varepsilon$ for all $r \in \mathbb{D}$, $r \leq r_0$. Hence, for all $x \in G$ and $r \in \mathbb{D}$, $r \leq r_0$ we have

$$\begin{aligned}
|f(x) - S_r f(x)| &\leq \int |f(x) - f(xy)| \, \mu_r(dy) \\
&= \int_U |f(x) - f(xy)| \, \mu_r(dy) + \int_{\complement U} |f(x) - f(xy)| \, \mu_r(dy) \\
&\leq \varepsilon \mu_r(U) + 2\|f\| \, \mu_r(\complement U) \leq \varepsilon(1 + 2\|f\|)
\end{aligned}$$

and thus $\lim_{\substack{r \downarrow 0 \\ r \in \mathbb{D}}} \|f - S_r f\| = 0$.

2. For $f \in \mathscr{C}_u(G)$ arbitrary we obtain $S_0 S_0 f = S_0 f$ and $S_r S_0 f = S_r f$. Therefore, $\lim_{\substack{r \downarrow 0 \\ r \in \mathbb{D}}} \|S_0 f - S_r f\| = 0$. □

1.5.8 Theorem. *Let G be a locally compact group and $S := (\mu_r)_{r \in \mathbb{D}}$ a convolution semigroup in $\mathscr{M}^1(G)$. The following statements are equivalent:*

(i) *S is 0-continuous;*
(ii) *S is continuous.*
In either case $\mu_0 = \lim_{\substack{r\downarrow 0\\ r\in\mathbb{D}}} \mu_r$ *is the identity of S.*

Proof. 1. (i) ⇒ (ii). Let S be 0-continuous with $\lim_{\substack{r\downarrow 0\\ r\in\mathbb{D}}} \mu_r = \mu_0$ and let $(r_\alpha)_{\alpha\in\mathbb{A}}$ be a net in $\mathbb{D}$ with $\lim_{\alpha\in\mathbb{A}} r_\alpha = 0$.

Since $\lim_{\alpha\in\mathbb{A}} \mu_{r_\alpha} = \mu_0$, Property (2) of 1.2.20 implies that $(\mu_{r_\alpha})_{\alpha\in\mathbb{A}}$ is tight.

Let $r\in\mathbb{D}$ and let without loss of generality $r_\alpha < r$ for all $\alpha\in\mathbb{A}$. From $\mu_r = \mu_{r_\alpha} * \mu_{r-r_\alpha}$ for all $\alpha\in\mathbb{A}$ one concludes that $(\mu_{r-r_\alpha})_{\alpha\in\mathbb{A}}$ is tight by Theorem 1.2.21. Hence, there exists a subnet $(\mu_{r-r_\beta})_{\beta\in\mathbb{B}}$ converging to some $\mu\in\mathcal{M}^1(G)$ by Property (1) of 1.2.20. In view of $\lim_{\beta\in\mathbb{B}} \mu_{r_\beta} = \mu_0$ the above equality yields $\mu_r = \mu_0 * \mu = \mu * \mu_0$ and hence $\mu_0 * \mu_r * \mu_0 = \mu_r$ for all $r\in\mathbb{D}$. This implies that μ_0 is the identity of S. By Theorem 1.5.7 the semigroup S is continuous.

2. (ii) ⇒ (i). For any net $(r_\alpha)_{\alpha\in\mathbb{A}}$ in $\mathbb{D}$ with $\lim_{\alpha\in\mathbb{A}} r_\alpha = 0$ and $r\in\mathbb{D}$ we have $\mu_{r+r_\alpha} = \mu_r * \mu_{r_\alpha}$ and $\lim_{\alpha\in\mathbb{A}} \mu_{r+r_\alpha} = \mu_r$. Therefore $(\mu_{r_\alpha})_{\alpha\in\mathbb{A}}$ is tight and hence again by Property (1) of 1.2.20 admits a subnet converging to $\mu\in\mathcal{M}^1(G)$. Without loss of generality we may assume that $\lim_{\alpha\in\mathbb{A}} \mu_{r_\alpha} = \mu$.

Let $(s_\beta)_{\beta\in\mathbb{B}}$ be another net in $\mathbb{D}$ with $\lim_{\beta\in\mathbb{B}} s_\beta = 0$ such that

$$\lim_{\beta\in\mathbb{B}} \mu_{s_\beta} = \mu' \in \mathcal{M}^1(G)$$

exists. By the continuity of the convolution mapping and of the semigroup S we obtain

$$\begin{aligned}\mu &= \lim_{\alpha\in\mathbb{A}} \mu_{r_\alpha} = \lim_{\alpha\in\mathbb{A}} \lim_{\beta\in\mathbb{B}} \mu_{r_\alpha + s_\beta}\\ &= \lim_{\alpha\in\mathbb{A}} (\mu_{r_\alpha} * \lim_{\beta\in\mathbb{B}} \mu_{s_\beta}) = \lim_{\alpha\in\mathbb{A}} \mu_{r_\alpha} * \mu' = \mu * \mu',\end{aligned}$$

and similarly, $\mu' = \mu * \mu'$. It therefore follows that $\mu' = \mu$ and hence

$$\mu_0 = \lim_{\substack{r\downarrow 0\\ r\in\mathbb{D}}} \mu_r. \quad \square$$

1.5.9 Theorem. *Let G be a locally compact group and* $S := (\mu_r)_{r\in\mathbb{D}_0}$ *a continuous convolution semigroup in* $\mathcal{M}^1(G)$. *There exists a unique continuous convolution semigroup* $(\nu_t)_{t\in\mathbb{R}_+}$ *in* $\mathcal{M}^1(G)$ *with* $\nu_r = \mu_r$ *for all* $r\in\mathbb{D}_0$.

Proof. 1. Put $\mathsf{X} := \mathscr{C}_u(G)$ and $S_r := \mathrm{Res}_{\mathsf{X}}\, T_{\mu_r}$ for all $r\in\mathbb{D}_0$ as above. Let $\mathscr{L}_s(\mathsf{X})$ denote the space $\mathscr{L}(\mathsf{X})$ equipped with the strong operator topology. By Theorem 1.5.7 the mapping $\phi: \mathbb{D}_0 \to \mathscr{L}_s(\mathsf{X})$ defined by $\phi(r) := S_r$ for all $r\in\mathbb{D}_0$ is a uniformly continuous homomorphism. Since X is barreled and complete, $\mathscr{L}_s(\mathsf{X})$ is quasi-complete ([40], p. 31). In view of $\|S_r f\| \le \|f\|$ for all $f\in\mathsf{X}$ and $r\in\mathbb{D}_0$ the set $\phi(\mathbb{D}_0)$ is bounded, so $\overline{\phi(\mathbb{D}_0)}$ is complete. As ϕ is uniformly continuous and $\mathbb{D}_0$ is dense in $\mathbb{R}_+$, ϕ is uniquely extendable to a continuous homomorphism $\bar{\phi}: \mathbb{R}_+ \to \mathscr{L}_s(\mathsf{X})$. Putting $S_t := \bar{\phi}(t)$ for all $t\in\mathbb{R}_+$ we obtain a strongly continuous semigroup $(S_t)_{t\in\mathbb{R}_+}$ in $\mathscr{L}_s(\mathsf{X})$ with $\|S_t\| \le 1$ for all $t\in\mathbb{R}_+$.

2. By assumption we conclude from Part 1 that there exists a strongly continuous operator semigroup $(T_t)_{t\in\mathbb{R}_+}$ on $\mathscr{C}^0(G)$ with

$$\|T_t\| = 1 \text{ for all } t\in\mathbb{R}_+ \quad\text{and}\quad T_r = \mathrm{Res}_{\mathscr{C}^0(G)}\, T_{\mu_r} \text{ for all } r\in\mathbb{D}_0.$$

By the Riesz representation theorem, for every $t\in\mathbb{R}_+$ there exists a measure $\nu_t\in\mathcal{M}^1(G)$ such that $T_t=\mathrm{Res}_{\mathscr{C}^0(G)}\,T_{\nu_t}$. The semigroup $(\nu_t)_{t\in\mathbb{R}_+}$ defined in this way satisfies the requirements of the theorem. □

1.5.10 Remark. For any continuous convolution semigroup $S:=(\mu_r)_{r\in\mathbb{D}_0}$ in $\mathcal{M}^1(G)$ one obtains the relative compactness of the set

$$\{\mu_r : r\in\mathbb{D}_0\cap[0,r_0]\}\qquad\text{for some } r_0\in\mathbb{D}.$$

In fact, by the continuity of S there exist an $r_1\in\mathbb{D}$ and a compact subset K of G such that $\mu_r(K)\geq\delta$ holds for all $r\in\mathbb{D}_0\cap[0,r_1]$ and some $\delta\in\mathbb{R}_+^*$. From $\mu_{r_1}=\mu_{r_1-r}*\mu_r$ we deduce via Theorem 1.2.21 (iii) the existence of a subset $(x_r)_{r\in\mathbb{D}_0\cap[0,r_1]}$ of G such that $\{\mu_r*\varepsilon_{x_r} : r\in\mathbb{D}_0\cap[0,r_1]\}$ is uniformly tight. Hence, for every $\varepsilon>0$ there exist a compact subset C of G and an $r_0\in\mathbb{D}$ with $\mu_r*\varepsilon_{x_r}(C)\geq1-\varepsilon$ for all $r\in\mathbb{D}_0\cap[0,r_0]$. From $\mu_r(K)\geq\delta$ we therefore obtain $x_r\in K^{-1}C$ (with the choice $\varepsilon<\delta$) and from Theorem 1.2.21 (ii) we conclude the uniform tightness. Hence, by Prohorov's theorem 1.1.11 the set $\{\mu_r : r\in\mathbb{D}_0\cap[0,r_0]\}$ is relatively compact in $\mathcal{M}^1(G)$.

An alternative proof of the statement of the theorem goes as follows:

Without loss of generality let $r_0=1$ and let $\mathscr{K}$ be a given compact subset of $\mathcal{M}^1(G)$ with $\mu_r\in\mathscr{K}$ for all $r\in\mathbb{D}_0\cap[0,1]$. For $t\in[0,1]$ there exists a sequence $(r_n)_{n\geq1}$ in $\mathbb{D}_0\cap[0,1]$ with $\lim_{n\to\infty}r_n=t$. The sequence $(\mu_{r_n})_{n\geq1}$ has an accumulation point $\nu_t\in\mathcal{M}^1(G)$. By Part 1 of the proof we have $\mathrm{Res}_{\mathscr{C}_u(G)}\,T_{\nu_t}=S_t$. Thus $\nu_s*\nu_t=\nu_{s+t}$ for all $s,t\in[0,1]$ with $s+t\leq1$. But $(\nu_t)_{t\in[0,1]}$ can be extended uniquely to a continuous convolution semigroup $(\nu_t)_{t\in\mathbb{R}_+}$ in $\mathcal{M}^1(G)$, for which $\nu_t=\mu_t$ for all $t\in\mathbb{D}_0$.

We are now ready to study the *representation of convolution semigroups* in $\mathcal{M}^1(G)$ via linear functionals on $\mathfrak{K}(G)$.

1.5.11. Let $S:=(\mu_t)_{t\in\mathbb{R}_+^*}$ be an $\{e\}$-continuous convolution semigroup in $\mathcal{M}^1(G)$. For every $f\in\mathfrak{K}(G)$ the limit

$$\psi(f):=\lim_{t\downarrow0}\frac{1}{t}\,(f(e)-\hat{\mu}_t(f))$$

exists, $\psi\in\mathfrak{K}(G)^*$ and ψ has the following properties:

(i) For every $D\in\mathrm{Rep}(G)$ there exists a $t_0:=t_0(D)\in\mathbb{R}_+^*$ with

$$\psi(D)=-\frac{1}{t}\log\hat{\mu}_t(D)\qquad\text{for all } t\in\,]0,t_0];$$

(ii) $\hat{\mu}_t(D)=\exp(-t\,\psi(D))$ for all $t\in\mathbb{R}_+^*$ and $D\in\mathrm{Rep}(G)$.

In fact, since S is an $\{e\}$-continuous semigroup, for every $D\in\mathrm{Rep}(G)$ there exists a $t_0:=t_0(D)\in\mathbb{R}_+^*$ satisfying $\|\hat{\mu}_t(D)-E_{n(D)}\|<1$ whenever $t\in\,]0,t_0]$. For all $t\in\,]0,t_0]$ the expression $\frac{1}{t}\log\hat{\mu}_t(D)$ is independent of t. Therefore we put

$$\psi(D):=-\frac{1}{t}\log\hat{\mu}_t(D)\qquad\text{for some } t\in\,]0,t_0]$$

and conclude that $\hat{\mu}_t(D)=\exp(-t\psi(D))$ for all $t\in\mathbb{R}_+^*$. The existence of

$$\psi(D)=\lim_{t\downarrow 0}\frac{1}{t}(E_{n(D)}-\hat{\mu}_t(D))$$

implies the assertions.

1.5.12. The linear functional ψ on $\mathfrak{R}(G)$ defined above admits the property of the following

Definition. A real linear functional ψ on $\mathfrak{R}(G)$ is said to be a *negative-definite form* (on $\mathfrak{R}(G)$) if
(i) $\psi(1)\geq 0$ and
(ii) $\psi(f)\leq 0$ for all $f\in\mathfrak{R}_+(G)$ with $f(e)=0$.

The negative-definite form of 1.5.11 is called the *negative-definite form corresponding to the semigroup S* (or the negative-definite form of S).

1.5.13 Theorem. *Let $G\in\mathbf{A}$, $\psi\in\mathfrak{R}(G)^*$, and for every $t\in\mathbb{R}_+^*$ let*

$$\phi_t := \mathrm{Exp}(-t\psi).$$

The following statements are equivalent:
(i) *ψ is negative-definite and satisfies $\psi(1)=0$;*
(ii) *ϕ_t is positive-definite and fulfills $\phi_t(1)=1$ for all $t\in\mathbb{R}_+^*$.*

The proof requires the following

1.5.14 Lemma. *Let E be a compact space and let F be in $\mathscr{C}(\mathbb{R}_+\times E)$. For every $(s,y)\in\mathbb{R}_+\times E$ with $s\in\mathbb{R}_+^*$ and $F(s,y)=\min_{x\in E}F(s,x)$ the partial derivative $\left.\frac{\partial}{\partial t}\right|_{t=s} F(t,y)$ is assumed to exist and to be nonnegative.*
Then $\min_{x\in E}F(t,x)\geq\min_{x\in E}F(0,x)$ holds for all $t\in\mathbb{R}_+$.

Proof. Without loss of generality we suppose that $\min_{x\in E}F(0,x)=1$. Let $c\in\mathbb{R}_+^*$ and define a function $Q\in\mathscr{C}(\mathbb{R}_+\times E)$ by $Q(t,x):=e^{ct}F(t,x)$ for all $(t,x)\in\mathbb{R}_+\times E$. Q attains its minimum at the same points of E as F does.

We shall show that $Q\geq 1$. Let $(t_0,x_0)\in\mathbb{R}_+^*\times E$ such that $Q(t_0,x_0)<1$ holds. Then there exists an $(s,y)\in\mathbb{R}_+^*\times E$ satisfying

$$q:=Q(s,y)=\min\{Q(t,x):t\in[0,t_0],x\in E\}<1,$$

where necessarily $s>0$. The subset

$$A:=\{(t,x)\in\mathbb{R}_+^*\times E:Q(t,x)=q\}\cap([0,t_0]\times E)$$

of $[0,t_0]\times E$ is compact so that its projection B onto $[0,t_0]$ is closed. Putting $s_0:=\inf B$ we thus have the existence of a $y_0\in E$ such that $Q(s_0,y_0)=q$ and for all $t\in[0,s_0[$ and $x\in E$ therefore $Q(t,x)>q$. So we get for $s:=s_0$ and $y:=y_0$

$$\left.\frac{\partial}{\partial t}\right|_{t=s} Q(t,y) \leq 0.$$

On the other hand, we have $Q(s,y)=\min_{x\in E} Q(s,x)$, which implies that

$$F(s,y)=\min_{x\in E} F(s,x) \quad \text{and}$$

$$\left.\frac{\partial}{\partial t}\right|_{t=s} Q(t,y)=e^{cs}\cdot\left.\frac{\partial}{\partial t}\right|_{t=s} F(t,y)+cQ(s,y)\geq cq>0,$$

which contradicts the preceding inequality. Thus $Q\geq 1$. For $c\in\mathbb{R}_+^*$ decreasing to 0 we obtain $F\geq 1$, and hence the result. □

We now start the *proof of the theorem*. First of all we observe that transition to the Bohr group allows us to assume the compactness of G.

1. (i) ⇒ (ii). Let $D\in\mathrm{Rep}(G)$ be fixed and let $N:=-\psi$. R_N is considered as a function on $\mathscr{V}(D)$. For every $t\in\mathbb{R}_+$ we define the bounded linear operator $T_t:=\exp t R_N$ on $\mathscr{V}(D)$ and hence obtain a strongly continuous semigroup of operators on $\mathscr{V}(D)$, whose infinitesimal generator $\lim_{t\downarrow 0}\frac{1}{t}(T_t-E)$ equals R_N. For $f\in\mathscr{V}(D)$ with $f=\bar{f}$ we introduce a function F on $\mathbb{R}_+\times G$ by means of

$$F(t,x):=(T_t f)(x) \qquad \text{for all } (t,x)\in\mathbb{R}_+\times G.$$

Plainly, $F\in\mathscr{C}(\mathbb{R}_+\times G)$. In fact, F is real, since ψ is. Furthermore, F is continuous, since by Property (4) of 1.3.6 there exist $D'\in\mathrm{Rep}(G)$ and $M\in\mathfrak{M}(n(D'),\mathbb{C})$ with $f=\mathrm{tr}(MD')$ and the trace is continuous.

For $s\in\mathbb{R}_+^*$ and $y\in G$ with $F(s,y)=\min_{x\in G} F(s,x)$ we compute

$$\begin{aligned} r(s) &:= \lim_{h\downarrow 0}\frac{1}{h}(F(s+h,y)-F(s,y)) \\ &= \lim_{h\downarrow 0}\frac{1}{h}((T_{s+h}f)(y)-(T_s f)(y)) \\ &= R_N(T_s f)(y)=N({}_y(T_s f))=-\psi({}_y(T_s f)). \end{aligned}$$

Thus r is continuous on $\mathbb{R}_+^*$, so that $F(\cdot,y)$ is differentiable in $s\in\mathbb{R}_+^*$ with derivative $r(s)$. Defining $g:=T_s f-(T_s f)(y)$ we observe that $g\in\mathscr{V}(D)$ and by the very choice of $y\in G$ also $g\geq 0$. Since by assumption ψ is negative-definite and satisfies $\psi(1)=0$, we have

$$0\geq\psi({}_y g)=\psi({}_y(T_s f))=-r(s)$$

and Lemma 1.5.14 yields $\min_{x\in G} T_t f(x)\geq\min_{x\in G} f(x)$. For $f\geq 0$ we therefore obtain

$$\begin{aligned} \phi_t(f) &= \mathrm{Exp}(-t\psi)(f)=\mathrm{Exp}(tN)(f)=((\exp tR_N)f)(e) \\ &= (T_t f)(e)\geq\min_{x\in G}(T_t f)(x)\geq\min_{x\in G} f(x)\geq 0. \end{aligned}$$

As every $f\in\mathfrak{R}_+(G)$ belongs to $\mathscr{V}(D)$ for some $D\in\mathrm{Rep}(G)$, ϕ_t is positive-definite.

2. (ii) ⇒ (i). From $1=\phi_t(1)=\exp(-t\psi(1))$ for all $t\in\mathbb{R}_+^*$ we conclude that $\psi(1)=0$. As a compact group G is in $\mathbf{B}$ by Theorem 1.4.9, for every $t\in\mathbb{R}_+^*$ there exists a $\mu_t\in\mathcal{M}^1(G)$ satisfying $\hat{\mu}_t=\phi_t=\operatorname{Exp}(-t\psi)$. By Theorems 1.3.8 and 1.4.5 $S:=(\mu_t)_{t\in\mathbb{R}_+^*}$ is an $\{e\}$-continuous semigroup whose negative-definite form is ψ. □

1.5.15 Definition. Let ψ be a real linear functional on $\mathfrak{R}(G)$.
(a) ψ is called a *primitive form* (on $\mathfrak{R}(G)$) if for all $f, g\in\mathfrak{R}(G)$ we have

$$\psi(fg^*)=\psi(f)\,g(e)-f(e)\,\psi(g).$$

(b) ψ is said to be a *quadratic form* if
(i) $\psi(D)$ is positive-semidefinite (in $\mathfrak{M}(n(D),\mathbb{C})$) for all $D\in\operatorname{Rep}(G)$ and
(ii) for all $f, g\in\mathfrak{R}(G)$ we have

$$\psi(fg)+\psi(fg^*)=2[\psi(f)\,g(e)+f(e)\,\psi(g)].$$

(c) A negative-definite form ψ (on $\mathfrak{R}(G)$) is called a *Poisson form* if there exists a positive-definite form ϕ on $\mathfrak{R}(G)$ with $\psi=\phi(1)\,\hat{\varepsilon}_e-\phi$.

1.5.16 Theorem. *Let $G\in\mathbf{A}$ and ψ a primitive from on $\mathfrak{R}(G)$. Then*
(i) *ψ is skew symmetric.*
(ii) $\operatorname{Exp}(-\psi)$ *is real and multiplicative.*
(iii) *ψ is negative-definite.*

Proof. (i) Choosing $f=g=1$ we see that the definition of a primitive form yields $\psi(1)=0$. For $f=1$ and $g\in\mathfrak{R}(G)$ arbitrary, however, we have $\psi^{\#}(g)=\psi(g^*)=-\psi(g)$, and thus $\psi^{\#}=-\psi$.

(ii) For any primitive form ψ on $\mathfrak{R}(G)$ and $D, D'\in\operatorname{Rep}(G)$ one has

$$\psi(D\otimes D')=\psi(D)\otimes E_{n(D')}+E_{n(D)}\otimes\psi(D').$$

Defining $N:=\operatorname{Exp}(-\psi)$ one immediately observes that $N\in\mathfrak{D}(G)$ and verifies that $N(D\otimes D')=N(D)\otimes N(D')$ for $D, D'\in\operatorname{Rep}(G)$, so that $N\in\mathfrak{Q}(G)$. Hence, by Theorem 1.3.5 N is real and multiplicative on $\mathfrak{R}(G)$.

(iii) By assumption, $t\psi$ is a primitive form on $\mathfrak{R}(G)$ for all $t\in\mathbb{R}_+^*$. (ii) now implies, that $\phi_t:=\operatorname{Exp}(-t\psi)$ is a real and multiplicative linear functional on $\mathfrak{R}(G)$ which satisfies

$$\phi_t(f\bar{f})=\phi_t(f)\,\phi_t(\bar{f})=\phi_t(f)\,\overline{\phi_t(f)}\geq 0 \qquad \text{for all } f\in\mathfrak{R}(G),$$

so that ϕ_t is a positive-definite form satisfying $\phi_t(1)=1$ for all $t\in\mathbb{R}_+^*$. Theorem 1.5.13 then yields the assertion. □

1.5.17 Theorem. *Let $G\in\mathbf{A}$ and ψ a quadratic form on $\mathfrak{R}(G)$. Then*
(i) *ψ is symmetric with $\psi(1)=0$.*
(ii) *ψ is a negative-definite form.*

Proof. (i) is proved analogously to (i) of Theorem 1.5.16.

(ii) Without loss of generality G is assumed to be compact. Since G is the projective limit of compact Lie groups we further assume that G is a compact Lie group.

Given a quadratic form ψ on $\mathfrak{K}(G)$ it suffices to prove that $\Psi := \operatorname{Exp}(-\psi)$ is positive-definite, since $t\psi$ is a quadratic form for every $t \in \mathbb{R}_+^*$ and the argument of the proof of (iii) in Theorem 1.5.16 can be re-applied. Finally, we assume $G = \mathfrak{U}(n)$ for some $n \geq 1$ since every compact Lie group can be viewed as a subgroup of $\mathfrak{U}(n)$ for some $n \geq 1$.

Let $D := (d_{ij})_{i,j=1,\dots,n}$ be the identity representation of $\mathfrak{U}(n)$.

By $D^{\check{}}$ we denote the *contragradient representation* $(\bar{d}_{ji})_{i,j=1,\dots,n}$ corresponding to D, which (for unitary representations) coincides with $D^* := (d_{ij}^*)_{i,j=1,\dots,n}$.

1. We first show that any $\Psi' \in \mathfrak{K}(G)^*$ satisfying

(a) $\Psi'(\bar{f}) = \overline{\Psi'(f)}$ for all $f \in \mathfrak{K}(G)$,

(b) $\Psi'(fg) + \Psi'(fg^*) = 2[\Psi'(f)\, g(e) + f(e)\, \Psi'(g)]$ for all $f, g \in \mathfrak{K}(G)$ and

(c) $\Psi'(D \otimes D) = \Psi'(D \otimes D^{\check{}})$

vanishes identically.

In fact, from (c) follows $\Psi'(d_{ij}\, d_{lk}) - \Psi'(d_{ij}\, \bar{d}_{kl}) = 0$ and from (b)

$$\Psi'(d_{ij}\, d_{kl}) + \Psi'(d_{ij}\, \bar{d}_{lk}) = 2\,[\delta_{kl}\, \Psi'(d_{ij}) + \delta_{ij}\, \Psi'(d_{kl})],$$

thus $\Psi'(d_{ij}\, d_{lk}) = \Psi'(d_{ij}\, \bar{d}_{kl}) = \delta_{kl}\, \Psi'(d_{ij}) + \delta_{ij}\, \Psi'(d_{lk})$ and so

$$\begin{aligned}
&\Psi'(d_{ll}\, d_{ll}) = \Psi'(d_{ll}\, \bar{d}_{ll}) = 2\Psi'(d_{ll}) \quad \text{for all } 1 \leq l \leq n,\\
&\Psi'(d_{ij}\, d_{ll}) = \Psi'(d_{ij}\, \bar{d}_{ll}) = \Psi'(d_{ij}) \quad \text{and}\\
&\Psi'(d_{ll}\, d_{ij}) = \Psi'(d_{ll}\, \bar{d}_{ji}) = \Psi'(d_{ij}) \quad \text{for all } 1 \leq i \neq j \leq n, 1 \leq l \leq n \quad \text{and}\\
&\Psi'(d_{ij}\, d_{lk}) = \Psi'(d_{ij}\, \bar{d}_{kl}) = 0 \quad \text{for all } 1 \leq i \neq j \leq n,\ 1 \leq l \neq k \leq n.
\end{aligned}$$

From this we obtain by the definition of $D = (d_{ij})_{i,j=1,\dots,n}$

$$2\Psi'(d_{ii}) = \Psi'(d_{ii}\, \bar{d}_{ii}) = -\textstyle\sum_{k=1, k \neq i}^{n} \Psi'(d_{ik}\, \bar{d}_{ik}) = 0$$

for $1 \leq i \leq n$ and

$$2\Psi'(d_{ij}) = \Psi'(d_{ii}\, \bar{d}_{ji}) + \Psi'(d_{ij}\, \bar{d}_{jj}) = -\textstyle\sum_{k=1, k \neq i, k \neq j}^{n} \Psi'(d_{ik}\, \bar{d}_{jk}) = 0$$

for $1 \leq i \neq j \leq n$.

Altogether we get $\Psi'(D) = 0$ and $\Psi'(D \otimes D) = \Psi'(D \otimes D^{\check{}}) = 0$. By (a) and (b) one concludes from the formula

$$\begin{aligned}
6fgh &= f(2g - g^*)(h + h^*) + g(2h - h^*)(f + f^*) + h(2f - f^*)(g + g^*)\\
&\quad + (f^* g h^* - f g^* h) + (f g^* h^* - f^* g h) + (f^* g^* h - f g h^*),
\end{aligned}$$

valid for all $f, g, h \in \mathfrak{K}(G)$, that

$$\Psi'(f) = \Psi'(fg) = \Psi'(fg^*) = \cdots = 0 \quad \text{implies} \quad \Psi'(fgh) = 0.$$

Thus Ψ' vanishes on the set of all polynomials of degree $k \geq 3$ (in d_{ij} and $\bar{d}_{kl}$ for $1 \leq i, j, k, l \leq n$) if it vanishes on the set of all polynomials of degree $k - 1$. From

$$\Psi'(D) = 0 \quad \text{and} \quad \Psi'(D \otimes D) = \Psi'(D \otimes D^{\check{}}) = 0$$

follows the assertion.

2. We retain the notation D for the identity representation of $G:=\mathfrak{U}(n)$ of dimension $n:=n(D)$. Let $\{X_{pq}: 1\leq p,q\leq n\}$ be a basis of the Lie algebra $\mathscr{L}(G)$ (of skew-Hermitian matrices) of G such that the following conditions hold:

$$A_{pp}D=(i\,\delta_{ps}\,\delta_{pr})_{i\leq s,r\leq n} \quad \text{for } 1\leq p\leq n,$$
$$A_{pq}D=(\delta_{pr}\,\delta_{qs}-\delta_{ps}\,\delta_{qr})_{1\leq r,s\leq n} \quad \text{and}$$
$$A_{qp}D=[i(\delta_{pr}\,\delta_{qs}+\delta_{ps}\,\delta_{qr})]_{1\leq r,s\leq n} \quad \text{for } 1\leq p<q\leq n.$$

Here we write

$$A_{pq}f:=(\tilde{X}_{pq}f)(e) \quad \text{and}$$
$$A_{pq}A_{kl}f:=(\tilde{X}_{pq}\tilde{X}_{kl}f)(e), \quad \text{where}$$
$$(\tilde{X}_{pq}f)(x):=\lim_{s\to 0}\frac{1}{s}[f(x\exp(s\cdot X_{pq}))-f(x)]$$

for all $f\in\mathfrak{K}(G)$, $x\in G$ and $1\leq p,q,k,l\leq n$.

Then $\{A_{pq}D: 1\leq p,q\leq n\}$ is a basis of the $\mathbb{R}$-vector space $\mathscr{L}(G)$, so that

$$\{A_{pq}D\otimes A_{kl}D: 1\leq p,q,k,l\leq n\}$$

is a basis of the $\mathbb{R}$-vector space of all Hermitian matrices in $\mathfrak{M}(n^2,\mathbb{C})$.

Let $\Psi\in\mathfrak{K}(G)^*$ have the properties (a) and (b) of Part 1. We shall show that there exists a family $\{a_{pq,kl}: 1\leq p,q,k,l\leq n\}$ of real numbers with $a_{pq,kl}=a_{kl,pq}$ for all $1\leq p,q,k,l\leq n$ such that

$$\Psi(f)=\sum_{p,q,k,l=1}^{n} a_{pq,kl}\,A_{pq}\,A_{kl}f$$

holds for all $f\in\mathfrak{K}(G)$.

First of all, there exists a family $\{a_{pq,kl}: 1\leq p,q,k,l\leq n\}$ of real numbers such that

$$4\sum_{p,q,k,l=1}^{n} a_{pq,kl}\,A_{pq}D\otimes A_{kl}D=\Psi(D\otimes D)-\Psi(D\otimes D^{\check{}}).$$

Property (b) of Ψ yields

$$\Psi(D\otimes D)-\Psi(D\otimes D^{\check{}})=2\Psi(D\otimes D)-2\Psi(D)\otimes E_n-2E_n\otimes\Psi(D).$$

Both relations together yield $a_{pq,kl}=a_{kl,pq}$ for all $1\leq p,q,k,l\leq n$, as a few computations show.

[It appears convenient to separate cases. For example, in the case $1\leq p<q<n$, $1\leq k<l\leq n$ we use the formulae

$$a_{pq,kl}-a_{qp,lk}=\tfrac{1}{2}\operatorname{Re}\Psi(d_{qp}\,d_{lk})=\tfrac{1}{2}\operatorname{Re}\Psi(d_{pq}\,d_{kl}) \quad \text{and}$$
$$a_{pq,kl}+a_{qp,lk}=-\tfrac{1}{2}\operatorname{Re}\Psi(d_{qp}\,d_{kl})=-\tfrac{1}{2}\operatorname{Re}\Psi(d_{pq}\,d_{lk}).]$$

Putting $\Psi''(f):=\sum_{p,q,k,l=1}^{n} a_{pq,kl}\,A_{pq}\,A_{kl}f$ for all $f\in\mathfrak{K}(G)$ and $\Psi':=\Psi-\Psi''$ we obtain a linear functional Ψ' on $\mathfrak{K}(G)$ satisfying (a) and (b) of Part 1. From the properties of the elements $a_{pq,kl}$ $(1\leq p,q,k,l\leq n)$ we deduce that

$$\begin{aligned}
&\Psi''(D\otimes D)-\Psi''(D\otimes D^{\sim})\\
&=\textstyle\sum_{p,q,k,l=1}^{n} a_{pq,kl}\{2A_{pq}D\otimes A_{kl}D+2A_{kl}D\otimes A_{pq}D\}\\
&=4\textstyle\sum_{p,q,k,l=1}^{n} a_{pq,kl}A_{pq}D\otimes A_{kl}D
\end{aligned}$$

and therefore $\Psi'(D\otimes D)-\Psi'(D\otimes D^{\sim})=0$. But then Part 1 implies that $\Psi'\equiv 0$ and hence, $\Psi=\Psi''$ on $\mathfrak{R}(G)$.

3. To prove the positive-definiteness of Ψ we have to show that

$$\textstyle\sum_{p,q,k,l=1}^{n} a_{pq,kl}h_{pq}h_{kl}\geq 0 \quad \text{for all } h_{pq}, h_{kl}\in\mathbb{R}\,(1\leq p,q,k,l\leq n).$$

Let $F\in\mathrm{Rep}(G)$ of the form $F:=\bigotimes_{i=1}^{t}F_i$ with $F_i\in\mathrm{Irr}(G)$ of degree

$$m_i:=n(F_i)\ (i=1,\dots,t) \quad\text{and}\quad \xi:=\bigotimes_{i=1}^{t}\xi_i\in\mathbb{C}^{\sum_{i=1}^{t}m_i}$$

with $\|\xi_i\|=1$ $(i=1,\dots,t)$.

From the representation of Ψ established in Part 2 we obtain that $\Psi(F)$ is negative-definite for all $F\in\mathrm{Rep}(G)$ and hence the chain of equations

$$\begin{aligned}
(5.1)\qquad &0\geq\langle\Psi(F)\xi,\xi\rangle=\textstyle\sum_{p,q,k,l=1}^{n}a_{pq,kl}\langle A_{pq}FA_{kl}F\xi,\xi\rangle\\
&=-\textstyle\sum_{p,q,k,l=1}^{n}a_{pq,kl}\langle\sum_{j=1}^{t}\xi_1\otimes\cdots\otimes\xi_{j-1}\otimes A_{kl}F_j\xi_j\otimes\xi_{j+1}\otimes\cdots\otimes\xi_t,\\
&\textstyle\sum_{j=1}^{t}\xi_1\otimes\cdots\otimes\xi_{j-1}\otimes A_{pq}F_j\xi_j\otimes\xi_{j+1}\otimes\cdots\otimes\xi_t\rangle\\
&=-\textstyle\sum_{p,q,k,l=1}^{n}a_{pq,kl}\{\sum_{j=1}^{t}\langle A_{kl}F_j\xi_j,A_{pq}F_j\xi_j\rangle\\
&+\textstyle\sum_{\substack{i,j=1\\ i\neq j}}^{t}\langle A_{kl}F_i\xi_i,\xi_i\rangle\langle\xi_j,A_{pq}F_j\xi_j\rangle\}\\
&=\textstyle\sum_{p,q,k,l=1}^{n}a_{pq,kl}\{[\sum_{j=1}^{t}\langle A_{kl}F_j\xi_j,\xi_j\rangle][\sum_{j=1}^{t}\langle A_{pq}F_j\xi_j,\xi_j\rangle]\\
&+\textstyle\sum_{j=1}^{t}\langle A_{pq}A_{kl}F_j\xi_j,\xi_j\rangle-\sum_{j=1}^{t}\langle A_{kl}F_j\xi_j,\xi_j\rangle\langle A_{pq}F_j\xi_j,\xi_j\rangle\}.
\end{aligned}$$

We now define for $1\leq p\leq n$

$$\begin{aligned}
&\xi_{pp}(k):=\delta_{kp},\\
&\eta_{pp}(k):=\delta_{kp}
\end{aligned}$$

and for $1\leq p<q\leq n$

$$\begin{aligned}
&\xi_{pq}(k):=\sqrt{\tfrac{1}{2}}(\delta_{kp}+i\,\delta_{kq}),\\
&\eta_{pq}(k):=\sqrt{\tfrac{1}{2}}(i\,\delta_{kp}+\delta_{kq}),\\
&\xi_{qp}(k):=\sqrt{\tfrac{1}{2}}(\delta_{kp}+\delta_{kq}) \quad\text{and}\\
&\eta_{qp}(k):=\sqrt{\tfrac{1}{2}}(\delta_{kp}-\delta_{kq}),
\end{aligned}$$

where $1\leq k\leq n$, and compute for the faithful representation D of Part 1 the relations

$$\begin{aligned}
(5.2)\qquad &\langle A_{pp}D\xi_{rs},\xi_{rs}\rangle=\langle A_{pp}D\eta_{rs},\eta_{rs}\rangle=\tfrac{1}{2}i(\delta_{pr}+\delta_{ps}) \qquad (1\leq p\leq n)\\
&\langle A_{pq}D\xi_{rs},\xi_{rs}\rangle=i\,\delta_{pr}\delta_{qs},\\
&\langle A_{pq}D\eta_{rs},\eta_{rs}\rangle=-\langle A_{pq}D\xi_{rs},\xi_{rs}\rangle \quad\text{and}\\
&\langle A_{pq}\bar{D}\xi_{ss},\xi_{ss}\rangle=\langle A_{pq}\bar{D}\eta_{ss},\eta_{ss}\rangle=0 \qquad (1\leq p\neq q\leq n)
\end{aligned}$$

as well as

$$\langle A_{pp} D \xi_{rs}, \xi_{rs}\rangle = \langle A_{pp} D \eta_{rs}, \eta_{rs}\rangle = -\langle A_{pp} \bar{D} \xi_{rs}, \xi_{rs}\rangle$$
$$= -\langle A_{pp} \bar{D} \eta_{rs}, \eta_{rs}\rangle \quad (1 \le p \le n) \text{ for all } 1 \le r, s \le n.$$

For $1 \le p,\, q \le n$ we now choose an even $h_{pq} \in \mathbb{Z}$ and set in the case $p \neq q$ and $h_{pq} \ge 0$ exactly h_{pq} of the vectors $\xi_1, \ldots, \xi_t$ equal to $\zeta_{pq} := \xi_{pq}$ and the corresponding $F_r := D$, and in the case $p \neq q$ and $h_{pq} \le 0$ exactly $|h_{pq}|$ of the vectors $\xi_1, \ldots, \xi_t$ equal to $\zeta_{pq} := \eta_{pq}$ and the corresponding $F_r := D$. Finally, we put

$$k_{pp} := h_{pp} - \tfrac{1}{2} \sum_{\substack{p,q=1\\ p \neq q}}^{n} (h_{pq} + h_{qp}).$$

Furthermore, we choose $|k_{pp}|$ of the vectors $\xi_1, \ldots, \xi_t$ equal to $\zeta_{pp} := \xi_{pp}$ $(1 \le p \le n)$. For the corresponding representations $F_r \in \operatorname{Irr}(G)$ we choose $F_r := D$, if $k_{pp} \ge 0$ and $F_r := \bar{D}$ whenever $k_{pp} < 0$. By this procedure we obtain

$$t = \sum_{\substack{p,q=1\\ p \neq q}}^{n} |h_{pq}| + \sum_{p=1}^{n} |k_{pp}|.$$

From (5.1) follows

$$\begin{aligned}
(5.3) \qquad 0 \ge \sum_{p,q,k,l=1}^{n} a_{pq,kl} \Big\{ \Big[&\sum_{\substack{r,s=1\\ r \neq s}}^{n} \sum_{j=1}^{|h_{rs}|} \langle A_{kl} D \zeta_{rs}, \zeta_{rs} \rangle \\
&+ \sum_{p=1}^{n} \sum_{j=1}^{|k_{pp}|} \langle A_{kl} F_r \zeta_{pp}, \zeta_{pp} \rangle \Big] \Big[\sum_{\substack{r,s=1\\ r \neq s}}^{n} \sum_{j=1}^{|h_{rs}|} \langle A_{pq} D \zeta_{rs}, \zeta_{rs} \rangle \\
&+ \sum_{l=1}^{n} \sum_{j=1}^{|k_{pp}|} \langle A_{pq} F_r \xi_{ll}, \xi_{ll} \rangle \Big] + \sum_{j=1}^{t} \langle A_{pq} A_{kl} F_j \xi_j, \xi_j \rangle \\
&- \sum_{j=1}^{t} \langle A_{kl} F_j \xi_j, \xi_j \rangle \langle A_{pq} F_j \xi_j, \xi_j \rangle \Big\}.
\end{aligned}$$

Applying the relations (5.2) to this inequality we get

$$0 \ge \sum_{p,q,k,l=1}^{n} a_{pq,kl} (i\, h_{pq})(i\, h_{kl}) + \alpha,$$

where α denotes some remainder. Replacing in (5.3) the numbers h_{pq} by $N h_{pq}$ with $N \in \mathbb{N}$ $(1 \le p, q \le n)$, we obtain

$$0 \ge -N^2 \sum_{p,q,k,l=1}^{n} a_{pq,kl}\, h_{pq}\, h_{kl} + N\alpha$$

and in the limit (as $N \to \infty$)

$$0 \le \sum_{p,q,k,l=1}^{n} a_{pq,kl}\, h_{pq}\, h_{kl},$$

where $h_{pq}, h_{kl} \in \mathbb{Z}$ $(1 \le p, q, k, l \le n)$. By the continuity of the quadratic form, this inequality holds also for $h_{pq}, h_{kl} \in \mathbb{R}$ $(1 \le p, q, k, l \le n)$. □

The correspondence between $\{e\}$-continuous convolution semigroups and continuous negative-definite forms is made precise in the following results.

1.5.18 Theorem. *Let $G \in \mathbf{A}$ and let $(\mu_t)_{t \in \mathbb{R}_+^*}$ be an $\{e\}$-continuous convolution semigroup in $\mathscr{M}^1(G)$. Then there exists exactly one continuous negative-definite form*

ψ on $\mathfrak{R}(G)$ with $\psi(1)=0$ such that

$$\hat{\mu}_t = \operatorname{Exp}(-t\psi)$$

holds for all $t \in \mathbb{R}_+^*$.

Proof. The fact that for $(\mu_t)_{t\in\mathbb{R}_+^*}$ there exists a negative-definite form ψ on $\mathfrak{R}(G)$ with the required properties follows directly from 1.5.11, where the negative-definite form corresponding to $(\mu_t)_{t\in\mathbb{R}_+^*}$ was introduced. It remains to prove the continuity of ψ. Let $D \in \operatorname{Rep}(G)$ and $\varepsilon \in]0, \frac{1}{2}]$ be given. The subset $\{\mu_t : t \in [0,1]\}$ of $\mathcal{M}^1(G)$ is $\mathcal{T}_w$-compact so that by Prohorov's theorem there exists a compact subset K of G with $\mu_t(\complement K) \leq \frac{\varepsilon}{4}$ for all $t \in [0,1]$. There is always a $t_0 \in]0,1[$ satisfying

$$\|\hat{\mu}_t(D) - E_{n(D)}\| < 1 - 2\varepsilon \quad \text{and} \quad \psi(D) = -\frac{1}{t}\log \hat{\mu}_t(D)$$

for all $t \in]0, t_0[$ (see 1.5.11). But for $t \in]0, t_0[$ and $D' \in W\left(D; K, \frac{\varepsilon}{2}\right)$ we have

$$\begin{aligned}
&\|\hat{\mu}_t(D') - \hat{\mu}_t(D)\|_s = \|\textstyle\int [D'(x) - D(x)]\, \mu_t(dx)\|_s \\
&\leq \textstyle\int_K \|D'(x) - D(x)\|_s\, \mu_t(dx) + \int_{\complement K} \|D'(x) - D(x)\|_s\, \mu_t(dx) \\
&\leq \frac{\varepsilon}{2} + 2\frac{\varepsilon}{4} = \varepsilon, \qquad \text{thus } \|\hat{\mu}_t(D') - E_{n(D')}\|_s \leq 1 - \varepsilon.
\end{aligned}$$

Hence, $\psi(D') = -\frac{1}{t}\log \hat{\mu}_t(D')$ for all $D' \in W$ ($t \in]0, t_0[$), and ψ is continuous on W, which implies the assertion. □

The converse of the statement in Theorem 1.5.18 is contained in

1.5.19 Theorem. *Let $G \in \mathbf{M}$ and let ψ be a continuous negative-definite form on $\mathfrak{R}(G)$ with $\psi(1)=0$. Then there exists exactly one $\{e\}$-continuous convolution semigroup $(\mu_t)_{t\in\mathbb{R}_+^*}$ in $\mathcal{M}^1(G)$ satisfying*

$$\hat{\mu}_t = \operatorname{Exp}(-t\psi) \quad \textit{for all } t \in \mathbb{R}_+^*.$$

Proof. Let ψ be a continuous negative-definite form on $\mathfrak{R}(G)$ with $\psi(1)=0$. For every $t \in \mathbb{R}_+^*$ the linear functional $\phi_t := \operatorname{Exp}(-t\psi)$ is a positive-definite form on $\mathfrak{R}(G)$ with $\phi_t(1)=1$, as follows from Theorem 1.5.13, and it is also continuous.

By Theorem 1.4.12 G is a B-group. Hence, for every $t \in \mathbb{R}_+^*$ there exists a $\mu_t \in \mathcal{M}^1(G)$ such that $\phi_t = \hat{\mu}_t$ holds. Plainly, for all $s, t \in \mathbb{R}_+^*$ we obtain

$$\begin{aligned}
&\hat{\mu}_{t+s} = \operatorname{Exp}(-(s+t)\psi) = \operatorname{Exp}(-s\psi) \odot \operatorname{Exp}(-t\psi) \\
&= \hat{\mu}_s \odot \hat{\mu}_t = (\mu_s * \mu_t)^\wedge
\end{aligned}$$

and by the injectivity of the Fourier transform we get the semigroup property of $(\mu_t)_{t\in\mathbb{R}_+^*}$. In order to show that $(\mu_t)_{t\in\mathbb{R}_+^*}$ is $\{e\}$-continuous we pick a sequence $(t_n)_{n\geq 1}$ in

$\mathbb{R}_+^*$ with $\lim_{n\to\infty} t_n = 0$ and a $D \in \operatorname{Rep}(G)$, and conclude

$$\lim_{n\to\infty} \hat{\mu}_{t_n}(D) = \hat{\varepsilon}_e(D)$$

which by Theorem 1.4.5 implies that $\mathscr{T}_w$-$\lim_{n\to\infty} \mu_{t_n} = \varepsilon_e$ and hence $\mathscr{T}_w$-$\lim_{t\downarrow 0} \mu_t = \varepsilon_e$. The uniqueness of the semigroup $(\mu_t)_{t\in\mathbb{R}_+^*}$ follows again from the injectivity of the Fourier transform. □

Continuous primitive forms on $\mathfrak{R}(G)$ are related to one-parameter subgroups of the group G.

1.5.20 Theorem. *Let $G \in \mathbf{A}$.*

(i) *For every one-parameter subgroup $(x_t)_{t\in\mathbb{R}}$ of G there exists a continuous primitive form ψ on $\mathfrak{R}(G)$ with*

$$\hat{\varepsilon}_{x_t} = \operatorname{Exp}(-t\psi) \qquad \text{for all } t \in \mathbb{R}.$$

(ii) *If $G \in \mathbf{M}$, then for every continuous primitive form ψ on $\mathfrak{R}(G)$ there exists a one-parameter subgroup $(x_t)_{t\in\mathbb{R}}$ of G with*

$$\hat{\varepsilon}_{x_t} = \operatorname{Exp}(-t\psi) \qquad \text{for all } t \in \mathbb{R}.$$

Proof. (i) follows directly from Theorem 1.5.18 if one considers the negative-definite form corresponding to the one-parameter convolution semigroup $(\varepsilon_{x_t})_{t\in\mathbb{R}_+^*}$ in $\mathscr{M}^1(G)$.

(ii) is derived from Theorem 1.5.19. One has only to note that, together with ψ, $-\psi$ is also a continuous primitive form on $\mathfrak{R}(G)$. □

1.5.21 Theorem. *Let $G \in \mathbf{A}$. For any $\psi \in \mathfrak{R}(G)^*$ the following statements are equivalent:*

(i) *ψ is a continuous negative-definite form;*

(ii) *there is a sequence $(\phi_n)_{n\geq 1}$ of continuous positive-definite forms on $\mathfrak{R}(G)$ such that*

$$\mathscr{T}_{co}\text{-}\lim_{n\to\infty} [(\psi(1) + \phi_n(1))\,\hat{\varepsilon}_e - \phi_n] = \psi$$

holds, and one has $\psi(1) \geq 0$.

Proof. 1. (i) ⇒ (ii). For every $n \geq 1$ we define

$$\chi_n := \frac{1}{n}(\psi(1)\,\hat{\varepsilon}_e - \psi)$$

and observe that

$$n\left\{\hat{\varepsilon}_e - \operatorname{Exp}\left[-\frac{1}{n}(\psi - \psi(1)\,\hat{\varepsilon}_e)\right]\right\} = n\left\{-\sum_{\nu\geq 1} \frac{1}{\nu!}\frac{1}{n^\nu}(\psi(1)\,\hat{\varepsilon}_e - \psi)^\nu\right\}$$

$$= -\psi(1)\,\hat{\varepsilon}_e + \psi - \frac{1}{n}(\psi(1)\,\hat{\varepsilon}_e - \psi)^2 \sum_{\nu\geq 2} \frac{1}{\nu!}\chi_n^{\nu-2}.$$

For $D \in \operatorname{Rep}(G)$ one has $\|\chi_n(D)\| \leq \|\psi(1)\,E_{n(D)} - \psi(D)\|$ $(n \geq 1)$.

By the continuity of ψ we thus conclude that $\sum_{\nu\geq 2}\frac{1}{\nu!}\chi_n^{\nu-2}$ is bounded uniformly in $n\geq 1$ on compact subsets of $\mathrm{Rep}(G)$. Therefore

$$\lim_{n\to\infty} n\left\{\hat{\varepsilon}_e - \mathrm{Exp}\left[-\frac{1}{n}(\psi-\psi(1)\,\hat{\varepsilon}_e)\right]\right\} = \psi - \psi(1)\,\hat{\varepsilon}_e$$

uniformly on compact subsets of $\mathrm{Rep}(G)$. Defining

$$\phi_n := n\,\mathrm{Exp}\left[-\frac{1}{n}(\psi-\psi(1)\,\hat{\varepsilon}_e)\right]$$

and deducing its positive-definiteness from Theorem 1.5.13 for every $n\geq 1$, one obtains

$$\mathscr{T}_{co}\text{-}\lim_{n\to\infty}[(\psi(1)+\phi_n(1))\,\hat{\varepsilon}_e - \phi_n] = \psi.$$

2. (ii) ⇒ (i). From Theorem 1.4.8 and $\psi(1)\geq 0$ it follows that for every $n\geq 1$ the linear functional $\psi_n := (\psi(1)+\phi_n(1))\,\hat{\varepsilon}_e - \phi_n$ is a negative-definite form on $\mathfrak{R}(G)$. This property carries over to $\psi := \lim_{n\to\infty}\psi_n$, and since the convergence of $(\psi_n)_{n\geq 1}$ is with respect to the topology $\mathscr{T}_{co}$, ψ is continuous on $\mathrm{Rep}(G)$. □

1.5.22 Corollary. *Every continuous negative-definite form ψ on $\mathfrak{R}(G)$ with $\psi(1)=0$ is the $\mathscr{T}_{co}$-limit of a sequence of continuous Poisson forms on $\mathfrak{R}(G)$.*

The *proof* is clear. □

References and Comments

R 1.1 Measures on a Locally Compact Space

This is a basic topic in any text book on measure and integration theory, which emphasizes the topological aspects of the theory. Examples of such books are Halmos [190] and Bauer [6]. Bauer [6] will remain our source for most of the abstract measure theory needed. A standard reference concerning the concrete theory of Radon measures on locally compact spaces is Livre VI of Bourbaki's Eléments de Mathématique [41], [42], [43] and [45]. In the last mentioned volume the theory has been extended to more general topological spaces. This need had been met previously by Varadarajan [488] and was taken up later by Topsøe [463]. Motivations for an extension of the notion of Radon measure to general topological spaces are numerous: Most of them have their roots in the theory of probability. See for example Billingsley's monograph [18], which is based on the pioneering works of Prohorov [392] and Le Cam [313].

In Prohorov's paper [392] the notion of an abstract (regular or Borel) measure on a metric space is predominant. A special emphasis on probability measures on a metric space is to be found in the useful reference book by Parthasarathy [377], in which the contributions of the Indian School to the theory have been included. In our presentation we first follow Bourbaki [41] or (equivalently) Dieudonné [114]. We only define the fundamental notions and cite the basic facts concerning Radon measures on a locally compact space in order to set up a consistent terminology. Naturally, the theory of L^p-spaces will be used in much more detail than that with which it is considered in this section. The interested reader is referred to Bourbaki [41] or Hewitt and Ross [218], Chapter III. For Borel measures on a locally compact space and their relationship to Radon measures (via the Riesz representation theorem) we follow the lecture notes of Courrège [93]. This reference should also be consulted concerning operations in the set of measures yielding measures with densities, images of measures under various kinds of mappings and induced measures. Projective limits of measures on a topological space are treated extensively in Bourbaki [45]. The discussion of the section is concentrated on the vague and weak topologies in the set $\mathscr{M}^1(E)$ of all probability measures on a locally compact space E. Theorems 1.1.1 to 1.1.7 contain elementary facts about the vague topology as collected in Choquet [73]. Theorems 1.1.8 to 1.1.12 provide a selection of facts on the weak topology with special emphasis on Prohorov's theorem 1.1.11, whose origin can be found in Prohorov [392] or Le Cam [313]. It should be noted that Theorem 1.1.9 on the comparison of vague and weak convergence can be proved in a more elementary way (see Heyer [222]).

Connections between weak convergence and uniform convergence of measures are discussed in detail in Kallianpur [282] and Ranga Rao [397], two references of particular interest. Just for completeness we add a few publications that will be useful in the development of our theory. In Tortrat [465], [475] and Csiszár [101] Borel measures on arbitrary topological spaces are studied. In Siebert [442] the tightness of a net of probability measures on a topological space is refined and the Prohorov property is slightly extended.

R 1.2 Convolution of Measures on a Locally Compact Group

Convolvable measures on locally compact groups and their convolutions have become standard notions for a modern approach to abstract harmonic analysis, in which unbounded measures are admitted. For the basic facts concerning convolvable measures one best consults Bourbaki [44] or Dieudonné [114]. Although our discussion will be concentrated on probability measures we have to include the general aspects in order to be able to at least indicate those developments which yield the most recent research in the field as for example work done in the potential theory of random walks on locally compact groups (see Guivarc'h [187]). The origin of the support formula appearing in Theorem 1.2.1 can be traced to Urbanik [482] (for probability measures on a compact group). Its generalization to probability measures on arbitrary locally compact groups appears in Heyer [220]. Extended studies of the measure algebra of a locally compact group are part of any treatise on abstract harmonic analysis as for example the encyclopedia by Hewitt and Ross [218]. See also Heyer [227].

By a previous convention we shall look at the convolution of two (bounded, positive) measures on a locally compact group G as Radon- and Borel measure as well. The note [456] by Stromberg deals for the first time with this aspect. Starting from the well-known fact that the sets $\mathscr{M}^b(G)$, $\mathscr{M}^b_+(G)$ and $\mathscr{M}^1(G)$ of bounded real, bounded positive and probability measures on G resp. are norm topological convolution semigroups, we turn our attention to the corresponding property for the weak topology $\mathscr{T}_w$ in $\mathscr{M}^b(G)$ and establish the property for $\mathscr{M}^b_+(G)$ and $\mathscr{M}^1(G)$. For compact groups G and $\mathscr{M}^1(G)$ the result has been known since the work of Glicksberg [168] and Rosenblatt [410]. The separate continuity with respect to $\mathscr{T}_w$ of the convolution in $\mathscr{M}^b(G)$ for compact G has been established by Glicksberg [168]. The second statement of Theorem 1.2.2 is shown (with a different proof) at various places; we simply cite Grenander [182], p. 188 and Kloss [307]. There are extensions of Theorem 1.2.2 in two directions. For arbitrary (not necessarily locally compact) topological groups G and τ-regular Borel measures the convolution operation is continuous with respect to the topology of pointwise convergence on the space of (right) uniformly continuous bounded functions on G. (Csiszár [101], [102] and Siebert [442]). Results for topological semigroups G, on the other hand, are due to Tortrat [473], [475] (in the completely regular case) and to Yuan [510] (in the locally compact case).

Invariant integrals and measures on a locally compact group G are central objects of discussion in the monographs of Hewitt and Ross [218] and Nachbin [368]. In our text we pick only a few properties for the sake of precise reference and aim directly at the problem of the convolution equation. The definition of the invariance subgroup of a measure $\mu \in \mathscr{M}_+(G)$ is due to Schmetterer, in whose work [428], [429] Theorem 1.2.4 appears. The (folklore) argument in the proof of this theorem appears in the paper of Heyer [220], which also contains Theorem 1.2.10 on the characterization of idempotent measures in $\mathscr{M}^b_+(G)$ as normed Haar measures of compact subgroups of G. In our presentation the characterization is derived from Theorem 1.2.7 on the solution of the convolution equation, which plays a special rôle in various branches of the theory. The determination of all (real and complex) idempotent measures on the torus goes back to the year 1939 and to the ingenuity of P. Lévy [320]. His results have been rediscovered in the case of probability measures by Kakehashi [281] and in the general case by Helson [215].

These first investigations were continued by Rudin [416] and completed in the work [85] of Cohen, who characterized all idempotent measures on a general locally compact Abelian group. See also Rudin [418] and [417], Chapter 3. For a not necessarily Abelian compact group idempotent probability measures on G were first described by Ito and Kawada [287], and later by Wendel [500] without the separability assumption made previously. A corresponding result for measures on an almost periodic group is contained in Kloss [306]. The complete solution to the problem for arbitrary locally compact groups can now be found also in Kelley [289] and Pym [393]. A short proof of the theorem for metric groups is due to Parthasarathy [378]. See also Parthasarathy's book [377]. The proof is close to that given in Heyer [220]. It is based on a special form of Numakura's theorem [373], extended later by Ellis, that a compact semigroup embedded in a group is a (compact sub)group. A proof of the characterization theorem based on representation theory has been given by Loynes [328]. See also Grenander [182], 5.2. We now mention a few more references concerning idempotents in $\mathscr{M}^1(G)$ for locally compact (semitopological) semigroups. The work done for compact semigroups can only be quoted selectively. See Kloss [304], Collins [87], Heble

[211], Heble and Rosenblatt [212]. Most of these contributions have been included in the book [412] by Rosenblatt. Among the authors who dealt with idempotents in $\mathcal{M}_+^b(G)$ for a locally compact semigroup G we mention Pym [393], Mukherjea [360], Tserpes and Mukherjea [477], Mukherjea and Tserpes [365] and Choy [79]. The study of idempotent probability measures on discrete semigroups was advanced by Martin-Löf [346]. His paper contains an extensive bibliography on the subject. Within the framework of semitopological semigroups as for example that chosen by Berglund and Hofmann in [13], papers by Pym [394] and Choy [80], [81] are of interest. One might in this context also consult the book [77] by Chow and Choy.

Probability measures of finite order on a compact semigroup G are defined as measures $\mu \in \mathcal{M}^1(G)$ with $\mu^{n+1} = \mu$ for some $n \geq 1$. For $n = 3$ the characterization of such measures has been discussed in Collins [88]. The more general problem is the subject of the papers [76] and [75] of Chow.

The convolution equation of Choquet and Deny [74], [109] for probability measures on topological groups has been Tortrat's main interest since 1964. His paper [464] contains a first contribution for Borel probability measures on a non-Abelian, completely regular topological group. The subsequent publications [465], [466] and [469] deepen these studies with respect to generalizations either to more general classes of bounded measures or to various kinds of topological semigroups. The paper [362] of Mukherjea contributes to the case of a cancellative locally compact semigroup. For topological groups the last word seems to have been said in the papers [101] and [102] by Csiszár, where the generalization reaches a desirable limitation.

There has appeared an additional interest in the extension of the Choquet-Deny convolution equation for unbounded measures to more general locally compact groups. The work in this direction has been founded on the classical theorem 1.2.12, which nowadays can be proved in various ways (see Revuz [399] for an operator-theoretic alternative). A far-reaching extension motivated by the theory of random walks on locally compact groups has been given by Azencott [2] and Guivarc'h [187], where the underlying groups are certain solvable Lie groups. Within the framework of locally compact semigroups very little is known: The interested reader is directed to Mukherjea [361]. Theorem 1.2.13 essentially due to Carnal [64] is proved in Schmetterer [429], where Remark 1.2.14 also appears. Theorem 1.2.15 indicated for compact groups in Carnal [64] is quoted in Siebert [439]. Part (iii) has been studied in more detail by Yuan [513], where the following result is shown: Let G be a locally compact semigroup and H a compact group in G with identity e. Then the space $\mathcal{M}^b(eGe/H)$ of all bounded regular Borel measures on eGe/H is a Banach space, the closed subspace $\mathcal{M}_H^b(eGe/H)$ of H-invariant measures in $\mathcal{M}^b(eGe/H)$ is a (convolution) Banach algebra and for any idempotent element ω of $\mathcal{M}^1(G)$ with compact support one has (algebraic and topological) isomorphisms

$$\omega * \mathcal{M}^b(G) * \omega \cong \omega_H * \mathcal{M}^b(G) * \omega_H \cong \mathcal{M}_H^b(eGe/H).$$

Theorem 1.2.17 is taken from Bourbaki [44]. The definitions and facts concerning convolvable functions on a locally compact group are designed to remind the reader of background material from abstract harmonic analysis. See Hewitt and

Ross [218] or Rudin [417]. These books are excellent sources for a basic knowledge of the group algebra.

In the rest of the section we present a choice of results on (tight) sets of translates of measures in $\mathcal{M}^1(G)$ for a locally compact group G. The notion of a tight net is due to Siebert [442]. The proof of Theorem 1.2.21 is similar to that given in Parthasarathy, Rao and Varadhan [382] or Parthasarathy [377], III.2. Theorem 1.2.22 on the σ-closedness of the convolution mapping appeared in a more general context in Heyer and Tortrat [237], where it was applied to factorization problems for probability measures on a locally compact group G. The theorem yields in particular a short proof of Theorem 1.2.23 due to Parthasarathy, Rao and Varadhan [382] that the set of all prime probability measures on G is a $\mathcal{T}_w$-G_δ set in $\mathcal{M}^1(G)$. In connection with the divisibility of measures in $\mathcal{M}^1(G)$ results of Štěpán [452], [453] on the $\mathcal{T}_w$-closed convex hull $\mathcal{N}_\nu$ of the set of translates of a given measure $\nu \in \mathcal{M}^1(G)$ are of interest. The reasoning in Štěpán [453] yields that $\mu \in \mathcal{N}_\nu$ iff $\mu = \nu * \lambda$ with $\lambda \in \mathcal{M}^1(G)$. But all this gives only little insight into the arithmetic for the semigroup $\mathcal{M}^1(G)$. More details from the classical theory can be found in the monographs of Dugué [125], Linnik [322] and Cuppens [104], which are certainly based on the famous contributions of Khintchine [299], Lévy [318] and their schools. For a recent account on the subject see also the expository paper [324] of Livšic, Ostrovskiĭ and Čistjakov. However, this topic lies outside the intentions of this book.

R 1.3 Fourier Transforms of Bounded Measures

The connection between the theory of probability and Fourier analysis is a discovery of P. Lévy, who in the early 1920's conceived the notion of characteristic function of a real-valued random variable or its probability distribution as a means of solving the classical central limit problem. In his early monograph Lévy [316] proved the one-to-one correspondence between distribution functions and their characteristic functions as well as a first formulation of the continuity theorem, which we shall discuss in greater detail in the following section. As useful references for the classical setup we mention the books of Lukacs [329] and Kawata [288], which contain a remarkable amount of material. Clearly, harmonic analysts have always worked with Fourier(-Stieltjes) transforms of measures and functions even on more general locally compact Abelian groups taking the real line as a particular example. Excellent accounts of the Abelian situation are the monographs of Rudin [417] and Hewitt and Ross [218]. What we need here is an extension of Fourier transform theory to arbitrary (not necessarily Abelian) locally compact groups. In order to build up a reasonable theory the characters appearing as basic objects in the case of a locally compact Abelian group have to be replaced by general infinite-dimensional representations in the case of an arbitrary locally compact group. See Heyer [227], in particular, Chapter IX. For the purpose of our special applications we restrict ourselves to finite-dimensional representations of the group G and pursue the corresponding duality theory. For compact groups rather complete results are available. The interested reader might consult Hewitt and Ross [218], preferably Chapter VII. For any locally compact group G we define in the text the dual Rep(G) of G on the basis of all finite-dimensional representations of G. The properties of Rep(G) are proved in Heyer [227]. Moreover, the coefficient algebra $\mathfrak{R}(G)$ of the given group is introduced only for finite-dimensional representations of G. An extensive discussion of $\mathfrak{K}(G)$, at least for compact G, is contained in Hewitt and Ross [218], §27. The equivalence of the statements (i) to (iii) following 1.3.1 is proved in Hewitt and Ross [218] (27.39). In the context of this result one also finds more details concerning the coefficient algebra $\mathfrak{K}(G)$ of G, in particular, a description of all real multiplicative linear functionals on $\mathfrak{K}(G)$, which yields the famous Tannaka-Krein duality theorem (Hewitt and Ross [218], §30 and Heyer [227], Chapter V). For the Fourier transform theory we are aiming at almost periodic (locally compact) groups as the suitable domain. The representation theory of this class of groups has already been indicated in the preliminaries. Here we add more precise information established with proofs in Heyer [227]. Pseudo- and quasirepresentations of Rep(G) are the building blocks of a duality theory for almost periodic groups in the spirit of Chu, who started an axiomatic treatment of the duality. See Heyer [231].

Theorems 1.3.3 to 1.3.5 are generalizations via Bohr compactification of corresponding results in Hewitt and Ross [218], §30, from compact to almost periodic groups. Properties 1.3.6 of the algebraic dual $\mathfrak{K}(G)^*$ of the coefficient algebra $\mathfrak{K}(G)$ of an almost periodic group G can be found in Siebert [437], where a new framework for the Fourier transform of bounded measures on an almost periodic group is set up. We continue with the presentation of Definition 1.3.7 of this Fourier transform and of Theorem 1.3.8 on its first fundamental properties. The most prominent property of any Fourier transformation to be considered is its injectivity. In Theorem 1.3.8 this injectivity was proved under the hypothesis that G is almost periodic. It should be noted that the Fourier transformation introduced on the basis of all (not necessarily finite-dimensional) representations of a locally compact group (Grenander [180], [181] and [182]) or via the Fourier algebra (Dixmier [118], Eymard [131]) admits similar properties, but seems to be less applicable for our purposes, although various results (in the context of probability theory) have been obtained (Heyer [223], Akemann and Walter [1]). In the last lines of the section topologies in $\mathfrak{K}(G)$ are studied. A topology of particular interest is the one which makes the Fourier transforms of bounded measures on G continuous as linear functionals on $\mathfrak{K}(G)$. Theorem 1.3.11 is taken from Siebert [436]. Similar problems have been discussed for the Fourier algebra of an arbitrary locally compact group. We mention the papers Eymard [131] and McKennon [353], Chapter 5.

R 1.4 The Theorems of Lévy and Bochner

Lévy's continuity theorem on the sequential bicontinuity of the Fourier transformation for probability measures on the group $\mathbb{R}$ goes back to the early 1920's (Lévy [316]). In 1936 Glivenko [170] strengthened part of the theorem. Further improvements were achieved by Cramér [96] and independently by Lévy [321]. In [96] Cramér noted that his proof of the theorem also yielded a generalization of Glivenko's result in the following sense: If a sequence $(\hat{\mu}_n)_{n\geq 1}$ of Fourier transforms of probability measures in $\mathscr{M}^1(\mathbb{R})$ converges pointwise to a limit function ϕ on $\mathbb{R}$ which is continuous at 0, then there exists some measure $\mu\in\mathscr{M}^1(\mathbb{R})$ such that $\phi=\hat{\mu}$ and $(\mu_n)_{n\geq 1}$ converges vaguely (weakly) to μ. It is essentially the same theorem which Bochner proved in [25] (within the framework of generalized distribution functions). The continuity theorem has become of great interest to probabilists (see for example Bauer [6], p. 250), and, in recent years, it has even appeared in some textbooks on Fourier analysis (see for example Katznelson [286], pp. 138–139, Exercises).

In our exposition we axiomatize the continuity theorem for almost periodic groups by introducing Lévy groups and then show (Theorem 1.4.5) that every Moore group is in fact a Lévy group. The proof carried out in analogy to that of Satz 5 in Siebert [437] is based on the fact (Theorem 1.4.2) that locally compact Abelian groups are Lévy groups. This result is known since about 1963. It is contained in Parthasarathy, Rao and Varadhan [383] (of course in Parthasarathy [377]), Rogalski [408] and Bingham and Parthasarathy [22] for probability measures, and in Lesca [314] for more general bounded measures. For compact not necessarily Abelian groups G the sequential bicontinuity of the Fourier transform on $\mathscr{M}^1(G)$ has already been shown by Ito and Kawada in [287]. See also Stromberg [457]. There are modifications of the continuity theorem stated in terms of nets of probability measures on locally compact Abelian groups. From Berg and Forst [12] we quote the following result, which by the method described in our text can be extended to general Moore groups: The Fourier transformation is a homeomorphism of the cone $\mathscr{M}_+^b(G)$ with the weak topology onto the cone of continuous positive-definite functions on $G^\wedge$ furnished with the compact open topology. The Cramér continuity theorem, however, cannot be generalized to nets of measures on a locally compact Abelian group, as has been shown in Berg and Forst [12], Example 3.15. In addition, one knows (for example from Kendall and Lamperti [293]) that in the case of a noncompact locally compact Abelian group G the open subsets of $\mathscr{M}^1(G)$ in the compact open topology do not in general contain neighborhoods in the topology of pointwise convergence of their elements.

There are other approaches aiming at a continuity theorem for measures on general locally compact groups G having a countable basis of their topology. In this context more general notions of Fourier transforms defined for general (not necessarily finite-dimensional) representations of G are applied. The fact that the weak convergence of sequences of measures in $\mathscr{M}^1(G)$ implies the strong convergence of the corresponding sequence of Fourier transforms was proved for the first time by Godement [173]. A restricted converse of this statement is due to Grenander [180]. A little later, Grenander's restriction was removed by Loynes [328], whose demonstration entered the book [182] of Grenander (See also Heyer [223] for a few applications of that theorem). Moreover, it should be mentioned that the continuity theorem for arbitrary locally compact groups G in the sense that a subset $\mathscr{N}$ of $\mathscr{M}_+^b(G)$ is weakly compact iff it is compact with respect to the Fourier topology (see Heyer [224]) has not yet reached its final generality. The proofs in Martin-Löf [345] and Heyer [224] depending on von Neumann's direct integration of Hilbert spaces rely heavily on the assumption of a countable basis of G. (Naimark's formulation of the direct integration theorem is stated more generally than has been proved). Finally references should be added concerning a continuity theorem stated in terms of positive-definite functions on an arbitrary locally compact group. In McKennon [353] the Krein algebra of the group has been exploited in a detailed study of various topologies. The paper of Akemann and

Walter [1] contains a W^*-algebraic approach to the comparison of topologies in the measure algebra of the group.

In the second part of the section Bochner's theorem on the one-to-one correspondence between bounded measures on a locally compact Abelian group G and continuous positive-definite functions on $G^\wedge$ (as can be found for example in Rudin [417]) is axiomatized for general almost periodic groups. One defines positive-definite (linear) forms on the coefficient algebra $\mathfrak{R}(G)$ of G and speaks of a Bochner group G, if every continuous positive-definite form on $\mathfrak{R}(G)$ is the Fourier transform of a measure in $\mathscr{M}_+^b(G)$. In analogy to the theory of Lévy groups it is shown in Theorem 1.4.12 that every Moore group is a Bochner group. Bochner groups have been introduced by Siebert [436], [437]. In order to show that any compact group is a Bochner group (Theorem 1.4.9), one needs the profound fact (Theorem 1.4.8) that for an almost periodic group G positive-definiteness of a linear functional on $\mathfrak{R}(G)$ implies positivity (and hence continuity). This auxiliary result is proved for compact groups in Hewitt and Ross [218], (30.2). Its origin can be traced in Krein [309], and the proof is due originally to Bochner [26]. For compact not necessarily Abelian groups a restricted Bochner theorem is proved in Hewitt and Ross [218] (34.10). The discussion on Bochner groups also shows that the class of all such groups is closed under the formation of certain projective limits and finite products. A comparison of Bochner groups with Lévy groups is still an open problem.

The connection of Bochner groups with groups that satisfy only the axiomatic duality theorem based on finite-dimensional representations (Chu groups) has been indicated in Theorem 1.4.15. For the theory of Chu groups the interested reader may study the relevant chapters of Heyer [227], and also the more specific results in Heyer [231]. Theorem 1.4.16 on a continuity property of the Fourier transform of a bounded measure is due to Siebert [436]. For a related theorem in the Abelian case see Hewitt and Ross [218], (33.20).

R 1.5 Convolution Semigroups and Negative-Definite Forms

An important tool for the analysis of probability measures on a locally compact group G is the mapping $\mu \to T_\mu$ from $\mathscr{M}^1(G)$ into $\mathscr{L}(\mathscr{C}^b(G))$ defined by

$$T_\mu f(x) := \int f(xy)\,\mu(dy)$$

for all $f \in \mathscr{C}^b(G)$ and $x \in G$. For a fixed $\mu \in \mathscr{M}^1(G)$ the linear operator T_μ on $\mathscr{C}^b(G)$ is called the probability operator of μ. More generally one considers the correspondence $\mu \to T_\mu$ from $\mathscr{M}_+^b(G)$ into $\mathscr{L}(\mathscr{C}^b(G))$ defined as above, which assigns to every measure $\mu \in \mathscr{M}_+^b(G)$ its convolution operator on $\mathscr{C}^b(G)$. The convolution operator T_μ of $\mu \in \mathscr{M}_+^b(G)$ enjoys properties similar to those of the Fourier transform $\hat{\mu}$ of μ introduced in any of the usual ways. The first aim of our discussion is to exploit this analogy in the direction of the bicontinuity of $\mu \to T_\mu$ from $\mathscr{M}^1(G)$ into $\mathscr{L}(\mathscr{C}^0(G))$, where $\mathscr{M}^1(G)$ and $\mathscr{L}(\mathscr{C}^0(G))$ are equipped with the weak topology and the strong operator topology resp. This is Theorem 1.5.5 due in the form presented to Siebert [442]. The preparations preceding the result are borrowed from Bourbaki [45], where a more general setup has been chosen. It is known (from harmonic analysis) that the convolution operator of a measure in $\mathscr{M}_+^b(G)$ can be extended to the Hilbert space $L^2_{\mathbb{C}}(G, \omega)$, where ω denotes a right Haar measure of G.

For the probability operator the possibility of this extension has been observed by Grenander in [179]. In Grenander [181] and [182] one also finds interesting properties of the extended probability operator, whose very definition makes it possible to apply Hilbert space methods to problems in the theory of probability on locally compact groups. Procedures for the extension of the convolution operator to the spaces $L^p_{\mathbb{C}}(G,\omega)$ with $p\in[1,\infty[$ have been described in the case of a locally compact Abelian group G in the monograph of Berg and Forst [12]. In [12] one finds for example the following result: Let T be a bounded operator on $L^p_{\mathbb{C}}(G,\omega)$ for $p\in[1,\infty[$ which is translation invariant and submarkovian. Then there exists a unique measure $\mu\in\mathscr{M}^{(1)}_+(G)$ satisfying $Tf=\mu*f$ for all $f\in L^p_{\mathbb{C}}(G,\omega)$.

Next we introduce the notion of (one-parameter) convolution semigroups in $\mathscr{M}^1(G)$ for a locally compact group G and the corresponding (one-parameter) semigroup of probability operators on $\mathscr{C}_u(G)$. The first task arising is the comparison of the various notions of continuity of a convolution semigroup in $\mathscr{M}^1(G)$. In Theorem 1.5.7 the continuity of the semigroup homomorphism $t\to\mu_t$ from $\mathbb{R}_+$ into $\mathscr{M}^1(G)$ is shown to be equivalent to the uniform continuity of the corresponding semigroup $(S_t)_{t\in\mathbb{R}^*_+}$ of probability operators $S_t=T_{\mu_t}$ on $\mathscr{C}_u(G)$. From Theorem 1.5.8 we obtain the fact that for convolution semigroups $(\mu_t)_{t\in\mathbb{R}_+}$ in $\mathscr{M}^1(G)$ continuity (at any point $t\in\mathbb{R}_+$) and continuity at the point $t=0$ are equivalent. In either case one obtains the existence of $\mu_0:=\lim_{t\downarrow 0}\mu_t$, and μ_0 is in fact an idempotent in $\mathscr{M}^1(G)$ (coming from a compact subgroup of G). In the text we treat the problem slightly more generally (following Siebert [442]), admitting convolution semigroups in $\mathscr{M}^1(G)$ which are indexed by a dense subsemigroup $\mathbb{D}$ of $\mathbb{R}^*_+$. In this case the extension of a convolution semigroup $(\mu_t)_{t\in\mathbb{D}}$ in $\mathscr{M}^1(G)$ to a convolution semigroup $(\nu_t)_{t\in\mathbb{R}_+}$ in $\mathscr{M}^1(G)$ has to be secured. This is done with the help of some functional analysis in Theorem 1.5.9. Remark 1.5.10 indicates the validity of the theorem also for general topological groups and measures.

At this point of the discussion a reference to continuous convolution semigroups in $\mathscr{M}^1(G)$ (or $\mathscr{M}^{(1)}(G)$) and strongly continuous semigroups of probability (or contraction convolution) operators on $\mathscr{C}_u(G)$ or $\mathscr{C}^0(G)$ would be desirable. We are referring to the theory of Hille-Yosida which was introduced in full power in the work [261] of Hunt and was analyzed in detail in various publications on the subject as for example Deny [110] and Berg and Forst [12], § 11. In both works the authors collect results from the Hille-Yosida theory on the infinitesimal generator and the resolvent of a contraction semigroup on a Banach space which find most publication and quote further references. Specific aspects of the theory concerning Feller semigroups are contained in Berg [8]. We shall take up these sources again when we study in more detail the canonical representations of continuous convolution semigroups in $\mathscr{M}^1(G)$, a problem which will dominate the discussion of Chapter IV. The formalism of differentiating a contraction semigroup will lead to the correspondence between continuous convolution semigroups in $\mathscr{M}^1(G)$ and certain (generating) linear functionals on the coefficient algebra $\mathfrak{R}(G)$ of G.

In 1.5.11 we ascribe to an $\{e\}$-continuous convolution semigroup $(\mu_t)_{t\in\mathbb{R}^*_+}$ in $\mathscr{M}^1(G)$ a negative-definite form $\psi\in\mathfrak{R}(G)^*$ defined by

$$\psi(f):=\lim_{t\downarrow 0}\frac{1}{t}(f(e)-\hat{\mu}_t(f))\qquad\text{for all } f\in\mathfrak{R}(G).$$

The subsequent presentation is based on Siebert [437]. In [437] the notions of negative-definite, primitive, quadratic and Poisson forms on $\mathfrak{R}(G)$ are introduced for at least almost periodic groups. These notions are based on the corresponding ones for functions on the dual $G^\wedge$ of a locally compact Abelian group. The pioneering work within the Abelian setup is due to Schoenberg [434]. About twenty years later Herz wrote his fundamental paper [217], which was the basis for studies in this context of Beurling and Deny. See Deny [110], [111]. Thus Theorem 1.5.13 on the duality of negative- and positive-definite forms on $\mathfrak{R}(G)$ is the first step in the more general approach to a characterization of convolution semigroups in $\mathcal{M}^1(G)$ in terms of negative-definite forms on $\mathfrak{R}(G)$. The proof of the theorem is based on Lemma 1.5.14 which is due to Hunt [261]. The full generalization to almost periodic groups of the Schoenberg representation is contained in our theorems 1.5.18 and 1.5.19. It should be noted that the restriction to Moore groups is required for the proof that for every continuous negative-definite form ψ on $\mathfrak{R}(G)$ with $\psi(1)=0$ there exists exactly one continuous convolution semigroup $(\mu_t)_{t\in\mathbb{R}_+^*}$ in $\mathcal{M}^1(G)$ satisfying $\hat{\mu}_t=\mathrm{Exp}(-t\psi)$ for all $t\in\mathbb{R}_+^*$.

For locally compact Abelian groups G the statement is also contained in Bochner [29], Rogalski [408] and Berg and Forst [12], Theorem 8.3. As primitive and quadratic forms on $\mathfrak{R}(G)$ were defined independently of negative-definiteness, it appears necessary to show that they in fact possess this property. The corresponding theorems 1.5.16 and 1.5.17 are not at all trivial. To show that a quadratic form ψ on $\mathfrak{R}(G)$ for an almost periodic group G is a negative-definite form or equivalently, that $\Psi:=\mathrm{Exp}(-\psi)$ is a positive-definite form on $\mathfrak{R}(G)$ requires the application of a method developed for compact groups G by Carnal in the proof of Hilfssatz 4.2 in [64]. This method is purely matrix-theoretic and reduces the solution of the problem to the framework of the Lie algebra of the unitary groups $G:=\mathfrak{U}(n)$ for $n\geq 1$. Clearly, the proof of Theorem 1.5.17 is far from being natural. But there exists as yet no better one.

Theorems 1.5.20 and 1.5.21 yield first applications of the characterization result incorporated in Theorems 1.5.16 and 1.5.17. In Theorem 1.5.20 the one-to-one correspondence between continuous primitive forms on $\mathfrak{R}(G)$ and one-parameter subgroups in G is established, Theorem 1.5.21 and Corollary 1.5.22 yield the sequential approximation by Poisson forms in the compact open topology of continuous negative-definite forms on $\mathfrak{R}(G)$. The Poisson forms on $\mathfrak{R}(G)$ will be analyzed in greater detail in Chapter VI.

We close the report on §5 of Chapter I with a few remarks on the more general setup of duality between continuous positive-definite and negative-definite functions for arbitrary topological groups. Here the book of Parthasarathy and Schmidt [385] and the previous work referred to in it becomes the relevant source. The authors of [385] define positive-definite and conditionally positive-definite kernels on an arbitrary set X and develop a reformulation of the central limit theorem of probability theory in terms of these kernels. In particular, they show in their lemma 1.7 that a Hermitian kernel L on X is conditionally positive-definite iff $\exp tL$ is positive-definite for all $t\in\mathbb{R}_+^*$.

Chapter II

Convergence of Convolution Sequences of Probability Measures

The study of the convergence of sequences of probability measures on a locally compact group is basic for the development of the central limit theorem. This chapter serves as a source of preparatory material as well as a presentation of the smooth part of the theory. Methodically the problems treated are chosen from the axiomatic point of view: Classical theorems are not only generalized to certain sufficiently large classes of locally compact groups, but in addition their domains of validity are determined. The fact that convergence of convolution sequences is a rare event is indicated in the first section. The analysis of conditions for the convergence of the sequence of powers of a probability measure makes evident all the difficulties which can arise in solving the general problem. The theorem of Ito and Kawada and its limitations are the main result in this direction. The theorem is valid only in the special case of a compact group. The next problem solved is the equivalence of stochastic, a.s. convergence and convergence in distribution for sequences of n-fold products of group-valued random variables. The domain of validity of this equivalence principle appears to be the class of locally compact groups without nontrivial compact subgroups. In a similar way we discuss the convergence principle in its various forms and analyze the convergence behavior of a sequence of convolution products after a certain shift. According to the particular shift one obtains probabilistic characterizations of the classes of all compact groups and of all totally disconnected compact groups. There follows the description of an axiomatic setup for the variance of a group-valued random variable. Although the axiomatization presented seems in some sense too restricted, one can apply the notion of variance to convergence problems of convolution sequences which are not directly treatable by the more elementary methods. Finally we present a number of results concerning asymptotic equidistribution of sequences of probability measures on a locally compact Abelian group. The restriction to Abelian groups seems natural from the point of view of existence of such sequences. Although the limiting procedure for an asymptotically equidistributed sequence is understood in the sense of the norm topology in the set of measures on the group, a comparison with limits in the sense of the vague topology will also be given. It appears that for compact Abelian groups the treatment has elements in common with the general theory of the first section. In the last section we concentrate on the analysis of asymptotically equidistributed powers. These can be completely characterized solely in terms of the underlying measure.

2.1 Convolution Powers on a Compact Group

In the course of this section we shall use some basic facts from the theory of compact semigroups. Whenever S denotes a semigroup we shall abbreviate by $E(S)$ the totality of idempotents in S and by $H(e)$ for $e \in E(S)$ the group of units in eSe. If S is a (Hausdorff) topological semigroup, then eSe is a closed subsemigroup of S with identity e. For any compact semigroup S and any $s \in S$ we introduce the set $P(s) := \{s^n : n \geq 1\}$ and its closure $Q(s) := P(s)^-$, which is the closed Abelian subsemigroup of S generated by s. It can be shown that $Q(s)$ contains a closed Abelian group

$$A(s) := \bigcap_{k \geq 1} \{s^k, s^{k+1}, \ldots\}^- .$$

Plainly, $E(S) \neq \emptyset$. Moreover, S admits a unique closed minimal ideal $M(S)$. Given a compact semigroup S, an element $s \in S$ and an idempotent e of $Q(s)$ we obtain that the equality $se = s$ implies $Q(s) = M(Q(s))$ so that $Q(s)$ is a group with identity e. In particular, if $E(S) = \{e\}$, then S is itself a group.

We are now going to apply these notions from the theory of compact semigroups to the $\mathscr{T}_v$-compact convolution semigroup $\mathscr{M}^1(G)$ for a compact group G. The aim of the discussion is the convergence behavior of the sequence $(\mu^n)_{n \geq 1}$ of n-th powers μ^n of a given measure $\mu \in \mathscr{M}^1(G)$. For $\mu \in \mathscr{M}^1(G)$ we define the set $\mathscr{P}(\mu) := \{\mu^n : n \geq 1\}$, its closure $\mathscr{Q}(\mu)$ and its accumulation set $\mathscr{A}(\mu)$. Clearly, $\mathscr{Q}(\mu)$ is a compact Abelian subsemigroup of $\mathscr{M}^1(G)$. Moreover, $\mathscr{A}(\mu)$ is the maximal group contained in $\mathscr{Q}(\mu)$ and the minimal ideal or kernel of $\mathscr{Q}(\mu)$. If $\lambda = \lambda^2$ is the unit element of $\mathscr{A}(\mu)$, then

$$\mathscr{Q}(\mu) * \lambda = \lambda * \mathscr{Q}(\mu) = \mathscr{A}(\mu) \quad \text{and} \quad \mathscr{A}(\mu) = \{\nu^n : \nu := \mu * \lambda = \lambda * \mu, n \geq 1\}^- .$$

Plainly, the sequence $(\mu^n)_{n \geq 1}$ converges iff $\mathscr{A}(\mu) = \{\lambda\}$.

2.1.1 Theorem. *Let G be a compact group, $\mathscr{N}$ a group of measures in $\mathscr{M}^1(G)$ with unit element λ and let G_1 denote the support* $\text{supp}(\mathscr{N})$ *of $\mathscr{N}$ defined as the subset* $(\bigcup \{\text{supp}(\nu) : \nu \in \mathscr{N}\})^-$ *of G and $H := \text{supp}(\lambda)$. Then*

(i) *G_1 is a closed subgroup of G;*
(ii) *H is a normal subgroup of G_1;*
(iii) *for any $\nu \in \mathscr{N}$ and every $x \in \text{supp}(\nu)$ one has $\nu = \lambda * \varepsilon_x = \varepsilon_x * \lambda$.*

Proof. (i) For $\nu_1, \nu_2 \in \mathscr{N}$ and $x_1 \in \text{supp}(\nu_1)$, $x_2 \in \text{supp}(\nu_2)$ we obtain $\nu_1 * \nu_2 \in \mathscr{N}$ and $x_1 x_2 \in \text{supp}(\nu_1 * \nu_2)$. Let $\nu \in \mathscr{N}$ with inverse ν^{-1} with respect to the group $\mathscr{N}$ and $x \in \text{supp}(\nu)$. For any element y of $\text{supp}(\nu^{-1})$ we obtain $z = xy \in \text{supp}(\lambda)$, and therefore

$$x^{-1} = yz^{-1} \in \text{supp}(\nu^{-1}) \, \text{supp}(\lambda) = \text{supp}(\nu^{-1}).$$

This implies that $\text{supp}(\nu)^{-1} \subset \text{supp}(\nu^{-1})$. Analogously

$$(\text{supp}(\nu^{-1}))^{-1} \subset \text{supp}(\nu) \quad \text{or} \quad \text{supp}(\nu^{-1}) \subset \text{supp}(\nu)^{-1}.$$

Hence, combining all of the previous results

$$\text{supp}(\nu)^{-1} = \text{supp}(\nu^{-1}), \quad \text{i.e.,} \quad \bigcup \{\text{supp}(\nu) : \nu \in \mathcal{N}\} \quad \text{is a group.}$$

Consequently, G_1 is a closed subgroup of G.

(ii) For $\nu \in \mathcal{N}$ and $x \in \text{supp}(\nu)$ we showed in (i) that $x^{-1} \in \text{supp}(\nu^{-1})$; hence,

$$x^{-1} H x \in \text{supp}(\nu^{-1}) H \,\text{supp}(\nu) = \text{supp}(\nu^{-1})\,\text{supp}(\lambda)\,\text{supp}(\nu) = H.$$

Since the set $\{x \in G : x^{-1} H x \subset H\}$ is clearly dense in G_1, we obtain $x^{-1} H x \subset H$ for all $x \in G_1$ and thus that H is a normal subgroup of G_1.

(iii) Let $\nu \in \mathcal{N}$. Since $\nu * \lambda = \nu$, $\text{supp}(\nu)$ is the union of a certain number of H-cosets of G_1. If two distinct H-cosets belong to $\text{supp}(\nu)$, then there exist $x_1, x_2 \in G_1$ such that

$$x_1 H \subset \text{supp}(\nu), \qquad x_2 H \subset \text{supp}(\nu) \quad \text{and} \quad x_1^{-1} x_2 H \cap H = \varnothing.$$

Since

$$x_1^{-1}(x_2 H) \subset \text{supp}(\nu)^{-1}\,\text{supp}(\nu) = \text{supp}(\nu^{-1})\,\text{supp}(\nu) = H,$$

$\text{supp}(\nu)$ is in fact an H-coset of G_1. Hence, $x \in \text{supp}(\nu)$ implies that H is the support of $\nu * \varepsilon_{x^{-1}}$ and of $\varepsilon_{x^{-1}} * \nu$. Moreover, if $y \in H$, then

$$\varepsilon_y * \nu = \varepsilon_y * (\lambda * \nu) = (\varepsilon_y * \lambda) * \nu = \lambda * \nu = \nu$$

and therefore

$$\varepsilon_y * (\nu * \varepsilon_{x^{-1}}) = (\varepsilon_y * \nu) * \varepsilon_{x^{-1}} = \nu * \varepsilon_{x^{-1}}.$$

Thus, $x \in \text{supp}(\nu)$ implies that $\nu * \varepsilon_{x^{-1}}$ is H-invariant and $\text{supp}(\nu * \varepsilon_{x^{-1}}) = H$. But then $\nu * \varepsilon_{x^{-1}} = \lambda$ or $\nu = \lambda * \varepsilon_x$ for all $x \in \text{supp}(\nu)$. Analogously one obtains $\nu = \varepsilon_x * \lambda$ for all $x \in \text{supp}(\nu)$ and hence, the complete statement. □

2.1.2 Corollary. *Let $\mu \in \mathcal{M}^1(G)$ and let λ be the unit element of the group $\mathscr{A}(\mu)$ defined above. Then* $\text{supp}(\mu)$ *is contained in an H-coset of the group* $\text{supp}(\mathscr{A}(\mu))$ *and hence, in an H-coset of the group G.*

Proof. Since $\mathscr{Q}(\mu) * \lambda = \mathscr{A}(\mu)$, we conclude that $\mu * \lambda = \nu$ for some $\nu \in \mathscr{A}(\mu)$ and hence by the theorem $\mu * \lambda = \lambda * \varepsilon_x$ for some $x \in \text{supp}(\mathscr{A}(\mu))$. Consequently,

$$\text{supp}(\mu) \subset \text{supp}(\mu)\,\text{supp}(\lambda) = Hx = xH \quad \text{for some } x \in \text{supp}(\mathscr{A}(\mu)).$$

The remaining statement follows again by the theorem. □

2.1.3. We first remark that for any sequence $(E_n)_{n \geq 1}$ of subsets of a topological space E the upper limit $\overline{\lim}_{n \geq 1} E_n$ is defined as the set of all points of E each neighborhood of which has nonempty intersection with infinitely many of the E_n and the lower limit $\underline{\lim}_{n \geq 1} E_n$ as the set of all points of E each neighborhood of which has a nonempty intersection with all but (the first) finitely many of the E_n. $(E_n)_{n \geq 1}$ is said to *converge* if $\overline{\lim}_{n \geq 1} E_n = \underline{\lim}_{n \geq 1} E_n$.

Theorem. *Let G be a compact group, $\mu\in\mathscr{M}^1(G)$ and $K:=[\mathrm{supp}(\mu)]^-$ the closed subgroup of G generated by* $\mathrm{supp}(\mu)$. *Then*
(i) $\mathrm{supp}(\mathscr{A}(\mu))=\overline{\lim}_{n\geq 1}\mathrm{supp}(\mu^n)=\mathrm{supp}(\mathscr{Q}(\mu))=K$.
(ii) $\underline{\lim}_{n\geq 1}\mathrm{supp}(\mu^n)\neq\emptyset$ *implies* $\mathrm{supp}(\mathscr{A}(\mu))\subset\underline{\lim}_{n\geq 1}\mathrm{supp}(\mu^n)$.

Proof. (i) First of all one notes that

$$\mathrm{supp}(\mathscr{A}(\mu))\subset\overline{\lim}_{n\geq 1}\mathrm{supp}(\mu^n)\subset\mathrm{supp}\,\mathscr{Q}(\mu)\subset K.$$

On the other hand one knows by Theorem 2.1.1 that $\mathrm{supp}(\mathscr{A}(\mu))$ is a closed subgroup of G and one has the inclusion $\mathrm{supp}(\mu)\subset\mathrm{supp}(\mathscr{A}(\mu))$ by Corollary 2.1.2. Therefore $K\subset\mathrm{supp}(\mathscr{A}(\mu))$, and the assertion is proved.

(ii) We first show that $\mathrm{supp}(\mathscr{A}(\mu))$ is the kernel of the semigroup $\mathrm{supp}(\mathscr{Q}(\mu))$. One has

$$\mathscr{Q}(\mu)*\mathscr{A}(\mu)\subset\mathscr{A}(\mu)\quad\text{and}\quad\mathscr{A}(\mu)*\mathscr{Q}(\mu)\subset\mathscr{A}(\mu).$$

Hence,

$$\mathrm{supp}(\mathscr{Q}(\mu))\,\mathrm{supp}(\mathscr{A}(\mu))\subset\mathrm{supp}(\mathscr{A}(\mu))\quad\text{and}$$

$$\mathrm{supp}(\mathscr{A}(\mu))\,\mathrm{supp}(\mathscr{Q}(\mu))\subset\mathrm{supp}(\mathscr{A}(\mu))\quad\text{resp.}$$

Thus $\mathrm{supp}(\mathscr{A}(\mu))$ is an ideal of $\mathrm{supp}(\mathscr{Q}(\mu))$. By Theorem 2.1.1 $\mathrm{supp}(\mathscr{A}(\mu))$ is a group and hence is a minimal ideal or the kernel of $\mathrm{supp}(\mathscr{Q}(\mu))$. We now assume that $\underline{\lim}_{n\geq 1}\mathrm{supp}(\mu^n)\neq\emptyset$ and deduce that $\underline{\lim}_{n\geq 1}\mathrm{supp}(\mu^n)$ is an ideal of $\mathrm{supp}(\mathscr{Q}(\mu))$. In fact, let

$$x_1\in\mathrm{supp}(\mu^m),\quad x_2\in\underline{\lim}_{n\geq 1}\mathrm{supp}(\mu^n)\quad\text{and}\quad U\in\mathfrak{B}(x_1x_2).$$

Then there exists a $V\in\mathfrak{B}(x_2)$ satisfying $x_1V\subset U$. Moreover, there is an $n_0\geq 1$ such that for every $n>n_0$ one has $V\cap\mathrm{supp}(\mu^n)\neq\emptyset$. Let $n-m>n_0$ and $x_3\in V\cap\mathrm{supp}(\mu^{n-m})$. Then

$$x_1x_3\in x_1V\subset U\quad\text{and also}\quad x_1x_3\in\mathrm{supp}(\mu^m)\,\mathrm{supp}(\mu^{n-m})=\mathrm{supp}(\mu^n).$$

From this we obtain

$$U\cap\mathrm{supp}(\mu^n)\neq\emptyset\ \text{for all}\ n>n_0+m\ \text{and thus}\ x_1x_2\in\underline{\lim}_{n\geq 1}\mathrm{supp}(\mu^n).$$

Since

$$\mathrm{supp}(\mathscr{Q}(\mu))\,\underline{\lim}_{n\geq 1}\mathrm{supp}(\mu^n)\subset\underline{\lim}_{n\geq 1}\mathrm{supp}(\mu^n),$$

$\underline{\lim}_{n\geq 1}\mathrm{supp}(\mu^n)$ is a left ideal of $\mathrm{supp}(\mathscr{Q}(\mu))$. Analogously one obtains that

$$\underline{\lim}_{n\geq 1}\mathrm{supp}(\mu^n)\quad\text{is a right ideal in}\quad\mathrm{supp}(\mathscr{Q}(\mu)).$$

Taking these facts together, we see that

$$\underline{\lim}_{n\geq 1}\mathrm{supp}(\mu^n)\quad\text{is an ideal in}\quad\mathrm{supp}(\mathscr{Q}(\mu)),$$

which was to be proved. □

2.1.4 Theorem. *Let G be a compact group, $\mu \in \mathscr{M}^1(G)$ and let λ be the unit element of the group $\mathscr{A}(\mu)$. We further put $H := \operatorname{supp}(\lambda)$ and set $K := [\operatorname{supp}(\mu)]^-$. Then the following conditions are equivalent:*

(i) $\lim_{n \to \infty} \mu^n$ *exists;*
(ii) $\underline{\lim}_{n \geq 1} \operatorname{supp}(\mu^n) \neq \emptyset$;
(iii) *the sequence* $(\operatorname{supp}(\mu^n))_{n \geq 1}$ *converges;*
(iv) $[\bigcup_{n \geq 1} \operatorname{supp}(\mu)^n \operatorname{supp}(\mu)^{-n}]^- = K$;
(v) $\operatorname{supp}(\mu)$ *is not contained in a proper coset of any closed normal subgroup of* K;
(vi) $\operatorname{supp}(\mu)$ *is not contained in any proper coset of* H *in* K;
(vii) $\lambda = \omega_K$.

Proof. 1. (i) $\Rightarrow$ (ii). Let $\lim_{n \to \infty} \mu^n = \nu$. We shall show that

$$\operatorname{supp}(\nu) \subset \underline{\lim}_{n \geq 1} \operatorname{supp}(\mu^n).$$

Assuming the contrary holds one obtains for every $x \in \operatorname{supp}(\nu)$ and every $U \in \mathfrak{B}(x)$ a subsequence $(n_i)_{i \geq 1}$ in $\mathbb{N}$ satisfying $U \cap \operatorname{supp}(\mu^{n_i}) = \emptyset$ for all $i \geq 1$. Therefore one has

$$x \in U \cap \operatorname{supp}(\nu) \subset U \cap \operatorname{supp}(\{\mu^{n_i} : i \geq 1\}^-)$$

and, on the other hand, $\operatorname{supp}(\{\mu^{n_i} : i \geq 1\}^-) \subset \complement U$, which is the desired contradiction.

2. (ii) $\Rightarrow$ (iii) is an immediate consequence of Theorem 2.1.3.

3. (iii) $\Rightarrow$ (iv). By Theorem 2.1.3 it suffices to show that

$$F := \underline{\lim}_{n \geq 1} \operatorname{supp}(\mu^n) \subset (\bigcup_{n \geq 1} \operatorname{supp}(\mu^n)(\operatorname{supp}(\mu^n))^{-1})^-.$$

Let $x_1 \in F$ and $U \in \mathfrak{B}(x_1)$. Since F is a group, we have $x_1 = x_2 x_3$ for some $x_2, x_3 \in F$. Choosing neighborhoods V, W in $\mathfrak{B}(x_1)$, $\mathfrak{B}(x_2)$ resp. such that $VW \subset U$ one obtains $V \cap \operatorname{supp}(\mu^n) \neq \emptyset$ and hence $W^{-1} \cap \operatorname{supp}(\mu^n) \neq \emptyset$ for $n \geq 1$ sufficiently large. We conclude that $\emptyset \neq W \cap (\operatorname{supp}(\mu^n))^{-1}$ and thus

$$\emptyset \neq VW \cap \operatorname{supp}(\mu^n)(\operatorname{supp}(\mu^n))^{-1} \subset U \cap \operatorname{supp}(\mu^n)(\operatorname{supp}(\mu^n))^{-1}.$$

This implies that $x_1 \in (\bigcup_{n \geq 1} \operatorname{supp}(\mu^n)(\operatorname{supp}(\mu^n))^{-1})^-$ and hence, the assertion.

4. (iv) $\Rightarrow$ (v). Let N be a proper closed normal subgroup of K and $x \in K$ such that $\operatorname{supp}(\mu) \subset Nx$ holds. Plainly

$$\operatorname{supp}(\mu^n) \subset N x^n, \qquad (\operatorname{supp}(\mu^n))^{-1} \subset x^{-n} N$$

and hence, $\operatorname{supp}(\mu^n)(\operatorname{supp}(\mu^n))^{-1} \subset N$. But then

$$[\bigcup_{n \geq 1} \operatorname{supp}(\mu^n)(\operatorname{supp}(\mu^n))^{-1}]^- \subset N,$$

which is a contradiction to the assumption that N is a proper subgroup of K.

5. (v) $\Rightarrow$ (vi) follows from the normality of H in K via Theorems 2.1.1 and 2.1.3.

6. (vi) $\Rightarrow$ (vii). By Corollary 2.1.2 and Theorem 2.1.3 (i) we have that $\operatorname{supp}(\mu)$ is contained in some H-coset of K. The assumption implies therefore that $\operatorname{supp}(\mu) \subset H$ or $K = H$. In this case $\lambda = \omega_K$.

7. (vii) $\Rightarrow$ (i). If $\lambda=\omega_K$, then $\lambda*\mu=\lambda$ and since $\mathscr{A}(\mu)=\{(\lambda*\mu)^n\colon n\geqq 1\}^-$, we conclude that $\mathscr{A}(\mu)=\{\lambda\}$ or $\lim_{n\to\infty}\mu^n=\lambda$. $\square$

2.1.5 Corollary. *For any $\mu\in\mathscr{M}^1(G)$ the following statements are equivalent:*
(i) $\lim_{n\to\infty}\mu^n=\omega_G$;
(ii) $\underline{\lim}_{n\geq 1}\operatorname{supp}(\mu^n)=G$.

Proof. 1. (i) $\Rightarrow$ (ii) holds since

$$\underline{\lim}_{n\geq 1}\operatorname{supp}(\mu^n)=\lim_{n\to\infty}\operatorname{supp}(\mu^n)=\operatorname{supp}(\omega_G)=G.$$

2. (ii) $\Rightarrow$ (i). From

$$G=\underline{\lim}_{n\geq 1}\operatorname{supp}(\mu^n)\subset\overline{\lim}_{n\geq 1}\operatorname{supp}(\mu^n)\subset G$$

one obtains $\underline{\lim}_{n\geq 1}\operatorname{supp}(\mu^n)=\overline{\lim}_{n\geq 1}\operatorname{supp}(\mu^n)$, and then from the theorem the existence of $\nu:=\lim_{n\to\infty}\mu^n$. But $\operatorname{supp}(\nu)=\lim_{n\to\infty}\operatorname{supp}(\mu^n)=G$ implies the assertion. $\square$

2.1.6 Corollary. *If G is connected and μ is a symmetric ω_G-absolutely continuous measure in $\mathscr{M}^1(G)$, then* $\lim_{n\to\infty}\mu^n=\omega_G$.

Proof. Since $\operatorname{supp}(\mu^{n+2})=\operatorname{supp}(\mu^n)\operatorname{supp}(\mu^2)$ and $\operatorname{supp}(\mu^2)$ contains the unit element e of G, one has $\underline{\lim}_{n\geq 1}\operatorname{supp}(\mu^{2n})=\overline{\lim}_{n\geq 1}\operatorname{supp}(\mu^{2n})$. Hence, by the theorem $\lim_{n\to\infty}\mu^{2n}$ exists. But as $\operatorname{supp}(\mu)\operatorname{supp}(\mu)^{-1}$ admits an interior point,

$$\underline{\lim}_{n\geq 1}\operatorname{supp}(\mu^{2n})=G.$$

In this case $\lim_{n\geq 1}\mu^{2n}=\omega_G$ by Corollary 2.1.5 and

$$\lim_{n\to\infty}\mu^{2n+1}=\lim_{n\to\infty}\mu^{2n}*\mu=\omega_G*\mu=\omega_G.\quad\square$$

2.1.7 Theorem. *Let G be a compact group, $\mu\in\mathscr{M}^1(G)$ and λ the unit element of the group $\mathscr{A}(\mu)$ with H as its support. Let $S(\mu)$ be the H-coset of G containing* $\operatorname{supp}(\mu)$. *Then for every $x\in S(\mu)$ the sequence $(\nu_n)_{n\geq 1}$ defined by $\nu_n:=\varepsilon_{x^{-n}}*\mu^n$ for all $n\geq 1$ and the sequence $(\pi_n)_{n\geq 1}$ defined by $\pi_n:=\mu^n*\varepsilon_{x^{-n}}$ for all $n\geq 1$ converge to λ.*

Proof. Let $\nu\in\mathscr{A}(\mu)$. From the compactness of $\mathscr{M}^1(G)$ and the continuity of the convolution mapping it follows that $\nu=\varepsilon_{x_1}*\nu'$, where x_1 and ν' are accumulation points of the sequences $(x^{-n})_{n\geq 1}$ and $(\mu^n)_{n\geq 1}$ resp. Since $\nu'\in\mathscr{A}(\mu)$, Theorem 2.1.1 implies that $\nu'=\varepsilon_{x_2}*\lambda$ for some $x_2\in G$ and hence $\nu=\varepsilon_{x_1}*\varepsilon_{x_2}*\lambda$. But $\operatorname{supp}(\nu_n)\subset H$ for all $n\geq 1$ implies $\operatorname{supp}(\nu)\subset H$. Consequently, $\nu=\lambda$ and hence $\lim_{n\to\infty}\nu_n=\lambda$. One establishes analogously the convergence to λ of $(\pi_n)_{n\geq 1}$. $\square$

2.1.8 Remark. The subgroup $H:=\operatorname{supp}(\lambda)$ appearing in the statement of the theorem is the smallest closed subgroup of G with the property that $\operatorname{supp}(\mu)$ lies in one of its two-sided cosets. In fact, if $\operatorname{supp}(\mu)$ is contained in a two-sided H'-coset of G with $H'\subset H$, $H'\neq H$, then for $\nu:=\lim_{n\to\infty}\varepsilon_{x^{-n}}*\mu^n$ (with $x\in\operatorname{supp}(\mu)$) one has $\operatorname{supp}(\nu)\subset H'$; but $\operatorname{supp}(\nu)=H$ since by Theorem 2.1.4 $\nu=\omega_H$.

In the following we shall discuss the limitations of the fundamental theorem of K. Ito and Y. Kawada. It can be generalized from sequences of powers of a measure to normal sequences (in the sense of Borel) of measures on a compact group; but it cannot be modified for an arbitrary locally compact group.

2.1.9. Definition. A sequence $(\mu_j)_{j\geq 1}$ of measures in $\mathscr{M}^1(G)$ is said to be *normal* if for every $n\geq 1$ there exists a sequence $(j_l)_{l\geq 1}$ in $\mathbb{N}$ with $j_1<j_2<\cdots$ such that $\mu_s=\mu_{j_l+s}$ for all $s=1,2,\ldots,n$; $l\geq 1$.

Plainly the sequence $(\mu_n)_{n\geq 1}$ with $\mu_n:=\mu\in\mathscr{M}^1(G)$ for all $n\geq 1$ is normal.

2.1.10 Theorem. *Let G be a compact group, $(\mu_j)_{j\geq 1}$ a normal sequence in $\mathscr{M}^1(G)$ and $(\nu_n)_{n\geq 1}$ the corresponding sequence of n-th convolution products $\nu_n:=\mu_1*\cdots*\mu_n$ $(n\geq 1)$. Then $\lim_{n\to\infty}\nu_n=:\nu\in\mathscr{M}^1(G)$ implies that $\nu=\omega_H$ for the compact subgroup $H:=[\bigcup_{j\geq 1}\operatorname{supp}(\mu_j)]^-$ generated by the set $\bigcup_{j\geq 1}\operatorname{supp}(\mu_j)$.*

Proof. Let $(\mu_j)_{j\geq 1}$ be a normal sequence in $\mathscr{M}^1(G)$ such that the corresponding sequence $(\nu_n)_{n\geq 1}$ converges to $\nu\in\mathscr{M}^1(G)$. Fix $n\in\mathbb{N}$ and let $(j_l)_{l\geq 1}$ be a sequence in $\mathbb{N}$ with $j_1<j_2<\cdots$ such that $\nu_n=\mu_{j_l+1}*\cdots*\mu_{j_l+n}$ holds for all $l\geq 1$. Then the equalities

$$\lim_{l\to\infty}\nu_{j_l+n}=\nu \quad\text{and}\quad \lim_{l\to\infty}\nu_{j_l+n}=\lim_{l\to\infty}\nu_{j_l}*\nu_n=\nu*\nu_n$$

imply that $\nu=\nu*\nu_n$ and hence $\nu=\nu^2$. Thus by Theorem 1.2.10 $\nu=\omega_H$ for some compact subgroup H of G. For every $k\geq 2$ and some corresponding sequence $(j_l)_{l\geq 1}$ in $\mathbb{N}$ with $j_1<j_2<\cdots$ one obtains $\nu_{j_l+k}=\nu_{j_l+k-1}*\mu_k$. Hence, for $l\to\infty$ the equality $\nu=\nu*\mu_k$ holds and according to Theorem 1.2.7 we have the inclusion $\operatorname{supp}(\mu_k)\subset H$, which implies that $\operatorname{supp}(\nu)\supset[\bigcup_{j\geq 1}\operatorname{supp}(\mu_j)]^-$.

The inverse inclusion is evident. Thus $H=[\bigcup_{j\geq 1}\operatorname{supp}(\mu_j)]^-$. □

2.1.11 Theorem. *Let G be a compact group, $(\mu_j)_{j\geq 1}$ a normal sequence in $\mathscr{M}^1(G)$ and $(\nu_n)_{n\geq 1}$ the corresponding sequence of n-th convolution products $\nu_n:=\mu_1*\cdots*\mu_n$ $(n\geq 1)$. Then the following statements are equivalent:*
(i) $\lim_{n\to\infty}\nu_n$ *exists;*
(ii) $[\bigcup_{j\geq 1}\operatorname{supp}(\mu_j)]^-$
$=[\bigcup_{n\geq 1}\operatorname{supp}(\mu_1)\cdot\cdots\cdot\operatorname{supp}(\mu_n)\cdot\operatorname{supp}(\mu_n)^{-1}\cdot\cdots\cdot\operatorname{supp}(\mu_1)^{-1}]^-$.

Proof. 1. We first observe that for any $n\geq 1$, $x_n\in\operatorname{supp}(\mu_1)\cdot\cdots\cdot\operatorname{supp}(\mu_n)$ and for $\pi_n:=\nu_n*\varepsilon_{x_n^{-1}}$ we obtain

$$\begin{aligned}&[\textstyle\bigcup_{n\geq 1}\operatorname{supp}(\pi_n)]^-\\&=[\textstyle\bigcup_{n\geq 1}\operatorname{supp}(\mu_1)\cdot\cdots\cdot\operatorname{supp}(\mu_n)\operatorname{supp}(\mu_n)^{-1}\cdot\cdots\cdot\operatorname{supp}(\mu_1)^{-1}]^-.\end{aligned}$$

In fact, the assumption implies

$$\begin{aligned}&\operatorname{supp}(\mu_1)\cdot\cdots\cdot\operatorname{supp}(\mu_n)x_n^{-1}\\&\subset\operatorname{supp}(\mu_1)\cdot\cdots\cdot\operatorname{supp}(\mu_n)\operatorname{supp}(\mu_n)^{-1}\cdot\cdots\cdot\operatorname{supp}(\mu_1)^{-1}\\&\subset[\operatorname{supp}(\mu_1)\cdot\cdots\cdot\operatorname{supp}(\mu_n)x_n^{-1}]^-\end{aligned}$$

for all $n \geq 1$ and hence

$$[\bigcup_{n\geq 1} \operatorname{supp}(\mu_1)\cdot \cdots \cdot \operatorname{supp}(\mu_n) x_n^{-1}]^-$$
$$=[\bigcup_{n\geq 1} \operatorname{supp}(\mu_1)\cdot \cdots \cdot \operatorname{supp}(\mu_n)\cdot \operatorname{supp}(\mu_n)^{-1}\cdot \cdots \cdot \operatorname{supp}(\mu_1)^{-1}]^-.$$

2. (i) ⇒ (ii). Let $\lim_{n\to\infty} \nu_n =: \nu$ exist. Then by Theorem 2.1.10 we obtain $\nu = \omega_H$ for $H := [\bigcup_{j\geq 1} \operatorname{supp}(\mu_j)]^-$. Moreover,

$$\lim_{n\to\infty} \hat{\nu}_n(D) = \hat{\nu}(D) = 0 \quad \text{holds for all } D \in \operatorname{Irr}_0(H) := \operatorname{Irr}(H)\setminus\{I\}.$$

Then by Part 1 we have $\lim_{n\to\infty} \hat{\pi}_n(D) = 0 = \hat{\nu}(D)$ for all $D \in \operatorname{Irr}_0(H)$, and using the Peter-Weyl theorem, we get $\lim_{n\to\infty} \hat{\pi}_n(D) = \hat{\nu}(D)$ for all $D \in \operatorname{Rep}(H)$. By the very definition of H we have finally $\lim_{n\to\infty} \hat{\pi}_n(D) = \hat{\nu}(D)$ for all $D \in \operatorname{Rep}(G)$. Since G is an L-group by Theorem 1.4.5, we obtain $\lim_{n\to\infty} \pi_n = \nu$. Hence,

$$\operatorname{supp}(\nu) \subset [\bigcup_{n\geq 1} \operatorname{supp}(\pi_n)]^-$$

and with the help of Part 1 we obtain the asserted equality.

3. Let $\mu \in \mathscr{M}^1(G)$, $A := [\operatorname{supp}(\mu)]^-$ and $e \in \operatorname{supp}(\mu)$. If $D \in \operatorname{Rep}(A)$ and $\|\hat{\mu}(D)\| = 1$, then $D \notin \operatorname{Irr}_0(A)$. Indeed, let $D \in \operatorname{Rep}(A)$ be of the form $D := (d_{ik})_{i,k=1,\ldots,n(D)}$. From $\|\hat{\mu}(D)\| = 1$ we deduce that

$$\sum_{i,k=1}^{n(D)} \left(\left|\int_{\operatorname{supp}(\mu)} d_{ik}\, d\mu\right|^2 - \int_{\operatorname{supp}(\mu)} |d_{ik}|^2\, d\mu\right) = 0.$$

Hence, $d_{ik}(x) = \text{const}$ or since $e \in \operatorname{supp}(\mu)$, we have more precisely

$$d_{ik}(x) = d_{ik}(e) = \delta_{ik}$$

for all $x \in \operatorname{supp}(\mu)$, $\quad i = 1, \ldots, n(D) \quad (k = 1, \ldots, n(D))$.

Then $(\delta_{1k}, \ldots, \delta_{n(D)k})$ is the $D(x)$-invariant vector of $\mathscr{H}(D)$ for $x \in \operatorname{supp}(\mu)$ and hence for $x \in [\operatorname{supp}(\mu)]^- = A$ $(k = 1, \ldots, n(D))$, which implies that $D \notin \operatorname{Irr}_0(A)$.

4. (ii) ⇒ (i). Let the equality of (ii) be satisfied. Then according to Part 1 we have

$$[\bigcup_{n\geq 1} \operatorname{supp}(\pi_n)]^- = [\bigcup_{j\geq 1} \operatorname{supp}(\mu_j)]^-.$$

Let $P := [\bigcup_{n\geq 1} \operatorname{supp}(\pi_n)]^-$. We pick $D \in \operatorname{Rep}(P)$. If $\|\hat{\pi}_n(D)\| = 1$ for every $n \geq 1$, then according to Part 3 we have $D \notin \operatorname{Irr}_0([\operatorname{supp}(\pi_n)]^-)$ for every $n \geq 1$. But

$$[\operatorname{supp}(\pi_{n+1})]^- \supset [\operatorname{supp}(\pi_n)]^- \text{ implies } D \notin \operatorname{Irr}_0([\bigcup_{j=1}^n \operatorname{supp}(\pi_j)]^-)$$

for every $n \geq 1$ and consequently, $D \notin \operatorname{Irr}_0(P)$. Hence, by the above equality of groups we have $D \notin \operatorname{Irr}_0(H)$. For $D \in \operatorname{Irr}_0([\bigcup_{j\geq n+1} \operatorname{supp}(\mu_j)]^-)$ there therefore exists a $k \geq 1$ such that $\|\hat{\pi}_k(D)\| < 1$ or $\|\hat{\nu}_k(D)\| < 1$ holds. Let now $(j_l)_{l\geq 1}$ be a sequence in $\mathbb{N}$ with $j_1 < j_2 < \ldots$ such that $\nu_n = \mu_{j_l+1} * \cdots * \mu_{j_l+n}$ and $j_l + n < j_{l+1}$ holds for all $l \geq 1$.

Plainly $\|\hat{\nu}_m(D)\| \leq \|\hat{\nu}_n(D)\|^{j_l}$ for all $j_l + n \leq m$. But by the above discussion we have $\lim_{n\to\infty} \hat{\nu}_n(D) = 0$ for all $D \in \operatorname{Irr}_0(P)$ or, using the Peter-Weyl theorem, $\lim_{n\to\infty} \hat{\nu}_n(D) = 0$ for all $D \in \operatorname{Rep}(G)$ such that $\operatorname{Res}_P D$ does not contain I.

Since by Theorem 1.4.5 G is an L-group, Statement (i) of the theorem follows. □

2.1.12. Let G be an arbitrary locally compact group and $(\tilde{G}, \beta)$ a Bohr compactification of G. Given a sequence $(\mu_n)_{n\geq 1}$ of measures in $\mathscr{M}^1(G)$, one might be interested in finding the limit measure in $\mathscr{M}^1(\tilde{G})$ of the sequence $(\tilde{\mu}_n)_{n\geq 1}$ of image measures $\tilde{\mu}_n := \beta(\mu_n)$ $(n\geq 1)$. If $(\mu_n)_{n\geq 1}$ converges (in $\mathscr{M}^1(G)$), then $(\tilde{\mu}_n)_{n\geq 1}$ converges in $\mathscr{M}^1(\tilde{G})$. The converse of this statement is, however, false. In this setup the Ito-Kawada theorem (the equivalence (i)$\Leftrightarrow$(v) of Theorem 2.1.4) can be rewritten as follows: Given a measure $\mu\in\mathscr{M}^1(G)$ the sequence $(\tilde{\mu}^n)_{n\geq 1}$ converges iff $\operatorname{supp}(\tilde{\mu}) = \overline{\beta(\operatorname{supp}(\mu))}$ is not contained in a proper coset of any closed normal subgroup of the compact group $\tilde{G}$.

Theorem. *Let $G\in\mathbf{A}$ and $\mu\in\mathscr{M}^1(G)$. The following statements are equivalent:*
(i) $\lim_{n\to\infty}\tilde{\mu}^n = \omega_{\tilde{G}}$;
(ii) $\operatorname{supp}(\mu)$ *is not contained in a proper coset of any closed normal subgroup* G_1 *of* G *satisfying* $G/G_1\in\mathbf{A}$.

Proof. 1. (i)$\Rightarrow$(ii). Let $\operatorname{supp}(\mu)$ be contained in a proper coset xG_1 of a closed normal subgroup G_1 of G satisfying $G/G_1\in\mathbf{A}$. Then $\beta(x)\notin\overline{\beta(G_1)}$ since $G_1=\beta^{-1}(\overline{\beta(G_1)})$, which is equivalent to $G/G_1\in\mathbf{A}$ (under the hypothesis $G\in\mathbf{A}$). Clearly

$$\operatorname{supp}(\tilde{\mu})\subset\beta(x)\,\overline{\beta(G_1)}=\overline{\beta(xG_1)}$$

and hence $\operatorname{supp}(\tilde{\mu})$ is contained in a proper coset of a closed normal subgroup of $\tilde{G}$. But by Theorem 2.1.4 this contradicts Statement (i).

2. (ii)$\Rightarrow$(i). By Theorem 2.1.4 the negation of Statement (i) yields the existence of a proper closed normal subgroup K of $\tilde{G}$ such that $\operatorname{supp}(\tilde{\mu})\subset zK$ for some $z\in\tilde{G}$. For $G_1 := \beta^{-1}(K)$ we get

$$G_1=\beta^{-1}(\overline{\beta(G_1)})\quad\text{and thus}\quad G/G_1\in\mathbf{A}.$$

Let $x\in\operatorname{supp}(\mu)$. Then $\beta(x)\in\operatorname{supp}(\tilde{\mu})\subset zK$ and $\beta(x)K=zK$. We conclude that for every $y\in\operatorname{supp}(\mu)$ one has $\beta(x)K=\beta(y)K$ and hence, $\beta(y)=\beta(x)k$ for some $k\in K$ and $y\in xG_1$ by the injectivity of β. But $K\neq\tilde{G}$ implies that $G_1\neq G$ and we arrive at the negation of Statement (ii). □

2.2 Equivalence of Types of Convergence

Let G be an arbitrary locally compact group, $(\mu_j)_{j\geq 1}$ a sequence of measures in $\mathscr{M}^1(G)$, and for every $k\geq 0$ let $(\nu_{k,n})_{n>k}$ be the *corresponding sequence of partial convolution products* $\nu_{k,n} := \mu_{k+1} * \cdots * \mu_n$. The preceding section suggested that even in the special case of a compact group G the sequence $(\nu_{k,n})_{n>k}$ does not in general converge, even for $k=0$ and a sequence $(\mu_j)_{j\geq 1}$ with $\mu_j := \mu\in\mathscr{M}^1(G)$ for all $j\geq 1$. Nevertheless the accumulation behavior of $(\nu_{k,n})_{n>k}$ can be studied in detail for an arbitrary locally compact group G.

2.2.1. For the following we shall assume that the family

$$\mathscr{P}((\mu_j)_{j\geq 1}) := \bigcup_{k\geq 0}\bigcup_{n>k}\{\nu_{k,n}\}$$

of partial convolution products corresponding to the sequence $(\mu_j)_{j\geq 1}$ in $\mathcal{M}^1(G)$ is ($\mathcal{T}_v$-) relatively compact. For every $0\leq k<n$ one defines the set $\mathcal{A}_k$ of all accumulation points $\nu^{(k)}(\in\mathcal{M}^1(G))$ of $(\nu_{k,n})_{n>k}$ and observes immediately that $\mathcal{A}_k=\mu_{k+1}*\mathcal{A}_{k+1}$ holds. Furthermore, we introduce the set $\mathcal{A}_\infty$ of all accumulation points $\nu^{(\infty)}$ of the sequence $(\nu^{(k)})_{k\geq 1}$ with $\nu^{(k)}\in\mathcal{A}_k$ for $k\geq 0$. Finally, we abbreviate by $\mathcal{L}$ the set of accumulation points of $(\nu_{l,m})_{m>l}$ in the sense that the filter $\{(l,m):m>l\}$ on $\mathbb{Z}\times\mathbb{N}$ is finer than the product of the Fréchet filters on $\mathbb{Z}$ and $\mathbb{N}$. That is to say, $\lambda\in\mathcal{L}$ iff there exist subnets $(l_\alpha)_{\alpha\in\mathbf{A}}$ and $(m_\alpha)_{\alpha\in\mathbf{A}}$ of $\mathbb{Z}$ and $\mathbb{N}$ resp. with $l_\alpha<m_\alpha$ for all $\alpha\in\mathbf{A}$ and $\lambda=\lim_{\alpha\in\mathbf{A}}\nu_{l_\alpha,m_\alpha}$.

2.2.2 Theorem. *Let G be a locally compact group and $(\mu_j)_{j\geq 1}$ a sequence in $\mathcal{M}^1(G)$ with relatively compact family $\mathcal{P}((\mu_j)_{j\geq 1})$.*

(i) *For $\nu,\nu'\in\mathcal{A}_k$ there exists an $x\in G$ such that $\nu'=\nu*\varepsilon_x$ holds, and for the right invariance subgroups $H(\nu)$ and $H(\nu')$ of ν and ν' resp. we obtain $H(\nu')=x^{-1}H(\nu)x$. More precisely,*

(ii) *there exist $\lambda'\in\mathcal{L}$ satisfying $H(\nu')=x^{-1}H(\nu)x$ for all $x\in(\operatorname{supp}(\lambda'))^{-1}$ and $\lambda\in\mathcal{L}$ satisfying $H(\nu)=x'^{-1}H(\nu')x'$ for all $x'\in(\operatorname{supp}(\lambda))^{-1}$.*

Proof. If $\nu:=\nu^{(k)}=\lim_{n\to\infty}\nu_{k,n}$ and $\nu':=\nu'^{(k)}=\lim_{n'\to\infty}\nu_{k,n'}$, then there exists a $\lambda\in\mathcal{L}$ satisfying $\lim_{n,n'\to\infty}\nu_{n,n'}=\lambda$ with the limit understood in the sense of the definition of $\mathcal{L}$. It follows that

$$\nu'=\lim_{n'\to\infty}\nu_{k,n'}=\lim_{n,n'\to\infty}\nu_{k,n}*\nu_{n,n'}=\nu*\lambda$$

and hence, $\nu'=\nu*\lambda$. Interchanging n and n' one obtains $\nu=\nu'*\lambda'$ for some $\lambda'\in\mathcal{L}$. Then by Theorem 1.2.7 $\nu*\lambda*\lambda'=\nu$ and $\nu'*\lambda'*\lambda=\nu'$ imply $\operatorname{supp}(\lambda*\lambda')\subset H(\nu)$ and $\operatorname{supp}(\lambda'*\lambda)\subset H(\nu')$ resp., and one obtains

$$\nu'=\nu*\lambda*\varepsilon_{x^{-1}}*\varepsilon_x=\nu*\varepsilon_x \qquad \text{for all } x\in(\operatorname{supp}(\lambda'))^{-1}$$

as well as

$$\nu=\nu'*\lambda'*\varepsilon_{x'^{-1}}*\varepsilon_{x'}=\nu'*\varepsilon_{x'} \qquad \text{for all } x'\in(\operatorname{supp}(\lambda))^{-1}.$$

But the right invariance subgroup $H(\nu')$ of $\nu'=\nu*\varepsilon_x$ for $x\in(\operatorname{supp}(\lambda'))^{-1}$ is just $x^{-1}H(\nu)x$ and the right invariance subgroup $H(\nu)$ of $\nu=\nu'*\varepsilon_{x'}$ for $x'\in(\operatorname{supp}(\lambda))^{-1}$ is $x'^{-1}H(\nu')x'$. □

2.2.3. By the general assumption, $\mathcal{P}((\mu_j)_{j\geq 1})$ is relatively compact or by Prohorov's theorem, uniformly tight in $\mathcal{M}^1(G)$. Hence, there exists a compact subset K of G, which we shall fix once and for all, satisfying $\nu_{k,n}(\complement K)\leq\frac{1}{3}$ for all $0\leq k<n$.

Consequently there exists for every $k\geq 0$ an $x_k\in K^{-1}K$ such that

$$H(\nu')=x_k^{-1}H(\nu)x_k \qquad \text{holds for all } \nu',\nu\in\mathcal{A}_k.$$

Indeed, for $k\geq 0$, $\nu^{(k)}$ and $\nu'^{(k)}$ in $\mathcal{A}_k$ we obtain by Theorem 2.2.2 the existence of an $x_k\in G$ such that $\nu'^{(k)}=\nu^{(k)}*\varepsilon_{x_k}$ holds, and by assumption we conclude from

$$\nu^{(k)}(K)\geq\tfrac{2}{3} \quad\text{and}\quad \nu^{(k)}(Kx_k^{-1})=\nu'^{(k)}(K)\geq\tfrac{2}{3}$$

that $K\cap Kx_k^{-1}\neq\emptyset$ so that $x_k\in K^{-1}K$ is satisfied.

2.2.4. Let $\mathfrak{D}$ be the set of all sequences $\mathscr{D}:=(\nu^{(k)})_{k\geq 0}$ with

$$\nu^{(k)}\in\mathscr{A}_k \quad \text{and} \quad \nu^{(k)}=\mu_{k+1}*\nu^{(k+1)} \qquad \text{for all } k\geq 0.$$

For any $\mathscr{D}\in\mathfrak{D}$ we denote the set of accumulation points of $\mathscr{D}$ by $\mathscr{A}(\mathscr{D})$. Clearly,

$$\bigcup_{\mathscr{D}\in\mathfrak{D}}\mathscr{A}(\mathscr{D})\subset\mathscr{A}_\infty\subset\mathscr{L}.$$

Furthermore, one observes that for $\mathscr{D}:=(\nu^{(k)})_{k\geq 0}\in\mathfrak{D}$ one has

$$H(\nu^{(k)})\supset H(\nu^{(k+1)}) \qquad \text{for all } k\geq 0.$$

The compact subgroup $H(\mathscr{D}):=\bigcap_{k\geq 0}H(\nu^{(k)})$ is called the *basis* of the sequence $\mathscr{D}\in\mathfrak{D}$. If $H(\mathscr{D})$ is independent of $\mathscr{D}\in\mathfrak{D}$, and hence, uniquely determined for the given sequence $(\mu_j)_{j\geq 1}$, it will be called the *basis of* $(\mu_j)_{j\geq 1}$ and abbreviated by $H((\mu_j)_{j\geq 1})$. For sequences $\mathscr{D}:=(\nu^{(k)})_{k\geq 0}$ and $\mathscr{D}':=(\nu'^{(k)})_{k\geq 0}$ in $\mathfrak{D}$ we finally put

$$A(\mathscr{D},\mathscr{D}'):=\{x\in G\colon x \text{ is accumulation point of a sequence } (x_k)_{k\geq 0} \text{ in } G \text{ such that } \nu'^{(k)}=\nu^{(k)}*\varepsilon_{x_k} \text{ holds for all } k\geq 0\}.$$

Obviously, $A(\mathscr{D},\mathscr{D}')=A(\mathscr{D}',\mathscr{D})^{-1}$. Moreover, one has $H(\mathscr{D}')=x^{-1}H(\mathscr{D})x$ for all $x\in A(\mathscr{D},\mathscr{D}')$. In fact, by Theorem 2.2.2 for every $k\geq 0$ there exists an $x_k\in G$ such that $H(\nu'^{(k)})=x_k^{-1}H(\nu^{(k)})x_k$ holds. For $y'\in H(\mathscr{D}')$ we have $y'=x_k^{-1}y^{(k)}x_k$ for some $y^{(k)}\in H(\nu^{(k)})$.

Furthermore, $(y^{(k)})_{k\geq 0}$ admits an accumulation point $y:=x\,y'\,x^{-1}$ for $x\in A(\mathscr{D},\mathscr{D}')$ since $\bigcup_{k\geq 0}\{x_k\}$ is relatively compact in G. Therefore $y\in H(\mathscr{D})$ and thus $H(\mathscr{D}')\subset x^{-1}H(\mathscr{D})x$. The definition of $A(\mathscr{D},\mathscr{D}')$ then yields the assertion.

2.2.5 Theorem. *Let G be a locally compact group, $(\mu_j)_{j\geq 1}$ a sequence in $\mathscr{M}^1(G)$ such that $\mathscr{P}((\mu_j)_{j\geq 1})$ is relatively compact and let $\mathfrak{D}$ be defined as above. For every $\mathscr{D}\in\mathfrak{D}$ the normed Haar measure $\omega_{H(\mathscr{D})}$ of the compact subgroup $H(\mathscr{D})$ of G is in $\mathscr{A}(\mathscr{D})$, and every $\nu\in\mathscr{A}(\mathscr{D})$ has the form $\nu=\varepsilon_x*\omega_{H(\mathscr{D})}$ for some $x\in G$.*

Proof. Let $\mathscr{D}:=(\nu^{(k)})_{k\geq 0}\in\mathfrak{D}$. For every $k\geq 0$ we have

$$\lim_{n\to\infty}\mu_1*\cdots*\mu_k*\nu_{k,n}=\mu_1*\cdots*\mu_k*\nu^{(k)},$$

where the limit is taken with respect to a filter $\mathfrak{F}_k$ on $\mathbb{N}$. Plainly $\mathfrak{F}_k<\mathfrak{F}_{k+1}$ for all $k\geq 0$. If one puts $\mathfrak{F}:=\sup_{k\geq 0}\mathfrak{F}_k$, one obtains with respect to this filter $\mathfrak{F}$ and for every $k\geq 0$ the limit relation $\lim_{n\to\infty}\nu_{k,n}=\nu^{(k)}$. For $0\leq k<k'<n$ we have

$$\nu_{k,n}=\nu_{k,k'}*\nu_{k',n} \quad \text{and thus} \quad \nu^{(k)}=\nu_{k,k'}*\nu^{(k')}.$$

If $\nu:=\nu^{(\infty)}$ is an accumulation point of $(\nu^{(k')})_{k'\geq 0}$ with respect to $\mathfrak{F}$, then $\nu^{(k)}=\nu^{(k)}*\nu$ for all $k\geq 0$ and hence $\nu=\nu*\nu$ as an idempotent in $\mathscr{M}^1(G)$ is of the form ω_H with $H:=\operatorname{supp}(\nu)$.

Let $\nu,\nu'\in\mathscr{A}(\mathscr{D})$. From $\nu^{(k)}=\nu_{k,k'}*\nu^{(k')}$ for all $0\leq k<k'<n$ one concludes that $\nu=\lambda*\nu'$ with $\lambda\in\mathscr{L}$ as well as $\nu'=\lambda'*\nu$ with $\lambda'\in\mathscr{L}$ as in the proof of Theorem 2.2.2. This yields the existence of $x\in G$ satisfying $\nu'=\varepsilon_x*\nu$.

We still have to show that $\operatorname{supp}(\nu)=H(\mathscr{D})$ holds. Indeed, since $H(\nu^{(k)})\supset H(\nu^{(k+1)})$ and $\nu^{(k)}*\varepsilon_x=\nu^{(k)}$ for all $x\in H(\nu^{(k)})$, $k\geq 0$, one concludes that $\operatorname{supp}(\nu)\subset H(\mathscr{D})$. On the other hand, for $x\in H(\mathscr{D})$ the above equality yields $\nu*\varepsilon_x=\nu$ and thus $H(\mathscr{D})\subset\operatorname{supp}(\nu)$. □

2.2.6 Theorem. *Let G be a locally compact group and $(\mu_j)_{j\geq 1}$ a sequence in $\mathcal{M}^1(G)$ with relatively compact family $\mathcal{P}((\mu_j)_{j\geq 1})$. For every $\lambda\in\mathcal{L}$ there exist sequences $\mathcal{D}:=(\nu^{(k)})_{k\geq 0}$ and $\mathcal{D}':=(\nu'^{(k)})_{k\geq 0}$ in $\mathfrak{D}$ satisfying the following conditions:*
(i) *For all $k\geq 0$ one has $\nu^{(k)}*\lambda=\nu'^{(k)}$;*
(ii) *for all $x\in A(\mathcal{D},\mathcal{D}')$ the inclusion $\operatorname{supp}(\lambda)\subset H(\mathcal{D})x=xH(\mathcal{D}')$ holds.*

Proof. By assumption $\lim_{l,m\to\infty}\nu_{l,m}=\lambda$ with respect to the filter $\mathfrak{F}$ in the sense of the definition of $\mathcal{L}$. As in the proof of Theorem 2.2.5 we construct for every $k\geq 0$ filters $\mathfrak{F}_k$ and $\mathfrak{F}'_k$ with $\mathfrak{F}_k<\mathfrak{F}_{k+1}$ and $\mathfrak{F}'_k<\mathfrak{F}'_{k+1}$ such that

$$\lim_{l\to\infty}\nu_{k,l}=\nu^{(k)} \quad\text{and}\quad \lim_{m\to\infty}\nu_{k,m}=\nu'^{(k)}$$

with respect to $\mathfrak{F}_k$ and $\mathfrak{F}'_k$ resp.

As in the proof of Theorem 2.2.2 we obtain for every $k\geq 0$ the equality

$$\nu^{(k)}*\lambda=\nu'^{(k)}$$

which is Assertion (i) of our theorem. Furthermore, one deduces as in Theorem 2.2.2 the equalities

$$\nu^{(k)}*\varepsilon_{x_k}=\nu'^{(k)} \quad\text{and}\quad H(\nu^{(k)})x_k=x_kH(\nu'^{(k)}) \quad\text{for } x_k\in G.$$

By the very definition of $A(\mathcal{D},\mathcal{D}')$ we obtain $\operatorname{supp}(\lambda)\subset H(\mathcal{D})x=xH(\mathcal{D}')$ for all $x\in A(\mathcal{D},\mathcal{D}')$ which proves (ii). □

2.2.7 Corollary. *The set $K_0:=\overline{\bigcup_{\lambda\in\mathcal{L}}\operatorname{supp}(\lambda)}$ is compact in G.*

Proof. Under the assumption of the relative compactness of $\mathcal{P}((\mu_j)_{j\geq 1})$ there exists a compact subset K of G satisfying $\nu_{k,n}(\complement K)\leq\frac{1}{3}$ for all $0\leq k<n$. Given $\lambda\in\mathcal{L}$ we obtain by the theorem $\operatorname{supp}(\lambda)\subset H(\nu')x$ for $\nu'\in\mathcal{A}_k$ and $x\in K^{-1}K$. We fix $\nu\in\mathcal{A}_k$ and get by Theorem 2.2.2 that $H(\nu')=y^{-1}H(\nu)y$ holds for some $y\in K^{-1}K$. Thus $\operatorname{supp}(\lambda)\subset KK^{-1}H(\nu)(K^{-1}K)^2=:L$. Since L is compact, so is K_0 as a closed subset of L. □

2.2.8 Application. For any $U\in\mathfrak{V}(e)$ we have $\lim_{k,n\to\infty}\nu_{k,n}(UK_0)=1$.

Proof. Assuming the negation of the statement one obtains the existence of a filter $\mathfrak{F}$ on $\mathbb{N}\times\mathbb{N}$ and of a neighborhood $U\in\mathfrak{V}(e)$ such that

$$\lim_{k,n\to\infty}\nu_{k,n}(UK_0)<1 \quad\text{and}\quad \lim_{k,n\to\infty}\nu_{k,n}=\lambda$$

with respect to the filter $\mathfrak{F}$.

On the other hand, there exists an $f\in\mathcal{K}(G)$ such that

$$f(x)=1 \text{ for all } x\in K_0 \quad\text{and}\quad f(x)=0 \text{ for all } x\in\complement UK_0.$$

Obviously $K_0\subset V:=\{x\in G: f(x)>\eta \text{ for } \eta\in]0,1[\}\subset UK_0$ and hence

$$\lambda(K_0)\leq\lambda(V)\leq\underline{\lim}_{n>k\geq 0}\nu_{k,n}(V)\leq\lim_{k,n\to\infty}\nu_{k,n}(UK_0)<1,$$

which is a contradiction of $\operatorname{supp}(\lambda)\subset K_0$. □

2.2.9 Application. If $\lim_{n\to\infty}\nu_{0,n}=\nu^{(0)}$, then K_0 is contained in the unique (maximal) compact subgroup $H:=H(\nu^{(0)})$, for which $\nu^{(0)}$ is right H-invariant.

Proof. By Theorem 2.2.6 one obtains from the hypothesis $\nu^{(0)}*\lambda=\nu^{(0)}$ or $\operatorname{supp}(\lambda)\subset H(\nu^{(0)})$ for all $\lambda\in\mathscr{L}$. But then $K_0\dot{\subset} H:=H(\nu^{(0)})$. The rest of the statement is clear. □

2.2.10 Remarks. (1) Let H be a compact subgroup of G, $\mu_1:=\omega_H$ and $\mu_j:=\varepsilon_{x_j}$ for $x_j\in H$, whenever $j\geq 2$. Then for every $k\geq 1$ the measures $\nu^{(k)}\in\mathscr{A}_k$ are precisely the Dirac measures supported by the points of the compact accumulation set $C^{(k)}$ of the sequence $(x_{k+1}\cdot\cdots\cdot x_n)_{n\geq k+1}$. One has $C^{(k)}=x_k^{-1}\cdot\cdots\cdot x_2^{-1}\,C^{(1)}$ for all $k\geq 2$, and $\mathscr{L}$ consists of the Dirac measures supported by the points of the compact subset K_0 of H. For every $\mathscr{D}:=(\nu^{(k)})_{k\geq 0}\in\mathfrak{D}$ we get $H(\mathscr{D})=\{e\}$, and $K_0\neq\{e\}$, if the sequence $(x_2\cdot\cdots\cdot x_n)_{n\geq 2}$ does not converge in G. So for $k\geq 1$, $\lim_{n\to\infty}\nu_{k,n}$ does not exist, but as $\mu_1=\omega_H$, $\lim_{n\to\infty}\nu_{0,n}=\omega_H$ does.

(2) Let H be the basis of the sequence $(\mu_j)_{j\geq 1}$ in $\mathscr{M}^1(G)$ and suppose that one has $xH=Hx$ for $x\in G$ and therefore $x^pH=Hx^p$ for all $p\geq 1$. Suppose also that the set $\{x^p: p\geq 1\}$ is relatively compact in G. Then the accumulation set C of $\{x^p: p\geq 1\}$ is a compact subgroup of G, and one has $cH=Hc$ for all $c\in C$. Taking now the particular sequence $(\mu_j)_{j\geq 1}$ defined by $\mu_{2p}:=\varepsilon_x$ for all $p\geq 1$ and $\mu_{2p+1}:=\omega_H$ for all $p\geq 0$, we see that $K_0=CH$ and every $\lambda\in\mathscr{L}$ is of the form $\lambda=\varepsilon_c*\omega_H$ for some $c\in C$.

(3) Let $H(\mathscr{D})$ and $H(\mathscr{D}')$ be the bases of sequences $\mathscr{D}$ and $\mathscr{D}'$ in $\mathfrak{D}$ for a given sequence $(\mu_j)_{j\geq 1}$ in $\mathscr{M}^1(G)$. Assume that $\omega_{H(\mathscr{D})}=\lambda*\lambda'$ with $\operatorname{supp}(\lambda)\subset H(\mathscr{D})$ and $\operatorname{supp}(\lambda')\subset H(\mathscr{D}')$. For the particular sequence $(\mu_j)_{j\geq 1}$ with

$$\mu_{2p+1}:=\lambda \quad\text{and}\quad \mu_{2p+2}:=\lambda' \qquad \text{for all } p\geq 0$$

we obtain $\nu_{k,n}=\omega_{H(\mathscr{D})}$, whenever $n\geq k+2$. Moreover, one has

$$\lim_{k\to\infty}\nu^{(k)}=\omega_{H(\mathscr{D})} \quad\text{and}\quad K_0=H(\mathscr{D}).$$

But $\mathscr{L}$ consists of the measures λ, λ' and $\omega_{H(\mathscr{D})}$.

In the following we maintain the assumption that G is a locally compact group admitting a countable basis of its topology and $(\mu_j)_{j\geq 1}$ is a sequence in $\mathscr{M}^1(G)$.

There exists a probability space $(\Omega,\mathfrak{A},P)$ consisting of a nonempty set Ω, a σ-algebra $\mathfrak{A}$ of subsets of Ω and an (abstract) probability measure P on $(\Omega,\mathfrak{A})$, and there exists for each $j\geq 1$ a measurable mapping $X_j:\Omega\to G$ such that $X_j(P)=\mu_j$. In fact, one chooses $\Omega:=G^{\mathbb{N}}$, $\mathfrak{A}:=\mathfrak{B}^{\mathbb{N}}$ the $\mathbb{N}$-tensor power of the Borel σ-algebra $\mathfrak{B}:=\mathfrak{B}(G)$ of G and X_j equal to the j-th projection of $G^{\mathbb{N}}$ onto its j-th component ($j\geq 1$). Moreover, the sequence $(X_j)_{j\geq 1}$ constructed is *stochastically independent* in the sense that for all $n\geq 1$ and sets $B_1,\ldots,B_n\in\mathfrak{B}$ the formula

$$P\bigcap_{j=1}^{n}[X_j\in B_j]=\prod_{j=1}^{n}P[X_j\in B_j]$$

holds.

For every $0\leq k<n$ the measures μ_k and $\nu_{k,n}$ are the distributions of X_k and

$$Y_{k,n}:=X_{k+1}\cdot\cdots\cdot X_n=Y_{0,k}^{-1}\,Y_{0,n} \text{ resp.}$$

For simplicity we further put $Y_n := Y_{0,n}$ for all $n \geq 1$. For this sequence $(Y_n)_{n\geq 1}$ the classical notions of stochastic and almost sure convergence can be extended in the following manner: Given any probability space $(\Omega, \mathfrak{A}, P)$ and a locally compact group G with a countable basis of its topology one introduces a *G-valued random variable* or G-random variable (on $(\Omega, \mathfrak{A}, P)$) as a measurable mapping $Z: \Omega \to G$ and its *distribution* as the image measure $Z(P)$ of P under Z.

2.2.11 Definition. A sequence $(Z_n)_{n\geq 1}$ of G-valued random variables Z_n is said to be *stochastically convergent* to (a G-valued random variable) Z, in symbols, st-$\lim_{n\to\infty} Z_n = Z$, if for all $U \in \mathfrak{B}(e)$ we have $\lim_{n\to\infty} P[Z_n^{-1} Z \in U] = 1$, and *almost surely convergent* to Z, in symbols, as-$\lim_{n\to\infty} Z_n = Z$, if for all $U \in \mathfrak{B}(e)$ one has $\lim_{n\to\infty} P \bigcap_{k\geq n} [Z_k^{-1} Z \in U] = 1$.

Clearly, as-$\lim_{n\to\infty} Z_n$ exists iff $\lim_{n\to\infty} P \bigcap_{k,l\geq n} [Z_k^{-1} Z_l \in U] = 1$ for every $U \in \mathfrak{B}(e)$.

2.2.12. It should be noted that st-$\lim_{n\to\infty} Z_n = Z$ signifies stochastic convergence with respect to the left uniform structure of G. But this is equivalent to the stochastic convergence with respect to the right uniform structure of G, as follows from the next

Lemma. *Let G be a locally compact group. For every compact subset K of G and all $U \in \mathfrak{B}(e)$ there are $U' := U'(U, K)$, $U'' := U''(U, K) \in \mathfrak{B}(e)$ such that*
(a) $x U' \subset U x$ *and*
(b) $U'' x \subset x U$ *hold for all* $x \in K$.

Proof. It suffices to prove (a). One chooses a symmetric $V \in \mathfrak{B}(e)$ such that $V^3 \subset U$ holds, a finite covering $\{V x_i : i = 1, \ldots, l\}$ of K and puts $U' := \bigcap_{i=1}^{l} x_i^{-1} V x_i$. Then for every $x \in K$ we obtain $x U' \subset V x_i U' \subset V^2 x_i$ for some $i \in \{1, \ldots, l\}$ and hence $x U' \subset V^3 x \subset U x$. □

The equivalence cited above is implied by the lemma since given a compact subset K of G and a $U \in \mathfrak{B}(e)$ one obtains a $U' := U'(U, K) \in \mathfrak{B}(e)$ satisfying $[Y \in K] \cap [Y_n \in Y U'] \subset [Y_n \in U Y]$ for all $n \geq 1$. Hence, from the assumptions $P[Y \in K] > 1 - \varepsilon$ and $P[Y_n \in Y U'] > 1 - \varepsilon$ for $\varepsilon > 0$ one gets the assertion $P[Y_n \in U Y] > 1 - 2\varepsilon$ for sufficiently large $n \geq 1$.

Let H be a compact subgroup of G, $\dot{G} := G/H$ and p the canonical mapping $G \to \dot{G}$. Neighborhoods of left cosets $\dot{x} := xH \in \dot{G}$ are left classes of $U \dot{x} = U x H$ with $U \in \mathfrak{B}(e)$. There is a natural left uniform structure in $\dot{G}$ defined by the system of vicinities $\{(\dot{x}, \dot{y}) \in \dot{G} \times \dot{G} : \dot{x} \in U \dot{y}\}$ $(U \in \mathfrak{B}(e))$. The following modes of convergence are understood with respect to this uniform structure.

2.2.13 Definition. The sequence $(Z_n)_{n\geq 1}$ (of G-valued random variables) converges *stochastically modulo H* to Z, in symbols, (H)st-$\lim_{n\to\infty} Z_n = Z$, if for all $U \in \mathfrak{B}(e)$ one has

$$\lim_{n\to\infty} P[p(Z_n) \in U\, p(Z)] = \lim_{n\to\infty} P[Z_n \in U Z H] = 1,$$

and it *converges a.s. modulo H* to Z, in symbols (H)as-$\lim_{n\to\infty} Z_n = Z$, if for all $U \in \mathfrak{B}(e)$ the relations

$$\lim_{n\to\infty} P \bigcap_{k\geq n} [p(Z_k) \in U p(Z)] = \lim_{n\to\infty} P \bigcap_{k\geq n} [Z_k \in UZH] = 1$$

are satisfied.

Plainly, (H)st-$\lim_{n\to\infty} Z_n = Z$ implies the Cauchy condition

(CC) $\qquad \lim_{n,m\to\infty} P[Z_n \in U Z_m H] = 1 \qquad$ for all $U \in \mathfrak{B}(e)$.

This follows directly from the inclusion

$$[Z \in U Z_m H] \cap [Z_n \in UZH] \subset [Z_n \in U^2 Z_m H].$$

If we consider the special case of a sequence $(X_j)_{j\geq 1}$ of G-valued random variables with corresponding sequence $(Y_{k,n})_{0\leq k<n}$ of partial products $Y_{k,n} := X_{k+1} \cdot \dots \cdot X_n$ and corresponding sequences $(\mu_j)_{j\geq 1}$ and $(\nu_{k,n})_{0\leq k<n}$ of distributions resp., then the relative compactness of $\mathscr{P}((\mu_j)_{j\geq 1})$ implies the equivalence of (CC) and

(CC′) $\qquad \lim_{k,n\to\infty} \nu_{k,n}(UH) = \lim_{k,n\to\infty} P[Y_k^{-1} Y_n \in UH] = 1 \qquad$ for all $U \in \mathfrak{B}(e)$.

In fact, by assumption for every $\eta > 0$ there exists a compact subset $K := K_\eta$ of G such that $\nu_{k,n}(K) > 1 - \eta$ holds for all $0 \leq k < n$.

Furthermore, for $U \in \mathfrak{B}(e)$ there are by Lemma 2.2.12 a neighborhood $U'' := U''(U, K) \in \mathfrak{B}(e)$ such that $P[Y_k^{-1} Y_n \in UH] \geq P[Y_n \in U'' Y_k H] - \eta$ holds and a neighborhood $U' := U'(U, K) \in \mathfrak{B}(e)$ such that $P[Y_k^{-1} Y_n \in U'H] \leq P[Y_n \in U Y_k H] + \eta$ is satisfied. Both inequalities then imply the result.

We summarize the results of the discussion in the following

2.2.14 Theorem. *Let $(X_j)_{j\geq 1}$ be a sequence of G-valued random variables X_j with distribution μ_j and $(Y_n)_{n\geq 1}$ the corresponding sequence of n-th partial products $Y_n := X_1 \cdot \dots \cdot X_n$. It is assumed that $\mathscr{P}((\mu_j)_{j\geq 1})$ is relatively compact. Let H be a compact subgroup of G. Then the following statements are equivalent:*
(i) *Condition* (CC) *holds with respect to H;*
(ii) *Condition* (CC′) *holds with respect to H;*
(iii) (H)st-$\lim_{n\to\infty} Y_n = Y$ *exists.*

Proof. It remains to prove the implication (ii)$\Rightarrow$(iii). Let $\{U_m : m \geq 1\}$ be a basis of $\mathfrak{B}(e)$ with the additional property that $U_l U_{l-1} \cdot \dots \cdot U_m \subset U_{m-1}$ for all $l \geq m \geq 1$. There exists a sequence $(n_m)_{m\geq 1}$ in $\mathbb{N}$ such that $P[Y_k \notin U_m Y_n H] < \frac{1}{2^m}$ for all $k, n \geq n_m$. With $Y'_m := Y_{n_m}$ for all $m \geq 1$ one obtains

$$P \bigcup_{m\geq n+1} [Y'_{m+1} \notin U_m Y'_m H] \leq \sum_{m\geq n+1} \frac{1}{2^m} = \frac{1}{2^n},$$

and hence

$$P \bigcap_{n\geq 1} \bigcup_{m\geq n+1} [Y'_{m+1} \notin U_m Y'_m H] = 0.$$

For every ω outside a P-null set $C \in \mathfrak{A}$ there exists an index $N := N_\omega \geq 1$ such that $Y'_{m+1}(\omega) \in U_m Y'_m(\omega)$ holds whenever $m \geq N$. Consequently, for such $\omega \in C$ one has

$$\begin{aligned} Y'_{l+1}(\omega) &\in U_l Y'_l(\omega) H \subset U_l U_{l-1} Y'_{l-1}(\omega) H \subset \cdots \\ &\subset U_l \cdot \cdots \cdot U_m Y'_m(\omega) H \subset U_{m-1} Y'_m(\omega) H \quad \text{if } l \geq m \geq N. \end{aligned}$$

But this together with the completeness of $\dot{G} := G/H$ implies the a.s. convergence of the sequence $(p(Y'_m))_{m \geq 1}$ to a $\dot{G}$-valued random variable $\dot{Y}$. The a.s. convergence of $(Y_{n_m})_{m \geq 1}$ implies the stochastic convergence, and the assertion follows from the inclusion

$$[Y_{n_m} \in UYH] \cap [Y_k \in U Y_{n_m} H] \subset [Y_k \in U^2 YH]$$

which is valid for all $k \geq n_m$, $m \geq 1$ and $U \in \mathfrak{B}(e)$. □

2.2.15 Theorem. *Let $(X_j)_{j \geq 1}$ be a sequence of independent G-valued random variables X_j with distribution μ_j and $(Y_n)_{n \geq 1}$ the corresponding sequence of n-th partial products $Y_n := X_1 \cdot \cdots \cdot X_n$ with distribution $\nu_{0,n} := \mu_1 * \cdots * \mu_n$. It is assumed that $\mathscr{P}((\mu_j)_{j \geq 1})$ is relatively compact in $\mathscr{M}^1(G)$. Let H be a compact subgroup of G. Then stochastic convergence modulo H of the sequence $(Y_n)_{n \geq 1}$ implies a.s. convergence modulo H.*

Proof. From the stochastic convergence modulo H of the sequence $(Y_n)_{n \geq 1}$ or equivalently from Condition (CC) by Theorem 2.2.14 follows the existence of a subsequence $(Y_{n_m})_{m \geq 1}$ of $(Y_n)_{n \geq 1}$ converging a.s. modulo H. Since the conditions

(a) $\lim_{m \to \infty} P \bigcup_{l \geq n_m} [Y_l \notin UYH] = 0$ for every $U \in \mathfrak{B}(e)$ and

(b) $P \bigcup_{n_m < k \leq n_{m+1}} [Y_k \notin U_m Y_{n_m} H] < \varepsilon_m$ for all $m \geq 1$,

where $\{U_m : m \geq 1\}$ is a basis of $\mathfrak{B}(e)$ and $(\varepsilon_m)_{m \geq 1}$ is a sequence in $\mathbb{R}_+^*$ with $\sum_{m \geq 1} \varepsilon_m < \infty$, imply the convergence a.s. modulo H of the sequence $(Y_n)_{n \geq 1}$, it remains to show (b). First of all, for every $m \geq 1$ there exists a $U'_m := U'_m(U_m) \in \mathfrak{B}(e)$ satisfying $P[Y_k^{-1} Y_n \notin U'_m H] \leq \frac{1}{2^m}$ for all $k, n \geq n_m$. Putting

$$Z_i := Y_{n_m}^{-1} Y_k \quad \text{for all } i := k - n_m \text{ with } k = n_m + 1, n_m + 2, \ldots, n_{m+1},$$

and accordingly

$$A_i := [Z_i \notin U_m'^2 H] \quad \text{and} \quad B_i := [Z_{i_0} \in Z_i U''_m H],$$

where $i_0 := n_{m+1} - n_m$ and $U''_m := U''_m(U'_m) \in \mathfrak{B}(e)$ such that $HU''_m \subset U'_m H$.

For $i, j \geq 1$ with $i \leq j$ one has $P(A_i \cap B_j) = P(A_i) P(B_j)$, and hence the well-known inequality of P. Lévy yields in this case

$$P[Z_{i_0} \notin U'_m H] \geq P\left(\bigcup_{i=1}^{i_0} (A_i \cap B_i)\right) \geq \inf_{1 \leq i \leq i_0} P(B_i) P\left(\bigcup_{i=1}^{i_0} A_i\right).$$

Applying $P(B_i) \geq 1 - \frac{1}{2^m}$ for all $i := k - n_m$ with $k = n_m + 1, n_m + 2, \ldots, n_{m+1}$ we obtain

$$P\left(\bigcup_{i=1}^{i_0} A_i\right) = P \bigcup_{k=n_m+1}^{n_{m+1}} [Y_k \notin Y_{n_m} U_m'^2 H] \leq \frac{1}{2^m - 1}.$$

But the relative compactness of $\mathscr{P}((\mu_j)_{j\geq 1})$ implies the existence of a basis $\{U_m: m\geq 1\}$ of $\mathfrak{B}(e)$ such that with $(\varepsilon_m)_{m\geq 1}$ in $\mathbb{R}_+^*$ defined by $\varepsilon_m := \frac{1}{2^m-1}$ for all $m\geq 1$ Condition (b) is satisfied. □

Summarizing the results of this section we obtain the following

2.2.16 Theorem. *Let G be a locally compact group having a countable basis of its topology and $(X_j)_{j\geq 1}$ a sequence of independent G-valued random variables X_j with distributions μ_j. It is assumed that the set $\mathscr{P}((\mu_j)_{j\geq 1})$ of partial convolution products*

$$\nu_{k,n} := \mu_{k+1} * \cdots * \mu_n \qquad (0\leq k<n)$$

is relatively compact. Finally, we let K_0 be the compact subset $\overline{\bigcup_{\lambda\in\mathscr{L}} \operatorname{supp}(\lambda)}$, where $\mathscr{L}$ denotes the set of all accumulation points of $(\nu_{k,n})_{n>k}$ (with respect to the filter $\{(l,m): m>l\}$ on $\mathbb{Z}\times\mathbb{N}$ which is finer than the product of the Fréchet filters on $\mathbb{Z}$ and $\mathbb{N}$).

(i) *If there exists a compact subgroup H of G with $K_0\subset H$, then the sequence $(Y_n)_{n\geq 1}$ of partial products $Y_n := X_1\cdot\cdots\cdot X_n$ converges a.s. modulo H.*
(ii) *If the sequence $(\nu_{0,n})_{n\geq 1}$ converges, then there exists a compact subgroup H of G such that the sequence $(Y_n)_{n\geq 1}$ converges a.s. modulo H.*

Proof. (i) By Application 2.2.8 we get for every $U\in\mathfrak{B}(e)$ that

$$\lim_{n,m\to\infty} \nu_{n,m}(UK_0)=1 \quad \text{and hence,} \quad \lim_{n,m\to\infty} \nu_{n,m}(UH)=1$$

holds for all compact subgroups H of G satisfying $K_0\subset H$. But this shows that Condition (CC′) prior to Theorem 2.2.14 holds, which by Theorem 2.2.14 implies that $(H)\text{st-}\lim_{n\to\infty} Y_n = Y$.

This, however, is equivalent to $(H)\text{as-}\lim_{n\to\infty} Y_n = Y$ by Theorem 2.2.15.

(ii) is a direct consequence of Application 2.2.9 together with (i). □

2.2.17 Definition. A locally compact group G having a countable basis of its topology is said to *satisfy the equivalence principle* (EP) if for every independent sequence $(X_j)_{j\geq 1}$ of G-valued random variables X_j with distribution μ_j the sequence $(Y_n)_{n\geq 1}$ of n-th partial products $Y_n := X_1\cdot\cdots\cdot X_n$ has the property that the sequence $(\nu_{0,n})_{n\geq 1}$ of n-th partial convolution products $\nu_{0,n} := \mu_1 * \cdots * \mu_n$ converges iff $(Y_n)_{n\geq 1}$ converges a.s.

Plainly, G satisfies (EP) iff for every independent sequence $(X_j)_{j\geq 1}$ convergence of $(\nu_{0,n})_{n\geq 1}$, stochastic convergence and a.s. convergence of $(Y_n)_{n\geq 1}$ are equivalent.

2.2.18 Definition. A locally compact group G is called *aperiodic* if the only compact subgroup of G is $\{e\}$.

2.2.19 Theorem. *For every locally compact group G having a countable basis of its topology the following statements are equivalent:*
(i) *G satisfies* (EP);
(ii) *G is aperiodic.*

Proof. 1. (i) ⇒ (ii) follows for the special case of a sequence $(X_j)_{j\geq 1}$ of G-valued random variables X_j with distribution $\mu_j := \omega_H$ for some compact subgroup $H \neq \{e\}$ by the Borel-Cantelli lemma ([6], p. 168).

2. (ii) ⇒ (i) is implied by (ii) of Theorem 2.2.16. □

2.2.20 Remark. The class of all aperiodic groups can be characterized using the solution of Hilbert's fifth problem (Theorem F) as the class of all torsion-free Lie groups. By Theorem 2.3 of [246], p. 138 it contains the connected, simply connected, solvable Lie groups, and in particular, the connected component of the group of all invertible upper triangular matrices in $\mathfrak{M}(2, \mathbb{R})$ (which is the simplest noncommutative example of the class), the torsion-free groups (among the discrete groups) and the Euclidean groups. We also note that the aperiodic Abelian groups (having a countable basis of their topology) are subgroups of (finite-dimensional) vector spaces.

2.3 The Normed Convergence Property

We shall now study the limiting behavior of convolution sequences $(\nu_{0,n} * \varepsilon_{x_n})_{n\geq 1}$ for sequences $(\mu_j)_{j\geq 1}$ of measures in $\mathscr{M}^1(G)$ and certain sequences $(x_j)_{j\geq 1}$ of elements in the underlying locally compact group G. The sequences $(x_j)_{j\geq 1}$ will be chosen such that the normed (or shifted) sequence $(\nu_{0,n} * \varepsilon_{x_n})_{n\geq 1}$ converges in $\mathscr{M}^1(G)$.

2.3.1 Definition. A locally compact group G is said to admit the *normed convergence property* (NCP) if for every sequence $(\mu_j)_{j\geq 1}$ in $\mathscr{M}^1(G)$ there exists a sequence $(x_j)_{j\geq 1}$ in G such that the normed sequence $(\nu_{0,n} * \varepsilon_{x_n})_{n\geq 1}$ converges (to a measure in $\mathscr{M}^1(G)$).

2.3.2 Theorem. *Let G be a locally compact group having a countable basis of its topology. Then the following statements are equivalent:*
(i) *G admits* (NCP);
(ii) *G is compact.*

Proof. 1. (ii) ⇒ (i). Let G be compact. For any sequence $(\mu_j)_{j\geq 1}$ in $\mathscr{M}^1(G)$ the set $\mathscr{P}((\mu_j)_{j\geq 1})$ is relatively compact and therefore, by Theorem 2.2.2, for measures $\nu, \nu' \in \mathscr{A}_0$ there exists an $x_\nu \in G$ such that $\nu' = \nu * \varepsilon_{x_\nu}$ holds. Since G admits a countable basis of its topology, there exists a countable neighborhood filter $(U_n)_{n\geq 1}$ of ν. As $\mathscr{A}_0$ is compact in $\mathscr{M}^1(G)$ by assumption and $\mathscr{A}_0 \subset \bigcup \{U_n * \varepsilon_{x_\nu} : \nu \in \mathscr{A}_0\}$, there exists for every $n \geq 1$ a finite subset $\{\varepsilon_{x_{\nu_1}}, \ldots, \varepsilon_{x_{\nu_r}}\}$ of $\{\varepsilon_{x_\nu} : \nu \in \mathscr{A}_0\}$ satisfying $\mathscr{A}_0 \subset \bigcup_{i=1}^r U_n * \varepsilon_{x_{\nu_i}}$. Furthermore, by the definition of $\mathscr{A}_0$ there is an $N := N(n) \geq 1$ such that $\nu_{0,k} \in \bigcup_{i=1}^r U_n * \varepsilon_{x_{\nu_i}}$ holds for all $k \geq N$. Choosing for $k \in [N(n), N(n+1)[$ a measure $\varepsilon_{x_{\nu_{i(k)}}} \in \mathscr{M}^1(G)$ with $\nu_{0,k} \in U_n * \varepsilon_{x_{\nu_{i(k)}}}$ and putting $\varepsilon_{y_k} := \varepsilon_{x^{-1}_{\nu_{i(k)}}}$ we obtain $\nu_{0,k} * \varepsilon_{y_k} \in U_n$ for all $k \geq N$, which proves the assertion.

2. (i) ⇒ (ii). Let G be a locally compact but noncompact group admitting (NCP). Clearly, there exists a symmetric measure $\mu \in \mathscr{M}^1(G)$ with noncompact support

$\operatorname{supp}(\mu)$. By assumption there is a sequence $(x_n)_{n\geq 1}$ in G such that the sequences $(\nu_n)_{n\geq 1}$ with $\nu_n := \mu^n * \varepsilon_{x_n}$ for all $n \geq 1$ and $(\tilde{\nu_n})_{n\geq 1}$ converge to limits ν and $\tilde{\nu}$ (in $\mathscr{M}^1(G)$) resp. Hence,

$$\lim_{n\to\infty} \nu_n * \tilde{\nu_n} = \lim_{n\to\infty} \mu^{2n} =: \lambda.$$

On the other hand,

$$\lim_{n\to\infty} \mu^{2n} = \lim_{n\to\infty} (\mu^2)^n = \lim_{n\to\infty} (\mu^2)^{2n} = \lambda^2,$$

and therefore, $\lambda = \omega_H$ for a compact subgroup H of G. In addition, one verifies that $\lambda = \mu^2 * \lambda$. But $\operatorname{supp}(\lambda) = \operatorname{supp}(\mu^2 * \lambda)$ is not compact by the construction of μ. Thus we have a contradiction. $\square$

In the sequel we shall need a strengthened form of the normed convergence property for a compact group admitting a countable basis of its topology. To this end we present the following definition, valid for any locally compact group G.

2.3.3 Definition. A sequence $(\mu_j)_{j\geq 1}$ in $\mathscr{M}^1(G)$ is called *composition convergent* if for every $k \geq 0$ the sequence $(\nu_{k,n})_{n>k}$ of partial convolution products corresponding to $(\mu_j)_{j\geq 1}$ converges.

We know from the preceding discussion that for composition convergent sequences $(\mu_j)_{j\geq 1}$ in $\mathscr{M}^1(G)$ one has $\lim_{n\to\infty} \nu_{k,n} = \nu^{(k)}$ for every $k \geq 0$, and $\lim_{k\to\infty} \nu^{(k)} = \nu^{(\infty)} = \omega_H$ for some compact subgroup H of G which is the basis $H((\mu_j)_{j\geq 1})$ of $(\mu_j)_{j\geq 1}$ in the sense of 2.2.4. In fact, H is the maximal compact subgroup of G with the property $\nu^{(k)} = \nu^{(k)} * \varepsilon_x$ for all $x \in H$, or $\nu^{(k)} = \nu^{(k)} * \omega_H$ whenever $k \geq 0$.

2.3.4 Theorem. *Let G be a compact group having a countable basis of its topology. For every sequence $(\mu_j)_{j\geq 1}$ in $\mathscr{M}^1(G)$ there exists a sequence $(x_j)_{j\geq 1}$ in G with the property that the sequence $(\mu_j')_{j\geq 1}$ defined by*

$$\mu_j' := \varepsilon_{x_{j-1}^{-1}} * \mu_j * \varepsilon_{x_j} \qquad \text{for all } j \geq 1 \ (\text{with } x_0 := e)$$

is composition convergent.

Proof. Choosing $\nu^{(0)}$ in the set $\mathscr{B}_0 := \mathscr{A}_0$ of all accumulation points of the sequence $(\nu_{0,n})_{n\geq 1}$, we see that Theorem 2.3.2 yields the existence of a sequence $(y_{0,n})_{n\geq 1}$ in G such that

$$\lim_{n\to\infty} \nu_{0,n} * \varepsilon_{y_{0,n}} = \nu^{(0)}$$

holds.

Denoting by $\mathscr{B}_1$ the set of all accumulation points of the sequence $(\nu_{1,n} * \varepsilon_{y_{0,n}})_{n\geq 2}$, we obtain $\mu_1 * \nu^{(1)} = \nu^{(0)}$ for all $\nu^{(1)} \in \mathscr{B}_1$, and, again by Theorem 2.3.2, a sequence $(y_{1,n})_{n\geq 2}$ in G such that

$$\lim_{n\to\infty} \nu_{1,n} * \varepsilon_{y_{0,n}} * \varepsilon_{y_{1,n}} = \nu^{(1)}$$

is satisfied. If $\mathscr{B}_2$ denotes the set of accumulation points of the sequence $(\nu_{2,n} * \varepsilon_{y_{0,n}} * \varepsilon_{y_{1,n}})_{n\geq 3}$, we obtain $\mu_2 * \nu^{(2)} = \nu^{(1)}$ for all $\nu^{(2)} \in \mathscr{B}_2$. This procedure can be continued to exhibit for every $k \geq 0$ a subset $\mathscr{B}_k$ of $\mathscr{M}^1(G)$, a measure $\nu^{(k)} \in \mathscr{B}_k$ and a sequence $(y_{k,n})_{n>k}$ in G satisfying

$$\lim_{n\to\infty} \nu_{k,n} * \varepsilon_{y_{0,n}} * \cdots * \varepsilon_{y_{k,n}} = \nu^{(k)} \quad \text{and} \quad \mu_{k+1} * \nu^{(k+1)} = \nu^{(k)}.$$

For every $k \geq 0$ let the family $\{U_i^{(k)}: i \geq 1\}$ be a (countable) neighbourhood filter of $\nu^{(k)}$ such that $\mu_{k+1} * U_i^{(k+1)} \subset U_{i+1}^{(k)}$ for all $i \geq 1$. Furthermore, let $(n_l)_{l \geq 0}$ denote a sequence in $\mathbb{N}$ with $l < n_l < n_{l+1}$ for all $l \geq 0$ and such that for all $n \geq n_k$ $(k \geq 0)$ one has

$$\nu_{k,n} * \varepsilon_{y_{0,n}} * \cdots * \varepsilon_{y_{k,n}} \in U_1^{(k)}.$$

The desired sequence $(x_n)_{n \geq 1}$ in G is constructed as follows: We put $x_n := y_{0,n}$ for $n < n_1$ and $x_n := y_{0,n} \cdot \cdots \cdot y_{k,n}$ for $n_{k-1} \leq n < n_k$, and show that $\lim_{n\to\infty} \nu_{k,n} * \varepsilon_{x_n} = \nu^{(k)}$ for every $k \geq 0$. In fact, if $n_k \leq n < n_{k+1}$, then

$$\nu_{k,n} * \varepsilon_{x_n} = \mu_{k+1} * \nu_{k+1,n} * \varepsilon_{y_{0,n}} * \cdots * \varepsilon_{y_{k+1,n}}.$$

By the choice of the integer n_{k+1} one gets

$$\nu_{k+1,n} * \varepsilon_{y_{0,n}} * \cdots * \varepsilon_{y_{k+1,n}} \in U_1^{(k+1)}.$$

By the definition of the neighborhood filter $\{U_i^{(k)}: i \geq 1\}$ for each $k \geq 0$ one has $\mu_{k+1} * U_i^{(k+1)} \subset U_{i+1}^{(k)}$ and hence, $\nu_{k,n} * \varepsilon_{x_n} \in \mu_{k+1} * U_1^{(k+1)} \subset U_2^{(k)}$.

Similarly, we conclude that $\nu_{k,n} * \varepsilon_{x_n} \in U_{i+2}^{(k)}$ for $n_{k+i} \leq n < n_{k+i+1}$. Again by the construction of $\{U_i^{(k)}: i \geq 1\}$ for $k \geq 0$ we finally obtain

$$\lim_{n\to\infty} \nu_{k,n} * \varepsilon_{x_n} = \nu^{(k)}, \text{ and thus}$$

$$\lim_{n\to\infty} \mu'_{k+1} * \cdots * \mu'_n = \lim_{n\to\infty} \varepsilon_{x_k^{-1}} * \nu_{k,n} * \varepsilon_{x_n} = \varepsilon_{x_k^{-1}} * \nu^{(k)}$$

for each $k \geq 1$. The sequence $(\mu'_j)_{j \geq 1}$ is therefore composition convergent. □

2.3.5 Theorem. *Let G be a finite group. Then a sequence $(\mu_j)_{j \geq 1}$ in $\mathscr{M}^1(G)$ composition converges if there exists an $a \in \mathbb{R}_+^*$ such that $\mu_j(\{e\}) \geq a$ holds for all $j \geq 1$.*

The *proof* of this theorem follows from a sequence of lemmata valid for a finite group G of order s.

2.3.6 Lemma. *If a sequence $(\mu_j)_{j \geq 1}$ in $\mathscr{M}^1(G)$ is composition convergent with a compact subgroup H of G as its basis, then there exists a $j_0 \geq 1$ such that $\prod_{j \geq j_0} \mu_j(H) > 0$.*

Proof. 1. First of all we note that the mappings

$$p \to pq + (1-p)(1-q) \quad \text{and} \quad q \to pq + (1-p)(1-q)$$

from $]\frac{1}{2}, \infty[$ into $\mathbb{R}_+$ are isotone. Moreover, let $\mathbb{Z}_2 := \{0, 1\}$ be the group of order 2 with neutral element 0 and let

$$\gamma_n := p_n \varepsilon_0 + (1-p_n)\varepsilon_1 \in \mathscr{M}^1(\mathbb{Z}_2) \quad \text{for every } n \geq 1.$$

One shows (with an argument from the theory of Markov chains with n-th step transition matrix $\left(\begin{pmatrix} p_n & 1-p_n \\ 1-p_n & p_n \end{pmatrix}\right)$ the validity of the following statement:

If $\lim_{n\to\infty} p_n = 1$ and $\prod_{n\geq n_0} p_n = 0$ for all $n_0 \geq 1$, then the sequence $(\gamma_{n_0, n_0+m})_{m\geq 1}$ of partial convolution products

$$\gamma_{n_0, n_0+m} := \gamma_{n_0+1} * \cdots * \gamma_{n_0+m} \quad \text{for } m \geq 1$$

converges for all $n_0 \geq 1$ to $\frac{1}{2}\varepsilon_0 + \frac{1}{2}\varepsilon_1$ as $m \to \infty$.

2. Given $(\mu_j)_{j\geq 1}$ in $\mathscr{M}^1(G)$ we consider the corresponding sequence $(\nu_{k,n})_{n>k}$. Since G is finite, Application 2.2.8 implies that

$$\lim_{k,n\to\infty} \nu_{k,n}(K_0) = 1.$$

Hence, by Application 2.2.9 one obtains

$$\lim_{k,n\to\infty} \nu_{k,n}(H(\nu^{(0)})) = 1.$$

But $(\mu_j)_{j\geq 1}$ is assumed to be composition convergent so that

$$\lim_{k,n\to\infty} \nu_{k,n}(H(\nu^{(j)})) = 1 \qquad \text{for all } j \geq 1.$$

Again by the finiteness of G we have $H(\nu^{(j)}) = H((\mu_j)_{j\geq 1}) =: H$ for sufficiently large $j \geq 1$, or $\lim_{k,n\to\infty} \nu_{k,n}(H) = 1$. In particular, $\lim_{j\to\infty} \mu_j(H) = 1$, and hence $\mu_j(H) > \frac{1}{2}$ for sufficiently large $j \geq 1$.

Clearly,

$$\nu_{k-1,k+1}(H) \leq \mu_k(H)\,\mu_{k+1}(H) + (1-\mu_k(H))(1-\mu_{k+1}(H))$$

and, using the first remark in Part 1, one gets

$$\begin{aligned} \nu_{k-1,k+2}(H) &\leq \nu_{k-1,k+1}(H)\,\mu_{k+2}(H) + (1-\nu_{k-1,k+1}(H))(1-\mu_{k+2}(H)) \\ &\leq \mu_{k+2}(H)[\mu_k(H)\,\mu_{k+1}(H) + (1-\mu_k(H))(1-\mu_{k+1}(H))] \\ &\quad + (1-\mu_{k+2}(H))[1-\mu_k(H)\,\mu_{k+1}(H) - (1-\mu_k(H))(1-\mu_{k+1}(H))]. \end{aligned}$$

For every $i \geq 0$ we consider the measure

$$\gamma_i := \mu_{k+i}(H)\,\varepsilon_0 + (1-\mu_{k+i}(H))\,\varepsilon_1 \in \mathscr{M}^1(\mathbb{Z}_2)$$

such that the above inequality reads $\nu_{k-1,k+2}(H) \leq \gamma_0 * \gamma_1 * \gamma_2(\{0\})$. More generally, one obtains for all $k \geq 1$, $m \geq 0$ the inequality

$$\nu_{k-1,k+m}(H) \leq \gamma_0 * \cdots * \gamma_m(\{0\}).$$

Assuming now that $\prod_{j\geq j_0} \mu_j(H) = 0$ for all $j_0 \geq 1$ we conclude from the preceding arguments together with the second remark in Part 1 that for every $k \geq 1$ there exists a sequence $(a_{k,m})_{m\geq 0}$ with $\lim_{m\to\infty} a_{k,m} = \frac{1}{2}$ satisfying $\nu_{k-1,k+m}(H) \leq a_{k,m}$ for all $m \geq 0$. On the other hand we have by the above argument $\lim_{k,m\to\infty} \nu_{k-1,k+m}(H) = 1$ and hence a contradiction. □

2.3.7 Lemma. *For any sequence $(\mu_j)_{j\geq 1}$ in $\mathscr{M}^1(G)$ and every compact subgroup H of G the following statements are equivalent:*
(i) *$(\mu_j)_{j\geq 1}$ is composition convergent with basis H;*
(ii) *there exists a $j_0\geq 1$ such that $\prod_{j\geq j_0}\mu_j(H)>0$, and the sequence $(\sigma_j)_{j\geq 1}$ of measures $\sigma_j\in\mathscr{M}^1(G)$ defined by $\sigma_j(A):=\dfrac{\mu_j(A\cap H)}{\mu_j(H)}$ for all $A\subset G$ is composition convergent to ω_H.*

Proof. 1. (ii) ⇒ (i). Let $(\varepsilon_k)_{k\geq 1}$ be a sequence in $\mathbb{R}_+^*$ with $\lim_{k\to\infty}\varepsilon_k=0$ such that $\prod_{j\geq k}\mu_j(H)>1-\varepsilon_k$ holds. For every $0\leq k<n$ we define as usual

$$\nu_{k,n}:=\mu_{k+1}*\cdots*\mu_n \quad\text{and}\quad \tau_{k,n}:=\sigma_{k+1}*\cdots*\sigma_n.$$

One immediately obtains

$$1-\varepsilon_k<\textstyle\prod_{j=k}^n\mu_j(H)\leq\nu_{k-1,n}(H)\leq 1 \qquad\text{for all } 1\leq k<n.$$

For fixed $k\geq 1$ we therefore get

$$\|\nu_{k-1,n}-\tau_{k-1,n}\|=\|\textstyle\mathop{\ast}_{j=k}^n\mu_j-\mathop{\ast}_{j=k}^n\sigma_j\|\leq 2\varepsilon_k$$

or the set $\mathscr{A}_{k-1}$ of accumulation points of $(\nu_{k-1,n})_{n\geq k}$ is contained in a neighborhood of ω_H of $\|\cdot\|$-radius $\leq 2\varepsilon_k$. Clearly, for any $\nu\in\mathscr{A}_0$ one has $\nu=\nu*\omega_H$. We shall show that $(\nu_{0,n})_{n\geq 1}$ converges with basis H. Indeed, if ν' is another element of $\mathscr{A}_0$, then for

$$\nu:=\lim_{k\to\infty}\nu_{0,n_k} \quad\text{and}\quad \nu':=\lim_{k\to\infty}\nu_{0,n'_k}$$

(with subsequences $(n_k)_{k\geq 1}$ and $(n'_k)_{k\geq 1}$ in $\mathbb{N}$ resp.) we obtain $\nu'=\nu*\lambda$, where λ is an accumulation point of the sequence $(\mu_{n_k+1}*\cdots*\mu_{n'_k})_{n\geq k}$. By assumption we have

$$\lim_{j_0\to\infty}\textstyle\prod_{j\geq j_0}\mu_j(H)=1 \quad\text{and hence,}\quad \lim_{k\to\infty}\prod_{j=n_k+1}^{n'_k}\mu_j(H)=1$$

or $\operatorname{supp}(\lambda)\subset H$ which implies $\nu=\nu'$. By the same reasoning we get the convergence of $(\nu_{k-1,n})_{n\geq 1}$ for every $k\geq 1$, and thus the composition convergence of $(\mu_j)_{j\geq 1}$ to ω_H.

2. (i) ⇒ (ii). From the assumption there follows via Lemma 2.3.6 the existence of a $j_0\geq 1$ such that $\prod_{j\geq j_0}\mu_j(H)>0$ holds. Suppose that $(\sigma_j)_{j\geq 1}$ is not composition convergent to ω_H. Then there exist a $\delta>0$ and a $k_1\geq 1$ satisfying $\|\tau_{k-1,n}-\omega_H\|\geq 2\delta$ for infinitely many $n\geq k\geq k_1$. Choosing $k_2\geq k_1$ such that $\prod_{j\geq k_2}\mu_j(H)\geq 1-\delta$ we obtain

$$\|\mu_k(H)\,\sigma_k*\cdots*\mu_n(H)\,\sigma_n-\tau_{k-1,n}\|\leq\delta \quad\text{and hence}$$
$$\|\textstyle\mathop{\ast}_{j=k}^n\mu_j(H)\,\sigma_j-\omega_H\|\geq\delta \qquad\text{for all } n\geq k\geq k_2.$$

We now choose $\varepsilon>0$ such that for all $\alpha,\beta\in\mathscr{M}^b(G)$ with $\|\alpha-\omega_H\|\geq\delta$ and

$$\beta:=\textstyle\sum_{i=1}^s p_i\,\varepsilon_{x_i}, \qquad x_i\in G,\ |p_i|<\varepsilon \qquad\text{for all } i=1,\dots,s$$

the inequality $\|\alpha+\beta-\omega_H\| \geq \frac{\delta}{2}$ holds, and we choose $k_0 := k_0(\varepsilon) \geq k_2$ such that $\prod_{j \geq k_0} \mu_j(H) \geq 1-\varepsilon$ is satisfied. Then for every $k \geq k_0$ we obtain

$$\nu_{k-1,k+m} = \ast_{j=0}^{m} \mu_{k+j}(H)\, \sigma_{k+j} + \mu$$

with $\mu \in \mathscr{M}_+^b(G)$ of the form $\sum_{i=1}^{s} p_i \varepsilon_{x_i}$, $x_i \in G$, $0 \leq p_i \leq \varepsilon$ for all $i=1,\dots,s$. For $k \geq k_0$ and $m \geq 0$ the measure $\nu_{k-1,k+m}$ therefore lies outside a neighborhood of ω_H of $\|\cdot\|$-radius $<\frac{\delta}{2}$, so that $(\mu_j)_{j\geq 1}$ is not composition convergent with basis H. □

2.3.8. A sequence $(\Omega_j)_{j\geq 1}$ of subsets of G is said to *fill* G, if for every $j \geq 1$ there exists an $m(j) \geq 0$ such that $\prod_{m=0}^{m(j)} \Omega_{j+m} = G$.

Lemma. *For any sequence $(\mu_j)_{j\geq 1}$ in $\mathscr{M}^1(G)$ the following statements are equivalent:*
(i) *$(\mu_j)_{j\geq 1}$ is composition convergent with basis G;*
(ii) *any sequence $(\Omega_j)_{j\geq 1}$ of subsets of G with $\prod_{j\geq j_0} \mu_j(\Omega_j) > 0$ for some $j_0 \geq 1$ fills G.*

Proof. 1. (ii)⇒(i). Let $(\mu_j)_{j\geq 1}$ be not composition convergent to ω_G. By Theorem 2.3.4 we know that the sequence $(\mu'_j)_{j\geq 1}$ of normed partial convolution products

$$\mu'_j := \varepsilon_{x_{j-1}^{-1}} * \mu_j * \varepsilon_{x_j}$$

for a properly choosen $x_j \in G$ $(j \geq 1)$ is composition convergent, but with a compact subgroup $H \neq G$ as its basis. Lemma 2.3.6 implies the existence of $j_0 \geq 1$ such that $\prod_{j\geq j_0} \mu'_j(H) > 0$ holds. The sequence $(\Omega_j)_{j\geq 1}$ with $\Omega_j := x_{j-1} H x_j^{-1}$ for all $j \geq 1$ (where $x_0 := e$) does not fill G, whereas

$$\prod_{j\geq j_0} \mu_j(\Omega_j) = \prod_{j\geq j_0} \mu'_j(H) > 0,$$

which is a contradiction.

2. (i)⇒(ii). Let $\lim_{n\to\infty} \nu_{k-1,n} = \omega_G$ for all $k \geq 1$ and let $(\Omega_j)_{j\geq 1}$ be a sequence of subsets of G with the properties in (ii) which does not fill G. Choosing $j_0 \geq 1$ such that

$$\prod_{j\geq j_0} \Omega_j =: \Omega \neq G \quad \text{and} \quad \prod_{j=j_0}^{m} \mu_j(\Omega_j) > \frac{|\Omega|}{s}$$

with $|\Omega| := \operatorname{card}(\Omega)$ for all $m \geq j_0$ one obtains the existence of a $\beta \in \mathbb{R}_+^*$ with $\nu_{j_0-1,m}(\Omega) > \beta > \frac{|\Omega|}{s}$ for all $m \geq j_0$; but

$$\lim_{n\to\infty} \nu_{j_0-1,m}(\Omega) = \omega_G(\Omega) = \frac{|\Omega|}{s},$$

which establishes the contradiction. □

2.3.9 Lemma. *For any sequence $(\mu_j)_{j\geq 1}$ in $\mathscr{M}^1(G)$ the following statements are equivalent:*
(i) *$(\mu_j)_{j\geq 1}$ is composition convergent with basis H;*

(ii) (a) $\prod_{j\geq j_0}\mu_j(H)>0$ *for some* $j_0\geq 1$.

(b) *Every sequence* $(\Omega_j)_{j\geq 1}$ *of subsets of* G *with* $\Omega_j\subset H$ *for all* $j\geq 1$ *and* $\prod_{j\geq j_0}\mu_j(\Omega_j)>0$ *for some* $j_0\geq 1$ *fills* H.

Proof. Since the implication (i)$\Rightarrow$(ii) (a) follows immediately from Lemma 2.3.6, it remains to show (i)$\Rightarrow$(ii)(b) and (ii)$\Rightarrow$(i).

1. (i)$\Rightarrow$(ii)(b). Let $(\sigma_j)_{j\geq 1}$ be the sequence corresponding to $(\mu_j)_{j\geq 1}$ introduced in Lemma 2.3.7. Then $(\sigma_j)_{j\geq 1}$ composition converges to ω_H and for any sequence $(\Omega_j)_{j\geq 1}$ of subsets Ω_j of H satisfying $\prod_{j\geq j_0}\mu_j(\Omega_j)>0$ for some $j_0\geq 1$ one obtains $\prod_{j\geq j_0}\sigma_j(\Omega_j)>0$. Thus Lemma 2.3.8 implies that $(\Omega_j)_{j\geq 1}$ fills H.

2. (ii)$\Rightarrow$(i). Let $(\sigma_j)_{j\geq 1}$ be as above. From (ii)(b) and Lemma 2.3.8 we conclude the composition convergence of $(\sigma_j)_{j\geq 1}$ to ω_H. But Lemma 2.3.7 implies (i). □

We are now ready to present the *proof of Theorem* 2.3.5. 1. Let $(\Omega_j)_{j\geq 1}$ be a sequence of subsets of G with $e\in\Omega_j$ for all $j\geq 1$. For every $j\geq 1$ there exists an $m(j)\geq 1$ such that

$$\prod_{m=j}^{l}\Omega_m=:M_j\supset M_{j+1}\supset\cdots\supset\{e\}\qquad\text{for all } l\geq m(j).$$

Therefore there exists a $j_0\geq 1$ such that $M_j=:M$ for all $j\geq j_0$. Obviously $M=M^2$ or M is a subgroup of G, called the *basis* of $(\Omega_j)_{j\geq 1}$. Clearly the sequence $(\Omega_j)_{j\geq 1}$ fills its basis M.

2. We now consider the nonempty family **G** of all subgroups H' of G satisfying $\prod_{j\geq j_1}\mu_j(H')>0$ for some $j_1\geq 1$ and a minimal element H of **G**. Then any sequence $(\Omega_j)_{j\geq 1}$ of subsets of G satisfying $\Omega_j\subset H$ for all $j\geq 1$ and $\prod_{j\geq j_2}\mu_j(\Omega_j)>0$ for some $j_2\geq 1$ fills H. In fact, given such a sequence $(\Omega_j)_{j\geq 1}$, we first observe that $e\in\Omega_j$ for all $j\geq j_2$.

Assuming $e\notin\Omega_j$ for infinitely many $j\geq 1$ we obtain $\mu_j(\{e\})\leq 1-\mu_j(\Omega_j)$ for such $j\geq 1$. But by assumption there exists an $\alpha\in\mathbb{R}_+^*$ satisfying $\mu_j(\{e\})\geq\alpha$ for all $j\geq 1$ and hence $\mu_j(\Omega_j)\leq 1-\alpha<1$ for infinitely many $j\geq 1$ or

$$\prod_{j\geq j_2}\mu_j(\Omega_j)=0\qquad\text{for all } j_2\geq 1,$$

which contradicts the hypothesis on $(\Omega_j)_{j\geq 1}$. By Part 1 we conclude that $(\Omega_j)_{j\geq j_2}$ admits a basis M. We next show $M=H$. Assuming $M\neq H$, we proceed as follows: Plainly

$$M^2=M=\prod_{i\geq k}\Omega_i\qquad\text{for some } k\geq j_2,$$

and for all $j\geq k$ one has

$$\Omega_j\subset\prod_{i\geq k}\Omega_i=M.$$

On the other hand, $\Omega_j\subset H$ for all $j\geq 1$, so that $\prod_{i\geq k}\Omega_i=M\subset H$ for all $k\geq 1$. The minimality of H implies that $\prod_{j\geq j_1}\mu_j(M)=0$ for all $j_1\geq 1$ and hence

$$\prod_{j\geq k}\mu_j(\Omega_j)\leq\prod_{j\geq k}\mu_j(M)=0\qquad\text{for all } k\geq j_2,$$

which contradicts the hypothesis. Thus $(\Omega_j)_{j\geq j_2}$ and so $(\Omega_j)_{j\geq 1}$ fills H.

3. From Part 2 one sees that the hypotheses of Lemma 2.3.9 are fulfilled. It follows that $(\mu_j)_{j\geq 1}$ is composition convergent (with basis H minimal in **G**). □

The normed convergence property for a compact group G having a countable basis of its topology yields for every sequence $(\mu_j)_{j\geq 1}$ in $\mathcal{M}^1(G)$ the existence of a sequence $(x_j)_{j\geq 1}$ in G depending on the entire sequence $(\mu_j)_{j\geq 1}$ such that $(\nu_{0,n} * \varepsilon_{x_n})_{n\geq 1}$ converges. There is a sharpening of (NCP) which stems from the following question. For which compact groups do the elements x_k of the sequence $(x_n)_{n\geq 1}$ depend only on the first k members of $(\mu_j)_{j\geq 1}$ (for $k\geq 1$)?

2.3.10 Definition. Given a locally compact group G we call a subset $\mathcal{N}$ of $\mathcal{M}^1(G)$ *full* (in $\mathcal{M}^1(G)$) if

(a) for every $\mu\in\mathcal{M}^1(G)$ there exists an $x\in G$ such that $\mu * \varepsilon_x \in \mathcal{N}$ holds and

(b) for every sequence $(\mu_j)_{j\geq 1}$ in $\mathcal{N}$ the sequence $(\nu_{0,n})_{n\geq 1}$ with

$$\nu_{0,n} := \mu_1 * \cdots * \mu_n \quad \text{for all } n\geq 1$$

converges.

For a compact group G having a countable basis of its topology the existence of a full subset $\mathcal{N}$ of $\mathcal{M}^1(G)$ implies the following property of the group G.

2.3.11 Definition. A locally compact group G is said to admit the *strong normed convergence property* (SNCP) if for every $(\mu_j)_{j\geq 1}$ in $\mathcal{M}^1(G)$ there exists a sequence $(x_j)_{j\geq 1}$ with x_k depending only on $\mu_1, \ldots, \mu_k$ $(k\geq 1)$ such that $(\nu_{0,n} * \varepsilon_{x_n})_{n\geq 1}$ converges.

In fact, for every $\mu\in\mathcal{M}^1(G)$ one chooses $x(\mu)\in G$ such that $\mu * \varepsilon_{x(\mu)}\in\mathcal{N}$. Let $(\mu_j)_{j\geq 1}$ be a sequence in $\mathcal{M}^1(G)$ and put

$$x_1 := x(\mu_1), \quad x_j := x(\varepsilon_{x_{j-1}^{-1}} * \mu_j) \quad \text{for all } j\geq 2$$

as well as

$$\mu_1' := \mu_1 * \varepsilon_{x_1}, \quad \mu_j' := \varepsilon_{x_{j-1}^{-1}} * \mu_j * \varepsilon_{x_j} \quad \text{for all } j\geq 2.$$

Then the fact that $\mathcal{N}$ is full in $\mathcal{M}^1(G)$ implies that the sequence

$$(\nu_{0,n}')_{n\geq 1} \quad \text{with} \quad \nu_{0,n}' := \mu_1' * \cdots * \mu_n' = \nu_{0,n} * \varepsilon_{x_n} \quad \text{for all } n\geq 1$$

converges.

2.3.12. Properties of full subsets of $\mathcal{M}^1(G)$ for a compact group G having a countable basis of its topology. We have

(1) For every full subset $\mathcal{N}$ of $\mathcal{M}^1(G)$ the semigroup $\langle\mathcal{N}\rangle$ generated by $\mathcal{N}$ is full in $\mathcal{M}^1(G)$.

(2) If $\mathcal{N}\subset\mathcal{M}^1(G)$ is full, then so is $\overline{\mathcal{N}}$.

(3) Let $\dot{G}$ be a compact group and ϕ a continuous homomorphism from G into $\dot{G}$. Then for any full subset $\mathcal{N}$ of $\mathcal{M}^1(G)$ the set $\phi(\mathcal{N})$ is full in $\mathcal{M}^1(\dot{G})$.

(4) Let $\mathcal{N}$ be full in $\mathcal{M}^1(G)$ and N a closed normal subgroup of G. If $\mu\in\mathcal{N}$ with $\operatorname{supp}(\mu)\subset Nx$ for some $x\in G$, then $\operatorname{supp}(\mu)\subset N$ and the set $\mathcal{B} := \{\mu\in\mathcal{N} : \operatorname{supp}(\mu)\subset N\}$ is a full subset of $\mathcal{M}^1(N)$.

The proofs of these properties can be given without difficulty by using only the definition of a full set (via the Fourier transformation).

2.3.13 Theorem. *Let G be a compact group having a countable basis of its topology. The following statements are equivalent:*
(i) *G admits* (SNCP);
(ii) *G is totally disconnected.*

The proof of the theorem will be based on a sequence of lemmata.

2.3.14 Lemma. *Let G be a finite group. Then there exists a full subset of* $\mathcal{M}^1(G)$.

Proof. Let $s := |G|$. For every $\mu \in \mathcal{M}^1(G)$ we consider the set

$$\mathcal{N}_\mu := \{\mu * \varepsilon_x : x \in G\}.$$

The system $\{\mathcal{N}_\mu : \mu \in \mathcal{M}^1(G)\}$ is a partition of $\mathcal{M}^1(G)$ containing classes of translations of measures $\mu \in \mathcal{M}^1(G)$ by elements $x \in G$. Since $\|\mu\| = 1$, there exists an $x \in G$ such that $\mu(\{x\}) \geq \frac{1}{s}$, hence a representative $\mu' \in \mathcal{N}_\mu$ with $\mu'(\{e\}) \geq \frac{1}{s}$. Putting

$$\mathcal{N} := \left\{\mu' \in \mathcal{N}_\mu : \mu'(\{e\}) \geq \frac{1}{s},\ \mu \in \mathcal{M}^1(G)\right\},$$

we see that Theorem 2.3.5 implies that $\mathcal{N}$ is a full subset of $\mathcal{M}^1(G)$. □

Proof of Theorem 2.3.13 *Part* 1. (ii)⇒(i). Let G be a totally disconnected compact group of the form $\varprojlim_{n \geq 1} G_n$ with finite groups G_n of order $s_n := |G_n|$ $(n \geq 1)$. Then

$$G \cong \{(x_n)_{n \geq 1} \in \textstyle\prod_{n \geq 1} G_n : p_{m,n}(x_n) = x_m \quad \text{for all } 1 \leq m \leq n\},$$

where $p_{m,n}$ denotes the canonical projection from G_n onto G_m satisfying for all $l \leq k \leq n$ the relation $p_{l,k} \circ p_{k,n} = p_{l,n}$. For $n \geq 1$ we abbreviate by p_n the canonical projection from G onto G_n. Without loss of generality we assume that $s_n < s_{n+1}$ for all $n \geq 1$. Let $\mu \in \mathcal{M}^1(G)$. We have to construct elements b_n of G_n such that $p_n(\mu)(\{b_n\}) \geq \frac{1}{s_n}$ for all $n \geq 1$. One starts by choosing $b_1 \in G_1$ such that

$$\mu_1(\{b_1\}) = \sup_{x \in G_1} p_1(\mu)(\{x\}).$$

Suppose $b_1, \ldots, b_k$ have been chosen in $G_1, \ldots, G_k$ resp. such that

$$p_k(\mu)(\{b_k\}) \geq \frac{1}{s_k} \quad \text{and} \quad p_{i,i+1}(\{b_{i+1}\}) = b_i \qquad \text{for all } i = 1, \ldots, k.$$

The set $p_{k,k+1}^{-1}(\{b_k\})$ contains $\frac{s_{k+1}}{s_k} =: r_k$ elements, and one has

$$p_k(\mu)(\{b_k\}) = p_{k+1}(\mu)(p_{k,k+1}^{-1}(\{b_k\})) \geq \frac{1}{s_k}.$$

Hence, there exists a $b_{k+1}\in p_{k,k+1}^{-1}(\{b_k\})$ satisfying

$$p_{k+1}(\mu)(\{b_{k+1}\})\geq\frac{1}{s_k r_k}=\frac{1}{s_{k+1}}.$$

Putting $x(\mu):=(b_1^{-1}, b_2^{-1}, \ldots)$ and choosing $\mu*\varepsilon_{x(\mu)}$ as a representative of μ we obtain a set $\mathcal{N}$ of representatives which is full in $\mathcal{M}^1(G)$. It suffices to show Property (b) of the definition of a full set. Let $(\mu_j)_{j\geq 1}$ be a sequence in $\mathcal{N}$ and let $(\nu_{0,m})_{m\geq 1}$ be the corresponding sequence of m-th partial convolution products. It remains to show that if N is an open normal subgroup of G and p is the canonical projection from G onto $\dot{G}:=G/N$, then the sequence $(p(\nu_{0,m}))_{m\geq 1}$ in $\mathcal{M}^1(\dot{G})$ converges. We know that there is a system $\{U_i: i=1,\ldots,n\}$ of elements $U_i\in\mathfrak{V}(e)$ $(i=1,\ldots,n)$ satisfying

$$\left(\prod_{i=1}^n U_i\times\prod_{i\geq n+1} G_i\right)\cap G\subset N.$$

In particular,

$$p_n^{-1}(\{e_n\})\subset N \quad\text{or}\quad \nu_{0,m}(N)\geq p_n(\nu_{0,m})(\{e_n\})\geq\frac{1}{s_n}\quad\text{for all } m\geq 1.$$

Theorem 2.3.5 then yields the convergence of $(p(\nu_{0,m}))_{m\geq 1}$. □

2.3.15 Lemma. *Let $\mathbb{T}$ be the one-dimensional torus. There exists a sequence $(\mu_n)_{n\geq 1}$ in $\mathcal{M}^1(\mathbb{T})$ such that for every sequence $(x_n)_{n\geq 1}$ in $\mathbb{T}$ and corresponding sequence $(\mu_n')_{n\geq 1}$ with $\mu_n':=\mu_n*\varepsilon_{x_n}$ $(n\geq 1)$ the following conditions hold:*

(i) *There is an infinite subset $\mathbb{N}_0$ of $\mathbb{N}$ such that the product $\bigast_{n\in\mathbb{N}_0}\mu_n'$ does not converge;*

(ii) *there is an $n\geq 1$ such that the sequence $(\mu_n'^k)_{k\geq 1}$ does not converge.*

Proof. Let f be a mapping $\mathbb{N}\to\mathbb{N}$ such that

$$\sum_{n\geq 1}\frac{1}{f(n)}=\infty \quad\text{and}\quad \sum_{n\geq 1}\frac{1}{f(n)^2}<\infty$$

hold. For every $n\geq 1$ we put $\mu_n:=\frac{1}{2}(\varepsilon_{\frac{2\pi}{f(n)}}+\varepsilon_0)$.

Assume that there exists a sequence $(x_n)_{n\geq 1}$ in $\mathbb{T}$ such that for any $\mathbb{N}_0\subset\mathbb{N}$ and $n\geq 1$ the products in Conditions (i) and (ii) converge. The negation of Condition (ii) together with Theorem 2.1.4 yields the existence of an $m(n)\in\mathbb{Z}_+$ with $m(n)<f(n)$ satisfying $x_n=\dfrac{2\pi}{f(n)}m(n)$.

Clearly, for every $\chi\in\mathbb{Z}\cong\mathbb{T}^\wedge$ one obtains

$$\hat{\mu}_n'(\chi)=\frac{1}{2}\exp\left(2\pi i\frac{m(n)}{f(n)}\right)\left[\exp\left(2\pi i\frac{1}{f(n)}\right)+1\right]$$

and concludes that

$$\arg\hat{\mu}_n'(1)=2\pi\left(\frac{m(n)}{f(n)}+\frac{1}{2f(n)}\right)=2\pi\frac{2m(n)+1}{2f(n)}$$

as well as

$$|\hat{\mu}_n(1)|^2 = \frac{1}{4}\left[\left(\cos 2\pi \frac{1}{f(n)} + 1\right)^2 + \sin^2 2\pi \frac{1}{f(n)}\right]$$
$$= \frac{1}{2}\left(1 + \cos 2\pi \frac{1}{f(n)}\right).$$

Firstly we have the convergence of the product

$$\ast_{n \geq k} (\mu_n * \tilde{\mu_n}) = \ast_{n \geq k} (\mu'_n * \tilde{\mu'_n})$$

to a measure $\lambda_k \in \mathscr{M}^1(\mathbb{T})$ for every $k \geq 1$.
[For fixed $k \geq 1$ and $\chi \in \mathbb{Z}$ the sequence $(\prod_{n=k}^{m} (\mu_n * \tilde{\mu_n})^\wedge(\chi))_{m \geq 1}$ in $\mathbb{R}_+$ is antitone and hence convergent. The assertion follows from the compactness of $\mathbb{T}$ by the continuity of the Fourier transformation.]
It thus follows that $\lim_{k \to \infty} \lambda_k = \omega_H$ for some compact subgroup H of $\mathbb{T}$. Furthermore, we have $\lim_{k \to \infty} \hat{\lambda}_k(1) = 1$.
[For every $k \geq 1$ one obtains

$$\hat{\lambda}_k(1) = \prod_{n \geq k} (\mu_n * \tilde{\mu_n})^\wedge(1) = \prod_{n \geq k} \left[1 - \frac{2\pi^2}{f(n)^2} \phi\left(\frac{2\pi}{f(n)}\right)\right]$$

with $\phi(x) := \frac{1}{2} - \frac{x^2}{4!} + \cdots \ (x \in \mathbb{R})$. But

$$\lim_{k \to \infty} \prod_{n \geq k} \left[1 - \frac{2\pi^2}{2f(n)^2} \phi\left(\frac{2\pi}{f(n)}\right)\right] = 1 \Leftrightarrow \sum_{n \geq 1} \frac{1}{f(n)^2} \phi\left(\frac{2\pi}{f(n)}\right) < \infty.$$

This, however, is fulfilled since $\phi(x) \leq e$ for all $|x| \leq 1$.]

Thus $\hat{\omega}_H(1) = 1$ or $\omega_H = \varepsilon_0$, i.e., $\lim_{k \to \infty} \ast_{n \geq k} (\mu_n * \tilde{\mu_n}) = \varepsilon_0$.

We now consider the sequence $(\nu'_{k,n})_{0 \leq k < n}$ with

$$\nu'_{k,n} := \mu'_{k+1} * \cdots * \mu'_n \qquad \text{for all } 0 \leq k < n.$$

By the negation of Condition (i) we have

$$\lim_{n \to \infty} \nu'_{k,n} = \sigma_k \in \mathscr{M}^1(\mathbb{T}) \qquad \text{for all } k \geq 0$$

and $\lim_{k \to \infty} \sigma_k = \omega_{H'}$ for some compact subgroup H' of G and hence

$$\lim_{k \to \infty} \prod_{n \geq k} \hat{\mu'_n}(\chi) = \hat{\omega}_{H'}(\chi) \qquad \text{for all } \chi \in \mathbb{Z}.$$

Since

$$|\hat{\mu'_n}(\chi)| \geq |\hat{\mu'_n}(\chi)|^2 = (\mu'_n * \tilde{\mu'_n})^\wedge(\chi) \qquad \text{for all } \chi \in \mathbb{Z},\ n \geq 1,$$

one has

$$|\hat{\omega}_{H'}(1)| = \lim_{k \to \infty} \prod_{n \geq k} |\hat{\mu'_n}(1)| = 1$$

and therefore again $\lim_{k \to \infty} \ast_{n \geq k} \mu'_n = \varepsilon_0$.

Let

$$\mathbb{N}_1 := \{n \in \mathbb{N} : 0 < \arg \widehat{\nu'_{0,n}}(1) \leq \pi\} \quad \text{and}$$
$$\mathbb{N}_2 := \{n \in \mathbb{N} : \pi < \arg \widehat{\nu'_{0,n}}(1) \leq 2\pi\}.$$

Then either

$$\sum_{n \in \mathbb{N}_1} \frac{1}{f(n)} = \infty \quad \text{or} \quad \sum_{n \in \mathbb{N}_2} \frac{1}{f(n)} = \infty.$$

Without loss of generality we assume that

$$\sum_{n \in \mathbb{N}_1} \frac{1}{f(n)} = \infty \quad \text{and put} \quad \mathbb{N}_1 := \{n_k : k \geq 1\}.$$

Then by the above discussion one obtains $\lim_{k\to\infty} \ast_{n \geq k} \mu'_n = \varepsilon_0$ and thus for a suitable choice of $m(k)$ $\lim_{k\to\infty} \prod_{\substack{n \in \mathbb{N}_1 \\ k \leq n \leq m(k)}} \widehat{\mu'_n}(1) = 1$. Therefore $\lim_{k\to\infty} \widehat{\mu'_{n_k}}(1) = 1$. Consequently, there exists a $k_0 \geq 1$ such that for all $k \geq k_0$ one has

$$0 \leq \arg \widehat{\mu'_{n_k}}(1) < \frac{\pi}{8} \quad \text{and} \quad -\frac{\pi}{8} \leq \sum_{\substack{n \in \mathbb{N}_1 \\ n_k \leq n \leq m(k)}} \arg \widehat{\mu'_n}(1) < \frac{\pi}{8} \quad (\operatorname{mod} 2\pi).$$

On the other hand, $\frac{\pi}{8} \geq \arg \widehat{\mu'_n}(1) \geq \frac{\pi}{f(n)}$ for all $n \in \mathbb{N}_1$. Since

$$\sum_{n \in \mathbb{N}_1} \frac{1}{f(n)} = \infty \quad \text{and} \quad \lim_{n\to\infty} f(n) = \infty,$$

for every $k \geq 1$ there exists an $m(k) \geq 1$ such that

$$\pi > \sum_{\substack{n \in \mathbb{N}_1 \\ n_k \leq n \leq m(k)}} \arg \widehat{\mu'_n}(1) > \frac{\pi}{8}$$

holds, which is the desired contradiction. □

2.3.16 Lemma. *Let G be a compact group having a countable basis of its topology, H a closed commutative subgroup of G and $\mathcal{N}$ a full subset of $\mathcal{M}^1(G)$. For every sequence $(\mu_n)_{n\geq 1}$ in $\mathcal{M}^1(G)$ with $\operatorname{supp}(\mu_n) \subset H$ for all $n \geq 1$ there exist a sequence $(x_n)_{n \geq 1}$ in H and an element y of G such that $\varepsilon_{y^{-1}} * \mu_n * \varepsilon_{x_n} * \varepsilon_y \in \mathcal{N}$ for all $n \geq 1$.*

Proof. 1. We first consider the special case of two measures $\mu_1, \mu_2 \in \mathcal{M}^1(G)$ and the corresponding sequence

$$\{\mu_1 * \varepsilon_{y_1}, \varepsilon_{y_1^{-1}} * \mu_2 * \varepsilon_{y_2}, \ldots, \varepsilon_{y_{2n-1}^{-1}} * \mu_2 * \varepsilon_{y_{2n}}\}.$$

For every $k \geq 1$ the symbol ν_k denotes the convolution product of the first k members of the sequence. This convolution is in $\mathcal{N}$ if one chooses

$$y_{2j} := x(\varepsilon_{y_{2j-1}^{-1}} * \mu_2) \quad \text{and} \quad y_{2j-1} := x(\varepsilon_{y_{2j-1}^{-1}} * \mu_1) \quad (j \geq 1).$$

Evidently,

$$\nu_{2k-1}=(\mu_1*\mu_2)^{k-1}*\mu_1*\varepsilon_{y_{2k-1}},\qquad \nu_{2k}=(\mu_1*\mu_2)^k*\varepsilon_{y_{2k}}$$

and $\lim_{k\to\infty}\nu_{2k-1}=\lim_{k\to\infty}\nu_{2k}=\mu$. By Theorem 2.3.2 there exist a sequence $(a_n)_{n\geq 1}$ in H and a closed subgroup H_1 of H such that

$$\lim_{n\to\infty}(\mu_1*\mu_2)^n*\varepsilon_{a_n}=\omega_{H_1}$$

holds. Since

$$\lim_{n\to\infty}(\mu_1*\mu_2)^n*\varepsilon_{a_n}*\varepsilon_{a_n^{-1}}*\varepsilon_{y_{2n}}=\mu,$$

μ is a translate of ω_{H_1}.

Hence, there are sequences $(h_{1,n})_{n\geq 1}$ in H and $(z_{1,n})_{n\geq 1}$ in G with

$$\lim_{n\to\infty}z_{1,n}=z_1\quad\text{and}\quad y_{2n}=h_{1,n}z_{1,n}\qquad\text{for all } n\geq 1.$$

Furthermore,

$$\mu_1*(\mu_1*\mu_2)^{n-1}*\varepsilon_{y_{2n-1}}=(\mu_1*\mu_2)^{n-1}*\mu_1*\varepsilon_{a_n}*\varepsilon_{a_{n-1}^{-1}}*\varepsilon_{y_{2n-1}}$$

converges to μ for $n\to\infty$ and $\operatorname{supp}(\mu_1)\subset H$ since H is commutative. Thus there exist sequences $(h_{2,n})_{n\geq 1}$ in H and $(z_{2,n})_{n\geq 1}$ in G with $\lim_{n\to\infty}z_{2,n}=z_1$ such that $h_{2,n}z_{2,n}=y_{2n-1}$ for all $n\geq 1$. By the compactness of G there are sequences $(h_{1,n_k})_{k\geq 1}$, $(h_{2,n_k})_{k\geq 1}$, $(h_{1,n_k-1})_{k\geq 1}$ and $(h_{2,n_k-1})_{k\geq 1}$ converging to q_1, q_2, q_1' and q_2' resp. Then

$$\lim_{k\to\infty}\varepsilon_{y_{2n_k-1}^{-1}}*\mu_2*\varepsilon_{y_{2n_k}}=\varepsilon_{z_1^{-1}}*\mu_2*\varepsilon_{q_1}*\varepsilon_{q_2^{-1}}*\varepsilon_{z_1}$$

and

$$\lim_{k\to\infty}\varepsilon_{y_{2(n_k-1)}^{-1}}*\mu_2*\varepsilon_{y_{2n_k-1}}=\varepsilon_{z_1^{-1}}*\mu_1*\varepsilon_{(q_1')^{-1}}*\varepsilon_{q_2'}*\varepsilon_{z_1}.$$

Putting

$$x_2^{(2)}:=q_1q_2^{-1}\in H\quad\text{and}\quad x_1^{(2)}:=(q_1')^{-1}q_2'\in H$$

one obtains

$$\varepsilon_{z_1^{-1}}*(\mu_2*\varepsilon_{x_2^{(2)}})*\varepsilon_{z_1}\in\overline{\mathcal{N}}\qquad\text{and}\qquad\varepsilon_{z_1^{-1}}*(\mu_1*\varepsilon_{x_1^{(2)}})*\varepsilon_{z_1}\in\overline{\mathcal{N}}\ .$$

2. The general case is handled analogously: For $\mu_1,\dots,\mu_n\in\mathcal{M}^1(G)$ one obtains $x_1^{(n)},\dots,x_n^{(n)}\in H$ and a $z_n\in G$ such that $\varepsilon_{z_n^{-1}}*\mu_k*\varepsilon_{x_k^{(n)}}*\varepsilon_{z_n}\in\overline{\mathcal{N}}$ for all $k=1,\dots,n$. Let z be an accumulation point of $(z_n)_{n\geq 1}$. For every $k\geq 1$ there exists an accumulation point (z,x_k) of the sequence $((z_n,x_k^{(n)}))_{n\geq 1}$. Hence $\varepsilon_{z^{-1}}*\mu_k*\varepsilon_{x_k}*\varepsilon_z\in\overline{\mathcal{N}}$ for all $k=1,\dots,n$. But by Property (2) of 2.3.12 $\mathcal{N}$ can be taken closed, and thus the lemma is proved. □

Proof of Theorem 2.3.13 *Part* 2. (i)⇒(ii). Let G admit (SNCP) and let G be not totally disconnected. Then there exists a $D\in\operatorname{Rep}(G)$ such that $D(G)$ is a compact group with connected component $G_0\neq\{e\}$. For any full subset $\mathcal{N}$ of $\mathcal{M}^1(G)$ the set $D(\mathcal{N})$ is full in $D(\mathcal{M}^1(G))$ by Property (3) of 2.3.12. The set

$$\mathcal{B}:=\{\mu\in D(\mathcal{N}):\operatorname{supp}(\mu)\subset G_0\}$$

is full in $\mathcal{M}^1(G_0)$ by Property (4) of 2.3.12 since G_0 is a closed normal subgroup of $D(G)$. Now G_0 contains a subgroup isomorphic to $\mathbb{T}$. We consider the sequence $(\mu_n)_{n\geq 1}$ of measures in $\mathcal{M}^1(\mathbb{T})$ constructed in Lemma 2.3.15. By Lemma 2.3.16 there exist a sequence $(x_n)_{n\geq 1}$ in $\mathbb{T}$ and a $z\in D(G)$ such that $\varepsilon_{z^{-1}} * \mu_n * \varepsilon_{x_n} * \varepsilon_z \in \mathcal{B}$ or $\mu'_n := \mu_n * \varepsilon_{x_n} \in \varepsilon_z * \mathcal{B} * \varepsilon_{z^{-1}}$ for all $n\geq 1$. Property (3) of 2.3.12 yields that $\varepsilon_z * \mathcal{B} * \varepsilon_{z^{-1}}$ is full, and hence for every subset $\mathbb{N}_0$ of $\mathbb{N}$ the product $\bigstar_{n\in\mathbb{N}_0} \mu'_n$ and for every $n\geq 1$ the sequence $(\mu_n'^k)_{k\geq 1}$ converge. This is a contradiction of the statement of Lemma 2.3.15. □

2.4 Convergence in Variance

The variance of a real-valued random variable X on a probability space $(\Omega, \mathfrak{A}, P)$ is defined as $V(X) := \int (X - E(X))^2\, dP = \int (x - E(X))^2\, P_X(dx)$, if $X \in L^2(\Omega, \mathfrak{A}, P)$, and ∞ otherwise, where $P_X = X(P)$ denotes, as usual, the distribution of X and $E(X) := \int X\, dP = \int x\, P_X(dx)$ signifies the expectation of X. Since $V(X)$ depends only on the probability measure $P_X \in \mathcal{M}^1(\mathbb{R})$, V can be viewed as a mapping of the set $\mathcal{M}^1(\mathbb{R})$ into the compactified half-line $\bar{\mathbb{R}}_+ := \mathbb{R}_+ \cup \{\infty\}$. Considering $\mathcal{M}^1(\mathbb{R})$ as a $\mathcal{T}_v$-topological convolution semigroup the following properties of V can be stated:

(1) $\quad V(\mu * \nu) = V(\mu) + V(\nu) \quad$ for all $\mu, \nu \in \mathcal{M}^1(\mathbb{R})$.

(2) $\quad V(\mu) = 0 \Leftrightarrow \mu = \varepsilon_x \quad$ for some $x \in \mathbb{R}$.

Replacing V by V' defined by $V'(\mu) := \exp(-V(\mu))$ for all $\mu \in \mathcal{M}^1(\mathbb{R})$ one notes that

(1′) $\quad V'(\mu * \nu) = V'(\mu)\, V'(\nu) \quad$ for all $\mu, \nu \in \mathcal{M}^1(\mathbb{R})$.

(2′) $\quad V'(\mu) = 1 \Leftrightarrow \mu = \varepsilon_x \quad$ for some $x \in \mathbb{R}$.

Thus V' is a semigroup homomorphism from $\mathcal{M}^1(\mathbb{R})$ into $\mathbb{R}_+$. Clearly, V can be retrieved from V' by the formula $V(\mu) = -\log V'(\mu)$, valid for all $\mu \in \mathcal{M}^1(\mathbb{R})$.

The semigroup homomorphisms V and V' are not in general continuous. The continuity is already violated on the subsemigroup $\mathcal{M}^1_K(\mathbb{R})$ of all measures in $\mathcal{M}^1(\mathbb{R})$ with compact support, as the following simple example shows. Choosing a sequence $(\mu_n)_{n\geq 1}$ in $\mathcal{M}^1_K(\mathbb{R})$ with

$$\mu_n := \left(1 - \frac{1}{n}\right) \varepsilon_0 + \frac{1}{n} \varepsilon_n \qquad \text{for all } n \geq 1$$

one obtains $\lim_{n\to\infty} \mu_n = \mu := \varepsilon_0$ with respect to $\mathcal{T}_v$, but not $\lim_{n\to\infty} V(\mu_n) = V(\mu)$ since $V(\mu_n) = n - 1$ for all $n \geq 1$ and $V(\mu) = V(\varepsilon_0) = 0$. It should be noted, however, that V is continuous on the set $\mathcal{M}^1(\mathbb{R}, K)$ of all measures $\mu \in \mathcal{M}^1(\mathbb{R})$ with $\operatorname{supp}(\mu) \subset K$ for some fixed compact subset K of $\mathbb{R}$.

We now generalize the concept of a variance for an arbitrary locally compact group G.

Let S be a subsemigroup of $\mathcal{M}^b(G)$.

2.4.1 Definition. A *weak variance on S* is any nonconstant continuous (semigroup) homomorphism V from S into $\mathbb{R}_+$ (endowed with multiplication as semigroup operation).

A weak variance V on S is called a *variance on S* if

(V1) $V(S)\subset[0,1]$ and

(V2) $V(\mu)=1\Leftrightarrow\mu\in\mathscr{D}(G)\cap S.$

In the special case $S:=\mathscr{M}^1(G)$ we talk of a *weak variance* or *variance V for G*. Let $\mathsf{S}(S)$ and $\mathsf{V}(S)$ denote the sets of all weak variances and variances on S resp.

For $V_1, V_2\in\mathsf{S}(S)$ one defines the composition $V_1\circ V_2$ by $V_1\circ V_2(\mu):=V_1(\mu)\,V_2(\mu)$ for all $\mu\in S$. Thus $\mathsf{S}(S)$ becomes a commutative semigroup with respect to composition $\circ$), which contains $\mathsf{V}(S)$ as a subsemigroup. If, in particular, $S:=\mathscr{M}^1(G)$, we write $\mathsf{S}(G)$ and $\mathsf{V}(G)$ instead of $\mathsf{S}(S)$ and $\mathsf{V}(S)$ resp.

2.4.2. Let G be a compact group and S a *shift closed* subsemigroup of $\mathscr{M}^1(G)$ in the sense that the semigroup S is closed in $\mathscr{M}^1(G)$ and such that for every $\mu\in S$ and $x\in G$ the shifts $\mu*\varepsilon_x$ and $\varepsilon_x*\mu$ belong to S.

Properties. (1) Axioms (V1) and (V2) are equivalent to Axiom

(V3) $V(\mu)=1$ implies $\mu\in\mathscr{D}(G)$.

In fact, since $V(S)$ is a compact subsemigroup of $\mathbb{R}_+$, it is contained in $[0,1]$. If $V(\varepsilon_x)<1$ held for some $x\in G$, then one would have $V(\varepsilon_{x_0})=0$ for every accumulation point x_0 of the sequence $(x^n)_{n\geq 1}$ and hence

$$V(\varepsilon_e)=V(\varepsilon_{x_0^{-1}}*\varepsilon_{x_0})=V(\varepsilon_{x_0^{-1}})\,V(\varepsilon_{x_0})=0,$$

which would contradict the nonconstancy of V.

(2) A weak variance V on S is a variance on S iff the following axiom holds:

(V4) $V(\mu)=0$ for all $\mu\in\mathscr{J}(G)\cap S$ with $\mu\neq\varepsilon_e$.

Indeed, let $\mu\in S\cap\complement\mathscr{D}(G)$ be given such that $V(\mu)=1$. By Theorem 2.1.7 there exist an element $x\in G$ and an idempotent $\nu\in\mathscr{J}(G)$ with the property

$$\lim_{n\to\infty}\mu^n*\varepsilon_{x^{-n}}=\nu.$$

In particular, $\nu\in S$. The properties of the weak variance together with (V4) imply that

$$\lim_{n\to\infty}V(\mu)^n=\lim_{n\to\infty}V(\mu^n)=\lim_{n\to\infty}V(\mu^n*\varepsilon_{x^{-n}})=V(\nu)=0$$

or $V(\mu)<1$. The proof of the converse is obvious.

2.4.3 Examples. Let G be a compact group and S a shift closed subsemigroup of $\mathscr{M}^1(G)$.

1. Given $D\in \operatorname{Rep}(G)$, one defines a weak variance V on S by

$$V(\mu)=V_D(\mu):=|\det \hat{\mu}(D)| \quad \text{for all } \mu\in S.$$

V is a variance on S iff D is faithful (injective).
[To be shown is that $V:=V_D$ satisfies (V4) iff D is faithful. For every idempotent $\mu\in S$ of the form $\mu=\omega_H$ for some compact subgroup $H\neq\{e\}$ of G and $D\in\operatorname{Rep}(G)$ one has $\hat{\mu}(D)=\hat{\mu}(D)D(x)$ for every $x\in H$. Thus $\hat{\mu}(D)$ is not invertible and therefore $|\det\hat{\mu}(D)|=0$ iff for $x\in G$ with $x\neq e$ one has $D(x)\neq E$.]

2. More generally, let $D_1,\dots,D_n\in\operatorname{Rep}(G)$ and $k_1,\dots,k_n\in\mathbb{R}_+^*$. One defines a weak variance V on S by putting

$$V(\mu):=V_{D_1,\dots,D_n}(\mu):=\prod_{i=1}^n|\det\hat{\mu}(D_i)|^{k_i} \quad \text{for all } \mu\in S.$$

V is a variance on S iff $D_1\oplus\cdots\oplus D_n$ is faithful. This assertion follows readily from

2.4.4 Theorem. *Let G be a compact group, $\sigma_1,\dots,\sigma_n\in\Sigma(G)$ and $k_1,\dots,k_n\in\mathbb{R}_+^*$. We define a weak variance V for G by putting*

$$V(\mu):=\prod_{i=1}^n|\det\hat{\mu}(\sigma_i)|^{k_i} \quad \textit{for all } \mu\in\mathscr{M}^1(G).$$

The following statements are equivalent:
(i) *V is a variance for G;*
(ii) $[\sigma_1,\dots,\sigma_n]=\Sigma(G)$.

Proof. We shall show that for the weak variance V for G Statement (ii) is equivalent to Axiom (V4). We note that for every closed subgroup H of G and $\sigma\in\Sigma(G)$ the operator $\hat{\omega}_H(\sigma)$ is the projector onto some Hilbert space $\mathscr{H}(\sigma)$ with orthonormal basis $\{\xi_1^{(\sigma)},\dots,\xi_n^{(\sigma)}\}$ such that there exists an $l_\sigma\in\{0,1,\dots,n(\sigma)\}$ satisfying

$$\int d_{ij}^{(\sigma)}\,d\omega_H=\begin{cases}\delta_{ij}, & \text{if } j\le l_\sigma\\ 0 & \text{otherwise}\end{cases},$$

where $(d_{ij}^{(\sigma)})_{i,j=1,\dots,n(\sigma)}=D^{(\sigma)}\in\sigma\in\Sigma(G)$. Furthermore, one plainly concludes for any closed subgroup H_1 of G and the closed normal subgroup H_2 generated by H_1 the identity of the sets $H_1^\perp, H_2^\perp$,

$$\{\sigma\in\Sigma(G):\hat{\omega}_{H_1}(\sigma)\neq 0\},\quad \{\sigma\in\Sigma(G):\hat{\omega}_{H_2}(\sigma)\neq 0\}\quad\text{and}$$
$$\{\sigma\in\Sigma(G):\hat{\omega}_{H_2}(\sigma)=E\}.$$

1. (ii) $\Rightarrow$ (i). Let V be defined as in the statement of the theorem and let $\sigma_1,\dots,\sigma_n\in\Sigma(G)$ be such that $[\sigma_1,\dots,\sigma_n]=\Sigma(G)$. By the preceding remarks it suffices to show (V4) for closed normal subgroups of G. Let H be a closed normal subgroup $\neq\{e\}$ of G. Clearly, $H^\perp\neq\Sigma(G)$ and hence there exists a $j\in\{1,\dots,n\}$ such that $\sigma_j\notin H^\perp$. Consequently, $\hat{\omega}_H(\sigma_j)=0$ and V is a variance for G.

2. (i) $\Rightarrow$ (ii). Let V be a variance for G and suppose that $P:=[\sigma_1,\dots,\sigma_n]\neq\Sigma(G)$. Then $F:=P^\perp:=\{x\in G: D^{(\sigma)}(x)=E \text{ for all } \sigma\in P\}$ is a closed normal subgroup $\neq\{e\}$

of G. We obtain

$$\hat{\omega}_F(\sigma) := \begin{cases} E & \text{for } \sigma \in P \\ 0 & \text{otherwise} \end{cases}$$

and hence $V(\omega_F) = 1$, which contradicts the hypothesis. □

2.4.5 Examples. 1. There exists a variance for any finite group since for a finite group G the dual $\Sigma(G)$ is finite.

2. For any compact Abelian Lie group of the form $G := \mathbb{T}^n \times F$, where $n \geq 0$ and F is a finite Abelian group, there exists a variance since for such a group one has the existence of a finite subset $\{\chi_1, \ldots, \chi_n\}$ of $G^\wedge$ satisfying $[\chi_1, \ldots, \chi_n] = G^\wedge$. In particular, for $G := \mathbb{T}$ the mappings $V_1, V_2, \ldots$ from $\mathscr{M}^1(G)$ into $\mathbb{R}_+$ defined by

$$V_1(\mu) = V_1^\alpha(\mu) := |\hat{\mu}(1)|^\alpha, \quad V_2(\mu) = V_2^{\alpha,\beta}(\mu) := |\hat{\mu}(2)|^\alpha \, |\hat{\mu}(3)|^\beta, \ldots$$

for all $\mu \in \mathscr{M}^1(G)$ $(\alpha, \beta, \ldots \in \mathbb{R}_+^*)$ are variances for G. More generally

3. There exists a variance on any compact Lie group G since G admits a faithful (finite-dimensional) representation.

The following discussion is devoted to a detailed analysis of the connection between weak variances and variances for G.

2.4.6 Theorem. *Let G be a compact group. For every $V \in \mathbf{S}(G)$ there exists a closed normal subgroup H of G with $H \neq G$ satisfying the following conditions:*
(i) $V(\omega_H) = 1$.
(ii) *For every closed subgroup K of G with $V(\omega_K) = 1$ one has $K \subset H$.*

Proof. If $V \in \mathbf{V}(G)$, then by Property (2) of 2.4.2 $H := \{e\}$ is the desired normal subgroup of G. Let V now be in $\mathbf{S}(G)$, but not in $\mathbf{V}(G)$.

The compact subsemigroup $S := \{\mu \in \mathscr{M}^1(G) : V(\mu) = 1\}$ of $\mathscr{M}^1(G)$ contains minimal idempotents since $E(S)$ is inductively ordered by the order relation

$$\mu < \nu : \Leftrightarrow \mu * \nu = \nu * \mu = \mu \quad (\text{for } \mu, \nu \in E(S))$$

([253], A1.2.1). As the idempotents in $\mathscr{M}^1(G)$ are exactly the Haar measures of compact subgroups of G by Theorem 1.2.10, we have for compact subgroups H_1 and H_2 of G the equivalences

$$\omega_{H_1} < \omega_{H_2} \Leftrightarrow \omega_{H_1} * \omega_{H_2} = \omega_{H_2} * \omega_{H_1} = \omega_{H_1} \Leftrightarrow H_2 \subset H_1.$$

But to minimal idempotents in S there correspond maximal compact subgroups of G. Hence, the system $\mathbf{K}$ of all closed subgroups K of G with $V(\omega_K) = 1$ contains a maximal element H. It remains to show that H is normal in G.

Let therefore $x \in G$. Then clearly, $x^{-1}Hx \in \mathbf{K}$ and also

$$H_1 := [H \cup x^{-1}Hx]^- \in \mathbf{K} \quad \text{since } \omega_{H_1} = \lim_{n \to \infty} (\omega_H * \omega_{x^{-1}Hx})^n$$

holds. But this implies that $H_1 = H$ and $H_1 = x^{-1}Hx$ and hence H is a closed normal subgroup of G with $H \neq G$ satisfying the conditions (i) and (ii). □

2.4.7 Definition. The closed normal subgroup H proved in Theorem 2.4.6 to be the unique maximal closed normal subgroup of G satisfying $V(\omega_H)=1$ is called the *kernel of the weak variance* V for G.

For any closed normal subgroup H of G we denote by V_H a weak variance for G with kernel H. $\mathbf{S}_H(G)$ denotes the totality of weak variances V_H for G with kernel H. Plainly $\mathbf{S}(G)=\bigcup\{\mathbf{S}_H(G)$: H is a closed normal subgroup of $G\}$ and $\mathbf{V}(G)=\mathbf{S}_{\{e\}}(G)$.

2.4.8 Properties. (1) For every closed normal subgroup H of G and every $\mu\in\mathcal{M}^1(G)$ with $\operatorname{supp}(\mu)\subset H$ one has $V_H(\mu)=1$. Indeed, $\mu*\omega_H=\omega_H$ holds and therefore by Theorem 2.4.6

$$1=V_H(\omega_H)=V_H(\mu*\omega_H)=V_H(\mu)\,V_H(\omega_H)=V_H(\mu).$$

(2) For closed normal subgroups H_1, H_2 of G

$$\mathbf{S}_{H_1}(G)\circ\mathbf{S}_{H_2}(G)\subset\mathbf{S}_{H_1\cap H_2}(G)$$

holds. In particular, $\mathbf{S}_H(G)$ is a commutative subsemigroup of $\mathbf{S}(G)$ for any closed normal subgroup H of G.

In fact, let H be a closed normal subgroup of G such that

$$V_{H_1}(\omega_H)\cdot V_{H_2}(\omega_H)=V_{H_1}\circ V_{H_2}(\omega_H)=1$$

holds. Then $V_{H_1}(\omega_H)=V_{H_2}(\omega_H)=1$. But by Theorem 2.4.6 $H\subset H_1$, and $H\subset H_2$, so that $H\subset H_1\cap H_2$. On the other hand, Property (1) implies that

$$V_{H_1}(\omega_{H_1\cap H_2})=V_{H_2}(\omega_{H_1\cap H_2})=1,$$

and therefore $V_{H_1}\circ V_{H_2}(\omega_{H_1\cap H_2})=1$. Thus $H_1\cap H_2$ is the kernel of the weak variance $V_{H_1}\circ V_{H_2}$.

(3) If H_1, H_2 are closed normal subgroups of G such that $H_1\cap H_2=\{e\}$, then $\mathbf{S}_{H_1}(G)\circ\mathbf{S}_{H_2}(G)\subset\mathbf{V}(G)$. In particular,

(4) $\mathbf{V}(G)\circ\mathbf{S}(G)\subset\mathbf{V}(G)$.

2.4.9 Theorem. *Let G be a compact group and H a closed normal subgroup of G. Then* $\mathbf{S}_H(G)\cong\mathbf{V}(G/H)$.

Proof. The canonical homomorphism p from G onto $\dot{G}:=G/H$ can be extended to an epimorphism $\bar{p}$ from $\mathcal{M}^1(G)$ onto $\mathcal{M}^1(\dot{G})$.

1. For every $V_H\in\mathbf{S}_H(G)$ the mapping $V:\mathcal{M}^1(\dot{G})\to\mathbb{R}_+$ defined by

$$V(\bar{p}(\mu)):=V_H(\mu)\qquad\text{for all }\mu\in\mathcal{M}^1(G)$$

is a weak variance for $\dot{G}$. We show that V is well-defined. Let $\mu,\nu\in\mathcal{M}^1(G)$ be such that $\bar{p}(\mu)=\bar{p}(\nu)$. Then for every $f\in\mathscr{C}(G)$ one defines the function

$$F:=F_f\in\mathscr{C}(G)\quad\text{by}\quad F(x):=\varepsilon_x*\omega_H(f)\qquad\text{for all }x\in G$$

and by the factorization $F=\bar{F}\circ p$ with $\bar{F}\in\mathscr{C}(\dot{G})$ it follows that

$$\mu(F)=\bar{p}(\mu)(\bar{F})=\bar{p}(\nu)(\bar{F})=\nu(F).$$

Thus $\mu*\omega_H=\nu*\omega_H$ and therefore $V_H(\mu)=V_H(\mu*\omega_H)=V_H(\nu*\omega_H)=V_H(\nu)$.

2. V is a variance for $\dot{G}$. Let K be a closed subgroup of $\dot{G}$ with $K\neq p(H)$. There is a closed subgroup L of G, $L\supset H$, $L\neq H$ such that $K=p(L)$ holds. By the definition of $\mathsf{S}_H(G)$ we conclude that $V_H(\omega_L)=0$ and thus $V(\omega_K)=0$.

3. Conversely, if V is a variance for $\dot{G}$, then the mapping $V_H:\mathscr{M}^1(G)\to\mathbb{R}_+$ defined by $V_H(\mu):=V(\bar{p}(\mu))$ for all $\mu\in\mathscr{M}^1(G)$ is an element of $\mathsf{S}_H(G)$. □

2.4.10 Corollary. *For any compact group G the following statements are equivalent:*
(i) $\mathsf{S}(G)=\mathsf{V}(G)$;
(ii) *G does not admit any proper normal subgroup* ($\neq\{e\}, G$).

Proof. 1. (ii) ⇒ (i). Let $V\in\mathsf{S}(G)$ not be a variance for G. Then there exists a closed subgroup $K\neq\{e\}$ of G with $V(\omega_K)=1$. By Theorem 2.4.6 there exists a closed normal subgroup $H\neq G$ of G satisfying $V(\omega_H)=1$ and $K\subset H$. We show that $H\neq\{e\},G$. In fact, $H\supset K\neq\{e\}$. Furthermore, H is a normal subgroup of G satisfying $V(\omega_H)=1$. If we had $H=G$, then $V(\mu*\omega_G)=V(\mu)$ and also $V(\mu*\omega_G)=V(\omega_G)$ and consequently $V(\mu)=1$ for all $\mu\in\mathscr{M}^1(G)$. Hence $H\neq G$, and (i) is established.

2. (i) ⇒ (ii). Let H be a closed normal subgroup of G with $H\neq\{e\}$. Since by the theorem $\mathsf{V}(G/H)=\varnothing$ iff $\mathsf{S}_H(G)=\varnothing$, we may assume that

$$\mathsf{V}(G/H)\cong\mathsf{S}_H(G)\neq\varnothing.$$

For every $V\in\mathsf{V}(G/H)$ one defines $\bar{V}\in\mathsf{S}_H(G)$ by

$$\bar{V}(\mu):=V(\bar{p}(\mu))\qquad\text{for all }\mu\in\mathscr{M}^1(G),$$

where as above $\bar{p}$ is the extension to $\mathscr{M}^1(G)$ of the canonical homomorphism p from G onto G/H. From $\bar{V}(\omega_H)=1$ one concludes that $\bar{V}$ is not in $\mathsf{V}(G)$. □

Let $\mathbb{K}$ denote one of the fields $\mathbb{R}$, $\mathbb{C}$ and $\mathbb{H}$ of real, complex numbers and quaternions resp. To every $h\in\mathbb{H}$ there corresponds a unique matrix $\begin{pmatrix} z & -\bar{w}\\ w & \bar{z}\end{pmatrix}$ with $z,w\in\mathbb{C}$ and to every $A\in\mathfrak{M}(n,\mathbb{H})$ a matrix $\bar{A}\in\mathfrak{M}(2n,\mathbb{C})$ (via replacement of a quaternion by its corresponding matrix in $\mathfrak{M}(2,\mathbb{C})$). Considering $\mathfrak{M}(n,\mathbb{H})$ as an algebra over $\mathbb{R}$ one establishes a bicontinuous algebra isomorphism $A\to\bar{A}$ from $\mathfrak{M}(n,\mathbb{H})$ onto a subalgebra of $\mathfrak{M}(2n,\mathbb{C})$. For $A\in\mathfrak{M}(n,\mathbb{H})$ we put $\det A:=\det\bar{A}$. As usual, $\mathbb{K}^*$ denotes the multiplicative group $\mathbb{K}\setminus\{0\}$ of the field $\mathbb{K}$.

2.4.11 Lemma. *Every nonconstant continuous semigroup homomorphism ϕ from the multiplicative semigroup $\mathfrak{M}(n,\mathbb{K})$ into the multiplicative semigroup $\mathbb{R}_+$ is of the form $\phi(A):=|\det A|^k$ for some $k\in\mathbb{R}_+^*$ and all $A\in\mathfrak{M}(n,\mathbb{K})$.*

Proof. Since $\mathfrak{GL}(n,\mathbb{K})$ is dense in $\mathfrak{M}(n,\mathbb{K})$, it suffices to show that the group homomorphisms from $\mathfrak{GL}(n,\mathbb{K})$ into $\mathbb{R}_+^*$ have the desired form. Plainly any nonconstant continuous homomorphism ϕ from $\mathfrak{G}\mathfrak{L}(n,\mathbb{K})$ into $\mathbb{R}_+^*$ is surjective since $\mathfrak{G}\mathfrak{L}(n,\mathbb{K})$ is connected for $\mathbb{K}:=\mathbb{C}$ or $\mathbb{H}$. Since there exists an isomorphism

$$\psi:\mathfrak{G}\mathfrak{L}(n,\mathbb{K})/\mathfrak{G}\mathfrak{L}(n,\mathbb{K})'\to\mathbb{K}^*/\mathbb{K}^{*\prime},$$

we conclude from $\mathfrak{G}\mathfrak{L}(n,\mathbb{K})'\subset\ker\phi$ that ϕ induces a homomorphism ϕ_0 from $\mathbb{K}^*/\mathbb{K}^{*\prime}$ into $\mathbb{R}_+^*$ satisfying $\phi=\phi_0\circ\psi\circ\psi_1$, where ψ_1 denotes the canonical homomorphism

$$\mathfrak{G}\mathfrak{L}(n,\mathbb{K})\to\mathfrak{G}\mathfrak{L}(n,\mathbb{K})/\mathfrak{G}\mathfrak{L}(n,\mathbb{K})'.$$

This implies that $\ker\phi=\psi_1^{-1}(\psi^{-1}(\ker\phi_0))$. For ϕ_0 there is exactly one homomorphism $\phi_1:\mathbb{K}^*\to\mathbb{R}_+^*$ whose kernel determines $\ker\phi_0$. Obviously ϕ_1 maps the elements of $\mathbb{K}$ of modulus 1 onto $\{1\}$ and thus $\phi_1(z)$ depends only on $|z|$ for $z\in\mathbb{K}^*$. Since the automorphisms of $\mathbb{R}_+^*$ are of the form $x\to x^k$ for $k\in\mathbb{R}^*$ we conclude that $\ker\phi_1=\{z\in\mathbb{K}^*:|z|=1\}$.

In particular, $\ker\phi$ is independent of ϕ for all nonconstant continuous homomorphisms ϕ from $\mathfrak{G}\mathfrak{L}(n,\mathbb{K})$ into $\mathbb{R}_+^*$. Defining $d(A):=|\det A|$ for all $A\in\mathfrak{G}\mathfrak{L}(n,\mathbb{K})$ one obtains a nonconstant continuous epimorphism from $\mathfrak{G}\mathfrak{L}(n,\mathbb{K})$ into $\mathbb{R}_+^*$ and $\ker\phi=\ker d$. Thus the mapping $d(A)\to\phi(A)$ defines for all $A\in\mathfrak{G}\mathfrak{L}(n,\mathbb{K})$ a homomorphism from $\mathbb{R}_+^*$ onto itself, and $\phi(A)=d(A)^k$ for $k\in\mathbb{R}_+^*$ and all $A\in\mathfrak{G}\mathfrak{L}(n,\mathbb{K})$. ϕ is continuously extendable to $\mathfrak{M}(n,\mathbb{K})$. □

2.4.12 Lemma. *Let* A *be a simple algebra over* $\mathbb{R}$ *with unit element and* ϕ *a nonconstant continuous homomorphism of the multiplicative semigroup* A *into* $\mathbb{R}_+$. *Then there exist a representation* D *of* A *and a* $k\in\mathbb{R}_+^*$ *such that* $\phi(a)=|\det D(a)|^k$ *for all* $a\in\mathsf{A}$.

Proof. Every simple algebra over $\mathbb{R}$ with unit is by the Wedderburn theorem isomorphic to a full matrix algebra with elements in a division algebra over $\mathbb{R}$, and any division algebra over $\mathbb{R}$ is isomorphic to one of the algebras $\mathbb{R}$, $\mathbb{C}$ or $\mathbb{H}$. The first mentioned isomorphism $a\to D(a)$ from A into $\mathfrak{M}(n,\mathbb{K})$ is a representation of A. Lemma 2.4.11 then implies the assertion. □

2.4.13 Lemma. *Let* $\mathsf{A}:=\sum_{i\in\mathbb{I}}^{\oplus}\mathsf{A}_i$ *be the (direct) Hilbert sum of a family* $(\mathsf{A}_i)_{i\in\mathbb{I}}$ *of simple algebras over* $\mathbb{R}$ *with unit and let* ϕ *be a nonconstant continuous semigroup homomorphism from* A *into* $\mathbb{R}_+$. *Then there exist a finite subset* $\mathbb{J}_0$ *of* $\mathbb{I}$ *and for every* $i\in\mathbb{J}_0$ *a representation* D_i *of* A_i *as well as a* $k_i\in\mathbb{R}_+^*$ *such that for all* $a:=(a_i)_{i\in\mathbb{I}}\in\mathsf{A}$ *one obtains* $\phi(a)=\prod_{i\in\mathbb{J}_0}|\det D_i(a_i)|^{k_i}$.

Proof. 1. We first show the following special case of the statement: Let $\mathsf{A}_1,\dots,\mathsf{A}_n$ be simple algebras over $\mathbb{R}$ with unit and $\mathsf{A}:=\sum_{i=1}^{\oplus n}\mathsf{A}_i$. Then for every nonconstant continuous semigroup homomorphism ϕ from A into $\mathbb{R}_+$ there exist representations $D_1,\dots,D_n$ of $\mathsf{A}_1,\dots,\mathsf{A}_n$ resp. and $k_1,\dots,k_n\in\mathbb{R}_+$ with $k_j>0$ for at least one $j\in\{1,\dots,n\}$ such that for all

$$a:=a_1\oplus\cdots\oplus a_n\in\mathsf{A}=\sum_{i=1}^{\oplus n}\mathsf{A}_i$$

we have $\phi(a)=\prod_{i=1}^{n}|\det D_i(a_i)|^{k_i}$.

In fact, for every $i=1,\dots,n$ the algebra A_i is naturally embedded in A. Since the A_i $(i=1,\dots,n)$ mutually annihilate themselves, ϕ induces on A_i a mapping ϕ_i with the properties of ϕ, and one obtains

$$\phi(a)=\prod_{i=1}^n \phi_i(a_i) \quad \text{for all } a:=a_1\oplus\cdots\oplus a_n\in\mathsf{A}=\sum_{i=1}^{\oplus n}\mathsf{A}_i.$$

The result follows from Lemma 2.4.1.

2. Let $\mathscr{F}(\mathbb{I})$ denote the system of all finite subsets of $\mathbb{I}$. For every $\mathbb{J}\in\mathscr{F}(\mathbb{I})$ we put $\mathsf{A}^{(\mathbb{J})}:=\sum_{i\in\mathbb{J}}^{\oplus}\mathsf{A}_i$, denote by $j_{\mathbb{J}}$ the natural embedding of $\mathsf{A}^{(\mathbb{J})}$ into A and write $a^{(\mathbb{J})}:=(a_i)_{i\in\mathbb{J}}$ for all $a:=(a_i)_{i\in\mathbb{I}}\in\mathsf{A}$. If $\mathbb{J}\subset\mathbb{J}'\in\mathscr{F}(\mathbb{I})$, then $j_{\mathbb{J},\mathbb{J}'}$ denotes the natural embedding from $\mathsf{A}^{(\mathbb{J})}$ into $\mathsf{A}^{(\mathbb{J}')}$.

Since $\bigcup_{\mathbb{J}\in\mathscr{F}(\mathbb{I})} j_{\mathbb{J}}(\mathsf{A}^{(\mathbb{J})})$ is dense in A, there exists a subset $\mathbb{J}\in\mathscr{F}(\mathbb{I})$ such that $\phi\circ j_{\mathbb{J}}$ is nonconstant. By Part 1 there are a subset $\mathbb{J}_0\subset\mathbb{J}$ and for every $i\in\mathbb{J}_0$ a representation D_i of A_i as well as a number $k_i\in\mathbb{R}_+^*$ satisfying

$$\phi\circ j_{\mathbb{J}}(a^{(\mathbb{J})})=\prod_{i\in\mathbb{J}_0}|\det D_i(a_i)|^{k_i} \quad \text{for all } a^{(\mathbb{J})}\in\mathsf{A}^{(\mathbb{J})}.$$

Let $\mathbb{J}'\in\mathscr{F}(\mathbb{I})$ be such that $\mathbb{J}'\supset\mathbb{J}$ and $\mathbb{J}'\neq\mathbb{J}$. Then there exist a subset $\mathbb{J}_0'$ of $\mathbb{J}'$ and for every $i\in\mathbb{J}_0'$ a representation D_i' of A_i as well as a number $k_i'\in\mathbb{R}_+^*$ satisfying

$$\phi\circ j_{\mathbb{J}'}(a^{(\mathbb{J}')})=\prod_{i\in\mathbb{J}_0'}|\det D_i'(a_i)|^{k_i'} \quad \text{for all } a^{(\mathbb{J}')}\in\mathsf{A}^{(\mathbb{J}')}.$$

Plainly $\mathbb{J}_0'\subset\mathbb{J}$ since otherwise for all $a^{(\mathbb{J})}\in\mathsf{A}^{(\mathbb{J})}$ the expression

$$\begin{aligned}\phi\circ j_{\mathbb{J}}(a^{(\mathbb{J})}) &= \phi\circ j_{\mathbb{J}'}(j_{\mathbb{J},\mathbb{J}'}(a^{(\mathbb{J})}))\\ &=\prod_{i\in\mathbb{J}_0'\setminus\mathbb{J}}|\det D_i'(a_i)|^{k_i'}\prod_{i\in\mathbb{J}_0'\cap\mathbb{J}}|\det D_i'(a_i)|^{k_i'}\end{aligned}$$

would be identically zero, which contradicts the hypothesis. Successive replacement of special entries yields $\mathbb{J}_0=\mathbb{J}_0'$ and $k_i=k_i'$ for all $i\in\mathbb{I}$, $|\det D_i|=|\det D_i'|$ for all $i\in\mathbb{J}_0$ or without loss of generality $D_i=D_i'$ for all $i\in\mathbb{J}_0$. Hence one has for all $a\in\bigcup_{\mathbb{J}\in\mathscr{F}(\mathbb{I})} j_{\mathbb{J}}(\mathsf{A}^{(\mathbb{J})})$ the equality

$$\phi(a)=\prod_{i\in\mathbb{J}_0}|\det D_i(a_i)|^{k_i},$$

which by the continuity requirement extends to all $a\in\mathsf{A}$. □

In order to establish the general form of weak variances on $\mathscr{M}^b(G)$ for a compact group G we apply the preceding discussion to the algebra $\mathsf{A}:=L^2_{\mathbb{R}}(G,\omega)$. A refined version of the Peter-Weyl theorem ([326], p. 158) is needed. Let G be a compact group. In the Hilbert algebra $L^2_{\mathbb{R}}(G,\omega)$ there exists a family $\{L_\sigma:\sigma\in\Sigma(G)\}$ of simple ideals

$$L_\sigma:=\{\chi^{(\sigma)}*f: f\in L^2_{\mathbb{R}}(G,\omega)\}, \quad \text{where } \chi^{(\sigma)}:=\operatorname{tr}D^{(\sigma)} \text{ for } D^{(\sigma)}\in\sigma\in\Sigma(G),$$

such that $L^2_{\mathbb{R}}(G,\omega)=\sum_{\sigma\in\Sigma(G)}^{\oplus} L_\sigma$ holds. Clearly, for each $\sigma\in\Sigma(G)$ the ideal L_σ has a unit, and for all $\sigma,\sigma'\in\Sigma(G)$ with $\sigma\neq\sigma'$ one has

$$L_\sigma * L_{\sigma'}=L_{\sigma'} * L_\sigma=\{0\}.$$

This result provides us with the determination of all $\|\cdot\|_2$-continuous homomorphisms from $L^2_{\mathbb{R}}(G,\omega)$ into $\mathbb{R}_+$.

2.4.14 Theorem. *Let G be a compact group and let V be a weak variance on $\mathcal{M}^b(G)$. Then there exist $D_1, \ldots, D_n \in \operatorname{Rep}(G)$ and $k_1, \ldots, k_n \in \mathbb{R}_+^*$ with the property that*

$$V(\mu) = \prod_{i=1}^n |\det \hat{\mu}(D_i)|^{k_i} \qquad \textit{for all } \mu \in \mathcal{M}^b(G).$$

Proof. Let D_0 be a finite-dimensional representation of $L^2(G) := L^2_{\mathbb{R}}(G, \omega)$. There exists exactly one $D \in \operatorname{Rep}(G)$ such that $D_0(f) = \hat{f}(D)$ holds for all $f \in L^2(G)$. The mapping $f \to \hat{f}(D)$ from $L^2(G)$ into the algebra $\mathcal{L}(\mathcal{H}(D))$ of bounded operators on $\mathcal{H}(D)$ is weakly continuous. Then the $\|\cdot\|_2$-continuous semigroup homomorphisms from $L^2(G)$ onto $\mathbb{R}_+$ are also weakly continuous. By Lemma 2.4.13 $\mathsf{S}(L^2(G))$ consists therefore of all homomorphisms V_0 from $L^2(G)$ into $\mathbb{R}_+$ of the form

$$V_0(f) := \prod_{i=1}^n |\det \hat{f}(D_i)|^{k_i}$$

for some $D_1, \ldots, D_n \in \operatorname{Rep}(G)$, $k_1, \ldots, k_n \in \mathbb{R}_+^*$ and all $f \in L^2(G)$. We now define for every $\mu \in \mathcal{M}^b(G)$ the real number

$$V(\mu) := \prod_{i=1}^n |\det \hat{\mu}(D_i)|^{k_i}.$$

Then $V \in \mathsf{S}(\mathcal{M}^b(G))$ with $\operatorname{Res}_{L^2(G)} V = V_0$. Since $L^2(G)$ is dense in $\mathcal{M}^b(G)$, V is the unique continuous extension of V_0. □

2.4.15 Theorem. *Let G be a compact group and $V \in \mathsf{S}(G)$. Then V is extendable to a weak variance $\bar{V}$ on $\mathcal{M}^b(G)$ of the form*

$$\bar{V}(\mu) = \prod_{i=1}^n |\det \hat{\mu}(D_i)|^{k_i} \qquad \textit{for all } \mu \in \mathcal{M}^b(G)$$

and some $D_1, \ldots, D_n \in \operatorname{Rep}(G)$, $k_1, \ldots, k_n \in \mathbb{R}_+^*$.

Proof. For any subsemigroup $\mathcal{N}$ of $\mathcal{M}^b := \mathcal{M}^b(G)$ we denote by $\mathcal{N}^0$ the set of all measures in $\mathcal{N}$ of total mass 0 and by $\mathcal{N}^1$ the set of all probability measures in $\mathcal{N}$. The spaces $\mathcal{C} := \mathcal{C}_{\mathbb{R}}(G)$ and $\mathcal{D} := \mathcal{C} \oplus \mathbb{R}\,\varepsilon_e$ are considered as subsemigroups of $\mathcal{M}^b$. Thus we can put $\mathcal{C}^{(1)} := \mathcal{C}^1 \ominus \omega_G$ and $\mathcal{D}^{(1)} := \mathcal{D}^1 \ominus \omega_G$.

1. Given a weak variance V for G we define the mapping

$$V_0 : \mathcal{D}^{(1)} \to \mathbb{R}_+ \quad \text{by} \quad V_0(\nu - \omega_G) := V(\nu) \qquad \text{for all } \nu \in \mathcal{D}^1.$$

Obviously $V_0 \in \mathsf{S}(\mathcal{D}^{(1)})$, $\varepsilon_e - \omega_G$ is the unit element of $\mathcal{D}^{(1)}$ and $t(\varepsilon_e - \omega_G) \in \mathcal{D}^{(1)}$ for all $t \in [0,1]$. Moreover, $t\,\nu \in \mathcal{D}^{(1)}$ for all $\nu \in \mathcal{D}^{(1)}$ and $t \in [0,1]$, since $t\,\nu = \nu * [t(\varepsilon_e - \omega_G)]$ holds.

2. For the mapping $\tilde{V}_0 : [0,1] \to \mathbb{R}$ defined by $\tilde{V}_0(t) := V_0(t(\varepsilon_e - \omega_G))$ for all $t \in [0,1]$ one has $\tilde{V}_0(t\,u) = \tilde{V}_0(t)\,\tilde{V}_0(u)$ for all $t, u \in [0,1]$ and therefore the existence of a $k := k(V) \in \mathbb{R}_+^*$ such that $\tilde{V}_0(t) = t^k$ holds for all $t \in [0,1]$.

3. For $\mu, \nu \in \mathcal{D}^{(1)}$ with $\mu = t\,\nu$ for some $t \in [0,1]$ one plainly obtains $V_0(\mu) = t^k\,V_0(\nu)$ since $t\,\nu = \nu * [t(\varepsilon_e - \omega_G)]$ holds.

4. If one defines for $g \in \mathcal{C}^0$, $h \in \mathcal{D}^{(1)}$ and $t \in \mathbb{R}_+$ with $g = t\,h$

$$\bar{V}_0(g) := t^k\,V_0(h),$$

then $\operatorname{Res}_{\mathscr{C}^{(1)}} \overline{V}_0 = \operatorname{Res}_{\mathscr{C}^{(1)}} V_0$ and $\overline{V}_0 \in \mathbf{S}(\mathscr{C}^0)$. This follows from the equality

$$\mathscr{C}^0 = \{t\,h : t \in \mathbb{R}_+,\ h \in \mathscr{D}^{(1)}\}.$$

5. Let μ be an element of the weak hull $\overline{\mathscr{C}^0}$ of $\mathscr{C}^0$ in $\mathscr{M}^b$. Then there exists $(f_\alpha)_{\alpha \in \mathbf{A}}$ in $\mathscr{C}^0$ with $\mu = \lim_{\alpha \in \mathbf{A}} f_\alpha$. Since $\overline{V}_0$ is nonconstant on $\mathscr{C}^0$, there exists an $h \in \mathscr{C}^0$ with the property $\overline{V}_0(h) > 0$. We have $\mu * h = \lim_{\alpha \in \mathbf{A}} (f_\alpha * h)$ and by the continuity of $\overline{V}_0$

$$\lim_{\alpha \in \mathbf{A}} \overline{V}_0(f_\alpha)\, \overline{V}_0(h) = \lim_{\alpha \in \mathbf{A}} \overline{V}_0(f_\alpha * h) = \overline{V}_0(\mu * h).$$

Thus $(\overline{V}_0(f_\alpha))_{\alpha \in \mathbf{A}}$ converges. We now define $\overline{\overline{V}}_0(\mu) := \lim_{\alpha \in \mathbf{A}} \overline{V}_0(f_\alpha)$ for all $\mu = \lim_{\alpha \in \mathbf{A}} f_\alpha$ with $(f_\alpha)_{\alpha \in \mathbf{A}}$ in $\mathscr{C}^0$ and obtain a continuous homomorphism $\overline{\overline{V}}_0$ on $\overline{\mathscr{C}^0}$ which extends $\overline{V}_0$ to a continuous homomorphism from $\mathscr{M}^{b0}$ into $\mathbb{R}_+$.

6. Plainly $\mathscr{M}^b = \mathscr{M}^{b0} \oplus \mathbb{R}\,\omega_G$. Let $r \in \mathbb{R}_+^*$. We put

$$\overline{V}(\mu + t\,\omega_G) := \overline{\overline{V}}_0(\mu)\,|t|^r \quad \text{for all } \mu \in \mathscr{M}^{b0},\ t \in \mathbb{R}.$$

Clearly $\overline{V} \in \mathbf{S}(\mathscr{M}^b)$, $\operatorname{Res}_{\mathscr{D}^1} \overline{V} = V$ and by $\overline{\mathscr{D}^1} = \mathscr{M}^1$ also $\operatorname{Res}_{\mathscr{M}^1} \overline{V} = V$. Thus $\overline{V}$ is the desired extension of V, which by Theorem 2.4.14 is of the required form. □

2.4.16 Corollary. *For $V_1, V_2 \in \mathbf{S}(\mathscr{M}^b(G))$ with $\operatorname{Res}_{\mathscr{M}^1(G)} V_1 = \operatorname{Res}_{\mathscr{M}^1(G)} V_2$ there exists an $s \in \mathbb{R}$ such that $V_1(\mu) = |\mu(G)|^s\, V_2(\mu)$ holds for all $\mu \in \mathscr{M}^b(G)$.*

We are now able to present the main existence result for variances on $\mathscr{M}^1(G)$.

2.4.17 Theorem. *Let G be an infinite compact group. The following statements are equivalent:*

(i) *There exists a variance for G;*

(ii) *G is a Lie group.*

Proof. By Theorems 2.4.4 and 2.4.14 it suffices to show that an infinite compact group is a Lie group iff there exists a finite subset $\{\sigma_1, \ldots, \sigma_n\}$ of $\Sigma(G)$ with $[\sigma_1, \ldots, \sigma_n] = \Sigma(G)$.

1. Let G be a compact Lie group. Then there exists an open $U \in \mathfrak{V}(e)$ such that $\{e\}$ is the only subgroup contained in U. For every $\sigma \in \Sigma(G)$ the set $\{\sigma\}^\perp$ is a closed normal subgroup of G.

Plainly (since $G \in \mathbf{A}$) one has

$$\bigcap_{\sigma \in \Sigma(G)} \{\sigma\}^\perp = \{e\} \quad \text{and hence} \quad \bigcap_{\sigma \in \Sigma(G)} \{\sigma\}^\perp \cap \complement U = \emptyset.$$

As G is compact, there exists a subset $\{\sigma_1, \ldots, \sigma_n\}$ of $\Sigma(G)$ satisfying

$$\bigcap_{i=1}^n \{\sigma_i\}^\perp \cap \complement U = \emptyset \quad \text{so that} \quad N := \bigcap_{i=1}^n \{\sigma_i\}^\perp \subset U \quad \text{or} \quad N = \{e\}.$$

From

$$\bigcap_{i=1}^n \{\sigma_i\}^\perp = \{[\sigma_1] \times [\sigma_2] \times \cdots \times [\sigma_n]\}^\perp = [\sigma_1, \ldots, \sigma_n]^\perp = \{e\}$$

one concludes that $[\sigma_1, \ldots, \sigma_n] = \Sigma(G)$.

2. Conversely, let there exist a subset $\{\sigma_1, \ldots, \sigma_n\}$ of $\Sigma(G)$ satisfying

$$[\sigma_1, \ldots, \sigma_n] = \Sigma(G).$$

Then

$$\{e\}=[\sigma_1,\ldots,\sigma_n]^{\perp}=\bigcap_{i=1}^{n}\{\sigma_i\}^{\perp}.$$

Thus the mapping $x\to D^{(\sigma_1)}(x)\oplus\cdots\oplus D^{(\sigma_n)}(x)$ is an injective representation of G and hence a topological isomorphism, and G is topologically isomorphic to a closed subgroup of $\mathfrak{U}(m)$ for some $m\geq 1$. Since a closed subgroup of a Lie group is itself a Lie group, G is a Lie group. □

2.4.18 Corollary. *Let G be a compact group and H the kernel of a weak variance for G. Then G/H is either finite or a Lie group.*

Proof. By Theorem 2.4.9 we have $\mathsf{V}(G/H)\neq\emptyset$, and G/H is either finite or a Lie group by the theorem. □

We shall now give a few applications of the concept of variance for a compact group. Let G be a compact group and $\mu\in\mathscr{M}^1(G)$. The measure μ is said to *admit a (bilateral) idempotent factor* ω_H for some compact subgroup $H\neq\{e\}$ of G if there exist measures $\nu,\lambda\in\mathscr{M}^1(G)$ satisfying $\mu=\nu*\omega_H*\lambda$. Clearly, if G is Abelian, μ admits an idempotent factor ω_H iff $\mu=\mu*\omega_H$. We furthermore agree on the notation that $\mu\in\mathscr{M}^1(G)$ is a *measure with finite variance* if there exists $V\in\mathsf{V}(G)$ such that $V(\mu)>0$ holds.

2.4.19 Theorem. *Let G be a compact Abelian group such that $\mathsf{V}(G)\neq\emptyset$. Then $\mu\in\mathscr{M}^1(G)$ is a measure with finite variance iff μ does not admit an idempotent factor.*

The *proof* of this theorem will be preceded by the following

2.4.20 Lemma. *Let G be a compact group, $\nu\in\mathscr{M}^1(G)$ and N a closed normal subgroup of G. Then the following assertions are equivalent:*

(i) $\nu=\nu*\omega_N=\omega_N*\nu$;

(ii) $N\subset\{\sigma\in\Sigma(G):\hat{\nu}(\sigma)\neq 0\}^{\perp}$.

Proof. The set $H:=\{\sigma\in\Sigma(G):\hat{\nu}(\sigma)\neq 0\}^{\perp}$ is a closed normal subgroup of G. Moreover, we have

$$\hat{\omega}_N(\sigma)=\begin{cases}E_{n(\sigma)}, & \text{if } \sigma\in N^{\perp}\\ 0 & \text{otherwise,}\end{cases}$$

and $N^{\perp\perp}=N$ holds for every closed normal subgroup N of G. Hence, we obtain the equivalences

$$\begin{aligned}\nu*\omega_N=\nu &\Leftrightarrow \hat{\nu}(\sigma)\,\hat{\omega}_N(\sigma)=\hat{\nu}(\sigma) \quad \text{for all } \sigma\in\Sigma(G)\\ &\Leftrightarrow \hat{\omega}_N(\sigma)=E_{n(\sigma)} \quad \text{for all } \sigma\in H^{\perp}\\ &\Leftrightarrow H^{\perp}\subset N^{\perp}\Leftrightarrow N\subset H,\end{aligned}$$

which yield the assertion. □

2.4.21 Corollary. *$\nu\in\mathscr{M}^1(G)$ does not admit an idempotent factor in the center of $\mathscr{M}^1(G)$ iff $[\{\sigma\in\Sigma(G):\hat{\nu}(\sigma)\neq 0\}]=\Sigma(G)$.*

The *proof* follows from the lemma together with the equality

$$[\{\sigma\in\Sigma(G):\hat{\nu}(\sigma)\neq 0\}]=\{\sigma\in\Sigma(G):\hat{\nu}(\sigma)\neq 0\}. \quad \square$$

The *proof of Theorem* 2.4.19 follows now from Corollary 2.4.21 together with Theorem 2.4.4. □

2.4.22 Remark. Theorem 2.4.19 is no longer true if G is not Abelian. In fact, we consider the group $G:=\mathfrak{S}_3$ of symmetries of degree 3 written as $\mathfrak{S}_3=\{(1), (12), (13), (23), (123), (132)\}$. Plainly, $\Sigma(G)=\{1,\chi,\sigma\}$, where 1 denotes the unit character of G, χ the character of G satisfying $\chi(f)=1$ for all f in the alternating subgroup $\mathfrak{A}_3$ of $\mathfrak{S}_3$ and $\chi(f)=-1$ otherwise and σ the unitary representation of G of dimension 2. Choosing a proper basis of $\mathscr{H}(D)$ we can give the representation D of G in the following form:

$$(1)\to\begin{pmatrix}1&0\\0&1\end{pmatrix},\quad (12)\to\begin{pmatrix}-1&0\\0&1\end{pmatrix},\quad (13)\to\begin{pmatrix}\frac{1}{2}&-\frac{3}{2}\\-\frac{3}{2}&-\frac{1}{2}\end{pmatrix},$$

$$(23)\to\begin{pmatrix}\frac{1}{2}&\frac{3}{2}\\\frac{3}{2}&-\frac{1}{2}\end{pmatrix},\quad (123)\to\begin{pmatrix}-\frac{1}{2}&-\frac{3}{2}\\\frac{3}{2}&-\frac{1}{2}\end{pmatrix},\quad (132)\to\begin{pmatrix}-\frac{1}{2}&\frac{3}{2}\\-\frac{3}{2}&-\frac{1}{2}\end{pmatrix}$$

(see [218], (27.61)).

Let

$$\mu:=\tfrac{1}{9}\varepsilon_{\{(1)\}}+\tfrac{2}{9}\varepsilon_{\{(12)\}}+\tfrac{1}{9}\varepsilon_{\{(13)\}}+\tfrac{3}{9}\varepsilon_{\{(123)\}}+\tfrac{2}{9}\varepsilon_{\{(132)\}}\in\mathscr{M}^1(G).$$

One computes

$$\hat{\mu}(\chi)=\tfrac{1}{9}+\tfrac{3}{9}+\tfrac{2}{9}-\tfrac{2}{9}-\tfrac{1}{9}=\tfrac{1}{3}\neq 0 \quad\text{and}$$

$$\hat{\mu}(\sigma)=\tfrac{1}{9}\begin{pmatrix}1&0\\0&1\end{pmatrix}+\tfrac{2}{9}\begin{pmatrix}-1&0\\0&1\end{pmatrix}+\tfrac{1}{9}\begin{pmatrix}\frac{1}{2}&-\frac{3}{2}\\-\frac{3}{2}&-\frac{1}{2}\end{pmatrix}$$

$$+\tfrac{3}{9}\begin{pmatrix}-\frac{1}{2}&\frac{3}{2}\\\frac{3}{2}&-\frac{1}{2}\end{pmatrix}+\tfrac{2}{9}\begin{pmatrix}-\frac{1}{2}&\frac{3}{2}\\-\frac{3}{2}&-\frac{1}{2}\end{pmatrix}=\begin{pmatrix}-\frac{1}{3}&-\frac{3}{9}\\0&0\end{pmatrix}.$$

Since $\{\sigma\}^\perp=\{(1)\}$, one has $[\sigma]=\{(1)\}^\perp=\Sigma(G)$, and hence the mapping V: $\mathscr{M}^1(G)\to\mathbb{R}_+$ defined by $V(\mu):=|\det\hat{\mu}(\sigma)|^\alpha$ for some $\alpha\in\mathbb{R}_+^*$ and all $\mu\in\mathscr{M}^1(G)$ is a variance for G. Moreover, every variance for G admits V as a component in the sense of Theorem 2.4.4. Clearly, μ is a measure with infinite variance. It remains to show that μ does not admit an idempotent factor.

(a) For any compact group G and any measure $\mu\in\mathscr{M}^1(G)$ the existence of an idempotent factor ω_H for some compact subgroup H of G implies that

$$[\{\sigma\in\Sigma(G):\det\hat{\mu}(\sigma)\neq 0\}]\subset H^\perp.$$

In fact, let ν, $\lambda\in\mathscr{M}^1(G)$ satisfy $\mu=\nu*\omega_H*\lambda$ and let $\sigma\notin H^\perp$. Then

$$\det\hat{\omega}_H(\sigma)=0 \quad\text{and}\quad \det\hat{\mu}(\sigma)=\det\hat{\nu}(\sigma)\det\hat{\omega}_H(\sigma)\det\hat{\lambda}(\sigma)=0.$$

(b) $\mathfrak{S}_3$ admits the following subgroups:

$$G_1:=\{(1),(12)\},\quad G_2:=\{(1),(13)\},\quad G_3:=\{(1),(23)\}\quad\text{and}$$
$$\mathfrak{A}_3:=\{(1),(123),(132)\}.$$

It should be noted that $\mathfrak{A}_3$ is the only normal subgroup of G. Moreover,

$$N(G_1)=N(G_2)=N(G_3)=\{(1)\}$$

holds.

(c) If μ admitted the idempotent factor $\omega_{\mathfrak{A}_3}$, then by Corollary 2.4.21 one would have

$$[\{\sigma\in\Sigma(G):\hat{\mu}(\sigma)\neq 0\}]=\mathfrak{A}_3^{\perp}=\{1,\chi\}.$$

But $[\{\sigma\in\Sigma(G):\hat{\mu}(\sigma)\neq 0\}]=\Sigma(G)$, which is a contradiction. If μ admitted any of the idempotent factors ω_{G_1}, ω_{G_2} or ω_{G_3}, then by (a) one would have

$$\{\sigma\in\Sigma(G):\det\hat{\mu}(\sigma)\neq 0\}\subset G_i^{\perp}=G^{\perp}=\{1\}\qquad\text{for all } i=1,2,3,$$

but $\hat{\mu}(\chi)=\frac{1}{3}\neq 0$. Thus μ does not admit any idempotent factor.

Let G be a compact Lie group (finite or infinite) which, as we know, implies $\mathsf{V}(G)\neq\emptyset$.

2.4.23 Definition. A sequence $(X_j)_{j\geq 1}$ of G-random variables X_j (on a probability space $(\Omega,\mathfrak{A},P)$) with distribution μ_j is said to be *convergent in variance* (*V-convergent*) if there exist a variance $V\in\mathsf{V}(G)$ and an $N\geq 1$ such that $\prod_{j\geq N}V(\mu_j)>0$ holds.

2.4.24 Theorem. *Let $(X_j)_{j\geq 1}$ be an independent sequence of G-random variables X_j with distribution $\mu_j (j\geq 1)$ and $\mathscr{A}_\infty$ the accumulation set of*

$$(\nu_{k,n})_{0\leq k<n}\quad\textit{with}\quad \nu_{k,n}:=\mu_{k+1}*\cdots*\mu_n\quad\textit{for every } 0\leq k<n$$

as introduced in 2.2.1. The following statements are equivalent:
(i) *$(X_j)_{j\geq 1}$ is V-convergent for some $V\in\mathsf{V}(G)$;*
(ii) $\mathscr{A}_\infty=\{\varepsilon_e\}$.

Proof. 1. Let $V\in\mathsf{V}(G)$ be such that $\prod_{j\geq N}V(\mu_j)>0$ holds for $N\geq 1$ and let $\nu\in\mathscr{A}_\infty$ satisfy $\nu=\lim_{k'\to\infty}\nu^{(k')}$, where $\nu^{(k)}:=\lim_{n'\to\infty}\nu_{k,n'}$ for $k\geq 0$ (the limits being taken along filters on $\mathbb{Z}$ and $\mathbb{N}$ resp.).

Then we have $\nu\in\mathscr{I}(G)$. Moreover, for every $\varepsilon>0$ there exists $N(\varepsilon)\geq 1$ satisfying for all k', $n'\geq N(\varepsilon)$ the inequality $V(\nu_{k',n'})>1-\varepsilon$. Hence, $V(\nu)=1$ or $\nu\in\mathscr{D}(G)$, but this implies $\nu=\varepsilon_e$.

2. Conversely, if we have $\varepsilon_e=\lim_{k'\to\infty}\nu^{(k')}$ with $\nu^{(k)}:=\lim_{n'\to\infty}\nu_{k,n'}$ for $k\geq 0$, then for any $\varepsilon>0$ there exists an $N(\varepsilon)\geq 1$ such that for k', $n'\geq N(\varepsilon)$ we obtain $V(\nu_{k',n'})>1-\varepsilon$ and hence

$$\prod_{j\geq k'}V(\mu_j)=\lim_{n'\to\infty}V(\nu_{k',n'})\geq 1-\varepsilon.\quad\square$$

2.4.25 Corollary. *Let $(X_j)_{j\geq 1}$ be an independent sequence of G-random variables X_j with distribution $\mu_j (j\geq 1)$ such that $(X_j)_{j\geq 1}$ V'-converges for some $V'\in \mathsf{V}(G)$. Then for every $V\in \mathsf{V}(G)$ there exists an $N\geq 1$ satisfying $\prod_{j\geq N} V(\mu_j)>0$.*

Proof. Let $\mathscr{A}_\infty$ be the accumulation set of $(\nu_{k,n})_{0\leq k<n}$ with

$$\nu_{k,n}:=\mu_{k+1}*\cdots*\mu_n \quad \text{for every } 0\leq k<n \text{ and } \nu\in\mathscr{A}_\infty.$$

Then we have $\nu=\lim_{k'\to\infty}\nu^{(k')}$, where $\nu^{(k)}:=\lim_{n'\to\infty}\nu_{k,n'}$ for $k\geq 0$ (the limits being taken as in the proof of 2.4.24). Let $V\in\mathsf{V}(G)$ be fixed (but arbitrary). Since $(X_j)_{j\geq 1}$ V'-converges for some $V'\in\mathsf{V}(G)$ the theorem yields $\nu=\varepsilon_e$ or $V(\nu)=1$ and hence for every $\varepsilon>0$ there exists an $N(\varepsilon)\geq 1$ satisfying

$$V(\nu_{k',n'})\geq 1-\varepsilon \quad \text{for all } n'>k'\geq N(\varepsilon).$$

It follows that

$$\prod_{j\geq k'} V(\mu_j)=\lim_{n'\to\infty} V(\nu_{k',n'})\geq 1-\varepsilon,$$

and thus the assertion is true. □

2.5 Asymptotic Equidistribution

Let G be a compact group with Haar measure $\omega:=\omega_G$ in $\mathscr{M}^1(G)$. A sequence $(x_j)_{j\geq 1}$ in G is called *equidistributed* if

$$\lim_{n\to\infty}\frac{1}{n}\sum_{j=1}^n f(x_j)=\omega(f)=\int_G f\,d\omega \quad \text{for all } f\in\mathscr{C}(G).$$

If for every $n\geq 1$ one introduces the measure $\mu_n:=\frac{1}{n}\sum_{j=1}^n \varepsilon_{x_j}$, then this limit relation reads $\lim_{n\to\infty}\mu_n(f)=\omega(f)$ or $\mathscr{T}_v\text{-}\lim_{n\to\infty}\mu_n=\omega$.

Attempting to generalize the concept of equidistribution to arbitrary locally compact groups G one faces the problem that under these general conditions the sequence $(\mu_n)_{n\geq 1}$ does not converge to a probability measure since in the noncompact case ω is unbounded. Thus a reformulation of the concept is needed which avoids the usage of the Haar measure and is hence applicable to general locally compact groups. In fact, $\mathscr{T}_v\text{-}\lim_{n\to\infty}\mu_n=\omega$ is equivalent to

$$\mathscr{T}_v\text{-}\lim_{n\to\infty}(\varepsilon_x*\mu_n-\mu_n)=0 \quad \text{for all } x\in G.$$

In order to get extensions of the theory, the special sequence $(\mu_n)_{n\geq 1}$ can be replaced by arbitrary sequences of measures in $\mathscr{M}^1(G)$ and the vague topology $\mathscr{T}_v$ can be replaced by any other topology in $\mathscr{M}^1(G)$. For the discussion of this and the next section we choose the norm topology in $\mathscr{M}^1(G)$ for a locally compact group G.

2.5.1 Definition. A sequence $(\mu_n)_{n\geq 1}$ in $\mathcal{M}^1(G)$ is called *asymptotically equidistributed* (or an a.e.-sequence) if for all $x\in G$ one has

$$\lim_{n\to\infty} \|\mu_n - \varepsilon_x * \mu_n\| = 0.$$

2.5.2 Remark. For any μ, $\nu\in\mathcal{M}^1(G)$ the inequality

$$\|\mu - \nu * \mu\| \leq \int_G \|\mu - \varepsilon_x * \mu\| \, \nu(dx)$$

obtains. From this it follows that a sequence $(\mu_n)_{n\geq 1}$ in $\mathcal{M}^1(G)$ is an a.e.-sequence iff for all ν_1, $\nu_2\in\mathcal{M}^1(G)$ the relation

$$\lim_{n\to\infty} \|\nu_1 * \mu_n - \nu_2 * \mu_n\| = 0$$

holds.

2.5.3 Definition. A sequence $(\mu_n)_{n\geq 1}$ in $\mathcal{M}^1(G)$ is said to be *asymptotically ω-continuous* if the sequence of the norms of the ω-continuous parts of μ_n tends to 1 as $n\to\infty$.

2.5.4 Theorem. *Every a.e.-sequence in $\mathcal{M}^1(G)$ is asymptotically ω-continuous.*
More generally, let $(\mu_n)_{n\geq 1}$ be a sequence in $\mathcal{M}^1(G)$ such that

$$\lim_{n\to\infty} \|\mu - \varepsilon_x * \mu_n\| = 0$$

for all x in a set $B\in\mathfrak{B}(G)$ with $\omega(B)>0$. Then $(\mu_n)_{n\geq 1}$ is asymptotically ω-continuous.

Proof. Without loss of generality we suppose that $\omega(B)<\infty$. For any such $B\in\mathfrak{B}(G)$ we abbreviate

$$\omega^B := \frac{1}{\omega(B)} \operatorname{Res}_B \omega.$$

Let $\varepsilon>0$, and for every $m\geq 1$ let

$$B_m := \{x\in B : \|\mu_n - \varepsilon_x * \mu_n\| < \varepsilon \text{ for all } n\geq m\}.$$

Plainly $(B_m)_{m\geq 1}$ is an isotone sequence in $\mathfrak{B}(G)$ with $B=\bigcup_{m\geq 1} B_m$, and hence $\omega(B_m)>0$ for all $m\geq m_0$. For all $n\geq m_0$ we define $\bar{\mu}_n := \omega^{B_n} * \mu_n$ and observe that $\bar{\mu}_n \ll \omega$. But for all $n\geq m_0$ we obtain

$$\begin{aligned}\|\bar{\mu}_n - \mu_n\| &\leq \int_G \|\varepsilon_x * \mu_n - \mu_n\| \, d\omega^{B_n} \\ &\leq \frac{1}{\omega(B_n)} \int_{B_n} \|\mu_n - \varepsilon_x * \mu_n\| \, \omega(dx) \leq \varepsilon,\end{aligned}$$

which implies the result. □

2.5.5 Theorem. *For any sequence $(\mu_n)_{n\geq 1}$ in $\mathcal{M}^1(G)$ the following statements are equivalent:*

(i) $(\mu_n)_{n\geq 1}$ *is an a.e.-sequence;*
(ii) $\lim_{n\to\infty} \sup_{x\in K} \|\varepsilon_x * \mu_n - \mu_n\| = 0$ *for all compact subsets K of G.*

Proof. We need only prove the implication (i) $\Rightarrow$ (ii).

Let K be a compact subset of G, $\varepsilon > 0$ and $\sigma \in \mathcal{M}^1_\omega(G) := L^1_{\mathbb{R}}(G,\omega) \cap \mathcal{M}^1(G)$. Since the mapping $x \to \varepsilon_x * \sigma$ from G into $\mathcal{M}^1_\omega(G)$ is continuous, there exists a neighborhood $U \in \mathfrak{B}(e)$ satisfying $\|\varepsilon_y * \sigma - \sigma\|_1 < \varepsilon$ for all $y \in U$. As K is assumed to be compact, there is a finite subset $\{x_1, \dots, x_N\}$ of G such that $K \subset \bigcup_{i=1}^N x_i U$ holds.

By Hypothesis (i), $(\mu_n)_{n\geq 1}$ is an a.e.-sequence and hence there exists an $n_0 \geq 1$ satisfying $\|\sigma * \mu_n - \mu_n\| < \varepsilon$ and

$$\|\varepsilon_{x_i} * \sigma * \mu_n - \mu_n\| < \varepsilon \qquad \text{for all } n \geq n_0 \text{ and } i = 1, \dots, N.$$

Consequently, for $x \in K$ of the form $x = x_i y$ for some $i \in \{1, \dots, N\}$ and $y \in U$ we obtain

$$\begin{aligned}
&\|\varepsilon_x * \mu_n - \mu_n\| \leq \|\varepsilon_x * \sigma * \mu_n - \varepsilon_x * \mu_n\| \\
&+ \|\varepsilon_x * \sigma * \mu_n - \varepsilon_{x_i} * \sigma * \mu_n\| + \|\varepsilon_{x_i} * \sigma * \mu_n - \mu_n\| \\
&\leq \|\sigma * \mu_n - \mu_n\| + \|\varepsilon_y * \sigma * \mu_n - \sigma * \mu_n\| + \|\varepsilon_{x_i} * \sigma * \mu_n - \mu_n\| \\
&\leq \|\sigma * \mu_n - \mu_n\| + \|\varepsilon_y * \sigma - \sigma\|_1 + \|\varepsilon_{x_i} * \sigma * \mu_n - \mu_n\| < 3\varepsilon,
\end{aligned}$$

whenever $n \geq n_0$. But this implies the assertion. □

For general non-Abelian locally compact groups G asymptotic equidistributed sequences do not necessarily exist. Thus the restriction to Abelian locally compact groups becomes a natural hypothesis. Although part of the subsequent discussion can be carried out for amenable groups, we shall restrict ourselves from now on for simplicity to the Abelian case.

2.5.6 Theorem. *There exists an a.e.-sequence in $\mathcal{M}^1(G)$ iff G is σ-compact.*

Proof. 1. Let G be a σ-compact locally compact Abelian group. By the well-known theorem (18.13) in [218] there exists a sequence $(H_n)_{n\geq 1}$ of open, relatively compact subsets H_n of G satisfying

$$H_1 \subset H_2 \subset \cdots, \quad \bigcup_{n\geq 1} H_n = G \quad \text{and}$$
$$\lim_{n\to\infty} \frac{\omega(xH_n \cap \complement H_n)}{\omega(H_n)} = 0 \quad \text{whenever} \quad x \in G.$$

Clearly, the sequence $(\omega_n)_{n\geq 1}$ of measures $\omega_n := \dfrac{1}{\omega(H_n)} \operatorname{Res}_{H_n} \omega \in \mathcal{M}^1(G)$ is an a.e.-sequence if the last mentioned limit relation is satisfied.

2. Conversely, let $(\mu_n)_{n\geq 1}$ be an a.e.-sequence in $\mathcal{M}^1(G)$. For every $n \geq 1$ there exists an isotone sequence $(K_m^{(n)})_{m\geq 1}$ of open, relatively compact subsets $K_m^{(n)}$ of G satisfying $\lim_{m\to\infty} \mu_n(K_m^{(n)}) = 1$. The subgroup $H := [\bigcup_{n,m\geq 1} K_m^{(n)}]$ is open and hence closed, and one has $\mu_n(H) = 1$ for all $n \geq 1$. For any $x \in G \cap \complement H$ one obtains

$\mu_n \perp \varepsilon_x * \mu_n$ $(n \geq 1)$, which is a contradiction to the asymptotic equidistribution of $(\mu_n)_{n \geq 1}$. Thus $G = H$ and G is σ-compact. □

2.5.7 Theorem. *Let G be compact. Then a sequence $(\mu_n)_{n \geq 1}$ in $\mathcal{M}^1(G)$ is an a.e.-sequence iff* $\lim_{n \to \infty} \|\mu_n - \omega\| = 0$.

Proof. Let $(\mu_n)_{n \geq 1}$ be an *a.e.*-sequence in $\mathcal{M}^1(G)$. Clearly, $\omega * \mu_n = \omega$ for all $n \geq 1$ and hence by the inequality of Remark 2.5.2 the assertion follows.

Conversely, if $\lim_{n \to \infty} \|\mu_n - \omega\| = 0$, then for all $x \in G$ and $n \geq 1$ one has

$$\|\mu_n - \varepsilon_x * \mu_n\| \leq \|\mu_n - \omega\| + \|\omega - \varepsilon_x * \omega\| + \|\varepsilon_x * \omega - \varepsilon_x * \mu_n\| = 2\|\mu_n - \omega\|,$$

which implies the result. □

We now define a mapping $N: \mathcal{M}^1(G) \times G \to \mathbb{R}_+$ by

$$N(\mu, x) := \|\mu - \varepsilon_x * \mu\| \quad \text{for all } \mu \in \mathcal{M}^1(G) \text{ and } x \in G.$$

For any $\mu, \nu \in \mathcal{M}^1(G)$ and $x, y \in G$ one has the following

2.5.8 Properties. (1) $N(\mu, e) = 0$ and $N(\mu, x) = N(\mu, x^{-1})$,
(2) $N(\mu, xy) \leq N(\mu, x) + N(\mu, y)$,
(3) $N(\mu * \nu, x) \leq N(\mu, x)$, in particular, $N(\mu * \varepsilon_y, x) = N(\mu, x)$.
(4) The mapping $x \to N(\mu, x)$ from G into $\mathbb{R}_+$ is lower semicontinuous.

The proofs are straightforward; only (4) requires an argument. For every $\mu \in \mathcal{M}^1(G)$ and $x \in G$ one has

$$N(\mu, x) = \sup \{|\int (f(xy) - f(y))\, \mu(dy)| : f \in \mathcal{K}(G), \|f\| \leq 1\}.$$

Since every $f \in \mathcal{K}(G)$ is uniformly continuous, the mapping $x \to \int f(xy)\, \mu(dy)$ from G into $\mathbb{R}$ is continuous. But then $N(\mu, \cdot)$ is lower semicontinuous as the upper envelope of a family of continuous functions on G.

For any sequence $(\mu_n)_{n \geq 1}$ in $\mathcal{M}^1(G)$ we introduce the set

$$T((\mu_n)_{n \geq 1}) := \{x \in G : \lim_{n \to \infty} N(\sigma * \mu_n, x) = 0 \text{ for all } \sigma \in \mathcal{M}^1_\omega(G)\}.$$

Obviously $T((\mu_n)_{n \geq 1})$ is a subgroup of G. It is called the *shift group* of the sequence $(\mu_n)_{n \geq 1}$.

2.5.9 Theorem. *Let $(\mu_n)_{n \geq 1}$ be a sequence in $\mathcal{M}^1(G)$ and $x_1, x_2 \in G$ with $x_1 \neq x_2$. The following two statements are equivalent:*
(i) $x_1 x_2^{-1} \in T((\mu_n)_{n \geq 1})$;
(ii) *for every open $U \in \mathfrak{B}(e)$ there exist measures $\nu_1, \nu_2 \in \mathcal{M}^1_\omega(G)$ satisfying*
(a) $\nu_1(x_1 U) = \nu_2(x_2 U) = 1$ *and*
(b) $\lim_{n \to \infty} \|\nu_1 * \mu_n - \nu_2 * \mu_n\| = 0$.

Proof. 1. (i) ⇒ (ii). Let $x_1 x_2^{-1} \in T((\mu_n)_{n \geq 1})$. Choosing $\sigma \in \mathcal{M}^1_\omega(G)$ such that $\sigma(U) = 1$ holds, one obtains

$$(\varepsilon_{x_1} * \sigma)(x_1 U) = (\varepsilon_{x_2} * \sigma)(x_2 U) = 1 \quad \text{and}$$

$$\lim_{n \to \infty} \|(\varepsilon_{x_1} * \sigma) * \mu_n - (\varepsilon_{x_2} * \sigma) * \mu_n\|$$
$$= \lim_{n \to \infty} \|\sigma * \mu_n - \varepsilon_{x_1 x_2^{-1}} * \sigma * \mu_n\| = 0.$$

Hence, the conditions (a) and (b) in (ii) are satisfied with the choice

$$\nu_1 := \varepsilon_{x_1} * \sigma \quad \text{and} \quad \nu_2 := \varepsilon_{x_2} * \sigma \qquad (\text{in } \mathscr{M}^1_\omega(G)).$$

2. (ii)⇒(i). Let $x_1, x_2 \in G$ with $x_1 \neq x_2$ be given such that the conditions (a) and (b) of (ii) are satisfied. For every $\sigma \in \mathscr{M}^1_\omega(G)$ the function $x \to \|\sigma - \varepsilon_x * \sigma\|$ on G is continuous. Hence, there exists for every $m \geq 1$ an open $U_m \in \mathfrak{B}(e)$ satisfying

$$\|\sigma - \varepsilon_x * \sigma\| < \frac{1}{m} \quad \text{for all } x \in U_m.$$

By assumption there are

$$\nu_1^{(m)}, \nu_2^{(m)} \in \mathscr{M}^1(G) \quad \text{with} \quad \nu_1^{(m)}(x_1 U_m) = \nu_2^{(m)}(x_2 U_m) = 1$$

and

$$\lim_{n\to\infty} \|\nu_1^{(m)} * \mu_n - \nu_2^{(m)} * \mu_n\| = 0 \qquad \text{for all } m \geq 1.$$

But then

$$\|\nu_1^{(m)} * \sigma - \varepsilon_{x_1} * \sigma\| = \|(\varepsilon_{x_1^{-1}} * \nu_1^{(m)}) * \sigma - \sigma\|$$

$$\leq \int_{U_m} \|\sigma - \varepsilon_y * \sigma\| (\varepsilon_{x_1^{-1}} * \nu_1^{(m)})(dy) \leq \frac{1}{m},$$

and analogously $\|\nu_2^{(m)} * \sigma - \varepsilon_{x_2} * \sigma\| \leq \frac{1}{m}$. Thus

$$\|\sigma * \mu_n - \varepsilon_{x_1 x_2^{-1}} * \sigma * \mu_n\| = \|\varepsilon_{x_1} * \sigma * \mu_n - \varepsilon_{x_2} * \sigma * \mu_n\|$$

$$\leq \|\varepsilon_{x_1} * \sigma - \nu_1^{(m)} * \sigma\| + \|\nu_1^{(m)} * \mu_n - \nu_2^{(m)} * \mu_n\| + \|\varepsilon_{x_2} * \sigma - \nu_2^{(m)} * \sigma\|$$

$$\leq \|\nu_1^{(m)} * \mu_n - \nu_2^{(m)} * \mu_n\| + \frac{2}{m}, \quad \text{or}$$

$$\overline{\lim}_{n \geq 1} \|\sigma * \mu_n - \varepsilon_{x_1 x_2^{-1}} * \sigma * \mu_n\| \leq \frac{2}{m} \qquad \text{for all } m \geq 1.$$

This implies that

$$\lim_{n\to\infty} \|\sigma * \mu_n - \varepsilon_{x_1 x_2^{-1}} * \sigma * \mu_n\| = 0, \quad \text{or} \quad x_1 x_2^{-1} \in T((\mu_n)_{n\geq 1}). \quad \square$$

2.5.10 Corollary. *The shift group $T((\mu_n)_{n\geq 1})$ of a sequence $(\mu_n)_{n\geq 1}$ in $\mathscr{M}^1(G)$ is closed in G.*

Proof. Let $a \neq e$ be an accumulation point of $T((\mu_n)_{n\geq 1})$ and let U be an open neighbourhood in $\mathfrak{B}(e)$. For every $x \in aU \cap T((\mu_n)_{n\geq 1})$ there is an open $V \in \mathfrak{B}(e)$ with $V \subset U$ and $xV \subset aU$. By the theorem there are $\nu_1, \nu_2 \in \mathscr{M}^1(G)$ satisfying

$$\nu_1(V) = \nu_2(xV) = 1 \quad \text{and} \quad \lim_{n\to\infty} \|\nu_1 * \mu_n - \nu_2 * \mu_n\| = 0.$$

But $\nu_1(U) = \nu_2(aU) = 1$ implies via the theorem that $a \in T((\mu_n)_{n\geq 1})$. $\square$

2.5.11 Definition. A sequence $(\mu_n)_{n\geq 1}$ in $\mathscr{M}^1(G)$ will be called *weakly asymptotically equidistributed* (or a w.a.e.-sequence) if $T((\mu_n)_{n\geq 1})=G$ or equivalently if

$$\lim_{n\to\infty} \|\varepsilon_x * \sigma * \mu_n - \sigma * \mu_n\| = 0 \qquad \text{for all } \sigma\in\mathscr{M}^1_\omega(G) \text{ and } x\in G.$$

2.5.12 Remark. A sequence $(\mu_n)_{n\geq 1}$ in $\mathscr{M}^1(G)$ is a w.a.e.-sequence iff for all $\sigma_1, \sigma_2 \in \mathscr{M}^1_\omega(G)$ the relation $\lim_{n\to\infty}\|\sigma_1 * \mu_n - \sigma_2 * \mu_n\| = 0$ holds.

By Theorem 2.5.9 it suffices to show that the weak asymptotic equidistribution of $(\mu_n)_{n\geq 1}$ implies the above limit relation: For $\sigma_1, \sigma_2\in\mathscr{M}^1_\omega(G)$ and $x\in G$ one has

$$\lim_{n\to\infty}\|\sigma_1 * \mu_n - \varepsilon_x * \sigma_1 * \mu_n\| = 0$$

which by Lebesgue's dominated convergence theorem yields

$$\lim_{n\to\infty}\int \|\sigma_1 * \mu_n - \varepsilon_x * \sigma_1 * \mu_n\|\, \sigma_2(dx) = 0$$

or

$$\lim_{n\to\infty}\|\sigma_1 * \mu_n - \sigma_2 * \sigma_1 * \mu_n\| = 0.$$

Analogously one has

$$\lim_{n\to\infty}\|\sigma_2 * \mu_n - \sigma_1 * \sigma_2 * \mu_n\| = 0.$$

The inequality

$$\|\sigma_1 * \mu_n - \sigma_2 * \mu_n\| \leq \|\sigma_1 * \mu_n - \sigma_1 * \sigma_2 * \mu_n\| + \|\sigma_2 * \mu_n - \sigma_1 * \sigma_2 * \mu_n\|,$$

valid for all $n\geq 1$, completes the proof. □

Concerning the notion of asymptotic equidistribution, we shall discuss the special case of a compact group also for weakly asymptotically distributed sequences. Some preparations are needed.

In $\mathscr{M}^b(G)$ for a locally compact group G one introduces the topologies $\mathscr{T}_t$ and $\mathscr{T}_a$ defined by the seminorms p_f (for $f\in\mathscr{K}(G)$) and q_σ (for $\sigma\in\mathscr{M}^1_\omega(G)$) given by

$$p_f(\mu) := \sup_{x\in G} \left|\int f(y^{-1}x)\,\mu(dy)\right| \quad \text{and}$$

$$q_\sigma(\mu) := \|\mu * \sigma\| \qquad \text{for all } \mu\in\mathscr{M}^b(G) \text{ resp.}$$

2.5.13 Lemma. *Let G be an arbitrary locally compact group. The restrictions to $\mathscr{M}^1(G)$ of the topologies $\mathscr{T}_v$, $\mathscr{T}_t$ and $\mathscr{T}_a$ coincide.*

Proof. For any $\nu\in\mathscr{M}^1(G)$ and $\sigma\in\mathscr{M}^1_\omega(G)$ of the form $\sigma := u\cdot\omega$ with $u\in L^1_{\mathbb{R}}(G,\omega)$ we have

$$q_\sigma(\nu) = \|\nu * \sigma\| = \|\nu * u\|_1.$$

We further note that given $f\in\mathscr{K}(G)$ and $\varepsilon>0$ there exists a function $u\in L^1_{\mathbb{R}}(G,\omega)$ satisfying $|(\nu * u)(f) - \nu(f)| < \varepsilon$ for all $\nu\in\mathscr{M}^1(G)$.

1. $\mathrm{Res}_{\mathscr{M}^1(G)}\mathscr{T}_a > \mathrm{Res}_{\mathscr{M}^1(G)}\mathscr{T}_v$. In fact, let $(\mu_\alpha)_{\alpha\in\mathbf{A}}$ be a net in $\mathscr{M}^1(G)$ with $\mathscr{T}_a$-$\lim_{\alpha\in\mathbf{A}}\mu_\alpha = \mu\in\mathscr{M}^1(G)$.

For any $f\in\mathscr{K}(G)$ and $\varepsilon>0$ one obtains with the function $u\in L^1_{\mathbb{R}}(G,\omega)$ introduced above

$$|\mu_\alpha(f)-\mu(f)|\leq|\mu_\alpha(f)-(\mu_\alpha*u)(f)|+\|\mu_\alpha*u-\mu*u\|_1\,\|f\|$$
$$+|(\mu*u)(f)-\mu(f)|<\varepsilon(2+\|f\|),$$

whenever $\alpha\geq\alpha_0$ for some $\alpha_0\in\mathbb{A}$, which shows that $\mathscr{T}_v$-$\lim_{\alpha\in\mathbb{A}}\mu_\alpha=\mu$.

2. $\mathrm{Res}_{\mathscr{M}^1(G)}\mathscr{T}_v>\mathrm{Res}_{\mathscr{M}^1(G)}\mathscr{T}_t$. We choose $f\in\mathscr{K}(G)$ with $\mathrm{supp}(f)$ contained in a compact subset L of G. Let $(\mu_\alpha)_{\alpha\in\mathbb{A}}$ be a net in $\mathscr{M}^1(G)$ which by assumption satisfies

$$\mathscr{T}_v\text{-}\lim_{\alpha\in\mathbb{A}}\mu_\alpha=\mu\in\mathscr{M}^1(G).$$

Then by Property (2) of 1.2.20 there exists a compact subset K of G such that $\mu_\alpha(K)\geq1-\varepsilon$ for large $\alpha\in\mathbb{A}$ and $\mu(K)\geq1-\varepsilon$. Since the mapping $x\to{}_xf$ from G into $\mathscr{K}(G)$ is continuous, the set $\{_xf\colon x\in LK^{-1}\}$ is totally bounded in $\mathscr{K}(G)$, and hence

$$\lim_{\alpha\in\mathbb{A}}\sup_{x\in LK^{-1}}|\mu_\alpha(_xf)-\mu(_xf)|=0.$$

But for $x\notin LK^{-1}$ we have

$$|\mu_\alpha(_xf)|\leq\varepsilon\|f\|\qquad\text{for large }\alpha\in\mathbb{A}\text{ and }|\mu(_xf)|\leq\varepsilon\|f\|$$

since $\mathrm{supp}(_xf)\subset\complement K$.

This implies that

$$\sup_{x\in G}|\mu_\alpha(_xf)-\mu(_xf)|\leq2\varepsilon\|f\|,$$

whenever $\alpha\geq\alpha_0$ for some $\alpha_0\in\mathbb{A}$, i.e., $\mathscr{T}_t$-$\lim_{\alpha\in\mathbb{A}}\mu_\alpha=\mu$.

3. $\mathrm{Res}_{\mathscr{M}^1(G)}\mathscr{T}_t>\mathrm{Res}_{\mathscr{M}^1(G)}\mathscr{T}_a$. Let $(\mu_\alpha)_{\alpha\in\mathbb{A}}$ be a net in $\mathscr{M}^1(G)$ satisfying $\mathscr{T}_t$-$\lim_{\alpha\in\mathbb{A}}\mu_\alpha=\mu\in\mathscr{M}^1(G)$. We first choose a function $\phi\in\mathscr{K}_+(G)$ with $\phi\leq1$ such that $\mu_\alpha(\phi)\geq1-\varepsilon$ holds for all $\alpha\geq\alpha_0$ for some $\alpha_0\in\mathbb{A}$. For $\alpha\in\mathbb{A}$ we introduce the measures $\mu'_\alpha:=\phi\cdot\mu_\alpha$. Then by assumption $\mathscr{T}_v$-$\lim_{\alpha\in\mathbb{A}}\mu'_\alpha=\mu'$ and moreover,

$$\lim_{\alpha\in\mathbb{A}}\sup_{x\in G}|\mu'_\alpha(_xf)-\mu'(_xf)|=0\qquad\text{for all }f\in\mathscr{K}(G).$$

For $f\in\mathscr{K}(G)$ we therefore get

$$\lim_{\alpha\in\mathbb{A}}\|\mu'_\alpha*f-\mu'*f\|_1$$
$$=\lim_{\alpha\in\mathbb{A}}\int\left|\int f(y^{-1}x)\,\phi(y)\,\mu_\alpha(dy)-\int f(y^{-1}x)\,\phi(y)\,\mu(dy)\right|\omega(dx)$$
$$=\lim_{\alpha\in\mathbb{A}}\int|\mu'_\alpha(_{x^{-1}}f^*)-\mu'(_{x^{-1}}f^*)|\,\omega(dx)=0$$

since the integrand converges to 0 uniformly in x.

But for all $\alpha\geq\alpha_0$ we have

$$\|\mu'_\alpha*f-\mu_\alpha*f\|_1\leq\iint|f(y^{-1}x)|\,|\phi(y)-1|\,\mu_\alpha(dy)\,\omega(dx)\leq\varepsilon\|f\|_1.$$

Hence, $\lim_{\alpha\in\mathbb{A}}\|\mu_\alpha*f-\mu*f\|_1=0$ for all $f\in\mathscr{K}(G)$. This implies that

$$\lim_{\alpha\in\mathbb{A}}\|\mu_\alpha*f-\mu*f\|_1=0\qquad\text{for all }f\in L^1_{\mathbb{R}}(G,\omega)$$

since $\overline{\mathscr{K}(G)}^{\|\cdot\|_1}=L^1_{\mathbb{R}}(G,\omega)$. Thus $\mathscr{T}_a$-$\lim_{\alpha\in\mathbb{A}}\mu_\alpha=\mu$. □

2.5.14 Theorem. *Let G be compact. Then a sequence $(\mu_n)_{n\geq 1}$ in $\mathscr{M}^1(G)$ is a w.a.e.-sequence iff $\mathscr{T}_v$-$\lim_{n\to\infty}\mu_n=\omega$.*

Proof. One first notes that $\sigma*\omega=\omega$ for all $\sigma\in\mathscr{M}^1(G)$. Thus Lemma 2.5.13 implies that $(\mu_n)_{n\geq 1}$ $\mathscr{T}_v$-converges to ω iff

$$\lim_{n\to\infty}\|\sigma*\mu_n-\omega\|=0 \qquad \text{for all } \sigma\in\mathscr{M}^1_\omega(G).$$

But by Theorem 2.5.7 this is equivalent to the asymptotic equidistribution of the sequence $(\sigma*\mu_n)_{n\geq 1}$ for all $\sigma\in\mathscr{M}^1_\omega(G)$, which proves the assertion. ▯

2.5.15 Remark. The proof of the preceding theorem implies that a sequence $(\mu_n)_{n\geq 1}$ in $\mathscr{M}^1(G)$ is a w.a.e.-sequence iff $(\sigma*\mu_n)_{n\geq 1}$ is an a.e.-sequence for all $\sigma\in\mathscr{M}^1_\omega(G)$.

In the noncompact case the asymptotic behavior of w.a.e.-sequences is of opposite character.

2.5.16 Theorem. *Let G be a noncompact locally compact Abelian group and $(\mu_n)_{n\geq 1}$ a w.a.e.-sequence in $\mathscr{M}^1(G)$. Then for every compact subset K of G one has*

$$\lim_{n\to\infty}\sup_{x\in G}\mu_n(Kx)=0.$$

Proof. 1. One observes that for compact subsets K_1, K_2 of G there exists a $y\in G$ such that $K_1\cap K_2 y=\emptyset$. Hence, by induction, for every compact subset K of G and for every $n\geq 1$ there exist $y_1,\ldots,y_n\in G$ such that the sets $K, Ky_1,\ldots,Ky_n$ are pairwise disjoint.

2. Without loss of generality we suppose that $(\mu_n)_{n\geq 1}$ is an a.e.-sequence in $\mathscr{M}^1(G)$. Indeed, if K is a compact subset of G and $(\mu_n)_{n\geq 1}$ a w.a.e.-sequence in $\mathscr{M}^1(G)$, then one can choose a compact $U\in\mathfrak{V}(e)$ and a measure $\sigma\in\mathscr{M}^1_\omega(G)$ such that $\sigma(U)=1$. Plainly then, KU is compact, $(\sigma*\mu_n)_{n\geq 1}$ is an a.e.-sequence in $\mathscr{M}^1(G)$ and for all $x\in G$ one has $\mu_n(Kx)\leq(\sigma*\mu_n)(KUx)$.

3. Let K be a compact subset of G and $(\mu_n)_{n\geq 1}$ an a.e.-sequence in $\mathscr{M}^1(G)$. By Part 1 there exists for every $k\geq 1$ a sequence $\{y_1,\ldots,y_k\}$ in G such that the sets $Kx, Kxy_1,\ldots,Kxy_k$ are pairwise disjoint for every $x\in G$. For $i=1,\ldots,k$ and $x\in G$ one has $\mu_n(Kx)\leq\mu_n(Kxy_i)+\|\mu_n-\varepsilon_{y_i}*\mu_n\|$ and hence

$$(k+1)\,\mu_n(Kx)\leq\mu_n((Kx)\cup\cdots\cup(Kxy_k))+\textstyle\sum_{i=1}^k\|\mu_n-\varepsilon_{y_i}*\mu_n\|.$$

It follows that

$$\sup_{x\in G}\mu_n(Kx)\leq\frac{1}{k+1}\left(1+\textstyle\sum_{i=1}^k\|\mu_n-\varepsilon_{y_i}*\mu_n\|\right)$$

and therefore

$$\overline{\lim}_{n\geq 1}\left(\sup_{x\in G}\mu_n(Kx)\right)\leq\frac{1}{k+1} \qquad \text{for all } k\geq 1.$$

This, however, implies the assertion. ▯

2.5.17 Lemma. *Let G_1, G_2 be locally compact Abelian groups and ϕ a homomorphism from G_1 into G_2. Then for any sequence $(\mu_n)_{n\geq 1}$ in $\mathscr{M}^1(G_1)$ one has*

$$\phi(T((\mu_n)_{n\geq 1}))\subset T((\phi(\mu_n))_{n\geq 1}).$$

In particular, for $G_1:=G$, $G_2:=\dot{G}:=G/H$ with a closed subgroup H of G, the canonical projection $p: G\to\dot{G}$ and a w.a.e.-sequence $(\mu_n)_{n\geq 1}$ in $\mathscr{M}^1(G)$, the sequence $(p(\mu_n))_{n\geq 1}$ is a w.a.e.-sequence in $\mathscr{M}^1(\dot{G})$.

Proof. Let $x\in T((\mu_n)_{n\geq 1})$. Then

$$\|\phi(\mu_n)-\varepsilon_{\phi(x)}*\phi(\mu_n)\|=\|\phi(\mu_n-\varepsilon_x*\mu_n)\|\leq\|\mu_n-\varepsilon_x*\mu_n\|$$

holds for all $n\geq 1$ so that

$$\overline{\lim}_{n\geq 1}\|\phi(\mu_n)-\varepsilon_{\phi(x)}*\phi(\mu_n)\|\leq\lim_{n\to\infty}\|\mu_n-\varepsilon_x*\mu_n\|=0,$$

which implies that $\phi(x)\in T((\phi(\mu_n))_{n\geq 1})$. □

2.5.18 Remark. The converse of the second statement of Lemma 2.5.17 is in general false. We consider the special case $G:=\mathbb{R}$ and show that not every sequence $(\mu_n)_{n\geq 1}$ in $\mathscr{M}^1(G)$ which is modulo $h\in\mathbb{R}_+^*$ weakly asymptotically equidistributed is itself weakly asymptotically equidistributed. Indeed, let

$$p(x):=a_k x^k+\cdots+a_0$$

be a real polynomial for which there exist indices $0<k_1<k_2\leq k$ such that $a_{k_1}\in\mathbb{R}_+^*$ and $\frac{a_{k_2}}{a_{k_1}}$ is irrational. Then by Theorem VI in [70] the sequence $(\mu_n)_{n\geq 1}$ in $\mathscr{M}^1(G)$ with

$$\mu_n:=\frac{1}{n}\sum_{j=1}^{n}\varepsilon_{p(j)}\qquad\text{for all } n\geq 1$$

is weakly asymptotically equidistributed modulo h for all $h\in\mathbb{R}_+^*$. But since $\lim_{j\to\infty}|p(j+1)-p(j)|=\infty$, $(\mu_n)_{n\geq 1}$ is not weakly asymptotically equidistributed.

In order to discuss a few examples for the special case $G:=\mathbb{R}$ we present two auxiliary facts. For every $h\in\mathbb{R}_+^*$ one defines a norm M^h in $\mathscr{M}^b(\mathbb{R})$ by

$$M^h(\mu):=\sup_{x\in[0,h[}\sum_{k=-\infty}^{\infty}|\mu([x+kh, x+(k+1)h[)|$$

for all $\mu\in\mathscr{M}^b(\mathbb{R})$.

Clearly, $M^h(\mu)\leq\|\mu\|$ for all $\mu\in\mathscr{M}^b(\mathbb{R})$, $h\in\mathbb{R}_+^*$.

2.5.19 Lemma. *Let I be a bounded interval in $\mathbb{R}$ with $\lambda^1(I)\geq d$ and*

$$\lambda_I:=\frac{1}{\lambda^1(I)}\operatorname{Res}_I\lambda^1.$$

Then for all $\nu_1, \nu_2 \in \mathscr{M}^1(\mathbb{R})$ *and* $m \geq 1$ *one has*

$$\|\nu_1 * \lambda_I - \nu_2 * \lambda_I\| \leq M^{\frac{d}{m}}(\nu_1 - \nu_2) + \frac{2}{m}.$$

Proof. For $x, y \in \left[0, \frac{d}{m}\right]$ we have $\|\varepsilon_x * \lambda_I - \varepsilon_y * \lambda_I\| \leq \frac{2}{m}$. Hence, for all $\gamma_1, \gamma_2 \in \mathscr{M}^1(\mathbb{R})$ with $\gamma_1\left(\left[0, \frac{d}{m}\right]\right) = \gamma_2\left(\left[0, \frac{d}{m}\right]\right) = 1$ one has the estimate

$$\|\gamma_1 * \lambda_I - \gamma_2 * \lambda_I\| \leq \frac{2}{m}.$$

Let $\rho_1, \rho_2 \in \mathscr{M}_+^b(\mathbb{R})$ be such that

$$\rho_1\left(\left[0, \frac{d}{m}\right]\right) = \rho_1(\mathbb{R}) \quad \text{and} \quad \rho_2\left(\left[0, \frac{d}{m}\right]\right) = \rho_2(\mathbb{R})$$

are fulfilled. Without loss of generality we suppose $\rho_1(\mathbb{R}) > 0$.

Then

$$\|\rho_1 * \lambda_I - \rho_2 * \lambda_I\| \leq \left\|\rho_1 * \lambda_I - \frac{\rho_2(\mathbb{R})}{\rho_1(\mathbb{R})} \rho_1 * \lambda_I\right\|$$

$$+ \left\|\frac{\rho_2(\mathbb{R})}{\rho_1(\mathbb{R})} \rho_1 * \lambda_I - \rho_2 * \lambda_I\right\| \leq |\rho_1(\mathbb{R}) - \rho_2(\mathbb{R})| + \frac{2}{m} \rho_2(\mathbb{R}).$$

Putting $I_k := \left[k\frac{d}{m}, (k+1)\frac{d}{m}\right[$ for all $k \geq 1$ we obtain

$$\|\nu_1 * \lambda_I - \nu_2 * \lambda_I\| \leq \sum_{k=-\infty}^{\infty} \|(\mathrm{Res}_{I_k} \nu_1) * \lambda_I - (\mathrm{Res}_{I_k} \nu_2) * \lambda_I\|$$

$$\leq \sum_{k=-\infty}^{\infty} \left(|\nu_1(I_k) - \nu_2(I_k)| + \frac{2}{m} \nu_2(I_k)\right) \leq M^{\frac{d}{m}}(\nu_1 - \nu_2) + \frac{2}{m}. \quad \square$$

2.5.20 Lemma. *A sequence* $(\mu_n)_{n \geq 1}$ *in* $\mathscr{M}^1(\mathbb{R})$ *is a w.a.e.-sequence iff*

$$\lim_{n \to \infty} M^h(\mu_n - \varepsilon_h * \mu_n) = 0 \qquad \textit{for all } h \in \mathbb{R}_+^*.$$

Proof. 1. Let $\lim_{n \to \infty} M^h(\mu_n - \varepsilon_h * \mu_n) = 0$ be satisfied for all $h \in \mathbb{R}_+^*$. By Lemma 2.5.19 we have for any bounded interval I in $\mathbb{R}$ with $\lambda^1(I) \geq h$ and all $k, m \geq 1$ the inequalities

$$\|\mu_n * \lambda_I - \mu_n * \varepsilon_{kh} * \lambda_I\| \leq M^{\frac{h}{m}}(\mu_n - \varepsilon_{kh} * \mu_n) + \frac{2}{m}$$

$$\leq kmM^{\frac{h}{m}}(\mu_n - \varepsilon_{\frac{h}{m}} * \mu_n) + \frac{2}{m},$$

whence

$$\overline{\lim}_{n \geq 1} \|\mu_n * \lambda_I - \mu_n * \varepsilon_{kh} * \lambda_I\| \leq \frac{2}{m} \quad \text{or}$$

$$\lim_{n \to \infty} \|\lambda_I * \mu_n - \lambda_{I+kh} * \mu_n\| = 0.$$

Theorem 2.5.9 implies that $\mathbb{R}_+^* \subset T((\mu_n)_{n \geq 1})$, i.e. $(\mu_n)_{n \geq 1}$ is a w.a.e.-sequence.

2. Let $(\mu_n)_{n\geq 1}$ be a w.a.e.-sequence in $\mathscr{M}^1(\mathbb{R})$ and let d be such that $0<2d<h$. By Lemma 2.5.17 $(\mu_n)_{n\geq 1}$ is weakly asymptotically equidistributed modulo h, and we have

$$\lim_{n\to\infty}\mu_n\left(\bigcup_{k=-\infty}^{\infty}[x+kh-d, x+kh+d]\right)=\frac{2d}{h} \quad \text{uniformly in } x\in\mathbb{R}.$$

Let $\sigma\in\mathscr{M}^1_\omega(G)$ be given with $\sigma([0,d])=1$. For every $k\geq 1$ we put

$$I_k:=[x+kh, x+(k+1)h[$$

and observe that

$$\begin{aligned}&|\mu_n(I_k)-(\sigma*\mu_n)(I_k)|\\ &\leq\mu_n([x+kh-d, x+kh+d])+\mu_n([x+(k+1)h-d, x+(k+1)h+d]),\end{aligned}$$

which implies

$$\begin{aligned}&\textstyle\sum_{k=-\infty}^{\infty}|\mu_n(I_k)-\mu_n(I_{k+1})|\\ &\leq\textstyle\sum_{k=-\infty}^{\infty}|(\sigma*\mu_n)(I_k)-(\sigma*\mu_n)(I_{k+1})|\\ &\quad+4\mu_n\left(\bigcup_{k=-\infty}^{\infty}[x+kh-d, x+kh+d]\right)\\ &\leq\|\sigma*\mu_n-\varepsilon_h*\sigma*\mu_n\|+4\mu_n\left(\bigcup_{k=-\infty}^{\infty}[x+kh-d, x+kh+d]\right)\\ &\quad(\text{all } n\geq 1).\end{aligned}$$

This implies that $\overline{\lim}_{n\geq 1} M^h(\mu_n-\varepsilon_h*\nu_n)\leq\frac{8d}{h}$ or $\lim_{n\to\infty} M^h(\mu_n-\varepsilon_h*\mu_n)=0$. □

2.5.21 Examples. 1. The sequence $(\mu_n)_{n\geq 1}$ in $\mathscr{M}^1(\mathbb{R})$ defined by

$$\mu_n:=\frac{1}{2n^2+1}\sum_{k=-n^2}^{n^2}\varepsilon_{\frac{k}{n}} \quad \text{for all } n\geq 1$$

is a w.a.e.-sequence by Lemma 2.5.20, but by Theorem 2.5.4 not an a.e.-sequence, since it is not asymptotically λ^1-continuous.

2. The sequence $(\sigma_n)_{n\geq 1}$ of measures

$$\sigma_n:=\mu_n*\lambda_{[0,\frac{1}{2^n}]}\in\mathscr{M}^1_{\lambda^1}(\mathbb{R})$$

with μ_n as in Example 1 and

$$\lambda_{[0,\frac{1}{2^n}]}:=2^n\operatorname{Res}_{[0,\frac{1}{2^n}]}\lambda^1 \quad \text{for } n\geq 1$$

is clearly a w.a.e.-sequence, but the fact that $\|\sigma_n-\varepsilon_{\frac{1}{2^n}}*\sigma_n\|=2$ for all $n\geq 1$ shows that $(\sigma_n)_{n\geq 1}$ cannot be an a.e.-sequence since the asymptotic equidistribution of $(\sigma_n)_{n\geq 1}$ implies that $\lim_{n\to\infty}\|\sigma_n-\varepsilon_x*\sigma_n\|=0$ holds uniformly on compact subsets of $\mathbb{R}$.

2.6 Shifting Iterated Convolutions

In this section we concentrate on a detailed study of asymptotic equidistribution for convolution powers of a measure which will complement the theory presented in § 1. Let G be a locally compact Abelian group. To every $\mu\in\mathcal{M}^1(G)$ one associates the system $\mathbf{H}_\mu$ of all closed subgroups H of G for which there exists an element x_H of G with the property that $\operatorname{supp}(\mu)\subset x_H H$. Plainly a closed subgroup H of G belongs to $\mathbf{H}_\mu$ iff $\operatorname{supp}(\mu)\operatorname{supp}(\mu)^{-1}\subset H$. This implies that

$$H_\mu := \bigcap\{H : H\in\mathbf{H}_\mu\}\in\mathbf{H}_\mu.$$

The closed subgroup H_μ is called the *lattice group* of μ.

2.6.1 Theorem. *Let* $\nu,\lambda,\rho\in\mathcal{M}^1(G)$ *and let* $p,q,r\in\mathbb{R}_+$ *with* $p,q>0$, $p+q+r=1$. *For the measure*

$$\mu := p\,\nu + q\,\lambda + r\,\rho\in\mathcal{M}^1(G)$$

we have the estimate

$$\|\nu*\mu^n-\lambda*\mu^n\|\le\sqrt{\frac{p+q}{pq}}\,\frac{1}{\sqrt{n+1}}\qquad \textit{for all } n\ge 1.$$

Proof. 1. In the course of the proof we shall use the following two elementary inequalities ([295], pp. 308–312):

(a) If $p\in]0,1]$ and $r\in[1,\infty[$, then for all $n\ge 1$ one has

$$\sum_{k=0}^n \frac{1}{\sqrt[r]{k+1}}\binom{n}{k}p^k(1-p)^{n-k}\le\frac{1}{\sqrt[r]{p}}\,\frac{1}{\sqrt[r]{n+1}}.$$

(b) $\sup\left\{\binom{n}{k}p^k(1-p)^{n-k}\sqrt{p(1-p)}\,\sqrt{n+1} : p\in]0,1[,\ 0\le k\le n\right\}\le\frac{1}{2}.$

2. Putting $\alpha := \frac{p}{p+q}$, $\beta := \frac{q}{p+q}$ and $\pi := \alpha\,\nu+\beta\,\lambda$ we obtain

$$\begin{aligned}&\|\nu*\pi^n-\lambda*\pi^n\|\\ &=\left\|\sum_{k=0}^n\binom{n}{k}\alpha^k\beta^{n-k}\nu^{k+1}*\lambda^{n-k}-\sum_{k=0}^n\binom{n}{k}\alpha^k\beta^{n-k}\nu^k*\lambda^{n-k+1}\right\|\\ &\le\sum_{k=0}^{n+1}\left|\binom{n}{k-1}\alpha^{k-1}\beta^{n-k+1}-\binom{n}{k}\alpha^k\beta^{n-k}\right|\end{aligned}$$

(with the conventions that

$$\nu^0=\lambda^0 := \varepsilon_e \quad\text{and}\quad \binom{n}{-1}=\binom{n}{n+1} := 0).$$

This implies for all $n \geq 1$ that

$$\|\nu * \pi^n - \lambda * \pi^n\| \leq 2 \max_{k=0,1,\ldots,n} \binom{n}{k} \alpha^k \beta^{n-k}$$

$$\leq \frac{1}{\sqrt{\alpha\beta}} \frac{1}{\sqrt{n+1}} = \frac{p+q}{\sqrt{pq}} \frac{1}{\sqrt{n+1}},$$

where the second inequality follows from Part 1(b).

But introducing $s := p+q$, which implies $\mu = s\pi + r\rho$, one gets

$$\|\nu * \mu^n - \lambda * \mu^n\| \leq \sum_{k=0}^{n} \binom{n}{k} s^k r^{n-k} \|\nu * \pi^k - \lambda * \pi^k\|$$

$$\leq \sum_{k=0}^{n} \binom{k}{n} s^k r^{n-k} \frac{s}{\sqrt{pq}} \frac{1}{\sqrt{k+1}}$$

or by Part 1(a) the final inequality

$$\|\nu * \mu^n - \lambda * \mu^n\| \leq \frac{s}{\sqrt{pq}} \frac{1}{\sqrt{s}} \frac{1}{\sqrt{n+1}} = \sqrt{\frac{p+q}{pq}} \frac{1}{\sqrt{n+1}}. \quad \square$$

2.6.2 Theorem. *For any* $\mu \in \mathcal{M}^1(G)$ *we have* $H_\mu = T((\mu^n)_{n \geq 1})$.

Proof. 1. Let $\mu \in \mathcal{M}^1(G)$ and let $x_1, x_2 \in \operatorname{supp}(\mu)$ with $x_1 \neq x_2$. For every open neighborhood $U \in \mathfrak{B}(e)$ we have $\mu(x_1 U)$, $\mu(x_2 U) \in \mathbb{R}_+^*$. Hence, there exist $\rho \in \mathcal{M}^1(G)$ and $p, q, r \in \mathbb{R}_+$ with $p, q > 0$, $p+q+r=1$ such that $\mu = p\,\mu_{x_1 U} + q\,\mu_{x_2 U} + r\rho$ is satisfied if $x_1 U \cap x_2 U = \emptyset$ holds.

Here as above

$$\mu_{x_i U} := \frac{1}{\mu(x_i U)} \operatorname{Res}_{x_i U} \mu \qquad (i = 1, 2).$$

Now Theorem 2.6.1 implies (in general)

$$\lim_{n \to \infty} \|\mu_{x_1 U} * \mu^n - \mu_{x_2 U} * \mu^n\| = 0$$

and hence one obtains with the aid of Theorem 2.5.9 that $x_1 x_2^{-1} \in T((\mu^n)_{n \geq 1})$. Consequently, $\operatorname{supp}(\mu)\operatorname{supp}(\mu)^{-1} \subset T((\mu^n)_{n \geq 1})$ or, since $T((\mu^n)_{n \geq 1})$ is closed by Corollary 2.5.10, the asserted inclusion $H_\mu \subset T((\mu^n)_{n \geq 1})$.

2. Let $x \notin H_\mu$. Then there exists a $U \in \mathfrak{B}(e)$ satisfying $xU \cap H_\mu = \emptyset$. Choosing an open neighborhood $V \in \mathfrak{B}(e)$ with $VV^{-1} \subset U$ one obtains $(xVH_\mu) \cap (H_\mu V) = \emptyset$. We now pick a $\nu \in \mathcal{M}_\omega^1(G)$ with $\nu(V) = 1$ and get for every $n \geq 1$ the equalities

$$(\nu * \mu^n)(x_{H_\mu}^n H_\mu V) = 1, \qquad (\varepsilon_x * \nu * \mu^n)(x\, x_{H_\mu}^n H_\mu V) = 1$$

and

$$(x_{H_\mu}^n H_\mu V) \cap (x\, x_{H_\mu}^n H_\mu V) = \emptyset.$$

From this follows

$$\|\nu * \mu^n - \varepsilon_x * \nu * \mu^n\| = 2 \qquad \text{for all } n \geq 1,$$

and thus $x \notin T((\mu^n)_{n \geq 1})$. □

2.6.3 Definition. A measure $\mu \in \mathscr{M}^1(G)$ is called *non-lattice* if $H_\mu = G$.

2.6.4 Corollary. *The sequence $(\mu^n)_{n \geq 1}$ is a w.a.e.-sequence iff μ is non-lattice.*

For measures μ, $\mu' \in \mathscr{M}^1(G)$ one writes $\mu' \prec \mu$ if there exist an $\alpha \in]0, 1]$ and a $\mu'' \in \mathscr{M}^1(G)$ such that $\mu = \alpha \mu' + (1-\alpha) \mu''$.

2.6.5 Remark. $\mu' \prec \mu$ holds iff $\mu' \ll \mu$ and the μ-density of μ' is μ-a.s. bounded.

In fact, let $\mu' \prec \mu$ and let $\mu = \alpha \mu' + (1-\alpha) \mu''$ be given with $\alpha \in]0, 1[$ and $\mu'' \in \mathscr{M}^1(G)$. From $\mu' \ll \mu$ and $\mu'' \ll \mu$ follows the existence of Borel measurable functions f' and $f'' \geq 0$ with $\mu' = f' \cdot \mu$ and $\mu'' = f'' \cdot \mu$ resp. But the above equality yields $\alpha f' + (1-\alpha) f'' = 1$ μ-a.s. and hence, $f' \leq \frac{1}{\alpha}$ μ-a.s. If, conversely, $\mu' = f' \cdot \mu$ with $0 \leq f' \leq \beta$ μ-a.s. (with $\beta \in \mathbb{R}_+^*$), then choosing $\alpha \in \mathbb{R}_+^*$ with $\alpha \leq \frac{1}{\beta}$ and putting $f'' := \frac{1 - \alpha f'}{1 - \alpha}$ one obtains $f'' \geq 0$ μ-a.e. and $\int f'' \, d\mu = 1$. Clearly, $\mu'' := f'' \cdot \mu \in \mathscr{M}^1(G)$ and μ'' satisfies $\mu = \alpha \mu' + (1-\alpha) \mu''$.

2.6.6 Theorem. *Let G be a locally compact Abelian group, $\mu \in \mathscr{M}^1(G)$ and $x \in G$. The following statements are equivalent:*
(i) $N(\mu, x) < 2$;
(ii) *there exists a measure $\mu_0 \in \mathscr{M}^1(G)$ such that $\frac{1}{2}(\varepsilon_e + \varepsilon_x) * \mu_0 \prec \mu$.*

Proof. 1. (ii) ⇒ (i). Let $\mu_0 \in \mathscr{M}^1(G)$ exist with $\frac{1}{2}(\varepsilon_e + \varepsilon_x) * \mu_0 \prec \mu$. Then there exist an $\alpha \in]0, 1[$ and a $\mu_1 \in \mathscr{M}^1(G)$ satisfying

$$\mu = \frac{\alpha}{2}(\varepsilon_e + \varepsilon_x) * \mu_0 + (1-\alpha) \mu_1,$$

and it follows that

$$\begin{aligned} N(\mu, x) &= \|\mu * (\varepsilon_e - \varepsilon_x)\| \\ &= \left\| \frac{\alpha}{2}(\varepsilon_e + \varepsilon_x) * (\varepsilon_e - \varepsilon_x) * \mu_0 + (1-\alpha) \mu_1 * (\varepsilon_e - \varepsilon_x) \right\| \\ &= \left\| \frac{\alpha}{2}(\varepsilon_e - \varepsilon_{x^2}) * \mu_0 + (1-\alpha) \mu_1 * (\varepsilon_e - \varepsilon_x) \right\| \leq \alpha + 2(1-\alpha) = 2 - \alpha < 2. \end{aligned}$$

2. (i) ⇒ (ii). Conversely, let $N(\mu, x) < 2$ and let G_+, G_- be the components of a Hahn decomposition of G with respect to the measure $\lambda := \mu - \varepsilon_x * \mu$. Putting

$$\nu := \varepsilon_x * \mu, \qquad \nu_1 := \mathrm{Res}_{G_+} \mu - \mathrm{Res}_{G_+} \nu, \qquad \nu_2 := \mathrm{Res}_{G_-} \nu - \mathrm{Res}_{G_-} \mu$$

and

$$\nu_0 := \mathrm{Res}_{G_-}\mu + \mathrm{Res}_{G_+}\nu$$

(i.e. $\nu_1 = \lambda_+$ and $\nu_2 = \lambda_-$) we obtain $\mu = \nu_1 + \nu_0$ and $\nu = \nu_2 + \nu_0$ with $\nu_0, \nu_1, \nu_2 \in \mathscr{M}_+^b(G)$. Clearly $\nu_0 \neq 0$ since $\nu_0 = 0$ implies

$$\mu + \nu = \nu_1 + \nu_2 = \lambda_+ + \lambda_- = |\lambda|$$

or

$$2 = \|\mu\| + \|\nu\| = \|\mu + \nu\| = \|\,|\lambda|\,\| = \|\lambda\| = N(\mu, x),$$

which contradicts the hypothesis. From

$$\mu = \nu_1 + \nu_0 \quad \text{and} \quad \mu = \varepsilon_{x^{-1}} * \nu = \varepsilon_{x^{-1}} * \nu_2 + \varepsilon_{x^{-1}} * \nu_0$$

we conclude that

$$\mu = \tfrac{1}{2}(\varepsilon_e + \varepsilon_x) * \varepsilon_{x^{-1}} * \nu_0 + \tfrac{1}{2}(\nu_1 + \varepsilon_{x^{-1}} * \nu_2).$$

Obviously,

$$\mu_0 := \frac{\varepsilon_{x^{-1}} * \nu_0}{\|\nu_0\|} \in \mathscr{M}^1(G) \quad \text{and} \quad 0 < \|\nu_0\| \leq 1$$

since

$$\|\nu_0\| \leq \|\nu_1\| + \|\nu_0\| = \|\mu_0\| = 1.$$

If, now, $N(\mu, x) = 0$, then $\lambda = 0$, and hence $\nu_1 + \nu_2 = |\lambda| = 0$. Thus

$$\nu_1 = \nu_2 = 0 \quad \text{or} \quad \mu = \tfrac{1}{2}(\varepsilon_e + \varepsilon_x) * \varepsilon_{x^{-1}} * \nu_0,$$

which is the asserted relationship as $\mu = \nu_0$ implies $\|\nu_0\| = 1$. If, however, $N(\mu, x) > 0$, then $\nu_1 \neq 0$ or $\nu_2 \neq 0$ and thus

$$\beta := \tfrac{1}{2}\|\nu_1 + \varepsilon_{x^{-1}} * \nu_2\| > 0.$$

Furthermore,

$$\|\nu_0\| + \beta = \|\nu_0\| + \tfrac{1}{2}\|\nu_1\| + \tfrac{1}{2}\|\nu_2\| = \tfrac{1}{2}\|\mu\| + \tfrac{1}{2}\|\nu\| = 1.$$

But

$$\mu = \|\nu_0\| \tfrac{1}{2}(\varepsilon_e + \varepsilon_x) * \mu_0 + \beta \frac{1}{2\beta}(\nu_1 + \varepsilon_{x^{-1}} * \nu_2)$$

yields the assertion in this case as well. □

2.6.7. For $\mu \in \mathscr{M}^1(G)$ one defines the set

$$S(\mu) := \{x \in G : \lim_{n \to \infty} N(\mu^n, x) = 0\}$$

which is a subgroup of G with the *properties:*
(1) $S(\mu) \subset S(\mu * \nu)$ if $\nu \in \mathscr{M}^1(G)$ and
(2) $S(\mu^n) = S(\mu)$ for all $n \geq 1$.

It remains to prove (2). First of all one notes that (1) implies $S(\mu) \subset S(\mu^n)$ for all $n \geq 1$. If, conversely, $x \in S(\mu^n)$ $(n \geq 1)$, then $\lim_{m \to \infty} N(\mu^{nm}, x) = 0$. But since the sequence $(N(\mu^n, x))_{n\geq 1}$ is antitone, we obtain $\lim_{n\to\infty} N(\mu^n, x)=0$, whence $x\in S(\mu)$. $S(\mu)$ will be called the *Stam group* of μ.

Plainly $S(\mu)\subset T((\mu^n)_{n\geq 1})=H_\mu$.

2.6.8 Theorem. *For any $\mu, \nu\in\mathcal{M}^1(G)$ with $\mu\prec\nu$ one has $S(\mu)\subset S(\nu)$.*

Proof. Let $\alpha\in]0,1[$ and $\mu'\in\mathcal{M}^1(G)$ be given such that $\nu=\alpha\mu+(1-\alpha)\mu'$ holds. Then by the properties of the function N one concludes that

$$N(\nu^n, x)\leq\sum_{k=0}^{n}\binom{n}{k}\alpha^k(1-\alpha)^{n-k}N(\mu^k, x)\qquad\text{for all } x\in G.$$

Let $(X_j)_{j\geq 1}$ be an independent sequence of integer-valued random variables X_j on a probability space $(\Omega, \mathfrak{A}, P)$ with common distribution

$$P_{X_j}:=\beta_1^\alpha:=\alpha\varepsilon_0+(1-\alpha)\varepsilon_1\qquad\text{for all } j\geq 1,$$

and let $a(k):=N(\mu^k, x)$ for all $k\geq 0$ (x being chosen fixed). Then

$$\begin{aligned}&E(a\circ(X_1+\cdots+X_n))=\int a\circ(X_1+\cdots+X_n)\,dP\\&=\sum_{k=0}^{n}N(\mu^k, x)P[X_{i_1}=\cdots=X_{i_k}=1,\ X_j=0\text{ for } j\leq n, j\notin\{i_1,\ldots,i_k\}]\\&=\sum_{k=0}^{n}N(\mu^k, x)\binom{n}{k}\alpha^k(1-\alpha)^{n-k}=\sum_{k=0}^{n}\binom{n}{k}\alpha^k(1-\alpha)^{n-k}a(k).\end{aligned}$$

Thus from the above inequality one gets

$$N(\nu^n, x)\leq E(a\circ(X_1+\cdots+X_n)).$$

Let $x\in S(\mu)$, i.e., $\lim_{k\to\infty}a(k)=\lim_{k\to\infty}N(\mu^k, x)=0$. For $n\geq 1$ we put

$$A_n:=[X_1+\cdots+X_n\leq k_0]\qquad\text{for some } k_0\geq 1,\quad A:=\bigcap_{n\geq 1}A_n,$$

$$B_n:=[X_n=1]\quad\text{and}\quad B:=\overline{\lim}_{n\geq 1}B_n.$$

Clearly, for every $n\geq 1$ we have $A_{n+1}\subset A_n$, and $A\subset\complement B$. Since the sequence $(B_n)_{n\geq 1}$ is independent and $P(B_n)=1-\alpha$ holds for all $n\geq 1$, the Borel-Cantelli lemma yields $P(B)=1$ or $P(A)=0$, and hence $\lim_{n\to\infty}P(A_n)=0$. For $\varepsilon>0$ there exists a $k_0\geq 1$ such that $a(k)<\varepsilon$ is satisfied for all $k>k_0$. It follows that

$$P(A_n)=\sum_{k=0}^{k_0}\binom{n}{k}\alpha^k(1-\alpha)^{n-k}\qquad\text{for all } n\geq 1,$$

and together with $a(k)\leq 2$ for all $k\geq 1$ we obtain

$$\begin{aligned}&E(a\circ(X_1+\cdots+X_n))\\&=\sum_{k=0}^{k_0}\binom{n}{k}\alpha^k(1-\alpha)^{n-k}a(k)+\sum_{k=k_0+1}^{n}\binom{n}{k}\alpha^k(1-\alpha)^{n-k}a(k)\\&\leq 2P(A_n)+\varepsilon\sum_{k=0}^{n}\binom{n}{k}\alpha^k(1-\alpha)^{n-k}=2P(A_n)+\varepsilon\qquad(n\geq 1).\end{aligned}$$

This implies that

$$\lim_{n\to\infty} E(a\circ(X_1+\cdots+X_n))=0 \quad\text{or}\quad \lim_{n\to\infty} N(\nu^n, x)=0,$$

i.e., $x\in S(\nu)$. □

2.6.9 Theorem. *For every $x\in G$ we have $x\in S(\frac{1}{2}(\varepsilon_e+\varepsilon_x))$.*

Proof. 1. Let $x\in G$ be of infinite order. Then

$$N((\tfrac{1}{2}(\varepsilon_e+\varepsilon_x))^n, x)=\left\|\frac{1}{2^n}(\varepsilon_e+\varepsilon_x)^n * (\varepsilon_e-\varepsilon_x)\right\|$$

equals the sum of absolute values of the coefficients of the polynomial

$$\frac{1}{2^n}(1+z)^n(1-z).$$

This sum asymptotically equals $\dfrac{2}{\sqrt{\pi n}}$. But then

$$\lim_{n\to\infty} N((\tfrac{1}{2}(\varepsilon_e+\varepsilon_x))^n, x)=0.$$

2. Let $x\in G$ be of finite order and such that $[x]$ is a finite subgroup H of G. Clearly, $\mu:=\frac{1}{2}(\varepsilon_e+\varepsilon_x)\in\mathscr{M}^1(H)$ and $\operatorname{supp}(\mu)$ is not contained in a coset of a proper normal subgroup of H. By the Ito-Kawada theorem 2.1.4 $\mathscr{T}_v$-$\lim_{n\to\infty}\mu^n=\omega_H$. Thus the sequence $(\mu^n*(\varepsilon_e-\varepsilon_x))_{n\geq 1}$ $\mathscr{T}_v$-converges to the zero measure. But since $\mathscr{T}_v$ coincides with the norm topology on $\mathscr{M}^b(H)$ and the embedding of $\mathscr{M}^b(H)$ in $\mathscr{M}^b(G)$ is an isometry, the asserted statement follows. □

2.6.10 Theorem. *For every measure $\mu\in\mathscr{M}^1(G)$ one has*

$$S(\mu)=\bigcup_{n\geq 1}\{x\in G: N(\mu^n, x)<2\}.$$

In other words, the antitone sequence $(N(\mu^n, x))_{n\geq 1}$ $(x\in G)$ is either constant and $\equiv 2$ or convergent to 0.

Proof. Let $x\in G$. Suppose there exists an $n\geq 1$ such that $N(\mu^n, x)<2$ holds. Then by Theorem 2.6.6 there is a measure $\mu_0\in\mathscr{M}^1(G)$ satisfying $\frac{1}{2}(\varepsilon_e+\varepsilon_x)*\mu_0\prec\mu^n$. But Theorems 2.6.9 and 2.6.8 together with Properties (1) and (2) of 2.6.7 yield

$$x\in S(\tfrac{1}{2}(\varepsilon_e+\varepsilon_x))\subset S(\tfrac{1}{2}(\varepsilon_e+\varepsilon_x)*\mu_0)\subset S(\mu^n)=S(\mu). \quad □$$

2.6.11 Corollary. *$S(\mu)$ is an F_σ-subset of G.*

Proof. The antitonicity of the sequence $(N(\mu^n, x))_{n\geq 1}$ $(x\in G)$ yields via the theorem the representation $S(\mu)=\bigcup_{n\geq 1}\{x\in G: N(\mu^n, x)\leq 1\}$. But for every $n\geq 1$ the mapping $x\to N(\mu^n, x)$ from G into $\mathbb{R}_+$ is lower semicontinuous. This yields the assertion. □

2.6.12 Theorem. *For any $\mu, \nu\in\mathscr{M}^1(G)$ with $\mu\ll\nu$ one has $S(\mu)\subset S(\nu)$.*

Proof. Let $\mu, \nu \in \mathscr{M}^1(G)$ be such that $\mu \ll \nu$ holds. We define f to be the ν-density of μ and put $\beta_n := \int \min(f(x), n)\, \nu(dx)$ for all $n \geq 1$. Clearly, $\beta_n \leq \beta_{n+1}$ for all $n \geq 1$ and $\lim_{n\to\infty} \beta_n = 1$. Hence, there exists an $n_0 \geq 1$ such that $\beta_n > 0$ for all $n \geq n_0$. For every $n \geq n_0$ we put $f_n := \frac{1}{\beta_n} \min(f, n)$ and $\mu_n := f_n \cdot \nu$. Since

$$\mu_n(1) = \int f_n \, d\nu = \frac{1}{\beta_n} \left(\int \min(f, n)\, d\nu\right) = 1,$$

one has $\mu_n \in \mathscr{M}^1(G)$ for all $n \geq n_0$. Without loss of generality let $\beta_n \geq \frac{1}{2}$ for all $n \geq n_0$. This implies that $f_n \leq 2f$ for all $n \geq n_0$. But now the sequence $(f_n)_{n \geq 1}$ converges ν-a.e. to f, and the Lebesgue dominated convergence theorem yields

$$\lim_{n\to\infty} \|\mu - \mu_n\| = \lim_{n\to\infty} \int |f - f_n| \, d\nu = 0.$$

By Remark 2.6.5 we obtain $\mu_n \prec \nu$ and therefore by Theorem 2.6.8 $S(\mu_n) \subset S(\nu)$ for all $n \geq n_0$. It remains to prove that $S(\mu) \subset \bigcup_{n \geq n_0} S(\mu_n)$. Let $x \in S(\mu)$. Then there is a $k \geq 1$ with $N(\mu^k, x) < 2$. Choosing $\varepsilon := 2 - N(\mu^k, x)$ we get for all $n \geq n_0$ the chain

$$N(\mu_n^k, x) = \|\mu_n^k * (\varepsilon_e - \varepsilon_x)\| \leq 2\,\|\mu_n^k - \mu^k\| + \|\mu^k * (\varepsilon_e - \varepsilon_x)\|.$$

Moreover, there exists an $n_1 \geq n_0$ satisfying $2\,\|\mu_n^k - \mu^k\| < \varepsilon$ for all $n \geq n_1$, since $\lim_{n\to\infty} \|\mu_n - \mu\| = 0$. Consequently, $N(\mu_n^k, x) < \varepsilon + N(\mu^k, x) = 2$. But then Theorem 2.6.10 implies that $x \in S(\mu_n)$ for all $n \geq n_1$. □

2.6.13 Theorem. *Let $\mu \in \mathscr{M}^1(G)$ be such that there exist an $n \geq 1$ and a measure $\nu \in \mathscr{M}^1_\omega(G)$ satisfying $\nu \prec \mu^n$. Then $S(\mu)$ is an open (hence closed) subgroup of G.*

Proof. By Property (2) of 2.6.7 we assume that $\nu \prec \mu$. Moreover, Theorem 2.6.8 implies that we can also assume that $\mu \ll \omega$ in such a way that $\mu = f \cdot \omega$ holds with $f \in L^1_{\mathbb{R}}(G, \omega)$. Without loss of generality we may even take $f \in \mathscr{C}_{\mathbb{R}}(G)$. Indeed, one chooses a bounded function f' on G such that

$$0 \leq f' \leq f \quad \text{and} \quad \alpha := \int f' \, d\omega \in]0, 1[$$

hold. For $\mu' := \frac{f'}{\alpha} \cdot \omega$ we obtain $\mu' \prec \mu$. By Theorem 2.6.8 one can replace μ by μ'. Thus we assume f to be bounded, get $f \in L^1_{\mathbb{R}}(G, \omega) \cap L^\infty_{\mathbb{R}}(G, \omega)$ and hence, find that $f * f \in \mathscr{C}_{\mathbb{R}}(G)$.

For every $n \geq 1$ let

$$f_n := \frac{d\mu^n}{d\omega} = \underbrace{f * \cdots * f}_{n\text{-times}} \in \mathscr{C}(G).$$

Then $x \to N(\mu^n, x) = \int |f_n(y) - f_n(yx)| \, \omega(dy)$ is continuous for every $n \geq 1$, and Theorem 2.6.10 yields that $S(\mu)$ is an open subgroup of G. □

We are now prepared to characterize the a.e.-sequences $(\mu^n)_{n \geq 1}$ in analogy to the w.a.e.-sequences $(\mu^n)_{n \geq 1}$ given in Corollary 2.6.4. One observes that for a given $\mu \in \mathscr{M}^1(G)$ the sequence $(\mu^n)_{n \geq 1}$ is an a.e.-sequence in $\mathscr{M}^1(G)$ iff $S(\mu) = G$.

2.6.14 Theorem. *Let G be a locally compact Abelian group and $\mu \in \mathcal{M}^1(G)$. The following statements are equivalent:*
(i) $S(\mu)=G$;
(ii) *μ is non-lattice, and there is an $n \geq 1$ such that μ^n is not ω-singular.*

Proof. 1. (i) $\Rightarrow$ (ii). The hypothesis together with Theorem 2.6.2 implies that

$$G=S(\mu) \subset T((\mu^n)_{n \geq 1})=H_\mu$$

and therefore μ is non-lattice. For every $n \geq 1$ the function $x \to N(\mu^n, x)$ is lower semicontinuous, hence Borel measurable, and for $\lambda \in \mathcal{M}^1(G)$ of the form $\lambda=f \cdot \omega$ with $f \in L^1_{\mathbb{R}}(G, \omega)$ we have

$$\|\mu^n - \lambda * \mu^n\| \leq \int f(y) \, \|\mu^n - \varepsilon_y * \mu^n\| \, \omega(dy),$$

which follows from Remark 2.5.2. Since by assumption

$$\lim_{n \to \infty} N(\mu^n, y)=0 \quad \text{for all } y \in G,$$

the Lebesgue dominated convergence theorem yields

$$\lim_{n \to \infty} \|\mu^n - \lambda * \mu^n\|=0.$$

But $\lambda * \mu^n \ll \omega$ for all $n \geq 1$. Thus, if $\mu^n \perp \omega$ for all $n \geq 1$, we would have

$$\|\mu^n - \lambda * \omega^n\| = \|\mu^n\| + \|\lambda * \mu^n\| = 2 \quad \text{for all } n \geq 1,$$

which would contradict the above limit relationship. Consequently, there exist an $n \geq 1$ and a $\nu \in \mathcal{M}^1_\omega(G)$ with $\nu \prec \mu^n$.

2. (ii) $\Rightarrow$ (i). By Theorem 2.6.13 $S(\mu)$ is an open subgroup of G. Hence, $H:=G/S(\mu)$ is discrete and $\mu':=p(\mu)$ is atomic, where $p\colon G \to H$ denotes the canonical epimorphism. We shall show that $|H|=1$ and assume for this purpose the contrary: $|H| \geq 2$. Since μ is non-lattice, there are $h_1, h_2 \in H$ with $h_1 \neq h_2$ and $\mu'(\{h_1\}) \neq 0$, $\mu'(\{h_2\}) \neq 0$. For every $h \in H$ with $\mu'(\{h\}) \neq 0$ we put

$$\mu_h := \operatorname{Res}_{p^{-1}(h)} \frac{1}{\mu'(\{h\})} \mu.$$

Since μ has a non-zero value for at most countably many H-cosets, we obtain $\mu=\sum_{h \in H} \mu'(\{h\}) \mu_h$. Clearly, $\mu \ll \omega$ implies $\mu_{h_1}, \mu_{h_2} \ll \omega$. Therefore there are $x_1 \in p^{-1}(\{h_1\})$ and $x_2 \in p^{-1}(\{h_2\})$ satisfying $\|\mu_{h_1} - \mu_{h_2} * \varepsilon_{x_1 x_2^{-1}}\| = 2-\alpha < 2$, where $\alpha \in \mathbb{R}^*_+$. In fact, let $\mu_{h_1} := f_1 \cdot \omega$ and $\mu_{h_2} := f_2 \cdot \omega$ with $f_1, f_2 \in L^1_{\mathbb{R}}(G, \omega)$. Then $\mu_{h_2} * \varepsilon_{x_1 x_2^{-1}} = (f_2)_{x_2 x_1^{-1}} \cdot \omega$. But there are $x_1 \in p^{-1}(\{h_1\})$ and $x_2 \in p^{-1}(\{h_2\})$ with

$$\int \inf(f_1, f_2') \, d\omega =: \alpha \in \mathbb{R}^*_+, \quad \text{where } f_2' := (f_2)_{x_2 x_1^{-1}}.$$

Hence, from

$$f_1 - f_2' + |f_1 - f_2'| = \sup(f_1, f_2') \quad \text{and}$$
$$f_1 + f_2' = \sup(f_1, f_2') + \inf(f_1, f_2')$$

one concludes that

$$\begin{aligned}\|\mu_{h_1}-\mu_{h_2}*\varepsilon_{x_1x_2^{-1}}\| &= \int |f_1-f_2'|\,d\omega\\ &= \int \sup(f_1,f_2')\,d\omega - \int (f_1-h')\,d\omega\\ &= 2-\int \inf(f_1,f_2')\,d\omega = 2-\alpha<2.\end{aligned}$$

For every $h\in H$ we have $\operatorname{supp}(\mu_h)\cap\operatorname{supp}(\mu_h*\varepsilon_{x_1x_2^{-1}})=\emptyset$, whence

$$\begin{aligned}&\|\mu'(\{h\})\,\mu_h-\mu'(\{h\})\,\mu_h*\varepsilon_{x_1x_2^{-1}}\|\\ &=\mu'(\{h\})\,\|\mu_h-\mu_h*\varepsilon_{x_1x_2^{-1}}\|=2\mu'(\{h\}).\end{aligned}$$

Without loss of generality we assume that $\mu'(\{h_1\})\le\mu'(\{h_2\})$. Then

$$\begin{aligned}N(\mu,x_1x_2^{-1}) \le &\sum_{\substack{h\in H\\ h\neq h_1,h_2}} \|\mu'(\{h\})\,\mu_h-\mu'(\{h\})\,\mu_h*\varepsilon_{x_1x_2^{-1}}\|\\ &+\|\mu'(\{h_1\})(\mu_{h_1}-\mu_{h_2}*\varepsilon_{x_1x_2^{-1}})+\mu'(\{h_2\})(\mu_{h_2}-\mu_{h_2}*\varepsilon_{x_1x_2^{-1}})\|\\ &=2(1-\mu'(\{h_1\})-\mu'(\{h_2\}))\\ &+\|\mu'(\{h_1\})(\mu_{h_1}-\mu_{h_2}*\varepsilon_{x_1x_2^{-1}})-(\mu'(\{h_2\})-\mu'(\{h_1\}))\,\mu_{h_2}*\varepsilon_{x_1x_2^{-1}}\\ &+\mu'(\{h_1\})(\mu_{h_2}-\mu_{h_1}*\varepsilon_{x_1x_2^{-1}})+(\mu'(\{h_2\})-\mu'(\{h_1\}))\,\mu_{h_2}\|\\ &\le 2(1-\mu'(\{h_1\})-\mu'(\{h_2\}))+\mu'(\{h_1\})(2-\alpha)\\ &+2(\mu'(\{h_2\})-\mu'(\{h_1\}))+2\mu'(\{h_1\})\\ &=2-\alpha\,\mu'(\{h_1\})<2.\end{aligned}$$

Theorem 2.6.10 yields $x_1x_2^{-1}\in S(\mu)$ or $h_1=h_2$, contradicting the assumption $h_1\neq h_2$ and thus the hypothesis $|H|\ge 2$. ▯

2.6.15 Corollary. *If G is connected, then $S(\mu)=G$ iff there exist an $n\ge 1$ and a $\nu\in\mathscr{M}^1_\omega(G)$ such that $\nu\prec\mu^n$ holds.*

The *proof* follows from the theorem and Theorem 2.6.13 together with the fact, that in a connected locally compact group there is no proper open subgroup. ▯

2.6.16 Corollary. *Let $\mu\in\mathscr{M}^1(G)$ be atomic and let $H\in\mathbf{H}_\mu$ be such that $x_H H$ is the smallest H-coset of this form containing the atoms of μ. Then $S(\mu)=H$.*

Proof. Let G_d be the given group G furnished with the discrete topology. The inclusion from $\mathscr{M}^b(G_d)$ into $\mathscr{M}^b(G)$ is norm preserving. From

$$N(\mu*\varepsilon_y,x_H)=N(\mu,x_H)\qquad\text{for all } y\in G$$

we conclude that

$$S(\mu)=S(\mu*\varepsilon_{x_H^{-1}})\subset H.$$

But $\mu*\varepsilon_{x_H^{-1}}$ is non-lattice (as a measure in $\mathscr{M}^1(H_d)$) and $\ll \operatorname{Res}_{H_d}\omega$. The theorem then yields the result. ▯

For any measure $\mu\in\mathcal{M}^1(G)$ we put $L_\mu:=[\mathrm{supp}(\mu)]^-$. Clearly, $L_\mu\supset H_\mu$, and $L_\mu=H_\mu$ if $e\in\mathrm{supp}(\mu)$.

2.6.17 Theorem. *Let G be a locally compact Abelian group, $\mu\in\mathcal{M}^1(G)$ and $(\rho_n)_{n\geq 1}$ an a.e.-sequence in $\mathcal{M}^1(\mathbb{Z})$ with $\rho_n(\mathbb{N})=1$ for all $n\geq 1$. Then for the sequence $(\mu_n)_{n\geq 1}$ in $\mathcal{M}^1(G)$ defined by*

$$\mu_n:=\sum_{k\geq 1}\rho_n(\{k\})\,\mu^k \quad \textit{for all } n\geq 1$$

we have $T((\mu_n)_{n\geq 1})=L_\mu$.

Proof. 1. For any $\sigma\in\mathcal{M}^1_\omega(G)$ and $\tau\in\mathcal{M}^1(G)$ with $\tau(H_\mu)=1$ one has

$$\|\sigma*\mu^n-\sigma*\tau*\mu^n\|\leq\int_{H_\mu}\|\sigma*\mu^n-\varepsilon_x*\sigma*\mu^n\|\,\tau(dx)$$

by the inequality of Remark 2.5.2. Theorem 2.6.2 then yields

$$\lim_{n\to\infty}\|\sigma*\mu^n-\sigma*\tau*\mu^n\|=0.$$

2. For all $\sigma\in\mathcal{M}^1_\omega(G)$ and $x\in\mathrm{supp}(\mu)$ one has

$$\lim_{n\to\infty}\|\sigma*\varepsilon_x*\mu^n-\sigma*\mu^{n+1}\|=0.$$

Indeed, the assumptions imply that $(\varepsilon_{x^{-1}}*\mu)(H_\mu)=1$, whence by

$$\|\sigma*\varepsilon_x*\mu^n-\sigma*\mu^{n+1}\|=\|\sigma*\mu^n-\sigma*(\varepsilon_{x^{-1}}*\mu)*\mu^n\|,$$

Part 1 of this proof becomes applicable, and the limit relationship follows.

3. We first prove that $L_\mu\subset T((\mu_n)_{n\geq 1})$. Since by Corollary 2.5.10 $T((\mu_n)_{n\geq 1})$ is a closed subgroup of G, it suffices to show $\mathrm{supp}(\mu)\subset T((\mu_n)_{n\geq 1})$. Let $\sigma\in\mathcal{M}^1(G)$ and $x\in\mathrm{supp}(\mu)$ be given. For $n\geq 1$ and $k\geq 1$ we get

$$\begin{aligned}
\|\sigma*\mu_n-\sigma*\varepsilon_x*\mu_n\| &\leq\rho_n(\{1,\ldots,k\})+\rho_n(\{1,\ldots,k-1\})\\
&+\sum_{j\geq k}\|\rho_n(\{j+1\})\,\sigma*\mu^{j+1}-\rho_n(\{j\})\,\sigma*\varepsilon_x*\mu^j\|\\
&\leq 2\rho_n(\{1,\ldots,k\})\\
&+\sum_{j\geq k}(|\rho_n(\{j+1\})-\rho_n(\{j\})|+\rho_n(\{j\})\|\,\sigma*\mu^{j+1}-\sigma*\varepsilon_x*\mu^j\|)\\
&\leq 2\rho_n(\{1,\ldots,k\})+\|\rho_n-\varepsilon_1*\rho_n\|+\sum_{j\geq k}\rho_n(\{j\})\,\|\sigma*\mu^{k+1}-\sigma*\varepsilon_x*\mu^k\|\\
&\leq 2\rho_n(\{1,\ldots,k\})+\|\rho_n-\varepsilon_1*\rho_n\|+\|\sigma*\mu^{k+1}-\sigma*\varepsilon_x*\mu^k\|,
\end{aligned}$$

and hence, for all $k\geq 1$

$$\overline{\lim}_{n\geq 1}\|\sigma*\mu_n-\sigma*\varepsilon_n*\mu_n\|\leq\|\sigma*\mu^{k+1}-\sigma*\varepsilon_x*\mu^k\|,$$

from which by Part 2 the desired limit relation follows.

4. We finally show that $T((\mu_n)_{n\geq 1})\subset L_\mu$. Let $x\notin L_\mu$. Then there exists a $U\in\mathfrak{V}(e)$ with $xU\cap L_\mu=\emptyset$, and for every open neighborhood $V\in\mathfrak{V}(e)$ with $VV^{-1}\subset U$ we have $(VL_\mu)\cap(x\,VL_\mu)=\emptyset$. Let $\sigma\in\mathcal{M}^1_\omega(G)$ with $\sigma(V)=1$ be given. For every $n\geq 1$ one obtains $(\sigma*\mu^n)(VL_\mu)=1$ and consequently, $(\sigma*\mu_n)(VL_\mu)=1$.

But this implies that $\|\sigma * \mu_n - \varepsilon_x * \sigma * \mu_n\| = 2$ for all $n \geq 1$, and thus $x \notin T((\mu_n)_{n \geq 1})$. □

2.6.18 Remark. For every $\mu \in \mathscr{M}^1(G)$ the sequence $(\mu_n)_{n \geq 1}$ in $\mathscr{M}^1(G)$ defined by $\mu_n := \frac{1}{n} \sum_{k=1}^{n} \mu^k$ for all $n \geq 1$ satisfies the assumptions of the theorem.

After those measures $\nu \in \mathscr{M}^1(G)$ for which $S(\nu) = G$ holds have been characterized, the question arises as to whether there exists a non-lattice measure $\mu \in \mathscr{M}^1(G)$ such that $S(\mu) = \{e\}$ is fulfilled.

Let $(\xi_n)_{n \geq 1}$ be a sequence in $]0, 1[$ which is dense in $[0, 1]$, and let $(a_n)_{n \geq 1}$ be a sequence in $\mathbb{R}_+$ such that $\sum_{n \geq 1} a_n = 1$. We define a measure

$$\nu := \sum_{n \geq 1} a_n \varepsilon_{\xi_n} \in \mathscr{M}^1(\mathbb{R})$$

with distribution function $F := F_\nu$ defined by $F_\nu(x) := \nu([0, x[)$ for all $x \in \mathbb{R}$. Clearly, for $A := F([0, 1])$ we have $\lambda^1(A) = 0$.

2.6.19 Theorem. *If* $\sum_{i > n} a_i < a_n$ *for all* $n \geq 1$, *then*

(i) $A \cap (A + x)$ *contains at most one element if* $x \neq 0$.

(ii) *Every diffuse measure* $\mu \in \mathscr{M}^1(\mathbb{R})$ *with* $\operatorname{supp}(\mu) = A$ *satisfies* $\mu \perp \mu * \varepsilon_x$ *for all* $x \in \mathbb{R} \setminus \{0\}$.

Proof. 1. Let $(\varepsilon_n)_{n \geq 1}$ be a sequence of elements from $\{-1, 0, 1\}$ such that $\sum_{n \geq 1} \varepsilon_n a_n = 0$. Then $\varepsilon_n = 0$ for all $n \geq 1$. In fact, let n_0 be the first integer such that $\varepsilon_{n_0} \neq 0$. Without loss of generality we suppose that $\varepsilon_{n_0} = 1$. Then

$$\sum_{n \geq 1} \varepsilon_n a_n = a_{n_0} + \sum_{n > n_0} \varepsilon_n a_n \geq a_{n_0} - \sum_{n > n_0} a_n > 0,$$

which is a contradiction.

2. Since (ii) follows immediately from (i), it suffices to show (i). Let $F(a), F(a') \in A \cap (x + A)$. Then there are $b, b' \in \mathbb{R}$ such that $F(a) = x + F(b)$, $F(a') = x + F(b')$ resp. Without loss of generality let $x > 0$. Then $x = \nu([b, a[) = \nu([b', a'[)$. The sequence $(\varepsilon_n)_{n \geq 1}$ defined by $\varepsilon_n := 1_{[b,a[}(\xi_n) - 1_{[b',a'[}(\xi_n)$ for all $n \geq 1$ satisfies the condition of Part 1 and thus $\varepsilon_n = 0$ for all $n \geq 1$. But for every $n \geq 1$ we have $\xi_n \in [b, a[$ iff $\xi_n \in [b', a'[$. Since $(\xi_n)_{n \geq 1}$ is dense in $[0, 1]$, this implies $a = a'$ or $F(a) = F(a')$. □

2.6.20 Theorem. *Let* $A_1 := A$ *and for every* $n \geq 1$ *let* $A_n := A_{n-1} + A$. *Suppose furthermore, that* $\frac{a_n}{a_{n+1}} \in \mathbb{N}$ *for all* $n \geq 1$ *and that* $\lim_{n \to \infty} \frac{1}{a_n} \sum_{i > n} a_i = 0$ *is satisfied. Then*

(i) $x \notin A_{n-1}$ *implies that* $(x + A) \cap A_n$ *contains at most one element.*

(ii) *For any diffuse measure* $\mu \in \mathscr{M}^1(\mathbb{R})$ *with* $\operatorname{supp}(\mu) = A$ *and every* $n \geq 1$ *we have* $\mu^n \perp \mu^n * \varepsilon_x$ *whenever* $x \in \mathbb{R} \setminus \{0\}$.

Proof. 1. Let $(x_n)_{n \geq 1}$ be a bounded sequence in $\mathbb{R}$ such that $\sum_{n \geq 1} x_n a_n = 0$ holds. Then there exists an $n_0 \geq 1$ with $x_n = 0$ for all $n \geq n_0$. In fact, let $M := \sup\{|x_n| : n \geq 1\}$ and let $n_0 \geq 1$ be such that $n \geq n_0$ implies $\frac{1}{a_n} \sum_{i > n} a_i < \frac{1}{M}$.

We define

$$u := \sum_{n \geq n_0} x_n a_n \quad \text{and} \quad v := \sum_{n > n_0} x_n a_n.$$

By assumption $|v| \leq M \sum_{i > n_0} a_i < a_{n_0}$, and since $\frac{u}{a_{n_0}} \in \mathbb{N}$, we get $u = v = 0$.

Let n_1 be the first integer $> n_0$ such that $x_n \neq 0$ holds. Without loss of generality we suppose $x_{n_1} \geq 1$. Then

$$v = x_{n_1} a_{n_1} + \sum_{n > n_1} x_n a_n \geq a_{n_1} - M \sum_{n > n_1} a_n > 0$$

which contradicts the hypothesis. Thus $x_n = 0$ for all $n > n_0$.

2. We proceed to the proof of the theorem.

(i) Suppose that there exist a, a', b_i, b_i' $(i = 1, \ldots, n)$ in $\mathbb{R}$ satisfying

$$F(a) + x = \sum_{i=1}^n F(b_i) \quad \text{and} \quad F(a') + x = \sum_{i=1}^n F(b_i').$$

Clearly, for every $i = 1, \ldots, n$ we have $a \neq b_i$ and $a' \neq b_i'$, which follows from the choice of $x \in A_{n-1}$. Defining

$$f := 1_{[0,a[} - \sum_{i=1}^n 1_{[0,b_i[} - 1_{[0,a'[} + \sum_{i=1}^n 1_{[0,b_i'[}$$

and $x_m := f(\xi_m)$ for all $m \geq 1$ we obtain a sequence $(x_m)_{m \geq 1}$ in $\mathbb{R}$ satisfying the hypotheses of Part 1 and hence the existence of an $m_0 \geq 1$ such that $m > m_0$ implies $x_m = 0$. The sequence $(\xi_m)_{m > m_0}$ being still dense in $[0, 1]$ and f being lower semicontinuous, one obtains $f = 0$, and thus $a = a'$ or $F(a) = F(a')$.

(ii) Plainly, $\operatorname{supp}(\mu^n) = A_n$ for all $n \geq 1$. We shall show that

$$\mu^n(x + A_n) = 0 \quad \text{for all } x \in \mathbb{R} \setminus \{0\}.$$

Let $x \in \mathbb{R} \setminus \{0\}$. Then, first of all, (i) implies $\mu(x + A_1) = 0$ (and consequently $x \notin A_0 := \{0\}$). Suppose we have proved that $\mu^{n-1}(x + A_{n-1}) = 0$ holds. We have

$$\mu^n(x + A_n) = \int \mu(x - y + A_n)\, \mu^{n-1}(dy).$$

But by (i) $y - x \notin A_{n-1}$ implies $\mu(x - y + A_n) = 0$ and hence

$$\mu^n(x + A_n) \leq \int 1_{x + A_{n-1}}(y)\, \mu^{n-1}(dy) \leq \mu^{n-1}(x + A_{n-1}) = 0,$$

which yields the desired result. □

2.6.21 Remark. Let $i \in \{1, \ldots, m\}$, $n \geq 1$, $\mu_i \in \mathscr{M}^1(\mathbb{R})$ and $\mu_i^n \perp \mu_i^n * \varepsilon_x$ for all $x \in \mathbb{R} \setminus \{0\}$. Then $\mu^n := \bigotimes_{i=1}^m \mu_i^n \in \mathscr{M}^1(\mathbb{R}^m)$ satisfies $\mu^n \perp \mu^n * \varepsilon_x$ for all $x \in \mathbb{R}^m \setminus \{0\}$. The proof follows immediately from the formula

$$\mu^n\left(\left(\prod_{i=1}^m A_{i,n}\right) \cap \left(x + \prod_{i=1}^m A_{i,n}\right)\right) = \prod_{i=1}^m \mu_i^n(A_{i,n} \cap (x_i + A_{i,n})),$$

where

$A_{i,n} := \operatorname{supp}(\mu_i^n)$ $(i = 1, \ldots, m;\ n \geq 1)$ and $x \in \mathbb{R}^m \setminus \{0\}$ is of the form $x = (x_i)_{i=1,\ldots,m}$.

Thus there exists a non-lattice measure in $\mathscr{M}^1(\mathbb{R}^m)$ whose Stam group is trivial, and in general, the Stam group of a non-lattice measure in $\mathscr{M}^1(\mathbb{R}^m)$ is not dense in $\mathbb{R}^m$.

References and Comments

R 2.1 Convolution Powers on a Compact Group

The general facts concerning compact semigroups which precede the first theorem of the section are taken from the standard literature such as Hofmann and Mostert [253] or Berglund and Hofmann [13]. In [13] one also finds an extensive discussion of the $\mathscr{T}_v$-compact convolution semigroup $\mathscr{M}^1(G)$ for a compact group G. A central problem pertaining to the analysis of $\mathscr{M}^1(G)$ is the study of the limiting behavior of the sequence $(\mu^n)_{n\geq 1}$ of convolution powers for a measure $\mu\in\mathscr{M}^1(G)$. This problem was treated for the first time by P. Lévy in [320] and Kakehashi in [281] for the torus group, by Horton and Smith [254], [255] for the two-element group and by Dvoretzky and Wolfowitz [126] and Gyires [189] for arbitrary finite cyclic groups. The case of a general compact group has been developed on the basis of the fundamental contribution of Ito and Kawada [287]. These authors proved the equivalence (i)$\Leftrightarrow$(v) of Theorem 2.1.4 under the additional hypothesis that G admits a countable basis of its topology. Extensions of the theorem to general compact groups have been given by Kloss [303], [304] and Stromberg [457]. In the papers of Prékopa, Rényi and Urbanik [391] and of Rivkind [406], in which the convergence ($\mathscr{T}_v$- and setwise resp.) of a sequence $(\mu^n)_{n\geq 1}$ in $\mathscr{M}^1(G)$ to the Haar measure ω_G is established under an additional assumption on the measure μ (it must be full or ω_G-absolutely continuous and symmetric resp.) for Abelian compact and connected compact groups with a countable basis of their topology resp. The separability condition in Rivkind's theorem was later removed by Kloss [304], [305]. The approach to the Ito-Kawada equivalence of Theorem 2.1.4 that is chosen for our presentation is purely semigroup-theoretic. It is due to Collins [88] (See also Sazonov and Tutubalin [423] and Roynette [415]). Rivkind's result [406] includes Corollary 2.1.6. The proof of Stromberg [457] reproduced in Grenander [182] uses some structure theory of locally compact Abelian groups, in particular, Fourier transforms (as did Ito and Kawada). A proof based on results from the theory of equidistribution on compact groups is due to Cigler [82].

The Ito-Kawada theorem can also be deduced from the work [187] of Guivarc'h. See also Derriennic [112] for an Abelian version. Finally Csiszár presented a demonstration in [99] which makes use of Rényi's notion of I-divergence.

There are numerous applications of the Ito-Kawada theorem. At this point, we mention only the following result used in the theory of random walks on Lie groups (Roynette [415], Chapitre II): Let $G:=\mathfrak{SO}(n)$ for $n\geq 3$, and μ be a measure in $\mathscr{M}^1(G)$ generating G. Then $\mathscr{T}_v\text{-}\lim_{n\to\infty}\mu^n=\omega_G$.

Here are some more references on the individual implications of Theorem 2.1.4. The equivalence (i)$\Leftrightarrow$(iv) is due to Urbanik [482] and Kloss [304]. Urbanik's method yields the more general theorem 2.1.11. The implications (v)$\Rightarrow$(vi)$\Rightarrow$(vii) depend on Theorem 2.1.1, which implies Theorem 2.1.7 containing a first result on the convergence of normalized powers of a measure in $\mathscr{M}^1(G)$, a problem that will be of further interest in the next section. See Tortrat

[468], [469] for a general approach to the problem under the condition that the normed sequence in $\mathscr{M}^1(G)$ for an arbitrary locally compact group G is relatively $\mathscr{T}_v$-compact in $\mathscr{M}^1(G)$. Although the convergence behavior of general sequences $(\mu_1 * \cdots * \mu_n)_{n \geq 1}$ of measures in $\mathscr{M}^1(G)$ will be studied for arbitrary locally compact groups G in the subsequent sections, we shall still comment on a few related questions here. In their paper Ullrich and Urbanik [481] study sequences $(\mu_j)_{j \geq 1}$ of symmetric measures in $\mathscr{M}^1(G)$ for a compact group G such that there exist a full measure λ on G and a sequence $(\alpha_n)_{n \geq 1}$ in $[0, 1]$ satisfying the condition $\mu_n \geq \alpha_n \lambda$ (setwise on $\mathfrak{B}(G)$). It is shown that

$$\mathscr{T}_v\text{-}\lim_{n \to \infty} \mu_1 * \cdots * \mu_n = \omega_G \quad \text{if } \sum_{n \geq 1} \alpha_n = \infty.$$

Moreover, the condition of divergence appears to be essential. Related results have been found by Ito and Kawada [287] and Kloss [303], [304]. A recent contribution is due to Šlosman [445]. In Kloss [304] and Grenander [182] one also finds a first attempt to obtain assertions on the speed of convergence once the convergence has been assured. A sample result is as follows:

Let $(\mu_j)_{j \geq 1}$ be a sequence in $\mathscr{M}^1(G)$ for a compact group G and let $(\alpha_n)_{n \geq 1}$ be a sequence in $\mathbb{R}_+^*$ such that $\mu_n \geq \alpha_n \omega_G$ holds. Then

$$|\mu_1 * \cdots * \mu_n - \omega_G| \leq \prod_{j=1}^{n} (1 - \alpha_j).$$

In particular, the condition $\sum_{j \geq 1} \alpha_j = \infty$ implies the (setwise) convergence of $(\mu_1 * \cdots * \mu_n)_{n \geq 1}$ to ω_G on $\mathfrak{B}(G)$. More recent research on the speed of convergence of sequences $(\mu^n)_{n \geq 1}$ in $\mathscr{M}^1(G)$ can be found in Bhattacharya [17]. Another setup for quantitative studies in this theory has been proposed by Egorov and Maksimov, in whose work [127] a general autoregression problem for measures on compact Abelian groups is discussed.

In connection with the central theorem 2.1.4 we finally mention the following strengthened form due to Maurer [349]: Let $\mu \in \mathscr{M}^1(G)$ and let ω_H be an accumulation point of $(\mu^n)_{n \geq 1}$ for a compact subgroup H of G. Then

(i) H is normal in $[\operatorname{supp}(\mu)]^-$,
(ii) $\operatorname{supp}(\mu) \subset aH$ for all $a \in \operatorname{supp}(\mu)$,
(iii) $\mu^n * \omega_H = \varepsilon_{a^n} * \omega_H$ for all $n \geq 1$ and
(iv) $\mathscr{T}_v\text{-}\lim_{n \to \infty} (\mu^n - \varepsilon_{a^n} * \omega_H) = 0$.

The rest of the section is devoted to the discussion of the limitations of the Ito-Kawada theorem encountered when one tries to generalize it to noncompact groups. We restrict ourselves to Theorem 2.1.12 which essentially is due to Klasa [302]. In this theorem the attempt is made to extend the Ito-Kawada statement to almost periodic groups. Further approaches to possible extensions to noncompact locally compact groups are closely related to the theory of probability measures on locally compact semigroups, a framework which we generally avoid in our presentation. For any noncompact, locally compact group G which is generated by the support of a measure $\mu \in \mathscr{M}^1(G)$ Mukherjea showed in [363] that $\mathscr{T}_v\text{-}\lim_{n \to \infty} \mu^n = 0$. For an alternative proof one may consult Derriennic [112], Théorème 8. This result has forerunners. We mention only the paper [346] of Martin-Löf and the previous research quoted therein. The methods of [346] are semigroup-theoretic. For compact semigroups an Ito-Kawada type result is as old as the papers [410] and [211] by Rosenblatt and Heble resp. See also Rosenblatt's monograph [412], Chapter V, where the convergence problem is set up starting from its historical origin and moving to the theory of random walks on compact semigroups. As an important reference to basic work in this field one has to mention the early papers of Bellman [7] and of Hewitt and Zuckerman [219] which together with [492] of Vorobev contain first discoveries in the direction of an arithmetic in the semigroup $\mathscr{M}^1(G)$ for a semigroup G. More recent work on the limiting behavior of convolution iterates in $\mathscr{M}^1(G)$ for a semigroup G has been done by Mukherjea [364]. See also the book of Mukherjea and Tserpes [366].

Motivated by the needs of ergodic theory one has studied the sequence of arithmetic means of convolution powers of a measure in $\mathscr{M}^1(G)$, where G is a locally compact group. For compact G the $\mathscr{T}_v$-convergence of the sequence

$$\left(\frac{1}{n}\sum_{j=1}^{n}\mu^j\right)_{n\geq 1}$$

for $\mu\in\mathscr{M}^1(G)$ to an idempotent in $\mathscr{M}^1(G)$ is derived in a most simple fashion. For general locally compact groups G the problem (related to the convolution equation) is solved in Grenander [182], 3.1. See also Tortrat [465] for further extensions concerning groups. Naturally the problem can be posed for measures on semigroups as well. A survey of the research done up to 1967 can be found in Williamson [503], Section 7. The central result is proved in detail in Rosenblatt [412], Section 5 of Chapter V. Some generalization has been achieved in Tortrat [469]. Interesting contributions rounding off the picture are to be ascribed to Chow [75] and Choy [78]. The authors' book [77] serves as a useful introduction to the theory of probability measures on compact (semitopological) semigroups.

For countable discrete groups G the asymptotic behavior of convolution powers of a measure $\mu\in\mathscr{M}^1(G)$ can be described in terms of additional properties of G. In Kesten [298] it is shown that for a symmetric measure $\mu\in\mathscr{M}^1(G)$ whose support generates G and which satisfies $\mu^n(\{e\})>0$ for sufficiently large $n\geq n_0$ the condition

$$\overline{\lim}_{n\geq 1}\mu^n(\{e\})^{\frac{1}{n}}=1$$

is equivalent to the amenability of G. Related studies are due to Gerl [158], [159]. In Gerl's paper [161] the relationship between the above condition and the existence of ratio limits

$$\lim_{n\to\infty}\frac{\mu^{n+1}(\{x\})}{\mu^n(\{x\})}\quad\text{for all } x\in G$$

is studied.

R 2.2 Equivalence of Types of Convergence

Our starting point is the general discussion of the limiting behavior of the sequence $(\nu_{k,n})_{0\leq k<n}$ defined by $\nu_{k,n}:=\mu_{k+1}*\cdots*\mu_n$ for a given sequence $(\mu_j)_{j\geq 1}$ of

measures in $\mathscr{M}^1(G)$, where G is a locally compact group. The aim of the presentation is a generalization of the classical equivalence theorem stating that for a series of independent $\mathbb{R}$-valued random variables convergence in distribution, stochastic convergence (convergence in probability) and almost sure convergence (convergence with probability 1) are equivalent (Tortrat [474], p. 185; Ito [263]). The structure of the section is adapted from Heyer [232], 2., 6. and 7. A reference for the various results is the article of Tortrat [472]. It was Tortrat in [472] who for the first time collected the work done in this area up to 1970 and put it in a general frame. Theorem 2.2.2 appears already in Csiszár [100] for the case of a locally compact group whose topology has a countable basis and in Tortrat [468] for more general topological groups (under additional assumptions for the sequence $(\nu_{0,n})_{n\geq 1}$). The introduction of the basis of a sequence $(\nu^{(k)})_{k\geq 0}$ of accumulation points $\nu^{(k)}\in\mathscr{M}^1(G)$ of the sequence $(\nu_{k,n})_{0\leq k<n}$ is due for a compact group G to Maksimov [338]. Corollary 2.2.7 contains the definition of a compact set K_0 of G which plays an important rôle in determining the convergence behavior of the sequence $(\nu_{k,n})_{0\leq k<n}$. It was introduced by Galmarino in [154], where an extended version of the equivalence theorem 2.2.16 is proved. Remarks 2.2.10 are borrowed from Tortrat [472]. The formulation of the equivalence theorem makes it necessary to extend the classical notions of random variable, stochastic independence, stochastic and almost sure convergence to general locally compact groups G having a countable basis of their topology. This has been achieved by Grenander [182] and Heyer [223]. In the more general setup of arbitrary topological groups, fundamental problems arise, as were pointed out in a more systematic way by Sazonov and Tutubalin in [423].

Even if G is compact, a product of random variables need not be a random variable. In fact, let the cardinal number of G be larger than that of the continuum. Then the Borel set $D := \{(x,y)\in G\times G : x=y\}$ does not belong to $\mathfrak{B}(G)\otimes\mathfrak{B}(G)$. For the underlying measurable space $(\Omega,\mathfrak{A})$ we choose $(G\times G, \mathfrak{B}(G)\otimes\mathfrak{B}(G))$, and we define G-valued random variables X and Y on $(\Omega,\mathfrak{A})$ by $X((x,y)) := x$ and $Y((x,y)) := y^{-1}$ resp. for all $(x,y)\in G\times G$. Clearly, $X\cdot Y$ is not a G-valued random variable. This example is due to Nedoma [369] and was inspired by the work of Hanš [192] on generalized random variables. Another problem concerns the various possibilities for defining stochastic and almost sure convergence for random variables taking values in a general topological group. In this context Loynes in [327] defines left and right stochastic convergence and studies their equivalence under special assumptions (regularity or semiregularity) on the probability measure involved. Loynes also shows in [327] that for Abelian or locally compact groups such complications do not usually occur. A general treatment of the notion of stochastic convergence (for random variables with values) in arbitrary uniform spaces has been given by Doss in [122]. See also Bourbaki [41], Chapitre IV, § 5, n° 11. The Cauchy criterion for stochastic convergence of G-random variables, which precedes 2.2.12, appears in Grenander [182], p. 108. Its direct proof relies on the fact that a locally compact group is complete. Lemma 2.2.12 containing a well-known fact from the elementary theory of topological groups is added for an immediate application. In Definition 2.2.13 the notions of stochastic and almost sure convergence are extended to random variables with values in a homogeneous space G/H with respect to a compact subgroup H of G (stochastic and almost sure convergence modulo H). See Galmarino [154], Bártfai [5] and Csiszár [100], or for a more general framework, [122] of Doss. The Cauchy-type criterion of Theorem 2.2.14 for sequences $(Y_n)_{n\geq 1}$ of n-th partial products $Y_n := X_1 \cdot \dots \cdot X_n$ for a sequence $(X_j)_{j\geq 1}$ of G-random variables is proved in analogy to the classical case. It yields Theorem 2.2.15 concerning the equivalence of stochastic and almost sure convergence modulo H. For proofs (which depend on Loève's lemma for events or Lévy's inequalities; see [325], pp. 246–247) see Loynes [327], Csiszár [100] (Proof of Theorem 3.2) or Tortrat [472] resp.

Theorems 2.2.16 and 2.2.19 establish the validity of the equivalence principle (defined in 2.2.17). An alternative approach to the solution based on Theorem 3.2 of Csiszár [100] is given in Heyer [225]. A version of the equivalence theorem for random variables with values in a finite group has been given by Maksimov in [330]. This paper will be an important reference for Section 4. More information on the class of aperiodic groups characterized by the equivalence principle can be found in Heyer [225]. Extensions to regular G-invariant Markov chains on homogeneous spaces of the form G/H have been achieved by Galmarino [154].

There are several papers dealing with the equivalence theorem for random variables taking values in locally compact or metric semigroups by Byczkowski [57] and Byczkowski and Wós [60], [61], [62]. The most far-reaching results have been found in the case of certain metric semigroups G. For every sequence $(Y_n)_{n\geq 1}$ of product random variables with values in G convergence in distribution implies almost sure convergence iff G has no nontrivial compact subgroup or right-zero subsemigroup.

There are generalizations to locally compact groups of the classical three series theorem (of Kolmogorov) by Pakshirajan [375] and Maksimov [340]. Relations with the purity law can be seen in the work of Brown and Moran [50], which also contains an approach to the equivalence theorem. The emphasis of these authors, however, is on the study of Bernoulli convolutions and their significance for questions in harmonic analysis. See Brown and Moran [47], [48], [49].

Finally, it should be noted that the study of sequences of compositions of independent G-random variables has suggested the idea of relaxing the hypothesis of independence. Thus Markov chains with values in groups and semigroups and their limiting behavior have become interesting. First results in this direction were proved by Koutský [308]. The theory was extended to finite groups by Cigler [83], [84] and to finite semigroups by Schmetterer [426], [427]. The case of more general semigroups is contained in the work of Muthsam [367]. A first contribution to the theory of Markov chains with values in a compact semigroup based on methods from Banach lattice theory is due to Wolff [504].

R 2.3 The Normed Convergence Property

Our treatment of the normed convergence property follows the arrangement of Heyer [232], Sections 3 to 5. For a locally compact group G having a countable basis of its topology this property states that given any sequence $(\mu_j)_{j\geq 1}$ in $\mathcal{M}^1(G)$ convergence of the corresponding sequence $(\nu_{0,n})_{n\geq 1}$ of n-th partial products can be achieved through right translation by a sequence $(\varepsilon_{x_n})_{n\geq 1}$ of Dirac measures. That the normed convergence property holds for compact groups was proved for the first time by Kloss in [304] using pure measure theory and in [305] via Fourier transforms for the special case of an Abelian group. In [306] Kloss shows under additional assumptions on G that the validity of the normed convergence property implies the compactness of G. The characterization of compact groups having a countable basis of their topology by the normed convergence property (convergence principle) appears in our presentation as Theorem 2.3.2. The proof due to Heyer [221] is based on Theorem 2.2.2. See also Kloss [307] for a generalization of the characterizing implication given previously in Kloss [306].

It is a puzzling fact that is has remained unknown whether the condition of the existence of a countable basis for the topology of G could be dropped without affecting the validity of the theorem. Kloss's early proofs of the unrestricted statement do not indicate how to overcome this shortcoming. Definition 2.3.3 of a composition convergent sequence of measures in $\mathcal{M}^1(G)$ is taken from Maksimov [338], where further material on the accumulation behavior of the sequences $(\nu_{k,n})_{n>k}$ $(k\geq 0)$ and their translates can be found. An example is Theorem 2.3.4 which states that for a compact group G admitting a countable basis of its topology and any sequence $(\mu_j)_{j\geq 1}$ in $\mathcal{M}^1(G)$ the corresponding sequence $(\nu_{k,n})_{n>k}$ converges after left translation by $\varepsilon_{x_k^{-1}}$ and right translation by ε_{x_n} $(n>k)$ for a sequence $(\varepsilon_{x_n})_{n\geq 1}$ of Dirac measures on G. Theorem 2.3.5 is a remarkable, though not fully understood contribution of Maksimov [332]. The theorem is of central importance for the proof of Theorem 2.3.13, again due to Maksimov [336], which contains the characterization of all totally disconnected compact groups having a countable basis of their topology by the strong normed convergence property defined in 2.3.11 (strong convergence principle). Various generalizations of the convergence principle have been studied. Tortrat in [468] and Csiszár [100] extended the theorem to more general topological groups. The efforts of Mukherjea in [363] yield results for semigroups. The work [363] of Mukherjea is grounded on the following general version of the convergence principle of Csiszár [100] which is cited only for reference: Let G be a locally compact group having a countable basis of its topology and $(\mu_j)_{j\geq 1}$ be a sequence in $\mathcal{M}^1(G)$ with corresponding sequences $(\nu_{k,n})_{n>k}$ $(k\geq 0)$. Then either $\lim_{n\to\infty} \sup_{y\in G} \nu_{0,n}$ $(Ky)=0$ for every compact subset K of G or there exists a sequence $(x_n)_{n\geq 1}$ in G such that the sequences $(\nu_{k,n} * \varepsilon_{x_n})_{n>k}$ $(k\geq 0)$ converge for $n\to\infty$.

R 2.4 Convergence in Variance

The origin of an axiomatic approach to the notion of variance for probability measures on a locally compact group is Lévy's idea of introducing the (circular) variance of a random variable taking values in the torus group [320]. Another idea contributing to the generalization of the classical variance of a (real-valued) random variable is due to Khintchine who defined in [299] a functional which now has his name and which led to the famous theorem (of Khintchine) that every measure in $\mathscr{M}^1(\mathbb{R})$ is the convolution product of a measure without prime factors and a countable product of prime measures. (See Lukacs [329] or Cuppens [104]).

From the latter theorem one derives the remarkable fact that any measure in $\mathscr{M}^1(\mathbb{R})$ having no prime factor is infinitely divisible (Cuppens [104]). While the notion of circular variance had been used by Bártfai [5] to study the limiting behavior of sequences $(\nu_n)_{n\geq 1}$ $(=\mu_1 * \cdots * \mu_n)$ for sequences $(\mu_j)_{j\geq 1}$ in $\mathscr{M}^1(\mathbb{T})$ to the Haar measure of $\mathbb{T}$ in terms of the almost sure divergence of a sequence of random variables corresponding to the sequence $(\mu_j)_{j\geq 1}$, the technique of Khintchine was extended to general locally compact Abelian groups having a countable basis of their topology by Parthasarathy, Rao and Varadhan in [383] (see also Parthasarathy [377]). More recent research in this direction is due to Fel'dman [139]. Both approaches, however, do not provide an algebraic-topological interpretation of what should be meant by the variance of a probability measure on a locally compact group or semigroup. The relevant point has been made by Grenander in [182], Section 3.3, where possible candidates for variances (or variance functionals) are shown to lack either the continuity (as is the variance of a real random variable) or the full additivity (in favor of super- or subadditivity).

Not necessarily continuous dispersions have been studied in Heyer [226], where the axiomatic formulation of a generalized variance is modelled after Maksimov, who in [330] introduced the notion for probability measures on a finite group. In [330] the aim of the discussion is the proof of an equivalence principle (in the sense of 2.2) and of a three series theorem (in the sense of Kolmogorov), both for random variables taking values in a finite group. Maksimov's notion of a variance proved to be useful also in the theory of additive processes with values in a finite group [337]. (For the precise connection see our discussion in §6 of Chapter VI.)

The central results presented in this section concern the existence of (generalized) variances for probability measures on a compact group G (Theorem 2.4.17) and an application leading to characterizations of sequences of G-random variables which converge in variance (Theorem 2.4.24). The notion of a weak variance given in Definition 2.4.1 and its discussion is due to Maksimov [335]. The proof of Theorem 2.4.4 is taken from Byczkowska, Byczkowski and Timoszyk [55]. The results following from the discussion of the kernel of a weak variance have been published in Maksimov [335], where one also finds a sketch of an approach yielding the general form of all weak variances on $\mathscr{M}^b(G)$. We achieve this basic result (as Theorem 2.4.15) by a slightly different, more algebraic method. Lemmata 2.4.11 to 2.4.13 are exercises from the elementary theory of algebras. Theorem 2.4.15 and its corollary show that weak variances can always be extended to variances and hence, in retrospect it motivates the notion of weak variance. As a consequence of Theorem 2.4.15 we obtain the existence theorem 2.4.17. Its proof is completed by the well-known argument that yields the characterization of compact Lie groups by the finite generation of their dual (see Hewitt and Ross [218], (28.61)). The illustrative result 2.4.19 for variances on $\mathscr{M}^1(G)$, where G is a compact Abelian group, and the subsequent remark are borrowed from Byczkowska, Byczkowski and Timoszyk [56]. In [55] the property (of G) that $\mu \in \mathcal{M}^1(G)$ is a measure with a finite variance iff μ does not admit an idempotent factor is discussed. It is shown that this property is not preserved under the formation of finite direct products. There exists an extension of the theory of variances for compact semigroups due to Byczkowski [58]. The main theorem of [58] culminates in the following generalization of Theorem 2.4.17: Let G be a compact semigroup admitting a variance (in a sense adapted to the semigroup). Then every compact simple subsemigroup of S is a Lie group.

Further references of interest concern the delphic theory invented by Kendall [291], advanced by Davidson [107], [108] and related to the previous work on generalized convolutions by Urbanik [485], [486] through the efforts of Gilewski (and Urbanik) [164] ([165]). In [291] Kendall introduced Abelian topological semigroups for which the central limit theorem for triangular systems holds. Gilewski in [164] proves that each generalized convolution admitting generating functions defines a delphic semigroup. The axiomatized generating function ϕ of Urbanik [485] corresponds to the delphic homomorphism Δ (from the delphic semigroup S into the semigroup $\mathbb{R}_+^*$ satisfying $\Delta(u)=0$ iff u equals the neutral element of S). More recent results on delphic theory (aiming at factorization problems and stability) have been published by N.H. Bingham [23]. A particular point of interest in this connection is the still unsolved problem of Davidson of "delphicizing" compactly generated locally compact Abelian groups G by reducing $\mathcal{M}^b(G)$ to "strongly delphic form". Davidson formulated this idea in a letter to F. Papangelou, which has appeared in print [292].

After having exploited the notion of a variance of a measure for the purpose of describing the convergence of sequences of random variables or measures one is inclined to consider an analogous theory for general moments, in particular, for the expectation. Ideas in this spirit have been introduced by Maksimov [340]. His axiomatization of an expectation on the semigroup $\mathcal{M}^1(G)$ for a compact group G is limited to compact groups with a special subgroup structure, and thus all the applications are merely to groups $\mathbb{T}^n \times \mathfrak{SU}(2, \mathbb{R})^m$ for $n, m \geq 1$. The main application cited in [340] is a three series criterion characterizing the almost sure convergence of G-random variables by conditions on their first and second moments, i.e., expectations and variances. These moment conditions also appear in the papers of Maksimov [333], [334] and [342]. In [334] it is shown that for a compact Lie group G of dimension k the almost sure convergence behavior of the sequence $(Y_n)_{n \geq 1}$ of G-random variables $Y_n := X_1 \cdot \dots \cdot X_n$ for a sequence $(X_j)_{j \geq 1}$ of G-random variables cannot in general be expressed in terms of the almost sure convergence of the corresponding sequence of local random variables taking values in a coordinate neighborhood $U \in \mathfrak{V}_{\mathbb{R}^k}(0)$. A correspondence between the almost sure convergence for G and U can, however, be established if $(Y_n)_{n \geq 1}$ is compared with the sequence $(Y_n')_{n \geq 1}$ of G-random variables

$$Y_n' := X_1' \cdot \dots \cdot X_n' \ (n \geq 1), \text{ where } X_j' := x_j^{-1} X_j x_{j+1}$$

for a suitable sequence $(x_j)_{j \geq 1}$ in G. In Maksimov [342] the underlying group is an arbitrary Lie group G, and the almost sure convergence or divergence is characterized in terms of moment conditions for the corresponding sequence of local random variables. A three series theorem for arbitrary Lie groups remains a conjecture.

For $\mathfrak{SO}(2,\mathbb{R})$-bi-invariant probability measures on $\mathfrak{SU}(2,\mathbb{R})$ dispersions have been introduced via spherical functions by Karpelevich, Tutubalin and Shur in [284] and by Tutubalin in [478] in order to establish a central limit theorem. The results of [284] can be generalized to K-bi-invariant probability measures on a semisimple Lie group G of noncompact type and rank 1, where K is a maximal compact subgroup of G.

R 2.5 Asymptotic Equidistribution

The theory of asymptotically equidistributed sequences in $\mathscr{M}^1(G)$ for a locally compact Abelian group G has been developed by Kerstan and Matthes in their papers [294], [295] and [296]; applications can be found in the monograph of Kerstan, Matthes and Mecke [297], and also in publications by H. Herrmann [216] and by Fichtner [141]. We restrict our discussion to the Abelian case since a famous example of Dieudonné [113] tells us that on arbitrary non-Abelian locally compact groups asymptotically equidistributed sequences of measures do not necessarily exist. Clearly, the existence problem can be solved for amenable groups as is indicated by Gerl [160] with the help of a result of Emerson [130] and a technique presented in the book of Greenleaf [178]. The paper of Gerl also contains extensions of asymptotic equidistribution to nets of measures in $\mathscr{M}^1(G)$ as well as a more detailed treatment of the topologies introduced in $\mathscr{M}^1(G)$ by the various definitions of asymptotically equidistributed families. In this context the reader is referred to the work in the theory of equidistribution on locally compact groups by Cigler [82], Rindler [405], Maxones and Rindler [351] and others mentioned in [405]. The definitions of asymptotically equidistributed and weakly asymptotically equidistributed sequences in $\mathscr{M}^1(G)$ as well as the main results are taken from Kerstan and Matthes [295]. The proof of Theorem 2.5.5 is due to Maxones. Lemma 2.5.13 concerning the comparison of topologies on $\mathscr{M}^1(G)$ is stated in [295]. The proof given in our text is shorter than that of [295]. The weakly asymptotically equidistributed sequence of Example 2.5.21 which is not asymptotically equidistributed is due to Dobrushin [120]. See also Stone [455]. Theorem 2.5.14 can be generalized to non-Abelian groups G, but is not necessarily true for arbitrary asymptotically equidistributed sequences in $\mathscr{M}^1(G)$, as is proved in Gerl [160]. Theorem 2.5.16 can be viewed in connection with Theorem 3.1 in the paper of Csiszár [100].

Finally we want to mention the monographs of Cassels [70] and of Kuipers and Niederreiter [311] as standard references for questions in classical equidistribution theory.

R 2.6 Shifting Iterated Convolutions

This section is the immediate continuation of the preceding one. The method of asymptotic equidistribution is applied to sequences of convolution powers of a given probability measure on a locally compact Abelian group G. The main ideas reproduced in the text are due to Stam [449], [450], who treated the case of measures on the group $\mathbb{R}^n$ (for $n \geq 1$). The work of Stam contains much more detailed information on shifts of iterated convolutions than we could possibly indicate here. Our aim, however, was the generalization of various classical results to general locally compact Abelian groups. This task is based on the paper of Letac [315], in which the fundamental notion of a Stam group has been introduced (as defined in 2.6.7). Following, but independent of Stam were the contributions to the theory by Kerstan and Matthes [295] and [296]. Their studies are the starting point of our presentation. The notion of the lattice group appears in Kerstan and Matthes [295], also Theorems 2.6.1 and 2.6.2 on the relationship between the lattice group of $\mu \in \mathcal{M}^1(G)$ and the shift group of the sequence $(\mu^n)_{n \geq 1}$ defined in Section 5. The lattice group of μ can be used to characterize the weak asymptotic equidistribution of $(\mu^n)_{n \geq 1}$. This is the contents of Corollary 2.6.4. Theorem 2.6.6 provides preparatory facts for the proof of the zero-two law stated as Theorem 2.6.10. A collection of properties of the Stam group is included in Theorems 2.6.8 and 2.6.9. Remarkably, the proofs of these theorems are purely probabilistic, the first one being based on the Borel-Cantelli lemma, and the second one (following Stam [449]) on the central limit theorem for binomial distributions. Consequences of the zero-two law follow as Corollary 2.6.11 and Theorem 2.6.13. The Stam group is an F_σ-subset of G, and for a large class of measures in $\mathcal{M}^1(G)$ it is open (hence closed) in G.

In Theorem 2.6.14 are characterized those measures in $\mathcal{M}^1(G)$ whose Stam group coincides with G or equivalently, whose sequence of powers is asymptotically equidistributed. The measures involved turn out to be non-lattice. An extension of the discussion to general (weakly) asymptotically equidistributed sequences $(\mu_n)_{n \geq 1}$ in $\mathcal{M}^1(G)$ requires the introduction of a system of lattice groups connected with the sequence $(\mu_n)_{n \geq 1}$. This has been accomplished in Section 4 of Kerstan and Matthes [295]. The origins of the theory are the fundamental contributions [121] and [446] of Doob and Smith to classical renewal theory. Theorem 2.6.17 concerning mixtures of asymptotically equidistributed sequences in $\mathcal{M}^1(G)$ is taken from Kerstan and Matthes [295]. The rest of the section is devoted to the question of whether non-lattice measures in $\mathcal{M}^1(G)$ exist whose Stam group is the degenerate group $\{e\}$. This problem has its origin with Wiener and Young, who gave in [502] an example of a probability measure on $\mathbb{R}$ singular with respect to all of its translates. Theorems 2.6.19 and 2.6.20 due to G. Fourt [151] answer the above cited question in the affirmative.

It remains to mention that zero-two laws for probability measures on a locally compact Abelian group have been proved in various places. The approach of Kerstan and Matthes [295] yields a related result (Satz 3.6). Within the more general framework of zero-two laws for Markov processes presented by Derriennic [112] the corresponding result (Proposition 5) is reduced to the Hewitt-Savage theorem. A first zero-two law for probability measures on a discrete Abelian semigroup was discovered by Kinzl [301]. For non-Abelian groups or semigroups G the theory is still in statu nascendi. The paper [187] of Guivarc'h contains starting points. They lead to the note [147] of Foguel, who investigated sufficient conditions for the limit relation

$$\lim_{n \to \infty} \|\nu * \mu^n\| = 0 \quad \text{for } \nu, \mu \in \mathcal{M}^1(G)$$

with $\nu \ll \omega_G$ to hold.

Chapter III

Embedding of Infinitely Divisible Probability Measures

One of the most prominent subclasses of probability measures on a locally compact group is the class of infinitely divisible probability measures. Its rôle for the solution of the central limit problem is well-known from the classical theory. The most important step on the way to the central limit theorem is the embedding of an infinitely divisible measure into a continuous one-parameter convolution semigroup. Since the embedding theorem does not hold for any general locally compact group, the question arises as to what classes of groups yield the validity of an embedding theorem. Establishing these classes of groups will be the aim of the following chapter. First of all, root compact groups which admit an algebraic version of the embedding theorem are studied. The class of root compact groups enables us to describe the domain of validity for the theorem asserting the closedness of the infinitely divisible probability measures in the semigroup of all probability measures and for the fact that every infinitely divisible measure is submonogeneously embeddable. This particular form of algebraic embedding is also studied without the condition of root compactness. For this purpose it appears necessary to introduce a special class of infinitely divisible measures, the Poisson measures, whose various characterizations are presented in detailed fashion. A deeper analysis of the submonogeneous embeddings for general locally compact Abelian groups has been initiated by considerations of the roots of divisible measures on certain free groups. The next problem on the way to continuous embeddings is the existence of one-parameter semigroups in general topological semigroups. This setting is contained in the literature pertaining to the structure theory of compact semigroups. We apply the general theory to the semigroup of probability measures on a locally compact group and obtain the general continuous embedding, which in the Abelian case leads to a most satisfying solution. There follows a discussion of the uniqueness of roots of infinitely divisible probability measures and its implications for the general embedding problem. Finally a treatment of injective algebraic and continuous embeddings illustrates the strong relationship between embedding and the structure of the underlying group. It is shown that the class of locally compact groups for which submonogeneous homomorphisms into the semigroup of probability measures are injective or trivial contains various types of groups which play an important part in representation theory. An extraordinarily deep construction yields an example which shows that there exist locally compact groups for which some homomorphism of the above definition is neither injective nor trivial. Although these considerations concerning noninjective embeddings are of marginal interest for applications, they indicate the complications arising from the general embedding problem.

3.1 Root Compact Groups

We start with the introduction of a class of locally compact groups, which will appear to be of major interest in connection with the embedding of probability measures.

3.1.1 Definition. Let $n\geq 1$. A locally compact group G is called *n-root compact* if for every compact subset C of G there exists a compact subset C_n of G such that all finite sequences $\{x_1,\ldots,x_n\}$ in G with $x_n=e$ satisfying

$$Cx_i\,Cx_j\cap Cx_{i+j}\neq\emptyset \qquad \text{for all } i+j\leq n$$

are contained in C_n.

G is called *root compact* if G is n-root compact for all $n\geq 1$.

For every $n\geq 1$ we abbreviate the class of all n-root compact groups by $\mathbf{R}_n$. The class of all root compact groups will be denoted by $\mathbf{R}$ so that clearly $\mathbf{R}=\bigcap_{n\geq 1}\mathbf{R}_n$ holds.

3.1.2 Properties of the classes $\mathbf{R}_n(n\geq 1)$ and $\mathbf{R}$. (1) If $G\in\mathbf{R}_n$ (or $\mathbf{R}$) and H is a closed subgroup of G, then $H\in\mathbf{R}_n$ (or $\mathbf{R}$).

(2) Let G be a locally compact group and K a compact normal subgroup of G. Then $G\in\mathbf{R}_n$ (or $\mathbf{R}$) iff $G/K\in\mathbf{R}_n$ (or $\mathbf{R}$).

Both properties are special cases of the following property:

(3) Let G and $\dot{G}$ be locally compact groups and ϕ a continuous proper homomorphism from G into $\dot{G}$. Then $G\in\mathbf{R}_n$ (or $\mathbf{R}$) iff $\phi(G)\in\mathbf{R}_n$ (or $\mathbf{R}$).

In fact, for a continuous and proper mapping $\phi\colon G\to\dot{G}$ the subset $\phi(G)$ of $\dot{G}$ is closed in $\dot{G}$, and for every compact subset $\dot{C}$ of $\phi(G)$ the set $\phi^{-1}(\dot{C})$ is compact in G. Let C be a compact subset of G and $\{x_1,\ldots,x_n\}$ with $x_n=e$ a finite sequence in G (or $\phi(G)$) satisfying

$$Cx_i\,Cx_j\cap Cx_{i+j}\neq\emptyset \qquad \text{for all } i+j\leq n.$$

Then this property remains true after application of ϕ (or ϕ^{-1}) also in $\phi(G)$ (or G).

One shows easily that

(4) locally compact direct products $\prod_{\alpha\in\mathbf{A}}G_\alpha$ of groups G_α in $\mathbf{R}_n$ (or $\mathbf{R}$) are again in $\mathbf{R}_n$ (or $\mathbf{R}$). Consequently,

(5) projective limits $\varprojlim_{\alpha\in\mathbf{A}}G_\alpha$ of projective families $(G_\alpha)_{\alpha\in\mathbf{A}}$ of groups $G_\alpha\in\mathbf{R}_n$ (or $\mathbf{R}$) with locally compact product $\prod_{\alpha\in\mathbf{A}}G_\alpha$ are in $\mathbf{R}_n$ (or $\mathbf{R}$).

This follows directly from (4) with the aid of (1).

3.1.3 First Examples. 1. Every compact group is in $\mathbf{R}$.

2. The additive groups $\mathbb{R}$ and $\mathbb{Z}$ are in $\mathbf{R}$.

We shall give the proof for $\mathbb{R}$ only. Let C be a compact subset of $\mathbb{R}$ and $r\in\mathbb{R}_+^*$ such that $C\subset\{x\in\mathbb{R}:|x|\leq r\}$. We suppose that for a given $n\geq 1$ and a sequence $\{x_1,\ldots,x_n\}$ in $\mathbb{R}$ with $x_n:=0$ we have $(C+x_i+C+x_j)\cap(C+x_{i+j})\neq\emptyset$ for all $i+j\leq n$.

Let

$$|x_j| = \max_{i=1,\dots,n-1} |x_i| \quad \text{and put}$$

$$k := j, \quad \text{if } j \le \frac{n}{2}, \quad \text{and} \quad k := n-j, \quad \text{if } j > \frac{n}{2}.$$

Since $(C+x_j+C+x_{n-j}) \cap C \neq \emptyset$, we get $|x_{n-j}| \ge |x_j| - 3r$, and hence $|x_k| \ge |x_j| - 3r$. Moreover, one has $(C+x_k+C+x_k) \cap (C+x_{2k}) \neq \emptyset$, whence

$$|x_j| \ge |x_{2k}| \ge 2|x_k| - 3r \ge 2|x_j| - 9r \quad \text{so that} \quad |x_j| \le 9r.$$

Choosing $C_n := \{x \in \mathbb{R} : |x| \le 9r\}$ (which in this case does not depend on $n \ge 1$) one obtains the assertion.

3. Every compactly generated locally compact Abelian group G is in **R**.

In fact, Theorem A implies that $G \cong \mathbb{R}^m \times \mathbb{Z}^n \times K$, where $m, n \ge 0$ and K is a compact Abelian group. Hence the examples 1 and 2 together with Property (4) of 3.1.2 yield the result.

More generally,

4. every compactly generated locally compact group G with compact commutator subgroup $K(G)$ is in **R**.

This follows immediately from the induction principle (Property (2) of 3.1.2) together with Example 3.

5. There are non-compactly generated locally compact Abelian groups belonging to **R**. One just takes the direct product of infinitely many copies of $\mathbb{Z}$, furnished with the discrete topology.

6. The group $G := \left\{ \begin{pmatrix} 1 & x \\ 0 & y \end{pmatrix} : x \in \mathbb{R}, y \in \mathbb{R}_+^* \right\}$ of proper affine mappings of $\mathbb{R}$ is in **R**. In fact, let $n \ge 1$ and C a compact subset of G. For $i = 1, \dots, n$ we assume that there are $\alpha, \beta, \gamma \in \mathbb{R}$ and $A, B, \Gamma \in]\varepsilon, \infty[$ for some $\varepsilon > 0$ satisfying

$$(*) \qquad \begin{pmatrix} 1 & \alpha \\ 0 & A \end{pmatrix} \begin{pmatrix} 1 & x_i \\ 0 & y_i \end{pmatrix} \begin{pmatrix} 1 & \beta \\ 0 & B \end{pmatrix} \begin{pmatrix} 1 & x_j \\ 0 & y_j \end{pmatrix} = \begin{pmatrix} 1 & \gamma \\ 0 & \Gamma \end{pmatrix} \begin{pmatrix} 1 & x_{i+j} \\ 0 & y_{i+j} \end{pmatrix}.$$

Furthermore, let $x_n = 0$ and $y_n = 1$. Then $(*)$ implies

$$(**) \qquad x_j + y_j \beta + y_j B x_i + y_j \alpha B y_i = x_{i+j} + \gamma y_{i+j} \quad \text{and}$$
$$A B y_i y_j = \Gamma y_{i+j} \quad \text{for } 1 \le i+j \le n.$$

Taking logarithms in the second equation one obtains a constant $m \in]0, M[$ with $m \le y_i \le M$ (for some $M \in \mathbb{R}_+^*$), whenever $i = 1, \dots, n$. But the first equation of $(**)$ implies that

$$c_{i+1} x_1 + x_i = x_{i+1} + \beta_{i+1} \qquad \text{for } i = 1, \dots, n-1,$$

where $0 < \varepsilon \le c_{i+1} \le \varepsilon'$, and $|\beta_{i+1}| \le \delta$ for $\varepsilon \le 1 \le \varepsilon'$ and $\delta \in \mathbb{R}_+^*$. This implies that

$$x_{i+1} \in c_{i+1} x_1 + x_i + \Delta \quad \text{with} \quad \Delta := [-\delta, \delta],$$

and an induction yields

$$x_i \in \left(\sum_{j=1}^{i} c_j\right) x_1 + i\Delta \qquad \text{for } i=1,\dots,n, \text{ where } c_1 := 1.$$

From $x_n = 0$ we obtain

$$\left(\sum_{j=1}^{n} c_j\right) x_1 \in n\Delta \quad \text{and} \quad n\varepsilon |x_1| \le \left(\sum_{j=1}^{n} c_j\right) |x_1| \le n\delta,$$

whence $|x_1| \le \frac{\delta}{\varepsilon}$. But the fact that

$$|x_i| \le \left(\sum_{j=1}^{i} c_j\right) |x_1| + i\delta \le n\varepsilon' \frac{\delta}{\varepsilon} + n\delta = n\delta \left(\frac{\varepsilon'}{\varepsilon} + 1\right)$$

for $i=1,\dots,n$ yields the assertion.

7. The group $G := \left\{ \begin{pmatrix} 1 & x \\ 0 & y \end{pmatrix} : x \in \mathbb{R}, y \in \mathbb{R}^* \right\}$ of general affine mappings of $\mathbb{R}$ is not in **R** since its subset of all elements of order 2 which contains the matrices $\begin{pmatrix} 1 & a \\ 0 & -1 \end{pmatrix}$ with $a \in \mathbb{R}$ is not compact.

With a computation as in Example 6 one can show that

8. the group

$$G = \left\{ \begin{pmatrix} 1 & x & y \\ 0 & 1 & z \\ 0 & 0 & 1 \end{pmatrix} : x, y, z \in \mathbb{Z} \right\}$$

belongs to **R**. This group and the group introduced in Example 6 are non-MAP Lie groups.

For every locally compact group G, any measure $\mu \in \mathscr{M}^1(G)$ and any $n \ge 1$ one defines the set $R(n,\mu) := \{\nu \in \mathscr{M}^1(G) : \nu^n = \mu\}$. Since the convolution in $\mathscr{M}^1(G)$ is continuous, $R(n,\mu)$ is $\mathscr{T}_v$-closed in $\mathscr{M}^1(G)$ for every $n \ge 1$. For any subset $\mathscr{N}$ of $\mathscr{M}^1(G)$ and every $n \ge 1$ we further define $R(n,\mathscr{N}) := \bigcup_{\mu \in \mathscr{N}} R(n,\mu)$. We shall now connect the notion of a root compact group with the structure of $\mathscr{M}^1(G)$.

3.1.4 Theorem. *For any locally compact group G and every $n \ge 1$ we consider the following statements:*

(i) $G \in \mathbf{R}_n$.

(ii) *Given any $\mathscr{T}_v$-relatively compact subset $\mathscr{N}$ of $\mathscr{M}^1(G)$ the set $R(n,\mathscr{N})$ is relatively compact.*

(iii) *Given any compact subset C of G the set $\{x \in G : x^n \in C\}$ is compact.*

Then one has the implications (i) ⇒ (ii) ⇒ (iii).

If, in addition, G is class compact *in the sense that every conjugacy class of G is relatively compact, then* (iii) ⇒ (i) *also holds.*

Proof. 1. (i) ⇒ (ii). Let $\mathscr{N}$ be a $\mathscr{T}_v$-relatively compact subset of $\mathscr{M}^1(G)$. Then by Prohorov's theorem there exists for $\varepsilon \in]0, \frac{1}{3}[$ a compact subset C of G satisfying $\mu(C) \ge 1 - \varepsilon$ for all $\mu \in \mathscr{N}$. Let $\nu \in R(n,\mu)$. Since

$$\nu^{n-k} * \nu^k(C) = \mu(C) \ge 1 - \varepsilon \qquad \text{for all } 1 \le k < n,$$

there exists an $x_k \in G$ such that $\nu^k(Cx_k) \geq 1-\varepsilon$ holds. Putting $x_n := e$ one obtains for all $l \geq 1$ with $k+l \leq n$ the relations

$$\nu^{k+l}(Cx_k\,Cx_l) \geq \nu^k(Cx_k)\,\nu^l(Cx_l) \geq (1-\varepsilon)^2, \quad \text{whence}$$
$$\nu^{k+l}(Cx_k\,Cx_l \cap Cx_{k+l}) \geq \nu^{k+l}(Cx_k\,Cx_l) - \nu^{k+l}(Cx_k\,Cx_l \cap (\complement C)\,x_{k+l})$$
$$\geq \nu^{k+l}(Cx_k\,Cx_l) - \nu^{k+l}((\complement C)\,x_{k+l}) \geq (1-\varepsilon)^2 - \varepsilon > 1 - 3\varepsilon > 0.$$

Therefore $Cx_k\,Cx_l \cap Cx_{k+l} \neq \emptyset$ for all $k+l \leq n$. Since by assumption $G \in \mathbf{R}_n$, there exists a compact subset C_n of G containing the sequence $\{x_1, \ldots, x_n\}$. Consequently, $\nu(CC_n) \geq \nu(Cx_1) \geq 1-\varepsilon$. But this implies that $\nu(CC_n) \geq 1-\varepsilon$ for all $\nu \in R(n, \mathcal{N})$, which by Prohorov's theorem shows the $\mathcal{T}_v$-relative compactness of $R(n, \mathcal{N})$ in $\mathcal{M}^1(G)$.

2. (ii) ⇒ (iii). Let C be a compact subset of G. Then $\mathcal{N} := \{\varepsilon_x : x \in C\}$ is $\mathcal{T}_v$-compact in $\mathcal{M}^1(G)$. Hence, by assumption $R(n, \mathcal{N}) = \{\varepsilon_y : y^n \in C\}$ is relatively compact in $\mathcal{M}^1(G)$. Thus the set $\{x \in G : x^n \in C\}$ is relatively compact in G and, since it is closed, even compact.

3. Let G be class compact. We prove the implication (iii) ⇒ (i). We first note that any class compact locally compact group is uniformly class compact in the sense that every compact subset of G is contained in a G-invariant compact subset of G ([407], Theorem 3D and [323], Theorem 2.2). Given a compact subset C of G we may suppose without loss of generality that $C = C^{-1}$, $e \in C$ and $xCx^{-1} = C$ for all $x \in G$ hold. Let $\{x_1, \ldots, x_n\}$ be a sequence in G with $x_n = e$ such that

$$Cx_i\,Cx_j \cap Cx_{i+j} \neq \emptyset \qquad \text{for all } i+j \leq n.$$

By induction one shows that

$$x_i \in C^{3(i-1)}\,x_1^i \qquad \text{for } i = 1, \ldots, n.$$

In particular, $e = x_n \in C^{3(n-1)}\,x_1^n$, and hence $x_1^n \in C^{3(n-1)}$. But by assumption the set

$$K_n := \{x \in G : x^n \in C^{3(n-1)}\}$$

is compact, and $x_1 \in K_n$. From $e \in C \cap K_n$ follows

$$x_i \in C_n := C^{3(n-1)}\,K_n^n \qquad \text{for all } i = 1, \ldots, n,$$

and, since C_n is compact, the proof of (i) is complete. □

In order to establish further classes of examples of root compact groups we first present a restricted induction principle. If G is a locally compact group and N a closed normal subgroup of G, then a subset F of G is called *N-invariant* if $xFx^{-1} = F$ for all $x \in N$.

3.1.5 Theorem. *Let G be a locally compact group, N a closed normal subgroup of G and let any compact subset of G be contained in an N-invariant compact subset of G. If N and G/N are in $\mathbf{R}_n$, then also $G \in \mathbf{R}_n$ (for $n \geq 1$).*

Proof. Let $\dot{G} := G/N$ and let p denote the canonical epimorphism from G onto $\dot{G}$. Let C be a compact subset of G, which without loss of generality can be assumed N-invariant. Since $\dot{G} \in \mathbf{R}_n$, for $\dot{C} := p(C)$ there exists a compact subset $\dot{C}_n$ of $\dot{G}$ such that the defining relationship of n-root compactness of $\dot{G}$ is satisfied. Moreover, there is a compact subset F of G satisfying $e \in F$ and $p(F) \supset \dot{C}_n$. Without loss of generality we assume F to be N-invariant. But $K := \{e\} \cup (N \cap (CF)^{-2} CF)$ is a compact subset of N. Since $N \in \mathbf{R}_n$, for K there exists a compact subset K_n of N satisfying the defining relationship of n-root compactness of N. Let $\{x_1, \ldots, x_n\}$ be a sequence in G with $x_n = e$ satisfying

$$C x_i \, C x_j \cap C x_{i+j} \neq \varnothing \quad \text{for all } i+j \leq n.$$

Clearly, $p(x_n) = \dot{e} \in \dot{G}$ and

$$p(C)\, p(x_i)\, p(C)\, p(x_j) \cap p(C)\, p(x_{i+j}) \neq \varnothing \quad \text{for all } i+j \leq n.$$

Thus $\{p(x_1), \ldots, p(x_n)\} \subset \dot{C}_n$ and there are $a_1, \ldots, a_n \in F$, $y_1, \ldots, y_n \in N$ satisfying $x_i = a_i y_i$ for $i = 1, \ldots, n$ and $a_n = y_n = e$. From the defining property of $\{x_1, \ldots, x_n\}$ we conclude that $CF y_i \, CF y_j \cap CF y_{i+j} \neq \varnothing$ for $i+j \leq n$. The N-invariance of C and F implies that of CF. Hence,

$$y_i \, y_j \, y_{i+j}^{-1} \in (CF)^{-2} CF \cap N \subset K \quad \text{for } i+j \leq n.$$

Consequently, $e \in K$ implies $K y_i \, K y_j \cap K y_{i+j} \neq \varnothing$ for all $i+j \leq n$ and $y_n = e$ and thus $\{y_1, \ldots, y_n\} \subset K_n$. But then $\{x_1, \ldots, x_n\} \subset F K_n =: C_n$, and C_n is compact. This completes the proof. □

3.1.6 Remark. The assumptions of Theorem 3.1.5 yielding the restricted induction principle are satisfied in the following special cases:

(i) N is a compact extension of a subgroup of $Z(G)$. In particular,
(ii) N is a subgroup of $Z(G)$ and
(iii) N is a compact subgroup of G.

Moreover,

(iv) G is class compact.

We shall prove (i) and (iv). As for (i), let C be a compact subset of G. There is a closed subgroup Z of $Z(G)$ such that $Z \subset N$ and N/Z is compact. We define a mapping $f \colon C \times (N/Z) \to G$ by $f(c, \dot{x}) := x c x^{-1}$ for all $c \in C$, $\dot{x} \in N/Z$. The map f is continuous, and $C' := f(C \times (N/Z))$ is an N-invariant compact subset of G with $C \subset C'$. Concerning (iv), one notes that G is in fact uniformly class compact by [323], Theorem 2.2. Hence, every compact subset of G is contained in a G-invariant compact set.

3.1.7 Theorem. *Let G be a class compact group and $\mathbf{N}$ an ascending system of open normal subgroups in $\mathbf{R}_n$ $(n \geq 1)$ with $G = \bigcup \{N : N \in \mathbf{N}\}$ such that for every $N \in \mathbf{N}$ there exists only a finite number of elements in G/N whose order divides n. Then $G \in \mathbf{R}_n$.*

Proof. As a class compact group G is in fact uniformly class compact (by [323], Theorem 2.2). Let C be a G-invariant compact subset of G and put

$$C_{(n)} := \{x \in G : x^n \in C\} \quad (n \geq 1).$$

By Theorem 3.1.4 we have to show that $C_{(n)}$ is compact in G. Let $N \in \mathbf{N}$ be given such that $C \subset N$, which follows from the assumption, and denote the canonical mapping $G \to G/N$ by p. Again by hypothesis there exists a sequence $\{x_1, \ldots, x_r\}$ in G such that $p(x_1), \ldots, p(x_r) \in G/N$ are the elements whose order divides n. Let $x \in C_{(n)}$. From $x^n \in C \subset N$ follows $p(x)^n = p(e)$ and hence there are a $j \in \{1, \ldots, r\}$ and an $a \in N$ satisfying $x = x_j a$. But there exists an invariant compact subset D of G with $\{x_1^{-1}, \ldots, x_r^{-1}\} \subset D$. From $a \in Dx$ follows $a^n \in D^n x^n \subset D^n C$. Since $D^n C$ is compact, there exists an $N' \in \mathbf{N}$ with $N \subset N'$ and $D^n C \subset N'$. The set

$$D_{(n)} := \{y \in N' : y^n \in D^n C\}$$

is compact by Theorem 3.1.4 since $N' \in \mathbf{R}_n$. But $a \in N \subset N'$ and $a^n \in D^n C$ imply that $a \in D_{(n)}$, which shows that $x = x_j a \in x_j D_{(n)}$. Hence,

$$C_{(n)} \subset \bigcup_{j=1}^{r} x_j D_{(n)}.$$

Thus $C_{(n)}$ is compact and consequently, $G \in \mathbf{R}_n$. □

An element a of an arbitrary topological group G is called *compact* if $[a]^-$ is compact in G. If G is a locally compact Abelian group, then the *set* $B(G)$ *of all compact elements* of G is a closed subgroup of G which by the structure theorem M admits compact open subgroups. It should be noted that in non-Abelian locally compact groups G the set $B(G)$ may fail to be a group or even a closed set.

3.1.8 Theorem. *For any locally compact Abelian group G the following statements are equivalent:*

(i) $G \in \mathbf{R}_n$ $(n \geq 1)$;

(ii) *for every* $\mathcal{T}_v$*-relatively compact subset* $\mathcal{N}$ *of* $\mathcal{M}^1(G)$ *the set* $R(n, \mathcal{N})$ *is relatively compact;*

(iii) *for every* $\mu \in \mathcal{M}^1(G)$ *the set* $R(n, \mu)$ *is* $\mathcal{T}_v$*-compact in* $\mathcal{M}^1(G)$;

(iv) *for every compact open subgroup K of* $B(G)$ *the set* $K_{(n)} := \{x \in G : x^n \in K\}$ *is compact in G;*

(v) *for some compact open subgroup K of* $B(G)$ *the set* $K_{(n)}$ *is compact in G.*

Proof. The implication (i)⇒(ii) follows from Theorem 3.1.4. The assertions (ii)⇒(iii) and (iv)⇒(v) are evident. It remains to prove (iii)⇒(iv) and (v)⇒(i).

1. (iii)⇒(iv). Let K be a compact open subgroup of $B(G)$. Then the subset $\{\varepsilon_x * \omega_K : x \in K_{(n)}\}$ of $R(n, \omega_K)$ is by assumption relatively compact, hence uniformly tight, and therefore there exists a compact subset C of G satisfying

$$\tfrac{1}{2} \leq \varepsilon_x * \omega_K(C) = \omega_K(x^{-1} C) \qquad \text{for all } x \in K_{(n)}.$$

But this yields $K \cap x^{-1} C \neq \emptyset$ and $K_{(n)}$ is contained in the compact set KC, and is thus itself compact.

2. (v)⇒(i). By the structure theorem M there exist a $k \geq 0$ and a locally compact Abelian group H containing a compact open subgroup K such that $G \cong \mathbb{R}^k \times H$ holds. By Example 3 of 3.1.3 we have $\mathbb{R}^k \in \mathbf{R}_n$. Hence, by Property (4) of 3.1.2 it suffices to show that $H \in \mathbf{R}_n$. Moreover, Remark 3.1.6 implies that it is sufficient to prove that the discrete group $D := H/K$ lies in $\mathbf{R}_n$. Finally, Theorem 3.1.4 reduces the proof of the desired implication to that of the statement that

every $x \in D$ admits at most finitely many n-th roots. Indeed, let $D_{(n)} := \{x \in D : x^n = e\}$. Then $D_{(n)}$ is finite since $K_{(n)}$ is compact and $K_{(n)}/K = D_{(n)}$ is discrete. If a and b are two n-th roots of $x \in D$, then $(ab^{-1})^n = xx^{-1} = e$ implies that $ab^{-1} \in D_{(n)}$. The finiteness of $D_{(n)}$ yields the assertion. □

3.1.9 Further examples. 9. A discrete Abelian group G is in $\mathbf{R}_n$ iff G admits for every divisor m of n at most a finite number of elements of m-th order.

This follows immediately from Theorem 3.1.8. In particular,

10. $\mathbb{Q}_d$ and every free Abelian group furnished with the discrete topology are in $\mathbf{R}$.

11. The free Abelian group G with generators $x_1, x_2, \ldots$ and relations $x_1 = x_n^n$ for all $n \geq 1$ is, however, not in $\mathbf{R}_n$ for all $n \geq 1$. In fact, the relationship $x_1 = (x_{mn}^m)^n$ for all $m \geq 1$ shows that G admits infinitely many elements of order n.

12. For every prime p the group Ω_p of p-adic numbers is in $\mathbf{R}$.

This follows from Theorem 3.1.7 since in Ω_p there is an increasing sequence of compact open normal subgroups Δ_m with $\Omega_p = \bigcup_{m \geqq 1} \Delta_m$ such that for every $m \geq 1$ and every $n \geq 1$ there are only finitely many elements in Ω_p/Δ_m of order n.

13. Let G be a locally compact group with compact commutator subgroup $K(G)$ in which case $B(G)$ is compact. Then $G \in \mathbf{R}$.

Indeed, Remark 3.1.6 permits us to assume without loss of generality that G is Abelian. But then the statement follows from Theorem 3.1.8.

14. The class $\mathbf{R}$ is not closed under the formation of finite extensions, as becomes clear from the following counterexample: The semidirect product $G := \mathbb{R} \times_\eta \mathbb{Z}_2$ with $\eta(\varepsilon)(x) := \varepsilon x$ for all $\varepsilon \in \mathbb{Z}_2$ and $x \in \mathbb{R}$ is a finite extension of $\mathbb{R}$, but is not in $\mathbf{R}$ since the set $G_2 := \{(x, -1) : x \in \mathbb{R}\}$ of elements of order 2 of G is not compact. Therefore by Theorem 3.1.4 G cannot lie in $\mathbf{R}$.

We now strengthen the definition of a root compact group in the sense that the compact set C_n appearing in its formulation is required to be independent of n.

3.1.10 Definition. A locally compact group G is called *strongly root compact* if for every compact subset C of G there exists a compact subset C_0 in G with the property that for every $n \geq 1$ the finite sequences $\{x_1, \ldots, x_n\}$ of G with $x_n = e$ satisfying $Cx_i Cx_j \cap Cx_{i+j} \neq \emptyset$ for all $i + j \leq n$ are contained in C_0.

The class of all strongly root compact groups will be denoted by $\mathbf{R}_0$. Clearly, $\mathbf{R}_0 \subset \mathbf{R}$. We collect the following

3.1.11 Properties of the class $\mathbf{R}_0$, which are obtained directly from 3.1.2 and 3.1.5: (1) The class $\mathbf{R}_0$ is closed under the formation of closed subgroups, locally compact direct products and locally compact projective limits (in the sense of Property (5) of 3.1.2).

(2) If G is a locally compact group, N a closed normal subgroup of G and if every compact subset of G is contained in an N-invariant compact set, then $N, G/N \in \mathbf{R}_0$ implies that $G \in \mathbf{R}_0$.

In particular,

(3) if G is a locally compact group and K is a compact normal subgroup of G, then $G \in \mathbf{R}_0$ iff $G/K \in \mathbf{R}_0$.

3.1.12 Examples. 1. Every compact group is in $\mathbf{R}_0$.

2. Every compactly generated locally compact Abelian group is in $\mathbf{R}_0$.

3. Every free Abelian group furnished with the discrete topology belongs to $\mathbf{R}_0$.

For every $\mu \in \mathscr{M}^1(G)$ the set

$$R(\mu) := \bigcup_{n \geq 1} \{\nu^m : \nu \in \mathscr{M}^1(G) \text{ with } \nu^n = \mu, 1 \leq m \leq n\}$$

is called the *root set* of μ. Analogously the root set of a subset $\mathscr{N}$ of $\mathscr{M}^1(G)$ is defined as the set $R(\mathscr{N}) := \bigcup_{\mu \in \mathscr{N}} R(\mu)$.

3.1.13 Theorem. *Let G be a locally compact group. We consider the following statements:*

(i) $G \in \mathbf{R}_0$,

(ii) *for every $\mathscr{T}_v$-relatively compact subset $\mathscr{N}$ of $\mathscr{M}^1(G)$ the root set $R(\mathscr{N})$ is itself relatively compact in $\mathscr{M}^1(G)$,*

(iii) *for every compact subset C of G the set $C_\infty := \{x \in G : x^n \in C$ for some $n \geq 1\}$ is relatively compact in G,*

(iv) *$B(G)$ is compact.*

Then one has the implications (i)$\Rightarrow$(ii)$\Rightarrow$(iii)$\Rightarrow$(iv).

Proof. 1. (i)$\Rightarrow$(ii) follows in analogy to the proof of (i)$\Rightarrow$(ii) in Theorem 3.1.4.

2. (ii)$\Rightarrow$(iii). Let C be a compact subset of G and $\mathscr{N} := \{\varepsilon_x : x \in C\}$. Then $\mathscr{N}$ is relatively compact and hence by assumption $R(\mathscr{N})$ is also relatively compact. By Prohorov's theorem there is a compact subset F of G satisfying $\mu(F) \geq \frac{1}{2}$ for all $\mu \in R(\mathscr{N})$. If $x \in C_\infty = \{x \in G : x^n \in C$ for some $n \geq 1\}$, then $\varepsilon_x \in R(\mathscr{N})$ or $\varepsilon_x(F) \geq \frac{1}{2}$ so that $x \in F$ and consequently, $C_\infty \subset F$, which shows the relative compactness of C_∞.

3. (iii)$\Rightarrow$(iv). Let U be a compact neighborhood in $\mathfrak{B}(e)$. Then

$$U_\infty := \{x \in G : x^n \in U \text{ for some } n \geq 1\}$$

is relatively compact in G. For every $x \in B(G)$ there exists an $n \geq 1$ satisfying $x^n \in U$. Thus, $B(G) \subset U_\infty$, and $B(G)$ is relatively compact. But $B(G)$ is also closed. Indeed, if $(x_\alpha)_{\alpha \in \mathbf{A}}$ is a net in $B(G)$ with $\lim_{\alpha \in \mathbf{A}} x_\alpha = x$, then x^n lies in the compact set $\overline{B(G)}$ for all $n \geq 1$. Hence, $[x]^- \subset \overline{B(G)}$ and therefore $[x]^-$ is compact, which implies that $x \in B(G)$. ▯

3.1.14 Remark. The preceding theorem implies that strongly root compact groups which are class compact admit largest compact normal subgroups.

3.1.15 Remark. $\mathbf{R} \neq \mathbf{R}_0$ since $\mathbb{Q}_d \in \mathbf{R}$ (by Example 10 of 3.1.9), and $\mathbb{Q}_d \notin \mathbf{R}_0$ (by Theorem 3.1.13). But $\mathbb{Q}_d$ satisfies Statement (iv) of Theorem 3.1.13 which shows that the statements in the theorem are not in general equivalent.

We are now going to add to our list of examples in $\mathbf{R}_0$. The next theorem contains Examples 6 and 8 of 3.1.3 as special cases.

3.1.16 Lemma. *Let G be a compactly generated locally compact group which is nilpotent of rank n_G in the sense that $Z_n(G)=\overline{C_n(G)}=\{e\}$ with a smallest $n:=n_G\geq 1$. Then $Z_{n_G-1}(G)$ is compactly generated.*

Proof. 1. If G is nilpotent of rank n_G and $N:=Z_{n_G-1}(G)$, then

$$\{e\}\neq N\subset Z(G).$$

Let p denote the canonical projection from G onto $\dot{G}:=G/N$. Then for every $n\geq 0$ we have $C_n(\dot{G})=p(C_n(G))$ and hence, $Z_n(\dot{G})\supset p(Z_n(G))$. This implies that $\dot{G}$ is also nilpotent of rank $n_{\dot{G}}=n_G-1$.

2. In order to prove the statement of the lemma we perform an induction with respect to the rank n_G of G. For $n_G=1$ one gets $N=G$ and the assertion becomes evident. Let the lemma be proved for all nilpotent compactly generated locally compact groups G with $n_G=r\geq 1$. We have to deal with the case $n_G=r+1$. Since N is a locally compact Abelian group, there exists an open compactly generated subgroup H of N with $H\neq N$. H is a subgroup of $Z(G)$. With the notation q for the canonical mapping from G onto G/H we obtain that $q(N)\cong N/H$ is discrete and closed in G/H. Then

$$Z_r(G/H)=\overline{C_r(G/H)}=\overline{q(C_r(G))}=q(\overline{C_r(G)})=q(N)$$

is discrete. But G/H is compactly generated and nilpotent of rank $n_{G/H}=n_G$. By the closure properties of the class of all compactly generated locally compact groups it suffices to show that $Z_r(G/H)$ is compactly generated, for in this case

$$Z_r(G/H)=q(N)\cong N/H$$

will yield that N is compactly generated.

3. Without loss of generality we assume that N is discrete. Then

$$N=Z_r(G)=C_r(G).$$

There exists a symmetric compact neighborhood $U\in\mathfrak{B}_G(e)$ with $G=\bigcup_{n\geq 1}U^n$. The canonical mapping from G onto G/N will be abbreviated by p. Since by Part 1 we have $n_{\dot{G}}=n_G-1=r$, the induction hypothesis implies that $Z_{r-1}(\dot{G})=p(Z_{r-1}(G))$ is compactly generated. Thus there is a symmetric compact $V\in\mathfrak{B}_{Z_{r-1}(G)}(e)$ satisfying

$$Z_{r-1}(\dot{G})=\bigcup_{n\geq 1}p(V)^n=p(\bigcup_{n\geq 1}V^n).$$

It remains to show that N is generated by the compact set

$$K:=\{[x,y]:x\in U,\ y\in V\}.$$

In fact, $N=C_r(G)$ is generated by the set

$$F:=\{[x,y]:x\in G,\ y\in Z_{r-1}(G)\}.$$

Moreover, $N \subset Z(G)$, and hence $[a,b] \in Z(G)$ for all $a \in G$, $b \in Z_{r-1}(G)$. Let now $x \in G$ and $y \in Z_{r-1}(G)$. There exist $x_1, \ldots, x_n \in U$ satisfying

$$[x,y] = [(x_1 \cdot \cdots \cdot x_{n-1})x_n, y] = x_n^{-1}[x_1 \cdot \cdots \cdot x_{n-1}, y]\, x_n [x_n, y]$$
$$= [x_1 \cdot \cdots \cdot x_{n-1}, y][x_n, y] = \cdots = [x_1, y] \cdot \cdots \cdot [x_n, y].$$

From the fact that $Z_{r-1}(\dot{G}) = p(\bigcup_{n\geq 1} V^n)$ we deduce the existence of $m \geq 1$, $y_1, \ldots, y_m \in V$ and $z \in Z_r(G)$ with $y = y_1 \cdot \cdots \cdot y_m z$. For every $j = 1, \ldots, n$ we further have

$$[x_j, y] = [x_j, y_1 \cdot \cdots \cdot y_m z] = [x_j, y_m][x_j, y_1 \cdot \cdots \cdot y_{m-1}]$$
$$= [x_j, y_m] \cdot \cdots \cdot [x_j, y_1].$$

This shows that $[x,y] = \prod_{\substack{1 \leq j \leq n \\ 1 \leq k \leq m}} [x_j, y_k]$.

For all $j = 1, \ldots, n$ and $k = 1, \ldots, m$ the elements $[x_j, y_k]$ belong to K, so that $F \subset [K]$ or $N \subset [K]$. But $K \subset N$ implies the assertion. □

3.1.17 Theorem. Every nilpotent, compactly generated locally compact group G belongs to $\mathbf{R}_0$.

Proof. Again we perform an induction with respect to the rank n_G of G. For $n_G = 1$ we obtain the special case of an Abelian group G, and Example 1 of 3.1.12 settles the problem. We assume the assertion to be true for all nilpotent, compactly generated locally compact groups G with $n_G = r \geq 1$ and show it for those G with $n_G = r+1$. It follows that $\dot{G} := G/Z_r(G)$ is a nilpotent, compactly generated locally compact group with $n_{\dot{G}} = n_G - 1 = r$. By the induction hypothesis $\dot{G}$ is in $\mathbf{R}_0$. We now infer from Lemma 3.1.16 that $Z_r(G)$ is compactly generated. As a central subgroup of G the group $Z_r(G)$ is Abelian and is thus in $\mathbf{R}_0$. Property (2) of 3.1.11 yields $G \in \mathbf{R}_0$. □

3.1.18 Remark. If the group G of the theorem is connected, then $Z_n(G)$ is connected for every $n \geq 1$, and the proof of the theorem does not require Lemma 3.1.16.

In fact, let H be a connected subgroup of G and

$$L := \{[x,y] : x \in G,\ y \in H\} = \bigcup_{x \in G} \{[x,y] : y \in H\}.$$

For every $x \in G$ the set $\{[x,y] : y \in H\}$ is connected (as a continuous image of H), and $\bigcap_{x \in G} \{[x,y] : y \in H\} = \{e\}$. Hence, L is connected. But then also $L_1 := L \cup L^{-1}$ and the sets L_1^n $(n \geq 1)$ are connected. This implies the connectedness of $[G,H] = \bigcup_{n \geq 1} L_1^n$.

In order to establish classes of solvable groups belonging to $\mathbf{R}_0$ the following

3.1.19 Preparations concerning complex Lie algebras are needed.

(1) Let $\mathfrak{L}$ be a solvable Lie algebra of finite dimension over $\mathbb{C}$. Furthermore, let M be an $\mathfrak{L}$-module of finite dimension over $\mathbb{C}$ such that there exists a *Jordan-Hölder series* of modules of the form

$$\mathsf{M} = \mathsf{M}_0 \supset \mathsf{M}_1 \supset \cdots \supset \mathsf{M}_{t-1} \supset \mathsf{M}_t = \{0\}.$$

Then the $\mathfrak{L}$-module $\mathsf{M}_k/\mathsf{M}_{k+1}$ is of dimension 1 over $\mathbb{C}$ for all $0 \le k \le t-1$, and for every element $X \in \mathfrak{L}$ the projection $X_{\mathsf{M}_k/\mathsf{M}_{k+1}}$ of X to $\mathsf{M}_k/\mathsf{M}_{k+1}$ is of the form $\rho_k(X) \cdot 1$, where ρ_k is a linear functional on $\mathfrak{L}$ which vanishes on the characteristic ideal $[\mathfrak{L}, \mathfrak{L}]$ of $\mathfrak{L}$. In particular, every irreducible (simple) $\mathfrak{L}$-module of finite dimension over $\mathbb{C}$ is one-dimensional. Finally we mention that a solvable Lie algebra $\mathfrak{L}$ over $\mathbb{C}$ of dimension n has the property that every ideal of $\mathfrak{L}$ is a member of a Jordan-Hölder series of ideals of the form

$$\mathfrak{L} = \mathfrak{L}'_0 \supset \mathfrak{L}'_1 \supset \cdots \supset \mathfrak{L}'_{n-1} \supset \mathfrak{L}'_n = \{0\}$$

such that $\mathfrak{L}'_k$ has dimension $n-k$ for $0 \le k \le n$.

We note an important special case.

(2) Let $\mathfrak{L}$ be a solvable Lie algebra over $\mathbb{R}$ (of finite dimension). Then every irreducible representation of $\mathfrak{L}$ is of dimension ≤ 2, and every ideal of $\mathfrak{L}$ is a member of a decreasing sequence of ideals in $\mathfrak{L}$ of the form

$$\mathfrak{L} = \mathfrak{L}_0 \supset \mathfrak{L}_1 \supset \cdots \supset \mathfrak{L}_{m-1} \supset \mathfrak{L}_m = \{0\}$$

such that the dimension of $\mathfrak{L}_k/\mathfrak{L}_{k+1}$ is ≤ 2 for all $0 \le k \le m-1$.

(3) We now assume G to be a connected solvable real Lie group of (real) dimension n with Lie algebra $\mathfrak{L}(G)$ such that $\{X_1, \ldots, X_n\}$ is a basis of

$$\mathfrak{L}(G) = \mathfrak{L}_0 \supset \mathfrak{L}_1 \supset \cdots \supset \mathfrak{L}_{n-1} \supset \mathfrak{L}_n = \{0\}$$

with $X_k \in \mathfrak{L}_k \setminus \mathfrak{L}_{k+1}$ for $0 \le k \le n-1$. We apply Theorem D. Let $\bar{G}$ be the simply connected universal covering group of G and p the corresponding covering homomorphism. Then $D := \ker p$ is a discrete subgroup with a finite number of generators in $Z(G)$. Moreover, there exists a subset $\{X_{k_1}, \ldots, X_{k_d}\}$ of $\{X_1, \ldots, X_n\}$ with $[X_{k_r}, X_{k_s}] = 0$ for $1 \le r, s \le d$, and $\{\exp X_{k_1}, \ldots, \exp X_{k_d}\}$ is a generator of D. Every $x \in G$ can be written uniquely in the form $\prod_{i=1}^n \exp s_i X_i$, where uniqueness is meant in the following sense:

$$\prod_{i=1}^n \exp s_i X_i = \prod_{i=1}^n \exp t_i X_i$$
$$\Leftrightarrow s_i = \begin{cases} t_i, & \text{if } i \notin \{k_1, \ldots, k_d\} \\ t_i + \mathbb{Z}, & \text{if } i \in \{k_1, \ldots, k_d\}. \end{cases}$$

One has a homeomorphism $G \approx \prod_{i=1}^n A_i$ with

$$A_i := \begin{cases} \mathbb{R}, & \text{if } i \notin \{k_1, \ldots, k_d\} \\ \mathbb{T} := \mathbb{R}/\mathbb{Z}, & \text{if } i \in \{k_1, \ldots, k_d\}, \end{cases}$$

the homeomorphism being defined as the mapping

$$\prod_{i=1}^n \exp s_i X_i \to (\bar{s}_i)_{i=1,\ldots,n},$$

where

$$\bar{s}_i := \begin{cases} s_i, & \text{if } i \notin \{k_1, \ldots, k_d\} \\ s_i + \mathbb{Z}, & \text{if } i \in \{k_1, \ldots, k_d\}. \end{cases}$$

(4) To a given Jordan-Hölder series $\mathfrak{L}(G) = \mathfrak{L}_0 \supset \mathfrak{L}_1 \supset \cdots \supset \mathfrak{L}_{m-1} \supset \mathfrak{L}_m = \{0\}$ of the Lie algebra $\mathfrak{L}(G)$ of a given connected solvable (real) Lie group G there

corresponds a chain of groups

$$G=G_0\supset G_1\supset\cdots\supset G_{m-1}\supset G_m=\{e\},$$

where G_{k+1} is a closed analytic normal subgroup of G_k for $0\le k\le m-1$. For a given $k=1,\dots,m$ one introduces the orthogonal complement $\mathfrak{M}_k$ of $\mathfrak{L}_k$ such that $\mathfrak{L}(G)=\mathfrak{L}_k\oplus\mathfrak{M}_k$. Using the canonical mappings, in particular the canonical surjection $p_k: G\to G/G_k$ and its differential

$$dp_k: \mathfrak{L}(G)=\mathfrak{L}_k\oplus\mathfrak{M}_k\to\mathfrak{L}(G/G_k)\cong\mathfrak{L}(G)/\mathfrak{L}_k \qquad \text{for } k=1,\dots,m,$$

one obtains that $\operatorname{Res}_{\mathfrak{M}_k} dp_k: \mathfrak{M}_k\to\mathfrak{L}(G/G_k)$ is an isomorphism of vector spaces. Introducing the number $r_k := 1+\dim G_k$ and the subset

$$E_k := \prod_{i=r_k}^{n} \exp\mathbb{R}X_i$$

of G one observes that $\operatorname{Res}_{E_k} p_k: E_k\to G/G_k$ is a homeomorphism.

(5) Let G be a connected solvable real Lie group of finite dimension with Lie algebra $\mathfrak{L}(G)$ and complexification $\tilde{\mathfrak{L}}(G) := \mathfrak{L}(G)\otimes_{\mathbb{R}}\mathbb{C}$ of $\mathfrak{L}(G)$. We assume the existence of a Jordan-Hölder series of the form

$$\mathfrak{L}(G)=\mathfrak{L}_0\supset\mathfrak{L}_1\supset\cdots\supset\mathfrak{L}_{m-1}\supset\mathfrak{L}_m=\{0\}.$$

Applying the discussion of (1) to the complex Lie algebra $\tilde{\mathfrak{L}}(G)$ and using the fact that every irreducible representation ρ of $\tilde{\mathfrak{L}}(G)$ is of complex dimension 1, one obtains complex linear functionals ρ_k on $\tilde{\mathfrak{L}}(G)$ (with values in the vector space $\mathsf{M}_k/\mathsf{M}_{k+1}$) vanishing on the characteristic ideal $[\tilde{\mathfrak{L}}(G),\tilde{\mathfrak{L}}(G)]$ of $\tilde{\mathfrak{L}}(G)$ for $0\le k\le m-1$. If ρ is taken to be the adjoint representation ad of $\tilde{\mathfrak{L}}(G)$, then the linear functionals ρ_k deduced from ρ in $\mathsf{M}_k/\mathsf{M}_{k+1}$ are called the *roots* of $\mathfrak{L}(G)$. These roots are of the form $\phi_1+i\phi_2$, where ϕ_1 and ϕ_2 are real linear functionals on $\mathfrak{L}(G)$. $\mathfrak{L}(G)$ admits roots of the form $\phi_1+i\phi_2$ for $\phi_2\neq 0$ iff there exists a $k\in\{0,\dots,m-1\}$ such that $\dim(\mathfrak{L}_k/\mathfrak{L}_{k+1})=2$ and

$$[Y, X_{k_r}+iX_{k_{r+1}}]=(\phi_1(Y)+i\phi_2(Y))(X_{k_r}+iX_{k_{r+1}}) \pmod{\mathfrak{L}_k}$$

for all $Y\in\mathfrak{L}(G)$. In particular, $\mathfrak{L}(G)$ admits only real roots iff

$$\dim(\mathfrak{L}_k/\mathfrak{L}_{k+1})=1 \qquad \text{for all } k=0,\dots,m-1.$$

(6) Let G be a connected nilpotent real Lie group with Lie algebra $\mathfrak{L}(G)$. Then $\mathfrak{L}(G)$ admits only real roots. This follows from the definition of a nilpotent Lie algebra.

3.1.20 Lemma. *Let K_1 and K_2 be two compact subsets of $\mathbb{R}_+^*$ and $\mathbb{R}$ resp. Then there exists a compact subset F of $\mathbb{R}$ depending only on K_1 and K_2 such that F contains all sequences $\{x_1,\dots,x_n\}$ with $x_n=0$ satisfying $x_{i+j}\in x_i+K_1x_j+K_2$ for every $1\le i$, $j\le i+j\le n$.*

Proof. We put $x_p := \max\{x_i: 1\le i\le n\}$ and $x_q := \min\{x_j: 1\le j\le n\}$. Without loss of generality we suppose that $K_2=-K_2$, and assume $x_p=\max\{|x_i|: 1\le i\le n\}$.

Let $m_1 := \min K_1$, $m_2 := \min K_2$, $M_1 := \max K_1$ and $M_2 := \max K_2$. Clearly, $m_2 = -M_2$. We discuss three cases.

1. $p=q$ is trivial.

2. $p>q$. For $i=1,\ldots,n-p$ and $j=1,\ldots,n-q$ one has

$$x_p \geq x_{p+i} = x_p + k_1(p,i)\,x_i + k_2(p,i)$$

and

$$x_q \leq x_{q+j} = x_q + k_1(q,j)\,x_j + k_2(q,j),$$

where $k_1(p,i)$, $k_1(q,j) \in K_1$ and $k_2(p,i)$, $k_2(q,j) \in K_2$, so that

$$x_i \leq -k_2(p,i)/k_1(p,i) \leq M_2/m_1 \quad \text{and}$$
$$x_j \geq -k_2(q,j)/k_1(q,j) \geq m_2/M_1.$$

In particular, $m_2/M_1 \leq x_{n-p} \leq M_2/m_1$. Moreover, we obtain

$$x_p \in -K_1 x_{n-p} - K_2 \quad \text{or} \quad x_p = -k_1 x_{n-p} - k_2$$

for $k_1 \in K_1$, $k_2 \in K_2$, and then

$$m_2\left(\frac{M_1}{m_1}+1\right) \leq \frac{k_1 m_2}{m_1} + m_2 \leq x_p \leq 2M_2.$$

3. $p<q$. In analogy to Part 2 one gets $m_2\left(1+\frac{M_1}{m_1}\right) \leq x_q \leq 2M_2$. It suffices to assume that $q-p<\frac{n}{2}$. Under the hypothesis $q-p \geq \frac{n}{2}$ one has $p \leq \frac{n}{2}$ and the bound for x_p is supplied by the inequality

$$x_p \geq x_{2p} = (1+k_1)\,x_p + k_2 \geq (1+m_1)\,x_p + m_2.$$

Thus there are $k_1 \in K_1$, $k_2 \in K_2$ such that

$$x_q \leq x_{2(q-p)} = (1+k_1)\,x_{q-p} + k_2.$$

Hence,

$$-x_{q-p} \leq -\frac{1}{1+k_1}x_q + \frac{k_2}{1+k_1} \leq -\frac{1}{1+k_1} m_2\left(1+\frac{M_1}{m_1}\right) + \frac{M_2}{1+m_1}$$
$$\leq \frac{1}{1+k_1} M_2\left(1+\frac{M_1}{m_1}\right) + \frac{M_2}{1+m_1} \leq \frac{M_2}{1+m_1}\left(2+\frac{M_1}{m_1}\right).$$

In addition, there are $k_1 \in K_1$, $k_2 \in K_2$ satisfying

$$x_p = x_q - k_1 x_{q-p} + k_2$$
$$\leq x_q + \frac{M_1 M_2}{1+m_1}\left(2+\frac{M_1}{m_1}\right) + M_2 \leq 2M_2 + \frac{M_1 M_2}{1+m_1}\left(2+\frac{M_1}{m_1}\right) + M_2$$
$$\leq M_2\left(3+\frac{M_1}{m_1}+\frac{M_1}{1+m_1}\left(2+\frac{M_1}{m_1}\right)\right).$$

This shows that x_p is bounded by a real number depending only on K_1 and K_2. □

3.1.21 Lemma. *Let G be a connected solvable Lie group of dimension $n \geq 1$ and G_1 the closed analytic normal subgroup corresponding to the ideal $\mathfrak{L}_1$ of dimension 1 in the Lie algebra $\mathfrak{L}(G)$ of G. One assumes that $\{s_1, \ldots, s_p\}$ is a sequence in $\mathbb{R}$ with $s_p = 0$ such that*

$$\exp s_{i+j} X_1 \in (\exp s_i X_1)\, K (\exp s_j X_1)\, K$$

for every $1 \leq i, j \leq i+j \leq p$, where K denotes a compact subset of G. Then there exists a compact subset C_1 of G_1 depending only on K and containing the sequence $\{\exp s_1 X_1, \ldots, \exp s_p X_1\}$.

Proof. We assume that G_1 is simply connected. Otherwise $G_1 \cong \mathbb{R}/\mathbb{Z}$ is compact and the proof is complete. Moreover, we may suppose that

$$K = \prod_{k=1}^{n} \exp T_k X_k,$$

where $T_k \subset \mathbb{R}$ is compact for every $k = 1, \ldots, n$. For any $X, Y \in \mathfrak{L}(G)$ such that $[X, Y] = \alpha X$ holds with $\alpha \in \mathbb{R}$ one has the formula

$$\exp t Y \exp s X \exp(-t Y) = \exp e^{\alpha t} s X$$

whenever $s, t \in \mathbb{R}$. This formula follows from

$$\exp t Y \exp s X \exp(-t Y) = \exp \operatorname{Ad}(\exp t Y)(s X)$$

$$= \exp(e^{\operatorname{ad} t Y})(s X) = \exp\left(\sum_{k \geq 0} \frac{t^k}{k!} (\operatorname{ad} Y)^k (s X)\right)$$

$$= \exp\left(\sum_{k \geq 0} \frac{(\alpha t)^k}{k!}\right)(s X) = \exp(e^{\alpha t} s X).$$

Successive application of this formula and the uniqueness of the representation in (3) of 3.1.19 implies $\exp s_{i+j} X_1 \in \exp(s_i + s_j e^{K_1} + K_2) X_1$, where K_1, K_2 are compact subsets of $\mathbb{R}$ depending only on K. Again by the uniqueness of the representation in (3) of 3.1.19 one deduces that

$$s_{i+j} \in s_i + s_j e^{K_1} + K_2.$$

But Lemma 3.1.20 yields the existence of a compact subset T_1 of $\mathbb{R}$ depending only on K_1 and K_2 such that $s_i \in T_1$ for all $1 \leq i \leq p$. Thus $C_1 := \exp T_1 X_1$ is the desired subset of G_1. ☐

3.1.22 Theorem. *Let G be a connected solvable Lie group, whose Lie algebra $\mathfrak{L}(G)$ admits an ideal $\mathfrak{L}_1$ of dimension 1. Let G_1 be the closed analytic normal subgroup of G corresponding to $\mathfrak{L}_1$. If $G/G_1 \in \mathbf{R}_0$, then $G \in \mathbf{R}_0$.*

Proof. Let

$$K = \prod_{i=1}^{n} \exp T_i X_i$$

(with compact subsets T_i of $\mathbb{R}$ for $i = 1, \ldots, n$) be a compact subset of G. We suppose that $\{x_1, \ldots, x_N\}$ with $x_N = e$ is a sequence in G satisfying $x_{i+j} \in x_i K x_j K$ for every

$1 \le i, j \le i+j \le N$. Let p denote the canonical mapping from G onto $\dot{G} := G/G_1$. One has

$$\dot{x}_{i+j} \in \dot{x}_i \dot{K} \dot{x}_j \dot{K} \quad \text{for all } 1 \le i, j \le i+j \le N,$$

and thus the existence of a compact subset

$$D_1 := \prod_{i=2}^{n} \exp A_i X_i^*$$

with compact subsets A_i of $\mathbb{R}$ $(i=2, \ldots, n)$ depending only on K such that $\dot{x}_i \in D_1$ for every $1 \le i \le N$. Here X^* denotes the differential $dp(X)$ of p evaluated at $X \in \mathfrak{L}(G)$. Preparation (4) in 3.1.19 yields that

$$D := \prod_{i=2}^{n} \exp A_i X_i$$

is compact in G. Without loss of generality we suppose that

$$x_i = \prod_{k=1}^{n} \exp s_k(i) X_k$$

with $s_k(N) = 0$ and $s_k(i) \in A_k$ for every $1 \le i < N$, $2 \le k \le n$. Hence,

$$x_{i+j} \in (\exp s_1(i) X_1) DK (\exp s_1(j) X_1) DK,$$

and thus

$$\exp s_1(i+j) X_1 \in (\exp s_1(i) X_1) K^* (\exp s_1(j) X_1) K^*$$

for every $1 \le i, j \le i+j \le N$, where $K^* := DK \cup DKD^{-1}$ is a compact subset of G depending only on K. Lemma 3.1.21 now implies the existence of a compact subset $C_1 = \exp A_1 X_1$ of G_1 depending only on K which contains the sequence $\{\exp s_1(1) X_1, \ldots, \exp s_1(N) X_1\}$. Consequently, $C_1 D$ is a compact subset of G depending only on K and containing $\{x_1, \ldots, x_N\}$. This shows that $G \in \mathbf{R}_0$. □

3.1.23 Theorem. *Every connected solvable Lie group G such that its Lie algebra $\mathfrak{L}(G)$ admits only real roots belongs to $\mathbf{R}_0$.*

Proof. Let G_1 be the closed analytic normal subgroup of G corresponding to the minimal ideal $\mathfrak{L}_1$ of $\mathfrak{L}(G)$. Then G/G_1 is also a connected solvable Lie group with the property that its Lie algebra $\mathfrak{L}(G/G_1) \cong \mathfrak{L}(G)/\mathfrak{L}_1$ admits only real roots. We suppose by induction that $G/G_1 \in \mathbf{R}_0$. Since $\mathfrak{L}_1$ is of dimension 1 by (5) of 3.1.19, the assertion follows from Theorem 3.1.22. □

3.1.24 Remark. By Preparation (6) of 3.1.19 the theorem implies that every connected nilpotent Lie group belongs to $\mathbf{R}_0$, a result which is contained in Theorem 3.1.17.

3.1.25. An **example** of a connected solvable Lie group whose Lie algebra admits only real roots is the group G of all proper affine mappings of $\mathbb{R}$, which by Example 6 of 3.1.3 is in $\mathbf{R}$. The above theorem strengthens this result in the sense that $G \in \mathbf{R}_0$.

3.1.26 Remark. The groups $G := \mathbb{C} \times_\eta \mathbb{R}$ defined by $\eta(x)(c) := c e^{ib\pi x}$ for all $x \in \mathbb{R}$, $c \in \mathbb{C}$ $(b \in \mathbb{R}^*)$ and $H := \mathbb{C} \times_\xi \mathbb{R}$ defined by

$$\xi(x)(c) := c e^{\left(1 - \frac{i\pi}{\log 2}\right) x} \quad \text{for all } x \in \mathbb{R},\ c \in \mathbb{C}$$

are connected solvable Lie groups not in $\mathbf{R}$. In fact, $G \notin \mathbf{R}_2$ and $H \notin \mathbf{R}_3$. We carry out the proof for the group H. Let

$$K := \{0\} \times \{\log 2, -\log 2\} \subset H,$$

$$x_1 := (c, 0),\ x_2 := (c(1-e^{-\log 2}), 0) \text{ and } x_3 := (0,0). \text{ Then}$$

$$x_2 = (c,0)(0, -\log 2)(c,0)(0, \log 2) \quad \text{and}$$

$$\begin{aligned} x_3 &= (c,0)(0,\log 2)(c(1-e^{-\log 2}),0)(0,-\log 2) \\ &= (c(1-e^{\log 2}),0)(0,-\log 2)(c,0)(0,\log 2). \end{aligned}$$

In particular, $x_{i+j} \in x_i K x_j K$ for all $c \in \mathbb{C}$ (appearing in the definition of x_1, x_2), $1 \le i$, $j \le i+j \le 3$. But there exists no compact subset K_3 of G satisfying $x_1 \in K_3$. That is to say, $H \notin \mathbf{R}_3$.

We now present a first application of the concept of root compactness.

3.1.27 Definition. Let G be any locally compact group. A measure $\mu \in \mathcal{M}^1(G)$ is called *infinitely divisible* if for every $n \ge 1$ there exists an n-th root μ_n such that $\mu_n^n = \mu$ holds.

The set of all infinitely divisible measures in $\mathcal{M}^1(G)$ will be denoted by $\mathcal{I}(G)$. If G is Abelian, then $\mathcal{I}(G)$ is a subsemigroup of $\mathcal{M}^1(G)$. But in general the semigroup property clearly does not hold. A large class of examples of measures in $\mathcal{I}(G)$ will be treated in detail in the following section. Here we content ourselves with a fundamental topological property of the set $\mathcal{I}(G)$.

3.1.28 Theorem. *For any root compact group G the set $\mathcal{I}(G)$ of infinitely divisible measures is a $\mathcal{T}_v$-closed subset of $\mathcal{M}^1(G)$.*

Proof. 1. Let G be a root compact group and $(\mu_\alpha)_{\alpha \in \mathbf{A}}$ a tight net in $\mathcal{M}^1(G)$. If for every $\alpha \in \mathbf{A}$ there exists a $\nu_\alpha \in \mathcal{M}^1(G)$ such that $\nu_\alpha^n = \mu_\alpha$ (for some fixed $n \ge 1$), then $(\nu_\alpha)_{\alpha \in \mathbf{A}}$ is a tight net in $\mathcal{M}^1(G)$. In fact, let $\varepsilon \in]0, \frac{1}{3}[$. There are an $\alpha_\varepsilon \in \mathbf{A}$ and a compact subset C of G satisfying $\mu_\alpha(C) > 1-\varepsilon$ for all $\alpha > \alpha_\varepsilon$. Since $\nu_\alpha^p * \nu_\alpha^{n-p} = \mu_\alpha$ for all $p = 1, \dots, n$, there is an $x_{\alpha,p} \in G$ (for $\alpha > \alpha_\varepsilon$ and $p = 1, \dots, n$) such that $\nu_\alpha^p(C x_{\alpha,p}) \ge 1-\varepsilon$ holds. Consequently, we have

$$\nu_\alpha^{p+q}(C x_{\alpha,p+q}) \ge 1-\varepsilon > \tfrac{2}{3}$$

and

$$\nu_\alpha^{p+q}(C x_{\alpha,p}\, C x_{\alpha,q}) \ge \nu_\alpha^p(C x_{\alpha,p})\, \nu_\alpha^q(C x_{\alpha,q}) \ge (1-\varepsilon)^2 > 1-2\varepsilon > \tfrac{1}{3}.$$

Hence, $C x_{\alpha,p} C x_{\alpha,q} \cap C x_{\alpha,p+q} \ne \emptyset$ for all $\alpha > \alpha_\varepsilon$ and $p, q \in \{1, \dots, n\}$ such that $p+q \le n$. Without loss of generality we assume that $x_{\alpha,n} := e$ for all $\alpha > \alpha_\varepsilon$. Since G is supposed to be root compact, there exists a compact subset C_n of G such that $x_{\alpha,p} \in C_n$ for all $\alpha > \alpha_\varepsilon$ and $p \in \{1, \dots, n\}$. Thus $\nu_\alpha(CC_n) \ge 1-\varepsilon$ for every $\alpha > \alpha_\varepsilon$, and the assertion is established.

2. Let $(\mu_\alpha)_{\alpha \in \mathbf{A}}$ be a net in $\mathcal{I}(G)$ with $\mathcal{T}_v$-$\lim_{\alpha \in \mathbf{A}} \mu_\alpha = \mu \in \mathcal{M}^1(G)$. Clearly, $(\mu_\alpha)_{\alpha \in \mathbf{A}}$ is a tight net. For every $\alpha \in \mathbf{A}$ let $\nu_\alpha \in \mathcal{M}^1(G)$ be an n-th root of μ_α (for $n \ge 1$). By Part 1 $(\nu_\alpha)_{\alpha \in \mathbf{A}}$ is a tight net and it therefore admits a subnet $(\nu_{\alpha(\beta)})_{\beta \in \mathbf{B}}$ converging (vaguely)

to some $\nu\in\mathcal{M}^1(G)$. But the continuity of the convolution mapping yields $\nu^n=\mu$, and thus $\mu\in\mathcal{I}(G)$. □

3.1.29 Remark. There are locally compact (non-root compact) groups G for which $\mathcal{I}(G)$ is not closed in $\mathcal{M}^1(G)$. As an example one might consider any locally compact Abelian group G with the property that the set $T(G)$ of divisible elements of G (which in the case of an Abelian group is a subgroup of G) is dense in G, but different from G. Such groups exist as is shown in [218], (24.44). For $x\in T(G)$ we have $\varepsilon_x\in\mathcal{I}(G)$. If $y\in G\setminus T(G)$, then $\varepsilon_y\notin\mathcal{I}(G)$, but ε_y clearly belongs to the $\mathcal{T}_v$-closure of $\mathcal{I}(G)$.

Let $(r_i)_{i\geq 0}$ and $(n_i)_{i\geq 0}$ be sequences in $\mathbb{R}_+^*$ and $\mathbb{N}$ resp. such that $r_i=n_i r_{i+1}$ holds for all $i\geq 0$. For $i\geq 0$ we denote by $\mathbb{M}_i$ the semigroup $\mathbb{M}(r_i)$ generated by r_i, and we put $\mathbb{M}:=\bigcup_{i\geq 0}\mathbb{M}_i$. The semigroup $\mathbb{M}$ (which is a subsemigroup of $\mathbb{R}_+^*$) is called a *real submonogeneous semigroup with generator* $(r_i)_{i\geq 0}$. Given a real submonogeneous semigroup $\mathbb{M}$, its images $f(\mathbb{M})$ under all homomorphisms f from $\mathbb{M}$ into some semigroup S are introduced as ($\mathbb{M}$-)*submonogeneous semigroups* (in S). Replacing $\mathbb{R}_+^*$ by $\mathbb{R}$ and the semigroup $f(\mathbb{M})$ by the subgroup generated by $f(\mathbb{M})$ one obtains the notion of an ($\mathbb{M}$-)*submonogeneous group*.

Moreover, one can start the discussion with an ($\mathbb{M}$-)submonogeneous group and later restrict oneself to the ($\mathbb{M}_+$-)submonogeneous semigroup (induced by the set $\mathbb{M}_+$ of nonnegative elements of $\mathbb{M}$). If one considers sequences $(r_i)_{i\geq 0}$ in $\mathbb{Q}_+^*$ of elements r_i of the form $\frac{1}{m_i}$ for $m_i\in\mathbb{N}$ $(i\geq 0)$, then one obtains the notion of a *rational submonogeneous semigroup* with generator $(r_i)_{i\geq 0}$ (satisfying the relation $m_{i+1}=n_i m_i$ for all $i\geq 0$) and as above also the notion of an ($\mathbb{M}$-)submonogeneous semigroup induced by the rational submonogeneous semigroup $\mathbb{M}:=\left\langle\left\{\frac{1}{m_i}:i\geq 0\right\}\right\rangle$. We note that the generator of a rational submonogeneous semigroup is not uniquely determined. The set $\left\{\frac{1}{(i+k)!}:i\geq 0\right\}$ is a generator of $\mathbb{Q}_+^*$ for every $k\geq 1$. Finally we observe that any noncyclic rational submonogeneous semigroup is dense in $\mathbb{R}_+$.

The following auxiliary fact is an immediate consequence of the definition of a submonogeneous semigroup.

3.1.30 Lemma. *Let $(m_i)_{i\geq 0}$ be a sequence in $\mathbb{N}$ and f a mapping from the set $\left\{\frac{1}{m_i}:i\geq 0\right\}$ into the semigroup S such that*

$$f\left(\frac{1}{m_i}\right)=f\left(\frac{1}{m_{i+1}}\right)^{n_i} \quad \textit{for } n_i\in\mathbb{N} \textit{ and every } i\geq 0.$$

Then f can be extended uniquely to a (semigroup) homomorphism from

$$\mathbb{M}:=\left\langle\left\{\frac{1}{m_i}:i\geq 0\right\}\right\rangle \quad \textit{into } S.$$

This lemma enables us to speak of a *submonogeneous semigroup generated* (in S) *by the sequence* $\left\{f\left(\frac{1}{m_i}\right): i\geq 0\right\}$.

Let G be a locally compact group. Replacing the above mentioned semigroup S by the semigroup $\mathscr{M}^1(G)$ we obtain the notion of an ($\mathbb{M}$-)*submonogeneous convolution semigroup in* $\mathscr{M}^1(G)$.

3.1.31 Definition. A measure $\mu\in\mathscr{M}^1(G)$ is called *submonogeneously embeddable* if there exists an $\mathbb{M}$-submonogeneous convolution semigroup $(\mu_r)_{r\in\mathbb{M}}$ in $\mathscr{M}^1(G)$ (where $\mathbb{M}$ is a rational submonogeneous semigroup with generator $(m_i)_{i\geq 1}$) such that $\mu_1=\mu$ holds.

If one wishes to make particular reference to the rational submonogeneous semigroup $\mathbb{M}$ chosen, one may speak of an $\mathbb{M}$-*(submonogeneous) embedding*. In the special case $\mathbb{M}:=\mathbb{Q}_+^*$ the notion of a *rational embedding* has entered the literature.

3.1.32 Theorem. *Let G be a locally compact group and $\mathbb{M}$ a rational submonogeneous semigroup with generator $(m_i)_{i\geq 1}$.*

(i) *If $\mu\in\mathscr{I}(G)$ admits the property that $R(m_i, \mu)$ is compact in $\mathscr{M}^1(G)$ for all $i\geq 1$, then μ is $\mathbb{M}$-embeddable.*

(ii) *If $G\in\bigcap_{i\geq 1}\mathbf{R}_{m_i}$ (in particular if $G\in\mathbf{R}$), then every $\mu\in\mathscr{I}(G)$ is $\mathbb{M}$-embeddable.*

Proof. (i) Since $\mu\in\mathscr{I}(G)$, we have $R(m_i, \mu)\neq\emptyset$ for all $i\geq 1$. For every $i\geq 1$ we denote by f_{ii} the identity on $R(m_i, \mu)$ and by f_{ij} for $j>i$ the mapping from $R(m_j, \mu)$ in $R(m_i, \mu)$ defined by

$$f_{ij}(\mu_j):=\mu_j^{n_i\cdot\ldots\cdot n_{j-1}} \qquad \text{for all } \mu_j\in R(m_j, \mu).$$

The family $(R(m_i, \mu), f_{ij}, j>i\geq 1)$ is a projective system of compact spaces together with continuous projections. Its projective limit R is not empty. Choosing a sequence $(\mu_i)_{i\geq 1}$ in R and putting $f(m_i^{-1}):=\mu_i$ for all $i\geq 1$ we obtain

$$f(m_i^{-1})=f(m_{i+1}^{-1})^{n_i}.$$

By Lemma 3.1.30 f can be uniquely extended to a homomorphism $\bar{f}$ from $\mathbb{M}$ into $\mathscr{M}^1(G)$ for which one clearly has $\bar{f}(1)=\mu$. Assertion (ii) is an immediate consequence of (i) and Theorem 3.1.4. □

3.2 Poisson Measures and Their Characterizations

The classical *Poisson distribution* π_α *with parameter* or mean value $\alpha\in\mathbb{R}$ can be redefined as the measure

$$\exp[\alpha(\varepsilon_1-\varepsilon_0)]=\varepsilon_0+\sum_{k\geq 1}\frac{\alpha^k}{k!}(\varepsilon_1-\varepsilon_0)^k \quad \text{in } \mathscr{M}^1(\mathbb{R}),$$

where the convergence of the series is understood in the sense of the norm. Here the measure $\nu := \alpha(\varepsilon_1 - \varepsilon_0) \in \mathscr{M}^b(\mathbb{R})$ satisfies the conditions $\nu(f) \geq 0$ for all $f \in \mathscr{C}_+^b(\mathbb{R})$ with $f(0)=0$ and $\nu(1)=0$. By considering measures $\nu \in \mathscr{M}^b(\mathbb{R})$ of the form $\alpha(\lambda - \varepsilon_0)$ for an arbitrary $\lambda \in \mathscr{M}^1(\mathbb{R})$ an important generalization is achieved. Furthermore, one notes that since $\{0\}$ is the only compact subgroup of $\mathbb{R}$, one might also define generalized Poisson measures on a locally compact group with respect to any given compact subgroup.

Let G be a locally compact group, H a compact subgroup of G and ω_H the normed Haar measure of H (in $\mathscr{M}^1(G)$). Clearly the set

$$\mathscr{M}_H^b(G) := \omega_H * \mathscr{M}^b(G) * \omega_H$$

is a Banach subalgebra of $\mathscr{M}^b(G)$ with multiplicative unit ω_H. We can introduce the *exponential* $\exp_H(\nu)$ of a measure $\nu \in \mathscr{M}_H^b(G)$ by

$$\exp_H(\nu) := \omega_H + \sum_{k \geq 1} \frac{\nu^k}{k!}$$

since the series on the right side of the defining equality converges in the norm of $\mathscr{M}^b(G)$. For every $\nu \in \mathscr{M}_H^b(G)$ satisfying $\|\nu - \omega_H\| < 1$ its *logarithm* $\log_H(\nu)$ is defined by

$$\log_H(\nu) := -\sum_{k \geq 1} \frac{(\omega_H - \nu)^k}{k}.$$

Plainly, for any $\nu \in \mathscr{M}_H^b(G)$ the measures $\exp_H(\nu)$ and $\log_H(\nu)$ are elements of $\mathscr{M}_H^b(G)$. We recall that the set $\mathscr{M}_H^1(G) := \omega_H * \mathscr{M}^1(G) * \omega_H$ is a subsemigroup of $\mathscr{M}^1(G)$. It was studied in some detail in §2 of Chapter I.

Let

$$\begin{aligned}\mathscr{N}_H(G) &:= \{\nu \in \mathscr{M}_H^b(G) : \nu = \nu_1 - \|\nu_1\|\, \omega_H \text{ for some } \nu_1 \in \mathscr{M}_H^b(G)_+\} \\ &= \{\nu \in \mathscr{M}_H^b(G) : \nu = \gamma(\lambda - \omega_H) \text{ for } \gamma \in \mathbb{R}_+, \lambda \in \mathscr{M}_H^1(G)\}.\end{aligned}$$

Clearly $\nu \in \mathscr{N}_H(G)$ iff $\nu \in \mathscr{M}_H^b(G)$ and $\mathrm{Res}_{\mathfrak{B}(G \setminus H)}\, \nu \geq 0$, $\nu(1)=0$ or iff $\nu \in \mathscr{M}_H^b(G)$ satisfying $\nu(f) \geq 0$ for all $f \in \mathscr{C}_+^b(G)$ such that $\mathrm{Res}_H f = 0$ and $\nu(1)=0$.

3.2.1 Definition. A measure $\mu \in \mathscr{M}^1(G)$ is called an *H-Poisson measure*, if there exists a measure $\nu \in \mathscr{N}_H(G)$ such that $\mu = \exp_H(\nu)$ holds.

$\mu \in \mathscr{M}^1(G)$ is said to be a *Poisson measure*, if μ is an H-Poisson measure for some compact subgroup H of G.

The set of all H-Poisson measures on G will be denoted by $\mathscr{P}_H(G)$. Naturally one puts $\mathscr{P}(G) := \bigcup \{\mathscr{P}_H(G) : H \text{ is a compact subgroup of } G\}$. Evidently $\mathscr{P}(G) \subset \mathscr{I}(G)$, so that we have established a large class of examples in $\mathscr{I}(G)$ for an arbitrary locally compact group G. We finally define for every compact subgroup H of G the set

$$\mathscr{L}_H(G) := \{\mu \in \mathscr{M}_H^b(G) : \text{There exists an inverse } \mu^{-1} \text{ of } \mu \text{ (in } \mathscr{M}_H^b(G))\},$$

where the inverse μ^{-1} of μ in $\mathscr{M}_H^b(G)$ is defined by

$$\mu * \mu^{-1} = \mu^{-1} * \mu = \omega_H.$$

3.2.2 Properties of exponential, logarithm and Poisson measure. Let μ, ν, π, $\rho \in \mathcal{M}_H^b(G)$. Then we have

(1) $$\exp_H \log_H(\mu) = \mu, \quad \text{if } \|\mu - \omega_H\| < 1.$$

(2) $$\log_H \exp_H(\pi) = \pi, \quad \text{if } \|\pi\| < \log 2.$$

(3) $$\exp_H(\pi + \rho) = \exp_H(\pi) * \exp_H(\rho), \quad \text{if } \pi * \rho = \rho * \pi.$$

(4) $$\log_H(\mu * \nu) = \log_H(\mu) + \log_H(\nu), \quad \text{if } \mu * \nu = \nu * \mu,$$

$$\|\mu - \omega_H\|, \quad \|\nu - \omega_H\| \quad \text{and} \quad \|\mu * \nu - \omega_H\| < 1.$$

If one assumes that $\mu \in \mathcal{M}_H^1(G)$ and the existence of an $n \geq 1$ such that $\|\mu^n - \omega_H\| + 2\|\mu - \omega_H\| < 1$ holds, then

(5) $$\|\mu^k - \omega_H\| + 2\|\mu - \omega_H\| < 1 \quad \text{and}$$

$$\log_H(\mu^k) = k \log_H(\mu) \quad \text{for all } 1 \leq k \leq n.$$

While the proofs of (1) to (4) are straightforward, the argument for (5) should be noted. We perform an induction. For $n = 1$ the statement is obviously true. Let it be established for an arbitrary $n > 1$. We abbreviate

$$\nu := \mu - \omega_H, \quad \lambda := \mu^n - \omega_H \quad \text{and} \quad \pi := \mu^{n+1} - \omega_H$$

and denote the norms of these measures by r, s and t resp. For $\kappa \in \mathcal{M}^b(G)$ with $\kappa(1) = 0$ we have

$$\|\kappa\| = |\kappa|(H) + |\kappa|(\complement H) \geq |\kappa(H)| + |\kappa(\complement H)| = 2|\kappa(\complement H)|,$$

and for $\kappa \in \mathcal{N}_H(G)$ we get

$$\|\kappa\| = \kappa_+(\complement H) + \kappa_-(H) = 2\kappa(\complement H).$$

From $\nu, \lambda, \pi \in \mathcal{N}_H(G)$ and $\nu * \lambda(1) = 0$ we obtain $t = 2\pi(\complement H)$ and from $\pi = \nu + \lambda + \nu * \lambda$ immediately

$$t = r + s + 2\nu * \lambda(\complement H) \geq r + s - \|\nu * \lambda\| \geq r + s - rs.$$

By assumption $t + 2r < 1$, so that $0 < 1 - r \leq 1$ and $0 < 1 - t \leq (1 - r)(1 - s) \leq 1 - s$. Hence, $s \leq t < 1$ and $r + s + rs \leq t + 2rs < 1$. Property (4) is therefore applicable and one has $\log_H(\mu^{n+1}) = \log_H(\mu) + \log_H(\mu^n)$.

From $s \leq t$ follows the (induction) assumption

$$\|\mu^k - \omega_H\| + 2\|\mu - \omega_H\| < 1, \quad \text{so that } \log_H(\mu^k) = k \log_H(\mu)$$

for all $k = 1, \ldots, n$, whence the assertion.

Let H and H' be two compact subgroups of G. Then

(6) $$\mathcal{L}_H(G) \cap \mathcal{M}_{H'}^b(G) \neq \emptyset \quad \text{implies } H' \subset H.$$

In fact, for $\mu \in \mathscr{L}_H(G) \cap \mathscr{M}^b_{H'}(G)$ with $\mu * \mu^{-1} = \omega_H$ we obtain

$$\omega_{H'} * \omega_H = \omega_{H'} * \mu * \mu^{-1} = \mu * \mu^{-1} = \omega_H$$

and thus $H' \subset H$.

(7) If $H \neq H'$, then $\mathscr{L}_H(G) \cap \mathscr{L}_{H'}(G) = \emptyset$.

(8) For $\mu \in \mathscr{L}_H(G)$ one has $I(\mu) = H$.

This follows from (6) if one notes that $H' = I(\mu) \supset H$.

3.2.3 Theorem. *Let G be a locally compact group, H a compact subgroup of G, $(\mu_r)_{r \in \mathbb{Q}^*_+}$ a rational convolution semigroup in $\mathscr{M}^1(G)$ and $(x_r)_{r \in \mathbb{Q}^*_+}$ a family in $N(H)$ with the property* $\lim_{r \downarrow 0} \|\mu_r - \omega_H * \varepsilon_{x_r}\| = 0$. *Then*

(i) $\mu_r \in \mathscr{L}_H(G)$ *and* $I(\mu_r) = H$ *for every* $r \in \mathbb{Q}^*_+$.

(ii) *There exists a unique homomorphism* $r \to H y_r$ *from* $\mathbb{Q}^*_+$ *into* $N(H)/H$ *such that* $\lim_{r \downarrow 0} \|\mu_r - \omega_H * \varepsilon_{y_r}\| = 0$ *holds.*

Proof. (i) For every $\delta > 0$ there exists an $r_0 := r_0(\delta)$ such that

$$\|\mu_r - \omega_H * \varepsilon_{x_r}\| \leq \delta \quad \text{holds for all } r \leq r_0.$$

Given $t \in \mathbb{Q}^*_+$ one obtains the existence of $r, s \in \mathbb{Q}^*_+$ with $r + s = t$ and $r \leq r_0$. From

$$\|\mu_t - \omega_H * \varepsilon_{x_r} * \mu_s\| = \|(\mu_r - \omega_H * \varepsilon_{x_r}) * \mu_s\| \leq \delta$$

we conclude that

$$\|\omega_H * \mu_t - \omega_H * \varepsilon_{x_r} * \mu_s\| \leq \delta \quad \text{and} \quad \|\mu_t - \omega_H * \mu_t\| \leq 2\delta,$$

whence $\mu_t = \omega_H * \mu_t$. Analogously one derives $\mu_t = \mu_t * \omega_H$, and thus altogether $\mu_t \in \mathscr{M}^1_H(G)$. Since we have $\|\mu_r * \varepsilon_{x_r^{-1}} - \omega_H\| < 1$ for all $r \leq r_0(\delta)$ with $\delta := \frac{1}{2}$ say, $\nu_r := \log_H(\mu_r * \varepsilon_{x_r^{-1}})$ is defined. From this follows

$$\mu_r^{-1} = \varepsilon_{x_r^{-1}} * \exp_H(-\nu_r) \in \mathscr{M}^b_H(G),$$

whence $\mu_r \in \mathscr{L}_H(G)$ for all $r \leq r_0(\delta)$. Since for every $t \in \mathbb{Q}^*_+$ there are an $n \geq 1$ and an $r \in \mathbb{Q}^*_+$ with $r \leq r_0(\delta)$ and $\mu_t = \mu_r^n$, we finally conclude that $\mu_t \in \mathscr{L}_H(G)$ for all $t \in \mathbb{Q}^*_+$. The remaining statement follows from Property (8) above.

(ii) Let $\delta := \frac{1}{2}$ and $r_0 := r_0(\delta)$. For every $r, s \in \mathbb{Q}^*_+$ with $r + s \leq r_0$ one obtains the inequalities

$$\|\omega_H * \varepsilon_{x_r} * \varepsilon_{x_s} - \omega_H * \varepsilon_{x_r} * \mu_s\| = \|\omega_H * \varepsilon_{x_r} * \omega_H * \varepsilon_{x_s} - \omega_H * \varepsilon_{x_r} * \mu_s\| \leq \tfrac{1}{2},$$

$$\|\omega_H * \varepsilon_{x_r} * \mu_s - \mu_r * \mu_s\| = \|\omega_H * \varepsilon_{x_r} * \mu_s - \mu_{r+s}\| \leq \tfrac{1}{2} \quad \text{and}$$

$$\|\mu_{r+s} - \omega_H * \varepsilon_{x_{r+s}}\| \leq \tfrac{1}{2}.$$

Hence, one concludes that

$$\|\omega_H * \varepsilon_{x_r} * \varepsilon_{x_s} - \omega_H * \varepsilon_{x_{r+s}}\| < 2 \quad \text{or}$$

$$H x_r H x_s = H x_r x_s = H x_{r+s}.$$

By a well-known extension procedure the mapping $r \to H x_r$ can be extended to a homomorphism from $\mathbb{Q}_+^*$ into $N(H)/H$. □

3.2.4 Theorem. *Let G be a locally compact group, H a compact subgroup of G and $\mu \in \mathcal{M}_H^1(G)$ such that $\|\mu - \omega_H\| < 1$ holds. The following statements are equivalent:*

(i) $\mu \in \mathcal{P}_H(G)$;

(ii) *there exists exactly one* $\nu \in \mathcal{N}_H(G)$ *of the form* $\nu = \log_H(\mu)$ *satisfying* $\mu = \exp_H(\nu)$;

(iii) *there exists a rational convolution semigroup* $(\mu_r)_{r \in \mathbb{Q}_+^*}$ *in* $\mathcal{M}_H^1(G)$ *with the properties*

(a) $\lim_{r \downarrow 0} \|\mu_r - \omega_H\| = 0$ *and*

(b) $\mu_1 = \mu$.

(iv) *For every $n \geq 1$ there exists a $\mu_n \in \mathcal{M}_H^1(G)$ with $\mu_n^n = \mu$, and we have*

$$\inf_{n \geq 1} \|\mu_n - \omega_H\| = 0.$$

Proof. 1. The implications (ii)⇒(i), (i)⇒(iii) and (iii)⇒(iv) are evident. We only have to prove

2. (iv)⇒(ii). From the hypothesis we obtain for sufficiently large $n \geq n_0$ a measure $\mu_n \in \mathcal{M}_H^1(G)$ such that $\mu_n^n = \mu$ and $\|\mu - \omega_H\| + 2\|\mu_n - \omega_H\| < 1$ holds. Property (5) of 3.2.2 yields for every $n \geq n_0$ the equality

$$\log_H(\mu_n) = \frac{1}{n} \log_H(\mu).$$

Putting $\nu := \log_H(\mu) \in \mathcal{M}_H^b(G)$ we conclude that $\mu_n = \exp_H\left(\frac{\nu}{n}\right)$ for all $n \geq n_0$.

From $0 \leq \mu_n = \omega_H + \sum_{k \geq 1} \left(\frac{\nu}{n}\right)^k \left(\frac{1}{k!}\right)$ one obtains

$$\nu(f) \geq -\frac{1}{n} \sum_{k \geq 2} \frac{\nu^k(f)}{n^{k-2}} \frac{1}{k!}$$

and hence, $\nu(f) \geq 0$ for all $f \in \mathcal{C}_+^b(G)$ with $\operatorname{Res}_H f = 0$. Since $1 = \mu(1) = \exp(\nu(1))$, we also have $\nu(1) = 0$ and thus $\nu \in \mathcal{N}_H(G)$.

It remains to prove the uniqueness of ν in the representation $\mu = \exp_H(\nu)$. Let $\nu' \in \mathcal{N}_H(G)$ be such that $\mu = \exp_H(\nu')$ holds. Choosing

$$\mu_n := \exp_H\left(\frac{\nu'}{n}\right) \quad \text{for all } n \geq 1$$

we obtain for sufficiently large n the inequalities

$$\|\mu - \omega_H\| + 2\|\mu_n - \omega_H\| < 1 \quad \text{and} \quad \left\|\frac{\nu'}{n}\right\| < \log 2.$$

The above discussion together with Property (2) of 3.2.2 shows that

$$\frac{\nu'}{n} = \log_H(\mu_n) = \frac{\nu}{n} \quad \text{or} \quad \nu' = \nu. \quad \square$$

3.2.5 Theorem. *Let G be a locally compact group, H a compact subgroup of G and* $(\mu_r)_{r\in\mathbb{Q}_+^*}$ *a rational convolution semigroup in* $\mathscr{M}^1(G)$. *The following statements are equivalent:*

(i) *There exists a* $\nu\in\mathscr{N}_H(G)$ *with* $\mu_r=\exp_H(r\nu)$ *for all* $r\in\mathbb{Q}_+^*$;

(ii) $\lim_{r\downarrow 0}\|\mu_r-\omega_H\|=0$.

Moreover, (i) *or* (ii) *implies* $\mu_r\in\mathscr{L}_H(G)$ *for all* $r\in\mathbb{Q}_+^*$.

Proof. We show (ii) ⇒ (i). First of all we note that Theorem 3.2.3 implies that $\mu_r\in\mathscr{M}_H^1(G)$ for all $r\in\mathbb{Q}_+^*$. By assumption there exists an $r_0>0$ such that $\|\mu_r-\omega_H\|<1$ holds for all $r\le r_0$. The family $(\mu_{rs})_{s\in\mathbb{Q}_+^*}$ is a convolution semigroup $(\mu_s')_{s\in\mathbb{Q}_+^*}$ satisfying

$$\lim_{s\downarrow 0}\|\mu_s'-\omega_H\|=0 \quad\text{and}\quad \mu_1'=\mu_r.$$

Therefore Theorem 3.2.4 yields for any $r\le r_0$ the existence of exactly one measure $\nu_r\in\mathscr{N}_H(G)$ with $\mu_r=\exp_H(\nu_r)$. For $r+s\le r_0$ we have

$$\mu_{r+s}=\mu_r*\mu_s=\exp_H(\nu_r+\nu_s)$$

and since $\nu_r+\nu_s\in\mathscr{N}_H(G)$ also $\nu_{r+s}=\nu_r+\nu_s$. Hence, there is exactly one $\nu\in\mathscr{N}_H(G)$ with $\nu_r=r\nu$ and $\mu_r=\exp_H(r\nu)$ for all $r\le r_0$. The mapping $r\to\mu_r$ can be extended uniquely to a homomorphism from $\mathbb{Q}_+^*$ into $\mathscr{M}^1(G)$ and the assertion follows (for all $r\in\mathbb{Q}_+^*$).

The remaining statement is deduced from the fact that

$$\mu_r^{-1}=\exp_H(-r\nu)\in\mathscr{M}_H^b(G)\qquad\text{for all } r\in\mathbb{Q}_+^*. \quad\square$$

3.2.6 Remark. The representation of a measure $\mu\in\mathscr{P}_H(G)$ in the form $\mu=\exp_H(\nu)$ for $\nu\in\mathscr{N}_H(G)$ is in general not unique. In fact, there exists a $\mu\in\mathscr{P}_H(G)$ admitting representations

$$\mu=\exp_H(\nu_1)=\exp_H(\nu_2)\quad\text{with}\quad \nu_1,\nu_2\in\mathscr{N}_H(G)\quad\text{and}\quad \nu_1\neq\nu_2.$$

1. Let G be a locally compact Abelian group and let there exist a measure $\lambda\in\mathscr{M}_H^b(G)$ with $\lambda\neq 0$ satisfying $\exp_H(\lambda)=\omega_H$. If $\lambda=\lambda_1-\lambda_2$ with $\lambda_1,\lambda_2\in\mathscr{M}_+^b(G)$ and $\lambda_1\neq\lambda_2$, then

$$\exp_H[\omega_H*(\lambda_1-\lambda_2)]=\exp_H(\lambda_1-\lambda_2)=\omega_H$$

and consequently

$$\exp_H(\omega_H*\lambda_1)=\exp_H(\omega_H*\lambda_2).$$

Putting $\pi_1:=\omega_H*\lambda_1$ and $\pi_2:=\omega_H*\lambda_2$ we obtain $\pi_1,\pi_2\in\mathscr{M}_H^b(G)$, $\pi_1-\pi_2=\lambda$ and so $\pi_1\neq\pi_2$. Since $(\pi_1-\pi_2)(1)=\lambda(1)=0$ we get $\|\pi_1\|=\|\pi_2\|$. Clearly, for $i=1,2$ one has $\nu_i:=\pi_i-\|\pi_i\|\,\omega_H\in\mathscr{N}_H(G)$ and therefore $\exp_H(\nu_1)=\exp_H(\nu_2)$. But by construction $\nu_1\neq\nu_2$.

2. To establish the hypothesis of Part 1 we proceed in the case of a finite group G of order $n\ge 1$ with at least one element x_0 of order >2 as follows. For every $\mu\in\mathscr{M}^b(G)$ of the form $\mu=\sum_{x\in G}\mu_x\,\varepsilon_x$ and every $\chi\in G^\wedge$ we define

$\hat{\mu}(\chi) := \sum_{x\in G} \mu_x \chi(x)$, where for every $x\in G$ one has the inversion formula

$$\mu_x = \frac{1}{n} \sum_{\chi\in G^\wedge} \chi^{-1}(x)\, \hat{\mu}(\chi)$$

and $\mu_x = \overline{\mu_x}$ iff $\bar{\hat{\mu}}(\chi) = \hat{\mu}(\chi^{-1})$ for all $\chi \in G^\wedge$.

Plainly, there exists a $\lambda \in \mathcal{M}^b(G)$ satisfying $\exp_{\{e\}}(\lambda) = \varepsilon_e$ iff $\exp \hat{\lambda}(\chi) = 1$ for all $\chi \in G^\wedge$. We want to show that $\lambda \in \mathcal{M}^b(G)$ can be chosen $\neq 0$. Recalling the existence of $x_0 \in G$ with $x_0^2 \neq e$ we choose $\chi_0 \in G^\wedge$ such that $\chi_0(x_0) \neq \chi_0^{-1}(x_0)$ holds. The measure $\lambda := \sum_{x\in G} \lambda_x \varepsilon_x$ with $\lambda_x = \frac{1}{n}(\chi_0(x)\, 2\pi i - \chi_0^{-1}(x)\, 2\pi i)$ for every $x\in G$ fulfills the requirements.

We are now going to discuss characterizations of measures in $\mathscr{P}_H(G)$ within $\mathscr{I}(G)$ by means of conditions on their roots.

3.2.7 Theorem. *Let G be a locally compact group, H a compact subgroup of G and $\mu \in \mathscr{I}(G)$. Then following statements are equivalent:*

(i) $\mu \in \mathscr{P}_H(G)$;

(ii) *there exist an isotone (infinite) sequence $(m_i)_{i\geq 1}$ in $\mathbb{N}$ and a sequence $(\mu_{m_i})_{i\geq 1}$ of roots of μ satisfying the following conditions:*

(a) $\mu_{m_i} \in \mathcal{M}^1_{H_i}(G)$ *for a compact subgroup H_i of G with $H_i \supset H$ for every $i \geq 1$,*

(b) $\{\mu_{m_i} : i \geq 1\}$ *generates a submonogeneous convolution semigroup (in $\mathcal{M}^1(G)$),*

(c) $\lim_{i\to\infty} \|\mu_{m_i} - \omega_H\| = 0$.

From either statement follows $I(\mu) = H$.

Proof. It suffices to prove the implication (ii)$\Rightarrow$(i) and $I(\mu) = H$.

1. Without loss of generality we assume that the sequence $(\mu_{m_i})_{i\geq 1}$ contains infinitely many elements; otherwise one obtains $\mu = \omega_H$. There exists a $j_0 \geq 1$ such that $\|\mu_{m_{j_0}} - \omega_H\| < 1$ holds. This inequality together with Condition (a) enables us to define

$$\nu_0 := \log_H(\mu_{m_{j_0}}) \quad \text{and} \quad \nu := m_{j_0} \nu_0,$$

and we obtain $\mu = \exp_H(\nu)$. It remains to show that $\nu \in \mathcal{N}_H(G)$. Obviously $\nu(1) = 0$. For every $i \geq 1$ there exists $n_i \geq 1$ such that $\mu_{m_i} = \mu_{m_{i+1}}^{n_i}$ by (b). Condition (c) together with Property (5) of 3.2.2 yields

$$\mu_{m_i} = \exp_H \frac{\nu_0}{n_{j_0} \cdot \dots \cdot n_{i-1}}$$

for sufficiently large $i \geq j_0$. Moreover,

$$\lim_{i\to\infty} N_i := \lim_{i\to\infty} n_{j_0} \cdot \dots \cdot n_{i-1} = \infty.$$

For any $f \in \mathscr{C}^b_+(G)$ with $\operatorname{Res}_H f = 0$ we have

$$0 \leq \mu_{m_i}(f) = \frac{\nu_0(f)}{N_i} + O\left(\frac{1}{N_i^2}\right),$$

whence the assumption $\nu_0(f)<0$ or equivalently $\nu(f)<0$ would lead to a contradiction. Thus $\nu(f)\geq 0$ for all $f\in\mathscr{C}_+^b(G)$ with $\mathrm{Res}_H f=0$, and $\nu\in\mathscr{N}_H(G)$.

2. For every $\nu\in\mathscr{N}_H(G)$ we have $\exp_H(\nu)*\exp_H(-\nu)=\omega_H$ and thus $\mu:=\exp_H(\nu)$ satisfies $I(\mu)=H$. □

3.2.8 Remark. If G is compact, then Conditions (b) and (c) in Statement (ii) of the theorem can be replaced by the condition

(d) $\sup_{i\geq 1} m_i\,\|\mu_{m_i}-\omega_H\|<\infty.$

Without loss of generality

$$\mathscr{T}_v\text{-}\lim_{i\to\infty} m_i(\mu_{m_i}-\omega_H)=:\nu\in\mathscr{M}_H(G) \quad \text{with } \mathrm{Res}_{\mathfrak{B}(G\setminus H)}\,\nu\geq 0.$$

Let $\alpha>0$ be such that $m_i\|\mu_{m_i}-\omega_H\|\leq\alpha$ for all $i\geq 1$. Defining for every $i\geq 1$ the measure

$$\lambda_{m_i}:=\sum_{k\geq 2}\frac{(\mu_{m_i}-\omega_H)^k}{k!}$$

we obtain

$$\|\lambda_{m_i}\|\leq\sum_{k\geq 2}\frac{1}{k!}\left(\frac{\alpha}{m_i}\right)^k\leq\frac{1}{m_i^2}\sum_{k\geq 2}\frac{\alpha^k}{k!},\quad\text{whence}$$

$$\lim_{m_i\to\infty} m_i\,\|\lambda_{m_i}\|=0 \quad\text{or}\quad \lim_{m_i\to\infty}(1+\|\lambda_{m_i}\|)^{m_i}=1.$$

On the other hand, we have

$$\|\mu-\exp_H[m_i(\mu_{m_i}-\omega_H)]\|\leq\sum_{k=1}^{m_i}\left\|\binom{m_i}{k}\mu_{m_i}^{m_i-k}\lambda_{m_i}^k\right\|$$

$$\leq\sum_{k=1}^{m_i}\binom{m_i}{k}\|\lambda_{m_i}\|^k=(1+\|\lambda_{m_i}\|)^{m_i}-1$$

which tends to 0 as $m_i\to\infty$. We therefore conclude that $\mu=\exp_H(\nu)$ and $\nu(1)=0$, so that $\mu\in\mathscr{P}_H(G)$.

3.2.9 Theorem. *Let G be a finite group. Then $\mathscr{I}(G)=\mathscr{P}(G)$.*

Proof. It remains to prove the inclusion $\mathscr{I}(G)\subset\mathscr{P}(G)$. Let $\mu\in\mathscr{I}(G)$. By $n\geq 2$ we denote the order of G. For every $i\geq 0$ there exists an i-th root $\bar\mu_i$ of μ satisfying $\mu=\bar\mu_0$ and $\bar\mu_{i-1}=\bar\mu_i^n$ for all $i\geq 1$. In fact, $\mathscr{M}^1(G)$ is compact in the vague topology. Hence, since G is finite, $\mathscr{M}^1(G)$ is also compact in the norm topology, and so is $\prod_{j\geq 0}\mathscr{M}_j^1(G)$ with $\mathscr{M}_j^1(G):=\mathscr{M}^1(G)$ for all $j\geq 0$. Moreover, for every $k\geq 1$ there exists an n^k-th root μ_{n^k} of μ satisfying $\mu_{n^k}^{n^k}=\mu$. Defining for every $k\geq 1$ the set

$$\mathscr{N}_k:=\bigcup_{j\geq k}\left[\{\mu_{n^j}^{n^j}\}\times\{\mu_{n^j}^{n^{j-1}}\}\times\cdots\times\{\mu_{n^j}\}\times\prod_{i>j}\mathscr{M}_i^1(G)\right]$$

one observes that $\{\mathscr{N}_k\colon k\geq 1\}$ is a filterbasis in $\prod_{j\geq 0}\mathscr{M}_j^1(G)$ which admits an accumulation point $(\bar\mu_0,\bar\mu_1,\ldots)\in\prod_{j\geq 0}\mathscr{M}_j^1(G)$. Clearly, the sequence $(\bar\mu_i)_{i\geq 0}$ in $\mathscr{M}^1(G)$ satisfies the required conditions.

By Lemma 3.1.30 $(\bar{\mu}_i)_{i\geq 0}$ generates a submonogeneous convolution semigroup in $\mathcal{M}^1(G)$. But there exists a subsequence $(\nu_j)_{j\geq 0}$ of $(\bar{\mu}_i)_{i\geq 0}$ with

$$\lim_{j\to\infty} \nu_j =: \bar{\mu} \in \mathcal{M}^1(G).$$

Moreover, for every $l\geq 0$ there exists a $k_0 := k_0(l) \geq 1$ such that for all $k\geq k_0$ and appropriately chosen $r_k\geq 2$ one gets $\bar{\mu}_l = \nu_k^{r_k}$. For fixed $l\geq 0$ we obtain from $\bar{\mu}_l = \nu_k * \nu_k^{r_k-1} = \nu_k^{r_k-1} * \nu_k$ in the limit for $k\to\infty$ the relationship $\bar{\mu}_l = \bar{\mu} * \lambda_l = \lambda_l * \bar{\mu}$ with $\lambda_l \in \mathcal{M}^1(G)$. But then

$$\bar{\mu}_l \in (\bar{\mu} * \mathcal{M}^1(G)) \cap (\mathcal{M}^1(G) * \bar{\mu}) \qquad \text{for all } l\geq 0,$$

which implies that

$$\nu_i \in (\bar{\mu} * \mathcal{M}^1(G)) \cap (\mathcal{M}^1(G) * \bar{\mu}) \qquad \text{for all } i\geq 0$$

and

$$\nu_k^{r_k-1} \in (\bar{\mu} * \mathcal{M}^1(G)) \cap (\mathcal{M}^1(G) * \bar{\mu}) \qquad \text{for all } k\geq 1.$$

Since the set $(\bar{\mu} * \mathcal{M}^1(G)) \cap (\mathcal{M}^1(G) * \bar{\mu})$ is compact, it contains λ_l for all $l\geq 0$. It now follows easily that $\bar{\mu} = \bar{\mu} * \lambda = \lambda * \bar{\mu}$ holds with $\lambda \in (\bar{\mu} * \mathcal{M}^1(G)) \cap (\mathcal{M}^1(G) * \bar{\mu})$ and that $\lambda \in \mathcal{I}(G)$, whence $\lambda = \omega_H$ for some subgroup H of G. Consequently, $\bar{\mu} \in \mathcal{M}_H^1(G)$ and thus $\mathcal{M}_H^1(G) = (\bar{\mu} * \mathcal{M}^1(G)) \cap (\mathcal{M}^1(G) * \bar{\mu})$. Theorem 1.2.13 now implies the existence of $x \in N(H)$ such that $\bar{\mu} = \varepsilon_x * \omega_H$ holds. Considering the subsequence $(\bar{\mu}_{m_k})_{k\geq 1}$ of $(\bar{\mu}_l)_{l\geq 0}$ defined by $\bar{\mu}_{m_k} := \nu_k^n$ for all $k\geq 1$ one observes that

$$\lim_{k\to\infty} \nu_k^n = (\varepsilon_x * \omega_H)^n = \omega_H, \quad \text{and hence,} \quad \lim_{k\to\infty} \|\bar{\mu}_{m_k} - \omega_H\| = 0.$$

The conditions of Theorem 3.2.7 are thus satisfied and the result $\mu \in \mathcal{P}_H(G)$ follows. ▯

We give an immediate application which contributes to our knowledge of the structure of the set $\mathcal{P}(G)$.

3.2.10 Remark. Raikov's classical theorem stating that Poisson measures on $\mathbb{R}$ do not admit non-Poisson factors fails to hold for measures on an arbitrary locally compact group. Let q be a prime number >2 and put

$$G := \mathbb{Z}_q := \{0, 1, \ldots, q-1\}$$

the group of residue classes modulo q. Moreover, let $\gamma \in \mathbb{R}_+^*$ and $x \in G \setminus \{0\}$. We show that the Poisson measure

$$\mu := \exp_{\{0\}}[\gamma(\varepsilon_x - \varepsilon_0)] \in \mathcal{M}^1(G)$$

does not admit only Poisson factors from $\mathcal{M}^1(G)$. For any $t \in]0, 1[$ we define

$$\nu_1 := t\varepsilon_0 + (1-t)\omega_G \quad \text{and} \quad \nu_2 := \frac{1}{t}\mu + \left(1 - \frac{1}{t}\right)\omega_G.$$

Evidently $\nu_1 \in \mathcal{M}^1(G)$ and $\nu_2 \in \mathcal{M}^b(G)$. Moreover, $\nu_1 * \nu_2 = \mu$. If t is chosen appropriately, ν_2 can be shown to be in $\mathcal{M}^1(G)$. In fact, $\mu \neq \omega_G$ and hence,

$$0 < M := \min_{k \in G} \mu(\{k\}) < \frac{1}{q}.$$

Choosing $t \in [1 - qM, 1[$ we obtain for all $l \in G$ the inequalities

$$\nu_2(\{l\}) \geq \frac{1}{t} M + \left(1 - \frac{1}{t}\right) \frac{1}{q} \geq \frac{1}{1 - qM} (qM - 1) \frac{1}{q} + \frac{1}{q} \geq 0.$$

Furthermore, $\nu_2(G) = 1$ and so $\nu_2 \in \mathcal{M}^1(G)$ by the choice of t. Clearly $\nu_1 \in \mathcal{I}(G)$ so that by Theorem 3.2.9, $\nu_1 \in \mathcal{P}(G)$. Finally,

$$\hat{\nu}_1(k) = t \qquad \text{for all } k \in G \setminus \{0\}$$

and

$$\hat{\mu}(j) = \exp\left[\gamma \left\{ \exp\left(\frac{2\pi i x j}{q} - 1\right) \right\}\right] \qquad \text{for all } j \in G,$$

so that ν_2 cannot be a Poisson factor of μ.

3.2.11 Theorem. *Let G be a locally compact group, H a compact subgroup of G and $\mu \in \mathcal{I}(G)$ such that for every $n \geq 1$ there exists an n-th root μ_n of μ and $I(\mu_n) = H$. The following statements are equivalent:*

(i) $\mu \in \mathcal{P}_H(G)$.

(ii) (a) *There exists a subsequence $(\mu_{m_i})_{i \geq 1}$ (with an isotone sequence $(m_i)_{i \geq 1}$ in $\mathbb{N}$) of $(\mu_n)_{n \geq 1}$ satisfying Conditions* (ii)(a) *and* (ii)(b) *of Theorem* 3.2.7.

(b) *There exist sequences $(k_n)_{n \geq 1}$ in $[0, 1]$ and $(w_n)_{n \geq 1}$ in $\mathbb{R}_+$ such that one has*

(b1) $\lim_{n \to \infty} w_n k_n = \infty$,

(b2) $\|\mu_n - \omega_H\| \leq 2(1 - k_n)$ *for all $n \geq 1$ and*

(b3) $\sup_{n \geq 1} w_n \|\mu_n * \mu_n^{\sim} - \omega_H\| < \infty$.

Proof. It remains to prove the implication (ii) ⇒ (i).

1. For any measure $\nu \in \mathcal{M}_H^1(G)$ one establishes immediately that

(α) $\|\nu - \omega_H\| = 2\nu(G \setminus H)$ and

(β) $\nu * \nu^{\sim}(G \setminus H) \geq \nu(G \setminus H)\, \nu(H)$.

2. For every $n \geq 1$ we conclude from these properties that

$$w_n \|\mu_n * \mu_n^{\sim} - \omega_H\| \geq 2 w_n \mu_n(G \setminus H)\, \mu_n(H) = w_n \|\mu_n - \omega_H\|\, \mu_n(H),$$

and with the aid of (ii)(b2), $\mu_n(H) \geq k_n$. Therefore

$$w_n \|\mu_n * \mu_n^{\sim} - \omega_H\| \geq w_n k_n \|\mu_n - \omega_H\|.$$

Assumptions (ii)(b1) and (b3) yield $\lim_{n \to \infty} \|\mu_n - \omega_H\| = 0$ and application of Theorem 3.2.7 leads to the assertion. □

3.2.12 Remark. The conditions of (ii)(b) of Theorem 3.2.11 cannot be weakened, as the following discussion shows. Let $G=\mathbb{R}$ and consider a family $(\alpha_p)_{p\in\mathbb{P}}$ of numbers in $\mathbb{R}_+^*$ satisfying $\sum_{p\in\mathbb{P}}\alpha_p=1$, where $\mathbb{P}$ denotes the set of prime numbers (in $\mathbb{N}$). We further introduce the measure

$$\nu:=\sum_{p\in\mathbb{P}}\alpha_p(\varepsilon_{\frac{1}{p}}-\varepsilon_0)=\sum_{p\in\mathbb{P}}\alpha_p\varepsilon_{\frac{1}{p}}-\varepsilon_0\in\mathscr{M}^1(\mathbb{R})$$

as well as for every $t\in\mathbb{R}_+^*$ the measures

$$\mu_t:=\exp_{\{0\}}(t\nu)\quad\text{and}\quad\lambda_t:=\varepsilon_{-t}*\mu_t\ \text{in}\ \mathscr{M}^1(\mathbb{R}).$$

Then the semigroup $(\lambda_t)_{t\in\mathbb{R}_+^*}$ is not Poisson, but we have

(i)
$$\|\lambda_{\frac{1}{p}}-\varepsilon_0\|=2\left(1-\mu_{\frac{1}{p}}\left(\left\{\frac{1}{p}\right\}\right)\right)$$
$$=2\left[1-\sum_{k\geq 0}e^{-\frac{1}{p}}\frac{1}{p^k}\frac{1}{k!}\left(\sum_{q\in\mathbb{P}}\alpha_q\varepsilon_{\frac{1}{q}}\right)^k\left(\left\{\frac{1}{p}\right\}\right)\right]$$
$$=2\left(1-e^{-\frac{1}{p}}\frac{1}{p}\alpha_p\right)<2\quad\text{for all } p\in\mathbb{P}.$$

(ii)
$$\|\lambda_{\frac{1}{p}}*\lambda_{\frac{1}{p}}^{\sim}-\varepsilon_0\|=2\lambda_{\frac{1}{p}}*\lambda_{\frac{1}{p}}^{\sim}(\complement\{0\})$$
$$\geq 2\lambda_{\frac{1}{p}}(\{0\})\,\lambda_{\frac{1}{p}}(\complement\{0\})=2\left[\mu_{\frac{1}{p}}\left(\left\{\frac{1}{p}\right\}\right)\left(1-\mu_{\frac{1}{p}}\left(\left\{\frac{1}{p}\right\}\right)\right)\right]$$
$$=2\left[e^{-\frac{1}{p}}\frac{\alpha_p}{p}\left(1-e^{-\frac{1}{p}}\frac{\alpha_p}{p}\right)\right]\quad\text{for all } p\in\mathbb{P}$$

or $\lim_{p\to\infty}\|\lambda_{\frac{1}{p}}*\lambda_{\frac{1}{p}}^{\sim}-\varepsilon_0\|=0$.

Choosing sequences $(k_n)_{n\geq 1}$ in $[0,1]$ and $(w_n)_{n\geq 1}$ in $\mathbb{R}_+$ such that

$$\|\lambda_{\frac{1}{n}}-\varepsilon_0\|\leq 2(1-k_n)\quad\text{holds for all } n\geq 1$$

and

$$\sup_{n\geq 1}w_n\|\lambda_{\frac{1}{n}}*\lambda_{\frac{1}{n}}^{\sim}-\varepsilon_0\|<\infty$$

is satisfied, we obtain

$$k_p\leq e^{-\frac{1}{p}}\frac{\alpha_p}{p}\quad\text{for all } p\in\mathbb{P}$$

and

$$\sup_{p\in\mathbb{P}}w_p\left[e^{-\frac{1}{p}}\frac{\alpha_p}{p}\left(1-e^{-\frac{1}{p}}\frac{\alpha_p}{p}\right)\right]<\infty,$$

whence for all $p \in \mathbb{P}$ the estimate

$$w_p k_p \leq w_p e^{-\frac{1}{p}\frac{\alpha_p}{p}} \leq c \sup_{p \in \mathbb{P}} w_p \left[e^{-\frac{1}{p}\frac{\alpha_p}{p}} \left(1 - e^{-\frac{1}{p}\frac{\alpha_p}{p}}\right)\right] < \infty$$

with a constant $c \in \mathbb{R}_+^*$.

Poisson measures on almost periodic groups can be characterized in a more useful way. The following result is a modification of Theorem 3.2.11 in the sense that Condition (ii)(a) of that theorem is replaced by the almost periodicity of G.

3.2.13 Theorem. *Let G be in* **A**, *H a compact subgroup of G and $\mu \in \mathscr{I}(G)$ such that for every $n \geq 1$ there exists an n-th root $\mu_n \in \mathscr{M}_H^1(G)$ with $\mu_n^n = \mu$. Then the following statements are equivalent:*
(i) $\mu \in \mathscr{P}_H(G)$;
(ii) (a) $\underline{\lim}_{n \geq 1} \|\mu_n - \omega_H\| < 2$,
(b) *there exists an $n_0 \geq 1$ such that $\|\mu_{n_0} * \tilde{\mu_{n_0}} - \omega_H\| < 1$ holds.*

Proof. To be proved is the sufficiency of the conditions under (ii). First of all, we may assume that G is σ-compact. This implies that given a Bohr compactification $(\tilde{G}, \beta)$ of G the image $\beta(G)$ is in $\mathfrak{B}(\tilde{G})$. Furthermore, we restrict ourselves without loss of generality to the proof of the assertion for a compact group G. In this case we consider $\mu \in \mathscr{I}(G)$ as a measure in $\mathscr{M}^1(\tilde{G})$ and apply the theorem for measures in $\mathscr{M}^1(\tilde{G})$, such that $\mu \in \mathscr{P}_H(\tilde{G})$, i.e., $\mu = \exp_H(\nu)$ for $\nu \in \mathscr{N}_H(\tilde{G})$ holds. But $\nu = \nu_1 + \nu_2$ with $\nu_1 := \operatorname{Res}_{\mathfrak{B}(\tilde{G} \setminus G)} \nu$ and $\nu_2 := \operatorname{Res}_{\mathfrak{B}(G)} \nu$. Since $\mu(\tilde{G} \setminus G) = 0$, we obtain $\nu_1 = 0$ and thus $\mu = \exp_H(\nu_2)$ with $\nu_2 \in \mathscr{M}^1(G)$ or $\mu \in \mathscr{P}_H(G)$.

In order to show the implication (ii) $\Rightarrow$ (i) for a compact group G we need two lemmata.

3.2.14 Lemma. *Let G be a compact group, H a closed subgroup of G, $(\mu_n)_{n \geq 1}$ a sequence of measures in $\mathscr{M}_H^1(G)$ and $\nu_n := \mu_n * \tilde{\mu_n}$ for all $n \geq 1$. If*
(α) $\mu_n^n = \mu_1$ *and*
(β) *there exists an $n_0 \geq 1$ such that $\|\mu_{n_0} * \tilde{\mu_{n_0}} - \omega_H\| < 1$,*
then $\sup_{n \geq 1} \sup_{D \in \mathscr{U}_H} n(1 - \operatorname{tr} \hat{\nu}_n(D)/n_H) < \infty$.

Here $\mathscr{U}_H := \mathscr{U}_H(G)$ denotes the set of all representations $D \in \operatorname{Irr}(G)$ with $n_H := n_H(D) > 0$ elements $\neq 0$ in the (diagonalized) matrix

$$\hat{\omega}_H(D) := \begin{pmatrix} 1 & 0 & \cdots & & 0 \\ 0 & \ddots & & & \\ \vdots & & 1 & 0 & \vdots \\ & & & \ddots & \\ 0 & & \cdots & & 0 \end{pmatrix}.$$

Proof. Given $n_0 \geq 1$ *and* $D \in \mathscr{U}_H$ the matrices $M_H := \hat{\omega}_H(D)$ and $\hat{\nu}_{n_0}(D)$ can be diagonalized simultaneously (after a possible modification of D within $\sigma(D)$) as one deduces from $\mu_{n_0} \in \mathscr{M}_H^1(G)$. For $1 \leq i \leq n_H$ we denote by $\lambda_i(D, n)$ the i-th eigenvalue

$\neq 0$ of $\hat{\nu}_n(D)$. Assumption (β) implies the existence of an $\alpha \in]0, 1[$ independent of D satisfying the condition $0 \leq 1 - \lambda_i(D, n_0) \leq \alpha < 1$.

Assumption (α) yields $\lambda_i(D, 1) \geq (1-\alpha)^{n_0} =: \beta > 0$ for $1 \leq i \leq n_H$ and thus

$$\left(\frac{\operatorname{tr}\hat{\nu}_n(D)}{n_H}\right)^{n_H n} \geq \left(\prod_{i=1}^{n_H} \lambda_i(D, n)\right)^n = |\det(\hat{\mu}_n(M_H))|^{2n}$$

$$= \prod_{i=1}^{n_H} \lambda_i(D, 1) \geq \beta^{n_H} \quad \text{or}$$

$$n\left(1 - \frac{1}{n_H} \operatorname{tr}\hat{\nu}_n(D)\right) \leq n(1 - \beta^{\frac{1}{n}})$$

which tends to $-\log\beta$ as $n \to \infty$ uniformly on $\mathscr{U}_H$. The assertion follows. □

3.2.15 Lemma. *Under the assumptions of the preceding lemma one has*

$$\sup_{n \geq 1} n\nu_n(G \setminus H) < \infty.$$

Proof. Let V be an open subset of G such that $V^{-1}V \in \mathfrak{V}(e)$. There exists always a function $f_V \in \mathscr{C}(G)$ with the following properties:
(1) $f_V \geq 0$,
(2) $\operatorname{Res}_{G \setminus V} f_V = 0$,
(3) $f_V(xy) = f_V(yx)$ for all $x, y \in G$ and
(4) $\int f_V \, d\omega_G = 1$.

For any given $D \in \operatorname{Irr}(G)$ Condition (3) together with Schur's lemma implies the existence of $\gamma_D \in \mathbb{C}^*$ with $\hat{f}_V(D) = \gamma_D E_{n(D)}$. We now define functions g_V and h_V on G by

$$g_V(x) := \int_G f_V(y) f_V(yx^{-1}) \omega_G(dy) \quad \text{and}$$

$$h_V(z) := \int g_V(y^{-1}z) \omega_H(dy)$$

for all $x, z \in G$ resp. One easily verifies that $h_V(x) = 0$ for all

$$x \in U := \complement V^{-1}VHV^{-1}V = U^{-1},$$

$$\hat{g}_V(D) = |\gamma_D|^2 E_{n(D)} \quad \text{and} \quad \hat{h}_V(D) = \hat{\omega}_H(D) |\gamma_D|^2$$

whenever $D \in \operatorname{Irr}(G)$. Here again we assume $\hat{\omega}_H(D)$ to be of the form given in Lemma 3.2.14. The function $h_V \in \mathscr{L}^2_{\mathbb{C}}(G, \omega)$ admits by the Peter-Weyl theorem an expansion of the form

$$h_V = \sum_{D \in \mathscr{V}} n(D) \sum_{i,j=1}^{n(D)} d_{ij} \overline{\hat{h}_V(d_{ij})} \qquad (D := (d_{ij})_{i,j=1,\dots,n(D)}),$$

where $\mathscr{V} := \mathscr{V}(G)$ denotes a complete system of representations in $\operatorname{Irr}(G)$ chosen in such a way that it contains exactly one representative of every class corresponding to the elements of $\mathscr{U}_H$. For every $z \in G$ we obtain

$$h_V(z) = \sum_{D \in \mathscr{V}} n(D) \sum_{i=1}^{n_H} |\gamma_D|^2 d_{ii}(z)$$

and, in particular,

$$h_V(e) = \sum_{D \in \mathscr{V}} n(D) n_H |\gamma_D|^2 > 0.$$

Lemma 3.2.14 provides the existence of $K\in\mathbb{R}_+^*$ such that

$$n\left(1-\frac{\operatorname{tr}\hat{\nu}_n(D)}{n_H}\right)\leq K$$

for all $n\geq 1$ and $D\in\mathscr{U}_H$. Multiplying this inequality by $h_V(e)$ and using the above expansions we deduce that

$$\begin{aligned}K h_V(e) &\geq n\int_U [h_V(e)-\textstyle\sum_{D\in\mathscr{U}_H} n(D)\sum_{i=1}^{n_H}|\gamma_D|^2\, d_{ii}(z)]\,\nu_n(dz)\\ &= n\int_U [h_V(e)-h_V(z)]\,\nu_n(dz) = n h_V(e)\,\nu_n(U)\end{aligned}$$

and hence $n\nu_n(U)\leq K$ for all $n\geq 1$. Since $\sup_{n\geq 1} n\nu_n(\complement V^{-1}VHV^{-1}V)\leq K$ for all $V\in\mathfrak{V}(e)$, the lemma is proved. □

We continue the proof of the theorem. Condition (ii) (a) of the theorem and the statement of Lemma 3.2.15 imply the existence of a sequence $(n_i)_{i\geq 1}$ in $\mathbb{N}$ such that $\sup_{i\geq 1} n_i\|\mu_{n_i}-\omega_H\|<\infty$ holds. Remark 3.2.8 yields the assertion. □

3.2.16 Remark. If $G\in\mathbf{A}$ and $H:=\{e\}$, the measures of $\mathscr{P}_{\{e\}}(G)$ can be characterized as measures $\mu\in\mathscr{I}(G)$ such that for every $n\geq 1$ there exists an n-th root $\mu_n\in\mathscr{M}^1(G)$ and that $\lim_{n\to\infty}\mu_n(\{e\})=1$ holds. In fact, one has the equality $\|\mu_n-\varepsilon_e\|=2(1-\mu_n(\{e\}))$ valid for all $n\geqq 1$. Thus

(a′) $\lim_{n\to\infty}\|\mu_n-\varepsilon_e\|=0$ and therefore

(b′) $\lim_{n\to\infty}\|\mu_n*\mu_n^\sim-\varepsilon_e\|$
$\leq\lim_{n\to\infty}\|\mu_n\|\,\|\mu_n^\sim-\varepsilon_e\|+\lim_{n\to\infty}\|\varepsilon_e\|\,\|\mu_n-\varepsilon_e\|=0.$

But (a′) and (b′) yield the conditions (a) and (b) of (ii) in Theorem 3.2.13 resp. Hence, $\mu\in\mathscr{P}_{\{e\}}(G)$.

3.3 Submonogeneous Embedding of Infinitely Divisible Measures

The final application of Section 1 implied that for root compact groups G every measure $\mu\in\mathscr{I}(G)$ is rationally embeddable. As yet we do not have any result concerning rational embedding of infinitely divisible measures on a non-root compact locally compact group. Such a group exists, however. The example 11 following 3.1.9 yields a deeper analysis of the submonogeneous embeddings of measures in $\mathscr{I}(G)$ for general locally compact Abelian groups G.

3.3.1. Let F_1 denote the free Abelian group with countable generator $\{e,x_1,x_2,\ldots,x_n,\ldots\}$. Introducing in F_1 the relations $x_1=x_n^n$ for all $n\geq 1$ one obtains the group G_1 previously proved to be in no $\mathbf{R}_n$ for all $n\geq 1$, where G_1 has to be considered as a discretely topologized locally compact group. Clearly, there exists no homomorphism $f:\mathbb{Q}_+^*\to G_1$ satisfying $f(1)=x_1$. This implies that

3.3.2. there exist a locally compact Abelian group G having a countable basis of its topology and an element $\mu \in \mathscr{I}(G)$, such that there is no homomorphism $f: \mathbb{Q}_+^* \to \mathscr{M}^1(G)$ satisfying $f(1)=\mu$ or equivalently, such that μ is not rationally embeddable.

3.3.3. We extend to our present framework a few notions introduced previously. If G is a semigroup with neutral element e, we define $x \in G$ to be *submonogeneously embeddable* if there exist a submonogeneous semigroup $\mathbb{M}$ with $1 \in \mathbb{M}$ and a homomorphism $f: \mathbb{M} \to G$ such that $f(1)=x$ holds. In the special case $\mathbb{M}:=\mathbb{Q}_+^*$ we call $x \in G$ *rationally embeddable.*

It should be noted that in general we assume $\mathbb{M}$ to be noncyclic and $f(\mathbb{M}) \neq \{e\}$.

3.3.4 Theorem. *Let G_1 be the free Abelian group defined above.*

(i) *The only (infinitely) divisible elements of G_1 are of the form x_1^β with $\beta \in \mathbb{Z}$. In particular,*

(ii) *for $i \geq 2$ the element $x_i \in G_1$ admits no m-th root if $(m, i) \geq 2$.*

(iii) *If p is a prime number and $m:=p^\kappa$ for $\kappa \in \mathbb{N}$, then for $i \geq 1$ the element $x_i \in G_1$ does not admit an m-th root of the form $x_1^{\alpha_1} x_2^{\alpha_2} \cdot \dots \cdot x_n^{\alpha_n}$ with $n<m$.*

(iv) *No element of G_1 is rationally embeddable.*

(v) *Every element y of G_1 is submonogeneously embeddable, and for the submonogeneous embedding of $y \in G_1$ it suffices to choose an arbitrary sequence $(s_i)_{i \geq 0}$ in $\mathbb{N}$ such that*
 (a) *$y^{s_0}=x_1^\beta$ for some $\beta \in \mathbb{Z}$ and*
 (b) *$(\prod_{l=0}^i s_l, s_{i+1})=1$ for all $i \geq 0$ hold.*

Proof. 1. We first consider the case of $\beta \in \mathbb{N}$ and $i \geq 2$. Let $x_1^{\alpha_1} x_2^{\alpha_2} \cdot \dots \cdot x_n^{\alpha_n}$ be an m-th root of x_i^β. By the definition of G_1 we may assume that $0 \leq \alpha_j \leq j-1$ for $j \geq 2$ and also that $1 \leq \beta \leq i-1$. From $x_i^\beta=(x_1^{\alpha_1} x_2^{\alpha_2} \cdot \dots \cdot x_n^{\alpha_n})^m$ we conclude that

(α) $\beta-\alpha_i m \equiv 0(i)$,

(β) $\alpha_j m \equiv 0(j)$ for all $2 \leq j \leq n$, $j \neq i$ and

(γ) $\dfrac{\beta-\alpha_i m}{i}=m\alpha_1+\sum_{\substack{2 \leq j \leq n \\ j \neq i}} \dfrac{\alpha_j m}{j}$.

Conversely, every $(\alpha_1, \dots, \alpha_n) \in \mathbb{Z}^n$ satisfying (α), (β), (γ) and the assumption $0 \leq \alpha_j \leq j-1$ for $j \geq 2$ defines an m-th root of x_i^β. Problem (α) is solvable iff $d_{m,i}:=(m, i)$ divides β, and in this case there are exactly $d_{m,i}$ different solutions. If $l_{m,j}:=(m, j)$, then (β) admits $l_{m,j}$ solutions of the form

$$\alpha_j:=\beta_j \frac{j}{l_{m,j}} \quad \text{for } 0 \leq \beta_j \leq l_{m,j}-1.$$

It should be noted that for $l_{m,j} \equiv 1$ one chooses $\alpha_j=\beta_j=0$. But writing $m_j:=\dfrac{m}{l_{m,j}}$

it remains to show by (γ) that there exists an $\alpha_1 \in \mathbb{Z}$ satisfying

$$\frac{\beta - \alpha_i m}{i} = m\alpha_1 + \sum_{\substack{2 \le j \le n \\ j \ne i \\ (m,j) > 1}} \beta_j m_j$$

for suitably chosen solutions α_i of (α) and (β). The existence of α_i, however, can be established iff the module generated by the elements m_j with $2 \le j \le n$, $j \ne i$, $(m,j) > 1$, contains the integer $\dfrac{\beta - \alpha_i m}{i}$. This occurs, in particular, if the greatest common divisor of the above set of elements m_j equals 1, and this can be achieved if $n \ge m$ and $i \ne m$.

2. For every $n \ge m$ there is at least one m-th root of x_i^β of the form $x_1^{\alpha_1} \cdot \ldots \cdot x_n^{\alpha_n}$, if $d_{m,i}$ divides β.

3. For $i = 1$ we obtain the relations

(β') $\qquad \alpha_j m \equiv 0(j) \quad$ for $2 \le j \le n$ and

(γ') $\qquad \beta = m\alpha_1 + \sum_{2 \le j \le n} \dfrac{\alpha_j m}{j}$,

and their discussion becomes a special case of the treatment in Parts 1 and 2. Hence for $n \ge m$ there exists also in this case an m-th root of x_i^β of the form $x_1^{\alpha_1} \cdot \ldots \cdot x_n^{\alpha_n}$.

4. Analogously one handles the case $\beta = 0$, and thus all roots of e are determined.

5. The discussion of the roots of an arbitrary element of G_1 runs as follows: Let $x_1^{\gamma_1} x_2^{\gamma_2} \cdot \ldots \cdot x_k^{\gamma_k} = (x_1^{\alpha_1} \cdot \ldots \cdot x_n^{\alpha_n})^m$ and assume $k \le n$. Then without loss of generality one supposes that $k = n$; otherwise one considers instead of $x_1^{\gamma_1} x_2^{\gamma_2} \cdot \ldots \cdot x_k^{\gamma_k}$ the element

$$x_1^{\gamma_1} \cdot \ldots \cdot x_k^{\gamma_k} x_{k+1}^{\gamma_{k+1}} \cdot \ldots \cdot x_n^{\gamma_n} \quad \text{with } \gamma_l = 0,\ k+1 \le l \le n.$$

The existence of the initial product representation is equivalent to

(β'') $\qquad -\gamma_j + \alpha_j m \equiv 0(j) \quad$ for $2 \le j \le k$ and

(γ'') $\qquad \gamma_1 = m\alpha_1 + \sum_{2 \le j \le k} \dfrac{-\gamma_j + \alpha_j m}{j}$

under the norming condition $0 \le \alpha_j \le j-1$ for $j \ge 2$. If $k > n$, an analoguous procedure yields the result.

Suppose that $x_1^{\gamma_1} \cdot \ldots \cdot x_k^{\gamma_k}$ is itself an r-th root of x_1^β for $\beta \in \mathbb{Z}$. Then the elements $\gamma_j (2 \le j \le k)$ satisfy (β'), where α_j, m and n are replaced by γ_j, r and k resp. But (β'') admits solutions α_j iff (m,j) divides γ_j for $2 \le j \le k$. If, in particular, $(m,r) = 1$ holds, then one has

$$\left(m, \frac{j}{(r,j)}\right) = (m,j)$$

and hence, the divisibility condition is satisfied. In the case $k=n\geq m$ the discussion of (γ) leads to the result.

Up to this point we have proved Statements (i) to (iv). We are left with

6. the proof of Statement (v). If $y\in G_1$ admits the initial product representation, then by Part 3 we have $y^{n!}=x_1^{\beta}$ for $\beta\in\mathbb{Z}$.

Let $y_i=y_{i+1}^{s_i}$ for $i\geq 1$ be such that $y_1=y$ holds. Then the semigroup generated in G_1 by $(y_i)_{i\geq 1}$ is a homomorphic image of the real submonogeneous semigroup $\mathbb{M}$ generated by the set

$$\left\{1, \frac{1}{s_1}, \ldots, \frac{1}{s_1\cdot\ldots\cdot s_k}, \ldots\right\}. \quad \square$$

3.3.5 Remark. For every sequence $(s_i)_{i\geq 0}$ in $\mathbb{N}$ satisfying $(\prod_{l=0}^{i} s_l, s_{i+1})=1$ there exists a homomorphism

$$f:\mathbb{M}:=\left\langle\left\{1, \frac{1}{s_1}, \ldots, \frac{1}{s_1\cdot\ldots\cdot s_k}, \ldots\right\}\right\rangle\to G_1$$

such that $f(1)=x_1$ and $f(\mathbb{M})\subset[\{x_{s_i}: i\geq 0\}]$ hold.

Let F_2 be the free Abelian group with countable generator

$$\{e, x_i\ (i\geq 1),\ x_{kj}\ (k,j\geq 2)\}.$$

Introducing in F_2 the relations $x_1=x_n^n$ for all $n\geq 1$ and $x_i=x_{ij}^j$ for $i,j\geq 2$ we obtain a group G_2 which again is not in $\mathbf{R}_n$ for all $n\geq 1$.

3.3.6 Theorem. *G_2 contains a divisible element x_1 which admits infinitely many divisible roots, but is not rationally embeddable.*

Proof. Clearly, for every $i\geq 1$ the element x_i of G_2 is (infinitely) divisible, and so is every element of G_2 of the form $x_1^{\alpha_1}\cdot\ldots\cdot x_n^{\alpha_n}$ with $\alpha_1,\ldots,\alpha_n\in\mathbb{Z}$. It remains to show that x_1 is not rationally embeddable. We shall establish the assertion that no element $y\in G_2$ of the above product form is rationally embeddable. Let f be a homomorphism from $\mathbb{Q}_+^*$ into G_2 satisfying $f(1)=y$. By Theorem 3.3.4 $f(\mathbb{Q}_+^*)$ is not contained in $[\{e, x_i\ (i\geq 1)\}]$ (generated in G_1). Thus there is an $r\in\mathbb{Q}_+^*$ such that

$$f(r)=x_1^{\alpha_1}\cdot\ldots\cdot x_{s_0}^{\alpha_{s_0}}\prod_{\substack{2\leq i\leq s_1\\ 2\leq j\leq s_2}} x_{ij}^{\beta_{ij}}$$

with $s_l\in\mathbb{N}$, $0\leq l\leq 2$, $\alpha_n\in\mathbb{Z}$, $1\leq n\leq s_0$ and $0\leq\beta_{ij}\leq j-1$, $2\leq i\leq s_1$, $2\leq j\leq s_2$, where one has $\beta_{i_0 j_0}>0$ at least once. Plainly $f(r)$ admits no j_0-th root. $\square$

Let G_3 be the (non-Abelian) free group with countable generator $\{e, x_1, x_2, \ldots\}$ and relations $x_1=x_n^n$ for all $n\geq 1$, G_4 the free group with countable generator $\{e, x_i\ (i\geq 1),\ x_{kj}\ (k,j\geq 2)\}$ and relations $x_1=x_n^n$ for all $n\geq 1$, $x_i=x_{ij}^j$ for $i,j\geq 2$.

3.3.7 Theorem. *For the groups G_3 and G_4 we obtain the following properties:*
(i) *$G_3\setminus Z(G_3)$ contains no divisible element,*
(ii) *G_3 contains no submonogeneously embeddable element,*
(iii) *the element x_1 of G_4 is divisible and admits for every $n\geq 1$ divisible n-th roots and*
(iv) *no element of G_4 is submonogeneously embeddable.*

Proof. We concentrate ourselves on the proof of (i). The rest follows in analogy to the proof of Theorem 3.3.6.

1. One observes that in G_3 the relation $x_n^k x_m^l = x_m^l x_n^k$ holds iff $m=n$ or $k\equiv 0(n)$ or $l\equiv 0(m)$ is satisfied. In particular, one concludes that $Z(G_3)$ is the cyclic group generated by x_1. Indeed, the initial equations are equivalent to $x_n^k x_m^l x_n^{-k} x_m^{-l}=e$. But if none of the conditions is satisfied, then $x_n^k x_m^l x_n^{-k} x_m^{-l}$ cannot be reduced.

2. Let $m\in\mathbb{N}$, $\gamma\in\mathbb{Z}$ and $u\in G_3$ be such that $u^m=x_1^\gamma$ holds. Then there exist $y\in G_3$, $\alpha,\beta\in\mathbb{Z}$ and $s\in\mathbb{N}$ satisfying

$$(\alpha) \qquad \beta m\equiv 0(s) \quad \text{and}$$

$$(\beta) \qquad \alpha m+\frac{m\beta}{s}=\gamma \quad \text{such that}$$

$u=x_1^\alpha y x_s^\beta y^{-1}$ holds. Conversely, every such element u is an m-th root of x_1^γ if (α) and (β) are satisfied. If $\beta=0$, then $u=x_1^\alpha$ with $\alpha m=\gamma$ and x_1^γ admits only finitely many roots of this form. Let $u\in G_3$ be such that $u^m=x_1^\gamma$ holds, be not of the form x_1^α and be represented in reduced form as $u=x_1^{\alpha_0}x_{i_1}^{\alpha_1}\cdot\dots\cdot x_{i_n}^{\alpha_n}$ with

$$(*) \qquad \alpha_0\in\mathbb{Z},\ \ 0<\alpha_j<i_j,\ \ 2\leq i_j,\ \ 1\leq j\leq n,\ \ i_j\neq i_{j+1},\ \ 1\leq j<n,\ \ n\geq 1.$$

Then $x_1^{\gamma-m\alpha_0}=(x_{i_1}^{\alpha_1}\cdot\dots\cdot x_{i_n}^{\alpha_n})^m$. But this representation can only hold if the right side is reducible. This implies by Part 1 the validity of the equations

$$i_{n-l}=i_{l+1},\quad \alpha_{n-l}+\alpha_{l+1}\equiv 0(i_{l+1}),\quad 0\leq l<\frac{n}{2}.$$

Clearly, $n=2k+1$ for $k\geq 0$ and hence

$$\begin{aligned} u&=x_1^{\alpha_0}(x_{i_1}^{\alpha_1}\cdot\dots\cdot x_{i_k}^{\alpha_k})\,x_{i_{k+1}}^{\alpha_{k+1}}(x_{i_k}^{-\alpha_k}\cdot\dots\cdot x_{i_1}^{-\alpha_1})\cdot(x_{i_1}^{\alpha_n+\alpha_1}\cdot\dots\cdot x_{i_k}^{\alpha_{k+2}+\alpha_k})\\ &=x_1^{\alpha_0+a}(x_{i_1}^{\alpha_1}\cdot\dots\cdot x_{i_k}^{\alpha_k})\,x_{i_{k+1}}^{\alpha_{k+1}}(x_{i_1}^{\alpha_1}\cdot\dots\cdot x_{i_k}^{\alpha_k})^{-1} \qquad \text{with}\\ a&:=\sum_{l=0}^{k-1}\frac{\alpha_{n-l}+\alpha_{l+1}}{i_{l+1}}.\end{aligned}$$

Defining $y:=x_{i_1}^{\alpha_1}\cdot\dots\cdot x_{i_k}^{\alpha_k}$ one obtains from $u^m=x_1^\gamma$ the relations

$$x_1^{\gamma-m(\alpha_0+a)}=y\,x_{i_{k+1}}^{m\alpha_{k+1}}\,y^{-1}$$

and therefore by Part 1

$$m\alpha_{k+1}\equiv 0(i_{k+1}),\qquad \gamma-m(\alpha_0+a)=m\frac{\alpha_{k+1}}{i_{k+1}}.$$

But these relations correspond to (α) and (β).

3. If $z \in G_3$ is not of the form $z = x_1^\gamma$ for $\gamma \in \mathbb{Z}$, then there exists an $m_z \geq 1$ such that z admits no m-th root for all $m \geq m_z$. Let

$$z = x_1^{\alpha_0} x_{i_1}^{\alpha_1} \cdot \ldots \cdot x_{i_n}^{\alpha_n}$$

and assume that the conditions $(*)$ of Part 2 are satisfied. One defines

$$\overline{m}_z := n \max_{1 \leq j \leq n}(i_j - 1) + |\alpha_0|.$$

Let $m > \overline{m}_z$ and let there exist an m-th root v of z of the form

$$v = x_1^{\delta_0} x_{k_1}^{\delta_1} \cdot \ldots \cdot x_{k_s}^{\delta_s}$$

with

$$(**) \qquad \delta_0 \in \mathbb{Z}, \quad 0 < \delta_i < k_i, \quad 1 \leq i \leq s, \quad k_i \neq k_{i+1}, \quad 1 \leq i < s, \quad s \geq 1.$$

From $(*)$ we conclude that $k_i \geq 2$, $1 \leq i \leq s$ also holds. But the right side of

$$x_1^{\alpha_0} x_{i_1}^{\alpha_1} \cdot \ldots \cdot x_{i_n}^{\alpha_n} = (x_1^{\delta_0} x_{k_1}^{\delta_1} \cdot \ldots \cdot x_{k_s}^{\delta_s})^m$$

is reducible.

As in Part 2 we obtain $s = 2k+1$, $k \geq 0$ and $k_{s-l} = k_{l+1}$ $\delta_{s-l} + \delta_{l+1} \equiv 0(k_{s-l})$ for all $0 \leq l < k$. Thus $v = x_1^{\delta'} y_1 x_t^{\gamma'} y_1^{-1}$ and so $z = x_1^{\delta''} y_1 x_t^{\gamma''} y_1^{-1}$ with $0 < \gamma^{(i)} < t$ $(i = 1, 2)$ satisfying

$$(\alpha') \qquad m\gamma' \equiv \gamma''(t) \quad \text{and}$$

$$(\beta') \qquad m\delta' + \frac{m\gamma' - \gamma''}{t} = \delta''.$$

These relations imply that for fixed $t \geq 2$, $\gamma'', \delta'' \in \mathbb{Z}$ there exist only finitely many m-th roots of z with $m > \overline{m}_z$. This, however, shows the existence of m_z. From Parts 2 and 3 follows the desired result. □

3.3.8 Corollary. *There exist a locally compact group G having a countable basis of its topology and a measure $\mu \in \mathscr{I}(G)$ which is not submonogeneously embeddable.*

3.3.9. From Remark 3.3.5 it follows immediately that for an Abelian group G and any divisible element x of G there exist a noncyclic submonogeneous group $\mathbb{M}$ and a homomorphism $f: \mathbb{M} \to G$ such that $f(1) = x$ holds. In fact, let $x \in G$ be divisible $\neq e$. For every $n \geq 1$ there exists an $x_n \in G$ such that $x_n^n = x$, and the subgroup of G generated by $(x_n)_{n \geq 1}$ is a homomorphic image of G_1. The group G_1 is *universal* for this problem in the following sense. Let G be an arbitrary Abelian group and $x \in G$ a divisible element. If the submonogeneous group $\mathbb{M}$ has the property that there is a homomorphism $f: \mathbb{M} \to G_1$ with $f(1) = x_1$, then there exists also a homomorphism $g: \mathbb{M} \to G$ satisfying $g(1) = x$. Thus it seems instructive to study the class **SM** of all noncyclic submonogeneous groups $\mathbb{M}$ with $1 \in \mathbb{M}$ such that there exists a homomorphism

$$f: \mathbb{M} \to G_1 \quad \text{satisfying} \quad f(1) = x_1.$$

3.3.10 Theorem. *A noncyclic submonogeneous group $\mathbb{M}$ with $1\in\mathbb{M}$ belongs to the class* **SM** *iff $\mathbb{M}/\mathbb{Z}$ can be embedded in a direct product of infinitely many cyclic groups.*

Proof. 1. Let $\mathbb{M}$ be a noncyclic submonogeneous group with $1\in\mathbb{M}$ and $f: \mathbb{M}\to G_1$ a homomorphism with $f(1)=x_1$. Since $[x_1]\cong\mathbb{Z}$, the homomorphism f is injective. On the other hand,

$$G_1/[x_1]\cong\prod^*_{i\geq 1}\mathbb{Z}_i \quad \text{with } \mathbb{Z}_i:=\mathbb{Z}/i\mathbb{Z} \ (i\geq 1)$$

as is easily seen. Hence,

$$\mathbb{M}/\mathbb{Z}\cong\{f(r): r\in\mathbb{M}\}/[x_1]\subset\prod_{i\geq 1}\mathbb{Z}_i.$$

2. Let $\mathbb{M}$ be a submonogeneous group with $1\in\mathbb{M}$ such that $\mathbb{M}/\mathbb{Z}\subset\prod_{i\geq 1}E_i$ with finite cyclic groups E_i of order $n_i\geq 1$ $(i\geq 1)$. We assume that

$$\mathbb{M}=\bigcup_{i\geq 1}\left[\left\{\frac{1}{k_1\cdot\dots\cdot k_i}\right\}\right]$$

with $k_i\in\mathbb{N}$ and define $m_i:=k_1\cdot\dots\cdot k_i$ for $i\geq 1$. Every prime number $p>1$ divides only finitely many of the k_i for $i\geq 1$. Indeed, if there were infinitely many indices i_j $(j\geq 1)$ such that p divided every element of $(k_{i_j})_{j\geq 1}$, then for any $n\geq 1$ there would exist an $i(n)\geq 1$ with $p^n\mid m_{i(n)}$. Hence

$$\mathbb{M}\supset\bigcup_{n\geq 1}\left[\left\{\frac{1}{p^n}\right\}\right]$$

and therefore $\mathbb{M}/\mathbb{Z}\supset\mathbb{Z}(p^\infty)$, which would lead to a contradiction. We shall construct a generator $\{q_i: i\geq 1\}$ for $\mathbb{M}$ satisfying the condition

$$\left(\prod_{l=1}^{i} q_l, q_{i+1}\right)=1 \qquad \text{for all } i\geq 1.$$

Let $(p_j)_{j\geq 1}$ be the sequence of common prime divisors of $(k_i)_{i\geq 1}$ and d_j the largest exponent such that for a suitable $i_0\geq 1$ (and hence for all $i>i_0$) one has $p_j^{d_j}\mid m_{i_0}$. Putting $q_j:=p_j^{d_j}$ for all $j\geq 1$ one obtains the desired generator $\{q_j: j\geq 1\}$ of $\mathbb{M}$. On one side $q_j\mid m_i$ for sufficiently large i, on the other side $m_j\mid\prod_{l=1}^{i} q_l$ since $\{q_j: j\geq 1\}$ runs through all prime powers in $\{m_i: i\geq 1\}$. By the condition achieved for the modified generator $\{q_i: i\geq 1\}$ of $\mathbb{M}$ the hypothesis of Theorem 3.3.4 is satisfied and hence $\mathbb{M}\in$ **SM**. □

3.3.11 Corollary. *Let G be an Abelian group, $x\in G$ divisible and $\mathbb{M}\in$* **SM**. *Then there exists a homomorphism $f:\mathbb{M}\to G$ with $f(1)=x$.*

The *proof* is immediate from the theorem by the universality of the group G_1 described in 3.3.9. □

3.3.12. Defining for any locally compact group G the class **SM**(G) of all noncyclic submonogeneous groups $\mathbb{M}$ such that for every $\mu\in\mathscr{I}(G)$ there exists a homomorphism $f: \mathbb{M}_+ \to \mathcal{M}^1(G)$ satisfying $f(1) = \mu$, one obtains the following results which are reformulations of preceding theorems:

(1) $\mathbb{Q} \notin \mathbf{SM}(G_1)$, but

(2) $\mathbf{SM}(G_1) \neq \emptyset$, and

(3) $\mathbf{SM}(G_3) = \emptyset$.

We are now ready to prove the main theorem of this section on the submonogeneous embedding of infinitely divisible measures.

3.3.13. Let G be an arbitrary locally compact group. For every measure $\mu \in \mathcal{M}^b(G)$ we denote by μ^s the discrete part $\sum_{x \in G} \mu(\{x\}) \varepsilon_x$ of μ. Furthermore, we introduce the real functions p, q and u on $\mathcal{M}^b(G)$ by means of

$$p(\mu) := \sum_{x \in G} |\mu(\{x\})| = \|\mu^s\|, \quad q(\mu) := \sum_{x \in G} |\mu(\{x\})|^2$$

and

$$u(\mu) := \sup_{x \in G} |\mu(\{x\})| \quad \text{for all } \mu \in \mathcal{M}^b(G) \text{ resp.}$$

Properties. (1) The mapping $\mu \to \mu^s$ is an involution invariant, (norm-) continuous homomorphism from $\mathcal{M}^b(G)$ into itself. In particular, for any $\mu \in \mathcal{M}^b(G)$ one has

$$(\mu^s)^n = (\mu^n)^s, \quad (\mu^s)^{-1} = (\mu^{-1})^s \quad \text{and} \quad (\mu^s)^{\sim} = (\mu^{\sim})^s$$

(where, of course, the inverse μ^{-1} of μ is taken in the algebra $\mathcal{M}^b(G)$).

(2) The functions p, q and u are invariant under the mapping $\mu \to \mu^s$.

(3) For every $\mu \in \mathcal{M}^b(G)$ there exists an element x_μ of G satisfying

$$|\mu(\{x_\mu\})| = u(\mu).$$

This element becomes uniquely determined if $u(\mu) > \frac{1}{2}\|\mu\|$.

(4) If $\mu \in \mathcal{M}_+^b(G)$ and $x \in G$, then $\mu^n(\{x^n\}) \geq [\mu(\{x\})]^n$ for all $n \geq 1$.

(5) For every $\mu \in \mathcal{M}^b(G)$ we have

$$\begin{aligned} q(\mu) &= \mu * \mu^{\sim}(\{e\}) = \mu^{\sim} * \mu(\{e\}) \\ &= \mu^s * (\mu^s)^{\sim}(\{e\}) = (\mu^s)^{\sim} * \mu^s(\{e\}). \end{aligned}$$

(6) If $\mu \in \mathcal{M}^{(1)}(G) := \{\mu \in \mathcal{M}^b(G) : \|\mu\| \leq 1\}$, then

$$1 \geq p(\mu) \geq u(\mu) \geq p(\mu)\, u(\mu) \geq q(\mu) \geq u(\mu)^2.$$

We restrict ourselves to the proof of Property (6). Clearly, for $\mu \in \mathcal{M}^{(1)}(G)$ one has $1 \geq p(\mu) \geq u(\mu)$ and hence $u(\mu) \geq p(\mu)\, u(\mu)$. The chains

$$q(\mu) = \sum_{x \in G} |\mu(\{x\})|^2 \leq \left(\sum_{x \in G} |\mu(\{x\})|\right) \sup_{x \in G} |\mu(\{x\})| = p(\mu)\, u(\mu)$$

and

$$\sum_{x \in G} |\mu(\{x\})|^2 \geq \sup_{x \in G} |\mu(\{x\})|^2 = u(\mu)^2,$$

valid for all $\mu \in \mathcal{M}^{(1)}(G)$, yield the assertion.

For every $\mu\in\mathscr{M}^b(G)$ the convolution operator T_μ will be considered as an operator on the Hilbert space $L^2_{\mathbb{C}}:=L^2_{\mathbb{C}}(G,\omega)$, where ω is chosen to be a right Haar measure on G. This choice enables us to obtain the formula $T_\mu f=\mu * f$ for all $f\in L^2_{\mathbb{C}}(G,\omega)$ which is consistent with the notions in § 2 of Chapter I.

3.3.14 Lemma. *The mapping $\mu\to T_\mu$ from $\mathscr{M}^b(G)$ into $\mathscr{L}(L^2_{\mathbb{C}})$ is an involution invariant continuous homomorphism with the following additional properties:*
(i) *$\mu\to T_{\mu^s}$ is an involution invariant continuous homomorphism from $\mathscr{M}^b(G)$ into the space $\mathscr{L}(L^2_{\mathbb{C}})$,*
(ii) *$\|T_{\mu^s}\|^2\geq q(\mu)$ for all $\mu\in\mathscr{M}^b(G)$,*
(iii) *if $\mu\in\mathscr{M}^b(G)$ and T_{μ^s} is invertible (in $\mathscr{L}(L^2_{\mathbb{C}})$), then $\|T_{\mu^s}^{-1}\|^{-2}\leq q(\mu)$.*

The *proof* of (i) is clear. It remains to show (ii) and (iii).

1. Let $\operatorname{supp}(\mu^s)$ be finite or $\mu^s:=\sum_{i=1}^m\alpha_i\varepsilon_{x_i}$ for $\alpha_1,\dots,\alpha_m\in\mathbb{R}$ and $x_1,\dots,x_m\in G$. Then there exists an open, relatively compact neighborhood $U\in\mathfrak{B}(e)$ with $U^{-1}=U$ satisfying $x_ix_j^{-1}U\cap U=\emptyset$ for all $i,j=1,\dots,m$, $i\neq j$.

For the function $f:=\dfrac{1}{\sqrt{\omega(U)}}1_U$ we have $f\in L^2_{\mathbb{R}}(G,\omega)$, $\|f\|_2=1$ and

$$\langle T_{(\mu^s)^{\sim}*\mu^s}f,f\rangle=\|T_{\mu^s}f\|_2^2=q(\mu).$$

Hence,

$$\begin{aligned}\|T_{\mu^s}^{-1}\|^{-2}&=\inf_{\|g\|_2=1}\|T_{\mu^s}g\|_2^2\leq q(\mu)\\ &\leq\sup_{\|g\|_2=1}\|T_{\mu^s}g\|_2^2=\|T_{\mu^s}\|^2.\end{aligned}$$

2. If $\operatorname{supp}(\mu^s)$ is not finite or $\mu^s:=\sum_{i\geq1}\alpha_i\varepsilon_{x_i}$ with $(\alpha_i)_{i\geq1}$ in $\mathbb{R}$ and $(x_i)_{i\geq1}$ in G, then one puts $\nu_n:=\sum_{i=1}^n\alpha_i\varepsilon_{x_i}$ for any $n\geq1$, observes that

$$\lim_{n\to\infty}\|\mu^s-\nu_n\|=0 \quad\text{and}\quad \lim_{n\to\infty}q(\nu_n)=q(\mu),$$

and concludes

$$\lim_{n\to\infty}\|T_{\mu^s}-T_{\nu_n}\|=0.$$

If, in addition, $T_{\mu^s}^{-1}$ exists, then $T_{\nu_n}^{-1}$ exists for sufficiently large $n\geq1$, and one obtains

$$\lim_{n\to\infty}\|T_{\mu^s}^{-1}-T_{\nu_n}^{-1}\|=0.$$

But

$$\|T_{\nu_n}^{-1}\|^{-2}\leq q(\nu_n)\leq\|T_{\nu_n}\|^2$$

for sufficiently large $n\geq1$. Letting $n\to\infty$, we see that this inequality yields the desired relations. □

3.3.15 Lemma. *Let $\mu\in\mathscr{I}(G)$ such that for every $n\geq1$ the n-th root μ_n of μ is normal (in the sense that $\mu_n*\mu_n^{\sim}=\mu_n^{\sim}*\mu_n$ is satisfied) and let T_{μ^s} be an invertible operator. Then there exist $\varepsilon,\eta\in\mathbb{R}_+^*$ with $\varepsilon<\eta$ such that $\sqrt[n]{\eta}\geq q(\mu_n)\geq\sqrt[n]{\varepsilon}$ holds. In particular, $\lim_{n\to\infty}q(\mu_n)=1$.*

Proof. One observes that, under the assumptions of the lemma, T_{μ^s} is invertible if either $T_{\mu_n^s}$ or μ_n^s or μ_n is invertible for some $n \geq 1$. Clearly for every $n \geq 1$ the operator $T_{\mu_n^s}$ is invertible. Hence, by the preceding lemma one gets

$$\eta_n := \|T_{\mu_n^s}\|^2 \geq q(\mu_n) \geq \|T_{\mu_n^s}^{-1}\|^{-2} =: \varepsilon_n.$$

For the normal operator $T_{\mu_n^s}$ the spectral radius $r(T_{\mu_n^s})$ satisfies

$$r(T_{\mu_n^s}) = \|T_{\mu_n^s}\| = \sqrt[n]{r(T_{\mu^s})},$$

from which $\eta_n = \sqrt[n]{\eta_1}$ and $\varepsilon_n = \sqrt[n]{\varepsilon_1}$ follow $(n \geq 1)$. This yields the assertion. □

3.3.16 Theorem. *Let G be a locally compact group and let $\mu \in \mathscr{I}(G)$ be such that for every $n \geq 1$ there exists an n-th root $\mu_n \in \mathscr{M}^1(G)$ satisfying $\mu_n^n = \mu$. We suppose for all $n \geq 1$ that*

(a) *μ_n is normal and*

(b) *μ_n is invertible in $\mathscr{M}^b(G)$*

Then there exists a sequence $(x_n)_{n \geq 1}$ in G such that

(i) *$(x_n^n)_{n \geq 1}$ is a finite set and*

(ii) *$\lim_{n \to \infty} \|\mu_n - \varepsilon_{x_n}\| = 0$.*

Proof. Condition (b) implies the invertibility of T_{μ^s}. Putting $\varepsilon := \|T_{\mu^s}^{-1}\|^{-2}$ we see from Lemma 3.3.15 that $q(\mu_n) \geq \sqrt[n]{\varepsilon}$ for $n \geq 1$. Moreover, by Property (6) of 3.3.13 one has $u(\mu_n) \geq \sqrt[n]{\varepsilon}$, and there exists an $x_n \in G$ satisfying $u(\mu_n) = \mu_n(\{x_n\})(n \geq 1)$. This implies $u(\mu_n) = (\mu_n * \varepsilon_{x_n^{-1}})(\{e\})$ and hence

$$\|\mu_n - \varepsilon_{x_n}\| = \|\mu_n * \varepsilon_{x_n^{-1}} - \varepsilon_e\| \leq 2(1 - u(\mu_n)).$$

Therefore $\|\mu_n - \varepsilon_{x_n}\| \leq 2(1 - \sqrt[n]{\varepsilon})$ for $n \geq 1$ or (ii). Again by Property (4) of 3.3.13 one obtains $\mu(\{x_n^n\}) \geq [\mu_n(\{x_n\})]^n \geq \varepsilon$ for every $n \geq 1$, i.e., there exists only a finite set of elements $y \in G$ of the form $y = x_k^k$ for $k \geq 1$, which implies Assertion (i). □

3.3.17 Theorem. *Let G be a discrete Abelian group and $\mathbb{M} \in \mathbf{SM}$. Then for every $\mu \in \mathscr{I}(G)$ there exists a homomorphism ϕ: $\mathbb{M}_+ \to \mathscr{M}^1(G)$ with $\phi(1) = \mu$,* i.e., *every $\mu \in \mathscr{I}(G)$ is $\mathbb{M}$-(submonogeneously) embeddable.*

The *proof* of the theorem will be preceded by the following

3.3.18 Lemma. *Let G be a discrete Abelian group and μ a measure in $\mathscr{I}(G)$ with n-th root μ_n for every $n \geq 1$. There exists a finite subgroup H of G such that μ_n is an invertible element of $\mathscr{M}_H^1(G)$ for all $n \geq 1$.*

Proof. Let $\nu_n := \mu_n * \mu_n^{\sim}$ for every $n \geq 1$. Then $\hat{\nu}_n \geq 0$ and so

$$\hat{\nu}_n = \hat{\nu}_1^{\frac{1}{n}} \quad \text{for all } n \geq 1.$$

Furthermore, $\hat{\nu}_{nk}^k = \hat{\nu}_n$, whence $\nu_{nk}^k = \nu_n$ for all $n, k \geq 1$. This implies that the mapping

$$r := \frac{k}{n} \to \nu_n^k =: \nu_r$$

is a homomorphism from $\mathbb{Q}_+^*$ into $\mathcal{M}^1(G)$. Plainly we have

$$\lim_{r\to 0} \hat{\nu}_r = \lim_{r\to 0} \hat{\nu}_1^r = 1_{[\hat{\nu}_1 > 0]}$$

pointwise on $G^\wedge$. Since the set

$$\mathcal{A} := \{f \in \mathcal{C}^0(G) : f = \hat{\hat{f}} \text{ with } \hat{f} \in L^1(G^\wedge, \omega_{G^\wedge})\}$$

is dense in $\mathcal{C}^0(G)$, the limit relations

$$\begin{aligned}\lim_{r\to 0} \nu_r(f) &= \lim_{r\to 0} \int_G \left(\int_{G^\wedge} \chi(x) \hat{f}(\chi)\, \omega_{G^\wedge}(d\chi)\right) \nu_r(dx)\\ &= \lim_{r\to 0} \int_{G^\wedge} \hat{f}(\chi)\, \hat{\nu}_r(\chi)\, \omega_{G^\wedge}(d\chi) = \int_{G^\wedge} \hat{f}\, 1_{[\hat{\nu}_1 > 0]}\, d\omega_{G^\wedge},\end{aligned}$$

valid for all $f \in \mathcal{A}$, imply the existence of $\lim_{r\to 0} \nu_r(f)$ for all $f \in \mathcal{C}^0(G)$.

On the other hand, by the $\mathcal{T}_v$-compactness of $\mathcal{M}^{(1)}(G)$ there exists a measure $\lambda \in \mathcal{M}_+^{(1)}(G)$ such that $\mathcal{T}_v\text{-}\lim_{r\to 0} \nu_r = \lambda$ holds. Thus

$$\mathcal{T}_v\text{-}\lim \nu_{r+s} = \nu_s * \lambda,$$

and since $\lim_{r\to 0} \hat{\nu}_{r+s} = \hat{\nu}_s$, we obtain $\nu_s * \lambda = \nu_s$ for all $s \in \mathbb{Q}_+^*$. It follows that $\lambda \in \mathcal{M}^1(G)$ or

$$\mathcal{T}_w\text{-}\lim_{r\to 0} \nu_r = \lambda = \omega_H$$

for some compact subgroup H of G. Clearly, $\nu_s = \omega_H * \nu_s = \nu_s * \omega_H$ for all $s \in \mathbb{Q}_+^*$. Since G is discrete, $\lim_{r\to 0} \nu_r = \omega_H$ holds in the norm topology of $\mathcal{M}^1(G)$; in particular,

$$\lim_{n\to\infty} \|\nu_n - \omega_H\| = 0 \quad \text{or} \quad \|\mu_n * \mu_n^\sim - \omega_H\| < 1$$

for sufficiently large $n \geq n_0$. □

Proof of the theorem. Let $\mu \in \mathcal{I}(G)$ be such that for every $n \geq 1$ there exists a $\mu_n \in \mathcal{M}^1(G)$ satisfying $\mu_n^n = \mu$. Without loss of generality we may assume that $I(\mu_n) = \{e\}$ for all $n \geq 1$. Then by Lemma 3.3.18, Theorem 3.3.16 is applicable, and there exists a sequence $(x_n)_{n\geq 1}$ in G such that $(x_n^n)_{n\geq 1}$ is finite and

$$\lim_{n\to\infty} \|\varepsilon_{x_n^{-1}} * \mu_n - \varepsilon_e\| = 0.$$

But then the sequence $(x_{n!}^{n!})_{n\geq 1}$ is also finite, and one can find a strictly isotone sequence $(n_k)_{k\geq 1}$ in $\mathbb{N}$ satisfying

$$x_{n_k!}^{n_k!} = x_{n_l!}^{n_l!} =: \bar{x} \in G \qquad \text{for all } k, l \geq 1.$$

Consequently, $\bar{x}$ is divisible with n-th root $\bar{x}_n = x_{n_k!}^{\frac{n_k!}{n}}$ for every $k \geq 1$ and

$$\lim_{k\to\infty} \|\varepsilon_{x_{n_k!}^{-1}} * \mu_{n_k!} - \varepsilon_e\| = 0.$$

An application of Theorem 3.2.13 now implies that the measure $\varepsilon_{\bar{x}^{-1}} * \mu$ belongs to $\mathcal{P}_{\{e\}}(G)$ or is of the form $\exp_{\{e\}}(\nu)$ with $\nu \in \mathcal{N}_{\{e\}}(G)$. By Corollary 3.3.11 there exists for every $\mathbb{M}\in\mathbf{SM}$ a homomorphism $f: \mathbb{M}\to G$ with $f(1)=\bar{x}$. Defining

$$\phi: \mathbb{M}_+ \to \mathcal{M}^1(G) \quad \text{by} \quad \phi(r) := \varepsilon_{f(r)} * \exp_{\{e\}}(rv)$$

for all $r\in\mathbb{M}_+$ (and $v\in\mathcal{N}_{\{e\}}(G)$) one obtains a homomorphism ϕ of the desired form satisfying $\phi(1)=\varepsilon_{\bar{x}} * \varepsilon_{\bar{x}^{-1}} * \mu=\mu$. □

3.3.19 Corollary. $\mathbf{SM}(G)=\mathbf{SM}$.

3.4 Existence of One-Parameter Semigroups

This section is devoted to the discussion of existence theorems for one-parameter semigroups in topological semigroups. These results are well-known but rarely applied, although fairly short proofs have been available for some time. Before stating the main result in the form needed in our further investigations we collect a few notions and facts from the theory of topological semigroups.

As in the introduction to § 1 of Chapter II we denote by S a topological semigroup and keep the notations $E(S)$ for the totality of idempotents in S, $H(e)$ with $e\in E(S)$ for the group of units in eSe and $M(S)$ for the minimal closed ideal of S, whose existence and uniqueness is garanteed, whenever S is assumed to be compact. Let S be a compact semigroup and I a closed ideal of S. Let Δ denote the diagonal of $S\times S$. Putting $R := \Delta \cup (I\times I)$ one observes that R is a closed congruence relation and S/R is a semigroup with zero. S/R is called the Rees quotient modulo I and is denoted by S/I. If S is a compact semigroup with identity 1, then $H(1)$ is a compact group. A compact semigroup S with identity 1 is said to be irreducible if S is connected and does not contain any proper compact connected subsemigroup T satisfying $1\in T$ and $T\cap M(S)\neq\emptyset$. In compact semigroups S with identity 1 one introduces the $\mathcal{H}$-order relation $\leq(\mathcal{H})$ by $x\leq(\mathcal{H})\,y$ for $x, y\in S$, if $Sx\subset Sy$ and $xS\subset yS$ hold. The $\mathcal{H}$-class of an element $x\in S$ will be denoted by $\mathcal{H}(x)$. One has $H(1)=\mathcal{H}(1)$, and if all idempotents $\neq 1$ of S are in $M(S)$, then

$$\mathcal{H}(x)=H(1)\,x\cap x\,H(1) \qquad \text{for all } x\notin M(S).$$

For a given compact semigroup S one defines the compact semigroup $\bar{S}$ of all nonempty closed subsets of S, the space $\bar{\bar{S}}$ of all closed Abelian subsemigroups of S and the space $\bar{\bar{S}}_0$ of all $T\in\bar{\bar{S}}$ which are totally ordered with respect to the $\mathcal{H}_T$-order relation, where $\mathcal{H}_T$ denotes the restriction of $\mathcal{H}$ to T. It is readily proved that $\bar{\bar{S}}_0$ is closed in $\bar{S}$ ([253], B.4.10).

We are now going to study the existence of one-parameter semigroups (groups) in a topological semigroup S defined as continuous semigroup homomorphisms from $\mathbb{R}_+(\mathbb{R})$ into S.

3.4.1 Theorem. *Let S be a compact semigroup with identity* 1. *We assume that*
(i) $H(1)$ *is not open in* S;
(ii) *there exists a neighborhood V of* 1 *such that* 1 *is the only idempotent of S belonging to V.*

Then there exists a one-parameter semigroup Φ in S satisfying $\Phi(t)\notin H(1)$ for all $t\in\mathbb{R}_+^*$.

Proof. Since 1 is isolated in $E(S)$, one infers that $I:=S(E(S)\setminus\{1\})S$ is a closed ideal in S containing all elements of $E(S)\setminus\{1\}$. Hence, for the compact semigroup $T:=S/I$ we have $E(T)=\{0,1\}$. We first show that if the desired one-parameter semigroup exists in T, it also exists in S. At a later stage it will be proved that every such compact semigroup T contains a compact subsemigroup I isomorphic to $[0, 1]$ such that $H_I(1):=H(1)\cap I=\{1\}$ holds. It then follows that the mapping $\phi:\mathbb{R}_+\to[0,1]$ defined by $\phi(t):=e^{-t}$ for all $t\in\mathbb{R}_+$ extends to a one-parameter semigroup Φ in S with the required property. Starting with the homomorphism ϕ we clearly obtain homomorphisms $\Phi_I:\mathbb{R}_+\to I$ and $\Phi_T:\mathbb{R}_+\to T$ related to each other as in the following diagram:

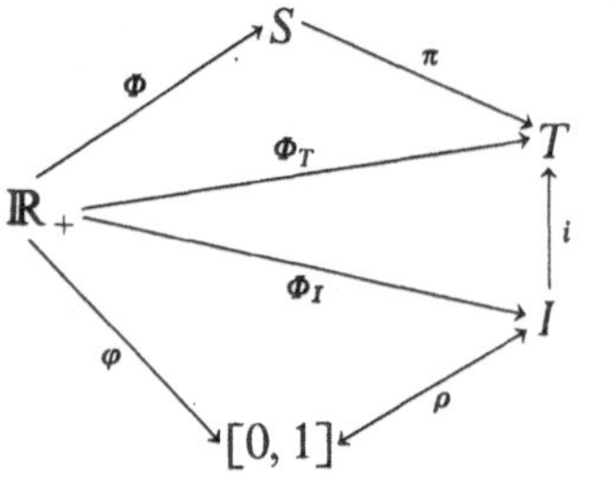

Here π denotes the canonical projection $S\to S/I=:T$, i the inclusion on I and ρ the isomorphism $I\to[0,1]$ (still to be established). It is easy to see that the existence of Φ_T implies that of Φ. In fact, we have $\phi(\mathbb{R}_+)\subset]0,1]$, i.e. $\Phi_I(\mathbb{R}_+)\subset I\setminus\{0\}$ or $\Phi_T(\mathbb{R}_+)\subset T\setminus\{0\}$. But now $\pi^{-1}(\{0\})=I$ and $\pi^{-1}(\{x\})=x$ for all $x\notin I$. The mapping $\Phi:\mathbb{R}_+\to S$ defined by $\Phi(t):=\Phi_T(t)$ for all $t\in\mathbb{R}_+$ is the desired one-parameter semigroup in S since $S\setminus I$ is open and identifiable with $T\setminus\{0\}$. We are now justified in assuming without loss of generality that S is a compact semigroup with identity 1 satisfying $E(S)=\{0,1\}$.

1. Let Δ be the diagonal of $S\times S$ and $\mathfrak{U}$ a neighborhood basis of Δ. For every $\mathcal{U}\in\mathfrak{U}$ we define $\mathcal{V}:=\mathcal{V}_{\mathcal{U}}$ as the set

$$\{(x,y)\in S\times S:\{(x,y)\}\cup(x,y)\Delta\cup\Delta(x,y)\cup\Delta(x,y)\Delta\subset\mathcal{U}\}$$

which is again a neighborhood of Δ with $\mathcal{V}\subset\mathcal{U}$. Let U be a neighborhood of 1 with $U\times U\subset\mathcal{V}$. By Assumption (i) $U\not\subset H(1)$ and hence there exists an element $x_{\mathcal{U}}$ of $U\setminus H(1)$. From the definition of $\mathcal{V}$ it follows that the set $\Gamma(x_{\mathcal{U}}):=\{x_{\mathcal{U}},x_{\mathcal{U}}^2,\ldots,x_{\mathcal{U}}^n,\ldots\}$ is $\mathcal{U}$-connected in the sense that for elements $x,y\in\Gamma(x_{\mathcal{U}})$ there are

$$x_0,x_1,\ldots,x_n\in\Gamma(x_{\mathcal{U}})\quad\text{with}\quad x_0=x,\ x_n=y$$

and satisfying $(x_k,x_{k-1})\in\mathcal{U}$ for all $k=1,\ldots,n$. This is an immediate consequence of

$$(x_{\mathcal{U}}^k,x_{\mathcal{U}}^{k-1})=(x_{\mathcal{U}},1)(x_{\mathcal{U}}^{k-1},x_{\mathcal{U}}^{k-1})\in\mathcal{U}$$

for all $k=1,\dots,n$. The $\mathscr{U}$-connectedness of $\Gamma(x_{\mathscr{U}})$ implies that of $\overline{\Gamma(x_{\mathscr{U}})}$. Since plainly $\Gamma(x_{\mathscr{U}})$ is an $\mathscr{H}$-chain in the sense that for any $x, y \in \Gamma(x_{\mathscr{U}})$ one has either $x \leq (\mathscr{H}) y$ or $y \leq (\mathscr{H}) x$, $\overline{\Gamma(x_{\mathscr{U}})}$ is also such an $\mathscr{H}$-chain. We consider the net $(\overline{\Gamma(x_{\mathscr{U}})})_{\mathscr{U} \in \mathfrak{U}}$ and pick a universal subnet $(\Gamma_\alpha)_{\alpha \in \mathbf{A}}$ of it. Since $x_{\mathscr{U}} \notin H(1)$, one has $0 \in \overline{\Gamma(x_{\mathscr{U}})}$ and therefore $\underline{\lim}_{\alpha \in \mathbf{A}} \Gamma_\alpha \neq \emptyset$. Hence, $(\Gamma_\alpha)_{\alpha \in \mathbf{A}}$ converges with $\lim_{\alpha \in \mathbf{A}} \Gamma_\alpha =: T$. In fact, let $x \in \overline{\lim}_{\alpha \in \mathbf{A}} \Gamma_\alpha$ and $\mathfrak{A} := \{B \subset S : x \in B\}$. For all $\alpha \in \mathbf{A}$ there exists a $\beta \in \mathbf{A}$ with $\beta \geq \alpha$ and $\Gamma_\beta \in \mathfrak{A}$. Since $(\Gamma_\alpha)_{\alpha \in \mathbf{A}}$ is universal, there is an $\alpha_0 \in \mathbf{A}$ such that for all $\beta \geq \alpha_0$ one has $\Gamma_\beta \in \mathfrak{A}$, therefore $x \in \bigcup_{\alpha \in \mathbf{A}} \bigcap_{\beta \geq \alpha} \Gamma_\beta = \underline{\lim}_{\alpha \in \mathbf{A}} \Gamma_\alpha$.

2. T is a compact Abelian subsemigroup of S and an $\mathscr{H}_T$-chain. Clearly, T is an Abelian subsemigroup of S. That T is also compact and an $\mathscr{H}_T$-chain follows from the remarks preceding the statement of the theorem. Furthermore, we obtain $1 \in T$, and since $0 \in \overline{\Gamma(x_{\mathscr{U}})}$, also $0 \in T$. From an obvious extension of Theorem 9.1 in [501] we infer that T is connected. Since $1 \in T$ and $T \cap M(S) = T \cap \{0\} = \{0\} \neq \emptyset$ holds, there exists an irreducible subsemigroup T' of T. We assume without loss of generality that T itself is irreducible. For every $n \geq 1$ we define the mapping $f_n : T \to T$ by $f_n(x) := x^n$ for all $x \in T$. Since T is irreducible and Abelian, f_n is a surjective homomorphism from T onto itself for every $n \geq 1$ and thus T is divisible.

3. $T \setminus H(1)$ is an ideal in T, where for simplicity $H(1) := H_T(1)$. For $y \in H(1)$ there exists an $x \in T$ with $x^n = y$. Moreover, $x \in H(1)$. Hence, $H(1)$ is divisible. As a divisible compact Abelian group $H(1)$ is therefore connected.

4. It remains to show that $H(1) = \{1\}$. If this has been verified, the proof will be complete. In fact, from $H(1) = \{1\}$ it follows by a remark preceding the statement of the theorem that

$$H(1) = \{x \in T : x \leq (\mathscr{H}_T)\, 1 \text{ and } 1 \leq (\mathscr{H}_T)\, x\} = \mathscr{H}(1)$$

as well as $\mathscr{H}(x) = x H(1)$, so that $\mathscr{H}(x) = \{x\}$ for all $x \in T$. T is totally ordered, connected and for every $x \in T$ the relationship $0 \leq (\mathscr{H}_T)\, x \leq (\mathscr{H}_T)\, 1$ holds. From [137], Theorem 1 we obtain that T is isomorphic to $[0, 1]$, and the theorem is established.

5. We therefore assume that $H(1) \neq \{1\}$ and derive a contradiction. Let χ be a nontrivial character of $H(1)$. Then

$$R := \{(x, y) \in T \times T : x \in y \ker \chi\}$$

is a closed congruence relation, and the quotient mapping $p : T \to T/R$ is open. Consequently, T/R is irreducible. Since $H(1)$ is connected and $\chi \neq 1$, we obtain

$$p(H(1)) = H_{T/R}(1) \cong H(1)/\ker \chi \cong \mathbb{T},$$

where $H_{T/R}(1)$ denotes in the obvious way the group of units of 1 in T/R. Also, T/R is again an $\mathscr{H}$-chain. Thus by the assumption $H(1) \neq \{1\}$ there exists an irreducible compact Abelian semigroup $T := T/R$ which is an $\mathscr{H}$-chain and satisfies

$$E(T) = \{0, 1\}, \qquad H(1) \cong \mathbb{T}.$$

This statement will now lead to a contradiction. Let U be an open neighborhood of 1 satisfying

$$\bar{U} \cap \left(\left[\frac{\pi}{2}, \frac{3\pi}{2}\right] \cup \{0\}\right) = \emptyset,$$

where $[a,b]$ is interpreted as the set

$$[a,b]_{H(1)} := \{h \in H(1) : a \le \arg h \le b\}.$$

Let $(x_\alpha)_{\alpha \in \mathbf{A}}$ be a net in T with $\lim_{\alpha \in \mathbf{A}} x_\alpha = 1$. Since T is divisible, there exists a net $(y_\alpha)_{\alpha \in \mathbf{A}}$ in T such that $y_\alpha^2 = x_\alpha$ holds for all $\alpha \in \mathbf{A}$.

The only possible accumulation points of $(y_\alpha)_{\alpha \in \mathbf{A}}$ are 1 and -1. Hence, one of the nets $(y_\alpha)_{\alpha \in \mathbf{A}}$ and $((-1)\, y_\alpha)_{\alpha \in \mathbf{A}}$ admits 1 as an accumulation point and so is cofinal in U. Consequently, $(x_\alpha)_{\alpha \in \mathbf{A}}$ is cofinal in $f_2(U)$ and one gets $1 \in \overset{\circ}{\widehat{f_2(U)}}$. Let W be an open neighborhood of 1 with

$$W \subset U \cap f_2(U), \qquad W \cap \mathbb{T} \supset]-\varepsilon, \varepsilon[,$$

(where $]a,b[:=]a,b[_{H(1)}$ in analogy to $[a,b]$) and $\overline{W} \cap [2\varepsilon, 2\pi - \varepsilon] = \emptyset$ for some $\varepsilon > 0$. For later purposes we choose W such that $\overline{W} \subset U$ and $W^2 \subset U$ hold.

For every proper closed ideal I in T let $W_I := W \setminus I$. We first show that there exists such an ideal I in T satisfying $W_I \subset f_2(W_I)$. Supposing that this is not the case one chooses $x_I \in W_I \setminus f_2(W_I)$. From $W_I \subset W \subset f_2(U)$ follows the existence of $y_I \in U$ with $f_2(y_I) = y_I^2 = x_I$. Since I is an ideal, $x_I \notin I$ implies $y_I \notin I$, and so $y_I \notin W$. The complements of proper closed ideals of T form a neighborhood basis of $\mathbb{T}$. Hence, the net $\{y_I : I$ is a proper closed ideal in $T\}$ admits an accumulation point $y \in (\bar{U} \setminus W) \cap \mathbb{T}$. Without loss of generality we assume that the net converges to y. But then the corresponding net $\{x_I : I$ is a proper closed ideal in $T\}$ converges to $x := y^2 \in \overline{W} \cap \mathbb{T}$. But $y \in (\bar{U} \setminus W) \cap \mathbb{T}$ implies that

$$\varepsilon \le |\arg y| < \frac{\pi}{2} \quad \text{or} \quad 2\varepsilon \le |\arg x| < \pi.$$

On the other hand, $0 \le |\arg x| < 2\varepsilon$ holds by the choice of W. This implies the desired contradiction.

Let I be a proper closed ideal of T with $W_I \subset f_2(W_I)$, $x_0 \in W_I$ and $x_n \in W_I$ with $x_n^2 = x_{n-1}$ $(n \ge 1)$. We have $T/\mathscr{H} \cong [0,1]$, whence all accumultion points of $(x_n)_{n \ge 1}$ belong to $\mathbb{T}$. Moreover, every accumulation point of $(x_n)_{n \ge 1}$ belongs to $\overline{W_I} \cap \mathbb{T} \subset \bar{U} \cap \mathbb{T} \subset \left]-\frac{\pi}{2}, \frac{\pi}{2}\right[$. If h is an accumulation point of $(x_n)_{n \ge 1}$, so also is h^{2^k} for all $k \ge 1$. But for $h \ne 1$ there exists a $k \ge 1$ with

$$h^{2^k} \in \left[\frac{\pi}{2}, \frac{3\pi}{2}\right] \quad \text{and hence} \quad \lim_{n \to \infty} x_n = 1.$$

From [358], Lemma 4.1.2 we infer the existence of a neighborhood V of 1 with $V \subset W$ such that if $x \in V$ and $x^k \in U$ for $1 \leq k \leq n$, $x^n \in V$, then $x^k \in W$ for $1 \leq k \leq n$.

Since $\lim_{n\to\infty} x_n = 1$, there exists an $n_0 \geq 1$ such that $x_n \in V$ for all $n \geq n_0$. Without loss of generality we may assume that $x_n \in V$ for all $n \geq 1$. One shows that $\{x_n^m : 1 \leq m \leq 2^n\} \subset W$. By $x_n^m = x_0^l$ for all $m \geq 2^n$ and a fixed $l \geq 1$ $\Gamma(x_n) \subset W \cap Tx_0$. Thus

$$T_0 := \overline{\bigcup_{n\geq 0} \Gamma(x_n)} \subset \overline{W} \cup Tx_0 \subset \overline{U} \cup Tx_0.$$

As in Part 2 one sees that T_0 is a compact connected subsemigroup of T which contains 1 and 0. Since T is irreducible, $T_0 = T$. On the other hand the definition of U implies that $T_0 \cap \mathbb{T} \subset \left]-\frac{\pi}{2}, \frac{\pi}{2}\right[$, which provides us with the desired contradiction. □

The following sequence of lemmata will yield the main result on the existence of one-parameter semigroups in noncompact topological semigroups.

3.4.2 Lemma. *Let S be a topological semigroup, $d \in \mathbb{R}_+^*$ and $f: \mathbb{Q}_+^* \to S$ an (algebraic) semigroup homomorphism such that $f(]0, d[\cap \mathbb{Q})$ is relatively compact in S. For each $x \in \mathbb{Q}_+^*$ we define $S(x) := \overline{f(]0, x[\cap Q)}$. Then*

(i) *$S(nd) = (S(d))^n$ is compact for every $n \geq 1$;*

(ii) *$S(x)$ is compact for every $x \in \mathbb{Q}_+^*$;*

(iii) *$S(x+y) = S(x)\,S(y)$ for all $x, y \in \mathbb{Q}_+^*$;*

(iv) *$\overline{f(\mathbb{Q}_+^*)}$ admits an identity e such that $K := \bigcap_{x\in\mathbb{Q}_+^*} S(x)$ is a compact connected Abelian subgroup of the group $H(e)$ of units in eSe ($e \in E(S)$); in particular*

(v) *$\overline{f(\mathbb{Q}_+^*)} \subset eSe$.*

Proof. (i) First of all one notes that

$$f(]0, 2d[\cap \mathbb{Q}_+^*) = f(]0, d[\cap \mathbb{Q}_+^*)\, f(]0, d[\cap \mathbb{Q}_+^*).$$

The continuity of the multiplication in S yields

$$S(2d) \supset S(d)\, S(d) \supset f(]0, 2d[\cap \mathbb{Q}_+^*),$$

and hence $S(2d) = (S(d))^2$.

(ii) For each $x \in \mathbb{Q}_+^*$ there exists an $m_0 \geq 1$ such that $x \leq m_0 d$ holds. This implies that $S(x) \subset S(m_0 d)$. Since by (i) $S(m_0 d)$ is compact, $S(x)$ is also compact.

(iii) follows as in the proof of (i).

(iv) and (v) K is the intersection of a descending family of compact subsets of $S(1)$ and hence $\neq \varnothing$. Moreover,

$$K^2 = \left(\bigcap_{x\in\mathbb{Q}_+^*} S(x)\right)\left(\bigcap_{y\in\mathbb{Q}_+^*} S(y)\right) \subset \bigcap_{x,y\in\mathbb{Q}_+^*} S(x+y) = \bigcap_{t\in\mathbb{Q}_+^*} S(t) = K,$$

whence K is a compact Abelian subsemigroup of S.

From the general theory we know that $E(K) \neq \emptyset$. Let $e \in E(K)$. Then e can be obtained as $\lim_{\alpha \in \mathbb{A}} f(x_\alpha)$, where $(x_\alpha)_{\alpha \in \mathbb{A}}$ is an appropriate net in $]0, 1[\cap \mathbb{Q}$ satisfying $\lim_{\alpha \in \mathbb{A}} x_\alpha = 0$. Hence, for every $t \in]0, 1[\cap \mathbb{Q}$ there exists an $\alpha_0 \in \mathbb{A}$ such that $x_\alpha < t$ for all $\alpha > \alpha_0$, so that

$$f(t) = f(t - x_\alpha) f(x_\alpha) = f(x_\alpha) f(t - x_\alpha) \qquad \text{for all } \alpha > \alpha_0.$$

Consequently, $f(t) = x e = e x$ for some $x \in S(1)$. In particular, $S(1) \subset e S e$, so that $f(]0, 1[\cap \mathbb{Q}) \subset e S e$ or $e \in K \subset \overline{f(\mathbb{Q}_+^*)} \subset e S e$. This implies that $\overline{f(\mathbb{Q}_+^*)}$ is a closed subsemigroup of $e S e$ with identity e for every $e \in E(K)$. But then $E(K) = \{e\}$ since the identity of a semigroup is uniquely determined.

It follows that K as a compact semigroup with identity and no other idempotent than e is already a group and hence a compact Abelian subgroup of $H(e)$.

We still have to show that K is connected or equivalently, that K is divisible. Let $x \in K$. There is a net $(x_\alpha)_{\alpha \in \mathbb{A}}$ in $]0, 1[\cap \mathbb{Q}$ such that

$$\lim_{\alpha \in \mathbb{A}} x_\alpha = 0 \quad \text{and} \quad x = \lim_{\alpha \in \mathbb{A}} f(x_\alpha)$$

hold. Given $n \geq 1$ one observes that $\left(\frac{x_\alpha}{n}\right)_{\alpha \in \mathbb{A}}$ is also a net in $]0, 1[\cap \mathbb{Q}$ satisfying $\lim_{\alpha \in \mathbb{A}} \frac{x_\alpha}{n} = 0$. For a suitable subnet $(x_\beta)_{\beta \in \mathbb{B}}$ of $(x_\alpha)_{\alpha \in \mathbb{A}}$ one obtains the existence of $y := \lim_{\beta \in \mathbb{B}} f\left(\frac{x_\beta}{n}\right) \in K$, so that $y^n = x$. □

3.4.3 Lemma. *Let S be a topological semigroup, $f: \mathbb{Q}_+^* \to S$ a semigroup homomorphism such that $f(]0, 1[\cap \mathbb{Q})$ is relatively compact in S and K the compact connected Abelian subgroup of $H(e)$ defined in the preceding lemma. Then f can be extended to $\mathbb{Q}_+$ such that $f(0) = e$ holds, and one obtains*

$$\overline{K f([x, y[\cap \mathbb{Q})} = K \overline{f([x, y[\cap \mathbb{Q})} = \overline{f([x, y[\cap \mathbb{Q})}$$

for all $x, y \in \mathbb{Q}_+^*$ with $x < y$.

Proof. We show the last statement. Clearly, for $x, y \in \mathbb{Q}_+^*$ with $x < y$ one has $\overline{f([x, y[\cap \mathbb{Q})} \subset K \overline{f([x, y[\cap \mathbb{Q})}$. For the reverse inclusion it suffices to show $K f([x, y[\cap \mathbb{Q}) \subset \overline{f([x, y[\cap \mathbb{Q})}$. Let $k \in K$. Then there exists a net $(x_\alpha)_{\alpha \in \mathbb{A}}$ in $]0, 1[\cap \mathbb{Q}$ satisfying $\lim_{\alpha \in \mathbb{A}} x_\alpha = 0$ and $k = \lim_{\alpha \in \mathbb{A}} f(x_\alpha)$. For any $z \in [x, y[\cap \mathbb{Q}$ there exists $\alpha_0 \in \mathbb{A}$ such that $x_\alpha + z \in [x, y[\cap \mathbb{Q}$ for all $\alpha > \alpha_0$. By the closedness of $\overline{f([x, y[\cap \mathbb{Q})}$ we then conclude that $k f(z) \in \overline{f([x, y[\cap \mathbb{Q})}$. □

3.4.4 Lemma. *Let S be a topological semigroup. For every semigroup homomorphism $f: \mathbb{Q}_+ \to S$ such that $f(]0, 1[\cap \mathbb{Q})$ is relatively compact in S the following statements are equivalent:*

(i) $K = \{f(0)\}$;

(ii) *f is continuous at* 0;

(iii) *f is continuous on* $[0, 1] \cap \mathbb{Q}$;

(iv) *f is continuous.*

Proof. 1. (i) $\Rightarrow$ (ii) follows immediately from the fact that a filterbasis of compact subsets of a Hausdorff space converges iff it intersects at a single element.
2. (ii) $\Rightarrow$ (iii). Let $p \in [0, 1] \cap \mathbb{Q}$ and $(x_\alpha)_{\alpha \in \mathbb{A}}$ a net in $[0, 1] \cap \mathbb{Q}$ satisfying

$$\lim_{\alpha \in \mathbb{A}} x_\alpha = p.$$

Without loss of generality we assume $x_\alpha \leq p$ for all $\alpha \in \mathbb{A}$. By assumption the set $f([0, 1] \cap \mathbb{Q})$ is relatively compact, whence it suffices to show that every limit y of a convergent subnet $(f(x_\beta))_{\beta \in \mathbb{B}}$ of $(f(x_\alpha))_{\alpha \in \mathbb{A}}$ in $f([0, 1] \cap \mathbb{Q})$ coincides with $f(p)$. But now

$$f(p) = f(p - x_\beta) f(x_\beta) \text{ for all } \beta \in \mathbb{B} \quad \text{and} \quad \lim_{\beta \in \mathbb{B}} f(p - x_\beta) = f(0),$$

since by hypothesis f is continuous at 0. Therefore

$$f(p) = \lim_{\beta \in \mathbb{B}} f(p - x_\beta) \lim_{\beta \in \mathbb{B}} f(x_\beta) = f(0)\, y = y,$$

which follows from $f(0)x = x$ for every $x \in \overline{f(\mathbb{Q}_+)}$. Thus every convergent subnet of $(f(x_\alpha))_{\alpha \in \mathbb{A}}$ has the same limit $f(p)$. This together with the compactness of the set $\overline{\{f(x_\alpha) : \alpha \in \mathbb{A}\} \cup \{f(p)\}}$ yields $\lim_{\alpha \in \mathbb{A}} f(x_\alpha) = f(p)$.

3. (iii) $\Rightarrow$ (iv). By Lemma 3.4.2 $S(x)$ is compact for every $x \in \mathbb{Q}_+^*$. The argument used in Part 2 of the proof then yields the assertion.

4. The implication (iv) $\Rightarrow$ (i) is obvious. □

Given an abstract semigroup S and a number $d \in \mathbb{R}_+^*$ we define a *partial semigroup homomorphism* from $[0, d]$ into S as a mapping $f: [0, d] \to S$ satisfying $f(p + q) = f(p) f(q)$ for all $p, q \in [0, d]$ with $p + q \in [0, d]$.

3.4.5 Lemma. *Let S be a topological semigroup and $f: \mathbb{Q}_+ \to S$ a continuous semigroup homomorphism such that $f(]0, 1[\cap \mathbb{Q})$ is relatively compact in S. There exists a unique continuous partial semigroup homomorphism $F: [0, 1] \to S$ satisfying*

$$\mathrm{Res}_{[0,1] \cap \mathbb{Q}} F = \mathrm{Res}_{[0,1] \cap \mathbb{Q}} f.$$

Proof. Given $p \in [0, 1]$ there is a net $(x_\alpha)_{\alpha \in \mathbb{A}}$ in $]0, 1] \cap \mathbb{Q}$ with $\lim_{\alpha \in \mathbb{A}} x_\alpha = p$. Since $(f(x_\alpha))_{\alpha \in \mathbb{A}}$ is a net in the compact set $S(1)$, we may assume without loss of generality that the limit

$$F(p) := \lim_{\alpha \in \mathbb{A}} f(x_\alpha)$$

exists. Let $(y_\alpha)_{\alpha \in \mathbb{A}}$ be another net in $]0, 1] \cap \mathbb{Q}$ satisfying

$$\lim_{\alpha \in \mathbb{A}} x_\alpha = \lim_{\alpha \in \mathbb{A}} y_\alpha = p.$$

It suffices to assume that $x_\alpha > y_\alpha$ holds for all $\alpha \in \mathbb{A}$ and that the limits

$$\lim_{\alpha \in \mathbb{A}} f(x_\alpha) \quad \text{and} \quad \lim_{\alpha \in \mathbb{A}} f(y_\alpha)$$

exist. Clearly, $(x_\alpha - y_\alpha)_{\alpha\in\mathbf{A}}$ is a net in $]0, 1] \cap \mathbb{Q}$ converging to 0, whence by the continuity of f one obtains $\lim_{\alpha\in\mathbf{A}} f(x_\alpha - y_\alpha) = f(0)$. It follows that

$$\begin{aligned}\lim_{\alpha\in\mathbf{A}} f(x_\alpha) &= \lim_{\alpha\in\mathbf{A}} f(x_\alpha - y_\alpha) \lim_{\alpha\in\mathbf{A}} f(y_\alpha)\\ &= f(0) \lim_{\alpha\in\mathbf{A}} f(y_\alpha) = \lim_{\alpha\in\mathbf{A}} f(y_\alpha),\end{aligned}$$

or F is independent of the choice of the prescribed nets. Now let $0 \leq p, q \leq p+q \leq 1$. There exist nets $(x_\alpha)_{\alpha\in\mathbf{A}}$ and $(y_\alpha)_{\alpha\in\mathbf{A}}$ in $]0, 1] \cap \mathbb{Q}$ satisfying

$$\begin{aligned}&\lim_{\alpha\in\mathbf{A}} x_\alpha = p, \quad \lim_{\alpha\in\mathbf{A}} y_\alpha = q,\\ &F(p) = \lim_{\alpha\in\mathbf{A}} f(x_\alpha) \quad \text{and} \quad F(q) = \lim_{\alpha\in\mathbf{A}} f(y_\alpha).\end{aligned}$$

Without loss of generality we now suppose that $x_\alpha - y_\alpha \in]0, 1] \cap \mathbb{Q}$ for all $\alpha\in\mathbf{A}$. Thus one obtains

$$F(p)\,F(q) = \lim_{\alpha\in\mathbf{A}} f(x_\alpha) \lim_{\alpha\in\mathbf{A}} f(y_\alpha) = \lim_{\alpha\in\mathbf{A}} f(x_\alpha + y_\alpha) = F(p+q),$$

and F is a partial semigroup homomorphism. The continuity of F on $[0, 1]$ is shown as in the proof of Lemma 3.4.4 with the aid of

$$\bigcap_{x\in\mathbb{R}_+^*} \overline{F(]0, 1[)} = \{f(0)\}. \quad \square$$

3.4.6 Theorem. *Let S be a topological semigroup and f a continuous semigroup homomorphism from $\mathbb{Q}_+$ into S such that $f(]0, 1[\cap \mathbb{Q})$ is relatively compact in S. There exists a unique one-parameter semigroup ϕ in S such that* $\operatorname{Res}_{\mathbb{Q}_+} \phi = f$.

Proof. By Lemma 3.4.5 it remains to show that for any partial semigroup homomorphism $F: [0, 1] \to S$ there exists a unique semigroup homomorphism $\phi: \mathbb{R}_+ \to S$ satisfying $\operatorname{Res}_{[0,1]} \phi = F$. In fact, for a given $r\in\mathbb{R}_+^*$ there is an $n \geq 1$ such that $\frac{r}{n} \in [0, 1]$. We define

$$\phi: \mathbb{R}_+^* \to S \quad \text{by} \quad \phi(r) := F\left(\frac{r}{n}\right)^n.$$

For any other integer $m \geq 1$ with $\frac{r}{m} \in [0, 1]$ we then obtain

$$F\left(\frac{r}{n}\right)^n = \left(F\left(\frac{r}{mn}\right)^m\right)^n = \left(F\left(\frac{r}{mn}\right)^n\right)^m = F\left(\frac{r}{m}\right)^m,$$

and thus ϕ is well-defined. Let $p, q\in\mathbb{R}_+$. Then there exists an $n \geq 1$ such that $\frac{p+q}{n} \in [0, 1]$ holds. Consequently,

$$\phi(p+q) = \left(F\left(\frac{p+q}{n}\right)\right)^n = \left(F\left(\frac{p}{n}\right)\right)^n \left(F\left(\frac{q}{n}\right)\right)^n = \phi(p)\,\phi(q).$$

The uniqueness of ϕ follows directly from the construction. $\square$

3.4.7 Lemma. *Let S be a topological semigroup and ϕ a nontrivial one-parameter semigroup in S. Then there exists a number $d \in]0,1]$ such that $\mathrm{Res}_{[0,d]}\,\phi$ is injective. Moreover, with a given $c \in \mathbb{R}_+^*$ one can reparametrize ϕ such that $\mathrm{Res}_{[0,c]}\,\phi$ remains injective.*

Proof. 1. We assume that $\mathrm{Res}_{[0,r]}\,\phi$ is not injective for all $r \in \mathbb{R}_+$. Then there exist nets $(x_\alpha)_{\alpha \in \mathbb{A}}$ and $(y_\alpha)_{\alpha \in \mathbb{A}}$ in $[0,r]$ such that $\lim_{\alpha \in \mathbb{A}} x_\alpha = \lim_{\alpha \in \mathbb{A}} y_\alpha = 0$ and $x_\alpha \neq y_\alpha$ but $\phi(x_\alpha) = \phi(y_\alpha)$ for all $\alpha \in \mathbb{A}$. Without loss of generality let $x_\alpha > y_\alpha$ for all $\alpha \in \mathbb{A}$. One has

$$\phi(x_\alpha) = \phi(x_\alpha - y_\alpha)\,\phi(y_\alpha) = \phi(x_\alpha - y_\alpha)\,\phi(x_\alpha)$$

and by induction

$$\phi(x_\alpha) = \phi(n(x_\alpha - y_\alpha))\,\phi(x_\alpha) \qquad \text{for all } n \geq 1.$$

Since $\lim_{\alpha \in \mathbb{A}}(x_\alpha - y_\alpha) = 0$, for $r_0 \in]0,1]$ one can choose a net $(n_\beta)_{\beta \in \mathbb{B}}$ in $\mathbb{N}$ and subnets $(x_\beta)_{\beta \in \mathbb{B}}$, $(y_\beta)_{\beta \in \mathbb{B}}$ of $(x_\alpha)_{\alpha \in \mathbb{A}}$, $(y_\alpha)_{\alpha \in \mathbb{A}}$ resp. such that

(a) $\frac{r_0}{2} \leq n_\beta(x_\beta - y_\beta) \leq r_0$ holds for all $\beta \in \mathbb{B}$ and

(b) $r = \lim_{\beta \in \mathbb{B}} n_\beta(x_\beta - y_\beta) \in \left[\frac{r_0}{2}, r_0\right]$.
It follows that

$$\begin{aligned}\phi(0) &= \lim_{\beta \in \mathbb{B}} \phi(x_\beta) = \lim_{\beta \in \mathbb{B}}(\phi(n_\beta(x_\beta - y_\beta))\,\phi(y_\beta)) \\ &= \lim_{\beta \in \mathbb{B}} \phi(n_\beta(x_\beta - y_\beta)) \lim_{\beta \in \mathbb{B}} \phi(y_\beta) = \phi(r)\,\phi(0) = \phi(r)\end{aligned}$$

and thus one obtains for each $r_0 \in]0,1[$ an $r \in \left[\frac{r_0}{2}, r_0\right]$ with $\phi(nr) = \phi(0)$ for all $n \geq 1$. Consequently, there is a dense subset $\mathbb{D}$ of $\mathbb{R}_+^*$ with $\phi(\mathbb{D}) = \phi(0)$ which yields a contradiction to the nontriviality of ϕ.

2. Let $c \in \mathbb{R}_+$. The mapping $\phi': \mathbb{R}_+^* \to S$ defined by $\phi'(x) = \phi\left(\left(\frac{d}{c}\right)x\right)$ for all $x \in \mathbb{R}_+^*$ is a one-parameter semigroup in S with the property that $\mathrm{Res}_{[0,c]}\,\phi'$ is injective. This proves the second statement of the lemma. □

3.4.8 Theorem. *Let S be a topological semigroup, f an (algebraic) semigroup homomorphism from $\mathbb{Q}_+$ into S such that $f(]0,1[\cap\mathbb{Q})$ is relatively compact in S and K the compact, connected Abelian subgroup of $H(e)$, where e is the identity of $\overline{f(\mathbb{Q}_+^*)}$, as in Lemma 3.4.2. Then*

(i) *K acts on the spaces $\overline{f(\mathbb{Q}_+)}$ and $\overline{f([x,y[\cap\mathbb{Q})}$ ($x, y \in \mathbb{Q}_+^*$ with $x<y$); the canonical mapping onto the corresponding quotient spaces will be denoted by π;*

(ii) *$\overline{f(\mathbb{Q}_+)}/K$ is a topological semigroup with identity under the multiplication $(xK, yK) \to xK \cdot yK := xyK$;*

(iii) *if* $f(\mathbb{Q}_+)\not\subset K$, *then the composite mapping* $\pi\circ f:\mathbb{Q}_+\to\overline{f(\mathbb{Q}_+)}/K$ *is a nontrivial continuous semigroup homomorphism such that*

$$\overline{\pi(f([x,y[\cap\mathbb{Q}))}=\overline{f([x,y[\cap\mathbb{Q})}/K$$

holds for all $x,y\in\mathbb{Q}_+^*$ *with* $x<y$.

Proof. 1. (i) is clear by Lemma 3.4.3.

2. (ii) follows immediately from the fact that $\{xK:x\in\overline{f(\mathbb{Q}_+)}\}$ is the family of equivalence classes of a closed congruence relation since K is compact. Hence, $\overline{f(\mathbb{Q}_+)}/K$ is a topological semigroup with identity with respect to the quotient topology.

3. Clearly $\pi\circ f$ is a nontrivial semigroup homomorphism. Its continuity follows from Lemma 3.4.4. Let $x,y\in\mathbb{Q}_+^*$ with $x<y$. Then

$$\pi\circ f([x,y[\cap\mathbb{Q})=\{f(r)K:r\in[x,y[\cap\mathbb{Q}\}$$

is contained in $\overline{f([x,y[\cap\mathbb{Q})}/K$. Moreover,

$$\pi(f([x,y[\cap\mathbb{Q}))\subset\pi(\overline{f([x,y[\cap\mathbb{Q})})\subset\overline{\pi f([x,y[\cap\mathbb{Q})}.$$

This implies the assertion. □

For the rest of the section we maintain the following hypotheses: S is a topological semigroup and f an algebraic semigroup homomorphism from $\mathbb{Q}_+$ into S. It will be assumed that

(H1) $S(x)=\overline{f(]0,x[\cap\mathbb{Q})}$ is compact for each $x\in\mathbb{Q}_+$,

(H2) $f(1)\notin K:=\bigcap_{x\in\mathbb{Q}_+^*}S(x)$,

(H3) $\mathrm{Res}_{]0,2]\cap\mathbb{Q}}\pi\circ f$ is injective, where π denotes the quotient mapping from $\overline{f(\mathbb{Q}_+)}$ onto $\overline{f(\mathbb{Q}_+)}/K$ or from $\overline{f([x,y[\cap\mathbb{Q})}$ onto $\overline{f([x,y[\cap\mathbb{Q})}/K$ for $x,y\in\mathbb{Q}_+$ with $x<y$.

From these hypotheses we obtain via Theorems 3.4.6, 3.4.8 and Lemma 3.4.7 the existence of a unique nontrivial one-parameter semigroup g in $\overline{f(\mathbb{Q}_+)}/K$ such that $\mathrm{Res}_{[0,2]}g$ is injective (after a possible reparametrization of g or f) such that the diagram below commutes:

$$\begin{array}{ccc}
]0,2[\cap\mathbb{Q} & \xrightarrow{\ f\ } & S(2)\\
\Big\downarrow i & & \Big\downarrow \pi\\
[0,2] & \overset{}{\underset{g}{\dashrightarrow}} & S(2)/K.
\end{array}$$

Here the mapping i denotes inclusion. It is easily derived that

$$\rho:=g^{-1}\circ\pi:S(2)\to[0,2]$$

is a continuous semigroup homomorphism satisfying

$$\mathrm{Res}_{[0,2]\cap\mathbb{Q}}\,\rho\circ f=\mathrm{id}_{[0,2]\cap\mathbb{Q}}.$$

3.4.9 Lemma. *Under the hypotheses* (H 1), (H 2) *and* (H 3) *with the above notation the following statements are valid:*

(i) $x\in Kf(r)$ *iff* $x\in\pi^{-1}(g(r))$ *for* $r\in\mathbb{Q}_+$;

(ii) $x\in S(2)$ *iff there is a unique* $t\in[0,2]$ *with* $x\in\pi^{-1}(g(t))$;

(iii) $\pi^{-1}(g([x,y])=\overline{Kf([x,y[\cap\mathbb{Q})}=\overline{f([x,y[\cap\mathbb{Q})}$ *for* $x,y\in\mathbb{Q}_+^*$ *with* $x<y$;

(iv) $S(1)\,Kf(1)\subset\overline{Kf([1,2[\cap\mathbb{Q})}$;

(v) $S(1)\backslash Kf(1)=S(2)\backslash\overline{Kf([1,2[\cap\mathbb{Q})}$;

(vi) $X:=(S(1)\backslash Kf(1))\cup\{Kf(1)\}$ *together with the multiplication*

$$(x,y)\to m_X(x,y)=\begin{cases}x\,y, & \text{if } x,y,x\,y\in S(1)\backslash Kf(1)\\ Kf(1) & \text{otherwise}\end{cases}$$

becomes a compact Abelian semigroup with identity.

Proof. While (i) and (ii) are evident it remains to prove (iii) to (vi).

1. (iii) For all $x,y\in\mathbb{Q}_+^*$ with $x<y$ we have

$$\mathrm{Res}_{[x,y[\cap\mathbb{Q}}\,g=\mathrm{Res}_{[x,y[\cap\mathbb{Q}}\,\pi\circ f$$

and hence

$$g([x,y])=\pi(\overline{f([x,y[\cap\mathbb{Q})}).$$

Moreover, we have proved that $\overline{Kf([x,y[\cap\mathbb{Q})}=\overline{f([x,y[\cap\mathbb{Q})}$, whence the assertion.

2. (iv) Let $x\in S(1)$ and $y\in Kf(1)$. There exists a unique $s\in[0,1]$ such that $\pi(x)=g(s)$ as well as $\pi(y)=g(1)$ hold. It follows that

$$\pi(x\,y)=\pi(x)\,\pi(y)=g(s)\,g(1)=g(s+1)\in S(2).$$

Since $s+1\in[0,2]$, one obtains $x\,y\in\overline{Kf([1,2[\cap\mathbb{Q})}$.

3. (v) is an immediate consequence of

$$\begin{aligned}S(1)\backslash Kf(1)&=g([0,1[)=g([0,2])-g([1,2])\\&=S(2)\backslash\overline{Kf([1,2]\cap\mathbb{Q})}.\end{aligned}$$

4. (vi) Let m denote the multiplication in $S(1)$. We define $\pi':S(2)\to X$ by

$$\mathrm{Res}_{S(1)\backslash Kf(1)}\,\pi'=\mathrm{Res}_{S(2)\backslash\overline{Kf([1,2[\cap\mathbb{Q})}}\,\pi\quad\text{and}$$

$$\pi'(\overline{Kf([1,2[\cap\mathbb{Q})})=\{Kf(1)\}.$$

The commutativity of the diagram below shows that m_X is a multiplication on X which is also continuous, since π' is closed.

$$\begin{array}{ccc} S(1)\times S(1) & \xrightarrow{\;m\;} & S(2) \\ \downarrow{\scriptstyle \pi'\times\pi'} & & \downarrow{\scriptstyle \pi'} \\ X\times X & \dashrightarrow_{m_X} & X. \end{array}$$

3.4.10 Theorem. *Under the hypotheses* (H1), (H2) *and* (H3) *the following statements hold:*

(i) $E(X)=\{e,0\}$ *with* $0=\{Kf(1)\}$;

(ii) $H_X(e)=K$;

(iii) K *is not open in* X;

(iv) $X\setminus\{Kf(1)\}\cong S(1)\setminus Kf(1)$.

Let $[0,1]_*$ *denote the compact Abelian semigroup* $[0,1]$ *(with identity) together with the multiplication* $(x,y)\to x*y:=\min\{1,x+y\}$.

(v) *There exists a continuous semigroup homomorphism* $\phi_*:[0,1]_*\to X$ *such that* $\phi_*(0)=e$ *and* $\phi_*(1)=0$.

Let $\tau:\mathbb{R}_+\to[0,1]_*\cong\mathbb{R}_+/[1,\infty[$ *be the canonical mapping and* $\rho_X:X\to[0,1]_*$ *the unique continuous semigroup homomorphism defined by the commutativity of the subsequent diagram*

$$\begin{array}{ccc} S(2) & \xrightarrow{\rho:=g^{-1}\circ\pi} & [0,2] \\ \downarrow{\scriptstyle \pi} & & \downarrow{\scriptstyle \tau} \\ X & \dashrightarrow_{\rho_X} & [0,1]_*. \end{array}$$

(vi) $\rho_X\circ\phi_*=\mathrm{id}_{[0,1]_*}$ *(after a suitable reparametrization of* ϕ_**)*;

(vii) *there is a one-parameter semigroup* ϕ *in* S *which is the unique extension of the partial semigroup homomorphism* $\mathrm{Res}_{[0,\frac{1}{2}]}\,\phi_*$ *from* $[0,\frac{1}{2}]$ *into* S *satisfying* $\phi(t)\in f(t)K$ *for all* $t\in\mathbb{Q}_+$.

Proof. Statements (i) and (iv) are obvious.

1. (ii) In order to show $H_X(e)=K$ we pick $x\in X\setminus K$ and argue as follows: If $x=0$, then x is surely not a unit of X. If, however, $x\neq 0$, then there is a net $(x_\alpha)_{\alpha\in\mathbf{A}}$ in $]0,1]\cap\mathbb{Q}$ with

$$\lim_{\alpha\in\mathbf{A}} x_\alpha=:p\in[0,1] \quad\text{and}\quad x=\lim_{\alpha\in\mathbf{A}} f(x_\alpha).$$

By Assumption (H3) we have the identification

$$0=Kf(1)=K\overline{f([1,2[\cap\mathbb{Q})}.$$

Since $np\in]1,2]$ for some $n\geq 1$, we therefore get $x^n=\lim_{\alpha\in\mathbf{A}}(f(x_\alpha))^n=0$ in X or that x is not a unit of X.

2. (iii) There exists a net $(x_\alpha)_{\alpha\in\mathbf{A}}$ in $]0,1[\cap\mathbb{Q}$ with

$$\lim_{\alpha\in\mathbf{A}} x_\alpha=0 \quad\text{and}\quad e=\lim_{\alpha\in\mathbf{A}} f(x_\alpha).$$

Thus there is no neighborhood U of e with $U\subset K$, i.e., K is not open in X.

3. (v) follows from Theorem 3.4.1 with the aid of (i) to (iv). In fact, by that theorem there exists a continuous semigroup homomorphism $\phi_*: [0,1]_* \to X$ with $\phi_*([0,1]_*) \cap K = \{e\}$ which implies $\phi_*(0) = e$. But for any $y \notin K$ one obtains $\lim_{n\to\infty} y^n = 0$. In fact, for the compact Abelian semigroup $P(y)$ we have $E(P(y)) \neq \emptyset$. But $e \in E(P(y))$ implies

$$e = \lim_{\alpha \in \mathbb{A}} y^{n_\alpha} = y \lim_{\beta \in \mathbb{B}} y^{n_\beta - 1} = (\lim_{\beta \in \mathbb{B}} y^{n_\beta - 1})\, y$$

for a subnet $(n_\beta)_{\beta \in \mathbb{B}}$ of $(n_\alpha)_{\alpha \in \mathbb{A}}$ in $\mathbb{N}$ and hence, $y \in H_X(e) =: K$, so that $e \notin E(P(y))$. Since $E(X) = \{e, 0\}$ by assumption, we conclude that $\lim_{n\to\infty} y^n = 0$. The assertion $\phi_*(1) = 0$ is an immediate consequence.

4. (vi) First one notes that $G := \rho_X \circ \phi_*: [0,1]_* \to [0,1]_*$ is a continuous semigroup homomorphism satisfying $G(0) = 0$ and $G(1) = 1$. By Lemma 3.4.5 and the proof of Theorem 3.4.6 there exists a unique continuous semigroup homomorphism $F: \mathbb{R}_+^* \to \mathbb{R}_+^*$ such that $\tau \circ F = G \circ \tau$ holds. Since F is nontrivial, we may assume without loss of generality (after a suitable reparametrization) that $\mathrm{Res}_{[0,2]} F$ is injective. In particular, $\rho_X \circ \phi_* = \mathrm{id}_{[0,1]_*}$ (after reparametrization).

5. (vii) By the preceding discussion the following diagram is commutative:

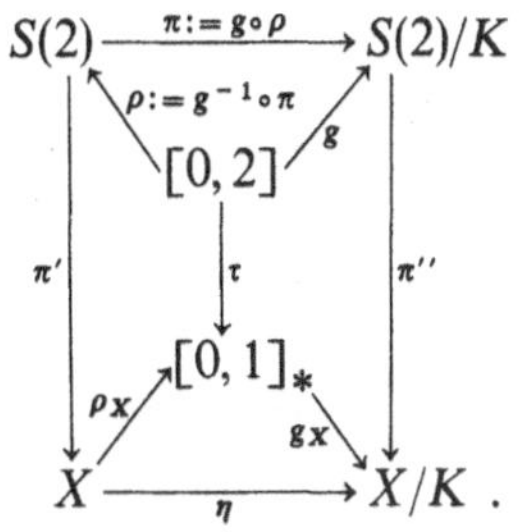

Here η denotes the quotient mapping from X onto X/K, and the mappings π'' and g_X are defined in the obvious manner. As a consequence one obtains the commutativity of the diagram

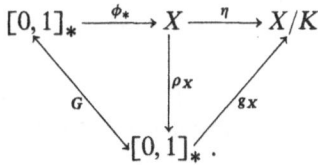

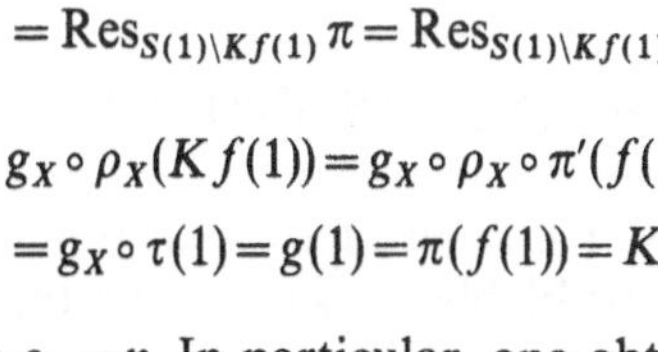

In fact,

$$\mathrm{Res}_{S(1)\backslash Kf(1)}\, g_X \circ \rho_X = \mathrm{Res}_{S(1)\backslash Kf(1)}\, g \circ \rho$$
$$= \mathrm{Res}_{S(1)\backslash Kf(1)}\, \pi = \mathrm{Res}_{S(1)\backslash Kf(1)}\, \eta$$

and

$$g_X \circ \rho_X(Kf(1)) = g_X \circ \rho_X \circ \pi'(f(1)) = g_X \circ \tau \circ \rho(f(1))$$
$$= g_X \circ \tau(1) = g(1) = \pi(f(1)) = Kf(1) = \eta(Kf(1)),$$

so that $g_X \circ \rho_X = \eta$. In particular, one obtains

$$\eta \circ \phi_*(t) = g(t) = \pi \circ f(t) \in f(t)K \quad \text{for all } t \in [0,1] \cap \mathbb{Q}.$$

The rest is obvious. □

3.5 The General Continuous Embedding

Basing ourselves on the existence of one-parameter semigroups we establish a general theorem concerning continuous embedding of infinitely divisible probability measures on a locally compact group. We first collect the results of the preceding paragraph and reformulate those which will be frequently used in résumé form.

3.5.1 Theorem. *Let T be a Hausdorff topological semigroup and f an abstract homomorphism from $\mathbb{Q}_+^*$ into T with the property that $f(]0,1]\cap\mathbb{Q})$ is relatively compact in T. Then there exist a compact connected Abelian group K in T and a one-parameter semigroup ϕ in T satisfying the following conditions:*
(i) *$\phi(0)$ is the identity e of K;*
(ii) *K is a subgroup of the group $H(e)$ of units of eTe;*
(iii) *$\phi(\mathbb{R}_+^*)$ lies in the centralizer $\{x\in eTe: xk=kx$ for all $k\in K\}$ of K in eTe;*
(iv) *for every $r\in\mathbb{Q}_+^*$ there is a $k_r\in K$ such that $\phi(r)=f(r)\,k_r$ holds.*

Proof. From the discussion of § 4 it is clear that

$$K:=\bigcap_{r\in]0,1[}\overline{f(]0,r[\cap\mathbb{Q})}$$

is the desired subgroup of $H(e)$. If now $f(1)\in K$, then by the properties of K the theorem is proved. If, conversely, $f(1)\notin K$, then by Theorem 3.4.10 there is a one-parameter semigroup ϕ in T such that $\phi(\mathbb{R}_+^*)\subset\overline{f(\mathbb{Q}_+^*)}\subset eTe$ and $\phi(r)\in f(r)\,K$ holds for all $r\in\mathbb{Q}_+$. In particular, $\phi(\mathbb{R}_+^*)$ lies in the centralizer of K in eTe. □

Applying the preceding theorem to $T:=\mathscr{M}^1(G)$ for a locally compact group G and a homomorphism $f:\mathbb{Q}_+^*\to\mathscr{M}^1(G)$ defined by $f(r):=\mu_r$ for all $r\in\mathbb{Q}_+^*$ one obtains a rational convolution semigroup $S:=f(\mathbb{Q}_+^*)=(\mu_r)_{r\in\mathbb{Q}_+^*}$ in $\mathscr{M}^1(G)$ with *accumulation group* $\mathscr{A}(S):=K$.

3.5.2 Definition. A measure $\mu\in\mathscr{I}(G)$ is said to be *root compact* if the root set $R(\mu)$ of μ is relatively compact in $\mathscr{M}^1(G)$.

Let **S** denote the class of all locally compact groups G such that any $\mu\in\mathscr{I}(G)$ is root compact. Clearly, **S** contains all compact groups, and **S** has the following

3.5.3 Properties. (1) Let G be a locally compact group and K a compact normal subgroup of G. Then $G\in\mathbf{S}$ iff $G/K\in\mathbf{S}$.

(2) Let $G\in\mathbf{S}$ and H a closed subgroup of G. Then $H\in\mathbf{S}$.

(3) Let $(G_i)_{i=1,\dots,n}$ be a finite sequence of groups in **S**. Then the product $\prod_{i=1}^n G_i$ is in **S**.

By Theorem 3.1.32 every root compact measure $\mu\in\mathscr{M}^1(G)$ (for an arbitrary locally compact group G) is rationally embeddable. This motivates the hypothesis of the following important embedding result.

3.5.4 Theorem. *Let G be a locally compact group, $\mu\in\mathscr{M}^1(G)$ a root compact measure and $S:=(\mu_r)_{r\in\mathbb{Q}_+^*}$ a rational convolution semigroup in $\mathscr{M}^1(G)$ such that $\mu_1=\mu$ holds.*

Then

(i) *the accumulation group* $\mathscr{A}(S)$ *of* S *is a compact, connected Abelian group in* $\mathscr{M}^1(G)$ *with unit element* ω_H *for some compact subgroup* H *of* G;

(ii) *for every* $r \in \mathbb{Q}_+^*$ *the measure* μ_r *has* ω_H *as an idempotent factor and satisfies* $\mu_r * \nu = \nu * \mu_r$ *for all* $\nu \in \mathscr{A}(S)$;

(iii) *there exists a compact subgroup* $H(S)$ *of* G *with*

$$H(S) \subset N(H) \quad \text{and} \quad H(S)_0\, H = H(S)$$

such that $H(S)/H$ *and* $\mathscr{A}(S)$ *are isomorphic (as topological groups)*;

(iv) *for every* $r \in \mathbb{Q}_+^*$ *there exists an* $x_r \in H(S)_0$ *such that* $(\varepsilon_{x_r} * \mu_r)_{r \in \mathbb{Q}_+^*}$ *is an* H*-continuous convolution semigroup in* $\mathscr{M}^1(G)$.

Proof. Considering the homomorphism $f: \mathbb{Q}_+^* \to \mathscr{M}^1(G)$ defined by $f(r) := \mu_r$ for all $r \in \mathbb{Q}_+^*$ one observes immediately that $f([0,1] \cap \mathbb{Q}_+^*) \subset R(\mu)$, and by assumption we obtain that the hypothesis of Theorem 3.5.1 with $T := \mathscr{M}^1(G)$ and f is satisfied. Thus the assertions (i) and (ii) of the theorem become immediate consequences of the previous result. One need only note that the unit of $\mathscr{A}(S)$ is the only idempotent of $\mathscr{A}(S)$ and therefore by Theorem 1.2.10, it is of the form ω_H for some compact subgroup H of G. Since Assertion (iv) follows directly from Theorem 3.5.1 together with (iii), it remains to show (iii). For every $\nu \in \mathscr{A}(S)$ there exists a $\nu' \in \mathscr{A}(S)$ satisfying $\nu * \nu' = \omega_H$, and hence an $x_\nu \in N(H)$ with $\nu = \varepsilon_{x_\nu} * \omega_H$ via Theorem 1.2.13. Clearly,

$$H(S) := \{x \in N(H) : \varepsilon_x * \omega_H \in \mathscr{A}(S)\}$$

is a closed subgroup of G, and the mapping $g: H(S) \to \mathscr{A}(S)$ defined by $g(x) := \varepsilon_x * \omega_H$ for all $x \in H(S)$ is a continuous epimorphism onto $\mathscr{A}(S)$ with kernel H. If p denotes the canonical mapping from $H(S)$ onto $H(S)/H$ and $g = h \circ p$ the factorization of g over $H(S)/H$, then $h: H(S)/H \to \mathscr{A}(S)$ is continuous and open. Indeed, embedding $\mathscr{A}(S)$ into $\mathscr{M}_H^1(H(S))$ and $H(S)/H$ into $\mathscr{M}^1(H(S)/H)$ and observing that by Theorem 1.3.15 the topological semigroups $\mathscr{M}_H^1(H(S))$ and $\mathscr{M}^1(H(S)/H)$ are isomorphic, one obtains $H(S)/H \cong \mathscr{A}(S)$. Since now H and $\mathscr{A}(S)$ are compact groups, so is $H(S)$. As $\mathscr{A}(S)$ is also connected, we further have

$$g(H(S)_0) = \mathscr{A}(S) \quad \text{and thus} \quad H(S)_0\, H = H(S). \quad \square$$

3.5.5 Corollary. *Every divisible group in* **S** *is connected.*

Proof. Let G be a divisible group and let $x \in G$ be such that $\varepsilon_x \in \mathscr{I}(G)$. Since $G \in \mathbf{S}$, the measure ε_x is root compact. Hence, by the theorem there exists a $y \in G_0$ such that $x\,y$ lies on a one-parameter subgroup of G. But this implies that $x\,y \in G_0$ or $x \in G_0$, thus $G_0 = G$. $\square$

3.5.6 Definition. A measure $\mu \in \mathscr{M}^1(G)$ is called *continuously embeddable* if there exists a continuous convolution semigroup $(\mu_r)_{r \in \mathbb{D}}$ in $\mathscr{M}^1(G)$ with $\mathbb{D} := \mathbb{Q}_+^*$ or $\mathbb{R}_+^*$ satisfying $\mu_1 = \mu$.

If $\mathscr{T}_v\text{-}\lim_{r \downarrow 0} \mu_r = \omega_H$ for some compact subgroup H of G, then μ is called *H-continuously embeddable.*

For any compact subgroup H of G let the symbols $\mathscr{E}_H(G)$ and $\mathscr{E}(G)$ denote the classes of all H-continuously embeddable and all continuously embeddable measures in $\mathscr{M}^1(G)$ resp. Clearly, $\mathscr{E}(G) := \bigcup \{\mathscr{E}_H(G): H$ is a compact subgroup of $G\}$. For a compact subgroup H of G we further denote by $\mathscr{I}_H(G)$ the class of all root compact measures in $\mathscr{M}^1(G)$ which are rationally embeddable into a (rational) convolution semigroup S in $\mathscr{M}^1(G)$ such that ω_H is the unit element of $\mathscr{A}(S)$. If $G \in \mathbf{S}$, then Theorem 3.5.4 yields the equality

$$\mathscr{I}(G) = \bigcup \{\mathscr{I}_H(G): H \text{ is a compact subgroup of } G\}.$$

Moreover, for any compact subgroup H of G we have $\mathscr{E}_H(G) \subset \mathscr{I}_H(G)$, but in general, $\mathscr{E}_H(G) \neq \mathscr{I}_H(G)$, as can be seen from the example of the group $G := \mathbb{Q}_d$, for which $\varepsilon_x \in \mathscr{I}_{\{0\}}(G) \setminus \mathscr{E}_{\{0\}}(G)$ for all $x \in G$.

3.5.7 Corollary. *For any compact subgroup H of G we have $\overline{\mathscr{E}_H(G)} = \mathscr{I}_H(G)$ and, if $G \in \mathbf{S}$, also $\overline{\mathscr{E}(G)} = \mathscr{I}(G)$.*

Proof. By the preceding discussion it suffices to prove the first assertion. Let $\mu \in \mathscr{I}_H(G)$ be rationally embedded into a convolution semigroup S such that ω_H is the unit element of $\mathscr{A}(S)$. But in the compact, connected Abelian group $\mathscr{A}(S)$ the union of all its one-parameter subgroups is dense ([218], (25.20)). The theorem implies the existence of a $\nu \in \mathscr{A}(S)$ with $\nu * \mu \in \mathscr{E}_H(G)$. Choosing a net $(\nu_\alpha)_{\alpha \in \mathbf{A}}$ in $\mathscr{A}(S)$ with $\lim_{\alpha \in \mathbf{A}} \nu_\alpha = \nu'$ such that $\nu * \nu' = \omega_H$ holds and ν_α lies on a one-parameter subgroup of $\mathscr{A}(S)$ for all $\alpha \in \mathbf{A}$ we obtain $\nu_\alpha * \nu * \mu \in \mathscr{E}_H(G)$ for all $\alpha \in \mathbf{A}$, and from

$$\lim_{\alpha \in \mathbf{A}} \nu_\alpha * \nu * \mu = \nu' * \nu * \mu = \omega_H * \mu = \mu$$

the assertion. □

The following result contains the consequences of Theorem 3.5.4 for the embedding problem.

3.5.8 Theorem. *Let G be a locally compact group in $\mathbf{S}$ such that every compact connected subgroup of G is arcwise connected. Then $\mathscr{I}(G) = \mathscr{E}(G)$.*

Proof. Let H be a compact subgroup of G. It clearly suffices to show that $\mathscr{I}_H(G) \subset \mathscr{E}_H(G)$. We apply Theorem 3.5.4 to a measure $\mu \in \mathscr{I}_H(G)$. Firstly, there exists a continuous epimorphism g from $H(S)$ onto $\mathscr{A}(S)$ with the property that

$$g(H(S)_0) = \mathscr{A}(S).$$

By assumption $H(S)_0$ is arcwise connected, and hence so is $\mathscr{A}(S)$. As an arcwise connected, compact Abelian group, $\mathscr{A}(S)$ is in fact the union of its one-parameter subgroups (Theorem 1 of [117]). In particular, the measure $\varepsilon_{x_1^{-1}} * \omega_H$ (with $x_1 \in H(S)_0$) lies on a one-parameter subgroup of $\mathscr{A}(S)$. But by Theorem 3.5.4 $\varepsilon_{x_1} * \mu_1$ with $\mu_1 = \mu \in \mathscr{I}_H(G)$ is H-embeddable. Thus so is μ or $\mu \in \mathscr{E}_H(G)$. □

3.5.9 Corollary. *Let $G \in \mathbf{S}$. Then we have $\mathscr{I}(G) = \mathscr{E}(G)$ in any of the following special cases:*

(i) *G_0 is a Lie group, in particular, if*
(ii) *G is a Lie group or*
(iii) *G is a totally disconnected group.*

Proof. It suffices to show the assertion in Case (i). This, however, follows from the fact that any compact connected subgroup K of G is a Lie group and is thus, in particular, arcwise connected. The theorem applies. □

3.5.10 Definition. A locally compact group G is said to *satisfy the embedding property* (EMP) if $\mathscr{I}(G)=\mathscr{E}(G)$.

The class of all groups satisfying (EMP) will be denoted by $\mathbf{E}$.

3.5.11 Examples of groups in the class $\mathbf{E}$ are:
1. All compactly generated Abelian Lie groups.
2. All compact Lie groups.
3. All totally disconnected compact groups.
4. All connected nilpotent Lie groups.
5. Every solvable connected Lie group whose Lie algebra admits only real roots.

In fact, groups of type 1 to 5 are in $\mathbf{R}_0$ by Examples 1 and 2 of 3.1.12 and Theorems 3.1.17 and 3.1.23. They are also in $\mathbf{S}$ since $\mathbf{R}_0\subset\mathbf{S}$ by Theorem 3.1.13. The conclusion is supplied by Corollary 3.5.9.

3.5.12 Theorem. *For any locally compact Abelian group* $G\in\mathbf{S}$ *the following statements are equivalent:*
(i) $G\in\mathbf{E}$;
(ii) G_0 *is locally arcwise connected.*

Proof. 1. (ii) ⇒ (i). Let $\mu\in\mathscr{I}(G)$. Since $G\in\mathbf{S}$, Theorem 3.5.4 yields the existence of an $x\in G_0$, a compact subgroup H of G and of an H-continuously embeddable measure $\nu\in\mathscr{M}^1(G)$ such that $\mu=\varepsilon_x*\nu$ holds. By Theorems 1 and 4 of [117] the group G_0, being locally arcwise connected, is the union of its one-parameter subgroups. Thus ε_x is continuously embeddable or μ is H-continuously embeddable, i.e., $\mu\in\mathscr{E}(G)$.

2. (i) ⇒ (ii). Since by (24.24) of [218] one has $G_0\subset T(G)$, we get $\varepsilon_x\in\mathscr{I}(G)$ for all $x\in G_0$. Consequently, G_0 is the union of the one-parameter subgroups of G and is hence arcwise connected. But then G_0 is also locally arcwise connected, as follows from the implications of Theorem H. □

3.5.13 Remark. The implication (ii) ⇒ (i) of Theorem 3.5.12 strengthens the statement of Theorem 3.5.8. By the structure theorem A we have $G_0\cong\mathbb{R}^n\times K$ for some $n\geq 0$ and a compact connected Abelian group K. But the hypothesis of Theorem 3.5.8 yields that K is arcwise connected and hence that G_0 is arcwise connected. But then, by Theorem H, G_0 is also locally arcwise connected.

In order to present a characterization of the locally compact Abelian groups in $\mathbf{S}$ we shall study in more detail the roots of a measure in $\mathscr{I}(G)$. A problem of special interest is the existence of unique roots.

3.5.14 Definition. A locally compact group G is said to *satisfy the uniqueness property* (UP) if for every $\mu\in\mathscr{I}(G)$ and every $n\geq 1$ the n-th root of μ is uniquely determined.

Clearly, the groups $\mathbb{R}$ and $\mathbb{Q}_d$ satisfy (UP).

3.5.15 Theorem. *Let G be a locally compact Abelian group. Then the following statements are equivalent:*
(i) *G satisfies* (UP);
(ii) *G is aperiodic.*

The proof of the theorem will be preceded by a useful

3.5.16 Lemma. *Let G be an aperiodic locally compact Abelian group. Then for every $\mu \in \mathscr{I}(G)$ one has $\hat{\mu}(\chi) \neq 0$ for $\chi \in G^\wedge$.*

Proof. By assumption $G^\wedge$ is a connected locally compact Abelian group. Let $\mu \in \mathscr{I}(G)$ and define $\phi := \hat{\mu}$. Then ϕ is a normed continuous positive-definite function on $G^\wedge$ such that for every $n \geq 1$ there exists a normed continuous positive-definite function ϕ_n on $G^\wedge$ satisfying $\phi_n^n = \phi$. Without loss of generality we may assume that ϕ and hence $\phi_n (n \geq 1)$ is real. Indeed, the properties of ϕ extend to $|\phi|$ and $|\phi|^2$, but $\phi(\chi) = 0$ iff $|\phi|^2(\chi) = 0$ for all $\chi \in G^\wedge$. One observes that

$$\lim_{n \to \infty} \phi_n = \lim_{n \to \infty} \phi^{\frac{1}{n}} = \psi$$

with $\psi(\chi) = 0$, if $\phi(\chi) = 0$, and $\psi(\chi) = 1$, if $\phi(\chi) \neq 0$, and ψ is positive-definite. But from $\phi(1) = 1$ one concludes that $\phi(\chi) \neq 0$ for all χ in a neighborhood $U \in \mathfrak{V}(1)$. Hence, $\psi(\chi) = 1$ for all $\chi \in U$. Thus ψ is continuous on U. The positive-definiteness of ψ implies its continuity on the whole of $G^\wedge$, as follows from the inequality

$$|\psi(\chi) - \psi(\eta)|^2 \leq 2\psi(1)[\psi(1) - \operatorname{Re} \psi(\chi^{-1}\eta)],$$

valid for any (possibly complex-valued) positive-definite function ψ on $G^\wedge$ and $\chi, \eta \in G^\wedge$. But now we infer that $G^\wedge$ is connected. Since ψ takes on only the values 0 or 1, it must necessarily be identical to 1 on G which shows that $\hat{\mu} = \phi \neq 0$ on $G^\wedge$. □

We now proceed to the *proof of the theorem.*

1. (ii) ⇒ (i). Let $\mu \in \mathscr{I}(G)$ and let ν, λ be two n-th roots of μ in $\mathscr{M}^1(G)$ (for $n \geq 1$). By Lemma 3.5.16 we obtain $\hat{\mu}(\chi) \neq 0$ for all $\chi \in G^\wedge$, so that $\phi(\chi) := \dfrac{\hat{\nu}(\chi)}{\hat{\lambda}(\chi)}$ for all $\chi \in G^\wedge$ defines a continuous function ϕ on $G^\wedge$ satisfying $\phi(\chi)^n = 1$ for all $\chi \in G^\wedge$. Let E_n denote the set of all n-th roots of unity. Then $\phi(G^\wedge) \subset E_n$. But by the hypothesis $G^\wedge$ is connected, so that $\phi(1) = 1$ together with the continuity of ϕ implies that $\phi \equiv 1$ since E_n is discrete. This yields $\hat{\nu} = \hat{\lambda}$ or $\nu = \lambda$.

2. (i) ⇒ (ii). The following argument does not depend on the commutativity property of G. We show that there exists a measure $\mu \in \mathscr{P}(G)$ whose n-th roots are not unique for all $n \geq 1$. Let H be a nontrivial compact subgroup of G. If H admits an element x of finite order $n \geq 3$, then by Remark 3.2.6 there exists a Poisson measure μ on $[x] = \{e, x, x^2, \ldots, x^{n-1}\}$ possessing the required property. In the case that H admits an element x of order 2 one puts

$$\mu := \exp_{\{e\}}(\varepsilon_x - \varepsilon_e) = \frac{1}{e}\left(\sum_{k \geq 0} \frac{1}{(2k+1)!}\right) \varepsilon_x + \frac{1}{e}\left(\sum_{k \geq 0} \frac{1}{(2k)!}\right) \varepsilon_e$$

(where Euler's e in $\frac{1}{e}$ must not be confused with the neutral element of G) and verifies that

$$\nu_2 := \exp_{\{e\}}[\tfrac{1}{2}(\varepsilon_x - \varepsilon_e)] \quad \text{and} \quad \lambda_2 := \varepsilon_x * \nu_2$$

are second roots of μ with $\nu_2 \neq \lambda_2$. Finally, if H is torsion-free and compact, then by the Lie projectivity of H there exists a compact normal subgroup K of H such that H/K is a Lie group. Thus we assume without loss of generality that H itself is a Lie group. But as a compact Lie group H is either finite or contains a torus and therefore elements of finite order. The proof is thus reduced to the first two special cases discussed. □

3.5.17 Remark. There exists a non-Abelian, aperiodic locally compact group G which does not satisfy (UP). One simply chooses $G := G_3$ of Theorem 3.3.7 as an example for which the roots of $\varepsilon_{x_1} \in \mathcal{M}^1(G)$ are not unique. More generally, if G is a non-Abelian, aperiodic locally compact group, $\mu \in \mathcal{I}(G)$ and if there exist an $x \in G$ and an $n > 1$ such that $\varepsilon_x * \mu * \varepsilon_{x^{-1}} = \mu$ and $\nu_n := \varepsilon_x * \mu_n * \varepsilon_{x^{-1}} \neq \mu_n$ hold, then ν_n and μ_n are different n-th roots of μ.

3.5.18 Theorem. *For any locally compact Abelian group G the following statements are equivalent:*

(i) *G satisfies* (UP);

(ii) *there exist a $d \geq 0$ and a discrete weak direct product $\mathbb{Q}_d^{\mathfrak{m}*}$ of a finite or infinite number of copies of $\mathbb{Q}_d$ such that G is continuously (monomorphically) embedded into $\mathbb{R}^d \times \mathbb{Q}_d^{\mathfrak{m}*}$*;

(iii) *there exist a locally convex vector space* E *(over* $\mathbb{R}$*) and a continuous monomorphism from G into* E.

Proof. 1. (i) ⇒ (ii). By the structure theorem M G is of the form $\mathbb{R}^d \times H$, where $d \geq 0$ and H is a locally compact Abelian group admitting a compact open subgroup. Let now G satisfy (UP). Then by Theorem 3.5.15 G is aperiodic. Hence, H is discrete and torsionfree. Theorems A14 and A16 of [218] imply that the smallest divisible extension of H is of the form $\mathbb{Q}_d^{\mathfrak{m}*}$, whence the continuous monomorphism from G into $\mathbb{R}^d \times \mathbb{Q}_d^{\mathfrak{m}*}$ is established.

2. (ii) ⇒ (iii) is clearly achieved with $\mathsf{E} := \mathbb{R}^d \times \mathbb{R}^{\mathfrak{m}}$.

3. (iii) ⇒ (i). By assumption there exists a continuous monomorphism from G into a locally convex vector space E which can be embedded into $\mathbb{R}^{\mathfrak{n}}$ for some cardinal $\mathfrak{n}$ and hence into the projective limit of a projective system of Euclidean groups $\mathbb{R}^n$ ($n \in \mathbb{N}$) with their natural projections. The fact that $\mathbb{R}^n$ for $n \geq 1$ satisfies (UP) together with the subsequent permanence property for (UP) implies the assertion. Let G be a locally compact group, $(G_\alpha, p_{\alpha\beta}, \mathbb{A})$ a projective system of locally compact groups and $(p_\alpha)_{\alpha \in \mathbb{A}}$ a system of continuous homomorphisms p_α: $G \to G_\alpha$ consistent and separating for $(G_\alpha, p_{\alpha\beta}, \mathbb{A})$. If G_α satisfies (UP) for all $\alpha \in \mathbb{A}$, then so does G. □

We are now ready to establish the desired characterization of the class of all Abelian groups in **S**.

3.5.19 Theorem. *For any locally compact Abelian group G the following statements are equivalent:*
(i) $G \in \mathbf{S}$;
(ii) $B(G)$ *is compact and* $T(G) = G_0$.

Proof. 1. (i) $\Rightarrow$ (ii). Let $G \in \mathbf{S}$. By Theorem M there exist a $d \geq 0$ and a locally compact Abelian group H admitting a compact open subgroup K such that $G \cong \mathbb{R}^d \times H$ holds. Property (1) of 3.5.3 implies that the discrete group $D := H/K$ is in $\mathbf{S}$. This yields the relative compactness of $R(\varepsilon_e) := \{\varepsilon_x : x \in B(D)\}$ and hence $B(D)$ is finite. Using $B(D) \cong B(H)/K$ and $B(H) \cong B(G)$ one obtains that $B(G)$ is compact. In view of $T(G) \cong \mathbb{R}^d \times T(H)$ and $T(H) \supset H_0$, which follows from (24.24) of [218], it remains to show that $T(H) \subset H_0$. But this is obtained from the proof of Corollary 3.5.5 since by Property (2) of 3.5.3 $H \in \mathbf{S}$.

2. (ii) $\Rightarrow$ (i). Let $B(G)$ be compact (or let G possess a largest compact subgroup) and $T(G) = G_0$. Firstly $T(G/B(G)) = (G/B(G))_0$. In fact, Theorem M implies that $G \cong \mathbb{R}^d \times H$, where H admits a compact open subgroup, whence $B(G) \cong B(H) \subset H$. Any $x \in H$ with $x^n \in B(G)$ for some $n \geq 1$ lies in $B(G)$, so that $B(G)$ is a pure compact subgroup of H in the sense that whenever $y \in H$ and $y^m \in B(G)$ for some $m \geq 1$, there is a $z \in B(G)$ with $z^m = y^m$. By (25.21) in [218] $B(G)$ is an algebraic direct factor of H, i.e., there is a subgroup N of H such that H and $B(G) \times N$ are isomorphic as abstract groups. But $T(H) = H_0 \subset B(G)$ and the algebraic isomorphism between N and $H/B(G)$ imply that $T(H/B(G)) = \{e\}$, whence $T(G/B(G)) = \mathbb{R}^d = (G/B(G))_0$.

On the basis of Property (1) of 3.5.3 we may assume that G is aperiodic. But then $G \cong \mathbb{R}^d \times D$ with a torsionfree discrete Abelian group D satisfying $T(D) = \{e\}$. We have to show that $G \in \mathbf{S}$. Since $\mathbb{R}^d \in \mathbf{R}_0 \subset \mathbf{S}$ and $\mathbf{S}$ is closed under the formation of direct products by Property (3) of 3.5.3, it remains to show that $D \in \mathbf{S}$ or that without loss of generality $G := D$ can be assumed. From Theorem 3.5.15 we infer that for any $\mu \in \mathscr{I}(G)$ there exists a unique rational convolution semigroup $(\mu_r)_{r \in \mathbb{Q}_+^*}$ in $\mathscr{M}^1(G)$ satisfying $\mu_1 = \mu$. Putting $\nu_r = \mu_r * \mu_r^{\sim}$ we obtain $\hat{\nu}_r = |\hat{\mu}_r|^2 = (\hat{\nu}_1)^r$ for every $r \in \mathbb{Q}_+^*$ and hence $\lim_{r \downarrow 0} \nu_r = \varepsilon_e$ or $\lim_{r \downarrow 0} \nu_r(\{e\}) = 1$. Using the technique introduced in the properties 3.3.13 we show that the convolution semigroup $(\mu_r)_{r \in \mathbb{Q}_+^*}$ is continuous. In fact, for every $r \in \mathbb{Q}_+^*$ one chooses $x_r \in G$ with $\mu_r(\{x_r\}) = \max_{x \in G} \mu_r(\{x\})$. From

$$\begin{aligned} \nu_r(\{e\}) &= (\mu_r * \mu_r^{\sim})(\{e\}) \\ &= \textstyle\sum_{x \in G} \mu_r(\{x\})^2 \leq \mu_r(\{x_r\}) \sum_{x \in G} \mu_r(\{x\}) = \mu_r(\{x_r\}) \leq 1 \end{aligned}$$

for all $r \in \mathbb{Q}_+^*$ we obtain $\lim_{r \downarrow 0} \mu_r(\{x_r\}) = 1$. Consequently, there exists an $r_0 \in \mathbb{Q}_+^*$ with $\mu_r(\{x_r\}) > \sqrt{\varepsilon}$ for all $r \leq r_0$, where $\varepsilon > \frac{1}{2}$. But then one has for all $r, s \in \mathbb{Q}_+^*$ with $r + s \leq r_0$ on the one hand $\mu_{r+s}(\{x_{r+s}\}) > \sqrt{\varepsilon} > \varepsilon$, and on the other hand

$$\begin{aligned} \mu_{r+s}(\{x_r x_s\}) &= \mu_r * \mu_s(\{x_r\}\{x_s\}) \\ &\geq \mu_r(\{x_r\}) \mu_s(\{x_s\}) > \sqrt{\varepsilon}\sqrt{\varepsilon} = \varepsilon. \end{aligned}$$

Thus $\{x_{r+s}\} \cap \{x_r x_s\} \neq \varnothing$ or $x_{r+s} = x_r x_s$. In particular, $x_r \in T(G)$ for all $r \leq r_0$. Since by assumption $T(G) = \{e\}$, we obtain $x_r = e$ for all $r \leq r_0$ and therefore

$\lim_{r\downarrow 0}\mu_r(\{e\})=1$. This shows the continuity of $(\mu_r)_{r\in\mathbb{Q}_+^*}$. Theorem 1.5.9 yields the extension of $(\mu_r)_{r\in\mathbb{Q}_+^*}$ to a continuous (real) convolution semigroup $(\mu_t)_{t\in\mathbb{R}_+^*}$ in $\mathscr{M}^1(G)$ with $\mu_1=\mu$.

We finally have $R(\mu)\subset\{\mu_t: t\in[0,1]\}=:\mathscr{N}$. But since $\mathscr{N}$ is compact as a continuous image of $[0,1]$, $R(\mu)$ is also compact in $\mathscr{M}^1(G)$, whence $G\in\mathbf{S}$. □

3.5.20 Remark. The preceding theorem enables us to establish the inequality $\mathbf{S}\neq\mathbf{R}_0$. Indeed, the discrete (additive) subgroup of $\mathbb{Q}$ generated by the sequence $\left(\frac{1}{2^n}\right)_{n\geq 1}$ is an element of $\mathbf{S}$ by Theorem 3.5.19, but it does not belong to $\mathbf{R}_0$ by Theorem 3.1.13.

3.5.21 Remark. In view of Theorem 3.5.19 one might conjecture the following extension of Theorem 3.5.12: A locally compact Abelian group G lies in $\mathbf{E}$ iff $B(G)$ is compact, $T(G)=G_0$ and G_0 is locally arcwise connected. Such a result, however, is false, as the group

$$G:=\prod_{n\geq 1}G_n \quad\text{with}\quad G_n:=\mathbb{Z}_2 \quad\text{for all } n\geq 1,$$

furnished with the discrete topology, shows. First of all one notes that G is a group of bounded order, whence $T(G)=\{e\}$. Moreover, $G_0=\{e\}$ implies that G_0 is locally arcwise connected, and $B(G)=G$ is certainly not compact. But it can be shown that $\mathscr{I}(G)\subset\mathscr{E}(G)$. Let $\dot{G}$ be the group $\prod_{n\geq 1}G_n$ furnished with the product topology and ϕ the canonical mapping from G onto $\dot{G}$. Given $\mu\in\mathscr{I}(G)$ we observe that $\phi(\mu)\in\mathscr{I}(\dot{G})$. But since $\dot{G}\in\mathbf{R}$ the measure $\phi(\mu)$ can be rationally embedded in a (rational) convolution semigroup $(\dot{\mu}_r)_{r\in\mathbb{Q}_+^*}$ in $\mathscr{M}^1(\dot{G})$, as follows from Theorem 3.1.32. Since $\phi(\mu)$ is discrete, $\dot{\mu}_r$ is also discrete for all $r\in\mathbb{Q}_+^*$. Therefore for every $r\in\mathbb{Q}_+^*$ there is a $\mu_r\in\mathscr{M}^1(G)$ with $\phi(\mu_r)=\dot{\mu}_r$, which shows that μ is rationally embeddable in the convolution semigroup $S:=(\mu_r)_{r\in\mathbb{Q}_+^*}$ in $\mathscr{M}^1(G)$. We shall show that S is in fact continuous. Since

$$\mu_r*\mu_{1-r}=\mu_1 \qquad\text{for all } r\in\mathbb{J}:=[0,1]\cap\mathbb{Q}_+^*$$

there exists by Theorem 1.3.21 for every $r\in\mathbb{J}$ an $x_r\in G$ such that the family $\{\mu_r*\varepsilon_{x_r}: r\in\mathbb{J}\}$ is $\mathscr{T}_v$-relatively compact in $\mathscr{M}^1(G)$. From $x^2=e$ for all $x\in G$ we infer that

$$(\mu_r*\varepsilon_{x_r})^2=\mu_r^2*\varepsilon_{x_r^2}=\mu_{2r} \qquad\text{for all } r\in\mathbb{Q}_+^*.$$

Therefore the set $\{\mu_{2r}: r\in\mathbb{J}\}$ is relatively compact in $\mathscr{M}^1(G)$. Thus the hypothesis of Theorem 3.5.4 is satisfied and the accumulation set $\mathscr{A}(S)$ of S exists. As G is discrete, Corollary 3.5.9 yields $\mathscr{A}(S)=\{e\}$, and the continuity of S is established.

3.6 Injective Submonogeneous Embeddings

Roughly speaking, injective embeddings are injective convolution semigroups of probability measures on a locally compact group G that can be used for the embedding of infinitely divisible probability measures on G. We shall discuss the wider problem of describing those locally compact groups G for which every homomorphism from $\mathbb{Q}_+^*$ into $\mathcal{M}^1(G)$ is injective (or trivial). Interestingly enough, the class of groups with this property contains a large number of families of locally compact groups known in harmonic analysis and representation theory. We add a few simple facts concerning the algebraic classification of submonogeneous semigroups.

As in §3 we start with real submonogeneous semigroups and groups. Clearly, any subgroup of $\mathbb{R}$ generated by a submonogeneous subsemigroup of $\mathbb{R}_+$ is itself submonogeneous. If a subsemigroup of $\mathbb{Q}_+^*$ is the intersection of a submonogeneous subgroup of $\mathbb{R}$ with $\mathbb{R}_+^*$, then it is submonogeneous. This follows from the fact that any subgroup of a submonogeneous group is submonogeneous. The following algebraic result will be the initial point for the discussion of injective embeddings.

3.6.1 Lemma. *Let $\mathbb{T}$ be a submonogeneous subsemigroup of $\mathbb{R}_+$, S an arbitrary semigroup and f a homomorphism from $\mathbb{T}$ into S. We define the subsets $\mathbb{T}_0$ and $\mathbb{T}_1$ of $\mathbb{T}$ by*

$$\mathbb{T}_0 := \{t\in\mathbb{T}: \text{There exists an } s\in\mathbb{T} \text{ with } t<s \text{ and } f(s)=f(t)\} \quad \text{and}$$

$$\mathbb{T}_1 := \{t\in\mathbb{T}: \text{There exists an } s\in\mathbb{T} \text{ with } 0\le s<t \text{ and } f(s)=f(t)\}$$

and correspondingly

$$t_0 := \inf\mathbb{T}_0, \text{ if } \mathbb{T}_0\neq\emptyset, \text{ and } t_0:=\infty, \text{ if } \mathbb{T}_0=\emptyset, \text{ and}$$

$$t_1 := \inf\mathbb{T}_1, \text{ if } \mathbb{T}_1\neq\emptyset, \text{ and } t_1:=\infty, \text{ if } \mathbb{T}_1=\emptyset, \text{ resp.}$$

Then

(i) *$t_0\le t_1$, and $t_0=\infty$ iff f is injective.*
Let $t_0<t_1$. Then, moreover,
(ii) *$f(\mathbb{T}_0)$ is a group isomorphic to a subgroup of the group of rationals modulo 1;*
(iii) *$\mathrm{Res}_{[0,t_0]\cap\mathbb{T}} f$ is injective;*
(iv) *$t_1-t_0\in\mathbb{T}$, there exists a smallest $n\ge 1$ with $(t_1-t_0)n\in\mathbb{T}_0$, and $f((t_1-t_0)n)$ is the unit element of the group $f(\mathbb{T}_0)$.*

Proof. It remains to prove the assertions (ii) and (iv).

1. We first show that $f(\mathbb{T}_0)$ is a group. For this purpose we have to prove the solvability of the equation $f(t)f(r)=f(s)$ with respect to $f(r)\in f(\mathbb{T}_0)(t,s\in\mathbb{T}_0)$. Since $s\in\mathbb{T}_0$, there exists an $h\in\mathbb{T}\setminus\{0\}$ satisfying $f(s)=f(s+h)=f(s)f(h)$. Therefore one obtains $f(s)=f(s+mh)$ for all $m\ge 1$, and we may suppose without loss of generality that $h\in\mathbb{T}_0$. Clearly, there exists a natural number $n\ge 1$ such that $r:=nh+s-t\in\mathbb{T}_0$. But then $f(t)f(r)=f(nh)f(s)=f(s)$ which was to be proved.

2. Let $\mathbb{V}:=\{v\in\mathbb{T}: f(t)f(v)=f(t) \text{ for all } t\in\mathbb{T}_0\}$ and let ε be the unit element of the group $f(\mathbb{T}_0)$. Then $\mathbb{V}\supset f^{-1}(\varepsilon)$. We observe that $v,w\in\mathbb{V}$ with $v>w$ implies $v-w\in\mathbb{V}$, since

$$f(t)f(v-w)=(f(t)f(w))f(v-w)=f(t)f(v)=f(t)$$

holds ($t \in \mathbb{T}_0$). Thus $\mathbb{V}$ is of the form $\mathbb{V}' \cap \mathbb{R}_+$, where $\mathbb{V}'$ denotes a subgroup of $\mathbb{R}$. $\mathbb{V}'$ is discrete and therefore $\mathbb{V}' = [l]$ with $l \geq t_1 - t_0$. But $l > t_1 - t_0$ yields a contradiction to the definition of t_1, whence $\mathbb{V}' = [l]$ with $l = t_1 - t_0$. There is then a uniquely determined smallest $n \geq 1$ such that $nl \in \mathbb{T}_0$, and $f(nl) = \varepsilon$ is the unit element of $f(\mathbb{T}_0)$. This proves Assertion (iv).

3. Finally we define a mapping

$$F: \frac{1}{l}\mathbb{T} \to f(\mathbb{T}_0) \quad \text{by} \quad F(r) := f(l(n+r)) \qquad \text{for all } r \in \frac{1}{l}\mathbb{T}.$$

F is a homomorphism of semigroups onto $f(\mathbb{T}_0)$ which is constant on the classes modulo 1. Since $f(\mathbb{T}_0)$ is a group, F can be extended to a homomorphism from the group $\frac{1}{l}\mathbb{T}' := \left[\frac{1}{l}\mathbb{T}\right]$ generated by $\frac{1}{l}\mathbb{T}$ in $\mathbb{R}$ onto the group $f(\mathbb{T}_0)$, whose kernel is the subgroup of $\mathbb{R}$ generated by 1. This completes the proof of Statement (ii). □

For the remaining discussion of this section we shall suppose that all submonogeneous subgroups are noncyclic.

3.6.2 Theorem. *Let G be a locally compact group, $\mathbb{M}$ a submonogeneous subgroup of $\mathbb{Q}$ with $1 \in \mathbb{M}$ and f a homomorphism from $\mathbb{M}_+$ into $\mathscr{M}^1(G)$ of the form*

$$f(r) := \mu_r \in \mathscr{M}^1(G) \quad \textit{for all } r \in \mathbb{M}_+.$$

We suppose that f is noninjective. Then

(i) *there exist an $r_0 \in \mathbb{M}$ and a compact subgroup H of G such that*
   (a) *$\{\mu_r : r \geq r_0\}$ is a group in $\mathscr{M}^1(G)$ with unit element*
   (b) *$\mu_{r_0} = \omega_H$, and*

(ii) *there exists a homomorphism $\phi: r \to x_r$ from $\mathbb{M}$ into $N(H)$ satisfying*
   (c) *$\mu_r = \varepsilon_{x_r} * \omega_H = \omega_H * \varepsilon_{x_r}$ for all $r \geq r_0$;*
   (d) *$\operatorname{supp}(\mu_r) \subset x_r H$ for all $r \in \mathbb{M}_+$.*

Proof. (i) Since by assumption f is noninjective, we obtain

$$\mathbb{M}_0 := \{r \in \mathbb{M}_+ : \text{There exists } s \in \mathbb{M}_+ \text{ with } r < s \text{ and } f(r) = f(s)\} \neq \emptyset,$$

whence by Lemma 3.6.1 $f(\mathbb{M}_+)$ contains a group $\{\mu_r : r \geq r_0\}$ for some $r_0 \in \mathbb{M}_+$ with unit μ_{r_0}, which as an idempotent in $\mathscr{M}^1(G)$ is of the form $\mu_{r_0} = \omega_H$ for some compact subgroup H of G. This proves the statements (a) and (b).

(ii) Let $r \in \mathbb{M}$ with $r > r_0$ and $k \geq 1$ be such that $(k+1) r_0 > r \geq k r_0$. Putting $r' := (k+2) r_0 - r$ one obtains

$$\mu_r * \mu_{r'} = \mu_{r'} * \mu_r = \mu_{(k+2) r_0} = \omega_H,$$

whence for all $y_r \in \operatorname{supp}(\mu_r)$ and $y_{r'} \in \operatorname{supp}(\mu_{r'})$ we have $y_r, y_{r'} \in N(H)$,

$$\mu_r = \varepsilon_{y_r} * \omega_H = \omega_H * \varepsilon_{y_r} \quad \text{and} \quad \mu_{r'} = \varepsilon_{y_{r'}} * \omega_H = \omega_H * \varepsilon_{y_{r'}},$$

as follow from Theorem 1.2.13. We now put

$$y_r := y_{r_0 - r} \text{ for all } 0<r<r_0 \quad \text{and} \quad y_{-r} := y_r^{-1} \text{ for all } r \in \mathbb{M}_+.$$

The desired homomorphism $\phi: r \to x_r$ from $\mathbb{M}$ into $N(H)$ is constructed as follows: Let

$$\mathbb{M} := \left[\left\{\frac{1}{n_k} : k \geq 1\right\}\right]$$

with a sequence $(n_k)_{k \geq 1}$ in $\mathbb{N}$. We put

$$r_k := \frac{1}{n_1 \cdot \ldots \cdot n_k} \quad \text{for all } k \geq 1.$$

For m and $k \geq 1$ one defines

$$x_{r_k}^{(m)} := \begin{cases} y_{r_k} & \text{for } k \geq m \\ y_{r_m}^{n_{k+1} \cdot \ldots \cdot n_m} & \text{for } k < m. \end{cases}$$

Since $\prod_{k \geq 1} x_{r_k} H$ is a compact set, one can choose an accumulation point $(x_{r_1}, x_{r_2}, \ldots, x_{r_k}, \ldots)$ of the sequence

$$((x_{r_1}^{(m)}, x_{r_2}^{(m)}, \ldots, x_{r_k}^{(m)}, \ldots))_{m \geq 1}.$$

From $(x_{r_{k+1}}^{(m)})^{n_{k+1}} = x_{r_k}^{(m)}$ for all $m > k+1$ one concludes that $x_{r_{k+1}}^{n_{k+1}} = x_{r_k}$ for all $k \geq 1$. Moreover, one has $x_{r_k} K = y_{r_k} K$ for all $k \geq 1$. Hence, by Lemma 3.1.30 there exists a homomorphism $\phi: r \to x_r$ from $\mathbb{M}$ into $N(H)$ such that

$$\mu_r = \varepsilon_{\phi(r)} * \omega_H = \varepsilon_{x_r} * \omega_H$$

holds for all $r \geq r_0$. The rest is clear. □

3.6.3. Let G be a locally compact group with topology $\mathcal{T}$, H a $\mathcal{T}$-compact subgroup of G and $\mathcal{T}_{N(H)}$ the restriction of $\mathcal{T}$ to $N(H)$.

Definition. By $\bar{\mathcal{T}}$ we denote the topology in $N(H)$ generated by the $\mathcal{T}_{N(H)}$-open subsets of $N(H)$ and all cosets xH with $x \in N(H)$. The group $N(H)$ furnished with the topology $\bar{\mathcal{T}}$ will be abbreviated by $\bar{G}_H$.

Clearly, $N(H)$ is a $\bar{\mathcal{T}}$-locally compact space since xH is a compact and open subset of $N(H)$ for all $x \in N(H)$. Moreover, $\bar{G}_H$ is a topological group since for all $x \in N(H)$ the mapping $\tau_x: h \to xhx^{-1}$ from H into itself is continuous. Hence, $\bar{G}_H$ becomes a locally compact group with the following properties:

(1) $\bar{G}_H$ contains H as a compact open normal subgroup.

(2) The embedding $\mathrm{Id}: \bar{G}_H := (N(H), \bar{\mathcal{T}}) \to (N(H), \mathcal{T}_{N(H)})$ is continuous.

(3) $\mathrm{Res}_{xH}\, \mathrm{Id}: (xH, \mathrm{Res}_{xH} \bar{\mathcal{T}}) \to (xH, \mathrm{Res}_{xH} \mathcal{T}_{N(H)})$ is a topological isomorphism.

3.6.4 Definition. Let G be a locally compact group, $\mathbb{M}$ a submonogeneous subgroup of $\mathbb{Q}$ and f a homomorphism $\mathbb{M}_+ \to \mathcal{M}^1(G)$ of the form $r \to \mu_r$. f is called *trivial* if there exist a compact subgroup H of G and for every $r \in \mathbb{M}_+$ an $x_r \in G$ satisfying $\mu_r = \varepsilon_{x_r} * \omega_H$.

3.6.5 Remark. Let G be a locally compact group, $\mathbb{M}$ a submonogeneous subgroup of $\mathbb{Q}$ and f a noninjective homomorphism from $\mathbb{M}_+$ into $\mathscr{M}^1(G)$ of the form $r \to \mu_r$ such that an $r_0 \in \mathbb{M}$, a compact subgroup H of G and a homomorphism $\phi: r \to x_r$ from $\mathbb{M}$ into $N(H)$ exist satisfying (a) to (d) of Theorem 3.6.2. Then $(\mu_r)_{r \in \mathbb{M}_+}$ is trivial iff $(\mathrm{Id}(\mu_r))_{r \in \mathbb{M}_+}$ is trivial. This follows immediately from the fact that $\mathrm{supp}(\mu_r) \subset x_r H$ implies $\mu_r \in \mathscr{M}^1(N(H))$ for all $r \in \mathbb{M}_+$.

We recall that the groups $A(H) := \mathrm{Aut}(H)$ and $I(H) := \mathrm{Int}(H)$ of all and of inner automorphisms of H resp. are topological groups with respect to the usual topology $\mathscr{T}_A$.

In the sequel we shall consider the subgroup $A^G(H) := \{\tau_x : x \in N(H)\}$ of $A(H)$, which clearly contains $I(H)$ as a subgroup.

3.6.6 Properties. Let H be a compact subgroup of a locally compact group G.

(1) The groups $A(H)/I(H)$ and $A^G(H)/I(H)$ are totally disconnected.

(2) The canonical homomorphisms $\xi: N(H) \to A^G(H)$ defined by $\xi(x) = \tau_x$ for all $x \in N(H)$ and $\zeta := p \circ \xi: N(H) \to A^G(H)/I(H)$ (with the canonical epimorphism $p: A^G(H) \to A^G(H)/I(H)$) possess $Z_G(H)$ and $Z_G(H)H$ resp. as their kernels.

(3) The factor group $N(H)/Z_G(H)H$ has a basis of compact open subgroups and

(4) $N(H)_0 \subset Z_G(H)H$.

Concerning the proofs of these properties we mention Theorem 1 of [270] as reference for (1), the statements in (2) can be easily verified, (3) follows from (7.7) of [218] since $N(H)/Z_G(H)H$ is totally disconnected, and (4) follows from (1) and (2) since $\xi((N(H))_0)$ is connected in $A^G(H)/I(H)$ and is therefore trivial.

3.6.7 Theorem. *Let G be a locally compact group, $\mathbb{M}$ a submonogeneous subgroup of $\mathbb{Q}$, f a noninjective homomorphism $r \to \mu_r$ from $\mathbb{M}_+$ into $\mathscr{M}^1(G)$, and let $r_0 \in \mathbb{M}_+$, the compact subgroup H of G and the homomorphism $\phi: r \to x_r$ from $\mathbb{M}$ into $N(H)$ be as in Theorem 3.6.2. We define a homomorphism $F: \mathbb{M}_+ \to A^G(H)/I(H)$ as the composition $\zeta \circ \phi = p \circ \xi \circ \phi$ of the mapping $\zeta := p \circ \xi$ introduced in Property (2) of 3.6.6 with ϕ. If $(\mu_r)_{r \in \mathbb{M}_+}$ is nontrivial, then F has a cyclic kernel.*

Proof. One first observes that $\mathbb{M}' := \ker F$ is submonogeneous or cyclic since every subgroup of a submonogeneous group is of this form. Moreover, $\mathbb{M}' = \{r \in \mathbb{M} : \tau_{x_r} \in I(H)\}$. In particular, $x_{r_0} \in H$ implies $r_0 \in \mathbb{M}'$, so that without loss of generality we assume that $r_0 = 1$. We now prove the negation of the statement in the formulation of the theorem. Let $\mathbb{M}'$ be submonogeneous and noncyclic and let $\{m_j : j \geq 1\}$ be a generator of $\mathbb{M}'$. For every $r \in \mathbb{M}'$ there exists a $y_r \in H$ with $\tau_{x_r} = \tau_{y_r}$ and hence, with $\tau_{x_r}(\lambda) = \tau_{y_r}(\lambda)$ for all $\lambda \in \mathscr{M}^1(H)$. In particular, we obtain for all $r, s \in \mathbb{M}'$ the equality

$$\tau_{x_r}(\mu_s * \varepsilon_{x_s^{-1}}) = \tau_{y_r}(\mu_s * \varepsilon_{x_s^{-1}})$$

and therefore for every $j \geq 1$ with $r_j := \dfrac{1}{m_1 \cdot \dots \cdot m_j}$ the following chain of equalities:

$$
\begin{aligned}
(\mu_{r_j} * \varepsilon_{x_{r_j}^{-1}} * \varepsilon_{y_{r_j}})^k &= (\mu_{r_j} * \varepsilon_{x_{r_j}^{-1}}) * \tau_{y_{r_j}}(\mu_{r_j} * \varepsilon_{x_{r_j}^{-1}}) * \cdots * \tau_{y_{r_j}}^{k-1}(\mu_{r_j} * \varepsilon_{x_{r_j}^{-1}}) * \varepsilon_{y_{r_j}^k} \\
&= (\mu_{r_j} * \varepsilon_{x_{r_j}^{-1}}) * \tau_{x_{r_j}}(\mu_{r_j} * \varepsilon_{x_{r_j}^{-1}}) * \cdots * \tau_{x_{r_j}}^{k-1}(\mu_{r_j} * \varepsilon_{x_{r_j}^{-1}}) * \varepsilon_{y_{r_j}^k} \\
&= (\mu_{r_j} * \varepsilon_{x_{r_j}^{-1}}) * \varepsilon_{x_{r_j}} * (\mu_{r_j} * \varepsilon_{x_{r_j}^{-1}}) * \varepsilon_{x_{r_j}^{-1}} * \varepsilon_{x_{r_j}^2} * (\mu_{r_j} * \varepsilon_{x_{r_j}^{-1}}) * \varepsilon_{x_{r_j}^{-2}} * \cdots \\
&\quad * \varepsilon_{x_{r_j}^{k-1}} * (\mu_{r_j} * \varepsilon_{x_{r_j}^{-1}}) * \varepsilon_{x_{r_j}^{-k+1}} * \varepsilon_{y_{r_j}^k} \\
&= \mu_{r_j}^k * \varepsilon_{x_{r_j}^{-k}} * \varepsilon_{y_{r_j}^k} = \mu_{kr_j} * \varepsilon_{x_{-kr_j}} * \varepsilon_{y_{r_j}^k}.
\end{aligned}
$$

Thus

$$
\begin{aligned}
&(\mu_{r_{j+1}} * \varepsilon_{x_{r_{j+1}}^{-1}} * \varepsilon_{y_{r_{j+1}}})^{m_{j+1}} = \mu_{r_j} * \varepsilon_{x_{r_j}^{-1}} * \varepsilon_{y_{r_{j+1}}^{m_{j+1}}} \quad \text{and} \\
&(\mu_{r_j} * \varepsilon_{x_{r_j}^{-1}} * \varepsilon_{y_{r_j}})^{m_1 \cdot \dots \cdot m_j} = \omega_H \qquad (j \geq 1).
\end{aligned}
$$

Putting

$$\alpha_{r_j} := (\mu_{r_j} * \varepsilon_{x_{r_j}^{-1}} * \varepsilon_{y_{r_j}}) - \omega_H$$

we immediately observe that $\alpha_{r_j}^{m_1 \cdot \dots \cdot m_j} = 0$ or for any $\sigma \in \Sigma(H)$, that the matrix $\hat{\alpha}_{r_j}(\sigma)$ in $\mathfrak{M}(n(\sigma), \mathbb{C})$ is nilpotent. The Cayley-Hamilton theorem yields $\hat{\alpha}_{r_j}(\sigma)^k = 0$ for all $k \geq n(\sigma)$. Let $j \geq 1$ and $j_1 > j$ be such that $m_{j+1} \cdot \dots \cdot m_{j_1} \geq n(\sigma)$. Then

$$
\begin{aligned}
&\hat{\alpha}_{r_{j_1}}(\sigma)^{m_{j+1} \cdot \dots \cdot m_{j_1}} = 0 \quad \text{or} \\
&[(\mu_{r_{j_1}} * \varepsilon_{x_{r_{j_1}}^{-1}} * \varepsilon_{y_{r_{j_1}}})\hat{\ }(\sigma) - \hat{\omega}_H(\sigma)]^{m_{j+1} \cdot \dots \cdot m_{j_1}} \\
&= [(\mu_{r_j} * \varepsilon_{x_{r_j}^{-1}} - \omega_H) * \varepsilon_{y_{r_{j_1}}}^{m_{j+1} \cdot \dots \cdot m_{j_1}}]\hat{\ }(\sigma) = 0.
\end{aligned}
$$

Consequently,

$$(\mu_{r_j} * \varepsilon_{x_{r_j}^{-1}})\hat{\ }(\sigma) = \hat{\omega}_H(\sigma) \qquad \text{for all } \sigma \in \Sigma(H)$$

and thus

$$\mu_{r_j} = \varepsilon_{x_{r_j}} * \omega_H \qquad \text{for all } j \geq 1.$$

We have therefore shown that if $\mathbb{M}'(=\ker F)$ is submonogeneous but noncyclic, then $\mu_r = \varepsilon_{x_r} * \omega_H$ for all $r \in \mathbb{M}'_+$. Let finally $r \in \mathbb{M}_+ \setminus \mathbb{M}'_+$. Then there exists an $s \in \mathbb{M}'_+$ with $0 < s < r$, and hence such that

$$
\begin{aligned}
\mu_r &= \mu_s * \mu_{r-s} = \omega_H * \varepsilon_{x_s} * \mu_{r-s} \\
&= \varepsilon_{x_r} * (\varepsilon_{x_{r-s}^{-1}} * \omega_H * \varepsilon_{x_{r-s}}) * (\varepsilon_{x_{r-s}^{-1}} * \mu_{r-s}) \\
&= \varepsilon_{x_r} * \omega_H * (\varepsilon_{x_{r-s}^{-1}} * \mu_{r-s}) = \varepsilon_{x_r} * \omega_H.
\end{aligned}
$$

This follows from the fact that $x_{r-s} \in N(H)$ and $\operatorname{supp}(\varepsilon_{x_{r-s}^{-1}} * \mu_{r-s}) \subset H$ for all $r, s \in \mathbb{M}_+$ with $0 < s < r$. Altogether, one has $\mu_r = \varepsilon_{x_r} * \omega_H$ for all $r \in \mathbb{M}_+$ which implies the triviality of $r \to \mu_r$. □

3.6.8 Definition. For any submonogeneous subgroup $\mathbb{M}$ of $\mathbb{Q}$ we introduce the class $\mathbf{G}_{\mathbb{M}}$ of all locally compact groups G such that every homomorphism

$f: \mathbb{M}_+ \to \mathcal{M}^1(G)$ (of the form $r \to \mu_r$) is injective or trivial. We put

$$\mathbf{G}^* := \bigcap \{\mathbf{G}_{\mathbb{M}} : \mathbb{M} \text{ is a submonogeneous subgroup of } \mathbb{Q}\} \quad \text{and}$$

$$\overline{\mathbf{G}^*} := \bigcup \{\complement\, \mathbf{G}_{\mathbb{M}} : \mathbb{M} \text{ is a submonogeneous subgroup of } \mathbb{Q}\}.$$

Moreover, for every submonogeneous subgroup $\mathbb{M}$ of $\mathbb{Q}$ we define the class $\mathbf{H}_{\mathbb{M}}$ of all locally compact groups G such that every homomorphism $F: \mathbb{M}_+ \to A^G(H)/I(H)$ of the form given in Theorem 3.6.7 has a noncyclic kernel, and put

$$\mathbf{H}^* := \bigcap \{\mathbf{H}_{\mathbb{M}} : \mathbb{M} \text{ is a submonogeneous subgroup of } \mathbb{Q}\}.$$

3.6.9. The following **properties** of the classes just defined are established:
(1) For all submonogeneous subgroups $\mathbb{M}$ of $\mathbb{Q}$ we have $\mathbf{H}_{\mathbb{M}} \subset \mathbf{G}_{\mathbb{M}}$.
(2) $\mathbf{H}^* \subset \mathbf{G}^*$.
(3) For all submonogeneous subgroups $\mathbb{M}, \mathbb{M}'$ of $\mathbb{Q}$ with $\mathbb{M} \subset \mathbb{M}'$ we have $\mathbf{H}_{\mathbb{M}} \subset \mathbf{H}_{\mathbb{M}'}$ and $\mathbf{G}_{\mathbb{M}} \subset \mathbf{G}_{\mathbb{M}'}$.

In fact, Property (3) is obvious, and Properties (1) and (2) follow from Theorem 3.6.7.

3.6.10 Remark. It follows immediately from Theorem 3.6.7 that every discrete group and every locally compact Abelian group G is in $\mathbf{H}^*$ and therefore in $\mathbf{G}^*$. If G is discrete, all compact subgroups H of G are finite and the assertion follows. In the case of an Abelian group G we obtain $A^G(H) = \{e\}$ and therefore the triviality of the homomorphism F (appearing in the definition of $\mathbf{H}^*$).

In order to establish larger classes of examples in $\mathbf{H}^*$ we begin with a few lemmata.

3.6.11 Lemma. *Let Γ be a compact subgroup of $\mathfrak{GL}(n, \mathbb{C})$ and h a homomorphism from a submonogeneous subgroup $\mathbb{M}$ of $\mathbb{Q}$ into $\mathfrak{GL}(n, \mathbb{C})$ satisfying $h(1) \in \Gamma$. Then $h(\mathbb{M})$ is a relatively compact subgroup of $\mathfrak{GL}(n, \mathbb{C})$.*

Proof. For every $B \in \Gamma$ there exists a (nonsingular) matrix $V \in \mathfrak{GL}(n, \mathbb{C})$ such that $VBV^{-1} \in \mathfrak{U}(n)$. Hence without loss of generality we assume that Γ is a compact subgroup of $\mathfrak{U}(n)$. Since $h(1) \in \Gamma$, there exist by the spectral theorem a pairwise orthogonal resolution of unity $(P_j)_{j=1,\dots,p}$ and a sequence $(l_j)_{j=1,\dots,p}$ in $\mathbb{N}$ satisfying

$$h(1) = \sum_{j=1}^{p} e^{2\pi i l_j} P_j.$$

For every $s \in \mathbb{M}$ we define

$$U_s := h(s)\left(\sum_{j=1}^{p} e^{-2\pi i l_j s} P_j\right).$$

Since for $s \in \mathbb{M}$ the operator $h(s)$ commutes with $h(1)$ and with $h(1)^* = h(-1)$, it also commutes with $\sum_{j=1}^{p} e^{-2\pi i l_j s} P_j$. Hence, the mapping $s \to U_s$ is a homomorphism from $\mathbb{M}$ into Γ with $U_1 = E$. Let $(n_k)_{k \ge 1}$ be a sequence in $\mathbb{N}$ such that

$$\mathbb{M} = \left[\left\{\frac{1}{n_k} : k \ge 1\right\}\right].$$

Then for every $k\geq 1$ we have $U_{\frac{1}{n_k}}^{n_k}=E$, and by Theorem 1 of [240] there exist a pairwise orthogonal resolution of unity $(Q_j^{(k)})_{j=1,\ldots,p'}$ with $p':=p'(k)$ and a sequence $(k_j)_{j=1,\ldots,p'}$ in $\mathbb{N}$ satisfying

$$U_{\frac{1}{n_k}}=\sum_{j=1}^{p'} e^{2\pi i\frac{k_j}{n_k}}\, Q_j^{(k)}.$$

It should be noted that one does not necessarily have $Q_j^{(k)*}=Q_j^{(k)}$ for $k,j\geq 1$. From

$$U^{n_{k+1}}_{\frac{1}{n_i\cdot\ldots\cdot n_{k+1}}}=U_{\frac{1}{n_i\cdot\ldots\cdot n_k}}$$

for all $k\geq 1$ it follows that there are only finitely many different projectors $Q_j^{(k)}$ $(j=1,\ldots,p'')$ such that $U_s=\sum_{j=1}^{p''} e^{2\pi i k_j s}\, Q_j^{(k)}$. Therefore

$$\|U_s\|\leq\sum_{j=1}^{p''}\|Q_j^{(k)}\|\qquad\text{for all } s\in\mathbb{M}.$$

Thus $\{U_s: s\in\mathbb{M}\}$ is uniformly bounded, hence relatively compact, and consequently, $h(\mathbb{M})$ relatively compact in $\mathfrak{GL}(n,\mathbb{C})$. □

3.6.12 Lemma. *Let H be a compact connected Lie group. Then $A(H)/I(H)$ is discrete, and there exists a continuous homomorphism π from $A(H)$ into $\mathfrak{GL}(n,\mathbb{C})$ (for some $n\geq 1$) such that $\pi(A(H))/\pi(I(H))$ is discrete and* ker π *is finite.*

Proof. 1. By the structure theorem E for compact Lie groups there exist a connected Abelian subgroup $Z_0:=Z(H)_0$ of H and a (maximal) connected, semisimple compact normal subgroup S of H such that $H=Z_0\cdot S$ holds. Every automorphism $\sigma\in A(H)$ induces automorphisms $\sigma_1\in A(Z_0)$ and $\sigma_2\in A(S)$ in Z_0 and S resp. The mapping $\phi: A(H)\to A(Z_0)\times A(S)$ defined by

$$\phi(\sigma):=(\sigma_1,\sigma_2)\qquad\text{for all } \sigma\in A(H)$$

is an embedding of $A(H)$ into $A(Z_0)\times A(S)$ such that

$$\phi(I(H))=I(Z_0)\times I(S)=\{e\}\times I(S)\quad\text{holds.}$$

The last equality follows from the fact that Z_0 admits only the trivial inner automorphism. We conclude that $\phi(A(H))/\phi(I(H))$ is isomorphic to a group of the form $K:=(A_1/I(Z_0))\times(A_2/I(S))$ with subgroups $A_1\subset A(Z_0)$ and $A_2\subset A(S)$. If $A(Z_0)/I(Z_0)$ and $A(S)/I(S)$ are discrete, then the group K is also discrete and consequently, $A(H)/I(H)$ is discrete. It therefore remains to show that $A(Z_0)/I(Z_0)$ and $A(S)/I(S)$ are discrete.

(a) If $H=Z_0$, then H is a connected compact Abelian Lie group which by the structure theorem C is a torus group of dimension $m\geq 1$ say. Moreover, there exist a topological isomorphism between $A(H)$ and the matrix group $\{B\in\mathbb{M}(m,\mathbb{Z})\colon \det B=\pm 1\}$. Since this matrix group is obviously discrete, so is $A(H)=A(H)/I(H)$ and the proof is complete.

(b) If $H=S$, then H is a connected semisimple Lie group with Lie algebra $\mathfrak{L}:=\mathfrak{L}(H)$. Let $A(\mathfrak{L}):=\mathrm{Aut}(\mathfrak{L})$. Clearly, $A(\mathfrak{L})$ is isomorphic to a subgroup of $\mathfrak{GL}(l,\mathbb{R})$

for some $l \geq 1$ and hence is a Lie group, whose connected component A_0 coincides with the adjoint group of H defined as the group of automorphism of $\mathfrak{L}$ that are induced by an inner automorphism of H. Thus to every $\sigma \in A(H)$ there corresponds the induced automorphism $\sigma' \in A(\mathfrak{L})$, and we obtain a continuous homomorphism $\psi: A(H) \to A(\mathfrak{L})$ defined by $\psi(\sigma) := \sigma'$ for all $\sigma \in A(H)$. Plainly, $I(H) = \psi^{-1}(A_0)$. As $A(\mathfrak{L})$ is a Lie group and $A(\mathfrak{L})/A_0$ is discrete, $A(H)/I(H)$ is discrete.

2. The mapping $\pi := (\mathrm{Id} \times \psi) \circ \phi$ from $A(H)$ into $A(Z_0) \times A(\mathfrak{L}) \subset \mathfrak{GL}(n, \mathbb{C})$ is a continuous homomorphism of groups. Since $\pi(I(H)) = A(\mathfrak{L})_0$ is open in $\pi(A(H))$, $\pi(A(H))/\pi(I(H))$ is discrete, and $\ker \pi$ is finite. □

3.6.13 Lemma. *Let H be a compact Lie group. There exist a subgroup A_1 of $A(H)$ with $[A(H):A_1] < \infty$, a subgroup I^* of $I(H)$ with $[I(H):I^*] < \infty$ and a homomorphism $\psi: A_1 \to A(H_0)$ with $[\psi^{-1}(I(H_0)):I^*] < \infty$. Moreover $A(H)/I(H)$ is discrete, and there is a $d \geq 1$ such that for all $\tau \in A(H)$ and $\sigma \in I(H)$ one has $\tau^d \in A_1$, $\sigma^d \in I^*$ and $\psi(\sigma^d) \in I(H_0)$.*

Proof. We define $A_1 := \{\tau \in A(H) : \tau$ leaves the H_0-cosets of H invariant$\}$ and $I^* := \{\tau_x \in I(H) : x \in H_0\}$. Clearly, A_1 is a closed normal subgroup of $A(H)$, and I^* is a group of finite index. Since $A(H)/A_1$ is isomorphic to a subgroup of $\mathrm{Aut}(H/H_0)$, A_1 is of finite index in $A(H)$. Analoguously one sees that I^* is of finite index in $I(H)$. Let ψ be the continuous homomorphism from A_1 into $A(H_0)$ defined by $\psi(\sigma) := \sigma'$ for all $\sigma \in A_1$, where for every $\sigma \in A_1$ the automorphism σ' is the one induced by σ. We have $\psi(I^*) = I(H_0)$ and therefore $[\psi^{-1}(I(H_0)) : I^*] < \infty$. One observes that I^* is at the same time closed and open in $\psi^{-1}(I(H_0))$. By Lemma 3.6.12 $A(H_0)/I(H_0)$ is discrete, whence $\psi(A_1)/I(H_0)$ is discrete and also $A_1/\psi^{-1}(I(H_0))$. It follows that A_1/I^* is discrete and consequently $A(H)/I(H)$. The remaining statements in the lemma are easily deduced. □

3.6.14 Theorem. *Every Lie group G lies in* $\mathbf{H}^*$ *(and hence in* $\mathbf{G}^*$*).*

Proof. Let G be a Lie group, $\mathbb{M}$ a submonogeneous subgroup of $\mathbb{Q}$, H a compact subgroup of G (which again is a Lie group) and $F: \mathbb{M} \to A^G(H)$ a homomorphism of the form $F = \xi \circ \phi$ composed of a homomorphism $\phi: r \to x_r$ from $\mathbb{M}$ into $N(H)$ and the homomorphism $\xi: x \to \tau_x$ from $N(H)$ into $A^G(H)$ such that

$$F(1) = \xi \circ \phi(1) = \xi(x_1) = \tau_{x_1} \in I(H).$$

By Theorem 3.6.7 it suffices to show that $\ker F$ is noncyclic. From Lemma 3.6.13 we infer the existence of an integer $d \geq 1$ such that for all $\tau \in A(H)$, $\sigma \in I(H)$ one has $\tau^d \in A_1$, $\sigma^d \in I^*$ and $\psi(\sigma^d) \in I(H_0)$. Let $\mathbb{M}^d := \{rd : r \in \mathbb{M}\}$. Using the notation of Lemmata 3.6.12 and 3.6.13 we observe $A^G(H) \subset A(H)$, $A(H)^d \subset A_1$, $I(H)^d \subset I^*$ and by the very statements of the lemmata we obtain a homomorphism

$$J := \pi \circ \psi \circ F \quad \text{from } \mathbb{M}^d \text{ into } \mathfrak{GL}(n, \mathbb{C}) \text{ with}$$
$$J(d) = \pi \circ \psi \circ F(d) = \pi \circ \psi(\tau_{x_d}) = \pi \circ \psi(\tau_{x_1}^d) \in \pi(I(H_0)).$$

It follows that $J(d)$ is contained in a compact subgroup of $\mathfrak{GL}(n, \mathbb{C})$, $J(\mathbb{M}^d)/\pi(I(H_0))$ is discrete and $\ker J$ finite. By Lemma 3.6.11 $J(\mathbb{M}^d)/\pi(I(H_0))$ is finite. But then

$J(\mathbb{M}^d)/I(H)$ is finite since $\ker J$ is finite. This, however, implies that the homomorphism $J':\mathbb{M}^d \to A(H)/I(H)$ of the form $J':=p\circ F$ (with corresponding canonical mapping p) is noncyclic, which is the desired result. □

3.6.15. Closure properties of the class $\mathbf{G}_{\mathbb{M}}$ for a submonogeneous subgroup $\mathbb{M}$ of $\mathbb{Q}$.

(1) Let $G\in\mathbf{G}_{\mathbb{M}}$ and H a locally compact group such that there exists a continuous injective homomorphism from H into G. Then $H\in\mathbf{G}_{\mathbb{M}}$. In particular,

(2) every closed subgroup of a group $G\in\mathbf{G}_{\mathbb{M}}$ is itself in $\mathbf{G}_{\mathbb{M}}$.

(3) Let $G\in\mathbf{G}_{\mathbb{M}}$ and H a locally compact group such that there exists a continuous open homomorphism from G onto H with a compact kernel. Then $H\in\mathbf{G}_{\mathbb{M}}$. In particular,

(4) every factor group G/N by a compact normal subgroup N of a group $G\in\mathbf{G}_{\mathbb{M}}$ lies in $\mathbf{G}_{\mathbb{M}}$.

(5) Let G_α and G $(\alpha\in\mathbb{A})$ be locally compact groups and $\phi_\alpha: G\to G_\alpha$ a continuous homomorphism $(\alpha\in\mathbb{A})$ such that the family $(\phi_\alpha)_{\alpha\in\mathbb{A}}$ is consistent for the projective system $(G_\alpha, p_{\alpha\beta}, \mathbb{A})$ and separating. If $G_\alpha\in\mathbf{G}_{\mathbb{M}}$ for all $\alpha\in\mathbb{A}$, then $G\in\mathbf{G}_{\mathbb{M}}$.

Concerning the proofs of these properties it should be noted that (1) to (4) are obvious consequences of the definition of $\mathbf{G}_{\mathbb{M}}$ since in these cases the semigroup of measures in $\mathscr{M}^1(H)$ with compact support can be embedded into $\mathscr{M}^1(G)$ by an injective semigroup homomorphism. It remains to show (5). Let G be a locally compact group and $f:\mathbb{M}_+\to\mathscr{M}^1(G)$ a noninjective homomorphism of the form $r\to\mu_r$ with $r_0\in\mathbb{M}_+$, a compact subgroup H of G and a homomorphism $\phi: r\to x_r$ from $\mathbb{M}$ into $N(H)$ as in Theorem 3.6.2.

We note that $(\mu_r)_{r\in\mathbb{M}_+}$ is trivial iff $\mu_r=\mu_{r+r_0}$ for all $r\in\mathbb{M}_+$.

Let $r\to\mu_r$ be a noninjective homomorphism from $\mathbb{M}_+$ into $\mathscr{M}^1(G)$. Then for every $\alpha\in\mathbb{A}$ the mapping $r\to p_\alpha(\mu_r)$ from $\mathbb{M}_+$ into $\mathscr{M}^1(G_\alpha)$ is a homomorphism of semigroups which by

$$p_\alpha(\mu_{r_0})^2=p_\alpha(\mu_{r_0}^2)=p_\alpha(\mu_{r_0})$$

is noninjective. Since $G_\alpha\in\mathbf{G}_{\mathbb{M}}$ for all $\alpha\in\mathbb{A}$, we obtain

$$p_\alpha(\mu_r)=p_\alpha(\mu_{r+r_0}) \quad \text{for all } r\in\mathbb{M}_+,\ \alpha\in\mathbb{A}.$$

But by Theorem 3.6.2 $\operatorname{supp}(\mu_r)$ is compact for all $r\in\mathbb{M}_+$, whence it follows from [41], Chapitre III, §4, no 5, that μ_r is uniquely determined by $p_\alpha(\mu_r)$ $(\alpha\in\mathbb{A})$ for every $r\in\mathbb{M}_+$ or $\mu_r=\mu_{r+r_0}$ for all $r\in\mathbb{M}_+$, which shows that $G\in\mathbf{G}_{\mathbb{M}}$.

3.6.16 Definition. A locally compact group G is called *of type* (L) (in the sense of Iwasawa) if there exist a family $(G_\alpha)_{\alpha\in\mathbb{A}}$ of Lie groups and a family $(\phi_\alpha)_{\alpha\in\mathbb{A}}$ of continuous homomorphisms ϕ_α from G into G_α which separates the points of G.

3.6.17 Remark. Since by the theorem of Cartan a locally compact group L such that there exists a continuous homomorphism from L into a Lie group which is injective at least on a neighborhood of e is itself a Lie group, groups of type (L) can also be defined as locally compact groups G containing a family $(K_\alpha)_{\alpha\in\mathbb{A}}$ of closed normal subgroups K_α such that G/K_α is a Lie group $(\alpha\in\mathbb{A})$ and $\bigcap_{\alpha\in\mathbb{A}}K_\alpha=\{e\}$. Indeed, under the hypothesis of a group of type (L) the groups $G/\ker\phi_\alpha$ $(\alpha\in\mathbb{A})$ are Lie groups.

3.6.18. Further examples of groups in $\mathbf{G}^*$.

1. Every group G of type (L) lies in $\mathbf{G}^*$, in particular,
2. every Lie projective (locally compact) group is in $\mathbf{G}^*$.
3. Every almost periodic (locally compact) group is in $\mathbf{G}^*$, so that
4. locally compact Abelian groups and arbitrary compact groups, and also arbitrary connected locally compact groups belong to $\mathbf{G}^*$.

The argument for Example 1 is based on Theorem 3.6.14, Property (5) of 3.6.15 and the fact that the system $(\phi_\alpha)_{\alpha\in\mathbb{A}}$ of continuous homomorphisms ϕ_α from G into a Lie group $G_\alpha\,(\alpha\in\mathbb{A})$ is in fact consistent and separating. It remains to show that the family of all closed normal subgroups K such that G/K is a Lie group becomes an ordered system with respect to inclusion. This can be done in the following manner. For $i=1,\dots,n$ let K_i be a closed normal subgroup of G such that G/K_i is a Lie group. The set $\{p_1,\dots,p_n\}$ of canonical projections $p_i: G\to G/K_i$ induces a continuous homomorphism from G into the Lie group $\prod_{i=1}^n G/K_i$ whose kernel is $\bigcap_{i=1}^n K_i$. Cartan's theorem implies that $G/\bigcap_{i=1}^n K_i$ is a Lie group.

The proof of Example 2 is trivial since Lie projective groups are clearly of type (L).

Almost periodic groups admit by 1.3.1 a separating system of representations in $\mathrm{Rep}(G)$ or $\Sigma(G)$, and hence are of type (L) and thus in $\mathbf{G}^*$. The groups mentioned in Example 4 are almost periodic and Lie projective resp., and are therefore in $\mathbf{G}^*$.

3.6.19 Theorem. *Let G be a locally compact group, $\mathbb{M}$ a submonogeneous subgroup of $\mathbb{Q}$, f a noninjective homomorphism $r\to\mu_r$ from $\mathbb{M}_+$ into $\mathcal{M}^1(G)$. Furthermore let $r_0\in\mathbb{M}_+$, the compact subgroup H of G and the homomorphism $\phi: r\to x_r$ from $\mathbb{M}$ into $N(H)$ be as in Theorem 3.6.2. We further have a homomorphism $F:\mathbb{M}\to A^G(H)/I(H)$ defined in Theorem 3.6.7. If $F(\mathbb{M})$ is precompact in $A^G(H)/I(H)$, then $(\mu_r)_{r\in\mathbb{M}_+}$ is trivial.*

Proof. 1. We first show the general statement that for any locally compact group G and a compact subgroup H of G with precompact $A^G(H)$ the group $\bar{G}_H$ introduced in 3.6.3 is Lie projective. In fact, one notes that the automorphism groups $A^G(H)$ and $A^{\bar{G}_H}(H)$ induced by $N(H)$ in H and $\bar{G}_H$ resp. are topologically isomorphic. Hence, both are precompact in $A(H)$. Since $\bar{G}_H$ is locally compact, the compact group H possesses small $A^G(H)$-invariant neighborhoods. Since H is an open normal subgroup in $\bar{G}_H$, we obtain that $\bar{G}_H$ has small invariant neighborhoods and is therefore Lie projective.

2. By Part 1 we assume without loss of generality that G is itself Lie projective. Putting

$$L:=\bigcup_{r\in\mathbb{M}} x_r H\subset\bar{G}_H$$

we observe that $\mathrm{supp}(\mu_r)\subset L$ and hence $\mu_r\in\mathcal{M}^1(L)$ for all $r\in\mathbb{M}_+$. But since $L=\bar{L}_H$ and $A^{\bar{L}_H}(H)$ is precompact, Part 1 of this proof implies the Lie projectivity of L. It follows $L\in\mathbf{G}^*$ by Example 2 of 3.6.18, and thus that $(\mu_r)_{r\in\mathbb{M}_+}$ is trivial. □

3.6.20 Theorem. *Let G be a locally compact group, $\mathbb{M}$ a submonogeneous subgroup of $\mathbb{Q}$ furnished with the topology induced from $\mathbb{R}$ and $(\mu_r)_{r\in\mathbb{M}_+}$ a continuous ($\mathbb{M}_+$-)*

submonogeneous convolution semigroup in $\mathscr{M}^1(G)$. Then $(\mu_r)_{r\in\mathbb{M}_+}$ is either injective or trivial.

Proof. Let $f:\mathbb{M}_+\to\mathscr{M}^1(G)$ be defined by $f(r):=\mu_r$ for all $r\in\mathbb{M}_+$. We suppose that f is noninjective and show that then f is trivial. By Theorem 3.6.2 there exist an $r_0:=1\in\mathbb{M}_+$, a compact subgroup H of G with $f(1)=\mu_1=\omega_H$ and a homomorphism $\phi:r\to x_r$ from $\mathbb{M}$ into $N(H)$ such that $f(r)=\mu_r=\varepsilon_{x_r}*\omega_H$ holds for all $r\geq r_0=1$. Obviously the mapping $\Delta:\mathbb{M}\to N(H)/H$ defined by $\Delta(r):=x_rH$ for all $r\in\mathbb{M}$ is continuous on $\mathbb{M}\cap[1,\infty[$. Since $\Delta(\mathbb{M})=\Delta([1,2])$, $\Delta(\mathbb{M})$ is relatively compact in $N(H)/H$.

Let $F:=p\circ\xi\circ\phi$ be the continuous homomorphism from $\mathbb{M}$ into $A^G(H)/I(H)$ composed of ϕ with $\xi:x\to\tau_x$ from $N(H)$ into $A^G(H)$ and the projection $p:\sigma\to\sigma I(H)$ from $A^G(H)$ into $A^G(H)/I(H)$. Then $F(\mathbb{M})$ is relatively compact in $A^G(H)/I(H)$. From Theorem 3.6.19 we infer that $(\mu_r)_{r\in\mathbb{M}_+}$ is trivial. □

3.6.21 Definition. A locally compact group G *admits the property* (M) if for every isotone sequence $(K_n)_{n\geq1}$ of compact subgroups of G the set $\bigcup_{n\geq1}K_n$ is precompact, the *property* (MA) if for every isotone sequence $(K_n)_{n\geq1}$ of compact Abelian subgroups of G the set $\bigcup_{n\geq1}K_n$ is precompact, the *property* (MI) if for every isotone sequence $(K_n)_{n\geq1}$ of compact Abelian subgroups of G the image of $\bigcup_{n\geq1}K_n$ in $I(G)$ is precompact, and the *property* (R) if for all compact subsets A of G the root set $R(A):=\{x\in G: x^n\in A \text{ for some } n\geq1\}$ of A is relatively compact in G.

The classes of groups admitting the properties (M), (MA), (MI) and (R) will be denoted by [M], [MA], [MI] and [R] resp. We have $[\mathrm{R}]\subset[\mathrm{M}]\subset[\mathrm{MA}]\subset[\mathrm{MI}]$ and $\mathbf{R}_0\subset[\mathrm{R}]$. It is sufficient to treat the first and the last inclusion. Concerning the first one we observe that for any compact neighborhood $U\in\mathfrak{V}(e)$ and each $x\in B(G)$ there exists a $k:=k(U,x)\geq1$ satisfying $x^k\in U$. This implies that $R(U)\supset B(G)$. Since by assumption $R(U)$ is relatively compact, the assertion follows. The last inclusion follows immediately from Theorem 3.1.4.

3.6.22 Examples. 1. Every compactly generated locally compact Abelian group belongs to [M].
2. Every connected almost periodic group is in [M].
3. Every closed subgroup of $\mathfrak{GL}(n,\mathbb{C})$ (for $n\geq1$) lies in [MA].
4. Any connected locally compact group belongs to [MI].

The proofs of Examples 1 and 2 consist of an application of the structure theorems A and L resp. Concerning Example 3 we note that by Corollary 4.2 of [494] it suffices to show that any Lie group G such that every discrete Abelian torsion subgroup of G is finite belongs to [MA]. But from the structure theorem A we conclude that an Abelian closed subgroup H of a Lie group of the form $\mathbb{R}^m\times\mathbb{T}^n\times D$ with $m,n\geq0$ and a discrete group D is compactly generated iff D is finitely generated. If the torsion part of D is finite, then the group H possesses a unique maximal compact subgroup, and so $H\in[\mathrm{MA}]$. As to Example 4 we restrict ourselves without loss of generality to connected Lie groups G. But then $A(G)$ is isomorphic to a closed subgroup of $\mathfrak{GL}(n,\mathbb{C})$ (for some $n\geq1$) (see [246], IX, Theorem 1.2), and Example 3 yields the assertion. □

3.6.23 Theorem. $[\mathrm{MI}] \subset \mathbf{G}^*$.

Proof. Let $f: r \to \mu_r$ be a noninjective homomorphism from the semigroup $\mathbb{M}_+$ of a submonogeneous subgroup $\mathbb{M}$ of $\mathbb{Q}$ into the semigroup $\mathcal{M}^1(G)$ for a locally compact group G. As in the proof of Theorem 3.6.20 we have $r_0 := 1 \in \mathbb{M}_+$, a compact subgroup H of G with $f(1) = \mu_1 = \omega_H$ and a homomorphism $\phi: r \to x_r$ from $\mathbb{M}$ into $N(H)$ satisfying

$$f(r) = \mu_r = \varepsilon_{x_r} * \omega_H \qquad \text{for all } r \geq r_0 = 1.$$

We define the homomorphism $F' := \xi \circ \phi : \mathbb{M} \to A^G(H)$ as the composition of ϕ with the homomorphism $\xi: x \to \tau_x$ from $N(H)$ into $A^G(H)$. If $F'(\mathbb{M})$ is precompact in $A^G(H)$, then Theorem 3.6.19 implies that $r \to \mu_r$ is trivial. It therefore remains to deduce from the assumption $G \in [\mathrm{MI}]$ that $F'(\mathbb{M})$ is in fact precompact in $A^G(H)$. The homomorphism $\xi: N(H) \to A^G(H)$ can be factorized via natural embeddings with homomorphisms $i: x \to \tau'_x$ from $N(H)$ into $I(G)$ and $j: \sigma' \to \sigma$ from $I(G)$ into $A^G(H)$ such that $\xi = j \circ i$ holds. Let $(n_k)_{k \geq 1}$ be a sequence in $\mathbb{N}$ satisfying

$$\mathbb{M} = \bigcup_{k \geq 1} \left[\left\{\frac{1}{n_1 \cdot \ldots \cdot n_k}\right\}\right].$$

For every $k \geq 1$ we put

$$\mathbb{M}_k := \left[\left\{\frac{1}{n_1 \cdot \ldots \cdot n_k}\right\}\right]$$

and observe that $(\mathbb{M}_k)_{k \geq 1}$ is an isotone sequence of sets with $\bigcup_{k \geq 1} \mathbb{M}_k = \mathbb{M}$. Plainly, for every $k \geq 1$ the set $\phi(\mathbb{M}_k)$ is a relatively compact Abelian subgroup and hence $K_k := \overline{\phi(\mathbb{M}_k)}$ a compact Abelian subgroup of G with $K_k \subset K_{k+1}$ for all $k \geq 1$. Since $G \in [\mathrm{MI}]$, $i(\bigcup_{k \geq 1} K_k)$ is precompact in $I(G)$ and hence also $i \circ \phi(\mathbb{M})$. But the restriction of j to the subgroup $\{\tau_x \in I(G): x \in N(H)\}$ of $I(G)$ is continuous. Therefore $j \circ i \circ \phi(\mathbb{M}) = \xi \circ \phi(\mathbb{M}) = F'(\mathbb{M})$ is precompact in $A^G(H)$. This was to be proved. ▯

3.6.24 Corollary. $[\mathrm{R}] \subset \mathbf{G}^*$.

The *proof* is clear since $[\mathrm{R}] \subset [\mathrm{MI}]$. ▯

After having studied in detail the class $\mathbf{G}^*$, it remains to comment on its complement $\overline{\mathbf{G}^*}$.

3.6.25 Example. Let $\mathbb{Q}_1 := \mathbb{Q}/\mathbb{Z}$ and let $H := \mathbb{T}^{\mathbb{Q}_1}$ be a compact torus group whose elements will be regarded as mappings $f: \mathbb{Q}_1 \to \mathbb{T}$.

For every $r \in \mathbb{Q}$ we define the automorphism $R_{-r} \in A(H)$ by

$$R_{-r} f(t) := f(t - r) \qquad \text{for all } t \in \mathbb{Q}_1 \text{ (with } t - r \in \mathbb{Q}_1).$$

Considering $\mathbb{Q}$ as a topological group with the discrete topology one obtains a continuous homomorphism $\eta: \mathbb{Q} \to A(H)$ defined by $\eta(r) := R_{-r}$ for all $r \in \mathbb{Q}$ and a topological group $G_1:=H\times_\eta \mathbb{Q}$ which is the semidirect product of H with $\mathbb{Q}$ with respect to η. Clearly, H can be viewed as a compact normal subgroup of G_1 and of G. Since $\mathbb{Z}$ can be regarded as a discrete normal subgroup of G_1, we obtain a group $G:=G_1/\mathbb{Z}$ serving our purposes. We shall construct a convolution semigroup $(\mu_r)_{r\in\mathbb{Q}_+^*}$ in $\mathscr{M}^1(G)$ with $\mu_1=\omega_H$, but such that for $r\in\mathbb{Q}_+^*\cap[0,1[$ the measure μ_r is not of the form $\varepsilon_{x_r}*\omega_H$ for some $x_r\in G$.

First of all we observe that for every $r\in\mathbb{Q}$ the automorphism R_{-r} of H induces an isomorphism in $\mathscr{M}^1(H)$ defined by

$$R_{-r}(\mu)(B):=\mu(R_{-r}(B)) \quad \text{for all } B\in\mathfrak{B}(H).$$

Given $r,s\in\mathbb{Q}\cap[0,1]$ with $r<s$ we define in H the compact subgroup

$$H_{r,s}:=\{f\in H: f(t)=1 \text{ for all } t\in\complement(\mathbb{Q}\cap[r,s[)\}$$

and in $\mathscr{M}^1(H)$ the measure $\lambda_{r,s}:=\omega_{H_{r,s}}$ with $\lambda_{0,1}:=\omega_H$. One easily sees that $R_{-r}(H_{0,t})=H_{r,r+t}$ and $R_{-r}(\lambda_{0,t})=\lambda_{r,r+t}$ for all $r,t\in\mathbb{Q}\cap[0,1]$ such that $r+t\in[0,1]$. From $H_{0,s}=H_{0,r}\times H_{r,s}$ it follows directly that

$$\lambda_{0,s}=\lambda_{0,r}*\lambda_{r,s}=\lambda_{r,s}*\lambda_{0,r} \quad (r,s\in\mathbb{Q}\cap[0,1] \text{ with } r<s).$$

We now define measures μ_r' in $\mathscr{M}^1(G_1)$ by

$$\mu_r':=\begin{cases}\varepsilon_r*\lambda_{0,r}, & \text{if } r\in\mathbb{Q}\cap]0,1]\\ \varepsilon_r*\lambda_{0,1}, & \text{if } r\in\mathbb{Q}\cap[1,\infty[.\end{cases}$$

For $s:=r+t\in\mathbb{Q}\cap]0,1]$ we obtain

$$\begin{aligned}\mu_t'*\mu_r'&=\varepsilon_t*\varepsilon_r*(\varepsilon_{-r}*\lambda_{0,t}*\varepsilon_r)*\lambda_{0,r}=\varepsilon_t*\varepsilon_r*R_{-r}(\lambda_{0,t})*\lambda_{0,r}\\&=\varepsilon_{t+r}*\lambda_{r,s}*\lambda_{0,r}=\varepsilon_s*\lambda_{0,s}=\mu_s' \quad \text{and}\end{aligned}$$

for $r,t\in\mathbb{Q}\cap]0,1[$ and $s:=r+t\in\mathbb{Q}\cap]1,\infty[$ similarly

$$\begin{aligned}\mu_t'*\mu_r'&=\mu_t'*\mu_{1-t}'*\mu_{s-1}'=\varepsilon_1*\lambda_{0,1}*\varepsilon_{s-1}*\lambda_{0,s-1}\\&=\varepsilon_s*R_{-(s-1)}(\lambda_{0,1})*\lambda_{0,s-1}=\varepsilon_s*\lambda_{0,1}*\lambda_{0,s-1}=\varepsilon_s*\lambda_{0,1}=\mu_s'.\end{aligned}$$

Analogously $\mu_t'*\mu_r'=\mu_{t+r}'$ is obtained also for $t,r\in\mathbb{Q}\cap[1,\infty[$. Putting all of this together $(\mu_r')_{r\in\mathbb{Q}_+^*}$ is shown to be a convolution semigroup in $\mathscr{M}^1(G_1)$. Denoting the canonical projection from G_1 onto G by p and identifying H with its isomorphic image $p(H)$ in G one obtains a convolution semigroup $(\mu_r)_{r\in\mathbb{Q}_+^*}$ in $\mathscr{M}^1(G)$ defined by $\mu_r:=p(\mu_r')$ for all $r\in\mathbb{Q}_+^*$ with the property $\mu_1=\omega_H$. But for $r\in\mathbb{Q}\cap]0,1[$ the measure μ_r is not of the form $\varepsilon_{x_r}*\omega_H$ with $x_r\in G$. Thus the assertion that $G\in\complement\mathbf{G}_{\mathbb{Q}}$ has been proved.

3.6.26 Remark. If G is the group constructed in the preceding example, then there exists a nontrivial, noninjective convolution semigroup $(\mu_r)_{r\in\mathbb{Q}_+^*}$ in $\mathscr{M}^1(G)$ such that $[\mathrm{supp}(\mu_r)]^-=x_rH$ for all $r\in\mathbb{Q}_+^*$. Let $(\mu_r)_{r\in\mathbb{Q}_+^*}$ be as defined above, $\alpha\in\mathbb{R}_+^*$ and

$$\mathring{\mu}_r:=e^{-\alpha r}\mu_r+(1-e^{-\alpha r})\mu_{r+1} \quad \text{for all } r\in\mathbb{Q}_+^*.$$

Then $(\mathring{\mu}_r)_{r\in\mathbb{Q}_+^*}$ is a nontrivial, noninjective convolution semigroup in $\mathscr{M}^1(G)$ with the properties

$$\mathring{\mu}_r = \mu_r \quad \text{for all } r\in\mathbb{Q}\cap[1,\infty[\quad \text{and}$$
$$\operatorname{supp}(\mathring{\mu}_r) = \operatorname{supp}(\mu_{r+1}) = H \quad \text{for all } r\in\mathbb{Q}_+^*.$$

3.6.27 Discussion of a generalization of Example 3.6.25. Let $\mathbb{M}$ be a noncyclic submonogeneous subgroup of $\mathbb{Q}$ with $1\in\mathbb{M}$, $M_1 := \mathbb{M}/\mathbb{Z}$, $H := \mathbb{T}^{M_1}$ and $\Gamma_1 := H \times_\eta \mathbb{M}$, where η is the continuous homomorphism $r \to R_{-r}$ from $\mathbb{M}$ into $A(H)$ with

$$R_{-r} f(t) := f(t-r) \quad \text{for all } t\in M_1 \quad (\text{and } t-r\in M_1)\ (r\in\mathbb{M}).$$

Plainly $H_1 := \{(f,0)\in\Gamma_1 : f\in H\}$ is a compact open normal subgroup of Γ_1 isomorphic to H and $\bar{\Gamma}_1 := \{(0,k)\in\Gamma_1 : k\in\mathbb{Z}\}$ is a discrete normal subgroup of Γ_1.

(1) $\Gamma_1\in[M]$, whence $\Gamma_1\in\mathbf{G}^*$.

We now define the desired group $\Gamma := \Gamma_1/\mathbb{Z}$, which is the generalization of G introduced in 3.6.25.

(2) Γ is locally compact since $\Gamma = \bar{\Gamma}_H$.

(3) One shows as in 3.6.25 that Γ is in $\complement\mathbf{G}_{\mathbb{M}}$, and hence is in $\overline{\mathbf{G}^*}$.

(4) Γ has invariant neighborhoods of the identity.

(5) Γ is unimodular.

The last two properties follow from the fact that Γ_1 contains a compact open subgroup.

(6) $\Gamma\in\mathbf{R}$.

This is a consequence of Property (2) of 3.1.2, since $\Gamma/H \cong M_1\in\mathbf{R}$.

(7) Γ is class compact and admits a compact commutator subgroup.

These properties are implied by the facts $(h,r)\circ(k,s)\circ(h,r)^{-1}\circ(k,s)^{-1}\in H$ and $(h,r)^{-1}\circ(k,s)\circ(h,r)\in(e,s)H$, valid for all

$$(h,r),(k,s)\in\Gamma = H\times_\eta \mathbb{M}/\mathbb{Z} \cong H\times_\eta M_1.$$

Finally (8). Since M_1 is Abelian and H a compact normal subgroup of G, we have that Γ is amenable ([178], Theorem 2.3.3) and of type I ([227], 11.3.5).

In order to complete our list of properties of the class $\overline{\mathbf{G}^*}$ it should be mentioned that by Example 1 of 3.6.18 Γ is not of type (L).

References and Comments

R 3.1 Root Compact Groups

Root compact locally compact groups are the discovery of Böge. In his paper [31] he gives the definitions of root compactness and strong root compactness, collects a few permanence properties of the respective classes, discusses first examples and indicates the relationship between root compactness and the embedding problem for probability measures, which is the topic of this chapter.

Since the work of Böge root compact groups have become more and more fundamental for the solution of the embedding problem. A fairly complete account on the subject comprising the research done up to the year 1971 has been published by Heyer [230]. The paper of Siebert [439] discusses details on root compact groups and the general embedding theorem even for arbitrary, not necessarily locally compact groups. Our presentation absorbs most of the results of Siebert [439], but is restricted to the locally compact case.

Example 2 of 3.1.3 has been computed in Böge [31]. Similarly, one can show that the group $\mathbb{Q}_p$ of p-adic rational numbers (for a prime number $p \geq 2$) is root compact. We derive this result from the more general framework of Theorem 3.1.7 (see Example 12 of 3.1.9). The computation of Example 6 has been carried out by Siebert in connection with [439]. The example is designed to show that there are root compact nilpotent Lie groups. In a similar fashion one establishes the assertion in Example 8 of 3.1.3, that there exists a non-almost periodic Lie group which is root compact. The proof of the implication (i) $\Rightarrow$ (ii) of Theorem 3.1.4 appears in Böge [31]. The full theorem has been shown by Siebert [439]. Theorem 3.1.5 is a generalization of a result of Burrell and McCrudden [52], [53], which has been incorporated as (ii) of Remark 3.1.6. In Burrell and McCrudden [52] convenient locally compact groups are introduced which appear to have properties similar to those of root compact groups. It turns out that root compact groups are necessarily convenient, so convenience is the weaker notion. The origin of Theorem 3.1.7 is in § 5 of Böge [31]. A complete proof of Böge's result was given by Schmetterer [429].

Theorem 3.1.8 permits one to discuss further examples of root or non-root compact groups. Example 11 of 3.1.9 of a non-root compact group was brought to the author's attention by Böge. It has been reproduced in Heyer [230] and Schmetterer [430]. From [210], Footnote 1, one learns that the example was known previously to Hofmann [251]. For a more complete analysis of the example one may read Hazod and Schmetterer [210] or the exposition in § 3. Definition 3.1.10 contains a strong form of root compactness, which will turn out later to be the proper condition for a full embedding theory. The chain of implications in Theorem 3.1.13 is a result in Siebert [439]. In general, one does not obtain equivalence, as is pointed out in Remark 3.1.15. The implication (iii) $\Rightarrow$ (iv) of Theorem 3.1.13 was communicated to Siebert by Hazod. It implies that strongly root compact and class compact groups admit largest compact normal subgroups (Remark 3.1.14). There is a large number of different notions (in the sense) of root compactness arising from the statements of Theorems 3.1.4 and 3.1.13. These notions have been efficiently analyzed by Yuan in Chapter III of [509]. The fact that any connected, nilpotent Lie group is root compact is due to Burrell and McCrudden [52], [53]. The result was strengthened to strong root compactness by Siebert in [439] using heavily the structure of nilpotent Lie groups. Theorem 3.1.17 contains a further generalization. Every nilpotent, compactly generated locally compact group is strongly root compact. The proof (especially of Lemma 3.1.16) inspired by an argument of Guivarc'h in [187], p. 349 has been given by Siebert. The following discussion aiming at Theorem 3.1.23, which claims that every connected solvable Lie group whose Lie algebra has only real roots is strongly root compact is contained in Yuan [509] and [512]. The preparations of 3.1.19

are gathered from the papers [72] and [116] by Chevalley and Dixmier resp., in which the structure theory of solvable Lie groups has been developed. Lemma 3.1.20 is the crucial step in the proof of Theorem 3.1.23. It can be used for a direct proof of the fact that the group of proper affine mappings of $\mathbb{R}$ is actually strongly root compact. In Remark 3.1.26 two examples from Yuan [512] show that there are connected solvable Lie groups which are not root compact.

The first example mentioned concerns a group G which is also simply connected, but not exponential in the sense that $\exp_G$ is surjective (see Hochschild [246], p. 140, Example 1).

In the sequel the set $\mathscr{I}(G)$ of infinitely divisible probability measures on a locally compact group G is studied with respect to closure in $\mathscr{M}^1(G)$ and rational embedding. These topics motivate the introduction of root compact groups. First of all it should be noted that the notion of an infinitely divisible probability measure on $\mathbb{R}$ was invented by de Finetti in his early papers [142], [143], [144]. The classical theory of infinitely divisible measures leads to the solution of the central limit problem as we know it today. Fairly modern books on the subject are Gnedenko and Kolmogorov [172], Loève [325] and also Cuppens [104], where de Finetti's work receives appropriate emphasis. For a finite group G the set $\mathscr{I}(G)$ was studied for the first time by Vorobev [492] and Böge [30], for locally compact Abelian groups by Kloss [305], [306], and for general compact groups by Carnal [64]. A survey of what has been done in the theory up to 1971 is contained in Heyer [230]. Theorem 3.1.28 states that for root compact groups G the set $\mathscr{I}(G)$ is $\mathscr{T}_v$-closed in $\mathscr{M}^1(G)$. Its proof in its final form is due to Siebert [442]. For σ-compact locally compact groups a proof can be found in Siebert [439]. A less direct proof of the general statement has been given by Hazod [202]. The next aim is Theorem 3.1.32 on the embedding of infinitely divisible measures in $\mathscr{M}^1(G)$ into rational convolution semigroups in $\mathscr{M}^1(G)$. This is the first step on the way (to be terminated in § 4) towards the general continuous embedding. In view of the generality needed later we start with the notions of a submonogeneous (convolution) semigroup or group in $\mathscr{M}^1(G)$. In a broader context these notions appear already in the paper of Hofmann [249]. Their application to problems in the theory of probability is due to Schmetterer [429]. Lemma 3.1.30 is taken from Hofmann [249]. Theorem 3.1.32 given in Siebert [439] summarizes all previous knowledge concerning rational embedding contained in Böge [31] and Schmetterer [429].

We mention finally that the example of Remark 3.1.29 of a non-root compact group G for which $\mathscr{I}(G)$ is not $\mathscr{T}_v$-closed in $\mathscr{M}^1(G)$ is worked out in Siebert [439]. A more sophisticated counterexample yielding the same conclusion is Example (1) at the end of Yuan [512].

R 3.2 Poisson Measures and their Characterizations

Poisson measures on groups as they are introduced in this section are the generalization of the classical compound Poisson distribution. The rôle of the Poisson distribution in the solution of the central limit problem is convincingly described by de Finetti's theorem mentioned in References and Comments of § 1. See also Cuppens [104], where in Section 2.9 exponentials of signed measures on $\mathbb{R}^n$ are introduced and Section 4.2 where they are applied to the study of infinitely divisible distributions. The background for the definition of exponential measures in $\mathscr{M}^b(\mathbb{R}^n)$ is the research done for Poisson measures on general locally compact groups. This research started with the work of Vorobev [492] and of Böge [30] on Poisson measures in $\mathscr{M}^1(G)$ for a finite group G. In a later paper Böge [31] extended his previous results to arbitrary locally compact groups. Although the exponential and logarithm can be defined in any Banach algebra as was done for algebras of operators much earlier by various authors and yielded the extensions of Hazod and Schmetterer [209], Part III, and Hazod [195], we restrict ourselves here to the special case of the Banach algebra $\mathscr{M}^b(G)$ for a locally compact group G and its subsemigroup $\mathscr{M}^1(G)$.

In our exposition of the theory we define the exponential and logarithm in $\mathscr{M}^b(G)$ and study their elementary properties. The definition 3.2.1 of an H-Poisson measure in $\mathscr{M}^1(G)$ for a compact subgroup H of G reflects our efforts to incorporate also those Poisson measures in $\mathscr{M}^1(G)$ whose exponential series expansion starts with a nontrivial idempotent in $\mathscr{M}^1(G)$. Theorems 3.2.3 to 3.2.5 on the continuous rational embedding of H-Poisson measures are taken from Böge [31]. Theorem 3.2.5 gives a characterization of rational Poisson semigroups in $\mathscr{M}^1(G)$ by their norm-continuity. Further characterizations of this sort are discussed in § 1 of Chapter VI. In Remark 3.2.6 Böge's result from [30] is reproduced which states that the exponential representation of an H-Poisson measure in $\mathscr{M}^1(G)$ is not in general unique. See also Example 4.2 and the preceding comments in Schmetterer [429].

Theorem 3.2.7 appears in Schmetterer [429]. It is the first attempt to characterize Poisson measures on G by the submonogeneity of certain sequences of roots (see also Schmetterer [428]). For compact groups the conditions on these sequences can be reformulated in a more applicable way as indicated in Remark 3.2.8. See also a related result by Hazod in [202], p. 299. The assertion of Theorem 3.2.9 stating that on a finite group the set $\mathscr{I}(G)$ is contained in the set $\mathscr{P}(G)$ of all Poisson measures on G is known from Böge [30]. The proof presented is borrowed from Schmetterer [429]. It utilizes the idea of submonogeneous semigroups. Clearly, the theorem can also be proved with the help of the embedding theorem to be established later and a result from Martin-Löf [346], p. 99. In [346] one finds an extension of Böge's result to finite semigroups.

For further aspects of the exponential principle (characterizing all locally compact groups G that satisfy $\mathscr{I}(G)=\mathscr{P}(G)$) see Heyer [232], Section 9, also our discussion in § 1 of Chapter VI. Raikov's classical theorem on the Poisson decomposability of Poisson measures on $\mathbb{R}$ published in [395] and its classical extensions (see Cuppens [104], Linnik [322] and Lukacs [329]) cannot be generalized to arbitrary locally compact groups. Remark 3.2.10 contains a counterexample due to Rihl [404] which is described in Schmetterer [429]. Further studies in the direction of rescuing Raikov's theorem at least for some classes of groups are due to Lévy [319] and Ruhin [419], [420]. In [419] Ruhin proved the following result including that of Remark 3.2.10: Let G be a locally compact Abelian group, μ an $\{e\}$-Poisson measure in $\mathscr{M}^1(G)$ with parameter $x_0 \in G$ (as introduced in Urbanik [483]) of the form

$$\mu = \exp_{\{e\}}[\gamma(\varepsilon_{x_0} - \varepsilon_e)] \quad \text{with } x_0 \in G \text{ and } \gamma \in \mathbb{R}_+ \text{ and}$$

$$\mu = \nu * \lambda \quad \text{with } \nu, \lambda \in \mathscr{M}^1(G).$$

Then (i) μ and ν are Poisson measures if x_0 is an element of infinite order or of order 2; (ii) μ and ν can be chosen to be non-Poisson if x_0 is of finite order >2.

Theorem 3.2.11 is due to Hazod and Schmetterer [209]. Remark 3.2.12 concerning a possible weakening of the conditions in the theorem has been communicated to the author by Hazod. The origin of the characterization theorem 3.2.13 is in Urbanik [483], where it was proved for compact Abelian groups. The generalization to arbitrary compact groups is due to Carnal [64]. See also Heyer [230] and Schmetterer [429] for a slightly modified proof. The final extension to almost periodic groups has been indicated in Hazod and Schmetterer [210] and explicitly stated in Heyer [232]. Theorem 3.2.13 will be of great value for the presentation of the material in the next sections.

We note that the theory of Poisson measures and semigroups has been generalized to large classes of locally compact semigroups. One generalization in the direction of discrete semigroups was pursued by Martin-Löf [346], another in the direction of at least compact semigroups has been pursued by Grenander [182] and more recently by Hazod [200]. While the emphasis of the studies in Grenander [182], Section 2.3, is on the convergence of sequences of probability measures to a Poisson measure, Hazod in [200] occupies himself mainly with the approximation of continuous convolution semigroups by (elementary) Poisson semigroups. In [200] the technique chosen is operator-theoretic, and most of the theorems are based on corresponding results on the approximation of contraction semigroups. For further references see the references and comments to § 1 of Chapter VI.

R 3.3 Submonogeneous Embedding of Infinitely Divisible Probability Measures

In § 1 it was shown that for root compact groups G every measure in $\mathscr{I}(G)$ is rationally embeddable. Moreover, it was established that there exist a locally compact Abelian (non-root compact) group G_1 with a countable basis of its topology and a measure $\mu\in\mathscr{I}(G_1)$ which is not rationally embeddable (see Example 11 of 3.1.9). In the section under discussion this example is exploited in more detail and this leads to the statement of Corollary 3.3.8 that there exist a locally compact group G admitting a countable basis of its topology and a measure $\mu\in\mathscr{I}(G)$ which is not even submonogeneously embeddable. On the other hand, it turns out in Theorem 3.3.17 that on a discrete Abelian group G every measure in $\mathscr{I}(G)$ is in fact $\mathbb{M}$-submonogeneously embeddable for every noncyclic submonogeneous group $\mathbb{M}$. Both results are the main contents of the paper by Hazod and Schmetterer [210], which is the leading source for our presentation in this section. In Theorem 3.3.4 the roots of infinitely divisible elements in the group G_1 are calculated, and their embedding properties are exhibited. Theorems 3.3.6 and 3.3.7 contain similar information for free groups G_2, G_3 and G_4 appearing as modifications of the group G_1. It is the group G_4 which proves the statement of Corollary 3.3.8. Theorem 3.3.10 taken from Hazod [202] characterizes all noncyclic submonogeneous groups $\mathbb{M}$ with $1\in\mathbb{M}$ such that there exists a homomorphism $f:\mathbb{M}\to G_1$ satisfying $f(1)=x_1$. The main results of the section up to Corollary 3.3.11 are summarized in 3.3.12.

In 3.3.13 to 3.3.16 preparatory material from Hazod [195] is presented. The principal auxiliary result is Theorem 3.3.16 on the convergence of normal and invertible roots of a measure in $\mathscr{I}(G)$ for an arbitrary locally compact group G. Now Theorem 3.3.17 on the submonogeneous embedding can be deduced. The proof of Lemma 3.3.18 makes use of the technique of Hazod developed in [195]. In the paper [195] various refinements of Theorem 3.3.16 on the (Poisson) embedding of the roots of an infinitely divisible probability measure on a locally compact group G are obtained. The corresponding theorems are stated in terms of arbitrary complex measures in $\mathscr{M}^b_{\mathbb{C}}(G)$ and are adjacent to the framework of a general Banach algebra as indicated in Hazod and Schmetterer [209], Part III. The statement of Theorem 3.3.17 remains true if G is a connected locally compact Abelian group admitting a countable basis of its topology. This is Satz 2.3 of Hazod [202]. Whether there exist a locally compact Abelian group G and a measure in $\mathscr{I}(G)$ which is not submonogeneously embeddable is still an open question.

R 3.4 Existence of One-Parameter Semigroups

The problem of the title lies outside the theory of probabilities on algebraic-topological structures, but it has recently attained remarkable importance by its application to the general embedding theorem, to be discussed in the next section. One-parameter semigroups are the invention of Gleason [166]. The problem of the existence of one-parameter semigroups in topological semigroups is a central topic of the theory of compact semigroups. The historical basis for the main theorem 3.4.1 are the papers [358] and [359] by Mostert and Shields. The theorem has been generalized in various directions, which are made explicit in the monograph [253] by Hofmann and Mostert. The general approach to Theorem III of [253] gives substantial contributions to the subject due to Hofmann. (See [253], Historical Comments to Section 3 of Chapter B.) For our presentation we have chosen a theorem and its proof from Carruth and Lawson [68], whose formulation and arguments seem more directly applicable to our purpose. It should be noted, however, that the proof in [68] does not yield the extended theory of Hofmann and Mostert [253], Section 3 of Chapter B, as has been pointed out by Carruth, Hofmann and Mislove [69], p. 303.

We start with a few basic facts from the theory of compact semigroups taken from Hofmann and Mostert [253] and then prove Theorem 3.4.1. The demonstration makes use of the structure theorem 1 of Faucett [137] and of the Yamabe-type lemma 4.1.2 of Mostert and Shields [358]. The following discussion has as its purpose the application of the one-parameter semigroup theorem to a special case needed for the continuous embedding. This discussion is essentially due to Hofmann, whose communication to the author appeared in Heyer [230], Appendix (A), and initiated the work of Siebert [439] and Heyer [234]. Theorem 3.4.10 in the form presented is due to Yuan [514]. More detailed proofs are contained in Yuan [509].

R 3.5 The General Continuous Embedding

Rational or more generally submonogeneous embedding of infinitely divisible probability measures on a locally compact group G was treated in Sections 1 and 3. Given a rationally embeddable measure in $\mathscr{I}(G)$ we now try to find a continuous convolution semigroup $(\mu_t)_{t\in\mathbb{R}_+^*}$ in $\mathscr{M}^1(G)$ such that $\mu=\mu_1$ holds. Here continuity of

$(\mu_t)_{t\in\mathbb{R}_+^*}$ means the existence of the $\mathscr{T}_v$-$\lim_{t\downarrow 0}\mu_t=\mu_0$, which implies that $\mu_0=\omega_H$ for some compact subgroup H of G (not necessarily equal to $\{e\}$). For locally compact groups G without nontrivial compact subgroup (for example $G:=\mathbb{R}^n$, $n\geq 1$) or for an arbitrary locally compact group G but a measure $\mu\in\mathscr{I}(G)$ without idempotent factor we can seek an $\{e\}$-embedding of μ. This is the framework of the classical theory, in which the full embedding theorem holds for any $\mu\in\mathscr{I}(G)$ with a uniquely determined embedding semigroup. The classical embedding result is due to P. Lévy [321]. A straightforward proof of it can be found in Bauer [6]. A more generalizable approach to the theorem has been proposed by Parthasarathy [379]. Parthasarathy's proof makes use of some theory from the central limit problem contained in Gnedenko and Kolmogorov [172], p. 73.

The first attempt to extend the result to the torus $\mathbb{T}^m$ $(m\geq 1)$ was made by Bochner [29], Theorem 3.6.2. See also [29], in particular the notes and references to Chapter 3 for further history on the subject. In [29] Bochner aims directly at the generalization (for the torus $\mathbb{T}^m$) of a Lévy-Khintchine type formula and does not explicitly distinguish between the embedding of an infinitely divisible measure on the one hand and the canonical representation of its Fourier transform on the other. This has been done by Kloss in his papers [305], [306], where he treats the continuous embedding for compact Abelian Lie groups and groups of the type $\mathbb{T}^m\times\mathbb{R}^n$ $(m,n\geq 1)$ resp. in his theorem 4. Kloss claims in [305], p. 370 that in the case of a non-Lie group the embedding theorem does not in general hold. We present a counterexample in the discussion preceding Corollary 3.5.7. See also Heyer [230], p. 140 or Schmetterer [430], p. 85, where the counterexample is attributed to Carnal. For general locally compact Abelian groups G with arcwise connected dual group $G^\wedge$ the embedding theorem has been proved by Rogalski in [408]. Rogalski's proof depends on the choice of a continuous version of the logarithm of the Fourier transform (on $G^\wedge$) of the given measure $\mu\in\mathscr{I}(G)$, a fact which will become crucial in the entire theory, for example in Theorem 3.5.12.

Following the historical development of the embedding problem the break-through to the non-Abelian case due to Carnal [63] deserves mention. Carnal's ideas carried out in [64] for compact groups seem to be limited to almost periodic groups. See also Heyer [230]. Therefore, a different setup appears desirable. This one was initiated by Hofmann (see Heyer [230], Appendix (A)) and carried out by Siebert [439], Heyer [234] and Yuan [514].

Theorem 3.5.1 is the general continuous embedding theorem based on the analysis of Section 4, which shows that abstract homomorphisms f from $\mathbb{Q}_+^*$ into a Hausdorff semigroup T with relatively compact image $f(]0,1]\cap\mathbb{Q})$ can be shifted such that they become continuous one-parameter semigroups in T. In addition to the statement proved, one can show that the shifting elements from a well-defined compact connected Abelian group K in T lie on a rational one-parameter semigroup in K. See Yuan [514], Proposition 3.5.

Application of the theorem to the topological semigroup $T:=\mathscr{M}^1(G)$ for a locally compact group G yields the special continuous embedding theorem 3.5.4, which is the fundamental step on the way towards the discussion of the embedding principle. The accumulation group $A(S)$ of a convolution semigroup $S:=(\mu_r)_{r\in\mathbb{Q}_+^*}$ in $\mathscr{M}^1(G)$ was introduced by Carnal in [64]. The notion of a root compact measure in $\mathscr{M}^1(G)$ goes back to Hofmann. The class **S** of locally compact groups G such that any $\mu \in \mathscr{I}(G)$ is root compact is an invention of Siebert [439]. The proof of Theorem 3.5.4 modeled after Carnal [64] is contained in Siebert [439], whose presentation we follow. Corollary 3.5.7 is the so-called density theorem. It asserts for groups $G \in \mathbf{S}$ that continuously embeddable measures in $\mathscr{M}^1(G)$ are dense in $\mathscr{I}(G)$. Both classes of measures coincide for groups in $\mathbf{S}$ if G_0 is a Lie group. Hence, the embedding principle can be shown to be valid in the series of examples 3.5.11 of groups which are known to be strongly root compact. These are the compactly generated Abelian Lie groups (Kloss), compact Lie groups and totally disconnected compact groups (Carnal), connected nilpotent Lie groups (Burrell and McCrudden) and connected solvable Lie groups with only real roots (Yuan). The most complete result in this area is Theorem 3.5.12 due to Siebert [439], which contains the description of the exact domain of the embedding principle at least for Abelian groups. It should be noted that the embedding theorem for measures in $\mathscr{I}(G)$ can hold independent of the strong root compactness or membership in the class $\mathbf{S}$. In [514] Yuan proves for the group $G := \mathfrak{SL}(2, \mathbb{R})$ that the divisible element $\begin{pmatrix} -1 & 0 \\ 0 & -1 \end{pmatrix}$ of G is not root compact, but lies on the one-parameter subgroup $f : \mathbb{R} \to G$ defined by

$$f(t) := \begin{pmatrix} \cos \pi t & \sin \pi t \\ -\sin \pi t & \cos \pi t \end{pmatrix} \quad \text{for all } t \in \mathbb{R}.$$

On the other hand, a reasonable conjecture would be that all exponential Lie groups admit the embedding property. Among the connected solvable groups, exponential Lie groups are those Lie groups G which satisfy the equivalent conditions (i) $\exp_G$ is bijective and (ii) every root of $\mathfrak{L}(G)$ is of the form $\phi + i\phi'$ with ϕ' proportional to ϕ. See Dixmier [116] or Bourbaki [39], pp. 278–279. These conditions define a class of Lie groups containing that introduced by Yuan in [512]. In Definition 3.5.14 groups G are introduced for which every measure in $\mathscr{I}(G)$ has uniquely determined roots. Again for Abelian groups the domain of validity of the uniqueness principle is characterized in Theorem 3.5.15. See Hazod [202], Section II, for related material and also for the origin of Remark 3.5.17. Theorem 3.5.18 is due to Hazod [202]. It indicates the relationship between the uniqueness property and order structures in G. The idea of studying this relationship was initiated by Schmetterer and might still be carried out with useful results. A limitation, however, seems to be set by the example 2.12 in Hazod [202] of a non-Abelian group admitting no ordering and not embeddable in a direct product of ordered groups, on which every infinitely divisible probability measure has unique roots. Theorem 3.5.19 gives a characterization of all Abelian groups in $\mathbf{S}$. The proof given here differs slightly from that in Siebert [439] and it avoids Parthasarathy's embedding result 7.1 in [377], Chapter IV. Remark 3.5.21 destroys the illusion that Condition (ii) of Theorem 3.5.19 could be used to give an alternative description of the domain of validity of the embedding principle in terms of the set $B(G)$ of compact elements of G.

We now make a few remarks concerning results that are connected with our theme but not explicitly incorporated. The embedding within a translate of infinitely divisible probability measures on a locally compact Abelian group was discussed at an early stage of the development by Parthasarathy, Rao and

Varadhan [383] and Parthasarathy and Sazonov [384]. See also Parthasarathy [377]. There are also the contributions by Parthasarathy [379] and [380] for compact not necessarily Abelian groups G. The slight error in Theorem 4.1 of Parthasarathy [379] was corrected in Parthasarathy [380], where the proof uses the theory of second order cocycles. In [380] only measures in $\mathscr{I}(G)$ without idempotent factors are considered. Subsequent papers of Parthasarathy and Schmidt contain extensive studies of the embedding within a translate of continuous infinitely divisible, normalized, positive-definite functions ϕ on a locally compact group G (having a countable basis of its topology) in the sense of the following sample result, which is Theorem 12.4 of [385]: Let ϕ have continuous roots and let $G_1 := \{x \in G : \phi(x) \neq 0\}$ have a locally connected (closed) commutator subgroup. Then there exist a continuous normalized, conditionally positive-definite function ψ on G_1 and a continuous homomorphism χ from G_1 into $\mathbb{T}$ such that $\phi(x) = \chi(x) \exp \psi(x)$, if $x \in G_1$, and $\phi(x) = 0$ otherwise. For connected, locally connected locally compact groups G one obtains $G_1 = G$ and one can choose χ identical to 1. In the context of homogeneous spaces this approach appears already in the papers [155] and [156] of Gangolli. For special classes of homogeneous spaces the embedding problem had been previously discussed by Tutubalin [478]. The case of (right) invariant infinitely divisible probability measures on a connected, semisimple Lie group G of noncompact type is treated in Parthasarathy [379]. Here the main embedding theorem (5.1) depends on the existence of a spherical function on G in the sense of Harish-Chandra. See [213], Chapter X for basic knowledge on the subject.

Finally we note that the embedding problem for infinitely divisible probability measures on a locally compact group G can be extended to embedding in Markov chains. This theory is as old as the papers [128] and [129] of Elfving and their successors [421] and [424] by Runnenburg and Scheffer resp.

A more structural analysis of the problem was set up by Kingman [300] and Johansen [278], [272], [276] and [273], [275]. We shall provide more precise information on this topic in Chapter IV, § 6, and Chapter VI, § 6.

R 3.6 Injective Submonogeneous Embeddings

In this section homomorphisms $r \to \mu_r$ from a noncyclic submonogeneous subsemigroup $\mathbb{M}_+$ of $\mathbb{Q}$ into the semigroup $\mathscr{M}^1(G)$ for a locally compact group G are studied which are either injective or trivial in the sense that there exist a compact subgroup H of G and for every $r \in \mathbb{M}_+$ an $x_r \in G$ satisfying $\mu_r = \varepsilon_{x_r} * \omega_H$. The idea of studying injective homomorphisms $r \to \mu_r$ from $\mathbb{M}_+$ into $\mathscr{M}^1(G)$ goes back to Schmetterer [428], [429], where Theorem 3.6.2 has been shown under the additional conditions that G be Abelian or compact. The argument which settles the compact case yields for an arbitrary locally compact group G the fact that a homomorphism $r \to \exp_H(r\nu)$ from $\mathbb{M}_+$ into $\mathscr{M}^1(G)$ (with a compact subgroup H of G and a measure $\nu \in \mathscr{N}_H(G)$) is injective iff $\nu \neq 0$.

The proof of Theorem 3.6.2 is based on a result of Hofmann [249], which in our presentation appears as Lemma 3.6.1. The generalization of the theorem to arbitrary locally compact groups is due to Hazod, in whose paper [198] a more general approach to the problem is taken. The analysis started in Hazod [198] was continued in Hazod [201]. Both papers remain the basic sources for the discussion of this section. The topology $\mathscr{T}$ on the normalizer $N(H)$ of a compact subgroup H of G was introduced by Hazod in [201]. It helps to study noninjective homomorphisms $r \to \mu_r$ from $\mathbb{M}_+$ into $\mathscr{M}^1(G)$ in terms of group homomorphisms ϕ from $\mathbb{M}$ into $N(H)$ (as in Theorem 3.6.2) and F from $\mathbb{M}$ into $A^G(H)/I(H)$, where $A^G(H)$ denotes the group of automorphisms $h \to xhx^{-1}$ $(x \in N(H))$ of H (as in Theorem 3.6.7). For any noncyclic submonogeneous subgroup $\mathbb{M}$ of $\mathbb{Q}$ one introduces the class $\mathbf{G}_{\mathbb{M}}$ of all locally compact groups G such that every homomorphism $r \to \mu_r$ from $\mathbb{M}_+$ into $\mathscr{M}^1(G)$ is either injective or trivial, and the intersection $\mathbf{G}^*$ of all such classes. The accomplishment of Hazod in [201] is a satisfactory description of the classes $\mathbf{G}^*$ and its complement $\overline{\mathbf{G}^*}$. The first major result in this direction is Theorem 3.6.14 which states that every Lie group is in $\mathbf{G}^*$. Its proof uses Lemma 3.6.11 which depends on functional analytic tools from Riesz and Nagy [403] and Hille [240]. It also uses Lemmata 3.6.12 and 3.6.13 which are essentially due to Iwasawa [270]. In 3.6.15 we collect closure properties of the class $\mathbf{G}_{\mathbb{M}}$ yielding further examples in 3.6.18. It can be deduced for example, that every group of type (L) in the sense of Iwasawa [270] lies in $\mathbf{G}^*$. Theorem 3.6.19 asserts that a noninjective homomorphism $r \to \mu_r$ from $\mathbb{M}_+$ into $\mathscr{M}^1(G)$ is trivial if for the corresponding homomorphism F (of Theorem 3.6.7) the image $F(\mathbb{M})$ of $\mathbb{M}$ is precompact in $A^G(H)/I(H)$. This result indicates the connection between the injectivity of convolution semigroups in $\mathscr{M}^1(G)$ and compactness properties of the underlying group G, which is discussed in detail in Hazod [201]. Theorem 3.6.20 on the injectivity of continuous convolution semigroups in $\mathscr{M}^1(G)$ can be proved more directly. Following a comment of Siebert, such a proof consists of reducing the problem (via Theorem 3.6.2) to the special case of a compact group, which is easier to handle (see Schmetterer [428], p. 317 for the method). Theorem 3.6.23 and Corollary 3.6.24 provide new classes of examples G in $\mathbf{G}^*$ which are defined by the compactness properties of G introduced in Definition 3.6.21. The related results remain valid for general topological groups, as has been shown in the work of Hazod [201]. In [201], p. 31 one also finds an alternative proof of Theorem 3.6.23 making use of the structure theorem for Abelian Lie groups as well as of the theorem of Wang [494] which states that a discrete Abelian torsion subgroup of $\mathfrak{G}\mathfrak{L}(n, \mathbb{C})$ is finite. Finally, some knowledge about the class $\overline{\mathbf{G}^*}$ is presented. The example 3.6.25 of a group in $\complement \mathbf{G}_{\mathbb{Q}}$ is an unpublished contribution of Carnal. A discussion of the example and the additional properties deduced in Remark 3.6.26 can be found in Hazod [198]. In the discussion 3.6.27 Carnal's example is generalized from rational to submonogeneous convolution semigroups. The group Γ constructed in $\mathbf{G}^*$ has invariant neighborhoods, is unimodular, root compact, class compact, admits a compact commutator subgroup, is amenable and is of type I, but is not of type (L). A proof of the last property is given at the end of Hazod [198]. Given an arbitrary locally compact group G and a noncyclic submonogeneous group $\mathbb{M}$ one can construct a connected (not necessarily locally compact) topological group $\Gamma_0 := \Gamma_{\mathbb{M}}$ such that G and Γ are injectively embedded into Γ_0, i.e., $\Gamma_0 \in \complement \mathbf{G}_{\mathbb{M}}$. One simply takes for Γ_0 the group $\mathfrak{U}(\mathscr{H})$ of all unitary operators on an infinite-dimensional Hilbert space furnished with the strong operator topology (Hazod [201], Satz 4.3).

Chapter IV

Canonical Representations of Convolution Semigroups

In the preceding chapter the problem of embedding an infinitely divisible probability measure in a continuous convolution semigroup was discussed in great detail. The next step towards a solution of the central limit theorem is the canonical representation of all continuous convolution semigroups in the sense of a Lévy-Khintchine formula. As in the classical theory this formula will enable us in subsequent chapters to characterize special classes of infinitely divisible measures and to illuminate the rôle of the Gauss measures. A systematic presentation of the theory of canonical representations of continuous convolution semigroups of probability measures on a locally compact group G requires some knowledge of the theory of positive semigroups and their infinitesimal generators or generating functionals.

The first section establishes a natural connection between convolution semigroups on G and contraction semigroups of (probability) operators on certain function spaces E over G such that the problem of generating a convolution semigroup becomes a problem of determining the existence of the infinitesimal generator of the corresponding contraction semigroup on E. This problem can be solved first of all for a Lie group G by the theory developed by G.A. Hunt. This is done in Section 2. The canonical representations obtained for convolution semigroups on Lie groups will be the building blocks for the case of more general locally compact groups G being treated in Sections 3 and 4. Here the domain of the analysis is the class of Lie projective almost periodic groups, which, as we worked out earlier, allow finite-dimensional Fourier analysis. The culmination of the treatment is the discussion of the representation problem for probability measures on arbitrary locally compact groups G.

Using the solution of Hilbert's fifth problem and the ideas of Bruhat a differentiable structure can be introduced in any locally compact group and the problem will be reduced via Lie projectivity to the Lie group case. The generating functionals of continuous convolution semigroups of probability measures on G can be characterized as almost positive and normed forms on the space $\mathfrak{D}(G)$ of all differentiable functions on G with compact support, i.e., as distributions on G.

Finally we add a section on convolution hemigroups of probability measures on a locally compact group G. These generalize the convolution semigroups by their dependence on two parameters rather than one. Since the results of the one-parameter theory can be applied also to the two-parameter case, first results concerning the generation and representation of convolution hemigroups can be achieved. The techniques used in the proofs, however, are matrix- and operator-theoretic resp. The theorems proved relate the probabilistic framework of convolution hemigroups to various aspects of the evolution equations for matrices and operators resp.

4.1 Positive Semigroups and Their Generating Functionals

Let G be a locally compact group and E one of the Banach spaces $\mathscr{C}_u(G)$ of all bounded, with respect to the left uniform structure of G uniformly continuous real functions on G or $\mathscr{C}^0(G)$ of all continuous real functions on G which vanish at infinity. Given an $\{e\}$-continuous convolution semigroup $(\mu_t)_{t\in\mathbb{R}_+^*}$ in $\mathscr{M}^1(G)$ we define the family $(S_t)_{t\in\mathbb{R}_+^*}$ of probability operators $S_t := T_{\mu_t}$ on E by $S_t f(y) := \mu_t({}_y f)$ for all $f\in\mathsf{E}$, $y\in G$ and $t\in\mathbb{R}_+^*$. $(S_t)_{t\in\mathbb{R}_+^*}$ is a strongly continuous semigroup of contraction operators on E which is *positive* in the sense that

$$\inf_{y\in G} S_t f(y) \geq \inf_{y\in G} f(y) \quad \text{holds for all } f\in\mathsf{E},\ t\in\mathbb{R}_+^*,$$

and *invariant* in the sense that

$$L_x S_t f = S_t L_x f \quad \text{is satisfied for all } f\in\mathsf{E},\ x\in G.$$

Here and in the sequel we use the notation L_x and R_x for the operators on $\mathscr{C}_u(G)$ induced by $L(x)$ and $R(x)$ $(x\in G)$ resp.

Resuming the above discussion we have the following

4.1.1 Lemma. *Let G be a locally compact group and E one of the spaces $\mathscr{C}_u(G)$ or $\mathscr{C}^0(G)$. Then*

(i) *For every $\{e\}$-continuous convolution semigroup $(\mu_t)_{t\in\mathbb{R}_+^*}$ in $\mathscr{M}^1(G)$ there exists a (positive, invariant) strongly continuous semigroup $(S_t)_{t\in\mathbb{R}_+^*}$ of contraction operators on E given by*

$$S_t f(y) := \mu_t({}_y f) \quad \textit{for all } f\in\mathsf{E},\ y\in G \quad \textit{and} \quad t\in\mathbb{R}_+^*.$$

(ii) *For $\mathsf{E} := \mathscr{C}^0(G)$ the correspondence $\mu_t \to S_t := T_{\mu_t}$ defined in* (i) *between $\{e\}$-continuous convolution semigroups $(\mu_t)_{t\in\mathbb{R}_+^*}$ in $\mathscr{M}^1(G)$ and (positive, invariant) strongly continuous contraction semigroups $(S_t)_{t\in\mathbb{R}_+^*}$ on E is one-to-one.*

The *proof* of Statement (ii) of the lemma is a direct consequence of the introductory discussion together with the fact that by the positivity of $(S_t)_{t\in\mathbb{R}_+^*}$, for any $t\in\mathbb{R}_+^*$ there exists a measure $\mu_t\in\mathscr{M}^1(G)$ satisfying

$$S_t f(e) = \mu_t(f) \quad \text{for all } f\in\mathscr{C}^0(G).$$

The invariance of $(S_t)_{t\in\mathbb{R}_+^*}$ yields

$$S_t f(y) = \mu_t({}_y f) \quad \text{for all } f\in\mathscr{C}^0(G) \text{ and } y\in G. \quad \square$$

The lemma enables us to speak of the *(contraction) semigroup* $(S_t)_{t\in\mathbb{R}_+^*}$ (on E) *corresponding to the convolution semigroup* $(\mu_t)_{t\in\mathbb{R}_+^*}$ and conversely.

Let $(S_t)_{t\in\mathbb{R}_+^*}$ be the contraction semigroup (on E) corresponding to the given convolution semigroup $(\mu_t)_{t\in\mathbb{R}_+^*}$ in $\mathcal{M}^1(G)$. The *infinitesimal generator* N of $(S_t)_{t\in\mathbb{R}_+^*}$ is defined by

$$Nf := \lim_{t\downarrow 0} \frac{1}{t}(S_t f - f)$$

for all f in its domain D_N, which by the Hille-Yosida theorem [241], p. 307 is a dense subspace of E. By the invariance of $(S_t)_{t\in\mathbb{R}_+^*}$ we have $L_x \mathsf{D}_N \subset \mathsf{D}_N$ for all $x\in G$ and $N(L_x f) = L_x(Nf)$ for every $x\in G$ and all $f\in \mathsf{D}_N$. In the following discussion we shall concentrate ourselves on the domain D_A of the *generating functional*

$$f \to Af := \lim_{t\downarrow 0} \frac{1}{t}(S_t f(e) - f(e))$$

corresponding to the given semigroup $(S_t)_{t\in\mathbb{R}_+^*}$ on E. Plainly, $Af = Nf(e)$ whenever $f\in\mathsf{D}_N$. Having established the existence of Af, the invariance property immediately implies the existence of

$$A(L_x f) = Nf(x) = \lim_{t\downarrow 0} \frac{1}{t}(S_t f(x) - f(x)) \qquad \text{for all } x\in G.$$

With the justification of Lemma 4.1.1 N and A will also be called the *infinitesimal generator* and *generating functional of the convolution semigroup* $(\mu_t)_{t\in\mathbb{R}_+^*}$ *resp.*

In order to describe the subspace D_N in more detail we are forced to use more extensively the structure of the underlying group G.

4.1.2 Lemma. *Let* B *be a real Banach space,* M *a dense linear subspace of* B, $(T_t)_{t\in\mathbb{R}_+^*}$ *a family of continuous linear functionals in* B' *and c a constant independent of t such that* $|T_t x| \le c\|x\|$ *holds for all* $x\in\mathsf{B}$. *If* $\lim_{t\downarrow 0} T_t x$ *exists for all* $x\in\mathsf{M}$, *then it also exists for all* $x\in\mathsf{B}$.

The *proof* is clear. □

4.1.3 Remark. Under the conditions of the lemma let $(T_t^\lambda)_{t\in\mathbb{R}_+^*}$ be a family in B' for any $\lambda\in\Lambda$ and let $c\in\mathbb{R}_+^*$ be a constant independent of t and $\lambda\in\Lambda$ such that $|T_t^\lambda x| \le c\|x\|$ holds for all $x\in\mathsf{B}$. Then the existence of $\lim_{t\downarrow 0} T_t^\lambda x$ for all $x\in\mathsf{M}$ implies the existence of $\lim_{t\downarrow 0} T_t^\lambda x$ for all $x\in\mathsf{B}$, where the limit is understood uniformly in $\lambda\in\Lambda$.

4.1.4 Lemma. *Let G be a locally compact group and* $(\mu_t)_{t\in\mathbb{R}_+^*}$ *a convolution semigroup in* $\mathcal{M}^1(G)$. *Then for every* $V\in\mathfrak{V}(e)$ *we have*

$$\sup_{t\in\mathbb{R}_+^*} \frac{1}{t}\mu_t(\complement V) < \infty.$$

Proof. Let $V \in \mathfrak{B}(e)$. We shall prove that there exist a function $\psi \in \mathsf{D}_N$ and a constant $c \in \mathbb{R}_+^*$ with the properties $\psi \geq 0$, $\psi(e)=0$ and $\psi(x) \geq c$ for all $x \in \complement V$. In this case we immediately conclude that

$$c\frac{1}{t}\mu_t(\complement V) \leq \frac{1}{t}\int_{\complement V} \psi\, d\mu_t \leq \frac{1}{t}\int_G \psi\, d\mu_t \quad \text{and also that}$$

$$\sup_{t \in \mathbb{R}_+^*} \frac{1}{t}\int \psi\, d\mu_t < \infty \qquad \text{since } \psi \in \mathsf{D}_N \text{ and } \psi(e)=0.$$

We are therefore left with the proof of the first statement. Let V_1 be a compact neighborhood in $\mathfrak{B}(e)$ such that $V_1^{-1}=V_1$ and $V_1^2 \subset V$, let $f \in \mathscr{C}_+(G)$ with $f \leq 1$, $f(e)=0$ and $f(x)=1$ for all $x \in \complement V_1$ and finally let $\varepsilon \in]0, \frac{1}{4}[$. Since D_N is dense in $\mathscr{C}_u(G)$, there is a function $g \in \mathsf{D}_N$ satisfying

$$\|f-g\| \leq \varepsilon \quad \text{or} \quad f(x)-\varepsilon \leq g(x) \leq f(x)+\varepsilon \qquad \text{for all } x \in G.$$

Putting $d := \inf_{x \in V_1} g(x)$ we obtain $d \leq g(e) \leq \varepsilon$, and since V_1 is compact, there is an $x_0 \in V_1$ satisfying $g(x_0)=d$.

If $x \in \complement V_1$, then $g(x) \geq f(x)-\varepsilon \geq 1-\varepsilon > \varepsilon$. Moreover, $d \leq \varepsilon$ implies that $d = \inf_{x \in G} g(x)$ or $\inf_{x \in V_1} g(x) = \inf_{x \in G} g(x)$. For $h := g-d$ we get $h \in \mathsf{D}_N$, $h \geq 0$ and $h(x_0)=0$. We define $\psi := L_{x_0} h = {}_{x_0}h$. As $h \in \mathsf{D}_N$ and $L_x \mathsf{D}_N \subset \mathsf{D}_N$ for all $x \in G$, one has $\psi \in \mathsf{D}_N$, but also $\psi \geq 0$ and $\psi(e)=0$. Since $y \in \complement V$ implies that $x_0 y \in \complement V_1$, one obtains for all $y \in \complement V$ the estimate

$$\psi(y) = {}_{x_0}h(y) = {}_{x_0}g(y) - d \geq 1-\varepsilon-d \geq 1-2\varepsilon > \tfrac{1}{2},$$

which is the desired conclusion. □

In the following we shall often use the notation $G^* := G \setminus \{e\}$.

4.1.5 Theorem. *Let G be a discrete group. Then for any convolution semigroup $(\mu_t)_{t \in \mathbb{R}_+^*}$ in $\mathscr{M}^1(G)$ with corresponding semigroup $(S_t)_{t \in \mathbb{R}_+^*}$ on $\mathscr{C}_u(G)$ and generator N on D_N we have the following properties:*

(i) *The family* $\left(\mathrm{Res}_{G^*} \frac{1}{t}\mu_t\right)_{t \in \mathbb{R}_+^*}$ $\mathscr{T}_w$*-converges (as $t \downarrow 0$) to a measure $\nu \in \mathscr{M}_+^b(G^*)$.*

(ii) $\mathsf{D}_N = \mathscr{C}_u(G)$.

(iii) *For all $f \in \mathscr{C}_u(G)$ and $x \in G$ the formula $Nf(x) = \int_{G^*}(f(xy)-f(x))\,\nu(dy)$ holds.*

Proof. (i) Let $f \in \mathscr{C}_u(G)$ be such that $f(e)=0$. Since G is discrete, Lemma 4.1.4 implies the existence of a constant $\gamma \in \mathbb{R}_+$ with

$$\sup_{t \in \mathbb{R}_+^*} \frac{1}{t}\mu_t(G^*) \leq \gamma.$$

$B_1 := \{f \in \mathscr{C}_u(G) : f(e)=0\}$ is a Banach subspace of $\mathscr{C}_u(G)$. We shall show that $D_1 := \{f \in \mathsf{D}_N : f(e)=0\}$ is dense in B_1. Given $f \in B_1$ and $\varepsilon > 0$ there exists a $g \in \mathsf{D}_N$ with $\|f-g\| \leq \frac{\varepsilon}{2}$. The function $h := g-g(e)$ belongs to D_1, and one obtains

$$\|f-h\| \leq \|f-g\| + |g(e)| \leq \varepsilon.$$

For every $f \in B_1$, $t \in \mathbb{R}_+^*$ we put $N_t f := \frac{1}{t} \int_{G^*} f \, d\mu_t$. Then for all $t \in \mathbb{R}_+^*$ the mapping $f \to N_t f$ is an element of B_1' and $|N_t f| \leq \gamma \|f\|$ for all $f \in B_1$. Since for $g \in D_1$ the limit $\lim_{t \downarrow 0} N_t g$ exists, Lemma 4.1.2 yields the existence of $\lim_{t \downarrow 0} N_t f$ for all $f \in B_1$, whence the assertion.

(ii) For any $f \in \mathscr{C}_u(G)$ and $x \in G$ we have

$$\left|\frac{1}{t}(S_t f(x) - f(x))\right| \leq \frac{1}{t} \int_{G^*} |f(xy) - f(x)| \, \mu_t(dy) \leq 2\gamma \|f\|.$$

Remark 4.1.3 applied to the family $(N_t^x)_{t \in \mathbb{R}_+^*}$ of continuous linear functionals $f \to N_t^x f := \frac{1}{t}(S_t f(x) - f(x))$ on $\mathscr{C}_u(G)$, where $x \in G$, yields that $(N_t^x f)_{t \in \mathbb{R}_+^*}$ converges for all $f \in \mathscr{C}_u(G)$ (as $t \downarrow 0$) uniformly in $x \in G$. This is the desired property.

(iii) For $f \in \mathscr{C}_u(G)$ and $x \in G$ we obtain from (i) and (ii) that

$$N f(x) = \lim_{t \downarrow 0} \frac{1}{t}(S_t f(x) - f(x)) = \lim_{t \downarrow 0} \frac{1}{t} \int_{G^*} (f(xy) - f(x)) \, \mu_t(dy)$$

$$= \int_{G^*} (f(xy) - f(x)) \, \nu(dy). \quad \square$$

4.1.6. Let G be a Lie group of dimension $n \geq 1$ with Lie algebra $\mathfrak{L}(G)$ having a basis $\{X_1, \ldots, X_n\}$. For any $Y \in \mathfrak{L}(G)$ and $s \in \mathbb{R}^*$ we put $\zeta(s) := \exp sY$ and for every $f \in \mathscr{C}_u(G)$ we define

$$Yf := \lim_{t \to 0} \frac{1}{t}(L_{\zeta(t)} f - f) \quad \text{as well as}$$

$$\tilde{Y}f := \lim_{t \to 0} \frac{1}{t}(R_{\zeta(t)} f - f),$$

if the limits exist in the norm of $\mathscr{C}_u(G)$. Yf is called the *left derivative* and $\tilde{Y}f$ the *right derivative of* Y *in* f. We define the sets

$$\mathscr{C}_1(G) := \{f \in \mathscr{C}_u(G) \colon Xf \text{ exists for all } X \in \mathfrak{L}(G)\},$$

$$\tilde{\mathscr{C}}_1(G) := \{f \in \mathscr{C}_u(G) \colon \tilde{X}f \text{ exists for all } X \in \mathfrak{L}(G)\},$$

$$\mathscr{C}_2(G) := \{f \in \mathscr{C}_1(G) \colon Y(Xf) \text{ exists for all } X, Y \in \mathfrak{L}(G)\} \quad \text{and}$$

$$\tilde{\mathscr{C}}_2(G) := \{f \in \tilde{\mathscr{C}}_1(G) \colon \tilde{Y}(\tilde{X}f) \text{ exists for all } X, Y \in \mathfrak{L}(G)\}.$$

One observes that the set of once continuously differentiable functions in $\mathscr{K}(G)$ is contained in $\mathscr{C}_1(G)$ and $\tilde{\mathscr{C}}_1(G)$ and that the set of twice continuously differentiable functions in $\mathscr{K}(G)$ is a subset of $\mathscr{C}_2(G)$ and of $\tilde{\mathscr{C}}_2(G)$. The sets $\mathscr{C}_1(G)$, $\tilde{\mathscr{C}}_1(G)$, $\mathscr{C}_2(G)$ and $\tilde{\mathscr{C}}_2(G)$ are dense subspaces of $\mathscr{C}_u(G)$. Defining for every $f \in \mathscr{C}_1(G)(f \in \tilde{\mathscr{C}}_1(G))$

$$\|f\|_1 := \|f\| + \sum_{i=1}^n \|X_i f\| \quad (\|f\|_{\tilde{1}} := \|f\| + \sum_{i=1}^n \|\tilde{X}_i f\|)$$

and for every $f \in \mathscr{C}_2(G)(f \in \tilde{\mathscr{C}}_2(G))$

$$\|f\|_2 := \|f\|_1 + \sum_{i=1}^n \sum_{j=1}^n \|X_i X_j f\|$$
$$(\|f\|_{\tilde{2}} := \|f\|_{\tilde{1}} + \sum_{i=1}^n \sum_{j=1}^n \|\tilde{X}_i \tilde{X}_j f\|)$$

we see that the sets $\mathscr{C}_1(G)$, $\tilde{\mathscr{C}}_1(G)$, $\mathscr{C}_2(G)$ and $\tilde{\mathscr{C}}_2(G)$ become Banach spaces with respect to the norms $\|\cdot\|_1$, $\|\cdot\|_{\tilde{1}}$, $\|\cdot\|_2$ and $\|\cdot\|_{\tilde{2}}$ resp.

Further *properties.* (1) $R_x \mathscr{C}_1(G) \subset \mathscr{C}_1(G)(R_x \mathscr{C}_2(G) \subset \mathscr{C}_2(G))$ and

$$\|R_x f\|_1 = \|f\|_1 \quad \text{for all } f \in \mathscr{C}_1(G)$$
$$(\|R_x f\|_2 = \|f\|_2 \text{ for all } f \in \mathscr{C}_2(G)), \quad x \in G.$$

(2) $L_x \tilde{\mathscr{C}}_1(G) \subset \tilde{\mathscr{C}}_1(G)$ $(L_x \tilde{\mathscr{C}}_2(G) \subset \tilde{\mathscr{C}}_2(G))$ and

$$\|L_x f\|_{\tilde{1}} = \|f\|_{\tilde{1}} \quad \text{for all } f \in \tilde{\mathscr{C}}_1(G)$$
$$(\|L_x f\|_{\tilde{2}} = \|f\|_{\tilde{2}} \text{ for all } f \in \tilde{\mathscr{C}}_2(G)), \quad x \in G.$$

4.1.7. For the Lie group G of dimension $n \geq 1$ introduced in 4.1.6 we consider a convolution semigroup $(\mu_t)_{t \in \mathbb{R}_+^*}$ in $\mathscr{M}^1(G)$ and its corresponding semigroup $(S_t)_{t \in \mathbb{R}_+^*}$. $(S_t)_{t \in \mathbb{R}_+^*}$ can plainly be regarded as a strongly continuous contraction semigroup on the Banach spaces $\mathscr{C}_1(G)$ and $\mathscr{C}_2(G)$. In fact,

(i) $S_t \mathscr{C}_1(G) \subset \mathscr{C}_1(G)$ $(S_t \mathscr{C}_2(G) \subset \mathscr{C}_2(G))$ for all $t \in \mathbb{R}_+^*$.
(ii) $X(S_t f) = S_t(Xf)$ for all $X \in \mathfrak{L}(G)$, $f \in \mathscr{C}_1(G)$
$(YX(S_t f) = S_t(YXf)$ for all X, $Y \in \mathfrak{L}(G)$, $f \in \mathscr{C}_2(G)$ $(t \in \mathbb{R}_+^*))$.
(iii) $\|S_t f\|_1 \leq \|f\|_1$ for all $f \in \mathscr{C}_1(G)$ $(\|S_t f\|_2 \leq \|f\|_2$ for all $f \in \mathscr{C}_2(G))$ and thus
(iv) $\lim_{t \downarrow 0} \|S_t f - f\|_1 = 0$ for all $f \in \mathscr{C}_1(G)$ $(\lim_{t \downarrow 0} \|S_t f - f\|_2 = 0$ for all $f \in \mathscr{C}_2(G))$.

Let D' be the domain of the infinitesimal generator N' of the semigroup $(S_t)_{t \in \mathbb{R}_+^*}$ considered as a semigroup of operators on $\mathscr{C}_2(G)$. Then D' is dense in $\mathscr{C}_2(G)$ by the Hille-Yosida theorem [241] p. 307. But clearly $\mathsf{D}' \subset \mathsf{D}_N$, whence $\mathsf{D}' \subset \mathsf{D}_N \cap \mathscr{C}_2(G) \subset \mathscr{C}_2(G)$. We conclude that $\mathsf{D}_N \cap \mathscr{C}_2(G)$ is dense in $\mathscr{C}_2(G)$.

4.1.8. Let $f \in \mathscr{C}_2(G)$. Then the function $g: t \to f(\zeta(t))$ is twice differentiable on $\mathbb{R}$ and admits therefore a Taylor expansion valid up to the second order. Translated into the Lie group G this shows that

$$f(x) = f(e) + \sum_{i=1}^n x_i(x) X_i f(e)$$
$$+ \frac{1}{2!} \sum_{i=1}^n \sum_{j=1}^n x_i(x) x_j(x) X_i X_j f(\zeta(x))$$

is a generalized *Taylor expansion* for all x in a suitable canonical neighborhood $U \in \mathfrak{B}(e)$ with system $\{x_1, \ldots, x_n\}$ of canonical coordinates with respect to the given basis $\{X_1, \ldots, X_n\}$ of $\mathfrak{L}(G)$. Here $\zeta(x)$ denotes some point of U (on the one-parameter subgroup through x).

4.1.9 Lemma. *Let G be a Lie group of dimension $n \geq 1$ with Lie algebra $\mathfrak{L}(G)$ and basis $\{X_1, \ldots, X_n\}$ of $\mathfrak{L}(G)$. Let $U_0 \in \mathfrak{B}(e)$ be a fixed open relatively compact canonical* *neighborhood of e and* $\{x_1, \ldots, x_n\}$ *a system of canonical coordinates (of any point* $x \in U_0$*) with respect to* $\{X_1, \ldots, X_n\}$. *Then there exist bounded functions*

$$y_1, \ldots, y_n \in \mathscr{C}^\infty(G) \cap \mathscr{K}(G) \quad \text{and} \quad \phi := \phi_{U_0} \in \mathscr{C}^\infty(G)$$

such that the following conditions are satisfied:

(i) $y_i(x) = x_i(x)$ *for all* $x \in U_0$ $(i = 1, \ldots, n)$,
(ii) $\phi(x) = \sum_{i=1}^n x_i^2(x)$ *for all* $x \in U_0$,
(iii) $\phi(y) > 0$ *for all* $y \in G^*$.

Proof. Let U be another open relatively compact canonical neighborhood in $\mathfrak{V}(e)$ such that $\bar{U}_0 \subset U$. Then $U \cup \complement \bar{U}_0 = G$, so that there exist functions $\phi_1, \phi_2 \in \mathscr{C}^\infty(G)$ satisfying $0 \leq \phi_1, \phi_2 \leq 1$ having supports in U, $\complement U_0$ resp. with the property $\phi_1 + \phi_2 = 1$. For $i = 1, \ldots, n$ we define functions $y_i := x_i \phi_1$. Clearly,

$$y_i \in \mathscr{C}^\infty(G) \cap \mathscr{K}(G) \quad \text{and} \quad y_i(x) = x_i(x)$$

for all $x \in U_0$ $(i = 1, \ldots, n)$. We now define

$$\phi := \phi_{U_0} := \left(\sum_{i=1}^n x_i^2\right)\phi_1 + \phi_2.$$

Plainly $\phi \in \mathscr{C}^\infty(G)$ and

$$\phi(x) = \sum_{i=1}^n x_i^2(x) \quad \text{for all } x \in U_0.$$

Moreover, $\phi(y) > 0$ for all $y \in G^*$ and $\phi(\complement U) = 1$. □

4.1.10. For $f \in \mathscr{C}_2(G)$ and $i, j = 1, \ldots, n$ we denote by $A_i f$ and $A_{ij} f$ the numbers $X_i f(e) = \tilde{X}_i f(e)$ and $X_i X_j f(e) = \tilde{X}_j \tilde{X}_i f(e)$ resp. Obviously $f \to A_i f$ and $f \to A_{ij} f$ are continuous linear functionals on $\mathscr{C}_2(G)$ for $i, j = 1, \ldots, n$. In connection with the lemma above we note that for all $i, j = 1, \ldots, n$ one has $y_i(e) = 0$, $A_j y_i = \delta_{ij}$ as well as $A_i \phi = 0$, $A_{ii} \phi = 2$ and $A_{ij} \phi = 0$, if $i \neq j$.

4.1.11 Lemma. *Let* E *be a locally convex (Hausdorff) vector space,* E_1 *a dense subspace of* E, F *a subspace of* E *of finite codimension,* $y \in \mathsf{E}$ *and* $\mathsf{M} := y + \mathsf{F}$. *Then* $\mathsf{M}_1 := \mathsf{M} \cap \mathsf{E}_1$ *is dense in* M.

Proof. Without loss of generality we suppose that $y = 0$, and we restrict ourselves to $\operatorname{codim} \mathsf{F} = 1$ and assume $\bar{\mathsf{F}} = \mathsf{F}$ as well as $\dim \mathsf{E} = \infty$. Since $\mathsf{E} \setminus \mathsf{F}$ is open $\neq \emptyset$, there exists an $x_1 \in \mathsf{E}_1 \cap (\mathsf{E} \setminus \mathsf{F})$. Letting F_1 be the linear subspace generated by x_1 we have $\mathsf{E} = \mathsf{F} \oplus \mathsf{F}_1$. The projection p_F onto F is continuous and thus $p_\mathsf{F}(\mathsf{E}_1)$ is dense in F. For $y \in \mathsf{E}_1$ one has $y = x + \lambda x_1$ and therefore $p_\mathsf{F}(y) = x = y - \lambda x_1 \in \mathsf{E}_1$. We conclude that $p_\mathsf{F}(\mathsf{E}_1) \subset \mathsf{F} \cap \mathsf{E}_1 \subset \mathsf{F}$ and finally $\overline{\mathsf{F} \cap \mathsf{E}_1} = \mathsf{F}$ since $p_\mathsf{F}(\mathsf{E}_1)$ is dense in F. □

4.1.12 Lemma. *We keep the notation of the preceding discussion. There exist functions* $z_1, \ldots, z_n$ *and* ψ *in* $\mathsf{D}_N \cap \mathscr{C}_2(G)$ *with the following properties:*

(i) $z_i(e) = 0$, $A_j z_i = \delta_{ij}$ *for* $i, j = 1, \ldots, n$;
(ii) $\psi(e) = 0$, $A_i \psi = 0$, $A_{ij} \psi = 2\delta_{ij}$ *for* $i, j = 1, \ldots, n$;

(iii) *there exist a neighborhood* $V\in\mathfrak{V}(e)$, $V\subset U_0$ *and a constant* $\delta\in\mathbb{R}_+^*$ *such that* $\psi(x)\geq\delta\sum_{i=1}^n x_i^2(x)$ *for all* $x\in V$.

Proof. Application of Lemma 4.1.11 to the Banach space $\mathsf{E}:=\mathscr{C}_2(G)$, to the subspace $\mathsf{E}_1:=\mathsf{D}_N\cap\mathscr{C}_2(G)$ and equally to elements $y:=y_i\in\mathsf{E}$ and $\phi\in\mathsf{E}$ with corresponding subspaces M_i and M of the form

$$\{f\in\mathscr{C}_2(G): f(e)=y_i(e), A_i f=A_i y_i\} \quad\text{and}$$
$$\{f\in\mathscr{C}_2(G): f(e)=\phi(e), A_i f=A_i\phi, A_{ij}f=A_{ij}\phi\}$$

yields the asserted properties (i) and (ii).

We are left with the proof of (iii). The generalized Taylor expansion of 4.1.8 applied to the function $\psi\in\mathsf{D}_N\cap\mathscr{C}_2(G)$ and to a neighborhood $V_1\in\mathfrak{V}(e)$ with $V_1\subset U_0$ yields

$$\psi(x)=\frac{1}{2!}\sum_{i=1}^n\sum_{j=1}^n x_i(x)\,x_j(x)\,X_i X_j\psi(\zeta(x))$$

for all $x\in V_1$ (with $\zeta(x)\in V_1$).

Since $A_{ii}\psi=2$ for all $i=1,\dots,n$ and $A_{ij}\psi=0$ for all $i,j=1,\dots,n$, $i\neq j$, for a given $\varepsilon>0$ such that $2-\varepsilon-\varepsilon(n-1)>0$ there exists a $V_2\in\mathfrak{V}(e)$ with the properties

$$-\varepsilon\leq X_i X_j\psi(x)\leq\varepsilon \qquad \text{for all } i,j=1,\dots,n,\ i\neq j,$$
$$2-\varepsilon\leq X_i X_i\psi(x)\leq 2+\varepsilon \qquad \text{for } i=1,\dots,n,$$

whenever $x\in V_2$. From

$$|x_i(x)\,x_j(x)\,X_i X_j\psi(\zeta(x))|\leq\varepsilon\,|x_i(x)\,x_j(x)|$$

for all $x\in V_1\cap V_2$, $i,j=1,\dots,n$, $i\neq j$, we obtain

$$\psi(x)\geq\frac{1}{2!}\Big\{(2-\varepsilon)\sum_{i=1}^n x_i^2(x)-\varepsilon\sum_{i=1}^n\sum_{\substack{j=1\\ j\neq i}}^n|x_i(x)\,x_j(x)|\Big\}$$
$$\geq\frac{1}{2!}(2-\varepsilon-\varepsilon(n-1))\sum_{i=1}^n x_i^2(x) \qquad \text{for all } x\in V_1\cap V_2.$$

Putting $V:=V_1\cap V_2$ and $\delta:=\frac{1}{2!}(2-\varepsilon-\varepsilon(n-1))$ we obtain

$$\psi(x)\geq\delta\sum_{i=1}^n x_i^2(x) \qquad \text{for all } x\in V,$$

which proves the assertion. □

4.1.13 Lemma. *For the initially given convolution semigroup in* $\mathscr{M}^1(G)$ *and the neighborhood* $U_0\in\mathfrak{V}(e)$ *introduced in Lemma* 4.1.9 *we get*

$$\sup_{t\in\mathbb{R}_+^*}\frac{1}{t}\int_{U_0}\Big(\sum_{i=1}^n x_i^2\Big)\,d\mu_t<\infty.$$

Proof. Let ψ and V be as in Lemma 4.1.12. Since $\psi \in \mathsf{D}_N$ and $\psi(e)=0$ one obtains immediately $\sup_{t\in\mathbb{R}_+^*} \frac{1}{t}\int \psi\, d\mu_t < \infty$. But ψ is bounded and by Lemma 4.1.4

$$\sup_{t\in\mathbb{R}_+^*} \frac{1}{t}\mu_t(\complement V) < \infty$$

holds. Therefore $\sup_{t\in\mathbb{R}_+^*} \frac{1}{t}\int_V \psi\, d\mu_t < \infty$. We now infer that $\psi(x) \geq \delta \sum_{i=1}^n x_i^2(x)$, which is valid for all $x \in V$ by Lemma 4.1.12, and obtain

$$\sup_{t\in\mathbb{R}_+^*} \frac{1}{t}\int_V \sum_{i=1}^n x_i^2\, d\mu_t < \infty.$$

But

$$\frac{1}{t}\int_{U_0} \sum_{i=1}^n x_i^2\, d\mu_t = \frac{1}{t}\int_V \sum_{i=1}^n x_i^2\, d\mu_t + \frac{1}{t}\int_{U_0 \cap \complement V} \sum_{i=1}^n x_i^2\, d\mu_t$$

$$= \frac{1}{t}\int_V \sum_{i=1}^n x_i^2\, d\mu_t + \frac{1}{t}\int_{U_0 \cap \complement V} \phi_{U_0}\, d\mu_t,$$

where ϕ_{U_0} has been proved to exist in Lemma 4.1.9. Since $\phi := \phi_{U_0}$ is bounded and $\sup_{t\in\mathbb{R}_+^*} \frac{1}{t}\mu_t(\complement V) < \infty$, we deduce that $\sup_{t\in\mathbb{R}_+^*} \frac{1}{t}\int_{U_0\cap\complement V} \phi\, d\mu_t < \infty$, whence

$$\sup_{t\in\mathbb{R}_+^*} \frac{1}{t}\int_{U_0} \sum_{i=1}^n x_i^2\, d\mu_t$$

$$\leq \sup_{t\in\mathbb{R}_+^*} \frac{1}{t}\int_V \sum_{i=1}^n x_i^2\, d\mu_t + \sup_{t\in\mathbb{R}_+^*} \frac{1}{t}\int_{U_0\cap\complement V} \phi\, d\mu_t < \infty. \quad \square$$

4.1.14 Theorem. *Let G be a Lie group and $(\mu_t)_{t\in\mathbb{R}_+^*}$ a convolution semigroup in $\mathscr{M}^1(G)$ with corresponding semigroup $(S_t)_{t\in\mathbb{R}_+^*}$. Then the generating functional A of $(S_t)_{t\in\mathbb{R}_+^*}$ exists on $\mathscr{C}_2(G)$.*

Proof. It suffices to show that for every $f \in \mathscr{C}_2(G)$ the limit

$$\lim_{t\downarrow 0} \frac{1}{t}(S_t f(e) - f(e))$$

exists. By Lemma 4.1.2 we have to prove the existence of a constant $c \in \mathbb{R}_+^*$ such that

$$\left|\frac{1}{t}(S_t f(e) - f(e))\right| \leq c\,\|f\|_2 \qquad \text{for all } f \in \mathscr{C}_2(G).$$

This can be established as follows: Given $f \in \mathscr{C}_2(G)$ we consider the function $g \in \mathscr{C}_2(G)$ defined by

$$g(x) := f(x) - f(e) - \sum_{i=1}^n (A_i f)\, z_i(x) \qquad \text{for all } x \in G,$$

where the functions $z_1, \ldots, z_n$ have been exhibited in Lemma 4.1.12. The generalized Taylor expansion of g in a neighborhood $W \in \mathfrak{B}(e)$ with $W \subset U_0$ yields

$$g(x) = \frac{1}{2!} \sum_{i=1}^{n} \sum_{j=1}^{n} x_i(x)\, x_j(x)\, X_i X_j g(\zeta(x))$$

for all $x \in W$ (with $\zeta(x) \in W$).

Hence, there exists a constant $c_1 \in \mathbb{R}_+^*$ depending only on the dimension n of G such that

$$|g(x)| \le c_1 \|g\|_2 \sum_{i=1}^{n} x_i^2(x) \qquad \text{for all } x \in W$$

and therefore

$$\sup_{t \in \mathbb{R}_+^*} \left| \frac{1}{t} \int_W g \, d\mu_t \right|$$

$$\le c_1 \|g\|_2 \sup_{t \in \mathbb{R}_+^*} \frac{1}{t} \int_W \left(\sum_{i=1}^{n} x_i^2(x)\right) \mu_t(dx) < \infty,$$

the last inequality being implied by Lemma 4.1.13. Thus there is a constant $c \in \mathbb{R}_+^*$ (independent of t) such that for all $t \in \mathbb{R}_+^*$ we have

$$\left| \frac{1}{t} \int_W \left[f(x) - f(e) - \sum_{i=1}^{n} (A_i f)\, z_i(x)\right] \mu_t(dx) \right| \le c_2 \|g\|_2 .$$

On the other hand, $\left| \frac{1}{t} \int_{\complement W} g \, d\mu_t \right| \le \|g\|_2 \frac{\mu_t(\complement W)}{t}$ holds for all $t \in \mathbb{R}_+^*$. By Lemma 4.1.4 we get $\sup_{t \in \mathbb{R}_+^*} \frac{\mu_t(\complement W)}{t} < \infty$, hence $\left| \frac{1}{t} \int_{\complement W} g \, d\mu_t \right| < \infty$ and thus there exists a constant $\bar{c}_2 \in \mathbb{R}_+^*$ (independent of t) such that for all $t \in \mathbb{R}_+^*$ we have

$$\left| \frac{1}{t} \int_{\complement W} \left[f(x) - f(e) - \sum_{i=1}^{n} (A_i f)\, z_i(x)\right] \mu_t(dx) \right| \le \bar{c}_2 \|g\|_2 .$$

Adding the two inequalities involving the constants c_2 and $\bar{c}_2$ resp. we get

$$\left| \frac{1}{t} (S_t f(e) - f(e)) - \frac{1}{t} \sum_{i=1}^{n} A_i f\, S_t z_i(e) \right| \le (c_2 + \bar{c}_2) \|g\|_2 .$$

Since $z_i \in \mathsf{D}_N$ and $z_i(e) = 0$, $\sup_{t \in \mathbb{R}_+^*} \frac{1}{t} S_t z_i(e) < \infty$ for all $i = 1, \ldots, n$.

Hence, $\left| \frac{1}{t} \sum_{i=1}^{n} A_i f\, S_t z_i(e) \right|$ and so $\left| \frac{1}{t} (S_t f(e) - f(e)) \right|$ is bounded above by $c \|f\|_2$ (with c independent of f) for all $t \in \mathbb{R}_+^*$. □

4.1.15 Remark. If G is Abelian (in particular, if $G := \mathbb{R}^n$ for $n \ge 1$), then clearly $\mathscr{C}_2(G) = \tilde{\mathscr{C}}_2(G)$ and $\|\cdot\|_2 = \|\cdot\|_{\tilde{2}}$, and by the proof of the theorem there exists a constant $c \in \mathbb{R}_+^*$ such that

$$\left|\frac{1}{t}(S_t f(e) - f(e))\right| \leq c \|f\|_2^{\sim}$$

holds for all $f \in \tilde{\mathscr{C}}_2(G)$. Consequently, A exists on $\tilde{\mathscr{C}}_2(G)$ and

$$\left|\frac{1}{t}(S_t f(x) - f(x))\right| \leq c \|f\|_2^{\sim}$$

holds for all $f \in \tilde{\mathscr{C}}_2(G)$ and $x \in G$. By Remark 4.1.3 this implies that $\tilde{\mathscr{C}}_2(G) \subset \mathbf{D}_N$.

In order to get the existence of A on $\tilde{\mathscr{C}}_2(G)$ also for a general Lie group G we proceed as follows:

4.1.16 Theorem. *Let G be a Lie group and $(\mu_t)_{t \in \mathbb{R}_+^*}$ a convolution semigroup in $\mathscr{M}^1(G)$ with corresponding semigroup $(S_t)_{t \in \mathbb{R}_+^*}$. Then the generating functional A of $(S_t)_{t \in \mathbb{R}_+^*}$ exists also on $\tilde{\mathscr{C}}_2(G)$.*

Proof. For $(\mu_t)_{t \in \mathbb{R}_+^*}$ one considers the adjoint convolution semigroup $(\mu_t^{\sim})_{t \in \mathbb{R}_+^*}$ in $\mathscr{M}^1(G)$ defined by $\mu_t^{\sim}(B) = \mu_t(B^{-1})$ for all $B \in \mathfrak{B}(G)$, $t \in \mathbb{R}_+^*$. Clearly, $(\mu_t^{\sim})_{t \in \mathbb{R}_+^*}$ is again an $\{e\}$-continuous convolution semigroup with corresponding semigroup $(S_t^{\sim})_{t \in \mathbb{R}_+^*}$ on $\mathscr{C}_u(G)$. For $f \in \tilde{\mathscr{C}}_2(G)$ we have $f^{\sim} := f^* \in \mathscr{C}_2(G)$ and $\|f^{\sim}\|_2 = \|f\|_2^{\sim}$.

From the proof of Theorem 4.1.14 we conclude that

$$\lim_{t \downarrow 0} \frac{1}{t}(S_t^{\sim} f^{\sim}(e) - f^{\sim}(e))$$

exists for all $f \in \tilde{\mathscr{C}}_2(G)$. But the equalities

$$S_t^{\sim} f^{\sim}(e) = S_t f(e) \quad \text{and} \quad f^{\sim}(e) = f(e)$$

yield the existence of $\lim_{t \downarrow 0} \frac{1}{t}(S_t f(e) - f(e))$. □

4.1.17 Remark. Applying the existence of A to functions $L_x f$ for $f \in \tilde{\mathscr{C}}_2(G)$ and $x \in G$ one arrives at the existence of

$$\lim_{t \downarrow 0} \frac{1}{t}(S_t f(x) - f(x)) \quad \text{for all } f \in \tilde{\mathscr{C}}_2(G),\ x \in G,$$

but not necessarily of N on $\tilde{\mathscr{C}}_2(G)$.

4.2 Hunt's Representation Theorem

Let G be a Lie group of dimension $n \geq 1$ with Lie algebra $\mathfrak{L}(G)$ having a basis $\{X_1, \ldots, X_n\}$. In analogy to the preceding section we shall study the generating functional and the infinitesimal generator of contraction semigroups $(S_t)_{t\in\mathbb{R}_+^*}$ corresponding to convolution semigroups $(\mu_t)_{t\in\mathbb{R}_+^*}$ in $\mathcal{M}^1(G)$ on the space of continuous functions on G which are continuously extendable to the one-point compactification of G. Hence, we introduce the subspace

$$\overline{\mathscr{C}^0}(G) := \mathscr{C}^0(G) \oplus \mathbb{R}$$

of the Banach space $\mathscr{C}_u(G)$ of all real-valued uniformly continuous functions on G, where, as usual, $\mathscr{C}^0(G)$ denotes the space of those elements of $\mathscr{C}_u(G)$ that vanish at infinity. In a most natural way we now modify the spaces $\mathscr{C}_2(G)$ and $\tilde{\mathscr{C}}_2(G)$ to contain just functions $f \in \overline{\mathscr{C}^0}(G)$ with existing XYf and $\tilde{X}\tilde{Y}f$ resp. for all $X, Y \in \mathfrak{L}(G)$. Introducing the norms $\|\cdot\|_2$ and $\|\cdot\|_{\tilde{2}}$ on $\mathscr{C}_2^0(G)$ and $\tilde{\mathscr{C}}_2^0(G)$ we obtain Banach spaces $(\mathscr{C}_2^0(G), \|\cdot\|_2)$ and $(\tilde{\mathscr{C}}_2^0(G)\ \|\cdot\|_{\tilde{2}})$ resp. which are dense subspaces of $\overline{\mathscr{C}^0}(G)$ and are right and left invariant resp. By Lemma 4.1.1 for every convolution semigroup $(\mu_t)_{t\in\mathbb{R}_+^*}$ in $\mathcal{M}^1(G)$ there exists a contraction semigroup $(S_t)_{t\in\mathbb{R}_+^*}$ on $\overline{\mathscr{C}^0}(G)$ consisting of the probability operators $S_t := T_{\mu_t}$ of μ_t (for all $t \in \mathbb{R}_+^*$).

To the basis $\{X_1, \ldots, X_n\}$ of $\mathfrak{L}(G)$ there are associated a *compact canonical neighborhood* $U_0 \in \mathfrak{V}(e)$, a *canonical coordinate system* $\{x_1, \ldots, x_n\}$ *in* $\tilde{\mathscr{C}}_2^0(G) \cap \mathscr{K}(G)$ such that x_i is skew-symmetric, and hence satisfies $(X_i x_j)(e) = \delta_{ij}$ for $i, j = 1, \ldots, n$, and a *Hunt function* $\phi := \phi_{U_0}$ *for* G defined by

$$\phi(x) := \textstyle\sum_{i=1}^n x_i(x)^2 \qquad \text{for all } x \in U_0,$$

extended to a function in $\tilde{\mathscr{C}}_2^0(G)$, with $0 < \phi(x) \leq 1$ for all $x \in G^*$ and $\phi(\complement V) = 1$ for some compact $V \in \mathfrak{V}(e)$. This all follows from the proof of Lemma 4.1.9. Introducing for $k \geq 1$ the spaces

$$\mathscr{E}_k(G) := \{f \in \mathscr{C}^0(G) : f \text{ is } k\text{-times continuously differentiable in a neighborhood of } e \in G\} \quad \text{and}$$

$$\overline{\mathscr{E}_k}(G) := \mathscr{E}_k(G) \oplus \mathbb{R}$$

we obtain in analogy to Theorems 4.1.14 or 4.1.16 the following result:

4.2.1 Theorem. *Let G be a Lie group and $(\mu_t)_{t\in\mathbb{R}_+^*}$ a convolution semigroup in $\mathcal{M}^1(G)$ with corresponding semigroup $(S_t)_{t\in\mathbb{R}_+^*}$ on $\overline{\mathscr{C}^0}(G)$. Then the generating functional A of $(S_t)_{t\in\mathbb{R}_+^*}$ exists on $\overline{\mathscr{E}_2}(G)\ (\supset \tilde{\mathscr{C}}_2^0(G))$.*

In the following we shall extend the discussion to the one-point compactification G_∞ of G and to the corresponding space $G_\infty^* := G_\infty \setminus \{e\}$. Measures defined on G can always be considered as measures on G_∞ (by trivial extension). Whenever integrals with respect to measures in $\mathcal{M}^b(G_\infty)$ or $\mathcal{M}^b(G_\infty^*)$ appear their integrands are viewed (by trivial extension) as functions on G_∞ and G_∞^* resp.

Let U_0 be a canonical neighborhood of $e \in G$ and $\phi := \phi_{U_0}$ a Hunt function for G. For every $t \in \mathbb{R}_+^*$ we define the measure $\nu_t := \left(\frac{1}{t}\phi\right) \cdot \mu_t \in \mathcal{M}_+^b(G)$. By Theorem 4.2.1

one has for every $f \in \overline{\mathscr{E}_2}(G)$ the existence of the limit

$$\nu(f) := \lim_{t \downarrow 0} \int_G f \, d\nu_t = \lim_{t \downarrow 0} \frac{1}{t} \int f \phi \, d\mu_t = A(f\phi).$$

ν is a positive linear functional on the dense subspace $\overline{\mathscr{E}_2}(G)$ of $\overline{\mathscr{C}^0}(G)$ and hence is extendable to a positive Radon measure ν on G_∞. Since

$$\lim_{t \downarrow 0} \nu_t(1) = \nu(1),$$

the net $(\nu_t)_{t \in \mathbb{R}_+^*}$ in $\mathscr{M}_+^b(G_\infty)$ $\mathscr{T}_w$-converges to $\nu \in \mathscr{M}_+^b(G_\infty)$.

Putting $\eta := \mathrm{Res}_{G_\infty^*} \frac{1}{\phi} \cdot \nu$ we observe that $\eta \in \mathscr{M}_+(G_\infty^*)$.

4.2.2 Lemma. *Let $f \in \overline{\mathscr{E}_2}(G)$. We have the following assertions:*

(i) *If $\frac{f}{\phi} \in \mathscr{C}^b(G_\infty^*)$, then $\int_{G_\infty^*} f \, d\eta$ exists.*

(ii) *If $h := \frac{f}{\phi}$ is continuously extended to G by $h(e) = 0$, then*

$$\int_{G_\infty^*} f \, d\eta = A(f) = \lim_{t \downarrow 0} \frac{1}{t} \int_G f \, d\mu_t.$$

Proof. (i) Clearly, the function $h := 1_{G^*} \frac{|f|}{\phi}$ on G is lower semicontinuous and bounded, whence

$$\int_{G_\infty^*} |f| \, d\eta = \int_{G_\infty^*} |f| \frac{1}{\phi} \, d\nu = \int_{G_\infty} h \, d\nu < \infty.$$

(ii) Given $h\phi = f$ and $f(e) = 0$ we obtain

$$\lim_{t \downarrow 0} \int_G h \, d\nu_t = \lim_{t \downarrow 0} \frac{1}{t} \int_G f \, d\mu_t = A(f).$$

On the other hand the assumption $h(e) = 0$ yields

$$\lim_{t \downarrow 0} \int_G h \, d\nu_t = \int_{G_\infty} h \, d\nu = \int_{G_\infty^*} h \, d\nu = \int_{G_\infty^*} \frac{f}{\phi} \, d\nu = \int_{G_\infty^*} f \, d\eta,$$

which is the desired conclusion. □

As in 4.1.10 one defines for any function $f \in \overline{\mathscr{E}_2}(G)$ and $i, j = 1, \ldots, n$ the numbers $A_i f$ and $A_{ij} f$ and observes that the restrictions of the mappings $A_i : f \to A_i f$ and $A_{ij} : f \to A_{ij} f$ to $\mathscr{C}_2^0(G)$ and $\tilde{\mathscr{C}}_2^0(G)$ are continuous linear functionals.

4.2.3 Remark. The hypothesis in (ii) of Lemma 4.2.2 is satisfied if

$$f(e) = A_i(f) = A_{ij}(f) = 0 \qquad \text{for all } 1 \leq i, j \leq n$$

holds. This is an immediate consequence of the generalized Taylor expansion in a neighborhood of $e \in G$ together with the fact that $A_{ii}\,\phi = 2$ for all $1 \leq i \leq n$. We shall now fix a canonical neighborhood $U_0 \in \mathfrak{B}(e)$, a system $\{x_1, \ldots, x_n\}$ of canonical coordinates in $\tilde{\mathscr{C}}_2^0(G)$ with $x_i(e) = 0$, $A_j x_i = \delta_{ij}$ for $1 \leq i, j \leq n$, a Hunt function $\phi := \phi_{U_0} \in \tilde{\mathscr{C}}_2(G)$ with $\phi(e) = A_i\,\phi = A_{ij}\,\phi = 0$ for $1 \leq i, j \leq n$ with $i \neq j$ and $A_{ii}\,\phi = 2$ for $1 \leq i \leq n$ such that for any $V \in \mathfrak{B}(e)$ there exists a $\delta > 0$ satisfying $\phi(x) \geq \delta$ for all $x \in \complement V$ and formulate the following

4.2.4 Theorem. *Let $(\mu_t)_{t \in \mathbb{R}_+^*}$ be a convolution semigroup in $\mathscr{M}^1(G)$ with generating functional A existing on the space $\overline{\mathscr{E}_2}(G)$. Then there exist numbers $a_1, \ldots, a_n \in \mathbb{R}$, a symmetric, positive-semidefinite matrix $(a_{ij})_{i,j=1,\ldots,n} \in \mathfrak{M}(n, \mathbb{R})$ and a measure $\eta \in \mathscr{M}_+(G_\infty^*)$ satisfying $\int_{G_\infty^*} \phi\, d\eta < \infty$ such that for all $f \in \overline{\mathscr{E}_2}(G)$ one has*

$$Af = \sum_{i=1}^n a_i(\tilde{X}_i f)(e) + \sum_{i,j=1}^n a_{ij}(\tilde{X}_i \tilde{X}_j f)(e)$$
$$+ \int_{G_\infty^*} [f(x) - f(e) - \sum_{i=1}^n (\tilde{X}_i f)(e)\, x_i(x)]\, \eta(dx).$$

Proof. Let N be the infinitesimal generator of $(\mu_t)_{t \in \mathbb{R}_+^*}$ with domain D_N. By Lemma 4.1.12 and its proof there exist functions x_i, x_{ij} $(1 \leq i, j \leq n)$ in $\mathsf{D}_N \cap \mathscr{C}_2(G)$ such that

(i) $x_i(e) = x_{ij}(e) = A_i x_{jl} = 0$,
(ii) $A_j x_i = \delta_{ij}$ and
(iii) $A_{ij} x_{lm} = \delta_{il}\delta_{jm} + \delta_{im}\delta_{jl}$ for $1 \leq i, j, l, m \leq n$.

Let $f \in \overline{\mathscr{E}_2}(G)$ and put

$$c := f(e), \quad c_i := A_i f \quad \text{and}$$
$$c_{ij} := \tfrac{1}{2} A_{ij}(f - \sum_{l=1}^n c_l x_l) \quad (i, j = 1, \ldots, n).$$

Clearly, $c_{ij} = c_{ji}$, since

$$A_{ij} f = \frac{\partial^2}{\partial x_i\, \partial x_j} f(e) + \sum_{l=1}^n (A_l f) A_{ij} x_l \qquad \text{for } i, j = 1, \ldots, n.$$

But then the function

$$g := f - c - \sum_{i=1}^n c_i x_i - \sum_{i,j=1}^n c_{ij} x_{ij}$$

lies in $\overline{\mathscr{E}_2}(G)$ and satisfies $g(e) = A_i g = A_{ij} g = 0$ for $1 \leq i, j \leq n$. Lemma 4.2.2 together with Remark 4.2.3 yields the representation

$$A(f) = A(c + \sum_{i=1}^n c_i x_i + \sum_{i,j=1}^n c_{ij} x_{ij}) + \int_{G_\infty^*} g\, d\eta$$
$$= \sum_{i=1}^n b_i c_i + \sum_{i,j=1}^n b_{ij} c_{ij}$$
$$+ \int_{G_\infty^*} [f - c - \sum_{i=1}^n c_i x_i - \sum_{i,j=1}^n c_{ij} x_{ij}]\, d\eta$$

for properly chosen b_i and b_{ij} $(1 \leq i, j \leq n)$.

But since $A_l x_{ij} = 0$ for $l = 1, \ldots, n$, we have $\dfrac{x_{ij}}{\phi} \in \mathscr{C}^b(G^*)$, whence the existence of $\int_{G_\infty^*} x_{ij}\, d\eta$ by Lemma 4.2.2 $(i, j = 1, \ldots, n)$. We put

$$a_i := b_i + \tfrac{1}{2} \sum_{j,l=1}^n A_{jl} x_i (\int_{G_\infty^*} x_{jl}\, d\eta) \quad \text{and}$$
$$a_{ij} := \tfrac{1}{2}(b_{ij} - \int_{G_\infty^*} x_{ij}\, d\eta) \quad (i, j = 1, \ldots, n).$$

Then

$$Af = \sum_{i=1}^{n} a_i A_i f + \sum_{i,j=1}^{n} a_{ij} A_{ij} f + \int_{G_\infty^*} [f - f(e) - \sum_{i=1}^{n} (A_i f) x_i]\, d\eta.$$

Without loss of generality we suppose $x_{ij} = x_{ji}$ and therefore get

$$a_{ij} = a_{ji} \quad \text{for all } i, j = 1, \ldots, n.$$

Finally, $(a_{ij})_{i,j=1,\ldots,n} \in \mathfrak{M}(n, \mathbb{R})$ is symmetric and positive-semidefinite. In fact, let $f \in \overline{\mathscr{E}}_2(G)$ with $f \geq 0$ and $f(e) = A_i f = 0$ for $1 \leq i \leq n$. Then from Lemma 4.2.2 we infer that

$$\sum_{i,j=1}^{n} a_{ij} A_{ij} f + \int_{G_\infty^*} f\, d\eta = A(f) = \lim_{t \downarrow 0} \frac{1}{t} \int_G f\, d\mu_t \geq 0.$$

Assuming that $(a_{ij})_{i,j=1,\ldots,n}$ were not positive-semidefinite one would get the existence of a function f of the above sort satisfying

$$\sum_{i,j=1}^{n} a_{ij} A_{ij} f < 0.$$

For any $k \geq 1$ let $f_k := f e^{-k\phi^2}$. Then

$$A_{ij} f_k = A_{ij} f \quad \text{for all } i, j = 1, \ldots, n;\ k \geq 1$$

and we get

$$\lim_{k \to \infty} \int_{G_\infty^*} f_k\, d\eta = 0.$$

For k sufficiently large we therefore obtain

$$\sum_{i,j=1}^{n} a_{ij} A_{ij} f_k + \int_{G_\infty^*} f_k\, d\eta < 0,$$

in contradiction to the above positivity. □

The following result is a certain converse of the preceding theorem.

4.2.5 Theorem. *Let G be a Lie group of dimension $n \geq 1$, U_0 a canonical neighborhood of e, $\{x_1, \ldots, x_n\}$ a canonical coordinate system and $\phi := \phi_{U_0}$ a Hunt function as defined above. Furthermore, let $a_1, \ldots, a_n \in \mathbb{R}$, $(a_{ij})_{i,j=1,\ldots,n} \in \mathfrak{M}(n, \mathbb{R})$ symmetric, positive-semidefinite and $\eta \in \mathscr{M}_+(G^*)$ with $\int_{G^*} \phi\, d\eta < \infty$ be given. For all $f \in \tilde{\mathscr{C}}_2^0(G)$ and $x \in G$ we define*

$$\text{(LK)} \qquad \begin{aligned} Nf(x) := &\sum_{i=1}^{n} a_i (\tilde{X}_i f)(x) + \sum_{i,j=1}^{n} a_{ij} (\tilde{X}_i \tilde{X}_j f)(x) \\ &+ \int_{G^*} [f(xy) - f(x) - \sum_{i=1}^{n} (\tilde{X}_i f)(x)\, x_i(y)]\, \eta(dy). \end{aligned}$$

Then there exists exactly one convolution semigroup in $\mathscr{M}^1(G)$ with corresponding semigroup $(S_t)_{t \in \mathbb{R}_+^*}$ and infinitesimal generator N' such that $\operatorname{Res}_{\tilde{\mathscr{C}}_2^0(G)} N' = N$.

Given a subspace E of $\mathscr{C}_u(G)$ we shall say that the operator N *admits on* E *a representation* $(a_i, a_{ij}, \eta)_{1 \leq i,j \leq n}$ if for any $f \in \mathsf{E}$ and $x \in G$ the *Lévy-Khintchine formula* (LK) is valid.

We prepare the proof of the theorem by two lemmata.

4.2.6 Lemma. *Let $(\mu_t)_{t\in\mathbb{R}_+^*}$ be a convolution semigroup in $\mathscr{M}^1(G)$, $(S_t)_{t\in\mathbb{R}_+^*}$ the corresponding semigroup and A the generating functional of $(S_t)_{t\in\mathbb{R}_+^*}$. Then $\mathrm{Res}_{\tilde{\mathscr{C}}_2^0(G)} A$ determines the semigroup $(S_t)_{t\in\mathbb{R}_+^*}$ uniquely.*

Proof. From the construction of the numbers $a_1, \ldots, a_n \in \mathbb{R}$, the matrix $(a_{ij})_{i,j=1,\ldots,n} \in \mathfrak{M}(n, \mathbb{R})$ and the measure $\eta \in \mathscr{M}_+(G_\infty^*)$ which appear in the representation of A given in Theorem 4.2.4 we see that they are determined by $\mathrm{Res}_{\tilde{\mathscr{C}}_2^0(G)} A$ and hence by $\mathrm{Res}_{\overline{\mathscr{E}}_2(G)} A$.

Putting $S_t' := \mathrm{Res}_{\mathscr{C}_2^0(G)} S_t$ for every $t \in \mathbb{R}_+^*$ and denoting the infinitesimal generator of the semigroup $(S_t')_{t\in\mathbb{R}_+^*}$ by N' we conclude from the invariance condition $(N'f)(x) = A(L_x f)$ for all $f \in \mathrm{D}_{N'} \subset \mathscr{C}_2^0(G)$ that N' is uniquely determined by $\mathrm{Res}_{\tilde{\mathscr{C}}_2^0(G)} A$. From the Hille-Yosida theory one deduces the uniqueness of the semigroup $(S_t')_{t\in\mathbb{R}_+^*}$. But since $\mathscr{C}_2^0(G)$ is dense in $\overline{\mathscr{C}^0}(G)$, the convolution semigroup $(\mu_t)_{t\in\mathbb{R}_+^*}$ in $\mathscr{M}^1(G)$ is uniquely determined by $\mathrm{Res}_{\tilde{\mathscr{C}}_2^0(G)} A$, and hence also $(S_t)_{t\in\mathbb{R}_+^*}$. □

4.2.7 Lemma. *For every $m \geq 1$ let $(\mu_t^{(m)})_{t\in\mathbb{R}_+^*}$ be a convolution semigroup in $\mathscr{M}^1(G)$ and $(S_t^{(m)})_{t\in\mathbb{R}_+^*}$ the corresponding contraction semigroup on $\overline{\mathscr{C}^0}(G)$ with infinitesimal generator $N^{(m)}$ such that $N^{(m)}$ admits on $\tilde{\mathscr{C}}_2^0(G)$ the representation $(a_i^{(m)}, a_{ij}^{(m)}, \eta^{(m)})_{1\leq i,j\leq n}$ with numbers $a_1^{(m)}, \ldots, a_n^{(m)} \in \mathbb{R}$, a symmetric, positive-semi-definite matrix $(a_{ij}^{(m)})_{i,j=1,\ldots,n} \in \mathfrak{M}(n, \mathbb{R})$ and a measure $\eta^{(m)} \in \mathscr{M}_+(G^*)$.*

For every $f \in \tilde{\mathscr{C}}_2^0(G)$ let the sequence $(N^{(m)} f)_{m\geq 1}$ converge uniformly to a function Nf such that N admits on $\tilde{\mathscr{C}}_2^0(G)$ a representation $(a_i, a_{ij}, \eta)_{1\leq i,j\leq n}$ as above. Then there exists a convolution semigroup $(\mu_t)_{t\in\mathbb{R}_+^*}$ in $\mathscr{M}^1(G)$ with corresponding contraction semigroup $(S_t)_{t\in\mathbb{R}_+^*}$ on $\overline{\mathscr{C}^0}(G)$, whose infinitesimal generator N' is defined on $\tilde{\mathscr{C}}_2^0(G)$, and one has

(i) *$N'f = Nf$ for all $f \in \tilde{\mathscr{C}}_2^0(G)$ and*

(ii) *$\lim_{m\to\infty} \|S_t^{(m)} f - S_t f\| = 0$ for all $f \in \overline{\mathscr{C}^0}(G)$ uniformly in t on compact subsets of $\mathbb{R}_+$.*

Proof. 1. For every $m \geq 1$ and $f \in \tilde{\mathscr{C}}_2^0(G)$ one obtains

$$\left\| \frac{d}{dt} S_t^{(m)} f \right\| = \lim_{h\to 0} \frac{1}{h} \|(S_{t+h}^{(m)} - S_t^{(m)}) f\| = \|S_t^{(m)} N^{(m)} f\| \leq \|N^m f\|$$

and $\lim_{m\to\infty} \|N^{(m)} f\| = \|Nf\|$. Hence, there exists a constant $c_f \in \mathbb{R}_+^*$ satisfying the inequality $\|S_s^{(m)} f - S_t^{(m)} f\| \leq c_f |s-t|$ whenever $s, t \in \mathbb{R}_+^*$.

2. Since there exists a σ-compact open subgroup H of G with

$$\mu_t^{(m)}(\complement H) = 0 \qquad \text{for all } t \in \mathbb{R}_+^* \text{ and } m \geq 1,$$

we can apply the Kakutani-Kodaira theorem ([218], (8.7) valid also for σ-compact groups) and we may assume without loss of generality that G admits a countable basis of its topology. Transition to an appropriate subsequence yields for every $t \in \mathbb{Q}_+^*$ the existence of $\mu_t \in \mathscr{M}_+^b(G)$ satisfying

$$\mathscr{T}_v\text{-}\lim_{k\to\infty} \mu_t^{(m_k)} = \mu_t.$$

By the last inequality in Part 1 we may furthermore assume that

$$\mathscr{T}_v\text{-}\lim_{k\to\infty} \mu_t^{(m_k)} = \mu_t \qquad \text{for all } t\in\mathbb{R}_+^*$$

since $\tilde{\mathscr{C}}_2^0(G)$ is norm-dense in $\overline{\mathscr{C}^0}(G)$.

3. For every $t\in\mathbb{R}_+^*$ let S_t be the probability operator corresponding to μ_t. For every $f\in\mathscr{C}^0(G)$ and $x\in G$ we thus obtain

$$S_t f(x) = \lim_{k\to\infty} S_t^{(m_k)} f(x)$$

and consequently for all $f\in\tilde{\mathscr{C}}_2^0(G)\cap\mathscr{C}^0(G)$ and $s,t\in\mathbb{R}_+^*$ the inequality

$$\|S_s f - S_t f\| \le c_f |s-t|$$

which implies $\lim_{t\downarrow 0} S_t f = f$ for all $f\in\mathscr{C}^0(G)$. Moreover, one has for each $f\in\tilde{\mathscr{C}}_2^0(G)$, $x\in G$, $t\in\mathbb{R}_+^*$ and $k\ge 1$ the inequalities

$$\begin{aligned}
&|S_t^{(m_k)} N^{(m_k)} f(x) - S_t N f(x)| \\
&\le \|S_t^{(m_k)} N^{(m_k)} f - S_t^{(m_k)} N f\| + |S_t^{(m_k)} N f(x) - S_t N f(x)| \\
&\le \|N^{(m_k)} f - N f\| + |S_t^{(m_k)}(N f)(x) - S_t(N f)(x)|,
\end{aligned}$$

which by assumption yield $\lim_{k\to\infty} S_t^{(m_k)} N^{(m_k)} f(x) = S_t N f(x)$. By the Hille-Yosida theory [241], p. 308 we get for every $k\ge 1$, $t\in\mathbb{R}_+^*$ and $f\in\tilde{\mathscr{C}}_2^0(G)$ the formula

$$S_t^{(m_k)} f - f = \int_0^t S_s^{(m_k)} N^{(m_k)} f\, ds$$

and thus by the Lebesgue convergence theorem we obtain $S_t f - f = \int_0^t S_s N f\, ds$, whence from

$$\begin{aligned}
&\left|\frac{1}{t}(S_t f(x) - f(x)) - N f(x)\right| \le \frac{1}{t}\int_0^t |S_s N f(x) - N f(x)|\, ds \\
&\le \sup_{0\le s\le t} \|S_s(N f) - N f\|
\end{aligned}$$

the relationship

$$\lim_{t\downarrow 0} \frac{1}{t}(S_t f - f) = \lim_{t\downarrow 0} \frac{1}{t}\int_0^t S_s N f\, ds = N f.$$

4. We now show that $\mu_t\in\mathscr{M}^1(G)$ and deduce that

$$\mathscr{T}_w\text{-}\lim_{k\to\infty} \mu_t^{(m_k)} = \mu_t \qquad \text{for all } t\in\mathbb{R}_+^*.$$

In fact, let U be a compact neighborhood in $\mathfrak{V}(e)$ and let $f\in\tilde{\mathscr{C}}_2^0(G)$ satisfy $0\le f\le 1$ and $f(U)=1$. From Part 3 we obtain the inequalities

$$\begin{aligned}
&\underline{\lim}_{t\downarrow 0} \frac{1}{t}(\mu_t(G) - 1) \ge \lim_{t\downarrow 0} \frac{1}{t}(\mu_t(f) - 1) = \lim_{t\downarrow 0} \frac{1}{t}(S_t f(e) - f(e)) \\
&= N f(e) = \int_{\complement U} (f(y) - 1)\, \eta(dy) \ge -\eta(\complement U) \qquad \text{for all } U\in\mathfrak{V}(e),
\end{aligned}$$

whence $\underline{\lim}_{t\downarrow 0} \frac{1}{t}(\mu_t(G) - 1) \ge 0.$

We assume that $\mu_s(G)=e^{-cs}$ with $c\in\mathbb{R}_+$ for some $s\in\mathbb{R}_+^*$ and choose for $f\in\mathscr{K}(G)$ with $1_{\{e\}}\leq f\leq 1_G$ a function $g\in\mathscr{K}(G)$ satisfying $1_{C^2}\leq g\leq 1_G$, where $C:=\operatorname{supp}(f)$. Then with $t:=\frac{s}{2}$ we conclude that

$$\mu_t^{(m_k)}(f)\leq\mu_t^{(m_k)}(C)\leq[\mu_s^{(m_k)}(C^2)]^{\frac{1}{2}}\leq[\mu_s^{(m_k)}(g)]^{\frac{1}{2}}.$$

For $k\to\infty$ this yields

$$\mu_t(f)\leq[\mu_s(g)]^{\frac{1}{2}}\leq[\mu_s(G)]^{\frac{1}{2}}=e^{-ct}$$

and with f approaching 1_G also $\mu_t(G)\leq e^{-ct}$. Iteration with $t(n):=\frac{s}{2^n}$ gives $\mu_{t(n)}(G)\leq e^{-ct(n)}$ for all $n\geq 1$ and

$$0\leq\underline{\lim}_{t\downarrow 0}\frac{1}{t}(\mu_t(G)-1)\leq\frac{1}{s}\underline{\lim}_{n\geq 1}2^n(\mu_{t(n)}(G)-1)$$

$$\leq\frac{1}{s}\lim_{n\to\infty}2^n(e^{-ct(n)}-1)=-c$$

implies that $c=0$ or $\mu_s(G)=1$ for $s\in\mathbb{R}_+^*$.

5. By Theorem 1.5.5 we conclude that

$$\lim_{k\to\infty}\|S_t^{(m_k)}f-S_tf\|=0\qquad\text{for all } f\in\overline{\mathscr{C}^0}(G)\text{ and } t\in\mathbb{R}_+^*.$$

Thus $(S_t)_{t\in\mathbb{R}_+^*}$ is a positive, invariant semigroup of operators on $\overline{\mathscr{C}^0}(G)$. By Part 3 this semigroup is strongly continuous on $\overline{\mathscr{C}^0}(G)$, and on $\tilde{\mathscr{C}}_2^0(G)$ its infinitesimal generator N' coincides with N.

6. By Lemma 4.2.6 the semigroup $(S_t)_{t\in\mathbb{R}_+^*}$ is uniquely determined by N. This implies that the discussion above is independent of the particular choice of the subsequence $(m_k)_{k\geq 1}$ of $\mathbb{N}$. Therefore

$$\lim_{m\to\infty}\|S_t^{(m)}f-S_tf\|=0\qquad\text{for all } f\in\overline{\mathscr{C}^0}(G)\text{ and } t\in\mathbb{R}_+^*.$$

Moreover, the convergence holds uniformly in t on compact subsets of $\mathbb{R}_+$, as can be seen from the last inequality of Part 1. This completes the proof. □

We are ready to give the *proof of Theorem* 4.2.5. 1. Let $a_1,\ldots,a_n\in\mathbb{R}$ and $(a_{ij})_{i,j=1,\ldots,n}$ a symmetric, positive-semidefinite matrix in $\mathfrak{M}(n,\mathbb{R})$. Then the operator

$$\Delta:=\sum_{i=1}^n a_iX_i+\sum_{i,j=1}^n a_{ij}X_iX_j\qquad\text{on } \tilde{\mathscr{C}}_2^0(G)$$

can be transformed into $Y+\sum_{i=1}^l Y_i^2$ by a change of the basis in $\mathfrak{L}(G)$ such that Y and $Y_1,\ldots,Y_l$ are elements of $\mathfrak{L}(G)$ with $1\leq l\leq n$. For every $s\in\mathbb{R}^*$ we define as usual

$$\zeta(s):=\exp(sY)\quad\text{and}\quad\zeta_i(s):=\exp(sY_i)\qquad(i=1,\ldots,l)$$

and also

$$\mu_m:=m\varepsilon_{\zeta(\frac{1}{m})}+m^2\sum_{i=1}^l(\varepsilon_{\zeta_i(\frac{1}{m})}+\varepsilon_{\zeta_i(-\frac{1}{m})})\qquad(m\geq 1).$$

Clearly, $\mu_m \in \mathscr{M}_+^b(G^*)$ for every $m \geq 1$, and for any $f \in \tilde{\mathscr{C}}_2^0(G)$ we obtain

$$\Delta f(x) = \lim_{m \to \infty} \int_{G^*} [f(xy) - f(x)]\, \mu_m(dy)$$

uniformly in $x \in G$.

2. Let us now start with an operator N on $\tilde{\mathscr{C}}_2^0(G)$ of the form

$$Nf(x) = \int_{G^*} [f(xy) - f(x)]\, \eta(dy)$$

for all $f \in \tilde{\mathscr{C}}_2^0(G)$, $x \in G$, where $\eta \in \mathscr{M}_+^b(G^*)$. Clearly, N can be extended uniquely to a bounded operator on $\overline{\mathscr{C}^0}(G)$, which will again be denoted by N. For every $t \in \mathbb{R}_+^*$ we put

$$S_t := \sum_{m \geq 0} \frac{t^m}{m!} N^m \quad \text{and} \quad \mu_t := \exp_{\{e\}}[t(\eta - \|\eta\|\, \varepsilon_e)].$$

Plainly $(\mu_t)_{t \in \mathbb{R}_+^*}$ is a convolution semigroup in $\mathscr{M}^1(G)$ with corresponding semigroup $(S_t)_{t \in \mathbb{R}_+^*}$ such that the infinitesimal generator of $(S_t)_{t \in \mathbb{R}_+^*}$ coincides on $\tilde{\mathscr{C}}_2^0(G)$ with N.

3. Let then

$$Nf(x) = \sum_{i=1}^n a_i (\tilde{X}_i f)(x) + \int_{G^*} [f(xy) - f(x) - \sum_{i=1}^n (\tilde{X}_i f)(x)\, x_i(y)]\, \eta(dy)$$

with $a_1, \ldots, a_n \in \mathbb{R}$ and $\eta \in \mathscr{M}_+^b(G^*)$ $(f \in \tilde{\mathscr{C}}_2^0(G), x \in G)$. We can choose $Y \in \mathfrak{L}(G)$ such that

$$Nf(x) = \tilde{Y} f(x) + \int_{G^*} [f(xy) - f(x)]\, \eta(dy) \qquad \text{for all } f \in \tilde{\mathscr{C}}_2^0(G)$$

and $x \in G$. By Part 1 there exists a sequence $(\mu_m)_{m \geq 1}$ in $\mathscr{M}_+^b(G^*)$ satisfying

$$\lim_{m \to \infty} \| \tilde{Y} f - \int_{G^*} [f_y - f]\, \mu_m(dy) \| = 0.$$

Defining for every $m \geq 1$

$$N_m f(x) := \int_{G^*} [f(xy) - f(x)] (\eta(dy) + \mu_m(dy)) \qquad (f \in \tilde{\mathscr{C}}_2^0(G), x \in G)$$

one concludes from Part 2 that to every N_m there corresponds an $\{e\}$-continuous convolution semigroup in $\mathscr{M}^1(G)$. We have

$$\lim_{m \to \infty} \|N_m f - Nf\| = 0 \qquad \text{for all } f \in \tilde{\mathscr{C}}_2^0(G).$$

Hence, by Lemma 4.2.7 to N there corresponds a convolution semigroup in $\mathscr{M}^1(G)$.

4. It remains to treat the general case. Let $(\mu_m)_{m \geq 1}$ be a sequence in $\mathscr{M}_+^b(G^*)$ satisfying

$$\lim_{m \to \infty} \| \sum_{i,j=1}^n a_{ij} \tilde{X}_i \tilde{X}_j f - \int_{G^*} [f_y - f]\, \mu_m(dy) \| = 0.$$

Such a sequence exists by Part 1. For $m \geq 1$, $f \in \tilde{\mathscr{C}}_2^0(G)$ and $x \in G$ we define

$$N_m f(x) := \sum_{i,j=1}^n a_i (\tilde{X}_i f)(x) + \int_{G^*} [f(xy) - f(x)]\, \mu_m(dy)$$
$$+ \int_{G^*} [f(xy) - f(x) - \sum_{i=1}^n (\tilde{X}_i f)(x)\, x_i(y)]\, \eta_m(dy),$$

where $(\eta_m)_{m \geq 1}$ is the sequence of measures $(1 - e^{-m\phi}) \cdot \eta$ in $\mathscr{M}_+^b(G^*)$ satisfying $\mathscr{T}_w$-$\lim_{m \to \infty} \eta_m = \eta$.

Following Part 3 we obtain for every N_m a convolution semigroup in $\mathscr{M}^1(G)$. Moreover,

$$\lim_{m\to\infty} \|N_m f - Nf\| = 0 \qquad \text{for all } f \in \tilde{\mathscr{C}}_2^0(G).$$

Thus by Lemma 4.2.7, to the operator N of the theorem there corresponds a convolution semigroup in $\mathscr{M}^1(G)$, whose uniqueness follows immediately from Lemma 4.2.6. □

We finish the discussion of this section with the complete formulation of Hunt's theorem including all statements to which we shall refer.

4.2.8 Theorem. *Let G be a Lie group of dimension $n \geq 1$ with Lie algebra $\mathfrak{L}(G)$ having a basis $\{X_1, \ldots, X_n\}$, canonical neighborhood $U_0 \in \mathfrak{V}(e)$, canonical coordinate system $\{x_1, \ldots, x_n\}$ (with respect to $\{X_1, \ldots, X_n\}$ and U_0) and Hunt function $\phi := \phi_{U_0}$.*

(A) *Let $(\mu_t)_{t\in\mathbb{R}_+^*}$ be a convolution semigroup in $\mathscr{M}^1(G)$ with corresponding semigroup $(S_t)_{t\in\mathbb{R}_+^*}$. Then*

(i) *the infinitesimal generator N of $(S_t)_{t\in\mathbb{R}_+^*}$ is defined at least on the space $\tilde{\mathscr{C}}_2^0(G)$ and*

(ii) *there are numbers $a_1, \ldots, a_n \in \mathbb{R}$, a symmetric, positive-semidefinite matrix $(a_{ij})_{i,j=1,\ldots,n} \in \mathfrak{M}(n, \mathbb{R})$, and a measure $\eta \in \mathscr{M}_+(G^*)$ with $\int_{G^*} \phi\, d\eta < \infty$ such that N admits on $\tilde{\mathscr{C}}_2^0(G)$ the representation $(a_i, a_{ij}, \eta)_{1\leq i,j\leq n}$. Moreover,*

(iii) *the convolution semigroup $(\mu_t)_{t\in\mathbb{R}_+^*}$ in $\mathscr{M}^1(G)$ is uniquely determined by* $\mathrm{Res}_{\tilde{\mathscr{C}}_2^0(G)} N$, *and one has*

(iv) $\int_{G^*} f\, d\eta = \lim_{t\downarrow 0} \frac{1}{t} \int_G f\, d\mu_t$ *for all $f \in \mathscr{K}(G^*)$.*

(B) *Given numbers $a_1, \ldots, a_n \in \mathbb{R}$, a symmetric, positive-semidefinite matrix $(a_{ij})_{i,j=1,\ldots,n} \in \mathfrak{M}(n, \mathbb{R})$ and a measure $\eta \in \mathscr{M}_+(G^*)$ satisfying $\int_{G^*} \phi\, d\eta < \infty$ there exists exactly one convolution semigroup $(\mu_t)_{t\in\mathbb{R}_+^*}$ in $\mathscr{M}^1(G)$ with corresponding $(S_t)_{t\in\mathbb{R}_+^*}$ and infinitesimal generator N' such that N' admits on $\tilde{\mathscr{C}}_2^0(G)$ the representation $(a_i, a_{ij}, \eta)_{1\leq i,j\leq n}$.*

Proof. We are given the statements of Theorems 4.2.4 and 4.2.5 which take care of the proofs of (B) and (iii) of (A).

1. We shall now show (i) of (A). To $(\mu_t)_{t\in\mathbb{R}_+^*}$ there corresponds the generating functional A defined on the space $\overline{\mathscr{E}}_2(G) \supset \tilde{\mathscr{C}}_2^0(G)$. We put

$$(Nf)(x) := A(L_x f) \qquad \text{for all } f \in \tilde{\mathscr{C}}_2^0(G) \text{ and } x \in G.$$

By Part (B) of the theorem there exists a convolution semigroup $(\mu_t')_{t\in\mathbb{R}_+^*}$ in $\mathscr{M}^1(G)$ with corresponding semigroup $(S_t')_{t\in\mathbb{R}_+^*}$ whose infinitesimal generator is defined on $\tilde{\mathscr{C}}_2^0(G)$ and coincides on $\tilde{\mathscr{C}}_2^0(G)$ with N. But (iii) of (A) implies that $\mu_t = \mu_t'$ for all $t \in \mathbb{R}_+^*$. Hence, on $\tilde{\mathscr{C}}_2^0(G)$ the operator N is also the infinitesimal generator of the semigroup $(S_t)_{t\in\mathbb{R}_+^*}$ corresponding to $(\mu_t)_{t\in\mathbb{R}_+^*}$. This implies that $\tilde{\mathscr{C}}_2^0(G) \subset \mathsf{D}_N$.

2. Next we extend the statement of Theorem 4.2.4 to (ii) (and (iv)) of (A): By Theorem 4.2.4 we have a representation of A of the form

$$\begin{aligned} Af &= \textstyle\sum_{i=1}^n a_i(\tilde{X}_i f)(e) + \sum_{i,j=1}^n a_{ij}(\tilde{X}_i \tilde{X}_j f)(e) \\ &\quad + \textstyle\int_{G_\infty^*} [f(x) - f(e) - \sum_{i=1}^n (\tilde{X}_i f)(e)\, x_i(x)]\, \eta(dx) \end{aligned}$$

for all $f \in \overline{\mathscr{E}_2}(G)$, where $a_1, \ldots, a_n \in \mathbb{R}$, $(a_{ij})_{i,j=1,\ldots,n}$ is a symmetric, positive-semidefinite matrix in $\mathfrak{M}(n, \mathbb{R})$ and $\eta \in \mathscr{M}_+(G^*_\infty)$.

Putting $\eta := \bar{\eta} + \eta(\{\infty\})\,\varepsilon_e$ we obtain

$$\begin{aligned} Af &= \textstyle\sum_{i=1}^n a_i(\tilde{X}_i f)(e) + \sum_{i,j=1}^n a_{ij}(\tilde{X}_i \tilde{X}_j f)(e) \\ &\quad + c\,\varepsilon_e(f) + \int_{G^*}[f(x) - f(e) - \textstyle\sum_{i=1}^n (\tilde{X}_i f)(e)\, x_i(x)]\,\bar{\eta}(dx) \end{aligned}$$

for all $f \in \overline{\mathscr{E}_2}(G)$, where $-c := \eta(\{\infty\}) \in \mathbb{R}_+$ and $\bar{\eta} \in \mathscr{M}_+(G^*)$.

Thus $A = \bar{A} + c\,\varepsilon_e$ with

$$\begin{aligned} \bar{A}f &:= \textstyle\sum_{i=1}^n a_i(\tilde{X}_i f)(e) + \sum_{i,j=1}^n a_{ij}(\tilde{X}_i \tilde{X}_j f)(e) \\ &\quad + \int_{G^*}[f(x) - f(e) - \textstyle\sum_{i=1}^n (\tilde{X}_i f)(e)\, x_i(x)]\,\bar{\eta}(dx) \end{aligned}$$

for all $f \in \overline{\mathscr{E}_2}(G)$.

From (B) we conclude that for $\bar{A}$ there exists exactly one continuous convolution semigroup $(\bar{\mu}_t)_{t \in \mathbb{R}^*_+}$ in $\mathscr{M}^1(G)$ with contraction semigroup $(\bar{S}_t)_{t \in \mathbb{R}^*_+}$ and generating functional $\bar{A}$. On the other hand for A there exists a unique contraction semigroup $(e^{ct}\,\bar{S}_t)_{t \in \mathbb{R}^*_+}$ of operators on $\tilde{\mathscr{C}}_2^0(G)$ with infinitesimal generator $S_A = S_{\bar{A} + c\varepsilon_e}$ defined by $(S_A f)(x) := A(L_x f)$ for all $f \in \tilde{\mathscr{C}}_2^0(G)$, $x \in G$.

Since S_A is the infinitesimal generator of the contraction semigroup $(S_t)_{t \in \mathbb{R}^*_+}$ on $\tilde{\mathscr{C}}_2^0(G)$, we obtain on $\tilde{\mathscr{C}}_2^0(G)$ the equality

$$S_t = e^{-\eta(\{\infty\})t}\,\bar{S}_t, \quad \text{whence}$$

$$\mu_t = e^{-\eta(\{\infty\})t}\,\bar{\mu}_t \quad \text{for all } t \in \mathbb{R}^*_+,$$

which implies $\eta(\{\infty\}) = 0$. □

4.3 The Lévy-Khintchine Formula for Almost Periodic Groups

We shall first extend Theorem 4.2.8 in the sense that the function space $\tilde{\mathscr{C}}_2^0(G)$ which is proved to be in the domain of the infinitesimal generator of a given $\{e\}$-continuous convolution semigroup in $\mathscr{M}^1(G)$ can be enlarged to the space $\mathscr{R}(G) := \tilde{\mathscr{C}}_2^0(G) \oplus \mathfrak{K}(G)$ generated in $\mathscr{C}_u(G)$ by the space $\tilde{\mathscr{C}}_2^0(G)$ of the preceding section and the coefficient algebra $\mathfrak{K}(G)$ of G.

Let G be a Lie group of dimension $n \geq 1$ with Lie algebra $\mathfrak{L}(G)$ admitting a basis $\{X_1, \ldots, X_n\}$. We observe that for each $D \in \mathrm{Rep}(G)$ of the form $D := (d_{ij}(D))_{i,j=1,\ldots,n(D)}$ we have $d_{ij}(D) \in \mathscr{C}^\infty(G)$ for all $i, j = 1, \ldots, n(D)$. In fact, there exists an $h \in \mathscr{C}^\infty(G) \cap \mathscr{K}(G)$ such that the matrix $M := \int D h\, d\omega$ is invertible. But then the equalities

$$D(x)\,M = \int D(xz)\,h(z)\,\omega(dz) = \int D(y)\,h(x^{-1}y)\,\omega(dy)$$

valid for all $x \in G$ yield the assertion. Furthermore, one notes that for every $D \in \mathrm{Rep}(G)$ and $X \in \mathfrak{L}(G)$ the limit

$$\tilde{X}D := \lim_{s \to 0} \frac{1}{s}(D_{\zeta(s)} - D)$$

exists (in the sup norm) and one obtains for X, $Y\in\mathfrak{L}(G)$ the useful formulae

$$\tilde{X}D=D(\tilde{X}D)(e) \quad \text{and} \quad \tilde{X}\tilde{Y}D=D(\tilde{X}D)(e)\,\tilde{Y}D(e).$$

These statements follow directly from the fact that

$$\frac{1}{s}(D(x\zeta(s))-D(x))=D(x)\frac{1}{s}(D(\zeta(s))-E_{n(D)}),$$

which is valid for all $x\in G$, $s\in\mathbb{R}^*$. The desired extension of Theorem 4.2.8 can now be stated in the following form.

4.3.1 Theorem. *Let G be a Lie group G of dimension $n\geq 1$ with Lie algebra $\mathfrak{L}(G)$ admitting a basis $\{X_1,\ldots,X_n\}$, $\{x_1,\ldots,x_n\}$ a canonical coordinate system with respect to $\{X_1,\ldots,X_n\}$ and let ϕ be a Hunt function for G.*

(A) *Let $S=(\mu_t)_{t\in\mathbb{R}_+^*}$ be a convolution semigroup in $\mathcal{M}^1(G)$ with infinitesimal generator N and corresponding negative-definite form ψ.*
Then

(i) *N is defined on $\mathscr{R}(G)$;*

(ii) *$Nf(e)=-\psi(f)$ for all $f\in\mathfrak{R}(G)$;*

(iii) *there exist numbers $a_1,\ldots,a_n\in\mathbb{R}$, a symmetric, positive-semidefinite matrix $(a_{ij})_{i,j=1,\ldots,n}\in\mathfrak{M}(n,\mathbb{R})$ and a measure $\eta\in\mathcal{M}_+(G^*)$ with $\int_{G^*}\phi\,d\eta<\infty$ such that N admits on $\mathscr{R}(G)$ the representation $(a_i,a_{ij},\eta)_{1\leq i,j\leq n}$;*

(iv) *the convolution semigroup S is uniquely determined by* $\mathrm{Res}_{\tilde{\mathscr{C}}_2^0(G)}N$ *and*

(v) *$\int_G f\,d\eta=\lim_{t\downarrow 0}\frac{1}{t}\int_G f\,d\mu_t$ for all $f\in\mathscr{K}(G^*)$.*

(B) *Given numbers $a_1,\ldots,a_n\in\mathbb{R}$, a symmetric, positive-semidefinite matrix $(a_{ij})_{i,j=1,\ldots,n}\in\mathfrak{M}(n,\mathbb{R})$ and a measure $\eta\in\mathcal{M}_+(G^*)$ satisfying $\int_{G^*}\phi\,d\eta<\infty$ there exists exactly one convolution semigroup S in $\mathcal{M}^1(G)$ with infinitesimal generator N such that N admits on $\mathscr{R}(G)$ the representation $(a_i,a_{ij},\eta)_{1\leq i,j\leq n}$.*

Proof. 1. By Theorem 4.2.8 N is defined on $\tilde{\mathscr{C}}_2^0(G)$. The definitions of N and of ψ (as in 1.5.11) yield for every $D\in\mathrm{Rep}(G)$ the relationship

$$N(D)(x)=-D(x)\,\psi(D)$$

whenever $x\in G$, whence the existence of

$$Nf=\lim_{t\downarrow 0}\frac{1}{t}(S_t f-f)$$

for every $f\in\mathfrak{R}(G)$ and $(Nf)(e)=-\psi(f)$.

This proves (A)(i) and (ii). Since for every $f\in\mathfrak{R}(G)$ the functions $\tilde{X}f$, $\tilde{X}\tilde{Y}f$ (for X, $Y\in\mathfrak{L}(G)$) and Nf exist, the proof of Part (A) of Theorem 4.2.8 can be applied to the space $\mathscr{R}(G)$ instead of $\tilde{\mathscr{C}}_2^0(G)$, and one obtains the assertions (iii) to (v).

2. Again by Theorem 4.2.8 there exists exactly one convolution semigroup S in $\mathcal{M}^1(G)$ with infinitesimal generator N such that N admits on $\tilde{\mathscr{C}}_2^0(G)$ the representation $(a_i,a_{ij},\eta)_{1\leq i,j\leq n}$. From Part (A) of this theorem it follows that N

admits on $\mathscr{R}(G)$ a representation $(a_i', a_{ij}', \eta')_{1\le i,j\le n}$ say. For a given basis $\{X_1, \ldots, X_n\}$ of $\mathfrak{L}(G)$ and a canonical coordinate system $\{x_1, \ldots, x_n\}$ with respect to this basis the representation of N on $\mathscr{R}(G)$ is uniquely determined by the representation of N on $\tilde{\mathscr{C}}_2^0(G)$. Therefore $a_i'=a_i$, $a_{ij}'=a_{ij}$ $(i,j=1,\ldots,n)$ and $\eta'=\eta$. So (B) is proved. □

4.3.2 Remark. The property $\int_{G^*}\phi\,d\eta<\infty$ of the measure $\eta\in\mathscr{M}_+(G^*)$ appearing in the theorem is independent of the choice of the basis of $\mathfrak{L}(G)$. In fact, if ϕ_1, ϕ_2 are two Hunt functions for G corresponding to different bases of $\mathfrak{L}(G)$, then there exist $c, c'\in\mathbb{R}_+^*$ satisfying $c\phi_1\le\phi_2\le c'\phi_1$.

By the discussion preceding Theorem 4.3.1 we have the existence of the linear functionals

$$f\to A_i(f):=(\tilde{X}_i f)(e),$$
$$f\to A_{ij}(f):=(\tilde{X}_i\tilde{X}_j f)(e)\ (\text{for } i,j=1,\ldots,n)\quad\text{and}$$
$$f\to\gamma(\cdot,f):=\textstyle\sum_{i=1}^n x_i(\cdot)A_i(f)\qquad\text{on } \mathfrak{K}(G).$$

4.3.3 Lemma. *We assume G to be a Lie group in* **A**. *Let $a_1,\ldots,a_n\in\mathbb{R}$, $(a_{ij})_{i,j=1,\ldots,n}$ be a symmetric, positive-semidefinite matrix in $\mathfrak{M}(n,\mathbb{R})$ and η a measure in $\mathscr{M}_+(G^*)$ with $\int_{G^*}\phi\,d\eta<\infty$. Then*
(i) $\psi_1=-\sum_{i=1}^n a_i A_i$ *is a continuous primitive form,*
(ii) $\psi_2=-\sum_{i,j=1}^n a_{ij}A_{ij}$ *is a continuous quadratic form and*
(iii) $f\to\psi_3(f):=-\int_{G^*}[f(x)-f(e)-\gamma(x,f)]\,\eta(dx)$ *is a continuous negative-definite form (on $\mathfrak{K}(G)$).*

Proof. 1. For the proof that the linear functionals ψ_1, ψ_2 and ψ_3 on $\mathfrak{K}(G)$ are in fact continuous negative-definite forms we assume without loss of generality $a_i=a_{ij}=0$ for $i,j=1,\ldots,n$ and concentrate the argument on ψ_3. In fact, by Theorem 4.3.1 (B) there exists an $\{e\}$-continuous convolution semigroup S in $\mathscr{M}^1(G)$ with infinitesimal generator N such that N admits on $\mathfrak{K}(G)$ the representation $(0,0,\eta)$. Let ψ be the continuous negative-definite form on $\mathfrak{K}(G)$ corresponding to S (via Theorem 1.5.18). Then we get $\psi(f)=-(Nf)(e)=\psi_3(f)$ for all $f\in\mathfrak{K}(G)$ and ψ_3 is a continuous negative-definite form on $\mathfrak{K}(G)$ as desired.

2. The primitivity of A_i $(i=1,\ldots,n)$ and of ψ_1 follows immediately from

$$\tilde{X}_i(fg)=(\tilde{X}_i f)g+f(\tilde{X}_i g)\qquad(\text{valid for all } f, g\in\mathfrak{K}(G))$$

and

$$\overline{A_i(D)}^T=A_i(D^*)=-A_i(D)\qquad(\text{valid for all } D\in\operatorname{Rep}(G)).$$

3. It remains to show that ψ_2 is a quadratic form on $\mathfrak{K}(G)$. Let $D\in\operatorname{Rep}(G)$ and let us perform a change of the basis of $\mathfrak{L}(G)$ denoted by adding a prime. Then one can obtain

$$\psi_2(D)=-\textstyle\sum_{i=1}^n b_i A_{ii}'(D)=-\sum_{i=1}^n b_i A_i'(D)^2=\sum_{i=1}^n b_i A_i'(D)\overline{A_i'(D)}^T$$

with suitable $b_i\in\mathbb{R}_+$ $(i=1,\ldots,n)$. Hence, $\psi_2(D)$ is a positive-semidefinite Hermitian matrix. Moreover, one computes $\tilde{X}_i\tilde{X}_j(fg)$ for f, $g\in\mathfrak{K}(G)$ and obtains via

$A_i(g^*) = -A_i(g)$ for all $g \in \mathfrak{K}(G)$ the formula

$$A_{ij}(fg) + A_{ij}(fg^*) = 2[A_{ij}(f)\,g(e) + f(e)\,A_{ij}(g)] \quad (i,j = 1, \ldots, n).$$

This finishes the proof that ψ_2 is a quadratic form on $\mathfrak{K}(G)$. □

The following definition will be given for an arbitrary locally compact group G.

4.3.4 Definition. A mapping Γ: $G \times \mathrm{Rep}(G) \to \bigcup_{n \geq 1} \mathfrak{M}(n, \mathbb{C})$ (or $G \times \mathfrak{K}(G) \to \mathbb{C}$) is called a *Lévy function for G* if it satisfies the following conditions:

(LF1) For every $n \geq 1$ the mapping Γ: $G \times \mathrm{Rep}_n(G) \to \mathfrak{M}(n, \mathbb{C})$ is continuous.

For every compact subset C of $\mathrm{Rep}(G)$ one has

(LF2) $\sup\{\|\Gamma(x, D)\| : x \in G, D \in C\} < \infty$ and

(LF3) $\lim_{x \to e} \sup_{D \in C} \|\Gamma(x, D)\| = 0$.

(LF4) For every $x \in G$ there exists a one-parameter subgroup $(x_t)_{t \in \mathbb{R}}$ of G such that $D(x_t) = \exp t\Gamma(x, D)$ holds for all $D \in \mathrm{Rep}(G)$ and $t \in \mathbb{R}$.

(LF5) For every compact $C \subset \mathrm{Rep}(G)$ there exists a neighborhood $U \in \mathfrak{V}(e)$ satisfying $D(x) = \exp \Gamma(x, D)$ for all $D \in C$ and $x \in U$.

We note that for any Lévy function Γ for G the mapping $f \to \Gamma(x, f)$ is a continuous primitive form on $\mathfrak{K}(G)$.

4.3.5 Lemma. *Let G be a Lie group in* **A**. *There exist a neighborhood $U \in \mathfrak{V}(e)$ and a representation D_0 in* $\mathrm{Rep}(G)$ *such that* $\mathrm{Res}_U D_0$ *is injective.*

Proof. It suffices to show that there exists a representation $D_0 \in \mathrm{Rep}(G)$ such that $\ker(D_0)$ is discrete. Let U be a symmetric, relatively compact, open neighborhood in $\mathfrak{V}(e)$ such that U contains no nontrivial subgroup of G. The set $C := \overline{U^2} \setminus U$ is compact, and $e \notin C$. Since $\bigcap_{D \in \mathrm{Rep}(G)} \ker(D) = \{e\}$ and since the system $(\ker(D))_{D \in \mathrm{Rep}(G)}$ is descending, there exists a representation $D_0 \in \mathrm{Rep}(G)$ satisfying $\ker(D_0) \cap C = \emptyset$. Let $H := \ker(D_0) \cap U$. Clearly $e \in H$, and from $x \in H$ follows $x^{-1} \in H$. Moreover, for $x, y \in H$ we get $xy \in \ker(D_0)$ and $xy \in U^2$. But $\ker(D_0) \cap C = \emptyset$ implies $xy \in U$, whence $xy \in H$. Thus H is a subgroup of G with $H \subset U$. By assumption $H = \{e\}$, and hence $\ker(D_0) \cap U = \{e\}$ which shows that $\ker(D_0)$ is discrete. □

4.3.6 Remark. Let $\{X_1, \ldots, X_n\}$ be a basis of the Lie algebra $\mathfrak{L}(G)$ of G. Then for the (locally faithful) representation D_0 of the lemma the set $\{(\tilde{X}_1 D_0)(e), \ldots, (\tilde{X}_n D_0)(e)\}$ of vectors (in $\mathfrak{M}(n(D_0), \mathbb{C})$) is linearly independent (over $\mathbb{R}$). In fact, let U be the neighborhood in $\mathfrak{V}(e)$ and D_0 the representation in $\mathrm{Rep}(G)$ constructed in the lemma. We choose a neighborhood $V \in \mathfrak{V}_{\mathfrak{L}(G)}(0)$ with $\exp V \subset U$ and suppose that $a_1, \ldots, a_n \in \mathbb{R}$ are given such that

$$\sum_{i=1}^n a_i X_i \in V \quad \text{and} \quad \sum_{i=1}^n a_i (\tilde{X}_i D_0)(e) = 0$$

holds.

Putting $x := \exp(\sum_{i=1}^n a_i X_i)$ we then obtain

$$D_0(x) = \exp(\sum_{i=1}^n a_i (\tilde{X}_i D_0)(e)) = E_{n(D_0)},$$

and thus $x = e$ or $\sum_{i=1}^n a_i X_i = 0$. But this implies that $a_1 = \cdots = a_n = 0$.

4.3.7 Theorem. *Let G be a Lie group G in $\mathbf{A}$ with Lie algebra $\mathfrak{L}(G)$ having a basis $\{X_1, \ldots, X_n\}$ and a canonical coordinate system $\{x_1, \ldots, x_n\}$ with respect to this basis.*

(i) *The mapping $\Gamma: G \times \operatorname{Rep}(G) \to \bigcup_{n \geq 1} \mathfrak{M}(n, \mathbb{C})$ defined by*

$$\Gamma(x, D) := \gamma(x, D) = \sum_{i=1}^{n} x_i(x) \, A_i(D)$$

for all $x \in G$ and $D \in \operatorname{Rep}(G)$ is a Lévy function for G.

Given a Lévy function Γ for G

(ii) *there exist functions $y_1, \ldots, y_n \in \mathscr{C}^b(G)$ and a neighborhood $U \in \mathfrak{B}(e)$ such that $y_i(x) = x_i(x)$ for all $x \in U$, $i = 1, \ldots, n$, and*

$$\Gamma(x, D) = \sum_{i=1}^{n} y_i(x) \, A_i(D) \qquad \textit{for all } x \in G, \ D \in \operatorname{Rep}(G).$$

Proof. 1. Since A_i is a continuous primitive form on $\mathfrak{R}(G)$, the properties (LF 1) to (LF 4) for Γ are fulfilled.

Now we choose $U \in \mathfrak{B}(e)$ canonically such that for every $x \in U$ one has

$$x = \exp_G X \quad \text{with} \quad X := \sum_{i=1}^{n} x_i(x) \, X_i.$$

Then for every $D \in \operatorname{Rep}(G)$ we get

$$\Gamma(x, D) = (\tilde{X} D)(e) = dD(X)$$

and hence,

$$D(x) = D(\exp_G X) = \exp(dD(X)) = \exp \Gamma(x, D).$$

Thus Γ satisfies (LF 5) and (i) is proved.

2. By Property (LF 4), for each $x \in G$ there exists a one-parameter subgroup $(x_t)_{t \in \mathbb{R}}$ in G such that $D(x_t) = \exp t\Gamma(x, D)$ holds for all $D \in \operatorname{Rep}(G)$ and $t \in \mathbb{R}$. On the other hand, there is a vector

$$X(x) \in \mathfrak{L}(G) \quad \text{with} \quad x_t = \exp(t X(x)) \qquad \text{for all } t \in \mathbb{R}.$$

Let $X(x) = \sum_{i=1}^{n} y_i(x) \, X_i$ with $y_1(x), \ldots, y_n(x) \in \mathbb{R}$.
Then

$$\exp t \tilde{X}(x)(D) = D(x_t) = \exp t\Gamma(x, D) \qquad \text{for all } D \in \operatorname{Rep}(G) \text{ and } t \in \mathbb{R},$$

whence

$$\Gamma(x, D) = \sum_{i=1}^{n} y_i(x) \, A_i(D).$$

We now infer from Remark 4.3.6 that for the representation $D := D_0$ constructed in Lemma 4.3.5 the set $\{A_1(D), \ldots, A_n(D)\}$ of vectors is linearly independent.

Let $\{\mathrm{e}_1, \ldots, \mathrm{e}_n\}$ be a canonical basis of $\mathbb{R}^n$. The mapping

$$\ell: \{A_i(D): i = 1, \ldots, n\} \to \mathbb{R}^n$$

defined by $\ell(A_i(D)) := \mathrm{e}_i$ for all $i = 1, \ldots, n$ can be extended to a continuous linear function ℓ on the linear subspace generated by $\{A_i(D): i = 1, \ldots, n\}$.

Since the composition $\ell \circ \Gamma(\cdot, D)=(y_1, \ldots, y_n)$ is continuous, we obtain $y_1, \ldots, y_n \in \mathscr{C}(G)$. By (LF 2) the set $\{\Gamma(x, D): x \in G\}$ is bounded. Hence, $\{(y_1(x), \ldots, y_n(x)): x \in G\}$ is bounded and so $y_1, \ldots, y_n \in \mathscr{C}^b(G)$. With a proper choice of $U \in \mathfrak{B}(e)$ one deduces from (LF 5) together with Part 1 of the proof the remaining statement. □

Let G again be an arbitrary locally compact group. For any $D \in \operatorname{Rep}(G)$ we introduce a function f_D on G defined by

$$f_D(x) := \operatorname{Re}[\operatorname{tr}(E_{n(D)} - D(x))] \quad \text{for all } x \in G.$$

Clearly, $f_D \in \mathfrak{K}_+(G)$ and $\ker D = f_D^{-1}(\{0\})$.

4.3.8 Definition. A measure $\eta \in \mathscr{M}_+(G^*)$ is called a *Lévy measure on G* if
(a) $\eta(\complement U) < \infty$ for every neighborhood $U \in \mathfrak{B}(e)$ and
(b) $\int_{G^*} f_D \, d\eta < \infty$.

If G is a Lie group and ϕ a Hunt function for G, then every measure $\eta \in \mathscr{M}_+(G^*)$ with $\int_{G^*} \phi \, d\eta < \infty$ is a Lévy measure on G.

In fact, for any $U \in \mathfrak{B}(e)$ one has $\inf \phi(\complement U) > 0$, whence $\eta(\complement U) < \infty$. Let $D \in \operatorname{Rep}(G)$. Then since $f_D(e) = A_i(f_D) = 0$ for $i = 1, \ldots, n$ there exist a $U \in \mathfrak{B}(e)$ and a $c \in \mathbb{R}_+^*$ satisfying $f_D(x) \leq c\,\phi(x)$ for all $x \in U$. This together with $\eta(\complement U) < \infty$ yields the assertion.

4.3.9 Lemma. *Let G be a Lie group in* **A** *and ϕ a Hunt function for G. There exist a neighborhood $U \in \mathfrak{B}(e)$, a representation $D \in \operatorname{Rep}(G)$ and a $c \in \mathbb{R}_+^*$ such that $\phi(x) \leq c f_D(x)$ for all $x \in U$.*

Proof. By Remark 4.3.2 the statement of the lemma is independent of the particular choice of ϕ. Therefore the basis of $\mathfrak{L}(G)$ and ϕ can be suitably chosen.

Let $\{X_1, \ldots, X_n\}$ be a basis of $\mathfrak{L}(G)$, $\{x_1, \ldots, x_n\}$ a canonical coordinate system with respect to $\mathfrak{L}(G)$ and ϕ the corresponding Hunt function. Let $D := D_0$ be the locally faithful representation of G constructed in Lemma 4.3.5. For $x \in G$ we put

$$\mathfrak{x}(x) := (x_1(x), \ldots, x_n(x)) \quad \text{and} \quad M(x) := ((\tilde{X}_i \tilde{X}_j f_D)(x))_{i,j=1,\ldots,n}.$$

Since f_D admits a local minimum in e, one obtains

$$M(e) = (-\operatorname{Re}[\operatorname{tr}(A_i(D)\, A_j(D))])_{i,j=1,\ldots,n},$$

which is a positive-definite matrix in $\mathfrak{M}(n, \mathbb{C})$.

Let $S := \{\mathfrak{x} \in \mathbb{R}^n : \|\mathfrak{x}\| = 1\}$. The mapping $\rho: G \times S \to \mathbb{R}$ defined by

$$\rho(x, \mathfrak{x}) := \langle M(x)\, \mathfrak{x}, \mathfrak{x} \rangle \quad \text{for all } x \in G,\ \mathfrak{x} \in S$$

is continuous. Since S is compact and $M(e)$ is positive-definite, there is a compact $V \in \mathfrak{B}(e)$ with $\rho(V \times S) \subset \mathbb{R}_+^*$. Consequently,

$$c_V := \inf\{\rho(x, \mathfrak{x}) : x \in V, \mathfrak{x} \in S\} > 0.$$

Choosing $U \in \mathfrak{B}(e)$ with $U \subset V$ such that

$$\phi(x) = \sum_{i=1}^{n} x_i(x)^2 \quad \text{and} \quad f_D(x) = \langle M(\zeta(x))\, \mathfrak{x}(x), \mathfrak{x}(x) \rangle$$

holds for all $x \in U$, where $\zeta(x) \in U$ lies on the one-parameter subgroup through x, and putting $c := c_V^{-1}$, one obtains for all $x \in U^* := U \setminus \{e\}$ the inequalities

$$\begin{aligned} \phi(x) &= \|\mathfrak{x}(x)\|^2 = c\, \|\mathfrak{x}(x)\|^2\, c_V \\ &\leq c\, \|\mathfrak{x}(x)\|^2\, \rho(\zeta(x), \|\mathfrak{x}(x)\|^{-1}\, \mathfrak{x}(x)) \\ &= c\, \langle M(\zeta(x))\, \mathfrak{x}(x), \mathfrak{x}(x) \rangle = c\, f_D(x). \quad \square \end{aligned}$$

4.3.10 Theorem. *Let G be a Lie group in* **A** *and Γ a Lévy function for G.*

(A) *Let $S := (\mu_t)_{t \in \mathbb{R}_+^*}$ be a convolution semigroup in $\mathscr{M}^1(G)$ and ψ its corresponding negative-definite form on $\mathfrak{R}(G)$. There exist*

(i) *a continuous primitive form ψ_1,*
(ii) *a continuous quadratic form ψ_2 on $\mathfrak{R}(G)$ and*
(iii) *a Lévy measure η on G*

such that ψ admits the representation (ψ_1, ψ_2, η) defined by

$$\text{(LK)} \qquad \psi(f) = \psi_1(f) + \psi_2(f) - \int_{G^*} [f(x) - f(e) - \Gamma(x, f)]\, \eta(dx)$$

for every $f \in \mathfrak{R}(G)$.

Moreover, ψ_2 and η are uniquely determined by S and

$$\int_{G^*} f\, d\eta = \lim_{t \downarrow 0} \frac{1}{t} \int_G f\, d\mu_t \qquad \text{for all } f \in \mathscr{K}(G^*).$$

(B) *Let $G \in$* **M** *and let ψ_1, ψ_2, η be given as in* (A). *Then* (LK) *defines a continuous negative-definite form ψ on G, and there exists a convolution semigroup $S := (\mu_t)_{t \in \mathbb{R}_+^*}$ in $\mathscr{M}^1(G)$ such that ψ is the negative-definite form corresponding to S.*

Proof. 1. From Theorem 4.3.1 and Lemma 4.3.3 it follows that there are a continuous primitive form ψ_1', a continuous quadratic form ψ_2 on $\mathfrak{R}(G)$ and a Lévy measure η on G such that for all $f \in \mathfrak{R}(G)$ one has

$$\psi(f) = \psi_1'(f) + \psi_2(f) - \int_{G^*} [f(x) - f(e) - \gamma(x, f)]\, \eta(dx).$$

Let $U \in \mathfrak{B}(e)$ be chosen as in Theorem 4.3.7 (ii). Since $\Gamma(x, f) = \gamma(x, f)$ for all $x \in U$, $f \in \mathfrak{R}(G)$ and $\eta(\complement U) < \infty$, the mapping

$$f \to \psi_1''(f) := \int_{G^*} [\gamma(\cdot, f) - \Gamma(\cdot, f)]\, d\eta$$

is a continuous primitive form ψ_1'' on $\mathfrak{R}(G)$ (by the properties of Γ, Theorem 4.3.7, and Lemma 4.3.3). Thus

$$\psi(f) = \psi_1'(f) + \psi_1''(f) + \psi_2(f) - \int_{G^*} [f(x) - f(e) - \Gamma(x, f)]\, \eta(dx)$$

for all $f \in \mathfrak{R}(G)$, so that $\psi_1 := \psi_1' + \psi_1''$ is a continuous primitive form on $\mathfrak{R}(G)$. The condition concerning η is an immediate consequence of Theorem 4.3.1. It follows that η is uniquely determined by S. In order to complete the proof of (A) it remains to show that ψ_2 is also uniquely determined by S. To do this let $\psi=\psi'+\psi''$ be the decomposition into a skew-symmetric and a symmetric part.

By Theorems 1.5.16, 1.5.17 and 4.3.7 we get

$$\psi''=\psi_2-\tfrac{1}{2}\int_{G^*}[f+f^*-2f(e)]\,d\eta,$$

and the assertion becomes clear.

2. We now prove Part (B) of the theorem. Let $U\in\mathfrak{V}(e)$, $D\in\mathrm{Rep}(G)$ and $c\in\mathbb{R}_+^*$ be chosen as in Lemma 4.3.9 such that for a given Hunt function ϕ for G one has $\phi(x)\leq c f_D(x)$ whenever $x\in U$. The properties of η yield

$$\int_{G^*}\phi\,d\eta\leq c\int_{U^*}f_D\,d\eta+\eta(\complement U)<\infty.$$

Without loss of generality let $\Gamma(x,f)=\gamma(x,f)$ for all $x\in U$, $f\in\mathfrak{K}(G)$. Defining ψ_1'' as in Part 1 we obtain by Lemma 4.3.3(iii) a continuous negative-definite form ψ_3 on $\mathfrak{K}(G)$ defined by

$$\begin{aligned}\psi_3(f):&=-\int_{G^*}[f(x)-f(e)-\gamma(x,f)]\,\eta(dx)-\psi_1''(f)\\&=-\int_{G^*}[f(x)-f(e)-\Gamma(x,f)]\,\eta(dx)\end{aligned}$$

for all $f\in\mathfrak{K}(G)$. Thus $\psi:=\psi_1+\psi_2+\psi_3$ is also a continuous negative-definite form on $\mathfrak{K}(G)$. Since $G\in\mathbf{M}$, Theorem 1.5.19 yields the assertion. □

In the following we shall extend Theorem 4.3.10 to Lie projective groups in $\mathbf{A}$. Some *preparations* are in order.

Let G and $\dot{G}$ be two Lie groups and $p:G\to\dot{G}$ an open continuous epimorphism such that $H:=\ker p$ is compact. By $dp=dp(e)$ we denote as usual the linear mapping $\mathfrak{L}(G)\to\mathfrak{L}(\dot{G})$ corresponding to p. If the groups G and $\dot{G}$ are connected, then $\mathfrak{L}(H)=\ker dp$ and $\mathfrak{L}(\dot{G})=\operatorname{im} dp$. This yields:

1. Let G be of dimension $n\geq 1$ and H of dimension $r\geq 1$. Then $\dot{G}$ is of dimension $s:=k-r$ and for every basis $\{\dot{X}_1,\dots,\dot{X}_s\}$ of $\mathfrak{L}(\dot{G})$ there exists a basis $\{X_1,\dots,X_n\}$ of $\mathfrak{L}(G)$ with $dp(X_i)=\dot{X}_i$ for $i=1,\dots,s$ and $dp(X_i)=0$ for $i=s+1,\dots,n$ and hence, $\{X_{s+1},\dots,X_n\}$ is a basis of $\mathfrak{L}(H)$.

2. Let the bases $\{X_1,\dots,X_n\}$ of $\mathfrak{L}(G)$ and $\{\dot{X}_1,\dots,\dot{X}_s\}$ of $\mathfrak{L}(\dot{G})$ be chosen as in Part 1. For every canonical coordinate system $\{\dot{x}_1,\dots,\dot{x}_s\}$ of $\dot{G}$ with respect to $\{\dot{X}_1,\dots,\dot{X}_s\}$ there exists a canonical coordinate system $\{x_1,\dots,x_n\}$ of G with respect to $\{X_1,\dots,X_n\}$ such that $x_i=\dot{x}_i\circ p$ for all $i=1,\dots,s$.

This is shown by an easy computation.

We now assume that G is a Lie projective group in $\mathbf{A}$, i.e., there exists a descending family $(K_\alpha)_{\alpha\in\mathbb{A}}$ of compact normal subgroups of G satisfying $\bigcap_{\alpha\in\mathbb{A}}K_\alpha=\{e\}$ and such that $G_\alpha:=G/K_\alpha$ is a Lie group for all $\alpha\in\mathbb{A}$. The index set becomes an ordered set; we simply put $\alpha<\beta$ for $\alpha,\beta\in\mathbb{A}$ if $K_\beta\subset K_\alpha$. Let $p_{\alpha,\beta}$ be the canonical mapping from G_β onto G_α and $K_{\alpha\beta}:=K_\alpha/K_\beta$ (such that $G_\beta/K_{\alpha\beta}\cong G_\alpha$) for $\alpha,\beta\in\mathbb{A}$ with $\alpha<\beta$. For $\alpha\in\mathbb{A}$ we denote by p_α the canonical projection from G onto G_α. This notation will be kept for the rest of the section.

We note a technical fact that will be useful for the subsequent discussion.

3. For coefficients $f^{(1)},\dots,f^{(m)}\in\mathfrak{R}(G)$ there exist an $\alpha\in\mathbb{A}$ and coefficients $f_\alpha^{(1)},\dots,f_\alpha^{(m)}\in\mathfrak{R}(G_\alpha)$ with $f^{(i)}=f_\alpha^{(i)}\circ p_\alpha$ for all $i=1,\dots,m$.

In fact, by Property (4) of 1.3.6 for every $i=1,\dots,m$ there are a $D^{(i)}\in\operatorname{Rep}(G)$ and an $M_i\in\mathfrak{M}(n(D^{(i)}),\mathbb{C})$ satisfying $f^{(i)}=\operatorname{tr}(M_i D^{(i)})$. From Property (3) of 1.3.2 we infer the existence of an $\alpha\in\mathbb{A}$ and a $D_\alpha^{(i)}\in\operatorname{Rep}(G_\alpha)$ with $D^{(i)}=D_\alpha^{(i)}\circ p_\alpha$ for $i=1,\dots,m$. The functions $f_\alpha^{(i)}:=\operatorname{tr}(M_i D_\alpha^{(i)})$ $(i=1,\dots,m)$ satisfy the requirements.

4.3.11 Theorem. *Let G be a Lie projective group in $\mathbf{A}$. Then there exists a Lévy function for G.*

We prepare the proof of the theorem with the following crucial

4.3.12 Lemma. *Let $G\in\mathbf{A}$, let K be a compact normal subgroup of G such that $\dot G:=G/K$ is a Lie group and let p be the canonical mapping from G onto $\dot G$. We are given a basis $\{X_1,\dots,X_n\}$ of the Lie algebra $\mathfrak{L}(\dot G)$ of $\dot G$, a canonical coordinate system $\{x_1,\dots,x_n\}$ with respect to this basis and a Lévy function Γ for G. For all $x\in G$ and $\dot f\in\mathfrak{R}(\dot G)$ we define*

$$\dot\Gamma(x,\dot f):=\Gamma(x,\dot f\circ p).$$

Then

(i) *there exist $y_1,\dots,y_n\in\mathscr{C}^b(G)$ with*

$$\dot\Gamma(x,\dot f)=\textstyle\sum_{i=1}^n y_i(x)(\tilde X_i\dot f)(\dot e)\qquad\text{for all } x\in G \text{ and } \dot f\in\mathfrak{R}(\dot G)\quad\text{and}$$

(ii) *there exists a $U\in\mathfrak{B}(e)$ such that $(x_i\circ p)(x)=y_i(x)$ holds for all $x\in U$ and $i=1,\dots,n$.*

Proof. 1. Let $x\in G$. Then Property (LF 4) of Definition 4.3.4 implies the existence of a one-parameter subgroup $(x_t)_{t\in\mathbb{R}}$ in G with

$$D(x_t)=\exp t\,\Gamma(x,D)\qquad\text{for all } D\in\operatorname{Rep}(G)\text{ and } t\in\mathbb{R}.$$

For every $t\in\mathbb{R}$ we put $\dot x_t:=p(x_t)$. For the one-parameter subgroup $(\dot x_t)_{t\in\mathbb{R}}$ in $\dot G$ there exists an $X(x)\in\mathfrak{L}(\dot G)$ with $\dot x_t=\exp_{\dot G}(tX(x))$ for all $t\in\mathbb{R}$. Let

$$X(x)=\textstyle\sum_{i=1}^n y_i(x)\,X_i.$$

From

$$\exp t\,\tilde X(x)(\dot D)(\dot e)=\dot D(\dot x_t)=\exp t\,\Gamma(x,\dot D\circ p)=\exp t\,\dot\Gamma(x,\dot D)$$

for all $t\in\mathbb{R}$ we obtain

$$\dot\Gamma(x,\dot D)=\textstyle\sum_{i=1}^n y_i(x)(\tilde X_i\dot D)(\dot e)$$

whenever $\dot D\in\operatorname{Rep}(\dot G)$.

By Remark 4.3.6 there exists a (locally faithful) representation $\dot D_0\in\operatorname{Rep}(\dot G)$ such that the set $\{(\tilde X_1\dot D_0)(\dot e),\dots,(\tilde X_n\dot D_0)(\dot e)\}$ is linearly independent. Hence, the properties (LF 1) and (LF 2) of $\dot\Gamma$ yield $y_1,\dots,y_n\in\mathscr{C}^b(G)$, which proves (i).

2. It remains to show (ii). By Property (LF 5) of Γ for the compact subset $C := \{\dot{D}_0 \circ p\}$ of Rep(G) there exists a $U \in \mathfrak{V}(e)$ satisfying

$$\dot{D}_0(p(x)) = \exp \Gamma(x, \dot{D}_0 \circ p) = \exp \dot{\Gamma}(x, \dot{D}_0) \quad \text{for all } x \in U.$$

A suitable modification of U implies that for all $\dot{x} \in p(U)$ we have

$$\dot{x} = \exp_{\dot{G}}(\textstyle\sum_{i=1}^n x_i(\dot{x}) X_i) \quad \text{or}$$

$$\dot{D}_0(p(x)) = \exp(\textstyle\sum_{i=1}^n x_i(p(x))(\tilde{X}_i \dot{D}_0)(\dot{e})) \quad \text{for all } x \in U.$$

Hence, Property (LF 3) of Γ enables us to choose U so small that for $x \in U$ the equalities

$$\textstyle\sum_{i=1}^n y_i(x)(\tilde{X}_i \dot{D}_0)(\dot{e}) = \dot{\Gamma}(x, \dot{D}_0)$$

$$\dot{D}_0(p(x)) = \log \dot{D}_0(p(x)) = \textstyle\sum_{i=1}^n x_i(p(x))(\tilde{X}_i \dot{D}_0)(\dot{e})$$

hold. Since $\{(\tilde{X}_1 \dot{D}_0)(\dot{e}), \ldots, (\tilde{X}_n \dot{D}_0)(\dot{e})\}$ is a linearly independent set of vectors, we arrive at $y_i(x) = x_i(p(x))$ for all $x \in U$ and $i = 1, \ldots, n$, which was to be proved. □

Proof of Theorem 4.3.11. 1. Let **H** denote the system of pairs $(G/K, \Gamma_K)$ with a compact normal subgroup K of G and a Lévy function Γ_K for G/K. Since G/K_α is a Lie group (in **A**), for which by Theorem 4.3.7 a Lévy function Γ_α exists (for all $\alpha \in \mathbf{A}$), **H** is not empty. If for $i = 1, 2$ $(G/K_i, \Gamma_i) \in \mathbf{H}$, we define $(G/K_1, \Gamma_1) < (G/K_2, \Gamma_2)$, if $K_2 \subset K_1$ and

$$\Gamma_1(p_{12}(x), D) = \Gamma_2(x, D \circ p_{12}) \quad \text{for all } x \in G/K_2$$

and $D \in \text{Rep}(G/K_1)$, where p_{12} denotes the canonical mapping from G/K_2 onto G/K_1. This definition introduces an order relation in **H**.

2. We show that **H** is inductively odered. Let

$$\mathbf{C} := \{(G/K_i, \Gamma_i) : i \in \mathbb{I}\}$$

be a chain in **H** with the properties that $\mathbb{I}$ is a (totally) ordered set and that for all $i, j \in \mathbb{I}$ satisfying $i < j$ we have $(G/K_i, \Gamma_i) < (G/K_j, \Gamma_j)$. Putting $K := \bigcap_{i \in \mathbb{I}} K_i$ and observing that $G/K_i \cong \dfrac{G/K}{K_i/K}$ for all $i \in \mathbb{I}$ we obtain $G/K = \varprojlim_{i \in \mathbb{I}} G/K_i$.

Given $x \in G/K$ and $f \in \mathfrak{R}(G/K)$ there exist by 3 of the preceding preparations an $i \in \mathbb{I}$ and an $f_i \in \mathfrak{R}(G/K_i)$ with $f = f_i \circ p_i$, where p_i is the canonical mapping $G/K \to G/K_i$ for $i \in \mathbb{I}$. We define Γ_K by

$$\Gamma_K(x, f) := \Gamma_i(p_i(x), f_i) \quad \text{for all } x \in G/K \text{ and } f \in \mathfrak{R}(G/K).$$

The definition is independent of the special choice of $i \in \mathbb{I}$ since **C** is totally ordered, and it can be shown that Γ_K satisfies (LF 1) to (LF 3) and (LF 5) in the definition of a Lévy function for G/K; one need merely apply 3 of the preparations.

It remains to show Property (LF 4) for Γ_K. For $x \in G/K$ and $i \in \mathbb{I}$ there exists by Property (LF4) for Γ_i a one-parameter subgroup $(x_t^i)_{t\in\mathbb{R}}$ in G/K_i with

$$D_i(x_t^i) = \exp t\,\Gamma_i(p_i(x), D_i) = \exp t\,\Gamma_K(x, D_i \circ p_i)$$
$$\text{for all } D_i \in \operatorname{Rep}(G/K_i) \ (t \in \mathbb{R}).$$

For $i, j \in \mathbb{I}$ with $i<j$ we introduce the canonical mapping $p_{ij}: G/K_j \to G/K_i$. $(G/K_i, \Gamma_i) < (G/K_j, \Gamma_j)$ implies $x_t^i = p_{ij}(x_t^j)$, and there exists an $x_t \in G$ with $p_i(x_t) = x_t^i$, whence

$$D_i \circ p_i(x_t) = \exp t\,\Gamma_K(x, D_i \circ p_i) \qquad \text{for all } i \in \mathbb{I} \ (t \in \mathbb{R}).$$

Plainly, $(x_t)_{t\in\mathbb{R}}$ is a one-parameter subgroup in G, so that the assertion is proved, and Γ_K is a Lévy function for G/K. By construction $(G/K_i, \Gamma_i) < (G/K, \Gamma_K)$ for all $i \in \mathbb{I}$, whence $(G/K, \Gamma_K)$ is an upper bound for **C**. We infer from Zorn's Lemma that **H** has maximal elements.

3. The proof of the theorem will be complete, if we have shown that for every $(G/K, \Gamma_K) \in \mathbf{H}$ with $K \neq \{e\}$ there exists a pair $(G/N, \Gamma_N) \in \mathbf{H}$ with $(G/K, \Gamma_K) < (G/N, \Gamma_N)$ and $N \neq K$.

Let $(G/K, \Gamma_K) \in \mathbf{H}$ be given, with $K \neq \{e\}$. There is a $\gamma \in \mathbb{A}$ with

$$N := K \cap K_\gamma \neq K.$$

From $K/N = K/(K \cap K_\gamma) \cong KK_\gamma/K_\gamma \subset G/K_\gamma$ we conclude that K/N is a Lie group. For all $\alpha > \gamma$ we have $K \cap K_\alpha \subset N$. Hence, we assume without loss of generality that $N = \{e\}$, $\mathbb{A} := \{\alpha : \alpha > \gamma\}$ and $K \cap K_\alpha = \{e\}$ for all $\alpha \in \mathbb{A}$.

Let $\{Y_1, \ldots, Y_r\}$ denote a basis of $\mathfrak{L}(K)$. We denote the canonical mapping from G onto $\dot G := G/K$ by p and abbreviate $G_\alpha := G/K_\alpha$, $\dot K_\alpha := p(K_\alpha)$ and $\dot G_\alpha := \dot G/\dot K_\alpha$. For $\alpha, \beta \in \mathbb{A}$ with $\alpha < \beta$ we have the following commutative diagram (with the corresponding mappings existing):

$$\begin{array}{ccccc}
G & \xrightarrow{p_\beta} & G_\beta & \xrightarrow{p_{\alpha\beta}} & G_\alpha \\
\downarrow p & & \downarrow q_\beta & & \downarrow q_\alpha \\
\dot G & \xrightarrow[\dot p_\beta]{} & \dot G_\beta & \xrightarrow[\dot p_{\alpha\beta}]{} & \dot G_\alpha .
\end{array}$$

By $K \cap K_\alpha = \{e\}$ the projection $p_\alpha: K \to KK_\alpha/K_\alpha = \ker q_\alpha =: N_\alpha$ is a topological isomorphism. Hence, by [213], p. 118 the Lie algebras $\mathfrak{L}(G_\alpha)$ and $\mathfrak{L}(K) \oplus \mathfrak{L}(\dot G_\alpha)$ can be identified. Let $\{x_1^\gamma, \ldots, x_r^\gamma\}$ be a canonical coordinate system in N_γ with respect to the basis $\{dp_\gamma(Y_1), \ldots, dp_\gamma(Y_r)\}$ of $\mathfrak{L}(N_\gamma)$. Furthermore, we choose bases $\{X_1^\alpha, \ldots, X_{k(\alpha)}^\alpha\}$ and $\{\dot X_1^\alpha, \ldots, \dot X_{s(\alpha)}^\alpha\}$ of $\mathfrak{L}(G_\alpha)$ and $\mathfrak{L}(\dot G_\alpha)$ resp. satisfying $X_j^\alpha = dp_\alpha(Y_j)$ for $j = 1, \ldots, r$ and $dq_\alpha(X_{r+j}^\alpha) = \dot X_j^\alpha$ for $j = 1, \ldots, s(\alpha)$. By Preparations 1 and 2 there exists a canonical coordinate system $\{\dot x_1^\alpha, \ldots, \dot x_{s(\alpha)}^\alpha\}$ in $\dot G_\alpha$ with respect to the basis $\{\dot X_1^\alpha, \ldots, \dot X_{s(\alpha)}^\alpha\}$ such that

$$\{x_1^\gamma \circ p_{\gamma\alpha}, \ldots, x_r^\gamma \circ p_{\gamma\alpha}, \dot x_1^\alpha \circ q_\alpha, \ldots, \dot x_{s(\alpha)}^\alpha \circ q_\alpha\}$$

is a canonical coordinate system in G_α with respect to the basis $\{X_1^\alpha, \ldots, X_{k(\alpha)}^\alpha\}$ of $\mathfrak{L}(G_\alpha)$. By Lemma 4.3.12 there exist functions

$$\dot y_1^\alpha, \ldots, \dot y_{s(\alpha)}^\alpha \in \mathscr{C}^b(\dot G)$$

such that (with $\dot{X}_\alpha(\dot{x}) := \sum_{i=1}^{s(\alpha)} \dot{y}_i^\alpha(\dot{x})\, \dot{X}_i^\alpha \in \mathfrak{L}(\dot{G}_\alpha)$) one has

$$\Gamma_K(\dot{x}, \dot{f}_\alpha \circ \dot{p}_\alpha) = (\dot{X}_\alpha(\dot{x}) \dot{f}_\alpha)(\dot{e}_\alpha) \quad \text{for all } \dot{f}_\alpha \in \mathfrak{R}(\dot{G}_\alpha).$$

But $\dot{X}_\alpha(\dot{x})$ is uniquely determined by $\Gamma_K(\dot{x}, \cdot)$, whence

$$d\dot{p}_{\alpha\beta}(\dot{X}_\beta(\dot{x})) = \dot{X}_\alpha(\dot{x}).$$

Let us now define for $x \in G$ the vectors

$$Y_\alpha(x) := \sum_{j=1}^r x_j^\gamma(p_\gamma(x))\, dp_\alpha(Y_j) = \sum_{j=1}^r (x_j^r \circ p_{\gamma\alpha})(p_\alpha(x))\, X_j^\alpha,$$

$$Z_\alpha(x) := \sum_{j=1}^{s(\alpha)} \dot{y}_j^\alpha(p(x))\, X_{r+j}^\alpha \quad \text{and}$$

$$X_\alpha(x) := Y_\alpha(x) + Z_\alpha(x) \qquad (\alpha \in \mathbb{A}).$$

By construction we obtain for $\alpha, \beta \in \mathbb{A}$ with $\alpha < \beta$ the equalities

$$dp_{\alpha\beta}(Y_\beta(x)) = Y_\alpha(x) \quad \text{and} \quad dq_\alpha(Z_\alpha(x)) = \dot{X}_\alpha(p(x)),$$

hence $dp_{\alpha\beta}(Z_\beta(x)) = Z_\alpha(x)$ and consequently, $dp_{\alpha\beta}(X_\beta(x)) = X_\alpha(x)$.

For $\alpha \in \mathbb{A}$, $f_\alpha \in \mathfrak{R}(G_\alpha)$ and $x \in G$ we now define

$$\Gamma(x, f_\alpha \circ p_\alpha) := (X_\alpha(x) f_\alpha)(e_\alpha).$$

By the above formulae Γ is well-defined.

For $\alpha \in \mathbb{A}$, $\dot{f}_\alpha \in \mathfrak{R}(\dot{G}_\alpha)$ and $x \in G$ we have

$$\begin{aligned} \Gamma_K(p(x), \dot{f}_\alpha \circ \dot{p}_\alpha) &= (\dot{X}_\alpha(p(x))\, \dot{f}_\alpha)(\dot{e}_\alpha) \\ &= (dq_\alpha(Z_\alpha(x))\, \dot{f}_\alpha)(\dot{e}_\alpha) = (Z_\alpha(x)\, (\dot{f}_\alpha \circ q_\alpha))(\dot{e}_\alpha) \\ &= (X_\alpha(x)(\dot{f}_\alpha \circ q_\alpha))(\dot{e}_\alpha) = \Gamma(x, \dot{f}_\alpha \circ q_\alpha \circ p_\alpha) \\ &= \Gamma(x, (\dot{f}_\alpha \circ \dot{p}_\alpha) \circ p). \end{aligned}$$

In order to establish $(G/K, \Gamma_K) < (G, \Gamma)$ it remains to show that Γ is a Lévy function for G.

For this purpose we let C be a compact subset of $\mathrm{Rep}(G)$. By Preparation 3 there exist an $\alpha \in \mathbb{A}$ and a compact subset C_α of $\mathrm{Rep}(G_\alpha)$ satisfying $C = C_\alpha \circ p_\alpha$. Thus Properties (LF 1) to (LF 3) for Γ follow.

We shall show (LF 5). By Lemma 4.3.12 there exists a $\dot{U} \in \mathfrak{B}(\dot{e})$ with

$$\dot{x}_i^\alpha \circ \dot{p}_\alpha(\dot{x}) = \dot{y}_i^\alpha(\dot{x}) \qquad \text{for all } \dot{x} \in \dot{U} \text{ (and } i = 1, \ldots, s(\alpha)).$$

For $x \in U := p^{-1}(\dot{U})$ we then obtain

$$\dot{y}_j^\alpha(p(x)) = \dot{x}_i^\alpha \circ \dot{p}_\alpha \circ p(x) = \dot{x}_i^\alpha \circ q_\alpha \circ p_\alpha(x), \quad \text{whence}$$

$$X_\alpha(x) = \sum_{j=1}^r (x_j^\gamma \circ p_{\gamma\alpha})(p_\alpha(x))\, X_j^\alpha + \sum_{j=1}^{s(\alpha)} (\dot{x}_i^\alpha \circ q_\alpha)(p_\alpha(x))\, X_{r+j}^\alpha.$$

We now infer from the canonical coordinate system chosen for G_α that with a modified neighborhood $U \in \mathfrak{B}(e)$ one gets

$$p_\alpha(x) = \exp_{G_\alpha} X_\alpha(x) \qquad \text{for all } x \in U.$$

But this implies with $D_\alpha \in C_\alpha$ that

$$D_\alpha(p_\alpha(x)) = \exp(X_\alpha(x)\, D_\alpha)(e_\alpha) = \exp \Gamma(x, D_\alpha \circ p_\alpha)$$

holds for all $x \in U$. Finally we have to show Property (LF 4) for Γ.

For $X_\alpha(x) \in \mathfrak{L}(G_\alpha)$ (for $\alpha \in \mathbb{A}$ and $x \in G$) there exists a one-parameter subgroup $(x_t^\alpha)_{t\in\mathbb{R}}$ with $x_t^\alpha = \exp_{G_\alpha}(t\, X_\alpha(x))$. For $\alpha, \beta \in \mathbb{A}$, $\alpha < \beta$ the equality $d p_{\alpha\beta}(X_\beta(x)) = X_\alpha(x)$ implies that $p_{\alpha\beta}(x_t^\beta) = x_t^\alpha$ $(t \in \mathbb{R})$. Hence, there exists a one-parameter subgroup $(x_t)_{t\in\mathbb{R}}$ with $p_\alpha(x_t) = x_t^\alpha$ for all $\alpha \in \mathbb{A}$. Finally, we obtain

$$(D_\alpha \circ p_\alpha)(x_t) = D_\alpha(x_t^\alpha) = \exp(X_\alpha(x)\, D_\alpha)(e_\alpha) = \exp t\, \Gamma(x, D_\alpha \circ p_\alpha)$$

for all $D_\alpha \in \mathrm{Rep}(G_\alpha)$ $(\alpha \in \mathbb{A})$ and $t \in \mathbb{R}$. Thus Γ is a Lévy function for G, and the proof is finished. □

4.3.13 Theorem. *Let G be a Lie projective group in $\mathbf{A}$ and Γ a Lévy function for G.*

(A) *Let $S := (\mu_t)_{t\in\mathbb{R}_+^*}$ be a convolution semigroup in $\mathcal{M}^1(G)$ and ψ its corresponding negative-definite form on $\mathfrak{K}(G)$. There exist*

(i) *a continuous primitive form ψ_1,*

(ii) *a continuous quadratic form ψ_2 on $\mathfrak{K}(G)$ and*

(iii) *a Lévy measure η on G*

such that ψ has the representation (ψ_1, ψ_2, η) defined by

$$\text{(LK)} \qquad \psi(f) = \psi_1(f) + \psi_2(f) - \int_{G^*} [f(x) - f(e) - \Gamma(x, f)]\, \eta(dx)$$

for every $f \in \mathfrak{K}(G)$.

Moreover, ψ_2 and η are uniquely determined by S and

$$\int_{G^*} f\, d\eta = \lim_{t\downarrow 0} \frac{1}{t} \int f\, d\mu_t \qquad \textit{for all } f \in \mathcal{K}(G^*).$$

(B) *Let $G \in \mathbf{M}$ and let ψ_1, ψ_2, η be given as in* (A). *Then* (LK) *defines a continuous negative-definite form ψ on G, and there exists a convolution semigroup $S := (\mu_t)_{t\in\mathbb{R}_+^*}$ in $\mathcal{M}^1(G)$ such that ψ is the negative-definite form corresponding to S.*

Proof. 1. We show (A). For $\alpha \in \mathbb{A}$ let Γ_α be a Lévy function for the Lie group $G_\alpha \in \mathbf{A}$. $S_\alpha := p_\alpha(S)$ is an $\{e_\alpha\}$-continuous convolution semigroup in $\mathcal{M}^1(G_\alpha)$ with corresponding negative-definite form $\psi^{(\alpha)}$ on $\mathfrak{K}(G_\alpha)$ satisfying

$$\psi^{(\alpha)}(f_\alpha) = \psi(f_\alpha \circ p_\alpha) \qquad \text{for all } f_\alpha \in \mathfrak{K}(G_\alpha).$$

By Theorem 4.3.10 (A) there exist a continuous primitive form $\psi_1^{(\alpha)}$, a continuous quadratic form $\psi_2^{(\alpha)}$ on $\mathfrak{K}(G_\alpha)$ and a Lévy measure η_α on G_α (all uniquely determined) such that

$$\psi^{(\alpha)}(f_\alpha) = \psi_1^{(\alpha)}(f_\alpha) + \psi_2^{(\alpha)}(f_\alpha) - \int_{G_\alpha^*} [f_\alpha(x_\alpha) - f_\alpha(e_\alpha) - \Gamma_\alpha(x_\alpha, f_\alpha)]\, \eta_\alpha(dx_\alpha)$$

holds for all $f_\alpha \in \mathfrak{K}(G_\alpha)$.

From Theorem 4.3.10 we also obtain

$$\int_{G_\alpha^*} f_\alpha \, d\eta_\alpha = \lim_{t\downarrow 0} \frac{1}{t} \int_G f_\alpha \circ p_\alpha \, d\mu_t \quad \text{for all } f_\alpha \in \mathscr{K}(G_\alpha^*).$$

Hence, the formula $\eta(f_\alpha \circ p_\alpha) := \eta_\alpha(f_\alpha)$ for all $f_\alpha \in \mathscr{K}(G_\alpha^*)$ $(\alpha \in \mathbb{A})$ defines a positive linear functional η on the vector space

$$\mathscr{V} := \{f_\alpha \circ p_\alpha : f_\alpha \in \mathscr{K}(G_\alpha^*), \alpha \in \mathbb{A}\}.$$

Since $\mathscr{V}$ is a dense subspace of $\mathscr{K}(G^*)$ by Lemma 1.2.18 and since for each compact subset K of G there is a function $f \in \mathscr{V}_+$ with $f(x) > 0$ for all $x \in K$, Proposition 9 of [41], p. 56 implies that η can be uniquely extended to a Radon measure $\eta \in \mathscr{M}_+(G^*)$. Moreover, η is a Lévy measure on G. For every $\alpha \in \mathbb{A}$ and $f_\alpha \in \mathfrak{R}(G_\alpha)$ we now have (again by Theorem 4.3.10)

$$\begin{aligned}
&\int_{G_\alpha^*} [f_\alpha - f_\alpha(e_\alpha) - \Gamma_\alpha(\cdot, f_\alpha)] \, d\eta_\alpha \\
&= \int_{G_\alpha^*} [f_\alpha - f_\alpha(e_\alpha) - \textstyle\sum_{i=1}^{n(\alpha)} x_i^\alpha \tilde{X}_i^\alpha f_\alpha(e_\alpha)] \, d\eta_\alpha \\
&= \int_{G^*} [f_\alpha \circ p_\alpha - (f_\alpha \circ p_\alpha)(e) - \textstyle\sum_{i=1}^{n(\alpha)} (x_i^\alpha \circ p_\alpha) \tilde{X}_i^\alpha f_\alpha(e_\alpha)] \, d\eta,
\end{aligned}$$

where $\{x_1^\alpha, \ldots, x_{n(\alpha)}^\alpha\}$ is a canonical coordinate system with respect to the basis $\{X_1^\alpha, \ldots, X_{n(\alpha)}^\alpha\}$ of $\mathfrak{L}(G_\alpha)$.

An application of Lemma 4.3.12 yields the existence of functions $y_1^\alpha, \ldots, y_{n(\alpha)}^\alpha \in \mathscr{C}^b(G)$ and of a neighborhood $U_\alpha \in \mathfrak{B}(e)$ such that $x_i^\alpha \circ p_\alpha(x) = y_i^\alpha$ for all $x \in U_\alpha$ $(i = 1, \ldots, n(\alpha))$. Hence

$$\begin{aligned}
&\int_{G^*} [f_\alpha \circ p_\alpha - (f_\alpha \circ p_\alpha)(e) - \textstyle\sum_{i=1}^{n(\alpha)} (x_i^\alpha \circ p_\alpha) \tilde{X}_i^\alpha f_\alpha(e_\alpha)] \, d\eta \\
&= \int_{G^*} [f_\alpha \circ p_\alpha - (f_\alpha \circ p_\alpha)(e) - \textstyle\sum_{i=1}^{n(\alpha)} y_i^\alpha \tilde{X}_i^\alpha f_\alpha(e_\alpha)] \, d\eta \\
&+ \int_{G \setminus U_\alpha} [\textstyle\sum_{i=1}^{n(\alpha)} (y_i^\alpha - x_i^\alpha \circ p_\alpha) \tilde{X}_i^\alpha f_\alpha(e_\alpha)] \, d\eta \\
&= \int_{G^*} [f_\alpha \circ p_\alpha - f_\alpha \circ p_\alpha(e) - \Gamma(\cdot, f_\alpha \circ p_\alpha)] \, d\eta + \bar{\psi}_1^{(\alpha)}(f_\alpha)
\end{aligned}$$

for all $\alpha \in \mathbb{A}$, $f_\alpha \in \mathfrak{R}(G_\alpha)$, where the linear functional $\bar{\psi}_1^{(\alpha)}$ on $\mathfrak{R}(G_\alpha)$ defined by

$$\bar{\psi}_1^{(\alpha)}(f_\alpha) := \int_{G \setminus U_\alpha} [\textstyle\sum_{i=1}^{n(\alpha)} (y_i^\alpha - x_i^\alpha \circ p_\alpha) \tilde{X}_i^\alpha f_\alpha(e_\alpha)] \, d\eta$$

for all $f_\alpha \in \mathfrak{R}(G_\alpha)$ is a continuous primitive form on $\mathfrak{R}(G_\alpha)$.

Thus we get

$$\psi(f_\alpha \circ p_\alpha) = (\psi_1^{(\alpha)} + \bar{\psi}_1^{(\alpha)})(f_\alpha) + \psi_2^{(\alpha)}(f_\alpha) + \psi_3(f_\alpha \circ p_\alpha) \text{ with}$$

$$\psi_3(f_\alpha \circ p_\alpha) := -\int_{G^*} [f_\alpha \circ p_\alpha - f_\alpha \circ p_\alpha(e) - \Gamma(\cdot, f_\alpha \circ p_\alpha)] \, d\eta$$

for all $f_\alpha \in \mathfrak{R}(G_\alpha)$ $(\alpha \in \mathbb{A})$.

Replacing in the representation of ψ the linear functional $\psi_1^{(\alpha)}$ by $\psi_1^{(\alpha)} + \bar{\psi}_1^{(\alpha)}$ we obtain

$$\psi_j^{(\alpha)}(f_\alpha) = \psi_j^{(\beta)}(f_\alpha \circ p_{\alpha\beta}) \quad \text{for all } \alpha, \beta \in \mathbb{A},\ \alpha < \beta,\ f_\alpha \in \mathfrak{R}(G_\alpha) \text{ and } j = 1, 2.$$

Putting $\psi_j(f_\alpha \circ p_\alpha) := \psi_j^{(\alpha)}(f_\alpha)$ for all $f_\alpha \in \mathfrak{K}(G_\alpha)$ ($\alpha \in \mathbb{A}$) and $j=1,2$, we obtain that ψ_1 is a continuous primitive form and ψ_2 is a continuous quadratic form on $\mathfrak{K}(G)$, and $\psi = \psi_1 + \psi_2 + \psi_3$ holds.

Finally, we show the relationship

$$\int_{G^*} f\, d\eta = \lim_{t \downarrow 0} \frac{1}{t} \int f\, d\mu_t \qquad \text{for all } f \in \mathscr{K}(G^*).$$

The net $\left(\frac{1}{t}\mu_t\right)_{t \in \mathbb{R}_+^*}$ of continuous linear functionals on the barreled space $\mathscr{K}(G^*)$ converges pointwise on the dense subspace $\mathscr{V}$ to a continuous linear functional η and is bounded (by Theorem 4.3.10). The Banach-Steinhaus theorem then yields the result. The rest of Statement (A) follows as in Theorem 4.3.10.

2. The proof of (B) is carried out in analogy to Theorem 4.3.10. With the aid of Theorems 1.5.16, 1.5.17 and 1.5.19 one sees that it suffices to show that the negative-definite form ψ_3 defined in Part 1 of this proof is continuous on $\mathfrak{K}(G)$. For $\alpha \in \mathbb{A}$, and $f_\alpha \in \mathfrak{K}(G_\alpha)$ we found by the reasoning above that

$$\psi_3(f_\alpha \circ p_\alpha) = -\int_{G^*} [f_\alpha \circ p_\alpha - f_\alpha \circ p_\alpha(e) - \Gamma(\cdot, f_\alpha \circ p_\alpha)]\, d\eta,$$

where $\Gamma(\cdot, f_\alpha \circ p_\alpha) = \sum_{i=1}^{n(\alpha)} (x_i^\alpha \circ p_\alpha) \tilde{X}_i^\alpha f_\alpha(e_\alpha)$ in a properly chosen neighborhood $U_\alpha \in \mathfrak{B}(e)$.

Since $\eta(\complement U_\alpha) < \infty$, it suffices to provide an estimate for the integral term with integration over U_α^* only.

Given a Hunt function ϕ_α for G_α we first obtain a constant $c_0 \in \mathbb{R}_+^*$ such that

$$\begin{aligned} &\left|\int_{U_\alpha^*} [f_\alpha \circ p_\alpha - f_\alpha \circ p_\alpha(e) - \textstyle\sum_{i=1}^{n(\alpha)} (x_i^\alpha \circ p_\alpha) \tilde{X}_i^\alpha f_\alpha(e_\alpha)]\, d\eta\right| \\ &\leq c_0 \int_{G^*} \phi_\alpha(p_\alpha(x))\, \eta(dx) \end{aligned}$$

holds.

But by Lemma 4.3.9 this bound is $\leq c \int_{G^*} f_{D_\alpha \circ p_\alpha} d\eta$ for a representation $D_\alpha \in \operatorname{Rep}(G_\alpha)$ and a constant $c \in \mathbb{R}_+^*$, which can be chosen uniformly for compact subsets of $\operatorname{Rep}(G)$. As $U_\alpha \downarrow \{e\}$ this constant tends to zero. By the definition of the Lévy measure η we obtain the continuity of ψ_3. □

4.4 The Canonical Representation of Almost Positive Functionals

The first topic of this section will be the introduction of the space $\mathfrak{D}(G)$ of infinitely differentiable functions on G with compact support for an arbitrary locally compact group G. We recall the notion of infinite differentiability for a Lie group G with Lie algebra $\mathfrak{L}(G)$ having a basis $\{X_1, \ldots, X_n\}$. For every $k \geq 1$ we introduce the spaces

$$\begin{aligned} \mathscr{C}_k^0(G) := \{f \in \mathscr{C}^0(G) : Y_{i_1} \cdot \cdots \cdot Y_{i_r} f \in \mathscr{C}^0(G) \\ \text{for all } Y_{i_1}, \ldots, Y_{i_r} \in \mathfrak{L}(G),\ r = 1, \ldots, k\} \quad \text{and} \end{aligned}$$

$$\mathscr{C}_k^0(G) := \{f \in \mathscr{C}^0(G) : \tilde{Y}_{i_1} \cdot \dots \cdot \tilde{Y}_{i_r} f \in \mathscr{C}^0(G)$$
$$\text{for all } Y_{i_1}, \dots, Y_{i_r} \in \mathfrak{L}(G),\ r = 1, \dots, k\}$$

which certainly contain the space

$$\mathfrak{D}(G) := \mathscr{C}^\infty(G) \cap \mathscr{K}(G)$$

as a linear subspace.

For any compact subset K of G we put

$$\mathfrak{D}(G, K) := \{f \in \mathfrak{D}(G) : \mathrm{supp}(f) \subset K\},$$

and for $f \in \mathfrak{D}(G, K)$ we define the number

$$p_{K,k}(f) := \sup_{x \in K}(|f(x)| + \textstyle\sum_{i=1}^n |X_i f(x)| + \cdots$$
$$+ \textstyle\sum_{i_1, \dots, i_k = 1}^n |X_{i_1} \cdot \dots \cdot X_{i_k} f(x)|).$$

Plainly $p_{K,k}$ is a seminorm on $\mathfrak{D}(G, K)$, and $\mathfrak{D}(G, K)$ becomes a locally convex vector space with respect to the family $(p_{K,k})_{k \geq 1}$. $\mathfrak{D}(G)$ will now be furnished with the topology of the inductive limit with respect to the family $\{\mathfrak{D}(G, K) : K \text{ compact} \subset G\}$.

Let G be an arbitrary locally compact group.

4.4.1 Definition. A *Lie system* (L-system) *for* G is a quintuple $\{G_1, H_\alpha, p_\alpha, p_{\alpha\beta}, \mathbb{A}\}$ (or briefly $\{G_1, (H_\alpha)_{\alpha \in \mathbb{A}}\}$) with the following properties:

(LS1) G_1 is an open subgroup of G.
(LS2) $(H_\alpha)_{\alpha \in \mathbb{A}}$ is a descending family of compact normal subgroups of G_1 satisfying $\bigcap_{\alpha \in \mathbb{A}} H_\alpha = \{e\}$.
(LS3) For every $\alpha \in \mathbb{A}$ the group G_1/H_α is a Lie group.
(LS4) The index set $\mathbb{A}$ is ordered by the relation $\alpha < \beta :\Leftrightarrow H_\beta \subset H_\alpha$.
(LS5) For $\alpha \in \mathbb{A}$ (or $\alpha, \beta \in \mathbb{A}$ with $\alpha < \beta$) one denotes by p_α (or $p_{\alpha\beta}$) the canonical mapping from G_1 onto G_1/H_α (or from G_1/H_β onto G_1/H_α).

Clearly, given an L-system $\{G_1, (H_\alpha)_{\alpha \in \mathbb{A}}\}$ for G one has $\varprojlim_{\alpha \in \mathbb{A}} G_1/H_\alpha = G_1$. The solution of Hilbert's fifth problem yields the profound result that every locally compact group G admits an L-system (see [356], p. 175 and [171], p. 91 or our Theorem G).

4.4.2. The construction of the space $\mathfrak{D}(G)$ of *infinitely differentiable functions in* $\mathscr{K}(G)$ *for an arbitrary locally compact group* G will be performed in two steps.

(α) Let G be a Lie projective group. One has an L-system $\{G_1, (H_\alpha)_{\alpha \in \mathbb{A}}\}$ for G with $G_1 = G$. For every $\alpha \in \mathbb{A}$ the mapping

$$\hat{p}_\alpha : \mathfrak{D}(G/H_\alpha) \to \mathscr{K}(G)$$

is defined by $\hat{p}_\alpha(f_\alpha) := f_\alpha \circ p_\alpha$ for all $f_\alpha \in \mathfrak{D}(G/H_\alpha)$ and

$$\mathfrak{D}_\alpha(G) := \hat{p}_\alpha(\mathfrak{D}(G/H_\alpha))$$

is furnished with the final topology with respect to the mapping $\hat{p}_\alpha$. Thus

$$\mathfrak{D}(G) := \varinjlim_{\alpha \in \mathbb{A}} \mathfrak{D}_\alpha(G)$$

is a locally convex Hausdorff space, independent of the special choice of the approximating L-system given for G.

(β) Let G be an arbitrary locally compact group. There exists an open Lie projective subgroup G_1 of G. For every $x \in G$ let the mappings $\rho(x): xG_1 \to G_1$ and $\sigma(x): G_1 x \to G_1$ be defined by

$$\rho(x)\, y := x^{-1} y \quad \text{and} \quad \sigma(x)\, y := y x^{-1}$$

for all $y \in xG_1$ and $y \in G_1 x$ resp. One introduces the subspaces

$$\mathfrak{D}(xG_1) := \{f \circ \rho(x) : f \in \mathfrak{D}(G_1)\} \quad \text{and} \quad \mathfrak{D}(G_1 x) := \{f \circ \sigma(x) : f \in \mathfrak{D}(G_1)\}$$

of $\mathscr{K}(xG_1)$ and $\mathscr{K}(G_1 x)$ resp. equipped with the final topology with respect to $f \to f \circ \rho(x)$ and $f \to f \circ \sigma(x)$ resp. Clearly, $xG_1 = yG_1$ implies that $\mathfrak{D}(xG_1) = \mathfrak{D}(yG_1)$ and $G_1 x = G_1 y$ implies that $\mathfrak{D}(G_1 x) = \mathfrak{D}(G_1 y)$. Let $\mathfrak{D}^\rho(G)$ (or $\mathfrak{D}^\sigma(G)$) denote the direct topological sum of the spaces $\mathfrak{D}(x_i G_1)$ (or $\mathfrak{D}(G_1 y_j)$), where the elements x_i (or y_j) form a representative system of left (or right) cosets of G_1.

One shows that $\mathfrak{D}^\rho(G) = \mathfrak{D}^\sigma(G)$ and that $\mathfrak{D}(G) := \mathfrak{D}^\rho(G)$ $(= \mathfrak{D}^\sigma(G))$ is independent of the special choice of the open Lie projective subgroup G_1 of G ([51], Proposition 4).

4.4.3 Properties of the space $\mathfrak{D}(G)$.

(1) Let G be a locally compact group and $\{G, (H_\alpha)_{\alpha \in \mathbb{A}}\}$ an L-system for G. For every $\alpha \in \mathbb{A}$ and $f \in \overline{\mathscr{C}^0}(G)$, $x \in G$ we introduce the number

$$P_\alpha f(x) := \int_{H_\alpha} f(yx)\, \omega_\alpha(dy) = \omega_\alpha(f_x),$$

where $\omega_\alpha := \omega_{H_\alpha}$ denotes the normed Haar measure of the compact subgroup H_α of G.

(i) P_α is an idempotent operator on $\overline{\mathscr{C}^0}(G)$.
(ii) $\mathfrak{D}_\alpha(G) := P_\alpha(\mathfrak{D}(G)) \subset \mathfrak{D}(G)$.
(iii) $P_\alpha(\mathfrak{D}_\alpha(G)) \subset \mathfrak{D}_\alpha(G)$ (for all $\alpha \in \mathbb{A}$).
(iv) $\mathfrak{D}_\alpha(G) \subset \mathfrak{D}_\beta(G)$ for all $\alpha, \beta \in \mathbb{A}$ with $\alpha < \beta$.
(v) $\mathfrak{D}(G) = \bigcup_{\alpha \in \mathbb{A}} \mathfrak{D}_\alpha(G)$.

We note that in the case of a Lie projective group G with $G = G_1$ one obtains

(vi) $\mathfrak{D}_\alpha(G) = \hat{p}_\alpha(\mathfrak{D}(G/H_\alpha))$ for all $\alpha \in \mathbb{A}$ in agreement with Step (α) of the above construction of $\mathfrak{D}(G)$.

(2) For every $U \in \mathfrak{V}(e)$ there exists a function $f \in \mathfrak{D}_+(G)$ with $\mathrm{supp}(f) \subset U$ such that $\omega_G(f) = 1$ holds. If K is a compact subset of G and $(U_i)_{i=1,\ldots,n}$ a finite covering of K, then for each $i = 1, \ldots, n$ there exists an $f_i \in \mathfrak{D}_+(G)$ with $\mathrm{supp}(f_i) \subset U_i$ satisfying $\sum_{i=1}^n f_i(x) = 1$ for all $x \in K$.

The proof of this property can be taken from [51], Proposition 2. It follows immediately that

(3) $\mathfrak{D}(G)$ is dense in $\mathscr{K}(G)$.

The following property contains an equivalent description of the space $\mathfrak{D}(G)$.

(4) Let $\mathbf{C}(G)$ denote the system of all compact subgroups H of G such that there is an open Lie projective subgroup G_1 of G with $H \subset G_1$ and G_1/H is a Lie group. Then $\mathfrak{D}(G)$ is the linear subspace of functions $f \in \mathscr{K}(G)$ satisfying the following conditions:

(i) There exists an $H \in \mathbf{C}(G)$ with a corresponding open Lie projective subgroup G_1 of G ($H \subset G_1$, G_1/H Lie group).

(ii) There exist $x_1, \ldots, x_n \in G$ and $g_1, \ldots, g_n \in \mathscr{K}(G)$ with $\operatorname{supp}(g_i) \subset G_1$ $(i=1, \ldots, n)$ and $f = \sum_{i=1}^{n} {}_{x_i}g_i$.

(iii) For every $i=1, \ldots, n$ one has $g_i(hxk) = g_i(x)$ whenever $x \in G_1$ and $h, k \in H$.

(iv) There exist $\dot{g}_1, \ldots, \dot{g}_n \in \mathfrak{D}(G_1/H)$ such that $g_i = \dot{g}_i \circ p$ holds for all $i=1, \ldots, n$, where p denotes the canonical projection from G_1 onto G_1/H.

(5) Let G be a totally disconnected locally compact group (in the sense that there is no nontrivial continuous homomorphism (from $\mathbb{R}$ into G). Then the system $\mathbf{C}(G)$ of (4) coincides with the family of all compact open subgroups of G. The space $\mathfrak{D}(G)$ consists in this case of all functions $f \in \mathscr{K}(G)$ with the property that there exists an $H \in \mathbf{C}(G)$ satisfying $f(x) = f(hxk)$ for all $x \in G$ and $h, k \in H$. A function $f \in \mathscr{K}(G)$ belongs to $\mathfrak{D}(G)$ iff $\operatorname{im} f$ is finite. This follows directly from the fact that for a totally disconnected group G the system $\mathbf{C}(G)$ is a basis of $\mathfrak{B}(e)$.

4.4.4 Lemma. *Let G be a locally compact group.*

(i) *If K is a compact normal subgroup of G and π denotes the canonical projection from G onto $\dot{G} := G/K$, then $\dot{f} \circ \pi \in \mathfrak{D}(G)$ for all $\dot{f} \in \mathfrak{D}(\dot{G})$.*

(ii) *For any descending family $(K_i)_{i \in \mathbb{I}}$ of compact normal subgroups of G with $\bigcap_{i \in \mathbb{I}} K_i = \{e\}$ (or equivalently $G = \varprojlim_{i \in \mathbb{I}} G/K_i$) one has*

$$\mathfrak{D}(G) = \varinjlim_{i \in \mathbb{I}} \mathfrak{D}(G/K_i).$$

Proof. 1. Let G be a Lie projective group with L-system $\{G, (H_\alpha)_{\alpha \in \mathbb{A}}\}$. For the proof of (i) we put $\dot{H}_\alpha := \pi(H_\alpha)$, denote by $\dot{p}_\alpha$ the canonical projection from $\dot{G}$ onto $\dot{G}/\dot{H}_\alpha$ and define

$$q_\alpha : G/H_\alpha \to \dot{G}/\dot{H}_\alpha \quad \text{by} \quad q_\alpha \circ p_\alpha = \dot{p}_\alpha \circ \pi \qquad \text{for all } \alpha \in \mathbb{A}.$$

Since for every $\alpha \in \mathbb{A}$ the mapping q_α is an open epimorphism, $\dot{G}/\dot{H}_\alpha$ is a Lie group, and $\dot{G} = \varprojlim_{\alpha \in \mathbb{A}} \dot{G}/\dot{H}_\alpha$. Let $\dot{f} \in \mathfrak{D}(\dot{G})$. There exist an $\alpha \in \mathbb{A}$ and an $\dot{f}_\alpha \in \mathfrak{D}(\dot{G}/\dot{H}_\alpha)$ with $\dot{f} = \dot{f}_\alpha \circ \dot{p}_\alpha$. Hence,

$$\dot{f} \circ \pi = \dot{f}_\alpha \circ (\dot{p}_\alpha \circ \pi) = \dot{f}_\alpha \circ (q_\alpha \circ p_\alpha) = (\dot{f}_\alpha \circ q_\alpha) \circ p_\alpha$$

for some $\alpha \in \mathbb{A}$. But $\ker q_\alpha = KH_\alpha/H_\alpha$ is compact and q_α is analytic, so that $\dot{f}_\alpha \circ q_\alpha \in \mathfrak{D}(G/H_\alpha)$ for some $\alpha \in \mathbb{A}$ and $\dot{f} \circ \pi \in \mathfrak{D}(G)$. This shows Assertion (i), which under the hypothesis of (ii) implies that

$$\varinjlim_{i \in \mathbb{I}} \mathfrak{D}(G/K_i) \subset \mathfrak{D}(G) \qquad (\text{for } G = \varprojlim_{i \in \mathbb{I}} G/K_i).$$

Let $f \in \mathfrak{D}(G)$ be of the form $f = f_\alpha \circ p_\alpha$ for some $\alpha \in \mathbb{A}$ and $f_\alpha \in \mathfrak{D}(G/H_\alpha)$. Since $\bigcap_{i \in \mathbb{I}} p_\alpha(K_i) = \{p_\alpha(e)\}$ and G/H_α is a Lie group for every $\alpha \in \mathbb{A}$, there exists an $i := i(\alpha) \in \mathbb{I}$ with $K_i \subset H_\alpha$. Defining for $i \in \mathbb{I}$ and $\alpha \in \mathbb{A}$ the canonical projections

$\pi_i: G \to G/K_i$ and $q_{\alpha i}: G/K_i \to G/H_\alpha$ such that $p_\alpha = q_{\alpha i} \circ \pi_i$ holds one obtains $f = (f_\alpha \circ q_{\alpha i}) \circ \pi_i$ and $f_\alpha \circ q_{\alpha i} \in \mathfrak{D}(G/K_i)$. But this implies that $f \in \varinjlim_{i \in \mathbb{I}} \mathfrak{D}(G/K_i)$, thus (ii).

2. We now assume G to be an arbitrary locally compact group with an open Lie projective subgroup G_1. We first show (i).

$$\dot{G}_1 := \pi(G_1) \cong G_1/G_1 \cap K$$

is an open Lie projective subgroup of $\dot{G}$. Let $\dot{x} \in \dot{G}$ and $\dot{g} \in \mathfrak{D}(\dot{x}\dot{G}_1)$. By definition there exists an $\dot{f} \in \mathfrak{D}(\dot{G}_1)$ with $\dot{g} = \dot{f} \circ \rho(\dot{x})$. Since $C := \pi^{-1}(\mathrm{supp}(\dot{g}))$ is compact, there exist $x_1, \ldots, x_n \in G$ such that

$$C \subset \bigcup_{i=1}^n x_i G_1 \quad \text{and} \quad x_i G_1 \cap x_j G_1 = \emptyset$$

holds for all $i \neq j$; $i, j = 1, \ldots, n$.

Without loss of generality we assume that $\pi(x_j) = \dot{x}$ for $j = 1, \ldots, n$. By Part 1 one obtains $f := \dot{f} \circ \pi' \in \mathfrak{D}(G_1)$, where $\pi' := \mathrm{Res}_{G_1} \pi$. Moreover,

$$\dot{g} \circ \pi(y) = \dot{f}(\dot{x}^{-1} \pi(y)) = \dot{f} \circ \pi'(x_j^{-1} y) = f \circ \rho(x_j)(y)$$

holds for all $y \in x_j G_1$, $j = 1, \ldots, n$. Consequently,

$$(\dot{g}\, 1_{\dot{x}\dot{G}_1}) \circ \pi = \sum_{j=1}^n (f \circ \rho(x_j))\, 1_{x_j G_1} \in \mathfrak{D}(G).$$

By the definition of $\mathfrak{D}(\dot{G})$ (as the direct sum of the spaces $\mathfrak{D}(\dot{x}\dot{G}_1)$) this shows (i).

To prove (ii) we assume without loss of generality that $K_i \subset G_1$ holds for all $i \in \mathbb{I}$, whence $G_1 = \varprojlim_{i \in \mathbb{I}} G_1/K_i$. Let $x \in G$, $g \in \mathfrak{D}(xG_1)$ and $f \in \mathfrak{D}(G_1)$ be such that $g = f \circ \rho(x)$ is fulfilled. By Part 1 there exist an $i \in \mathbb{I}$ and an $f_i \in \mathfrak{D}(G_1/K_i)$ with $f = f_i \circ \pi_i'$, where $\pi_i' := \mathrm{Res}_{G_1} \pi_i$.

Defining $H(x) := \pi_i(x)\, G_1/K_i$ one gets

$$g_i := (f_i \circ \rho(\pi_i(x)))\, 1_{H(x)} \in \mathfrak{D}(G/K_i).$$

For $y \in xG_1$ the equalities

$$\begin{aligned} g_i \circ \pi_i(y) &= f_i(\pi_i(x)^{-1} \pi_i(y)) = f_i(\pi_i'(x^{-1} y)) \\ &= (f_i \circ \pi_i') \circ \rho(x)(y) = (f \circ \rho(x))(y) = g(y) \end{aligned}$$

hold. They imply that

$$g \in \varinjlim_{i \in \mathbb{I}} \mathfrak{D}(G/K_i) \quad \text{or} \quad \mathfrak{D}(G) \subset \varinjlim_{i \in \mathbb{I}} \mathfrak{D}(G/K_i),$$

which yields Statement (ii). □

4.4.5. Let G be an arbitrary locally compact group and $\{G_1, (H_\alpha)_{\alpha \in \mathbb{A}}\}$ an L-system for G. For every $\alpha \in \mathbb{A}$ we denote by $\mathfrak{g}_\alpha$ the Lie algebra $\mathfrak{L}(G_1/H_\alpha)$ of the Lie group G_1/H_α and by $\exp_\alpha$ the exponential mapping $\exp_{G_1/H_\alpha}$ from $\mathfrak{g}_\alpha$ into G_1/H_α. One shows that $(\mathfrak{g}_\alpha, dp_{\alpha\beta})$ is a projective system of Lie algebras. Its projective limit $\mathfrak{g} := \varprojlim_{\alpha \in \mathbb{A}} \mathfrak{g}_\alpha$ will be called the *Lie algebra of the locally compact group G*. By [312] there is also an exponential mapping $\exp := \exp_G : \mathfrak{g} \to G$ and for every $\alpha \in \mathbb{A}$

a homomorphism $dp_\alpha: \mathfrak{g} \to \mathfrak{g}_\alpha$ satisfying $p_\alpha \circ \exp = \exp_\alpha \circ dp_\alpha$ as in classical Lie group theory.

Moreover, one has $dp_{\alpha\beta} \circ dp_\beta = dp_\alpha$ for all $\alpha, \beta \in \mathbb{A}$ with $\alpha < \beta$. $\mathfrak{g}$ and exp can be shown to be independent of the special choice of the L-system $\{G_1, (H_\alpha)_{\alpha \in \mathbb{A}}\}$ for G. We finally add that a family $\{X_1, \ldots, X_n\}$ of elements of $\mathfrak{g}$ is called a $\mathfrak{g}_\alpha$-*basis* if $\{dp_\alpha(X_1), \ldots, dp_\alpha(X_n)\}$ is a basis of the Lie algebra $\mathfrak{g}_\alpha$ ($\alpha \in \mathbb{A}$). In analogy to 1.5.15 we now introduce various types of real linear functionals on $\mathfrak{D}(G)$ for a locally compact group G. For abbreviation we put

$$\mathfrak{D}_0(G) := \{f \in \mathfrak{D}_+(G) : f(e) = 0\}.$$

4.4.6 Definition. A real linear functional L on $\mathfrak{D}(G)$ is called

(a) *almost positive* if $L(f) \geq 0$ for all $f \in \mathfrak{D}_0(G)$,

(b) a *primitive form* (on $\mathfrak{D}(G)$) if it is almost positive and for all $f, g \in \mathfrak{D}(G)$ one has $L(f g^*) = L(f)\, g(e) - f(e)\, L(g)$,

(c) a *quadratic form* if L is almost positive and the relationship $L(f g) + L(f g^*) = 2[L(f)\, g(e) + f(e)\, L(g)]$ holds for all $f, g \in \mathfrak{D}(G)$,

(d) *concentrated* (at the origin) if for all $f \in \mathfrak{D}(G)$ and every $g \in \mathfrak{D}(G)$ which equals 1 in a neighborhood of e one has $L(f) = L(f g)$ and

(e) *normed* if for some $U \in \mathfrak{B}(e)$ with $\mathscr{H}(U) := \{f \in \mathfrak{D}(G) : 1_U \leq f \leq 1_G\}$ the equality $\sup\{L(f) : f \in \mathscr{H}(U)\} = 0$ is satisfied.

4.4.7. We collect a few basic **properties.** (1) $L \in \mathfrak{D}(G)^*$ is concentrated iff $L(f) = 0$ for all $f \in \mathfrak{D}(G)$ which vanish in a neighborhood of e. In fact, let L be concentrated, $U \in \mathfrak{B}(e)$ and $f \in \mathfrak{D}(G)$ with $f(U) = 0$. Then there exist a $V \in \mathfrak{B}(e)$ with $\bar{V} \subset \mathring{U}$ and a $g \in \mathfrak{D}(G)$ satisfying $g(V) = 1$, $g(\complement U) = 0$. This implies that $f g = 0$, whence $L(f) = L(f g) = 0$. For the converse let $f, g \in \mathfrak{D}(G)$ with $g(U) = 1$ for some $U \in \mathfrak{B}(e)$. Then $h := f - f g \in \mathfrak{D}(G)$ and h satisfies $h(U) = 0$. Thus

$$0 = L(h) = L(f) - L(f g).$$

(2) Every primitive form and every quadratic form L on $\mathfrak{D}(G)$ is concentrated. In fact, let L be a primitive form and let $f \in \mathfrak{D}(G)$ with $f(U) = 0$ for some $U \in \mathfrak{B}(e)$. There exists a $g \in \mathfrak{D}(G)$ satisfying

$$g = g^*, \quad g(e) = 0 \quad \text{and} \quad g(\operatorname{supp}(f)) = 1,$$

so that $f g = f$. The primitivity of L implies that

$$L(f) = L(f g) = L(f)\, g(e) - f(e)\, L(g) = 0.$$

The second statement is proved in the same manner.

(3) Given $f \in \mathfrak{D}(G)$ we obtain $L(f^*) = -L(f)$ for any primitive form and $L(f^*) = L(f)$ for every quadratic form L on $\mathfrak{D}(G)$. The argument is easy. Let $g \in \mathfrak{D}(G)$ with $g = g^*$ and $g(U) = 1$ for some $U \in \mathfrak{B}(e)$. If L is a primitive form, its defining equation together with (2) yields $L(g) = L(g g^*) = 0$. Again by the primitivity of L one obtains $L(f^*) = L(g f^*) = -L(f)$. The assertion concerning quadratic forms follows similarly.

(4) Let K be a compact normal subgroup of G and π the canonical projection from G onto $\dot{G} := G/K$. Then to every almost positive and normed $L \in \mathfrak{D}(G)^*$ there corresponds an almost positive and normed $\pi(L) \in \mathfrak{D}(\dot{G})^*$ defined by $\pi(L)(\dot{f}) := L(\dot{f} \circ \pi)$ for all $\dot{f} \in \mathfrak{D}(\dot{G})$.

4.4.8 Examples. Let G be a Lie group of dimension $n \geq 1$ with Lie algebra $\mathfrak{L}(G)$ having a basis $\{X_1, \ldots, X_n\}$.

1. Let $a_1, \ldots, a_n \in \mathbb{R}$ and define the linear functional ψ on $\mathfrak{D}(G)$ by

$$\psi(f) := \sum_{i=1}^{n} a_i (X_i f)(e) \quad \text{for all } f \in \mathfrak{D}(G).$$

Then ψ is a primitive form on $\mathfrak{D}(G)$. This follows from the formulae

$$(X_i f^*)(e) = -(X_i f)(e) \quad \text{and}$$
$$(X_i fg)(e) = (X_i f)(e)\, g(e) + f(e)(X_i g)(e),$$

valid for all $f, g \in \mathfrak{D}(G)$ and $i = 1, \ldots, n$.

2. Let $(a_{ij})_{i,j=1,\ldots,n} \in \mathfrak{M}(n, \mathbb{R})$ be a symmetric, positive-semidefinite matrix and define the linear functional ψ on $\mathfrak{D}(G)$ by

$$\psi(f) := \sum_{i,j=1}^{n} a_{ij} (X_i X_j f)(e) \quad \text{for all } f \in \mathfrak{D}(G).$$

Then ψ is a quadratic form on $\mathfrak{D}(G)$.

We content ourselves with the proof of the almost positivity of ψ. Let $f \in \mathfrak{D}_+(G)$ with $f(e) = 0$. For $i, j = 1, \ldots, n$ we obtain

$$\frac{\partial^2}{\partial x_i\, \partial x_j} f(e) = (X_i X_j f)(e) =: b_{ij}.$$

Since f attains a minimum in e, the matrix $(b_{ij})_{i,j=1,\ldots,n} \in \mathfrak{M}(n, \mathbb{R})$ is positive-semidefinite. But then the matrix $(c_{ij})_{i,j=1,\ldots,n}$ with $c_{ij} = a_{ij} b_{ij}$ for $i, j = 1, \ldots, n$ also has this property. The fact that $\psi(f) = \sum_{i,j=1}^{n} c_{ij} \geq 0$ then yields the assertion.

4.4.9 Lemma. *Let G be a Lie group of dimension $n \geq 1$ with Lie algebra $\mathfrak{L}(G)$ admitting a basis $\{X_1, \ldots, X_n\}$ and $\psi \in \mathfrak{D}(G)^*$.*

(i) *If ψ is a primitive form on $\mathfrak{D}(G)$, then there exist numbers $a_1, \ldots, a_n \in \mathbb{R}$ such that for all $f \in \mathfrak{D}(G)$ one has*

$$\psi(f) = \sum_{i=1}^{n} a_i (X_i f)(e) = \left[\left(\sum_{i=1}^{n} a_i X_i\right) f\right](e).$$

The vector $\sum_{i=1}^{n} a_i X_i$ is uniquely determined by ψ.

(ii) *If ψ is a quadratic form on $\mathfrak{D}(G)$, then there is a symmetric, positive-semidefinite matrix $(a_{ij})_{i,j=1,\ldots,n} \in \mathfrak{M}(n, \mathbb{R})$ such that for all $f \in \mathfrak{D}(G)$ the formula*

$$\psi(f) = \sum_{i,j=1}^{n} a_{ij} (X_i X_j f)(e)$$

obtains.

Proof. 1. Let $f \in \mathfrak{D}(G)$. There exist a $U \in \mathfrak{V}(e)$ and $g_1, \ldots, g_n \in \mathfrak{D}(G)$ such that

$$f(x) = f(e) + \sum_{i=1}^{n} x_i(x)\, g_i(x) \quad \text{for all } x \in U,$$

where $x_1(x), \ldots, x_n(x)$ are local coordinates of x and

$$g_i(e) = (X_i f)(e) = \frac{\partial}{\partial x_i} f(e) \quad \text{for } i = 1, \ldots, n.$$

Plainly

$$g := \sum_{i=1}^n x_i g_i \in \mathfrak{D}(G).$$

We further choose $h \in \mathfrak{D}(G)$ with $h = h^*$, $h(\complement U) = 0$ and $h(V) = 1$ for some $V \in \mathfrak{B}(e)$, whence $(f-g)h = f(e)h$. Let ψ be a primitive form on $\mathfrak{D}(G)$. Properties (2) and (3) of 4.4.7 imply that

$$\psi(f) - \psi(g) = \psi(f-g) = \psi((f-g)h) = \psi(f(e)h) = f(e)\psi(h) = 0,$$

whence

$$\begin{aligned}\psi(f) = \psi(g) &= \sum_{i=1}^n \psi(x_i g_i)\\ &= \sum_{i=1}^n (\psi(x_i) g_i(e) + x_i(e)\psi(g_i)) = \sum_{i=1}^n \psi(x_i)(X_i f)(e).\end{aligned}$$

Putting $a_i := \psi(x_i)$ for $i = 1, \ldots, n$ one obtains the first assertion of (i). The second one is evident.

2. We shall prove (ii). For $f \in \mathfrak{D}(G)$ there exist a $U \in \mathfrak{B}(e)$ and $g_{ij} \in \mathfrak{D}(G)$ $(i, j = 1, \ldots, n)$ satisfying

$$f(x) = f(e) + \sum_{i=1}^n x_i(x) \frac{\partial}{\partial x_i} f(e) + \tfrac{1}{2} \sum_{i=1}^n x_i(x) x_j(x) g_{ij}(x)$$

for all $x \in U$, where

$$g_{ij}(e) := \frac{\partial^2}{\partial x_i \partial x_j} f(e) \quad \text{for } i, j = 1, \ldots, n.$$

Clearly,

$$g := \sum_{i=1}^n x_i \frac{\partial}{\partial x_i} f(e) + \tfrac{1}{2} \sum_{i,j=1}^n x_i x_j g_{ij} \in \mathfrak{D}(G).$$

Let h be in $\mathfrak{D}(G)$ with $h = h^*$, $h(\complement U) = 0$ and $h(V) = 1$ for some $V \in \mathfrak{B}(e)$. By hypothesis we are given a quadratic form ψ on $\mathfrak{D}(G)$. Property (2) of 4.4.7 together with the defining formula for ψ yields

$$\psi(h) = \psi(h^2) = 2\psi(h), \quad \text{whence } \psi(h) = 0.$$

As in Part 1 we obtain

$$\psi(f) - \psi(g) = \psi(f-g) = \psi((f-g)h) = \psi(f(e)h) = f(e)\psi(h) = 0$$

and consequently,

$$\psi(f) = \psi(g) = \sum_{i=1}^n \psi(x_i) \frac{\partial}{\partial x_i} f(e) + \tfrac{1}{2} \sum_{i=1}^n \psi(x_i x_j g_{ij}).$$

But $\psi(x_i) = \psi(x_i^*) = -\psi(x_i)$ or $\psi(x_i) = 0$ as well as

$$\begin{aligned}\psi(x_i x_j g_{ij}) &= \psi(x_i x_j) g_{ij}(e) + (x_i x_j)(e)\psi(g_{ij})\\ &= \psi(x_i x_j) g_{ij}(e) \quad \text{for } i, j = 1, \ldots, n\end{aligned}$$

implies that

$$\psi(f)=\tfrac{1}{2}\sum_{i,j=1}^{n}\psi(x_i x_j)\frac{\partial^2}{\partial x_i\,\partial x_j}f(e).$$

The matrix $(a_{ij})_{i,j=1,\dots,n}\in\mathfrak{M}(n,\mathbb{R})$ with $a_{ij}:=\tfrac{1}{2}\psi(x_i x_j)$ for $i,j=1,\dots,n$ is clearly symmetric. Its positive-semidefiniteness results from the almost positivity of ψ applied to the function $h^2\in\mathfrak{D}_0(G)$ with

$$h:=\sum_{i=1}^{n}a_i x_i \qquad (a_1,\dots,a_n\in\mathbb{R}).$$

For all $f\in\mathfrak{D}(G)$ and $i,j=1,\dots,n$ one has

$$(X_i X_j f)(e)=\frac{\partial^2}{\partial x_i\,\partial x_j}f(e)+\sum_{l=1}^{n}(X_i X_j x_l)(e)\frac{\partial}{\partial x_l}f(e),$$

whence $\psi(f)=\psi_1(f)-\psi_2(f)$ with

$$\psi_1(f):=\sum_{i,j=1}^{n}a_{ij}(X_i X_j f)(e)$$

and

$$\psi_2(f)=\sum_{i,j=1}^{n}a_{ij}\sum_{l=1}^{n}(X_i X_j X_l)(e)(X_l f)(e).$$

ψ_2 is a primitive form on $\mathfrak{D}(G)$ by (i) of the lemma. ψ_1 satisfies

$$\psi_1(f^*)=\psi_1(f) \qquad \text{for all } f\in\mathfrak{D}(G).$$

Consequently,

$$\psi_1(f)-\psi_2(f)=\psi(f)=\psi(f^*)=\psi_1(f)+\psi_2(f) \quad \text{or} \quad \psi_2=0.$$

Thus $\psi=\psi_1$ yields the result. □

We return to the general situation of an arbitrary locally compact group G.

4.4.10 Definition. A linear mapping $\Gamma:\mathfrak{D}(G)\to\mathfrak{D}(G)$ is said to be a *Lévy mapping for* G if the following conditions hold:

(LM1) For every primitive form L on $\mathfrak{D}(G)$ and every $f\in\mathfrak{D}(G)$ one has $L(f-\Gamma(f))=0$.

(LM2) $\Gamma(f)^*=-\Gamma(f)$ for all $f\in\mathfrak{D}(G)$.

(LM3) For every $x\in G$ the function $f\to\Gamma(f)(x)$ is a primitive form on $\mathfrak{D}(G)$.

4.4.11 Example. Let G be a Lie group of dimension $n\geq 1$ with Lie algebra $\mathfrak{L}(G)$ having a basis $\{Y_1,\dots,Y_n\}$. A *coordinate system* (on G) *with respect to the basis* $\{Y_1,\dots,Y_n\}$ is a set $\{y_1,\dots,y_n\}$ in $\mathfrak{D}(G)$ with $y_i^*=-y_i$ and $(Y_i y_j)(e)=\delta_{ij}$ for all $i,j=1,\dots,n$. Clearly, every canonical coordinate system with respect to $\{Y_1,\dots,Y_n\}$ is a coordinate system on G with respect to $\{Y_1,\dots,Y_n\}$. Given a coordinate system $\{y_1,\dots,y_n\}$ with respect to $\{Y_1,\dots,Y_n\}$ we put

$$\Gamma(f):=\sum_{i=1}^{n}y_i(Y_i f)(e) \qquad \text{for all } f\in\mathfrak{D}(G).$$

Then Γ is a Lévy mapping for G. In fact, let ψ be a primitive form on $\mathfrak{D}(G)$. By Lemma 4.4.9(i) there exist $a_1, \ldots, a_n \in \mathbb{R}$ such that

$$\psi(f) = \sum_{i=1}^{n} a_i (Y_i f)(e)$$

holds for all $f \in \mathfrak{D}(G)$. But for $f \in \mathfrak{D}(G)$ we have

$$\psi(\Gamma(f)) = \sum_{i,j=1}^{n} a_i (Y_j f)(e)(Y_i y_j)(e)$$
$$= \sum_{i=1}^{n} a_i (Y_i f)(e) = \psi(f),$$

i.e., (LM1) of the above definition. (LM2) follows from the definition of a coordinate system and (LM3) from Example 1 of 4.4.8.

4.4.12 Lemma. *Let G be a Lie group of dimension $n \geq 1$ with Lie algebra $\mathfrak{L}(G)$ admitting a basis $\{Y_1, \ldots, Y_n\}$. For every Lévy mapping Γ for G there exists a coordinate system $\{y_1, \ldots, y_n\}$ with respect to $\{Y_1, \ldots, Y_n\}$ such that*

$$\Gamma(f) = \sum_{i=1}^{n} y_i (Y_i f)(e) \qquad \textit{for all } f \in \mathfrak{D}(G).$$

Proof. Given $x \in G$ we infer from Property (LM3) of the Lévy mapping and (i) of Lemma 4.4.9 that there are numbers $y_1(x), \ldots, y_n(x) \in \mathbb{R}$ such that

$$\Gamma(f)(x) = \sum_{i=1}^{n} y_i(x)(Y_i f)(e)$$

holds for all $f \in \mathfrak{D}(G)$.

For any coordinate system $\{x_1, \ldots, x_n\}$ with respect to $\{Y_1, \ldots, Y_n\}$ and for $i = 1, \ldots, n$ we have $y_i := \Gamma(x_i) \in \mathfrak{D}(G)$, and (LM2) implies that

$$y_i^* = \Gamma(x_i)^* = -\Gamma(x_i) = -y_i.$$

By Example 1 of 4.4.8 the mapping $f \to (Y_i f)(e)$ is a primitive form on $\mathfrak{D}(G)$, whence from (LM1) we obtain

$$0 = Y_i(x_j - \Gamma(x_j))(e) = (Y_i x_j)(e) - (Y_i y_j)(e) = \delta_{ij} - (Y_i y_j)(e).$$

But then $\{y_1, \ldots, y_n\}$ is a coordinate system with respect to $\{Y_1, \ldots, Y_n\}$ and the assertion is proved. □

4.4.13 Theorem. *Let G be a locally compact group. Then there always exists a Lévy mapping for G.*

Proof. 1. We reduce the proof of the statement to the case of a Lie projective group: Any locally compact group G admits a Lie projective open subgroup G_1. For every $f \in \mathfrak{D}(G)$ we have $f_1 := \mathrm{Res}_{G_1} f \in \mathfrak{D}(G_1)$. Let Γ_1 be a Lévy mapping for G_1. Then the mapping $\Gamma : \mathfrak{D}(G) \to \mathfrak{D}(G)$ defined by

$$\Gamma(f)(x) := \begin{cases} \Gamma_1(f_1)(x), & \text{if } x \in G_1 \\ 0 & \text{otherwise} \end{cases}$$

for all $f \in \mathfrak{D}(G)$ is a Lévy mapping for G. In fact, let ψ be a primitive form on $\mathfrak{D}(G)$. For $f \in \mathfrak{D}(G_1)$ we introduce the function f' in $\mathfrak{D}(G)$ by means of

$$f'(x) := \begin{cases} f(x), & \text{if } x \in G_1 \\ 0 & \text{otherwise.} \end{cases}$$

Then $\psi_1(f) := \psi(f')$ for all $f \in \mathfrak{D}(G_1)$ defines a primitive form ψ_1 on $\mathfrak{D}(G_1)$. By Property (2) of 4.4.7 ψ_1 is concentrated, and $\psi(f) = \psi_1(f_1)$ holds for all $f \in \mathfrak{D}(G)$. Hence, Γ satisfies (LM1). Properties (LM2) and (LM3) of Γ are evident.

For the rest of the proof we may now assume that G is Lie projective with an L-system $\{G, (H_\alpha)_{\alpha \in \mathbb{A}}\}$. If K is a compact normal subgroup of G, then $\{G/K, (H_\alpha K/K)_{\alpha \in \mathbb{A}}\}$ is an L-system for G/K.

2. As in the proof of Theorem 4.3.11 we carry out a transfinite induction. Let $\mathbf{G}$ be the system of pairs $(G/K, \Gamma_K)$, where K is a compact normal subgroup of G and Γ_K denotes a Lévy mapping for G/K. Since G/K is a Lie group for some compact subgroup K, we infer from Example 4.4.11 that $\mathbf{G} \neq \varnothing$. On $\mathbf{G}$ an order relation $<$ is defined by

$$(G/K, \Gamma_K) < (G/L, \Gamma_L) \quad \text{if } L \subset K \text{ and } \Gamma_L(f \circ p_{KL}) = \Gamma_K(f) \circ p_{KL}$$

for all $f \in \mathfrak{D}(G/K)$, where p_{KL} denotes the canonical mapping from G/Linto G/K. $\mathbf{G}$ is inductively ordered with respect to the relation $<$. In fact, let

$$\mathbf{C} := \{(G/K_i, \Gamma_i) : i \in \mathbb{I}\}$$

be a chain in $\mathbf{G}$. Without loss of generality we may assume that

$$G := \varprojlim_{i \in \mathbb{I}} G/K_i,$$

which by (ii) of Lemma 4.4.4 implies that

$$\mathfrak{D}(G) = \varinjlim_{i \in \mathbb{I}} \mathfrak{D}(G/K_i).$$

As usual, p_i denotes the canonical mapping from G onto G/K_i, and we put

$$\Gamma(f \circ p_i) := \Gamma_i(f) \circ p_i \qquad \text{for all } f \in \mathfrak{D}(G/K_i) \ (i \in \mathbb{I}).$$

Since $\mathbf{C}$ is a chain, Γ is well defined, and Γ is clearly a linear mapping $\mathfrak{D}(G) \to \mathfrak{D}(G)$ satisfying (LM2) and (LM3) of Definition 4.4.10. (LM1) follows from the fact that for any primitive form ψ on $\mathfrak{D}(G)$ the function

$$f \to \psi_i(f) := \psi(f \circ p_i)$$

is a primitive form on $\mathfrak{D}(G/K_i)$ $(i \in \mathbb{I})$. Hence, Γ is a Lévy mapping for G satisfying $(G/K_i, \Gamma_i) < (G, \Gamma)$ for all $i \in \mathbb{I}$, and $\mathbf{C}$ admits an upper bound in $\mathbf{G}$. Thus Zorn's lemma provides us with maximal elements in $\mathbf{G}$.

Plainly it remains to show that if $(G/K, \Gamma_K) \in \mathbf{G}$ with $K \neq \{e\}$, then there exists a pair $(G/L, \Gamma_L) \in \mathbf{G}$ satisfying $(G/K, \Gamma_K) < (G/L, \Gamma_L)$ with $L \neq K$. In order to prove this we proceed as follows:

3. From $\bigcap_{\alpha \in \mathbb{A}} H_\alpha = \{e\}$ one deduces the existence of $\gamma \in \mathbb{A}$ with

$$L := K \cap H_\gamma \neq K.$$

Now $K/L = K/(K \cap H_\gamma) \cong KH_\gamma/H_\gamma$ implies that K/L is a Lie group, and for all $\alpha \in \mathbb{A}$ with $\alpha > \gamma$ one has $K \cap H_\alpha \subset L$. Hence we may suppose without loss of generality that $L = \{e\}$, $\alpha > \gamma$ for all $\alpha \in \mathbb{A}$ and so $K \cap H_\alpha = \{e\}$ for all $\alpha \in \mathbb{A}$.

Let $\{Y_1, \ldots, Y_r\}$ be a basis of $\mathfrak{L}(K)$. In the usual manner we denote by p the canonical mapping from G onto $\dot{G} := G/K$, and define for every $\alpha \in \mathbb{A}$ the groups $G_\alpha := G/H_\alpha$, $\dot{H}_\alpha := p(H_\alpha)$ and $\dot{G}_\alpha := \dot{G}/\dot{H}_\alpha$ as well as the canonical mapping q_α from G_α onto $\dot{G}_\alpha$. We recall the diagram of the proof of Theorem 4.3.11 and the corresponding notations.

Since $K \cap H_\beta = \{e\}$, $\mathrm{Res}_K\, p_\beta$ is a topological isomorphism from K onto $KH_\beta/H_\beta = \ker q_\beta$ ($\beta \in \mathbb{A}$). We therefore obtain the identification

$$\mathfrak{L}(G_\beta) \cong \mathfrak{L}(K) \oplus \mathfrak{L}(\dot{G}_\beta).$$

We now choose functions $y_1, \ldots, y_r \in \mathfrak{D}(G_\gamma)$ satisfying

$$(4.1) \qquad y_j^* = -y_j, \quad (dp_\gamma(Y_i)\, y_j)(e_\gamma) = \delta_{ij}, \quad (X y_j)(e_\gamma) = 0$$
$$\text{for all } X \in \mathfrak{L}(\dot{G}_\gamma);\ i, j = 1, \ldots, r,$$

and for every $\alpha \in \mathbb{A}$ bases $\{X_1^\alpha, \ldots, X_{k(\alpha)}^\alpha\}$ and $\{\dot{X}_1^\alpha, \ldots, \dot{X}_{s(\alpha)}^\alpha\}$ of $\mathfrak{L}(G_\alpha)$ and $\mathfrak{L}(\dot{G}_\alpha)$ resp. satisfying

$$(4.2) \qquad X_j^\alpha = dp_\alpha(Y_j) \quad \text{for all } j = 1, \ldots, r;$$
$$dq_\alpha(X_{r+j}^\alpha) = \dot{X}_j^\alpha \quad \text{for all } j = 1, \ldots, s(\alpha).$$

For every $\dot{x} \in \dot{G}$ and $\alpha \in \mathbb{A}$ the mapping $\dot{f} \to \dot{\psi}_\alpha(\dot{x})(\dot{f}) := \Gamma_K(\dot{f} \circ \dot{p}_\alpha)(\dot{x})$ is a primitive form $\dot{\psi}_\alpha(\dot{x})$ on $\mathfrak{D}(\dot{G}_\alpha)$ by Property (LM 3) of Γ_K. Hence, we get from (i) of Lemma 4.4.9 the existence of unique vectors

$$\dot{X}_\alpha(\dot{x}) = \sum_{i=1}^{s(\alpha)} a_i^\alpha(\dot{x})\, \dot{X}_i^\alpha \quad \text{in } \mathfrak{L}(\dot{G}_\alpha)$$

(with $a_i^\alpha(\dot{x}) \in \mathbb{R}$ for $i = 1, \ldots, s(\alpha)$) satisfying

$$(4.3) \qquad \dot{\psi}_\alpha(\dot{x})(\dot{f}) = (\dot{X}_\alpha(\dot{x})\dot{f})(\dot{e}_\alpha) \quad \text{and} \quad d\dot{p}_{\alpha\beta}(\dot{X}_\beta(\dot{x})) = \dot{X}_\alpha(\dot{x}) \quad (\beta \in \mathbb{A},\ \beta > \alpha).$$

Clearly, the functions $a_1^\alpha, \ldots, a_{s(\alpha)}^\alpha$ belong to $\mathfrak{D}(\dot{G})$.

Let $\Gamma : \mathfrak{D}(G) \to \mathfrak{D}(G)$ be defined by

$$\Gamma(g_\alpha \circ p_\alpha)(x) := \sum_{j=1}^r y_j(p_\gamma(x))(dp_\alpha(Y_j)\, g_\alpha)(e_\alpha)$$
$$+ \sum_{j=1}^{s(\alpha)} a_j^\alpha(p(x))(X_{r+j}^\alpha\, g_\alpha)(e_\alpha)$$

for all $g_\alpha \in \mathfrak{D}(G_\alpha)$, $x \in G$ and $\alpha \in \mathbb{A}$.

With the notation above we introduce the mappings

$$Y_\alpha(\cdot) := \sum_{j=1}^r y_j \circ p_{\gamma\alpha}\, dp_\alpha(Y_j),$$
$$Z_\alpha(\cdot) := \sum_{j=1}^{s(\alpha)} a_j^\alpha \circ p\, X_{r+j}^\alpha \quad \text{and}$$
$$X_\alpha(\cdot) := Y_\alpha(\cdot) + Z_\alpha(\cdot) \quad \text{on } G \quad (\alpha \in \mathbb{A}).$$

We then get

$$dp_{\alpha\beta}(Y_\beta(x))=Y_\alpha(x) \quad \text{and} \quad dq_\alpha(Z_\alpha(x))=\dot{X}_\alpha(p(x)),$$

whence

$$dp_{\alpha\beta}(Z_\beta(x))=Z_\alpha(x) \quad \text{and} \quad dp_{\alpha\beta}(X_\beta(x))=X_\alpha(x)$$

for all $x\in G$, $\alpha,\beta\in\mathbb{A}$ with $\alpha<\beta$. This shows that Γ is well-defined. Moreover,

$$\Gamma(\dot{g}_\alpha\circ\dot{p}_\alpha\circ p)=\Gamma_K(\dot{g}_\alpha\circ\dot{p}_\alpha)\circ p$$

holds for all $\dot{g}_\alpha\in\mathfrak{D}(\dot{G}_\alpha)$, $\alpha\in\mathbb{A}$, and it is clear that Γ is a linear mapping $\mathfrak{D}(G)\to\mathfrak{D}(G)$ satisfying (LM2) and (LM3) of Definition 4.4.10. It therefore remains to show (LM1) for Γ. Suppose given a primitive form ψ on $\mathfrak{D}(G)$ and observe that the functions

$$\dot{f}\to\dot{\psi}(\dot{f}):=\psi(\dot{f}\circ p) \quad \text{and} \quad g_\alpha\to\psi_\alpha(g_\alpha):=\psi(g_\alpha\circ p_\alpha) \quad (\alpha\in\mathbb{A})$$

are primitive forms $\dot{\psi}$ and ψ_α on $\mathfrak{D}(\dot{G})$ and $\mathfrak{D}(G_\alpha)$ resp. Application of Lemma 4.4.9 (i) yields

$$\psi_\alpha(g_\alpha)=\textstyle\sum_{i=1}^{k(\alpha)} b_i^\alpha(X_i^\alpha g_\alpha)(e_\alpha) \quad \text{with } b_i^\alpha\in\mathbb{R}\ (g_\alpha\in\mathfrak{D}(G_\alpha),\ \alpha\in\mathbb{A}). \tag{4.4}$$

From (4.1) and (4.2) one concludes that $X_i^\alpha(y_j\circ p_{\gamma\alpha})(e_\alpha)=\delta_{ij}$ for $i=1,\dots,k(\alpha)$; $j=1,\dots,r$.

From (4.4) one obtains for every $g_\alpha\in\mathfrak{D}(G_\alpha)$ $(\alpha\in\mathbb{A})$

$$\begin{aligned}(4.5)\qquad &\psi((Y_\alpha(\cdot)\,g_\alpha)(e_\alpha))\\ &=\psi_\alpha[\textstyle\sum_{j=1}^r(y_j\circ p_{\gamma\alpha})(dp_\alpha(Y_j)\,g_\alpha)(e_\alpha)]\\ &=\textstyle\sum_{i=1}^r b_i^\alpha(dp_\alpha(Y_i)\,g_\alpha)(e_\alpha)=\sum_{i=1}^r b_i^\alpha(X_i^\alpha g_\alpha)(e_\alpha).\end{aligned}$$

Choosing $\dot{g}_\alpha\in\mathfrak{D}(\dot{G}_\alpha)$ $(\alpha\in\mathbb{A})$ with

$$(X_{r+j}^\alpha g_\alpha)(e_\alpha)=(\dot{X}_j^\alpha\,\dot{g}_\alpha)(\dot{e}_\alpha) \qquad (j=1,\dots,s(\alpha)),$$

we get by (4.2)

$$(Z_\alpha(\cdot)\,g_\alpha)(e_\alpha)=(\dot{X}_\alpha(p(\cdot))\,\dot{g}_\alpha)(\dot{e}_\alpha)=\Gamma_K(\dot{g}_\alpha\circ\dot{p}_\alpha)\circ p$$

and hence, by Property (LM1) for Γ_K together with (4.3) the equalities

$$\begin{aligned}&\psi((Z_\alpha(\cdot)\,g_\alpha)(e_\alpha))=\dot{\psi}(\Gamma_K(\dot{g}_\alpha\circ\dot{p}_\alpha))\\ &=\dot{\psi}(\dot{g}_\alpha\circ\dot{p}_\alpha)=\psi(\dot{g}_\alpha\circ\dot{p}_\alpha\circ p)\\ &=\psi(\dot{g}_\alpha\circ q_\alpha\circ p_\alpha)=\psi_\alpha(\dot{g}_\alpha\circ f_\alpha)=\textstyle\sum_{i=1}^{k(\alpha)} b_i^\alpha(X_i^\alpha\,\dot{g}_\alpha\circ q_\alpha)(e_\alpha)\\ &=\textstyle\sum_{i=1}^{s(\alpha)} b_{i+r}^\alpha(\dot{X}_i^\alpha\,\dot{g}_\alpha)(\dot{e}_\alpha)=\sum_{i=r+1}^{k(\alpha)} b_i^\alpha(X_i^\alpha\,g_\alpha)(e_\alpha).\end{aligned}$$

But from these we conclude via (4.4) and (4.5) that

$$\begin{aligned}&\psi(\Gamma(g_\alpha\circ p_\alpha))=\psi((Y_\alpha(\cdot)\,g_\alpha)(e_\alpha))+\psi((Z_\alpha(\cdot)\,g_\alpha)(e_\alpha))\\ &=\textstyle\sum_{i=1}^{k(\alpha)} b_i^\alpha(X_i^\alpha\,g_\alpha)(e_\alpha)=\psi_\alpha(g_\alpha)=\psi(g_\alpha\circ p_\alpha),\end{aligned}$$

which completes the proof of (LM 1) for Γ and hence, the demonstration of the theorem. □

4.4.14 Definition. A measure $\eta \in \mathscr{M}_+(G^*)$ is called a *Lévy measure on G* if
(i) $\eta(\complement U) < \infty$ for all $U \in \mathfrak{B}(e)$ and
(ii) $\int_{G^*} f\, d\eta < \infty$ for all $f \in \mathfrak{D}_0(G)$.

4.4.15 Example. Let G be a Lie group of dimension $n \geq 1$ with Lie algebra $\mathfrak{L}(G)$ admitting a basis $\{X_1, \ldots, X_n\}$ and a coordinate system $\{x_1, \ldots, x_n\}$ with respect to $\{X_1, \ldots, X_n\}$. By ϕ we denote a Hunt function for G defined for the coordinate system $\{x_1, \ldots, x_n\}$ and a compact neighborhood $U \in \mathfrak{B}(e)$. For a measure $\eta \in \mathscr{M}_+(G^*)$ the following two statements are equivalent:
(i) η is a Lévy measure on G;
(ii) $\int_{G^*} \phi\, d\eta < \infty$.

For the proof of the implication (i) ⇒ (ii) we define $q := \sum_{i=1}^n x_i^2$, observe that $q \in \mathfrak{D}(G)$ and $q(e) = 0$ and conclude that

$$\int_{G^*} \phi\, d\eta \leq \int_{G^*} q\, d\eta + \int_{\complement U} \phi\, d\eta \leq \int_{G^*} q\, d\eta + \|\phi\|\, \eta(\complement U) < \infty.$$

In order to show (ii) ⇒ (i) we consider a function $f \in \mathfrak{D}_0(G)$ and infer that $(X_i f)(e) = 0$ for all $i = 1, \ldots, n$. Hence, there exist a $V \in \mathfrak{B}(e)$ and a $c \in \mathbb{R}_+^*$ with $f(x) \leq c\,\phi(x)$ for all $x \in V$, as one deduces from a Taylor expansion of f in the neighborhood $V \in \mathfrak{B}(e)$ with $V \subset U$. On the other hand, there exists a $d \in \mathbb{R}_+^*$ with $\phi(x) \geq d$ for all $x \in \complement V$. This implies that $\eta(\complement V) < \infty$ and $\int_{G^*} f\, d\eta < \infty$.

At this point a reformulation of Theorem 4.2.8 can be given which will be useful for future discussions.

4.4.16 Theorem. *Let G be a Lie group of dimension $n \geq 1$ and Γ a Lévy mapping for G.*

(A) *Given a convolution semigroup $(\mu_t)_{t \in \mathbb{R}_+^*}$ in $\mathscr{M}^1(G)$, its generating functional A exists on $\mathfrak{D}(G)$, and there are uniquely determined a primitive form ψ_1, a quadratic form ψ_2 on $\mathfrak{D}(G)$ and a Lévy measure η on G such that for all $f \in \mathfrak{D}(G)$ we have the canonical representation*

$$\text{(CR)} \qquad A(f) = \psi_1(f) + \psi_2(f) + \int_{G^*} [f - f(e) - \Gamma(f)]\, d\eta.$$

Moreover, one has

$$\int_{G^*} f\, d\eta = \lim_{t \downarrow 0} \frac{1}{t} \int_G f\, d\mu_t \qquad \text{for all } f \in \mathscr{K}(G^*).$$

(B) *Let ψ_1 be a primitive form, ψ_2 a quadratic form on $\mathfrak{D}(G)$, η a Lévy measure on G and let A be defined as in* (CR) *of* (A). *Then there exists a unique convolution semigroup $(\mu_t)_{t \in \mathbb{R}_+^*}$ in $\mathscr{M}^1(G)$ such that A is the restriction to $\mathfrak{D}(G)$ of the generating functional of $(\mu_t)_{t \in \mathbb{R}_+^*}$.*

Proof. 1. The assertions of (A) follow directly from Theorem 4.2.8 (A) since $\mathfrak{D}(G) \subset \mathscr{C}_2^0(G)$. One simply uses Lemmata 4.4.9, 4.4.12 and Example 4.4.15.

2. By these lemmata there exist numbers $a_1, \ldots, a_n \in \mathbb{R}$, a symmetric, positive-semidefinite matrix $(a_{ij})_{i,j=1,\ldots,n} \in \mathfrak{M}(n, \mathbb{R})$, a coordinate system $\{x_1, \ldots, x_n\}$ with respect to the basis $\{X_1, \ldots, X_n\}$ of $\mathfrak{L}(G)$ and a Lévy measure η on G such that for all $f \in \mathfrak{D}(G)$ one has

$$A(f) = \sum_{i=1}^n a_i(\tilde{X}_i f)(e) + \sum_{i,j=1}^n a_{ij}(\tilde{X}_i \tilde{X}_j f)(e)$$
$$+ \int_{G^*} [f(y) - f(e) - \sum_{i=1}^n x_i(y)(\tilde{X}_i f)(e)]\, \eta(dy).$$

But $A(f)$ is defined for all $f \in \tilde{\mathscr{C}}_2^0(G)$. Since for all $X \in \mathfrak{L}(G)$, $f \in \tilde{\mathscr{C}}_2^0(G)$ and $y \in G$ the invariance property ${}_y(\tilde{X} f) = \tilde{X}({}_y f)$ holds, we obtain a mapping $N: \tilde{\mathscr{C}}_2^0(G) \to \mathscr{C}^0(G)$ defined by $(Nf)(x) := A({}_x f)$ for all $f \in \tilde{\mathscr{C}}_2^0(G)$, $x \in G$. But then Theorem 4.2.8(B) implies the result. □

We proceed under the hypothesis that G is an arbitrary locally compact group.

4.4.17 Properties. Let η be a Lévy measure on G.

(1) If Γ is a Lévy mapping for G, then for every $f \in \mathfrak{D}(G)$ the function

$$g := f - f(e) - \Gamma(f) \quad \text{is } \eta\text{-integrable.}$$

For the proof let $\{G_1, (H_\alpha)_{\alpha \in \mathbb{A}}\}$ be an L-system for G. By Property (1) of 4.4.3 there exists an $\alpha \in \mathbb{A}$ with f and $\Gamma(f)$ in $\mathfrak{D}_\alpha(G)$. We fix a $\mathfrak{g}_\alpha$-basis $\{X_1^\alpha, \ldots, X_n^\alpha\}$ of the Lie algebra $\mathfrak{g}$ of G. Property (LM1) of Γ implies that

$$(X_j^\alpha(f - \Gamma(f)))(e) = 0 \qquad \text{for } j = 1, \ldots, n.$$

Since $g(e) = 0$, there are an $h \in \mathfrak{D}_\alpha(G)$ with $h \geq 0$, $h(e) = 0$ and a $U \in \mathfrak{V}(e)$ satisfying $|g(x)| \leq h(x)$ for all $x \in U$. But g is bounded, whence the η-integrability of g follows from Example 4.4.15.

(2) Let Γ_1, Γ_2 be two Lévy mappings for G. Then

$$f \to \psi(f) := \int_{G^*} (\Gamma_1(f) - \Gamma_2(f))\, d\eta$$

is a primitive form ψ on $\mathfrak{D}(G)$.

Clearly, for every $f \in \mathfrak{D}(G)$ the function $\Gamma_1(f) - \Gamma_2(f)$ is η-integrable by (1), whence ψ is well-defined. (LM2) and (LM3) yield

$$\psi(f g^*) = \psi(f)\, g(e) - f(e)\, \psi(g) \qquad \text{for all } f, g \in \mathfrak{D}(G).$$

Since ψ is linear, this proves the assertion.

4.4.18 Theorem. *Let G be a locally compact group and Γ a Lévy mapping for G. For every linear functional $L \in \mathfrak{D}(G)^*$ the following statements are equivalent:*

(i) *L is almost positive and normed;*

(ii) *there exist a primitive form ψ_1, a quadratic form ψ_2 on $\mathfrak{D}(G)$ and a Lévy measure η on G such that L admits a canonical representation (ψ_1, ψ_2, η) defined by*

$$\text{(CR)} \qquad L(f) = \psi_1(f) + \psi_2(f) + \int_{G^*} [f - f(e) - \Gamma(f)]\, d\eta$$

for all $f \in \mathfrak{D}(G)$.

The proof of the theorem will be prepared by a

4.4.19 Lemma. *Let V be an open neighborhood in $\mathfrak{V}_{\mathbb{R}^n}(0)$ and*

$$\mathfrak{D}(V) := \mathscr{C}^\infty(V) \cap \mathscr{K}(V).$$

We consider a linear functional T on $\mathfrak{D}(V)$ with the following properties:

(a) *$T(f) \geq 0$ for all $f \in \mathfrak{D}_+(V)$ with $f(0)=0$,*

(b) *for all $f \in \mathfrak{D}(V)$ and every $g \in \mathfrak{D}(V)$ with $g(U)=1$ for some U in the neighborhood filter $\mathfrak{V}_V(0)$ of $0 \in V$ one has $T(f) = T(fg)$,*

(c) *there exists a $U \in \mathfrak{V}_V(0)$ with $\sup\{T(f): f \in \mathfrak{D}(V),\ 1_U \leq f \leq 1_V\} = 0$.*

Then there exist $a_1, \ldots, a_n \in \mathbb{R}$ and a symmetric, positive-semidefinite matrix $(a_{ij})_{i,j=1,\ldots,n} \in \mathfrak{M}(n, \mathbb{R})$ such that for all $f \in \mathfrak{D}(V)$ one gets

$$T(f) = \sum_{i=1}^n a_i \frac{\partial}{\partial x_i} f(0) + \sum_{i,j=1}^n a_{ij} \frac{\partial^2}{\partial x_i \partial x_j} f(0).$$

Proof. 1. Let $x_1, \ldots, x_n$ be functions in $\mathfrak{D}(V)$ that coincide in a neighborhood of 0 with the coordinate functions in $\mathbb{R}^n$, and let $g_{ijl} \in \mathfrak{D}(V)$ for $i, j, l = 1, \ldots, n$. For every $\alpha \in \mathbb{R}_+^*$ and $m = 1, 2$ we define

$$h_\alpha^{(m)} := \alpha \sum_{i,j=1}^n x_i x_j + (-1)^m \sum_{i,j,l=1}^n x_i x_j x_l g_{ijl}.$$

Since $h_\alpha^{(m)}(0) = 0$ and

$$\frac{\partial^2}{\partial x_i \partial x_j} h_\alpha^{(m)}(0) = \alpha \delta_{ij} \qquad (i, j = 1, \ldots, n),$$

$h_\alpha^{(m)}$ has a relative minimum in 0 ($\alpha \in \mathbb{R}_+^*$, $m = 1, 2$). But then Assumption (a) implies that

$$T(h_\alpha^{(m)}) \geq 0 \quad \text{or} \quad \alpha T\Big(\sum_{i,j=1}^n x_i x_j\Big) \geq \pm T\Big(\sum_{i,j,l=1}^n x_i x_j x_l g_{ijl}\Big)$$

for all $\alpha \in \mathbb{R}_+^*$, whence $T(\sum_{i,j,l=1}^n x_i x_j x_l g_{ijl}) = 0$.

2. Let now $f \in \mathfrak{D}(V)$. There exist functions $g_{ijl} \in \mathfrak{D}(V) (i, j, l = 1, \ldots, n)$ such that in some neighborhood of 0 one obtains

$$f = f(0) + \sum_{i=1}^n x_i \frac{\partial}{\partial x_i} f(0) + \sum_{i,j=1}^n x_i x_j \frac{\partial^2}{\partial x_i \partial x_j} f(0) + \sum_{i,j,l=1}^n x_i x_j x_l g_{ijl}.$$

By Assumption (b) we may infer from Part 1 that

$$T(f) = f(0)\, T(g) + \sum_{i=1}^n T(x_i) \frac{\partial}{\partial x_i} f(0) + \sum_{i,j=1}^n T(x_i x_j) \frac{\partial^2}{\partial x_i \partial x_j} f(0),$$

where g denotes a function in $\mathfrak{D}(V)$ which equals 1 in some neighborhood of 0. Putting $a_i := T(x_i)$ and $a_{ij} := T(x_i x_j)$ for $i, j = 1, \ldots, n$ we obtain again from (a)

$$\sum_{i,j=1}^n a_{ij} \alpha_i \alpha_j = T\Big(\Big(\sum_{i=1}^n \alpha_i x_i\Big)^2\Big) \geq 0$$

whenever $\alpha_1, \dots, \alpha_n \in \mathbb{R}$, and thus that $(a_{ij})_{i,j=1,\dots,n}$ is a positive-semidefinite matrix in $\mathfrak{M}(n, \mathbb{R})$. Finally, Assumption (c) yields the last statement. □

Proof of Theorem 4.4.18. The proof of the implication (i) ⇒ (ii) will be broken up in three parts.

1. Let $\mathscr{A} := \{f \in \mathfrak{D}(G): e \notin \operatorname{supp}(f)\}$. Since L is almost positive, the mapping $f \to \eta(f) := L(f)$ is a positive linear functional on $\mathscr{A}$. Plainly, $\mathscr{A}$ is a dense subspace of $\mathscr{K}(G^*)$ satisfying Property (P) of Proposition 9 in [41], p. 56. Hence, η can be uniquely extended to a positive measure on G^*. Let $f \in \mathfrak{D}_0(G)$. Since L is almost positive, we get $L(g) \leq L(f)$ for all $g \in \mathscr{A}_+$ with $g \leq f$, and thus

$$\begin{aligned}&\sup\{\textstyle\int g\,d\eta : g \in \mathscr{A}_+, g \leq f\}\\ &= \sup\{L(g): g \in \mathscr{A}_+, g \leq f\} \leq L(f) < \infty.\end{aligned}$$

Consequently, f is η-integrable and η satisfies (ii) of Definition 4.4.14. But L is assumed to be normed, whence η also satisfies (i) of the same definition and so becomes a Lévy measure on G. By Property (1) of 4.4.17 the mapping

$$f \to L_1(f) := \textstyle\int_{G^*} [f - f(e) - \Gamma(f)]\, d\eta$$

thus defines a linear functional L_1 on $\mathfrak{D}(G)$. We show that $\psi := L - L_1$ is an almost positive and concentrated form on $\mathfrak{D}(G)$. For $f \in \mathfrak{D}_0(G)$ we derive from the above inequalities that

$$L_1(f) = \textstyle\int_{G^*} f\, d\eta \leq L(f), \quad \text{whence} \quad \psi(f) \geq 0.$$

Let $f \in \mathfrak{D}(G)$ with $f(W) = 0$ for some $W \in \mathfrak{V}(e)$. Then $f \in \mathscr{A}$, $f(e) = \Gamma(f) = 0$ and therefore

$$\psi(f) = L(f) - L_1(f) = L(f) - \textstyle\int_{G^*} f\, d\eta = L(f) - \eta(f) = 0.$$

Moreover, ψ is normed. In conclusion we may assume as in Part 1 of the proof of Theorem 4.4.13 that G is a Lie projective group.

2. Let G first be a Lie group of dimension $n \geq 1$ with Lie algebra $\mathfrak{L}(G)$ having a basis $\{X_1, \dots, X_n\}$ and let $V \in \mathfrak{V}(e)$ be a neighborhood which is homeomorphic to an open neighborhood in $\mathfrak{V}_{\mathbb{R}^n}(0)$. Let

$$\mathfrak{D}(V) := \{f \in \mathfrak{D}(G): \operatorname{supp}(f) \subset V\}.$$

Then $T := \operatorname{Res}_{\mathfrak{D}(V)} \psi$ satisfies the conditions of Lemma 4.4.9. By the same lemma there are $a_1, \dots, a_n \in \mathbb{R}$ and a symmetric, positive-semidefinite matrix $(a_{ij})_{i,j=1,\dots,n} \in \mathfrak{M}(n, \mathbb{R})$ such that for all $f \in \mathfrak{D}(V)$ one has

$$T(f) = \textstyle\sum_{i=1}^n a_i \dfrac{\partial}{\partial x_i} f(e) + \textstyle\sum_{i,j=1}^n a_{ij} \dfrac{\partial^2}{\partial x_i\, \partial x_j} f(e).$$

After a suitable change of the $a_1, \dots, a_n$ one then obtains as in Part 2 of the proof of Lemma 4.4.9

$$T(f) = \textstyle\sum_{i=1}^n a_i (X_i f)(e) + \textstyle\sum_{i,j=1}^n a_{ij} (X_i X_j f)(e) \qquad \text{for all } f \in \mathfrak{D}(G).$$

If $g \in \mathfrak{D}(V)$ with $g(W)=1$ for some $W \in \mathfrak{B}(e)$, then we get

$$\psi(f)=\psi(fg)=T(fg)=T(f) \quad \text{for all } f \in \mathfrak{D}(G).$$

And again by Lemma 4.4.9 there are a primitive form ψ_1 and a quadratic form ψ_2 on $\mathfrak{D}(G)$ satisfying $\psi=\psi_1+\psi_2$. This proves the assertion for a Lie group.

3. Now let G be a Lie projective group and $\{G,(H_\alpha)_{\alpha\in\mathbb{A}}\}$ an L-system for G. For every $\alpha\in\mathbb{A}$ we put $\psi^\alpha(f_\alpha):=\psi(f_\alpha\circ p_\alpha)$ for all $f_\alpha\in\mathfrak{D}(G/H_\alpha)$ and obtain an almost positive, concentrated and normed linear functional on $\mathfrak{D}(G/H_\alpha)$. While the first two assertions are obvious, the latter follows from Property (4) of 4.4.7. By Part 2 there exist for $\alpha\in\mathbb{A}$ a primitive form ψ_1^α and a quadratic from ψ_2^α on $\mathfrak{D}(G/H_\alpha)$ satisfying $\psi^\alpha=\psi_1^\alpha+\psi_2^\alpha$. But ψ has a unique decomposition $\psi=\psi'+\psi''$ with $(\psi')^\#=-\psi'$ and $(\psi'')^\#=\psi''$, where $\psi^\#(f):=\psi(f^*)$ for all $f\in\mathfrak{D}(G)$. For every $\alpha\in\mathbb{A}$ and $f_\alpha\in\mathfrak{D}(G/H_\alpha)$ we obtain

$$\psi'(f_\alpha\circ p_\alpha)=(\psi^\alpha)'(f_\alpha)=\psi_1^\alpha(f_\alpha) \quad \text{and}$$
$$\psi''(f_\alpha\circ p_\alpha)=(\psi^\alpha)''(f_\alpha)=\psi_2^\alpha(f_\alpha).$$

Hence, $\psi_1:=\psi'$ is a primitive form, $\psi_2:=\psi''$ is a quadratic form on $\mathfrak{D}(G)$ and $\psi=\psi_1+\psi_2$ is the desired decomposition of ψ.

Finally we show the implication (ii) $\Rightarrow$ (i).

4. First we note that by Property (1) of 4.4.17 the integral term in (CR) exists. By definition ψ_1 and ψ_2 are almost positive functionals on $\mathfrak{D}(G)$. For $f\in\mathfrak{D}_0(G)$ we obtain $\Gamma(f)=0$ via (LM3) and hence,

$$L(f)=\psi_1(f)+\psi_2(f)+\int_{G^*} f\,d\eta\geq 0,$$

which shows the almost positivity of L. That L is also normed becomes clear as follows. Let $U\in\mathfrak{B}(e)$. Then for every $f\in\mathscr{H}(U)$ we have $f(U)=1$. (CR) then implies that $L(f)=\int_{G^*}(f-1)\,d\eta$. But

$$\mathscr{H}'(U):=\{f-1: f\in\mathscr{H}(U)\}$$

is an ascending family in $\mathfrak{D}(G)$ with $\sup\mathscr{H}'(U)=0$. From this it follows that

$$\sup\{L(f): f\in\mathscr{H}(U)\}=\sup\{\int_{G^*} g\,d\eta: g\in\mathscr{H}'(U)\}=0,$$

and the assertion is proved. □

4.5 The Lévy-Khintchine Formula for General Locally Compact Groups

For the discussion in this section we assume given an arbitrary locally compact group G with an L-system $\{G_1,(H_\alpha)_{\alpha\in\mathbb{A}}\}$ and a Lie algebra $\underline{\mathfrak{g}}$ in the sense of 4.4.1 and 4.4.5 resp. Given $\alpha\in\mathbb{A}$ the idempotent operator P_α on $\mathscr{C}^0(G)$ defined in 4.4.3

obviously satisfies

$$P_\alpha f(kx) = P_\alpha f(x) \qquad \text{for all } f \in \overline{\mathscr{C}^0}(G),\ k \in H_\alpha \text{ and } x \in G.$$

We put $\mathscr{C}_\alpha^0(G) := P_\alpha(\mathscr{C}^0(G))$ for all $\alpha \in \mathbb{A}$ and define for every $X \in \mathfrak{g}$ and $f \in \bigcup_{\alpha \in \mathbb{A}} \mathscr{C}_\alpha^0(G)$ the function

$$Xf := \lim_{t \to 0} \frac{1}{t}(L_{\zeta(t)} f - f),$$

if the limit exists. Here the one-parameter subgroup ζ of G is given by $\zeta(t) := \exp_G tX$ for all $t \in \mathbb{R}^*$. Xf is called (as in the case of a Lie group G) the *left derivative of X in f.* Moreover, one introduces for every $\alpha \in \mathbb{A}$ and $k \geq 1$ the function space

$$\mathscr{C}_k^\alpha(G) := \{ f \in \mathscr{C}_\alpha^0(G) : X_{i_1} \cdot \dots \cdot X_{i_r} f \in \mathscr{C}_\alpha^0(G) \text{ for all } X_{i_1}, \dots, X_{i_r} \in \mathfrak{g},\ r = 1, \dots, k \}$$

and observes that

(1) $\quad R_x \mathscr{C}_k^\alpha(G) \subset \mathscr{C}_k^\alpha(G) \quad (x \in G).$

(2) $\quad X R_x = R_x X \quad (X \in \mathfrak{g}, x \in G).$

(3) $\quad P_\alpha R_x = R_x P_\alpha \quad (\text{all } x \in G).$

(4) $\quad (f_\alpha \circ p_\alpha) 1_{G_1} \in \mathscr{C}_k^\alpha(G) \quad \text{whenever} \quad f_\alpha \in \mathscr{C}_\alpha^0(G_1/H_\alpha).$

Again for every $\alpha \in \mathbb{A}$ and $k \geq 1$ one needs the spaces

$$\mathscr{E}_k^\alpha(G) := \{ f \in \mathscr{C}_\alpha^0(G) : \text{There is an } f_\alpha \in \mathscr{E}_k(G_1/H_\alpha) \text{ with } \mathrm{Res}_{G_1} f = f_\alpha \circ p_\alpha \},$$

$$\overline{\mathscr{E}_k^\alpha}(G) := \mathscr{E}_k^\alpha(G) \oplus \mathbb{R}$$

and for $k \geq 1$ the spaces

$$\mathscr{E}_k(G) := \bigcup_{\alpha \in \mathbb{A}} \mathscr{E}_k^\alpha(G), \qquad \overline{\mathscr{E}_k}(G) := \mathscr{E}_k(G) \oplus \mathbb{R}.$$

4.5.1 Properties. (1) Let α be in $\mathbb{A}$, $k \geq 1$ be fixed and let $\{X_1, \dots, X_{k(\alpha)}\}$ be a $\mathfrak{g}_\alpha$-basis. For every $f \in \mathscr{C}_k^\alpha(G)$ we define

$$\|f\|_k^\alpha := \|f\| + \sum_{i=1}^{k(\alpha)} \|X_i f\| + \dots + \sum_{i_1, \dots, i_k = 1}^{k(\alpha)} \|X_{i_1} \cdot \dots \cdot X_{i_k} f\|.$$

Then $\|\cdot\|_k^\alpha$ is a norm on $\mathscr{C}_k^\alpha(G)$ and $(\mathscr{C}_k^\alpha(G), \|\cdot\|_k^\alpha)$ becomes a Banach space. In addition $\mathscr{C}_k^\alpha(G)$ is $\|\cdot\|$-dense in $\mathscr{C}_\alpha^0(G)$. In fact, the first assertion follows from the equality

$$\mathscr{C}_k^\alpha(G) = \{ f \in \mathscr{C}_\alpha^0(G) : X_{i_1} \cdot \dots \cdot X_{i_r} f \in \mathscr{C}_\alpha^0(G) \text{ for all } r = 1, \dots, k(\alpha) \},$$

which can easily be checked. The second statement is implied by $X\mathfrak{D}(G) \subset \mathfrak{D}(G)$ for all $X \in \mathfrak{g}$.

(2) Let $\{G_1, (H_\alpha)_{\alpha \in \mathbb{A}}\}$ and $\{G_2, (K_\beta)_{\beta \in \mathbb{B}}\}$ be two L-systems for G. Then there exist an $\alpha_0 \in \mathbb{A}$ and a $\beta_0 \in \mathbb{B}$ such that $\{G_1 \cap G_2, (H_\alpha)_{\alpha > \alpha_0}\}$ and $\{G_1 \cap G_2, (K_\beta)_{\beta > \beta_0}\}$ are L-

systems for G. Moreover, there is for every $\beta>\beta_0$ an $\alpha>\alpha_0$ satisfying $H_\alpha\subset K_\beta$ and for every $\alpha'>\alpha_0$ a $\beta'>\beta_0$ with $K_{\beta'}\subset H_{\alpha'}$. Consequently, the space $\mathscr{E}_k(G)$ is independent of the special choice of the L-system (for $k\geq 1$). The statements concerning L-systems are known from [312], the rest appears as a direct consequence.

For $\mu\in\mathscr{M}_+^b(G)$ we now consider the convolution operator T_μ as an operator on the space $\mathscr{C}^0(G)$. Clearly,

(α) $\quad XT_\mu=T_\mu X \quad (X\in\mathfrak{g})$ and

(β) $\quad P_\alpha T_\mu=T_\mu P_\alpha \quad$ (all $\alpha\in\mathbb{A}$).

4.5.2 Lemma. *Let G be a locally compact group, $\{G_1,(H_\alpha)_{\alpha\in\mathbb{A}}\}$ a Lie system for G and $S:=(\mu_t)_{t\in\mathbb{R}_+^*}$ a convolution semigroup in $\mathscr{M}^1(G)$ with corresponding semigroup $S^0:=(S_t)_{t\in\mathbb{R}_+^*}$ on $\mathscr{C}^0(G)$ and. infinitesimal generator N. For every $\alpha\in\mathbb{A}$ and $k\geq 1$ we consider the families*

$$S^\alpha:=(\operatorname{Res}_{\mathscr{C}_\alpha^0(G)}S_t)_{t\in\mathbb{R}_+^*} \quad \textit{and} \quad S_k^\alpha:=(\operatorname{Res}_{\mathscr{C}_k^\alpha(G)}S_t)_{t\in\mathbb{R}_+^*}.$$

Then

(i) S^α and S_k^α are strongly continuous contraction semigroups on $\mathscr{C}_\alpha^0(G)$ and $\mathscr{C}_k^\alpha(G)$ resp.

(ii) For the infinitesimal generators N^α and N_k^α of S^α and S_k^α resp. one has $\mathsf{D}_{N_k^\alpha}\subset\mathsf{D}_{N^\alpha}\subset\mathsf{D}_N$ $(\alpha\in\mathbb{A}, k\geq 1)$.

Proof. From Lemma 4.1.1 we obtain that S^0 is a strongly continuous semigroup on $\mathscr{C}^0(G)$. By the above property (β) we have $P_\alpha S_t=S_tP_\alpha$ for all $\alpha\in\mathbb{A}$, $t\in\mathbb{R}_+^*$, which implies that S^α is also a strongly continuous contraction semigroup. Let $f\in\mathscr{C}_k^\alpha(G)$ and let $(t_n)_{n\geq 1}$ be a sequence in $\mathbb{R}_+$ with $\lim_{n\to\infty}t_n=:t_0$. Then the functions f and $X_{i_1}\cdot\dots\cdot X_{i_r}f$ for all $X_{i_1},\dots,X_{i_r}\in\mathfrak{g}(r=1,\dots,k)$ are in $\mathscr{C}_\alpha^0(G)$. From the above property (α) we infer that

$$X_{i_1}\cdot\dots\cdot X_{i_r}S_t=S_tX_{i_1}\cdot\dots\cdot X_{i_r}$$

for all $X_{i_1},\dots,X_{i_r}\in\mathfrak{g}\,(r=1,\dots,k)$ whenever $t\in\mathbb{R}_+^*$.

Since S^α has been shown to be strongly continuous, we obtain

$$\lim_{n\to\infty}S_{t_n}f=S_{t_0}f \quad \text{and}$$
$$\lim_{n\to\infty}X_{i_1}\cdot\dots\cdot X_{i_r}S_{t_n}f$$
$$=\lim_{n\to\infty}S_{t_n}X_{i_1}\cdot\dots\cdot X_{i_r}f=S_{t_0}X_{i_1}\cdot\dots\cdot X_{i_r}f$$
$$=X_{i_1}\cdot\dots\cdot X_{i_r}S_{t_0}f \quad (r=1,\dots,k).$$

Hence, by the definition of the norm $\|\cdot\|_k^\alpha$ the limit relationship

$$\lim_{n\to\infty}\|(S_{t_n}-S_{t_0})f\|_k^\alpha=0$$

holds. Since f was chosen arbitrarily from $\mathscr{C}_k^\alpha(G)$, the strong continuity of S_k^α is proved. This proves (i). The rest is clear. □

For the given convolution semigroup $S:=(\mu_t)_{t\in\mathbb{R}_+^*}$ in $\mathcal{M}^1(G)$ we now consider its corresponding semigroup $\underline{S^0}:=(S_t)_{t\in\mathbb{R}_+^*}$, its infinitesimal generator N and its generating functional A on $\mathscr{C}^0(G)$ with domains of definition D_N and D_A resp.

4.5.3 Lemma. *Let the assumptions be the same as in the preceding lemma. Let $\alpha\in\mathbb{A}$ be fixed and let $\{X_1^\alpha,\dots,X_{k(\alpha)}^\alpha\}$ be a $\mathfrak{g}_\alpha$-basis. There exists a function $\Phi_\alpha\in\mathsf{D}_{N_2^\alpha}$ satisfying the following conditions:*

(i) $\Phi_\alpha(H_\alpha)=0$, $\Phi_\alpha(G\setminus H_\alpha)>0$.

(ii) $A_i^\alpha\Phi_\alpha:=(X_i^\alpha\Phi_\alpha)(e)=0$, $A_{ij}^\alpha\Phi_\alpha:=(X_i^\alpha X_j^\alpha\Phi_\alpha)(e)=2\delta_{ij}$ *for* $i,j=1,\dots,k(\alpha)$.

(iii) *For every $U\in\mathfrak{B}(e)$ with $H_\alpha\subset U$ there is a $c\in\mathbb{R}_+^*$ with $\Phi_\alpha(\complement U)\geq c$.*

Proof. First we note that there exists a function $\phi_\alpha\in\mathscr{C}_2^0(G_1/H_\alpha)$ satisfying the conditions

$$0\leq\phi_\alpha\leq 1,\quad \phi_\alpha(e_\alpha)=0,\quad \phi_\alpha(x_\alpha)>0\quad\text{for all } x_\alpha\in G_1/H_\alpha,\ x_\alpha\neq e_\alpha,$$

$$\phi_\alpha(\complement U_\alpha)=1 \text{ for some compact neighborhood } U_\alpha\in\mathfrak{B}_{G_1/H_\alpha}(e_\alpha),$$

$$dp_\alpha(X_i)\,\phi_\alpha(e_\alpha)=0 \quad\text{and}$$

$$dp_\alpha(X_i)\,dp_\alpha(X_j)\,\phi_\alpha(e_\alpha)=2\delta_{ij}\qquad (i,j=1,\dots,k(\alpha)).$$

Putting

$$\psi_\alpha(x):=\begin{cases}\phi_\alpha\circ p_\alpha(x) & \text{for } x\in G_1\\ 1 & \text{for } x\in G\setminus G_1\end{cases}$$

we observe that ψ_α belongs to $\mathscr{C}_2^\alpha(G)$ and fulfills the conditions (i), (ii) and (iii). Let $\varepsilon>0$ be given. By Lemma 4.1.11 there exists a $\Phi_\alpha\in\mathsf{D}_{N_2^\alpha}$ with $\|\psi_\alpha-\Phi_\alpha\|_2^\alpha<\varepsilon$; Φ_α satisfies Property (ii) and $\Phi_\alpha(H_\alpha)=\Phi_\alpha(e)=0$. Moreover, for $\varepsilon\in]0,\tfrac{1}{2}[$ there is a compact neighborhood $U\in\mathfrak{B}(e)$ with $\Phi_\alpha(\complement U)\geq\tfrac{1}{2}$. Since $\Phi_\alpha\in\mathsf{D}_{N_2^\alpha}$, one has a $V\in\mathfrak{B}(e)$ satisfying $H_\alpha\subset V$ and $\Phi_\alpha(V\setminus H_\alpha)>0$. On the other hand, ψ_α has Property (iii), so that ε can be chosen such that Φ_α also has Properties (i) and (iii). □

4.5.4 Lemma. *We keep the notation of the preceding lemma. In particular, we are given Φ_α for a fixed $\alpha\in\mathbb{A}$. Furthermore, let $\delta\in\mathbb{R}_+$ and $f\in\mathscr{E}_2^\alpha(G)$. Then there exists $g\in\mathsf{D}_{N_2^\alpha}$ with $|f-g|\leq\delta\Phi_\alpha$.*

Proof. Let $f\in\mathscr{E}_2^\alpha(G)$. For each $\varepsilon>0$ there exist a neighborhood U of H_α and a function $h\in\mathscr{C}_2^\alpha(G)$ with $h(x)=f(x)$ for all $x\in U$ and $\|f-h\|<\varepsilon$. Lemma 4.1.11 provides us with the existence of a $g\in\mathsf{D}_{N_2^\alpha}$ satisfying

$$g(e)=h(e),\quad A_i^\alpha g=A_i^\alpha h,\quad A_{ij}^\alpha g=A_{ij}^\alpha h$$

for all $i,j=1,\dots,k(\alpha)$ and $\|g-h\|_2^\alpha<\varepsilon$. But this implies the existence of another neighborhood V of H_α with $V\subset U$ such that $|f(x)-g(x)|\leq\delta\Phi_\alpha(x)$ holds for all $x\in V$ if ε is sufficiently small. Since by Lemma 4.5.3 there exists a $c\in\mathbb{R}_+^*$ with $\Phi_\alpha(\complement V)\geq c$, we obtain for an even smaller $\varepsilon>0$ and for all $x\in\complement V$ the estimates

$$|f(x)-g(x)|\leq\|f-h\|+\|h-g\|<2\varepsilon<\delta\Phi_\alpha(x).$$

This together with the above inequality for $x\in V$ yields the result. □

4.5.5 Lemma. *Under the hypotheses of the preceding lemmata one has*
(i) $\overline{\mathscr{E}_2}(G) \subset \mathsf{D}_A$ *and*
(ii) $Af \leq 0$ *for all* $f \in \mathsf{D}_A$ *with* $f \leq f(e)$.

Proof. Clearly for all $f \in \mathsf{D}_A$ with $f \leq f(e)$ we get

$$Af = \lim_{t \downarrow 0} \frac{1}{t} (\mu_t(f) - f(e)) \leq 0,$$

whence (ii).

Let $\delta \in \mathbb{R}_+^*$ and $f \in \mathscr{E}_2(G)$ be given. By the definition of $\mathscr{E}_2(G)$ there exists an $\alpha \in \mathbb{A}$ with $f \in \mathscr{E}_2^\alpha(G)$. But from Lemma 4.5.4 follows the existence of a $g \in \mathsf{D}_{N_2^\alpha}$ satisfying $|f - g| \leq \delta \Phi_\alpha$, where Φ_α is the function constructed in Lemma 4.5.3. Since g and Φ_α belong to D_A, the estimate

$$\left| \frac{1}{t} (\mu_t(f) - f(e)) - \frac{1}{t} (\mu_t(g) - g(e)) \right| = \left| \frac{1}{t} \int_G (f - g) \, d\mu_t \right| \leq \delta \left(\frac{1}{t} \int_G \Phi_\alpha \, d\mu_t \right),$$

valid for all $t \in \mathbb{R}_+^*$, yields the result. □

In the following lemma we shall admit $\{e\}$-continuous convolution semigroups of measures in $\mathscr{M}^{(1)}(G)$. Their generating functionals and infinitesimal generators will be defined exactly as for convolution semigroups in $\mathscr{M}^1(G)$, and in such a way that the corresponding facts from the Hille-Yosida theory remain applicable.

4.5.6 Lemma. *Let G be a locally compact group, $(\mu_t)_{t \in \mathbb{R}_+^*}$ a convolution semigroup in $\mathscr{M}^1(G)$ and $(\nu_t)_{t \in \mathbb{R}_+^*}$ a convolution semigroup in $\mathscr{M}^{(1)}(G)$ with generating functionals A and B resp. such that $A(f) = B(f)$ for all $f \in \mathfrak{D}(G)$. Then $\mu_t = \nu_t$ for all $t \in \mathbb{R}_+^*$.*

Proof. We first note that Lemma 4.5.5 implies the existence of A at least on the space $\mathscr{E}_2(G)$. From Lemma 4.5.4 we infer that A is uniquely determined on $\mathscr{E}_2(G)$ by $\mathrm{Res}_{\mathfrak{D}(G)} A$, whence by hypothesis $A(f) = B(f)$ holds for all $f \in \mathscr{E}_2(G)$. Given a Lie system $(G_1, (H_\alpha)_{\alpha \in \mathbb{A}})$ for G we fix an $\alpha \in \mathbb{A}$ and consider the space $\mathscr{C}_2^\alpha(G)$ introduced above.

Let $(S_t)_{t \in \mathbb{R}_+^*}$ and $(\bar{S}_t)_{t \in \mathbb{R}_+^*}$ be the contraction semigroups corresponding to $(\mu_t)_{t \in \mathbb{R}_+^*}$ and $(\nu_t)_{t \in \mathbb{R}_+^*}$ considered as semigroups on $\mathscr{C}_2^\alpha(G)$ with infinitesimal generators N and $\bar{N}$ and domains of definition D_N and $\mathsf{D}_{\bar{N}}$ resp. From Lemma 4.5.5 we conclude that

$$Nf = S_A(f) := A(L_x f) \quad \text{and} \quad \bar{N} f = S_B(f)$$

for all f in D_N and $\mathsf{D}_{\bar{N}}$ resp. Since

$$(N - \lambda E) \, \mathsf{D}_N = (\bar{N} - \lambda E) \, \mathsf{D}_{\bar{N}} = \mathscr{C}_2^\alpha(G) \qquad \text{for all } \lambda \in \mathbb{R}_+^* \quad \text{and}$$

$$\lim_{t \downarrow 0} \frac{1}{t} (S_t f(x) - f(x)) = A(L_x f) = \lim_{t \downarrow 0} \frac{1}{t} (\bar{S}_t f(x) - f(x))$$

for all $x \in G$, one has $N = \bar{N}$. Hence, by the Hille-Yosida theory $S_t = \bar{S}_t$ for all $t \in \mathbb{R}_+^*$. We now conclude that $S_t = \bar{S}_t$ on $\bigcup_{\alpha \in \mathbb{A}} \mathscr{C}_2^\alpha(G)$, so that from $\|S_t\| \leq 1$ and $\|\bar{S}_t\| \leq 1$ for all $t \in \mathbb{R}_+^*$ we obtain that $S_t = \bar{S}_t$ on $\mathscr{C}^0(G)$. This implies $\mu_t = \nu_t$ for all $t \in \mathbb{R}_+^*$. □

4.5.7 Lemma. *Let G be a locally compact group and $(S^{(m)})_{m\geq 1}$ a sequence of convolution semigroups $S^{(m)}:=(\mu_t^{(m)})_{t\in\mathbb{R}_+^*}$ in $\mathscr{M}^1(G)$ whose infinitesimal generators $N^{(m)}$ are defined at least on $\mathfrak{D}(G)$. Let us further assume that the strong limit $Nf:=\lim_{m\to\infty} N^{(m)}f$ exists for all $f\in\mathfrak{D}(G)$ and that the almost positive functional ψ on $\mathfrak{D}(G)$ defined by $\psi(f):=Nf(e)$ for all $f\in\mathfrak{D}(G)$ is normed. Then there exists a (unique) convolution semigroup $(\mu_t)_{t\in\mathbb{R}_+^*}$ in $\mathscr{M}^1(G)$ with infinitesimal generator N' such that $\mathrm{Res}_{\mathfrak{D}(G)} N'=N$.*

In addition, for every $t\in\mathbb{R}_+^*$ one has $\mu_t=\mathscr{T}_w\text{-}\lim_{m\to\infty}\mu_t^{(m)}$ uniformly in t for compact subsets of $\mathbb{R}_+$.

Proof. 1. Let G be a locally compact group having a countable basis of its topology. Then the proof of Lemma 4.2.7 with the space $\mathscr{C}_2(G)$ replaced by $\mathfrak{D}(G)$ yields the result; in Part 4 of that proof one just uses the hypothesis that ψ is normed instead of the explicit representation of N.

2. In the case of an arbitrary locally compact group G we may assume as in Part 2 of the proof of Lemma 4.2.7 that G is σ-compact. But then by [218], (8.7) we obtain $G=\varprojlim_{i\in\mathbb{I}} G_i$, where $G_i:=G/K_i$ is a locally compact group having a countable basis of its topology and K_i is a compact normal subgroup of G (for all $i\in\mathbb{I}$). For every $i\in\mathbb{I}$ the canonical projection from G onto G_i will be denoted by π_i. Then for $i\in\mathbb{I}$ the family $(S_i^{(m)})_{m\geq 1}$ of convolution semigroups

$$S_i^{(m)}:=(\pi_i(\mu_t^{(m)}))_{t\in\mathbb{R}_+^*}$$

in $\mathscr{M}^1(G_i)$ with infinitesimal generators $N_i^{(m)}$ satisfies the hypothesis of the lemma since by Lemma 4.4.4 and Property (4) of 4.4.7 we have

$$\begin{aligned}
&(N_i^{(m)}f_i)\circ\pi_i=N^{(m)}(f_i\circ\pi_i),\\
&(N_i f_i)\circ\pi_i:=(\lim_{m\to\infty} N_i^{(m)}f_i)\circ\pi_i\\
&=\lim_{m\to\infty} N^{(m)}(f_i\circ\pi_i)=N(f_i\circ\pi_i)\quad\text{and}\\
&\psi_i(f_i):=(N_i f_i)(e_i)=\pi_i(\psi)(f_i)
\end{aligned}$$

for all $f_i\in\mathfrak{D}(G_i)$. We now infer from Part 1 that for every $i\in\mathbb{I}$ there exists a convolution semigroup $S_i:=(v_t^{(i)})_{t\in\mathbb{R}_+^*}$ in $\mathscr{M}^1(G_i)$ with infinitesimal generator N_i' such that

$$\mathrm{Res}_{\mathfrak{D}(G_i)} N_i'=N_i \quad\text{and}\quad v_t^{(i)}=\mathscr{T}_w\text{-}\lim_{m\to\infty}\pi_i(\mu_t^{(m)})\qquad (t\in\mathbb{R}_+^*)\text{ hold.}$$

Since $\{f_i\circ\pi_i: f_i\in\mathscr{K}(G_i), i\in\mathbb{I}\}$ is a dense subset of $\mathscr{K}(G)$, one gets the existence of $\mu_t:=\mathscr{T}_w\text{-}\lim_{m\to\infty}\mu_t^{(m)}$ in $\mathscr{M}^1(G)$ and with the aid of Lemma 4.4.4 (ii) that $(\mu_t)_{t\in\mathbb{R}_+^*}$ is a convolution semigroup in $\mathscr{M}^1(G)$ with infinitesimal generator N' satisfying $\mathrm{Res}_{\mathfrak{D}(G)} N'=N$, as desired. Clearly, the convergence in the defining limit relation for μ_t is uniform in t for compact subsets of $\mathbb{R}_+$. □

4.5.8 Theorem. *Let G be a locally compact group.*

(A) *If $(\mu_t)_{t\in\mathbb{R}_+^*}$ is a convolution semigroup in $\mathscr{M}^1(G)$ with generating functional A, then*

(i) $\mathfrak{D}(G)\subset \mathsf{D}_A$.

(ii) *A is almost positive and normed on* $\mathfrak{D}(G)$.

(B) *For every almost positive and normed linear functional L on* $\mathfrak{D}(G)$ *there exists exactly one convolution semigroup* $(\mu_t)_{t\in\mathbb{R}_+^*}$ *in* $\mathcal{M}^1(G)$ *with generating functional A and infinitesimal generator N such that*
(iii) $L=\mathrm{Res}_{\mathfrak{D}(G)}A$,
(iv) $\mathfrak{D}(G)\subset \mathbf{D}_N$.

Proof. 1. Statement (A)(i) and the almost positivity of A follow directly from Lemma 4.5.5. In order to prove (B) we denote by $\mathbf{H}$ the collection of all locally compact groups G with the property that any almost positive and normed linear functional on $\mathfrak{D}(G)$ is the (restriction to $\mathfrak{D}(G)$ of the) generating functional of a convolution semigroup in $\mathcal{M}^1(G)$.

2. We show that if G is the (locally compact) projective limit of a family $(G_\alpha)_{\alpha\in\mathbf{A}}$ of locally compact groups $G_\alpha:=G/H_\alpha\in\mathbf{H}$ with compact normal subgroups H_α of G such that $\bigcap_{\alpha\in\mathbf{A}}H_\alpha=\{e\}$ and canonical mappings $p_\alpha:G\to G_\alpha$ and $p_{\alpha\beta}:G_\beta\to G_\alpha$ $(\alpha,\beta\in\mathbf{A},\alpha<\beta)$, then $G\in\mathbf{H}$. In fact, let ψ be an almost positive and normed functional on $\mathfrak{D}(G)$. By Property (4) of 4.4.7 we get for every $\alpha\in\mathbf{A}$ an almost positive and normed functional $\psi_\alpha:=p_\alpha(\psi)$ on G_α, whence by assumption a unique convolution semigroup $(\mu_t^{(\alpha)})_{t\in\mathbb{R}_+^*}$ in $\mathcal{M}^1(G_\alpha)$ with generating functional ψ_α. Let $\alpha,\beta\in\mathbf{A}$ with $\alpha<\beta$. Then $p_{\alpha\beta}(\psi_\beta)=\psi_\alpha$ and so $p_{\alpha\beta}(\mu_t^{(\beta)})=\mu_t^{(\alpha)}$ for all $t\in\mathbb{R}_+^*$. Consequently, $(\mu_t^{(\alpha)})_{\alpha\in\mathbf{A}}$ is a projective system of measures $\mu_t^{(\alpha)}\in\mathcal{M}^1(G_\alpha)$, and its projective limit $\mu_t\in\mathcal{M}^1(G)$ exists and is such that $p_\alpha(\mu_t)=\mu_t^{(\alpha)}$ holds for all $t\in\mathbb{R}_+^*$ $(\alpha\in\mathbf{A})$. Since $\{f_\alpha\circ p_\alpha:f_\alpha\in\mathcal{K}(G_\alpha),\alpha\in\mathbf{A}\}$ is a dense subset of $\mathcal{K}(G)$, $(\mu_t)_{t\in\mathbb{R}_+^*}$ is easily shown to be a convolution semigroup in $\mathcal{M}^1(G)$. From Lemma 4.4.4 we know that $\mathfrak{D}(G)=\varinjlim_{\alpha\in\mathbf{A}}\mathfrak{D}(G_\alpha)$. Thus ψ is the generating functional of $(\mu_t)_{t\in\mathbb{R}_+^*}$, and this implies the assertion.

3. For a Lie projective group G and a convolution semigroup $(\mu_t)_{t\in\mathbb{R}_+^*}$ in $\mathcal{M}^1(G)$ with infinitesimal generator N we have $\mathfrak{D}(G)\subset\mathbf{D}_N$. In fact, let

$$G=\varprojlim_{\alpha\in\mathbf{A}}G_\alpha$$

with Lie groups G_α and N_α be the infinitesimal generator of the convolution semigroup $(p_\alpha(\mu_t))_{t\in\mathbb{R}_+^*}$ in $\mathcal{M}^1(G_\alpha)$ (for $\alpha\in\mathbf{A}$). Plainly we have $f_\alpha\circ p_\alpha\in\mathbf{D}_N$ for all $f_\alpha\in\mathbf{D}_{N_\alpha}$ and $N_\alpha(f_\alpha)\circ p_\alpha=N(f_\alpha\circ p_\alpha)$ $(\alpha\in\mathbf{A})$. By Theorem 4.2.8 we have $\mathfrak{D}(G_\alpha)\subset\mathbf{D}_{N_\alpha}$, whence by the definition of $\mathfrak{D}(G)$ also $\mathfrak{D}(G)\subset\mathbf{D}_N$.

4. We show finally that any locally compact group G belongs to $\mathbf{H}$. Let $\{G_1,(H_\alpha)_{\alpha\in\mathbf{A}}\}$ be an L-system for G. We know that G_1 is an open Lie projective subgroup of G, for which there exists (by Theorem 4.4.13) a Lévy-mapping Γ_1. Defining $\Gamma(f):=\Gamma_1(\mathrm{Res}_{G_1}f)1_{G_1}$ for all $f\in\mathfrak{D}(G)$ one obtains a Lévy mapping Γ for G. Any given almost positive and normed functional ψ on $\mathfrak{D}(G)$ admits by Theorem 4.4.18 a canonical representation of the form

$$\psi(f)=\psi_1(f)+\psi_2(f)+\int_{G^*}[f-f(e)-\Gamma(f)]\,d\eta$$

with the data ψ_1,ψ_2 and η as described in that theorem. The functionals $\eta':=1_{G_1^*}\cdot\eta$ and $\bar\eta':=\mathrm{Res}_{G_1^*}\eta$ are Lévy measures on G and G_1 resp., and $\eta'':=1_{\complement G_1}\cdot\eta$ is a measure in $\mathcal{M}_+^b(G)$. For every $f\in\mathfrak{D}(G_1)$ we put

$$\bar\psi'(f):=\psi_1(1_{G_1}f)+\psi_2(1_{G_1}f)+\int_{G_1^*}[f-f(e)-\Gamma_1(f)]\,d\bar\eta'$$

and observe that $\bar{\psi}'$ is an almost positive and normed functional on G_1. Since by Part 2 $G_1 \in \mathbf{H}$, there exists a convolution semigroup $(\bar{\mu}'_t)_{t\in\mathbb{R}_+^*}$ in $\mathscr{M}^1(G_1)$ with generating functional $\bar{\psi}'$. We now consider $(\bar{\mu}'_t)_{t\in\mathbb{R}_+^*}$ as a convolution semigroup $(\mu'_t)_{t\in\mathbb{R}_+^*}$ in $\mathscr{M}^1(G)$ with infinitesimal generator N' and generating functional ψ'. By Property (2) of 4.4.7 ψ_1 and ψ_2 are concentrated linear functionals on $\mathfrak{D}(G)$, and an easy computation shows that

$$\psi'(f)=\bar{\psi}'(\mathrm{Res}_{G_1} f)=\psi_1(f)+\psi_2(f)+\int_{G^*}[f-f(e)-\Gamma(f)]\,d\eta'$$

for all $f\in\mathfrak{D}(G)$. For $k\geq 1$ we introduce the linear operator N'_k on $\mathfrak{D}(G)$ defined by

$$N'_k f(x):=k\int_G [f(xy)-f(x)]\,\mu'_{\frac{1}{k}}(dy)$$

for all $f\in\mathfrak{D}(G)$ and $x\in G$ and observe that N'_k is a Poisson generator in the sense that it corresponds to a convolution semigroup of Poisson measures in $\mathscr{M}^1(G)$. (See Part 2 of the proof of Theorem 4.2.5.) Recalling the inductive limit representation of $\mathfrak{D}(G)$ (given in 4.4.2) we obtain from Part 3 that $\lim_{k\to\infty} N'_k f=N'f$ holds for all $f\in\mathfrak{D}(G)$. Furthermore, we define the operator N'' on $\mathfrak{D}(G)$ by

$$N''f(x):=\int_G [f(xy)-f(x)]\,\eta''(dy)$$

for all $f\in\mathfrak{D}(G)$ and $x\in G$ and the functional ψ'' on $\mathfrak{D}(G)$ by

$$\psi''(f):=N''f(e) \qquad \text{for all } f\in\mathfrak{D}(G).$$

Clearly N'' and $N_k:=N'_k+N''$ $(k\geq 1)$ are Poisson generators, and for $f\in\mathfrak{D}(G)$ one has

$$\lim_{k\to\infty} N_k f=N'f+N''f=:Nf$$

as well as

$$Nf(e)=\psi'(f)+\psi''(f)=\psi(f).$$

Hence, Lemma 4.5.7 can be applied, and there exists a convolution semigroup in $\mathscr{M}^1(G)$ whose generating functional is precisely ψ.

5. Finally, we infer from the Hille-Yosida theory that the convolution semigroup constructed for ψ (in Part 4) is uniquely determined by ψ. Thus Statement (B) of the theorem has been proved.

6. We are left with the proof that the generating functional A of the convolution semigroup $(\mu_t)_{t\in\mathbb{R}_+^*}$ is normed. From Lemma 4.5.5 (ii) one concludes that $c:=\sup\{Af: 1_U\leq f\leq 1_G, f\in\mathfrak{D}(G)\}\leq 0$ for every $U\in\mathfrak{V}(e)$. Hence, the linear functional $f\to A_1(f):=Af+cf(e)$ on $\mathfrak{D}(G)$ is almost positive and normed. Parts 4 and 5 of this proof (or Statement (B)(iii) of the theorem) yield the existence of exactly one convolution semigroup $(\mu'_t)_{t\in\mathbb{R}_+^*}$ in $\mathscr{M}^1(G)$ whose generating functional coincides on $\mathfrak{D}(G)$ with $\mathrm{Res}_{\mathfrak{D}(G)} A_1$. But $(e^{-ct}\mu'_t)_{t\in\mathbb{R}_+^*}$ is a convolution semigroup in $\mathscr{M}^{(1)}(G)$ with generating functional A on $\mathfrak{D}(G)$. By the uniqueness lemma 4.5.6 one obtains $\mu_t=e^{-ct}\mu'_t$ for all $t\in\mathbb{R}_+^*$, which implies that $c=0$, i.e., A is normed. Thus the proof of Statement (A) of the theorem (and so of the entire theorem) is completed. □

The combination of Theorems 4.4.18 and 4.5.8 yields the following final form of the Lévy-Khintchine theorem for arbitrary locally compact groups.

4.5.9 Theorem. *Let G be a locally compact group.*

(A) *If $(\mu_t)_{t\in\mathbb{R}_+^*}$ is a continuous convolution semigroup in $\mathscr{M}^1(G)$ with generating functional A, then*

(i) $\mathfrak{D}(G)\subset \mathsf{D}_A$;

(ii) *A is almost positive and normed on $\mathfrak{D}(G)$;*

(iii) *given a Lévy mapping Γ for G, there exist a primitive form ψ_1, a quadratic form ψ_2 on $\mathfrak{D}(G)$ and a Lévy measure η on G such that A admits a canonical representation (ψ_1, ψ_2, η) defined by*

(CR) $$A(f)=\psi_1(f)+\psi_2(f)+\int_{G^*}[f-f(e)-\Gamma(f)]\,d\eta$$

for all $f\in\mathfrak{D}(G)$;

(iv) *η and ψ_2 are uniquely determined by $(\mu_t)_{t\in\mathbb{R}_+^*}$ and*

(v) $\int_{G^*} f\,d\eta=\lim_{t\downarrow 0}\frac{1}{t}\int_G f\,d\mu_t$ *for all $f\in\mathscr{K}(G^*)$.*

(B) *If conversely, for a given Lévy mapping Γ for G, a primitive from ψ_1, a quadratic form ψ_2 on $\mathfrak{D}(G)$ and a Lévy measure η on G the linear functional $L\in\mathfrak{D}(G)^*$ is defined by* (CR) *of* (iii) *in* (A), *then*

(vi) *L is almost positive and normed on $\mathfrak{D}(G)$;*

(vii) *there exists a unique continuous convolution semigroup $(\mu_t)_{t\in\mathbb{R}_+^*}$ in $\mathscr{M}^1(G)$ with generating functional A and infinitesimal generator N such that $\operatorname{Res}_{\mathfrak{D}(G)} A=L$ and*

(viii) $\mathfrak{D}(G)\subset \mathsf{D}_N$.

4.6 Convolution Hemigroups. Generation and Representation

A useful extension of the theory of continuous convolution semigroups leads in the direction of two-parameter families of measures on a group. The analysis of such families requires auxiliary studies from the theory of two-parameter families of matrices and of operators as well. It should be noted that the double index $s, t\in\mathbb{R}_+$ attached to a two-parameter family is always understood in the sense of the natural ordering $s\le t$ in $\mathbb{R}_+$.

Let G be a locally compact group.

4.6.1 Definition. A (continuous) *convolution hemigroup* (evolution family) in $\mathscr{M}^1(G)$ is a family $(\mu_{s,t})_{s,t\in\mathbb{R}_+}$ of measures in $\mathscr{M}^1(G)$ satisfying the following properties:

(HG1) $\mu_{s,r}*\mu_{r,t}=\mu_{s,t}$ for all $s, r, t\in\mathbb{R}_+$ with $s\le r\le t$.

(HG2) $\mu_{s,s}=\varepsilon_e$ for all $s\in\mathbb{R}_+$.

(HG3) The mapping $(s, t)\to\mu_{s,t}$ from $\mathbb{R}_+^2:=\{(s,t)\in\mathbb{R}^2: 0\le s\le t\}$ into $\mathscr{M}^1(G)$ is $\mathscr{T}_v$-continuous.

Clearly, for any continuous convolution semigroup $(\mu_t)_{t\in\mathbb{R}_+}$ in $\mathscr{M}^1(G)$ one defines a continuous convolution hemigroup $(\mu_{s,t})_{s,t\in\mathbb{R}_+}$ in $\mathscr{M}^1(G)$ by $\mu_{s,t}:=\mu_{t-s}$ for all $s, t\in\mathbb{R}_+$ with $s\le t$. Conversely, to a continuous convolution hemigroup

$(\mu_{s,t})_{s,t\in\mathbb{R}_+}$ in $\mathscr{M}^1(G)$ satisfying $\mu_{s+h,t+h}=\mu_{s,t}$ for all $h\in\mathbb{R}_+$ there corresponds a continuous convolution semigroup $(\mu_r)_{r\in\mathbb{R}_+}$ in $\mathscr{M}^1(G)$ defined by $\mu_r:=\mu_{0,r}$ for all $r\in\mathbb{R}_+$.

4.6.2 Definition. Let E be a linear subspace of $\mathscr{C}_u(G)$. A continuous convolution hemigroup $(\mu_{s,t})_{s,t\in\mathbb{R}_+}$ in $\mathscr{M}^1(G)$ is called *weakly differentiable on* E if for every $s,t\in\mathbb{R}_+$ with $s\le t$ the weak derivatives

$$\frac{\partial^+}{\partial t}\mu_{s,t}:=\lim_{h\downarrow 0}\frac{1}{h}(\mu_{s,t+h}-\mu_{s,t}) \quad \text{and}$$

$$\frac{\partial^-}{\partial s}\mu_{s,t}:=\lim_{h\downarrow 0}\left[-\frac{1}{h}(\mu_{s-h,t}-\mu_{s,t})\right] \quad \text{exist on } \mathsf{E}.$$

$(\mu_{s,t})_{s,t\in\mathbb{R}_+}$ is called *strongly differentiable on* E if for all $s,t\in\mathbb{R}_+$ with $s\le t$ the strong derivatives

$$\frac{\partial^+}{\partial t}T_{\mu_{s,t}}=\lim_{h\downarrow 0}\frac{1}{h}(T_{\mu_{s,t+h}}-T_{\mu_{s,t}}) \quad \text{and}$$

$$\frac{\partial^-}{\partial s}T_{\mu_{s,t}}=\lim_{h\downarrow 0}\left[-\frac{1}{h}(T_{\mu_{s-h,t}}-T_{\mu_{s,t}})\right]$$

of the probability operators $T_{\mu_{s,t}}$ of $\mu_{s,t}$ exist on E in the norm topology of $\mathscr{C}_u(G)$.

If $(\mu_{s,t})_{s,t\in\mathbb{R}_+}$ is weakly differentiable on E, one introduces for any $s\in\mathbb{R}_+$ the linear functional $A(s)$ on E defined by

$$A(s)(f):=\frac{\partial^+}{\partial t}\mu_{s,t}(f)\bigg|_{t=s}=\lim_{h\downarrow 0}\frac{1}{h}(\mu_{s,s+h}-\varepsilon_e)(f) \qquad \text{for all } f\in\mathsf{E}.$$

The family $(A(s))_{s\in\mathbb{R}_+}$ so attached to the hemigroup $(\mu_{s,t})_{s,t\in\mathbb{R}_+}$ will be called the *generating family* of $(\mu_{s,t})_{s,t\in\mathbb{R}_+}$ (on E).

Examples of subspaces E of $\mathscr{C}_u(G)$ for which we shall need the concepts of differentiability of a convolution hemigroup are the spaces $\mathfrak{R}(G)$, $\tilde{\mathscr{C}}_2^0(G)$ and $\mathfrak{D}(G)$ introduced in the preceding sections.

One notes that any continuous convolution *semi*group $(\mu_t)_{t\in\mathbb{R}_+}$ in $\mathscr{M}^1(G)$ admits a corresponding *hemi*group $(\mu_{s,t})_{s,t\in\mathbb{R}_+}$ which is differentiable on $\mathfrak{R}(G)$, $\tilde{\mathscr{C}}_2^0(G)$ and $\mathfrak{D}(G)$ by Theorems 4.3.13, 4.2.8 and 4.5.8 for Lie projective almost periodic groups, Lie groups and arbitrary locally compact groups G resp. In this case the generating family of $(\mu_{s,t})_{s,t\in\mathbb{R}_+}$ contains just one linear functional A, the generating functional of the underlying convolution semigroup $(\mu_t)_{t\in\mathbb{R}_+}$.

4.6.3 Remark. While in the case of a convolution semigroup in $\mathscr{M}^1(G)$ the weak differentiability on $\mathfrak{D}(G)$ follows from the continuity, we have to add this property when dealing with general convolution hemigroups. In fact, let $(\mu_{s,t})_{s,t\in\mathbb{R}_+}$ be a continuous convolution hemigroup in $\mathscr{M}^1(G)$ and $\phi:\mathbb{R}_+\to\mathbb{R}_+$ a strictly increasing, continuous mapping. Then $(\mu_{\phi(s),\phi(t)})_{s,t\in\mathbb{R}_+}$ is a continuous convolution hemigroup in $\mathscr{M}^1(G)$ which by a special choice of ϕ lacks the differentiability property. One need only pick an admissible function such that ϕ^{-1} has a derivative vanishing a.e. If the hemigroups $(\mu_{s,t})_{s,t\in\mathbb{R}_+}$ and $(\mu_{\phi(s),\phi(t)})_{s,t\in\mathbb{R}_+}$ were differentiable on $\mathfrak{D}(G)$, then we would have

$$A(\phi(s))(f)=A(s)(f)(\phi^{-1})'(t)=0 \quad \text{a.e.}$$

for all $s\in\mathbb{R}_+$ and $f\in\mathfrak{D}(G)$.

In the following we shall discuss for a given family $(A(s))_{s\in\mathbb{R}_+}$ of linear functionals $A(s)$ on E the existence and uniqueness of a convolution hemigroup $(\mu_{s,t})_{s,t\in\mathbb{R}_+}$ in $\mathscr{M}^1(G)$ satisfying on E the *evolution equations*

$$\text{(EE)}\qquad \begin{aligned} &\frac{\partial^+}{\partial t} T_{\mu_{s,t}}(f)=T_{\mu_{s,t}} S_{A(t)}(f) \quad \text{and}\\ &\frac{\partial^-}{\partial s} T_{\mu_{s,t}}(f)=-S_{A(s)}\, T_{\mu_{s,t}}(f) \qquad \text{for } f\in\mathsf{E}, \end{aligned}$$

where for any linear functional A on E the operator S_A on E is defined by

$$S_A(f)(x):=A({}_xf) \qquad \text{for all } f\in\mathsf{E} \text{ and } x\in G.$$

We start by solving the problem in the case $\mathsf{E}:=\mathfrak{K}(G)$ for a group $G\in\mathbf{A}$. Under these hypotheses one easily verifies that every weakly differentiable convolution hemigroup in $\mathscr{M}^1(G)$ is also strongly differentiable (on $\mathfrak{K}(G)$).

4.6.4 Definition. Let $n\geq 1$. A family $(m(s,t))_{s,t\in\mathbb{R}_+}$ in $\mathfrak{M}(n,\mathbb{C})$ is called a *hemigroup of matrices* if the following conditions hold:

(HGM1) $m(s,r)\,m(r,t)=m(s,t)$ for all $s,r,t\in\mathbb{R}_+$ with $s\leq r\leq t$.

(HGM2) $m(s,s)=E:=E_n$ for all $s\in\mathbb{R}_+$.

(HGM3) The mapping $(s,t)\to m(s,t)$ from $\mathbb{R}_+^2$ into $\mathfrak{M}(n,\mathbb{C})$ is continuous (with respect to the natural topologies in $\mathbb{R}_+^2$ and $\mathfrak{M}(n,\mathbb{C})$).

$(m(s,t))_{s,t\in\mathbb{R}_+}$ is said to be *differentiable* if for every $s,t\in\mathbb{R}_+$
(i) the limits

$$a(s):=\frac{\partial^+}{\partial t} m(s,t)\bigg|_{t=s}=\lim_{h\downarrow 0}\frac{1}{h}(m(s,s+h)-E)$$

and

$$\frac{\partial^-}{\partial s} m(s,t)\bigg|_{s=t}=\lim_{h\downarrow 0}\left[-\frac{1}{h}(m(t-h,t)-E)\right]$$

exist and
(ii)

$$\frac{\partial^-}{\partial s} m(s,t)\bigg|_{s=t}=-a(t)$$

holds.

4.6.5 Properties. (1) Let $(m(s,t))_{s,t\in\mathbb{R}_+}$ be a differentiable hemigroup of matrices. Then the mappings $s\to m(s,t)$ and $t\to m(s,t)$ are differentiable (for all $s<t$ and $t>s$

resp.), and one has the *evolution equations (for matrices)* in the form

$$\text{(EEM)}\qquad \begin{aligned} &\frac{\partial^+}{\partial t} m(s,t) = m(s,t)\, a(t) \quad \text{and} \\ &\frac{\partial^-}{\partial s} m(s,t) = -a(s)\, m(s,t) \qquad (s,t \in \mathbb{R}_+,\ s \leq t). \end{aligned}$$

The proof follows directly from the invertibility of the matrices $m(s,t)$ for $s,t \in \mathbb{R}_+$ with $s \leq t$ together with the equalities

$$\frac{1}{h}(m(s,t+h) - m(s,t)) = m(s,t)\left[\frac{1}{h}(m(t,t+h) - E)\right]$$

and

$$\begin{aligned} -\frac{1}{h}(m(s-h,t) - m(s,t)) &= \left[-\frac{1}{h}(m(s,s-h)^{-1} - E)\right] m(s,t) \\ &= m(s,s-h)^{-1}\frac{1}{h}(m(s,s-h) - E)\, m(s,t), \end{aligned}$$

valid for all $s,t \in \mathbb{R}_+$, $s \leq t$ and $h \in \mathbb{R}_+^*$.

(2) Let $(m(s,t))_{s,t \in \mathbb{R}_+}$ be a hemigroup of matrices. We put

$$m(s,t) := m(t,s)^{-1}$$

for all $s,t \in \mathbb{R}_+$ with $s > t$ and suppose the existence of

$$\left.\frac{\partial^+}{\partial t} m(s,t)\right|_{t=s} =: a(s)$$

for all $s \in \mathbb{R}_+$. Then the mappings $s \to m(s,t)$ and $t \to m(s,t)$ are differentiable and the evolution equations (EEM) are satisfied.

The proof of this statement is a consequence of Property (1) if one notes the equalities

$$m(s,t+h) - m(s,t) = m(s,t)(m(t,t+h) - E)$$

and

$$m(s-h,t) - m(s,t) = (m(s-h,s) - E)\, m(s,t),$$

which are valid for all $s,t \in \mathbb{R}_+$, $s \leq t$.

4.6.6 Lemma. *Let a be a piecewise continuous mapping from $\mathbb{R}_+$ into $\mathfrak{M}(n,\mathbb{C})$. Then there exists at most one hemigroup of matrices $(m(s,t))_{s,t \in \mathbb{R}_+}$ satisfying the evolution equations* (EEM).

Proof. For every $s,t \in \mathbb{R}_+$, $s \leq t$, we put

$$m(s,t) := (m_{ij}(s,t))_{i,j=1,\ldots,n} \quad \text{and} \quad a(t) := (a_{ij}(t))_{i,j=1,\ldots,n}.$$

Then

$$\frac{\partial^+}{\partial t} m_{ij}(s,t) = m_{ij}(s,t)\, a_{ij}(t)$$

with $m_{ij}(s,s)=\delta_{ij}$ $(i,j=1,\ldots,n)$ is a system of linear differential equations of first order in n^2 variables with n^2 initial conditions. This initial value problem has a unique solution. □

The construction of convolution hemigroups in $\mathscr{M}^1(G)$ with prescribed generating families is based on the notion of *product-integrability*. We need some

4.6.7 Preparations. 1. Let q be a Riemann integrable real-valued function on an interval $[s,t]$ of $\mathbb{R}_+$ and $\mathfrak{Z}$ the filterbasis of decompositions

$$Z:=\{s=\tau_0<\tau_1<\cdots<\tau_n=t\} \quad \text{of } [s,t]$$

in the sense of refinement, i.e., with

$$|Z|:=\max_{1\leq i\leq n}|\tau_i-\tau_{i-1}|\to 0.$$

Then we have the following properties of the exponential function:

$$e^{\int_s^t q(\tau)d\tau}=\lim_{Z\in\mathfrak{Z}} e^{\sum_{i=1}^n q(\tau_i)(\tau_i-\tau_{i-1})}$$

$$(\alpha) \qquad =\lim_{Z\in\mathfrak{Z}}\prod_{i=1}^n e^{q(\tau_i)(\tau_i-\tau_{i-1})}$$

$$(\beta) \qquad =\lim_{Z\in\mathfrak{Z}}\prod_{i=1}^n(1+q(\tau_i)(\tau_i-\tau_{i-1}))$$

$$(\gamma) \qquad =\lim_{Z\in\mathfrak{Z}}\prod_{i=1}^n(1-q(\tau_i)(\tau_i-\tau_{i-1}))^{-1}.$$

Each of the properties (α), (β) and (γ) gives rise to the definition of a multiplicative integral for a piecewise continuous mapping q from $I:=[s,t]$ into $\mathfrak{M}(n,\mathbb{C})$. According to the properties of the exponential function one defines
(A) the *exponential integral* of q by

$$\int_s^{t\,\frown}\exp(q(\tau)\,d\tau):=\lim_{Z\in\mathfrak{Z}}\prod_{i=1}^n\exp[q(\tau_i)(\tau_i-\tau_{i-1})];$$

(B) the *product integral* of q by

$$\int_s^{t\,(P)}(E+q(\tau)\,d\tau):=\lim_{Z\in\mathfrak{Z}}\prod_{i=1}^n[E+q(\tau_i)(\tau_i-\tau_{i-1})] \quad \text{and}$$

(C) the *resolvent integral* of q by

$$\int_s^{t\,(R)}q(\tau)\,d\tau:=\lim_{Z\in\mathfrak{Z}}\prod_{i=1}^n[E-q(\tau_i)(\tau_i-\tau_{i-1})]^{-1}.$$

The existence of either type of product integral yields the corresponding notion of integrability. Plainly the above multiplicative integrals form families $(m(s,t))_{s,t\in\mathbb{R}_+}$ of matrices in $\mathfrak{M}(n,\mathbb{C})$ satisfying (HGM 1) and (HGM 2) of Definition 4.6.4.

2. If the integrals (A), (B) and (C) for Riemann integrable mappings $q:I\to\mathfrak{M}(n,\mathbb{C})$ exist, then they coincide. In fact, first let q be a piecewise constant mapping of the form

$$q:=\sum_{i=0}^{n-1}1_{[\tau_i,\tau_{i+1}[}\,q_i,$$

where $\{s=\tau_0<\tau_1<\cdots<\tau_n=t\}$ is a decomposition of $[s,t]$ and $q_i\in\mathfrak{M}(n,\mathbb{C})$ for $i=0,\ldots,n-1$. Then the multiplicative integrals of q coincide, since they are the same for the constant mapping. This follows directly from the properties of the exponential function. The case of general piecewise continuous functions is handled via approximation by piecewise constant functions: One shows as in [347], § VII that for any pair $s,t\in\mathbb{R}_+$ with $s\le t$ one has

$$\int_s^{t(P)}(E+q(\tau)\,d\tau)=E+\int_s^t q(\tau)\,d\tau+\int_s^t\int_s^{\tau_2} q(\tau_1)\,q(\tau_2)\,d\tau_1\,d\tau_2$$

$$+\int_s^t\int_s^{\tau_2}\int_s^{\tau_3} q(\tau_1)\,q(\tau_2)\,q(\tau_3)\,d\tau_1\,d\tau_2\,d\tau_3+\cdots$$

which is an absolutely convergent (Peano) series with each of its terms bounded by

$$(\sup_{s\le\tau\le t}\|q(\tau)\|^k)\frac{(t-s)^k}{k!}\qquad(k\ge 1).$$

Consequently, summation and limit are exchangeable. Hence, in view of the series representation above the approximation by piecewise constant functions yields the coincidence of the multiplicative integrals.

3. From Part 2 one obtains the possibility of defining the matrix

$$m(s,t):=\int_s^{t\,\frown}\exp(q(\tau)\,d\tau)$$

for all $s,t\in\mathbb{R}_+$ with $s\le t$, which can be shown to be the solution of the integral equation

$$m(s,t)=E+\textstyle\int_s^t m(s,\tau)\,q(\tau)\,d\tau$$

or of the corresponding differential equations

$$\frac{\partial^+}{\partial t}m(s,t)=m(s,t)\,q(t)\quad\text{and}\quad\frac{\partial^-}{\partial s}m(s,t)=-q(s)\,m(s,t)$$

under the initial condition $m(s,s)=E$.

The following result is of auxiliary interest.

4.6.8 Lemma. *For every piecewise continuous mapping* $q:\mathbb{R}_+\to\mathfrak{M}(n,\mathbb{C})$ *and all* $s,t\in\mathbb{R}_+$ *with* $s\le t$ *the exponential integral* $\int_s^{t\,\frown}\exp(q(\tau)\,d\tau)$ *exists.*

Proof. Given a continuous mapping $q:\mathbb{R}_+\to\mathfrak{M}(n,\mathbb{C})$ and $t,h\in\mathbb{R}_+$ we introduce the notation

$$T(t,h):=\exp[q(t)\,h]$$

and assume without loss of generality that

$$\|T(t,h)\|\le 1\qquad\text{for all }t,h\in\mathbb{R}_+.$$

Let

$$Z := \{s = \tau_0 < \tau_1 < \cdots < \tau_n = t\}, \qquad Z' := \{s = \sigma_0 < \sigma_1 < \cdots < \sigma_m = t\}$$

be two decompositions of $[s,t]$ (in $\mathfrak{Z}$) such that Z' is finer than Z and $(u_i)_{i=0,\ldots,n}$ a sequence in $\mathbb{N}$ defined by $\sigma_{u_i} := \tau_i$ for all $i = 0, 1, \ldots, n$. We furthermore put

$$J_i := \prod_{j=u_{i-1}+1}^{u_i} T(\sigma_j, \sigma_j - \sigma_{j-1}) \quad \text{and} \quad K_i := T(\tau_i, \tau_i - \tau_{i-1})$$

for $i = 1, \ldots, n$. Then one obtains

$$\begin{aligned} \Delta(Z, Z') &:= \left\| \prod_{j=1}^{m} T(\sigma_j, \sigma_j - \sigma_{j-1}) - \prod_{i=1}^{n} T(\tau_i, \tau_i - \tau_{i-1}) \right\| \\ &\le \sum_{i=1}^{n} \left\| \prod_{k=1}^{i-1} K_k \prod_{r=i}^{n} J_r - \prod_{k=1}^{i} K_k \prod_{r=i+1}^{n} J_r \right\| \\ &\le \sum_{i=1}^{n} \left\| \prod_{k=1}^{i-1} K_k J_i - \prod_{k=1}^{i-1} K_k K_i \right\| \le \sum_{i=1}^{n} \|J_i - K_i\|. \end{aligned}$$

On the other hand,

$$\begin{aligned} &\|J_i - K_i\| \\ &= \left\| \prod_{j=u_{i-1}+1}^{u_i} T(\sigma_j, \sigma_j - \sigma_{j-1}) - \prod_{j=u_{i-1}+1}^{u_i} T(\tau_i, \sigma_j - \sigma_{j-1}) \right\| \\ &\le \sum_{j=u_{i-1}+1}^{u_i} \left\| \prod_{h=u_{i-1}+1}^{j-1} T(\sigma_h, \sigma_h - \sigma_{h-1}) \prod_{r=j}^{u_i} T(\tau_i, \sigma_r - \sigma_{r-1}) \right. \\ &\quad \left. - \prod_{h=u_{i-1}+1}^{j} T(\sigma_h, \sigma_h - \sigma_{h-1}) \prod_{r=j+1}^{u_i} T(\tau_i, \sigma_r - \sigma_{r-1}) \right\| \\ &\le \sum_{j=u_{i-1}+1}^{u_i} \left\| \prod_{h=u_{i-1}+1}^{j-1} T(\sigma_h, \sigma_h - \sigma_{h-1})\, T(\tau_i, \sigma_j - \sigma_{j-1}) \right. \\ &\quad \left. - \prod_{h=u_{i-1}+1}^{j-1} T(\sigma_h, \sigma_h - \sigma_{h-1})\, T(\sigma_j, \sigma_j - \sigma_{j-1}) \right\| \\ &\le \sum_{j=u_{i-1}+1}^{u_i} \| T(\tau_i, \sigma_j - \sigma_{j-1}) - T(\sigma_j, \sigma_j - \sigma_{j-1}) \|. \end{aligned}$$

But the piecewise continuity of q implies $\|T(\tau, h) - T(\sigma, h)\| < \varepsilon h$ whenever $|\tau - \sigma| < \delta(\varepsilon)$. Hence,

$$\Delta(Z, Z') < (b - a)\varepsilon \quad \text{whenever } |Z| := \max_{1 \le i \le n} |\tau_i - \tau_{i-1}| < \delta(\varepsilon).$$

Since this inequality holds for every $\varepsilon > 0$, the assertion is established. ☐

4.6.9 Lemma. *For every piecewise continuous mapping* $q: \mathbb{R}_+ \to \mathfrak{M}(n, \mathbb{C})$ *there exists exactly one hemigroup of matrices* $(m(s,t))_{s,t \in \mathbb{R}_+}$ *satisfying the evolution equations* (EEM). *Moreover, one has for* $s, t \in \mathbb{R}_+$ *with* $s \le t$ *the product integral representation*

$$m(s,t) = \int_s^{t\,\frown} \exp(q(\tau)\, d\tau).$$

The *proof* follows from Lemmata 4.6.8 and 4.6.6 with the aid of Property (2) of 4.6.5. ☐

4.6.10 Theorem. *Let* $G \in \mathbf{A}$ *and* $(\mu_{s,t})_{s,t \in \mathbb{R}_+}$ *be a convolution hemigroup in* $\mathcal{M}^1(G)$ *such that for all* $s, t \in \mathbb{R}_+$ *the linear functionals*

$$A(s) := \frac{\partial^+}{\partial t} \mu_{s,t} \Big|_{t=s} \quad \text{and} \quad \frac{\partial^-}{\partial s} \mu_{s,t} \Big|_{s=t}$$

exist on $\mathfrak{K}(G)$ *and*

$$\frac{\partial^-}{\partial s}\mu_{s,t}\bigg|_{s=t} = -A(t)$$

holds on $\mathfrak{K}(G)$. *Then*

(a) *for every* $s\in\mathbb{R}_+$ *the linear functional* $\psi(s) := -A(s)$ *is a negative-definite form on* $\mathfrak{K}(G)$ *satisfying* $\psi(s)(1)=0$,

(b) $(\mu_{s,t})_{s,t\in\mathbb{R}_+}$ *is weakly and hence strongly differentiable on* $\mathfrak{K}(G)$ *and*

(c) $(\mu_{s,t})_{s,t\in\mathbb{R}_+}$ *satisfies on* Rep(G) *the weak evolution equations*

$$\text{(WEE)}\qquad \begin{aligned} &\frac{\partial^+}{\partial t}\hat{\mu}_{s,t}(D)=\hat{\mu}_{s,t}(D)\,A(t)(D) \quad \textit{and} \\ &\frac{\partial^-}{\partial s}\hat{\mu}_{s,t}(D)=-A(s)(D)\,\hat{\mu}_{s,t}(D) \qquad \textit{for all } D\in\mathrm{Rep}(G), \end{aligned}$$

and hence the evolution equations (EE) *on* $\mathfrak{K}(G)$.

Proof. For $s, h\in\mathbb{R}_+$ the linear functionals

$$-A_h(s) := -\frac{1}{h}(\mu_{s,s+h}-\varepsilon_e)$$

are negative-definite forms on $\mathfrak{K}(G)$. Thus

$$\psi(s) := -A(s) = -\lim_{h\downarrow 0}\frac{1}{h}(\mu_{s,s+h}-\varepsilon_e)$$

is a negative-definite form on $\mathfrak{K}(G)$. Moreover, one has for $s, h\in\mathbb{R}_+$ the equality

$$-\frac{1}{h}(\mu_{s,s+h}-\varepsilon_e)(1)=0,$$

whence $\psi(s)(1)=0$. For the rest of the proof we put

$$m(s,t) := \hat{\mu}_{s,t}(D) \quad\text{and}\quad a(t) := A(t)(D)$$

for any fixed representation $D\in\mathrm{Rep}(G)$ and all $s,t\in\mathbb{R}_+$, $s\leq t$. But then (b) and (c) follow directly from Property (1) of 4.6.5 (together with the assertion preceding Definition 4.6.4). □

4.6.11 Corollary. *If, in addition,* $G\in\mathbf{M}$ *and for* $s\in\mathbb{R}_+$ *we assume that* $\psi(s)\in\mathscr{S}(G)$, *then* $\psi(s)$ *is the (continuous) negative-definite form corresponding to a unique* $\{e\}$*-continuous convolution semigroup in* $\mathscr{M}^1(G)$.

The *proof* is a direct application of Theorem 1.5.19. □

4.6.12 Theorem. *Let* $G\in\mathbf{M}$ *and let* $\psi(s) := -A(s)$ *for every* $s\in\mathbb{R}_+$ *be a continuous negative-definite form on* $\mathfrak{K}(G)$ *with* $\psi(s)(1)=0$. *We furthermore suppose that for* *every* $f \in \mathfrak{K}(G)$ *the mapping* $s \to \psi(s)(f)$ *from* $\mathbb{R}_+$ *into* $\mathbb{R}$ *is continuous and that for every* $D \in \operatorname{Rep}(G)$ *there exists a neighborhood* $U := U(D)$ *of* D *satisfying for all* $s, t \in \mathbb{R}_+$ *with* $s < t$ *the inequality*

$$\sup\{\|\psi(\tau)(D')\| : \tau \in]s, t],\ D' \in U\} < \infty.$$

Then

(i) *there exists exactly one convolution hemigroup* $(\mu_{s,t})_{s,t \in \mathbb{R}_+}$ *in* $\mathcal{M}^1(G)$ *satisfying the weak evolution equations* (WEE) *on* $\operatorname{Rep}(G)$;

(ii) *for all* $s, t \in \mathbb{R}_+$ *with* $s \leq t$ *and* $D \in \operatorname{Rep}(G)$ *one has the product integral representation* $\hat{\mu}_{s,t}(D) = \int\limits_s^{t\,\frown} \exp(A(\tau)(D)\, d\tau)$.

Proof. 1. The uniqueness statement of (i) follows immediately from Lemma 4.6.6 if one puts, as in the proof of the preceding theorem,

$$m(s,t) := \hat{\mu}_{s,t}(D) \quad \text{and} \quad a(t) := A(t)(D)$$

for all $s, t \in \mathbb{R}_+$ with $s \leq t$ and every $D \in \operatorname{Rep}(G)$.

2. In order to show the remaining assertions in (i) and (ii) we apply Lemma 4.6.9 to $q : s \to \psi(s)(D) = -A(s)(D)$, which maps $\mathbb{R}_+$ into $\mathfrak{M}(n, \mathbb{C})$ (with $D \in \operatorname{Rep}_n(G)$), as follows: For a given $D \in \operatorname{Rep}(G)$ and $s, t \in \mathbb{R}_+$ with $s \leq t$ we define

$$u(s,t)(D) := \int\limits_s^{t\,\frown} \exp(A(\tau)(D)\, d\tau).$$

The family $(u(s,t)(D))_{s,t \in \mathbb{R}_+}$ is a hemigroup of matrices in $\mathfrak{M}(n, \mathbb{C})$ satisfying the evolution equations (EEM). For $s, t \in \mathbb{R}_+$ with $s \leq t$ the mapping $D \to u(s,t)(D)$ from $\operatorname{Rep}(G)$ into $\bigcup_{n \geq 1} \mathfrak{M}(n, \mathbb{C})$ defines a positive-definite form $u(s,t)$ on $\mathfrak{K}(G)$ with $u(s,t)(1) = 1$. This is deduced from the definition of the product integral above. Once we have shown that for $s, t \in \mathbb{R}_+$ with $s \leq t$ the linear functional $u(s,t)$ belongs in fact to $\mathcal{S}(G)$, then by Theorem 1.4.12 one obtains the existence of a measure $\mu_{s,t} \in \mathcal{M}^1(G)$ satisfying $\hat{\mu}_{s,t}(D) = u(s,t)(D)$ for all $D \in \operatorname{Rep}(G)$ and hence the full desired result.

3. The proof of the continuity of $u(s,t) \in \mathfrak{K}(G)^*$ for $s, t \in \mathbb{R}_+$, $s \leq t$ will be carried out in analogy to the proof of Lemma 4.5.7. First of all one notes that by Preparation 2 of 4.6.7 we have the Peano series representation

$$u(s,t)(D) = \int\limits_s^{t\,\frown} \exp(A(\tau)(D)\, d\tau) = E_{n(D)} + \int\limits_s^t A(\tau)(D)\, d\tau + \cdots$$

$$+ \sum_{k \geq 1} \int\limits_s^t \int\limits_s^{\tau_2} \cdots\cdots \int\limits_s^{\tau_{k+1}} A(\tau_1)(D) \cdot \cdots \cdot A(\tau_{k+1})(D)\, d\tau_1 \cdot \cdots \cdot d\tau_{k+1}$$

whenever $s, t \in \mathbb{R}_+$, $s \leq t$, and $D \in \operatorname{Rep}(G)$.

(α) Let G admit a countable basis of its topology. Then by Property (3) of 1.3.1 $\operatorname{Rep}(G)$ also has a countable basis of its topology. For every sequence $(D_j)_{j \geq 1}$ in $\operatorname{Rep}_n(G)$ and $D \in \operatorname{Rep}_n(G)$ ($n \geq 1$) with $\lim_{j \to \infty} D_j = D$ one has by assumption that

$$\lim_{j \to \infty} A(t)(D_j) = A(t)(D) \qquad (t \in \mathbb{R}_+).$$

By the boundedness assumption Lebesgue's dominated convergence theorem is applicable.

Hence, the limit relations

$$\lim_{j\to\infty} \int_s^t A(\tau)(D_j)\,d\tau = \int_s^t A(\tau)(D)\,d\tau$$

and

$$\lim_{j\to\infty} \int_s^t \int_s^{\tau_2} \cdot \dots \cdot \int_s^{\tau_{k+1}} A(\tau_1)(D_j) \cdot \dots \cdot A(\tau_{k+1})(D_j)\,d\tau_1 \cdot \dots \cdot d\tau_{k+1}$$
$$= \int_s^t \int_s^{\tau_2} \cdot \dots \cdot \int_s^{\tau_{k+1}} A(\tau_1)(D) \cdot \dots \cdot A(\tau_{k+1})(D)\,d\tau_1 \cdot \dots \cdot d\tau_{k+1} \qquad (k \geq 1)$$

hold, so that the absolute convergence of the Peano series representation yields

$$\lim_{j\to\infty} u(s,t)(D_j) = u(s,t)(D).$$

This, however, shows that $u(s,t)$ is continuous on $\operatorname{Rep}(G)$ for $s,t\in\mathbb{R}_+$, $s\leq t$.

(β) Let G be an arbitrary group in **M**. From Theorem O we infer that any closed subgroup of G belongs to **M**. Therefore one may suppose without loss of generality that G is σ-compact, whence by [218], (8.7),

$$G = \varprojlim_{\alpha\in\mathbb{A}} G_\alpha,$$

where G_α is a group in **M** having a countable basis of its topology ($\alpha\in\mathbb{A}$). Moreover, for every $D\in\operatorname{Rep}(G)$ there exist an $\alpha\in\mathbb{A}$ and a $D'_\alpha\in\operatorname{Rep}(G_\alpha)$ such that $D = D'_\alpha \circ p_\alpha$ holds, where p_α as usual denotes the canonical projection from G onto G_α. Consequently,

$$p_\alpha(A(t))(D'_\alpha) = A(t)(D'_\alpha \circ p_\alpha) = A(t)(D) \qquad \text{for all } t\in\mathbb{R}_+.$$

Since by (α) $u(s,t)$ is continuous on $\operatorname{Rep}(G_\alpha)\circ p_\alpha$, we obtain the existence of a

$$\mu_{s,t}^{(\alpha)} \in \mathcal{M}^1(G_\alpha) \qquad \text{with } u(s,t)(D) = \widehat{\mu_{s,t}^{(\alpha)}}(D)$$

for all $D\in\operatorname{Rep}(G_\alpha)$ ($s,t\in\mathbb{R}_+$, $s\leq t$). From the product integral representation of $u(s,t)(D)$ it follows that $(\mu_{s,t}^{(\alpha)})_{\alpha\in\mathbb{A}}$ is a projective system of measures $\mu_{s,t}^{(\alpha)}\in\mathcal{M}^1(G_\alpha)$ ($\alpha\in\mathbb{A}$) whose projective limit

$$\mu_{s,t} := \varprojlim_{\alpha\in\mathbb{A}} \mu_{s,t}^{(\alpha)}$$

in $\mathcal{M}^1(G)$ exists ($s,t\in\mathbb{R}_+$, $s\leq t$). Clearly, $(\mu_{s,t})_{s,t\in\mathbb{R}_+}$ is the desired convolution hemigroup in $\mathcal{M}^1(G)$ satisfying (i) and (ii). □

In the following we shall treat the problem of the solution of the evolution equations (EE) on the spaces $\mathsf{E} := \tilde{\mathscr{C}}_2^0(G)$ and $\mathfrak{D}(G)$ for a Lie group and for an arbitrary locally compact group G resp.

4.6.13. Let B be a Banach space. A family $(U_{s,t})_{s,t\in\mathbb{R}_+}$ of contraction operators in $\mathscr{L}(\mathsf{B})$ is called a *hemigroup of operators* if the following conditions hold:

(HGO1) $U_{s,r}U_{r,t} = U_{s,t}$ for all $s,r,t\in\mathbb{R}_+$ with $s\leq r\leq t$.

(HGO2) $U_{s,s} = E$ for all $s\in\mathbb{R}_+$.

(HGO3) The mapping $(s,t)\to U_{s,t}$ from $\mathbb{R}_+^2$ into $\mathscr{L}_s(\mathsf{B})$ is continuous.

$(U_{s,t})_{s,t\in\mathbb{R}_+}$ is said to be *differentiable on* B with respect to s, t *at* $s_0, t_0 \in \mathbb{R}_+$ if for all $x \in$ B the mappings $s \to U_{s,t}x$ and $t \to U_{s,t}x$ are differentiable (with respect to the norm) at s_0, t_0 for all $s, t \in \mathbb{R}_+$ resp.

$(U_{s,t})_{s,t\in\mathbb{R}_+}$ is called *differentiable on* B if the limits

$$N(s)\,x := \frac{\partial^+}{\partial t} U_{s,t}\,x\bigg|_{t=s} := \lim_{h\downarrow 0} \frac{1}{h}(U_{s,s+h} - E)\,x$$

and

$$\frac{\partial^-}{\partial s} U_{s,t}\,x\bigg|_{s=t} := \lim_{h\downarrow 0} \left[-\frac{1}{h}(U_{t-h,t} - E)\right] x$$

exist (in the norm topology of B) and if

$$\frac{\partial^-}{\partial s} U_{s,t}\,x\bigg|_{s=t} = -N(t)\,x \qquad \text{for all } x \in \text{B}.$$

We now establish some lemmata.

4.6.14 Lemma. *Let* $(U_{s,t})_{s,t\in\mathbb{R}_+}$ *be a hemigroup of operators in* $\mathscr{L}$(B) *which is differentiable on a dense linear subspace* C *of* B. *Then* $(U_{s,t})_{s,t\in\mathbb{R}_+}$ *is differentiable on* C *with respect to* $s, t \in \mathbb{R}_+$ *at* s_0, t_0 *resp. and satisfies on* C *the* evolution equations for operators

$$\text{(EEO)} \qquad \begin{aligned} &\frac{\partial^+}{\partial t} U_{s,t}\,x = U_{s,t}\,N(t)\,x \\ &\frac{\partial^-}{\partial s} U_{s,t}\,x = -N(s)\,U_{s,t}\,x \end{aligned}$$

(*for all* $x \in$ C).

Proof. For all $s, t \in \mathbb{R}_+$, $s < t$, $h \in \mathbb{R}_+^*$ and $x \in$ C we have

$$\lim_{h\downarrow 0} \frac{1}{h}(U_{s,t+h} - U_{s,t})\,x = \lim_{h\downarrow 0} U_{s,t}\left[\frac{1}{h}(U_{t,t+h} - E)\,x\right]$$

$$= U_{s,t} \lim_{h\downarrow 0} \frac{1}{h}(U_{t,t+h} - E)\,x = U_{s,t}\,N(t)\,x$$

and

$$\lim_{h\downarrow 0}\left[-\frac{1}{h}(U_{s-h,t} - U_{s,t})\right] x$$

$$= \lim_{h\downarrow 0}\left[-\frac{1}{h}(U_{s-h,s} - E)\,U_{s,t}\,x\right] = -N(s)\,U_{s,t}\,x.$$

This shows the existence of

$$\frac{\partial^+}{\partial t} U_{s,t}\,x \quad \text{and} \quad \frac{\partial^-}{\partial s} U_{s,t}\,x$$

and establishes the equations of (EEO) for all $x \in$ C. $\square$

In the sequel we consider strongly continuous contraction semigroups $(T_t)_{t\in\mathbb{R}_+}$ in $\mathscr{L}(\mathsf{B})$ and the set $\mathsf{G}(\mathsf{B})$ of their infinitesimal generators. Given $N\in\mathsf{G}(\mathsf{B})$ there always exists a unique contraction semigroup $(T_t)_{t\in\mathbb{R}_+}$ in $\mathscr{L}(\mathsf{B})$ with infinitesimal generator N. We shall write in this case $\operatorname{Exp}(tN)=T_t$ for all $t\in\mathbb{R}_+$ and observe that for all $x\in\mathsf{D}_N$ one has

$$\frac{d}{dt}T_t x = N T_t x = T_t N x.$$

Let $s\to N(s)$ be a mapping from $\mathbb{R}_+$ into $\mathsf{G}(\mathsf{B})$, where C is as in 4.6.14 and $\mathsf{D}_{N(s)}\supset\mathsf{C}$ for all $s\in\mathbb{R}_+$.

The following assumptions will be made:

(1) There exists a norm $\|\cdot\|_1$ on C such that C becomes $\|\cdot\|_1$-complete and $\|\cdot\|_1\geq\|\cdot\|$.

(2) For all $s,t\in\mathbb{R}_+$ one has $\operatorname{Exp}(tN(s))\,\mathsf{C}\subset\mathsf{C}$ and for every $x\in\mathsf{C}$ the inequality $\|\operatorname{Exp}(tN(s))x\|_1\leq\|x\|_1$ holds.

(3) For every $s\in\mathbb{R}_+$ the mapping $N(s)\colon\mathsf{C}\to\mathsf{B}$ is $\|\cdot\|_1$-$\|\cdot\|$-continuous.

(4) The mapping $s\to N(s)$ from $\mathbb{R}_+$ into $\mathsf{G}(\mathsf{B})$ is piecewise $|||\cdot|||$-continuous, where for every $s\in\mathbb{R}_+$ the norm $|||N(s)|||$ of $N(s)$ is defined to be

$$\inf\{c\in\mathbb{R}_+^*:\|N(s)\,x\|\leq c\,\|x\|_1 \text{ for all } x\in\mathsf{C}\}.$$

4.6.15 Lemma. *Let s_0, $t_0\in\mathbb{R}_+$ with $s_0<t_0$ and let $\mathfrak{Z}:=\{Z_i: i\in\mathbb{N}\}$ be a filterbasis of decompositions $Z_i:=\{s_0=\tau_0^{(i)}<\tau_1^{(i)}<\cdots<\tau_{n_i}^{(i)}=t_0\}$ of $[s_0,t_0]$ in the sense of refinement, i.e., with*

$$|Z_i|:=\max_{1\leq j\leq n_i}|\tau_j^{(i)}-\tau_{j-1}^{(i)}|\to 0.$$

For every $i\geq 1$ and $t\in\mathbb{R}_+$ we put

$$N^{(i)}(t):=\sum_{j=1}^{n_i}N(\tau_j^{(i)})\,1_{[\tau_{j-1}^{(i)},\tau_j^{(i)}]}.$$

Then

(i) *given $0<s_0\leq s_1<t_1\leq t_0$ one has*

(a) $$\lim_{i\to\infty}\int_{s_1}^{t_1}|||N^{(i)}(t)-N(t)|||\,dt=0,$$

(b) $$\sup\{|||N^{(i)}(t)|||:s_0\leq t\leq t_0, i\in\mathbb{N}\}<\infty \quad \textit{and}$$

(c) $$\lim_{h\downarrow 0}\frac{1}{h}\int_t^{t+h}|||N(t)-N(r)|||\,dr=0 \quad \textit{for all } t\in\mathbb{R}_+;$$

(ii) *for every $i\geq 1$ there exists a unique solution $(U_{s,t}^{(i)})_{s,t\in\mathbb{R}_+}$ of the evolution equations*

(EEOi) $$\left.\frac{\partial^+}{\partial t}U_{s,t}^{(i)}x\right|_{t=r}=U_{s,r}^{(i)}N^{(i)}(r)\,x$$
$$\left.\frac{\partial^-}{\partial s}U_{s,t}^{(i)}x\right|_{s=r}=-N^{(i)}(r)\,U_{r,t}^{(i)}x$$

for all $x\in\mathsf{B}$, whenever $s_0\leq s\leq t\leq t_0$ holds.

Proof. First of all we note that (i) follows immediately from the above assumption (4). To prove (ii) we define for all decompositions

$$Z_i := \{s_0 = \tau_0^{(i)} < \tau_1^{(i)} < \cdots < \tau_{n_i}^{(i)} = t_0\}$$

of $[s_0, t_0]$ $(i \geq 1)$, $s_0 \leq s < t \leq t_0$ and

$$\tau_{j-1}^{(i)} \leq s < \tau_j^{(i)} < \cdots < \tau_k^{(i)} \leq t < \tau_{k+1}^{(i)} \qquad (j, k = 1, \ldots, n_i - 1)$$

the operator

$$U_{s,t}^{(i)} := \operatorname{Exp}(N(\tau_{j-1}^{(i)})(\tau_j^{(i)} - s)) \operatorname{Exp}(N(\tau_j^{(i)})(\tau_{j+1}^{(i)} - \tau_j^{(i)})) \cdot \cdots$$
$$\cdot \operatorname{Exp}(N(\tau_{k-1}^{(i)})(\tau_k^{(i)} - \tau_{k-1}^{(i)})) \operatorname{Exp}(N(\tau_k^{(i)})(t - \tau_k^{(i)})).$$

Plainly $(U_{s,t}^{(i)})_{s,t \in \mathbb{R}_+}$ solves (EEOi) uniquely. □

4.6.16 Lemma. *With the notations of the preceding lemma one has the existence of the strong limit $U_{s,t} := \lim_{i \to \infty} U_{s,t}^{(i)}$ uniformly in $s, t \in \mathbb{R}_+$ with $s \leq t$ and that the family $(U_{s,t})_{s,t \in \mathbb{R}_+}$ so defined is a hemigroup of operators on* C.

Proof. From (ii) of Lemma 4.6.15 we obtain for all $s, t \in \mathbb{R}_+$ with $s_0 \leq s \leq t \leq t_0$, $n, m \geq 1$ and $x \in \mathsf{C}$

$$(U_{s,t}^{(n)} - U_{s,t}^{(m)})\, x = \int_s^t U_{s,r}^{(n)} (N^{(n)}(r) - N^{(m)}(r))\, U_{r,t}^{(m)}\, x\, dr,$$

whence

$$\begin{aligned}
&\|(U_{s,t}^{(n)} - U_{s,t}^{(m)})\, x\| \\
&\leq \int_s^t \| U_{s,r}^{(n)} (N^{(n)}(r) - N^{(m)}(r))\, U_{r,t}^{(m)}\, x\|\, dr \\
&\leq \|x\|_1 \int_s^t |||U_{s,r}^{(n)} (N^{(n)}(r) - N^{(m)}(r))\, U_{r,t}^{(m)}|||\, dr \\
&\leq \|x\|_1 \int_s^t \|U_{s,r}^{(n)}\|_1\, |||N^{(n)}(r) - N^{(m)}(r)|||\, \|U_{r,t}^{(m)}\|_1\, dr \\
&\leq \|x\|_1 \int_{s_0}^{t_0} |||N^{(n)}(r) - N^{(m)}(r)|||\, dr,
\end{aligned}$$

where the last inequality requires the assumption (2). We further conclude that

$$\overline{\lim}_{m \geq 1} \|(U_{s,t}^{(n)} - U_{s,t}^{(m)})x\| \leq 2 \int_{s_0}^{t_0} |||N^{(n)}(r) - N(r)|||\, dr$$

and with the help of (i) of Lemma 4.6.15 also that

$$\lim_{n,m \to \infty} \|(U_{s,t}^{(n)} - U_{s,t}^{(m)})\, x\| = 0$$

uniformly in $s_0 \leq s \leq t \leq t_0$.

Thus $U_{s,t} := \lim_{n \to \infty} U_{s,t}^{(n)}$ exists strongly for $s, t \in \mathbb{R}_+$ with $s \leq t$. The uniformity of the estimates established implies that $(U_{s,t})_{s,t \in \mathbb{R}_+}$ is in fact a hemigroup of operators on C. □

4.6.17 Lemma. *The hemigroup of operators $(U_{s,t})_{s,t \in \mathbb{R}_+}$ defined in the preceding lemma is differentiable with respect to s, t in s_0, t_0 for all $s_0, t_0 \in \mathbb{R}_+$ with $s_0 < t_0$ satisfying on* C *the evolution equations* (EEO).

Proof. By Lemma 4.6.14 it suffices to show the existence of

$$N(s)\,x := \frac{\partial^+}{\partial t} U_{s,t}\,x \Big|_{t=s} \quad \text{and} \quad \frac{\partial^-}{\partial s} U_{s,t}\,x \Big|_{s=t}$$

as well as the equality

$$\frac{\partial^-}{\partial s} U_{s,t}\,x \Big|_{s=t} = -N(t)\,x \qquad \text{for all } s,t\in\mathbb{R}_+,\ s\le t \text{ and } x\in\mathbf{C}.$$

For this we first note that the mapping

$$r \to \operatorname{Exp}(N(t)(r-t))\, U^{(n)}_{r,t+h}x$$

is differentiable for all $x\in\mathbf{C}$, provided that $t<r<t+h$, $h\in\mathbb{R}^*_+$ and r does not appear in the decomposition $Z := \{s_0=\tau_0<\cdots<\tau_n=t_0\}$. In fact,

$$\frac{d}{dr}[\operatorname{Exp}(N(t)(r-t))\, U^{(i)}_{r,t+h}x]$$
$$=\operatorname{Exp}(N(t)(r-t))\,N(t)\,U^{(i)}_{r,t+h}x-\operatorname{Exp}(N(t)(r-t))\,N^{(i)}(r)\,U^{(i)}_{r,t+h}x.$$

Integration from $r=t$ to $t+h$ then yields

$$\|\operatorname{Exp}(hN(t))x-U^{(i)}_{t,t+h}x\|$$
$$\le \|x\|_1 \textstyle\int_t^{t+h} |||N(t)-N^{(i)}(r)|||\,dr.$$

Furthermore, Assumption (4) together with Lemma 4.6.15 (i) and Lemma 4.6.16 implies that

$$\|\operatorname{Exp}(hN(t))x-U_{t,t+h}x\| \le \|x\|_1 \textstyle\int_t^{t+h} |||N(t)-N(r)|||\,dr,$$

whence

$$\left\|\frac{1}{h} U_{t,t+h}\,x-N(t)\,x\right\| \le \|x\|_1 \frac{1}{h}\textstyle\int_t^{t+h} |||N(t)-N(r)|||\,dr$$
$$+\left\|\frac{1}{h}\operatorname{Exp}(hN(t))\,x-N(t)\,x\right\| = \|x\|_1 \frac{1}{h}\textstyle\int_t^{t+h} |||N(t)-N(r)|||\,dr$$

whenever $x\in\mathbf{C}$.
But now (i) of Lemma 4.6.15 can be applied to obtain

$$\lim_{h\downarrow 0}\left\|\frac{1}{h} U_{t,t+h}\,x-N(t)\,x\right\|=0 \quad\text{or}\quad \frac{\partial^+}{\partial v} U_{t,v}\,x\Big|_{v=t}=N(t)\,x$$

for all $x\in\mathbf{C}$, as desired. The rest of the assertions can be proved in a similar way. □

4.6.18 Lemma. *The hemigroup of operators* $(U_{s,t})_{s,t\in\mathbb{R}_+}$ *solving* (EEO) *is uniquely determined.*

Proof. Let $(V_{s,t})_{s,t\in\mathbb{R}_+}$ be another hemigroup of operators solving (EEO). Then for all $s<r<t$, $n\geq 1$ and $x\in \mathsf{C}$ the mapping $r\to V_{s,r}\,U_{r,t}^{(n)}x$ is differentiable, and one has

$$\frac{d}{dr}(V_{s,r}\,U_{r,t}^{(n)}\,x)=V_{s,r}N(r)\,U_{r,t}^{(n)}\,x-V_{s,r}N^{(n)}(r)\,U_{r,t}^{(n)}\,x.$$

Integration with respect to r from s to t yields

$$V_{s,t}\,x-U_{s,t}^{(n)}\,x=\int_s^t V_{s,r}(N(r)-N^{(n)}(r))\,U_{r,t}^{(n)}\,x\,dr$$

and so

$$\|(V_{s,t}-U_{s,t}^{(n)})\,x\|\leq\|x\|_1\int_s^t|||N(r)-N^{(n)}(r)|||\,dr.$$

Since $\lim_{n\to\infty}U_{s,t}^{(n)}=U_{s,t}$, we get $V_{s,t}=U_{s,t}$ on C for all $s,t\in\mathbb{R}_+$, $s\leq t$. □

The preceding results will be collected in the following main

4.6.19 Lemma. *Under the assumptions* (1) *to* (4) *on a given mapping $s\to N(s)$ from $\mathbb{R}_+$ into $\mathsf{G}(\mathsf{B})$ (with $\mathsf{D}_{N(s)}\supset\mathsf{C}$ for all $s\in\mathbb{R}_+$) there exists exactly one hemigroup of operators $(U_{s,t})_{s,t\in\mathbb{R}_+}$ satisfying on C the evolution equations for operators* (EEO). *Moreover, $(U_{s,t})_{s,t\in\mathbb{R}_+}$ admits a product integral representation (with respect to the strong operator topology) in the following sense: If $s_0,t_0\in\mathbb{R}_+$, $s_0<t_0$ and $\mathfrak{Z}$ denotes a filterbasis of decompositions $Z:=\{s_0=\tau_0<\tau_1<\cdots<\tau_n=t_0\}$ of $[s_0,t_0]$, then for all $s,t\in\mathbb{R}_+$, $s\leq t$ we have*

$$U_{s,t}=\int_s^{t}{}^{\frown}\operatorname{Exp}(N(\tau)\,d\tau):=\lim_{Z\in\mathfrak{Z}}\prod_{i=1}^{n}\operatorname{Exp}(N(\tau_i)(\tau_i-\tau_{i-1}))$$

(with respect to the strong operator topology).

Proof. It remains to show the last statement. This follows from the uniqueness of $(U_{s,t})_{s,t\in\mathbb{R}_+}$ (Lemma 4.6.18) since

$$U_{s,t}=\lim_{Z\in\mathfrak{Z}}U_{s,t}^{(n)},$$

where Z is defined as above and $U_{s,t}^{(n)}$ corresponds to Z. □

The following result is a generalization of the preceding one.

4.6.20 Lemma. *Let B be a Banach space, $(P_\alpha)_{\alpha\in\mathbb{A}}$ a net of projections on B with $\|P_\alpha\|=1$ for all $\alpha\in\mathbb{A}$ and $\lim_{\alpha\in\mathbb{A}}P_\alpha=E$ with respect to the strong operator topology. For every $\alpha\in\mathbb{A}$ we introduce the Banach space $\mathsf{B}_\alpha:=P_\alpha\mathsf{B}$ and given a dense subspace C of B the spaces $\mathsf{C}_\alpha:=P_\alpha\mathsf{C}$ such that $\mathsf{C}=\bigcup_{\alpha\in\mathbb{A}}\mathsf{C}_\alpha$ holds. Let $\mathsf{G}(\mathsf{B})$ be the set of infinitesimal generators of strongly continuous contraction semigroups in $\mathscr{L}(\mathsf{B})$ and $s\to N(s)$ a mapping $\mathbb{R}_+\to\mathsf{G}(\mathsf{B})$ with $\mathsf{D}_{N(s)}\supset\mathsf{C}$ for all $s\in\mathbb{R}_+$. We assume that*

(i) $\operatorname{Exp}(tN(s))\,P_\alpha=P_\alpha\operatorname{Exp}(tN(s))$ *and* $\operatorname{Exp}(tN(s))\,\mathsf{C}_\alpha\subset\mathsf{C}_\alpha$ *for all $t,s\in\mathbb{R}_+$, $\alpha\in\mathbb{A}$;*

(ii) *the mapping $s\to N_\alpha(s):=P_\alpha N(s)$ from $\mathbb{R}_+$ into $\mathsf{G}(\mathsf{B}_\alpha)$ with $\mathsf{D}_{N_\alpha(s)}\supset\mathsf{C}_\alpha$ for all $s\in\mathbb{R}_+$ satisfies the hypotheses* (1) *to* (4) *preceding Lemma* 4.6.15 *(for $\alpha\in\mathbb{A}$).*

Then there exists a unique hemigroup of operators $(U_{s,t})_{s,t\in\mathbb{R}_+}$ *satisfying the evolution equations* (EEO) *on* C, *and* $(U_{s,t})_{s,t\in\mathbb{R}_+}$ *admits a product integral representation in the sense of Lemma* 4.6.19.

Proof. One need only note that for every $\alpha\in\mathbb{A}$ one has a unique solution $(U_{s,t}^{(\alpha)})_{s,t\in\mathbb{R}_+}$ of (EEO) on B_α such that for $s,t\in\mathbb{R}_+$ with $s\leq t$

$$\begin{aligned} U_{s,t}^{(\alpha)} &= \lim_{Z\in\mathfrak{Z}} \prod_{i=1}^n P_\alpha \operatorname{Exp}(N(\tau_i)(\tau_i-\tau_{i-1})) \\ &= \lim_{Z\in\mathfrak{Z}} P_\alpha\left(\prod_{i=1}^n \operatorname{Exp}(N(\tau_i)(\tau_i-\tau_{i-1}))\right) \end{aligned}$$

holds. This follows from Lemma 4.6.19 together with the assumptions. One then concludes that $((U_{s,t}^{(\alpha)})_{s,t\in\mathbb{R}_+})_{\alpha\in\mathbb{A}}$ is a projective system of hemigroups of operators. Thus its projective limit $(U_{s,t})_{s,t\in\mathbb{R}_+}$ defined on C by $U_{s,t}:=\varprojlim_{\alpha\in\mathbb{A}} U_{s,t}^{(\alpha)}$ for $s,t\in\mathbb{R}_+$, $s\leq t$ exists. Since C is dense in B, the proof is accomplished. ☐

Lemmata 4.6.19 and 4.6.20 will now be applied to our initial problem concerning convolution hemigroups in $\mathcal{M}^1(G)$ for Lie groups and arbitrary locally compact groups G resp.

4.6.21 Theorem. *Let G be a Lie group of dimension $n\geq 1$ and $t\to A(t)$ a mapping from* $\mathbb{R}_+$ *into the set* $\mathsf{H}(G)$ *of generating functionals of* $\{e\}$*-continuous convolution semigroups in* $\mathcal{M}^1(G)$ *such that the mapping*

$$t\to S_{A(t)}=N(t)$$

is piecewise continuous with respect to the norm in $\mathcal{L}(\tilde{\mathscr{C}}_2^0(G),\mathscr{C}^0(G))$. *Then there exists a unique convolution hemigroup* $(\mu_{s,t})_{s,t\in\mathbb{R}_+}$ *in* $\mathcal{M}^1(G)$ *satisfying on* $\tilde{\mathscr{C}}_2^0(G)$ *the evolution equations* (EE) (*following Remark* 4.6.3). *Moreover,* $(\mu_{s,t})_{s,t\in\mathbb{R}_+}$ *admits a weak product integral representation in the sense that for all* $s,t\in\mathbb{R}_+$, $s\leq t$,

$$T_{\mu_{s,t}}=\int_s^{t\,\frown} \operatorname{Exp}(N(\tau)\,d\tau) \qquad \textit{holds on } \tilde{\mathscr{C}}_2^0(G).$$

The *proof* can be deduced from Lemma 4.6.19 if one substitutes

$$\mathsf{B}:=\mathscr{C}^0(G), \qquad \mathsf{C}:=\tilde{\mathscr{C}}_2^0(G)$$

and takes $|||\cdot|||$ equal to the norm of $\mathcal{L}(\tilde{\mathscr{C}}_2^0(G),\mathscr{C}^0(G))$. The assumptions of the lemma are clearly satisfied. One need only observe that by Theorem 4.2.8 for any $t\in\mathbb{R}_+$ the operator

$$S_{A(t)}=N(t):\tilde{\mathscr{C}}_2^0(G)\to\mathscr{C}^0(G)$$

is defined and continuous. Then $(U_{s,t})_{s,t\in\mathbb{R}_+}$ with $U_{s,t}:=T_{\mu_{s,t}}$ for all $s,t\in\mathbb{R}_+$, $s\leq t$, yields the assertions. ☐

4.6.22 Corollary. *Let the generating functional $A(t)$ for every $t\in\mathbb{R}_+$ have on* $\tilde{\mathscr{C}}_2^0(G)$ *the representation* $(a_i(t),a_{ij}(t),\eta_t)_{i,j=1,\dots,n}$ *with real numbers* $a_1(t),\dots,a_n(t)$, *a symmetric, positive-semidefinite matrix* $(a_{ij}(t))_{i,j=1,\dots,n}$ *in* $\mathfrak{M}(n,\mathbb{R})$ *and a Lévy measure*

$\eta_t \in \mathcal{M}_+(G^*)$. Suppose that the functions $t \to a_i(t)$ $(i=1,\ldots,n)$, $t \to a_{ij}(t)$ $(i,j=1,\ldots,n)$ are piecewise continuous on $\mathbb{R}_+$ and that $\lim_{t\to s}\|\eta_t - \eta_s\| = 0$ holds for all $s \in \mathbb{R}_+$ with the exception of a discrete subset without accumulation points. Then $t \to S_{A(t)} = N(t)$ is piecewise continuous and the statements of the theorem hold.

Proof. Let $\{X_1,\ldots,X_n\}$ be a basis of the Lie algebra $\mathfrak{L}(G)$ of G and $\{x_1,\ldots,x_n\}$ a canonical coordinate system with respect to $\{X_1,\ldots,X_n\}$. Theorem 4.2.8 then implies for every $t \in \mathbb{R}_+$ the canonical representation

$$A(t) = \psi_1(t) + \psi_2(t) + \psi_3(t),$$

where

$$\psi_1(t)(f) := \sum_{i=1}^n a_i(t)(\tilde{X}_i f)(e),$$
$$\psi_2(t)(f) := \sum_{i,j=1}^n a_{ij}(t)(\tilde{X}_i \tilde{X}_j f)(e),$$
$$\psi_3(t)(f) := \int_{G^*} [f - f(e) - \Gamma(f)]\, d\eta_t$$

with a Lévy measure η_t on G and a Lévy mapping Γ for G defined by

$$\Gamma(f) := \sum_{i=1}^n (\tilde{X}_i f)(e)\, x_i \qquad (t \in \mathbb{R}_+,\ f \in \tilde{\mathscr{C}}_2^0(G)).$$

By hypothesis the operator $S_{\psi_1(t)+\psi_2(t)}: \tilde{\mathscr{C}}_2^0(G) \to \mathscr{C}^0(G)$ is continuous ($t \in \mathbb{R}_+$). The inequalities

$$\left\| \int_{G^*} [{}_x f - f - \Gamma({}_x f)]\, d(\eta_t - \eta_s) \right\| \le \|\eta_t - \eta_s\|\, \|{}_x f - f - \Gamma({}_x f)\|$$
$$\le \|\eta_t - \eta_s\| (2\|f\| + c\|f\|_{\tilde{1}}) \le \|\eta_t - \eta_s\|\, c\, \|f\|_{\tilde{2}}$$

with $c \ge 2$, valid for all $f \in \tilde{\mathscr{C}}_2^0(G)$, $t, s \in \mathbb{R}_+$, together with the last mentioned hypothesis then yield the result. □

As an application of Lemma 4.6.20 we obtain

4.6.23 Theorem. *Let G be an arbitrary locally compact group and $s \to A(s)$ a mapping from $\mathbb{R}_+$ into the set $\mathsf{H}(G)$ of generating functionals of $\{e\}$-continuous convolution semigroups in $\mathcal{M}^1(G)$. Given a Lévy mapping Γ for G we then have for every $t \in \mathbb{R}_+$ a canonical representation of $A(t)$ on $\mathfrak{D}(G)$ of the form*

$$A(t) = \psi_1(t) + \psi_2(t) + \psi_3(t)$$

with a primitive form $\psi_1(t)$, a quadratic form $\psi_2(t)$ on $\mathfrak{D}(G)$ and an integral form $\psi_3(t)$ on $\mathfrak{D}(G)$ defined by

$$\psi_3(t)(f) = \int_{G^*} [f - f(e) - \Gamma(f)]\, d\eta_t \qquad \text{for all } f \in \mathfrak{D}(G),$$

where η_t denotes a Lévy measure on G. We suppose that for every $f \in \mathfrak{D}(G)$ the functions $t \to \psi_1(t)(f)$ and $t \to \psi_2(t)(f)$ are piecewise continuous and that

$$\lim_{t\to s} \|\eta_t - \eta_s\| = 0$$

holds for all $s \in \mathbb{R}_+$ with the exception of a discrete subset without accumulation points. Then the mapping $t \to S_{A(t)} = N(t)$ from $\mathbb{R}_+$ into $\mathsf{G}(\mathscr{C}^0(G))$ is piecewise *continuous and there exists exactly one convolution hemigroup* $(\mu_{s,t})_{s,t\in\mathbb{R}_+}$ *in* $\mathcal{M}^1(G)$ *satisfying on* $\mathfrak{D}(G)$ *the evolution equations* (EE).

Moreover, $(\mu_{s,t})_{s,t\in\mathbb{R}_+}$ *admits a weak product integral representation in the sense that for all* $s,t\in\mathbb{R}_+$, $s\leq t$ *one has*

$$T_{\mu_{s,t}}=\int_s^{t\,\frown} \mathrm{Exp}(N(\tau)\,d\tau) \quad \text{on } \mathfrak{D}(G).$$

The *proof* follows from Lemma 4.6.20 together with Theorem 4.4.16. □

References and Comments

R 4.1 Positive Semigroups and Their Generating Functionals

In the course of preparing the proof of the general Lévy-Khintchine formula for infinitely divisible probability measures on an arbitrary locally compact group G we must describe the infinitesimal generator or generating functional of a continuous convolution semigroup $(\mu_t)_{t\in\mathbb{R}_+^*}$ in $\mathcal{M}^1(G)$. These objects are defined as the infinitesimal generator or generating functional of the strongly continuous contraction semigroup $(T_t)_{t\in\mathbb{R}_+^*}$ of convolution operators corresponding to $(\mu_t)_{t\in\mathbb{R}_+^*}$. While the first approach to the Lévy-Khintchine formula originally due to Lévy [317] argues in a probabilistic fashion (see Lévy [321], Gnedenko and Kolmogorov [172] or Loève [325]), the techniques of more recent treatments are either via radial limits as in Rogalski [408] or via extreme point methods as developed in Kendall [290] and Johansen [271]. In all of these results the canonical form of the Fourier transform of an infinitely divisible probability measure on $\mathbb{R}$ is established without explicitly separating the following two steps of the proof: First the given infinitely divisible measure is embedded in a continuous convolution semigroup, secondly the infinitesimal generator of the embedding semigroup is represented. The first to perform this separation was Rogalski [408]. The embedding problem which can be solved without difficulty in the case of the real line has been extensively studied for general locally compact groups in Chapter III. Here we continue with the canonical representation which we shall discuss from various points of view.

The first step in the direction of a general canonical representation of convolution semigroups on a locally compact group G was taken by G.A. Hunt in his pioneering work [261]. Hunt established a canonical representation for convolution semigroups on a Lie group. His ideas lead Courrège in [94] to a proof of great clarity for the classical Lévy-Khintchine formula. In the meantime other approaches to the classical representation theorem have been found. We mention only the approach via helical varieties taken up recently by Masani [348].

The method of our exposition will be the one discovered by Hunt. A slight modification of his framework will direct us to the final results on almost positive or dissipative distributions on the test functions over the group.

We start with preparations for Theorems 4.1.14 and 4.1.15 in which we have proved that for a Lie group G the generating functional of a convolution semigroup in $\mathscr{M}^1(G)$ exists on the spaces $\mathscr{C}_2(G)$ and $\tilde{\mathscr{C}}_2(G)$ of twice continuously left and right differentiable functions on G resp. The proofs in Hunt [261] of these results have been redone by Wehn in [496], [497] and Ramaswamy in [396]. Ramaswamy added the illustration of the discrete group case which appears as Theorem 4.1.5. The operator-theoretic basis for the problem of existence of the infinitesimal generator or generating functional of a convolution semigroup is the theory of Hille and Yosida, on which the entire presentation of this chapter will rely.

Aside from Yosida's book [508] we recommend Neveu [372] and Deny [110] as useful references. The Lie-theoretic tools are taken from Helgason [213]. The construction of the functions in Lemma 4.1.12 can be done with the help of Yamabe [507]. A more general argument is given in Lemma 4.1.11 which has already appeared in Heyer [230].

A comment on Remark 4.1.15 might be added. In the case of an Abelian Lie group G (for example for $G := \mathbb{R}^n$ $(n \geq 1)$) the reasoning implies the existence of the infinitesimal generator (rather than the generating functional) of the underlying semigroup on $\mathscr{C}_2(G) = \tilde{\mathscr{C}}_2(G)$. This can be seen as follows: If $f \in \mathscr{C}_2(G)$, then $L_x f \in \mathscr{C}_2(G)$ and $\|L_x f\|_2 = \|f\|_2$. Hence,

$$\left|\frac{1}{t}(S_t(L_x f)(e) - L_x f(e))\right| \leq c\,\|L_x f\|_2,$$

whence

$$\left|\frac{1}{t}(L_x(S_t f)(e) - L_x f(e))\right| \leq c\,\|L_x f\|_2,$$

i.e.,

$$\left|\frac{1}{t}(S_t f(x) - f(x))\right| \leq c\,\|f\|_2,$$

which by Remark 4.1.3 yields the existence of

$$\lim_{t \downarrow 0} \frac{1}{t}(S_t f(x) - f(x))$$

for every $f \in \mathscr{C}_2(G)$ and uniformly in $x \in G$.

R 4.2 Hunt's Representation Theorem

The program of Hunt [261], §§1–5, will be continued. In the preceding section we discussed

I. the existence of the generating functional A on the space $\tilde{\mathscr{C}}_2(G)$ of the given convolution semigroup $(\mu_t)_{t \in \mathbb{R}_+^*}$ in $\mathscr{M}^1(G)$ for a Lie group G.

The following steps are still to be taken:

II. The representation of A on $\tilde{\mathscr{C}}_2(G)$ in the form

$$Af = \sum_{i=1}^n a_i(\tilde{X}_i f)(e) + \sum_{i,j=1}^n a_{ij}(\tilde{X}_i \tilde{X}_j f)(e)$$
$$+ \int_{G^*} [f(x) - f(e) - \sum_{i=1}^n (\tilde{X}_i f)(e)\, x_i(x)]\, \eta(dx),$$

where n is the dimension of G, $(a_i)_{i=1,\dots,n}$ is a sequence of constants in $\mathbb{R}$, $(a_{ij})_{i,j=1,\dots,n}$ is a symmetric, positive-semidefinite matrix in $\mathfrak{M}(n,\mathbb{R})$ and η is a Lévy measure on $G^*:=G\setminus\{e\}$ depending on the basis $\{X_1,\dots,X_n\}$ of the Lie algebra $\mathfrak{L}(G)$ of G and on a canonical coordinate system $\{x_1,\dots,x_n\}$ chosen with respect to $\{X_1,\dots,X_n\}$.

III. The linear functional A on $\tilde{\mathscr{C}}_2(G)$ determines the convolution semigroup $(\mu_t)_{t\in\mathbb{R}_+^*}$ completely in the sense that if $(\mu_t)_{t\in\mathbb{R}_+^*}$ and $(\mu_t')_{t\in\mathbb{R}_+^*}$ are two convolution semigroups in $\mathscr{M}^1(G)$ (or $(T_t)_{t\in\mathbb{R}_+^*}$ and $(T_t')_{t\in\mathbb{R}_+^*}$ are two positive (invariant) semigroups of contraction operators on $\mathscr{C}_u(G)$) with generating functionals A and A' resp. satisfying $Af=A'f$ for all $f\in\tilde{\mathscr{C}}_2(G)$, then $\mu_t=\mu_t'$ (or $T_t=T_t'$) for all $t\in\mathbb{R}_+^*$.

IV. Given a Lévy measure η on G^* and the data $(a_i)_{i=1,\dots,n}$ and $(a_{ij})_{i,j=1,\dots,n}$ as in II and letting the linear mapping N on $\tilde{\mathscr{C}}_2(G)$ be defined by

$$Nf(x)=\sum_{i=1}^n a_i(\tilde{X}_i f)(x)+\sum_{i,j=1}^n a_{ij}(\tilde{X}_i\tilde{X}_j f)(x)$$
$$+\int_{G^*}[f(xy)-f(x)-\sum_{i=1}^n \tilde{X}_i f(x)\,x_i(y)]\,\eta(dy)$$

for all $f\in\tilde{\mathscr{C}}_2(G)$ and $x\in G$, we see that $\operatorname{Res}_{\tilde{\mathscr{C}}_2(G)}N$ is in fact the infinitesimal generator of a (unique) convolution semigroup in $\mathscr{M}^1(G)$.

V. The infinitesimal generator of the convolution semigroup $(\mu_t)_{t\in\mathbb{R}_+^*}$ in $\mathscr{M}^1(G)$ is defined at least on $\tilde{\mathscr{C}}_2(G)$.

In the section under discussion we follow Hunt's program [261], but choose instead of $\tilde{\mathscr{C}}_2(G)$ a slightly modified function space $\tilde{\mathscr{C}}_2^0(G)$ which is built from $\overline{\mathscr{C}^0}(G):=\mathscr{C}^0(G)\oplus\mathbb{R}$ rather than $\mathscr{C}_u(G)$ with the aim of reducing problems that arise at infinity (Hunt uses the one-point compactification of G) or when trying to specialize to $\mathbb{R}^n (n\geq 1)$ (Hunt's results do not directly contain the classical ones as special cases). In addition, the choice of the function space $\tilde{\mathscr{C}}_2^0(G)$ corrects an error in Hunt [261] already pointed out by Wehn [496], [497] and Grenander [182].

Theorem 4.2.1 is the modified version of Theorem 4.1.16 and Lemma 4.2.2 contains the construction of the Lévy measure and prepares Theorem 4.2.4, which yields the solution to the problem of Step II above. Lemma 4.2.6 takes care of Step III. Step IV is the contents of Theorem 4.2.5, whose proof relies on the fundamental Lemma 4.2.7 of Hunt ([261], Lemma 4.1). In Theorem 4.2.8 we present the complete formulation of the canonical representation of a convolution semigroup $(\mu_t)_{t\in\mathbb{R}_+^*}$ on a Lie group. This theorem also contains Step V of Hunt's program concerning the domain D_N of the infinitesimal generator N of $(\mu_t)_{t\in\mathbb{R}_+^*}$. In fact, $\mathsf{D}_N\supset\tilde{\mathscr{C}}_2^0(G)$.

Extensions of Hunt's theorem 4.2.8 to homogeneous spaces $M:=G/H$, where G is a Lie group and H is a compact subgroup of G, are given already in Hunt [261], §§8–9. Again a modified demonstration was desirable. For a compact Lie group a complete proof of the theorem was given in Heyer [230]. Specialization to the homogeneous space of the three-dimensional sphere yields a result of Bochner [27]. For arbitrary Abelian Lie groups the Hunt representation also appears as a basic tool in the papers of Kloss [305] and [306]. The main contributions in these papers concern the Lévy-Khintchine formula for measures on the torus $\mathbb{T}^n (n\geq 1)$. They can also be derived from Theorem 3.6.2 of Bochner [29].

For symmetric spaces $M=G/H$ with a noncompact, semisimple, connected Lie group G with finite center and a compact subgroup H of G Lévy-Khintchine formulae have been proposed by Gangolli [155]. Gangolli proceeds in a way different from ours. He establishes first a Lévy-Khintchine formula for the given H-invariant infinitely divisible probability measure μ on M and then proves the continuous embeddability of μ. The main tools for this approach are the Iwasawa-Mostow decomposition of G and the Fourier transform (via spherical functions) on the commutative Banach algebra $\mathscr{M}^{\#}(M)$ of all bounded H-invariant measures on M. In a later paper [156] Gangolli presents a wider framework which contains his previous work as a special case. He achieves a fairly complete classification and characterization of all normalized, continuous, spherical, infinitely divisible, positive-definite functions on a homogeneous space $M=G/H$ with a locally compact group G and a closed subgroup H of G. This work is based on the paper [434] of Schoenberg and has been enhanced in some aspects by Faraut and Harzallah in [136]. We shall present more details on recent developments in this direction in the references and comments to §4 of this chapter.

R 4.3 The Lévy-Khintchine Formula for Almost Periodic Groups

The main theorem 4.3.13 of this section contains the general canonical form of continuous negative-definite forms on the coefficient algebra $\mathfrak{K}(G)$ of a Lie projective Moore group. The theorem is due to Siebert [437] and generalizes previous results in this direction established for locally compact Abelian groups by Parthasarathy, Rao and Varadhan [383], Parthasarathy and Sazonov [384] and Forst [150], and for compact, not necessarily abelian groups by Carnal [63], [64] and [66]. The discussion leading to the proof of Theorem 4.3.13 starts with the choice of the proper function space

$$\mathscr{R}(G):=\tilde{\mathscr{C}}_2^0(G)\oplus\mathfrak{K}(G) \qquad \text{for a Lie group } G.$$

Given a continuous convolution semigroup $(\mu_t)_{t\in\mathbb{R}_+^*}$ in $\mathscr{M}^1(G)$ its infinitesimal generator N is defined on $\tilde{\mathscr{C}}_2^0(G)$ and on $\mathfrak{K}(G)$, hence on $\mathscr{R}(G)$, and N has canonical form. Since, on the other hand, for the semigroup $(\mu_t)_{t\in\mathbb{R}_+^*}$ there exists a corresponding negative-definite form ψ on $\mathfrak{K}(G)$, we obtain for the generating functional A of $(\mu_t)_{t\in\mathbb{R}_+^*}$ the identity

$$Af=Nf(e)=-\psi(f), \qquad \text{whenever } f\in\mathfrak{K}(G).$$

This yields Part (A) of Theorem 4.3.1. Part (B) of this theorem concerns the converse of the canonical representation prescribing the form of the infinitesimal generator and concluding the existence of a corresponding convolution semigroup in $\mathscr{M}^1(G)$. The Lévy function introduced in Definition 4.3.4 is the generalization of Parthasarathy's g-function (see Parthasarathy [383] or [377]), which we shall call local inner product in Chapter V. The axioms for the Lévy function are due in a slightly modified form to Siebert [437]. Siebert's construction of such a function has been given in Theorem 4.3.7. The method will remain crucial in all of our canonical representations presented in this chapter. The axiomatic approach to the theorem of Lévy-Khintchine type is continued with Definition 4.3.8 of a Lévy measure. The construction of a Lévy measure for a given continuous convolution semigroup on a Lie-Moore group is borrowed from Hunt [261] and explicitly given in Part (A) of Theorem 4.3.10.

The next and last step in the representation theorem for Lie projective almost periodic groups is a suitable refinement of the construction of a Lévy function. This refinement is made precise in the proof of Theorem 4.3.11. The proof uses transfinite induction which extends the existence of a Lévy function from the projective system of the Lie quotients to its limit. The technical preparations preceding Theorem 4.3.11 make use of some Lie group theory adapted from Helgason [213] and a few facts on the lifting of finite-dimensional representations and their coefficient functions taken from Heyer [227].

So far canonical representations have been found for continuous convolution semigroups on the class of Lie projective almost periodic groups which of course comprises the classes of locally compact Abelian groups and of arbitrary compact groups. Moreover, it comprises the class of almost periodic groups with small invariant neighborhoods (Murakami groups) and hence the subclass of Moore groups and its subclass of almost periodic groups G with compact quotient group G/G_0 (Kuranishi groups) fall in the domain of validity of the representation theorem (Part (A)). In order to obtain genuine Lévy-Khintchine formulae for infinitely divisible measures in $\mathscr{M}^1(G)$ we have to look in the intersection of this domain with the class of locally compact groups for which the embedding principle (in the strong sense of an $\{e\}$-continuous embedding) is fulfilled (see § 4 of Chapter III).

It turns out that every infinitely divisible measure in $\mathscr{M}^1(G)$ admits a Lévy-Khintchine representation if G is a compactly generated Abelian Lie group with connected dual group. More restrictively, we have that any infinitely divisible measure in $\mathscr{M}^1(G)$ without idempotent factor admits a Lévy-Khintchine representation if G satisfies the general embedding property in the sense of Definition 3.5.10. All compact Lie groups and all totally disconnected compact groups appear as examples of this case.

We add a few bibliographical remarks. The results of Parthasarathy, Rao and Varadhan [383] on the representation of convolution semigroups on locally compact Abelian groups have been reproduced in Sazonov and Tutubalin [423]. They also appear as consequences of more general approaches to the study of negative-definite (conditionally positive-definite) functions on groups in the work of Guichardet [186], § 4.3. and of Drumm [123]. In [123] extreme point methods have been used to establish a Lévy-Khintchine formula. Forerunners of this approach are Kendall [290] and Johansen [271]. The special problem of characterizing (à la Lévy-Khintchine) all real negative-definite functions on an Abelian locally compact group has been solved by Harzallah [194]. Embedding and canonical representation of convolution semigroups on almost periodic groups (and more general locally compact groups) are also the main subject of the papers [234] and [235] of Heyer. We finally mention that Part (A) of Theorem 4.3.13 remains true for all almost periodic groups G. In [444] Siebert deduces this extension from the Lévy-Khintchine formula for general locally compact groups, which shall be dealt with in Section 5.

R 4.4 The Canonical Representation of Almost Positive Functionals

In order to extend Hunt's theorem, the canonical representation of continuous convolution semigroups and the Lévy-Khintchine formula from Lie groups (§ 2) or Lie projective, almost periodic groups (§ 3) to general locally compact groups we first redo the Lie group case. We then study the case of a Lie projective group and proceed to the case of a general locally compact group. The first problem to be overcome is the definition of the notion of an infinitely differentiable function with compact support on a locally compact group. This problem has been solved by Bruhat in [51] on the basis of the solution of Hilbert's fifth problem (Montgomery and Zippin [356]). For a useful account of Hilbert's fifth problem and some of its implications one may read Gluškov [171]. A new approach to the solution of the problem related to previous work of Lashof [312] is due to Boseck [35].

The axiomatics of a Lie system introduced as Definition 4.4.1 is contained in Lashof [312] and explicitly formulated in Hazod [197]. The space $\mathfrak{D}(G)$ of test functions on a locally compact group G is introduced following Bruhat [51]. Lemma 4.4.4 appears in Siebert [443]. Definition 4.4.6 containing the notion of almost positive linear functionals on $\mathfrak{D}(G)$ is due to Siebert [438]. Almost positive linear functionals on $\mathfrak{D}(G)$ replace in this general setup the negative-definite forms introduced for almost periodic groups G on the coefficient algebra $\mathfrak{K}(G)$ in § 5 of Chapter I. In the case of a Lie group G special kinds of almost positive linear functionals such as primitive and quadratic ones can be characterized to fit Hunt's theory. This is done in Examples 4.4.8 and Lemma 4.4.9. The definition (4.4.10) and the construction (Theorem 4.4.13) of Lévy mappings for an arbitrary locally compact group are the basis for the proofs of Theorems 4.4.16 and 4.4.18 which are taken from Siebert [438]. In Theorem 4.4.16 the preparatory step towards Theorem 4.4.18 is taken. In particular, the canonical representation of the generating functional A of a continuous convolution semigroup on a Lie group is established.

Its proof is essentially that of Hunt. In Theorem 4.4.18 a more general framework is chosen: All almost positive and normed linear functionals on $\mathfrak{D}(G)$ for an arbitrary locally compact group G are canonically represented. The connection between this result and the generators of continuous convolution semigroups in $\mathscr{M}^1(G)$ will be postponed to the subsequent section.

Theorem 4.4.18 has been inspired by the work of Hazod [197], in which an implicit canonical representation depending on the data of the underlying Lie system can be found. The proof of the theorem given by Siebert depends on a result (Proposition I.1.2) of Bony, Courrège and Priouret [34], the one given here is self-contained (as is the relevant Lemma 4.4.19).

In this connection we wish to add a few references concerning almost positive linear functionals on general differentiable structures. The pioneering work in this direction has been done by von Waldenfels [493]. The results of Bony, Courrège and Priouret [34] center on an extended analysis of Feller semigroups on a differentiable manifold with compact boundary. In this analysis the integro-differential operators of Lévy arising from the classical theory of the Lévy-Khintchine formula, and of von Waldenfels satisfying a certain maximum principle are given a special importance. Further research on diffusion operators with variable coefficients is contained in Ito [264] and Stroock and Varadhan [458], [459].

More material on the canonical representation of convolution semigroups in $\mathcal{M}^1(G)$ for a locally compact group G is contained in Hazod [197] and Siebert [438]. We mention the fact that any almost positive linear functional on $\mathfrak{D}(G)$ is continuous with respect to the natural topology in $\mathfrak{D}(G)$ and is hence a distribution in the sense of L. Schwartz. A general distribution-theoretic treatise on the generating functionals of continuous convolution semigroups in $\mathcal{M}^1(G)$ is [206] of Hazod.

R 4.5 The Lévy-Khintchine Formula for General Locally Compact Groups

All efforts in this section are directed towards the Lévy-Khintchine theorem for arbitrary locally compact groups G which appears as Theorem 4.5.9. This theorem, due to Hazod [197] and Siebert [438], is a combination of Theorem 4.4.18 on the canonical form of almost positive and normed linear functionals on $\mathfrak{D}(G)$ with Theorem 4.5.8 in which the bijective correspondence between almost positive and normed linear functionals and continuous convolution semigroups in $\mathcal{M}^1(G)$ is established. The correspondence result is proved with the help of some technical tools from Hazod [197] (Lemmata 4.5.2 to 4.5.6) and a general convergence theorem for infinitesimal generators taken from Siebert [443] (Lemma 4.5.7). Lemma 4.5.7 has its origin in the work of Hunt [261], where it is shown for Lie groups. The original version has been reproduced in our presentation as Lemma 4.2.7. The generalization to arbitrary locally compact groups makes use of the facts that G can be assumed to be σ-compact and hence is the projective limit of locally compact groups admitting a countable basis of their topology. There are other demonstrations available for the correspondence theorem 4.5.8. In Heyer [235] one finds arguments for the Lie projective case, which are more or less straightforward. One simply uses more strongly the results of the Hille-Yosida theory (Yosida [508], also Ditzian [115]). The general locally compact case has been proved in Heyer [235] essentially along the lines of Hazod [197]. An additional point of interest is the proof of Part (B) of theorem 4.5.8. Here one can also use an approximation result of Seidman [435] or a theorem of Hirsch and Roth [243], [244] which opens the way to further generalizations to be discussed below.

It should be emphasized that our analysis concerns exclusively convolution semigroups of probability measures in $\mathcal{M}^1(G)$ and their real generating functionals. In the paper of M.J. Fisher [145] a first generalization in the direction of more general bounded measures on $\mathbb{R}$ is achieved. Fisher obtains in [145] a Lévy-Khintchine formula of the classical type for invertible measures. His method, however, seems to be limited to the group of the real line. Another kind of generalization in the classical spirit of the canonical representation has been given by Faraut. Faraut studies generalized Laplacians on $\mathbb{R}^n$ defined as real linear functionals L on $\mathfrak{D}(\mathbb{R}^n)$ with the property that $L(f) \leq 0$ for all $f \in \mathfrak{D}(\mathbb{R}^n)$ satisfying the maximum property

$$f(0) = \sup_{x \in \mathbb{R}^n} f(x) \geq 0.$$

The set of generalized Laplacians coincides with the dual cone of the convex cone of all functions $f \in \mathfrak{D}(\mathbb{R}^n)$ with the maximum property. In Faraut [134] the general form of any generalized Laplacian L on $\mathbb{R}^n$ is obtained, and it is shown that L is a distribution on $\mathbb{R}^n$ and that there exist a sequence $(\mu_k)_{k \geq 1}$ of measures in $\mathscr{M}_+^{(1)}(\mathbb{R}^n)$ and a sequence $(a_k)_{k \geq 1}$ in $\mathbb{R}_+^*$ such that

$$L(f) = \lim_{k \to \infty} a_k \left[\int f \, d\mu_k - f(0)\right] \quad \text{holds for all } f \in \mathfrak{D}(\mathbb{R}^n).$$

Turning to continuous convolution semigroups of complex measures we are lead in a natural way to the general framework of semigroups of invariant measures or operators resp.

Recent results in this direction are summarized in Faraut [133]. Let G be a locally compact group and K a compact subgroup of G. We denote by M the homogeneous space G/K. An operator T on $\mathscr{C}_{\mathbb{C}}^0(M)$ is said to be invariant if it commutes with the translations L_x for all $x \in G$. For a strongly continuous contraction semigroup $(T_t)_{t \in \mathbb{R}_+^*}$ (with $T_0 = E$) on $\mathscr{C}_{\mathbb{C}}^0(M)$ such that T_t is invariant for all $t \in \mathbb{R}_+^*$ (a) the infinitesimal generator (N, D_N) of $(T_t)_{t \in \mathbb{R}_+^*}$ is closed and its domain D_N is dense (Hille-Yosida) and (b) N is dissipative in the sense that $\operatorname{Re} Nf(x) \leq 0$ for all $f \in \mathsf{D}_N$ and $x \in M$ with $f(x) = \|f\|$. Moreover, (c) (N, D_N) is invariant in the natural sense. The problem whether any operator L on $\mathscr{C}_{\mathbb{C}}^0(M)$ satisfying the properties (a), (b) and (c) is in fact the infinitesimal generator of a contraction semigroup on $\mathscr{C}_{\mathbb{C}}^0(M)$ has been solved in various special cases. In particular, if G is compact (Faraut in [133]), if the algebra $\mathscr{M}^{\#}(G)$ of all bounded K-bi-invariant (complex) measures on G is commutative (Hirsch and Roth in [243], [244], see also Hirsch [242] and Roth [413]) and if G is a Lie group and $\mathsf{D}_L \supset \mathfrak{D}(M)$ (Faraut in [132] for the translation group of $\mathbb{R}^n$ and Duflo in [124] for the general case).

The second mentioned situation contains the case of a locally compact Abelian group and the case of a Riemann symmetric space treated earlier by Faraut and Harzallah [135].

The theorem of Hirsch and Roth proved (as Theorem 8) in [244] has been formulated in terms of continuous convolution semigroups of measures in Hazod [206]. The final step of the development seems to be the paper of Duflo [124]. In [124] a bijection is established between the set of continuous convolution semigroups $(\mu_t)_{t \in \mathbb{R}_+^*}$ of contraction measures in $\mathscr{M}_{\mathbb{C}}(G)$ with $\mathscr{T}_v\text{-}\lim_{t \downarrow 0} \mu_t = \mu_0$ and the set of μ_0-dissipative distributions L on $\mathfrak{D}(G)$ defined by the properties

$$\mu_0 * L * \mu_0 = L \quad \text{and} \quad \operatorname{Re} L(f) \leq 0$$

for all μ_0-invariant $f \in \mathfrak{D}(G)$ with $f(e) = \sup_{x \in G} |f(x)|$. This bijection is given by the formula

$$\lim_{t \downarrow 0} \frac{1}{t} \langle \mu_t - \mu_0, f \rangle = \langle L, f \rangle \quad \text{valid for all } f \in \mathfrak{D}(G).$$

R 4.6 Convolution Hemigroups. Generation and Representation

Hemigroups (or generalized semigroups) of contraction operators appear in connection with the theory of probability as early as the work of Neveu [372]. Their importance for the analysis of stochastic processes with independent increments has grown enormously since then. The measure-theoretic implications of contraction hemigroups became evident with the work of Wehn [496], [497]. Wehn studied continuous convolution hemigroups of probability measures on a Lie group and indicated their relationship to the evolution equations. The exposition of this section has been announced in Heyer [235]. It follows and absorbs the results of Guth [188].

We start with the definition 4.6.1 of a continuous convolution hemigroup of probability measures on a locally compact group G and aim at the solution of evolution equations (in terms of the corresponding probability operator semigroup) introduced after Remark 4.6.3 for the two function spaces $\mathfrak{K}(G)$ for a Moore group G and $\mathfrak{D}(G)$ for an arbitrary locally compact group G. The corresponding results containing the existence of a convolution hemigroup in $\mathscr{M}^1(G)$ for a given generating family on $\mathfrak{K}(G)$ and $\mathfrak{D}(G)$ as well as its product integral representation are Theorems 4.6.12 and 4.6.23 resp., which are essentially due to Guth [188]. In analogy to the two-way approach to the general Lévy-Khintchine formulae presented in §§ 3 and 5 we again separate the two cases of function spaces under discussion since the methods are different.

The differentiability conditions introduced in Definition 4.6.2 are modeled after Maksimov [341], who solved the evolution equations in the case of a compact Lie group basing his arguments on the work of Wehn [496]. Remark 4.6.3 contains an example showing that for general groups G weak differentiability on $\mathfrak{D}(G)$ of a convolution hemigroup in $\mathscr{M}^1(G)$ does not necessarily follow from its continuity (see Goodman [175]).

We now start with a discussion of the main problem posed for $\mathfrak{K}(G)$ with a Moore group G. Differentiability of hemigroups of matrices and the corresponding evolution equations are taken up. References for the Preparations 4.6.7 on product integration are Schlesinger [425] (for historical purposes), Masani [347] and the work of Johansen [276], in which an appendix (§ 4) takes care of the special preparations of this aspect of the theory. The Peano series expansion in Part 2 of the preparations 4.6.7 has been deduced in Masani [347]. Lemma 4.6.8 is well-known and its proof is taken from Guth [188]. In the development of the results 4.6.9 to 4.6.12 we make essential use of the theory of negative-definite forms on $\mathfrak{K}(G)$ as exposed in § 5 of Chapter I.

Our next aim is the solution of the evolution equations for the space $\mathfrak{D}(G)$ with an arbitrary locally compact group G. The approach is purely operator-theoretic. Basic references concerning the theory of product integration in Banach spaces are E.J.P. Georg Schmidt [431] and Webb [495]. The general form of the evolution equations for hemigroups of operators on a Banach space has been solved in Heyn [238] and Kato [285] (see in particular Theorem 4.1). It is Kato's technique which leads us to the desired results. There is an extension of the Hille-Yosida theory to certain classes of hemigroups of contraction operators on a Banach space due to Da Prato ([105], [106]). The generation of such hemigroups has been discussed in Mayer [350]. First of all, in Lemma 4.6.18 it is shown that the operator hemigroup solving the evolution equations is uniquely determined. Lemma 4.6.19 contains the existence of such a hemigroup for a given family of infinitesimal generators as well as its product representation. Then, in Theorem 4.6.21, the preparatory results are applied to a family of generating functionals of $\{e\}$-continuous convolution semigroups in $\mathcal{M}^1(G)$ for a Lie group G. By a projective limit argument this result is extended in Theorem 4.6.23 to arbitrary locally compact groups.

At this point the question of embedding (infinitely divisible) measures on a locally compact group G into continuous convolution hemigroups in $\mathcal{M}^1(G)$ arises. This question generalizes in a natural way the problem treated in Chapter III. Not much has yet been done in this direction. We mention the papers [276], [273] and [274] of Johansen on the embedding problem for Markov chains, which are based on previous work of Kingman [300]. In Hazod [202] H-continuous convolution hemigroups of measures in $\mathcal{M}^1(G)$ are studied for an arbitrary compact subgroup H of G. The results obtained concern the approximation of measures in $\mathcal{M}^1(G)$ that are embeddable into convolution hemigroups by certain Poisson measures. The conditions appearing in Hazod's theorems are the requirements of differentiability in the sense of Masani [347], Heyn [238] and Maksimov [341] resp. For the generalized embeddability see also § 6 of Chapter VI. Another series of results is available on the behavior at infinity of continuous convolution hemigroups $(\mu_{s,t})_{s,t\in\mathbb{R}_+}$ in $\mathcal{M}^1(G)$ in the sense that for any $s\in\mathbb{R}_+$ the limit

$$\mu_{s,\infty} := \lim_{t\to\infty} \mu_{s,t}$$

exists. For finite groups the problem has been affirmatively settled by Maksimov in [339]. Generalizations to totally disconnected compact groups are proved in Hazod [205]. In [205] one also finds some facts on the discrete analogues of convolution hemigroups, which are sequences $(\mu_{s,t})_{s,t\in\mathbb{Z}_+}$ in $\mathcal{M}^1(G)$ of n-th partial products

$$\nu_{k,n} := \mu_{k+1} * \cdots * \mu_n$$

of measures $\mu_j\in\mathcal{M}^1(G)$ $(j\geq 1)$ as discussed in § 2 of Chapter II.

In order to report on process-theoretic aspects of the theory of convolution hemigroups in $\mathcal{M}^1(G)$ for a locally compact group G having a countable basis we have to give a few basic facts well-known in the classical setup (see e.g. Hida [239], § 3 or further references to be noted in the references and comments to § 6 of Chapter V). Let $\mathfrak{X} := (X_t)_{t\in\mathbb{R}_+}$ be a process (on a probability space $(\Omega, \mathfrak{A}, P)$) with values in G. We assume that $\mathfrak{X}$ has independent increments in the sense that for every sequence $0\leq t_1\leq t_2\leq\cdots\leq t_n$ in $\mathbb{R}_+$ $(n\geq 1)$ the G-random variables $X_{t_1}^{-1}X_{t_2}, \ldots, X_{t_{n-1}}^{-1}X_{t_n}$ are stochastically independent. Then the corresponding family $(\mu_{s,t})_{s,t\in\mathbb{R}_+}$ of distributions

$$\mu_{s,t} := P_{X_s^{-1}X_t} = X_s^{-1}X_t(P) \quad (s\leq t)$$

satisfies Property (HG 1) of Definition 4.6.1. If, in addition, the process $\mathfrak{X}$ is normed, i.e., it satisfies $X_0 = e$, then Property (HG 2) of 4.6.1 is also satisfied. Finally, if $\mathfrak{X}$ is regular in the sense that for every $\omega\in\Omega$ the path $t\to X_t(\omega)$ is right continuous, has left-hand limits $X_{s-0}(\omega)$ and $P[X_{s-0}\neq X_s]=0$ holds for every $s\in\mathbb{R}_+$, then $(\mu_{s,t})_{s,t\in\mathbb{R}_+}$ also fulfills Property (HG 3) of 4.6.1 and is therefore a continuous convolution hemigroup in $\mathcal{M}^1(G)$. Moreover, one has a one-to-one correspondence between normed regular processes (additive processes) with values in G and continuous convolution hemigroups in $\mathcal{M}^1(G)$ given by $P_{X_s^{-1}X_t} = \mu_{s,t}$, valid for all $s, t\in\mathbb{R}_+$, $s\leq t$ (see Cuculescu [103]).

In particular, there is a one-to-one correspondence between stationary additive processes, for which by definition $P_{X_s^{-1}X_t}$ depends only on $t-s$ for all $s, t \in \mathbb{R}_+$, $s \leq t$, and continuous convolution semigroups $(\mu_r)_{r\in\mathbb{R}_+}$ in $\mathscr{M}^1(G)$ given by

$$P_{X_s^{-1}X_t} = \mu_{s,t} =: \mu_{t-s} \quad \text{for all } s, t \in \mathbb{R}_+ \text{ with } s \leq t.$$

A detailed study of additive processes with values in a locally compact Abelian group having a countable basis of its topology will be carried out in §6 of Chapter V.

We are left with the task of referring to recent advances in the theory of additive processes for not necessarily Abelian locally compact groups. The case of a finite group has been treated by Maksimov in [337]. Contributions to the problem for a differentiable manifold are due to Stroock and Varadhan [458], [459] and Sobko [447], [448]. On the basis of their previous work, Stroock and Varadhan solved part of the problem for a Lie group G in [460]. Their work has been extended to the complete solution of the problem for G in Feinsilver [138]. In Stroock and Varadhan [460] and Feinsilver [138] the construction of the solution of the evolution equation

$$\frac{\partial^+}{\partial t} T_{\mu_{s,t}} = T_{\mu_{s,t}} S_{A(t)}$$

for the given hemigroup $(\mu_{s,t})_{s,t\in\mathbb{R}_+}$ in $\mathscr{M}^1(G)$ is reduced to the solution of a corresponding martingale problem, i.e., to the construction of a process $\mathscr{X} = (X_t)_{t\in\mathbb{R}_+}$ such that for all $f \in \mathfrak{D}(G)$ the family $(g_t^{(f)})_{t\in\mathbb{R}_+}$ of functions

$$g_t^{(f)} := f \circ X_t - \int_0^t (S_{A(\tau)} f) \circ X_\tau \, d\tau$$

forms a martingale with respect to the family $(\mathfrak{A}_t)_{t\in\mathbb{R}_+}$ of σ-algebras

$$\mathfrak{A}_t := \mathfrak{A}(X_s : s \leq t).$$

Additive processes on a homogeneous space are still a challenging topic. First contributions inspired by Bochner have been made by Woll in [505].

Chapter V

The Central Limit Problem in the Abelian Case

The aim of this chapter is twofold: to motivate in the special case of an Abelian group the broad discussion of the central limit theorem that will follow in the subsequent chapter, and to give certain auxiliary results which will be needed for the treatment of the problems for more general groups.

The introduction of the notion of weak infinite divisibility yields the first theorem on the accumulation points of sequences of Poisson measures. Defining the local inner product of a group in analogy to the Lévy function studied in Chapter IV and proving their existence, the principal tool for the analysis of infinitesimal systems is established. The main result in this direction is the accompanying laws theorem. Generalizing the classical notion of a Gauss measure on the real line $\mathbb{R}$ and on the torus $\mathbb{T}$ one obtains a definition valid for any locally compact Abelian group. Its characterization by the special form of its Fourier transform is the basic result on the way to existence and convergence problems.

Further characterizations of the Gauss measure indicate its relationship to the historical background of the notion. In the case of an Abelian group the discussion of symmetric Gauss semigroups can be carried out on a technically less sophisticated level. Absolutely continuous Gauss measures will be characterized and their densities will be computed under special conditions.

Most of this material is applied to a detailed treatment of additive stochastic processes with values in a locally compact Abelian group having a countable basis of its topology. Here the corresponding jump process is analyzed, and it is shown that there exists an analogue of the Lévy decomposition of real-valued processes with independent increments. The decomposition splits the given process into a Gauss process with continuous paths and a remainder process, which takes care of the jumps. The distributions of the G-valued random variables of the additive process admit a Lévy-Khintchine representation with the jump measure as the Lévy measure and are thus weakly infinitely divisible without idempotent factors.

5.1 Convergence of Infinitesimal Systems

In the case of a locally compact Abelian group G the notion of shift compactness introduced in §2 of Chapter I can be reformulated as follows:

Two measures $\mu, \nu \in \mathcal{M}^1(G)$ are called *shift equivalent* (in symbols $\mu \sim \nu$) if there exists an $x \in G$ with $\mu = \nu * \varepsilon_x$. By $\tilde{\mathcal{M}}^1(G)$ we denote the collection of all shift equivalence classes of $\mathscr{M}^1(G)$. The canonical mapping $\mu \to \tilde{\mu}$ from $\mathscr{M}^1(G)$ onto $\tilde{\mathscr{M}}^1(G)$ yields the quotient topology in $\tilde{\mathscr{M}}^1(G)$ such that $\tilde{\mathscr{M}}^1(G)$ becomes again a commutative ($\mathscr{T}_v$-) topological semigroup.

A subset $\mathscr{N}$ of $\mathscr{M}^1(G)$ is said to be *shift compact* if $\tilde{\mathscr{N}}$ is relatively compact in $\tilde{\mathscr{M}}^1(G)$. It follows directly from Theorem 1.2.21 that for any shift compact subset $\mathscr{N}$ of $\mathscr{M}^1(G)$ the set of factors of elements in $\mathscr{N}$ is shift compact. Consequently, $\mathscr{N}$ is shift compact iff the set

$$|\mathscr{N}|^2 := \{|\mu|^2 := \mu * \mu^\sim : \mu \in \mathscr{N}\}$$

is relatively compact.

5.1.1 Lemma. *Let $\lambda \in \mathscr{M}^1(G)$ such that for some sequence $(x_n)_{n \geq 1}$ in G the sequence $(\lambda^k * \varepsilon_{x_k})_{k \geq 1}$ is relatively compact. Then there exists a compact subgroup H of G with the property that any accumulation point of $(\lambda^k * \varepsilon_{x_k})_{k \geq 1}$ is of the form $\omega_H * \varepsilon_x$ for some $x \in G$.*

Proof. Let $v', v \in \mathscr{M}^1(G)$ be two accumulation points of $(\lambda^k * \varepsilon_{x_k})_{k \geq 1}$.

By Theorem 2.2.2 there exists an $x \in G$ with $v' = v * \varepsilon_x$. For two nets $(n_\alpha)_{\alpha \in \mathbb{A}}$ and $(m_\alpha)_{\alpha \in \mathbb{A}}$ in $\mathbb{N}$ satisfying $n_\alpha > m_\alpha$ for all $\alpha \in \mathbb{A}$ such that the nets

$$(\lambda^{n_\alpha} * \varepsilon_{x_{n_\alpha}})_{\alpha \in \mathbb{A}}, \quad (\lambda^{m_\alpha} * \varepsilon_{x_{m_\alpha}})_{\alpha \in \mathbb{A}} \quad \text{and} \quad (\lambda^{n_\alpha - m_\alpha} * \varepsilon_{x_{n_\alpha - m_\alpha}})_{\alpha \in \mathbb{A}}$$

in $\mathscr{M}^1(G)$ converge to $v * \varepsilon_{x_1}$, $v * \varepsilon_{x_2}$ and $v * \varepsilon_{x_3}$ for $x_1, x_2, x_3 \in G$ resp., the relationship

$$\lambda^{n_\alpha} * \varepsilon_{x_{n_\alpha}} = (\lambda^{m_\alpha} * \varepsilon_{x_{m_\alpha}}) * (\lambda^{n_\alpha - m_\alpha} * \varepsilon_{x_{n_\alpha - m_\alpha}}) * \varepsilon_{x_{n_\alpha} x_{m_\alpha}^{-1} x_{n_\alpha - m_\alpha}^{-1}}$$

valid for all $\alpha \in \mathbb{A}$, implies the relative compactness of $(\varepsilon_{x_{n_\alpha} x_{m_\alpha}^{-1} x_{n_\alpha - m_\alpha}^{-1}})_{\alpha \in \mathbb{A}}$ and hence the existence of an $x_4 \in G$ with

$$v * \varepsilon_{x_1} = (v * \varepsilon_{x_2}) * (v * \varepsilon_{x_3}) * \varepsilon_{x_4}.$$

Putting $x := x_2 x_3 x_4 x_1^{-1}$ we obtain an idempotent $\mu := v * \varepsilon_x \in \mathscr{M}^1(G)$ which yields the assertion. □

5.1.2 Theorem. *Let $(\mu_n)_{n \geq 1}$ be a sequence in $\mathscr{M}^1(G)$ and $(x_n)_{n \geq 1}$ a sequence in G such that $(\mu_n^n * \varepsilon_{x_n})_{n \geq 1}$ is relatively compact and no accumulation point of $(\mu_n^n * \varepsilon_{x_n})_{n \geq 1}$ admits an idempotent factor. Then*

$$\lim_{n \to \infty} |\mu_n|^2 = \varepsilon_e.$$

Proof. 1. Since $(\mu_n^n * \varepsilon_{x_n})_{n \geq 1}$ is assumed to be relatively compact, there is a sequence $(y_n)_{n \geq 1}$ in G such that $(\mu_n * \varepsilon_{y_n})_{n \geq 1}$ is relatively compact. If μ is an accumulation point of $(\mu_n * \varepsilon_{y_n})_{n \geq 1}$, then $|\mu|^2$ is an accumulation point of $(|\mu_n|^2)_{n \geq 1}$. Conversely, for every accumulation point λ of $(|\mu_n|^2)_{n \geq 1}$ there exists an accumulation point μ of $(\mu_n * \varepsilon_{y_n})_{n \geq 1}$ with $\lambda = |\mu|^2$.

2. We show that $(|\mu_n|^2)_{n \geq 1}$ admits a unique accumulation point ε_e. Let $\mu \in \mathscr{M}^1(G)$ be an accumulation point of the sequence $(\mu_n * \varepsilon_{y_n})_{n \geq 1}$ constructed in

Part 1. Since for $k, n \geq 1$ with $k < n$ we have

$$\mu_n^n * \varepsilon_{x_n} = (\mu_n * \varepsilon_{y_n})^k * \mu_n^{n-k} * \varepsilon_{y_n^{-k} x_n},$$

there exists (for every k) an accumulation point ρ_k of the sequence $(\mu_n^{n-k} * \varepsilon_{y_n^{-k} x_n})_{n \geq 1}$ such that $\mu^k * \rho_k$ is an accumulation point of $(\mu_n^n * \varepsilon_{x_n})_{n \geq 1}$. Since the accumulation set of a relatively compact sequence in $\mathcal{M}^1(G)$ is itself relatively compact, there is a sequence $(z_n)_{n \geq 1}$ in G with relatively compact sequence $(\mu^k * \varepsilon_{z_k})_{k \geq 1}$. Lemma 5.1.1 implies the existence of a compact subgroup H of G such that every accumulation point of $(\mu^k * \varepsilon_{z_k})_{k \geq 1}$ is of the form $\omega_H * \varepsilon_x$ for an $x \in G$. For an accumulation point ρ' of $(\rho_k * \varepsilon_{z_k^{-1}})_{k \geq 1}$ the measure $\omega_H * \varepsilon_x * \rho'$ is an accumulation point of $(\mu_n^n * \varepsilon_{x_n})_{n \geq 1}$. By assumption we obtain $\omega_H = \varepsilon_e$ (or $H = \{e\}$). From

$$\mu^k * \varepsilon_{z_k} = \mu * (\mu^{k-1} * \varepsilon_{z_{k-1}}) * \varepsilon_{z_k z_{k-1}^{-1}}$$

for all $k \geq 2$ one deduces that $\varepsilon_e = \mu * \varepsilon_{x_0}$ for some $x_0 \in G$, i.e., $\mu \in \mathcal{D}(G)$ or

$$\lambda := |\mu|^2 = \mu * \mu^{\sim} = \varepsilon_e$$

is the unique accumulation point of $(|\mu_n|^2)_{n \geq 1}$. □

5.1.3 Definition. A measure $\mu \in \mathcal{M}^1(G)$ is called *weakly infinitely divisible* if for each $n \geq 1$ there exist a measure $\mu_n \in \mathcal{M}^1(G)$ and an $x_n \in G$ such that $\mu = \mu_n^n * \varepsilon_{x_n}$.

The collection of all weakly infinitely divisible measures in $\mathcal{M}^1(G)$ will be denoted by $\mathcal{I}_0(G)$. Obviously $\mathcal{I}(G) \subset \mathcal{I}_0(G)$. The reverse inclusion is not true unless G is divisible. Evident examples of subclasses of $\mathcal{I}_0(G)$ are the subclasses $\mathcal{J}(G)$ and $\mathcal{P}(G)$ of idempotent and Poisson measures on G resp.

5.1.4 Theorem. *$\mathcal{I}_0(G)$ is a sequentially closed subsemigroup of $\mathcal{M}^1(G)$.*

Proof. It remains to be proved that $\mathcal{I}_0(G)$ is sequentially closed in $\mathcal{M}^1(G)$.

Let $(\mu_k)_{k \geq 1}$ be a sequence in $\mathcal{I}_0(G)$ with $\lim_{k \to \infty} \mu_k = \mu \in \mathcal{M}^1(G)$. Clearly, $(\mu_k)_{k \geq 1}$ is relatively compact, and for all $k, n \geq 1$ we have

$$\mu_k = \lambda_{k,n}^n * \varepsilon_{x_{k,n}} \quad \text{with } \lambda_{k,n} \in \mathcal{M}^1(G) \quad \text{and} \quad x_{k,n} \in G.$$

Now $(\lambda_{k,n})_{k \geq 1}$ is shift compact for every $n \geq 1$, i.e., there exists a sequence $(y_{k,n})_{k \geq 1}$ in G such that $(\lambda_{k,n} * \varepsilon_{y_{k,n}})_{k \geq 1}$ is relatively compact for every $n \geq 1$. For any accumulation point λ_n of $(\lambda_{k,n} * \varepsilon_{y_{k,n}})_{k \geq 1}$ we obtain from

$$\mu_k = (\lambda_{k,n} * \varepsilon_{y_{k,n}})^n * \varepsilon_{x_{k,n} y_{k,n}^{-n}}$$

the existence of an $x_n \in G$ (where ε_{x_n} is an accumulation point of the relatively compact sequence $(\varepsilon_{x_{k,n} y_{k,n}^{-n}})_{k \geq 1}$) satisfying $\mu = \lambda_n^n * \varepsilon_{x_n}$ (for all $n \geq 1$). This implies that $\mu \in \mathcal{I}_0(G)$. □

5.1.5 Theorem. *For any $\mu \in \mathscr{I}_0(G)$ the following statements are equivalent:*
(i) *μ admits a nontrivial idempotent factor;*
(ii) *there exists a $\chi_0 \in G^\wedge$ with $\hat{\mu}(\chi_0)=0$.*

Proof. It remains to prove the implication (ii) $\Rightarrow$ (i). Let $\chi_0 \in G^\wedge$ with $\hat{\mu}(\chi_0)=0$. For every $n \geq 1$ there exist a $\mu_n \in \mathscr{M}^1(G)$ and an $x_n \in G$ satisfying $\mu = \mu_n^n * \varepsilon_{x_n}$. This implies that $\hat{\mu}_n(\chi_0)=0$ for all $n \geq 1$. Moreover, there is a sequence $(y_n)_{n \geq 1}$ in G such that $(\mu_n * \varepsilon_{y_n})_{n \geq 1}$ is relatively compact. For any accumulation point λ of $(\mu_n * \varepsilon_{y_n})_{n \geq 1}$ one obtains $\hat{\lambda}(\chi_0)=0$, and for all $k \geq 1$ the measure λ^k is a factor of μ. Hence, there exists a sequence $(z_k)_{k \geq 1}$ in G such that $(\lambda^k * \varepsilon_{z_k})_{k \geq 1}$ is relatively compact. For every accumulation point λ_0 of $(\lambda^k * \varepsilon_{z_k})_{k \geq 1}$ one has the following properties:
(α) λ_0 is a factor of μ,
(β) $\hat{\lambda}_0(\chi_0)=0$ and
(γ) there is a compact subgroup H of G such that $\lambda_0 = \omega_H * \varepsilon_z$ holds for some $z \in G$.
[The last statement follows from Lemma 5.1.1.]
Consequently, ω_H is an idempotent factor of μ with $\hat{\omega}_H(\chi_0)=0$ or $\omega_H \neq \varepsilon_e$. □

With the help of the preceding theorem we can prove a compactness result for sequences of Poisson measures which will be basic for the further development.

For any measure $\nu \in \mathscr{M}_+^b(G)$ we introduce the notation

$$\exp(\nu) := \exp_{\{e\}}(\nu - \|\nu\| \, \varepsilon_e).$$

5.1.6 Theorem. *Let $(\nu_n)_{n \geq 1}$ be a sequence in $\mathscr{M}_+^b(G)$ and $(\mu_n)_{n \geq 1}$ the sequence of Poisson measures $\mu_n := \exp(\nu_n)$ $(n \geq 1)$. We assume that*
(a) *$(\mu_n)_{n \geq 1}$ is shift compact,*
(b) *no accumulation point μ of any sequence of shifts of $(\mu_n)_{n \geq 1}$ admits an idempotent factor.*
Then
(i) *For every $U \in \mathfrak{B}(e)$ the sequence $(\mathrm{Res}_{\complement U}\, \nu_n)_{n \geq 1}$ is relatively compact.*
(ii) *For all $\chi \in G^\wedge$ one has $\sup_{n \geq 1} \int (1\text{-}\mathrm{Re}\,\chi(x))\, \nu_n(dx) < \infty$.*

Proof. 1. Without loss of generality we choose a symmetric $U \in \mathfrak{B}(e)$ and show that the sequence $(\nu_n(\complement U))_{n \geq 1}$ is bounded. If it were not bounded, then there would exist a subsequence $(\nu_{n_k})_{k \geq 1}$ of $(\nu_n)_{n \geq 1}$ with the property $\nu_{n_k}(\complement U) \geq k$ for all $k \geq 1$. For every $k \geq 1$ one defines a measure

$$\lambda_k := \frac{1}{k} \mathrm{Res}_{\complement U}\, \nu_{n_k} \quad \text{in } \mathscr{M}_+^b(G).$$

Clearly, $\rho_k^k := \exp(k\,\lambda_k)$ is a factor of μ_{n_k} for $k \geq 1$, and the shift compactness of $(\mu_n)_{n \geq 1}$ yields the shift compactness of $(\rho_k^k)_{k \geq 1}$. Assumption (b) implies that no accumulation point of a sequence of shifts of $(\rho_k^k)_{k \geq 1}$ admits an idempotent factor, and thus from Theorem 5.1.2 follows that $(|\rho_k|^2)_{k \geq 1}$ converges to ε_e and

$$\lim_{k \to \infty} \exp(\lambda_k + \tilde{\lambda}_k)(\complement U) = 0.$$

But

$$\exp(\lambda_k+\lambda_k^{\sim})(\complement U)\geq e^{-2}\lambda_k(\complement U)=e^{-2}\frac{1}{k}\nu_{n_k}(\complement U)\geq e^{-2}$$

for all $k\geq 1$ implies a contradiction.

2. Let $U\in\mathfrak{V}(e)$ be fixed and $a:=\sup_{n\geq 1}\nu_n(\complement U)$. For every $n\geq 1$ we put $\tau_n:=\mathrm{Res}_{\complement U}\,\nu_n$. Then $\exp(\tau_n)$ is a factor of $\exp(\nu_n)$ for every $n\geq 1$. Since $(\mu_n)_{n\geq 1}$ is shift compact, $(\exp(\tau_n))_{n\geq 1}$ is also shift compact and hence, $(\exp(\tau_n+\tau_n^{\sim}))_{n\geq 1}$ is relatively compact. For every $n\geq 1$ we put $\sigma_n:=\tau_n+\tau_n^{\sim}$. Prohorov's theorem yields for every $\varepsilon>0$ the existence of a compact set $K:=K_\varepsilon\subset G$ such that $\exp(\sigma_n)(\complement K)\leq\varepsilon$ holds for all $n\geq 1$, and thus, the validity of

$$e^{-2a}\tau_n(\complement K)\leq e^{-\|\sigma_n\|}\sigma_n(\complement K)\leq\exp(\sigma_n)(\complement K)\leq\varepsilon$$

for all $n\geq 1$. Therefore $(\tau_n)_{n\geq 1}$ is uniformly tight and again by Prohorov's theorem relatively compact. This proves (i) of the theorem.

3. Statement (ii) of the theorem, which is still to be proved, can be written as

$$\sup_{n\geq 1}(\hat{\nu}_n(1)-\mathrm{Re}\,\hat{\nu}_n(\chi))<\infty \qquad \text{for all } \chi\in G^{\wedge}.$$

If this were not true, then we would have a subsequence $(\nu_{n_k})_{k\geq 1}$ of $(\nu_n)_{n\geq 1}$ and a character $\chi_0\in G^{\wedge}$ with

$$\hat{\nu}_{n_k}(1)-\mathrm{Re}\,\hat{\nu}_{n_k}(\chi_0)\geq k.$$

Putting again $\sigma_n:=\nu_n+\nu_n^{\sim}$ for all $n\geq 1$, one concludes that the sequence $(\exp(\sigma_{n_k}))_{k\geq 1}$ is uniformly tight. Therefore, by Prohorov's theorem it is relatively compact with accumulation point $\lambda\in\mathscr{M}^1(G)$. By Theorem 5.1.4 $\lambda\in\mathscr{I}_0(G)$ and by Theorem 5.1.5 we have $|\hat{\lambda}(\chi_0)|\neq 0$. But this contradicts the above inequality since

$$\lim_{k\to\infty}\exp(\sigma_{n_k})^{\wedge}(\chi_0)=\lim_{k\to\infty}\exp(-2(\hat{\nu}_{n_k}(1)-\mathrm{Re}\,\hat{\nu}_{n_k}(\chi_0)))=0. \quad \square$$

We are now going to adapt the notion of a Lévy function introduced in § 3 of Chapter IV to the case of an Abelian group. In the classical situation it is the truncated expectation which makes the study of limit theorems possible if the expectations are infinite. An appropriate technique of centering measures on an Abelian group involves the following

5.1.7 Definition. A function $g\in\mathscr{C}(G\times G^{\wedge})$ is called a *local inner product (centering function) for* G if the following conditions are satisfied:

(LI 1) $\sup_{x\in G}\sup_{\chi\in K}|g(x,\chi)|<\infty$ for all compact subsets K of $G^{\wedge}$.

(LI 2) $g(x,\chi\eta)=g(x,\chi)+g(x,\eta)$ and $g(x^{-1},\chi)=-g(x,\chi)$ for all $x\in G$, $\chi,\eta\in G^{\wedge}$.

(LI 3) For every compact subset K of $G^{\wedge}$ there exists a $U:=U_K\in\mathfrak{V}(e)$ such that $\chi(x)=\exp i g(x,\chi)$ holds for all $x\in U$, $\chi\in K$.

(LI 4) For every compact subset K of $G^{\wedge}$ one has $\lim_{x\to e}\sup_{\chi\in K}g(x,\chi)=0$.

5.1.8 Remark. Any Lévy function for G in the sense of 4.3.4 is a local inner product for G. In fact, any Lévy function γ for an almost periodic group G has the property that for every $x \in G$ there exists a one-parameter subgroup $(x_t)_{t \in \mathbb{R}}$ of G such that

$$D(x_t) = \exp t\, \gamma(x, D)$$

holds for every $D \in \operatorname{Rep}(G)$ and $t \in \mathbb{R}$.

Let ψ be the primitive form corresponding to the one-parameter semigroup $(\varepsilon_{x_t})_{t \in \mathbb{R}_+^*}$ in $\mathscr{M}^1(G)$ (in the sense of 1.5.20). From

$$\psi(D) = \lim_{t \downarrow 0} \frac{1}{t}(D(e) - \hat{\varepsilon}_{x_t}(D))$$

$$= \lim_{t \downarrow 0} \frac{1}{t}(E_{n(D)} - \exp t\, \gamma(x, D)) = -\gamma(x, D)$$

for all $D \in \operatorname{Rep}(G)$ and every $x \in G$ we conclude that $\gamma(x, \cdot)$ is a primitive form for every $x \in G$.

In the special case of an Abelian group G we therefore obtain

$$\gamma(x, \chi\eta) = \gamma(x, \chi)\, \eta(e) + \chi(e)\, \gamma(x, \eta) = \gamma(x, \chi) + \gamma(x, \eta)$$

for all $x \in G$, $\chi, \eta \in G^\wedge$ which is the first mentioned property of Condition (LI2). The second property of (LI2) as well as the conditions (LI1), (LI3) and (LI4) follow directly from the definition of a Lévy function.

5.1.9 Examples. 1. If $G := \mathbb{R}^d$ with $d \geq 1$ and if for every $i = 1, \ldots, d$ a function $\zeta_i \in \mathscr{C}^b(\mathbb{R})$ is given with the properties $\zeta_i(t) := t$ for all $t \in U \in \mathfrak{V}(0)$ and

$$\zeta_i(-t) = -\zeta_i(t) \qquad \text{for all } t \in \mathbb{R},$$

then the function $g: \mathbb{R}^d \times \mathbb{R}^d \to \mathbb{R}$ defined by

$$g(x, y) := \sum_{i=1}^d \zeta_i(x_i)\, y_i$$

for all $x := (x_1, \ldots, x_d)$ and $y := (y_1, \ldots, y_d)$ in $\mathbb{R}^d$ is a local inner product for G.

2. Let $G := \mathbb{T}^d$ for $d \geq 1$. Then G can be viewed as the set

$$\{(x_1, \ldots, x_d) \in \mathbb{R}^d : x_i \in]-1, 1] \text{ for all } i = 1, \ldots, d\}$$

with addition modulo 2. Using for every $i = 1, \ldots, d$ the function ζ_i defined in Example 1 we obtain a local inner product g for G by means of putting

$$g(x, \mathfrak{m}) := \sum_{i=1}^d \zeta_i(x_i)\, m_i$$

for all $x := (x_1, \ldots, x_d) \in \mathbb{T}^d$ and $\mathfrak{m} := (m_1, \ldots, m_d) \in \mathbb{Z}^d$.

3. Let $G := \mathbb{Q}_d^\wedge$, $\zeta \in \mathscr{C}^b(G)$ and $\chi_0 \in G^\wedge\ (= \mathbb{Q}_d)$ with $\chi_0 \neq 1$ be such that

$$\exp(i\, \zeta(x)) = \chi_0(x) \qquad \text{for all } x \in \mathfrak{B}(1).$$

Then the mapping $g: G \times G^{\wedge} \to \mathbb{R}$ defined by

$$g(x, \chi) := \zeta(x) \frac{\chi}{\chi_0}$$

for all $x \in G$, $\chi \in G^{\wedge}$ is a local inner product for G.

4. If G is totally disconnected, then the zero function on $G \times G^{\wedge}$ is a local inner product for G since every homomorphism from $G^{\wedge}$ into $\mathbb{R}$ is trivial.

5.1.10 Theorem. *There exists a local inner product for G.*

Proof. The argument is taken from the proof of Theorem 4.3.11.

1. If $G = G_1 \times G_2$ is a product of two Abelian groups such that for G_1 and G_2 local inner products g_1 and g_2 resp. exist, then there also exists a local inner product g for G. One simply defines g by

$$g(x, \chi) := g_1(x_1, \chi_1) + g_2(x_2, \chi_2)$$

for all $x := (x_1, x_2) \in G \times G$ and $\chi := (\chi_1, \chi_2) \in G_1^{\wedge} \times G_2^{\wedge}$.

2. By the structure theorem M for Abelian groups we have $G \cong \mathbb{R}^n \times H$, where $n \geq 0$ and H admits a compact open subgroup K. On the basis of Part 1 we may assume that $G = H$ with a compact open normal subgroup K of H (such that G/K is discrete).

3. Let $\mathbf{M} := \{(G/K, g_K) : K$ a compact normal subgroup of G, g_K a local inner product for $G/K\}$. By Example 4 of 5.1.9 $\mathbf{M} \neq \emptyset$. For $(G/K_i, g_i) \in \mathbf{M}$ with $i = 1,2$ we introduce an order relation $<$ in $\mathbf{M}$ by the definition

$$(G/K_1, g_1) < (G/K_2, g_2) \quad \text{if } K_2 \subset K_1$$

and

$$[p_{12}(g_1)](p_{12}(x_2), \chi_1) = g_2(x_2, \chi_1 \circ p_{12})$$

for all $x_2 \in G/K_2$, $\chi_1 \in (G/K_1)^{\wedge}$, where p_{12} denotes the canonical mapping from G/K_2 onto G/K_1.

$\mathbf{M}$ is inductively ordered by $<$ as one observes from the proof of Theorem 4.3.11. Thus there exist maximal elements in $\mathbf{M}$ by Zorn's lemma.

4. Let $(G/K, g_K) \in \mathbf{M}$ with $K \neq \{e\}$. Then there exists a compact normal subgroup N of G such that $N \subset K$, $N \neq K$ and K/N is a Lie group. We show that for G/N there exists a local inner product g_N such that $(G/K, g_K) < (G/N, g_N)$ holds in the sense of Part 3.

In fact, we may assume that $N = \{e\}$, and hence that K is a Lie group.

(α) There is a local inner product g_{K_0} for G/K_0 with $(G/K, g_K) < (G/K_0, g_{K_0})$. Since K/K_0 is finite, we have

$$\frac{G/K_0}{K/K_0} \cong G/K =: H.$$

Without loss of generality $K_0 = \{e\}$, whence K is finite. Clearly, $H^{\wedge} = K^{\perp}$ has finite index in $G^{\wedge}$. We put $n := [G^{\wedge} : H^{\wedge}]$ and conclude that $\chi^n \in H^{\wedge}$ for all $\chi \in G^{\wedge}$. Let p denote the canonical mapping from G onto H and $g \in \mathscr{C}(G \times G^{\wedge})$ be defined by

$$g(x,\chi):=\frac{1}{n}g_K(p(x),\chi^n)$$

for all $x\in G$, $\chi\in G^\wedge$. Then g is a local inner product for G. In fact, (LI1), (LI2) and (LI4) are clearly satisfied. In order to obtain the validity of (LI3) one uses the fact that G and G/K are locally isomorphic and chooses the neighborhood $U\in\mathfrak{V}(e)$ appearing in (LI3) appropriately small.

(β) Since K_0 is a compact, connected Abelian Lie group, by Theorem C it is of the form $\mathbb{T}^n$ for some $n\in\mathbb{Z}_+$. By [218], (25.31)(b) the group K_0 is a topological direct factor of G, i.e.,

$$G\cong(G/K_0)\times K_0\cong(G/K_0)\times\mathbb{T}^n.$$

Since we have established the existence of a local inner product for G/K_0 in (α) and for $\mathbb{T}$ in Example 2 of 5.1.9, Part 1 of this proof yields the existence of a local inner product g for G such that

$$(G,g):=((G/K_0)\times K_0,g)>(G/K_0,g_{K_0}).\quad \square$$

We are now ready to treat the limits of infinitesimal triangular systems of measures introduced in the following

5.1.11 Definition. A triangular system $(\mu_{nj})_{j=1,\ldots,k_n;n\geq 1}$ of measures in $\mathscr{M}^1(G)$ is called *infinitesimal*, if for each $U\in\mathfrak{V}(e)$ one has

$$\lim_{n\to\infty}\sup_{1\leq j\leq k_n}\mu_{nj}(\complement U)=0.$$

5.1.12. By the properties of the Fourier transform one obtains that $(\mu_{nj})_{j=1,\ldots,k_n;n\geq 1}$ is infinitesimal iff

$$\lim_{n\to\infty}\sup_{1\leq j\leq k_n}|\hat{\mu}_{nj}-1|=0$$

with respect to the topology $\mathscr{T}_{co}$ in $G^\wedge$.

5.1.13 Lemma. *For each compact $K\subset G^\wedge$ there exist $U:=U_K\in\mathfrak{V}(e)$ and a finite subset $F:=F_K$ of $G^\wedge$ with the property*

$$\sup_{\chi\in K}(1-\operatorname{Re}\chi(x))\leq c_K\sup_{\chi\in F}(1-\operatorname{Re}\chi(x))$$

for all $x\in U$, where c_K is a constant in $\mathbb{R}_+^*$ depending only on K.

Proof. One establishes the result in analogy to the preceding theorem via structure theorem M. Since $G\cong\mathbb{R}^n\times H$, where $n\geq 0$ and H admits a compact open normal subgroup, it suffices to prove the statement of the lemma for $\mathbb{R}$, $\mathbb{T}$ and any discrete Abelian group and to show that if the statement is true for Abelian groups G_1 and G_2, then it is also true for $G_1\times G_2$. While the special cases are treated without difficulty, the assertion for the product $G_1\times G_2$ follows from the inequality

$$1-\operatorname{Re}\chi(x_1x_2)\leq 2[(1-\operatorname{Re}\chi(x_1))+(1-\operatorname{Re}\chi(x_2))]$$

valid for all $x_1, x_2\in G$ and $\chi\in G^\wedge$. This inequality is itself a consequence of the inequality

$$1-\cos(\alpha+\beta)\leq 2[(1-\cos\alpha)+(1-\cos\beta)],$$

valid for all $\alpha,\beta\in[0,2\pi]$. □

5.1.14 Corollary. *Let g be a local inner product for G. For each compact $K\subset G^\wedge$ there exist a $U:=U_K\in\mathfrak{V}(e)$ and a finite $F:=F_K\subset G^\wedge$ with the property*

$$\sup_{\chi\in K} g(x,\chi)^2\leq c_K \sup_{\chi\in F} g(x,\chi)^2$$

for all $x\in U$, where c_K is a constant in $\mathbb{R}_+^*$ depending only on K.

The *proof* follows immediately from the statement of the lemma if one notes that by (LI4) and the inequality $1-\cos\xi>\frac{1}{4}\xi^2$, valid for all $\xi\in\mathbb{R}$ sufficiently close to 0, there exists a $V:=V_K\in\mathfrak{V}(e)$ with $V\subset U$ such that

$$\tfrac{1}{4}g(x,\chi)^2\leq 1-\operatorname{Re}\chi(x)\leq\tfrac{1}{2}g(x,\chi)^2$$

for all $x\in V$, $\chi\in K\cup F$. □

5.1.15 Lemma. *Let $(\mu_{nj})_{j=1,\ldots,k_n;n\geq 1}$ be an infinitesimal triangular system in $\mathscr{M}^1(G)$, put*

$$\mu_n:=\ast_{j=1}^{k_n}\mu_{nj}$$

for all $n\geq 1$ and let μ be an accumulation point of a shift of the sequence $(\mu_n)_{n\geq 1}$. Then
(i) $H:=\{\chi\in G^\wedge:\hat{\mu}(\chi)\neq 0\}$ is an open subgroup of $G^\wedge$ and
(ii) $\omega_{H^\perp}$ is an idempotent factor of μ.

Proof. It suffices to prove (i). By assumption there exists a subsequence $(\mu_{n_i})_{i\geq 1}$ of $(\mu_n)_{n\geq 1}$ with the property $\lim_{i\to\infty}|\mu_{n_i}|^2=|\mu|^2$, and therefore

$$\lim_{i\to\infty}\prod_{j=1}^{k_{n_i}}|\hat{\mu}_{n_ij}|^2=|\hat{\mu}|^2$$

with respect to the topology $\mathscr{T}_{co}$ in $G^\wedge$.

For simplicity we denote the subsequence $(\mu_{n_i})_{i\geq 1}$ again by $(\mu_n)_{n\geq 1}$. Thus we have in the modified notation

$$\lim_{n\to\infty}\prod_{j=1}^{k_n}|\hat{\mu}_{nj}|^2=|\hat{\mu}|^2.$$

Therefore $\hat{\mu}(\chi)\neq 0$ for all $\chi\in G^\wedge$ iff

$$\sup_{n\geq 1}\sum_{j=1}^{k_n}(1-|\hat{\mu}_{nj}(\chi)|^2)<\infty.$$

Since for every $\nu\in\mathcal{M}^1(G)$ and $\chi,\eta\in G^\wedge$ one has

$$1-\hat{\nu}(\chi\eta)\leq 2[(1-\hat{\nu}(\chi))+(1-\hat{\nu}(\eta))]$$

(as in the proof of Lemma 5.1.13) we obtain for $\chi,\eta\in G^\wedge$ the inequality

$$\sum_{j=1}^{k_n}(1-|\hat{\mu}_{nj}(\chi\eta)|^2)\leq 2[\sum_{j=1}^{k_n}(1-|\hat{\mu}_{nj}(\chi)|^2)+\sum_{j=1}^{k_n}(1-|\hat{\mu}_{nj}(\eta)|^2)],$$

whence $\hat{\mu}(\chi\eta)\neq 0$ if $\hat{\mu}(\chi)\neq 0$ and $\hat{\mu}(\eta)\neq 0$. Consequently, H is a subgroup of $G^\wedge$. The continuity of the Fourier transform implies that H is open. □

5.1.16 Theorem. *Let g be a local inner product for G. For an infinitesimal triangular system $(\mu_{nj})_{j=1,\ldots,k_n;n\geq 1}$ in $\mathcal{M}^1(G)$ we define a triangular system $(x_{nj})_{j=1,\ldots,k_n;n\geq 1}$ in G by*

$$\chi(x_{nj}):=\exp(-i\textstyle\int g(x,\chi)\,\mu_{nj}(dx))$$

for all $\chi\in G^\wedge$ and a triangular system $(\nu_{nj})_{j=1,\ldots,k_n;n\geq 1}$ in $\mathcal{M}^1(G)$ by

$$\nu_{nj}:=\exp(\mu_{nj}*\varepsilon_{x_{nj}})\qquad \textit{for all } j=1,\ldots,k_n;\ n\geq 1.$$

Furthermore, for every $n\geq 1$ we put

$$\mu_n:=\ast_{j=1}^{k_n}\mu_{nj},\qquad x_n:=(\textstyle\prod_{j=1}^{k_n}x_{nj})^{-1}\quad \textit{and}\quad \lambda_n:=(\ast_{j=1}^{k_n}\nu_{nj})*\varepsilon_{x_n}.$$

Assume also for one of the sequences $(\lambda_n)_{n\geq 1}$ or $(\mu_n)_{n\geq 1}$ that
(i) *it is shift compact and that*
(ii) *no accumulation point of its shifts admits an idempotent factor.*

Then $\lim_{n\to\infty}(\hat{\lambda}_n-\hat{\mu}_n)=0$ with respect to $\mathcal{T}_{co}$.

Proof. 1. We first observe that the sequence $(x_{nj})_{j=1,\ldots,k_n;n\geq 1}$ in G is well-defined by Property (LI2) of the given local inner product g for G. Furthermore, we note that for any $U\in\mathfrak{B}(e)$ there is an $n_0\geq 1$ such that $x_{nj}\in U$ for all $j=1,\ldots,k_n$; $n\geq n_0$. Thus the infinitesimality of $(\mu_{nj})_{j=1,\ldots,k_n;n\geq 1}$ implies the infinitesimality of $(\nu_{nj})_{j=1,\ldots,k_n;n\geq 1}$. From this we conclude that given a compact subset K of $G^\wedge$ there exists an $n_0\geq 1$ such that for all $n\geq n_0$ one has $\hat{\lambda}_n(\chi)\neq 0$ and $\hat{\mu}_n(\chi)\neq 0$ for all $\chi\in K$.

2. In the following steps of the proof we assume the conditions (i) and (ii) of the theorem to be valid for the sequence $(\mu_n)_{n\geq 1}$. Since $(\mu_n)_{n\geq 1}$ is shift compact and no accumulation point of shifts of $(\mu_n)_{n\geq 1}$ admits an idempotent factor, Lemma 5.1.15 yields $|\hat{\mu}_n(\chi)|\geq\alpha>0$ for all $\chi\in K$ and $n\geq n_0$. Hence, it suffices to prove that

$$\lim_{n\to\infty}\sup_{\chi\in K}|\log\hat{\lambda}_n(\chi)-\log\hat{\mu}_n(\chi)|=0.$$

3. A further reduction of the statement to be proved appears as follows: For every $n\geq n_0$ and $\chi\in K$ we have

$$\begin{aligned}\log\hat{\lambda}_n(\chi)&=\textstyle\sum_{j=1}^{k_n}\log\hat{\nu}_{nj}(\chi)-\sum_{j=1}^{k_n}\log\chi(x_{nj})\\&=\textstyle\sum_{j=1}^{k_n}\log\hat{\nu}_{nj}(\chi)+i\sum_{j=1}^{k_n}\int g(x,\chi)\,\mu_{nj}(dx)\\&=\textstyle\sum_{j=1}^{k_n}[(\mu_{nj}*\varepsilon_{x_{nj}})^\wedge(\chi)-1]+i\sum_{j=1}^{k_n}\int g(x,\chi)\,\mu_{nj}(dx)\end{aligned}$$

and

$$\log \hat{\mu}_n(\chi) = \sum_{j=1}^{k_n} \log \hat{\mu}_{nj}(\chi).$$

Introducing measures $\sigma_{nj} := \mu_{nj} * \varepsilon_{x_{nj}}$ for $j=1,\ldots,k_n$; $n \geq 1$ and observing the validity of the inequality

$$|\log(1-z)+z| \leq |z|^2 \quad \text{for all } |z| \leq \tfrac{1}{2}$$

one obtains for all $n \geq n_0$ and $\chi \in K$

$$\begin{aligned}
&|\log \hat{\lambda}_n(\chi) - \log \hat{\mu}_n(\chi)| \\
&= |\textstyle\sum_{j=1}^{k_n} (\hat{\sigma}_{nj}(\chi)-1) + i \sum_{j=1}^{k_n} \int g(x,\chi)\, \mu_{nj}(dx) \\
&\quad - \textstyle\sum_{j=1}^{k_n} \log \hat{\sigma}_{nj}(\chi) + \sum_{j=1}^{k_n} \log \chi(x_{nj})| \\
&= |\textstyle\sum_{j=1}^{k_n} (\hat{\sigma}_{nj}(\chi)-1) - \sum_{j=1}^{k_n} \log \hat{\sigma}_{nj}(\chi)| \\
&\leq [\textstyle\sum_{j=1}^{k_n} |1-\hat{\sigma}_{nj}(\chi)|] \sup_{1 \leq j \leq k_n} |1-\hat{\sigma}_{nj}(\chi)|.
\end{aligned}$$

Since $(\sigma_{nj})_{j=1,\ldots,k_n; n \geq 1}$ is infinitesimal, it thus suffices to prove that

$$\sup_{n \geq 1} \sup_{\chi \in K} \sum_{j=1}^{k_n} |1-\hat{\sigma}_{nj}(\chi)| < \infty.$$

4. For every $U \in \mathfrak{B}(e)$, $\chi \in G^\wedge$ and $j=1,\ldots,k_n$; $n \geq 1$ we have

$$\begin{aligned}
|1-\hat{\sigma}_{nj}(\chi)| &\leq |\textstyle\int_U (1-\chi(x))\, \sigma_{nj}(dx)| + |\int_{\complement U} (1-\chi(x))\, \sigma_{nj}(dx)| \\
&\leq |\textstyle\int_U (1-\chi(x))\, \sigma_{nj}(dx)| + 2\sigma_{nj}(\complement U).
\end{aligned}$$

By Property (LI3) of the local inner product g for G there exists for any compact $K \subset G^\wedge$ a neighborhood $U := U_K \in \mathfrak{B}(e)$ such that

$$\chi(x) = \exp i\, g(x,\chi) \quad \text{for all } x \in U,\ \chi \in K.$$

For all $x \in U$, $\chi \in K$ we also have

$$|1-\chi(x)+i\, g(x,\chi)| \leq c^{(1)} g^2(x,\chi),$$

where $c^{(1)} := c^{(1)}(K)$ is a constant in $\mathbb{R}_+^*$. We therefore obtain for all $\chi \in K$, $n \geq n_0$

$$\begin{aligned}
&|1-\hat{\sigma}_{nj}(\chi)| \\
&\leq |\textstyle\int_U g(x,\chi)\, \sigma_{nj}(dx)| + c^{(1)} \int_U g^2(x,\chi)\, \sigma_{nj}(dx) + 2\sigma_{nj}(\complement U).
\end{aligned} \tag{1.1}$$

Property (LI1) of g implies the existence of a $c^{(2)} := c^{(2)}(K) \in \mathbb{R}_+^*$ such that

$$\begin{aligned}
|\textstyle\int g(x,\chi)\, \sigma_{nj}(dx)| &= |\textstyle\int g(x x_{nj},\chi)\, \mu_{nj}(dx)| \\
&\leq |\textstyle\int_U g(x x_{nj},\chi)\, \mu_{nj}(dx)| + c^{(2)} \mu_{nj}(\complement U)
\end{aligned} \tag{1.2}$$

holds for all $\chi \in K$, $j=1,\ldots,k_n$; $n \geq n_0$.

Since $x_{nj} \in U$ for all $j=1,\ldots,k_n$; $n \geq n_0$ and

$$\chi(x) = \exp i\, g(x,\chi) \quad \text{for all } x \in U,\ \chi \in K,$$

Property (LI4) of g implies $g(x\,x_{nj}, \chi)=g(x, \chi)+g(x_{nj}, \chi)$ for all $x\in U$, $\chi\in K$ and $n\geq n_0$. Furthermore,

$$\exp i\,g(x_{nj}, \chi)=\chi(x_{nj})=\exp(-i\int g(x, \chi)\,\mu_{nj}(dx))$$

for all $\chi\in K$, $j=1, \ldots, k_n$; $n\geq n_0$. Again by Property (LI4) of g we obtain

$$g(x_{nj}, \chi)=-\int g(x\ \chi)\,\mu_{nj}(dx)$$

for all $\chi\in K$, $j=1, \ldots, k_n$; $n\geq n_0$.

Thus there exists a constant $c^{(3)}:=c^{(3)}(K)\in\mathbb{R}_+^*$ such that the following relations hold for all $\chi\in K$, $j=1, \ldots, k_n$; $n\geq n_0$:

$$\begin{aligned}&|\int_U g(x\,x_{nj}, \chi)\,\mu_{nj}(dx)|=|\int_U (g(x, \chi)+g(x_{nj}, \chi))\,\mu_{nj}(dx)|\\&=|\int_U g(x, \chi)\,\mu_{nj}(dx)-\mu_{nj}(U)\int g(x, \chi)\,\mu_{nj}(dx)|\\&=|\mu_{nj}(\complement\, U)\int_U g(x, \chi)\,\mu_{nj}(dx)-\mu_{nj}(U)\int_{\complement U} g(x, \chi)\,\mu_{nj}(dx)|\\&\leq c^{(3)}\,\mu_{nj}(\complement\, U).\end{aligned}$$

This together with (1.2) yields

$$|\int g(x, \chi)\,\sigma_{nj}(dx)|\leq c^{(4)}\,\mu_{nj}(\complement\, U)$$

for all $\chi\in K$, $j=1, \ldots, k_n$; $n\geq n_0$, where $c^{(4)}:=c^{(4)}(K)$ is a constant in $\mathbb{R}_+^*$. Thus (1.1) and Property (LI1) of g imply

$$|1-\hat{\sigma}_{nj}(\chi)|\leq c^{(1)}\int g^2(x, \chi)\,\sigma_{nj}(dx)+c^{(5)}\,\sigma_{nj}(\complement\, U)+c^{(6)}\,\mu_{nj}(\complement\, U)$$

for all $\chi\in K$, $j=1, \ldots, k_n$; $n\geq n_0$, where $c^{(5)}:=c^{(5)}(K)$ and $c^{(6)}:=c^{(6)}(K)$ are constants in $\mathbb{R}_+^*$.

It therefore suffices to show the following three finiteness conditions:

(α) $\overline{\lim}_{n\geq 1}\sum_{j=1}^{k_n}\sigma_{nj}(\complement\, U)<\infty$,

(β) $\overline{\lim}_{n\geq 1}\sum_{j=1}^{k_n}\mu_{nj}(\complement\, U)<\infty$ and

(γ) $\overline{\lim}_{n\geq 1}\sup_{\chi\in K}\sum_{j=1}^{k_n} g^2(x, \chi)\,\sigma_{nj}(dx)<\infty$.

5. We shall first verify Condition (β) of Part 4. Since $(|\mu_n|^2)_{n\geq 1}$ is a relatively compact sequence such that none of its accumulation points admits an idempotent factor, we obtain by Lemma 5.1.15 $|\hat{\mu}_n(\chi)|^2\geq\alpha>0$ for all $\chi\in K$ and $n\geq n_0$. Hence,

$$\overline{\lim}_{n\geq 1}\sup_{\chi\in K}\sum_{j=1}^{k_n}(1-|\hat{\mu}_{nj}(\chi)|^2)<\infty.$$

By the reduction step 3 we get

$$\lim_{n\to\infty}\sup_{\chi\in K}|\exp[\sum_{j=1}^{k_n}(|\hat{\mu}_{nj}(\chi)|^2-1)]-|\hat{\mu}_n(\chi)|^2|=0.$$

Therefore the sequence $(\exp(\sum_{j=1}^{k_n}|\mu_{nj}|^2))_{n\geq 1}$ is relatively compact.

Theorem 5.1.6 implies that for every $U\in\mathfrak{B}(e)$

(1.3) $\qquad \overline{\lim}_{n\geq 1}\sum_{j=1}^{k_n}|\mu_{nj}|^2(\complement U)<\infty$ and

(1.4) $\qquad \overline{\lim}_{n\geq 1}\sum_{j=1}^{k_n}\int(1-\operatorname{Re}\chi(x))\,|\mu_{nj}|^2(dx)<\infty \qquad$ for all $\chi\in G^\wedge$

holds. We now choose $V\in\mathfrak{B}(e)$ with $V^2\subset U$. Then for every $n\geq 1$ we have

$$\begin{aligned}&\sum_{j=1}^{k_n}\mu_{nj}(\complement U)\leq\sum_{j=1}^{k_n}\mu_{nj}(\complement V^2)\leq\sum_{j=1}^{k_n}\inf_{x\in V}\mu_{nj}(\complement(Vx))\\ &=\sum_{j=1}^{k_n}\inf_{x\in V}\mu_{nj}((\complement V)\,x)\\ &\leq\sum_{j=1}^{k_n}\mu_{nj}(V)^{-1}\int_V\mu_{nj}((\complement V)\,x)\,\mu_{nj}(dx)\\ &\leq\sum_{j=1}^{k_n}\mu_{nj}(V)^{-1}\int_G\mu_{nj}((\complement V)\,x)\,\mu_{nj}(dx)\\ &\leq\sup_{1\leq j\leq k_n}\mu_{nj}(V)^{-1}\sum_{j=1}^{k_n}|\mu_{nj}|^2(\complement V).\end{aligned}$$

The infinitesimality of $(\mu_{nj})_{j=1,\ldots,k_n;\,n\geq 1}$ together with (1.3) yields (β).

6. Since $|\mu_{nj}|^2=|\sigma_{nj}|^2$ for all $j=1,\ldots,k_n$; $n\geq 1$ and $(\sigma_{nj})_{j=1,\ldots,k_n;\,n\geq 1}$ is infinitesimal, an analogous conclusion leads to (α).

7. We now show (γ). From (1.4) in Part 5 and Lemma 5.1.13 follows

$$\overline{\lim}_{n\geq 1}\sup_{\chi\in K}\sum_{j=1}^{k_n}\int(1-\operatorname{Re}\chi(x))\,|\sigma_{nj}|^2\,(dx)<\infty.$$

Together with (1.3) of Part 5 this implies

(1.5) $\qquad \overline{\lim}_{n\geq 1}\sup_{\chi\in K}\sum_{j=1}^{k_n}\int_V\int_V(1-\operatorname{Re}\chi(x_1x_2^{-1}))\,\sigma_{nj}(dx_1)\,\sigma_{nj}(dx_2)<\infty$

for any $V\in\mathfrak{B}(e)$ with $V^2\subset U$. Using the argument in the proof of Corollary 5.1.14 one gets for all $\chi\in K$, $x_1,x_2\in V$

$$1-\operatorname{Re}\chi(x_1x_2^{-1})\geq\tfrac{1}{4}(g^2(x_1,\chi)+g^2(x_2,\chi)-2g(x_1,\chi)\,g(x_2,\chi))$$

and thus from the inequality (1.5)

$$\overline{\lim}_{n\geq 1}\sup_{\chi\in K}\sum_{j=1}^{k_n}\left[\int_V g^2(x,\chi)\,\sigma_{nj}(dx)-\left(\int_V g(x,\chi)\,\sigma_{nj}(dx)\right)^2\right]<\infty,$$

which by the results of Part 5 implies the assertion.

8. Let $(\lambda_n)_{n\geq 1}$ satisfy the conditions (i) and (ii) of the theorem. Then by Theorem 5.1.6 and Lemma 5.1.13 we obtain for every $U\in\mathfrak{B}(e)$ and every compact $K\subset G^\wedge$

(1.6) $\qquad \sum_{j=1}^{k_n}\sigma_{nj}(\complement U)<\infty$ and

(1.7) $\qquad \sup_{n\geq 1}\sup_{\chi\in K}\sum_{j=1}^{k_n}\int(1-\operatorname{Re}\chi(x))\,\sigma_{nj}(dx)<\infty.$

Let $U\in\mathfrak{B}(e)$ be fixed and let $V\in\mathfrak{B}(e)$ be chosen such that $VV^{-1}\subset U$ holds. For all $n\geq n_0$ we have $\complement U\subset(\complement V)\,x_{nj}^{-1}$ $(j=1,\ldots,k_n)$, whence

$$\sum_{j=1}^{k_n}\mu_{nj}(\complement U)\leq\sum_{j=1}^{k_n}\mu_{nj}((\complement V)\,x_{nj}^{-1}),\quad\text{or}$$

$$\sup_{n\geq 1}\sum_{j=1}^{k_n}\mu_{nj}(\complement U)<\infty.$$

Using again the argument in the proof of Corollary 5.1.14 we obtain for any compact $K \subset G^\wedge$ a neighborhood $U := U_K \in \mathfrak{V}(e)$ such that

$$\overline{\lim}_{n\geq 1} \sup_{\chi\in K} \sum_{j=1}^{k_n} \int_U g^2(x,\chi)\, \sigma_{nj}(dx) < \infty$$

holds. This, however, implies the conditions (α), (β) and (γ) of Part 4 and thus the reduced assertion in Part 3 which yields the end of the proof. □

5.1.17 Theorem. *Let $(\mu_{nj})_{j=1,\ldots,k_n;\, n\geq 1}$ be an infinitesimal triangular system in $\mathcal{M}^1(G)$ and $(\mu_n)_{n\geq 1}$ the corresponding sequence of measures*

$$\mu_n := \ast_{j=1}^{k_n} \mu_{nj}$$

(for $n\geq 1$) such that $\mu := \lim_{n\to\infty} \mu_n$ exists. Then $\mu \in \mathscr{I}_0(G)$.

Proof. 1. If $\mu := \lim_{n\to\infty} \mu_n$ admits no idempotent factor, then $\mu = \lim_{n\to\infty} \lambda_n$, where for every $n\geq 1$ the measure $\lambda_n \in \mathcal{M}^1(G)$ has been defined in the statement of Theorem 5.1.16. Application of Theorem 5.1.4 yields $\mu \in \mathscr{I}_0(G)$.

2. Let $\mu := \lim_{n\to\infty} \mu_n$ be an arbitrary measure in $\mathcal{M}^1(G)$. By Lemma 5.1.15 the set $H := \{\chi \in G^\wedge : \hat{\mu}(\chi) \neq 0\}$ is an open subgroup of $G^\wedge$ and $\omega_{H^\perp}$ is an idempotent factor of μ. Denoting the canonical mapping from G onto $G/H^\perp$ by p we conclude that the triangular system $(p(\mu_{nj}))_{j=1,\ldots,k_n;\, n\geq 1}$ in $\mathcal{M}^1(G/H^\perp)$ is infinitesimal and $\lim_{n\to\infty} p(\mu_n) = p(\mu)$. Now $p(\mu)$ admits no idempotent factor, so by Part 1 $p(\mu) \in \mathscr{I}_0(G/H^\perp)$. Thus for every $n\geq 1$ there exist a $\nu_n \in \mathcal{M}^1(G/H^\perp)$ and a $z_n \in G/H^\perp$ such that $p(\mu) = \nu_n^n * \varepsilon_{z_n}$ holds. But then there exists for every $n\geq 1$ a measure $\bar{\nu}_n \in \mathcal{M}^1(G)$ with the properties

$$p(\bar{\nu}_n) = \nu_n \quad \text{and} \quad \hat{\bar{\nu}}_n(\chi) = \begin{cases} \hat{\nu}_n(\chi) & \text{if } \chi \in H \\ 0 & \text{if } \chi \notin H. \end{cases}$$

Choosing for every $n\geq 1$ an element x_n in the H-coset of z_n one concludes that $\mu = \bar{\nu}_n^n * \varepsilon_{x_n}$, and thus $\mu \in \mathscr{I}_0(G)$. □

5.2 Gauss Measures in the Sense of Parthasarathy

Of major interest for the solution of the central limit problem and its ramifications is the convergence of infinitesimal families in $\mathcal{M}^1(G)$ to Gauss measures.

We start with a review of the notion of *normal distribution* for the groups $\mathbb{R}$ and $\mathbb{T}$.

(α) $\mu \in \mathcal{M}^1(\mathbb{R})$ is called a normal distribution if

$$\mu = \nu_{a,\sigma^2} = n_{a,\sigma^2} \cdot \lambda^1$$

for some $a \in \mathbb{R}$ and $\sigma^2 \in \mathbb{R}_+^*$, where n_{a,σ^2} is defined by

$$n_{a,\sigma^2}(x) := \frac{1}{\sqrt{2\pi}\,\sigma} e^{-\frac{1}{2\sigma^2}(x-a)^2}$$

for all $x \in \mathbb{R}$. In the language of the Fourier transform this normal distribution can be characterized by $\hat{\mu}(x) = e^{iax - \frac{1}{2}\sigma^2 x^2}$ for all $x \in \mathbb{R}$.

(β) Let τ be the continuous homomorphism $t \to e^{it}$ from $\mathbb{R}$ into the torus $\mathbb{T}$. $\mu \in \mathscr{M}^1(\mathbb{T})$ is called a normal distribution if $\mu = \tau(\nu_{a,\sigma^2})$ for some $a \in \mathbb{R}$ and $\sigma^2 \in \mathbb{R}_+^*$. One computes its measure $\mu(B)$ of a set $B \in \mathfrak{B}(\mathbb{T})$ as

$$\mu(B) = \int_{\arg B} \sum_{n \in \mathbb{Z}} \frac{1}{\sqrt{2\pi}\,\sigma} e^{-\frac{1}{2\sigma^2}(u-a+2n\pi)^2} du$$

and notes that μ is normal iff $\hat{\mu}(k) = e^{iak - \frac{1}{2}\sigma^2 k^2}$ for all $k \in \mathbb{Z}$.

In the more general case $d \geq 1$ one introduces *d-dimensional normal distributions* on $\mathbb{R}^d$ (as affine linear images of $\nu_0 := \underbrace{\nu_{0,1} \otimes \cdots \otimes \nu_{0,1}}_{d\text{-times}}$) and on $\mathbb{T}^d$ (preferably via Fourier transforms).

For every $d \geq 1$ the collection of normal distributions on $\mathbb{R}^d$ and $\mathbb{T}^d$ will be abbreviated by $\mathscr{N}(\mathbb{R}^d)$ and $\mathscr{N}(\mathbb{T}^d)$ resp. It is known that normal distributions on $\mathbb{R}^d$ and $\mathbb{T}^d$ do not admit Poisson factors. This motivates the following

5.2.1 Definition. Let G be a locally compact Abelian group. A measure $\mu \in \mathscr{M}^1(G)$ is called a *Gauss measure in the sense of Parthasarathy* (P-Gauss), if
(i) $\mu \in \mathscr{I}_0(G)$ and
(ii) for any factorization of μ of the form $\mu = \exp(\tau) * \lambda$ with $\tau \in \mathscr{M}_+^b(G)$ and $\lambda \in \mathscr{I}_0(G)$ one has $\tau = a\varepsilon_e$ for some $a \in \mathbb{R}_+$.

The class of P-Gauss measures in $\mathscr{M}^1(G)$ will be abbreviated by $\mathscr{G}_P(G)$ and the subclass of its symmetric elements by $\mathscr{G}_P^s(G)$.

5.2.2 Remark. Elements of $\mathscr{G}_P(G)$ do not admit idempotent factors. In fact, let $\mu \in \mathscr{G}_P(G)$ admit an idempotent factor σ of the form ω_H for some compact subgroup $H \neq \{e\}$ of G.

By assumption $\mu = \sigma * \rho$ with $\rho \in \mathscr{M}^1(G)$ and therefore

$$\mu = \omega_H * \rho = \omega_H * \omega_H * \rho = \omega_H * \mu.$$

Thus $\mu = \omega_H^n * \mu$ for every $n \geq 1$. This implies

$$\exp(\omega_H) * \mu = \left(\sum_{k \geq 0} \frac{\omega_H^k}{k!}\right) * \mu = \sum_{k \geq 0} \frac{\omega_H^k * \mu}{k!} = \mu,$$

which contradicts the hypothesis.

The characterization via Fourier transforms of P-Gauss measures requires some preparation concerning quadratic forms on Abelian groups.

5.2.3. A *continuous* mapping ψ from G into $\mathbb{R}$ is called a *quadratic form* on G if for all $x, y \in G$ we have

$$\psi(xy) + \psi(xy^{-1}) = 2(\psi(x) + \psi(y)),$$

and *positive* if $\psi(x) \geq 0$ for all $x \in G$.

The set of all quadratic forms on G will be denoted by $\mathsf{Q}(G)$, and the subset of its positive elements by $\mathsf{Q}_+(G)$. We further recall that a *continuous* mapping Ψ from $G \times G$ into $\mathbb{R}$ is said to be a (symmetric) *bilinear form on* G if for all $x, y, z \in G$ one has
(i) $\Psi(x, z) = \Psi(z, x)$ and
(ii) $\Psi(xy, z) = \Psi(x, z) + \Psi(y, z)$,
and *positive* if $\Psi(x, x) \geq 0$ for all $x \in G$.

The collection $\mathsf{B}(G)$ of all symmetric bilinear forms on G is a linear subspace of $\mathscr{C}(G \times G)$. The subset of positive elements of $\mathsf{B}(G)$ will be denoted by $\mathsf{B}_+(G)$.

5.2.4 Lemma. *There are one-to-one correspondences between the sets* $\mathsf{Q}(G)$ *and* $\mathsf{B}(G)$ *and between the sets* $\mathsf{Q}_+(G)$ *and* $\mathsf{B}_+(G)$.

Proof. 1. Given $\Psi \in \mathsf{B}(G)$ we define a function ψ on G by $\psi(x) := \Psi(x, x)$ for all $x \in G$. Ψ is uniquely determined by ψ. In fact, for all $x, y \in G$ we obtain

(2.1) $$\psi(xy) = \Psi(xy, xy) = \psi(x) + 2\Psi(x, y) + \psi(y), \quad \text{whence}$$

(2.2) $$\Psi(x, y) = \tfrac{1}{2}(\psi(xy) - \psi(x) - \psi(y)).$$

This equation shows that different bilinear forms Ψ on G yield different functions ψ on G.

Replacing $y \in G$ by y^{-1} in (2.1) we find that

$$\psi(xy^{-1}) = \psi(x) - 2\Psi(x, y) + \psi(y).$$

Addition of the last two equations implies

$$\psi(xy) + \psi(xy^{-1}) = 2(\psi(x) + \psi(y)) \qquad \text{for all } x, y \in G$$

so that $\psi \in \mathsf{Q}(G)$.

2. For any $\psi \in \mathsf{Q}(G)$ one immediately obtains

$$\psi(e) = 0 \quad \text{and} \quad \psi(y^{-1}) = \psi(y) \qquad \text{for all } y \in G.$$

If there exists a bilinear form $\Psi \in \mathsf{B}(G)$ such that $\Psi(x, x) = \psi(x)$ holds for all $x \in G$, then Ψ defined by (2.2) is an element of $\mathsf{B}(G)$. We content ourselves with the proof of the second defining relationship. First of all we obtain for all $x, y, z \in G$ the formulae

$$2\Psi(xy, z) = \psi(xyz) - \psi(xy) - \psi(z),$$

$$2\Psi(x, z) = \psi(xz) - \psi(x) - \psi(z) \quad \text{and}$$

$$2\Psi(y, z) = \psi(yz) - \psi(y) - \psi(z), \quad \text{whence}$$

(2.3) $$\begin{aligned} &2(\Psi(xy, z) - \Psi(x, z) - \Psi(y, z)) \\ &= (\psi(xyz) + \psi(z)) - (\psi(xz) + \psi(yz)) - (\psi(xy) - \psi(x) - \psi(y)). \end{aligned}$$

Furthermore, since $\psi \in \mathsf{Q}(G)$, we find that

$$\psi(xyz) + \psi(z) = \tfrac{1}{2}(\psi(xyz^2) + \psi(xy)) \quad \text{and}$$
$$\psi(xz) + \psi(yz) = \tfrac{1}{2}(\psi(xyz^2) + \psi(xy^{-1})), \quad \text{where} \quad x, y, z \in G.$$

From this (subtracting these two equations and using the defining property of ψ) follows

$$(\psi(xyz)+\psi(z))-(\psi(xz)+\psi(yz))=\tfrac{1}{2}(\psi(xy)-\psi(xy^{-1}))$$
$$=-\psi(x)-\psi(y)+\psi(xy).$$

This equation together with (2.3) yields

$$\Psi(xy,z)=\Psi(x,z)+\Psi(y,z) \qquad \text{for all } x,y,z\in G$$

such that the one-to-one correspondence between $\mathsf{Q}(G)$ and $\mathsf{B}(G)$ has been established. The correspondence between $\mathsf{Q}_+(G)$ and $\mathsf{B}_+(G)$ is deduced without difficulty. □

5.2.5 Lemma. *Let H be an open subgroup of G and $\psi_0\in\mathsf{Q}_+(H)$. Then there exists a $\psi\in\mathsf{Q}_+(G)$ with $\psi_0=\operatorname{Res}_H\psi$.*

Proof. Let $\mathbf{H}$ denote the system of all open subgroups H' of G with $H'\supset H$, for which there exists a quadratic form $\psi_{H'}\in\mathsf{Q}_+(H')$ satisfying $\psi_0=\operatorname{Res}_H\psi_{H'}$. We consider the system

$$\mathbf{H}_{\mathsf{Q}}:=\{(H',\psi_{H'})\colon H'\in\mathbf{H},\ \psi_{H'}\in\mathsf{Q}_+(H') \text{ with } \psi_0=\operatorname{Res}_H\psi_{H'}\}$$

which is naturally ordered by the relation $(H_1,\psi_{H_1})<(H_2,\psi_{H_2})$ if

$$H_1\subset H_2 \quad\text{and}\quad \psi_{H_1}=\operatorname{Res}_{H_1}\psi_{H_2}$$

(for $(H_1,\psi_{H_1}),(H_2,\psi_{H_2})\in\mathbf{H}_{\mathsf{Q}}$).

Let $\{(H_i,\psi_{H_i})\colon i\in\mathbb{I}\}$ be a chain in $\mathbf{H}_{\mathsf{Q}}$. Putting $H^*:=\bigcup_{i\in\mathbb{I}}H_i$ and defining $\psi_{H^*}\in\mathsf{Q}_+(H^*)$ by $\psi_{H^*}(x):=\psi_{H_i}(x)$, if $x\in H_i$ $(i\in\mathbb{I})$ we observe that ψ_{H^*} is well-defined and continuous. Thus (H^*,ψ_{H^*}) is an upper bound of the family $\{(H_i,\psi_{H_i})\colon i\in\mathbb{I}\}$. Zorn's Lemma yields the existence of maximal elements in $\mathbf{H}_{\mathsf{Q}}$. Let $(\bar{H},\bar{\psi})\in\mathbf{H}_{\mathsf{Q}}$ be maximal. We shall show that $\bar{H}=G$. If $\bar{H}\neq G$, then there exists an $x_1\notin\bar{H}$. We prove that $\bar{\psi}$ can be extended to the open subgroup $H_1:=[x_1]\bar{H}$.

(α) Let $[x_1]\cap\bar{H}=\{e\}$. One defines a real-valued function ψ_1 on H_1 by

$$\psi_1(x):=\psi_1(x_1^m x_0):=\bar{\psi}(x_0)$$
$$\text{for all } x:=x_1^m x_0 \quad\text{with}\quad m\in\mathbb{Z} \quad\text{and}\quad x_0\in\bar{H}.$$

The function ψ_1 is a nonnegative extension of $\bar{\psi}$ and satisfies

$$\psi_1(xy)+\psi_1(xy^{-1})=2(\psi_1(x)+\psi_1(y)) \qquad \text{for all } x,y\in H_1.$$

(β) Let $[x_1]\cap\bar{H}\neq\{e\}$. In this case there is a smallest number $n_0\in\mathbb{N}$ with $x_1^{n_0}\in\bar{H}$. For arbitrary $x\in H_1$ we have $x^{n_0}\in\bar{H}$. In fact, if $x\in H_1$ admits the representation $x=x_1^m x_0$ with $m\in\mathbb{Z}$ and $x_0\in\bar{H}$, then $x^{n_0}=x_1^{mn_0}x_0^{n_0}$ implies $x^{n_0}\in\bar{H}$. One now defines a real function ψ_1 on H_1 by

$$\psi_1(x):=\frac{1}{n_0^2}\bar{\psi}(x^{n_0}) \qquad \text{for all } x\in H_1.$$

Since $\bar{\psi}(y^m)=m^2\bar{\psi}(y)$ for all $y\in\bar{H}$, $m\in\mathbb{Z}$, we obtain for all $x\in\bar{H}$ the equalities

$$\psi_1(x)=\frac{1}{n_0^2}\bar{\psi}(x^{n_0})=\frac{n_0^2}{n_0^2}\bar{\psi}(x)=\bar{\psi}(x),$$

i.e., ψ_1 is a nonnegative continuous extension of $\bar{\psi}$ to H_1. For $x, y\in H_1$ we further have

$$\begin{aligned}\psi_1(xy)+\psi_1(xy^{-1})&=\frac{1}{n_0^2}[\bar{\psi}((xy)^{n_0})+\bar{\psi}((xy^{-1})^{n_0})]\\&=\frac{1}{n_0^2}[\bar{\psi}(x^{n_0}y^{n_0})+\bar{\psi}(x^{n_0}y^{-n_0})]=\frac{2}{n_0^2}[\bar{\psi}(x^{n_0})+\bar{\psi}(y^{n_0})]\\&=2(\psi_1(x)+\psi_1(y)),\quad\text{whence}\quad\psi_1\in\mathrm{Q}_+(H_1).\end{aligned}$$

But the existence of the achieved extension contradicts the maximality of $(\bar{H},\bar{\psi})$. □

5.2.6 Application. Let $G:=\mathbb{R}^d$ and let $H:=\mathbb{R}^m\times\mathbb{Z}^n$ be a (closed) subgroup of G with $m+n=d\in\mathbb{N}$. For every $\psi_0\in\mathrm{Q}_+(H)$ there exists exactly one $\psi\in\mathrm{Q}_+(G)$ with $\mathrm{Res}_H\psi=\psi_0$.

Proof. By Lemma 5.2.5 the quadratic form $\psi_1:=\mathrm{Res}_{\mathbb{Z}^d}\psi_0\in\mathrm{Q}_+(\mathbb{Z}^d)$ can be extended to $\psi_2\in\mathrm{Q}_+(\mathbb{Q}_d^d)$. Let Ψ_2 be the bilinear form corresponding to ψ_2 as in Lemma 5.2.4.

One shows that $\Psi_2(rx, y)=r\Psi_2(x, y)$ for all $r\in\mathbb{Q}$ and $x, y\in\mathbb{Q}^d$, so that there exists a symmetric matrix $(a_{ij})_{i,j=1,\dots,d}\in\mathfrak{M}(d,\mathbb{R})$ with

$$\Psi_2(x, y)=\textstyle\sum_{i,j=1}^d a_{ij}x_iy_j$$
for $x:=(x_1,\dots,x_d)$, $y:=(y_1,\dots,y_d)\in\mathbb{Q}^d$.

Putting

$$\psi(x):=\textstyle\sum_{i,j=1}^d a_{ij}x_ix_j\qquad\text{for all } x:=(x_1,\dots,x_d)\in\mathbb{R}^d$$

we obtain a quadratic form ψ in $\mathrm{Q}_+(\mathbb{R}^d)$. Obviously $\mathrm{Res}_{\mathbb{Z}^d}\psi=\psi_1=\mathrm{Res}_{\mathbb{Z}^d}\psi_0$.

For $x\in\mathbb{Q}^m\times\mathbb{Z}^n$ there exists a $k\in\mathbb{Z}^*$ with $kx\in\mathbb{Z}^d$, and one gets

$$\psi(x)=\frac{1}{k^2}\psi_1(kx)=\psi_0(x),$$

whence $\mathrm{Res}_{\mathbb{Q}^m\times\mathbb{Z}^n}\psi=\mathrm{Res}_{\mathbb{Q}^m\times\mathbb{Z}^n}\psi_0$. The continuity of ψ and ψ_0 yields $\mathrm{Res}_H\psi=\psi_0$. □

5.2.7 Theorem. *For every $\mu\in\mathscr{G}_P(G)$ there exist unique elements $x_0\in G$ and $\phi\in\mathrm{Q}_+(G^\wedge)$ such that $\hat{\mu}(\chi)=\chi(x_0)\exp(-\phi(\chi))$ holds for all $\chi\in G^\wedge$.*

Proof. 1. For $\mu\in\mathscr{G}_P(G)$ and $n\geq 1$ there exist $\mu_n\in\mathscr{M}^1(G)$ and $x_n\in G$ such that $\mu=\mu_n^n*\varepsilon_{x_n}$. By Remark 5.2.2 and Theorem 5.1.2 we may assume without loss of generality that $\lim_{n\to\infty}\mu_n=\varepsilon_e$ holds. Then the family $(\mu_{nj})_{j=1,\dots,n;n\geq 1}$ with $\mu_{nj}:=\mu_n$ for all $j=1,\dots,n$ is an infinitesimal triangular system in $\mathscr{M}^1(G)$. For every $n\geq 1$ we define $\nu_n := \mu_n^n$, $y_n \in G$ by

$$\chi(y_n) := \exp(-i \int g(x, \chi)\, \mu_n(dx))$$

for all $\chi \in G^\wedge$ (where g is a local inner product for G),

$$\alpha_n := \mu_n * \varepsilon_{y_n}, \quad \beta_n := \exp(\alpha_n) \quad \text{and} \quad \lambda_n := \beta_n^n * \varepsilon_{y_n^{-n}}.$$

One notes that accumulation points of shifts of $(\mu_n^n)_{n \geq 1}$ differ only by shifts of μ. Therefore, no accumulation point of any shift of $(\mu_n^n)_{n \geq 1}$ admits an idempotent factor. Theorem 5.1.16 implies for every compact subset K of $G^\wedge$

$$(2.4) \qquad \begin{aligned} 0 &= \lim_{n\to\infty} \sup_{\chi \in K} |(\beta_n^n * \varepsilon_{y_n^{-n}})^\wedge(\chi) - \mu_n^{n\wedge}(\chi)| \\ &= \lim_{n\to\infty} \sup_{\chi \in K} |(\lambda_n * \varepsilon_{x_n})^\wedge(\chi) - \hat{\mu}(\chi)|. \end{aligned}$$

In particular, for every $\chi \in G^\wedge$

$$\begin{aligned} |\hat{\mu}(\chi)| &= \lim_{n\to\infty} |(\lambda_n * \varepsilon_{x_n})^\wedge(\chi)| = \lim_{n\to\infty} |\beta_n^{n\wedge}(\chi)| \\ &= \lim_{n\to\infty} |\exp(\alpha_n)^\wedge(\chi)|^n = \lim_{n\to\infty} \exp(n \int (\operatorname{Re} \chi(x) - 1)\, \alpha_n(dx)). \end{aligned}$$

2. The real function ϕ on $G^\wedge$ defined by

$$\phi(\chi) := \lim_{n\to\infty} n \int (1 - \operatorname{Re} \chi(x))\, \alpha_n(dx)$$

for all $\chi \in G^\wedge$ is an element of $\mathbf{Q}_+(G^\wedge)$. In fact, for every $n \geq 1$ we define measures $\tau_n := n\alpha_n \in \mathscr{M}_+^b(G)$ such that $\exp(\tau_n)$ is a shift of $\lambda_n * \varepsilon_{x_n}$ $(n \geq 1)$. By (2.4) of Part 1 $\lim_{n\to\infty} \lambda_n * \varepsilon_{x_n} = \mu$. Thus the sequence $(\exp(\tau_n))_{n \geq 1}$ is shift compact and the sequence $(1_{\complement U} \cdot \tau_n)_{n \geq 1}$ is relatively compact for every $U \in \mathfrak{V}(e)$, as follows from Theorem 5.1.6. For any accumulation point $\tau := \lim_{k\to\infty} 1_{\complement U} \cdot \tau_{n_k}$ of the sequence $(1_{\complement U} \cdot \tau_n)_{n \geq 1}$ there exists an $\alpha \in \mathscr{I}_0(G)$ such that $\mu = \exp(\tau) * \alpha$ holds.
[There exists a sequence $(x'_{n_k})_{k \geq 1}$ in G satisfying

$$\begin{aligned} \mu &= \lim_{k\to\infty} \lambda_{n_k} * \varepsilon_{x_{n_k}} = \lim_{k\to\infty} \exp(\tau_{n_k}) * \varepsilon_{x'_{n_k}} \\ &= \lim_{k\to\infty} \exp(1_{\complement U} \cdot \tau_{n_k}) * \exp(1_U \cdot \tau_{n_k}) * \varepsilon_{x'_{n_k}} \end{aligned}$$

and therefore an accumulation point α of the sequence

$$(\exp(1_U \cdot \tau_{n_k}) * \varepsilon_{x'_{n_k}})_{k \geq 1}$$

satisfying $\mu = \exp(\tau) * \alpha$. This sequence lies in $\mathscr{I}_0(G)$ and the sequential closedness of $\mathscr{I}_0(G)$ in $\mathscr{M}^1(G)$ yields $\alpha \in \mathscr{I}_0(G)$.]

By assumption we have $\tau = \|\tau\|\, \varepsilon_e$. Without loss of generality we assume that

$$\lim_{n\to\infty} (1_{\complement U} \cdot \tau_n) = \|\tau\|\, \varepsilon_e.$$

Let $f \in \mathscr{C}^b(G)$ be such that $f(\complement U) = 1$ and $f(e) = 0$. Then

$$\begin{aligned} \|\tau\| &= \lim_{n\to\infty} (1_{\complement U} \cdot \tau_n)(G) = \lim_{n\to\infty} (1_{\complement U} \cdot \tau_n)(f) \\ &= \|\tau\| f(e) = 0. \end{aligned}$$

This implies that $\phi(\chi)=\lim_{n\to\infty}\int_U(1-\operatorname{Re}\chi(x))\tau_n(dx)$ for all $\chi\in G^\wedge$ whenever $U\in\mathfrak{B}(e)$.
The power series expansion of the cosine function together with the continuity of characters of G implies that for every $\varepsilon>0$ and all $\chi_1,\chi_2\in G^\wedge$ there exists a $U\in\mathfrak{B}(e)$ such that

$$\begin{aligned}&2(1-\varepsilon)[(1-\operatorname{Re}\chi_1(x))+(1-\operatorname{Re}\chi_2(x))]\\&\leq(1-\operatorname{Re}(\chi_1\chi_2)(x))+(1-\operatorname{Re}(\chi_1\chi_2^{-1})(x))\\&\leq 2(1+\varepsilon)[(1-\operatorname{Re}\chi_1(x))+(1-\operatorname{Re}\chi_2(x))]\end{aligned}$$

holds for all $x\in U$. Integration over U with respect to τ_n (for $n\geq 1$) and passage to the limit as $n\to\infty$ yields via the representation of ϕ that

$$\phi(\chi_1\chi_2)+\phi(\chi_1\chi_2^{-1})=2(\phi(\chi_1)+\phi(\chi_2))\quad\text{for } \chi_1,\chi_2\in G^\wedge$$

and hence that $\phi\in\mathbf{Q}_+(G^\wedge)$. Using Part 1 we have shown that $-\phi=\log|\hat{\mu}|$ with $\phi\in\mathbf{Q}_+(G^\wedge)$.

3. Since $\mu\in\mathscr{G}_P(G)$ does not admit an idempotent factor, Theorem 5.1.5 justifies the definition $\psi:=\frac{\hat{\mu}}{|\hat{\mu}|}$. We shall show that $\psi\in(G^\wedge)^\wedge$, which by Pontryagin's duality theorem yields the existence of an $x_0\in G$ with $\psi(\chi)=\chi(x_0)$ for all $\chi\in G^\wedge$ and therefore the desired result. First of all we note that for any $U\in\mathfrak{B}(e)$ and $\chi_1,\chi_2\in G^\wedge$ the following relation holds:

$$\frac{\psi(\chi_1\chi_2)}{\psi(\chi_1)\psi(\chi_2)}=\lim_{n\to\infty}\exp\int_U(\operatorname{Im}\chi_1\chi_2(x)-\operatorname{Im}\chi_1(x)-\operatorname{Im}\chi_2(x))\tau_n(dx).$$

It remains to be proved that the right side of this equation is 0. In fact, for any $x\in G$ and $\chi_1,\chi_2\in G^\wedge$ some computation yields

$$\begin{aligned}&|\operatorname{Im}\chi_1\chi_2(x)-\operatorname{Im}\chi_1(x)-\operatorname{Im}\chi_2(x)|\\&\leq|\operatorname{Im}\chi_1(x)|\,|1-\operatorname{Re}\chi_2(x)|+|\operatorname{Im}\chi_2(x)|\,|1-\operatorname{Re}\chi_1(x)|.\end{aligned}$$

Furthermore, Theorem 5.1.6 implies for all $\chi\in G^\wedge$ the condition

$$\sup_{n\geq 1}\int(1-\operatorname{Re}\chi(x))\tau_n(dx)=:c<\infty$$

with c independent of $n\geq 1$. Choosing $U\in\mathfrak{B}(e)$ such that $|\operatorname{Im}\chi_1(x)|$ and $|\operatorname{Im}\chi_2(x)|$ are smaller than $\varepsilon>0$ for all $x\in U$ one obtains

$$\left|\int_U(\operatorname{Im}\chi_1\chi_2(x)-\operatorname{Im}\chi_1(x)-\operatorname{Im}\chi_2(x))\tau_n(dx)\right|\leq 2c\varepsilon$$

and thus the desired assertion.

The uniqueness of $x_0\in G$ and $\phi\in\mathbf{Q}_+(G^\wedge)$ in the representation

$$\hat{\mu}(\chi)=\psi(\chi)|\hat{\mu}(\chi)|=\chi(x_0)\exp(-\phi(\chi)),$$

valid for $\chi\in G^\wedge$, is evident. □

5.2.8 Theorem. *Let $x_0 \in G$ and $\phi \in \mathbf{Q}_+(G^\wedge)$. Then there exists a measure $\mu \in \mathscr{G}_P(G)$ such that $\hat{\mu}(\chi) = \chi(x_0) \exp(-\phi(\chi))$ holds for all $\chi \in G^\wedge$.*

Proof. 1. We first show that for $x_0 \in G$ and $\phi \in \mathbf{Q}_+(G^\wedge)$ there exists a measure $\mu \in \mathscr{I}_0(G)$ such that the desired representation holds. To this end it suffices to establish the negative-definiteness of ϕ. For if ϕ is negative-definite, then $\exp(-\phi)$ is positive-definite. Thus by Bochner's theorem there exists a measure $\nu \in \mathscr{M}^1(G)$ with the property $\hat{\nu} = \exp(-\phi)$. Since for $n \geq 1$ one also has $\frac{1}{n}\phi \in \mathbf{Q}_+(G^\wedge)$, there is a measure $\nu_n \in \mathscr{M}^1(G)$ with $\hat{\nu}_n = \exp\left(-\frac{1}{n}\phi\right)$ and therefore $\hat{\nu} = (\hat{\nu}_n)^n = (\nu_n^n)^\wedge$. Thus $\nu \in \mathscr{I}(G)$ and $\mu := \varepsilon_{x_0} * \nu \in \mathscr{I}_0(G)$ such that

$$\chi(x_0) \exp(-\phi(\chi)) = \hat{\varepsilon}_{x_0}(\chi)\, \hat{\nu}(\chi) = \hat{\mu}(\chi)$$

for all $\chi \in G^\wedge$ holds.

Concerning the negative-definiteness of ϕ we have to show that for every $n \geq 1$, $\chi_1, \ldots, \chi_n \in G^\wedge$ and $c_1, \ldots, c_n \in \mathbb{C}$ one has

$$\textstyle\sum_{i,j=1}^n c_i \bar{c}_j (\phi(\chi_i) + \phi(\chi_j) - \phi(\chi_i \chi_j^{-1})) \geq 0.$$

Defining a real function ψ on $\mathbb{Z}^n$ by

$$\psi(\mathfrak{m}) := \phi(\textstyle\sum_{i=1}^n m_i \chi_i) \qquad \text{for all } \mathfrak{m} := (m_1, \ldots, m_n) \in \mathbb{Z}^n$$

one immediately verifies that $\psi(u+v) + \psi(u-v) = 2(\psi(u) + \psi(v))$ whenever $u, v \in \mathbb{Z}^n$. By Lemma 5.2.4 there exists a unique function $\Psi \in \mathbf{B}_+(\mathbb{Z}^n)$ with $\Psi(u,u) = \psi(u)$ for all $u \in \mathbb{Z}^n$ which is uniquely determined by the matrix

$$A := (\Psi(e_i, e_j))_{i,j=1,\ldots,n}$$

with $e_i := (0, \ldots, 0, \underset{(i)}{1}, 0, \ldots, 0) \in \mathbb{Z}^n$ for all $i = 1, \ldots, n$. Ψ can be extended to a bilinear form $\tilde{\Psi}$ on $\mathbb{R}^n$ for which

$$\tilde{\Psi}(\xi u, \xi u) = \xi^2 \tilde{\Psi}(u,u) = \xi^2 \Psi(u,u) = \xi^2 \psi(u) \geq 0$$

holds whenever $\xi \in \mathbb{R}$ and $u \in \mathbb{Z}^n$. Writing

$$D := \{x \in \mathbb{R}^n : x = \xi u \text{ for } \xi \in \mathbb{R} \text{ and } u \in \mathbb{Z}^n\}$$

we achieve $\tilde{\Psi}(z,z) \geq 0$ for all $z \in D$ and thus for all $z \in \mathbb{R}^n$, since D is dense in $\mathbb{R}^n$. This implies that $\tilde{\Psi}$ is a positive bilinear form on $\mathbb{R}^n$. As $\phi \in \mathbf{Q}_+(G^\wedge)$, one obtains for $i, j = 1, \ldots, n$ the equalities

$$\begin{aligned}
&\phi(\chi_i) + \phi(\chi_j) - \phi(\chi_i \chi_j^{-1}) = \psi(e_i) + \psi(e_j) - \psi(e_i e_j^{-1}) \\
&= \Psi(e_i, e_i) + \Psi(e_j, e_j) - \Psi(e_i e_j^{-1}, e_i e_j^{-1}) = 2\Psi(e_i, e_j) \\
&= 2\tilde{\Psi}(e_i, e_j)
\end{aligned}$$

and therefore

$$\sum_{i,j=1}^{n} c_i\bar{c}_j(\phi(\chi_i)+\phi(\chi_j)-\phi(\chi_i\chi_j^{-1}))=\sum_{i,j=1}^{n} c_i\bar{c}_j 2\tilde{\Psi}(e_i,e_j)$$
$$=2\Psi(\textstyle\sum_{i=1}^{n} c_i e_i, \sum_{i=1}^{n} c_i e_i)\geq 0.$$

2. For every $\tau\in\mathcal{M}_+^b(G)$ and $\chi_1,\chi_2\in G^\wedge$ one has

(2.5)
$$-\log|\exp(\tau)^\wedge(\chi_1\chi_2)|-\log|\exp(\tau)^\wedge(\chi_1\chi_2^{-1})|$$
$$\leq 2(-\log|\exp(\tau)^\wedge(\chi_1)|-\log|\exp(\tau)^\wedge(\chi_2)|).$$

This relation follows from the elementary inequality

$$1-ab\leq(1-a)+(1-b)$$

valid for all $a,b\in\mathbb{R}$ with $|a|,|b|\leq 1$. In fact, we have

$$\log|\exp(\tau)^\wedge(\chi)|=\int(\operatorname{Re}\chi(x)-1)\,\tau(dx)$$

for all $\chi\in G^\wedge$ and consequently,

$$-\log|\exp(\tau)^\wedge(\chi_1\chi_2)|-\log|\exp(\tau)^\wedge(\chi_1\chi_2^{-1})|$$
$$=2\int(1-\operatorname{Re}\chi_1(x)\operatorname{Re}\chi_2(x))\,\tau(dx).$$

Thus using the above inequality, one gets

$$2\int(1-\operatorname{Re}\chi_1(x)\operatorname{Re}\chi_2(x))\,\tau(dx)$$
$$\leq 2(-\log|\exp(\tau)^\wedge(\chi_1)|-\log|\exp(\tau)^\wedge(\chi_2)|)\qquad\text{for all } \chi_1,\chi_2\in G^\wedge.$$

3. Given any measure $\tau\in\mathcal{M}_+^b(G)$ one has the inequality

(2.6)
$$|\hat{\tau}(1)-\hat{\tau}(\chi)|^2\leq 2\hat{\tau}(1)(\hat{\tau}(1)-\operatorname{Re}\hat{\tau}(\chi))$$

valid for all $\chi\in G^\wedge$. From this it follows immediately that

$$\int(1-\operatorname{Re}\chi(x))\,\tau(dx)=0$$

for all $\chi\in G^\wedge$ implies $\hat{\tau}(\chi)=\hat{\tau}(1)=\|\tau\|$ for all $\chi\in G^\wedge$, and hence, $\tau=\|\tau\|\,\varepsilon_e$.

It remains to be shown that $\mu=\exp(\tau)*\lambda$ with $\tau\in\mathcal{M}_+^b(G)$ and $\lambda\in\mathcal{I}_0(G)$ implies that $\tau=\|\tau\|\,\varepsilon_e$. Let $\tau_0\in\mathcal{M}_+^b(G)$ and $\lambda\in\mathcal{I}_0(G)$ with $\mu=\exp(\tau_0)*\lambda$. Since $\hat{\mu}$ admits no zero by Theorem 5.1.5, $\hat{\lambda}$ does not either, so that λ admits no idempotent factor. Theorem 5.1.16 implies that λ is the limit of a sequence of shifts of measures $\exp(\tau)$ with $\tau\in\mathcal{M}_+^b(G)$. It follows from (2.5) in Part 2 that for all $\chi_1,\chi_2\in G^\wedge$ one has

$$-\log|\hat{\lambda}(\chi_1\chi_2)|-\log|\hat{\lambda}(\chi_1\chi_2^{-1})|$$
$$\leq 2(-\log|\hat{\lambda}(\chi_1)|-\log|\hat{\lambda}(\chi_2)|).$$

Addition of this inequality to the inequality (2.5) in Part 2 yields

$$\phi(\chi_1\chi_2)+\phi(\chi_1\chi_2^{-1})\leq 2(\phi(\chi_1)+\phi(\chi_2))\qquad\text{for all } \chi_1,\chi_2\in G^\wedge.$$

Since $\phi \in \mathbf{Q}_+(G^\wedge)$, we obtain equality in (2.5) of Part 2 if τ is replaced by τ_0. Using

$$|\exp(\tau_0)^\wedge(\chi)| = \int (1 - \operatorname{Re}\chi(x))\,\tau_0(dx) \qquad \text{for all } \chi \in G^\wedge$$

we arrive at

$$\begin{aligned}&\int (1 - \operatorname{Re}\chi_1\chi_2(x) + 1 - \operatorname{Re}\chi_1\chi_2^{-1}(x))\,\tau_0(dx)\\ &= 2\int (1 - \operatorname{Re}\chi_1(x) + 1 - \operatorname{Re}\chi_2(x))\,\tau_0(dx),\end{aligned}$$

and therefore

$$\int (1 - \operatorname{Re}\chi_1(x))(1 - \operatorname{Re}\chi_2(x))\,\tau_0(dx) = 0 \qquad \text{for all } \chi_1, \chi_2 \in G^\wedge.$$

If one chooses $\chi_2 := \chi_1$, inequality (2.6) implies that $\tau_0 = \|\tau_0\|\,\varepsilon_e$, which is the desired result. □

5.2.9 Theorem (*Résumé* of Theorems 5.2.7 and 5.2.8). *For any complex function f on $G^\wedge$ the following statements are equivalent:*
(i) *There exists a $\mu \in \mathscr{G}_P(G)$ with $f = \hat{\mu}$;*
(ii) *there exist an $x_0 \in G$ and a $\phi \in \mathbf{Q}_+(G^\wedge)$ such that*
$f(\chi) = \chi(x_0)\exp(-\phi(\chi))$ holds for all $\chi \in G^\wedge$.

5.2.10 Properties of the class $\mathscr{G}_P(G)$.
(1) $\mathscr{G}_P(G)$ is a subsemigroup of $\mathscr{I}_0(G)$.
This follows immediately from the characterization theorem 5.2.9.
(2) $\mathscr{G}_P(G)$ is closed in $\mathscr{M}^1(G)$.
In fact, let $(\mu_\alpha)_{\alpha \in \mathbf{A}}$ be a net in $\mathscr{G}_P(G)$ with

$$\lim_{\alpha \in \mathbf{A}} \mu_\alpha =: \mu \in \mathscr{M}^1(G).$$

By Theorem 5.2.9 there exist nets $(x_\alpha)_{\alpha \in \mathbf{A}}$ in G and $(\phi_\alpha)_{\alpha \in \mathbf{A}}$ in $\mathbf{Q}_+(G^\wedge)$ such that $\hat{\mu}_\alpha = \hat{\varepsilon}_{x_\alpha}\exp(-\phi_\alpha)$ for all $\alpha \in \mathbf{A}$. Thus by the continuity of the Fourier transform for all $\chi \in G^\wedge$

$$\log|\hat{\mu}(\chi)| = \lim_{\alpha \in \mathbf{A}} \log|\hat{\mu}_\alpha(\chi)| = \lim_{\alpha \in \mathbf{A}}(-\phi_\alpha(\chi)).$$

The function $\phi := -\log|\hat{\mu}|$ is in $\mathbf{Q}_+(G^\wedge)$. Putting

$$\Psi(\chi) := \frac{\hat{\mu}(\chi)}{|\hat{\mu}(\chi)|} = \lim_{\alpha \in \mathbf{A}} \frac{\hat{\mu}_\alpha(\chi)}{|\hat{\mu}_\alpha(\chi)|} = \lim_{\alpha \in \mathbf{A}} \chi(x_\alpha)$$

for all $\chi \in G^\wedge$ one gets $\Psi \in (G^\wedge)^\wedge$. By Pontryagin's duality theorem there exists an $x_0 \in G$ with $\Psi(\chi) = \chi(x_0)$ for all $\chi \in G^\wedge$ and hence, $\hat{\mu} = \hat{\varepsilon}_x \exp(-\phi)$. Another application of Theorem 5.2.9 yields $\mu \in \mathscr{G}_P(G)$.

(3) Let H be a compact subgroup of G and p the canonical mapping from G onto G/H. For every $\dot{\mu} \in \mathscr{G}_P(G/H)$ (or $\mathscr{G}_P^s(G/H)$) there is a measure $\mu \in \mathscr{G}_P(G)$ (or $\mathscr{G}_P^s(G)$) with $p(\mu) = \dot{\mu}$.

We show the assertion in brackets. Since H is compact, $G^\wedge/H^\perp$ is discrete and therefore $H^\perp$ is an open subgroup of $G^\wedge$. Let $\dot{\mu} \in \mathscr{G}_P^s(G/H)$ be given. By Theorem 5.2.9 there exists a $\psi_0 \in \mathbf{Q}_+(H^\perp)$ with $\hat{\dot{\mu}} = \exp(-\psi_0)$. By Lemma 5.2.5 ψ_0 can be extended to $\psi \in \mathbf{Q}_+(G^\wedge)$. But then there exists (again by Theorem 5.2.9) a $\mu \in \mathscr{G}_p^s(G)$ with $\hat{\mu} = \exp(-\psi)$. From

$$p(\mu)^\wedge = \operatorname{Res}_{H^\perp} \hat{\mu} = \exp(-\psi_0) = \hat{\hat{\mu}}$$

follows the assertion.

(4) For any $d \geq 1$ we have

$$\mathscr{G}_P(\mathbb{R}^d) \cap \complement\, \mathscr{D}(\mathbb{R}^d) = \mathscr{N}(\mathbb{R}^d) \quad \text{and}$$

$$\mathscr{G}_P(\mathbb{T}^d) \cap \complement\, \mathscr{D}(\mathbb{T}^d) = \mathscr{N}(\mathbb{T}^d).$$

We content ourselves with the proof of the first statement. Let $\mu \in \mathscr{N}(\mathbb{R}^d)$. There exists an affine linear mapping $T: \mathbb{R}^d \to \mathbb{R}^d$ of the form $T := L + a$ with $L \in \mathscr{L}(\mathbb{R}^d)$, $L \neq 0$ and $a \in \mathbb{R}^d$ such that $\mu = T(\nu_0)$ for $\nu_0 := \underbrace{\nu_{0,1} \otimes \cdots \otimes \nu_{0,1}}_{d\text{-times}}$. Clearly $\hat{\mu} = \hat{\varepsilon}_a \hat{\nu}_0 \circ L^T$ and hence

$$\hat{\mu}(x) = \hat{\varepsilon}_a(x) \exp\left(-\tfrac{1}{2}\langle L^T(x), L^T(x)\rangle\right) \qquad \text{for all } x \in \mathbb{R}^d.$$

The real function ϕ on $\mathbb{R}^d$ defined by $\phi(x) := \langle L^T(x), L^T(x)\rangle$ for all $x \in \mathbb{R}^d$ is an element $\neq 0$ in $\mathbf{Q}_+(\mathbb{R}^d)$. Theorem 5.2.9 yields $\mu \in \mathscr{G}_P(\mathbb{R}^d) \cap \complement\, \mathscr{D}(\mathbb{R}^d)$.

Conversely, let $\mu \in \mathscr{G}_P(\mathbb{R}^d) \cap \complement\, \mathscr{D}(\mathbb{R}^d)$. Then by Theorem 5.2.9 we have

$$\hat{\mu} = \hat{\varepsilon}_a \exp\left(-\tfrac{1}{2}\phi\right)$$

for $a \in \mathbb{R}^d$ and $\phi \in \mathbf{Q}_+(\mathbb{R}^d)$ with $\phi \neq 0$. Since ϕ is a positive quadratic form on $\mathbb{R}^d$, there is an $L \neq 0$ in $\mathscr{L}(\mathbb{R}^d)$ with $L = L^T$ such that $\phi(x) = \langle L^T(x), L^T(x)\rangle$ for all $x \in \mathbb{R}^d$. Putting $T := L + a$ one obtains $\mu = T(\nu_0) \in \mathscr{N}(\mathbb{R}^d)$.

(5) Cramér's theorem on normal classes is in general false. More precisely, for $G := \mathbb{T}$ there exists a $\mu \in \mathscr{G}_P(G)$ with $\mu = \sigma * \rho$ such that σ and ρ are not in $\mathscr{G}_P(G)$.

In fact, let

$$\mu := \tau(\nu_{0,\sigma^2}) = n_{0,\sigma^2} \cdot \lambda_{\mathbb{T}} \quad \text{with}$$

$$n_{0,\sigma^2}(u) = \frac{1}{\sqrt{2\pi}\,\sigma} \sum_{n \in \mathbb{Z}} e^{-\frac{1}{2\sigma^2}(u - 2n\pi)^2}$$

for all $u \in \mathbb{T}$ (and $\sigma^2 \in \mathbb{R}_+^*$). For all $\dot{u} \in \mathbb{T}$ and $v \in \mathbb{R}$ we define

$$n_1(u) := n_{0,\frac{\sigma^2}{2}}(u) + v \cos u \quad \text{and}$$

$$n_2(u) := n_{0,\frac{\sigma^2}{2}}(u) - 2\frac{v\alpha}{v + 2\alpha} \cos u,$$

where $\alpha := \int_{\mathbb{T}} n_{0,\frac{\sigma^2}{2}}(u) \cos u \, du$.

Then for sufficiently small $v \in \mathbb{R}$ one obtains that n_1 and n_2 are $\lambda_{\mathbb{T}}$-densities. But

$$n_1 * n_2 = n_{0,\frac{\sigma^2}{2}} * n_{0,\frac{\sigma^2}{2}} + \left(1 - \frac{2\alpha}{v+2\alpha}\right) n_{0,\frac{\sigma^2}{2}} * \cos$$

$$-2\frac{v^2\alpha}{v+2\alpha} \cos * \cos = n_{0,\sigma^2}$$

since $\cos * \cos = \frac{1}{2}\cos$ and $n_{0,\frac{\sigma^2}{2}} * \cos = \alpha \cos$.

Thus μ can be factorized into nonnormal factors $\sigma := n_1 \cdot \lambda_{\mathbb{T}}$ and $\rho := n_2 \cdot \lambda_{\mathbb{T}}$.

The following results aim at the existence of P-Gauss measures in the Abelian case.

5.2.11 Lemma. *Let A be a closed subgroup of $G^\wedge$ and $\mu \in \mathcal{M}^1(G)$. If $\hat{\mu}(\chi) = 1$ for all $\chi \in A$, then* $\operatorname{supp}(\mu) \subset A^\perp$.

Proof. Let K be a compact subset of $G \setminus A^\perp$. For every $x \in K$ there exists a $\chi \in A$ with $\chi(x) \neq 1$. Since χ is continuous, there is a neighborhood $U_x \in \mathfrak{B}(x)$ such that $\chi(y) \neq 1$ for all $y \in U_x$. We have $K \subset \bigcup_{x \in K} U_x$ and that K is compact. Hence, there exist $x_1, \ldots, x_n \in K$, $\chi_1, \ldots, \chi_n \in A$ and $U_{x_i} \in \mathfrak{B}(x_i)$ $(i = 1, \ldots, n)$ with

$$K \subset \bigcup_{i=1}^n U_{x_i} \quad \text{and} \quad \sum_{i=1}^n (1 - \operatorname{Re} \chi_i(x)) > 0 \qquad \text{for all } x \in K.$$

By assumption $\int (1 - \operatorname{Re} \chi(x))\, \mu(dx) = 0$ for all $\chi \in A$; in particular,

$$\sum_{i=1}^n (1 - \operatorname{Re} \chi_i(x)) = 0 \qquad \text{for all } x \in G.$$

But then

$$0 \geq \int_K \sum_{i=1}^n (1 - \operatorname{Re} \chi_i(x))\, \mu(dx) \geq \mu(K) \inf_{x \in K} \sum_{i=1}^n (1 - \operatorname{Re} \chi_i(x))$$

and thus $\mu(K) = 0$. This is true for all compact subsets K of $G \setminus A^\perp$, which implies that $\mu(G \setminus A^\perp) = 0$ or $\operatorname{supp}(\mu) \subset A^\perp$. □

5.2.12 Corollary. *Let $\mu \in \mathcal{G}_P^s(G)$. Then* $\operatorname{supp}(\mu) \subset G_0$.

Proof. By assumption $\hat{\mu} = \exp(-\phi)$ with $\phi \in \mathbf{Q}_+(G^\wedge)$. Every $\chi \in G_0^\perp$ is compact, i.e., $[\chi]^-$ is compact. For fixed $\chi_0 \in G_0^\perp$ and all $n \geq 1$ we have $\phi(\chi_0^n) = n^2 \phi(\chi_0)$. Since ϕ is bounded on $[\chi_0]^-$, we obtain $\phi(\chi_0) = 0$ and hence

$$\hat{\mu}(\chi) = \exp(-\phi(\chi)) = 1 \qquad \text{for all } \chi \in G_0^\perp.$$

The lemma implies that $\operatorname{supp}(\mu) \subset (G_0^\perp)^\perp = G_0$. □

5.2.13 Corollary. *Every $\mu \in \mathcal{G}_P(G)$ is a shift of a measure $\nu \in \mathcal{G}_P(G)$ with* $\operatorname{supp}(\nu) \subset G_0$.

The *proof* is immediate from Corollary 5.2.12. □

5.2.14 Theorem. *The set* $\mathscr{G}_P(G, G_0) := \{\mu \in \mathscr{G}_P(G) : \operatorname{supp}(\mu) \subset G_0\}$ *can be identified with* $\mathscr{G}_P(G_0)$.

Proof. 1. Let $\mu \in \mathscr{G}_P(G, G_0)$. We have to show that $\mu \in \mathscr{I}_0(G_0)$. Theorem 5.2.9 provides us with the existence of an $x_0 \in G$ and a $\phi \in \mathbf{Q}_+(G^\wedge)$ such that

$$\mu = \varepsilon_{x_0} * \nu \quad \text{with } \hat{\nu} = \exp(-\phi)$$

holds. By Corollary 5.2.12 we have $\operatorname{supp}(\nu) \subset G_0$. The assumption $\operatorname{supp}(\mu) \subset G_0$ implies that $x_0 \in G_0$. Since G_0 is divisible, for every $n \geq 1$ there is an

$$x_0^{\frac{1}{n}} \in G_0 \quad \text{with } (x_0^{\frac{1}{n}})^n = x_0.$$

For any $n \geq 1$ the measure $\nu_n \in \mathscr{M}^1(G)$ defined by $\hat{\nu}_n := \exp\left(-\frac{1}{n}\phi\right)$ satisfies $\operatorname{supp}(\nu_n) \subset G_0$. Thus ν_n can be considered as a measure in $\mathscr{M}^1(G_0)$ and $\mu = (\varepsilon_{x_0^{\frac{1}{n}}} * \nu_n)^n$ holds for every $n \geq 1$, so that $\mu \in \mathscr{I}_0(G_0)$.

2. Let $\mu \in \mathscr{G}_P(G_0)$ be extended to $\tilde{\mu} \in \mathscr{M}^1(G)$ with $\operatorname{supp}(\tilde{\mu}) \subset G_0$. By translation in G_0 one can achieve $\hat{\mu} = \exp(-\phi)$ with $\phi \in \mathbf{Q}_+(G_0^\wedge)$. As $G_0^\wedge = G^\wedge / G_0^\perp$, transformation of integrals yields for every $\chi \in G^\wedge$ the equalities

$$\begin{aligned} \tilde{\mu}^\wedge(\chi) &= \int_G \chi(x)\,\tilde{\mu}(dx) = \int_{G_0} \chi(x)\,\tilde{\mu}(dx) = \int_{G_0} (\chi G_0^\perp)(x)\,\mu(dx) \\ &= \hat{\mu}(\chi G_0^\perp) = \exp(-\phi(\chi G_0^\perp)). \end{aligned}$$

Putting $\tilde{\phi}(\chi) := \phi(\chi G_0^\perp)$ for all $\chi \in G^\wedge$ one obtains $\tilde{\phi} \in \mathbf{Q}_+(G^\wedge)$ and $\tilde{\mu}^\wedge = \exp(-\tilde{\phi})$ which by Theorem 5.2.9 yields $\tilde{\mu} \in \mathscr{G}_P(G)$. □

5.2.15 Corollary. $\mathscr{G}_P(G) = \mathscr{D}(G) * \mathscr{G}_P(G_0)$.

The *proof* is immediate from the theorem together with Corollary 5.2.13. □

5.2.16 Theorem. $\mathscr{G}_P(G) \cap \complement \mathscr{D}(G) \neq \emptyset$ iff $G_0 \neq \{e\}$.

Proof. 1. If there exists a nondegenerate measure in $\mathscr{G}_P(G)$, then by Corollary 5.2.12 we have $G_0 \neq \{e\}$.

2. Let $G_0 \neq \{e\}$. By Corollary 5.2.15 we may assume without loss of generality $G = G_0$ and therefore by the structure theorem L that $G = \mathbb{R}^d \times K$, where $d \geq 0$ and K is a connected compact Abelian group. But Property (4) of 5.2.10 yields $\mathscr{G}_P(\mathbb{R}^d) \cap \complement \mathscr{D}(\mathbb{R}^d) \neq \emptyset$ for all $d \geq 1$. In the case $d = 0$, however, we have $K^\wedge = G^\wedge$, and $K^\wedge$ is a torsion-free discrete Abelian group such that for $\chi_0 \in G^\wedge$ with $\chi_0 \neq 1$ the group $[\chi_0]$ is isomorphic to $\mathbb{Z}$. We now define real functions $\Psi \in \mathbf{B}_+(G^\wedge)$ by

$$\Psi(\xi, \eta) := \begin{cases} nm, & \text{if } \xi := \chi_0^n \text{ and } \eta := \chi_0^m \ (n, m \in \mathbb{Z}) \\ 0, & \text{if } \xi \text{ or } \eta \text{ are not in } [\chi_0] \end{cases}$$

and $\phi \in \mathbf{Q}_+(G^\wedge)$ by $\phi(\chi) := \Psi(\chi, \chi)$ for all $\chi \in G^\wedge$. Clearly $\phi \neq 0$, and $\exp(-\phi)$ is by Theorem 5.2.9 the Fourier transform of a nondegenerate measure in $\mathscr{G}_P(G)$. □

5.3 Gauss Measures in the Sense of Bernstein

From the large variety of characterizations of normal distributions we choose an idea of Bernstein for a generalization to arbitrary locally compact Abelian groups. This characterization admits a direct probabilistic interpretation.

5.3.1. Let G be an Abelian group and let $n: G\times G\to G\times G$ be the mapping defined by $n(x,y):=(xy, xy^{-1})$ for all $x, y\in G$.

Definition. A measure $\mu\in\mathscr{M}^1(G)$ is said to be a *Gauss measure in the sense of Bernstein* (B-Gauss) if the formula

$$n(\mu\otimes\mu)=(\mu*\mu)\otimes(\mu*\mu^{\sim})$$

is valid.

The class of all B-Gauss measures on G will be denoted by $\mathscr{G}_B(G)$, the subclass of symmetric elements by $\mathscr{G}_B^s(G)$.

5.3.2 Remark. If G has a countable basis of its topology, then $\mu\in\mathscr{M}^1(G)$ is a B-Gauss measure iff there exist a probability space $(\Omega,\mathfrak{A},P)$ and independent G-random variables X, Y on $(\Omega,\mathfrak{A},P)$ satisfying the following conditions:
(i) $X(P)=Y(P)=\mu$.
(ii) XY and XY^{-1} are independent.

In fact, for the G-random variables X, Y on $(\Omega,\mathfrak{A},P)$ we have

$$((XY)\otimes(XY^{-1}))(P)=(XY)(P)\otimes(XY^{-1})(P) \quad\text{and}$$
$$(X\otimes Y)(P)=X(P)\otimes Y(P).$$

By the definition of the mapping n we obtain $n\circ(X\otimes Y)=(XY)\otimes(XY^{-1})$ and hence,

$$\begin{aligned}n(\mu\otimes\mu)&=n(X(P)\otimes Y(P))\\&=(n\circ(X\otimes Y))(P)=(XY)(P)\otimes(XY^{-1})(P)\end{aligned}$$

and the independence of X and Y^{-1} yields

$$n(\mu\otimes\mu)=(\mu*\mu)\otimes(\mu*\mu^{\sim}).$$

Conversely, let $\mu\in\mathscr{M}^1(G)$ be given such that $n(\mu\otimes\mu)=\nu_1\otimes\nu_2$ for $\nu_1,\nu_2\in\mathscr{M}^1(G)$. We define a probability space $(\Omega,\mathfrak{A},P)$ by putting

$$\Omega:=G\times G,\qquad \mathfrak{A}:=\mathfrak{B}(G)\otimes\mathfrak{B}(G),\qquad P:=\mu\otimes\mu$$

and G-random variables X, Y on $(\Omega,\mathfrak{A},P)$ by $X:=pr_1$ and $Y:=pr_2$ resp. Obviously X, Y are independent and $X(P)=Y(P)=\mu$ holds. By assumption one has

$$((XY)\otimes(XY^{-1}))(P)=n\circ(X\otimes Y)(P)=n(\mu\otimes\mu)=\nu_1\otimes\nu_2$$

and by construction of the product measure we get

$$(XY)(P)=\nu_1 \quad\text{and}\quad (XY^{-1})(P)=\nu_2.$$

The independence of X, Y implies $\nu_1 = \mu * \mu$ and $\nu_2 = \mu * \mu^{\sim}$, whence

$$((XY) \otimes (XY^{-1}))(P) = (XY)(P) \otimes (XY^{-1})(P)$$

or the independence of XY and XY^{-1}.

5.3.3 Properties of B-Gauss measures.

(1) A measure $\mu \in \mathscr{M}^1(G)$ is B-Gauss iff for its Fourier transform $\phi = \hat{\mu}$ one has

$$\phi(\chi\eta)\phi(\chi\eta^{-1}) = \phi(\chi)^2 \phi(\eta)\phi(\eta^{-1}) \tag{3.1}$$

for all $\chi, \eta \in G^{\wedge}$.

This follows immediately from the equalities

$$\begin{aligned}
[n(\mu \otimes \mu)]^{\wedge}(\chi, \eta) &= \iint (\chi, \eta)(x, y)\, n(\mu \otimes \mu)(d(x, y)) \\
&= \iint (\chi\eta)(x)(\chi\eta^{-1})(y)(\mu \otimes \mu)(d(x, y)) = \int (\chi\eta)(x)\, \mu(dx) \int (\chi\eta^{-1})(y)\, \mu(dy) \\
&= \phi(\chi\eta)\phi(\chi\eta^{-1}) \quad \text{and} \\
\phi(\chi)^2 \phi(\eta)\phi(\eta^{-1}) &= (\mu * \mu)^{\wedge}(\chi)(\mu * \mu^{\sim})^{\wedge}(\eta) = [(\mu * \mu) \otimes (\mu * \mu^{\sim})]^{\wedge}(\chi, \eta),
\end{aligned}$$

valid for all $\chi, \eta \in G^{\wedge}$.

(2) The classes $\mathscr{G}_B(G)$ and $\mathscr{G}_B^s(G)$ are closed subsemigroups of $\mathscr{M}^1(G)$ and $\mathscr{D}(G) \subset \mathscr{G}_B(G)$.

This is a direct consequence of (1).

An important difference between B-Gauss measures and P-Gauss measures becomes apparent when we look at the class $\mathscr{J}_B(G)$ of idempotent elements of $\mathscr{G}_B(G)$.

5.3.4 Theorem. *For any compact subgroup H of G the following statements are equivalent:*

(i) $\omega_H \in \mathscr{G}_B(G)$;

(ii) *for all* $\chi \in \complement H^{\perp}$ *we have* $\chi^2 \in \complement H^{\perp}$.

Proof. 1. (i) $\Rightarrow$ (ii). Let $\omega_H \in \mathscr{G}_B(G)$ with Fourier transform $\phi = \hat{\omega}_H = 1_{H^{\perp}}$ satisfying the equation (3.1) of 5.3.3. For $\chi \in \complement H^{\perp}$ and $\eta := \chi$ this equation yields

$$\phi(\chi^2) = \phi(\chi^2)\phi(\chi\chi^{-1}) = \phi(\chi)^2 \phi(\chi)\phi(\chi^{-1}) = 0,$$

whence $\chi^2 \in \complement H^{\perp}$.

2. (ii) $\Rightarrow$ (i). Since $H^{\perp}$ is a subgroup of $G^{\wedge}$, $\chi, \eta \in H^{\perp}$ implies $\chi\eta, \chi\eta^{-1} \in H^{\perp}$ and thus for $\phi := \hat{\omega}_H = 1_{H^{\perp}}$ clearly

$$1 = \phi(\chi\eta)\phi(\chi\eta^{-1}) = \phi(\chi)^2 \phi(\eta)\phi(\eta^{-1}).$$

If $\chi \in H^{\perp}$ and $\eta \in \complement H^{\perp}$, then $\chi\eta \in \complement H^{\perp}$ and thus

$$\phi(\chi\eta)\phi(\chi\eta^{-1}) = 0 = \phi(\chi)^2 \phi(\eta)\phi(\eta^{-1}).$$

But if $\chi \in \complement H^\perp$ and $\eta \in \complement H^\perp$, then $\chi\eta \in \complement H^\perp$ or $\chi\eta^{-1} \in \complement H^\perp$ by Condition (ii). Hence

$$\phi(\chi\eta)\,\phi(\chi\eta^{-1}) = \phi(\chi)^2\,\phi(\eta)\,\phi(\eta^{-1})$$

also holds in this case. Thus altogether $\omega_H \in \mathscr{G}_B(G)$ by Property (1) of 5.3.3. □

5.3.5 Remark. $\mathscr{G}_P(G) \subset \mathscr{G}_B(G)$, but $\mathscr{G}_P(G) \neq \mathscr{G}_B(G)$.
In fact, the inclusion is an immediate consequence of Theorem 5.2.9 if one uses Property (1) of 5.3.3. The inequality follows from the fact that in $G := \mathbb{T}$ there is a sequence $(H_n)_{n \geq 1}$ of nontrivial compact subgroups

$$H_n := \left\{ \exp \frac{2\pi i l}{2n+1} : l = 0, 1, \ldots, n \right\}$$

and therefore measures $\omega_{H_n} \in \mathscr{G}_B(G)\,(n \geq 1)$ which cannot be in $\mathscr{G}_P(G)$, since elements of $\mathscr{G}_P(G)$ do not admit idempotent factors.

5.3.6 Definition. A locally compact Abelian group G is called a *Corwin group* (C-group) if the homomorphism $\zeta := \zeta_G$ from G into itself defined by $\zeta(x) := x^2$ for all $x \in G$ is an epimorphism.

G is said to be a *strong Corwin group* (SC-group) if ζ is an automorphism of G.

5.3.7 Properties.

(1) Let $\zeta^\wedge := \zeta_{G^\wedge}$. Then ζ is a monomorphism iff $H := \overline{\zeta^\wedge(G^\wedge)} = G^\wedge$.

In fact, if ζ is a monomorphism and $H \neq G^\wedge$, then $H^\perp = (G^\wedge/H)^\wedge \neq \{e\}$. For every $y \in H^\perp \setminus \{e\}$ we have $\eta(y) = 1$ whenever $\eta \in H$. In particular,

$$\chi(y)^2 = \chi(y^2) = 1 \qquad \text{for all } \chi \in G^\wedge,$$

whence $y^2 = e$, contradicting the injectivity of ζ. Conversely, if $H = G^\wedge$ holds, then for every $x \in G$ with $x^2 = e$ one has $\chi(x)^2 = \chi(x^2) = 1$ for all $\chi \in G^\wedge$, and since the elements of $G^\wedge$ separate the points of G, we obtain $x = e$.

(2) If the homomorphisms ζ and $\zeta^\wedge$ are surjective, then they are also bijective. This follows from Pontryagin's duality theorem with the help of (1).

(3) If G is strongly Corwin, then $G^\wedge$ is also strongly Corwin.

(4) The example of the torus $\mathbb{T}$ shows that the surjectivity properties of ζ and $\zeta^\wedge$ are in general independent of each other.

(5) The class of strong Corwin groups whose structure has been analyzed quite explicitly is closed under the formation of (locally compact) direct products, weak direct products and projective limits.

(6) Examples for non-Corwin groups are the groups $\mathbb{Z}_{2m}$ for $m \geq 1$ and $\mathbb{Z}$.

5.3.8 Theorem. *Let G be a C-group. Then $\mathscr{G}_B(G) = \mathscr{D}(G) * \mathscr{G}_B^s(G)$.*

Proof. In view of Property (2) of 5.3.3 it suffices to show the inclusion

$$\mathscr{G}_B(G) \subset \mathscr{D}(G) * \mathscr{G}_B^s(G).$$

Let μ be in $\mathscr{G}_B(G)$ with $\phi:=\hat{\mu}$ satisfying

$$\phi(\chi\eta)\phi(\chi\eta^{-1})=\phi(\chi)^2\phi(\eta)\phi(\eta^{-1}) \quad \text{for all } \chi,\eta\in G^\wedge.$$

The set $A:=A_\phi:=\{\chi\in G^\wedge:\phi(\chi)\neq 0\}$ is an open subgroup of $G^\wedge$. Defining the mapping $\Delta_0: A\to\mathbb{T}$ by $\Delta_0:=\frac{\phi}{\bar{\phi}}$ one verifies easily that Δ_0 is a continuous character of A extendable to a continuous character Δ of $G^\wedge$. By Pontryagin's duality theorem there exists an $s\in G$ with $s(\chi)=\Delta(\chi)$ for all $\chi\in G^\wedge$. Since by assumption G is a C-group and thus ζ is surjective, there exists for s an element y of G with $y^2=s$. Putting $x:=y^{-1}$ we obtain $\nu:=\varepsilon_x*\mu\in\mathscr{G}_B(G)$. We shall show that ν is symmetric. Define $\Phi:=\hat{\nu}$. Since A is a subgroup of $G^\wedge$, $\Phi(\chi)=\Phi(\chi^{-1})=0$, and thus $\hat{\nu}(\chi)=\nu^{\sim\wedge}(\chi)$ for all $\chi\in\complement A$. Moreover, for all $\chi\in A$ we obtain

$$\begin{aligned}\hat{\nu}(\chi)&=\Phi(\chi)=\chi(x)\phi(\chi)=\chi(y^{-1})\phi(\chi)\\&=\Delta_0(\chi)\frac{\phi(\chi^{-1})}{\chi(y)}=\frac{\chi(y^2)\phi(\chi^{-1})}{\chi(y)}=\frac{\chi(y)^2\phi(\chi^{-1})}{\chi(y)}\\&=\chi(y)\phi(\chi^{-1})=\chi(x^{-1})\phi(\chi^{-1})=\Phi(\chi^{-1})=\nu^{\sim\wedge}(\chi).\end{aligned}$$

Hence, altogether $\hat{\nu}(\chi)=\nu^{\sim\wedge}(\chi)$ for all $\chi\in G^\wedge$ or $\nu=\nu^\sim$. Since $\mu=\varepsilon_{x^{-1}}*\nu$ with $\nu\in\mathscr{G}_B^s(G)$, we have arrived at the desired result. □

5.3.9 Theorem. *If* $[G^\wedge:\overline{G^{\wedge 2}}]\leq 2$, *then* $\mathscr{G}_B^s(G)=\mathscr{G}_B(G)*\mathscr{G}_P^s(G)$.

The *proof* of the theorem will be prepared by three lemmata.

5.3.10 Lemma. *Let* $\mu\in\mathscr{G}_B^s(G)$ *with* $\phi:=\hat{\mu}\geq 0$,

$$A:=A_\phi=\{\chi\in G^\wedge:\phi(\chi)\neq 0\} \quad \textit{and} \quad H:=A^\perp.$$

Then $\omega_H\in\mathscr{G}_B(G)$ *and there exists a* $\nu\in\mathscr{G}_P^s(G)$ *with* $\mu=\omega_H*\nu$.

Proof. 1. $H:=A^\perp$ is a compact subgroup of G. In order to verify for H Condition (ii) of Theorem 5.3.4 we pick $\chi\in G^\wedge$ with $\chi^2\in\complement A$ and show $\chi\in\complement A$. But this follows from the symmetry of μ since Property (1) of 5.3.3 in this case yields $\phi(\chi^2)=\phi(\chi)^4$ for all $\chi\in G^\wedge$.

2. For $\sigma:=\mathrm{Res}_A\phi$ there exists by Bochner's theorem a measure $\nu_0\in\mathscr{M}^1(G/H)$ with $\hat{\nu}_0=\sigma$. Plainly $\sigma(\chi)\in]0,1]$ for all $\chi\in A$. The mapping $\psi: A\to\mathbb{R}_+$ defined by $\psi(\chi):=-\log\sigma(\chi)$ for all $\chi\in G$ is continuous and satisfies

$$\psi(\chi\eta)+\psi(\chi\eta^{-1})=2(\psi(\chi)+\psi(\eta)) \quad \text{for all } \chi,\eta\in A.$$

Hence, $\psi\in\mathrm{Q}_+(A)$ with $\hat{\nu}_0=\exp(-\psi)$, and Property (3) of 5.2.10 provides a measure $\nu\in\mathscr{G}_P^s(G)$ with $\mathrm{Res}_A\hat{\nu}=\hat{\nu}_0$. Using Part 1 we obtain $\hat{\mu}=\hat{\omega}_H\hat{\nu}$ and thus $\mu=\omega_H*\nu$. □

5.3.11 Lemma. *Suppose* $[G^\wedge:\overline{G^{\wedge 2}}]\leq 2$ *and let* $\Phi\in\mathscr{C}(G^\wedge)$ *satisfy the following conditions:*

(i) $\Phi(G^\wedge)\subset\{-1,1\}$ *and* $\Phi(1)=1$;

(ii) *Φ is constant on the cosets of the subgroup $\overline{G^{\wedge 2}}$ of $G^{\wedge}$.*
Then $\Phi \in (G^{\wedge})^{\wedge} \cong G$.

Proof. If $\overline{G^{\wedge 2}} = G^{\wedge}$, then Φ is obviously the unit character of $G^{\wedge}$. We therefore suppose $\overline{G^{\wedge 2}} \neq G^{\wedge}$. Without loss of generality it is assumed that

$$\Phi(\complement \overline{G^{\wedge 2}}) = \{-1\}$$

holds. Since $(\complement \overline{G^{\wedge 2}})^2 \subset \overline{G^{\wedge 2}}$ and $\overline{G^{\wedge 2}}$ is a subgroup of $G^{\wedge}$, one obtains

$$\Phi(\chi\eta) = \Phi(\chi)\,\Phi(\eta) \qquad \text{for all } \chi, \eta \in G^{\wedge}.$$

Thus $\Phi \in (G^{\wedge})^{\wedge}$, which by Pontryagin's duality theorem yields the assertion. □

5.3.12 Lemma. *Let $[G^{\wedge} : \overline{G^{\wedge 2}}] \leq 2$, $\mu \in \mathscr{G}_B^s(G)$ with $\phi := \hat{\mu} \geq 0$ and $A := A_\phi$ as in Lemma 5.3.10. Then $[A : \overline{A^2}] \leq 2$.*

Proof. As in Lemma 5.3.10 we establish that $\chi \in A$ iff $\chi^2 \in A$. Moreover, we show that $\overline{G^{\wedge 2}} \cap A = \overline{A^2}$. In fact, since A is a closed subgroup of $G^{\wedge}$, we have $\overline{A^2} \subset A$ and hence $\overline{A^2} \subset \overline{G^{\wedge 2}} \cap A$. Conversely, there exists for every $\chi \in \overline{G^{\wedge 2}} \cap A$ a net $(\chi_i)_{i \in \mathrm{I}}$ in $G^{\wedge 2}$ with $\lim_{i \in \mathrm{I}} \chi_i = \chi \in A$. For every $i \in \mathrm{I}$ we have $\chi_i = \eta_i^2$ for $\eta_i \in G^{\wedge}$. Since A is open, we assume without loss of generality that $\chi_i \in A$ for all $i \in \mathrm{I}$. But for $i \in \mathrm{I}$ we also have $\eta_i \in A$ and hence, $\chi_i = \eta_i^2 \in A^2$. Therefore $\lim_{i \in \mathrm{I}} \chi_i = \chi \in \overline{A^2}$ and thus $\overline{G^{\wedge 2}} \cap A \subset \overline{A^2}$.

Let $\chi, \eta \in A$ with $\chi, \eta \notin \overline{A^2}$. By assumption $\chi \in \eta \overline{G^{\wedge 2}}$ and as $\chi\eta^{-1} \in A$ one has $\chi \in \eta \overline{G^{\wedge 2}} \cap A = \eta \overline{A^2}$, which yields the assertion. □

Proof of Theorem 5.3.9. By Remark 5.3.5 it suffices to prove the inclusion $\mathscr{G}_B^s(G) \subset \mathscr{I}_B(G) * \mathscr{G}_P^s(G)$. Let $\mu \in \mathscr{G}_B^s(G)$. Lemma 5.3.10 applied to the measure

$$\mu * \mu \in \mathscr{G}_B^s(G) \quad \text{with}$$
$$\phi_0 := (\mu * \mu)^{\wedge} \geq 0, \qquad A := A_{\phi_0} \quad \text{and} \quad H := A^{\perp}$$

yields the existence of a $\nu \in \mathscr{G}_P^s(G)$ with the property $\mu * \mu = \omega_H * \nu$, where $\omega_H \in \mathscr{I}_B(G)$, or the existence of

$$\psi \in \mathbf{Q}_+(G^{\wedge}) \qquad \text{with } \phi_0 = 1_A \exp(-\psi).$$

We shall show the existence of an $x \in A^{\perp}$ with $x^2 = e$ such that for $\phi := \hat{\mu}$ we obtain $\phi = \hat{\varepsilon}_x 1_A \exp\left(-\frac{\psi}{2}\right)$.

Let $\Pi : A \to \mathbb{C}$ be given by $\phi = \Pi \exp\left(-\frac{\psi}{2}\right)$. Π maps A into $\{-1, 1\}$ and satisfies $\Pi(1) = 1$. Furthermore, for any $\chi, \eta \in A$ one has

$$\Pi(\chi\eta)\,\Pi(\chi\eta^{-1}) = (\Pi(\chi)\,\Pi(\eta))^2 = 1.$$

For $\chi=\chi'\eta^2\in A$ with $\chi',\eta\in A$ one deduces that

$$\Pi(\chi)=\Pi(\chi'\eta^2)=\Pi(\chi'\eta\eta^{-1})=\Pi(\chi'),$$

whence Π is constant on the cosets of the subgroup $\overline{A^2}$ of A. By Lemmata 5.3.12 and 5.3.11, $\Pi\in A^\wedge$. Moreover, Π can be extended to a function $\tilde{\Pi}\in\mathscr{C}(G^\wedge)$ by putting

$$\tilde{\Pi}(\chi):=1 \quad \text{for all } \chi\in G^\wedge, \text{ if } \Pi\equiv 1,$$

and

$$\tilde{\Pi}(\chi):=\begin{cases} 1 & \text{for all } \chi\in\overline{G^{\wedge 2}} \\ -1 & \text{otherwise} \end{cases}, \quad \text{if } \Pi\not\equiv 1.$$

Thus $\tilde{\Pi}\in(G^\wedge)^\wedge$ or by Pontryagin's duality theorem there exists an $x\in G$ with $\hat{\varepsilon}_x=\tilde{\Pi}$. We have therefore obtained

$$\phi=\hat{\mu}=\hat{\varepsilon}_x 1_A \exp\left(-\frac{\psi}{2}\right) \quad \text{with } \hat{\varepsilon}_x^2=1 \text{ or } x^2=e,$$

and thus $\hat{\varepsilon}_x\exp\left(-\frac{\psi}{2}\right)$ is the Fourier transform of a measure $\tau\in\mathscr{G}_P^s(G)$ by Theorem 5.2.9. Hence $\mu=\omega_H*\tau$ is the desired representation of μ with $\omega_H\in\mathscr{I}_B(G)$. □

5.3.13 Theorem. *Let G be a C-group such that* $[G^\wedge:\overline{G^{\wedge 2}}]\leq 2$ *holds. Then*

$$\mathscr{G}_B(G)=\mathscr{I}_B(G)*\mathscr{G}_P(G).$$

The *proof* is an immediate consequence of Theorems 5.3.9 and 5.3.8. □

5.3.14 Corollary. *If G is a C-group such that* $[G^\wedge:\overline{G^{\wedge 2}}]\leq 2$ *holds and* $G^\wedge$ *is connected, then*

$$\mathscr{G}_B(G)=\mathscr{G}_P(G).$$

Proof. By assumption G admits no nontrivial compact subgroup, whence $\mathscr{I}_B(G)=\{\varepsilon_e\}$. □

5.3.15 Corollary. *If G is an SC-group, then*

$$\mathscr{G}_B(G)=\mathscr{I}_B(G)*\mathscr{G}_P(G). \tag{3.2}$$

The *proof* follows from the property of an SC-group G that $G^{\wedge 2}=G^\wedge$. □

5.3.16 Remark. The representation (3.2) is valid for the C- (but not SC-) group $G:=\mathbb{T}$ since

$$\mathbb{T}^\wedge/\overline{\mathbb{T}^{\wedge 2}}\cong\mathbb{Z}/2\mathbb{Z}, \quad \text{and thus} \quad [\mathbb{T}^\wedge:\overline{\mathbb{T}^{\wedge 2}}]=2 \text{ holds.}$$

Theorem 5.3.4 yields in this case

$$\mathscr{I}_B(G)=\{\varepsilon_1,\omega_{\mathbb{T}}\}\cup\{\omega_{\mathbb{T}_n}:n\geq 1\} \quad \text{with } \omega_{\mathbb{T}_n}:=\frac{1}{2n+1}\sum_{k=0}^{2n}\varepsilon_{a_k},$$

where $a_k:=\exp\left(\frac{2\pi ik}{2n+1}\right)$ for $k=0,\dots,2n;\ n\geq 1$.

5.4 Convergence to Gauss Measures

The central limit theorem contains necessary and sufficient conditions for the convergence of infinitesimal triangular systems in $\mathscr{M}^1(\mathbb{R})$ to the normal distribution. We want to generalize this result to measures in $\mathscr{M}^1(G)$ for a locally compact Abelian group G.

5.4.1 Theorem. *Let $(\nu_n)_{n\geq 1}$ be a sequence in $\mathscr{M}_+^b(G)$ and $(x_n)_{n\geq 1}$ a sequence in G such that $\lim_{n\to\infty}(\exp(\nu_n) * \varepsilon_{x_n}) =: \mu \in \mathscr{M}^1(G)$. Then the following statements are equivalent:*
(i) $\mu \in \mathscr{G}_P(G)$;
(ii) (a) $\lim_{n\to\infty} \nu_n(\complement U) = 0$ *for all* $U \in \mathfrak{B}(e)$,
(b) $\sup_{n\geq 1} \int (1 - \operatorname{Re}\chi(x))\, \nu_n(dx) < \infty$ *for all* $\chi \in G^\wedge$.

Proof. 1. (i)⇒(ii). Let $\mu \in \mathscr{G}_P(G)$. By Remark 5.2.2 μ does not admit an idempotent factor. Then Theorem 5.1.6 yields (b) of Statement (ii) as well as the relative compactness of the sequence $(\operatorname{Res}_{\complement U} \nu_n)_{n\geq 1}$ for every $U \in \mathfrak{B}(e)$. Let $\nu := \nu_{\complement U}$ be an accumulation point of this sequence. Then $\exp(\nu)$ is a factor of μ, and since μ is Gauss, one necessarily has $\nu = 0$, i.e., $\lim_{n\to\infty} \operatorname{Res}_{\complement U} \nu_n = 0$ for $U \in \mathfrak{B}(e)$ or (a) of (ii) is satisfied.

2. (ii)⇒(i). Let $\lim_{n\to\infty} \nu_n(\complement U) = 0$ for every $U \in \mathfrak{B}(e)$. We shall show that $|\mu|^2 \in \mathscr{G}_P(G)$. Without loss of generality we may assume that ν_n is symmetric for all $n \geq 1$ and that $\lim_{n\to\infty} \exp(\nu_n) = \mu$ holds. By Theorem 5.2.9 it suffices to establish the existence of a quadratic form $\phi \in \mathrm{Q}_+(G^\wedge)$ such that $\hat{\mu} = \exp(-\phi)$ is fulfilled. Plainly

$$\hat{\mu}(\chi) = \lim_{n\to\infty} \exp \int (\operatorname{Re}\chi(x) - 1)\, \nu_n(dx)$$

or equivalently,

$$\phi(\chi) = -\log \hat{\mu}(\chi) = \lim_{n\to\infty} \int (1 - \operatorname{Re}\chi(x))\, \nu_n(dx)$$

for all $\chi \in G^\wedge$.

Since $\int_{\complement U} (1 - \operatorname{Re}\chi(x))\, \nu_n(dx) \leq 2\nu_n(\complement U)$ and by Hypothesis (a) of (ii)

$$\lim_{n\to\infty} \nu_n(\complement U) = 0,$$

we obtain

$$\phi(\chi) = \lim_{n\to\infty} \int_U (1 - \operatorname{Re}\chi(x))\, \nu_n(dx)$$

for all $\chi \in G^\wedge$ and $U \in \mathfrak{B}(e)$. The characteristic property of ϕ follows as in Part 2 of the proof of Theorem 5.2.7. □

5.4.2 Theorem. *Let $(\mu_{nj})_{j=1,\ldots,k_n; n\geq 1}$ be an infinitesimal triangular system and $(x_n)_{n\geq 1}$ a sequence in G such that for the sequence $(\mu_n)_{n\geq 1}$ with*

$$\mu_n := \mathop{\ast}_{j=1}^{k_n} \mu_{nj} * \varepsilon_{x_n}$$

(for $n \geq 1$) one has $\lim_{n\to\infty} \mu_n = \mu \in \mathscr{M}^1(G)$.

The following statements are equivalent:
(i) $\mu \in \mathscr{G}_P(G)$;
(ii) (a) $\lim_{n\to\infty} \sum_{j=1}^{k_n} \mu_{nj}(\complement U)=0$ *for all* $U \in \mathfrak{V}(e)$ *and*
(b) $\sup_{n\geq 1} \sum_{j=1}^{k_n} \int (1-\operatorname{Re}\chi(x))\,|\mu_{nj}|^2(dx)<\infty$ *for all* $\chi \in G^\wedge$.

Proof. 1. (i)$\Rightarrow$(ii). Let $\mu \in \mathscr{G}_P(G)$. By Theorem 5.2.9 we obtain $|\mu|^2 \in \mathscr{G}_P(G)$, and we have $\lim_{n\to\infty} \ast_{j=1}^{k_n} |\mu_{nj}|^2=|\mu|^2$. Since $|\mu|^2$ does not admit an idempotent factor, one has $(|\mu|^2)^\wedge(\chi)>0$ for all $\chi \in G^\wedge$ by Theorem 5.1.5. Hence,

$$\begin{aligned}(4.1)\qquad &\sup_{n\geq 1} \sum_{j=1}^{k_n} \int (1-\operatorname{Re}\chi(x))\,|\mu_{nj}|^2(dx)\\ &=\sup_{n\geq 1} \sum_{j=1}^{k_n} (1-|\hat{\mu}_{nj}(\chi)|^2)<\infty.\end{aligned}$$

In analogy to Part 3 of the proof of Theorem 5.1.16 we get

$$\begin{aligned}&|\log \exp(\textstyle\sum_{j=1}^{k_n} |\mu_{nj}|^2)^\wedge(\chi)-\log \prod_{j=1}^{k_n} (|\mu_{nj}|^2)^\wedge(\chi)|\\ &\leq \sup_{1\leq j\leq k_n} |1-(|\mu_{nj}|^2)^\wedge(\chi)|\,(\textstyle\sum_{j=1}^{k_n} (1-(|\mu_{nj}|^2)^\wedge(\chi)))\end{aligned}$$

for all $\chi \in G^\wedge$. Since $(|\mu_{nj}|^2)_{j=1,\dots,k_n;\, n\geq 1}$ is infinitesimal, (4.1) leads to

$$\lim_{n\to\infty} \exp(\textstyle\sum_{j=1}^{k_n} |\mu_{nj}|^2)=|\mu|^2.$$

But Theorem 5.4.1 implies that

$$\lim_{n\to\infty} \textstyle\sum_{j=1}^{k_n} |\mu_{nj}|^2(\complement U)=0,$$

and with the argument of Part 5 of the proof of Theorem 5.1.16 we have finally

$$\lim_{n\to\infty} \textstyle\sum_{j=1}^{k_n} \mu_{nj}(\complement U)=0$$

for every $U \in \mathfrak{V}(e)$. This completes the proof of Condition (a) in Statement (ii).

2. (ii)$\Rightarrow$(i). Condition (ii) (b) implies that $|\mu|^2$ admits no idempotent factor and that

$$\lim_{n\to\infty} \exp(\textstyle\sum_{j=1}^{k_n} |\mu_{nj}|^2)=|\mu|^2.$$

Let $U, V \in \mathfrak{V}(e)$ with $V^{-1}=V$ and $V^2 \subset U$. From

$$\begin{aligned}&|\nu|^2(\complement U)=\int \nu((\complement U)x)\,\nu(dx)\\ &=\int_V \nu((\complement U)x)\,\nu(dx)+\int_{\complement V} \nu((\complement U)x)\,\nu(dx)\leq 2\,\nu(\complement V)\end{aligned}$$

for every $\nu \in \mathscr{M}^1(G)$ we conclude that

$$\lim_{n\to\infty} \textstyle\sum_{j=1}^{k_n} |\mu_{nj}|^2(\complement U)=0$$

by Condition (ii) (a). Theorem 5.4.1 yields the result $|\mu|^2 \in \mathscr{G}_P(G)$. □

5.4.3 Theorem. *Let* $\mu \in \mathscr{M}^1(G)$ *and suppose that for each* $n\geq 1$ *there exist measures* $\mu_{nj} \in \mathscr{M}^1(G)$ *and* $z_{nj} \in G$ $(j=1,\dots,k_n)$ *with the properties*
(i) $\mu=\mu_{n1} * \cdots * \mu_{nk_n}$ *and*
(ii) $\lim_{n\to\infty} \sum_{j=1}^{k_n} (\mu_{nj} * \varepsilon_{z_{nj}})(\complement U)=0$ *for every* $U \in \mathfrak{V}(e)$.
Then $\mu \in \mathscr{G}_P(G)$.

Proof. Let g be a local inner product for G. For every $n \geq 1$ and $j = 1, \ldots, k_n$ we define $\lambda_{nj} := \mu_{nj} * \varepsilon_{z_{nj}}$, $x_{nj} \in G$ by

$$\chi(x_{nj}) := \exp(-i \int g(x, \chi) \lambda_{nj}(dx))$$

for all $\chi \in G^\wedge$, $\nu_{nj} := \lambda_{nj} * \varepsilon_{x_{nj}}$, $x_n := (\prod_{j=1}^{k_n} x_{nj})^{-1}$, $z_n := (\prod_{j=1}^{k_n} z_{nj})^{-1}$ and

$$\mu_n := \exp(\textstyle\sum_{j=1}^{k_n} \nu_{nj}) * \varepsilon_{x_n} * \varepsilon_{z_n}.$$

Evidently the infinitesimality of $(\lambda_{nj})_{j=1,\ldots,k_n; n \geq 1}$ implies that of $(\nu_{nj})_{j=1,\ldots,k_n; n \geq 1}$.

1. If $\sup_{n \geq 1} \sum_{j=1}^{k_n} |1 - \hat{\nu}_{nj}(\chi)| < \infty$ for all $\chi \in G^\wedge$, then $\lim_{n\to\infty} \mu_n = \mu$.

Obviously it suffices to show that $\overline{\lim}_{n \geq 1} |\hat{\mu}_n(\chi) - \hat{\mu}(\chi)| = 0$ holds for all $\chi \in G^\wedge$. For every $n \geq 1$ and $\chi \in G^\wedge$ we have

$$\begin{aligned}\hat{\mu}_n(\chi) &= \chi(x_n z_n) \exp(\textstyle\int (\chi(x) - 1)(\sum_{j=1}^{k_n} \nu_{nj})(dx)) \\ &= \textstyle\prod_{j=1}^{k_n} \chi(x_{nj} z_{nj})^{-1} \exp(\hat{\nu}_{nj}(\chi) - 1) \quad \text{and} \quad \hat{\mu}(\chi) = \prod_{j=1}^{k_n} \hat{\mu}_{nj}(\chi).\end{aligned}$$

An easy computation shows that

$$\begin{aligned}|\hat{\mu}_n(\chi) - \hat{\mu}(\chi)| &\leq \textstyle\sum_{j=1}^{k_n} |\chi(x_{nj} z_{nj})^{-1} \exp(\hat{\nu}_{nj}(\chi) - 1) - \hat{\mu}_{nj}(\chi)| \\ &= \textstyle\sum_{j=1}^{k_n} |\exp(\hat{\nu}_{nj}(\chi) - 1) - \hat{\nu}_{nj}(\chi)| \qquad \text{for all } \chi \in G^\wedge.\end{aligned}$$

Since $(\nu_{nj})_{j=1,\ldots,k_n; n \geq 1}$ is infinitesimal, we have

$$|1 - \hat{\nu}_{nj}(\chi)| \leq \tfrac{1}{2} \qquad \text{for all } n \geq n_0 := n_0(\chi) \text{ and } j = 1, \ldots, k_n.$$

Using the inequality $|\exp(z-1) - z| \leq |1 - z|^2$ for $|1 - z| \leq \frac{1}{2}$ one concludes from the above that

$$\begin{aligned}&\overline{\lim}_{n \geq 1} |\hat{\mu}_n(\chi) - \hat{\mu}(\chi)| \leq \overline{\lim}_{n \geq 1} \textstyle\sum_{j=1}^{k_n} |1 - \hat{\nu}_{nj}(\chi)|^2 \\ &\leq (\overline{\lim}_{n \geq 1} \sup_{1 \leq j \leq k_n} |1 - \hat{\nu}_{nj}(\chi)|)(\sup_{n \geq 1} \textstyle\sum_{j=1}^{k_n} |1 - \hat{\nu}_{nj}(\chi)|)\end{aligned}$$

for all $\chi \in G^\wedge$, which together with

$$\overline{\lim}_{n \geq 1} \sup_{1 \leq j \leq k_n} |1 - \hat{\nu}_{nj}(\chi)| = 0$$

for all $\chi \in G^\wedge$ implies the assertion.

2. $\lim_{n\to\infty} \exp(\sum_{j=1}^{k_n} |\mu_{nj}|^2) = |\mu|^2$.

In fact, from the infinitesimality of $(\lambda_{nj})_{j=1,\ldots,k_n; n \geq 1}$ follows that of $(|\mu_{nj}|^2)_{j=1,\ldots,k_n; n \geq 1}$ and in particular, the existence of an $n_0 := n_0(\chi) \geq 1$ such that $|\hat{\mu}_{nj}(\chi)|^2 > 0$ holds for all $n \geq n_0$, $j = 1, \ldots, k_n$. Since

$$|\hat{\mu}(\chi)|^2 = \textstyle\prod_{j=1}^{k_n} |\hat{\mu}_{nj}(\chi)|^2 \qquad \text{for all } n \geq 1,$$

we obtain

$$|\hat{\mu}(\chi)|^2 > 0 \quad \text{and} \quad \overline{\lim}_{n \geq 1} \textstyle\sum_{j=1}^{k_n} (1 - |\hat{\mu}_{nj}(\chi)|^2) < \infty \qquad \text{for all } \chi \in G^\wedge.$$

By Property (LI 2) of the local inner product g for G we get

$$\int g(x, \chi) |\mu_{nj}|^2 (dx) = \int (\int g(xy^{-1}, \chi) \mu_{nj}(dx)) \mu_{nj}(dy) = 0$$

for all $\chi \in G^\wedge$. Therefore Part 1 applied to $(|\mu_{nj}|^2)_{j=1,\ldots,k_n;n\geq 1}$ in place of $(\lambda_{nj})_{j=1,\ldots,k_n;n\geq 1}$ and to $|\mu|^2$ in place of μ together with the above finiteness condition implies the assertion.

3. Now Condition (ii) of the theorem implies Condition (ii) (a) of Theorem 5.4.1 by a well-known argument and Condition (ii) (b) of that theorem via Part 2. Thus Theorem 5.4.1 yields the assertion. □

5.4.4 Remark. Limit theorems in the sense of de Moivre-Laplace can hardly be established for general Abelian groups, since the general setting lacks norming techniques using second moments. Only in very special cases can the analogy to the classical framework be carried out. Let $G := \prod_{\alpha \in \mathbb{A}} G_\alpha$ be a product of solenoidal groups $G_\alpha := \mathbb{Q}_d^\wedge$ $(\alpha \in \mathbb{A})$.

For every $n \geq 1$ the mapping $\psi_n : G \to G$ defined by

$$\psi_n(x) := x^{\frac{1}{n}} \quad \text{for all } x \in G$$

is an automorphism of G. Let $\mu \in \mathscr{M}^1(G)$ with $\operatorname{supp}(\mu) = G$. Then there exists a $y \in G$ such that with

$$\mu_n := \psi_{[\sqrt{n}]}(\mu * \varepsilon_{y^{-1}})$$

for all $n \geq 1$ we have $\lim_{n\to\infty} \mu_n^n =: \nu \in \mathscr{G}_P^s(G)$.

In fact, for $\alpha \in \mathbb{A}$ and $\chi_\alpha \in G_\alpha^\wedge$ there exists an $r \in \mathbb{Q}$ and for $x \in G_\alpha$ there is a $\tau_\alpha(x) \in \mathbb{R}$ such that $\chi_\alpha(x) = e^{i\tau_\alpha(x)r}$. For every $\alpha \in \mathbb{A}$ the function τ_α can be chosen to be continuous. Moreover, every $\chi \in G^\wedge$ is of the form

$$\chi(x) = \exp\{i \textstyle\sum_{j=1}^s r_j \tau_{\alpha_j}(x_{\alpha_j})\}, \tag{4.2}$$

where $x := (x_\alpha)_{\alpha \in \mathbb{A}} \in G$, $r_1, \ldots, r_s \in \mathbb{Q}$, $\alpha_1, \ldots, \alpha_s \in \mathbb{A}$, $\tau_{\alpha_1}, \ldots, \tau_{\alpha_s}$ as above and $s \geq 1$.

Since for $\alpha \in \mathbb{A}$ the function τ_α is continuous, there exists a $y_\alpha \in G_\alpha$ satisfying

$$\int_G \tau_\alpha(p_\alpha(x))\, \mu(dx) = \tau_\alpha(y_\alpha),$$

where p_α denotes the projection from G onto G_α. In fact, since $\operatorname{supp}(\mu) = G$, $\|\mu\| = 1$ and G is compact and connected, $\tau_\alpha(p_\alpha(G))$ is a compact interval of $\mathbb{R}$, whence

$$\int \tau_\alpha(p_\alpha(x))\, \mu(dx) \in \tau_\alpha(p_\alpha(G)).$$

Let $y := (y_\alpha)_{\alpha \in \mathbb{A}}$ and $\chi \in G^\wedge$ be of the form (4.2). We have

$$\begin{aligned} \mu_n^{n\wedge}(\chi^k) &= [\hat{\mu}_n(\chi^k)]^n \\ &= [\textstyle\int \exp(ik[\sqrt{n}]^{-1} \sum_{j=1}^s r_j \tau_{\alpha_j}(x_{\alpha_j} y_{\alpha_j}^{-1}))\, \mu(dx)]^n. \end{aligned}$$

Expanding the exponential under the integral sign we obtain

$$\begin{aligned} &\textstyle\int \exp\{ik[\sqrt{n}]^{-1} \sum_{j=1}^s r_j \tau_{\alpha_j}(x_{\alpha_j} y_{\alpha_j}^{-1})\}\, \mu(dx) \\ &= 1 - \frac{k^2}{2n} \textstyle\int \{\sum_{j=1}^s r_j \tau_{\alpha_j}(x_{\alpha_j} y_{\alpha_j}^{-1})\}^2\, \mu(dx) + o\left(\frac{1}{n}\right). \end{aligned}$$

Putting

$$d(\chi) := \tfrac{1}{2}[\int \{\sum_{j=1}^{s} r_j \tau_{\alpha_j}(x_{\alpha_j} y_{\alpha_j}^{-1})\}^2 \mu(dx)]$$

one gets

$$\lim_{n\to\infty} \mu_n^{n\wedge}(\chi^k) = e^{-k^2 d(\chi)} \quad \text{for all } \chi \in G^\wedge \text{ and } k \in \mathbb{Z}.$$

If $d(\chi)=0$, it follows that $\sum_{j=1}^{s} r_j \tau_{\alpha_j} = 0\,[\mu]$ and since $\operatorname{supp}(\mu)=G$ we even have equality everywhere, whence $\chi=1$ or $d(\chi)>0$ for all $\chi \in G^\wedge$, $\chi \neq 1$.

The mapping $\chi^k \to e^{-k^2 d(\chi)}$ is the Fourier transform of a measure $\nu \in \mathscr{G}_P^s(G)$ by Theorem 5.2.9. The continuity theorem 1.4.2 yields $\lim_{n\to\infty} \mu_n^n = \nu$.

5.5 Symmetric Gauss Semigroups

In this section the generality of the definition of P-Gauss measures on a locally compact Abelian group G will be illustrated by looking at those P-Gauss measures which are absolutely continuous with respect to the Haar measure on G. Let us define the *special d-dimensional normal distribution* $\nu_{a,A} := n_{a,A} \cdot \lambda^d$ for $a \in \mathbb{R}^d$ and a positive-definite matrix $A \in \mathfrak{M}(d, \mathbb{R})$ with λ^d-density $n_{a,A}$ defined by

$$n_{a,A}(x) := \frac{1}{\sqrt{(2\pi)^d \det A}} \exp(-\tfrac{1}{2}\langle A^{-1}(x-a), x-a\rangle)$$

for all $x \in \mathbb{R}^d$. This distribution has a Fourier transform $\hat{\nu}_{a,A}$ given by

$$\hat{\nu}_{a,A}(t) = \exp(i\langle t, a\rangle - \tfrac{1}{2}\langle At, t\rangle)$$

for all $t \in \mathbb{R}^d$, and it shows the restrictive approach to Gauss measures via densities $n_{a,A}$ and motivates the more general introduction via Fourier transforms $\hat{\nu}_{a,A}$, where the matrix $A \in \mathfrak{M}(d, \mathbb{R})$ is allowed to be positive-semidefinite. These general d-dimensional normal distributions turn out to be precisely the nondegenerate P-Gauss measures on $\mathbb{R}^d$.

The restrictive rôle of the special d-dimensional normal distribution becomes apparent when we look at the following *properties of general d-dimensional normal distributions:*

(1) If $\nu \in \mathscr{N}(\mathbb{R}^d)$ is a special d-dimensional normal distribution, then $\nu = f \cdot \lambda^d$ with an analytic λ^d-density $f > 0$.

(2) If $\nu \in \mathscr{N}(\mathbb{R}^d)$ arbitrary, then $\nu \ll \lambda^d$ iff $\operatorname{supp}(\nu) = \mathbb{R}^d$.

(3) If $\nu \in \mathscr{N}(\mathbb{R}^d)$, then either $\nu \ll \lambda^d$ or $\nu \perp \lambda^d$. More precisely, one has that

(4) for any $\nu \in \mathscr{N}(\mathbb{R}^d)$ there exists a linear subspace V of $\mathbb{R}^d$ of dimension k with $\operatorname{supp}(\nu) = V$;

(5) considered as a measure on V the measure ν is a special k-dimensional normal distribution on V and hence admits an analytic ω_V-density $f > 0$;

(6) if $k < d$, then $\nu \perp \lambda^d$.

Let now G be an arbitrary locally compact Abelian group.

5.5.1 Definition. A convolution semigroup $(\nu_t)_{t\in\mathbb{R}_+^*}$ in $\mathscr{M}^1(G)$ is called a (symmetric) *Gauss semigroup in* $\mathscr{M}^1(G)$ *with corresponding quadratic form* ψ if there exists a positive quadratic form $\psi\in\mathrm{Q}_+(G^\wedge)$ such that $\hat{\nu}_t=\exp(-t\psi)$ holds for all $t\in\mathbb{R}_+^*$.

5.5.2 Remark. By Theorem 5.2.9 every measure ν_t of a Gauss semigroup $(\nu_t)_{t\in\mathbb{R}_+^*}$ in $\mathscr{M}^1(G)$ is an element of $\mathscr{G}_P^s(G)$.

5.5.3 Remark. Referring to the embedding problem discussed in Chapter III we note that the measures in $\mathscr{G}_P^s(G)$ are exactly those which are embeddable with a (symmetric) Gauss embedding semigroup. This again follows from an application of Theorem 5.2.9.

Let G now be a connected Abelian Lie group of the form

$$G:=\mathbb{R}^m\times\mathbb{T}^n \quad \text{with } m+n=:d.$$

We define a continuous epimorphism $p:\mathbb{R}^d:=\mathbb{R}^{m+n}\to G$ by

$$p(x_1,\dots,x_m,x_{m+1},\dots,x_d):=(x_1,\dots,x_m,\exp(2\pi i x_{m+1}),\dots,\exp(2\pi i x_d))$$

for all $(x_1,\dots,x_m,x_{m+1},\dots,x_d)\in\mathbb{R}^d$.

Putting

$$I:=\{(x_1,\dots,x_n)\in\mathbb{R}^n: x_j\in[0,1[\text{ for all } j=1,\dots,n\}$$

and $\lambda_I^n:=\mathrm{Res}_I\,\lambda^n$ we assume without loss of generality that ω_G is of the form $p(\lambda^m\otimes\lambda_I^n)$, and observe that $B\in\mathfrak{B}(\mathbb{R}^d)$ is a λ^d-null set iff $p(B)$ is an ω_G-null set.

5.5.4 Lemma. *For every Gauss semigroup* $(\nu_t)_{t\in\mathbb{R}_+^*}$ *in* $\mathscr{M}^1(G)$ *there exists exactly one Gauss semigroup* $(\mu_t)_{t\in\mathbb{R}_+^*}$ *in* $\mathscr{M}^1(\mathbb{R}^d)$ *such that* $p(\mu_t)=\nu_t$ *holds for all* $t\in\mathbb{R}_+^*$.

Proof. Since by assumption there exists a $\psi\in\mathrm{Q}_+(G^\wedge)$ such that

$$\hat{\nu}_t=\exp(-t\psi) \quad \text{for all } t\in\mathbb{R}_+^*,$$

Application 5.2.6 implies the existence of a unique $\bar{\psi}\in\mathrm{Q}_+(\mathbb{R}^d)$ with the property

$$\mathrm{Res}_{G^\wedge}\,\bar{\psi}=\mathrm{Res}_{\mathbb{R}^m\times\mathbb{Z}^n}\,\bar{\psi}=\psi.$$

Let $(\mu_t)_{t\in\mathbb{R}_+^*}$ be the Gauss semigroup in $\mathscr{M}^1(G)$ with corresponding quadratic form $\bar{\psi}$. From $\hat{\mu}_t(\chi\circ p)=\hat{\nu}_t(\chi)$ for all $\chi\in G^\wedge$ one concludes that $p(\mu_t)=\nu_t$ for every $t\in\mathbb{R}_+^*$. □

5.5.5 Lemma. *Let* $(\nu_t)_{t\in\mathbb{R}_+^*}$ *be a Gauss semigroup in* $\mathscr{M}^1(G)$ *with corresponding quadratic from* ψ *of the form* $\psi(\xi):=\langle A\xi,\xi\rangle$ *for some symmetric, positive-semidefinite matrix* $A\in\mathfrak{M}(d,\mathbb{R})$ *and all* $\xi\in\mathbb{R}^m\times\mathbb{Z}^n$. *Then the following statements are equivalent:*

(i) $\mathrm{supp}(\nu_t)=G$ *for all* $t\in\mathbb{R}_+^*$;

(ii) $\psi(\chi)>0$ *for all* $\chi\in G^\wedge$ *with* $\chi\neq 1$;

(iii) $\langle A\xi, \xi\rangle > 0$ *for all* $\xi \in \mathbb{R}^m \times \mathbb{Z}^n$ *with* $\xi \neq 0$;
(iv) $\operatorname{supp}(\chi(\nu_t)) = \mathbb{T}$ *for all* $\chi \in G^\wedge$ *with* $\chi \neq 1$ *and* $t \in \mathbb{R}_+^*$.

Proof. It suffices to show the chain (i) $\Rightarrow$ (iv) $\Rightarrow$ (ii) $\Rightarrow$ (i).

1. (i) $\Rightarrow$ (iv) follows from the equalities

$$\operatorname{supp}(\chi(\nu_t)) = \overline{\chi(\operatorname{supp}(\nu_t))} = \overline{\chi(G)} = \mathbb{T},$$

valid for all $\chi \in G^\wedge$ with $\chi \neq 1$.

2. If G is any locally compact Abelian group, $\mu \in \mathscr{M}^1(G)$ and if there exists a $\chi \in G^\wedge$ with $\chi \neq 1$ such that $\hat{\mu}(\chi) = e^{i\vartheta}$ holds for some $\vartheta \in \mathbb{R}$, then $\operatorname{supp}(\mu)$ is contained in a coset of a proper closed subgroup of G. This follows immediately from the assumption since

$$\int |\chi(x) - e^{i\vartheta}|^2 \, \mu(dx) = 0 \quad \text{implies}$$

$$\operatorname{supp}(\mu) \subset \{x \in G : \chi(x) = e^{i\vartheta}\} =: H_\vartheta = x_0 H_0$$

for some $x_0 \in H_\vartheta$ and the proper closed subgroup $H_0 := \{x \in G : \chi(x) = 1\}$ of G.

3. (iv) $\Rightarrow$ (ii). By Part 2 we conclude from the assumption that

$$1 \neq |\chi(\nu_t)^\wedge(\mathrm{Id}_{\mathbb{T}})| = \hat{\nu}_t(\chi) = \exp(-t\,\psi(\chi))$$

for all $\chi \in G^\wedge$ with $\chi \neq 1$. But this implies $\psi(\chi) > 0$ for all $\chi \in G^\wedge$, $\chi \neq 1$.

4. (ii) $\Rightarrow$ (i). By Lemma 5.5.4 there exists a Gauss semigroup $(\mu_t)_{t \in \mathbb{R}_+^*}$ in $\mathscr{M}^1(\mathbb{R}^d)$ such that $p(\mu_t) = \nu_t$ for all $t \in \mathbb{R}_+^*$.

Furthermore, there is a vector subgroup V of $\mathbb{R}^d$ with $\operatorname{supp}(\mu_t) = V$. Putting $H := p(V)$ one obtains $\operatorname{supp}(\nu_t) = \overline{H}$ for all $t \in \mathbb{R}_+^*$. If $\overline{H} \neq G$, then there exists a $\chi \in G^\wedge$ with $\chi \neq 1$ and $\chi(\overline{H}) = 1$, which implies that

$$1 = \int_{\overline{H}} \chi \, d\nu_t = \hat{\nu}_t(\chi) = \exp(-t\,\psi(\chi))$$

and thus $\psi(\chi) = 0$, contradicting the hypothesis. □

5.5.6 Theorem. *Let* $G := \mathbb{R}^m \times \mathbb{T}^n$ *be a connected Abelian Lie group of dimension* $d := m + n$ *and* $(\nu_t)_{t \in \mathbb{R}_+^*}$ *a Gauss semigroup in* $\mathscr{M}^1(G)$ with corresponding matrix $A \in \mathfrak{M}(d, \mathbb{R})$ *in the sense of Lemma* 5.5.5. *If* A *is positive-definite (not positive-definite), then*

$$\nu_t \ll \omega_G \quad (\nu_t \perp \omega_G) \quad \textit{for all } t \in \mathbb{R}_+^*.$$

Proof. By Lemma 5.5.4 there exists a Gauss semigroup $(\mu_t)_{t \in \mathbb{R}_+^*}$ in $\mathscr{M}^1(\mathbb{R}^d)$ with $p(\mu_t) = \nu_t$ for all $t \in \mathbb{R}_+^*$. Furthermore, for every $t \in \mathbb{R}_+^*$ and all $\xi \in \mathbb{R}^d$ we have $\hat{\mu}_t(\xi) = \exp(-t\langle A\xi, \xi\rangle)$.

Clearly $\mu_t \ll \lambda^d$ $(\mu_t \perp \lambda^d)$ iff $A \in \mathfrak{M}(d, \mathbb{R})$ is positive-definite (not positive-definite). The assertion follows from the fact that $B \in \mathfrak{B}(\mathbb{R}^d)$ is a λ^d-null set iff $p(B) \in \mathfrak{B}(G)$ is an ω_G-null set. □

5.5.7 Corollary. *For every Gauss measure* $\nu \in \mathscr{M}^1(G)$ *with* $\nu \ll \omega_G$ *we have* $\operatorname{supp}(\nu) = G$.

The *proof* follows from the theorem together with Lemma 5.5.5. □

5.5.8 Remark. The converse of the statement in Corollary 5.5.7 is in general false. In fact, let $G:=\mathbb{T}^2$ and $f:\mathbb{R}\to G$ be defined by

$$f(t):=(e^{it}, e^{it\alpha}) \qquad \text{for all } t\in\mathbb{R},$$

where $\alpha\in\mathbb{R}$ is irrational. Denoting $f(\mathbb{R})$ by H we obtain $\bar{H}=G$, but $H\neq G$, and thus $\omega_G(H)=0$. Let $\mu\in\mathcal{N}(\mathbb{R})$. Then $\nu:=f(\mu)$ is a Gauss measure in $\mathcal{M}^1(G)$ with

$$\operatorname{supp}(\nu)=\overline{f(\operatorname{supp}(\mu))}=\overline{f(\mathbb{R})}=\bar{H}=G.$$

But by $\nu(H)=1$ and $\omega_G(H)=0$ we have $\nu\perp\omega_G$.

5.5.9 Example. Let $G:=\mathbb{T}^d$ for $d\geq 1$ and $(\nu_t)_{t\in\mathbb{R}_+^*}$ a Gauss semigroup in $\mathcal{M}^1(G)$ with positive-definite matrix $A\in\mathfrak{M}(d,\mathbb{R})$. If $\nu_t\ll\omega_G$, then the function f_t on $\mathbb{T}^d$ defined by

$$(5.1) \qquad f_t(z_1,\ldots,z_d):=\sum_{\mathfrak{m}:=(m_1,\ldots,m_d)\in\mathbb{Z}^d}\exp(-t\langle A\,\mathfrak{m},\mathfrak{m}\rangle)\, z_1^{m_1}\cdot\ldots\cdot z_d^{m_d}$$

for all $(z_1,\ldots,z_d)\in\mathbb{T}^d$ is an ω_G-density of ν_t for all $t\in\mathbb{R}_+^*$.

Indeed, let $\|A\|_s$ denote the spectral norm of $A\in\mathfrak{M}(d,\mathbb{R})$. Then

$$\|A^{-1}\|_s^{-1}\,\|\mathfrak{m}\|^2\leq\langle A\mathfrak{m},\mathfrak{m}\rangle\leq\|A\|_s\,\|\mathfrak{m}\|^2 \quad \text{for all } \mathfrak{m}\in\mathbb{Z}^d.$$

Putting $\gamma:=\|A^{-1}\|_s^{-1}$ we obtain for all $(z_1,\ldots,z_d)\in\mathbb{C}^d$ such that $|z_1|=\cdots=|z_d|=1$ and $t\in\mathbb{R}_+^*$ the inequalities

$$\begin{aligned}&\sum_{\mathfrak{m}:=(m_1,\ldots,m_d)\in\mathbb{Z}^d}|\exp(-t\langle A\,\mathfrak{m},\mathfrak{m}\rangle)\, z_1^{m_1}\cdot\ldots\cdot z_d^{m_d}|\\ &\leq\Big(\sum_{m\in\mathbb{Z}}\exp(-t\gamma m^2)\Big)^d<\infty,\end{aligned}$$

and thus the absolute and uniform convergence of the series in (6.1) to f_t. That $\nu_t=f_t\cdot\omega_G$ for all $t\in\mathbb{R}_+^*$ follows from [218] (31.13).

In the special case $d=1$ with $A:=(a)\in\mathfrak{M}(1,\mathbb{R})$ we obtain for every $t\in\mathbb{R}_+^*$ and $x\in\mathbb{R}$ the representation

$$f_t(\exp(2\pi\, ix))=\vartheta\left(\frac{t\,a}{\pi}, x\right),$$

where ϑ is the theta function defined by

$$\vartheta(\alpha, x):=\sqrt{\frac{1}{\alpha}}\sum_{m=-\infty}^{\infty}\exp\left(-\frac{\pi}{\alpha}(x-m)^2\right)$$

for $\alpha\in\mathbb{R}_+^*$ and $x\in\mathbb{R}$.

In fact, for $t\in\mathbb{R}_+^*$ one has $\nu_t=p(\mu_t)$ with $\mu_t=g_t\cdot\lambda^1\in\mathcal{N}(\mathbb{R})$, where

$$g_t(x):=\sqrt{\frac{\pi}{t\,a}}\exp\left(-\frac{\pi^2}{t\,a}x^2\right)=\sqrt{\frac{1}{\alpha}}\exp\left(-\frac{\pi}{\alpha}x^2\right)$$

for all $x \in \mathbb{R}$ with the notation $\alpha := \frac{ta}{\pi}$. The formula

$$f_t(\exp(2\pi i x)) = \sum_{m=-\infty}^{\infty} g_t(x-m),$$

valid for all $t \in \mathbb{R}_+^*$ and $x \in \mathbb{R}$, yields the result.

We proceed with the discussion of absolutely continuous Gauss semigroups to arbitrary connected locally compact Abelian groups G. Those groups G are the projective limits $\varprojlim_{\alpha \in \mathbb{A}} G_\alpha$ of Lie groups $G_\alpha := G/K_\alpha$ of the form $G_\alpha := \mathbb{R}^{m_\alpha} \times \mathbb{T}^{n_\alpha}$ with $m_\alpha, n_\alpha \in \mathbb{Z}_+$ $(\alpha \in \mathbb{A})$ for a descending family $(K_\alpha)_{\alpha \in \mathbb{A}}$ of compact subgroups K_α of G with $\bigcap_{\alpha \in \mathbb{A}} K_\alpha = \{e\}$. As usual p_α denotes for every $\alpha \in \mathbb{A}$ the canonical mapping from G onto G_α.

5.5.10 Lemma. *Let $(\nu_t)_{t \in \mathbb{R}_+^*}$ be a Gauss semigroup in $\mathscr{M}^1(G)$ with corresponding quadratic form ψ. Then the following statements are equivalent:*
(i) $\operatorname{supp}(\nu_t) = G$ *for all* $t \in \mathbb{R}_+^*$;
(ii) $\psi(\chi) > 0$ *for all* $\chi \in G^\wedge$ *with* $\chi \neq 1$;
(iii) $\operatorname{supp}(\chi(\nu_t)) = \mathbb{T}$ *for all* $\chi \in G^\wedge$ *with* $\chi \neq 1$.

Proof. We have $G = \varprojlim_{\alpha \in \mathbb{A}} G_\alpha$. For every $\alpha \in \mathbb{A}$ the family $(p_\alpha(\nu_t))_{t \in \mathbb{R}_+^*}$ is a Gauss semigroup in $\mathscr{M}^1(G_\alpha)$ with corresponding quadratic form ψ_α defined by

$$\psi_\alpha(\chi_\alpha) := \psi(\chi_\alpha \circ p_\alpha) \qquad \text{for all } \chi_\alpha \in G_\alpha^\wedge \ (\alpha \in \mathbb{A}).$$

For every $\chi \in G^\wedge$ there exist an $\alpha \in \mathbb{A}$ and a $\chi_\alpha \in G_\alpha^\wedge$ such that $\chi = \chi_\alpha \circ p_\alpha$ is satisfied. One also notes that

$$\operatorname{supp}(\nu_t) = G \quad \text{iff} \quad \operatorname{supp}(p_\alpha(\nu_t)) = G_\alpha$$

for all $\alpha \in \mathbb{A}$ (and $t \in \mathbb{R}_+^*$). Lemma 5.5.5 yields the assertion. □

The following result extends Corollary 5.2.12.

5.5.11 Theorem. *For any Gauss semigroup $(\nu_t)_{t \in \mathbb{R}_+^*}$ in $\mathscr{M}^1(G)$ there exists a connected closed subgroup H of G such that* $\operatorname{supp}(\nu_t) = H$ *for all* $t \in \mathbb{R}_+^*$.

Proof. Let ψ be the quadratic form corresponding to $(\nu_t)_{t \in \mathbb{R}_+^*}$. The set

$$B := B_\psi := \{\chi \in G^\wedge : \psi(\chi) = 0\}$$

is a closed subgroup of $G^\wedge$. For every $\chi \in G^\wedge$ and $\eta \in B$ one has $\psi(\chi\eta) = \psi(\chi)$. Indeed, let Ψ be the bilinear form corresponding to ψ according to Lemma 5.2.4. Since $G^\wedge = \bigcup_{\alpha \in \mathbb{A}} G_\alpha^\wedge$, Application 5.2.6 yields $\Psi(\chi, \eta)^2 \leq \psi(\chi)\psi(\eta)$ for all $\chi, \eta \in G^\wedge$, and thus $\Psi(\chi, \eta) = 0$ for all $\chi \in G^\wedge$, $\eta \in B$.

Consequently, there exists a $\bar{\psi} \in \mathrm{Q}_+(G^\wedge/B)$ with $\psi = \bar{\psi} \circ p$, where p denotes the canonical mapping from $G^\wedge$ onto $G^\wedge/B$. Obviously $\bar{\psi}(\bar{\chi}) > 0$ for all $\bar{\chi} \in G^\wedge/B$ with $\bar{\chi} \neq 1$.

The subgroup $H:=B^{\perp}$ of G is closed, and $H^{\wedge}\cong G^{\wedge}/B$. We show that H is connected or equivalently, that any compact subgroup K of $G^{\wedge}/B$ is trivial. Let K be such a subgroup. There exists an $\alpha\in\mathbb{R}_{+}^{*}$ with $0\leq\bar{\psi}(\bar{\chi})\leq\alpha$ for all $\bar{\chi}\in K$. But $n^2\,\bar{\psi}(\bar{\chi})=\bar{\psi}(\bar{\chi}^n)\leq\alpha$ for all $\bar{\chi}\in K$, $n\geq 1$ implies that $\bar{\psi}(\bar{\chi})=0$, whence $\bar{\chi}=1$ and therefore $K=\{1\}$. Finally, $\bar{\psi}$ defines a Gauss semigroup $(\bar{\nu}_t)_{t\in\mathbb{R}_+^*}$ in $\mathscr{M}^1(H)$.

But $\bar{\nu}_t=\nu_t$ for all $t\in\mathbb{R}_{+}^{*}$. Hence, Lemma 5.5.10 implies that supp $(\nu_t)=H$ for all $t\in\mathbb{R}_{+}^{*}$. □

5.5.12 Example. Let $G:=\prod_{n\geq 1}G_n$ with $G_n:=\mathbb{T}$ for all $n\geq 1$. Given a sequence $(a_n)_{n\geq 1}$ in $\mathbb{R}_{+}^{*}$ we define a sequence $(\psi_n)_{n\geq 1}$ in $\mathrm{Q}_{+}(G_n^{\wedge})$ by $\psi_n(m):=a_n\,m^2$ for all $m\in G_n^{\wedge}\cong\mathbb{Z}$.

For every $t\in\mathbb{R}_{+}^{*}$ the measure $\nu_t^{(n)}$ defined by $\nu_t^{(n)\wedge}:=\exp(-t\,\psi_n)$ is in $\mathscr{G}_P^s(G_n)$ and $\nu_t:=\bigotimes_{n\geq 1}\nu_t^{(n)}\in\mathscr{G}_P^s(G)$ with $\hat{\nu}_t=\exp(-t\,\psi)$ for $\psi\in\mathrm{Q}_{+}(G^{\wedge})$ given by

$$\psi(\chi):=\sum_{n\geq 1}\psi_n(\chi_n)\qquad\text{for all }\chi:=(\chi_n)_{n\geq 1}\in G^{\wedge}.$$

$(\nu_t)_{t\in\mathbb{R}_+^*}$ is a Gauss semigroup in $\mathscr{M}^1(G)$.

(1) Given $t\in\mathbb{R}_{+}^{*}$ we have $\nu_t^{(n)}\sim\omega_{G_n}$ for all $n\geq 1$.

(2) By Kakutani's theorem, either $\nu_t\sim\omega_G$ or $\nu_t\perp\omega_G$ for all $t\in\mathbb{R}_{+}^{*}$.

(3) For $t\in\mathbb{R}_{+}^{*}$ the measure ν_t admits a square ω_G-integrable density iff $\sum_{n\geq 1}\exp(-2t\,a_n)<\infty$.

For every $n\geq 1$ one defines the closed subgroup $K_n:=\{(z_k)_{k\geq 1}\in G\colon z_k=1$ for all $k<n\}$ of G and on $K_n^{\perp}$ a positive quadratic form ψ by $\psi(\mathrm{m}):=\sum_{k<n}a_k\,m_k^2$ for all $\mathrm{m}:=(m_k)_{k\geq 1}\in K_n^{\perp}$.

From

$$\begin{aligned}&\int\hat{\nu}_t(\chi)^2\,\omega_{G^{\wedge}}(d\chi)=\lim_{n\to\infty}\int_{K_n^{\perp}}\hat{\nu}_t(\chi)^2\,\omega_{G^{\wedge}}(d\chi)\\&=\lim_{n\to\infty}\sum_{\mathrm{m}\in K_n^{\perp}}\exp(-2t\,\psi(\mathrm{m}))=\lim_{n\to\infty}\prod_{k<n}\Big(\sum_{m\in\mathbb{Z}}\exp(-2t\,a_k\,m^2)\Big)\end{aligned}$$

follow the equivalences

$$\begin{aligned}&\sum_{n\geq 1}\sum_{m\geq 1}\exp(-2t\,a_n\,m^2)<\infty\\&\Leftrightarrow\prod_{n\geq 1}\sum_{m\in\mathbb{Z}}\exp(-2t\,a_n\,m^2)<\infty\\&\Leftrightarrow\int\hat{\nu}_t(\chi)^2\,\omega_{G^{\wedge}}(d\chi)<\infty\end{aligned}$$

which by (31.18) and (31.33) of [218] are equivalent to the existence of a square ω_G-integrable density for ν_t (for all $t\in\mathbb{R}_{+}^{*}$).

Noting that $\sum_{n\geq 1}\exp(-a_n)<\infty$, we see that if there exists a $k\geq 1$ with $\sum_{n\geq 1}\frac{1}{a_n^k}<\infty$, then we obtain immediately for the choice $a_n:=n^2$ for all $n\geq 1$ $\nu_t\sim\omega_G$, hence $\nu_s\sim\nu_t$ for all $s,t\in\mathbb{R}_{+}^{*}$, and for the choice $a_n:=1$ for all $n\geq 1$ the relation $\nu_t\perp\omega_G$ for all $t\in\mathbb{R}_{+}^{*}$, hence $\nu_s\perp\nu_t$ for all $s,t\in\mathbb{R}_{+}^{*}$ with $s\neq t$.

5.5.13 Theorem. *Let G be a connected locally compact Abelian group. The following statements are equivalent:*

(i) *There exists a Gauss measure $\nu\in\mathscr{G}_P^s(G)$ with $\nu\ll\omega_G$;*

(ii) *G has a countable basis of its topology and is locally connected.*

Proof. 1. (i) $\Rightarrow$ (ii). By assumption G is of the form $\mathbb{R}^n \times K$ with $n \geq 0$ and a connected, compact Abelian group K. Let $\nu \in \mathscr{G}_P^s(G)$ with $\nu \ll \omega_G$. Denoting by p the canonical projection from G onto K we obtain $p(\nu) \in \mathscr{G}_P^s(K)$ with $p(\nu) \ll \omega_K$. Thus we may assume without loss of generality that G is compact. Let f be the ω_G-density of ν and ψ the quadratic form corresponding to ν. Since

$$\hat{f} = \hat{\nu} \in \mathscr{C}^0(G^\wedge) \quad \text{and} \quad \hat{\nu} = \exp(-\psi) \neq 0,$$

the discrete group $G^\wedge$ is necessarily countable, which implies that G admits a countable basis of its topology.

We now infer from the statements following Theorem H that G is locally connected iff every finite-dimensional factor group of G is locally connected. We therefore assume without loss of generality that G is finite-dimensional of dimension $n \geq 1$. Then by [218], (24.25) the group $H := G^\wedge$ is discrete, torsion free and of rank d. Since by [218], (A.16) the minimal divisible extension of H is torsion-free and of rank $d \geq 1$ it is isomorphic to $\mathbb{Q}_d^n$, and we may assume H to be a subgroup of $\mathbb{Q}_d^n$.

The quadratic form ψ corresponding to the given Gauss measure $\nu \in \mathscr{G}_P^s(H)$ can be extended by Lemma 5.2.5 to a positive quadratic form on $\mathbb{Q}_d^n$. Thus there exists a symmetric, positive-semidefinite matrix $A \in \mathfrak{M}(n, \mathbb{R})$ with $A \neq 0$ such that $\psi(\xi) = \langle A\xi, \xi \rangle$ holds for all $\xi \in H$. Putting $\gamma := \|A\|_s$ we conclude that there exist only finitely many $\xi \in H$ with $\|\xi\|^2 < \frac{\log 2}{\gamma}$. In fact, from

$$\psi(\xi) = \langle A\xi, \xi \rangle \leq \gamma \|\xi\|^2 \quad \text{and} \quad \|\xi\|^2 < \frac{\log 2}{\gamma}$$

follows $\psi(\xi) < \log 2$ and hence

$$\tfrac{1}{2} < \exp(-\psi(\xi)) = \hat{\nu}(\xi) \quad \text{for all } \xi \in H.$$

Since $\hat{\nu} = \hat{f} \in \mathscr{C}^0(H)$, there exist only finitely many $\xi \in H$ satisfying $\hat{\nu}(\xi) > \frac{1}{2}$, and the assertion is proved. We now have the existence of a $\delta > 0$ such that

$$H \cap \{\xi \in \mathbb{R}^n : \|\xi\|^2 < \delta\} = \{0\};$$

thus H is closed in $\mathbb{R}^n$ and therefore isomorphic to $\mathbb{Z}$. But then $G \cong \mathbb{T}^n$ is locally connected.

2. (ii) $\Rightarrow$ (i). By the structure theorem B there exist $n \geq 0$ and $\mathfrak{a} \geq 0$ or $\mathfrak{a} = \mathbb{N}$ such that $G \cong \mathbb{R}^n \times \mathbb{T}^{\mathfrak{a}}$ holds. If $\mathfrak{a}$ is finite, then G is a Lie group and the assertion has been proved in Theorem 5.5.6. It remains to carry out the proof for $\mathfrak{a} = \mathbb{N}$. Without loss of generality we assume that $G := \mathbb{T}^{\mathfrak{a}}$ since in the case $G := G_1 \times G_2$ with $G_1 := \mathbb{R}^n$ and $G_2 := \mathbb{T}^{\mathfrak{a}}$ the product $\nu := \nu_1 \otimes \nu_2$ of Gauss measures $\nu_i \in \mathscr{G}_P^s(G_i)$ with ω_G-densities f_i $(i = 1, 2)$ is a Gauss measure $\nu \in \mathscr{G}_P^s(G)$ with $\omega_G (= \omega_{G_1} \otimes \omega_{G_2})$-density f defined by

$$f(x_1, x_2) := f_1(x_1) f_2(x_2) \quad \text{for all } (x_1, x_2) \in G_1 \times G_2.$$

The proof of the statement for $G := \mathbb{T}^{\mathfrak{a}}$ with $\mathfrak{a} := \mathbb{N}$ follows from Example 5.5.12. $\square$

5.6 Additive Processes and Their Decomposition

In Chapter II we discussed sequences $(X_n)_{n\geq 1}$ of random variables on a given probability space $(\Omega, \mathfrak{A}, P)$ with values in a locally compact group G. These sequences were introduced in order to describe the corresponding convolution sequences of distributions in $\mathscr{M}^1(G)$. Generalizing the concept of a sequence of G-random variables one arrives at the notion of a (stochastic) process with values in G defined as a family $(X_t)_{t\in\mathbb{R}_+}$ of G-random variables X_t on $(\Omega, \mathfrak{A}, P)$. Such processes enable us to analyze the probabilistic meaning of convolution semi- and hemigroups in $\mathscr{M}^1(G)$. We develop part of the theory for continuous hemigroups $(\mu_{s,t})_{s,t\in[0,1]}$ in $\mathscr{M}^1(G)$ within the framework of a locally compact Abelian group G admitting a countable basis of its topology.

Let G be a locally compact Abelian group having a countable basis of its topology and hence an invariant metric ρ. By $\mathsf{D}_G[0,1]$ we denote the set of all functions from $[0,1]$ into G which are right continuous and admit left-hand limits. Functions in $\mathsf{D}_G[0,1]$ therefore allow only jumps as points of discontinuity. For every $f\in\mathsf{D}_G[0,1]$ and $t\in[0,1]$ one puts

$$f(t+) := \lim_{s\downarrow t} f(s) \quad \text{and} \quad f(t-) := \lim_{s\uparrow t} f(s).$$

Given an $\varepsilon>0$ there exist at most finitely many points t in $[0,1]$ satisfying $\rho(f(t), f(t-))>\varepsilon$.

5.6.1 Definition. A stochastic process $(X_t)_{t\in[0,1]}$ on $(\Omega, \mathfrak{A}, P)$ with values in G is called an *additive process* (or a process with independent increments) if the following conditions are satisfied:

(AP1) (Norming property.) $X_0 = e$.

(AP2) (Independence of increments.) For every sequence $t_0<t_1<\cdots<t_n$ in $[0,1]$ the G-random variables $X_{t_0}, X_{t_0}^{-1}X_{t_1}, \ldots, X_{t_{n-1}}^{-1}X_{t_n}$ are stochastically independent.

(AP3) (Stochastic continuity.) The process $(X_t)_{t\in[0,1]}$ is stochastically continuous in the sense that for every $\varepsilon>0$ one has $\lim_{t\to t_0} P[\rho(X_t, X_{t_0})>\varepsilon]=0$ whenever $t_0\in[0,1]$.

(AP4) For every $\omega\in\Omega$ the path $t\to X_t(\omega)$ from $[0,1]$ into G belongs to $\mathsf{D}_G[0,1]$.

It should be noted that by a well-known result of J.R. Kinney (see Theorem 3 in [474], p. 238) for any stochastic process $(X_t)_{t\in[0,1]}$ with values in G satisfying Conditions (AP1) to (AP3) above there exists an additive process with values in G which is stochastically equivalent to $(X_t)_{t\in[0,1]}$.

5.6.2 Remark. To every additive process $(X_t)_{t\in[0,1]}$ with values in G there corresponds a continuous convolution hemigroup $(\mu_{s,t})_{s,t\in[0,1]}$ in $\mathscr{M}^1(G)$ defined by $\mu_{s,t} := (X_s^{-1}X_t)(P)$ for all $s,t\in[0,1]$ with $s\leq t$. Conversely, every continuous convolution hemigroup $(\mu_{s,t})_{s,t\in[0,1]}$ in $\mathscr{M}^1(G)$ corresponds in this way to an additive process $(X_t)_{t\in[0,1]}$ with values in G.

If, in particular, the distribution $\mu_{s,t}$ of the increment $X_s^{-1}X_t$ depends only on the difference $t-s$ for $s,t\in[0,1]$ with $s\leq t$, then the continuous convolution hemigroup corresponding to the additive process $(X_t)_{t\in[0,1]}$ with values in G is in fact an $\{e\}$-continuous convolution *semi*group. In other words, $\{e\}$-continuous convolution semigroups in $\mathscr{M}^1(G)$ correspond to so-called time-homogeneous additive processes with values in G.

5.6.3 Definition. An additive process $(X_t)_{t\in[0,1]}$ with values in G is called a *Gauss process* (with values in G) if for every $t\in[0,1]$ the distribution $\mu_t:=P_{X_t}$ of X_t is a Gauss measure in $\mathscr{G}_P(G)$.

A homogeneous Gauss process $(X_t)_{t\in[0,1]}$ with values in G, for which $\mu_t\in\mathscr{G}_P^s(G)$ whenever $t\in[0,1]$ is called a *Wiener process* (with values in G).

Gauss processes with values in G can be characterized by the properties of their paths, as the following result shows.

5.6.4 Theorem. *Let G be a locally compact Abelian group having a countable basis of its topology (governed by an invariant metric ρ) and $(X_t)_{t\in[0,1]}$ an additive process on $(\Omega,\mathfrak{A},P)$ with values in G. If for every $\omega\in\Omega$ the path $t\to X_t(\omega)$ is continuous, then $(X_t)_{t\in[0,1]}$ is a Gauss process.*

Proof. Using the metric ρ defining the topology of G we introduce for every $n\geq 1$ the $\mathbb{R}$-random variable

$$S_n:=\sup\left\{\rho(X_t,X_s): t,s\in[0,1], |t-s|\leq\frac{1}{n}\right\} \qquad (\text{on } (\Omega,\mathfrak{A},P)).$$

Clearly S_n is determined by the G-random variables X_t for $t\in\mathbb{Q}\cap[0,1]$. Since for every $\omega\in\Omega$ the path $t\to X_t(\omega)$ is a uniformly continuous function on $[0,1]$ we get $\lim_{n\to\infty}S_n=0$ and hence for $\varepsilon>0$ the limit relation $\lim_{n\to\infty}P[S_n>\varepsilon]=0$. We now choose $t\in[0,1]$ and a sequence $0=:t_{n0}<t_{n1}<\cdots<t_{nn}:=t$ such that $\max_{1\leq j\leq n}(t_{nj}-t_{n,j-1})\leq\frac{1}{n}$. Then the set inclusions

$$\bigcup_{j=1}^n[\rho(X_{t_{nj}},X_{t_{n,j-1}})>\varepsilon]\subset[S_n>\varepsilon] \quad \text{and}$$
$$[\rho(X_{t_{nj}},X_{t_{n,j-1}})\leq\varepsilon \text{ for } j=1,\dots,k-1]\supset[S_n\leq\varepsilon]$$

imply

$$\begin{aligned}&P[S_n>\varepsilon]\\ &\geq P\Big(\bigcup_{k=1}^n[\rho(X_{t_{nj}},X_{t_{n,j-1}})\leq\varepsilon \text{ for } j=1,\dots,k-1 \text{ and } \rho(X_{t_{nk}},X_{t_{n,k-1}})>\varepsilon]\Big)\\ &=\sum_{k=1}^n P[\rho(X_{t_{nj}},X_{t_{n,j-1}})\leq\varepsilon \text{ for } j=1,\dots,k-1]\,P[\rho(X_{t_{nk}},X_{t_{n,k-1}})>\varepsilon]\\ &\geq\sum_{k=1}^n P[S_n\leq\varepsilon]\,P[\rho(X_{t_{nk}},X_{t_{n,k-1}})>\varepsilon],\end{aligned}$$

whence

$$\sum_{k=1}^n P[\rho(X_{t_{nk}},X_{t_{n,k-1}})>\varepsilon]\leq\frac{P[S_n>\varepsilon]}{1-P[S_n>\varepsilon]}$$

or

$$\lim_{n\to\infty}\sum_{k=1}^n P[\rho(X_{t_{nk}},X_{t_{n,k-1}})>\varepsilon]=0.$$

Thus for any neighborhood $U\in\mathfrak{B}(e)$ we obtain

$$\lim_{n\to\infty}\sum_{k=1}^{n}P[X_{t_{nk}}X_{t_{n,k-1}}^{-1}\in\complement U]=0.$$

Moreover, we have

$$X_t=\prod_{k=1}^{n}X_{t_{nk}}X_{t_{n,k-1}}^{-1}.$$

If we denote the distribution of $X_{t_{nk}}X_{t_{n,k-1}}^{-1}$ by μ_{nk} and observe that the G-random variables $X_{t_{nk}}X_{t_{n,k-1}}^{-1}$ $(k=1,\dots,n)$ are stochastically independent, then the representation of X_t yields

$$\mu_t:=P_{X_t}=\ast_{k=1}^{n}\mu_{nk}$$

and the limit relation above reads

$$\lim_{n\to\infty}\sum_{k=1}^{n}\mu_{nk}(\complement U)=0.$$

But then Theorem 5.4.3 applies and we get $\mu_t\in\mathscr{G}_P(G)$. This is valid for every $t\in[0,1]$, and the assertion is proved. □

In the following discussion we aim at the decomposition of a general additive process with values in G with one factor being a Gauss process.

5.6.5 Preparations. Let $U\in\mathfrak{B}(e)$ and $\mathfrak{B}_{\complement U}:=\mathfrak{B}(G)\cap\complement U$. We are given $f\in\mathsf{D}_G[0,1]$ and recall that for every $t\in[0,1]$ and $B\in\mathfrak{B}_{\complement U}$ there are at most finitely many discontinuities $t_0\in[0,t]$ of f, whose jumps $f(t_0)f(t_0-)^{-1}$ lie in B.

For $t\in[0,1]$, $B\in\mathfrak{B}_{\complement U}$ and for these jumps we define

$$f(t,B):=\prod_{t_0\in[0,t]}(f(t_0)\,f(t_0-)^{-1}).$$

Moreover, $f^{\complement B}(t):=f(t)\,f(t,B)^{-1}$.

Clearly, $f(t,B)$ is an element of G and the mapping $t\to f^{\complement B}(t)$ is a G-valued function on $[0,1]$ admitting no jumps which lie in B. For every $n\geq 1$ let a sequence $(t_{nj})_{j=0,\dots,n}$ in $[0,1]$ be given satisfying

$$0=:t_{n0}<t_{n1}<\dots<t_{nn}:=t$$

and such that $(t_{n+1,j})_{j=1,\dots,n+1}$ contains all points of $(t_{nj})_{j=1,\dots,n}$ and that $\lim_{n\to\infty}\max_{1\leq j\leq n}(t_{nj}-t_{n,j-1})=0$. We furthermore define for every $n\geq 1$ and $j=1,\dots,n$ elements f_{nj} of G by

$$f_{nj}:=\psi_U(f^U(t_{nj})(f^U(t_{n,j-1}))^{-1}),\quad\text{where}$$

$$\psi_U(x):=\begin{cases}x, & \text{if } x\in U\\ e, & \text{if } x\notin U.\end{cases}$$

(1) One has $\lim_{n\to\infty}\prod_{j=1}^{n}f_{nj}=f^U(t)$.

Indeed, let $V\in\mathfrak{B}(e)$ be an open neighborhood of e with $VV^{-1}\subset U$. Since f^V has no jumps lying in $\complement V$, we conclude from $f^V\in\mathsf{D}_G[0,1]$ that for every $s_0\in[0,1]$

there exists a $\delta(s_0)>0$ such that

$$f^V(s)f^V(s_0)^{-1}\in V \quad \text{for all } s\in]s_0-\delta(s_0), s_0+\delta(s_0)[\cap[0,t].$$

By the compactness of $[0,t]$ there exists a $\delta'>0$ satisfying

$$f^V(s)f^V(s')^{-1}\in VV^{-1}\subset U \quad \text{for all } s-s'\in]0,\delta'[\cap[0,t].$$

Let $t_1,\dots,t_m$ denote the discontinuities of f in $[0,t]$ whose jumps lie in $U\setminus V$. Since U is open, there is a $\delta''>0$ such that for all $j=1,\dots,m$ the inequalities

$$t_j+\delta''>s\geq t_j\geq s'>t_j-\delta'' \quad \text{imply } f^U(s)f^U(s')^{-1}\in U.$$

Let $0<\delta'''<\frac{1}{2}\min_{2\leq j\leq m}(t_j-t_{j-1})$ and $\delta:=\min(\delta',\delta'',\delta''')$. If $s-s'<\delta$ and $s,s'\in[0,t]$, then there is at most one element $t_j\in\{t_1,\dots,t_m\}\cap[s',s]$. But then the choice of δ''' implies that $f^U(s)f^U(s')^{-1}\in U$.

If there is no $t_j\in\{t_1,\dots,t_m\}\cap[s',s]$, then the choice of δ' implies

$$f^U(s)f^U(s')^{-1}\in VV^{-1}\subset U.$$

Thus we have $f^U(s)f^U(s')^{-1}\in U$ for all $s,s'\in[0,t]$ with $s-s'<\delta$ and if $\max_{1\leq j\leq n}(t_{nj}-t_{n,j-1})<\delta$, then

$$\prod_{j=1}^n f_{nj}=\prod_{j=1}^n f^U(t_{nj})(f^U(t_{n,j-1}))^{-1}=f^U(t)f^U(0)^{-1}=f^U(t).$$

(2) Let g be a local inner product for G in the sense of Definition 5.1.7 and $\chi\in G^\wedge$. Then there exists a neighborhood $U\in\mathfrak{V}(e)$ and for U there are the elements f_{nj} of G constructed above such that the sequence $(\sum_{j=1}^n g(f_{nj},\chi))_{n\geq 1}$ converges.

Indeed, for a properly chosen neighborhood $U\in\mathfrak{V}(e)$ we get by Property (LI3) of the local inner product g for G the equalities

$$\chi(f^U(t))=\lim_{n\to\infty}\chi\Big(\prod_{j=1}^n f_{nj}\Big)=\lim_{n\to\infty}\exp\Big(i\sum_{j=1}^n g(f_{nj},\chi)\Big)$$

and hence that $(\exp(i\sum_{j=1}^n g(f_{nj},\chi)))_{n\geq 1}$ is a Cauchy sequence in $\mathbb{T}$. Given $\varepsilon\in]0,1[$ this implies the existence of an $n_0:=n_0(\chi,f)\geq 1$ satisfying

$$\Big|\sum_{j=1}^n g(f_{nj},\chi)-\sum_{j=1}^{n_0} g(f_{n_0j},\chi)-2\pi\,\alpha(n)\Big|<\varepsilon$$

whenever $n\geq n_0$, where $\alpha(n)\in\mathbb{N}$ is suitably chosen. Consequently,

$$\Big|\sum_{j=1}^{n+1} g(f_{n+1,j},\chi)-\sum_{j=1}^n g(f_{nj},\chi)-2\pi(\alpha(n+1)-\alpha(n))\Big|<2\varepsilon$$

for all $n\geq n_0$, or

$$2\pi\,|\alpha(n+1)-\alpha(n)|\leq\Big|\sum_{j=1}^{n+1} g(f_{n+1,j},\chi)-\sum_{j=1}^n g(f_{nj},\chi)\Big|+2\varepsilon.$$

From the definition of $f_{n+1,j}$ we obtain for some $j_0\in\{1,\dots,n\}$ the equality

$$\begin{aligned}&\sum_{j=1}^{n+1} g(f_{n+1,j},\chi)-\sum_{j=1}^n g(f_{nj},\chi)\\ &=g(f_{n+1,j_0+1},\chi)+g(f_{n+1,j_0},\chi)-g(f_{n,j_0},\chi),\end{aligned}$$

whence

$$|\sum_{j=1}^{n+1} g(f_{n+1,j},\chi)-\sum_{j=1}^{n} g(f_{nj},\chi)| \leq 3\sup_{x\in U}|g(x,\chi)|.$$

Property (LI4) of g enables us to choose U such that

$$3\sup_{x\in U}|g(x,\chi)|<\varepsilon$$

holds, and we get $|\alpha(n+1)-\alpha(n)|<1$ or $\alpha(n)=\alpha(n+1)$ for all $n\geq n_0$. Thus without loss of generality $\alpha(n_0)=0$. The convergence of the Cauchy sequence $(\sum_{j=1}^{n} g(f_{nj},\chi))_{n\geq 1}$ completes the proof.

Given $t\in[0,1]$ and $B\in\mathfrak{B}_{\complement U}$ we now define the number $\nu(t,B)$ of discontinuities whose jumps lie in B of the process $(X_s)_{s\in[0,1]}$ in the interval $[0,t]$ and the product

$$X(t,B):=\prod_{t_0\in[0,t]} X_{t_0}(X_{t_0-})^{-1}$$

of these jumps.

By the remark preceding Definition 5.6.1 $\nu(t,B)$ and $X(t,B)$ are well-defined quantities. The families $(\nu(t,B))_{t\in[0,1]}$ and $(X(t,B))_{t\in[0,1]}$ are processes on $(\Omega,\mathfrak{A},P)$ with values in $\mathbb{Z}$ and G resp.

Finally, we introduce for $t\in[0,1]$ and $B\in\mathfrak{B}_{\complement U}$ the mathematical expectation $\pi(t,B):=E_P[\nu(t,B)]$ of the random variable $\nu(t,B)$ with respect to P. By its definition the mapping $B\to\pi(t,B)$ is a (positive) measure on $(\complement U,\mathfrak{B}_{\complement U})$ for any $t\in[0,1]$.

(3) For a given $B\in\mathfrak{B}_{\complement U}$ the process $(\nu(t,B))_{t\in[0,1]}$ is a stochastically continuous process with independent increments (taking values in $\mathbb{Z}$), and hence is a *Poisson process* with values in $\mathbb{Z}$ in the sense that for all $s,t\in[0,1]$, $s\leq t$ and $l\in\mathbb{Z}_+$ we have

$$P[\nu(t,B)-\nu(s,B)=l]=\frac{|\pi(s,t,B)|^l}{l!}\exp(-\pi(s,t,B)),$$

where

$$\pi(s,t;B):=E_P[\nu(t,B)-\nu(s,B)]=\pi(t,B)-\pi(s,B).$$

The proof of the first mentioned property is an immediate consequence of the assumption on $(X_t)_{t\in[0,1]}$, the conclusion stated is a computation known from the classical theory ([163], p. 263, Theorem 2).

(4) For every $B\in\mathfrak{B}_{\complement U}$ the process $(X(t,B))_{t\in[0,1]}$ is a stochastically continuous process with independent increments (taking values in G).

This follows directly from the assumptions on $(X_t)_{t\in[0,1]}$ with the help of Preparation (3).

(5) For every $B\in\mathfrak{B}_{\complement U}$ the processes $(X(t,B))_{t\in[0,1]}$ and $(Y(t,B))_{t\in[0,1]}$ defined by $Y(t,B):=X_t\cdot X(t,B)^{-1}$ for all $t\in[0,1]$ (both with values in G) are stochastically independent. The proof is a translation of classical arguments ([163], p. 264, Theorem 3) to the locally compact Abelian group G.

(6) Let $\{B_1,\ldots,B_k\}$ be a sequence of pairwise disjoint sets in $\mathfrak{B}_{\complement U}$ and $B:=\bigcup_{j=1}^k B_j$. We define processes

$$(Y_t^{(0)})_{t\in[0,1]},(Y_t^{(1)})_{t\in[0,1]},\ldots,(Y_t^{(k)})_{t\in[0,1]}$$

(with values in G) by putting

$$Y_t^{(0)} := Y(t, B),\ Y_t^{(1)} := X(t, B_1), \dots, Y_t^{(k)} := X(t, B_k) \quad \text{resp.}$$

(all $t \in [0, 1]$). Then

$$(Y_t^{(0)})_{t\in[0,1]}, (Y_t^{(1)})_{t\in[0,1]}, \dots, (Y_t^{(k)})_{t\in[0,1]}$$

are stochastically independent.

In fact, the processes $(Y_t^{(j)})_{t\in[0,1]}$ $(j=1, \dots, k)$ are determined by the process $(X(t, \complement B))_{t\in[0,1]}$, which is independent of the process $(Y_t^{(0)})_{t\in[0,1]}$ by Preparation (5). Hence, the process $(Y_t^{(0)})_{t\in[0,1]}$ does not depend on the processes

$$(Y_t^{(1)})_{t\in[0,1]}, \dots, (Y_t^{(k)})_{t\in[0,1]}.$$

But the process $(Y_t^{(j)})_{t\in[0,1]}$ is independent of $(Y_t^{(l)})_{t\in[0,1]}$ for $j, l = 1, \dots, k$, $j \neq l$, since $(Y_t^{(l)})_{t\in[0,1]}$ is independent of

$$(X(t, \textstyle\bigcup_{\substack{i=1\\ i\neq l}}^k B_i))_{t\in[0,1]}$$

and $(Y_t^{(j)})_{t\in[0,1]}$ is completely determined by

$$(X(t, \textstyle\bigcup_{\substack{i=1\\ i\neq l}}^k B_i))_{t\in[0,1]} \quad \text{for } j \neq l.$$

For an arbitrary but fixed $t \in [0, 1]$ and $\omega \in \Omega$ the mapping $B \to \nu(t, B)(\omega)$ is a (positive) measure on $(\complement U, \mathfrak{B}_{\complement U})$. Given a bounded measurable real-valued function f on G with $f(U) = \{0\}$ we consider the function

$$\omega \to g(\omega) := \int f(x)\, \nu(t, dx)(\omega) \quad \text{on } \Omega,$$

which is again measurable.

If g is a local inner product for G, we introduce the function

$$(\chi, \omega) \to \int_B g(x, \chi)\, \nu(t, dx)(\omega) \quad \text{on } G^\wedge \times \Omega$$

(depending on $t \in [0, 1]$ and $B \in \mathfrak{B}_{\complement U}$). Clearly this function is $\omega_{G^\wedge} \otimes P$-measurable. If f is $\pi(t, \cdot)$-integrable, we get the formulae

$$(7) \qquad \begin{aligned} &E_P[\textstyle\int f(x)\, \nu(t, dx)] = \int f(x)\, \pi(t, dx) \quad \text{and} \\ &V_P[\textstyle\int f(x)\, \nu(t, dx)] = \int f^2(x)\, \pi(t, dx), \end{aligned}$$

where E_P and V_P denote the (classical) expectation and variance of an $\mathbb{R}$-random variable on $(\Omega, \mathfrak{A}, P)$ with respect to P. We present the proof of the second formula only. Clearly it suffices to show the equality

$$E_P[(\textstyle\int f(x)\, \nu(t, dx))^2] - [E_P(\int f(x)\, \nu(t, dx))]^2 = \int f^2(x)\, \pi(t, dx)$$

for all functions $f = \sum_{i=1}^k \alpha_i 1_{B_i}$ with mutually disjoint sets $B_i \in \mathfrak{B}_{\complement U}$ and $\alpha_i \in \mathbb{R}_+$ $(i = 1, \dots, k)$.

Preparation (6) above yields the independence of the $\mathbb{Z}$-random variables $\nu(t,B_1),\ldots,\nu(t,B_k)$. Moreover, by Preparation (3) we obtain for every $i=1,\ldots,k$ the formula

$$E_P[\nu(t,B_i)^2]=\pi(t,B_i)^2+\pi(t,B_i).$$

Both facts together yield

$$\begin{aligned}
&E_P[(\int f(x)\,\nu(t,dx))^2]-[E_P(\int f(x)\,\nu(t,dx))]^2\\
&=E_P[\textstyle\sum_{\substack{i,j=1\\ i\neq j}}^k \alpha_i\alpha_j\,\nu(t,B_i)\,\nu(t,B_j)+\sum_{i=1}^k\alpha_i^2\,\nu(t,B_i)^2]\\
&\quad-\textstyle\sum_{i,j=1}^k\alpha_i\alpha_j\,\pi(t,B_i)\,\pi(t,B_j)\\
&=\textstyle\sum_{\substack{i,j=1\\ i\neq j}}^k\alpha_i\alpha_j\,\pi(t,B_i)\,\pi(t,B_j)+\sum_{i=1}^k\alpha_i^2(\pi(t,B_i)^2+\pi(t,B_i))\\
&\quad-\textstyle\sum_{i,j=1}^k\alpha_i\alpha_j\,\pi(t,B_i)\,\pi(t,B_j)\\
&=\textstyle\sum_{i=1}^k\alpha_i^2\,\pi(t,B_i)=\int f^2(x)\,\pi(t,dx).
\end{aligned}$$

5.6.6. We now proceed to the first step of the desired decomposition. Let G continue to be a locally compact Abelian group having a countable basis of its topology, g a local inner product for G and $(X_t)_{t\in[0,1]}$ a given additive process with values in G. To $(X_t)_{t\in[0,1]}$ there correspond the processes $(\nu(t,B))_{t\in[0,1]}$ and $(X(t,B))_{t\in[0,1]}$ as well as the family $(\pi(t,B))_{t\in[0,1]}$ of numbers

$$\pi(t,B):=E_P[\nu(t,B)]\qquad(t\in[0,1]).$$

Let $(U_m)_{m\geq 0}$ be a basis of $\mathfrak{B}(e)$ of open sets satisfying $U_m\subset U_n$ for all $m\geq n\geq 0$. We define $V_0:=\complement U_0$ and $V_j:=U_{j-1}\setminus U_j$ for $j\geq 1$.

Given $\chi\in G^\wedge$, by Property (LI 3) of g there exists a $j_0:=j_0(\chi)\geq 1$ satisfying $\chi(x)=\exp i\,g(x,\chi)$ for all $x\in U_j$ with $j\geq j_0$. For each $\omega\in\Omega$ and an open $U\in\mathfrak{B}(e)$ the range of the function $B\to\nu(t,B)(\omega)$ on $\mathfrak{B}_{\complement U}$ has only finitely many values. Hence, for every $t\in[0,1]$ and $j\geq 0$ there exist jumps $x_1,\ldots,x_r\in G$ of $s\to X_s(\omega)$ in $[0,t]$ that lie in V_j, and

$$X(t,V_j)(\omega)=\textstyle\prod_{i=1}^r x_i^{\nu(t,\{x_i\})(\omega)}=:\int_{V_j}x\,\nu(t,dx)(\omega)$$

has the property that for $\chi\in G^\wedge$ and $j\geq j_0$

$$\begin{aligned}
&\chi(X(t,V_j)(\omega))=\textstyle\prod_{i=1}^r\chi(x_i)^{\nu(t,\{x_i\})(\omega)}\\
&=\textstyle\prod_{i=1}^r(\exp i\,g(x_i,\chi))^{\nu(t,\{x_i\})(\omega)}\\
&=\exp i(\textstyle\sum_{i=1}^r g(x_i,\chi)\,\nu(t,\{x_i\})(\omega))\\
&=\exp i\textstyle\int_{V_j}g(x,\chi)\,\nu(t,dx)(\omega).
\end{aligned}$$

Preparation (7) of 5.6.5 yields

$$E_P(\textstyle\int_{V_j}g(x,\chi)\,\nu(t,dx))=\int_{V_j}g(x,\chi)\,\pi(t,dx).$$

This formula is the motivation for the notion of a generalized expectation of the G-random variable $X(t,V_j)$ ($t\in[0,1]$, $j\geq 0$). For $t\in[0,1]$ and $j\geq 0$ we consider the mapping $\phi: G^\wedge \to \mathbb{R}$ defined by

$$\phi(\chi) := \int_{V_j} g(x, \chi)\, \pi(t, dx) \quad \text{for all } \chi \in G^\wedge.$$

ϕ is a continuous homomorphism from $G^\wedge$ into $\mathbb{R}$. The multiplicativity follows from Property (LI2) of g together with the linearity of the integral. The continuity of ϕ is deduced as follows: Let $(\chi_n)_{n\geq 1}$ be a sequence in $G^\wedge$ with $\lim_{n\to\infty} \chi_n = \chi \in G^\wedge$. Then there exists a compact neighborhood $W \in \mathfrak{V}(e)$ with $\chi_n \chi^{-1} \in W$ for $n \geq n_0$. But Property (LI1) of g implies

$$|g(x, \chi_n) - g(x, \chi)| \leq \sup_{x\in G} \sup_{\chi \in W} |g(x, \chi)| < \infty$$

for all $x \in G$.

Since $\chi \to g(\cdot, \chi)$ is a continuous function on $G^\wedge$ and $\pi(t, V_j) < \infty$ we obtain by Lebesgue's convergence theorem

$$\lim_{n\to\infty} \int_{V_j} |g(x, \chi_n) - g(x, \chi)|\, \pi(t, dx) = 0,$$

whence $\lim_{n\to\infty} \phi(\chi_n) = \phi(\chi)$. We conclude that the mapping $\chi \to \exp i\phi(\chi)$ is a continuous character of $G^\wedge$. By the Pontryagin duality theorem there exists therefore a unique element $EX(t, V_j)$ of G satisfying

$$\chi(EX(t, V_j)) = \exp\left(i \int_{V_j} g(x, \chi)\, \pi(t, dx)\right)$$

for all $\chi \in G^\wedge$.

5.6.7 Definition. For $t \in [0, 1]$ and $j \geq 0$ the group element $EX(t, V_j)$ is said to be the *local expectation* of $X(t, V_j)$.

Our next aim will be the decomposition of the process $(X_t)_{t\in[0,1]}$ with a factor $(\Lambda_t)_{t\in[0,1]}$ defined by

$$\Lambda_t := \prod_{j\geq 0} X(t, V_j)(EX(t, V_j))^{-1} \quad \text{for all } t \in [0, 1],$$

where the convergence of the product of G-random variables is understood in the topology of G.

For $t \in]0, 1]$ and $n \geq 1$ we consider a sequence $(t_{nj})_{j=0,\dots,n}$ with $0 =: t_{n0} < t_{n1} < \cdots < t_{nn} := t$ and $\lim_{n\to\infty} \max_{1\leq j\leq n} (t_{nj} - t_{n,j-1}) = 0$.

As in Preparations 5.6.5 we introduce for an open $U \in \mathfrak{V}(e)$ and every $j = 1, \dots, n$ the G-random variables

$$X_{nj} := \psi_U \circ (X^U_{t_{nj}} (X^U_{t_{n,j-1}})^{-1}).$$

Clearly $X_{n1}, \dots, X_{nn}$ are stochastically independent.

5.6.8 Lemma. *Let $\chi \in G^\wedge$. Then for all sufficiently small open neighborhoods $U := U(\chi) \in \mathfrak{V}(e)$ and corresponding G-random variables X_{nj} $(j = 1, \dots, n;\ n \geq 1)$ one has*

$$V_P\left[\lim_{n\to\infty} \sum_{j=1}^n g(X_{nj}, \chi)\right] < \infty.$$

Proof. 1. We first show that under the hypothesis made we have

$$\overline{\lim}_{n\geq 1} \sum_{j=1}^{n} V_P[g(X_{nj},\chi)]<\infty.$$

Suppose this statement is false. Then there exists for every $k\geq 1$ an index $n_k\geq 1$ such that

$$\sigma_k^2:=\sum_{j=1}^{n_k} V_P[g(X_{n_k,j},\chi)]>k^2.$$

The $\mathbb{R}$-random variables

$$Y_{kj}:=\frac{g(X_{n_k,j},\chi)-E_P[g(X_{n_k,j},\chi)]}{\sigma_k} \qquad (j=1,\ldots,n_k)$$

are stochastically independent. Moreover, by Property (LI1) of g we have

$$\varepsilon:=\sup_{x\in U}|g(x,\chi)|<\infty.$$

Clearly, $E_P(Y_{kj})=0$ $(j=1,\ldots,n_k)$ and $V_P(\sum_{j=1}^{n_k} Y_{kj})=1$. For $\delta>0$ we choose $k_0\geq 1$ such that $\delta>\frac{2\varepsilon}{k}$ holds for all $k\geq k_0$. Then

$$|Y_{kj}|\leq\frac{2\varepsilon}{\sigma_k}\leq\frac{2\varepsilon}{k}<\delta \quad \text{implies} \quad P[|Y_{kj}|>\delta]=0$$

for all $j=1,\ldots,n_k$ and hence the validity of the classical Lindeberg-Feller condition

$$\sum_{j=1}^{n_k}\int_{|x|\geq\delta} x^2\, P_{Y_{kj}}(dx)=0 \qquad \text{for all } k\geq k_0.$$

From the classical central limit theorem ([6], p. 270) it follows that

$$\mathscr{T}_v\text{-}\lim_{k\to\infty} P_{\sum_{j=1}^{n_k} Y_{kj}}=\nu_{0,1}.$$

On the other hand, we obtained in Preparation (2) of 5.6.5 that for a sufficiently small $U:=U(\chi)\in\mathfrak{B}(e)$ the sequence $(\sum_{j=1}^{n} g(X_{nj},\chi))_{n\geq 1}$ converges, and by the classical equivalence principle (Theorem 2.2.19 for $G:=\mathbb{R}$) we obtain the a.s. convergence of the sequence $(\sum_{j=1}^{n_k} Y_{kj})_{k\geq 1}$ with

$$\sum_{j=1}^{n_k} Y_{kj}=\frac{1}{\sigma_k}\left(\sum_{j=1}^{n_k} g(X_{n_k,j},\chi)-E_P\left[\sum_{j=1}^{n_k}(g(X_{n_k,j},\chi))\right]\right)$$

for all $k\geq k_0$. But since $(E_P[\sum_{j=1}^{n} g(X_{nj},\chi)])_{n\geq 1}$ is a sequence in $\mathbb{R}$ and $\lim_{k\to\infty}\sigma_k=\infty$, the sequence $(\sum_{j=1}^{n_k} Y_{kj})_{k\geq 1}$ converges a.s. to a constant random variable, which is a contradiction of the $\mathscr{T}_v$-convergence above.

2. It remains to show the following classical result: If $(Z_n)_{n\geq 1}$ is a sequence of $\mathbb{R}$-random variables (on $(\Omega,\mathfrak{A},P)$) converging stochastically to an $\mathbb{R}$-random variable Z and if

$$\overline{\lim}_{n\geq 1} V_P(Z_n)=:c<\infty, \quad \text{then} \quad V_P(Z)<\infty.$$

In fact, let $\alpha\in\mathbb{R}_+^*$ with $\alpha^2>2c$ be chosen. From

$$[|E_P(Z_m)-E_P(Z_n)|>3\alpha]\subset[|E_P(Z_m)-Z_m|>\alpha]$$
$$\cup\,[|Z_m-Z_n|>\alpha]\cup[|Z_n-E_P(Z_n)|>\alpha] \quad \text{for } m,n\geq 1 \text{ follows}$$

$$\overline{\lim}_{n,m\geq 1} P[|Z_m-Z_n|>\alpha]\leq\frac{2}{\alpha^2}\overline{\lim}_{n\geq 1} V_P(Z_n)=\frac{2c}{\alpha^2}<1$$

(where the stochastic convergence of $(Z_n)_{n\geq 1}$ is applied). Since the expectations are constant, there exists an $n_0\geq 1$ with

$$|E_P(Z_m)-E_P(Z_n)|\leq 3\alpha \quad \text{for all } n,m\geq n_0.$$

Thus $\overline{\lim}_{n\geq 1}|E_P(Z_n)|<\infty$, whence by assumption $\overline{\lim}_{n\geq 1} E_P(Z_n^2)<\infty$.
Fatou's lemma finally implies that

$$E_P[(\text{st-lim}_{n\to\infty} Z_n)^2]\leq\underline{\lim}_{n\geq 1} E_P(Z_n^2)\leq\overline{\lim}_{n\geq 1} E_P(Z_n^2)<\infty,$$

and the assertion follows. □

5.6.9 Lemma. *For $t\in[0,1]$, $\chi\in G^\wedge$ and $j\geq 0$ we define the $\mathbb{R}$-random variable*

$$Y_j(t,\chi):=\int_{V_j} g(x,\chi)\,\nu(t,dx)-E_P\left(\int_{V_j} g(x,\chi)\,\nu(t,dx)\right)$$
$$=\int_{V_j} g(x,\chi)\,\nu(t,dx)-\int_{V_j} g(x,\chi)\,\pi(t,dx)$$

(on the basis of the discussion of 5.6.6).

Then for each $t\in[0,1]$ and $\chi\in G^\wedge$ the series $\sum_{j\geq 0} Y_j(t,\chi)$ converges with probability one.

Proof. 1. We show that for every $t\in[0,1]$ and $\chi\in G^\wedge$ we have

$$\sum_{j\geq 0} V_P(Y_j(t,\chi))<\infty.$$

Indeed, since

$$E_P\left[\left(\int_{V_j} g(x,\chi)\,\nu(t,dx)\right)^2\right]\leq\sup_{x\in V_j}|g(x,\chi)|^2\, E_P[\nu(t,V_j)^2]<\infty$$

for every $j\geq 0$, it suffices to show that $\sum_{j\geq k+1} V_P[Y_j(t,\chi)]<\infty$, where $k\geq 1$ is chosen such that $U_k\subset U$ holds and $U\in\mathfrak{B}(e)$ is chosen such that the assertion

$$V_P\left[\lim_{n\to\infty}\sum_{j=1}^n g(X_{nj},\chi)\right]<\infty$$

of Lemma 5.6.8 is satisfied.

For all $l,k\geq 0$ with $l>k$ one obtains

$$h:=\lim_{n\to\infty}\sum_{j=1}^n g(\psi_{U_k}\circ(X_{t_{nj}}^{U_k}(X_{t_{n,j-1}}^{U_l})^{-1}),\chi)$$
$$=\lim_{n\to\infty}\sum_{j=1}^n g(\psi_{U_l}\circ(X_{t_{nj}}^{U_l}(X_{t_{n,j-1}}^{U_l})^{-1}),\chi)+\int_{U_k\setminus U_l} g(x,\chi)\,\nu(t,dx).$$

But the two random variables on the right-hand side of this equality are stochastically independent by Preparation (6) of 5.6.5. Hence,

$$V_P(h) \geq V_P\left[\int_{U_k \setminus U_l} g(x,\chi)\, \nu(t,dx)\right] = \sum_{j=k+1}^{l} V_P[Y_j(t,\chi)].$$

By the choice of U we have $V_P(h) < \infty$, and thus

$$\sum_{j=k+1}^{l} V_P[Y_j(t,\chi)] < \infty.$$

As $l \to \infty$ we also have the assertion $\sum_{j \geq k+1} V_P[Y_j(t,\chi)] < \infty$.

2. Given $t \in [0,1]$ and $\chi \in G^\wedge$ we observe that the sequence $(Y_j(t,\chi))_{j \geq 0}$ of $\mathbb{R}$-random variables is stochastically independent. Let $\varepsilon > 0$. For $n, m \geq 1$ with $n \geq m$ Chebyshev's inequality together with the independence yields

$$P\left[\left|\sum_{j=m}^{n} Y_j(t,\chi)\right| > \varepsilon\right] \leq \frac{1}{\varepsilon^2} \sum_{j=m}^{n} V_P[Y_j(t,\chi)],$$

and by Part 1 the sequence $\left(\sum_{j=0}^{n} Y_j(t,\chi)\right)_{n \geq 1}$ is a Cauchy sequence with respect to the topology of stochastic convergence, and it converges stochastically. An application of the equivalence principle (Theorem 2.2.19 for $G := \mathbb{R}$) yields the assertion. □

5.6.10 Lemma. *Let G be a locally compact Abelian group having a countable basis of its topology and f a complex-valued $\omega_{G^\wedge}$-measurable function on $G^\wedge$ satisfying*

(i) $|f(\chi)| = 1$ *for all* $\chi \in G^\wedge$ *and*

(ii) $f(\chi\chi') = f(\chi) f(\chi')$ *for* $(\omega_{G^\wedge} \otimes \omega_{G^\wedge})$*-a.a.* $(\chi, \chi') \in G^\wedge \times G^\wedge$.

Then there exists a unique element $x \in G$ such that $f(\chi) = \chi(x)$ for $\omega_{G^\wedge}$-a.a. $\chi \in G^\wedge$.

Proof. For every relatively compact, $\omega_{G^\wedge}$-measurable subset C of $G^\wedge$ we put

$$\mu(C) := \int_C f(\chi)\, \omega_{G^\wedge}(d\chi).$$

By the translation invariance of $\omega_{G^\wedge}$ one obtains for $\omega_{G^\wedge}$-a.a. $\chi \in G^\wedge$ the equalities

$$\mu(C\chi^{-1}) = \int_C (\chi\chi')\, \omega_{G^\wedge}(d\chi') = f(\chi)\, \mu(C).$$

Let C be compact with $\mu(C) \neq 0$. Then

$$f(\chi) = \frac{\mu(C\chi^{-1})}{\mu(C)} \quad \text{for } \omega_{G^\wedge}\text{-a.a. } \chi \in G^\wedge, \text{ and by}$$

$$\begin{aligned} &|\mu(C\chi_1^{-1}) - \mu(C\chi_2^{-1})| \\ &\leq \int |1_{C\chi_1^{-1}}(\chi') - 1_{C\chi_2^{-1}}(\chi')|\, |f(\chi')|\, \omega_{G^\wedge}(d\chi') \\ &= \int |1_C(\chi'\chi_2^{-1}\chi_1) - 1_C(\chi')|\, \omega_{G^\wedge}(d\chi'), \end{aligned}$$

valid for all $\chi_1, \chi_2 \in G^\wedge$, the mapping $g: G^\wedge \to \mathbb{R}_+$ defined by

$$g(\chi) := \frac{\mu(C\chi^{-1})}{\mu(C)} \quad \text{for all } \chi \in G^\wedge$$

is a (continuous) character of $G^\wedge$.

By Pontryagin's duality theorem there exists an $x\in G$ with $g(\chi)=\chi(x)$ for all $\chi\in G^\wedge$, and by definition of g we get $f(\chi)=\chi(x)$ for $\omega_{G^\wedge}$-a.a. $\chi\in G^\wedge$. □

5.6.11 Theorem. *Let G be a locally compact Abelian group having a countable basis of its topology and let $(X_t)_{t\in[0,1]}$ be an additive process (on $(\Omega, \mathfrak{A}, P)$) with values in G. With the notation of the preceding discussion we put*

$$\Delta(t, V_j) := X(t, V_j)(EX(t, V_j))^{-1} \quad \text{for } j\geq 0.$$

Then there exists a process $(Z_t)_{t\in[0,1]}$ with values in G such that for each $t\in[0,1]$ we have with probability one the representation

$$X_t = \Lambda_t Z_t \quad \text{with} \quad \Lambda_t := \prod_{j\geq 0} \Delta(t, V_j),$$

where the convergence of the product of G-random variables is understood in the topology of G.

Proof. We fix $t\in[0,1]$.

1. The subset $A := \{(\chi, \omega)\in G^\wedge\times\Omega : \sum_{j\geq 0} Y_j(t,\chi)(\omega) < \infty\}$ of $G^\wedge\times\Omega$ is $\omega_{G^\wedge}\otimes P$-measurable since the random variable $Y_j(t,\chi)$ is for every $j\geq 0$. Moreover,

$$(\omega_{G^\wedge}\otimes P)((G^\wedge\times\Omega)\setminus A) = \iint 1_{(G^\wedge\times\Omega)\setminus A}(\chi, \omega)\, P(d\omega)\, \omega_{G^\wedge}(d\chi) = 0$$

since for all $\chi\in G^\wedge$ the series $\sum_{j\geq 0} Y_j(t,\chi)$ converges with probability one by Lemma 5.6.9.

Let $\chi\in G^\wedge$. From Property (LI3) of g we deduce the existence of a $j_0 := j(\chi)\geq 1$ such that

$$\chi\circ\Delta(t, V_j) = \exp i\, Y_j(t,\chi) \qquad \text{for all } j\geq j_0.$$

We now define

$$\xi(t,\chi)(\omega) := \begin{cases} \chi\left(\prod_{j=0}^{j_0} \Delta(t, V_j)(\omega)\right) \exp\left(i \sum_{j\geq j_0+1} Y_j(t,\chi)(\omega)\right), & \text{if } (\chi,\omega)\in A \\ 1 \quad \text{otherwise.} \end{cases}$$

Clearly, $\xi(t,\cdot)(\cdot)$ is an $(\omega_{G^\wedge}\otimes P)$-measurable function on $G^\wedge\times\Omega$ with

$$|\xi(t,\chi)(\omega)| = 1 \qquad \text{for all } (\chi,\omega)\in G^\wedge\times\Omega.$$

Moreover, by the very definition of $\xi(t,\cdot)(\cdot)$ we obtain

$$\xi(t, \chi\chi')(\cdot) = \xi(t,\chi)(\cdot)\,\xi(t,\chi')(\cdot)\ [P] \quad \text{for all } \chi, \chi'\in G^\wedge,$$

where the P-null set involved generally depends on $\chi, \chi'\in G^\wedge$. An application of Fubini's theorem yields

$$\xi(t,\chi\chi')(\omega) = \xi(t,\chi)(\omega)\,\xi(t,\chi')(\omega)$$

for $(\omega_{G^\wedge}\otimes\omega_{G^\wedge})$-a.a. $(\chi,\chi')\in G^\wedge\times G^\wedge$ and for all ω in a set $B\in\mathfrak{A}$ with $P(B)=0$.

Lemma 5.6.10 then implies the existence of an element $\xi(t)(\omega)$ of G satisfying $\chi(\xi(t)(\omega)) = \xi(t,\chi)(\omega)$ for $\omega_{G^\wedge}$-a.a. $\chi\in G^\wedge$ whenever $\omega\in B$.

2. We now show that $\chi(\xi(t)(\omega))=\xi(t,\chi)(\omega)$ for $(\omega_{G^\wedge}\otimes P)$-a.a. $(\chi,\omega)\in G^\wedge\times\Omega$. From the continuity of the mapping $\chi\to\chi(\xi(t)(\omega))$ for all $\omega\in B$ (introduced in Part 1) we deduce that

$$\frac{1}{\omega_{G^\wedge}(W)}\int_W \chi(\xi(t)(\omega))\omega_{G^\wedge}(d\chi)=\frac{1}{\omega_{G^\wedge}(W)}\int \xi(t,\chi)(\omega)\omega_{G^\wedge}(d\chi)$$

whenever $W\subset G^\wedge$ with $\omega_{G^\wedge}(W)>0$.

Let $(W_n)_{n\geq 1}$ be a decreasing sequence of compact neighborhoods in $\mathfrak{B}(\chi)$ satisfying $\bigcap_{n\geq 1} W_n=\{\chi\}$. By the continuity of the function $\chi\to\chi(\xi(t)(\omega))$ on $G^\wedge$ we obtain

$$\chi(\xi(t)(\omega))=\lim_{n\to\infty}\frac{1}{\omega_{G^\wedge}(W_n)}\int_{W_n}\chi(\xi(t)(\omega))\,\omega_{G^\wedge}(d\chi)$$

$$=\lim_{n\to\infty}\frac{1}{\omega_{G^\wedge}(W_n)}\int_{W_n}\xi(t,\chi)(\omega)\,\omega_{G^\wedge}(d\chi),\quad \text{if } \omega\in B.$$

In addition, we set $\xi(t)(\omega)=0$ if $\omega\notin B$. Then it becomes obvious that the function $\omega\to\chi(\xi(t)(\omega))$ on Ω is P-measurable. Since the σ-algebra $\mathfrak{A}(G^\wedge)$ generated by $G^\wedge$ is equal to $\mathfrak{B}(G)$, the G-random variable $\xi(t)$ is P-measurable and so the mapping $(\chi,\omega)\to\chi(\xi(t)(\omega))$ from $G^\wedge\times\Omega$ into $\mathbb{T}$ is $(\omega_{G^\wedge}\otimes P)$-measurable. Application of Fubini's theorem to the $(\omega_{G^\wedge}\otimes P)$-measurable function

$$(\chi,\omega)\to|\chi(\xi(t)(\omega))-\xi(t,\chi)(\omega)|$$

yields the assertion.

3. We put $\Lambda_t^{(n)}:=\prod_{j=0}^{n}\Delta(t,V_j)$ and obtain from Part 2 the limit relation

$$\lim_{n\to\infty}\chi(\xi(t)(\omega)\,\Lambda_t^{(n)}(\omega)^{-1})=\lim_{n\to\infty}\xi(t,\chi)(\omega)\,\chi(\Lambda_t^{(n)}(\omega)^{-1})$$
$$=\lim_{n\to\infty}\exp\Big(i\sum_{j\geq n+1}Y_j(t,\chi)(\omega)\Big)=1$$
for $(\omega_{G^\wedge}\otimes P)$-a.a. $(\chi,\omega)\in G^\wedge\times\Omega$.

We now apply Fubini's theorem to the $(\omega_{G^\wedge}\otimes P)$-measurable function

$$(\chi,\omega)\to\lim_{n\to\infty}|\chi(\xi(t)(\omega)\,\Lambda_t^{(n)}(\omega)^{-1})-1|.$$

Then there exists a $C\subset\Omega$ with $P(C)=1$ satisfying

$$\lim_{n\to\infty}\chi(\xi(t)(\omega)\,\Lambda_t^{(n)}(\omega)^{-1})=1\ [\omega_{G^\wedge}]\quad \text{for all } \omega\in C.$$

For each $\omega\in C$ the set

$$L(\omega):=\{\chi\in G^\wedge:\lim_{n\to\infty}\chi(\xi(t)(\omega)\,\Lambda_t^{(n)}(\omega)^{-1})=1\}$$

is a subgroup of $G^\wedge$ with $\omega_{G^\wedge}(G^\wedge\setminus L(\omega))=0$ and hence $L(\omega)$ equals $G^\wedge$. This shows that

$$\lim_{n\to\infty}\chi(\xi(t)(\omega)\,\Lambda_t^{(n)}(\omega)^{-1})=1\quad \text{for all } (\chi,\omega)\in G^\wedge\times C.$$

The continuity theorem 1.4.2 together with Theorem 1.1.3 then implies

$$\lim_{n\to\infty} \Lambda_t^{(n)}(\omega)=\xi(t)(\omega) \quad \text{for all } \omega\in C,$$

and the theorem has been proved. □

The previous analysis was based on the choice of an arbitrary basis $(U_m)_{m\geq 0}$ of open neighborhoods in $\mathfrak{V}(e)$. In the subsequent discussion we shall choose $(U_m)_{m\geq 0}$ more specifically in order to get the desired refinement of Theorem 5.6.11.

5.6.12 Lemma. *For arbitrary but fixed* $t\in[0,1]$ *and every compact subset* K *of* $G^\wedge$ *we have*

$$\int_G \sup_{\chi\in K} g(x,\chi)^2\, \pi(t,dx)<\infty.$$

Proof. 1. We first note that $\int g(x,\chi)^2\,\pi(t,dx)<\infty$ holds for every $\chi\in G^\wedge$. By definition of the G-random variables $Y_j(t,\chi)$ and the sets V_j $(j\geq 0)$ constructed in 5.6.6 we have via Preparation (7) of 5.6.5 the formula

$$V_P[Y_j(t,\chi)]=\int_{V_j} g(x,\chi)^2\,\pi(t,dx).$$

From $\bigcup_{j\geq 0} V_j=G\setminus\{e\}$ and $g(e,\chi)=0$ we conclude that

$$\sum_{j\geq 0} V_P[Y_j(t,\chi)]=\int_G g(x,\chi)^2\,\pi(t,dx) \quad \text{for all } \chi\in G^\wedge,$$

so that Part 1 of the proof of Lemma 5.6.9 yields the assertion.

2. Let $U\in\mathfrak{V}(e)$. Then Property (LI1) of g implies

$$\int_{\complement U} \sup_{\chi\in K} g(x,\chi)^2\,\pi(t,dx)<\sup_{x\in G}\sup_{\chi\in K} g(x,\chi)^2\,\pi(t,\complement U)<\infty.$$

It therefore suffices to show the existence of a neighborhood $U\in\mathfrak{V}(e)$ satisfying $\int_U \sup_{\chi\in K} g(x,\chi)^2\,\pi(t,dx)<\infty$. By Corollary 5.1.14 there are (for the given compact subset K of $G^\wedge$) a neighborhood $U_K\in\mathfrak{V}(e)$, a finite set $F_K\subset G^\wedge$ and a constant $c_K\in\mathbb{R}_+^*$ with

$$\sup_{\chi\in K} g(x,\chi)^2\leq c_K \sup_{\chi\in F_K} g(x,\chi)^2 \quad \text{whenever } x\in U_K.$$

Part 1 then implies

$$\int_{U_K} \sup_{\chi\in K} g(x,\chi)^2\,\pi(t,dx)\leq c_K\int_{U_K}\sum_{\chi\in F_K} g(x,\chi)^2\,\pi(t,dx)$$
$$\leq c_K \sum_{\chi\in F_K}\int_G g(x,\chi)^2\,\pi(t,dx)<\infty,$$

and the proof is complete. □

The statement of the lemma enables us to choose a sequence $(K_n)_{n\geq 1}$ of compact subsets K_n *of* $G^\wedge$ with $\bigcup_{n\geq 1} K_n=G^\wedge$ such that

$$\int_G \sup_{\chi\in K_n} g(x,\chi)^2\,\pi(t,dx)<\infty \quad \text{for all } n\geq 1.$$

But then there exists an increasing sequence $(j_0(n,k))_{k\geq 1}$ in $\mathbb{N}$ satisfying

$$\begin{aligned}&\sup_{\chi\in K_n}\sum_{j=j_0(n,k)+1}^{j_0(n,k+1)} V_P[Y_j(1,\chi)]\\&=\sup_{\chi\in K_n}\sum_{j=j_0(n,k)+1}^{j_0(n,k+1)}\int_{V_j} g(x,\chi)^2\,\pi(t,dx)\\&\leq\sum_{j=j_0(n,k)+1}^{j_0(n,k+1)}\int_{V_j}\sup_{\chi\in K_n} g(x,\chi)^2\,\pi(t,dx)<\frac{1}{k^6}.\end{aligned}$$

We now choose the sequence $(j_0(n,k))_{k\geq 1}$ such that $(j_0(n+1,k))_{k\geq 1}$ is a subsequence of $(j_0(n,k))_{k\geq 1}$. Since the processes $(Y_j(t,\chi))_{t\in[0,1]}$ have independent increments, we get

$$V_P[Y_j(t,\chi)]\leq V_P[Y_j(1,\chi)]\quad\text{for all } t\in[0,1]\ (j\geq 0,\ \chi\in G^\wedge).$$

Putting $n(0):=-1$ and $n(k):=j(k,k)$ for $k\geq 1$, we obtain for every $\chi\in G^\wedge$ the existence of an $n:=n(\chi)\geq 1$ satisfying $\chi\in K_n$. So for all $k>n$ we get

$$\sum_{j=n(k)+1}^{n(k+1)} V_P[Y_j(1,\chi)]<\frac{1}{k^6},$$

an inequality which we shall need later.

For every $t\in[0,1]$, $\chi\in G^\wedge$ and $k\geq 0$ we introduce the $\mathbb{R}$-random variable

$$W_k(t,\chi):=\sum_{j=n(k)+1}^{n(k+1)} Y_j(t,\chi).$$

5.6.13 Lemma. *For every $k\geq 0$, $\chi\in G^\wedge$ and $\omega\in\Omega$ the path $t\to W_k(t,\chi)(\omega)$ of the process $(W_k(t,\chi))_{t\in[0,1]}$ is right continuous.*

Proof. By the definition of the $\mathbb{R}$-random variable $Y_j(t,\chi)$ it suffices to show the right continuity of the functions

$$t\to\int_{V_j} g(x,\chi)\,\nu(t,dx)\quad\text{and}\quad t\to\int_{V_j} g(x,\chi)\,\pi(t,dx)$$
$$(j=n(k)+1,\ldots,n(k+1),\ \chi\in G^\wedge).$$

Let $U\in\mathfrak{V}(e)$ and $B\in\mathfrak{B}_{\complement U}$. Then for $t,s\in[0,1]$, $t\geq s$ we have

$$|\pi(t,B)-\pi(s,B)|\leq\int_\Omega|\nu(t,B)-\nu(s,B)|\,dP\leq 2\nu(1,B).$$

The stochastic continuity of the process $(\nu(t,B))_{t\in[0,1]}$ (Preparation (3) of 5.6.5) together with Lebesgue's convergence theorem yield the continuity of the function $t\to\pi(t,B)$ on $[0,1]$. Moreover, the path $t\to\nu(t,B)(\omega)$ is right continuous on $[0,1]$. The inequalities

$$\begin{aligned}&\left|\int_{V_j} g(x,\chi)\,\nu(t+h,dx)-\int_{V_j} g(x,\chi)\,\nu(t,dx)\right|\\&\leq\sup_{x\in G}|g(x,\chi)|\,|\nu(t+h,V_j)-\nu(t,V_j)|\end{aligned}$$

and

$$\begin{aligned}&\left|\int_{V_j} g(x,\chi)\,\pi(t+h,dx)-\int_{V_j} g(x,\chi)\,\pi(t,dx)\right|\\&\leq\sup_{x\in G}|g(x,\chi)|\,|\pi(t+h,V_j)-\pi(t,V_j)|\end{aligned}$$

($h\in\mathbb{R}_+$ such that $t+h\in[0,1]$) together with $\sup_{x\in G}|g(x,\chi)|<\infty$ then imply the assertion. □

5.6.14 Lemma. *For each $\chi\in G^\wedge$ the series $\sum_{k\geq 0} W_k(t,\chi)$ of $\mathbb{R}$-random variables converges with probability one uniformly in $t\in[0,1]$.*

Proof. Let $\chi\in G^\wedge$. By Lemma 5.6.13 the processes $(W_k(t,\chi))_{t\in[0,1]}$ are determined by their values at points $t\in\mathbb{Q}\cap[0,1]$. Hence, the functions $\sup_{0\leq t\leq 1}|W_k(t,\chi)|$ on Ω are P-measurable ($k\geq 0$). For $m\geq l\geq 0$ we obtain

$$P[\sup_{0\leq t\leq 1}|W_k(t,\chi)|>\frac{1}{k^2}]$$

$$=P\left[\lim_{m\to\infty}\sup_{0\leq l\leq m}\left|W_k\left(\frac{l}{m},\chi\right)\right|>\frac{1}{k^2}\right]$$

$$\leq\underline{\lim}_{m\geq 1} P\left[\sup_{0\leq l\leq m}\left|\sum_{p=1}^{l}\left(W_k\left(\frac{p}{m},\chi\right)-W_k\left(\frac{p-1}{m},\chi\right)\right)\right|>\frac{1}{k^2}\right]$$

$$\leq\underline{\lim}_{m\geq 1} k^4\sum_{j=1}^{m} V_P\left[W_k\left(\frac{j}{m},\chi\right)-W_k\left(\frac{j-1}{m},\chi\right)\right]$$

$$=\underline{\lim}_{m\geq 1} k^4 V_P\left[\sum_{j=1}^{m}\left(W_k\left(\frac{j}{m},\chi\right)-W_k\left(\frac{j-1}{m},\chi\right)\right)\right]$$

$$=\underline{\lim}_{m\geq 1} k^4 V_P[W_k(1,\chi)]\leq k^4\sum_{j=n(k)+1}^{n(k+1)} V_P[Y_j(1,\chi)]<\frac{1}{k^2}.$$

Here we have used Kolmogorov's inequality ([6], p. 174) in the third estimate and the discussion preceding Lemma 5.6.13 in the last one. Thus we have obtained

$$\sum_{k\geq 0} P\left[\sup_{0\leq t\leq 1}|W_k(t,\chi)|>\frac{1}{k^2}\right]<\infty$$

and by the Borel-Cantelli lemma ([6], p. 168)

$$P\left(\bigcup_{n\geq 1}\bigcap_{k\geq n}\left[\sup_{0\leq t\leq 1}|W_k(t,\chi)|\leq\frac{1}{k^2}\right]\right)=1. \quad \square$$

5.6.15 Theorem. *Let G be a locally compact Abelian group having a countable basis of its topology and $(X_t)_{t\in[0,1]}$ an additive process (on $(\Omega,\mathfrak{A},P)$) with values in G. For every $k\geq 0$ we consider the process $(\Lambda_t^k)_{t\in[0,1]}$ of G-random variables*

$$\Lambda_t^k:=\prod_{j=n(k)+1}^{n(k+1)}\Delta(t,V_j) \quad \text{on } (\Omega,\mathfrak{A},P).$$

Then uniformly in $t\in[0,1]$ the product $\prod_{k\geq 0}\Lambda_t^k$ converges with probability one.

The *proof* will be carried out in analogy to that of Theorem 5.6.11.

1. The set $A:=\{(\chi,\omega)\in G^\wedge\times\Omega:\sum_{k\geq 0} W_k(t,\chi)$ converges uniformly in $t\in[0,1]\}$ is $(\omega_{G^\wedge}\otimes P)$-measurable, and the path $t\to W_k(t,\chi)(\omega)$ is right continuous for $\omega\in\Omega$ $(k\geq 0,\ \chi\in G^\wedge)$ by Lemma 5.6.13.

Hence, for $n, m\geq 1$ the function

$$(\chi,\omega)\to\sup_{0\leq t\leq 1}\left|\sum_{k=0}^{n} W_k(t,\chi)-\sum_{k=0}^{m} W_k(t,\chi)\right|$$

on $G^\wedge\times\Omega$ is $(\omega_{G^\wedge}\otimes P)$-measurable, and

$$A=\{(\chi,\omega)\in G^\wedge\times\Omega:$$
$$\lim_{n,m\to\infty}\sup_{0\leq t\leq 1}\left|\sum_{k=0}^{n} W_k(t,\chi)-\sum_{k=0}^{m} W_k(t,\chi)\right|=0\}.$$

From Lemma 5.6.14 we deduce via Fubini's theorem that

$$(\omega_{G^\wedge}\otimes P)((G^\wedge\times\Omega)\setminus A)=0$$

holds. Property (LI3) of g and the definitions of the processes

$$(A_t^k)_{t\in[0,1]}\quad\text{and}\quad(W_k(t,\chi))_{t\in[0,1]}\qquad(\chi\in G^\wedge)$$

for $k\geq 0$ imply for a given $\chi\in G^\wedge$ the existence of a $j_0:=j(\chi)\geq 1$ with

$$\chi\circ A_t^k=\exp i\,W_k(t,\chi)\qquad\text{for all } k>j_0\qquad(t\in[0,1]).$$

We now define for $t\in[0,1]$

$$\zeta(t,\chi)(\omega):=\begin{cases}\chi\left(\prod_{k=0}^{j_0}A_t^k(\omega)\right)\exp\left(i\sum_{k\geq j_0+1} W_k(t,\chi)(\omega)\right), & \text{if } (\chi,\omega)\in A\\ 1 & \text{otherwise.}\end{cases}$$

Obviously $(\chi,\omega)\to\zeta(t,\chi)(\omega)$ is $(\omega_{G^\wedge}\otimes P)$-measurable with values of modulus 1 for every $t\in[0,1]$. Moreover, for $k\geq 0$ and $\omega\in\Omega$ the path $t\to A_t^k(\omega)$ is right continuous since $t\to X(t,V_j)(\omega)$ is right continuous and $t\to EX(t,V_j)$ is continuous for all $j\geq 0$ (as follows from the continuity of $t\to\int_{V_j}g(x,\chi)\,\pi(t,dx)$ shown in the proof of Lemma 5.6.13).

Hence, $t\to\zeta(t,\chi)(\omega)$ is right continuous on $[0,1]$, so that

$$(\chi,\chi',\omega)\to\sup_{0\leq t\leq 1}|\zeta(t,\chi\chi')(\omega)-\zeta(t,\chi)(\omega)\,\zeta(t,\chi')(\omega)|$$

is $(\omega_{G^\wedge}\otimes\omega_{G^\wedge}\otimes P)$-measurable on $G^\wedge\times G^\wedge\times\Omega$. Application of Lemma 5.6.14 yields $\zeta(t,\chi\chi')(\cdot)=\zeta(t,\chi)(\cdot)\,\zeta(t,\chi')(\cdot)$ for all $t\in[0,1]$, $[P]$, and for all $\chi,\chi'\in G^\wedge$.

Thus there exists a $B\in\mathfrak{A}$ with $P(B)=1$ such that

$$\zeta(t,\chi\chi')(\omega)=\zeta(t,\chi)(\omega)\,\zeta(t,\chi')(\omega)\qquad\text{for all } t\in[0,1],$$
$$(\omega_{G^\wedge}\otimes\omega_{G^\wedge})\text{-a.a. } (\chi,\chi')\in G^\wedge\times G^\wedge\quad\text{and all } \omega\in B.$$

By Lemma 5.6.10 there exists for each $t\in[0,1]$ and $\omega\in B$ an element $\zeta(t)(\omega)\in G$ such that $\chi(\zeta(t)(\omega))=\zeta(t,\chi)(\omega)$ for all $t\in[0,1]$, $\omega_{G^\wedge}$-a.a. $\chi\in G^\wedge$ and all $\omega\in B$.

2. As in the proof of Theorem 5.6.11 it can be shown that $\omega\to\zeta(t)(\omega)$ is a P-measurable mapping from Ω into G, whence $(\chi,\omega)\to\chi(\zeta(t)(\omega))$ is $(\omega_{G^\wedge}\otimes P)$-

measurable on $G^\wedge \times \Omega$. For each $\omega \in B$ the set

$$L(\omega) := \{\chi \in G^\wedge : t \to \chi(\zeta(t)(\omega)) \text{ is right continuous}\}$$

is a subgroup of $G^\wedge$ with $\omega_{G^\wedge}(G^\wedge \setminus L(\omega)) = 0$, whence $L(\omega) = G^\wedge$. That is to say, $t \to \chi(\zeta(t)(\omega))$ is right continuous for all $\chi \in G^\wedge$ and $\omega \in B$. Therefore the function $(\chi, \omega) \to \sup_{0 \le t \le 1} |\chi(\zeta(t)(\omega)) - \zeta(t, \chi)|$ on $G^\wedge \times \Omega$ is $(\omega_{G^\wedge} \otimes P)$-measurable, and

$$\chi(\zeta(t)(\omega)) = \zeta(t, \chi)(\omega) \quad \text{for all } t \in [0, 1],$$
$$(\omega_{G^\wedge} \otimes P)\text{-a.a } (\chi, \omega) \in G^\wedge \times \Omega.$$

Plainly,

$$\chi(\zeta(t)(\omega)) = \chi \circ \left(\prod_{k=0}^{j_0} \Lambda_t^k\right) \exp\left(i \sum_{k \ge j_0+1} W_k(t, \chi)\right)$$

for all $t \in [0, 1]$, $(\omega_{G^\wedge} \otimes P)$-a.a. $(\chi, \omega) \in G^\wedge \times \Omega$. We define the set

$$C := \{(\chi, \omega) \in G^\wedge \times \Omega :$$
$$\lim_{n \to \infty} \sup_{0 \le t \le 1} |\chi(\zeta(t)(\omega)(\prod_{k=0}^{n} \Lambda_t^k(\omega))^{-1}) - 1| = 0\}$$

and derive in a known manner $(\omega_{G^\wedge} \otimes P)((G^\wedge \times \Omega) \setminus C) = 0$.

Hence, there exists a set $D \in \mathfrak{A}$ with $P(D) = 1$ such that for the subgroup

$$C(\omega) := \{\chi \in G^\wedge : (\chi, \omega) \in C\}$$

of $G^\wedge$ one has $\omega_{G^\wedge}(G^\wedge \setminus C(\omega)) = 0$ for all $\omega \in D$. Again one concludes that $C(\omega) = G^\wedge$ whenever $\omega \in D$, or

$$\lim_{n \to \infty} \sup_{0 \le t \le 1} |\chi(\zeta(t)(\omega)(\prod_{k=0}^{n} \Lambda_t^k(\omega))^{-1}) - 1| = 0$$

for all $\chi \in G^\wedge$, $\omega \in D$. From the continuity theorem 1.4.2 it follows that

$$\lim_{n \to \infty} \prod_{k=0}^{n} \Lambda_t^k(\omega) = \zeta(t)(\omega)$$

uniformly in $t \in [0, 1]$, whenever $\omega \in D$. This completes the proof of the theorem. □

5.6.16 Theorem. *Let G be a locally compact Abelian group having a countable basis of its topology and $(X_t)_{t \in [0,1]}$ an additive process (on $(\Omega, \mathfrak{A}, P)$) with values in G. Then there exist processes $(\Lambda_t^k)_{t \in [0,1]}$ $(k \ge 0)$ and $(Y_t)_{t \in [0,1]}$ all with values in G such that $X_t = (\prod_{k \ge 0} \Lambda_t^k)\, Y_t$ uniformly in $t \in [0, 1]$ with probability one.*

Moreover,

(i) *the processes $(\Lambda_t^k)_{t \in [0,1]}$ $(k \ge 0)$ and $(Y_t)_{t \in [0,1]}$ have independent increments and are mutually independent;*

(ii) *the process $(Y_t)_{t \in [0,1]}$ is a Gauss process such that for P-a.a. $\omega \in \Omega$ the path $t \to Y_t(\omega)$ is continuous;*

(iii) *for every $k \ge 0$ and P-a.a. $\omega \in \Omega$ the path $t \to \Lambda_t^k(\omega)$ is continuous except at a finite number of points in $[0, 1]$ which are the discontinuities of a stochastically continuous Poisson process with values in $\mathbb{R}$.*

The *proof* relies on the preceding discussion; in particular, on the statements of Theorems 5.6.11 and 5.6.15. Let $D \in \mathfrak{A}$ be the set introduced in Part 2 of the proof of Theorem 5.6.15, where the existence with probability one of the process $(\prod_{k\geq 0} A_t^k)_{t\in[0,1]}$ was established. We define the process $(Y_t)_{t\in[0,1]}$ by

$$Y_t(\omega) := \begin{cases} X_t(\omega)(\prod_{k\geq 0} A_t^k(\omega))^{-1}, & \text{if } \omega\in D \\ e & \text{otherwise.} \end{cases}$$

What remains to be shown is that for every $\omega\in\Omega$ the path $t\to Y_t(\omega)$ of $(Y_t)_{t\in[0,1]}$ is continuous. If this has been proved, Theorem 5.6.4 implies that $(Y_t)_{t\in[0,1]}$ is a Gauss process.

First of all we note that for every $\omega\notin D$ the path $t\to Y_t(\omega)$ is obviously continuous. Let $\omega\in D$. Then by the uniform convergence in $t\in[0,1]$ of $\prod_{k\geq 0} A_t^k$ we get

$$\begin{aligned} &\overline{\lim}_{\substack{t\to s\\ t,s\in[0,1]}} \rho(Y_t(\omega), Y_s(\omega)) \\ &= \overline{\lim}_{t\to s} \lim_{m\to\infty} \rho(X_t(\omega)(\textstyle\prod_{k=0}^m A_t^k(\omega))^{-1}, X_s(\omega)(\prod_{k=0}^m A_s^k(\omega))^{-1}) \\ &= \lim_{m\to\infty} \overline{\lim}_{t\to s} \rho(X_t(\omega)(\textstyle\prod_{k=0}^m A_t^k(\omega))^{-1}, X_s(\omega)(\prod_{k=0}^m A_s^k(\omega))^{-1}). \end{aligned}$$

We assume that the path $t\to X_t(\omega)$ has a discontinuity at $s\in[0,1]$. Then the corresponding jump lies in $\bigcup_{j=0}^{n(m+1)} V_j$ for all $m\geq m_0$ (sufficiently large), and hence

$$\overline{\lim}_{t\to s} \rho(X_t(\omega)(\textstyle\prod_{k=0}^m A_t^k(\omega))^{-1}, X_s(\omega)(\prod_{k=0}^m A_s^k(\omega))^{-1})=0.$$

But this limit relation also obtains for the path $t\to X_t(\omega)$ which is continuous at s. Thus $\omega\in D$ implies that

$$\overline{\lim}_{t\to s} \rho(Y_t(\omega), Y_s(\omega))=0 \qquad \text{for every } s\in[0,1],$$

and the theorem is proved. □

5.6.17 Theorem. *Let G be a locally compact Abelian group having a countable basis of its topology and $(X_t)_{t\in[0,1]}$ an additive process (on $(\Omega, \mathfrak{A}, P)$) with values in G. We choose a fixed local inner product g for G. Then for each $t\in[0,1]$ and the distribution $\mu_t := P_{X_t}$ of X_t there exist an element x_t of G and a positive quadratic form ϕ_t on $G^\wedge$ such that*

$$\hat{\mu}_t(\chi)=\chi(x_t)\exp(-\phi_t(\chi)+\textstyle\int_G [\chi(x)-1-i\,g(x,\chi)]\,\pi(t,dx))$$

holds for all $\chi\in G^\wedge$.

Proof. By Theorem 5.6.16 we obtain for every $t\in[0,1]$ the product representation $\hat{\mu}_t=\hat{\alpha}_t\hat{\beta}_t$, where

$$\alpha_t := P_{\prod_{k\geq 0} A_t^k} \quad \text{and} \quad \beta_t := P_{Y_t}$$

is such that by Theorem 5.2.7 there exist $x_t\in G$ and $\phi_t\in \mathbf{Q}_+(G^\wedge)$ satisfying

$$\hat{\beta}_t(\chi)=\chi(x_t)\exp(-\phi_t(\chi)) \qquad \text{for all } \chi\in G^\wedge.$$

In order to prove the theorem it suffices to show that

$$\hat{\alpha}_t(\chi)=\exp(\int_G[\chi(x)-1-i\,g(x,\chi)]\,\pi(t,dx))$$

holds for all $\chi\in G^\wedge$.

Let ρ be an invariant metric generating the topology of G. For $B\subset G$ we denote the ρ-diameter of B by $\delta(B)$. Clearly $\delta(BB^{-1})\leq 2\delta(B)$. It follows that for any $U\in\mathfrak{B}(e)$ and $B\subset G$ with sufficiently small $\delta(B)$ we get $BB^{-1}\subset U$.

Let $U\in\mathfrak{B}(e)$ and $B\in\mathfrak{B}_{\complement U}$. We fix $\chi\in G^\wedge$. Then for each $n\geq 1$ there exists a (measurable) partition $\{B_{nj}: j\geq 1\}$ of B satisfying

$$\sup_{j\geq 1}\sup_{x\in B_{nj}B_{nj}^{-1}}|\chi(x)-1|<\frac{1}{n}.$$

For each $n\geq 1$ and $j\geq 1$ we choose $x_{nj}\in B_{nj}$. The inequality

$$\overline{\lim}_{m\geq 1}|\prod_{j=1}^m a_j-\prod_{j=1}^m b_j|\leq\sum_{j\geq 1}|a_j-b_j|,$$

valid for complex numbers a_j, b_j with $|a_j|,|b_j|\leq 1$ $(j=1,\dots,m)$, enables us to conclude that

$$\begin{aligned}&|\chi\circ(\prod_{j\geq 1}X(t,B_{nj})(\prod_{j\geq 1}x_{nj}^{\nu(t,B_{nj})})^{-1})-1|\\ &\leq\sum_{j\geq 1}|\chi\circ(X(t,B_{nj})\,x_{nj}^{-\nu(t,B_{nj})})-1|\\ &\leq\sum_{j\geq 1}\nu(t,B_{nj})\sup_{x\in B_{nj}B_{nj}^{-1}}|\chi(x)-1|\leq\frac{1}{n}\nu(t,B)\end{aligned}$$

for all $n\geq 1$, which implies that

$$\chi\circ X(t,B)=\lim_{n\to\infty}\chi\circ(\prod_{j\geq 1}x_{nj}^{\nu(t,B_{nj})}).$$

But $(\nu(t,B))_{t\in[0,1]}$ is a Poisson process with $E_P[\nu(t,B)]=\pi(t,B)$ by Preparation (3) of 5.6.5, and the sequence $(\nu(t,B_{nj}))_{j\geq 1}$ is stochastically independent for all $t\in[0,1]$, $n\geq 1$. Therefore

$$\begin{aligned}&E_P[\chi\circ X(t,B)]\\ &=\lim_{n\to\infty}E_P[\chi\circ(\prod_{j\geq 1}x_{nj}^{\nu(t,B_{nj})})]\\ &=\lim_{n\to\infty}E_P[\prod_{j\geq 1}\chi\circ x_{nj}^{\nu(t,B_{nj})}]\\ &=\lim_{n\to\infty}\prod_{j\geq 1}E_P[\chi\circ x_{nj}^{\nu(t,B_{nj})}]\\ &=\lim_{n\to\infty}\prod_{j\geq 1}\exp[\pi(t,B_{nj})(\chi(x_{nj})-1)]\\ &=\lim_{n\to\infty}\exp\sum_{j\geq 1}[\pi(t,B_{nj})(\chi(x_{nj})-1)]\\ &=\exp[\lim_{n\to\infty}\sum_{j\geq 1}(\chi(x_{nj})-1)\,\pi(t,B_{nj})]\\ &=\exp(\int_B(\chi(x)-1)\,\pi(t,dx))\qquad\text{for all } t\in[0,1].\end{aligned}$$

Moreover,

$$\begin{aligned}&\hat{P}_{EX(t,V_j)}(\chi)=\hat{\varepsilon}_{EX(t,V_j)}(\chi)=\chi(EX(t,V_j))\\ &=\exp(i\int_{V_j}g(x,\chi)\,\pi(t,dx))\end{aligned}$$

and so

$$\hat{P}_{\Delta(t,V_j)}(\chi)=\hat{P}_{X(t,V_j)(EX(t,V_j))^{-1}}(\chi)$$
$$=\exp\left(\int_{V_j}[\chi(x)-1-i\,g(x,\chi)]\,\pi(t,dx)\right)$$

for all $t\in[0,1]$, $j\geq 0$. From the P-a.e.-representation

$$\prod_{k\geq 0}\Lambda_t^k=\prod_{j\geq 0}\Delta(t,V_j)$$

and the stochastic independence of the sequence $(\Delta(t,V_j))_{j\geq 1}$ we finally conclude that

$$\hat{\alpha}_t(\chi)=\hat{P}_{\prod_{k\geq 0}\Lambda_t^k}(\chi)=\prod_{j\geq 0}\hat{P}_{\Delta(t,V_j)}(\chi)$$
$$=\exp\left(\sum_{j\geq 0}\int_{V_j}[\chi(x)-1-i\,g(x,\chi)]\,\pi(t,dx)\right)$$
$$=\exp\left(\int_G[\chi(x)-1-i\,g(x,\chi)]\,\pi(t,dx)\right)$$

for all $t\in[0,1]$, and the theorem is proved. □

5.6.18 Remark. Since for each $t\in[0,1]$ we have $\pi(t,\complement U)<\infty$ for all $U\in\mathfrak{B}(e)$, the measure $\pi(t,\cdot)$ on $(G,\mathfrak{B}(G))$ is σ-finite. Moreover, the condition $\int_G g(x,\chi)^2\,\pi(t,dx)<\infty$ appearing in Part 1 of the proof of Lemma 5.6.12 is equivalent to $\int_G(1-\operatorname{Re}\chi(x))\,\pi(t,dx)<\infty$, which can be deduced from the inequality

$$\tfrac{1}{4}g(x,\chi)^2\leq 1-\operatorname{Re}\chi(x)\leq\tfrac{1}{2}g(x,\chi)^2,$$

valid for all x in a sufficiently small neighborhood $U\in\mathfrak{B}(e)$. Hence, the formula of Theorem 5.6.17 is the Lévy-Khintchine representation of the measure μ_t ($t\in[0,1]$) which obviously belongs to $\mathscr{I}_0(G)$ and does not admit idempotent factors.

5.6.19 Theorem. *Let G be a locally compact Abelian group having a countable basis of its topology and g a local inner product for G. Then any continuous convolution hemigroup $(\mu_{s,t})_{s,t\in\mathbb{R}_+}$ (corresponding to an additive process $(X_t)_{t\in\mathbb{R}_+}$ with values in G) admits a Lévy-Khintchine representation of the form*

$$\hat{\mu}_{s,t}(\chi)=\chi(x_{s,t})\exp\{-\phi_{s,t}(\chi)+\int_G[\chi(x)-1-i\,g(x,\chi)]\,\pi(s,t;dx)\}$$

for all $\chi\in G^\wedge$. Here $x_{s,t}:=x_s^{-1}x_t\in G$, $\phi_{s,t}:=\phi_t-\phi_s$ with positive quadratic forms ϕ_s,ϕ_t on $G^\wedge$ satisfying $\phi_s\leq\phi_t$ and $\pi(s,t;\cdot):=\pi(t,\cdot)-\pi(s,\cdot)$.

Proof. Without loss of generality we restrict our attention to the family $(\mu_{s,t})_{s,t\in[0,1]}$ with index set $[0,1]$ instead of $\mathbb{R}_+$ keeping the convention $s\leq t$ for all $s,t\in[0,1]$. Let $(X_t)_{t\in[0,1]}$ be the additive process corresponding to $(\mu_{s,t})_{s,t\in[0,1]}$. Then by Theorem 5.6.16 we have

$$X_s^{-1}X_t=\left(\prod_{k\geq 0}(\Lambda_s^k)^{-1}\Lambda_t^k\right)Y_s^{-1}Y_t$$

uniformly in $s,t\in[0,1]$, $s\leq t$, with probability one, where the auxiliary processes appearing in the decomposition are as in the theorem quoted. For fixed $s\in[0,1]$

we consider the process $(X_s^{-1} X_t)_{t\in[s,1]}$. The process $(Y_s^{-1} Y_t)_{t\in[s,1]}$ appearing in the decomposition has continuous paths, independent increments and starts from e at $t=s$. By Theorem 5.6.4 we obtain that $(Y_s^{-1} Y_t)_{t\in[s,1]}$ is a Gauss process such that the distribution of $Y_s^{-1} Y_t$ admits a Fourier transform defined by

$$\chi(x_{s,t})\exp[-\phi_{s,t}(\chi)] \quad \text{for all } \chi\in G^\wedge;$$

where $x_{s,t}\in G$ and $\phi_{s,t}$ is a positive quadratic form on $G^\wedge$. By Theorem 5.6.17 the Fourier transform of the distribution $\mu_{s,t}$ of $\prod_{k\geq 0}(\Lambda_s^k)^{-1}\Lambda_t^k$ is of the form

$$\hat{\mu}_{s,t}(\chi)=\chi(x_{s,t})\exp\{-\phi_{s,t}(\chi) + \int_G [\chi(x)-1-i\,g(x,\chi)](\pi(t,dx)-\pi(s,dx))\}$$

for all $\chi\in G^\wedge (s,t\in[0,1], s\leq t)$. On the other hand, we have

$$\hat{\mu}_{s,t}(\chi)=\chi(x_s^{-1}x_t)\exp\{-(\phi_t(\chi)-\phi_s(\chi)) + \int_G [\chi(x)-1-i\,g(x,\chi)](\pi(t,dx)-\pi(s,dx))\}$$

for all $\chi\in G^\wedge (s,t\in[0,1], s\leq t)$. Comparison of the two formulae yields

$$\chi(x_{s,t})\exp(-\phi_{s,t}(\chi))=\chi(x_s^{-1}x_t)\exp[-(\phi_t(\chi)-\phi_s(\chi))]$$

for all $\chi\in G^\wedge$, and taking the logarithm of the absolute value of both sides gives us $\phi_{s,t}(\chi)=\phi_t(\chi)-\phi_s(\chi)$ for all $\chi\in G^\wedge$, whence $x_{s,t}=x_s^{-1}x_t$, and in particular, $\phi_t\geq\phi_s$ for all $s,t\in[0,1]$, $s\leq t$. □

References and Comments

R 5.1 Convergence of Infinitesimal Systems

The classical theory of infinitesimal triangular systems of probability measures on ℝ goes back to G.M. Bawli and A.J. Khintchine. A standard reference for the fundamental limit theorems remains the monograph of Gnedenko and Kolmogorov [172]. One might also consult Loève [325]. The generalization of the classical ideas to the theory of probability for locally compact Abelian groups is the work of Parthasarathy, Rao and Varadhan [383], which has been incorporated into the book of Parthasarathy [377]. Chapter IV of [377], in particular §§4–5, is the basic reference for our exposition. We emphasize, however, that our presentation is independent of the hypothesis of a countable basis for the underlying group. This emphasis is also the motivation for the work of Bossard [36], [37]. The measure-theoretic contribution which made the general idea possible is the notion of shift compactness, treated for general locally compact groups in §2 of Chapter I. Within the framework of Abelian groups this notion becomes even easier to handle and more directly applicable. Theorem 5.1.2 enables us to achieve convergence of a sequence of squares of measures on a locally compact Abelian group G if the corresponding sequence of powers is shift compact and its accumulation points do not have idempotent factors. This result helps to prove in Theorem 5.1.6 a compactness result for sequences of Poisson measures. The definition of weakly infinitely divisible measures has been introduced in [383] in accordance with shift compactness. Theorem 5.1.4 on the structure of the set of weakly infinitely divisible measures in $\mathcal{M}^1(G)$ still depends on a countability condition for G. The approach to Theorem 5.1.6 is inspired by ideas of Tortrat [467], [470] and [471]. The local inner product or g-function for G and its first construction are due to Parthasarathy. Our proof of Theorem 5.1.10 is different from the one given in Parthasarathy [377] or Bossard [37]. Corollary 5.1.14 is taken from M.S. Bingham [21]; it will be used in § 6. The aim of the section is the proof of the accompanying laws theorem 5.1.16 (due in the classical theory to Bawli) for arbitrary locally compact Abelian groups. Its proof follows from Parthasarathy [377]. A slight extension of Theorem 5.1.16 has been indicated in Sazonov and Tutubalin [423].

Previous to Parthasarathy infinitesimal systems of measures in $\mathcal{M}^1(G)$ were studied by Kloss [305], [306] for $G:=\mathbb{T}^s$ $(s\geq 1)$ and $G:=\mathbb{T}\times\mathbb{R}$ resp. As was first pointed out in Sazonov and Tutubalin [423], p. 21, the proofs of the accompanying laws theorem established are incomplete. Kloss' paper [305] also contains an assertion of accompanying laws type for totally disconnected, compact Abelian groups. The aim of Bossard [37] is to reprove the results mentioned for locally compact Abelian groups without using the Prohorov theorem and considering infinitely divisible measures rather than weakly infinitely divisible ones. As a consequence of his new proof Bossard introduces the notion of circle convergence for sequences $(\mu_n)_{n\geq 1}$ in $\mathcal{M}^1(G)$ by putting $\lim_{n\to\infty}\mu_n(f)=\mu(f)$ for all functions $f\in\mathscr{C}^b(G)$ which vanish in some neighborhood of e and he studies its relation to weak convergence. He illustrates his studies with a number of examples for special groups such as $\mathbb{T}$, $\mathbb{Z}_2\times\mathbb{Z}_2$ and the Klein 4-group.

R 5.2 Gauss Measures in the Sense of Parthasarathy

There are various ways to introduce normal distributions on $\mathbb{R}$ and $\mathbb{R}^n$ for $n>1$. In order to incorporate also the normal distributions on $\mathbb{R}^n$ whose support is a proper linear subspace of $\mathbb{R}^n$ one preferably chooses the definition via affine linear mappings or Fourier transforms. See for example Richter [400]. The first attempts to study normal distributions on the torus groups $\mathbb{T}^n$ for $n\geq 1$ were made by von Mises [355] and P. Lévy [320]. Notably, von Mises discovered the necessity of an analogue of the normal distribution for $\mathbb{T}$ in connection with physical studies concerning the weights of atoms. His circular normal distribution became more applicable later, also in the theory of statistics. For an extended account of this subject the interested reader might read the book of Mardia [344]. There has always been the intuitive idea that normal distributions are those infinitely divisible measures in $\mathcal{M}^1(\mathbb{R})$ which do not admit Poisson factors. This idea was made precise for general locally compact Abelian groups by the pioneering work of Parthasarathy, Rao and Varadhan [383] which entered Parthasarathy's book [377], Chapter IV. Previous to the work of the Indian school, Urbanik in [484] presented a less intrinsic definition of Gauss distributions for locally compact Abelian groups which nowadays appears to be even more suggestive and generalizable, as we shall point out in Chapter VI.

Our exposition starts with the introduction of Gauss measures in the sense of Parthasarathy (P-Gauss measures) and aims at a characterization of these measures through their Fourier transforms. Theorems 5.2.7, 5.2.8 and the résumé 5.2.9 containing the extension of [377], p. 97, Theorem 6.1 to locally compact Abelian groups, which do not necessarily admit a countable basis of their topology, establish the desired result. Since the characterization 5.2.9 is given in terms of quadratic forms on the character group $G^\wedge$ of G, the repetition of some folklore on positive bilinear forms on a group and their extensions seems to be in order. Lemma 5.2.5 is the extension result implicitly applied but not proved in Parthasarathy [377]. Application 5.2.6 is due to Siebert [440]. A further nonmetrizable version of Theorem 5.2.7 has been proved by Bossard in [36] and published in [37]. But in Bossard's work, normal laws in $\mathscr{M}^1(G)$ are defined as limits of certain infinitesimal triangular systems which circle converge in the sense of References and Comments 5.1.

Next, properties of the class $\mathscr{G}_P(G)$ of all P-Gauss measures on the locally compact Abelian group G are discussed. The above mentioned characterization theorem yields that $\mathscr{G}_P(G)$ is a closed subsemigroup of the class $\mathscr{I}_0(G)$ of all weakly infinitely divisible probability measures on G. Property (4) of 5.2.10 makes precise the relation between $\mathscr{G}_P(\mathbb{R}^n)$ and the classical notion of an n-dimensional normal distribution. It is observed in Property (5) that Cramér's theorem [95] on normal classes is no longer valid for the torus $\mathbb{T}$ (in place of $\mathbb{R}$). Counterexamples are due to Carnal [65] and to Martin-Löf ([182], p. 198). We do not discuss the various decomposition theorems for the classical normal distribution which have been established by Raikov [395] and the Linnik school (see for example [322] and [419]). A contribution of Ibragimov [262] will be discussed in the references and comments to Chapter VI. Corollary 5.2.15 makes precise the fact that P-Gauss measures on G are supported by the connected component G_0 of G. This is indicated in Parthasarathy [377], p. 101, Remark 2. Theorem 5.2.16 assures the existence of P-Gauss measures on G. The proof appears in Sazonov and Tutubalin [423] and Heyer [233].

Finally we mention the relationship between P-Gauss measures and generalizations of statistical characterizations (in terms of the sufficiency of the sample mean) of normal distributions to locally compact Abelian groups. Results in this direction are due to Ruhin [419], whose paper also contains a discussion of Raikov's theorem on Poisson classes. See also P. Lévy [319] and Marcinkiewicz [343].

R 5.3 Gauss Measures in the Sense of Bernstein

The classical normal distribution on $\mathbb{R}$ can be characterized as the only probability measure in $\mathscr{M}^1(\mathbb{R})$ which is invariant under the rotations in $\mathbb{R}^2$ in the following sense: Let $\xi_\theta : \mathbb{R}^2 \to \mathbb{R}^2$ rotate the plane through an angle θ which is not a multiple of $\frac{\pi}{2}$. Then, given a measure $\mu \in \mathscr{M}^1(\mathbb{R})$ and defining a measure $\sigma \in \mathscr{M}^1(\mathbb{R}^2)$ by $\sigma(B) := \mu \otimes \mu(\xi_\theta(B))$ for all $B \in \mathfrak{B}(\mathbb{R}^2)$, we find that the product representation $\sigma = \nu \otimes \nu$ for some $\nu \in \mathcal{M}^1(\mathbb{R})$ implies that ν is a normal distribution on $\mathbb{R}$. This way of characterizing the normal distribution is as old as Maxwell's investigations on the velocity distribution of molecules. Probabilistic studies in this direction were initiated by Kac [280] and S.N. Bernstein [16]. Their theorems can be found in the modern textbook literature, for example in Feller [140] and in Richter [400], where more general results are presented in detail. It should be noted that the studies centered on Bernstein's theorem are nowadays part of the so-called characterization theory of multivariate analysis. The statistical implications of this approach to the normal distribution can be deduced from papers by C.R. Rao [398], B.L.S. Prakasa Rao [390] and Flusser [146], in whose work the theory has already been generalized from $\mathbb{R}$ to arbitrary locally compact Abelian groups, a framework which we also choose for our presentation. The definition of a Gauss measure in the sense of Bernstein as given in 5.3.1 is due to Corwin [89], who studies the Bernstein property for general complex measures on a locally compact Abelian group. See his extended studies in [90], [91] and [92].

Gauss measures in the sense of Bernstein (*B*-Gauss measures) differ from *P*-Gauss measures by the fact that they admit idempotent factors (Remark 5.3.5). The main theorems proved in the text are taken from Heyer and Rall [236]. These authors aim at a complete comparison of the two types of Gauss measures at least for a certain class of locally compact Abelian groups. The groups appearing here in a natural way are called Corwin(*C*-)groups with regard to the structure theory developed by Corwin in [90]. They are defined by the property that the homomorphism $\zeta: x \to x^2$ from the group G into itself is in fact an epimorphism.

Corwin proves in [90] the following structure theorem, which contains a description of strong *C*-groups: Let G be a locally compact Abelian group and p a prime number such that the mapping $x \to x^p$ from G into itself is an automorphism of G. Then G contains an open subgroup of the form $\mathbb{R}^m \times \mathbb{Q}_p^n \times K$, where $m, n \geq 0$, $\mathbb{Q}_p$ denotes the group of p-adic (rational) numbers, K is a compact (Abelian) group and $x \to x^p$ is an automorphism of K. The first result concerning the relations between *P*- and *B*-Gauss measures on G is Theorem 5.3.9, which contains a sufficient condition on $G^\wedge$ for a symmetric *B*-Gauss measure to be a convolution product of an idempotent *B*-Gauss measure and a symmetric *P*-Gauss measure. Theorem 5.3.13 and its corollaries solve the problem posed for *C*-groups.

The conditions under which the assertions of Theorems 5.3.9 and 5.3.13 are valid are similar to those discovered by Ruhin in [419], Chapter III. Ruhin's work is motivated by and yields most interesting applications to estimation problems for probability measures on a locally compact Abelian group.

Within the theory of positive-definite functions the Bernstein property has been discussed by Schmidt, following Rao [398], in [432] and later on the basis of Corwin [89] and Heyer and Rall [236] in [433].

R 5.4 Convergence to Gauss Measures

Convergence of infinitesimal triangular systems of probability measures to the normal distribution is considered to be the most important part of the central limit problem. This claim has been supported with special authority by the work of Bochner [28], [29]. The classical results in this direction due to S.N. Bernstein, Khintchine and Feller are completely proved in Gnedenko and Kolmogorov [172], §26. For the multidimensional case see Takano [461]. These results are the core of our discussion. As in the preceding sections we extend the corresponding theorems of Parthasarathy [377] to locally compact Abelian groups which do not necessarily have a countable basis of their topology. So Theorem 5.4.2 appears in modified form in Parthasarathy [377], p. 115. Its proof based on Theorem 5.4.1 is new. Theorem 5.4.3, which will be applied in 5.6, is quoted in Bingham [21]. Remark 5.4.4 on a version of the de Moivre-Laplace theorem for a product solenoidal group contains a result of Urbanik [484]. The proof shows that only in groups with divisibility properties can the norming procedure of classical central limit theory be modeled. The limiting Gauss measure cannot be absolutely continuous with respect to Haar measure, as will be pointed out more precisely in § 5 following the investigations of Siebert [440]. In this context it seems appropriate to mention the generalized central limit theorems of Grenander [182] established for certain divisible, locally compact groups. Grenander's Theorem 5.3.2 presents moment conditions on the measures under which convergence to Gauss measures (defined by their generalized Fourier transforms) takes place. A slightly persuasive aspect of this approach to central limit theory is the illustration 5.5.1 of [182], where the special case of the torus group is treated.

Equally incomplete are the contributions [409] and [411] of Rosenblatt, which appear in condensed form in Chapter VII of his book [412]. Rosenblatt presents some central limit theory for measures on a compact Abelian group under mixing conditions. His corollary 1 in [412], p. 218 can be considered as a starting point for further studies in this direction.

R 5.5 Symmetric Gauss Semigroups

The results concerning symmetric Gauss semigroups which we chose to develop provide a first glance into the construction of absolutely continuous Gauss measures and their densities on arbitrary locally compact Abelian groups. The idea of considering Gauss measures on a locally compact Abelian group G, whose support coincides with G, comes from the paper of Urbanik [484], where it is shown that Gauss measures in the sense of Urbanik are full in the sense that they are positive on nonempty open subsets of G. The corresponding construction is contained in Lemma 5.5.5. The first class of groups treated is the class of connected Abelian Lie groups G which by the structure theory are of the form $\mathbb{R}^m \times \mathbb{T}^n$ for $m, n \geq 0$. Theorem 5.5.6 contains a sufficient condition for a symmetric Gauss semigroup in $\mathcal{M}^1(G)$ to be (ω_G-)absolutely continuous. From this result one obtains in Corollary 5.5.7 that absolutely continuous Gauss measures on G have the entire group G as their support. In the case $G := \mathbb{T}^d$ for $d \geq 1$ the density of a symmetric Gauss measure on G is represented in terms of theta functions. This calculus is due to Siebert [440], whose presentation will lead us through the section. More detailed studies for the infinite-dimensional torus group $\mathbb{T}^\infty$ are contained in Example 5.5.12. They have been enhanced in the meantime by profound contributions of Ch. Berg [10] in connection with his counterexample to the conjecture that harmonic groups are necessarily Lie groups.

Berg introduces in [10] the Brownian semigroup $(\mu_t^{\mathfrak{a}})_{t\in\mathbb{R}_+^*}$ (with vector $\mathfrak{a}:=(a_n)_{n\geq 1}\in(\mathbb{R}_+^*)^\infty$) on the (infinite-dimensional) torus

$$G:=\prod_{n\geq 1} G_n \quad \text{with} \quad G_n:=\mathbb{T} \quad \text{for all } n\geq 1$$

by

$$\widehat{\mu_t^{\mathfrak{a}}}(\mathfrak{m})=e^{-t\psi(\mathfrak{m})}$$

with

$$\psi(\mathfrak{m}):=\sum_{n\geq 1} a_n m_n^2 \quad \text{for all } \mathfrak{m}:=(m_n)_{n\geq 1}\in G^\wedge.$$

Clearly, Brownian semigroups $(\mu_t^{\mathfrak{a}})_{t\in\mathbb{R}_+^*}$ are symmetric Gauss semigroups in $\mathcal{M}^1(G)$ and they satisfy $\text{supp}(\mu_t^{\mathfrak{a}})=G$ for all $t\in\mathbb{R}_+^*$. The main results of [10], Part I, concern the absolute continuity of $(\mu_t^{\mathfrak{a}})_{t\in\mathbb{R}_+^*}$. In particular, the following statements are shown to be equivalent:

(i) $\mu_t^{\mathfrak{a}}\ll\omega$;

(ii) $\sum_{n\geq 1} e^{-2ta_n}<\infty$;

(iii) $\mu_t^{\mathfrak{a}}\ll\omega$, and there exists a square-integrable ω-density of $\mu_t^{\mathfrak{a}}$ (for all $t\in\mathbb{R}_+^*$).

Putting $t_0:=\inf\{t\in\mathbb{R}_+^*:\sum_{n\geq 1} e^{-2ta_n}<\infty\}$, we observe that $\mu_t^{\mathfrak{a}}\perp\omega$ for all $t\in]0,t_0[$ and $\mu_t^{\mathfrak{a}}\ll\omega$ for all $t\in]t_0,\infty[$.

Moreover, one has the equivalence of the following statements:

(i) $\mu_t^{\mathfrak{a}}\ll\omega$;

(ii) $\sum_{\mathfrak{m}\in G^\wedge} \widehat{\mu_t^{\mathfrak{a}}}(\mathfrak{m})<\infty$;

(iii) $\sum_{n\geq 1} e^{-ta_n}<\infty$;

(iv) $\mu_t^{\mathfrak{a}}\ll\omega$, and there exists a bounded ω-density of $\mu_t^{\mathfrak{a}}$ (for all $t\in\mathbb{R}_+^*$).

Hence, $\mu_t^{\mathfrak{a}}\ll\omega$ but without continuous ω-density of $\mu_t^{\mathfrak{a}}$ for $t\in]t_0,2t_0[$ and $\mu_t^{\mathfrak{a}}\ll\omega$ with continuous ω-density whenever $t\in]2t_0,\infty[$.

Lemma 5.5.10 and Theorem 5.5.11 contain full information on the supports of the measures of a symmetric Gauss semigroup on an arbitrary connected locally compact Abelian group. For such groups one can show (Theorem 5.5.13) that absolutely continuous symmetric Gauss measures (in the sense of Parthasarathy) exist iff G admits a countable basis of its topology and is locally connected. This existence theorem clarifies the special rôle of the de Moivre-Laplace type central limit theorem discussed in Remark 5.4.4.

This is only a short list of results on absolutely continuous Gauss semigroups, which without loss of generality we chose to be symmetric.

The full theory of continuous convolution semigroups on a locally compact Abelian group G has been laid out by Berg and Forst in their monograph [12]. This book aims at a complete description of recurrent and transient convolution semigroups in $\mathcal{M}^1(G)$ and only touches on Gauss semigroups and their importance for the central limit problem. In Chapter VI we shall introduce the definition of a Gauss semigroup due to Courrège [94] for an arbitrary locally compact group. This definition coincides with that of a Gauss semigroup in the sense of Parthasarathy if only symmetric measures are involved. There are papers by Forst [149] and Berg [9], supplementing the exposition of [12], which study Gauss semigroups in the sense of Courrège (semigroups of local type) for a locally compact Abelian group. In these papers the properties of Gauss semigroups are related to properties of the corresponding negative-definite functions on the character group

$G^\wedge$ of G, which in turn are connected with the potential operator and potential kernel of the semigroup. This is proved in Berg's paper [8]. See also Berg and Forst [12], Theorem 18.27.

Finally, we like to indicate two more directions of research on Gauss semigroups on locally compact Abelian groups G. The first direction concerns the diffusion of certain Gauss semigroups in $\mathscr{M}^1(G)$ established within a more general framework by Hazod [199]. We shall discuss Hazod's results in more detail in the references and comments to § 3 of Chapter VI.

Previous relevant work has been done by Hartmann and Wintner [193], Blum and Rosenblatt [24] and Huff [258]. The other direction of research culminates in the recent classical result of Hudson and Tucker [257] that for $G := \mathbb{R}$ any absolutely continuous, infinitely divisible measure $\mu \in \mathscr{M}^1(G)$ without Gauss component has a density which is strictly positive on supp(μ), and that supp(μ) is an unbounded interval. The theorem has predecessors in the work of Hudson and Mason [256]. A generalization of parts of this theorem is due to Berg, who in [11] proved the following: Let G be a locally compact Abelian group and $(\mu_t)_{t \in \mathbb{R}_+^*}$ a (continuous) symmetric convolution semigroup in $\mathscr{M}^1(G)$. Then there exists a closed subgroup H of G satisfying supp$(\mu_t) = H$ for all $t \in \mathbb{R}_+^*$. In particular, the support of a symmetric infinitely divisible probability measure on $\mathbb{R}^n$ is a closed subgroup of $\mathbb{R}^n$. Further descriptions of the supports of infinitely divisible probability measures on a locally compact Abelian group have been proposed by Yuan and Liang [515].

R 5.6 Additive Processes and their Decomposition

We discuss the results of M.S. Bingham, announced in [20] and proved completely in [21]. These results generalize from Euclidean groups to arbitrary locally compact Abelian groups having a countable basis of their topology the classical theory on the Lévy decomposition of additive processes (or processes with independent increments) as treated in Ito [263], [266] or Gihman and Skorohod [163]. For a locally compact Abelian group G having a countable basis of its topology the space $\mathsf{D}_G[0,1]$ of all G-valued functions on $[0,1]$ which are right continuous and admit left-hand limits is introduced. Detailed facts about this function space can be found in Parthasarathy [377], Chapter VII. Theorem 5.6.4 claims that an additive process with values in G is a Gauss process if it has continuous paths. Its proof is based on results in § 4. A related result characterizing Gauss processes $(X_t)_{t \in [0,1]}$ with a.s. continuous paths by properties on the data of the characteristic function of the P-Gauss distributions P_{X_t} $(t \in [0,1])$ has been established by Byczkowska in [54]. It follows from Byczkowska's result that any Wiener process with values in G is continuous in probability. The generalized jump process corresponding to a given additive process can be analyzed in analogy to the classical situation.

A useful treatment of the classical theory has been presented by Tortrat [474], where one finds also proofs of Preparations 5.6.5. Their generalizations to locally compact Abelian groups are proved in Bingham [19]. In the classical framework the jump process corresponding to an additive process has been studied extensively by Ito [263]. Extensions of his studies to arbitrary topological groups are due to Cuculescu [103] and Maksimov [331]. In 5.6.6 we begin with the discussion of the

Lévy decomposition for additive processes along the lines of Bingham [21]. Lemma 5.6.10 is taken from Bingham and Parthasarathy [22]. Theorems 5.6.11 and 5.6.15 contain the construction which yields the Lévy decomposition theorem 5.6.16. Every additive process with values in G is a product of a product of additive processes whose paths are a.s. continuous except at a finite number of points and a Gauss process. A result of particular interest is Theorem 5.6.17 which gives an explicit representation of the Fourier transform $\hat{\mu}_t$ of the distribution $\mu_t := P_{X_t}$ of the G-random variables X_t ($t \in [0, 1]$) of the given additive process. The representation is the canonical one in the sense of § 3 of Chapter IV for the family $(\mu_t)_{t \in \mathbb{R}_+^*}$ accompanying the process. It separates the quadratic part defined by the quadratic form ϕ_t and the integral part, whose Lévy measure appears as the jump measure $\pi(t, \cdot)$ introduced in (2) of Preparations 5.6.5. This is made more precise in Remark 5.6.18. The distributions μ_t ($t \in [0, 1]$) are weakly infinitely divisible and have no idempotent factors.

We note that, conversely, any weakly infinitely divisible measure $\mu \in \mathscr{M}^1(G)$ without any idempotent factor has a canonical representation of the form stated in Theorem 5.6.17. This is a result which has been generalized considerably in Chapter IV. Here we only note that for locally compact Abelian groups the theorem is proved in Parthasarathy [377] and Parthasarathy and Sazonov [384]. The canonical representations of continuous negative-definite functions on the character group $G^\wedge$ of G can be found in Herz [217] and Harzallah [194] for real functions and in Forst [150] as an appendix to Berg and Forst [12] and in Drumm [123] for arbitrary complex functions. Concerning additive processes with values in groups the following reference will be discussed in detail at a later point: The case of a finite group has been thoroughly studied in Maksimov [337], where product integration methods in the sense of § 6 of Chapter IV are applied. A less deep analysis for locally compact Abelian groups is contained in Pakshirajan [374]. A most recent contribution to the subject is the paper [138] of Feinsilver and its forerunner [460] of Stroock and Varadhan.

On the basis of Theorems 5.6.16 and 5.6.17 we finally obtain a Lévy-Khintchine formula for a continuous convolution hemigroup in $\mathscr{M}^1(G)$. In fact, Theorem 5.6.19 admits an additional interpretation of the Lévy measure and illustrates in the special case of an Abelian locally compact group having a countable basis of its topology the intentions of § 6 of Chapter IV. Relations between additive processes and the Markov property are of classical origin. See for example Bauer [6], § 66 or P.A. Meyer [354], Chapitre XIII, § 4. Their generalizations from the Euclidean group $\mathbb{R}^n$ ($n \geq 1$) to arbitrary locally compact Abelian groups G are indicated in Forst [148] and [149]. Let $\mathscr{X}$ be a Hunt process in the sense of Meyer [354] with state space $(G, \mathfrak{B}(G))$ and transition probabilities P_t defined by $P_t(x, B) := \mu_t(Bx^{-1})$ for all $t \in \mathbb{R}_+^*$, $x \in G$ and $B \in \mathfrak{B}(G)$. Then the continuous convolution semigroup $(\mu_t)_{t \in \mathbb{R}_+^*}$ in $\mathscr{M}^1(G)$ is of local type or is Gaussian iff $\mathscr{X}$ has continuous paths. Here again we observe the fact that continuity of the paths of a process is equivalent to the Wiener property, which in turn is equivalent with the vanishing of the jump measure in the integral part of the canonical representation of $(\mu_t)_{t \in \mathbb{R}_+^*}$.

The problem of the existence of the infinitesimal generator of a homogeneous Markov process taking values in a locally compact Abelian group has been treated already in 1959 by Woll [505]. See also Berg and Forst [12], § 12.

Chapter VI

The Central Limit Problem in the General Case

After the treatment of the embedding problem in Chapter III and the problem of canonical representation of continuous convolution semigroups in Chapter IV, the presentation of the central limit problem can be considered as the culmination of the development within the framework of our theory. Many facts discussed at an earlier stage will be combined here for a detailed study of Poisson and Gauss measures on an arbitrary locally compact group as well as for the study of the convergence behavior of triangular systems of probability measures in the sense of a Lindeberg-Feller central limit theorem.

The first section contains a deeper analysis of Poisson measures supplementing Section 2 of Chapter III, in which the first properties and some basic characterizations of Poisson measures were given. Poisson semigroups and measures are then characterized by their corresponding negative-definite forms or generating functionals; next, groups G for which the only H-continuous convolution semigroups in $\mathscr{M}^1(G)$ are the H-Poisson semigroups are characterized, and the density of various classes of Poisson measures among the H-embeddable ones is studied. Finally, emphasis is put on the description of Poisson measures with a parameter which will appear in a most natural way in connection with the central limit theorem.

Section 2 is devoted to Gauss semigroups and Gauss measures with special concentration on their characterizations (in the non-Abelian case). Moreover, Gauss measures are constructed by various general devices and Gauss semigroups are characterized by their generating functionals. It is shown by an operator-theoretic method that the set of symmetric Gauss measures on a locally compact group is closed in the set of embeddable measures.

Sections 3 and 4 are devoted to special aspects of the theory of Gauss distributions. In Section 3 the existence of Gauss measures whose support equals the whole group and of Haar-absolutely continuous Gauss measures is settled. The main result is a necessary and sufficient condition on the basis of the Lie group G for a Gauss semigroup on G to be absolutely continuous. Finally, diffuse or atomless Gauss semigroups are taken up. The discussion of central Gauss semigroups in Section 4 yields in the main theorem the characterization of locally connected, connected, almost periodic groups by the existence of an absolutely continuous Gauss measure. Special emphasis is given to particular properties of densities arising from absolutely continuous Gauss measures.

The last two sections (5 and 6) contain the theory of infinitesimal triangular systems of measures on an almost periodic group under restrictive conditions on the system and on the underlying group resp.

First of all, the accompanying laws theorem is treated in the strong form (without the admission of translates as was done in the Abelian case). Then convergence of infinitesimal systems to Dirac, Poisson and Gauss measures is studied.

As to the restrictions for the class of groups involved it is proved that for totally disconnected compact groups the limits of commutative infinitesimal systems are embeddable into continuous convolution semigroups, while the limits of noncommutative infinitesimal systems can only be shown to be immersible into continuous convolution hemigroups.

Here the natural limitations of the central limit theorem for general locally compact groups become apparent.

6.1 Poisson Embedding and Approximation

Let G be a locally compact group and H a compact subgroup of G. In §2 of Chapter III the set $\mathscr{P}_H(G)$ of all H-Poisson measures in $\mathscr{M}^1(G)$ was introduced and characterized in various ways. Clearly, $\mathscr{P}_H(G) \subset \mathscr{E}_H(G)$, so that H-Poisson measures fall within the class of H-continuously embeddable measures studied in detail in §5 of Chapter III. Moreover, H-Poisson measures in $\mathscr{M}^1(G)$ are embeddable into H-Poisson convolution semigroups in the sense of the following

6.1.1 Definition. An H-continuous convolution semigroup $(\mu_t)_{t\in\mathbb{R}_+^*}$ in $\mathscr{M}^1(G)$ is called an *H-Poisson semigroup* (in $\mathscr{M}^1(G)$), if there exists a measure $\nu\in\mathscr{N}_H(G)$ such that $\mu_t = \exp_H(t\nu)$ for all $t\in\mathbb{R}_+^*$.

By Theorem 3.2.5 a convolution semigroup $(\mu_t)_{t\in\mathbb{R}_+^*}$ in $\mathscr{M}^1(G)$ is an H-Poisson semigroup iff $(\mu_t)_{t\in\mathbb{R}_+^*}$ is norm-continuous at ω_H, which means that

$$\lim_{t\downarrow 0}\|\mu_t - \omega_H\| = 0$$

holds.

The results of Chapter IV enable us to characterize $\{e\}$-Poisson measures and $\{e\}$-Poisson semigroups in $\mathscr{M}^1(G)$ by means of negative-definite or almost positive forms on the spaces $\mathfrak{K}(G)$ or $\mathfrak{D}(G)$ for a Lie projective, almost periodic group G or for an arbitrary locally compact group G resp.

Let G be a locally compact group.

We recall that a negative-definite form ψ on $\mathfrak{K}(G)$ is said to be a Poisson form (on $\mathfrak{K}(G)$), if there exists a positive-definite form ϕ on $\mathfrak{K}(G)$ satisfying $\psi = \phi(1)\hat{\varepsilon}_e - \phi$.

For almost periodic groups G Poisson forms on $\mathfrak{K}(G)$ can be characterized and related naturally to $\{e\}$-Poisson measures in $\mathscr{M}^1(G)$.

6.1.2 Theorem. *Let $G\in\mathbf{A}$ and $\psi\in\mathfrak{K}(G)^*$. The following statements are equivalent:*
(i) *ψ is a Poisson form on $\mathfrak{K}(G)$;*
(ii) *ψ is a bounded negative-definite form on $\mathfrak{K}(G)$ with $\psi(1)=0$.*

Proof. 1. (i)$\Rightarrow$(ii). Let the Poisson form ψ on $\mathfrak{K}(G)$ be of the form $\psi = \phi(1)\hat{\varepsilon}_e - \phi$ for some positive-definite form ϕ on $\mathfrak{K}(G)$. By Theorem 1.4.8 ϕ and therefore ψ is real and bounded. Moreover, $\psi(1)=0$. But $\psi(f)=-\phi(f)\leq 0$ for all $f\in\mathfrak{R}_+(G)$ with $f(e)=0$, and thus ψ is a negative-definite form.

2. (ii) $\Rightarrow$ (i). Let ψ be a bounded negative-definite form on $\mathfrak{R}(G)$ with $\psi(1)=0$. Without loss of generality we assume that G is compact and is hence a B-group. By Theorems 1.5.13 and 1.4.8 there exists a semigroup homomorphism $t\to\mu_t$ from $\mathbb{R}_+^*$ into $\mathscr{M}^1(G)$ with

$$\hat{\mu}_t=\operatorname{Exp}(-t\psi)\quad \text{for all } t\in\mathbb{R}_+^*.$$

Since by hypothesis ψ is bounded, there exists a $\delta>0$ such that $\|\hat{\mu}_t-\hat{\varepsilon}_e\|\leq\delta t$, and by the norm-denseness of $\mathfrak{R}(G)$ in $\mathscr{C}^b(G)$ the inequality $\|\mu_t-\varepsilon_e\|\leq\delta t$ holds for all $t\in]0,1[$. Hence, $\lim_{t\downarrow 0}\|\mu_t-\varepsilon_e\|=0$ and therefore, by Theorem 3.2.5, there exists a measure $\kappa\in\mathscr{M}_+^b(G)$ with

$$\mu_t=\exp_{\{e\}}[-t(\|\kappa\|\,\varepsilon_e-\kappa)]\quad \text{for all } t\in\mathbb{R}_+^*.$$

But then $\psi=\|\kappa\|\,\hat{\varepsilon}_e-\hat{\kappa}$ and $\phi:=\hat{\kappa}$ is the desired positive-definite form on $\mathfrak{R}(G)$. □

6.1.3 Theorem. *Let $G\in\mathbf{A}$.*

(i) *For every $\{e\}$-Poisson semigroup $(\mu_t)_{t\in\mathbb{R}_+^*}$ in $\mathscr{M}^1(G)$ there exists a continuous Poisson form ψ on $\mathfrak{R}(G)$* with

$$\hat{\mu}_t=\operatorname{Exp}(-t\psi)\quad \textit{for all } t\in\mathbb{R}_+^*.$$

(ii) *If, in addition, $G\in\mathbf{M}$, then for every continuous Poisson form ψ on $\mathfrak{R}(G)$ there exists an $\{e\}$-Poisson semigroup $(\mu_t)_{t\in\mathbb{R}_+^*}$ in $\mathscr{M}^1(G)$ satisfying*

$$\hat{\mu}_t=\operatorname{Exp}(-t\psi)\quad \textit{for all } t\in\mathbb{R}_+^*.$$

Proof. (i) follows directly from the definition of an $\{e\}$-Poisson semigroup in $\mathscr{M}^1(G)$. There exists a $\nu\in\mathscr{N}_H(G)$ such that for every $t\in\mathbb{R}_+^*$ we have

$$\mu_t=\exp_{\{e\}}(t\nu).$$

But ν can be written as $\kappa-\|\kappa\|\,\varepsilon_e$ for $\kappa\in\mathscr{M}_+^b(G)$. Therefore

$$\mu_t=\exp_{\{e\}}[-t(\|\kappa\|\,\varepsilon_e-\kappa)]\quad\text{or}$$
$$\hat{\mu}_t=\operatorname{Exp}(-t\psi),\quad\text{where}\quad \psi:=\|\kappa\|\,\hat{\varepsilon}_e-\hat{\kappa}=\kappa(1)\,\hat{\varepsilon}_e-\hat{\kappa}$$

is a Poisson form on $\mathfrak{R}(G)$.

(ii) For a given continuous Poisson form ψ on $\mathfrak{R}(G)$ there is by Theorem 1.5.19 an $\{e\}$-continuous convolution semigroup $(\mu_t)_{t\in\mathbb{R}_+^*}$ in $\mathscr{M}^1(G)$ satisfying

$$\hat{\mu}_t=\operatorname{Exp}(-t\psi)\quad \text{for all } t\in\mathbb{R}_+^*.$$

Moreover, there exists a continuous positive-definite form ϕ on $\mathfrak{R}(G)$ with $\psi=\phi(1)\,\hat{\varepsilon}_e-\phi$. Since by Theorem 1.4.12 G is a B-group, we obtain the existence of a $\kappa\in\mathscr{M}_+^b(G)$ satisfying $\phi=\hat{\kappa}$.

Putting $\nu:=\kappa-\|\kappa\|\,\varepsilon_e$ we deduce that $\hat{\nu}=-\psi$ and

$$\hat{\mu}_t=\operatorname{Exp}(t\hat{\nu})=[\exp_{\{e\}}(t\nu)]^\wedge\quad \text{for all } t\in\mathbb{R}_+^*.$$

The uniqueness of the Fourier transform yields $\mu_t = \exp_{\{e\}}(tv) \in \mathscr{P}_{\{e\}}(G)$ for all $t \in \mathbb{R}_+^*$. □

The following theorem gives the characterizations of Poisson semigroups that are available in the case of a Lie projective, almost periodic group.

6.1.4 Theorem. *Let G be a Lie projective almost periodic group and $S := (\mu_t)_{t \in \mathbb{R}_+^*}$ an $\{e\}$-continuous convolution semigroup in $\mathscr{M}^1(G)$. Let Γ be a Lévy function for G and let ψ be the negative-definite form on $\mathfrak{R}(G)$ corresponding to S (and Γ) of the form (ψ_1, ψ_2, η) with a primitive form ψ_1, a quadratic form ψ_2 on $\mathfrak{R}(G)$ and a Lévy measure η on G. The following statements are equivalent:*
(i) $(\mu_t)_{t \in \mathbb{R}_+^*}$ *is an $\{e\}$-Poisson semigroup;*
(ii) ψ *is a Poisson form;*
(iii) ψ *is bounded;*
(iv) η *is bounded,* $\psi_2 = 0$ *and* $\psi_1 = \int_{G^*} \Gamma(x, \cdot)\, d\eta$;
(v) η *is bounded and* $\psi(f) = \int_{G^*} (f - f(e))\, d\eta$ *for all* $f \in \mathfrak{R}(G)$.

Proof. The equivalences (i)⇔(ii)⇔(iii) follow directly from Theorems 6.1.2 and 6.1.3; (iii)⇔(iv)⇔(v) can be deduced from Theorem 4.3.13 in a way similar to that chosen for the corresponding assertions in the more general setup below. □

6.1.5 Theorem. *Let G be an arbitrary locally compact group and $S := (\mu_t)_{t \in \mathbb{R}_+^*}$ an $\{e\}$-continuous convolution semigroup in $\mathscr{M}^1(G)$ whose corresponding contraction semigroup $(S_t)_{t \in \mathbb{R}_+^*}$ has generating functional A and infinitesimal generator N. Furthermore, let Γ be a Lévy mapping for G and let (ψ_1, ψ_2, η) be the canonical representation of A on $\mathfrak{D}(G)$ with a primitive form ψ_1, a quadratic form ψ_2 on $\mathfrak{D}(G)$ and a Lévy measure η on G. The following statements are equivalent:*
(i) $(\mu_t)_{t \in \mathbb{R}_+^*}$ *is an $\{e\}$-Poisson semigroup;*
(ii) $(\mu_t)_{t \in \mathbb{R}_+^*}$ *is norm-continuous,* i.e., $\lim_{t \downarrow 0} \|\mu_t - \varepsilon_e\| = 0$;
(iii) $(S_t)_{t \in \mathbb{R}_+^*}$ *is uniformly continuous;*
(iv) N *is a bounded operator on* $\mathfrak{D}(G)$;
(v) A *is bounded on* $\mathfrak{D}(G)$ *in the sense* $A(f) \le c \|f\|_\infty$ *for all* $f \in \mathfrak{D}(G)$ *and some* $c \in \mathbb{R}_+^*$;
(vi) η *is bounded,* $\psi_2 = 0$ *and* $\psi_1 = \int_{G^*} \Gamma(\cdot)(x)\, \eta(dx)$;
(vii) η *is bounded and* $A(f) = \int_{G^*} (f - f(e))\, d\eta$ *for all* $f \in \mathfrak{D}(G)$.

Proof. 1. The equivalence (i)⇔(ii) follows from Theorem 3.2.5, (i)⇔(iii)⇔(iv)⇔(v) is a consequence of the Hille-Yosida theory [241], X. The equivalence (vi)⇔(vii) is obvious. Similarly, the implication (vii)⇒(v) is clear. We must still show

2. the implication (v)⇒(vi): Let therefore A, the bounded generating functional of the semigroup $(\mu_t)_{t \in \mathbb{R}_+^*}$ in $\mathscr{M}^1(G)$, be of the canonical form (ψ_1, ψ_2, η). We assume the existence of a constant $c \in \mathbb{R}_+^*$ such that

$$A(f) \le c \|f\|_\infty \quad \text{for all } f \in \mathfrak{D}(G).$$

Then (a) η is bounded. Let

$$\mathscr{A} := \{f \in \mathfrak{D}_+(G) : f \le 1,\ f \text{ vanishes in a neighborhood of } e\}.$$

$\mathscr{A}$ is an ascending subset of $\mathfrak{D}(G)$ with $\sup \mathscr{A} = 1_{G^*}$. But then

$$\lim_{f\in\mathscr{A}} \eta(f) = \eta(G^*), \quad \text{and} \quad \eta(G^*) = \|\eta\| \leq c,$$

since from the canonical representation of A we obtain

$$\eta(f) = A(f) \leq c \quad \text{for all } f \in \mathscr{A}.$$

(b) We show that any primitive or quadratic form $L \neq 0$ on $\mathfrak{D}(G)$ is necessarily unbounded. It suffices to carry out the argument for a primitive form $L \neq 0$ on $\mathfrak{D}(G)$. Let L be a bounded primitive form $\neq 0$ on $\mathfrak{D}(G)$ and $a := \|L\| > 0$. Then the set

$$\mathscr{V} := \{f \in \mathfrak{D}(G) : \|f\| \leq 1, |L(f)| \geq \tfrac{2}{3} a\}$$

is nonempty.

From the defining functional equation of the primitive form L one concludes

$$L(f^2) = 2f(e)\, L(f) \quad \text{for all } f \in \mathfrak{D}(G).$$

In particular, for $f \in \mathscr{V}$ we get

$$\begin{aligned} 2|f(e)|\tfrac{2}{3}a &\leq |2f(e)\, L(f)| = |L(f^2)| \\ &\leq a\|f^2\| = a\|f\|^2 \leq a, \end{aligned}$$

and hence $|f(e)| \leq \frac{3}{4}$. Without loss of generality we assume $f(e) \geq 0$. Then there exists a $b \in \mathbb{R}_+^*$ satisfying $\frac{3}{4} < f(e) + b < 1$, and so there is an open neighborhood $U \in \mathfrak{V}(e)$ with $\frac{3}{4} < |f(x)| + b < 1$ for all $x \in U$. We now choose $V \in \mathfrak{V}(e)$ with $\overline{V} \subset U$ and $g \in \mathfrak{D}(G)$ such that $g = g^*$ with $b 1_V \leq g \leq b 1_U$, and put $h := f + g$. But g is constant in a neighborhood of e. Thus we get $L(g) = 0$ since for any $g' \in \mathfrak{D}(G)$ with $g' = 1$ in a neighborhood of e, the defining functional equation of L implies

$$L(g') = L(g'^2) = 2L(g'),$$

and we conclude that $|L(h)| = |L(f)| \geq \frac{2}{3}a$. For $x \in U$ we have

$$|h(x)| = |f(x) + g(x)| \leq |f(x)| + b < 1$$

and for $x \in \complement U$ clearly $h(x) = f(x)$. Thus, $\|h\| \leq 1$ and so $h \in \mathscr{V}$. But

$$h(e) = f(e) + g(e) = f(e) + b > \tfrac{3}{4}$$

yields a contradiction of the above assumption that $|f(e)| \leq \frac{3}{4}$.

The statements (a) and (b) together imply the assertion. □

We are now going to investigate the question, under what conditions on the locally compact group G we obtain the inclusion $\mathscr{I}(G) \subset \mathscr{P}(G)$, which clearly implies that $\mathscr{I}(G) = \mathscr{P}(G)$. Theorem 3.2.9 says that finite groups satisfy this property.

6.1.6 Theorem. *Let G be a locally compact group. The following statements are equivalent:*

(i) *For any given compact subgroup H of G every H-continuous convolution semigroup in $\mathscr{M}^1(G)$ is an H-Poisson semigroup;*
(ii) *G is discrete.*

Proof. 1. (ii) $\Rightarrow$ (i). Let $(\mu_t)_{t\in\mathbb{R}_+^*}$ be an H-continuous convolution semigroup in $\mathscr{M}^1(G)$, where G is assumed to be discrete. Then $(\mu_t)_{t\in\mathbb{R}_+^*}$ is norm-continuous at ω_H in the sense that

$$\lim_{t\downarrow 0}\|\mu_t-\omega_H\|=0.$$

For a given $\delta\in\mathbb{R}_+^*$ there thus exists a $t_0:=t_0(\delta)\in\mathbb{R}_+^*$ such that $\|\mu_t-\omega_H\|\leq\delta$ for all $t\leq t_0$. For $\delta<1/3$ the measure $\log_H(\mu_t)$ is defined, and $t\to\log_H(\mu_t)$ is a continuous mapping from $]0,t_0]$ into $\mathscr{N}_H(G)$. We have

$$\log_H(\mu_{t_0})=n\log_H(\mu_{\frac{t_0}{n}})\quad\text{and}\quad\log_H(\mu_{\frac{mt_0}{n}})=\frac{m}{n}\log_H(\mu_{t_0})$$

for any $\frac{m}{n}\in\mathbb{Q}_+\cap]0,1]$.

From the continuity of $t\to\log_H(\mu_t)$ and Property (1) of 3.2.2 there follows at once $\log_H(\mu_t)=t\,\frac{1}{t_0}\log(\mu_{t_0})=:t\nu$ and by a method applied in the proof of Theorem 3.2.7 $\mu_t=\exp_H(t\nu)$ for $\nu\in\mathscr{N}_H(G)$, provided that $t\leq t_0$. If $t\in\mathbb{R}_+^*$ is arbitrary, one chooses $n\geq 1$ such that $\frac{t}{n}\leq t_0$ and concludes again that

$$\mu_t=\mu_{\frac{t}{n}}^n=\exp_H(t\nu)\qquad\text{for}\quad\nu\in\mathscr{N}_H(G).$$

2. (i) $\Rightarrow$ (ii). Let G be nondiscrete. We shall have to treat two cases. (α) Let G be not totally disconnected. By [312] this is equivalent to the fact that there exists a nontrivial continuous homomorphism ψ from $\mathbb{R}$ into G. In this case we define an $\{e\}$-continuous convolution semigroup $(\mu_t)_{t\in\mathbb{R}_+^*}$ in $\mathscr{M}^1(G)$ with $\mu_t:=\varepsilon_{\psi(t)}$ for all $t\in\mathbb{R}_+^*$ which is easily seen not to be an $\{e\}$-Poisson semigroup.
(β) Let G be a nondiscrete, totally disconnected, compact group. There exists a decreasing sequence $(L_n)_{n\geq 1}$ of open, compact normal subgroups of G such that

$$L:=\bigcap_{n\geq 1}L_n$$

has nonvoid interior. It is known that the group $\dot{G}:=G/L$ is totally disconnected and metrizable.

Assume, now, that there exists a K-continuous convolution semigroup $(\dot{\mu}_t)_{t\in\mathbb{R}_+^*}$ in $\mathscr{M}^1(\dot{G})$ which is not a Poisson semigroup. Since by Theorem 1.2.15 $\mathscr{M}_L^1(G)$ and $\mathscr{M}^1(\dot{G})$ are isomorphic topological semigroups, there exists an H-continuous convolution semigroup $(\mu_t)_{t\in\mathbb{R}_+^*}$ in $\mathscr{M}_L^1(G)$ with $p(\mu_t)=\dot{\mu}_t$ for all $t\in\mathbb{R}_+^*$ and $p(\omega_H)=\omega_K$, where p denotes the canonical mapping from G onto $\dot{G}:=G/L$. But then $(\mu_t)_{t\in\mathbb{R}_+^*}$ cannot be a Poisson semigroup. Thus without loss of generality we may assume that G is metrizable. Under this condition there exists a decreasing sequence $(K_n)_{n\geq 1}$ of open normal subgroups K_n with $K_{n+1}\neq K_n$ for all $n\geq 1$ and $\bigcap_{n\geq 1}K_n=\{e\}$. We choose for every $n\geq 1$ an $x_n\in K_n\setminus K_{n+1}$ and put $\eta:=\sum_{n\geq 1}\varepsilon_{x_n}$. Clearly η is an unbounded Lévy measure on G. □

6.1.7 Corollary. *If G is discrete and H a finite subgroup of G, then* $\mathscr{E}_H(G)=\mathscr{P}_H(G)$.

The *proof* is clear. □

6.1.8 Definition. A locally compact group G is said to satisfy the *exponential principle* (EXP) if $\mathscr{I}(G)=\mathscr{P}(G)$.

6.1.9 Theorem. *Let G be a locally compact Abelian group. The following statements are equivalent:*
(i) *G satisfies* (EXP);
(ii) *G is discrete and completely nondivisible in the sense that* $T(G)=\{e\}$ *holds.*

Proof. 1. (i) ⇒ (ii). If G is nondiscrete, then by Theorem 6.1.6 there exists a compact subgroup H of G such that $\mathscr{E}_H(G)\backslash\mathscr{P}_H(G)\neq\emptyset$, and hence $\mathscr{I}(G)\backslash\mathscr{P}(G)\neq\emptyset$. In the case of a group G which satisfies $T(G)\neq\{e\}$ we have $\varepsilon_x\in\mathscr{I}(G)\backslash\mathscr{P}(G)$ for all $x\in T(G)$, and hence again, $\mathscr{I}(G)\backslash\mathscr{P}(G)\neq\emptyset$.

2. (ii) ⇒ (i). Now let G be a discrete Abelian group with $T(G)=\{e\}$ and $\mu\in\mathscr{I}(G)$. From the proof of Theorem 3.3.17 we obtain the existence of a divisible element $\bar{x}$ of G and of a measure $\lambda\in\mathscr{P}_H(G)$ satisfying $\mu=\varepsilon_{\bar{x}}*\lambda$. The hypothesis $T(G)=\{e\}$ implies that $\bar{x}=e$ and hence, that $\mu\in\mathscr{P}_{\{e\}}(G)$. □

In the non-Abelian case sufficient conditions for the validity of (EXP) can be given in various ways. A typical example is contained in the following

6.1.10 Theorem. *Let G be a discrete group in the class* **S**. *Then G satisfies* (EXP).

Proof. Let $\mu\in\mathscr{I}(G)$. By the definition of **S** there exists a rational one-parameter semigroup $(\mu_r)_{r\in\mathbb{Q}_+^*}$ in $\mathscr{M}^1(G)$ such that the set $\{\mu_r: r\in]0,1]\}$ is relatively compact in $\mathscr{M}^1(G)$ with the property $\mu_1=\mu$. Let $\mathscr{N}$ be the accumulation set of

$$\bigcap_{\varepsilon>0}\{\mu_r: r\in]0,\varepsilon[\}^-.$$

Then $\mathscr{N}$ is a group with unit ω_H for some finite subgroup H of G, and $\mathscr{N}$ is uniformly tight. Since G is discrete, we conclude that the set

$$\mathscr{N}\subset\{\varepsilon_x*\omega_H=\omega_H*\varepsilon_x: x\in G\}$$

is finite. Hence, there exists a sequence $(n_k)_{k\geq 1}$ in $\mathbb{N}$ with $\lim_{k\to\infty}n_k=\infty$ such that

$$\mathscr{T}_v\text{-}\lim_{k\to\infty}\mu_{\frac{1}{n_k!}}=\varepsilon_{x_0}*\omega_H \quad\text{or}\quad \lim_{k\to\infty}\|\mu_{\frac{1}{n_k!}}-\varepsilon_{x_0}*\omega_H\|=0$$

holds. Let

$$m_0:=|\mathscr{N}| \quad\text{and}\quad m_k:=\frac{n_k!}{m_0} \quad \text{for } n_k\geq m_0 \ (k\geq 1).$$

Then

$$\lim_{k\to\infty}\|\mu_{\frac{1}{m_k}}-\omega_H\|=0.$$

Theorem 3.2.7 implies that $(\mu_{\frac{1}{m_k}})_{k\geq 1}$ and so $(\mu_r)_{r\in\mathbb{Q}_+^*}$ is a family in $\mathscr{P}_H(G)$. □

6.1.11 Remark. For the validity of the theorem the condition of complete nondivisibility is redundant since $T(G) \neq \{e\}$ contradicts the fact that G belongs to **S**. Moreover, there exists a completely nondivisible locally compact group $G \notin \mathbf{S}$ for which (EXP) holds. One need only take the group $G := \prod_{n \geq 1}^{*} \mathbb{Z}_n$ with $\mathbb{Z}_n := \mathbb{Z}/n\mathbb{Z}$ for all $n \geq 1$ furnished with the discrete topology.

6.1.12 Properties of the set $\mathscr{P}(G)$ of all Poisson measures on a locally compact group G.

(1) Let H be a compact subgroup of G. Then the set $\mathscr{P}_H(G)$ is not in general a subsemigroup of $\mathscr{M}^1(G)$.

In fact, let G be a finite group with elements x, y such that $xy \neq yx$ holds. The measures $\lambda := \varepsilon_x - \varepsilon_e$ and $\pi := \varepsilon_y - \varepsilon_e$ are elements of $\mathscr{N}_{\{e\}}(G)$ satisfying

$$\lambda * \pi - \pi * \lambda = \varepsilon_x * \varepsilon_y - \varepsilon_y * \varepsilon_x \notin \mathscr{N}_{\{e\}}(G).$$

For any $\varepsilon > 0$ we define $\mu := \exp_{\{e\}}(\varepsilon\lambda)$ and $\nu := \exp_{\{e\}}(\varepsilon\pi)$. Obviously $\mu, \nu \in \mathscr{P}_{\{e\}}(G)$. For sufficiently small $\varepsilon > 0$ we have

$$\mu * \nu = \exp_{\{e\}} \log_{\{e\}}(\mu * \nu)$$

and in addition $\|\mu * \nu - \varepsilon_e\| < 1$. But by construction

$$\log_{\{e\}}(\mu * \nu) = \varepsilon(\lambda + \pi) + \tfrac{1}{2}\varepsilon^2 [\lambda * \pi - \pi * \lambda + O(\varepsilon)] \notin \mathscr{N}_{\{e\}}(G)$$

for sufficiently small $\varepsilon > 0$ so that $\mu * \nu \notin \mathscr{P}_{\{e\}}(G)$.

As an immediate consequence of the proof of (1) we obtain the following assertion.

(2) If G is a finite group, then $\mathscr{P}(G)$ is a subsemigroup of $\mathscr{M}^1(G)$ iff G is commutative.

(3) Let $(G_i)_{i=1,\ldots,n}$ be a finite family of locally compact groups and

$$G := \prod_{i=1}^{n} G_i.$$

If for every $i = 1, \ldots, n$ the measure μ_i is in $\mathscr{P}(G_i)$, then

$$\bigotimes_{i=1}^{n} \mu_i \in \mathscr{P}(G).$$

(4) For arbitrary families $(G_\alpha)_{\alpha \in \mathbf{A}}$ of even compact groups G_α and measures $\mu_\alpha \in \mathscr{P}(G_\alpha)$ for $\alpha \in \mathbf{A}$ the assertion of Property (3) for $G := \prod_{\alpha \in \mathbf{A}} G_\alpha$ is in general false.

In fact, for every $\alpha \in \mathbf{A}$ let a measure $\lambda_\alpha \in \mathscr{M}^1(G_\alpha)$ be given with $\operatorname{supp}(\lambda_\alpha) \subset G_\alpha$, $\lambda_\alpha(\{e\}) = 0$ and also a number $\gamma_\alpha \in \mathbb{R}_+$ such that

$$\mu^{(\alpha)} := \exp_{\{e\}}[\gamma_\alpha(\lambda_\alpha - \varepsilon_e)] \in \mathscr{P}_{\{e\}}(G_\alpha).$$

We define $\mu := \bigotimes_{\alpha \in \mathbf{A}} \mu^{(\alpha)}$. If the family $(\gamma_\alpha)_{\alpha \in \mathbf{A}}$ is not bounded, then the series $\sum_{\alpha \in \mathbf{A}} \gamma_\alpha(\lambda_\alpha - \varepsilon_e)$ (understood in the sense of a limit of finite partial sums) is not convergent. Hence, μ is not in $\mathscr{P}_{\{e\}}(G)$.

The following permanence properties expressed in terms of Poisson semigroups are easily established.

(5) Let G and G' be locally compact groups, H a compact subgroup of G and $\phi: G \to G'$ a continuous homomorphism. Then $\phi(\mathscr{P}_H(G)) \subset \mathscr{P}_{\phi(H)}(G')$.

More precisely, if $(\mu_t)_{t\in\mathbb{R}_+^*}$ is an H-Poisson semigroup in $\mathcal{M}^1(G)$, then $(\phi(\mu_t))_{t\in\mathbb{R}_+^*}$ is a $\phi(H)$-Poisson semigroup in $\mathcal{M}^1(G')$ of the form

$$\phi(\mu_t)=\exp_{\phi(H)}[t\gamma(\phi(\lambda)-\omega_{\phi(H)})]$$

with $\phi(\lambda)\in\mathcal{M}^1_{\phi(H)}(G')$.

(6) Let G and G' be locally compact groups, H a compact subgroup and ψ a $\mathcal{T}_v$-continuous algebra homomorphism from $\mathcal{M}^b(G)$ into $\mathcal{M}^b(G')$ satisfying $\psi(\mathcal{M}^1(G))\subset\mathcal{M}^1(G')$. If $(\mu_t)_{t\in\mathbb{R}_+^*}$ is an H-Poisson semigroup in $\mathcal{M}^1(G)$, then $(\psi(\mu_t))_{t\in\mathbb{R}_+^*}$ is an H'-Poisson semigroup in $\mathcal{M}^1(G')$, where the compact subgroup H' of G' is defined by $\omega_{H'}=\psi(\omega_H)$.

(7) Let G and G' be locally compact groups and ϕ a continuous epimorphism from G onto G' such that for every $\mu'\in\mathcal{M}^1(G')$ there exists a $\mu\in\mathcal{M}^1(G)$ satisfying $\phi(\mu)=\mu'$. Then for any $\{e'\}$-Poisson semigroup $(\mu_t')_{t\in\mathbb{R}_+^*}$ in $\mathcal{M}^1(G')$ there exists an $\{e\}$-Poisson semigroup $(\mu_t)_{t\in\mathbb{R}_+^*}$ in $\mathcal{M}^1(G)$ such that $\phi(\mu_t)=\mu_t'$ holds for all $t\in\mathbb{R}_+^*$.

In particular, given

$$\mu_t'=\exp_{\{e'\}}[t\gamma(\lambda'-\varepsilon_{e'})] \quad \text{with} \quad \lambda'\in\mathcal{M}^1(G'), \quad \gamma\in\mathbb{R}_+,$$

we obtain $\mu_t=\exp_{\{e\}}[t\gamma(\lambda-\varepsilon_e)]$ for $\lambda\in\mathcal{M}^1(G)$ satisfying $\phi(\lambda)=\lambda'$ whenever $t\in\mathbb{R}_+^*$.

(8) Let G be a locally compact group, K a compact normal subgroup of G and ϕ the canonical homomorphism from G onto $\dot{G}:=G/K$. For every $\dot{H}$-Poisson semigroup $(\dot{\mu}_t)_{t\in\mathbb{R}_+^*}$ in $\mathcal{M}^1(\dot{G})$ for a compact subgroup $\dot{H}$ of $\dot{G}$ there exists an H-Poisson semigroup $(\mu_t)_{t\in\mathbb{R}_+^*}$ in $\mathcal{M}^1(G)$ with $\phi(\mu_t)=\dot{\mu}_t$ for all $t\in\mathbb{R}_+^*$, where $H:=\dot{H}K$. This assertion follows from Property (7).

(9) Let G be a locally compact group, H a closed normal subgroup of G, $\dot{G}:=G/H$ and ϕ the canonical homomorphism from G onto $\dot{G}$. Then for every $\{\dot{e}\}$-Poisson semigroup $(\dot{\mu}_t)_{t\in\mathbb{R}_+^*}$ in $\mathcal{M}^1(\dot{G})$ there exists an $\{e\}$-Poisson semigroup in $\mathcal{M}^1(G)$ such that $\phi(\mu_t)=\dot{\mu}_t$ for all $t\in\mathbb{R}_+^*$.

The proof of this statement is based on the fact that by Theorem 1.2.15 the homomorphism ϕ induces a surjective mapping from $\mathcal{M}^1(G)$ onto $\mathcal{M}^1(\dot{G})$, and the statement hence follows from Property (7).

(10) In general, $\mathcal{P}_H(G)$ is far from being $\mathcal{T}_v$-closed in $\mathcal{E}_H(G)$. This can be seen from the following discussion concerning the approximation of embeddable measures by Poisson measures.

6.1.13 Theorem. *Let G be a locally compact group and H a compact subgroup of G. Then $\overline{\mathcal{P}_H(G)}=\mathcal{E}_H(G)$. Moreover, for every $\mu\in\mathcal{E}_H(G)$ there exists a sequence $(\sigma_n)_{n\geq 1}$ in $\mathcal{P}_H(G)$ satisfying $\mu=\mathcal{T}_v\text{-}\lim_{n\to\infty}\sigma_n$.*

The *proof* is a direct application of the Hille-Yosida theory of strongly continuous contraction semigroups $(T_t)_{t\in\mathbb{R}_+^*}$ on a Banach space B. In fact one has the formula

$$T_t=\lim_{s\downarrow 0}\exp tA_s, \text{ which is valid for all } t\in\mathbb{R}_+^*, \text{ where}$$

$$A_s:=\frac{1}{s}(T_s-E) \quad \text{for } s\in\mathbb{R}_+^* \quad ([241], \text{p. } 312).$$

Let now $\mu\in\mathscr{E}_H(G)$ be embeddable in an H-continuous convolution semigroup $(\mu_t)_{t\in\mathbb{R}_+^*}$ in $\mathscr{M}^1(G)$. Applying the above formula to the Banach space

$$\mathsf{B}:=\{f\in\mathscr{C}^0(G): f=\omega_H * f\}$$

and to the contraction semigroup $(T_t)_{t\in\mathbb{R}_+^*}$ corresponding to the semigroup $(\mu_t)_{t\in\mathbb{R}_+^*}$ we obtain

$$\mu=\lim_{n\to\infty}\sigma_n \quad \text{with} \quad \sigma_n:=\exp_H[n(\mu_{\frac{1}{n}}-\omega_H)]\in\mathscr{P}_H(G). \quad \square$$

6.1.14 Corollary. $\overline{\mathscr{P}_{\{e\}}(G)}=\overline{\mathscr{E}(G)}$.

Proof. It remains to show that for any compact subgroup H of G one has $\mathscr{E}_H(G)\subset\overline{\mathscr{P}_{\{e\}}(G)}$. This, however, follows from the approximation

$$\omega_H=\lim_{k\to\infty}\exp_{\{e\}}[k(\omega_H-\varepsilon_e)]$$

or from the more general considerations below. $\square$

6.1.15 Remark. The statement of the corollary generalizes in two directions the assertion of Corollary 1.5.22, a fact, which can be seen with the help of equivalence (i)$\Leftrightarrow$(ii) of Theorem 6.1.4.

We are keeping the hypotheses that G is a locally compact group and H denotes a compact subgroup of G.

6.1.16 Definition. A measure $\mu\in\mathscr{P}_H(G)$ of the form

$$\mu=\exp_H[\gamma(\lambda-\omega_H)] \qquad \text{with } \lambda\in\mathscr{M}_H^1(G) \text{ and } \gamma\in\mathbb{R}_+$$

is called *elementary* if there exists an $x_0\in G$ satisfying

$$\lambda=\omega_H*\varepsilon_{x_0}*\omega_H.$$

The set of all elementary H-Poisson measures on G will be abbreviated by $\mathscr{P}_H^e(G)$ and the totality of their finite (convolution) products by $\mathscr{F}\mathscr{P}_H^e(G)$.

Clearly, $\mathscr{F}\mathscr{P}_H^e(G)$ is a subsemigroup of $\mathscr{M}^1(G)$ and we have $\mathscr{P}_H^e(G)\subset\mathscr{F}\mathscr{P}_H^e(G)$ as well as $\mathscr{P}_H^e(G)\subset\mathscr{P}_H(G)$.

In order to establish refinements of the approximation results in 6.1.13 and 6.1.14 we prove a useful

6.1.17 Lemma. *Let* B *be a Banach algebra with unit* u, $\varepsilon\in\mathbb{R}_+^*$ *and* ρ *a continuous mapping from* $[0,\varepsilon[$ *into* B *satisfying the conditions*

$$\rho(0)=u \quad \text{and} \quad \|\rho(t)\|\le 1 \qquad \text{for all } t\in[0,\varepsilon[.$$

In B *the exponential of an element a is defined by*

$$\exp(a)=u+\sum_{k\geq 1}\frac{a^k}{k!},$$

where the convergence of the series is understood in the sense of the norm of B.

We now suppose that $b:=\lim_{t\downarrow 0}\frac{1}{t}(\rho(t)-u)$ *exists (in* B*). Then for all* $t\in[0,\varepsilon[$ *we have* $\exp(t\,b)=\lim_{n\to\infty}\rho\left(\frac{t}{n}\right)^n$.

Proof. 1. For any $a\in$ B with $\|a\|\leq 1$ and every $k\geq 1$ the inequality

$$\|a^k-\exp[k(a-u)]\|\leq\sqrt{k}\,\|a-u\|$$

holds. In fact, for $k\geq 1$ one obtains

$$\|\exp[k(a-u)]-a^k\|=e^{-k}\left\|\sum_{l\geq 0}\frac{k^l}{l!}(a^l-a^k)\right\|$$

$$\leq e^{-k}\sum_{l\geq 0}\frac{k^l}{l!}\|a^l-a^k\|\leq e^{-k}\sum_{l\geq 0}\frac{k^l}{l!}\|a^{|l-k|}-u\|$$

$$\leq e^{-k}\sum_{l\geq 0}\frac{k^l}{l!}|l-k|\,\|a-u\| \quad\text{and}$$

$$\left(\sum_{l\geq 0}\frac{k^l}{l!}|l-k|\right)^2\leq\left(\sum_{l\geq 0}\frac{k^l}{l!}\right)\left(\sum_{l\geq 0}\frac{k^l}{l!}(l-k)^2\right)$$

$$=e^k\left(\sum_{l\geq 0}\frac{k^l}{l!}(l^2-2lk+k^2)\right)$$

$$=e^k\left[k^2e^k-2k\sum_{l\geq 1}\frac{k^l}{(l-1)!}+\sum_{l\geq 1}\frac{k^l}{(l-1)!}l\right]$$

$$=e^k\left(k^2e^k-2k^2e^k+\sum_{l\geq 1}\frac{k^l}{(l-1)!}(l-1)+\sum_{l\geq 1}\frac{k^l}{(l-1)!}\right)$$

$$=e^k(k^2e^k-2k^2e^k+k^2e^k+ke^k)=ke^{2k},$$

whence the assertion.

2. Now let b be as assumed. For all $s,t\in\mathbb{R}_+$ we plainly have

$$\lim_{n\to\infty}\left\|\exp(s\,b)-\exp\left[s\frac{n}{t}\left(\rho\left(\frac{t}{n}\right)-u\right)\right]\right\|=0 \quad\text{and}$$

$$\lim_{n\to\infty}\left\|\exp\left[t\frac{n}{t}\left(\rho\left(\frac{t}{n}\right)-u\right)\right]-\rho\left(\frac{t}{n}\right)^n\right\|$$

$$=\lim_{n\to\infty}\left\|\exp\left[n\left(\rho\left(\frac{t}{n}\right)-u\right)\right]-\rho\left(\frac{t}{n}\right)^n\right\|$$

$$\leq\lim_{n\to\infty}\sqrt{n}\left\|\rho\left(\frac{t}{n}\right)-u\right\|,$$

where the last inequality follows from Part 1. Thus,

$$\lim_{n\to\infty} \left\| \exp\left[n\left(\rho\left(\frac{t}{n}\right)-u\right)\right]-\rho\left(\frac{t}{n}\right)^n\right\|=0$$

since

$$\lim_{n\to\infty} n \left\|\rho\left(\frac{t}{n}\right)-u\right\|=t\,\|b\|.$$

But this implies the statement of the lemma. □

6.1.18 Application of the lemma to the Banach algebra

$$\mathsf{B}:=\mathcal{M}_H^b(G):=\omega_H * \mathcal{M}^b(G) * \omega_H$$

for a locally compact group G and a compact subgroup H of G yields the following result: Given a $\|\cdot\|$-continuous mapping

$$\rho:[0,\varepsilon[\,\to \mathcal{M}_H^1(G) \quad \text{with} \quad \rho(0)=\omega_H \quad \text{and}$$

$$\lim_{t\downarrow 0} \frac{1}{t}(\rho(t)-\omega_H)=:\nu\in\mathcal{M}^b(G),$$

we first conclude that

(a) $$\exp_H(t\,\nu)=\lim_{n\to\infty} \rho\left(\frac{t}{n}\right)^n \quad \text{for all } t\in[0,\varepsilon[.$$

Putting $\rho(t):=\prod_{i=1}^k \exp_H[t(\lambda_i-\omega_H)]$ for all $t\in[0,\varepsilon[$ and measures $\lambda_i\in\mathcal{M}_H^1(G)$ $(i=1,\ldots,k)$, one obtains

(b) $$\exp_H\left[t\sum_{i=1}^k(\lambda_i-\omega_H)\right]=\lim_{n\to\infty}\left(\prod_{i=1}^k \exp_H\left(\frac{t}{n}\lambda_i\right)\right)^n.$$

Defining $\rho(t):=\omega_H * \exp_{\{e\}}[t(\lambda-\varepsilon_e)] * \omega_H$ for all $t\in[0,\varepsilon[$ and a measure $\lambda\in\mathcal{M}_H^1(G)$, one deduces that

(c) $$\exp_H[t(\omega_H*(\lambda-\varepsilon_e)*\omega_H)]$$
$$=\lim_{n\to\infty}\left[\omega_H*\exp_{\{e\}}\left(\frac{t}{n}(\lambda-\varepsilon_e)\right)*\omega_H\right]^n.$$

6.1.19 Theorem. *Let G be a locally compact group and H a compact subgroup of G. Then $\overline{\mathcal{F}\mathcal{P}_{\{e\}}^e(G)}\supset\mathcal{E}_H(G)$.*

By Corollary 6.1.14 the *proof* can be reduced to the demonstration of $\overline{\mathcal{F}\mathcal{P}_{\{e\}}^e(G)}\supset\mathcal{P}_{\{e\}}(G)$. Let $\mu\in\mathcal{P}_{\{e\}}(G)$ be of the form

$$\mu=\exp_{\{e\}}(\nu) \quad \text{with} \quad \nu:=\gamma(\lambda-\varepsilon_e)\in\mathcal{N}_{\{e\}}(G)$$

for $\lambda \in \mathscr{M}^1(G)$ and $\gamma \in \mathbb{R}_+$. For $\lambda \in \mathscr{M}^1(G)$ there exists by Theorem 1.1.7 a net $(\lambda_\alpha)_{\alpha \in \mathbf{A}}$ of measures $\lambda_\alpha \in \mathscr{M}^1(G)$ with finite support such that

$$\mathscr{T}_v\text{-}\lim_{\alpha \in \mathbf{A}} \lambda_\alpha = \lambda.$$

But then

$$\lim_{\alpha \in \mathbf{A}} \exp_{\{e\}} [\gamma(\lambda_\alpha - \varepsilon_e)] = \exp_{\{e\}} [\gamma(\lambda - \varepsilon_e)] = \mu,$$

where as a consequence of Formula (b) in 6.1.18 with

$$\lambda_\alpha := \sum_{i=1}^k c_i^\alpha \, \varepsilon_{x_i^\alpha} \in \mathscr{M}^1(G) \ (c_1^\alpha, \ldots, c_k^\alpha \in \mathbb{R}_+, x_1^\alpha, \ldots, x_k^\alpha \in G)$$

we have

$$\exp_{\{e\}} [\gamma(\lambda_\alpha - \varepsilon_e)] = \exp_{\{e\}} [\textstyle\sum_{i=1}^k \gamma \, c_i^\alpha (\varepsilon_{x_i^\alpha} - \varepsilon_e)]$$

$$= \lim_{n \to \infty} \left(\prod_{i=1}^k \exp_{\{e\}} \left[\gamma \frac{c_i^\alpha}{n} (\varepsilon_{x_i^\alpha} - \varepsilon_e) \right] \right)^n \in \overline{\mathscr{F}\mathscr{P}_{\{e\}}^e(G)} \qquad (\alpha \in \mathbf{A}). \quad \square$$

6.1.20 Corollary. *For all locally compact groups satisfying the embedding principle* (EMP) *of* 3.5.10 *we have* $\overline{\mathscr{F}\mathscr{P}_{\{e\}}^e(G)} \supset \mathscr{I}(G)$.

The *proof* is clear. $\square$

6.1.21 Remark. In general, the density statement of the corollary is false. In fact, there exists the discrete, torsionfree Abelian group $G := \mathbb{Q}_d$ satisfying

(i) $\mathscr{F}\mathscr{P}_{\{e\}}^e(G) \subset \mathscr{P}_{\{e\}}(G) = \mathscr{P}(G) = \mathscr{E}(G) \subset \mathscr{I}(G)$

such that

(ii) $\mathscr{F}\mathscr{P}_{\{e\}}^e(G)$ is not dense in $\mathscr{I}(G)$.

Let $\mu := \varepsilon_1 \in \mathscr{I}(G)$ and assume that (ii) does not hold. Then there exists a sequence $(\mu^{(n)})_{n \geq 1}$ in $\mathscr{P}_{\{e\}}(G)$ with $\lim_{n \to \infty} \mu^{(n)} = \varepsilon_1$ or $\lim_{n \to \infty} \|\mu^{(n)} - \varepsilon_1\| = 0$. Since $G \in \mathbf{R}$ and $\mathscr{N} := \{\mu^{(n)} : n \geq 1\} \cup \{\varepsilon_1\}$ is a compact subset of $\mathscr{M}^1(G)$, by Theorem 3.1.8 the root sets

$$R(k, \mathscr{N}) := \{\mu_k^{(n)} : n \geq 1\} \cup \{\varepsilon_{\frac{1}{k}}\} \quad (k \geq 1)$$

of $\mathscr{N}$ are relatively compact in $\mathscr{M}^1(G)$. Hence, one can find a subsequence $(n_j)_{j \geq 1}$ of $\mathbb{N}$ such that $\lim_{j \to \infty} \mu_k^{(n_j)} = \varepsilon_{\frac{1}{k}}$ holds for all $k \geq 1$. Representing $\mu^{(n)} \in \mathscr{P}_{\{e\}}(G)$ in the form

$$\exp_{\{e\}} [\gamma_n(\lambda_n - \varepsilon_e)] \qquad \text{for all } n \geq 1$$

one obtains

$$\mu_k^{(n_j)} = \exp_{\{e\}} \left[\frac{\gamma_{n_j}}{k} (\lambda_{n_j} - \varepsilon_e) \right] \qquad \text{for all } j, k \geq 1$$

and hence

$$\lim_{j \to \infty} \exp\left(-\frac{\gamma_{n_j}}{k}\right) \left(\varepsilon_e + \sum_{i \geq 1} \left(\frac{\gamma_{n_j}}{k}\right)^i \frac{1}{i!} \lambda_{n_j}^i\right)(\{r\}) = \begin{cases} 0, & \text{if } r \neq \frac{1}{k} \\ 1 & \text{otherwise.} \end{cases}$$

One deduces that $\lim_{j\to\infty}\gamma_{n_j}=\infty$ and thus that

$$\lim_{j\to\infty}\exp\left[\left(-\frac{\gamma_{n_j}}{k}\right)\left(\sum_{i\geq 1}\left(\frac{\gamma_{n_j}}{k}\right)^i\frac{1}{i!}\right)\right]$$

$$=\lim_{j\to\infty}\left[\exp\left(-\frac{\gamma_{n_j}}{k}\right)\left(\exp\left(\frac{\gamma_{n_j}}{k}\right)-1\right)\right]=1.$$

From $0\leq\lambda_n^i(\{r\})\leq 1$ for all $r\in\mathbb{Q}$; $i,n\geq 1$ one obtains

$$\lim_{j\to\infty}\lambda_{n_j}^i\left(\left\{\frac{1}{k}\right\}\right)=1 \qquad \text{for all } k,i\geq 1.$$

But this implies that

$$\lim_{j\to\infty}\|\lambda_{n_j}\|\geq\lim_{j\to\infty}\sum_{k\geq 1}\lambda_{n_j}\left(\left\{\frac{1}{k}\right\}\right)=\infty,$$

which is a contradiction of $\lambda_n\in\mathcal{M}^1(G)$ for all $n\geq 1$.

To complete the discussion of Poisson measures we shall present various characterizations of the measures in $\mathcal{P}^e_{\{e\}}(G)$. Such measures $\mu\in\mathcal{P}^e_{\{e\}}(G)$ of the form $\mu=\exp_{\{e\}}[\gamma(\varepsilon_{x_0}-\varepsilon_e)]$ with $\gamma\in\mathbb{R}_+$ and $x_0\in G$ are often called *Poisson distributions* (on G) *with parameter* x_0. They are equivalently described as measures $\mu\in\mathcal{M}^1(G)$ such that

$$\mu(B)=e^{-\gamma}\sum_{k\in K(B)}\frac{\gamma^k}{k!}$$

holds for all $B\in\mathfrak{B}(G)$, where $K(B):=\{k\geq 1: x_0^k\in B\}$.

6.1.22 Theorem. *Let $G\in\mathbf{A}$, $x_0\in G$ and $\mu\in\mathcal{M}^1(G)$ with $\mu\neq\varepsilon_e$. The following conditions are equivalent:*

(i) *μ is a Poisson measure (in $\mathcal{P}^e_{\{e\}}(G)$) with parameter x_0;*

(ii) *there exists a sequence $(\mu_n)_{n\geq 1}$ of n-th roots μ_n of μ such that*

(a) $\overline{\lim}_{n\geq 1}\mu_n(\{e\})>0$ *(or equivalently* $\overline{\lim}_{n\geq 1}\|\mu_n-\varepsilon_e\|<2$),

(b) *there exists an $n_0\geq 1$ such that $\|\mu_{n_0}*\mu_{n_0}^{\sim}-\varepsilon_e\|<1$ holds,*

(c) $\lim_{n\to\infty} n\,\mu_n(G\setminus\{x_0,e\})=0$.

Proof. Clearly one need only show the implication (ii) ⇒ (i). Let μ be in $\mathcal{M}^1(G)$, $\mu\neq\varepsilon_e$, with a sequence $(\mu_n)_{n\geq 1}$ of roots satisfying the conditions (a), (b) and (c) of (ii). Since $G\in\mathbf{A}$, the conditions (a) and (b) yield by Theorem 3.2.13 the representation

$$\mu=\exp_{\{e\}}[\gamma'(\lambda-\varepsilon_e)]\in\mathcal{P}_{\{e\}}(G)$$

with $\gamma'\in\mathbb{R}_+$ and $\lambda\in\mathcal{M}^1(G)$. Furthermore, one has the existence of a sequence $(n_k)_{k\geq 1}$ in $\mathbb{N}$ such that

$$\lim_{k\to\infty}n_k(\mu_{n_k}-\varepsilon_e)=\gamma'(\lambda-\varepsilon_e)$$

is satisfied. Hence, (c) implies

$$\gamma'(\lambda-\varepsilon_e)(G\setminus\{x_0,e\})=\gamma'\,\lambda(G\setminus\{x_0,e\})=0,$$

from which we get $\mu=\exp_{\{e\}}[\gamma(\varepsilon_{x_0}-\varepsilon_e)]$ with $\gamma\in\mathbb{R}_+$ or (i). □

For the proof of the following characterization of Poisson measures with parameter on a locally compact group G we are going to apply two auxiliary results, which in their general form will be useful also for the future development of the theory.

On $\mathscr{M}_+^{(1)}(G)$ we consider the topologies $\mathscr{T}_v$ and $\mathscr{T}_w$ as well as the weak operator topology $\mathscr{T}_{wo}$ defined by the seminorms

$$\mu\to p_{f,g}(\mu):=|\langle T_\mu f, g\rangle| \quad \text{for } f,g\in L^2_{\mathbb{C}}(G,\omega)$$

and the strong operator topology $\mathscr{T}_{so}$ defined by the seminorms

$$\mu\to q_f(\mu):=\|T_\mu f\|_2 \quad \text{for } f\in L^2_{\mathbb{C}}(G,\omega),$$

where, as usual, T_μ denotes the convolution operator of $\mu\in\mathscr{M}_+^{(1)}(G)$ extended to $L^2_{\mathbb{C}}(G,\omega)$ with a right Haar measure ω of G.

We note that by Theorem 1.2.2 $(\mathscr{M}^1(G),\mathscr{T}_w)$ is a topological semigroup. Let $\mathscr{T}'_{so}$ denote the topology in $\mathscr{M}^1(G)$ defined by the seminorms

$$\mu\to q'_f(\mu):=\|T_\mu f\| \quad \text{for all } f\in\mathscr{C}^0(G).$$

Then Theorem 1.5.5 shows that the topological semigroup $(\mathscr{M}^1(G),\mathscr{T}_w)$ is homeomorphic to $(\mathscr{M}^1(G),\mathscr{T}'_{so})$.

6.1.23 Lemma. *Let G be a locally compact group. Then*
(i) *on $\mathscr{M}_+^{(1)}(G)$ the topologies $\mathscr{T}_v$ and $\mathscr{T}_{wo}$ coincide;*
(ii) *on $\mathscr{M}^1(G)$ all topologies $\mathscr{T}_v$, $\mathscr{T}_w$, $\mathscr{T}_{wo}$ and $\mathscr{T}_{so}$ coincide.*

Proof. 1. Let $(\mu_\alpha)_{\alpha\in\mathbf{A}}$ be a net in $\mathscr{M}_+^{(1)}(G)$ with

$$\mathscr{T}_{wo}\text{-}\lim_{\alpha\in\mathbf{A}}\mu_\alpha=:\mu\in\mathscr{M}_+^{(1)}(G).$$

We shall show

$$\mathscr{T}_v\text{-}\lim_{\alpha\in\mathbf{A}}\mu_\alpha=\mu.$$

In fact, for $f,g\in\mathscr{K}(G)$ one defines the function $\phi_{f,g}$ on G by

$$\phi_{f,g}(x)=\int f g_x\, d\omega \quad \text{for all } x\in G.$$

Clearly the system $\mathscr{A}:=\{\phi_{f,g}: f,g\in\mathscr{K}(G)\}$ is norm-dense in $\mathscr{C}^0(G)$ since for an approximate identity $(\psi_\beta)_{\beta\in\mathbf{B}}$ in $\mathscr{C}^0(G)$ of functions $\psi_\beta\in\mathscr{K}_+(G)$ with $\int\psi_\beta\, d\omega=1$ satisfying

$$\mathscr{T}_w\text{-}\lim_{\beta\in\mathbf{B}}\psi_\beta\cdot\omega=\varepsilon_e$$

one obtains $\lim_{\beta\in\mathbf{B}}\|\phi_{\psi_\beta,g}-g\|=0$ whenever $g\in\mathscr{K}(G)$.

For all $\alpha \in \mathbf{A}$ and $f, g \in \mathscr{K}(G)$ we have

$$\mu_\alpha(\phi_{f,g}) = \int (\int f g_x \, d\omega) \, \mu_\alpha(dx) = \int (\int f g_x \, \mu_\alpha(dx)) \, d\omega = \langle T_{\mu_\alpha} g, f \rangle.$$

Hence, $\lim_{\alpha \in \mathbf{A}} \mu_\alpha(\phi_{f,g}) = \lim_{\alpha \in \mathbf{A}} \langle T_{\mu_\alpha} g, f \rangle = \langle T_\mu g, f \rangle = \mu(\phi_{f,g})$ by hypothesis, and the result is established.

2. $\mathscr{M}_+^{(1)}(G)$ is $\mathscr{T}_v$-compact by Corollary 1.1.2 and is $\mathscr{T}_{wo}$-compact as a closed subset of the unit ball of $\mathscr{L}(L^2_{\mathbb{C}}(G, \omega))$ of bounded operators on $L^2_{\mathbb{C}}(G, \omega)$ (furnished with the weak operator topology). Moreover, both topologies $\mathscr{T}_v$ and $\mathscr{T}_{wo}$ are Hausdorff. By Part 1 we have on $\mathscr{M}_+^{(1)}(G)$ the relation $\mathscr{T}_{wo} > \mathscr{T}_v$. Hence, the mapping

$$\mathrm{id} : (\mathscr{M}_+^{(1)}(G), \mathscr{T}_{wo}) \to (\mathscr{M}_+^{(1)}(G), \mathscr{T}_v)$$

is continuous. Thus, $(\mathscr{M}_+^{(1)}(G), \mathscr{T}_{wo})$ and $(\mathscr{M}_+^{(1)}(G), \mathscr{T}_v)$ are homeomorphic spaces, so that (i) is proved.

3. On $\mathscr{M}^1(G)$ we have $\mathscr{T}_w > \mathscr{T}_{wo}$. Let $(\mu_\alpha)_{\alpha \in \mathbf{A}}$ be a net in $\mathscr{M}^1(G)$ with

$$\mathscr{T}_w\text{-}\lim_{\alpha \in \mathbf{A}} \mu_\alpha =: \mu \in \mathscr{M}^1(G).$$

Then by the discussion preceding the lemma,

$$\lim_{\alpha \in \mathbf{A}} \| T_{\mu_\alpha} f - T_\mu f \| = 0 \qquad \text{for all } f \in \mathscr{K}(G),$$

so that

$$\lim_{\alpha \in \mathbf{A}} |\langle (T_{\mu_\alpha} - T_\mu) f, g \rangle| \le \lim_{\alpha \in \mathbf{A}} \| (T_{\mu_\alpha} - T_\mu) f \| \int |g| \, d\omega = 0$$

for all $f, g \in \mathscr{K}(G)$.

4. On $\mathscr{M}^1(G)$ we have $\mathscr{T}_w > \mathscr{T}_{so}$. Let $(\mu_\alpha)_{\alpha \in \mathbf{A}}$ be a net in $\mathscr{M}^1(G)$ with

$$\mathscr{T}_w\text{-}\lim_{\alpha \in \mathbf{A}} \mu_\alpha = \mu \in \mathscr{M}^1(G).$$

Clearly,

$$\mathscr{T}_w\text{-}\lim_{\alpha \in \mathbf{A}} \mu_\alpha^\sim * \mu_\alpha = \mu^\sim * \mu \quad \text{and} \quad \mathscr{T}_w\text{-}\lim_{\alpha \in \mathbf{A}} \mu_\alpha^\sim * \mu = \mu^\sim * \mu$$

since $(\mathscr{M}^1(G), \mathscr{T}_w)$ is a topological semigroup and the involution is continuous.

For all $\alpha \in \mathbf{A}$ and $f \in \mathscr{C}^0(G) \cap L^2_{\mathbb{C}}(G, \omega)$ the equality

$$\begin{aligned} \| (T_{\mu_\alpha} - T_\mu) f \|_2^2 &= \| T_{\mu_\alpha} f \|_2^2 + \| T_\mu f \|_2^2 - 2 \operatorname{Re} \langle T_\mu f, T_{\mu_\alpha} f \rangle \\ &= \langle T_{\mu_\alpha^\sim * \mu_\alpha} f, f \rangle + \langle T_{\mu^\sim * \mu} f, f \rangle - 2 \operatorname{Re} \langle T_{\mu_\alpha^\sim * \mu} f, f \rangle \end{aligned}$$

obtains, hence by Part 3

$$\begin{aligned} &\lim_{\alpha \in \mathbf{A}} \| (T_{\mu_\alpha} - T_\mu) f \|_2^2 \\ &= \langle T_{\mu^\sim * \mu} f, f \rangle + \langle T_{\mu^\sim * \mu} f, f \rangle - 2 \operatorname{Re} \langle T_{\mu^\sim * \mu} f, f \rangle = 0. \end{aligned}$$

5. The relation $\mathscr{T}_{wo} > \mathscr{T}_w$, which is valid on $\mathscr{M}^1(G)$ follows from Part 1 since on $\mathscr{M}^1(G)$ the topologies $\mathscr{T}_v$ and $\mathscr{T}_w$ coincide.

The union of the statements proved in Parts 3 to 5 yield Assertion (ii) of the lemma. □

6.1.24 Lemma. *Let G be a locally compact group, μ a measure in $\mathcal{M}^1(G)$ and $(\mu_{n_k})_{k\geq 1}$ a sequence of normal roots of μ such that $\mu_{n_k}^{n_k}=\mu$ for all $k\geq 1$. For every $k\geq 1$ we define the (symmetric) measure $\lambda_{n_k}:=\mu_{n_k}*\tilde{\mu}_{n_k}$ in $\mathcal{M}^1(G)$. Then there exist a compact group H of G and an H-continuous convolution semigroup $(\sigma_t)_{t\in\mathbb{R}_+^*}$ in $\mathcal{M}^1(G)$ uniquely determined by $n_1=1$ such that $\sigma_1=\lambda_1$ and $\sigma_{\frac{1}{n_k}}=\lambda_{n_k}$ for all $k\geq 1$.*

Proof. For every $k\geq 1$ we consider the convolution operators $T_{\mu_{n_k}}$ and $T_{\lambda_{n_k}}$ on $L^2_{\mathbb{C}}(G,\omega)$. $T_{\lambda_{n_k}}$ is positive-semidefinite and satisfies $T_{\lambda_{n_k}}^{n_k}=T_{\lambda_1}$ for all $k\geq 1$. We know that T_{λ_1} admits a spectral decomposition $T_{\lambda_1}=\int_0^1\rho\,dE_\rho$, where $(E_\rho)_{\rho\in[0,1]}$ denotes a spectral resolution corresponding to T_{λ_1}. Consequently

$$T_{\lambda_{n_k}}=\int_0^1\rho^{\frac{1}{n_k}}\,dE_\rho \qquad \text{for all } k\geq 1.$$

Let $\mathbb{D}$ denote the dense semigroup generated in $\mathbb{R}_+$ by $\left\{\frac{1}{n_k}:k\geq 1\right\}$. Then there exists a semigroup $(T_{\sigma_r})_{r\in\mathbb{D}}$ of operators

$$T_{\sigma_r}:=\int_0^1\rho^r\,dE_\rho \quad \text{on} \quad L^2_{\mathbb{C}}(G,\omega) \qquad \text{such that } T_{\sigma_{\frac{1}{n_k}}}=T_{\lambda_{n_k}}$$

holds for all $k\geq 1$, and $(T_{\sigma_r})_{r\in\mathbb{D}}$ is continuous with respect to the weak operator topology. It follows from Lemma 6.1.23 that the corresponding semigroup $(\sigma_r)_{r\in\mathbb{D}}$ in $\mathcal{M}^1(G)$ satisfying $\sigma_{\frac{1}{n_k}}=\lambda_{n_k}$ for all $k\geq 1$ is H-continuous for some compact subgroup H of G, and it is uniquely determined by $n_1=1$. By Theorem 1.5.9 $(\sigma_r)_{r\in\mathbb{D}}$ can be extended to an H-continuous convolution semigroup $(\sigma_t)_{t\in\mathbb{R}_+^*}$ in $\mathcal{M}^1(G)$. □

6.1.25 Theorem. *Let G be a locally compact group, $\mu\in\mathcal{M}^1(G)$ with $\mu\neq\varepsilon_e$ and $x_0\in G$ satisfying $x_0^2\neq e$.*
The following statements are equivalent:
(i) *μ is a Poisson measure with parameter x_0;*
(ii) *for every $n\geq 1$ there exists an n-th root $\mu_n\in\mathcal{M}^1(G)$ of μ with the properties*
(a) $\overline{\lim}_{n\geq 1}\,\mu_n(\{e\})>0$ *and*
(b) $\lim_{n\to\infty} n\,\mu_n(G\setminus\{e,x_0\})=0$.

Proof. It suffices to show the implication (ii) ⇒ (i). Let $\mu\in\mathcal{M}^1(G)$, $\mu\neq\varepsilon_e$, with a sequence $(\mu_n)_{n\geq 1}$ of roots satisfying Assumption (ii).

1. By K we denote the subgroup of G generated by x_0. Furthermore we put

$$\varepsilon(n):=n\,\mu_n(G\setminus\{x_0,e\}).$$

By assumption we have $\lim_{n\to\infty}\varepsilon(n)=0$. Moreover, we obtain

$$\mu_n(K)\geq\mu_n(\{e,x_0\})=1-\frac{\varepsilon(n)}{n},$$

whence

$$\mu(K)=\mu_n^n(K^n)\geq\mu_n(K)^n\geq\left(1-\frac{\varepsilon(n)}{n}\right)^n \quad \text{for all } n\geq 1.$$

Clearly there exists an $n_0\geq 1$ with $\frac{\varepsilon(n)}{n}<1$ for all $n\geq n_0$. This implies

$$1\geq\left(1-\frac{\varepsilon(n)}{n}\right)^n\geq 1-\varepsilon(n) \quad \text{for all } n\geq n_0$$

and thus

$$\lim_{n\to\infty}\left(1-\frac{\varepsilon(n)}{n}\right)^n=1.$$

Hence, $\mu(K)=1$. Since K is countable, the measure is discrete. Consequently, μ_n is discrete for every $n\geq 1$. We therefore may assume without loss of generality that G is discrete.

2. There exists an $n_0\geq 1$ such that $\mu_n(\{e\})>0$ for all $n\geq n_0$. In fact, if this were not the case, then there would exist a strictly isotone sequence $(n_k)_{k\geq 1}$ in $\mathbb{N}$ with $\lim_{k\to\infty} n_k=\infty$ and $\mu_{n_k}(\{e\})=0$ for all $k\geq 1$. But this implies

$$\mu_{n_k}(\{x_0\})=1-\frac{\varepsilon(n_k)}{n_k}.$$

Hence, without loss of generality we assume that

$$\left(1-\frac{\varepsilon(n_k)}{n_k}\right)^{n_k}\geq\frac{3}{4} \quad \text{for all } k\geq 1,$$

and we get

$$\mu(\{x_0^{n_k}\})=\mu_{n_k}^{n_k}(\{x_0\}^{n_k})\geq(\mu_{n_k}(\{x_0\}))^{n_k}=\left(1-\frac{\varepsilon(n_k)}{n_k}\right)^{n_k}\geq\frac{3}{4}$$

for all $k\geq 1$. But this yields that $x_0^{n_k}=x_0^{n_1}$ for all $k\geq 1$, and we have

$$\mu(\{x_0^{n_1}\})=\lim_{k\to\infty}\mu(\{x_0^{n_k}\})\geq\lim_{k\to\infty}\left(1-\frac{\varepsilon(n_k)}{n_k}\right)^{n_k}=1,$$

i.e., $\mu=\varepsilon_{x_0^{n_1}}$. From this we obtain $\mu_n=\varepsilon_{y_n}$ with $y_n\in G$ for all $n\geq 1$. Without loss of generality let

$$\frac{\varepsilon(n_k)}{n_k}<1 \quad \text{and hence} \quad \mu_{n_k}(\{x_0\})>0 \quad \text{for all } k\geq 1.$$

Then $y_{n_k}=x_0$ for all $k\geq 1$. If there exists an $n\geq 1$ with $\mu_n(\{e\})>0$, then $\mu_n=\varepsilon_e$, whence $\mu=\varepsilon_e$, which is impossible. Thus $\mu_n(\{e\})=0$ for all $n\geq 1$. But this contradicts the hypothesis (a) of Statement (ii).

3. Let $n\geq n_0$, where n_0 was chosen in Part 2. Then $e\in\text{supp}(\mu_n)$, whence

$$\text{supp}(\mu_n)\subset(\text{supp}(\mu_n))^n=\text{supp}(\mu_n^n)=\text{supp}(\mu)\subset K.$$

Since $(\mu_{nn_0}^{n_0})^n = \mu$ and $\operatorname{supp}(\mu_{nn_0}^{n_0}) \subset K$ for $nn_0 \geq n_0$ one can replace μ_n by μ_{nn_0} for $1 < n < n_0$. Clearly one may assume without loss of generality that G equals the discrete group generated by x_0. In particular, we may let G be Abelian.

4. For every $n \geq 1$ we introduce the measure $\lambda_n := \mu_n * \mu_n^\sim$. By Lemma 6.1.24 there exists a compact subgroup H of G such that

$$\lim_{n\to\infty} \lambda_n = \omega_H$$

holds. Moreover, we have

$$\lambda_n(\{x_0^{-1}, e, x_0\}) = \lambda_n(\{e, x_0\}\{e, x_0\}^{-1}) \geq \mu_n(\{e, x_0\})^2$$

for all $n \geq 1$. By Property (b) of Statement (ii) we get

$$\lim_{n\to\infty} \mu_n(\{e, x_0\}) = 1,$$

therefore

$$\lim_{n\to\infty} \lambda_n(\{x_0^{-1}, e, x_0\}) = 1,$$

and hence

$$\omega_H(\{x_0^{-1}, e, x_0\}) = 1.$$

This shows $H \subset \{x_0^{-1}, e, x_0\}$. By $x_0^2 \neq e$ this implies $H = \{e\}$ or $\omega_H = \varepsilon_e$.

5. From Part 4 we know that

$$\lim_{n\to\infty} \mu_n * \mu_n^\sim = \varepsilon_e.$$

Since G is discrete,

$$\lim_{n\to\infty} \|\mu_n * \mu_n^\sim - \varepsilon_e\| = 0.$$

Consequently, there exists an $n_0 \geq 1$ with

$$\|\mu_{n_0} * \mu_{n_0}^\sim - \varepsilon_e\| < 1.$$

But now the assumptions of Theorem 6.1.22 are satisfied, and the assertion follows. □

6.1.26 Remark. It can be shown that in Statement (ii) of the theorem Condition (b) implies Condition (a). In fact, one simply has to replace Part 5 of the proof by the following reasoning: Since G is a discrete Abelian group, Example 4 of 5.1.9 shows that the function $g \equiv 0$ is a local inner product for G. From

$$\lim_{n\to\infty} \mu_n(\{e, x_0\}) = 1 \quad \text{and} \quad \lim_{n\to\infty} \mu_n * \mu_n^\sim = \varepsilon_e$$

follows, that the sequence $(\mu_n)_{n\geq 1}$ admits as accumulation points at most ε_e and ε_{x_0}. In order to show that the case $\lim_{n\to\infty} \mu_n = \varepsilon_e$ does not occur, one applies Theorems 5.1.16 (with $g \equiv 0$ and $x_{nj} = e$ for all $n, j \geq 1$) and 5.1.6.

6.2 Gauss Measures and Their Characterizations

The section is devoted to the definition of Gauss semigroups for an arbitrary locally compact group and their first properties. For locally compact Abelian groups G a definition of a Gauss measure on G due to K.R. Parthasarathy has been given in §2 of Chapter V. In Parthasarathy's approach the set $\mathscr{G}_P(G)$ of Gauss measures was introduced as a subset of the set $\mathscr{I}_0(G)$ of all weakly infinitely divisible measures in $\mathscr{M}^1(G)$. Although by Theorem 5.2.7 every $\mu\in\mathscr{G}_P(G)$ admits a representation of its Fourier transform of the form

$$\hat{\mu}(\chi)=\chi(x_0)\exp(-\phi(\chi)) \quad \text{for all } \chi\in G^\wedge,$$

where $x_0\in G$ and $\phi\in \mathrm{Q}_+(G^\wedge)$ are uniquely determined, μ does not necessarily belong to the set $\mathscr{E}_{\{e\}}(G)$ and hence does not permit us to establish the canonical representation developed in §3 of Chapter IV.

In order to fill this gap we shall specify Gauss measures on an arbitrary locally compact group G within the set $\mathscr{E}_{\{e\}}(G)$ by a condition on the embedding semigroups.

6.2.1 Definition. An $\{e\}$-continuous convolution semigroup $(\mu_t)_{t\in\mathbb{R}_+^*}$ in $\mathscr{M}^1(G)\setminus\mathscr{D}(G)$ is called a *Gauss semigroup* if

$$\text{(G)} \qquad \lim_{t\downarrow 0}\frac{1}{t}\mu_t(\complement U)=0$$

holds for every (Borel) neighborhood $U\in\mathfrak{B}(e)$.

A measure $\mu\in\mathscr{E}_{\{e\}}(G)$ with embedding semigroup $(\mu_t)_{t\in\mathbb{R}_+^*}$ in $\mathscr{M}^1(G)$ is said to be a *Gauss measure* in $\mathscr{M}^1(G)$ *with embedding semigroup* $(\mu_t)_{t\in\mathbb{R}_+^*}$ if $(\mu_t)_{t\in\mathbb{R}_+^*}$ is a Gauss semigroup.

The study of the set $\mathscr{G}(G)$ of all Gauss measures in $\mathscr{M}^1(G)$ will be the central aim of this section.

6.2.2 Remark. The emphasis in the definition on the embedding semigroup of a Gauss measure stems from the fact that for general locally compact groups G one does not know whether the embedding of a Gauss measure on G is unique. Clearly, this defect does not appear in the classical situation $G:=\mathbb{R}^p$ $(p\geq 1)$, which was extensively discussed for the first time by Courrège in [94].

The comparison of the class $\mathscr{G}(G)$ with the previously introduced subclasses of $\mathscr{M}^1(G)$ immediately yields $\mathscr{G}(G)\subset\mathscr{E}(G)\subset\mathscr{I}(G)$.

6.2.3 Theorem. *Let G be a locally compact group and $(\mu_t)_{t\in\mathbb{R}_+^*}$ a Gauss semigroup in $\mathscr{M}^1(G)$. Then for every $t\in\mathbb{R}_+^*$ we have* $\operatorname{supp}(\mu_t)\subset G_0$.

Proof. Since G_0 is the intersection of all open subgroups of G, it suffices to prove that given an open subgroup H of G, one has $\operatorname{supp}(\mu_t)\subset H$ for all $t\in\mathbb{R}_+^*$. Let $(S_t)_{t\in\mathbb{R}_+^*}$ denote the contraction semigroup on $\mathscr{C}_u(G)$ corresponding to $(\mu_t)_{t\in\mathbb{R}_+^*}$ and let N be its infinitesimal generator with domain D_N. We define $M:=G/H$ and denote the canonical epimorphism $G \to M$ by p. For $\check{f} \in \mathscr{K}(M)$ one obviously has $f:=\check{f} \circ p \in \mathscr{C}_u(G)$ and $f_h = f$ for all $h \in H$. Hence, for all $t \in \mathbb{R}_+^*$ and $y \in G$ the following inequalities hold:

$$\left|\frac{1}{t}(S_t f(y) - f(y))\right| = \left|\frac{1}{t}\int [f(yx) - f(y)]\, \mu_t(dx)\right|$$

$$\leq \frac{1}{t}\left|\int_H [f(yx) - f(y)]\, \mu_t(dx)\right| + \frac{1}{t}\left|\int_{\complement H} [f(yx) - f(y)]\, \mu_t(dx)\right|$$

$$\leq 2\|f\|\frac{1}{t}\mu_t(\complement H).$$

Since $H \in \mathfrak{V}(e)$, Condition (G) of the above definition yields $f \in \mathsf{D}_N$ and $Nf = 0$. For every $t \in \mathbb{R}_+^*$ we also have $S_t f - f = \int_0^t S_s N f\, ds$ by the Hille-Yosida theory ([241], p. 308) and hence, $S_t f = f$. In particular, it follows for all $t \in \mathbb{R}_+^*$ and $\check{f} \in \mathscr{K}(M)$

$$p(\mu_t)(\check{f}) = \mu_t(\check{f} \circ p) = (S_t(\check{f} \circ p))(e) = (\check{f} \circ p)(e) = \varepsilon_{p(e)}(\check{f})$$

and so $p(\mu_t) = \varepsilon_{p(e)}$, which proves the assertion. □

From the preceding theorem we obtain the consequence that when dealing with Gauss semigroups it suffices to assume without loss of generality that the underlying locally compact group G is connected. A first application of this reduction concerns the construction of Gauss measures on G, which requires the following

6.2.4 Properties of the set $\mathscr{G}(G)$ of Gauss measures on G.

(1) Let μ be a Gauss measure in $\mathscr{G}(G)$ with embedding semigroup $(\mu_t)_{t \in \mathbb{R}_+^*}$. Then $\mu_t \in \mathscr{G}(G)$ for every $t \in \mathbb{R}_+^*$.

(2) Let G and G' be two locally compact groups and p a continuous homomorphism from G into G'. If $\mu \in \mathscr{G}(G)$ is a Gauss measure with embedding semigroup $S:=(\mu_t)_{t \in \mathbb{R}_+^*}$, then $p(\mu)$ is a Gauss measure in $\mathscr{G}(G')$ with embedding semigroup $p(S):=(p(\mu_t))_{t \in \mathbb{R}_+^*}$.

(3) Let $(G_\alpha)_{\alpha \in \mathbb{A}}$ be a family of locally compact groups such that

$$G := \prod_{\alpha \in \mathbb{A}} G_\alpha$$

is locally compact, and let $\mu^\alpha \in \mathscr{G}(G_\alpha)$ with embedding semigroup

$$S_\alpha := (\mu_t^\alpha)_{t \in \mathbb{R}_+^*} \qquad \text{for all } \alpha \in \mathbb{A}.$$

Then $\bigotimes_{\alpha \in \mathbb{A}} \mu^\alpha \in \mathscr{G}(G)$ with embedding semigroup

$$\bigotimes_{\alpha \in \mathbb{A}} S_\alpha := \left(\bigotimes_{\alpha \in \mathbb{A}} \mu_t^\alpha\right)_{t \in \mathbb{R}_+^*}.$$

The proofs of the previous properties are elementary. We proceed to a nontrivial property.

(4) Let G be a locally compact group of the form $\varprojlim_{K \in \mathbf{N}} G/K$ where $\mathbf{N}$ denotes a descending family of compact normal subgroups K of G with $\bigcap_{K \in \mathbf{N}} K = \{e\}$. Let $(\mu^K)_{K \in \mathbf{N}}$ be a projective system of measures $\mu^K \in \mathscr{G}(G/K)$ with embedding semigroup $S_K := (\mu_t^K)_{t \in \mathbb{R}_+^*}$ such that the system $(S_K)_{K \in \mathbf{N}}$ of semigroups is projective. Then the measure

$$\mu := \varprojlim_{K \in \mathbf{N}} \mu^K$$

is a Gauss measure in $\mathscr{G}(G)$ with embedding semigroup

$$S := \varprojlim_{K \in \mathbf{N}} S_K.$$

In fact, let p_L and p_{KL} (for $K, L \in \mathbf{N}$ with $L \subset K$) denote the canonical projections $G \to G/L$ and $G/L \to G/K$ resp. and let

$$S_L := (\mu_t^L)_{t \in \mathbb{R}_+^*}$$

be the embedding semigroup of the Gauss measures μ^L in $\mathscr{G}(G/L)$. By assumption we have $p_{KL}(S_L) = S_K$, i.e., the family $(\mu_t^K)_{K \in \mathbf{N}}$ is a projective system of Gauss measures for every $t \in \mathbb{R}_+^*$. Theorem 1.2.17 implies that for every $t \in \mathbb{R}_+^*$ there exists a unique projective limit measure

$$\mu_t := \varprojlim_{K \in \mathbf{N}} \mu_t^K \in \mathscr{M}^1(G) \quad \text{with } p_K(\mu_t) = \mu_t^K \quad \text{for all } K \in \mathbf{N}.$$

From $\mu^K = \mu_1^K \notin \mathscr{D}(G/K)$ we obtain $\mu = \mu_1 \notin \mathscr{D}(G)$. It remains to show that $(\mu_t)_{t \in \mathbb{R}_+^*}$ is a continuous convolution semigroup in $\mathscr{M}^1(G)$ satisfying Condition (G) of the definition of a Gauss semigroup. First of all we observe that the set

$$\mathscr{V} := \{f_K \circ p_K : f_K \in \mathscr{K}(G/K),\ K \in \mathbf{N}\}$$

is a dense subspace of $\mathscr{K}(G)$. For all $s, t \in \mathbb{R}_+^*$ and $K \in \mathbf{N}$ one obtains

$$p_K(\mu_s * \mu_t) = p_K(\mu_s) * p_K(\mu_t) = \mu_s^K * \mu_t^K = \mu_{s+t}^K = p_K(\mu_{s+t}),$$

whence $\mu_s * \mu_t = \mu_{s+t}$. Furthermore for $f := f_K \circ p_K \in \mathscr{V}$ one has

$$\lim_{t \downarrow 0} \mu_t(f) = \lim_{t \downarrow 0} \mu_t^K(f_K) = \varepsilon_{p_K(e)}(f_K) = \varepsilon_e(f)$$

since S_K is a continuous convolution semigroup in $\mathscr{M}^1(G/K)$. Therefore, the simply bounded net $(\mu_t)_{t \in \mathbb{R}_+^*}$ of continuous linear functionals μ_t on $\mathscr{K}(G)$ converges pointwise to ε_e on the dense subspace $\mathscr{V}$ of $\mathscr{K}(G)$. The Banach-Steinhaus theorem yields $\lim_{t \downarrow 0} \mu_t = \varepsilon_e$ or the continuity of the semigroup $S := (\mu_t)_{t \in \mathbb{R}_+^*}$ in $\mathscr{M}^1(G)$. Finally, for $U \in \mathfrak{B}(e)$ there exist $K \in \mathbf{N}$ and $V \in \mathfrak{B}(e)$ satisfying $VK \subset U$. Obviously $p_K(V) \in \mathfrak{B}_{G/K}(\dot{e})$, and for every $t \in \mathbb{R}_+^*$ we get

$$\mu_t^K(\complement p_K(V)) = p_K(\mu_t)(\complement p_K(V)) = \mu_t(\complement VK) \geq \mu_t(\complement U);$$

thus, $\lim_{t \downarrow 0} \frac{1}{t} \mu_t(\complement U) = 0$ since for $K \in \mathbf{N}$ the measure μ^K is a Gauss measure in $\mathscr{G}(G/K)$ with embedding semigroup S_K. The desired result $\mu \in \mathscr{G}(G)$ has been established.

At this point we shall establish the first existence results for measures in $\mathscr{G}(G)$. The one below contains a reformulation of Property (4).

6.2.5 Lemma. *Let G be a locally compact group of the form*

$$G:=\varprojlim_{K\in\mathbf{N}} G/K$$

as in Property (4) *above. For* $K_1\in\mathbf{N}$ *we introduce the cofinal subsystems* $\mathbf{N}_1:=\{K\in\mathbf{N}: K\subset K_1\}$ *of* $\mathbf{N}$. *The following statements are equivalent:*

(i) *There exists a measure* $\mu\in\mathscr{G}(G)$ *with embedding semigroup* $S:=(\mu_t)_{t\in\mathbb{R}_+^*}$;

(ii) *there is a* $K_1\in\mathbf{N}$ *such that for each* $K\in\mathbf{N}_1$ *there exists a* $\mu^K\in\mathscr{G}(G/K)$ *with embedding semigroup* $S_K:=(\mu_t^K)_{t\in\mathbb{R}_+^*}$ *for which the system* $(S_K)_{K\in\mathbf{N}_1}$ *is projective.*

Proof. It remains to supply an argument for the implication (i)$\Rightarrow$(ii). Since $\mu\notin\mathscr{D}(G)$, there is $K_1\in\mathbf{N}$ with $\operatorname{supp}(\mu)^{-1}\operatorname{supp}(\mu)\not\subset K_1$. For every $K\in\mathbf{N}_1$ we obtain $\mu^K:=p_K(\mu)\notin\mathscr{D}(G/K)$. Moreover, μ^K is a Gauss measure in $\mathscr{G}(G/K)$ by Property (2) of 6.2.4, it has embedding semigroup $S_K:=p_K(S)$ for $K\in\mathbf{N}_1$ and the system $(S_K)_{K\in\mathbf{N}_1}$ is obviously projective. □

6.2.6 Lemma. *Let G be a Lie group of dimension* $n\geq 1$. *Let* μ *be an embeddable measure in* $\mathscr{E}_{\{e\}}(G)$ *with embedding semigroup* $(\mu_t)_{t\in\mathbb{R}_+^*}$ *whose infinitesimal generator N admits on* $\tilde{\mathscr{C}}_2^0(G)$ *the representation* $(a_i, a_{ij}, \eta)_{i,j=1,\dots,n}$ *introduced in* §2 *of Chapter IV. The numbers* $a_1,\dots,a_n$ *are in* $\mathbb{R}$, $(a_{ij})_{i,j=1,\dots,n}\in\mathfrak{M}(n,\mathbb{R})$ *is a symmetric, positive-semidefinite matrix and* η *is a Lévy measure on G. The following statements are equivalent:*

(i) μ *is a Gauss measure in* $\mathscr{G}(G)$ *with embedding semigroup* $(\mu_t)_{t\in\mathbb{R}_+^*}$;

(ii) $\eta=0$ *and* $(a_{ij})_{i,j=1,\dots,n}\neq 0$.

Thus, for a Lie group G one always has $\mathscr{G}(G)\neq\emptyset$.

Proof. 1. (i)$\Rightarrow$(ii). For every $f\in\mathscr{K}(G^*)$ there exists a $U\in\mathfrak{V}(e)$ with $f(U)=0$. Using Condition (iv) of Part (A) of Theorem 4.2.8 we conclude that

$$|\textstyle\int_{G^*} f\,d\eta| = \lim_{t\downarrow 0}\frac{1}{t}|\int_{\complement U} f\,d\mu_t| \leq \|f\| \lim_{t\downarrow 0}\frac{1}{t}\mu_t(\complement U)=0,$$

which implies that $\eta=0$. Since $\mu\notin\mathscr{D}(G)$, plainly $(a_{ij})_{i,j=1,\dots,n}\neq 0$.

2. (ii)$\Rightarrow$(i). The assumption $(a_{ij})_{i,j=1,\dots,n}\neq 0$ implies that $\mu\notin\mathscr{D}(G)$. For $U\in\mathfrak{V}(e)$ there exist a $V\in\mathfrak{V}(e)$ with $\overline{V}\subset U$ and an $f\in\tilde{\mathscr{C}}_2^0(G)$ satisfying $0\leq f\leq 1$, $f(V)=0$ and $f(\complement U)=1$. From this we obtain

$$\lim_{t\downarrow 0}\frac{1}{t}\mu_t(\complement U)\leq \lim_{t\downarrow 0}\frac{1}{t}\textstyle\int f\,d\mu_t=(Nf)(e)=\int_{G^*} f\,d\eta=0,$$

which yields Condition (G) of Definition 6.2.1. □

6.2.7 Theorem. *Let G and* $\dot{G}$ *be two connected locally compact groups, p a continuous epimorphism from G onto* $\dot{G}$ *and* $K:=\ker p$ *a Lie group. Then for every Gauss measure* $\dot{\mu}\in\mathscr{G}(\dot{G})$ *with embedding semigroup* $\dot{S}:=(\dot{\mu}_t)_{t\in\mathbb{R}_+^*}$ *there exists a Gauss* *measure* $\mu \in \mathscr{G}(G)$ *with embedding semigroup* S *such that* $p(S)=\dot{S}$ *and hence* $p(\mu)=\dot{\mu}$ *holds.*

Proof. 1. G as a connected locally compact group admits the representation $G=\varprojlim_{\alpha\in\mathbb{A}} G_\alpha$ with Lie groups $G_\alpha := G/K_\alpha$ ($\alpha\in\mathbb{A}$), where $(K_\alpha)_{\alpha\in\mathbb{A}}$ is a descending family of compact normal subgroups in G with $\bigcap_{\alpha\in\mathbb{A}} K_\alpha=\{e\}$. Since K is a Lie group, we may assume without loss of generality that $K\cap K_\alpha=\{e\}$ for all $\alpha\in\mathbb{A}$. By p_α we denote as usual the canonical epimorphism from G onto G_α.

We also introduce $\dot{K}_\alpha := p(K_\alpha)$ and the canonical epimorphisms

$$\dot{p}_\alpha : \dot{G}\to\dot{G}_\alpha := \dot{G}/\dot{K}_\alpha \quad\text{and}\quad f_\alpha : G_\alpha\to\dot{G}_\alpha \qquad \text{for } \alpha\in\mathbb{A}.$$

Since K is σ-compact and K_α is compact for $\alpha\in\mathbb{A}$ one establishes as in the proof of Theorem 4.3.11 the isomorphism

$$KK_\alpha/K_\alpha \cong K/K\cap K_\alpha = K.$$

Finally, we assume given a basis $\{Y_1,\dots,Y_r\}$ in $\mathfrak{L}(K)$ and a symmetric, positive-definite matrix $(a_{ij})_{i,j=1,\dots,r}\in\mathfrak{M}(r,\mathbb{R})$.

2. For any $\alpha\in\mathbb{A}$ and a basis $\{\dot{X}_1^\alpha,\dots,\dot{X}_{s(\alpha)}^\alpha\}$ of $\mathfrak{L}(\dot{G}_\alpha)$ there exists a basis $\{X_1^\alpha,\dots,X_{k(\alpha)}^\alpha\}$ of $\mathfrak{L}(G_\alpha)$ with

$$df_\alpha(X_j^\alpha)=\dot{X}_j^\alpha \qquad \text{for } 1\le j\le s(\alpha)$$

and

$$X_j^\alpha = dp_\alpha(Y_{j-s(\alpha)}) \qquad \text{for } s(\alpha)+1\le j\le k(\alpha).$$

This follows from Preparation 1 preceding Theorem 4.3.11. Lemma 6.2.5 enables us to assume without loss of generality that for every $\alpha\in\mathbb{A}$ the measure $\dot{p}_\alpha(\dot{\mu}_1)$ is a Gauss measure in $\mathscr{G}(\dot{G}_\alpha)$ with embedding semigroup $\dot{S}_\alpha=\dot{p}_\alpha(\dot{S})$. Lemma 6.2.6 implies that the infinitesimal generator $\dot{N}_\alpha$ of $\dot{S}_\alpha$ admits on $\mathscr{C}_2^0(\dot{G}_\alpha)$ the representation $(\dot{a}_i^\alpha, \dot{a}_{ij}^\alpha, 0)_{i,j=1,\dots,s(\alpha)}$. Now let

$$(a_{ij}^\alpha)_{i,j=1,\dots,k(\alpha)} := (\dot{a}_{ij}^\alpha)_{i,j=1,\dots,s(\alpha)} \oplus (a_{ij})_{i,j=1,\dots,r}$$

and

$$N_\alpha := \sum_{i=1}^{s(\alpha)} \dot{a}_i^\alpha X_i^\alpha + \sum_{i,j=1}^{k(\alpha)} a_{ij}^\alpha X_i^\alpha X_j^\alpha.$$

Then by Theorem 4.2.8 there exists exactly one continuous semigroup

$$S^\alpha := (\mu_t^\alpha)_{\alpha\in\mathbb{R}_+^*}$$

in $\mathscr{M}^1(G_\alpha)$ with infinitesimal generator N'_α such that

$$\operatorname{Res}_{\mathscr{C}_2^0(G_\alpha)} N'_\alpha = N_\alpha.$$

On the other hand, μ_1^α is a Gauss measure in $\mathscr{G}(G_\alpha)$ with embedding semigroup S_α, which follows from Lemma 6.2.6. Since $(\dot{N}_\alpha \dot{f})\circ f_\alpha = N_\alpha(\dot{f}\circ f_\alpha)$ holds for every $\dot{f}\in\mathscr{C}_2^0(\dot{G}_\alpha)$, we conclude that $\dot{S}_\alpha = f_\alpha(S_\alpha)$. Moreover, the definition of N_α can be shown to be independent of the basis chosen in $\mathfrak{L}(\dot{G}_\alpha)$ (for all $\alpha\in\mathbb{A}$).

3. Now let α,β be in $\mathbb{A}$ with $\alpha<\beta$. We shall first show that $p_{\alpha\beta}(S_\beta)=S_\alpha$. In fact, the symbol $\dot{p}_{\alpha\beta}$ denotes as usual the canonical mapping from $\dot{G}_\beta$ onto $\dot{G}_\alpha$.

By Part 2 the bases $\{\dot{X}_1^\alpha, \ldots, \dot{X}_{s(\alpha)}^\alpha\}$ and $\{\dot{X}_1^\beta, \ldots, \dot{X}_{s(\beta)}^\beta\}$ of $\mathfrak{L}(\dot{G}_\alpha)$ and $\mathfrak{L}(\dot{G}_\beta)$ resp. can be chosen as in Preparation 1 preceding Theorem 4.3.11. From

$$\dot{p}_{\alpha\beta}(\dot{S}_\beta) = \dot{S}_\alpha \quad \text{follow}$$
$$\dot{a}_i^\beta = \dot{a}_i^\alpha \quad \text{and} \quad \dot{a}_{ij}^\beta = \dot{a}_{ij}^\alpha \qquad \text{whenever } i,j = 1, \ldots, s(\alpha).$$

Again from Part 2 we obtain

$$d p_{\alpha\beta}(X_j^\beta) = X_j^\alpha \qquad \text{for } j = 1, \ldots, s(\alpha)$$

and

$$d p_{\alpha\beta}(X_{s(\beta)+j}^\beta) = X_{s(\alpha)+j}^\alpha \qquad \text{for } j = 1, \ldots, k(\beta) - s(\beta) = r.$$

This shows that

$$(N_\alpha f) \circ p_{\alpha\beta} = N_\beta (f \circ p_{\alpha\beta}) \qquad \text{for all } f \in \tilde{\mathscr{C}}_2^0(G_\alpha)$$

and hence, $p_{\alpha\beta}(S_\beta) = S_\alpha$.

We now apply Lemma 6.2.5 to establish the existence of a measure $\mu \in \mathscr{G}(G)$ with embedding semigroup S such that $p_\alpha(S) = S_\alpha$ for all $\alpha \in \mathbf{A}$.

Since $\dot{p}_\alpha(\dot{S}) = \dot{S}_\alpha = f_\alpha(S_\alpha) = (f_\alpha \circ p_\alpha)(S) = (\dot{p}_\alpha \circ p)(S) = \dot{p}_\alpha(p(S))$ for all $\alpha \in \mathbf{A}$, one concludes from Theorem 1.2.17 that $\dot{S} = p(S)$; in particular, $\dot{\mu} = p(\mu)$, as desired. □

6.2.8 Theorem. *Let G be a connected locally compact group $\neq \{e\}$. Then $\mathscr{G}(G) \neq \emptyset$.*

Proof. Let $\mathbf{R}$ be the system of all pairs $(G/K, S_K)$, where K is a compact normal subgroup of G and $S_K := (\mu_t^K)_{t \in \mathbb{R}_+^*}$ is a continuous semigroup in $\mathscr{M}^1(G/K)$ such that μ_1^K is a Gauss measure in $\mathscr{G}(G/K)$ with corresponding semigroup S_K.

1. $\mathbf{R} \neq \emptyset$ since G is Lie projective, and Theorem 4.2.8 together with Lemma 6.2.6 implies the existence of Gauss measures on nondiscrete Lie groups.

2. For $i = 1, 2$ and $(G/K_i, S_{K_i}) \in \mathbf{R}$ one defines $(G/K_1, S_{K_1}) < (G/K_2, S_{K_2})$ if $K_2 \subset K_1$ and $p_{12}(S_{K_2}) = S_{K_1}$ and obtains and order relation in $\mathbf{R}$, which is inductive. In fact, if

$$\mathbf{S} := \{(G/K_i, S_{K_i}) : i \in \mathbb{I}\}$$

is a chain in $\mathbf{R}$ with $(G/K_i, S_{K_i}) < (G/K_j, S_{K_j})$ for all $i, j \in \mathbb{I}$, $i < j$, then putting $K := \bigcap_{i \in \mathbb{I}} K_i$, one obtains $G/K = \varprojlim_{i \in \mathbb{I}} G/K_i$.

Lemma 6.2.6 yields that $(G/K, S_K)$ is an upper bound of $\mathbf{S}$ and thus, by Zorn's Lemma, $\mathbf{R}$ contains maximal elements.

3. Let $(G/K, S_K)$ be a maximal element in $\mathbf{R}$. Without loss of generality we assume that $K \neq \{e\}$. Then there is a normal subgroup N of K with $N \neq K$ such that K/N is Lie. One shows that N is also normal in G.

In fact, if q is the canonical epimorphism from K onto K/N, then there is a $V \in \mathfrak{V}_K(e)$ such that $q(V)$ admits no nontrivial subgroup. For all $x \in G$ we have $x N x^{-1} \subset x K x^{-1} = K$. Hence, there exists a $U \in \mathfrak{V}_G(e)$ satisfying $q(x N x^{-1}) \subset q(V)$ for all $x \in U$, and thus $x N x^{-1} = N$ for all $x \in U$.

Since G is connected, one has $G=\bigcup_{n\geq 1} U^n$ and therefore $xNx^{-1}=N$ for all $x\in G$.

Denoting the canonical epimorphism from G/N onto G/K by p we obtain a compact Lie group $\ker p=K/N$. Theorem 6.2.7 implies the existence of a Gauss measure in $\mathscr{G}(G/N)$ with embedding semigroup S_N such that $(G/K, S_K)<(G/N, S_N)$ holds. This, however, is a contradiction of the maximality of $(G/K, S_K)$ since $N\neq K$. It follows that $K=\{e\}$, which completes the proof. □

6.2.9 Remark. There are various ways of proving Theorem 6.2.8. If one wants to avoid the construction presented above, one can rely on more profound structural results in order to get the existence statement right away. This can be done as follows:

(a) One uses the existence of Gauss measures on Lie groups and on compact connected Abelian groups. In the first case Lemma 6.2.6 provides Gauss measures. In the second case one has nontrivial one-parameter subgroups ([218], (25.20)) which via Property (2) of 6.2.4 yields Gauss measures as images of normal distributions on $\mathbb{R}$. For the general connected locally compact group G one proceeds as follows: If G admits a connected compact subgroup $H\neq\{e\}$, then it suffices to establish the existence of a Gauss measure in $\mathscr{M}^1(H)$. By Theorem E we have $H\cong(L\times A)/N$, where L is a compact connected Lie group, A is a compact connected Abelian group and N is a normal subgroup of $L\times A$. The above remarks together with Properties (2) and (3) of 6.2.4 yield $\mathscr{G}(H)\neq\emptyset$ and hence, $\mathscr{G}(G)\neq\emptyset$.

In case there exists no connected compact subgroup $H\neq\{e\}$ the group G is finite-dimensional. Then by Theorem H there exist a connected Lie group L and a continuous monomorphism f from L into G with $\overline{f(L)}=G$. Lemma 6.2.6 yields the existence of a measure $\mu\in\mathscr{G}(L)$. By Property (2) of 6.2.4 there follows $f(\mu)\in\mathscr{G}(G)$.

(b) More directly one can use the fact established in [356], pp. 118 and 192 that in any connected locally compact group $G\neq\{e\}$ there exists a nontrivial one-parameter subgroup. Then the normal distribution on $\mathbb{R}$ can be mapped via Property (2) of 6.2.4 to a Gauss measure in $\mathscr{G}(G)$ and $\mathscr{G}(G)\neq\emptyset$.

6.2.10 Theorem. *Let G be a Lie projective group in* **A** *and let μ be in $\mathscr{E}_{\{e\}}(G)\setminus\mathscr{D}(G)$ with embedding semigroup $S:=(\mu_t)_{t\in\mathbb{R}_+^*}$ in $\mathscr{M}^1(G)$. Let ψ be the negative-definite form on $\mathfrak{R}(G)$ corresponding to S (and a given Lévy mapping Γ for G) of the form (ψ_1, ψ_2, η), where ψ_1 is a primitive, ψ_2 a quadratic form on $\mathfrak{R}(G)$ and η the Lévy measure on G corresponding to S. The following statements are equivalent:*

(i) $\mu\in\mathscr{G}(G)$;

(ii) $\eta=0$;

(iii) *if $\psi=\psi'+\psi''$ is the representation of ψ in its skew-symmetric and symmetric part, then ψ' is primitive and ψ'' is quadratic;*

(iv) $\psi(f_D^2)=0$ *for all $D\in\operatorname{Irr}(G)$;*

(v) *for all $D\in\operatorname{Irr}(G)$ we have the (Gauss) condition*

$$\text{(GC)}\qquad |\det(\hat{\mu}(D\otimes D))|\cdot|\det(\hat{\mu}(D\otimes\overline{D}))|=|\det(\hat{\mu}(D))|^{4n(D)}.$$

Proof. 1. (i)$\Rightarrow$(ii) follows from Theorem 4.3.13 as in the proof of Lemma 6.2.6.

2. (ii)$\Rightarrow$(iii). By assumption we have $\psi = \psi_1 + \psi_2$ with a skew-symmetric part ψ_1 and a symmetric part ψ_2 on $\mathfrak{K}(G)$ as follows from Theorems 1.5.16 and 1.5.17 resp. But the uniqueness of the representation in skew-symmetric and symmetric part yields the assertion.

3. (iii)$\Rightarrow$(iv) follows immediately from the definitions of primitive and quadratic forms on $\mathfrak{K}(G)$ since for all $D \in \mathrm{Irr}(G)$ we have

$$\psi(f_D^2) = \psi'(f_D^2) + \psi''(f_D^2) = 0.$$

4. (iv)$\Leftrightarrow$(v). For every $D \in \mathrm{Irr}(G)$ one has the identities

$$f_{D\otimes D} + f_{D\otimes \bar{D}} = 4n(D) f_D - 2f_D^2 \quad \text{and} \quad |\det(\hat{\mu}(D))| = \exp(\psi(f_D)).$$

Therefore (GC) is equivalent to

$$-\infty < \psi(f_{D\otimes D}) + \psi(f_{D\otimes \bar{D}}) = 4n(D)\,\psi(f_D),$$

which implies the assertion.

5. (iv)$\Rightarrow$(ii). For every $D \in \mathrm{Irr}(G)$ we have by Theorem 4.3.13

$$\begin{aligned} 0 = \psi(f_D^2) &= \psi_1(f_D^2) + \psi_2(f_D^2) - \int_{G^*} [f_D^2(x) - f_D^2(e) - \Gamma(x, f_D^2)]\,\eta(dx) \\ &= -\int_{G^*} f_D^2\, d\eta. \end{aligned}$$

Since $\{x \in G : f_D^2(x) = 0\} = \ker D$ for every $D \in \mathrm{Irr}(G)$ and $\bigcap_{D \in \mathrm{Irr}(G)} \ker D = \{e\}$, we conclude that $\eta = 0$.

6. (ii)$\Rightarrow$(i). Let $G := \varprojlim_{K \in \mathbf{N}} G/K$, where $\mathbf{N}$ is a descending family of compact normal subgroups of G with $\bigcap_{K \in \mathbf{N}} K = \{e\}$ such that G/K is a Lie group for all $K \in \mathbf{N}$. Let $K \in \mathbf{N}$. By η_K we denote the Lévy measure corresponding to the canonical representation of $p_K(S)$ by Theorem 4.3.10. Let $f_K \in \mathscr{K}((G/K)^*)$.

Then $f := f_K \circ p_K \in \mathscr{K}(G^*)$, and the assumption together with Theorem 4.3.10 yields

$$\int f_K \, d\eta_K = \lim_{t \downarrow 0} \frac{1}{t} \int f_K \, d p_K(\mu_t) = \lim_{t \downarrow 0} \frac{1}{t} \int f \, d\mu_t = 0,$$

hence $\eta_K = 0$. Since $\mu \notin \mathscr{D}(G)$, there is a $K_1 \in \mathbf{N}$ such that $p_K(\mu_1) \notin \mathscr{D}(G/K)$ for all $K \in \mathbf{N}_1$ as in the proof of Lemma 6.2.5. For every $K \in \mathbf{N}_1$ the measure $p_K(\mu_1)$ is a Gauss measure in $\mathscr{G}(G/K)$ with embedding semigroup $p_K(S)$ by Lemma 6.2.6. But then Lemma 6.2.5 yields the assertion. □

6.2.11 Corollary. *The definition of a Gauss measure $\mu \in \mathscr{G}(G)$ is independent of its embedding semigroup.*

Proof. For $\mu \in \mathscr{G}(G)$ Condition (GC) of (v) of the theorem holds. This condition, however, is already independent of the embedding semigroup. □

6.2.12 Corollary. *The set $\mathscr{G}(G)$ is $\mathscr{T}_v$-closed in $\mathscr{E}_{\{e\}}(G) \setminus \mathscr{D}(G)$.*

The *proof* follows directly from Condition (GC) of (v) of the theorem. □

6.2.13 Corollary. *Let* $\mu, \nu \in \mathscr{G}(G)$ *be such that* $\mu * \nu \in \mathscr{E}_{\{e\}}(G)$. *Then* $\mu * \nu \in \mathscr{G}(G)$.

The *proof* follows obviously from Condition (GC) of (v) of the theorem. □

6.2.14 Remark. In general, $\mathscr{G}(G)$ is not a subsemigroup of $\mathscr{M}^1(G)$. We shall discuss the compact connected Lie group $G := \mathfrak{SO}(3)$ as a counterexample. Let $\{X_1, X_2, X_3\}$ be a basis of the Lie algebra $\mathfrak{L}(G)$ of G such that $\exp tX_i$ is the rotation about the x_i-axis by the angle $t \in \mathbb{R}_+^*$ ($i=1, 2, 3$). Further suppose that μ and ν are Gauss measures in $\mathscr{G}(G)$ with embedding semigroups $(\mu_t)_{t\in\mathbb{R}_+^*}$ and $(\nu_t)_{t\in\mathbb{R}_+^*}$ admitting infinitesimal generators N and M of the form

$$(a_i, a_{ij}, 0)_{i,j=1,2,3} \quad \text{and} \quad (b_i, b_{ij}, 0)_{i,j=1,2,3} \text{ resp.,}$$

with

$$a_1 = a_2 = a_3 = a_{22} = a_{33} = 0, \qquad a_{11} := a;$$
$$b_1 = b_2 = b_3 = b_{11} = b_{33} = 0, \qquad b_{22} := b \qquad (a, b \in \mathbb{R}_+^*,\ a \neq b)$$

and $a_{ij} = b_{ij} = 0$ for $i \neq j$ ($i, j = 1, 2, 3$). Here we have used Lemma 6.2.6. Let I denote the identity representation in $\operatorname{Rep}(G)$. For every $t \in \mathbb{R}_+^*$ one computes

$$\hat{\mu}_t(I) = \int_G I(x)\,\mu_t(dx) = \begin{pmatrix} 1 & 0 & 0 \\ 0 & e^{-at} & 0 \\ 0 & 0 & e^{-at} \end{pmatrix} \quad \text{and}$$

$$\hat{\nu}_t(I) = \begin{pmatrix} e^{-bt} & 0 & 0 \\ 0 & 1 & 0 \\ 0 & 0 & e^{-bt} \end{pmatrix}.$$

We further put $\lambda_t := \mu_t * \nu_t$ and consider $\hat{\lambda}_t = \hat{\mu}_t \cdot \hat{\nu}_t$ for all $t \in \mathbb{R}_+^*$. Clearly, for every $t \in \mathbb{R}_+^*$ the matrix

$$\hat{\lambda}_t(I) = \begin{pmatrix} e^{-bt} & 0 & 0 \\ 0 & e^{-at} & 0 \\ 0 & 0 & e^{-(a+b)t} \end{pmatrix} \in \mathfrak{M}(3, \mathbb{R})$$

is of diagonal form with positive eigenvalues, so that there exists exactly one (logarithm)

$$L_t := \begin{pmatrix} -bt & 0 & 0 \\ 0 & -at & 0 \\ 0 & 0 & -(a+b)t \end{pmatrix} \in \mathfrak{M}(3, \mathbb{R})$$

with $\exp L_t = \hat{\lambda}_t(I)$. That is, if $(\lambda_t)_{t\in\mathbb{R}_+^*}$ is a semigroup in $\mathscr{M}^1(G)$, then it is the unique Gauss semigroup in $\mathscr{M}^1(G)$ with infinitesimal generator of the form $(c_i, c_{ij}, 0)_{i,j=1,2,3}$, where

$$c_1 = c_2 = c_3 = 0, \quad c_{11} = c_{33} = a, \quad c_{22} = b \quad \text{and}$$
$$c_{ij} = 0 \qquad \text{for } i \neq j \ (i, j = 1, 2, 3).$$

Now let $(\pi_t)_{t\in\mathbb{R}_+^*}$ be the Gauss semigroup in $\mathscr{M}^1(G)$ with infinitesimal generator of the form $(d_i, d_{ij}, 0)_{i,j=1,2,3}$, where

$$\begin{aligned} &d_1=d_2=d_3=d_{33}=0, \quad d_{11}=a, \quad d_{22}=b \quad \text{and} \\ &d_{ij}=0 \quad \text{for } i\neq j \quad (i,j=1,2,3). \end{aligned}$$

If we can show that there exist numbers $t\in\mathbb{R}_+^*$ with $\lambda_t \neq \pi_t$, then $\mu_t * \nu_t = \lambda_t$ is not a Gauss measure in $\mathscr{G}(G)$.

For this purpose we first compute the infinitesimal generators N, M and Q of the semigroups $(\hat{\mu}_t(D))_{t\in\mathbb{R}_+^*}$, $(\hat{\nu}_t(D))_{t\in\mathbb{R}_+^*}$ and $(\hat{\pi}_t(D))_{t\in\mathbb{R}_+^*}$ resp., where $D:=I\otimes I$. These generators have the form

$$N=\left(\begin{array}{ccc|ccc|ccc} 0 & & & & & & & & \\ & -1 & & & & & & & \\ & & -1 & & & & & & \\ \hline & & & -1 & & & & & 0 \\ & & & & -2 & & & & 2 \\ & & & & & -2 & 0 & -2 & 0 \\ \hline & & & & & 0 & -1 & & \\ & & & & & -2 & & -2 & \\ & & & 0 & 2 & 0 & & & -2 \end{array}\right)\cdot a,$$

$$M=b\cdot\left(\begin{array}{ccc|ccc|ccc} -2 & & & & & & & & 2 \\ & -1 & & & & & & 0 & \\ & & -2 & & & & -2 & & \\ \hline & & & -1 & & & & & \\ & & & & 0 & & & & \\ & & & & & -1 & & & \\ \hline & & -2 & & & & -2 & & \\ & 0 & & & & & & -1 & \\ 2 & & & & & & & & -2 \end{array}\right) \quad \text{and}$$

$$Q=N+M.$$

(In these matrices all nonspecified entries are 0.)

Hence, the 9-dimensional vector space under discussion is a direct sum of subspaces that are invariant with respect to N and M. One of these subspaces is generated by the basis vector in the first, fifth and nineth place for which the exponentials are

$$n_t := \exp tN = \exp at\begin{pmatrix} 0 & 0 & 0 \\ 0 & -2 & 2 \\ 0 & 2 & -2 \end{pmatrix},$$

$$m_t := \exp bt \begin{pmatrix} -2 & 0 & 2 \\ 0 & 0 & 0 \\ 2 & 0 & -2 \end{pmatrix} \quad \text{and}$$

$$q_t := \exp t \begin{pmatrix} -2b & 0 & 2b \\ 0 & -2a & 2a \\ 2b & 2a & -2(a+b) \end{pmatrix} \quad (t \in \mathbb{R}_+^*).$$

Taking second derivatives with respect to t in the equation

$$n_t \cdot m_t = q_t$$

one concludes $NM = MM$ contradicting the obvious computation $NM \neq MN$. Thus

$$\hat{\mu}_t(D)\,\hat{\nu}_t(D) \neq \hat{\pi}_t(D),$$

whence $\mu_t * \nu_t \neq \pi_t$ for some $t \in \mathbb{R}_+^*$. One notes that the semigroups $(\mu_t)_{t\in\mathbb{R}_+^*}$ and $(\nu_t)_{t\in\mathbb{R}_+^*}$ in $\mathscr{M}^1(G)$ constructed here are clearly degenerate.

A slight modification yields nondegenerate ones. In fact, since by Corollary 6.2.12 $\mathscr{G}(G)$ is $\mathscr{T}_v$-closed in $\mathscr{E}_{\{e\}}(G) \setminus \mathscr{D}(G)$, for $t \geq t_0$ there exists a $\mathscr{T}_v$-neighborhood $W \in \mathfrak{V}(\lambda_t)$ which does not contain a Gauss measure. Hence, there exist neighborhoods $U \in \mathfrak{V}(\mu_t)$ and $V \in \mathfrak{V}(\nu_t)$ such that for $\mu \in U$ and $\nu \in W$ the product $\mu * \nu$ is not a Gauss measure. But there do exist (nondegenerate) Gauss measures $\mu \in U$ and $\nu \in V$, as one can assure by enlarging the data in the representations of the embedding semigroups of μ and ν resp.

6.2.15 Theorem. *Let G be a Lie projective group in* **A**. *For any $\mu \in \mathscr{E}_{\{e\}}(G) \setminus \mathscr{D}(G)$ the following statements are equivalent:*
(i) $\mu \in \mathscr{G}(G)$;
(ii) *for every $D \in \mathrm{Irr}(G)$ the measure $D(\mu)$ is either in $\mathscr{G}(\overline{D(G)})$ or in $\mathscr{D}(\overline{D(G)})$.*

Proof. 1. (i) $\Rightarrow$ (ii). Let D be in $\mathrm{Irr}(G)$ and put $G' := \overline{D(G)}$. For every $R \in \mathrm{Irr}(G')$ we have $D_1 := R \circ D \in \mathrm{Irr}(G)$ with $n(D_1) = n(R)$. Since $\hat{\mu}(D_1) = D(\mu)^\wedge(R)$, $D(\mu)$ satisfies Condition (GC) of Theorem 6.2.10. Thus, if $D(\mu) \notin \mathscr{D}(G')$, then $D(\mu) \in \mathscr{G}(G')$.

2. (ii) $\Rightarrow$ (i). Denoting for every $D \in \mathrm{Irr}(G)$ the identity on $G' := \overline{D(G)}$ by I_D, we obtain $D(\mu)^\wedge(I_D) = \hat{\mu}(D)$. Let $D(\mu) \in \mathscr{G}(G')$. Then Statement (v) of Theorem 6.2.10 applied to G' and $D(\mu)$ instead of G and μ yields the validity of Condition (GC) for D. But for degenerate $D(\mu)$ this condition is trivially fulfilled. □

6.2.16 Theorem. *Let G be a Lie projective group in* **A** *and let μ be in $\mathscr{E}_{\{e\}}(G) \setminus \mathscr{D}(G)$ with embedding semigroup S and corresponding negative-definite form ψ on $\mathfrak{R}(G)$. The following statements are equivalent:*

(i) $\mu \in \mathscr{G}(G)$;

(ii) *if $\psi = \psi^{(1)} + \psi^{(2)}$ is a representation of ψ with a continuous negative-definite form $\psi^{(1)}$ and a Poisson form $\psi^{(2)}$ of the form $\psi^{(2)} := \|\kappa\|\,\hat{\varepsilon}_e - \hat{\kappa}$ for some $\kappa \in \mathscr{M}_+^b(G)$, then $\psi^{(2)} = 0$.*

Proof. 1. (i) ⇒ (ii). By Theorem 6.2.10 we obtain for all $D\in\mathrm{Irr}(G)$ the relations

$$0=\psi(f_D^2)=\psi^{(1)}(f_D^2)+\psi^{(2)}(f_D^2)$$
$$\leq\psi^{(2)}(f_D^2)=-\hat{\kappa}(f_D^2)=-\int f_D^2\,d\kappa\leq 0.$$

Thus, $\int f_D^2\,d\kappa=0$ and so $\mathrm{Res}_{G^*}\kappa=0$, which implies that $\psi^{(2)}=0$.

2. (ii) ⇒ (i). Given a Lévy function Γ for G we have by Theorem 4.3.13 the formula

$$\psi(f)=\psi_1(f)+\psi_2(f)-\int_{G^*}[f(x)-f(e)-\Gamma(x,f)]\,\eta(dx),$$

which is valid for all $f\in\mathfrak{K}(G)$, where η is a Lévy measure on G. For $D\in\mathrm{Irr}(G)$ with $D\neq I$ and $g_D:=\dfrac{f_D}{\|f_D\|}$ the integral $\int_{G^*}g_D\,d\eta$ exists,

$$\sigma:=g_D\cdot\eta\in\mathscr{M}_+^b(G),\qquad \tau:=(1-g_D)\cdot\eta\in\mathscr{M}_+(G^*),$$

and we have $\int_{G^*}f_{D'}\,d\tau\leq\int_{G^*}f_{D'}\,d\eta<\infty$ for all $D'\in\mathrm{Rep}(G)$ as well as $\tau(\complement U)\leq\eta(\complement U)<\infty$ for all $U\in\mathfrak{B}(e)$. Consequently, τ is a Lévy measure on G.

From the properties of the Lévy function Γ for G we conclude that the function $\psi_1^{(1)}$ on $\mathfrak{K}(G)$ defined by

$$\psi_1^{(1)}(f):=\int_{G^*}\Gamma(x,f)\,\sigma(dx)\qquad\text{for all } f\in\mathfrak{K}(G)$$

is a continuous primitive form. We have for all $f\in\mathfrak{K}(G)$ the desired representation

$$\psi(f)=\psi_1(f)+\psi_1^{(1)}(f)+\psi_2(f)$$
$$-\int_{G^*}[f(x)-f(e)-\Gamma(x,f)]\,\tau(dx)+\int_G[f(e)-f(x)]\,\sigma(dx).$$

Therefore, $\int_G[f(x)-f(e)]\,\sigma(dx)=0$ for all $f\in\mathfrak{K}(G)$. Choosing $f:=f_{D'}$ for $D'\in\mathrm{Rep}(G)$, we obtain $g_D\cdot\eta=\sigma=0$ for all $D\in\mathrm{Irr}(G)$ and hence, $\eta=0$. Theorem 6.2.10 implies the result. □

6.2.17 Theorem. *Let G be a locally compact Abelian group and let μ be in $\mathscr{E}_{\{e\}}(G)\setminus\mathscr{D}(G)$ with embedding semigroup S. The following statements are equivalent:*

(i) $\nu\in\mathscr{G}(G)$;

(ii) *if $\mu=\sigma*\tau$ with $\sigma\in\mathscr{P}_{\{e\}}(G)$ and $\tau\in\mathscr{E}_{\{e\}}(G)$, then $\sigma=\varepsilon_e$;*

(iii) *there exist an $x\in G$, a one-parameter subgroup $(x_t)_{t\in\mathbb{R}}$ of G with $x=x_1$, and a quadratic form ψ_2 on $\mathfrak{K}(G)$, continuous on $G^\wedge$, such that for all $\chi\in G^\wedge$ the relation $\hat{\mu}(\chi)=\chi(x)\exp(-\psi_2(\chi))$ holds.*

Proof. 1. (i) ⇔ (ii). Since G is Abelian, the algebra $\mathscr{S}(G)$ is commutative. The equivalence under discussion follows immediately from Theorem 6.2.16 since $\sigma\in\mathscr{M}^1(G)$ is a Poisson measure iff there exists a measure $\kappa\in\mathscr{M}_+^b(G)$ with the property $\hat{\sigma}=\mathrm{Exp}(\hat{\kappa}-\|\kappa\|\,\hat{\varepsilon}_e)$.

2. (i) ⇒ (iii). Let ψ be the negative-definite form corresponding to the embedding semigroup S of μ. By Theorem 6.2.10 we have $\psi=\psi_1+\psi_2$ with a continuous primitive form ψ_1 and a continuous quadratic form ψ_2 on $\mathfrak{K}(G)$.

Since G is a Moore group, we conclude from Theorem 1.5.20 (ii) that there exists a one-parameter subgroup $(x_t)_{t\in\mathbb{R}}$ of G with $\hat{\varepsilon}_{x_t}=\operatorname{Exp}(-t\psi_1)$ for all $t\in\mathbb{R}$. Putting $x:=x_1$ we find that this implies

$$\hat{\mu}(\chi)=\exp(-\psi(\chi))=\chi(x)\exp(-\psi_2(\chi)) \quad \text{for all } \chi\in G^\wedge.$$

3. (iii) $\Rightarrow$ (i). Conversely, let $(x_t)_{t\in\mathbb{R}}$ be a one-parameter subgroup of G and put $x:=x_1$. By Theorem 1.5.20 (i) there exists a continuous primitive form ψ_1 on $\mathfrak{K}(G)$ satisfying

$$\hat{\varepsilon}_{x_t}=\operatorname{Exp}(-t\psi_1) \quad \text{for all } t\in\mathbb{R}.$$

Plainly, ψ_2 is also continuous on $\operatorname{Rep}(G)$, and we have

$$\hat{\mu}(D)=\exp(-\psi_1(D)-\psi_2(D)) \quad \text{for all } D\in\operatorname{Rep}(G),$$

or $\hat{\mu}=\operatorname{Exp}(-(\psi_1+\psi_2))$. Theorem 6.2.10 yields the assertion. □

6.2.18 Corollary. *A measure $\mu\in\mathscr{E}_{\{e\}}(G)\setminus\mathscr{D}(G)$ is in $\mathscr{G}_p(G)$ iff it belongs to $\mathscr{G}(G)$.*

Proof. 1. If $\mu\in\mathscr{G}_p(G)$, then obviously (ii) of the theorem is fulfilled, so that $\mu\in\mathscr{G}(G)$.

2. Conversely, if $\mu\in\mathscr{G}(G)$, then by the theorem there exist an $x\in G$ and a $\psi\in Q_+(G^\wedge)$ such that $\hat{\mu}(\chi)=\chi(x)\exp(-\psi(\chi))$ holds for all $\chi\in G^\wedge$.

This implies by Theorem 5.2.8 that $\mu\in\mathscr{G}_p(G)$. □

6.2.19 Remark. If G is a compact Lie group, then the class $\mathscr{G}(G)$ coincides with the class of normal measures in the sense of Carnal ([64]) by Lemma 6.2.6.

For arbitrary compact groups G, measures $\mu\in\mathscr{E}_{\{e\}}(G)$ which are Gauss in the sense of Carnal ([64]) are elements of $\mathscr{G}(G)$. This follows from Theorem 6.2.15.

In the case of a locally compact Abelian group G the Gauss measures in the sense of Urbanik, defined as measures $\mu\in\mathscr{E}_{\{e\}}(G)$ such that $\chi(\mu)\in\mathscr{N}(\mathbb{T})=\mathscr{G}(\mathbb{T})$ for all $\chi\in G^\wedge$, $\chi\neq 1$, are in $\mathscr{G}(G)$, as is again implied by Theorem 6.2.15.

Urbanik defined Gauss measures within the set $\mathscr{E}_{\{e\}}(G)$ rather than $\mathscr{I}(G)$ on the grounds of the following fact: There are non-Gauss measures μ in $\mathscr{M}^1(\mathbb{T}^2)$ such that $\chi(\mu)\in\mathscr{N}(\mathbb{T})$ for all $\chi\in(\mathbb{T}^2)^\wedge\cong\mathbb{Z}^2$. In fact, the measures μ to be constructed are not in $\mathscr{I}(G)$, and are hence not in $\mathscr{E}_{\{e\}}(G)$. Consider the function $g\in\mathscr{C}(\mathbb{T}^2)$ defined by

$$g(u,v):=\sum\nolimits_{m,n\in\mathbb{Z}} c_{mn}\, e^{-i(mu+nv)}=\sum\nolimits_{m,n\in\mathbb{Z}} c_{mn}\cos(mu+nv)$$

for all $(u,v)\in\mathbb{T}^2$, where

$$c_{mn}:=\begin{cases}\dfrac{1}{10^{n^2}} & \text{for } m=n\\[2ex] \dfrac{1}{10^{m^2+n^2}} & \text{for } m\neq n\ (m,n\in\mathbb{Z})\end{cases}$$

For every $(u,v)\in\mathbb{T}^2$ one computes

$$g(u,v)=\sum_{(m,n)\in\mathbb{Z}^2\setminus\{(0,0)\}} c_{mn}\cos(mu+nv)+1$$
$$\geq 1-\sum_{(m,n)\in\mathbb{Z}^2\setminus\{(0,0)\}} c_{mn}\geq 1-2\sum_{n\geq 1}\frac{1}{10^{n^2}}-4\left(\sum_{n\geq 1}\frac{1}{10^{n^2}}\right)^2$$
$$\geq 1-2\sum_{n\geq 1}\frac{1}{10^n}-4\left(\sum_{n\geq 1}\frac{1}{10^n}\right)^2=\frac{59}{81}.$$

Hence $g\geq 0$. Furthermore, it can be shown easily that

$$\int g\,d\lambda_{\mathbb{T}^2}=1 \quad\text{so that}\quad \mu:=g\cdot\lambda_{\mathbb{T}^2}\in\mathcal{M}^1(\mathbb{T}^2).$$

(a) For every nontrivial character χ of $\mathbb{T}^2$ of the form $\chi_{r,s}=(r,s)\in\mathbb{Z}^2$ we get $\chi(\mu)\in\mathcal{N}(\mathbb{T})$. In fact, for $(r,s)\in\mathbb{Z}^2$ one has

$$\hat{\mu}(r,s)=\frac{1}{4\pi^2}\int_0^{2\pi}\int_0^{2\pi} e^{iru}\,e^{isv}\sum_{m,\,n\in\mathbb{Z}} c_{mn}\,e^{-i(mu+nv)}\,du\,dv.$$

Hence,

$$\chi_{r,s}(\mu)^\wedge(k)=\hat{\mu}(\chi_{r,s}^k)=\hat{\mu}(kr,ks)=e^{-d(\chi)k^2}$$

for all $k\in\mathbb{Z}$, where

$$d(\chi):=\begin{cases} r^2\log 10 & \text{for } \chi=\chi_{r,r}=(r,r)\\ (r^2+s^2)\log 10 & \text{for } \chi=\chi_{r,s}=(r,s) \text{ with } r\neq s.\end{cases}$$

But this shows that $\chi_{r,s}(\mu)\in\mathcal{N}(\mathbb{T})$ by the discussion opening §2 of Chapter V.

(b) $\mu\notin\mathcal{G}(\mathbb{T}^2)$.

Suppose μ were in $\mathcal{G}(\mathbb{T}^2)$. Then there would exist $\alpha,\beta,\gamma\in\mathbb{R}$ such that

$$c_{mn}=e^{-(\alpha m^2+\beta mn+\gamma n^2)}$$

and thus

$$\alpha m^2+\beta mn+\gamma n^2=\begin{cases}(m^2+n^2)\log 10 & \text{for } m\neq n\\ n^2\log 10 & \text{for } m=n\ (m,n\in\mathbb{Z})\end{cases}$$

would hold. We would therefore obtain $\alpha+\beta+\gamma=\log 10$ for $m=n$ and $\alpha=\gamma=\log 10$, $\beta=0$ for $m\neq n$, or $2\log 10=\log 10$, which is a contradiction.

We are now ready to give general characterizations of Gauss semigroups on arbitrary locally compact groups in terms of their generating functionals as described in Theorems 4.4.18 and 4.5.8.

6.2.20 Theorem. *Let G be a locally compact group and $(\mu_t)_{t\in\mathbb{R}_+^*}$ an $\{e\}$-continuous convolution semigroup in $\mathcal{M}^1(G)$ with generating functional A defined at least on $\mathfrak{D}(G)$ and admitting (for a given Lévy mapping Γ for G) a canonical representation (ψ_1,ψ_2,η), where ψ_1 is a primitive form, ψ_2 is a quadratic form on $\mathfrak{D}(G)$ and η is a Lévy measure on G.*

The following statements are equivalent:
(i) $(\mu_t)_{t\in\mathbb{R}_+^*}$ *is a Gauss semigroup in* $\mathcal{M}^1(G)$;
(ii) A *is concentrated and* $\psi_2 \neq 0$;
(iii) A *is concentrated and there is an* $f\in\mathfrak{D}(G)$ *with* $f=f^*$ *satisfying* $A(f)\neq 0$;
(iv) $\eta=0$ *and* $\psi_2\neq 0$;
(v) $A=\psi_1+\psi_2$ *and* $\psi_2\neq 0$.

Proof. 1. (i)$\Rightarrow$(ii) follows as in Lemma 6.2.6 since for every $f\in\mathfrak{D}(G)$ with $f(U)=0$ for a neighborhood $U\in\mathfrak{V}(e)$ the inequalities

$$|Af| = \lim_{t\downarrow 0}\left|\frac{1}{t}(\mu_t-\varepsilon_e)(f)\right|$$
$$=\lim_{t\downarrow 0}\frac{1}{t}\left|\int_{\complement U} f\,d\mu_t\right| \le \|f\| \lim_{t\downarrow 0}\frac{1}{t}\mu_t(\complement U)$$

are valid.

2. (ii) $\Leftrightarrow$ (iii) is evident.

3. (ii) $\Rightarrow$ (iv). By assumption, the canonical representation of A as

$$Af=\psi_1(f)+\psi_2(f)+\int_{G^*}[f-f(e)-\Gamma(f)]\,d\eta \qquad \text{for all } f\in\mathfrak{D}(G)$$

yields $\int_{G^*} f\,d\eta=0$ for all $f\in\mathfrak{D}(G)$ with $f(U)=0$ for a neighborhood $U\in\mathfrak{V}(e)$ since ψ_1,ψ_2 and $f\to\Gamma(f)(x)$ $(x\in G)$ are concentrated by Properties (1) and (2) of 4.4.7. But then $\int_{G^*} f\,d\eta=0$ for all f in the dense subspace

$$\mathcal{A} := \{g\in\mathfrak{D}(G): e\notin \operatorname{supp}(g)\}$$

of $\mathcal{K}(G^*)$ satisfying Property (P) of [41], p. 56 and hence $\eta=0$.

4. (iv) $\Leftrightarrow$ (v) is again evident.

5. (iv)$\Rightarrow$(i) is deduced as in Lemma 6.2.6: For $U\in\mathfrak{V}(e)$ there exist a $V\in\mathfrak{V}(e)$ with $\bar{V}\subset U$ and an $f\in\mathfrak{D}(G)$, $0\le f\le 1$, with $f(V)=1$ and $f(\complement U)=0$. But then by the properties of the Lévy measure η (established in Lemma 4.5.5) we obtain

$$\lim_{t\downarrow 0}\frac{1}{t}\mu_t(\complement U)\le \lim_{t\downarrow 0}\frac{1}{t}\int_G (1-f)\,d\mu_t = A(1-f)=\eta(1-f)=0,$$

hence, Condition (G) of Definition 6.2.1 is satisfied. □

In the subsequent discussion a generalization of Corollary 6.2.12 for general locally compact groups will be established. A few *facts from the analytic theory of operator semigroups* are needed. Let B be a Banach space, $(T_t^{(\alpha)})_{t\in\mathbb{R}_+^*}$ $(\alpha\in\mathbb{A})$ and $(T_t)_{t\in\mathbb{R}_+^*}$ strongly continuous contraction semigroups of operators on B with infinitesimal generators $N^{(\alpha)}$ and N having domains of definition $\mathsf{D}_{N^{(\alpha)}}$ and D_N resp. For every $\lambda\in\mathbb{R}_+^*$ we define the operators $I_\lambda^{(\alpha)}$ $(\alpha\in\mathbb{A})$ and I_λ on B by

$$I_\lambda^{(\alpha)} := \int_0^\infty e^{-\lambda t}\, T_t^{(\alpha)}\,dt \quad\text{and}\quad I_\lambda := \int_0^\infty e^{-\lambda t}\, T_t\,dt$$

and the λ-resolvents $R_\lambda^{(\alpha)}$ $(\alpha\in\mathbb{A})$ and R_λ on B by

$$R_\lambda^{(\alpha)} := (\lambda E - N^{(\alpha)})^{-1} \quad\text{and}\quad R_\lambda := (\lambda E - N)^{-1} \quad \text{resp.}$$

We know that for semigroups $(T_t^{(\alpha)})_{t\in\mathbb{R}_+^*}$ $(\alpha\in\mathbb{A})$ and $(T_t)_{t\in\mathbb{R}_+^*}$ with the norming properties $T_0^{(\alpha)}=E$ and $T_0=E$ one has $R_\lambda^{(\alpha)}=I_\lambda^{(\alpha)}$ and $R_\lambda=I_\lambda$ resp.

Now if $((T_t^{(\alpha)})_{t\in\mathbb{R}_+^*})_{\alpha\in\mathbb{A}}$ is a net and the semigroups $(T_t^{(\alpha)})_{t\in\mathbb{R}_+^*}$ $(\alpha\in\mathbb{A})$ and $(T_t)_{t\in\mathbb{R}_+^*}$ have the norming property, then the Trotter-Kato theorem ([508], IX, 12) yields the equivalence of the following two statements valid in the sense of the strong operator topology:

(i) $\lim_{\alpha\in\mathbb{A}} R_\lambda^{(\alpha)}=R_\lambda$ for some $\lambda\in\mathbb{R}_+^*$;

(ii) $\lim_{\alpha\in\mathbb{A}} T_t^{(\alpha)}=T_t$ uniformly in t on compact subsets of $\mathbb{R}_+$.

Moreover, (i) or (ii) imply the useful property.

(iii) To every $x\in\mathsf{D}_N$ there exists a net $(x_\alpha)_{\alpha\in\mathbb{A}}$ in B with $x_\alpha\in\mathsf{D}_{N^{(\alpha)}}$ for all $\alpha\in\mathbb{A}$ satisfying $\lim_{\alpha\in\mathbb{A}} x_\alpha=x$ and $\lim_{\alpha\in\mathbb{A}} N^{(\alpha)}x_\alpha=Nx$.

In fact, from the Hille-Yosida theory [508] we obtain for all $\lambda\in\mathbb{R}_+^*$ the formulae

$$R_\lambda^{(\alpha)}\mathsf{B}=\mathsf{D}_{N^{(\alpha)}},\qquad R_\lambda\mathsf{B}=\mathsf{D}_N \quad\text{and}$$
$$(\lambda E-N^{(\alpha)})\,\mathsf{D}_{N^{(\alpha)}}=\mathsf{B}=(\lambda E-N)\,\mathsf{D}_N.$$

Now let x be in D_N and for all $\alpha\in\mathbb{A}$ (and a fixed $\lambda\in\mathbb{R}_+^*$) put

$$x_\alpha:=R_\lambda^{(\alpha)}(\lambda E-N)\,x=[(\lambda E-N^{(\alpha)})^{-1}(\lambda E-N)]\,x.$$

Then (i) implies $\lim_{\alpha\in\mathbb{A}} x_\alpha=x$. Furthermore, we have for all $\alpha\in\mathbb{A}$ (and $\lambda\in\mathbb{R}_+^*$) the equalities

$$N^{(\alpha)}x_\alpha=-(\lambda E-N^{(\alpha)})\,x_\alpha+\lambda x_\alpha=-(\lambda E-N)\,x+\lambda x_\alpha.$$

Hence, $\lim_{\alpha\in\mathbb{A}} N^{(\alpha)}x_\alpha=Nx$, and the assertion is established.

6.2.21 Lemma. *Let* B *be a Hilbert space and let* $(T^{(\alpha)})_{\alpha\in\mathbb{A}}$ *be a net of positive-semidefinite contraction operators on* B *converging in the strong operator topology to a (positive-semidefinite) contraction operator* T *on* B. *Let* $T^{(\alpha)}$ $(\alpha\in\mathbb{A})$ *and* T *admit spectral decompositions* $\int_0^1\rho\,dE_\rho^{(\alpha)}$ *and* $\int_0^1\rho\,dE_\rho$ *with spectral resolutions* $(E_\rho^{(\alpha)})_{\rho\in[0,1]}$ *and* $(E_\rho)_{\rho\in[0,1]}$ *resp.*

We define for every $t\in\mathbb{R}_+^*$ *the operators*

$$T_t^{(\alpha)}:=\int_0^1\rho^t\,dE_\rho^{(\alpha)}\ (\alpha\in\mathbb{A})\quad\text{and}\quad T_t:=\int_0^1\rho^t\,dE_\rho.$$

Then

(i) $\lim_{\alpha\in\mathbb{A}} T_t^{(\alpha)}=T_t$ *for all* $t\in\mathbb{R}_+^*$ *and*

(ii) $\lim_{\alpha\in\mathbb{A}} I_\lambda^{(\alpha)}=I_\lambda$ *for all* $\lambda\in\mathbb{R}_+^*$ *(in the strong operator topology).*

If, moreover, $(E_{0+}^{(\alpha)}-E_{0-}^{(\alpha)})\,\mathsf{B}=\{0\}=(E_{0+}-E_{0-})\,\mathsf{B}$ *for all* $\alpha\in\mathbb{A}$, *then*

(iii) $\lim_{\alpha\in\mathbb{A}} R_\lambda^{(\alpha)}=R_\lambda$ *for all* $\lambda\in\mathbb{R}_+^*$ *(in the strong operator topology).*

Proof. Since Statement (iii) is obvious, we need to show (i) and (ii). These two assertions, however, follow from the equality

$$\lim_{\alpha\in\mathbb{A}} f(T^{(\alpha)}):=\lim_{\alpha\in\mathbb{A}}\int_0^1 f(\rho)\,dE_\rho^{(\alpha)}=\int_0^1 f(\rho)\,dE_\rho=:f(T)$$

valid for all $f\in\mathscr{C}([0,1])$. This is because the operators $T_t^{(\alpha)}$ $(\alpha\in\mathbb{A})$ and T_t for $t\in\mathbb{R}_+^*$ as well as $I_\lambda^{(\alpha)}$ and I_λ for $\lambda\in\mathbb{R}_+^*$ are representable as functions of $T^{(\alpha)}$ and T

resp. if one chooses in the above equality the functions

$$\rho \to f(\rho) := \rho^t \ (t \in \mathbb{R}_+^*) \quad \text{and} \quad \rho \to f(\rho) := \frac{1}{-\lambda + \log \rho} \ (\lambda \in \mathbb{R}_+^*) \ \text{resp.}$$

Clearly, the above equality is valid for all polynomials in $\mathscr{C}([0,1])$. Let $f \in \mathscr{C}([0,1])$ be arbitrary. For every $\varepsilon > 0$ there exists a polynomial $p := p_\varepsilon$ satisfying $\|f - p\| < \varepsilon$. For every $x \in \mathsf{B}$ we obtain

$$\begin{aligned} &\|f(T^{(\alpha)})x - f(T)x\| \\ &\leq \|f(T^{(\alpha)})x - p(T^{(\alpha)})x\| + \|f(T)x - p(T)x\| + \|p(T^{(\alpha)})x - p(T)x\|. \end{aligned}$$

The first two summands of the right hand side of the inequality can be majorized by

$$\int_0^1 \|f - p\| \, d\|E_\rho^{(\alpha)} x\| \ (\alpha \in \mathbb{A}) \quad \text{and} \quad \int_0^1 \|f - p\| \, d\|E_\rho x\|,$$

hence by 2ε, and the last summand converges to 0 with respect to $\alpha \in \mathbb{A}$, since p is a polynomial. This proves the assertion. □

6.2.22 Theorem. *Let G be a Lie group with Lie algebra $\mathfrak{L}(G)$ and basis $\{X_1, \ldots, X_d\}$ and let $((\mu_t^{(n)})_{t \in \mathbb{R}_+^*})_{n \geq 1}$ be a sequence of Gauss semigroups $(\mu_t^{(n)})_{t \in \mathbb{R}_+^*}$ in $\mathscr{M}^1(G)$ with an infinitesimal generator $N^{(n)}$ admitting (with respect to $\{X_1, \ldots, X_d\}$) on $\mathfrak{D}(G)$ a representation $(a_i^{(n)}, a_{ij}^{(n)}, 0)_{i,j=1,\ldots,d}$ with $a_i^{(n)} \in \mathbb{R}$, $i = 1, \ldots, d$, and a symmetric, positive-semidefinite matrix $(a_{ij}^{(n)})_{i,j=1,\ldots,d} \in \mathfrak{M}(d, \mathbb{R})$.*

Let $(\mu_t)_{t \in \mathbb{R}_+^*}$ with $\mu_1 \notin \mathscr{D}(G)$ be an $\{e\}$-continuous convolution semigroup in $\mathscr{M}^1(G)$ satisfying $\lim_{n \to \infty} \mu_t^{(n)} = \mu_t$ uniformly in t on compact subsets of $\mathbb{R}_+$. Then $(\mu_t)_{t \in \mathbb{R}_+^*}$ is a Gauss semigroup in $\mathscr{M}^1(G)$ with an infinitesimal generator N of the form $(a_i, a_{ij}, 0)_{i,j=1,\ldots,d}$ with $a_i \in \mathbb{R}$, $i = 1, \ldots, d$, and a symmetric, positive-semidefinite matrix $(a_{ij})_{i,j=1,\ldots,d} \in \mathfrak{M}(d, \mathbb{R})$.

Moreover, one has for all $i, j = 1, \ldots, d$
(α) $\lim_{n \to \infty} a_i^{(n)} = a_i$ and
(β) $\lim_{n \to \infty} a_{ij}^{(n)} = a_{ij}$.

Proof. We apply the theory developed in Chapter IV, §1 and 2 (especially Theorem 4.2.8) and consider the convolution operators (of the measures appearing in our theorem) on the space $\tilde{\mathscr{C}}_2^0(G)$ furnished with the norm $\|\cdot\|_{\tilde{2}}$.

First of all one notes that for any sequence $(\mu^{(n)})_{n \geq 1}$ of measures in $\mathscr{M}^1(G)$ and a measure $\mu \in \mathscr{M}^1(G)$ the limit relationship

$$\mathscr{T}_w\text{-}\lim_{n \to \infty} \mu^{(n)} = \mu \quad \text{implies} \quad \lim_{n \to \infty} T_{\mu^{(n)}} = T_\mu$$

in the strong operator topology on $\tilde{\mathscr{C}}_2^0(G)$.

Now let $(T_t^{(n)})_{t \in \mathbb{R}_+^*}$ $(n \geq 1)$ and $(T_t)_{t \in \mathbb{R}_+^*}$ be the contraction semigroups $(S_{\mu_t^{(n)}})_{t \in \mathbb{R}_+^*}$ and $(S_{\mu_t})_{t \in \mathbb{R}_+^*}$ corresponding to $(\mu_t^{(n)})_{t \in \mathbb{R}_+^*}$ and $(\mu_t)_{t \in \mathbb{R}_+^*}$ with infinitesimal generators $N^{(n)}$ and N and generating functionals $A^{(n)}$ and A on $\tilde{\mathscr{C}}_2^0(G)$ resp.

By Property (iii) preceding Lemma 6.2.21, for any $f \in \mathsf{D}_N$ there exists a sequence $(f_n)_{n \geq 1}$ in $\mathsf{D}_{N^{(n)}}$ satisfying

$$\lim_{n \to \infty} \|f_n - f\|_{\tilde{2}} = 0 \quad \text{and} \quad \lim_{n \to \infty} \|N^{(n)}(f_n) - N(f)\|_{\tilde{2}} = 0.$$

1. Let $C:=\sup_{n\geq 1}[\sum_{i=1}^{d}|a_i^{(n)}|+\sum_{i,j=1}^{d}|a_{ij}^{(n)}|]<\infty$. Then for all $f\in \mathsf{D}_N$ we get

$$|A^{(n)}(f)-A(f)|\leq|A^{(n)}(f_n)-A(f)|+|A^{(n)}(f)-A^{(n)}(f_n)|$$
$$\leq|A^{(n)}(f_n)-A(f)|+\|f-f_n\|_2^{\sim} C$$

and hence

$$\lim_{n\to\infty}|A^{(n)}(f)-A(f)|=0.$$

Choosing for f the modified canonical coordinates x_i and x_{ij} $(i,j=1,\ldots,d)$ in D_N constructed in Lemma 4.1.12 we arrive at the assertion.

2. Let $(k_l)_{l\geq 1}$ be an isotone sequence in $\mathbb{R}_+$ with $\lim_{l\to\infty}k_l=\infty$ such that for some subsequence $(n_l)_{l\geq 1}$ of $\mathbb{N}$ with $\lim_{l\to\infty}n_l=\infty$ one has

$$\sum_{i=1}^{d}|a_i^{(n_l)}|+\sum_{i,j=1}^{d}|a_{ij}^{(n_l)}|=k_l.$$

For every $l\geq 1$ we define the linear functional $\bar{A}_l$ on $\tilde{\mathscr{C}}_2^0(G)$ by

$$\bar{A}_l(f):=\frac{1}{k_l}(N^{(n_l)}(f))(e) \qquad \text{for all } f\in\tilde{\mathscr{C}}_2^0(G),$$

Without loss of generality we assume the existence of an accumulation point $\bar{A}$ of the sequence $(\bar{A}_l)_{l\geq 1}$ of the form

$$\bar{A}f:=\sum_{i=1}^{d}\bar{a}_i(\tilde{X}_i f)(e)+\sum_{i,j=1}^{d}\bar{a}_{ij}(\tilde{X}_i\tilde{X}_j f)(e) \qquad \text{for } f\in\tilde{\mathscr{C}}_2^0(G),$$

where as above $\bar{a}_i\in\mathbb{R}$, $i=1,\ldots,d$ and $(\bar{a}_{ij})_{i,j=1,\ldots,d}$ is a symmetric, positive-semidefinite matrix in $\mathfrak{M}(d,\mathbb{R})$ such that

$$\lim_{l\to\infty}\frac{1}{k_l}a_i^{(n_l)}=\bar{a}_i \quad \text{and}$$

$$\lim\frac{1}{k_l}a_{ij}^{(n_l)}=\bar{a}_{ij} \quad (i,j=1,\ldots,d) \text{ hold.}$$

Let $(\nu_t)_{t\in\mathbb{R}_+^*}$ be the convolution semigroup corresponding to $\bar{A}$. Then by Lemma 4.2.7 we get

$$\mathscr{T}_w\text{-}\lim_{l\to\infty}\mu_{t\frac{1}{k_l}}^{(n_l)}=\nu_t$$

uniformly in t on compact subsets of $\mathbb{R}_+$. On the other hand, the hypothesis yields

$$\mathscr{T}_w\text{-}\lim_{l\to\infty}\mu_{t\frac{1}{k_l}}^{(n_l)}=\varepsilon_e.$$

Consequently, $\nu_t=\varepsilon_e$ for all $t\in\mathbb{R}_+^*$ or $\bar{A}=0$, which contradicts the construction. □

We have now arrived at our aim:

6.2.23 Theorem. *Let G be a locally compact group, $((\mu_t^{(\alpha)})_{t\in\mathbb{R}_+^*})_{\alpha\in\mathbb{A}}$ a net of Gauss semigroups in $\mathcal{M}^1(G)$ and $(\mu_t)_{t\in\mathbb{R}_+^*}$ with $\mu_1\notin\mathcal{D}(G)$ an $\{e\}$-continuous convolution semigroup in $\mathcal{M}^1(G)$ satisfying $\lim_{\alpha\in\mathbb{A}}\mu_t^{(\alpha)}=\mu_t$ uniformly in t on compact subsets of $\mathbb{R}_+$. Then $(\mu_t)_{t\in\mathbb{R}_+^*}$ is a Gauss semigroup in $\mathcal{M}^1(G)$.*

Proof. By Theorem 6.2.3 $\operatorname{supp}(\mu_t^{(\alpha)})\subset G_0$ for all $t\in\mathbb{R}_+^*$ and $\alpha\in\mathbb{A}$. Therefore we can assume without loss of generality that $G=G_0$, and hence, that G is Lie projective. But then the proof reduces to the special case of a connected Lie group admitting a countable basis of its topology. Thus it is enough to consider sequences of Gauss semigroups. The assertion then follows from Theorem 6.2.22. □

6.2.24 Theorem. *Let G be a locally compact group, $((\mu_t^{(\alpha)})_{t\in\mathbb{R}_+^*})_{\alpha\in\mathbb{A}}$ a net of Gauss semigroups in $\mathcal{M}^1(G)$ and $\mu\in\mathcal{M}^1(G)\setminus\mathcal{D}(G)$ such that $\lim_{\alpha\in\mathbb{A}}\mu_1^{(\alpha)}=\mu$ holds. We further suppose that the measures $\mu_t^{(\alpha)}$ ($t\in\mathbb{R}_+^*$, $\alpha\in\mathbb{A}$) are symmetric and that μ does not admit an idempotent factor. Then there exists a Gauss semigroup $(\mu_t)_{t\in\mathbb{R}_+^*}$ in $\mathcal{M}^1(G)$ with $\mu_1=\mu$, and for the generating functionals $A^{(\alpha)}$ ($\alpha\in\mathbb{A}$) and A of the semigroups $(\mu_t^{(\alpha)})_{t\in\mathbb{R}_+^*}$ and $(\mu_t)_{t\in\mathbb{R}_+^*}$ resp. we have*
(i) $\lim_{\alpha\in\mathbb{A}}\mu_t^{(\alpha)}=\mu_t$ uniformly in t on compact subsets of $\mathbb{R}_+$, so that
(ii) $\lim_{\alpha\in\mathbb{A}}A^{(\alpha)}(f)=A(f)$ for all $f\in\mathfrak{D}(G)$.

Proof. As in the proof of Theorem 6.2.23 we assume without loss of generality that G is a connected Lie group and that the given net of Gauss semigroups in $\mathcal{M}^1(G)$ is in fact a sequence $((\mu_t^{(n)})_{t\in\mathbb{R}_+^*})_{n\geq 1}$. We consider the convolution operators $T_t^{(n)}$ ($t\in\mathbb{R}_+^*$, $n\geq 1$) and T on the Hilbert space $L^2_{\mathbb{C}}(G,\omega)$ (with respect to a right Haar measure on G) corresponding to the measures $\mu_t^{(n)}$ and μ resp. Since $\mu_t^{(n)}$ is symmetric for all $t\in\mathbb{R}_+^*$ and $n\geq 1$, $T_t^{(n)}$ is a positive-semidefinite contraction operator on $L^2_{\mathbb{C}}(G,\omega)$ satisfying $\lim_{n\to\infty}T_1^{(n)}=T$ in the strong operator topology.

But then Lemma 6.2.21 yields for every $t\in\mathbb{R}_+^*$ the existence of an operator T_t defined by the spectral resolution of T with $\|T_t\|=1$ and such that

$$\lim_{n\to\infty}T_t^{(n)}=T_t \quad \text{for all } t\in\mathbb{R}_+^* \text{ and}$$

$$\lim_{n\to\infty}I_\lambda^{(n)}=I_\lambda \quad \text{for all } \lambda\in\mathbb{R}_+^*$$

(in the strong operator topology). Since $\lim_{n\to\infty}T_t^{(n)}=T_t$ for all $t\in\mathbb{R}_+^*$ holds also for the weak operator topology, we can conclude that there exists for every $t\in\mathbb{R}_+^*$ a measure $\mu_t\in\mathcal{M}^1(G)$ satisfying $\lim_{n\to\infty}\mu_t^{(n)}=\mu_t$. In fact, by Lemma 6.1.23 and the $\mathcal{T}_v$-compactness of $\mathcal{M}_+^{(1)}(G)$ there exists for every $t\in\mathbb{R}_+^*$ a measure $\mu_t\in\mathcal{M}_+^{(1)}(G)$ with $T_t=T_{\mu_t}$ and $\mathcal{T}_v\text{-}\lim_{n\to\infty}\mu_t^{(n)}=\mu_t$. But from

$$1=\|T_t\|=\|T_{\mu_t}\|\leq\|\mu_t\|\leq 1$$

we conclude that $\mu_t\in\mathcal{M}^1(G)$ and hence that $\mathcal{T}_w\text{-}\lim_{n\to\infty}\mu_t^{(n)}=\mu_t$ for all $t\in\mathbb{R}_+^*$.

Hence, μ is embeddable in an H-continuous convolution semigroup $(\mu_t)_{t\in\mathbb{R}_+^*}$ in $\mathcal{M}^1(G)$ for some compact subgroup H of G. Since μ does not admit an idempotent factor, $H=\{e\}$ or $\mu\in\mathcal{E}_{\{e\}}(G)$, and by Lemma 6.2.21 (iii)

$$\lim_{n\to\infty}R_\lambda^{(n)}=R_\lambda \quad \text{for all } \lambda\in\mathbb{R}_+^*.$$

The Trotter-Kato theorem now implies that $\lim_{n\to\infty} T_t^{(n)} = T_t$ (in the strong operator topology) uniformly in t on compact subsets of $\mathbb{R}_+$, which yields

$$\mathcal{T}_w\text{-}\lim_{n\to\infty} \mu_t^{(n)} = \mu_t$$

uniformly in t on compact subsets of $\mathbb{R}_+$. Now Theorem 6.2.22 becomes applicable, and the assertions stated in the theorem are proved. □

6.2.25 Corollary. *The set $\mathcal{G}^s(G)$ of (symmetric) Gauss measures on G which are embeddable in symmetric Gauss semigroups is $\mathcal{T}_v$-closed in the subset of all measures in $\mathcal{M}^1(G)$ without idempotent factors.*

The *proof* is clear. □

6.3 Absolute Continuity and Diffusion of Gauss Semigroups

For locally compact Abelian groups absolutely continuous Gauss semigroups have been discussed in §5 of Chapter V. The aim of that discussion was the existence problem. In the following we shall extend the theory of absolutely continuous and diffuse Gauss semigroups to general locally compact groups which by Theorem 6.2.3 will be assumed without loss of generality to be connected. We start with the special case of a Lie group.

Let G be a connected Lie group of dimension $n \geq 1$, $\{X_1, \ldots, X_n\}$ a basis of its Lie algebra $\mathfrak{L}(G)$ and $\{x_1, \ldots, x_n\}$ a system of canonical coordinates in $\mathfrak{D}(G)$ with respect to the basis $\{X_1, \ldots, X_n\}$. Given a vector $(a_1, \ldots, a_n) \in \mathbb{R}^n$ and a symmetric, positive-semidefinite matrix $(a_{ij})_{i,j=1,\ldots,n} \in \mathfrak{M}(n, \mathbb{R})$ with $(a_{ij})_{i,j=1,\ldots,n} \neq 0$ the mapping $N: \mathfrak{D}(G) \to \mathfrak{D}(G)$ defined by

$$N := \sum_{i=1}^n a_i X_i + \sum_{i,j=1}^n a_{ij} X_i X_j$$

is a linear operator on $\mathfrak{D}(G)$, and the function A on $\mathfrak{D}(G)$ defined by $A(f) := (Nf)(e)$ for all $f \in \mathfrak{D}(G)$ is a linear functional on $\mathfrak{D}(G)$. By Theorem 4.4.16 there exists exactly one $\{e\}$-continuous convolution semigroup $(\mu_t)_{t\in\mathbb{R}_+^*}$ in $\mathcal{M}^1(G)$, whose corresponding generating functional coincides with A, and by Lemma 6.2.6 $(\mu_t)_{t\in\mathbb{R}_+^*}$ is a Gauss semigroup. We shall speak of the *Gauss semigroup with vector $(a_1, \ldots, a_n)$ and matrix $(a_{ij})_{i,j=1,\ldots,n}$*.

In the subsequent presentation absolute continuity or singularity of an $\{e\}$-continuous convolution semigroup $(\mu_t)_{t\in\mathbb{R}_+^*}$ in $\mathcal{M}^1(G)$ will always be understood with respect to a fixed (left) Haar measure $\omega := \omega_G$ on G in the sense that $\mu_t \ll \omega$ for all $t \in \mathbb{R}_+^*$ or $\mu_t \perp \omega$ for all $t \in \mathbb{R}_+^*$ resp.

6.3.1 Theorem. *Let G be a connected Lie group of dimension $n \geq 1$ and $(\mu_t)_{t\in\mathbb{R}_+^*}$ a Gauss semigroup in $\mathcal{M}^1(G)$ with vector $(a_1, \ldots, a_n)$ and matrix $(a_{ij})_{i,j=1,\ldots,n}$.*

If $(a_{ij})_{i,j=1,\ldots,n}$ is positive-definite, then for every $t \in \mathbb{R}_+^*$ there exists a (strictly) positive analytic ω-density f_t of μ_t such that $\mu_t = f_t \cdot \omega$.

Proof. 1. Let $\{X_1, \ldots, X_n\}$ be a basis of $\mathfrak{L}(G)$. Then the operator

$$N := \sum_{i=1}^n a_i X_i + \sum_{i,j=1}^n a_{ij} X_i X_j$$

on $\mathfrak{D}(G)$ is a left invariant elliptic operator on G with analytic coefficients. In fact, the left invariance of N is obvious. By [213], p. 94 there exist a canonical coordinate system $\{x_1, \ldots, x_n\}$ with respect to $\{X_1, \ldots, X_n\}$ and a neighborhood $V \in \mathfrak{V}(e)$ such that on V the functions x_i are analytic and such that one has

$$X_i = \sum_{j=1}^n (X_i x_j) \frac{\partial}{\partial x_j} \quad \text{and}$$

$$X_i X_j = \sum_{k,l=1}^n (X_i x_k)(X_j x_l) \frac{\partial^2}{\partial x_k \partial x_l}$$

$$+ \sum_{k=1}^n (X_i X_j x_k) \frac{\partial}{\partial x_k} \qquad (i, j = 1, \ldots, n).$$

Thus on V the operator N admits the representation

$$N = \sum_{k=1}^n b_k \frac{\partial}{\partial x_k} + \sum_{k,l=1}^n b_{kl} \frac{\partial^2}{\partial x_k \partial x_l} \quad \text{with}$$

$$b_k := \sum_{i=1}^n a_i (X_i x_k) + \sum_{i,j=1}^n a_{ij} (X_i X_j x_k),$$

$$b_{kl} := \sum_{i,j=1}^n a_{ij} (X_i x_k)(X_j x_l) \qquad (k, l = 1, \ldots, n).$$

Clearly, the functions b_{kl}, b_k are analytic on V and satisfy $b_{kl}(e) = a_{kl}$ for $k, l = 1, \ldots, n$. Hence, the matrix $(b_{kl}(x))_{k,l=1,\ldots,n}$ is positive-definite for all x in some $W \in \mathfrak{V}(e)$. This implies the ellipticity of N.

2. By the discussion in [268], p. 78 the parabolic differential operator

$$L := N - \frac{\partial}{\partial t}$$

on $\mathbb{R} \times G$ admits exactly one fundamental solution in the following sense: There exists exactly one function $q > 0$ on

$$D := \{(t, x, s, y) \in \mathbb{R} \times G \times \mathbb{R} \times G : s < t\}$$

with $\int q(t, x, s, y)\, \omega(dy) = 1$ for all $x \in G$ and $s, t \in \mathbb{R}$, $s < t$, such that for every function $f \in \mathscr{C}_u(G)$ and for all $s \in \mathbb{R}$ one has the following properties: If

$$\bar{f}(t, x) := \int q(t, x, s, y) f(y)\, \omega(dy) \quad \text{for } t \in \mathbb{R},\ t > s,\ x \in G, \text{ then}$$

(α) $t \to \bar{f}(t, x)$ is once and $x \to \bar{f}(t, x)$ is twice continuously differentiable, and $L\bar{f} \equiv 0$;

(β) $\lim_{t \downarrow s} \bar{f}(t, \cdot) = f$;

(γ) $\bar{f}$ and $\frac{\partial}{\partial t} \bar{f}$ are bounded on $]s_1, t_1[\times G$ for all $s_1, t_1 \in \mathbb{R}$ with $s < s_1 < t_1$.

3. Defining for $a \in G$ a function $v := v_a$ on D by

$$v(t, x, s, y) := q(t, a x, s, a y)$$

for all $(t, x, s, y) \in D$ and defining for every $f \in \mathscr{C}_u(G)$

$$g(t, x) := \int v(t, x, s, y) f(y) \, \omega(dy)$$

for all $t \in \mathbb{R}$, $t > s$ and $x \in G$, one obtains

$$g(t, x) = \overline{({}_{a^{-1}}f)}(t, ax)$$

for all $t \in \mathbb{R}$, $t > s$ and $x \in G$, so that hence g satisfies the conditions (α), (β) and (γ) of Part 2. Since the fundamental solution of the operator L is uniquely determined, the solution v coincides with q and therefore $q(t, x, s, y)$ only depends on $x^{-1} y$. Analogously one shows that $q(t, x, s, y)$ depends only on $t - s$. We thus define

$$p(t, y) := q(t, e, 0, y) \quad \text{for all } t \in \mathbb{R}_+^* \text{ and } y \in G.$$

4. By Part 3 of the proof we conclude from

$$q(t, x, s, y) = \int q(t, x, r, z) \, q(r, z, s, y) \, \omega(dz) \quad \text{for all } s < r < t,$$

that for all $s, t \in \mathbb{R}_+^*$ the equality $p(s, \cdot) * p(t, \cdot) = p(s + t, \cdot)$ holds. Again by [268], p. 78 the functions $t \to p(t, \cdot)$ and $y \to p(\cdot, y)$ are once and twice continuously differentiable resp., and one has $Lp \equiv 0$. As a solution of a parabolic differential equation with analytic coefficients the function p is itself analytic. This can be seen from [152], Theorem 1.

5. By Part 2 we have $p \geq 0$. We now show that $p(t, x) > 0$ for all $t \in \mathbb{R}_+^*$ and $x \in G$.

Assuming the existence of $t \in \mathbb{R}_+^*$ and $x \in G$ with $p(t, x) = 0$ we conclude from Part 4 with the choice $s := \frac{t}{2}$ the equality

$$p(t, x) = \int p(s, y^{-1} x) \, p(s, y) \, \omega(dy)$$

and thus that $p(s, y^{-1} x) \, p(s, y) = 0$ for all $y \in G$, since p is continuous.

From $\int p(s, y) \, \omega_G(dy) = 1$ one gets $p(s, \cdot) \not\equiv 0$ and hence the existence of $y_0 \in G$ and $U \in \mathfrak{V}(y_0)$ with $p(s, y) > 0$ for all $y \in U$.

Consequently, $p(s, y^{-1} x) = 0$ for all $y \in U$. But this contradicts the hypothesis, as G is connected and so $p(s, \cdot)$ admits only isolated zeros.

6. For every $t \in \mathbb{R}_+^*$ we define $f_t := p(t, \cdot)$ and $v_t := f_t \cdot \omega$.

Parts 2 and 4 of this proof imply that $(v_t)_{t \in \mathbb{R}_+^*}$ is an $\{e\}$-continuous convolution semigroup in $\mathscr{M}^1(G)$, and that $\int {}_x f \, dv_t = \bar{f}(t, x)$ holds for all $f \in \mathfrak{D}(G)$, $x \in G$.

By Part 2 we obtain $\frac{\partial}{\partial t} \bar{f}(t, e) = N \bar{f}(t, e)$ for all $t \in \mathbb{R}_+^*$. If B denotes the generating functional of $(v_t)_{t \in \mathbb{R}_+^*}$, then an easy computation yields

$$\frac{\partial}{\partial t} \bar{f}(t, e) = \int B({}_y f) \, v_t(dy) \quad \text{and}$$

$$N \bar{f}(t, e) = \int N(f_y)(e) \, v_t(dy), \quad \text{so that}$$

$$\int B({}_y f) \, v_t(dy) = \int N(f_y)(e) \, v_t(dy)$$

for all $t \in \mathbb{R}_+^*$. Since $\lim_{t \downarrow 0} \nu_t = \varepsilon_e$, we conclude that

$$B(f) = (Nf)(e) = A(f) \quad \text{for all } f \in \mathfrak{D}(G)$$

and thus $\mu_t = \nu_t = f_t \cdot \omega_G$ for all $t \in \mathbb{R}_+^*$ by Theorem 4.4.16. For every $t \in \mathbb{R}_+^*$ the ω-density f_t is analytic by Part 4 and positive by Part 5. □

6.3.2 Corollary. *For every Gauss semigroup $(\mu_t)_{t \in \mathbb{R}_+^*}$ in $\mathscr{M}^1(G)$ with vector $(a_1, \ldots, a_n)$ and matrix $(a_{ij})_{i,j=1,\ldots,n}$ one has* $\operatorname{supp}(\mu_t) = G$ *whenever* $t \in \mathbb{R}_+^*$.

The *proof* is clear. □

6.3.3 Corollary. *Let $(\mu_t)_{t \in \mathbb{R}_+^*}$ be a Gauss semigroup in $\mathscr{M}^1(G)$ with vector* $(0, \ldots, 0)$ *and matrix $(a_{ij})_{i,j=1,\ldots,n}$ such that $\mu_t = f_t \cdot \omega$ holds with f_t as in the theorem for every $t \in \mathbb{R}_+^*$. Then*

(i) *$f_t^* = \Delta \cdot f_t$ for all $t \in \mathbb{R}_+^*$,*

(ii) *$g_t := \Delta^{\frac{1}{2}} \cdot f_t$ is symmetric, ω-square integrable, positive-definite, positive and analytic for all $t \in \mathbb{R}_+^*$ and*

(iii) *$g_s * g_t = g_{s+t}$ for all $s, t \in \mathbb{R}_+^*$.*

If, in particular, G is unimodular, then f_t is bounded for every $t \in \mathbb{R}_+^*$.

Proof. By assumption, one has $Nf^* = Nf$ for all $f \in \mathfrak{D}(G)$. Hence, A is also the generating functional of the convolution semigroup $(\tilde{\mu_t})_{t \in \mathbb{R}_+^*}$ or $\tilde{\mu_t} = \mu_t$ for all $t \in \mathbb{R}_+^*$. From $\mu_t = f_t \cdot \omega$ follows $\tilde{\mu_t} = \Delta^{-1} f_t^* \cdot \omega$, hence $f_t^* = \Delta \cdot f_t$ for all $t \in \mathbb{R}_+^*$, i.e., (i). From $\Delta^* = \Delta^{-1}$ we deduce that $g_t^* = g_t$; the formula

$$f_{2t}(e) = \int (\Delta^{\frac{1}{2}} \cdot f_t)^2 \, d\omega$$

implies that g_t is square integrable (for all $t \in \mathbb{R}_+^*$). Moreover, an easy computation yields (iii). As an ω-square integrable function satisfying $g_{2t} = g_t * g_t^*$ the function g_t is positive-definite for all $t \in \mathbb{R}_+^*$, as follows from [218], (32.43). $\Delta > 0$ and $f_t > 0$ imply $g_t > 0$ for all $t \in \mathbb{R}_+^*$. The continuous homomorphism $\Delta^{\frac{1}{2}}$ between the Lie groups G and $\mathbb{R}_+^*$ is analytic so that g_t is analytic for every $t \in \mathbb{R}_+^*$, which implies (ii).

If G is unimodular, i.e., $\Delta = 1$, then for every $t \in \mathbb{R}_+^*$ the density f_t is positive-definite, and hence bounded. □

The preceding theorem contains a sufficient condition for the absolute continuity of a Gauss semigroup on a Lie group. In order to obtain a necessary and sufficient condition we need deeper analytic tools.

6.3.4 Lemma. *Let G be an arbitrary locally compact group and μ a measure in $\mathscr{M}^1(G)$. The following conditions are equivalent:*

(i) *For every compact subset K in G with $\omega(K) = 0$ the function $x \to \mu(xK)$ on G is continuous in e;*

(ii) $\mu \ll \omega$;

(iii) *for every $B \in \mathfrak{B}(G)$ the function $x \to \mu(xB)$ is right uniformly continuous on G.*

Proof. 1. The implication (iii)$\Rightarrow$(i) is clear.

2. (i)$\Rightarrow$(ii). By assumption the function $x \to \mu(x^{-1}K)$ is continuous in e for every compact $K \subset G$ with $\omega(K)=0$. Let U be an open neighborhood in $\mathfrak{V}(e)$ with $\omega(U)<\infty$. Then $\nu := 1_U \cdot \omega \in \mathcal{M}_+^b(G)$ and $\nu \ll \omega$. But this implies $\nu * \mu \ll \omega$ and hence

$$\begin{aligned} 0 &= \nu * \mu(K) = \int_G \mu(x^{-1}K)\,\nu(dx) \\ &= \int_U \mu(x^{-1}K)\,\omega(dx) \quad \text{or} \quad \mu(x^{-1}K)=0 \end{aligned}$$

for ω-a.a. $x \in U$. Therefore $\mu(e^{-1}K)=\mu(K)=0$, and the regularity of the measures ω and μ yields the assertion.

3. (ii) $\Rightarrow$ (iii). We have $\mu = f \cdot \omega$ with $f \in \mathcal{L}^1(G,\omega)$ or $\mu(xB)=\int_B {}_xf\,d\omega$ for all $x \in G$, $B \in \mathfrak{B}(G)$. But this implies that

$$\begin{aligned} |\mu(xB)-\mu(yB)| &\le \int_B |{}_xf - {}_yf|\,d\omega \\ &\le \|{}_xf - {}_yf\|_1 \qquad \text{for all } B \in \mathfrak{B}(G),\ x, y \in G. \end{aligned}$$

Since by [218], (20.4) the mapping $z \to {}_zf$ from G into $\mathcal{L}^1(G,\omega)$ is right uniformly continuous, the assertion follows. □

6.3.5 Lemma. *Let G be a locally compact group, let μ be in $\mathcal{M}^1(G)$ and let $T := T_\mu$ be the corresponding probability operator defined on $\mathscr{C}^0(G)$. The following statements are equivalent:*

(i) $\mu \ll \omega$;

(ii) *T is a compact operator from $\mathscr{C}^0(G)$ into $(\mathscr{C}^0(G), \mathcal{T}_{co})$.*

Proof. 1. (i) $\Rightarrow$ (ii). By hypothesis $\mu = g \cdot \omega$ with $g \in \mathcal{L}^1(G,\omega)$. Now let f be in $\mathscr{C}^0(G)$. Then $|Tf(x)| \le \|f\|$ and

$$\begin{aligned} &|Tf(x)-Tf(y)| \\ &= \left|\int f(xz)\,g(z)\,\omega(dz) - \int f(yz)\,g(z)\,\omega(dz)\right| \\ &= \left|\int f(z)\,g(x^{-1}z)\,\omega(dz) - \int f(z)\,g(y^{-1}z)\,\omega(dz)\right| \\ &\le \|f\|\,\|{}_{x^{-1}}g - {}_{y^{-1}}g\|_1 \qquad \text{for all } x, y \in G. \end{aligned}$$

The Arzela-Ascoli theorem implies the compactness of T.

2. (ii) $\Rightarrow$ (i). Let K be a compact subset of G with $\omega(K)=0$. There exists a net $(f_\alpha)_{\alpha \in \mathbb{A}}$ in $\mathcal{K}(G)$ satisfying $f_\alpha \le 1_G$ for all $\alpha \in \mathbb{A}$ and $\inf_{\alpha \in \mathbb{A}} f_\alpha = 1_K$. But this yields $\inf_{\alpha \in \mathbb{A}} Tf_\alpha = T1_K$. Since the set $\{f_\alpha : \alpha \in \mathbb{A}\}$ is bounded and T is a compact operator, there is a $\mathcal{T}_{co}$-convergent subnet of $(Tf_\alpha)_{\alpha \in \mathbb{A}}$. Consequently, $T1_K$ is continuous. But $T1_K(x)=\mu(x^{-1}K)$ for all $x \in G$. Thus Lemma 6.3.4 yields the result. □

6.3.6 Theorem. *Let G be a Lie group (of dimension $n \ge 1$), $\mathscr{E}_1(G)$ the space of all functions of $\mathscr{C}^0(G)$ which are continuously differentiable in a neighborhood of $e \in G$ and μ a measure in $\mathcal{M}^1(G)$ with probability operator $T := T_\mu$. If $\mathscr{C}^0(G) \subset \mathscr{E}_1(G)$, then $\mu \ll \omega$.*

Proof. Let $\mathfrak{L}(G)$ be the Lie algebra of G with basis $\{X_1, \ldots, X_n\}$. Clearly the set

$$B := \{\sum_{i=1}^n \alpha_i X_i : \alpha_i \in \mathbb{R} \ (i=1,\ldots,n), \ \sum_{i=1}^n \alpha_i^2 \leq 1\}$$

is a compact neighborhood of 0 in $\mathfrak{L}(G)$. Hence, $U := \exp B$ is a compact neighborhood in $\mathfrak{B}(e)$. For $Y \in B$ and $t \in \mathbb{R}$ we put $\zeta_Y(t) := \exp tY$. Now let $x \in G$ be fixed. For every $m \geq 1$ we introduce

$$W_m(x) := \{f \in \mathscr{C}^0(G) : |Tf(x\zeta_Y(t)) - Tf(x)| \leq mt \text{ for all } Y \in B \text{ and all } t \in [0,1]\}.$$

We first show that for any given $f \in \mathscr{C}^0(G)$ there exists an $m \geq 1$ with $f \in W_m(x)$. In fact, let

$$m(f) := \sup_{y \in xU} [\sum_{i=1}^n |\tilde{X}_i(Tf)(y)|^2]^{\frac{1}{2}}.$$

By assumption, $m(f) < \infty$. For $Y := \sum_{i=1}^n \alpha_i X_i \in B$ we have

$$|\tilde{Y}(Tf)(y)| = |\sum_{i=1}^n \alpha_i \tilde{X}_i(Tf)(y)| \leq \sum_{i=1}^n |\alpha_i| \, |\tilde{X}_i(Tf)(y)|$$
$$\leq [\sum_{i=1}^n \alpha_i^2]^{\frac{1}{2}} [\sum_{i=1}^n |\tilde{X}_i(Tf)(y)|^2]^{\frac{1}{2}} \leq m(f)$$

whenever $y \in xU$. Hence,

$$|Tf(x\zeta_Y(t)) - Tf(x)| \leq t \sup_{0 \leq s_0 \leq t} \left| \frac{d}{ds}\right|_{s=s_0} Tf(x\zeta_Y(s))|$$
$$= t \sup_{0 \leq s_0 \leq t} |\tilde{Y}(Tf)(x\zeta_Y(s_0))| \leq t \sup_{y \in xU} |\tilde{Y}(Tf)(y)| \leq t\, m(f)$$

for all $Y \in B$ and $t \in [0,1]$. This proves the above statement. We have thus obtained $\bigcup_{m \geq 1} W_m(x) = \mathscr{C}^0(G)$. Moreover, $W_m(x)$ is a closed subset of $\mathscr{C}^0(G)$ for all $m \geq 1$. Since $\mathscr{C}^0(G)$ is a complete metric space, there exists an $m_0 \geq 1$ with $\overset{\circ}{\widehat{W_{m_0}(x)}} \neq \emptyset$, i.e., there exist $f_0 \in W_{m_0}(x)$ and $\varepsilon > 0$ such that for all $f \in \mathscr{C}^0(G)$ satisfying $\|f - f_0\| \leq \varepsilon$ one has $f \in W_{m_0}(x)$. Consequently,

$$|Tf(x\zeta_Y(t)) - Tf(x)| \leq \frac{2m_0 t}{\varepsilon} \|f\|$$

holds for all $f \in \mathscr{C}^0(G)$, $Y \in B$ and $t \in [0,1]$. Since $\{\exp tB : t \in]0,1]\}$ is a basis of $\mathfrak{B}(e)$, for every bounded subset $\mathscr{A}$ of $\mathscr{C}^0(G)$ the set $\{Tf : f \in \mathscr{A}\}$ is equicontinuous and bounded. The Arzela-Ascoli theorem yields the compactness of T and Lemma 6.3.5 the assertion. ☐

6.3.7 Preparations from the theory of partial differential operators. Let Ω be a nonempty open subset of $\mathbb{R}^m$ $(m \geq 1)$, $\mathscr{C}^\infty(\Omega)$ the space of all infinitely often continuously differentiable functions on Ω and

$$\mathfrak{D}(\Omega) := \mathscr{K}(\Omega) \cap \mathscr{C}^\infty(\Omega).$$

The space of distributions on Ω will be denoted by $\mathfrak{D}'(\Omega)$. Finally, we shall work with the space $L^1_{loc}(\Omega)$ of all locally λ^m-integrable (classes of) functions on Ω. The following examples of distributions on Ω will be of interest for the sequel:

(1) Given $u \in L^1_{loc}(\Omega)$ we define $\tilde{u}$ on $\mathfrak{D}(\Omega)$ by

$$\tilde{u}(f) := \int f u \, dx \quad (= \int f u \, d\lambda^m)$$

for all $f \in \mathfrak{D}(\Omega)$ and obtain $\tilde{u} \in \mathfrak{D}'(\Omega)$. In general, $\tilde{u}$ will also be denoted by u signifying that $L^1_{loc}(\Omega)$ can be viewed as a subspace of $\mathfrak{D}'(\Omega)$.

(2) For $u \in \mathfrak{D}'(\Omega)$ and $j = 1, \ldots, m$ one defines a linear functional $\frac{\partial}{\partial x_j} u$ on $\mathfrak{D}(\Omega)$ by

$$\left(\frac{\partial}{\partial x_j} u\right)(f) := -u\left(\frac{\partial}{\partial x_j} f\right)$$

for all $f \in \mathfrak{D}(\Omega)$ and observes that $\frac{\partial}{\partial x_j} u$ is an element of $\mathfrak{D}'(\Omega)$.

(3) Let $u \in \mathfrak{D}'(\Omega)$ and $g \in \mathscr{C}^\infty(\Omega)$. Then $(g u)(f) := u(f g)$ for all $f \in \mathfrak{D}(\Omega)$ defines a distribution $g u \in \mathfrak{D}'(\Omega)$.

(4) If Ω_1 is a nonempty open subset of Ω and $u \in \mathfrak{D}'(\Omega)$, then

$$\mathrm{Res}_{\Omega_1} u := \mathrm{Res}_{\mathfrak{D}(\Omega_1)} u \in \mathfrak{D}'(\Omega_1).$$

For $u \in \mathfrak{D}'(\Omega)$ one defines the *singular support* $\operatorname{sing\,supp}(u)$ of u as the set $\{x \in \Omega$: There exists no $U \in \mathfrak{B}(x)$ such that $\mathrm{Res}_U u \in \mathscr{C}^\infty(U)\}$. Clearly, $\operatorname{sing\,supp}(u) \subset \operatorname{supp}(u)$, and $\operatorname{sing\,supp}(u)$ is a closed subset of Ω.

This and a few more facts about distributions and partial differential equations can be found in Chapter I of [247].

Now let P be a linear differential operator with $\mathscr{C}^\infty$-coefficients on Ω. Then Pu is defined for all $u \in \mathfrak{D}'(\Omega)$ and $Pu \in \mathfrak{D}'(\Omega)$. A linear differential operator P on Ω with $\mathscr{C}^\infty$-coefficients is called *hypoelliptic*, if for every $u \in \mathfrak{D}'(\Omega)$ one has

$$\operatorname{sing\,supp}(u) = \operatorname{sing\,supp}(Pu).$$

Obviously every elliptic differential operator is hypoelliptic. The following very useful result on hypoelliptic differential operators (Hörmander's *hypoellipticity criterion*) is proved in [248]. Let $X_0, X_1, \ldots, X_r$ be homogeneous differential operators of order 1 on Ω with $\mathscr{C}^\infty$-coefficients and $c \in \mathscr{C}^\infty(\Omega)$. We introduce the differential operator

$$P := \sum_{i=1}^r X_i^2 + X_0 + c$$

on Ω and assume that among the operators

$$X_{j_1}, [X_{j_1}, X_{j_2}], [X_{j_1}, [X_{j_2}, X_{j_3}]], \ldots,$$
$$[X_{j_1}, [X_{j_2}, [X_{j_3}, [\ldots, [X_{j_{k-1}}, X_{j_k}] \ldots]]]]$$

for all $j_1, j_2, \ldots, j_k \in \{0, 1, \ldots, r\}$ there are m operators which are linearly independent at every point of Ω. Then P is hypoelliptic.

6.3.8 Theorem. *Let G be a Lie group of dimension $n \geq 1$ with Lie algebra $\mathfrak{L}(G)$ admitting a basis $\{X_1, \ldots, X_n\}$ and let $(\mu_t)_{t \in \mathbb{R}_+^*}$ be a symmetric Gauss semigroup in*

$\mathscr{M}^1(G)$ with infinitesimal generator N, which by the choice of the basis of $\mathfrak{L}(G)$ is of the form $N := \sum_{i=1}^r X_i^2$ with $r \leq n$.

The Lie algebra generated by $\{X_1, \ldots, X_r\}$ will be denoted by $\mathfrak{G}$. Then
(i) *$(\mu_t)_{t\in\mathbb{R}_+^*}$ is either absolutely continuous or singular and*
(ii) *$(\mu_t)_{t\in\mathbb{R}_+^*}$ is absolutely continuous iff $\mathfrak{G} = \mathfrak{L}(G)$.*

Proof. Both assertions of the theorem will have been proved when we have established that
(α) if $\mathfrak{G} = \mathfrak{L}(G)$, then $(\mu_t)_{t\in\mathbb{R}_+^*}$ is absolutely continuous and
(β) if $\mathfrak{G} \neq \mathfrak{L}(G)$, then $(\mu_t)_{t\in\mathbb{R}_+^*}$ is singular.

1. We consider the differential operator $P := \frac{\partial}{\partial t} - N$ on $\mathbb{R}_+^* \times G$. Let $(T_t)_{t\in\mathbb{R}_+^*}$ be the contraction semigroup on $\mathscr{C}^0(G)$ corresponding to $(\mu_t)_{t\in\mathbb{R}_+^*}$. For a fixed $f \in \mathscr{C}^0(G)$ we define a function u on $\mathbb{R}_+^* \times G$ by

$$u(t, x) := (T_t f)(x) \qquad \text{for all } t \in \mathbb{R}_+^*,\ x \in G.$$

Plainly, $u \in \mathscr{C}(\mathbb{R}_+^* \times G)$. Furthermore, let Ω be a chart of G and $\tilde{u}$ the distribution on $\mathbb{R}_+^* \times \Omega$ defined by

$$\tilde{u}(h) := \iint u(t, x)\, h(t, x)\, dt\, dx \qquad \text{for all } h \in \mathfrak{D}(\mathbb{R}_+^* \times \Omega).$$

We shall show that $P\tilde{u} = 0$ holds, i.e., that $\tilde{u}$ is a weak (distribution) solution of the differential equation $Pu = 0$. In fact, let $h \in \mathfrak{D}(\mathbb{R}_+^* \times \Omega)$ be of the form $h := h_1 \times h_2$ with $h_1 \in \mathfrak{D}(\mathbb{R}_+^*)$ and $h_2 \in \mathfrak{D}(\Omega)$. First of all we deduce from the symmetry of $(\mu_t)_{t\in\mathbb{R}_+^*}$ the equations

$$\begin{aligned}(N\tilde{u})(h) &= \int \left[\int u(t, x)(N h_2)(x)\, dx\right] h_1(t)\, dt \\ &= \int \left[\int f(x)(T_t N)\, h_2(x)\, dx\right] h_1(t)\, dt \\ &= \int \left[\int f(x)\, N(T_t h_2)(x)\, dx\right] h_1(t)\, dt.\end{aligned}$$

On the other hand, we get

$$\begin{aligned}\left(\frac{\partial}{\partial t}\tilde{u}\right)(h) &= -\int \left[\int u(t, x) \frac{\partial}{\partial t} h_1(t)\, dt\right] h_2(x)\, dx \\ &= -\int \left[\int (T_t f)(x)\, h_2(x)\, dx\right] \frac{\partial}{\partial t} h_1(t)\, dt \\ &= -\int \left[\int f(x)(T_t h_2)(x)\, dx\right] \frac{\partial}{\partial t} h_1(t)\, dt \\ &= -\int \left\{\int f(x) \frac{\partial}{\partial t} [(T_t h_2)(x)\, h_1(t)]\, dx\right\} dt \\ &\quad + \int \left\{\int f(x) \frac{\partial}{\partial t} [(T_t h_2)(x)]\, dx\right\} h_1(t)\, dt.\end{aligned}$$

But there exist $a, b \in \mathbb{R}_+^*$ with $\operatorname{supp}(h_1) \subset]a, b[$, whence

$$\int_0^\infty \frac{\partial}{\partial t}[(T_t h_2)(x) h_1(t)]\,dt = \int_a^b \frac{\partial}{\partial t}[(T_t h_2)(x) h_1(t)]\,dt$$
$$= (T_b h_2)(x) h_1(b) - (T_a h_2)(x) h_1(a) = 0$$

or

$$\int \left\{ \int f(x) \frac{\partial}{\partial t}[(T_t h_2)(x) h_1(t)]\,dx \right\} dt$$
$$= \int \left\{ \int_0^\infty \frac{\partial}{\partial t}[(T_t h_2)(x) h_1(t)]\,dt \right\} f(x)\,dx = 0.$$

This implies that $\left(\frac{\partial}{\partial t}\tilde{u}\right)(h) = \int \left\{ \int f(x) \frac{\partial}{\partial t}[(T_t h_2)(x)]\,dx \right\} h_1(t)\,dt$. But now we have $P(T_t h_2(x)) = 0$ since $h_2 \in \mathfrak{D}(G)$. Thus, the expressions for $(N\tilde{u})(h)$ and $\left(\frac{\partial}{\partial t}\tilde{u}\right)(h)$ just computed yield

$$\left[\left(\frac{\partial}{\partial t} - N\right)\tilde{u}\right](h_1 \times h_2) = (P\tilde{u})(h_1 \times h_2) = 0.$$

As $\mathfrak{D}(\mathbb{R}_+^*) \otimes \mathfrak{D}(\Omega)$ is dense in $\mathfrak{D}(\mathbb{R}_+^* \times \Omega)$ and $P\tilde{u}$ is continuous on $\mathfrak{D}(\mathbb{R}_+^* \times \Omega)$, i.e., in $\mathfrak{D}'(\mathbb{R}_+^* \times \Omega)$, we obtain $P\tilde{u} = 0$.

2. Let us now assume that $\mathfrak{G} = \mathfrak{L}(G)$. Applying Hörmander's hypoellipticity criterion (quoted above) to the differential operator

$$-P = N + X_0 \quad \text{with } X_0 = -\frac{\partial}{\partial t}$$

we deduce from $\left[\frac{\partial}{\partial t}, X\right] = 0$ for all $X \in \mathfrak{L}(G)$ that P is hypoelliptic. Since $\operatorname{sing\,supp}(0) = \emptyset$, one gets

$$\operatorname{sing\,supp}(\tilde{u}) = \operatorname{sing\,supp}(P\tilde{u}) = \operatorname{sing\,supp}(0) = \emptyset,$$

and so $\tilde{u}$ is a function in $\mathscr{C}^\infty(G)$. In particular, $T_t f$ is in $\mathscr{C}^\infty(G)$ and hence is in $\mathscr{E}_1(G)$ for all $t \in \mathbb{R}_+^*$. But then Theorem 6.3.6 implies that $\mu_t \ll \omega$ for all $t \in \mathbb{R}_+^*$, whence the absolute continuity of the semigroup $(\mu_t)_{t \in \mathbb{R}_+^*}$.

3. For the Lie subalgebra $\mathfrak{G}$ of $\mathfrak{L}(G)$ there exist a connected Lie group H and a continuous monomorphism m from H into G. Clearly, $m(H)$ is a Borel subset of G. Let N' be the given infinitesimal operator N viewed as an operator on $\mathfrak{D}(H)$ and let $(\mu_t')_{t \in \mathbb{R}_+^*}$ be the Gauss semigroup in $\mathscr{M}^1(H)$ corresponding to N'. Then $m(\mu_t') = \mu_t$ and $\mu_t' \ll \omega_H$ for all $t \in \mathbb{R}_+^*$. If now $\mathfrak{G} \neq \mathfrak{L}(G)$, then $m(H) \neq G$ and hence $\omega(m(H)) = 0$, which shows that $\mu_t \perp \omega$ for all $t \in \mathbb{R}_+^*$, and the semigroup $(\mu_t)_{t \in \mathbb{R}_+^*}$ is singular. □

6.3.9 Remark. Let us retain the assumptions of the preceding theorem and denote by $\mathfrak{G}(N)$ the smallest subalgebra of $\mathfrak{L}(G)$ whose universal enveloping algebra contains the operator N. Then $(\mu_t)_{t \in \mathbb{R}_+^*}$ is absolutely continuous iff

$\mathfrak{G}(N)=\mathfrak{L}(G)$. In order to show this we prove that $\mathfrak{G}(N)$ coincides with the smallest subalgebra $\mathfrak{G}_1(N)$ of $\mathfrak{L}(G)$ which contains the set $\{X_1, \ldots, X_r\}$.

Plainly, $\mathfrak{G}(N) \subset \mathfrak{G}_1(N)$. We note that for any Lie subgroup H of G the universal enveloping algebra of $\mathfrak{L}(H)$ is isomorphic to the algebra $D(H)$ of left invariant differential operators on H. If $\{Y_1, \ldots, X_d\}$ is a basis of $\mathfrak{L}(H)$, then $\{Y_1^{l_1} \cdot \ldots \cdot Y_d^{l_d} : l_1, \ldots, l_d \in \mathbb{Z}_+\}$ is a basis of $D(H)$.

Now, let $\{Y_1, \ldots, Y_d\}$ be a basis of $\mathfrak{G}(N)$. Then $N=\sum_{i,j=1}^d a_{ij} Y_i Y_j$ with a symmetric, positive-semidefinite matrix $(a_{ij})_{i,j=1,\ldots,d} \neq 0$ in $\mathfrak{M}(d, \mathbb{R})$. Let H and H_1 be the connected Lie groups corresponding to $\mathfrak{G}(N)$ and $\mathfrak{G}_1(N)$ resp. and let m be a continuous monomorphism from H into H_1. Then $\operatorname{supp}(m(\mu_t)) \subset m(H)$ for all $t \in \mathbb{R}_+^*$. But since $m(\mu_t) \ll \omega_{H_1}$ for all $t \in \mathbb{R}_+^*$, we obtain $m(H)=H_1$. This implies that H is isomorphic to H_1 and so $\mathfrak{G}(N)=\mathfrak{G}_1(N)$.

Let G be a connected locally compact group and put

$$\mathfrak{B}(G):=\{f \in \mathfrak{D}_0(G): \text{There are no } g \in \mathfrak{D}_0(G) \text{ and } U \in \mathfrak{V}(e) \text{ such that } f(x) \leq g(x)^2 \text{ for all } x \in U\}.$$

Plainly $\mathfrak{B}(G)$ is a convex cone in $\mathfrak{D}(G)$.

6.3.10 Definition. A quadratic form ψ on $\mathfrak{D}(G)$ is called *strict* if $\psi(f)>0$ for all $f \in \mathfrak{B}(G)$.

6.3.11 Properties. Let G be a Lie group of dimension $n \geq 1$, $\{X_1, \ldots, X_n\}$ a basis of its Lie algebra $\mathfrak{L}(G)$ and $\{x_1, \ldots, x_n\}$ a system of canonical coordinates in $\mathfrak{D}(G)$ with respect to $\{X_1, \ldots, X_n\}$.

(1) For every $f \in \mathfrak{D}_0(G)$ the following statements are equivalent:

(i) $f \in \mathfrak{B}(G)$;

(ii) $\alpha:=\sum_{i=1}^n (X_i X_i f)(e)>0$;

(iii) $M:=\left(\dfrac{\partial^2}{\partial x_i \partial x_j} f(e)\right)_{i,j=1,\ldots,n} \neq 0$.

[We observed in the proof of Theorem 6.3.1 that for each $f \in \mathfrak{D}_0(G)$ and $i,j=1, \ldots, n$ one has

$$(X_i X_j f)(e)=\frac{\partial^2}{\partial x_i \partial x_j} f(e).$$

1. (i)$\Rightarrow$(iii). Let $M=0$. Then there exists a $U \in \mathfrak{V}(e)$ such that f admits for all $x \in U$ a Taylor expansion

$$f(x)=\sum_{i,j,k,l=1}^n x_i(x) x_j(x) x_k(x) x_l(x)(X_i X_j X_k X_l f)(\zeta(x))$$

with $\zeta(x) \in U$. It follows for all $x \in U$ that

$$f(x) \leq c\left(\sum_{1 \leq i \leq j \leq n}(x_i^2+x_j^2)\right)^2 \quad \text{with } c \in \mathbb{R}_+^*.$$

Defining

$$g:=c^{\frac{1}{2}} \sum_{1 \leq i \leq j \leq n}(x_i^2+x_j^2) \in \mathfrak{D}_0(G)$$

one concludes that $f \notin \mathfrak{B}(G)$.

2. (iii)$\Rightarrow$(ii). Since f has a minimum at e, M is symmetric and positive-semidefinite. From $M \neq 0$ it thus follows that $\alpha = \operatorname{tr} M > 0$.

3. (ii)$\Rightarrow$(i). By setting $\psi(h) := \sum_{i=1}^{n} (X_i X_i h)(e)$ for all $h \in \mathfrak{D}(G)$ one defines a quadratic form on $\mathfrak{D}(G)$, as follows from Example 2 of 4.4.8. Let $f \notin \mathfrak{B}(G)$. Then there are a function $g \in \mathfrak{D}_0(G)$ with $g = g^*$ and a $U \in \mathfrak{V}(e)$ such that $f(x) \leq g(x)^2$ holds for all $x \in U$. But

$$0 \leq \psi(f) \leq \psi(g^2) = 2 g(e) \psi(g) = 0$$

implies that $\alpha = \psi(f) = 0$.]

(2) For every quadratic form ψ on $\mathfrak{D}(G)$ of the form

$$\psi(f) := \sum_{i,j=1}^{n} a_{ij} (X_i X_j f)(e) \quad \text{for all } f \in \mathfrak{D}(G)$$

the following statements are equivalent:

(i) ψ is strict;

(ii) the matrix $(a_{ij})_{i,j=1,\ldots,n}$ is positive-definite.

[Without loss of generality one may assume that $a_{ij} = a_i \delta_{ij}$ with $a_i \in \mathbb{R}_+$ for all $i, j = 1, \ldots, n$.

1. (i)$\Rightarrow$(ii). Property (1) yields $x_i^2 \in \mathfrak{B}(G)$ and so

$$0 < \psi(x_i^2) = \sum_{j=1}^{n} a_j (X_j X_j x_i^2)(e) = 2 a_i \quad \text{for all } i = 1, \ldots, n.$$

Thus, $(a_{ij})_{i,j=1,\ldots,n}$ is positive-definite.

2. (ii)$\Rightarrow$(i). By Property (1) there exists for $f \in \mathfrak{B}(G)$ an index $i \in \{1, \ldots, n\}$ with $(X_i X_i f)(e) > 0$. Since $a_1, \ldots, a_n \in \mathbb{R}_+^*$, we obtain

$$\psi(f) = \sum_{j=1}^{n} a_j (X_j X_j f)(e) \geq a_i (X_i X_i f)(e) > 0$$

and hence that ψ is strict.]

Let G, H be two Lie projective locally compact groups and p a continuous homomorphism from G into H with compact kernel. For every $f \in \mathfrak{D}(H)$ we have $f \circ p \in \mathfrak{D}(G)$. And for every $L \in \mathfrak{D}(G)^*$ one defines an element $p(L) \in \mathfrak{D}(H)^*$ by $p(L)(f) := L(f \circ p)$ for all $f \in \mathfrak{D}(H)$.

Let G be a locally compact connected group. By the Lie projectivity of G there exists a descending family $(K_\alpha)_{\alpha \in \mathbb{A}}$ of compact normal subgroups of G with $\bigcap_{\alpha \in \mathbb{A}} K_\alpha = \{e\}$ such that $G_\alpha := G/K_\alpha$ is a Lie group for every $\alpha \in \mathbb{A}$.

For $\alpha \in \mathbb{A}$ (or $\alpha, \beta \in \mathbb{A}$ with $K_\beta \subset K_\alpha$) we denote by p_α (or $p_{\alpha\beta}$) the canonical mapping from G onto G_α (or from G_β onto G_α). We know that

$$\mathfrak{D}(G) = \bigcup_{\alpha \in \mathbb{A}} \{f_\alpha \circ p_\alpha : f_\alpha \in \mathfrak{D}(G_\alpha)\}$$

and verify easily that

$$\mathfrak{B}(G) = \bigcup_{\alpha \in \mathbb{A}} \{f_\alpha \circ p_\alpha : f_\alpha \in \mathfrak{B}(G_\alpha)\}.$$

From this there follows directly another property.

(3) For every $\alpha \in \mathbb{A}$ let ψ_α be a quadratic form on $\mathfrak{D}(G_\alpha)$ with the property that for all $\alpha, \beta \in \mathbb{A}$ the inclusion $K_\beta \subset K_\alpha$ implies $p_{\alpha\beta}(\psi_\beta) = \psi_\alpha$. Then

(i) there exists exactly one quadratic form $\psi := \varprojlim_{\alpha \in \mathbb{A}} \psi_\alpha$ on $\mathfrak{D}(G)$ with $p_\alpha(\psi) = \psi_\alpha$ for all $\alpha \in \mathbb{A}$ and
(ii) ψ is strict iff ψ_α is strict for all $\alpha \in \mathbb{A}$.

6.3.12 Lemma. *Let G and $\dot{G}$ be two connected locally compact groups and p a continuous epimorphism from G onto $\dot{G}$ such that $K := \ker p$ is a compact Lie group. If $\dot{\psi}$ is a strict quadratic form on $\mathfrak{D}(\dot{G})$, then there exists a strict quadratic form ψ on $\mathfrak{D}(G)$ such that $p(\psi) = \dot{\psi}$.*

The *proof* is carried out in analogy to that of Theorem 6.2.7. For every $\alpha \in \mathbb{A}$ we denote by p_α the canonical epimorphism from G onto G_α. Since K is a Lie group, we may assume without loss of generality that $K \cap K_\alpha = \{e\}$ for all $\alpha \in \mathbb{A}$. Furthermore, let $\dot{K}_\alpha := p(K_\alpha)$, $\dot{G}_\alpha := \dot{G}/\dot{K}_\alpha$ and let $\dot{p}_\alpha$, f_α be the canonical epimorphisms $\dot{G} \to \dot{G}_\alpha$, $G_\alpha \to \dot{G}_\alpha$ resp. for all $\alpha \in \mathbb{A}$.

Let $\{Y_1, \ldots, Y_r\}$ be a basis of $\mathfrak{L}(K)$. For every $\alpha \in \mathbb{A}$ and every basis $\{\dot{X}_1^\alpha, \ldots, \dot{X}_{s(\alpha)}^\alpha\}$ of $\mathfrak{L}(\dot{G}_\alpha)$ there exists a basis $\{X_1^\alpha, \ldots, X_{k(\alpha)}^\alpha\}$ of $\mathfrak{L}(G_\alpha)$ with

$$df_\alpha(X_j^\alpha) = \dot{X}_j^\alpha \quad \text{for } 1 \le j \le s(\alpha) \text{ and}$$
$$X_j^\alpha = dp_\alpha(Y_{j-s(\alpha)}) \quad \text{for } s(\alpha)+1 \le j \le k(\alpha).$$

For every $\alpha \in \mathbb{A}$ the linear functional $\dot{\psi}_\alpha := \dot{p}_\alpha(\dot{\psi})$ is a strict quadratic form on $\mathfrak{D}(\dot{G}_\alpha)$. Thus, by Lemma 4.4.9 and Property (2) of 6.3.11 there is a real symmetric, positive-definite matrix $(\dot{a}_{ij}^\alpha)_{i,j=1,\ldots,s(\alpha)}$ such that

$$\dot{\psi}_\alpha(\dot{f}) = \sum_{i,j=1}^{s(\alpha)} \dot{a}_{ij}^\alpha (\dot{X}_i^\alpha \dot{X}_j^\alpha \dot{f})(\dot{e}_\alpha) \quad \text{for all } \dot{f} \in \mathfrak{D}(\dot{G}_\alpha).$$

Now let $(a_{ij})_{i,j=1,\ldots,r}$ be a real symmetric, positive-definite matrix. Then the matrix

$$(a_{ij}^\alpha)_{i,j=1,\ldots,k(\alpha)} := (\dot{a}_{ij}^\alpha)_{i,j=1,\ldots,s(\alpha)} \oplus (a_{ij})_{i,j=1,\ldots,r}$$

is again real, symmetric and positive-definite. Defining

$$\psi_\alpha(f) := \sum_{i,j=1}^{k(\alpha)} a_{ij}^\alpha (X_i^\alpha X_j^\alpha f)(e_\alpha) \quad \text{for all } f \in \mathfrak{D}(G_\alpha)$$

one obtains a strict quadratic form ψ_α on $\mathfrak{D}(G_\alpha)$ satisfying

$$f_\alpha(\psi_\alpha) = \dot{\psi}_\alpha \quad \text{for all } \alpha \in \mathbb{A} \text{ and}$$
$$p_{\alpha\beta}(\psi_\beta) = \psi_\alpha \quad \text{for all } \alpha, \beta \in \mathbb{A} \text{ with } \alpha < \beta.$$

It follows that $\psi := \varprojlim_{\alpha \in \mathbb{A}} \psi_\alpha$ is a strict quadratic form on $\mathfrak{D}(G)$ with $p(\psi) = \dot{\psi}$. □

6.3.13 Theorem. *Let G be a connected locally compact group. Then there exists a strict quadratic form on $\mathfrak{D}(G)$.*

Proof. We proceed in analogy to the proof of Theorem 6.2.8. Let $\mathbf{G}$ be the system of all pairs $(G/K, \psi_K)$, where K is a compact normal subgroup of G and ψ_K is a strict quadratic form on $\mathfrak{D}(G/K)$. By Property (2) of 6.3.11 $\mathbf{G} \neq \emptyset$. For compact normal subgroups K and L of G with $L \subset K$ the canonical epimorphism from G/L onto G/K is denoted by p_{KL}. In $\mathbf{G}$ the order relation $<$ is defined by $(G/K, \psi_K) < (G/L, \psi_L)$ if $p_{KL}(\psi_L) = \psi_K$.

By Property (3) of 6.3.11 $\mathbf{G}$ is inductively ordered with respect to the order relation $<$. Zorn's lemma thus provides maximal elements of $\mathbf{G}$. It remains to be shown that if $(G/K, \psi_K) \in \mathbf{G}$ with $K \neq \{e\}$, then there exists a $(G/L, \psi_L) \in \mathbf{G}$ such that $(G/K, \psi_K) < (G/L, \psi_L)$ and $L \neq K$ hold.

In fact, given $(G/K, \psi_K) \in \mathbf{G}$ with $K \neq \{e\}$ there is a compact normal subgroup L of K with $L \neq K$ such that K/L is a Lie group. Following the argument in Part 3 of the proof of Theorem 6.2.8 we find that L is also normal in G. Denoting the canonical mapping from G/L onto G/K by p, we conclude that $\ker p = K/L$ is a compact Lie group. Applying Lemma 6.3.12 one obtains the existence of a strict quadratic form ψ_L on $\mathfrak{D}(G/L)$ with $(G/K, \psi_K) < (G/L, \psi_L)$ which shows the assertion. □

6.3.14 Theorem. *Let G be a connected locally compact group. Then there exists a Gauss semigroup $(\mu_t)_{t \in \mathbb{R}_+^*}$ in $\mathscr{M}^1(G)$ with* $\operatorname{supp}(\mu_t) = G$ *for all $t \in \mathbb{R}_+^*$ and hence, a measure $\mu \in \mathscr{G}(G)$ with* $\operatorname{supp}(\mu) = G$.

Proof. 1. Let $(\mu_t)_{t \in \mathbb{R}_+^*}$ be a symmetric Gauss semigroup in $\mathscr{M}^1(G)$ with corresponding generating functional ψ on $\mathfrak{D}(G)$. Then ψ is a quadratic form and if ψ is strict, then $\operatorname{supp}(\mu_t) = G$ for all $t \in \mathbb{R}_+^*$. In fact, as in the preceding discussion $G = \varprojlim_{\alpha \in \mathbf{A}} G_\alpha$ and p_α denotes the canonical epimorphism from G onto G_α $(\alpha \in \mathbf{A})$. One observes that

$$\operatorname{supp}(\nu_t) = G \text{ iff } \operatorname{supp}(p_\alpha(\nu_t)) = G_\alpha \text{ for all } \alpha \in \mathbf{A} \, (t \in \mathbb{R}_+^*).$$

Property (3) of 6.3.11 permits us to assume without loss of generality that G is a Lie group. But in this case the assertion follows from Property (2) of 6.3.11 together with Corollary 6.3.2.

2. By Theorem 6.3.13 there exists a strict quadratic form on $\mathfrak{D}(G)$ and by Part 1 this implies that for the symmetric Gauss semigroup $(\mu_t)_{t \in \mathbb{R}_+^*}$ in $\mathscr{M}^1(G)$ corresponding to ψ one has $\operatorname{supp}(\mu_t) = G$ for all $t \in \mathbb{R}_+^*$ and $\mu := \mu_1$ is a Gauss measure in $\mathscr{G}(G)$ with $\operatorname{supp}(\mu) = G$. □

Let G be a connected locally compact group, $(\mu_t)_{t \in \mathbb{R}_+^*}$ a symmetric Gauss semigroup in $\mathscr{M}^1(G)$ and ψ the quadratic form corresponding to $(\mu_t)_{t \in \mathbb{R}_+^*}$.

6.3.15 Remark. If ψ is strict and G is a Lie group, then $\mu_t \ll \omega_G$ for all $t \in \mathbb{R}_+^*$.

This follows from Theorem 6.3.1 together with Property (2) of 6.3.11.

For non-Lie groups the assertion of the statement is in general false. As in Example 5.5.12, let G be equal to $\prod_{n \geq 1} G_n$ with $G_n := \mathbb{T}$ for all $n \geq 1$, let $\mu_n \in \mathscr{G}(G_n)$ be defined by $\hat{\mu}_n(m) := \exp(-m^2)$ for all $m \in \mathbb{Z}$ $(n \geq 1)$ and put

$$\mu := \bigotimes_{n \geq 1} \mu_n \in \mathscr{G}(G).$$

Then $\mu \perp \omega_G$, but the quadratic form ψ corresponding to μ is strict, since $\psi = \varprojlim_{n \geq 1} \psi_n$, where ψ_n is the strict quadratic form corresponding to

$$\bigotimes_{i=1}^n \mu_i \in \mathscr{G}(\mathbb{T}^n) \qquad (n \geq 1).$$

6.3.16 Remark. From the proof of Theorem 6.3.14 we find that for strict ψ one has $\operatorname{supp}(\mu_t)=G$ whenever $t\in\mathbb{R}_+^*$. The relation $\operatorname{supp}(\mu_t)=G$ for all $t\in\mathbb{R}_+^*$, however, does not generally imply that ψ is strict. This follows from the example in Remark 5.5.8, since the quadratic form corresponding to the Gauss semigroup constructed there is not ω_G-absolutely continuous and hence not strict.

The study of diffuse or atomless Gauss semigroups on a locally compact group will be preceded by the following

6.3.17 Lemma. *Let G be a locally compact group and $t\to\mu_t$ a homomorphism (not necessarily continuous) from $\mathbb{R}_+^*$ into $\mathscr{M}^1(G)$ such that μ_t is normal for every $t\in\mathbb{R}_+^*$. For every $t\in\mathbb{Q}_+^*$ we define measures*

$$\lambda_t := \mu_{\frac{t}{2}} * \mu_{\frac{t}{2}}^{\sim} \quad \textit{and} \quad \nu_t := \mu_{\frac{t}{2}}^s * (\mu_{\frac{t}{2}}^s)^{\sim}$$

in $\mathscr{M}^1(G)$ and $\mathscr{M}_+^b(G_d)$ resp., where μ_t^s denotes as usual the discrete part of μ_t, and assume $\mu_t^s \neq 0$ for all $t\in\mathbb{R}_+^*$. Then

(i) *$t\to\lambda_t$ is a continuous homomorphism from $\mathbb{Q}_+^*$ into $\mathscr{M}^1(G)$ and*

(ii) *there exist a finite subgroup H_1 of G and an H_1-Poisson semigroup $(\tau_t)_{t\in\mathbb{R}_+^*}$ in $\mathscr{M}^1(G_d)$ such that $\nu_t = \|\nu_1\|^t\,\tau_t$ holds for all $t\in\mathbb{Q}_+^*$* (and $\tau_0 := \lim_{t\downarrow 0}\nu_t = \omega_{H_1}$).

Proof. From Lemma 6.1.24 we conclude that there exist a compact subgroup H of G and an H-continuous convolution semigroup $(\sigma_t)_{t\in\mathbb{R}_+^*}$ in $\mathscr{M}^1(G)$ uniquely determined by λ_1 such that

$$\sigma_1 = \lambda_1 \quad \text{and} \quad \sigma_{\frac{1}{n}} = \lambda_{\frac{1}{n}} \qquad \text{for all } n \geq 1.$$

Consequently, the homomorphisms

$$t \to \lambda_t \quad \text{and} \quad t \to \tau_t := \nu_t\,\|\nu_1\|^{-t} \ (\text{for } \nu_1 \neq 0)$$

are continuous from $\mathbb{Q}_+^*$ into $\mathscr{M}^1(G)$ and $\mathscr{M}^1(G_d)$ resp., and thus by Theorem 1.5.9 extendable to continuous convolution semigroups. Moreover, $(\tau_t)_{t\in\mathbb{R}_+^*}$ is an H_1-continuous convolution semigroup in $\mathscr{M}^1(G_d)$ for the finite subgroup H_1 of G determined by the idempotent

$$\tau_0 := \lim_{t\downarrow 0} \nu_t \in \mathscr{M}^1(G_d).$$

Hence, by Theorem 6.1.6 it is an H_1-Poisson semigroup in $\mathscr{M}^1(G_d)$. □

6.3.18 Corollary. *There exist a measure $\kappa\in\mathscr{M}_{H_1}^1(G_d)$ and an $\alpha\in\mathbb{R}_+^*$ such that*

$$\nu_t = \exp_{H_1}[t(\alpha\kappa - (\alpha - \log\|\nu_1\|)\,\omega_{H_1})]$$

holds for all $t\in\mathbb{R}_+^*$.

The *proof* is clear. □

6.3.19 Corollary. *If $(\mu_t)_{t\in\mathbb{R}_+^*}$ is an $\{e\}$-continuous convolution semigroup in $\mathscr{M}^1(G)$, then so is $(\tau_t)_{t\in\mathbb{R}_+^*}$, and we have either $\mu_t^s = 0$ for all $t\in\mathbb{R}_+^*$ or $H_1 = \{e\}$.*

The *proof* follows immediately from the limit relations

$$\lim_{t\downarrow 0}\|\lambda_t-\nu_t\|=\lim_{t\downarrow 0}(1-\|\nu_t\|)$$
$$=\lim_{t\downarrow 0}(1-\|\nu_1\|^t)=0 \quad\text{and}\quad \lim_{t\downarrow 0}\|\nu_t-\omega_{H_1}\|=0$$

together with the $\{e\}$-continuity of $(\lambda_t)_{t\in\mathbb{R}_+^*}$. □

6.3.20 Corollary. *For any $\{e\}$-continuous convolution semigroup $(\mu_t)_{t\in\mathbb{R}_+^*}$ of normal measures in $\mathcal{M}^1(G)$ one has either $\mu_t^s=0$ for all $t\in\mathbb{R}_+^*$ or that $(\lambda_t)_{t\in\mathbb{R}_+^*}$ is an $\{e\}$-Poisson semigroup in $\mathcal{M}^1(G)$.*

Proof. For every $t\in\mathbb{R}_+^*$ we have $\nu_{2t}=\mu_t^s*(\mu_t^s)^\sim=(\mu_t^s)^\sim*\mu_t^s=0$ iff $\mu_t^s=0$. But $\nu_t\neq 0$ for every $t\in\mathbb{R}_+^*$ implies by the preceding corollaries that $\lim_{t\downarrow 0}\|\lambda_t-\varepsilon_e\|=0$. As a norm-continuous convolution semigroup $(\lambda_t)_{t\in\mathbb{R}_+^*}$ is $\{e\}$-Poisson by Theorem 6.1.5. □

6.3.21 Theorem. *Let G be a locally compact group and $(\mu_t)_{t\in\mathbb{R}_+^*}$ an $\{e\}$-continuous convolution semigroup of normal measures in $\mathcal{M}^1(G)$ with generating functional A on $\mathfrak{D}(G)$ admitting a unique decomposition $A=A'+A''$ into its skew symmetric and symmetric parts A' and $A'':=\frac{1}{2}(A+A^\sim)$ resp. with $A^\sim:=A^\#=A\circ S$. By $(\lambda_t)_{t\in\mathbb{R}_+^*}$ we denote the semigroup of measures*

$$\lambda_t:=\mu_{\frac{t}{2}}*\mu_{\frac{t}{2}}^\sim \qquad (t\in\mathbb{R}_+^*).$$

The following statements are equivalent:
(i) *$(\mu_t)_{t\in\mathbb{R}_+^*}$ is diffuse in the sense that $\mu_t^s=0$ for all $t\in\mathbb{R}_+^*$;*
(ii) *$(\lambda_t)_{t\in\mathbb{R}_+^*}$ is not an $\{e\}$-Poisson semigroup;*
(iii) *A'' is not a Poisson generator.*

Proof. It suffices to show the equivalence (i)⇔(iii).

1. We first observe that the generating functional of $(\lambda_t)_{t\in\mathbb{R}_+^*}$ equals A''.

Plainly $(\mu_t)_{t\in\mathbb{R}_+^*}$ is diffuse iff $(\lambda_t)_{t\in\mathbb{R}_+^*}$ is, so that we may restrict ourselves without loss of generality to the discussion of the semigroup $(\lambda_t)_{t\in\mathbb{R}_+^*}$.

Let $\lambda_t^s\neq 0$ for all $t\in\mathbb{R}_+^*$. Then Corollary 6.3.20 implies that $(\lambda_t)_{t\in\mathbb{R}_+^*}$ is an $\{e\}$-Poisson semigroup in $\mathcal{M}^1(G)$, and hence that A'' is a Poisson generator.

2. Conversely, let A'' be a Poisson generator on $\mathfrak{D}(G)$. Then the Poisson semigroup $(\lambda_t)_{t\in\mathbb{R}_+^*}$ satisfies $\lambda_t(\{e\})>0$ for all $t\in\mathbb{R}_+^*$, which implies that $\lambda_t^s\neq 0$ for all $t\in\mathbb{R}_+^*$. □

6.3.22 Corollary. *Let $(\mu_t)_{t\in\mathbb{R}_+^*}$ be a Gauss semigroup in $\mathcal{M}^1(G)$ with generating functional A on $\mathfrak{D}(G)$. If μ_t is normal for every $t\in\mathbb{R}_+^*$, then $(\mu_t)_{t\in\mathbb{R}_+^*}$ is diffuse. In particular, every symmetric Gauss semigroup is diffuse.*

The *proof* is clear. □

6.3.23 Remark. Let $G\in\mathbf{M}$ and let $(\mu_t)_{t\in\mathbb{R}_+^*}$ be an $\{e\}$-continuous convolution semigroup in $\mathcal{M}^1(G)$ with corresponding negative-definite form ψ on $\mathfrak{R}(G)$

admitting a decomposition $\psi=\psi'+\psi''$ into a skew-symmetric and a symmetric part ψ' and ψ'' resp.

The following statements are equivalent:

(i) $(\mu_t)_{t\in\mathbb{R}_+^*}$ is diffuse;

(ii) there exists a $t_0\in\mathbb{R}_+^*$ satisfying $\inf_{\substack{f\in\mathfrak{K}(G)\\ f\neq 0}}|\hat{\mu}_{t_0}(f)|=0$;

(iii) ψ'' is unbounded.

The proof of this result relies on Theorems 4.3.13 and 6.3.21. One need only use the facts that quadratic negative-definite forms on $\mathfrak{K}(G)$ are necessarily unbounded and that the boundedness of ψ'' implies the boundedness of the Lévy measure corresponding to ψ'' (see the proof of Theorem 6.1.5).

6.3.24 Remark. In general, the Fourier transform of a diffuse measure in $\mathscr{M}^1(G)$ is not in $\mathscr{C}^0(G)$.

The following example shows the limiting behavior of a diffuse semigroup in a neighborhood of infinity.

Consider $G:=\mathbb{T}^2$, an irrational $\alpha\in[0,1]$ and a mapping $\phi:\mathbb{R}\to G$ defined by $\phi(z):=(e^{2\pi i z},e^{2\pi i\alpha z})$ for all $z\in\mathbb{R}$. Then $\phi(\mathbb{R})$ is dense in G, and hence, for every $t\in\mathbb{R}_+^*$ the measure $\mu_t:=\phi(\nu_{0,t})$ is a Gauss measure on G with $\mathrm{supp}(\mu_t)=G$.

Since $G^\wedge=\mathbb{Z}^2$, for every k, $n\in\mathbb{Z}$, the character $\chi_{k,n}$ of G is defined by

$$\chi_{k,n}((e^{ix},e^{iy})):=e^{2\pi i(kx+ny)}\qquad (x,y\in[0,2\pi[).$$

Evidently, for all $t\in\mathbb{R}_+^*$ and k, $n\in\mathbb{Z}$ we have

$$\hat{\mu}_t(\chi_{k,n})=\hat{\nu}_{0,t}(k+\alpha n).$$

Let $(k_j,n_j)_{j\geq 1}$ be a sequence in $\mathbb{Z}^2$ with $\lim_{j\to\infty}(k_j,n_j)=(\infty,\infty)$. Then

$$\lim_{j\to\infty}\hat{\mu}_t(\chi_{k_j,n_j})=0\quad\text{iff}\quad \lim_{j\to\infty}|k_j+\alpha n_j|=\infty.$$

Thus, for $\sup_{j\geq 1}|k_j+\alpha n_j|=:c<\infty$ we obtain

$$\inf_{j\geq 1}\hat{\mu}_t(\chi_{k_j,n_j})=\inf_{j\geq 1}\hat{\nu}_{0,t}(k_j+\alpha n_j)=e^{-\frac{tc^2}{2}}>0.$$

For every $\alpha\in]0,1[$ there exists at least one sequence $(k_j,n_j)_{j\geq 1}$ converging to (∞,∞) such that

$$\inf_{\chi\in G^\wedge}|\hat{\mu}_t(\chi)|=\lim_{j\geq 1}|\hat{\mu}_t(\chi_{k_j,n_j})|=0$$

holds for all $t\in\mathbb{R}_+^*$.

Consequently, by Remark 6.3.23 μ_t is diffuse for all $t\in\mathbb{R}_+^*$, but $\hat{\mu}_t$ does not vanish at infinity for $t\in\mathbb{R}_+^*$. In fact, if $\alpha\in[0,1]$ is irrational, then the accumulation points for $\chi\to\infty$ of the set $\{\hat{\mu}_t(\chi):t\in\mathbb{R}_+^*\}$ exhaust $[0,1]$.

6.4 Central Gauss Semigroups

The preceding studies concerning the existence of absolutely continuous Gauss semigroups will be extended to those Gauss semigroups which are invariant under inner automorphisms of the underlying group. Let G be a locally compact group and μ a measure in $\mathscr{M}^1(G)$. μ is called central if $\alpha(\mu)=\mu$ for all inner automorphisms $\alpha\in\operatorname{Int}(G)$. More generally, an $\{e\}$-continuous convolution semigroup $S:=(\mu_t)_{t\in\mathbb{R}_+^*}$ in $\mathscr{M}^1(G)$ is said to be *central* if μ_t is central for all $t\in\mathbb{R}_+^*$. If, for $\alpha\in\operatorname{Int}(G)$ we define $\alpha(S):=(\alpha(\mu_t))_{t\in\mathbb{R}_+^*}$, then the centrality of S is equivalent to $\alpha(S)=S$ for all $\alpha\in\operatorname{Int}(G)$.

Clearly, a convolution semigroup $(\mu_t)_{t\in\mathbb{R}_+^*}$ in $\mathscr{M}^1(G)$ is central iff

$$\varepsilon_x*\mu_t*\varepsilon_{x^{-1}}=\mu_t \qquad \text{for all } x\in G \text{ and } t\in\mathbb{R}_+^*.$$

In this case the measure μ_t belongs to the center of $\mathscr{M}^1(G)$ for every $t\in\mathbb{R}_+^*$.

Concerning the existence of central convolution semigroups we prove the following

6.4.1 Theorem. *Let G be a σ-compact, locally compact group with compact commutator subgroup $K:=K(G)$. Then there exists a central $\{e\}$-continuous convolution semigroup $(\mu_t)_{t\in\mathbb{R}_+^*}$ in $\mathscr{M}^1(G)$ with* $\operatorname{supp}(\mu_t)=G$ *for all $t\in\mathbb{R}_+^*$.*

Proof. It suffices to construct a central measure μ in $\mathscr{M}^1(G)$ satisfying $\operatorname{supp}(\mu)=G$. The definition $\mu_t:=\exp_{\{e\}}[t(\mu-\varepsilon_e)]$ for all $t\in\mathbb{R}_+^*$ then yields the desired semigroup.

Let $\dot{G}:=G/K$ and p be the canonical epimorphism from G onto $\dot{G}$. For every $f\in\mathscr{K}(G)$ we introduce the function $f^\#\in\mathscr{K}(\dot{G})$ defined by

$$f^\#(p(x)):=\int_K f(xk)\,\omega_K(dk) \qquad \text{for all } x\in G.$$

Since $\dot{G}$ is σ-compact, there exists a measure $\dot{\mu}\in\mathscr{M}^1(\dot{G})$ with $\operatorname{supp}(\dot{\mu})=\dot{G}$. The measure μ on G defined by $\mu(f):=\dot{\mu}(f^\#)$ for all $f\in\mathscr{K}(G)$ belongs to $\mathscr{M}^1(G)$. Moreover, we have $\operatorname{supp}(\mu)=G$. In fact, $p(\mu)=\dot{\mu}$ implies $p(\operatorname{supp}(\mu))=\operatorname{supp}(\dot{\mu})=\dot{G}$ since p is a closed mapping. On the other hand, one has $\mu*\omega_K=\mu$ so that $\operatorname{supp}(\mu)\cdot K=\operatorname{supp}(\mu)$. But then the fact that $\operatorname{supp}(\mu)=G$ becomes obvious.

It remains to show that μ is central. For $f\in\mathscr{K}(G)$ and $y\in G$ we introduce the function f^y on G by setting $f^y(x):=f(yxy^{-1})$ for all $x\in G$. Plainly

$$\begin{aligned}(f^y)^\#(p(x))&=\int_K f^y(xk)\,\omega_K(dk)=\int_K f(yxky^{-1})\omega_K(dk)\\&=\int_K f((yxy^{-1})(yky^{-1}))\,\omega_K(dk)\\&=\int_K f((yxy^{-1})k)\varepsilon_y*\omega_K*\varepsilon_{y^{-1}}(dk)\\&=\int f((yxy^{-1})k)\omega_K(dk)=f^\#(p(yxy^{-1}))=f^\#(p(x))\end{aligned}$$

since $\dot{G}$ is Abelian. This implies that $(f^y)^\#=f^\#$, and we get

$$\varepsilon_y*\mu*\varepsilon_{y^{-1}}(f)=\mu(f^y)=\dot{\mu}((f^y)^\#)=\dot{\mu}(f^\#)=\mu(f)$$

for all $y\in G$, which proves the assertion. □

6.4.2 Remark. At a later stage we shall need to know whether there exist noncentral, connected, locally compact groups G with compact commutator subgroup admitting central convolution semigroups $(\mu_t)_{t\in\mathbb{R}_+^*}$ with $\operatorname{supp}(\mu_t)=G$

for all $t \in \mathbb{R}_+^*$. Let $H := \mathfrak{SU}(2)$ and $\eta: \mathbb{R} \to H$ a nontrivial one-parameter subgroup in H. The definition $\eta(r)(x) := \eta(r) x \eta(r)^{-1}$ for all $r \in \mathbb{R}$ and $x \in H$ establishes a continuous homomorphism η from $\mathbb{R}$ into $\mathrm{Aut}(H)$. Since $Z(H)$ consists of two elements, η is nontrivial.

The (nondirect) semidirect product $G := H \times_\eta \mathbb{R}$ is not almost periodic by Theorem L, and hence is not central. But G is connected, and since $G/H \cong \mathbb{R}$ and H is compact, the commutator subgroup $K(G)$ is compact. But Theorem 6.4.1 yields the existence of the desired convolution semigroup in $\mathcal{M}^1(G)$.

Now let G be a connected locally compact group and let $S := (\mu_t)_{t \in \mathbb{R}_+^*}$ be a symmetric Gauss semigroup in $\mathcal{M}^1(G)$. For any automorphism $\alpha \in \mathrm{Int}(G)$ the transformed semigroup $\alpha(S)$ is again a symmetric Gauss semigroup in $\mathcal{M}^1(G)$. Thus, central (symmetric) Gauss semigroups in $\mathcal{M}^1(G)$ are well-defined.

6.4.3. Let $S := (\mu_t)_{t \in \mathbb{R}_+^*}$ be a symmetric Gauss semigroup in $\mathcal{M}^1(G)$ and ψ its generating functional. For every $f \in \mathfrak{D}(G)$ and $x \in G$ the function f^x defined above is an element of $\mathfrak{D}(G)$. For every $x \in G$ we define the linear form ψ^x on $\mathfrak{D}(G)$ by putting $\psi^x(f) := \psi(f^x)$ for all $f \in \mathfrak{D}(G)$ and list the following **properties**:

(1) ψ^x is the quadratic form corresponding to the transformed Gauss semigroup $(\varepsilon_x * \mu_t * \varepsilon_{x^{-1}})_{t \in \mathbb{R}_+^*}$ in $\mathcal{M}^1(G)$;

(2) ψ^x is strict iff ψ is strict.

Let G be a Lie group of dimension $n > 0$ with basis $\{X_1, \ldots, X_n\}$ of its Lie algebra $\mathfrak{L}(G)$. By Lemma 4.4.9 there exists a real symmetric, positive-semidefinite matrix $A^x := (a_{ij}^x)_{i,j=1,\ldots,n}$ such that

$$\psi^x(f) = \sum_{i,j=1}^n a_{ij}^x (X_i X_j f)(e) \quad \text{holds for all } f \in \mathfrak{D}(G).$$

We put $A := A^e$ and $a_{ij} := a_{ij}^e$ for all $i, j = 1, \ldots, n$.

(3) Let the adjoint representation Ad of G be given with respect to the basis $\{X_1, \ldots, X_n\}$ by the matrix $B(x) := (b_{ij}(x))_{i,j=1,\ldots,n}$. Then for every $f \in \mathfrak{D}(G)$ we have

$$\psi^x(f) = \sum_{i,j=1}^n a_{ij} (\mathrm{Ad}(x) X_i \, \mathrm{Ad}(x) X_j f)(e) \quad \text{and}$$
$$A^x = B(x) A B(x)^T.$$

This follows directly from the relations $X f^x = (\mathrm{Ad}(x) X f)^x$ (with $X \in \mathfrak{L}(G)$) and

$$\mathrm{Ad}(x) X_i = \sum_{j=1}^n b_{ji}(x) X_j$$

for all $f \in \mathfrak{D}(G)$ and $i = 1, \ldots, n$ resp.

Let now G be an arbitrary connected locally compact group.

(4) For every $f \in \mathfrak{D}(G)$ the mapping $x \to \psi^x(f)$ from G into $\mathbb{R}$ is continuous.

Without loss of generality one assumes that G is a Lie group. Since Ad is an analytic homomorphism, the function $x \to a_{ij}^x$ on G is continuous for all $i, j = 1, \ldots, n$ by Property (3) and hence $x \to \psi^x(f)$ is continuous for all $f \in \mathfrak{D}(G)$.

6.4.4 Definition. A quadratic form ψ on $\mathfrak{D}(G)$ is called *G-invariant* if $\psi^x = \psi$ for all $x \in G$.

6.4.5 Remark. Let ψ be a quadratic form ψ on $\mathfrak{D}(G)$ and $S := (\mu_t)_{t \in \mathbb{R}_+^*}$ the symmetric Gauss semigroup corresponding to ψ. Then S is central iff ψ is G-invariant.

The set of all strict G-invariant quadratic forms on $\mathfrak{D}(G)$ will be denoted by $\mathscr{Q}(G)$.

6.4.6 Theorem. *$\mathscr{Q}(G) \neq \emptyset$ iff G is a Z-group.*

Proof. 1. Let $\mathscr{Q}(G) \neq \emptyset$ and hence $\psi \in \mathscr{Q}(G)$. We are going to show that G is a Z-group. It suffices to prove this statement for the special case of a Lie group G. In fact, if G is an arbitrary connected locally compact group and K a compact normal subgroup of G, then the projection of ψ on $\mathfrak{D}(G/K)$ is a strict (G/K)-invariant quadratic form. Given the assertion for Lie groups, the Lie projectivity of G implies that G is a projective limit of Z-groups, hence itself a Z-group (by 9.1.9 and 9.1.10 of [227]).

From Property (3) of 6.4.3 we conclude that $A = B(x) A B(x)^T$ for all $x \in G$. Since ψ is strict, A is positive-definite by Property (2) of 6.4.3. Consequently,

$$A^{-1} = B(x)^T A^{-1} B(x)$$

and so

$$\langle A^{-1} X, X \rangle = \langle A^{-1} \operatorname{Ad}(x) X, \operatorname{Ad}(x) X \rangle$$

for all $x \in G$ and $X \in \mathfrak{L}(G)$. Thus, Ad leaves a positive, nondegenerate, symmetric bilinear form invariant and the closure of $\operatorname{Ad}(G)$ in the full linear group of $\mathfrak{L}(G)$ is compact. By [246], p. 153, Exercise 1, this implies that G is a Z-group.

2. Let G be a Z-group and ψ a strict quadratic form on $\mathfrak{D}(G)$ which exists by Theorem 6.3.13. For all $x \in G$ and $z \in Z(G)$ we have $\psi^{xz} = \psi^x$. Let p denote the canonical homomorphism from G onto $\dot{G} := G/Z(G)$ and let $\dot{x} := p(x)$ for $x \in G$. Then $\psi^{\dot{x}}(f) := \psi^x(f)$ is well-defined for every $f \in \mathfrak{D}(G)$ and $\dot{x} \to \psi^{\dot{x}}(f)$ is a continuous mapping from $\dot{G}$ into $\mathbb{R}$ for all $f \in \mathfrak{D}(G)$ by Property (4) of 6.4.3. Therefore, the function $\bar{\psi}$ on $\mathfrak{D}(G)$ defined by

$$\bar{\psi}(f) := \int \psi^{\dot{x}}(f)\, \omega_{\dot{G}}(d\dot{x}) \qquad \text{for all } f \in \mathfrak{D}(G)$$

is a strict quadratic form on $\mathfrak{D}(G)$ (by Properties (1) and (2) of 6.4.3). Since $\omega_{\dot{G}}$ is invariant on $\dot{G}$, the G-invariance of $\bar{\psi}$ follows. □

6.4.7 Corollary. *On every connected locally compact Z-group there exists a central Gauss semigroup $(\mu_t)_{t \in \mathbb{R}_+^*}$ with $\operatorname{supp}(\mu_t) = G$ for all $t \in \mathbb{R}_+^*$.*

The *proof* follows immediately from the theorem with the aid of the proof of Theorem 6.3.14. □

Let G be an arbitrary locally compact group.

6.4.8 Definition. Two quadratic forms ψ and $\psi' \in \mathscr{Q}(G)$ are called *equivalent* if there exists a number $c \in \mathbb{R}_+^*$ with $\psi' = c\psi$.

If ψ and $\psi' \in \mathscr{Q}(G)$ are the generating functionals of Gauss semigroups $(\mu_t)_{t \in \mathbb{R}_+^*}$ and $(\mu_t')_{t \in \mathbb{R}_+^*}$ in $\mathscr{M}^1(G)$ resp., then ψ and ψ' are equivalent in the sense of the definition iff $\mu_t' = \mu_{ct}$ for all $t \in \mathbb{R}_+^*$.

6.4.9 Theorem. *Let G be a connected central Lie group of dimension $n \geq 1$ with Lie algebra $\mathfrak{L}(G)$. The following statements are equivalent:*
(i) *On $\mathfrak{D}(G)$ there exists exactly one equivalence class of strict G-invariant quadratic forms;*
(ii) *$\mathfrak{L}(G)$ does not admit a proper ideal.*

Proof. 1. By assumption $G/Z(G)$ is compact. Hence, by [213], p. 119, Corollary 5.2, $\mathfrak{L}(G)$ is compact in the sense that the adjoint group Int $\mathfrak{L}(G)$ corresponding to the Lie algebra ad $\mathfrak{L}(G)$ is compact (in $\mathfrak{GL}(\mathfrak{L}(G))$). Let us denote $\mathfrak{L}(G)$ and its center by $\mathfrak{g}$ and $\mathfrak{c}$ resp. There exist simple ideals $\mathfrak{g}_1, \ldots, \mathfrak{g}_r$ of $\mathfrak{g}$ such that

$$\mathfrak{g} = \mathfrak{c} \oplus \sum_{i=1}^{r}{}^{\oplus} \mathfrak{g}_i$$

holds ([213], p. 122, Proposition 6.6 and Corollary 6.3).

We know that the invariant subspaces of the adjoint representation Ad of G are exactly the ideals of $\mathfrak{g}$. Hence, one obtains

$$\mathrm{Ad} = I_k \oplus \sum_{i=1}^{r}{}^{\oplus} \mathrm{Ad}_i,$$

where I_k denotes the identity on the k-dimensional Lie algebra $\mathfrak{c}$ generated by $\{X_1, \ldots, X_k\}$. Moreover, Ad_i is irreducible for every $i = 1, \ldots, r$. Let B be the Killing form of $\mathfrak{g}$ defined by

$$B(X, Y) := \mathrm{tr}(\mathrm{ad}\, X\, \mathrm{ad}\, Y) \qquad \text{for all } X, Y \in \mathfrak{g}.$$

Then for any $i = 1, \ldots, r$ we can choose a basis $\{X_1^{(i)}, \ldots, X_{s(i)}^{(i)}\}$ of $\mathfrak{g}_i$ satisfying

$$B(X_\alpha^{(i)}, X_\beta^{(i)}) = -\delta_{\alpha\beta} \qquad \text{where } \alpha, \beta \in \{1, \ldots, s(i)\}.$$

But then

$$\{X_1, \ldots, X_k, X_1^{(1)}, \ldots, X_{s(1)}^{(1)}, \ldots, X_1^{(r)}, \ldots, X_{s(r)}^{(r)}\}$$

is a basis $\{Y_1, \ldots, Y_n\}$ of $\mathfrak{g}$, and Ad is described with respect to this basis by the orthogonal matrices

$$B(x) := E_k \oplus \sum_{i=1}^{r}{}^{\oplus} B_i(x) \qquad (x \in G).$$

For $a_1, \ldots, a_k; t_1, \ldots, t_r \in \mathbb{R}_+^*$ we now define a linear functional ψ on $\mathfrak{D}(G)$ by

$$\psi(f) := \sum_{j=1}^{k} a_j (X_j X_j f)(e) + \sum_{i=1}^{r} t_i \Big(\sum_{j=1}^{s(i)} (X_j^{(i)} X_j^{(i)} f)(e)\Big)$$

for all $f \in \mathfrak{D}(G)$.

By Example 2 of 4.4.8 ψ is a quadratic form on $\mathfrak{D}(G)$. Property (2) of 6.3.11 yields that ψ is strict. If we write

$$\psi(f) = \sum_{i,j=1}^{n} a_{ij} (Y_i Y_j f)(e) \qquad \text{for all } f \in \mathfrak{D}(G),$$

where the parameters $a_1, \ldots, a_k; t_1, \ldots, t_r$ are absorbed in a matrix

$$A := (a_{ij})_{i,j=1,\ldots,n} \in \mathfrak{M}(n, \mathbb{R}),$$

then by the choice of the basis $\{Y_1, \ldots, Y_n\}$ of $\mathfrak{g}$ we get $A = B(x)AB(x)^T$ for all $x \in G$, whence the G-invariance of ψ.

2. (i) $\Rightarrow$ (ii). Let there exist just one equivalence class in $\mathcal{Q}(G)$. Then either $k=1, r=0$ or $k=0, r=1$ holds. In both cases $\mathfrak{g}$ admits no proper ideal.

3. (ii) $\Rightarrow$ (i). By assumption, Ad is irreducible. Let $\psi \in \mathcal{Q}(G)$ be given with

$$\psi(f) = \sum_{i,j=1}^{n} a_{ij}(Y_i Y_j f)(e) \quad \text{for all } f \in \mathfrak{D}(G)$$

and

$$A := (a_{ij})_{i,j=1,\ldots,n} \in \mathfrak{M}(n, \mathbb{R}).$$

Then $A = B(x)AB(x)^T$ for all $x \in G$ by Property (3) of 6.4.3. But for $x \in G$ the matrix $B(x)$ is orthogonal, so that by Schur's lemma there exists an $\alpha \in \mathbb{R}_+^*$ satisfying $A = \alpha E_n$. This implies that

$$\psi(f) = \alpha \sum_{i=1}^{n} (Y_i Y_i f)(e) \quad \text{for all } f \in \mathfrak{D}(G),$$

which yields the assertion. □

Let G be a connected compact group and $(\mu_t)_{t \in \mathbb{R}_+^*}$ an arbitrary Gauss semigroup in $\mathcal{M}^1(G)$ with generating functional ψ on $\mathfrak{D}(G)$. Clearly, ψ can be extended to $\mathfrak{R}(G)$ since for every $\sigma \in \Sigma(G)$ and $i,j = 1, \ldots, n(\sigma)$ the functions $\operatorname{Re} d_{ij}^{(\sigma)}$ and $\operatorname{Im} d_{ij}^{(\sigma)}$ are in $\mathfrak{D}(G)$ and are such that

$$\psi(\sigma) = \psi(D^{(\sigma)}) = (\psi(d_{ij}^{(\sigma)}))_{i,j=1,\ldots,n(\sigma)}$$

is defined. By Theorem 1.5.18 we obtain for every $t \in \mathbb{R}_+^*$

$$\hat{\mu}_t(\sigma) = \exp(-t\,\psi(\sigma)) \quad \text{whenever } \sigma \in \Sigma(G).$$

6.4.10 Properties. (1) If $\mu := \mu_1 \ll \omega_G$, then G admits a countable basis of its topology.

[If $\mu = f \cdot \omega_G$ for $f \in \mathcal{L}^1(G, \omega)$, then $\hat{f}(\sigma) := \hat{\mu}(\sigma) = \exp(-\psi(\sigma)) \neq 0$ for all $\sigma \in \Sigma(G)$. But $\{\sigma \in \Sigma(G) : \hat{f}(\sigma) \neq 0\}$ is countable by [218], (28.41), thus $\Sigma(G)$ is countable and therefore G admits a countable basis of its topology.]

(2) If the semigroup $(\mu_t)_{t \in \mathbb{R}_+^*}$ in $\mathcal{M}^1(G)$ is symmetric, then for every $\sigma \in \Sigma(G)$ the matrix $\psi(\sigma)$ is Hermitian positive-semidefinite by Theorem 6.2.10, and is thus equivalent to a diagonal matrix $(m_i^{(\sigma)} \delta_{ij})_{i,j=1,\ldots,n(\sigma)}$ with eigenvalues $m_i^{(\sigma)} \geq 0$ $(i = 1, \ldots, n(\sigma))$.

(3) If in addition $[\operatorname{supp}(\mu)]^- = G$, then the following statements hold:
(i) $m_i^{(\sigma)} > 0$ for all $i = 1, \ldots, n(\sigma)$ and $\sigma \in \Sigma(G) \setminus \{1\}$.
(ii) $\lim_{t \to \infty} \mu_t = \omega_G$.
[In fact, (i) follows from the Ito-Kawada theorem 2.1.4 since

$$\lim_{n \to \infty} \exp(-n\,\psi(\sigma)) = \lim_{n \to \infty} \hat{\mu}_n(\sigma) = \hat{\omega}_G(\sigma) = 0$$

for all $\sigma \in \Sigma(G) \setminus \{1\}$.
The implication (i) $\Rightarrow$ (ii) is a direct consequence of the continuity theorem 1.4.5 since (i) implies $\lim_{t \to \infty} \hat{\mu}_t(\sigma) = 0 = \hat{\omega}_G(\sigma)$ for all $\sigma \in \Sigma(G) \setminus \{1\}$.]

(4) For $\nu = f \cdot \omega_G$ with $f \in \mathscr{L}^2(G, \omega)$ the following statements hold true:

(i) $f = \sum_{\sigma \in \Sigma(G)} n(\sigma) \operatorname{tr}(\exp(-\psi(\sigma)) D^{(\sigma)})$ (in the $\mathscr{L}^2(G, \omega)$-seminorm $\|\cdot\|_2$);

(ii) $\|f\|_2^2 = \sum_{\sigma \in \Sigma(G)} n(\sigma) \operatorname{tr}(\exp(-2\psi(\sigma)))$;

(iii) if $\mu_{\frac{1}{2}} = g \cdot \omega_G$ with $g \in \mathscr{L}^2(G, \omega)$, then

$$\sum_{\sigma \in \Sigma(G)} n(\sigma) \operatorname{tr}(\exp(-\psi(\sigma)) D^{(\sigma)})$$

converges uniformly on G to a continuous function which ω_G-a.e. equals f. [(i) is clear from [218], (28.43). (ii) follows directly from the orthogonality relations if we use the formula

$$\operatorname{tr}(\exp(-\psi(\sigma)) D^{(\sigma)}) = \sum_{i=1}^{n(\sigma)} \exp(-m_i^{(\sigma)}) d_{ii}^{(\sigma)},$$

which is valid for all $\sigma \in \Sigma(G)$. (iii) is obtained via (ii) from

$$\begin{aligned} &\Big\|\sum_{\sigma \in P} n(\sigma) \sum_{i=1}^{n(\sigma)} \exp(-m_i^{(\sigma)}) d_{ii}^{(\sigma)}\Big\| \\ &\leq \sum_{\sigma \in P} n(\sigma) \sum_{i=1}^{n(\sigma)} \exp(-m_i^{(\sigma)}), \end{aligned}$$

where P is a finite subset of $\Sigma(G)$, and

$$\|g\|_2^2 = \sum_{\sigma \in \Sigma(G)} n(\sigma) \sum_{i=1}^{n(\sigma)} \exp(-m_i^{(\sigma)}).]$$

(5) Conversely, if $\sum_{\sigma \in \Sigma(G)} n(\sigma) \operatorname{tr}(\exp(-2\psi(\sigma)))$ converges, then there exists an $f \in \mathscr{L}^2(G, \omega)$ with $\mu = f \cdot \omega_G$. This follows again from [218], (28.43).

Let $\bar{\psi}$ be the G-invariant quadratic form on $\mathfrak{D}(G)$ defined by

$$\bar{\psi}(f) := \int \psi^x(f)\, \omega_G(dx)$$

for all $f \in \mathfrak{D}(G)$ and $(\bar{\mu}_t)_{t \in \mathbb{R}_+^*}$ the central Gauss semigroup corresponding to $\bar{\psi}$.

(6) Putting for every $\sigma \in \Sigma(G)$

$$\bar{m}^{(\sigma)} := \frac{\operatorname{tr}(\psi(\sigma))}{n(\sigma)} = \frac{1}{n(\sigma)} \sum_{i=1}^{n(\sigma)} m_i^{(\sigma)},$$

we obtain $\bar{\psi}(\sigma) = \bar{m}^{(\sigma)} E_{n(\sigma)}$ and thus

$$\bar{\mu}_t^\wedge(\sigma) = \exp(-t\, \bar{m}^{(\sigma)}) E_{n(\sigma)} \qquad \text{for all } t \in \mathbb{R}_+^*.$$

This is an immediate consequence of the orthogonality relations.

Now let G be a compact connected Lie group with basis $\{X_1, \ldots, X_n\}$ of its Lie algebra $\mathfrak{L}(G)$.

(7) Put $\psi(f) := \sum_{i=1}^n a_i (X_i X_i f)(e)$ for all $f \in \mathfrak{D}(G)$ with $a_i \in \mathbb{R}_+$ for all $i = 1, \ldots, n$. Then for every $\sigma \in \Sigma(G)$ we have

$$\bar{m}^{(\sigma)} = \frac{1}{n(\sigma)} \sum_{j=1}^n a_j \|(X_j D^{(\sigma)})(e)\|^2.$$

Indeed, for every $\sigma\in\Sigma(G)$ one computes

$$\operatorname{tr}(\psi(D^{(\sigma)}))=\sum_{j=1}^{n} a_j\,\|(X_j D^{(\sigma)})(e)\|^2.$$

Property (6) then implies the assertion.

6.4.11 Theorem. *Let G be a connected, locally connected compact group whose topology has a countable basis. Then there exists a central Gauss semigroup $(\mu_t)_{t\in\mathbb{R}_+^*}$ in $\mathcal{M}^1(G)$ with the following properties:*
(i) $\operatorname{supp}(\mu_t)=G$ *for all* $t\in\mathbb{R}_+^*$ *and*
(ii) $\mu_t=f_t\cdot\omega_G$ *with* $f_t\in\mathscr{C}(G)$ *for all* $t\in[2,\infty[$.

Proof. Without loss of generality we may assume that G is not finite-dimensional. If it were, the local connectedness would imply that G is a Lie group by Theorem H. In this case, however, the assertion follows from Theorems 6.4.6 and 6.3.1 with the aid of Property (2) of 6.4.3.

1. Since G is metrizable, there exists a decreasing sequence $(K_n)_{n\geq 1}$ of compact subgroups of G with $\bigcap_{n\geq 1} K_n=\{e\}$ such that $G_n:=G/K_n$ is a Lie group of dimension $d(n)$ for all $n\geq 1$. Without loss of generality we assume K_n to be connected for every $n\geq 1$. This can be done since $G/(K_n)_0$ is finite-dimensional and locally connected, and hence is a Lie group by Theorem H.

As G is not finite-dimensional, the sequence $(K_n)_{n\geq 1}$ can be chosen such that $d(n)<d(n+1)$ for all $n\geq 1$. With the standard notation p_n and q_n for the canonical mappings from G onto G_n and from G_{n+1} onto G_n resp., we have given $\Sigma(G_n)\cong K_n^{\perp}$, a basis $\{X_1^{(n)},\dots,X_{d(n)}^{(n)}\}$ (chosen consistently) of the Lie algebra $\mathfrak{L}(G_n)$ of G_n and a Haar-measure $\omega_n:=p_n(\omega_G)\in\mathcal{M}^1(G_n)$ for all $n\geq 1$.

2. Let $(a_m)_{m\geq 1}$ be a properly chosen sequence in $\mathbb{R}_+^*$, ψ_n the strict quadratic form on $\mathfrak{D}(G_n)$ defined by

$$\psi_n(f_n):=\sum_{i=1}^{d(n)} a_i(X_i^{(n)}X_i^{(n)}f_n)(e)\qquad\text{for all } f_n\in\mathfrak{D}(G_n)$$

and $\bar\psi_n:=\int\psi_n^x\,\omega_n(dx)$ the corresponding G_n-invariant strict quadratic form on $\mathfrak{D}(G_n)$.

By Theorem 6.3.1, for every $n\geq 1$ there exists a central Gauss semigroup $(\mu_t^{(n)})_{t\in\mathbb{R}_+^*}$ in $\mathcal{M}^1(G_n)$ with generating functional $\bar\psi_n$ such that $\mu_t^{(n)}=f_t^{(n)}\cdot\omega_n$ for $f_t^{(n)}\in\mathscr{C}(G_n)$ (all $t\in\mathbb{R}_+^*$).

It follows that $q_n(\bar\psi_{n+1})=\bar\psi_n$ for all $n\geq 1$ so that by Property (3) of 6.3.11 there exists a strict quadratic form $\bar\psi$ on $\mathfrak{D}(G)$ with $p_n(\bar\psi)=\bar\psi_n$ for all $n\geq 1$. $\bar\psi$ is G-invariant so that the Gauss semigroup $(\mu_t)_{t\in\mathbb{R}_+^*}$ corresponding to $\bar\psi$ is central and by the proof of Theorem 6.3.14, $\operatorname{supp}(\mu_t)=G$ for all $t\in\mathbb{R}_+^*$.

3. By Properties (6) and (7) of 6.4.10 we have for all $\sigma\in\Sigma(G_n)\cong K_n^{\perp}$ the formula

$$(*)\qquad \bar\psi(\sigma)=\bar m^{(\sigma)}E_{n(\sigma)}\quad\text{with}$$

$$\bar m^{(\sigma)}:=\frac{1}{n(\sigma)}\sum_{j=1}^{d(n)} a_j\,\|(X_j^{(n)}D^{(\sigma)})(e)\|^2,$$

and by Property (4)

$$\|f_t^{(n)}\|_2^2=\sum_{\sigma\in K_n^{\perp}} n(\sigma)^2\exp(-2t\,\bar m^{(\sigma)})\quad(n\geq 1).$$

Putting $K_0^\perp := \emptyset$, $R_n := K_n^\perp \setminus K_{n-1}^\perp$ and $\alpha_n := \sum_{\sigma\in R_n} n(\sigma)^2 \exp(-2\overline{m}^{(\sigma)})$ one concludes that

$$\|f_t^{(n)}\|_2^2 \leq \|f_1^{(n)}\|_2^2 = \sum_{j=1}^n \alpha_j \quad \text{for all } t\in\mathbb{R}_+^*,\ t\geq 1\ (n\geq 1).$$

For every $n\geq 1$ there is a finite subset F_n of R_n such that

$$\sum_{\sigma\in S_n := R_n\setminus F_n} n(\sigma)^2 \exp(-2\overline{m}^{(\sigma)}) \leq \frac{1}{2^n} \quad \text{for all } n\geq 1.$$

Furthermore, for every $\sigma\in R_n$ there exists a

$$j\in\{c(n) := d(n-1)+1, c(n)+1, \ldots, d(n)\}$$

with the property $\|(X_j^{(n)} D^{(\sigma)})(e)\| \neq 0$. In fact, let

$$\|(X_j^{(n)} D^{(\sigma)})(e)\| = 0 \quad \text{for } j=c(n), \ldots, d(n)$$

and define

$$X := x_{c(n)} X_{c(n)}^{(n)} + \cdots + x_{d(n)} X_{d(n)}^{(n)} \quad \text{with} \quad x_{c(n)}, \ldots, x_{d(n)} \in \mathbb{R}.$$

Then

$$(XD^{(\sigma)})(e) = 0 \quad \text{so that} \quad D^{(\sigma)}(\exp_{G_n}(X)) = \exp((XD^{(\sigma)})(e)) = E_{n(\sigma)}.$$

But $\{X_{c(n)}^{(n)}, \ldots, X_{d(n)}^{(n)}\}$ is a basis of the Lie algebra of the connected Lie group $\ker q_{n-1} = K_{n-1}/K_n$; thus, $D^{(\sigma)}(x) = E_{n(\sigma)}$ for all $x\in\ker q_n$ or $\sigma\in K_{n-1}^\perp$.

4. From (*) in Part 3 it follows that for every $n\geq 2$ there exists a $b_n\geq 1$ such that after replacing a_j by $b_n a_j$ for $j=c(n), \ldots, d(n)$ we obtain

$$\sum_{\sigma\in F_n} n(\sigma)^2 \exp(-2\overline{m}^{(\sigma)}) \leq \frac{1}{2^n} \quad \text{for all } n\geq 2.$$

Hence, $\alpha_n \leq \frac{1}{2^{n+1}}$ for all $n\geq 2$, and so

$$\|f_t^{(n)}\|_2^2 \leq \sum_{j=1}^n \alpha_j \leq \alpha_1 + \sum_{j=2}^n \frac{1}{2^{j+1}} \leq \alpha_1 + \sum_{j\geq 2} \frac{1}{2^{j+1}} \leq \alpha_1 + 1$$

for all $n\geq 1$ (and $t\in[1, \infty[$).

Since $K_n^\perp \subset K_{n+1}^\perp$ for all $n\geq 1$ and $\Sigma(G) = \bigcup_{n\geq 1} K_n^\perp$, one obtains

$$\begin{aligned}
&\sum_{\sigma\in\Sigma(G)} n(\sigma)\,\mathrm{tr}(\exp(-2t\,\overline{\psi}(\sigma)))\\
&= \sup_{n\geq 1} \sum_{\sigma\in K_n^\perp} n(\sigma)^2 \exp(-2t\,\overline{m}^{(\sigma)})\\
&= \sup_{n\geq 1} \|f_t^{(n)}\|_2^2 \leq \alpha_1 + 1 \quad \text{for all } t\in[1, \infty[.
\end{aligned}$$

From Property (5) of 6.4.10 it follows that for every $t\in[1, \infty[$ there exists an $f_t\in\mathscr{L}^2(G,\omega)$ with $\mu_t = f_t\cdot\omega_G$. Property (4) of 6.4.10 implies that for $t\in[2, \infty[$ f_t can be chosen in $\mathscr{C}(G)$. □

6.4.12 Theorem. *Let G be a connected almost periodic group having a countable basis of its topology.*

The following statements are equivalent:
(i) *There exists a* $\mu \in \mathscr{G}(G)$ *with* $\mu \ll \omega_G$;
(ii) *G is locally connected.*

The proof of the theorem will be preceded by three lemmata.

6.4.13 Lemma. *Let G be a Lie projective group and D a discrete normal subgroup of G. By p we denote the canonical epimorphism from G onto* $\dot{G} := G/D$.

Then for every quadratic (primitive) form $\dot{\psi}$ *on* $\mathfrak{D}(\dot{G})$ *there exists exactly one quadratic (primitive) form* ψ *on* $\mathfrak{D}(G)$ *such that* $p(\psi) = \dot{\psi}$ *holds.*

If, moreover, $\dot{\psi}$ *is strict, then so is* ψ.

Proof. 1. Since G and $\dot{G}$ are locally isomorphic, there exist symmetric open neighborhoods $U \in \mathfrak{B}(e)$ and $\dot{U} \in \mathfrak{B}(\dot{e})$ such that $\chi := \mathrm{Res}_U\, p$ is an isomorphism from U onto $\dot{U}$. Let $\phi := \chi^{-1}$. For every $f \in \mathfrak{D}(G)$ with $\mathrm{supp}(f) \subset U$ we have $f \circ \phi \in \mathfrak{D}(\dot{G})$.

Let $V \in \mathfrak{B}(e)$ with $\bar{V} \subset U$ and $h \in \mathfrak{D}_+(G)$ with $h = h^*$, $\mathrm{supp}(h) \subset U$ and $h(x) = 1$ for all $x \in V$. Then $\mathrm{supp}(fh) \subset U$, so that $(fh) \circ \phi \in \mathfrak{D}(\dot{G})$ for all $f \in \mathfrak{D}(G)$.

Defining a mapping $\alpha : \mathfrak{D}(G) \to \mathfrak{D}(\dot{G})$ by

$$\alpha(f) := (fh) \circ \phi \qquad \text{for all } f \in \mathfrak{D}(G)$$

we obtain an algebra homomorphism from $\mathfrak{D}(G)$ into $\mathfrak{D}(\dot{G})$ with $\alpha(f^*) = \alpha(f)^*$ for all $f \in \mathfrak{D}(G)$ with $\alpha(f) \in \mathfrak{D}_0(\dot{G})$ whenever $f \in \mathfrak{D}_0(G)$. Thus for every quadratic (primitive) form $\dot{\psi}$ on $\mathfrak{D}(\dot{G})$ there exists a quadratic (primitive) form $\psi := \dot{\psi} \circ \alpha$ on $\mathfrak{D}(G)$.

2. For $\dot{f} \in \mathfrak{D}(\dot{G})$ one has $\alpha(\dot{f} \circ p) = \dot{f}(h \circ \phi)$ and $(h \circ \phi)(\dot{x}) = 1$ for all $\dot{x} \in \chi(V)$. Since $\dot{\psi}$ is concentrated, we obtain

$$\psi(\dot{f} \circ p) = \dot{\psi}(\alpha(\dot{f} \circ p)) = \dot{\psi}(\dot{f}(h \circ \phi)) = \dot{\psi}(\dot{f}) \quad \text{or} \quad p(\psi) = \dot{\psi}.$$

The uniqueness of ψ follows from Property (2) of 4.4.7, the statement concerning strictness from $\alpha(\mathfrak{B}(G)) \subset \mathfrak{B}(\dot{G})$. □

6.4.14 Lemma. *Let G be a finite-dimensional, connected, locally compact group. Then there exist a connected Lie group L, a dense Borel measurable subgroup M of G and a continuous monomorphism m from L onto M with the following property: For every Gauss semigroup* $(\mu_t)_{t \in \mathbb{R}_+^*}$ *in* $\mathscr{M}^1(G)$ *there is a Gauss semigroup* $(\mu_t')_{t \in \mathbb{R}_+^*}$ *in* $\mathscr{M}^1(L)$ *with* $m(\mu_t') = \mu_t$ *for all* $t \in \mathbb{R}_+^*$.

Proof. 1. Since G is finite-dimensional, there exists by Theorem H a totally disconnected, compact Abelian group K, a connected Lie group L and a discrete normal subgroup D of $H := K \times L$ with $G = H/D$. Let p denote the canonical mapping from H onto G. Also by Theorem H, $m := \mathrm{Res}_L\, p$ is a continuous monomorphism from L onto a dense subgroup M of G. Since L is σ-compact, so is M. Thus, M is Borel measurable.

2. The generating functional ψ of the Gauss semigroup $(\mu_t)_{t \in \mathbb{R}_+^*}$ in $\mathscr{M}^1(G)$ admits by Theorem 6.2.20 a representation $\psi = \psi_1 + \psi_2$ with a primitive form ψ_1 and a quadratic form ψ_2 on $\mathfrak{D}(G)$.

The lemma can be shown by reduction to the case of a Lie group. Since H is Lie projective, Lemma 6.4.13 provides a primitive form ψ'_1 and a quadratic form ψ'_2 on $\mathfrak{D}(H)$ with $p(\psi'_j)=\psi_j$ for $j=1,2$. For the Gauss semigroup $(\mu'_t)_{t\in\mathbb{R}_+^*}$ corresponding to $\psi':=\psi'_1+\psi'_2$ one has $p(\mu'_t)=\mu_t$ for all $t\in\mathbb{R}_+^*$. By Theorem 6.2.3 we have $\operatorname{supp}(\mu'_t)\subset H_0=L$ for all $t\in\mathbb{R}_+^*$ which implies that $(\mu'_t)_{t\in\mathbb{R}_+^*}$ can be considered as a Gauss semigroup in $\mathscr{M}^1(L)$. □

6.4.15 Lemma. *Let G be a connected locally compact group which is not locally connected. Then there is a dense, Borel measurable subgroup M of G with* $M\neq G$ *such that every Gauss semigroup* $(\mu_t)_{t\in\mathbb{R}_+^*}$ *in* $\mathscr{M}^1(G)$ *satisfies* $\mu_t(M)=1$ *for all* $t\in\mathbb{R}_+^*$.

Proof. 1. Let G be also finite-dimensional. Applying the notation of the preceding lemma we note that M is a dense, Borel measurable subgroup of G with

$$\mu_t(M)=m(\mu'_t)(m(L))=\mu'_t(L)=1 \qquad \text{for all } t\in\mathbb{R}_+^*.$$

If $M=G$, then G is topologically isomorphic to L and hence locally connected. Thus $M\neq G$.

2. Suppose that G is not necessarily finite-dimensional. Since G is not locally connected, there exists by Theorem H a compact normal subgroup K of G such that $H:=G/K$ is finite-dimensional and not locally connected. If p denotes the canonical mapping from G onto H, then $(p(\mu_t))_{t\in\mathbb{R}_+^*}$ can be considered as a Gauss semigroup in $\mathscr{M}^1(H)$ (by a proper choice of K). By Part 1 there exists a dense Borel measurable subgroup M_1 of H with $M_1\neq H$ such that $p(\mu_t)(M_1)=1$ holds for all $t\in\mathbb{R}_+^*$. The subgroup $M:=p^{-1}(M_1)$ of G has the desired properties. □

Proof of Theorem 6.4.12. 1. (i) ⇒ (ii). Let μ be in $\mathscr{G}(G)$ with absolutely continuous part $\mu^a\neq 0$ and suppose that G is not locally connected. Then by Lemma 6.4.15 there exists a dense, Borel measurable subgroup M of G with $M\neq G$ such that $\mu(M)=1$ holds. Since G is connected, one gets $\omega_G(M)=0$ by [218] (20.17), which contradicts $\mu^a\neq 0$.

2. (ii) ⇒ (i). By assumption, $G=\mathbb{R}^n\times K$ for $n\geq 0$ and a connected compact group K. Since G is locally connected, so is K. Therefore, by Theorem 6.4.11 there exists a $\nu\in\mathscr{G}(K)$ with $\nu\ll\omega_K$. Taking an n-dimensional normal distribution μ on $\mathbb{R}^n$ (if $n\geq 1$) we obtain an ω_G-absolutely continuous measure $\mu\otimes\nu$ in $\mathscr{G}(G)$. □

6.5 Convergence of Triangular Systems of Probability Measures

In this section we shall analyze connections between the classes $\mathscr{P}(G)$ and $\mathscr{G}(G)$ of Poisson and Gauss measures on a locally compact group G and the central limit theorem.

We start with a preparatory result concerning the embeddability of sequences of Poisson measures in $\mathscr{P}_{\{e\}}(G)$.

Let G be a Moore group and $(\nu_n)_{n\geq 1}$ a sequence of measures $\nu_n\in\mathscr{P}_{\{e\}}(G)$ of the form $\nu_n=\exp_{\{e\}}(-\lambda_n)$ with $\lambda_n:=\|\kappa_n\|\,\varepsilon_e-\kappa_n$ for $\kappa_n\in\mathscr{M}_+^b(G)$ $(n\geq 1)$.

For every $n \geq 1$ the measure ν_n can be embedded in an $\{e\}$-continuous convolution semigroup $(\nu_{n,t})_{t \in \mathbb{R}_+^*}$ in $\mathcal{M}^1(G)$ of the form

$$\nu_{n,t} = \exp_{\{e\}}(-t\lambda_n) \quad \text{for all } t \in \mathbb{R}_+^*.$$

It is known that for every $n \geq 1$ the infinitesimal generator N_n of the semigroup $(\nu_{n,t})_{t \in \mathbb{R}_+^*}$ is defined on $\mathscr{C}_u(G)$, and one has

$$(N_n f)(x) = \int (f(xy) - f(x))\, \kappa_n(dy) \quad \text{for all } f \in \mathscr{C}_u(G) \text{ and } x \in G.$$

6.5.1 Theorem. *Let G be a root compact Moore group, i.e., $G \in \mathbf{R} \cap \mathbf{M}$, let $(\nu_n)_{n \geq 1}$ be a sequence of measures $\nu_n \in \mathscr{P}_{\{e\}}(G)$ of the form*

$$\nu_n = \exp_{\{e\}}(-\lambda_n) \quad \textit{with}$$

$$\lambda_n := \|\kappa_n\|\, \varepsilon_e - \kappa_n \quad \textit{for } \kappa_n \in \mathcal{M}_+^b(G) \quad (n \geq 1) \quad \textit{and}$$

$$\nu := \lim_{n \to \infty} \nu_n.$$

We further assume the following condition to be satisfied:

(A) *For every $D \in \operatorname{Rep}(G)$ one has $\overline{\lim}_{n \geq 1} \|\int (D(y) - E_{n(D)})\, \kappa_n(dy)\| < \infty$.*

Then ν is embeddable in an $\{e\}$-continuous (rational) convolution semigroup $(\sigma_r)_{r \in \mathbb{Q}_+^*}$, and there exists a subnet $(n_i)_{i \in \mathbb{I}}$ of $\mathbb{N}$ such that for every $r \in \mathbb{Q}_+^*$ one has $\lim_{i \to \infty} \nu_{n_i, r} = \sigma_r$.

Proof. 1. Since the set $\mathcal{N} := \{\nu_n : n \geq 1\}$ is relatively $\mathcal{T}_v$-compact, by Theorem 3.1.4 the same is true for $R(m, \mathcal{N})$ for every $m \geq 1$ and hence, for $R(r, \mathcal{N})$ for every $r \in \mathbb{Q}_+^*$. Thus, since $\{\nu_{n,r} : n \geq 1\} \subset R(r, \mathcal{N})$ there exists a convergent subnet $(\nu_{n_i, r})_{i \in \mathbb{I}}$ with

$$\lim_{i \in \mathbb{I}} \nu_{n_i, r} =: \sigma_r \in \mathcal{M}^1(G).$$

Plainly, one can choose the subnet such that $\lim_{i \in \mathbb{I}} \nu_{n_i, r} = \sigma_r$ for all $r \in \mathbb{Q}_+^*$, and hence $S := (\sigma_r)_{r \in \mathbb{Q}_+^*}$ is a rational convolution semigroup in $\mathcal{M}^1(G)$ with $\sigma_1 = \nu$.

2. Let $D \in \operatorname{Rep}(G)$. Then Condition (A) yields

$$c := c(D) := \overline{\lim}_{n \geq 1} \|\int (D(y) - E_{n(D)})\, \kappa_n(dy)\| < \infty.$$

Thus

$$\begin{aligned} |\int D(y)\, \nu_{n,r}(dy) - D(e)| &= |\int_0^r \int_G (N_n D)(y)\, \nu_{n,s}(dy)\, ds| \\ &\leq r\, \|N_n D\| \leq r \|\int (D(y) - E_{n(D)})\, \kappa_n(dy)\| \leq rc, \end{aligned}$$

and in the limit along $\mathbb{I}$ we get

$$|\int D(y)\, \sigma_r(dy) - D(e)| \leq rc \quad \text{for all } r \in \mathbb{Q}_+^*.$$

This shows that

$$\lim_{r \downarrow 0} \int D(y)\, \sigma_r(dy) = D(e) = E_{n(D)} \quad \text{for all } D \in \operatorname{Rep}(G).$$

Hence, by Theorem 1.4.5 $\lim_{r\downarrow 0}\sigma_r=\varepsilon_e$, and the embedding semigroup $S=(\sigma_r)_{r\in\mathbb{Q}_+^*}$ is continuous. □

The central limit problem for probability measures on a locally compact group G concerns the limiting behavior of triangular systems $(\mu_{nj})_{j=1,\dots,k_n;\, n\geq 1}$ in $\mathcal{M}^1(G)$.

6.5.2 Definition. A triangular system $(\mu_{nj})_{j=1,\dots,k_n;\, n\geq 1}$ in $\mathcal{M}^1(G)$ is called *infinitesimal* if for each $U\in\mathfrak{V}(e)$ one has

$$\lim_{n\to\infty}\sup_{1\leq j\leq k_n}\mu_{nj}(\complement U)=0.$$

The system $(\mu_{nj})_{j=1,\dots,k_n;\, n\geq 1}$ is said to be *commutative* if for all $n\geq 1, 1\leq j,k\leq k_n$ the equality $\mu_{nj}*\mu_{nk}=\mu_{nk}*\mu_{nj}$ holds, and *convergent with limit* μ if for the *sequence* $(\mu_n)_{n\geq 1}$ *of* (n-th) *row products*

$$\mu_n:=\mathop{\ast}_{j=1}^{k_n}\mu_{nj}\qquad (n\geq 1)$$

there exists a $\mu\in\mathcal{M}^1(G)$ with $\lim_{n\to\infty}\mu_n=\mu$.

6.5.3 Remark. If $G\in\mathbf{M}$, then a triangular system $(\mu_{nj})_{j=1,\dots,k_n;\, n\geq 1}$ in $\mathcal{M}^1(G)$ is infinitesimal iff for any $D\in\operatorname{Rep}(G)$ the relation

$$\lim_{n\to\infty}\sup_{1<j<k_n}\|\hat{\mu}_{nj}(D)-E_{n(D)}\|=0$$

obtains.

Indeed, let $(\mu_{nj})_{j=1,\dots,k_n;\, n\geq 1}$ be infinitesimal. Then for every $D\in\operatorname{Rep}(G)$ and $\varepsilon>0$ there exists a $U\in\mathfrak{V}(e)$ with

$$\|D(x)-E_{n(D)}\|\leq\varepsilon\qquad\text{for all } x\in U.$$

Consequently, we get

$$\begin{aligned}\|\hat{\mu}_{nj}(D)-E_{n(D)}\| &= \|\textstyle\int (D(x)-E_{n(D)})\,\mu_{nj}(dx)\|\\ &\leq \textstyle\int \|D(x)-E_{n(D)}\|\,\mu_{nj}(dx)\leq\varepsilon\,\mu_{nj}(U)+2\mu_{nj}(\complement U)\\ &\leq\varepsilon+2\mu_{nj}(\complement U)\end{aligned}$$

for all $j=1,\dots,k_n$; $n\geq 1$, and hence

$$\lim_{n\to\infty}\sup_{1\leq j\leq k_n}\|\hat{\mu}_{nj}(D)-E_{n(D)}\|=0\qquad\text{for all } D\in\operatorname{Rep}(G).$$

Conversely, let the last limit relationship be given for $D\in\operatorname{Rep}(G)$. Then for any $U\in\mathfrak{V}(e)$ there exists a $j_n\in\{1,\dots,k_n\}$ defined by

$$\mu_{nj_n}(\complement U)=\sup_{1\leq j\leq k_n}\mu_{nj}(\complement U).$$

By assumption we obtain

$$\lim_{n\to\infty}\|\hat{\mu}_{nj_n}(D)-E_{n(D)}\|=0,$$

whence by Theorem 1.4.5 $\lim_{n\to\infty}\mu_{nj_n}=\varepsilon_e$ and thus $\lim_{n\to\infty}\mu_{nj_n}(\complement U)=0$. This shows

$$\lim_{n\to\infty}\sup_{1\leq j\leq k_n}\mu_{nj}(\complement U)=0.$$

6.5.4 Definition. Given a triangular system $(\mu_{nj})_{j=1,\ldots,k_n;\, n\geq 1}$ in $\mathcal{M}^1(G)$ we introduce the corresponding *accompanying* (triangular) *system* $(\nu_{nj})_{j=1,\ldots,k_n;\, n\geq 1}$ of measures

$$\nu_{nj} := \exp_{\{e\}}(\mu_{nj} - \varepsilon_e) \in \mathcal{P}_{\{e\}}(G) \qquad (j=1,\ldots,k_n;\ n\geq 1).$$

A routine argument shows that a triangular system $(\mu_{nj})_{j=1,\ldots,k_n;\, n\geq 1}$ is infinitesimal iff the corresponding accompanying system $(\nu_{nj})_{j=1,\ldots,k_n;\, n\geq 1}$ is.

6.5.5 Theorem. *Let $G\in\mathbf{M}$ and let $(\mu_{nj})_{j=1,\ldots,k_n;\, n\geq 1}$ be an infinitesimal system of measures in $\mathcal{M}^1(G)$ with corresponding accompanying system $(\nu_{nj})_{j=1,\ldots,k_n;\, n\geq 1}$. By $(\mu_n)_{n\geq 1}$ and $(\nu_n)_{n\geq 1}$ we denote the sequences of* n-th *row products of $(\mu_{nj})_{j=1,\ldots,k_n;\, n\geq 1}$ and $(\nu_{nj})_{j=1,\ldots,k_n;\, n\geq 1}$ resp.*

We further assume the validity of condition

(B) $\qquad \overline{\lim}_{n\geq 1} \sum_{j=1}^{k_n} \|\hat{\mu}_{nj}(D) - E_{n(D)}\| < \infty \qquad$ *for all* $D\in \operatorname{Rep}(G)$.

Then

$$\lim_{n\to\infty} \|\hat{\mu}_n(D) - \hat{\nu}_n(D)\| = 0 \qquad \textit{for all } D\in\operatorname{Rep}(G).$$

In particular, we have that $(\mu_{nj})_{j=1,\ldots,k_n;\, n\geq 1}$ converges with limit $\mu\in\mathcal{M}^1(G)$ iff $(\nu_{nj})_{j=1,\ldots,k_n;\, n\geq 1}$ converges with the same limit.

Proof. First of all we note that for any $D\in\operatorname{Rep}(G)$ and every $n\geq 1$ the following chain of inequalities holds:

$$\begin{aligned}
&\|\hat{\mu}_n(D) - \hat{\nu}_n(D)\| = \left\|\prod_{j=1}^{k_n} \hat{\mu}_{nj}(D) - \prod_{j=1}^{k_n} \hat{\nu}_{nj}(D)\right\| \\
&= \left\|\sum_{j=1}^{k_n} \left\{\prod_{l=1}^{j-1} \hat{\mu}_{nl}(D)\,[\hat{\mu}_{nj}(D) - \hat{\nu}_{nj}(D)] \prod_{m=j+1}^{k_n} \hat{\nu}_{nm}(D)\right\}\right\| \\
&\leq \sum_{j=1}^{k_n} \left\|\prod_{l=1}^{j-1} \hat{\mu}_{nl}(D)\,[\hat{\mu}_{nj}(D) - \hat{\nu}_{nj}(D)] \prod_{m=j+1}^{k_n} \hat{\nu}_{nm}(D)\right\| \\
&\leq \sum_{j=1}^{k_n} \|\hat{\mu}_{nj}(D) - \hat{\nu}_{nj}(D)\| \\
&= \sum_{j=1}^{k_n} \|\hat{\mu}_{nj}(D) - \exp(\hat{\mu}_{nj}(D) - E_{n(D)})\| \\
&\leq 2 \sup_{1\leq j\leq k_n} \|\hat{\mu}_{nj}(D) - E_{n(D)}\| \sum_{j=1}^{k_n} \|\hat{\mu}_{nj}(D) - E_{n(D)}\|.
\end{aligned}$$

Applying Condition (B) and the infinitesimality of $(\mu_{nj})_{j=1,\ldots,k_n;\, n\geq 1}$ in the equivalent form established in Remark 6.5.3 we obtain the validity of the first assertion. The second assertion follows from Theorem 1.4.5 since G is assumed to be in **M**. □

6.5.6 Corollary. *Let $G\in\mathbf{R}\cap\mathbf{M}$ and let the commutative infinitesimal system*

$$(\mu_{nj})_{j=1,\ldots,k_n;\, n\geq 1}$$

in $\mathcal{M}^1(G)$ converge to a measure $\mu\in\mathcal{M}^1(G)$. If $(\mu_{nj})_{j=1,\ldots,k_n;\, n\geq 1}$ satisfies Condition (B) *of* 6.5.5, *then $\mu\in\mathcal{E}_{\{e\}}(G)$.*

Proof. Let $D\in\operatorname{Rep}(G)$. Then by Theorem 1.4.5 we conclude that

$$\lim_{n\to\infty} \|\hat{\mu}_n(D) - \hat{\mu}(D)\| = 0,$$

and hence by the preceding theorem

$$\lim_{n\to\infty} \|\hat{\nu}_n(D) - \hat{\mu}(D)\| = 0.$$

Hence, again by Theorem 1.4.5, $\lim_{n\to\infty} \nu_n = \mu$. The assumptions imply Condition (A) of Theorem 6.5.1 (with $\kappa_n := \sum_{j=1}^{k_n} \mu_{nj}$ for all $n \geq 1$), and consequently, the assertion. □

6.5.7 Remark. In order to illustrate Condition (B) for a group $G \in \mathbf{M}$ we consider two special cases.

(1) Let $k_n := n$ and $\mu_{n1} = \cdots = \mu_{nn} =: \alpha_n$ for all $n \geq 1$. If the triangular system $(\mu_{nj})_{j=1,\ldots,n;\, n\geq 1}$ satisfies Condition (B), then it is infinitesimal by Remark 6.5.3. For every $D \in \operatorname{Rep}(G)$ and $n \geq 1$ one obtains $1 - |\det \hat{\alpha}_n(D)| \leq \hat{\alpha}_n(f_D)$.

Hence, if $(\mu_{nj})_{j=1,\ldots,n;\, n\geq 1}$ satisfies Condition (B), then for all $D \in \operatorname{Rep}(G)$ the condition $\overline{\lim}_{n\geq 1}\, n(1 - |\det \hat{\alpha}_n(D)|) < \infty$ holds.

One can show that this condition implies the absence of idempotent factors of the limit measure α of $(\mu_{nj})_{j=1,\ldots,n;\, n\geq 1}$.

(2) Let G be a Lie group in $\mathbf{M}$ of dimension $m \geq 1$ with Lie algebra $\mathfrak{L}(G)$, a system $\{x_1, \ldots, x_m\}$ of canonical coordinates of G with respect to a given basis $\{X_1, \ldots, X_m\}$ of $\mathfrak{L}(G)$ and a Hunt function ϕ. For any commutative, infinitesimal system $(\mu_{nj})_{j=1,\ldots,k_n;\, n\geq 1}$ in $\mathscr{M}^1(G)$ the following two conditions imply Condition (B) above:

(W 1) $\quad \overline{\lim}_{n\geq 1} \sum_{j=1}^{k_n} \int \phi \, d\mu_{nj} < \infty;$

(W 2) $\quad$ there exists a $U \in \mathfrak{V}(e)$ with

$$\overline{\lim}_{n\geq 1} \sum_{j=1}^{k_n} \left|\int_U x_i \, d\mu_{nj}\right| < \infty \qquad \text{for all } i = 1, \ldots, m.$$

In fact, from (W 1) follows the existence of $U \in \mathfrak{V}(e)$ and $\gamma := \gamma(U) \in \mathbb{R}_+^*$ with

$$\sum_{j=1}^{k_n} \mu_{nj}(\complement U) \leq \gamma(U) \qquad \text{for all } n \geq 1.$$

It therefore remains to show that (W 1) and (W 2) imply that for every $U \in \mathfrak{V}(e)$ and $D \in \operatorname{Rep}(G)$ there exists a $\delta := \delta(U, D) \in \mathbb{R}_+^*$ satisfying

$$\sum_{j=1}^{k_n} \left\|\int_U (D(x) - E_{n(D)})\, \mu_{nj}(dx)\right\| \leq \delta \quad \text{whenever} \quad n \geq 1.$$

We first choose $U \in \mathfrak{V}(e)$ small enough that the Taylor expansion obtains in the following form: For D in $\operatorname{Rep}(G)$ and $x \in U$ one has

$$D(x) = E_{n(D)} + \sum_{i=1}^{m} x_i(x)\, A_i(D)$$
$$+ \tfrac{1}{2} \sum_{i,l=1}^{m} x_i(x)\, x_l(x)\, T(D)(x)\, A_i(D)\, A_l(D)$$

with matrices $A_i(D) := \tilde{X}_i(D)(e)$ $(i = 1, \ldots, m)$ and $T(D)(x) \in \mathfrak{M}(n(D), \mathbb{C})$ satisfying

$$\|T(D)(x)\| \leq 1.$$

The formula is derived from the integral form of the remainder term of the expansion. If $x \in U$ and $X := \sum_{i=1}^m x_i(x) X_i$, then

$$T(D)(x) = 2\int_0^1 (1-t)\, D(\exp tX)\, dt.$$

Using the expansion of $D(x)$ we get for all $j=1, \dots, k_n$; $n \geq 1$ the inequalities

$$
\begin{aligned}
&\| \textstyle\int_U (D(x) - E_{n(D)})\, \mu_{nj}(dx) \| \\
&\leq \textstyle\int_U \| \sum_{i=1}^m x_i(x)\, A_i(D) \\
(*)\qquad &+ \tfrac{1}{2} \textstyle\sum_{i,l=1}^m x_i(x)\, x_l(x)\, T(D)(x)\, A_i(D)\, A_l(D) \| \, \mu_{nj}(dx) \\
&\leq \textstyle\sum_{i=1}^m \| A_i(D) \| \, | \int_U x_i(x)\, \mu_{nj}(dx) | \\
&+ \tfrac{1}{2} \textstyle\sum_{i,l=1}^m \| T(D)(x) \| \, \| A_i(D) \| \, \| A_l(D) \| \, | \int_U x_i(x)\, x_l(x)\, \mu_{nj}(dx) |.
\end{aligned}
$$

By (W2) the first summand on the right hand side of $(*)$ becomes bounded uniformly in $n \geq 1$. Let now U and $\{x_1, \dots, x_m\}$ be such that for all $x \in U$ one has $\phi(x) = \sum_{i=1}^m x_i(x)^2$.

Then one obtains for all $j=1, \dots, k_n$; $n \geq 1$:

$$
\begin{aligned}
&\textstyle\sum_{i,l=1}^m | \int_U x_i(x)\, x_l(x)\, \mu_{nj}(dx) | \\
&\leq \textstyle\sum_{i,l=1}^m \int_U (x_i(x)^2 + x_l(x)^2)\, \mu_{nj}(dx) \\
&= 2m \textstyle\int_U \sum_{i=1}^m x_i(x)^2\, \mu_{nj}(dx) = 2m \int_U \phi(x)\, \mu_{nj}(dx) \\
&\leq 2m \textstyle\int \phi(x)\, \mu_{nj}(dx).
\end{aligned}
$$

But then (W1) yields the boundedness uniformly in $n \geq 1$ of

$$\textstyle\sum_{i,l=1}^m | \int x_i(x)\, x_l(x)\, \mu_{nj}(dx) |,$$

and hence the boundedness of the second summand on the right hand side of $(*)$.

6.5.8 Corollary. *Let $G \in \mathbf{R} \cap \mathbf{M}$ and let $\mu \in \mathcal{M}^1(G)$ satisfy the following condition*
(S) *For every $n \geq 1$ there exists an n-th root $\alpha_n \in \mathcal{M}^1(G)$ of μ, and for all $D \in \mathrm{Rep}(G)$ one has $\overline{\lim}_{n \geq 1}\, n \| \hat{\alpha}_n(D) - E_{n(D)} \| < \infty$.*
Then $\mu \in \mathscr{E}_{\{e\}}(G)$.

Proof. For every $n \geq 1$ one puts $k_n := n$ and $\mu_{nj} := \alpha_n$ $(j=1, \dots, k_n)$. Then

$$\mu_n := \mu_{n1} * \cdots * \mu_{nk_n} = \alpha_n^n = \mu$$

and thus $(\mu_{nj})_{j=1,\dots,k_n;\, n \geq 1}$ is a commutative triangular system in $\mathcal{M}^1(G)$ converging to μ. From (S) it follows that $\lim_{n \to \infty} \| \hat{\alpha}_n(D) - E_{n(D)} \| = 0$, whenever $D \in \mathrm{Rep}(G)$.

Hence, by Remark 6.5.3 $(\mu_{nj})_{j=1,\dots,k_n;\, n \geq 1}$ is infinitesimal. In addition, (S) implies (B) of Theorem 6.5.5 for $(\mu_{nj})_{j=1,\dots,k_n;\, n \geq 1}$. Corollary 6.5.6 then implies the assertion. □

6.5.9 Corollary. *Let* $G \in \mathbf{R} \cap \mathbf{M}$ *and* $\mu \in \mathcal{M}^1(G)$. *The following statements are equivalent:*
(i) $\mu \in \mathscr{E}_{\{e\}}(G)$;
(ii) μ *satisfies Condition* (S) *of Corollary* 6.5.8.

Proof. It remains to prove the implication (i) $\Rightarrow$ (ii). Let $(\mu_t)_{t \in \mathbb{R}_+^*}$ be an $\{e\}$-continuous convolution semigroup in $\mathcal{M}^1(G)$ with corresponding negative-definite form ψ on $\mathfrak{R}(G)$ such that $\mu_1 = \mu$. From 1.5.11 we conclude that for all $D \in \operatorname{Rep}(G)$ the equalities

$$\psi(D) = \lim_{t \downarrow 0} \frac{1}{t}(E_{n(D)} - \hat{\mu}_t(D)) = \lim_{n \to \infty} n(E_{n(D)} - \hat{\mu}_{\frac{1}{n}}(D))$$

hold. With $\alpha_n := \mu_{\frac{1}{n}}$ for all $n \geq 1$ Condition (S) of Corollary 6.5.8 is satisfied. □

6.5.10 Remark. Let $G \in \mathbf{M}$ and let $(\mu_{nj})_{j=1,\ldots,k_n;\, n \geq 1}$ be a commutative triangular system in $\mathcal{M}^1(G)$ satisfying

$$\lim_{n \to \infty} \sum_{j=1}^{k_n} \|\hat{\mu}_{nj}(D) - E_{n(D)}\| = 0$$

for all $D \in \operatorname{Rep}(G)$.

Then $(\mu_{nj})_{j=1,\ldots,k_n;\, n \geq 1}$ converges to ε_e.

We now consider the convergence of triangular systems to Poisson measures in $\mathscr{P}^e_{\{e\}}(G)$ with parameter x_0 which were studied in § 1.

6.5.11 Theorem. *Let* $G \in \mathbf{M}$ and let $(\nu_n)_{n \geq 1}$ be a sequence of Poisson measures $\nu_n \in \mathcal{M}^1(G)$ of the form $\nu_n := \exp_{\{e\}}(-\lambda_n)$ *with*

$$\lambda_n := \|\kappa_n\| \varepsilon_e - \kappa_n \quad \textit{for } \kappa_n \in \mathcal{M}^b_+(G) \ (n \geq 1),$$

$x_0 \in G^*$ *and* $U_0 \in \mathfrak{V}(e)$ *symmetric with* $x_0 \notin U_0$. *For all* $U \in \mathfrak{V}(e)$ *with* $U \subset U_0$ *the following conditions are assumed to hold true.*
(i) $\lim_{n \to \infty} \kappa_n(\complement(U \cup U x_0)) = 0$.
(ii) $\lim_{n \to \infty} \|\int_U (D(x) - E_{n(D)}) \kappa_n(dx)\| = 0$ *for all* $D \in \operatorname{Rep}(G)$.
(iii) $\lim_{n \to \infty} \kappa_n(U x_0) =: c \in \mathbb{R}_+^*$.

Then $\lim_{n \to \infty} \nu_n = \nu := \exp_{\{e\}}(-\lambda)$ *with* $\lambda := c(\varepsilon_e - \varepsilon_{x_0})$.

Proof. Let $D \in \operatorname{Rep}(G)$. For every $n \geq 1$ we define $A_n := \hat{\lambda}_n(D)$ and $A := \hat{\lambda}(D)$.

1. We shall show that $\lim_{n \to \infty} \|A_n - A\| = 0$.

In fact, for an arbitrary $\varepsilon > 0$ there exists a $U \in \mathfrak{V}(e)$ with $U \subset U_0$ and $\|D(x) - D(x_0)\| < \varepsilon$ for all $x \in U x_0$. Furthermore, one has

$$\begin{aligned}
\|A - A_n\| &= \|\textstyle\int (D(y) - E_{n(D)}) \kappa_n(dy) + A\| \\
&\leq \|\textstyle\int_U (D(y) - E_{n(D)}) \kappa_n(dy)\| \\
&\quad + \|\textstyle\int_{\complement(U \cup U x_0)} (D(y) - E_{n(D)}) \kappa_n(dy)\| \\
&\quad + \|\textstyle\int_{U x_0} (D(y) - E_{n(D)}) \kappa_n(dy) - c(D(x_0) - E_{n(D)})\|.
\end{aligned}$$

By Assumptions (ii) and (i) the first and second summands on the right side of the inequality tend to 0. The third summand can be majorized by

$$\|\int_{Ux_0}(D(y)-D(x_0))\,\kappa_n(dy)\| + \|(\kappa_n(Ux_0)-c)(D(x_0)-E_{n(D)})\|$$

whose second summand tends to 0 by Assumption (iii).

This yields

$$\|\int_{Ux_0}(D(y)-D(x_0))\,\kappa_n(dy)\| \le \varepsilon\kappa_n(Ux_0).$$

From $\lim_{n\to\infty}\varepsilon\kappa_n(Ux_0)=\varepsilon c$ one derives the assertion.

2. Plainly, by Part 1 there exists a constant $a := a(D)\in\mathbb{R}_+^*$ such that $\|A_n\|\le a$ holds for all $n\ge 1$. Therefore

$$\|\hat{v}_n(D)-v(D)\| \le \|A_n - A\| + \frac{1}{2!}\|A_n^2 - A^2\| + \cdots$$

for every $n\ge 1$.

For every $\delta>0$ there exists an $m_0\ge 1$ such that $\sum_{m>m_0}\frac{1}{m!}\|A_n^m - A^m\| < \delta$ for all $n\ge 1$. On the other hand, we have for every $m\ge 1$

$$\begin{aligned}\|A_n^m - A^m\| &= \|\textstyle\sum_{i=1}^m A_n^{i-1}(A_n - A)\,A^{m-i}\|\\ &\le \textstyle\sum_{i=1}^m a^{m-1}\|A_n - A\| \qquad (n\ge 1).\end{aligned}$$

Thus, by Part 1

$$\lim_{n\to\infty}\|A_n^m - A^m\| = 0 \qquad \text{for } m=1,\dots,m_0,$$

and so $\overline{\lim}_{n\ge 1}\|\hat{v}_n(D)-\hat{v}(D)\|<\delta$. This implies that $\lim_{n\to\infty}v_n = v$, the asserted statement. □

6.5.12 Theorem. *Let $G\in\mathbf{M}$ and let $(\mu_{nj})_{j=1,\dots,k_n;n\ge 1}$ be an infinitesimal and commutative triangular system in $\mathcal{M}^1(G)$. Let $x_0\in G^*$ and let U_0 be a symmetric neighborhood in $\mathfrak{B}(e)$ with $x_0\notin U_0$ such that for all $U\in\mathfrak{B}(e)$ with $U\subset U_0$ the following conditions hold:*

(i) $\lim_{n\to\infty}\sum_{j=1}^{k_n}\mu_{nj}(\complement(U\cup Ux_0))=0$.

(ii) $\lim_{n\to\infty}\sum_{j=1}^{k_n}\|\int_U(D(x)-E_{n(D)})\,\mu_{nj}(dx)\|=0$ *for all* $D\in\mathrm{Rep}(G)$.

(iii) $\lim_{n\to\infty}\sum_{j=1}^{k_n}\mu_{nj}(Ux_0) =: c\in\mathbb{R}_+^*$.

Then $(\mu_{nj})_{j=1,\dots,k_n;n\ge 1}$ is convergent with limit $v=\exp_{\{e\}}(-c(\varepsilon_e-\varepsilon_{x_0}))$.

Proof. Let $D\in\mathrm{Rep}(G)$. From the conditions (i) to (iii) we conclude that

$$\overline{\lim}_{n\ge 1}\sum_{j=1}^{k_n}\|\hat{\mu}_{nj}(D)-E_{n(D)}\|<\infty.$$

Let $(v_{nj})_{j=1,\dots,k_n;n\ge 1}$ denote the accompanying system corresponding to

$$(\mu_{nj})_{j=1,\dots,k_n;n\ge 1}$$

and $(v_n)_{n\ge 1}$ the sequence of its row products.

By Theorem 6.5.5 it suffices to show that $\lim_{n\to\infty} \nu_n = \nu$. But the sequence $(\kappa_n)_{n\geq 1}$ defined by $\kappa_n := \sum_{j=1}^{k_n} \mu_{nj}$ for all $n \geq 1$ satisfies the conditions (i) to (iii) of Theorem 6.5.11. This implies the assertion. □

6.5.13 Theorem. *Let $G \in \mathbf{R} \cap \mathbf{M}$ and let $(\nu_n)_{n\geq 1}$ be a sequence of Poisson measures $\nu_n \in \mathscr{M}^1(G)$ of the form $\nu_n = \exp_{\{e\}}(-\lambda_n)$ with*

$$\lambda_n := \|\kappa_n\| \, \varepsilon_e - \kappa_n$$

for $\kappa_n \in \mathscr{M}_+^b(G)$ (all $n \geq 1$). The following conditions are assumed to be satisfied:
(i) $\lim_{n\to\infty} \kappa_n(\complement U) = 0$ *for all $U \in \mathfrak{B}(e)$.*
(ii) *For every $D \in \mathrm{Irr}(G)$ there exists a $V \in \mathfrak{B}(e)$ with*

$$\overline{\lim}_{n\geq 1} \|\int_V (D(x) - E_{n(D)}) \kappa_n(dx)\| < \infty.$$

Then if $\lim_{n\to\infty} \nu_n =: \nu \in \mathscr{M}^1(G) \setminus \mathscr{D}(G)$, one has $\nu \in \mathscr{G}(G)$.

Proof. 1. By Theorem 6.5.1 we obtain that ν is embeddable in an $\{e\}$-continuous one-parameter semigroup $(\sigma_r)_{r\in\mathbb{Q}_+^*}$ in $\mathscr{M}^1(G)$, which by Theorem 1.5.9 can be uniquely extended to an $\{e\}$-continuous one-parameter semigroup $(\sigma_t)_{t\in\mathbb{R}_+^*}$ in $\mathscr{M}^1(G)$ with corresponding negative-definite form ψ on $\mathfrak{R}(G)$.

2. For every $D \in \mathrm{Irr}(G)$ there exists an $r \in \mathbb{Q}_+^*$ with

$$\|r\hat{\lambda}_n(D)\| < \log \tfrac{3}{2} \quad \text{for all } n \geq 1,$$

which is a consequence of the assumptions (i) and (ii). Thus, by Property (2) of 3.2.2 we have

$$-r\hat{\lambda}_n(D) = \log(\exp(-r\hat{\lambda}_n(D))) \quad \text{and}$$
$$\|\exp(-r\hat{\lambda}_n(D)) - E_{n(D)}\| \leq \tfrac{1}{2} \quad \text{for all } n \geq 1.$$

An application of Theorem 6.5.1 now yields for a suitable subnet $(\hat{\lambda}_{n_\alpha}(D))_{\alpha\in\mathbf{A}}$ of $(\hat{\lambda}_n(D))_{n\geq 1}$ the limit relation

$$\lim_{\alpha\in\mathbf{A}} \exp(-r\hat{\lambda}_{n_\alpha}(D)) = \hat{\sigma}_r(D), \quad \text{hence} \quad \|\hat{\sigma}_r(D) - E_{n(D)}\| \leq \tfrac{1}{2}$$

and hence, by 1.5.11, $-r\psi(D) = \log \hat{\sigma}_r(D)$. We conclude that $\psi(D) = \lim_{\alpha\in\mathbf{A}} \hat{\lambda}_{n_\alpha}(D)$.

3. Given $D \in \mathrm{Irr}(G)$ we deduce from the assumptions (i) and (ii) the existence of an $a \in \mathbb{R}_+^*$ with $\|\hat{\lambda}_n(D)\| \leq a$ for all $n \geq 1$. Consequently,

$$\int f_D \, d\kappa_n = \mathrm{Re}\{\mathrm{tr}(\int [E_{n(D)} - D(x)] \, \kappa_n(dx))\}$$
$$= \mathrm{Re}\{\mathrm{tr}\, \hat{\lambda}_n(D)\} \leq n(D) a \quad \text{for all } n \geq 1.$$

For $\varepsilon > 0$ there is a $U \in \mathfrak{B}(e)$ with $f_D(x) \leq \varepsilon$ for all $x \in U$. This implies that

$$|\hat{\lambda}_n(f_D^2)| = |\int f_D^2 \, d\kappa_n| \leq \int_U f_D^2 \, d\kappa_n + \int_{\complement U} f_D^2 \, d\kappa_n$$
$$\leq (\sup_{x\in U} f_D(x)) \int f_D \, d\kappa_n + \|f_D\|^2 \kappa_n(\complement U)$$
$$\leq \varepsilon n(D) a + \|f_D\|^2 \kappa_n(\complement U) \quad \text{for all } n \geq 1.$$

From (i) follows $\overline{\lim}_{n\geq 1} |\hat{\lambda}_n(f_D^2)| \leq \varepsilon n(D) a$, so that $\lim_{n\to\infty} \hat{\lambda}_n(f_D^2) = 0$.

On the other hand, one has $\lim_{\alpha\in A}\hat{\lambda}_{n_\alpha}(f_D^2)=\psi(f_D^2)$. Hence, $\psi(f_D^2)=0$. This implies that $\nu\in\mathscr{G}(G)$ by Theorem 6.2.10. □

6.5.14 Theorem. *Let $G\in\mathbf{R}\cap\mathbf{M}$ and $\nu\in\mathscr{M}^1(G)\setminus\mathscr{D}(G)$. The following statements are equivalent:*
(i) $\nu\in\mathscr{G}(G)$;
(ii) *there exists an infinitesimal and commutative triangular system*

$$(\mu_{nj})_{j=1,\ldots,k_n;n\geq 1} \quad \textit{in } \mathscr{M}^1(G)$$

convergent with limit ν and satisfying the conditions:
(a) $\lim_{n\to\infty}\sum_{j=1}^{k_n}\mu_{nj}(\complement U)=0$ *for all* $U\in\mathfrak{B}(e)$;
(b) *for every* $D\in\mathrm{Irr}(G)$ *there is a* $V\in\mathfrak{B}(e)$ *with*

$$\overline{\lim}_{n\geq 1}\sum_{j=1}^{k_n}\|\int_V(D(x)-E_{n(D)})\,\mu_{nj}(dx)\|<\infty.$$

Proof. 1. (i) ⇒ (ii). Let $\nu\in\mathscr{G}(G)$ be such that there exists an $\{e\}$-continuous semigroup $(\nu_t)_{t\in\mathbb{R}_+^*}$ in $\mathscr{M}^1(G)$ with $\nu_1=\nu$ and

$$\lim_{t\downarrow 0}\frac{1}{t}\nu_t(\complement U)=0 \qquad \text{for all } U\in\mathfrak{B}(e).$$

For every $n\geq 1$ we put $\mu_{nj}:=\nu_{\frac{1}{n}}$ $(j=1,\ldots,k_n:=n)$. Then the triangular system $(\mu_{nj})_{j=1,\ldots,n;\,n\geq 1}$ is commutative and convergent with limit ν. Since

$$\lim_{n\to\infty}\nu_{\frac{1}{n}}=\varepsilon_e,$$

the triangular system $(\mu_{nj})_{j=1,\ldots,n;\,n\geq 1}$ is also infinitesimal. Finally, for every $U\in\mathfrak{B}(e)$ we have

$$\lim_{n\to\infty} n\nu_{\frac{1}{n}}(\complement U)=0,$$

which implies Condition (a). Let ψ be the negative-definite form corresponding to $(\nu_t)_{t\in\mathbb{R}_+^*}$ and let $D\in\mathrm{Irr}(G)$. Then

$$\lim_{n\to\infty} n\,\|\int(D(x)-E_{n(D)})\,\nu_{\frac{1}{n}}(dx)\|=\|\psi(D)\|.$$

This implies Condition (b), again by (a).

2. (ii) ⇒ (i). For every $n\geq 1$ we consider the measures $\kappa_n:=\sum_{j=1}^{k_n}\mu_{nj}$ and λ_n, ν_n as in the statement of Theorem 6.5.13.

Applying Theorem 6.5.5 we obtain $\lim_{n\to\infty}\nu_n=\nu$. But Theorem 6.5.13 applied to the sequence $(\kappa_n)_{n\geq 1}$ defined above yields the assertion. □

6.5.15 Theorem. *Let $G\in\mathbf{R}\cap\mathbf{M}$ and let $(\mu_{nj})_{j=1,\ldots,k_n;n\geq 1}$ be an infinitesimal triangular system in $\mathscr{M}^1(G)$ convergent with limit $\nu\in\mathscr{M}^1(G)$.*

If $\nu\in\mathscr{G}(G)$, then for all $U\in\mathfrak{B}(e)$ the relation

$$\lim_{n\to\infty}\sum_{j=1}^{k_n}\mu_{nj}(\complement U)=0 \quad \textit{holds.}$$

Proof. Since by Theorem 6.2.3 the measure ν is always supported by G_0, we may assume without loss of generality that G is connected. In this case there exist an $n \geq 0$ and a compact connected group K such that $G = \mathbb{R}^n \times K$ by the Freudenthal-Weil theorem L.

1. In the case $G := \mathbb{R}^n$ the theorem follows from Theorem 5.4.2.

2. The validity of the asserted limit relation is preserved under the formation of direct products of groups. Let $G := G_1 \times G_2$ be a locally compact group and p_i the canonical projection from G onto G_i for $i = 1, 2$. We choose $U \in \mathfrak{B}(e)$ which without loss of generality is of the form $U := U_1 \times U_2$ with $U_i \in \mathfrak{B}(e_i)$ for $i = 1, 2$. But for every $\mu \in \mathscr{M}^1(G)$ one has

$$\mu(\complement U) \leq p_1(\mu)(\complement U_1) + p_2(\mu)(\complement U_2).$$

Thus, if $\nu \in \mathscr{G}(G)$ and if the asserted limit relation is valid for the measures $p_i(\nu) \in \mathscr{G}(G_i)$ for $i = 1, 2$, then it is also valid for ν.

3. By Part 2 it suffices to prove the theorem in the case of a compact (connected) group G $(= K)$.

For a given $U \in \mathfrak{B}(e)$ there exists a $D \in \operatorname{Rep}(G)$ such that

$$f_D(x) = \operatorname{Re}[\operatorname{tr}(E_{n(D)} - D(x))] \geq a > 0$$

for all $x \in \complement U$, where $a \in \mathbb{R}_+^*$.

For every $j = 1, \ldots, k_n$; $n \geq 1$ we put

$$\pi_{nj} := \mu_{nj} * \mu_{nj}^{\sim} \quad \text{and} \quad \pi_n := \mu_n * \mu_n^{\sim}.$$

Hence, $\det \hat{\pi}_n(D) = |\det \hat{\mu}_n(D)|^2 = \prod_{j=1}^{k_n} \det \hat{\pi}_{nj}(D)$.

Since for every $j = 1, \ldots, k_n$; $n \geq 1$ the matrix $\hat{\pi}_{nj}(D)$ is positive Hermitian, the infinitesimality of $(\mu_{nj})_{j=1,\ldots,k_n; n \geq 1}$ implies that the eigenvalues of $\hat{\pi}_{nj}(D)$ are near 1 for sufficiently large $n \geq 1$ and therefore that the logarithm of $\det \hat{\pi}_n(D)$ is arbitrarily small.

Consequently, one obtains for fixed $\varepsilon > 0$, $n \geq n(\varepsilon)$ and for all $j = 1, \ldots, k_n$

$$\begin{aligned} &\int_G f_D(x)\, \pi_{nj}(dx) = \operatorname{tr}(E_{n(D)} - \hat{\pi}_{nj}(D)) \\ &\leq -\log(\det \hat{\pi}_{nj}(D)) \leq (1+\varepsilon) \operatorname{tr}(E_{n(D)} - \hat{\pi}_{nj}(D)). \end{aligned}$$

On the other hand, one has $4n(D) f_D - f_{D \otimes D} - f_{D \otimes \bar{D}} = 2 f_D^2$, which together with the above inequalities implies that

$$\begin{aligned} &-4n(D) \log(\det \hat{\pi}_n(D)) \\ &+ \frac{1}{1+\varepsilon} \{\log(\det \hat{\pi}_n(D \otimes \bar{D})) + \log(\det \hat{\pi}_n(D \otimes D))\} \\ &\geq 2 \sum_{j=1}^{k_n} \int_G f_D^2 \, d\pi_{nj} \geq 2a^2 \sum_{j=1}^{k_n} \pi_{nj}(\complement U). \end{aligned}$$

The left side of the first inequality converges for $n \to \infty$ to the analogous expression with π_n replaced by $\pi := \nu * \nu^{\sim}$. Hence, the sequence

$$\left(\sum_{j=1}^{k_n} \pi_{nj}(\complement U)\right)_{n \geq 1}$$

is bounded. Replacing the neighborhood U by a symmetric one $V\in\mathfrak{B}(e)$ such that $V^2\subset U$ holds, one obtains $V(\complement U)\subset\complement V$. Thus,

$$\pi_{nj}(\complement V)\geq\mu_{nj}(\complement U)\,\mu_{nj}^{\sim}(V)$$

and since by assumption $\mu_{nj}^{\sim}(V)\geq\frac{1}{2}$ for sufficiently large n $(j=1,\dots,k_n)$, there exists a constant $c:=c(U)$ with the property

$$\overline{\lim}_{n\geq 1}\sum_{j=1}^{k_n}\mu_{nj}(\complement U)\leq c<\infty.$$

Since $\nu\in\mathscr{G}(G)$, by Theorem 6.2.10 the (Gauss) condition

$$-4n(D)\log(\det\hat{\pi}(D))+\log(\det\hat{\pi}(D\otimes D))$$
$$+\log(\det\hat{\pi}(D\otimes\bar{D}))=0$$

is satisfied. Then the above chain of inequalities yields

$$\lim_{n\to\infty}\sum_{j=1}^{k_n}\pi_{nj}(\complement U)=0\qquad\text{for every } U\in\mathfrak{B}(e).$$

The preceding discussion implies the assertion. □

6.6 Central Limit Theorems for Totally Disconnected Groups

For the class of compact groups the main aspects of the central limit theorem can be studied more explicitly. We shall take up in this section the detailed analysis of the accompanying laws theorem established in the general framework of Moore groups in the preceding section as well as the embeddability of the limits of infinitesimal triangular systems into continuous convolution semi- or hemigroups.

First of all we shall settle the problem of the infinite divisibility of limits of commutative, infinitesimal triangular systems of probability measures on a compact group.

6.6.1 Theorem. *Let G be a locally compact group in $\mathbf{R}$ which is the projective limit of a projective system $(G_\alpha, p_{\alpha\beta}, \mathbb{A})$ of the groups $G_\alpha:=G/K_\alpha$ $(\alpha\in\mathbb{A})$, where $(K_\alpha)_{\alpha\in\mathbb{A}}$ is a descending system of compact normal subgroups of G satisfying $\bigcap_{\alpha\in\mathbb{A}}K_\alpha=\{e\}$. Then for any measure $\mu\in\mathscr{M}^1(G)$ the following statements are equivalent:*
(i) $\mu\in\mathscr{I}(G)$;
(ii) $p_\alpha(\mu)\in\mathscr{I}(G_\alpha)$ *for all* $\alpha\in\mathbb{A}$.

Proof. 1. (i)⇒(ii) is trivial. We need only show

2. (ii)⇒(i). Let μ be in $\mathscr{M}^1(G)$ such that $p_\alpha(\mu)\in\mathscr{I}(G_\alpha)$ for all $\alpha\in\mathbb{A}$. We fix $\alpha\in\mathbb{A}$ and observe that $G_\alpha\in\mathbf{R}$ by Property (3) of 3.1.2. Hence, for all $n\geq 1$ the root set $R(n,p_\alpha(\mu))\neq\varnothing$ is $\mathscr{T}_v$-compact in $\mathscr{M}^1(G_\alpha)$ by Theorem 3.1.4. But then the set

$$R_\alpha:=p_\alpha^{-1}(R(n,p_\alpha(\mu)))$$

is nonempty, and $\mathscr{T}_v$-compact by the compactness of K_α. For $\alpha, \beta \in \mathbb{A}$ with $\alpha < \beta$ we now have $R_\beta \subset R_\alpha$. Therefore,

$$R := \bigcap_{\alpha \in \mathbb{A}} R_\alpha \neq \emptyset.$$

We pick $v \in R$ and get $p_\alpha(v^n) = (p_\alpha(v))^n = p_\alpha(\mu)$ for all $\alpha \in \mathbb{A}$, which implies that $v^n = \mu$. This can be done for every $n \geq 1$, so that $\mu \in \mathscr{I}(G)$. $\square$

The following theorem will be preceded by a technical

6.6.2 Lemma. *Let $\{x_1, \dots, x_m\}$ be a set of vectors in $\mathbb{R}^n$ satisfying for $\varepsilon > 0$ $\|x_i\| < \varepsilon$. Let $a := \sum_{i=1}^m x_i$ and $\lambda \in]0, 1[$. Then there exists a subset $\mathbb{I}$ of $\{1, \dots, m\}$ such that $\|\sum_{i \in \mathbb{I}} x_i - \lambda a\| < n\varepsilon$.*

Proof. We introduce the mapping $\psi : t := (t_1, \dots, t_m) \to \sum_{i=1}^m t_i x_i$ from $\mathbb{R}^m$ into $\mathbb{R}^n$ and the set

$$E := \psi([0, 1]^m).$$

For a given $b \in E$ the set

$$F := \{t \in [0, 1]^m : \psi(t) = b\}$$

is a nonempty convex compact subset of $\mathbb{R}^m$ with a nonempty set F_e of extreme points. We shall show that any $f \in F_e$ has at most n coordinates different from 0 and 1. In fact, let $f_{i_1}, \dots, f_{i_k}$ be the coordinates of f which equal 0 or 1. The system of equations $t_{i_j} = 0$ for $j = 1, \dots, k$ and $\psi(t) = 0$ admits a nontrivial solution $t_0 \in \mathbb{R}^m$ if $k + n < m$. But for a sufficiently small $\alpha \in \mathbb{R}$ (with $|\alpha| < \delta$ say) we get $f + \alpha t_0 \in F$, which contradicts the extremality of f. Let $\mathbb{I}$ denote the set of all $i \in \{i_1, \dots, i_k\}$ satisfying $f_i = 1$. Then one obtains

$$\|\sum_{i \in \mathbb{I}} x_i - b\| = \|\sum_{i \in \mathbb{I}} x_i - \psi(f)\| = \|\sum_{\substack{i \in \complement \mathbb{I} \\ f_i \in]0, 1[}} f_i x_i\| < n\varepsilon. \quad \square$$

6.6.3 Corollary. *For fixed $p \in \mathbb{N}$, $p \leq m$, the set $\{1, \dots, m\}$ can be decomposed into p disjoint classes $\mathbb{I}_1, \dots, \mathbb{I}_p$ such that for some constant $C := C(n, p) \in \mathbb{R}_+^*$ one has*

$$\|\sum_{i \in \mathbb{I}_r} x_i - \sum_{i \in \mathbb{I}_s} x_i\| < C\varepsilon$$

whenever $r, s = 1, \dots, p$.

The *proof* follows from a successive application of the lemma. $\square$

6.6.4 Theorem. *Let G be a compact group and $(\mu_{nj})_{j=1,\dots,k_n; n \geq 1}$ a commutative infinitesimal system in $\mathscr{M}^1(G)$ which converges to a measure $\mu \in \mathscr{M}^1(G)$. Then $\mu \in \mathscr{I}(G)$.*

Proof. 1. Being a compact group G is the projective limit of (a projective family of) compact Lie groups whose topology has a countable basis. By Theorem 6.6.1 we may therefore assume without loss of generality that G is a compact (Lie) group having a countable basis of its topology. Let $\Sigma(G)=\{\sigma_q : q\geq 1\}$ be the (countable) system $\mathrm{Irr}(G)/\sim$ of equivalence classes of all irreducible (unitary, continuous) representations in $\mathrm{Rep}(G)$ (introduced in §3 of Chapter I). For every $\sigma\in\Sigma(G)$ we shall choose a representative $D^{(\sigma)}\in\sigma$.

Given the triangular system $(\mu_{nj})_{j=1,\dots,k_n;n\geq 1}$ in $\mathscr{M}^1(G)$ and a $\sigma\in\Sigma(G)$ we introduce the corresponding system $(M_{nj}^{(\sigma)})_{j=1,\dots,k_n;n\geq 1}$ of matrices

$$M_{nj}^{(\sigma)} := \hat{\mu}_{nj}(D^{(\sigma)})\in\mathfrak{M}(n(\sigma),\mathbb{C}).$$

2. Let $p,Q\geq 1$ and $\varepsilon>0$. We shall show that there exists an $n_0:=n_0(Q,\varepsilon)$ such that for every $n\geq n_0$ one can decompose the set $\{1,\dots,k_n\}$ into p disjoint classes $\mathrm{II}_1^{(n)},\dots,\mathrm{II}_p^{(n)}$ satisfying

$$\|\prod_{j\in\mathrm{II}_r^{(n)}} M_{nj}^{(\sigma_q)} - \prod_{j\in\mathrm{II}_s^{(n)}} M_{nj}^{(\sigma_q)}\| < \varepsilon \quad \text{whenever } q=1,\dots,Q$$

and $r,s=1,\dots,p$. In fact, by Remark 6.5.3 the infinitesimality of $(\mu_{nj})_{j=1,\dots,k_n;n\geq 1}$ implies that for every $\sigma\in\Sigma(G)$

$$\lim_{n\to\infty}\sup_{1\leq j\leq k_n}\|M_{nj}^{(\sigma)}-E_{n(\sigma)}\|=0.$$

Hence, the matrices $N_{nj}^{(\sigma)} := \log M_{nj}^{(\sigma)}$ are well-defined for $j=1,\dots,k_n$ and sufficiently large n, and

$$\lim_{n\to\infty}\sup_{1\leq j\leq k_n}\|N_{nj}^{(\sigma)}\|=0.$$

We now consider the Euclidean space $\mathbb{R}^m$ of dimension

$$m := 2(m_1^2+\dots+m_Q^2),$$

where $m_q := n(\sigma_q)$ for $q=1,\dots,Q$, and the vectors $x_1,\dots,x_{k_n}\in\mathbb{R}^m$ whose components are the real and the imaginary parts of the coefficients of $N_{nj}^{(\sigma_q)}$ for $q=1,\dots,Q$. These vectors have arbitrary small norms for sufficiently large $n\geq n_0$ and hence satisfy the hypothesis of Corollary 6.6.3. We can therefore decompose $\{1,\dots,k_n\}$ into p disjoint classes $\mathrm{II}_1^{(n)},\dots,\mathrm{II}_p^{(n)}$ such that

$$\|\sum_{i\in\mathrm{II}_r^{(n)}} x_i - \sum_{i\in\mathrm{II}_s^{(n)}} x_i\|,$$

whence

$$\|\sum_{j\in\mathrm{II}_r^{(n)}} N_{nj}^{(\sigma_q)} - \sum_{j\in\mathrm{II}_s^{(n)}} N_{nj}^{(\sigma_q)}\|$$

becomes arbitrarily small for all $q=1,\dots,Q$ and $r,s=1,\dots,p$. The commutativity of $(\mu_{nj})_{j=1,\dots,k_n;n\geq 1}$ implies the equality

$$\prod_{j\in\mathrm{II}_r^{(n)}} M_{nj}^{(\sigma_q)} - \prod_{j\in\mathrm{II}_s^{(n)}} M_{nj}^{(\sigma_q)}$$
$$=\exp\big(\sum_{j\in\mathrm{II}_s^{(n)}} N_{nj}^{(\sigma_q)}\big)\big[\exp\big(\sum_{j\in\mathrm{II}_r^{(n)}} N_{nj}^{(\sigma_q)} - \sum_{j\in\mathrm{II}_s^{(n)}} N_{nj}^{(\sigma_q)}\big) - E_{n(\sigma_q)}\big],$$

which is valid for all $q=1,\dots,Q$ and $r,s=1,\dots,p$ whence the assertion.

3. We shall now apply the convergence of the triangular system

$$(\mu_{nj})_{j=1,\dots,k_n;n\geq 1}$$

in order to complete the proof of the theorem. Let $p, Q \geq 1$ be given as in Part 2. We put

$$\varepsilon_Q := \frac{1}{Q}, \quad n_Q := n_0(Q, \varepsilon_Q) \quad \text{and}$$

$$\mu_{n_Q, r} := \ast_{j \in \mathbb{I}_r^{(n_Q)}} \mu_{n_Q j} \quad \text{for } r = 1, \ldots, p.$$

We then obtain

$$\|\hat{\mu}_{n_Q, r}(D^{(\sigma_q)}) - \hat{\mu}_{n_Q, s}(D^{(\sigma_q)})\| < \frac{1}{Q}$$

for all $q = 1, \ldots, Q$; $r, s = 1, \ldots, p$ and $\mu_{n_Q} = \ast_{r=1}^{p} \mu_{n_Q, r}$. Going over to a suitable subsequence yields the existence of $\nu_r := \mathcal{T}_w\text{-}\lim_{Q \to \infty} \mu_{n_Q, r}$ in $\mathcal{M}^1(G)$. Plainly,

$$\nu_1 = \nu_2 = \cdots = \nu_p =: \nu \quad \text{and hence} \quad \mu = \nu^p.$$

But this shows that for every $p \geq 1$ there exists a $\lambda_p \in \mathcal{M}^1(G)$ satisfying $\mu = \lambda_p^p$ or $\mu \in \mathcal{I}(G)$. □

6.6.5 Theorem. *Let G be a compact group with the property that every connected compact subgroup of G is arcwise connected (for example, a totally disconnected compact group). Then every convergent commutative, infinitesimal system in $\mathcal{M}^1(G)$ has a limit measure in $\mathcal{E}(G)$.*

The *proof* follows from Theorem 6.6.4 together with Theorem 3.5.8. □

We are now returning to the case of an arbitrary locally compact group G. A measure $\mu \in \mathcal{M}^1(G)$ is called *regular* if the convolution operator T_μ (on $\mathscr{C}^0(G)$) of μ is injective. For compact groups G the regularity of $\mu \in \mathcal{M}^1(G)$ is equivalent with the condition $\det \hat{\mu}(D) \neq 0$ for all $D \in \mathrm{Rep}(G)$.

6.6.6 Remark. Let G be a non-totally disconnected, locally compact group such that G_0 is not arcwise connected. Then there exists an infinitesimal system $(\mu_{nj})_{j=1,\ldots,k_n; n \geq 1}$ in $\mathcal{M}^1(G)$ which converges to a regular limit $\mu \notin \mathcal{E}(G)$.

In fact, applying Theorem K to the connected group G_0 we obtain the existence of a compact connected group K and of an $m \geq 0$ such that $\mathbb{R}^m \times K$ is homeomorphic to G_0 under the mapping $(x, k) \to xk$. Since by assumption G_0 is not arcwise connected, K is also not arcwise connected. Therefore, there exists a subgroup of K which is isomorphic to a non-arcwise connected solenoidal group $\mathbb{S}$.

In other words, there exists a continuous homomorphism $\xi: \mathbb{R} \to K$ such that $\overline{\xi(\mathbb{R})} = \mathbb{S}$, and $\mathbb{S}$ contains elements which lie on none of its one-parameter subgroups. Thus without loss of generality it can be assumed that $G := \mathbb{S} := \mathbb{Q}_d^{\wedge}$.

We shall now construct an infinitesimal system $(\mu_{nj})_{j=1,\ldots,k_n; n \geq 1}$ in $\mathcal{M}^1(G)$ which converges to a regular limit $\mu \notin \mathcal{E}(G)$. Given an $x \in G$ such that $\varepsilon_x \notin \mathcal{E}(G)$ we choose a sequence $(r_n)_{n \geq 1}$ in $\mathbb{R}$ satisfying

$$\lim_{n \to \infty} y^{(n)} = \lim_{n \to \infty} \xi(r_n) = x.$$

For a fixed $n \geq 1$ we have

$$\lim_{k\to\infty} y_k^{(n)} := \lim_{k\to\infty} \xi\left(\frac{r_n}{k}\right) = e$$

and $(y_k^{(n)})^k = y^{(n)}$ for all $k \geq 1$. Let $(U_n)_{n\geq 1}$ be a basis of $\mathfrak{V}(e)$. Then for every $n \geq 1$ there exists a $k_n \geq 1$ with $y_{k_n}^{(n)} \in U_n$. Consequently, the triangular system $(\mu_{nj})_{j=1,\ldots,k_n;\, n\geq 1}$ in $\mathscr{M}^1(G)$ defined by

$$\mu_{nj} := \varepsilon_{y_{k_n}^{(n)}}$$

for all $j = 1, \ldots, k_n; n \geq 1$ is infinitesimal and the corresponding sequence $(\mu_n)_{n\geq 1}$ of n-th row products

$$\ast_{j=1}^{k_n} \mu_{nj} = \varepsilon_{y^{(n)}} \qquad (n \geq 1)$$

satisfies $\lim_{n\to\infty} \mu_n = \varepsilon_x$ with a regular limit $\varepsilon_x \notin \mathscr{E}(G)$.

In order to find a triangular system of nondegenerate measures serving our purpose we modify the above construction as follows: For every $t \in \mathbb{R}_+$ we introduce

$$\nu_t := \exp_{\{e\}}[t(\omega_G - \varepsilon_e)] = (1 - e^{-t})\omega_G + e^{-t}\varepsilon_e$$

and define

$$\bar{\mu}_{nj} := \varepsilon_{y_{k_n}^{(n)}} * \nu_{\frac{1}{k_n}} \qquad \text{for all } j = 1, \ldots, k_n;\ n \geq 1.$$

The infinitesimal system $(\bar{\mu}_{nj})_{j=1,\ldots,k_n;\, n\geq 1}$ in $\mathscr{M}^1(G)$ satisfies $\operatorname{supp}(\bar{\mu}_{nj}) = G$ for all $j = 1, \ldots, k_n;\ n \geq 1$, and for the corresponding sequence $(\bar{\mu}_n)_{n\geq 1}$ of n-th row products $\bar{\mu}_n = \varepsilon_{y^{(n)}} * \nu_1$ $(n \geq 1)$ we get

$$\lim_{n\to\infty} \bar{\mu}_n = \varepsilon_x * \nu_1,$$

where $\varepsilon_x * \nu_1$ is a regular limit $\notin \mathscr{E}(G)$.

The connection between the convergence of infinitesimal systems $(\mu_{nj})_{j=1,\ldots,k_n;\, n\geq 1}$ in $\mathscr{M}^1(G)$ and their corresponding accompanying systems $(\nu_{nj})_{j=1,\ldots,k_n;\, n\geq 1}$ will be discussed in two steps.

First we settle the problem for the case of a finite group and later we extend the result to totally disconnected compact groups by the usual projective limit argument. We shall need various

6.6.7 Elementary Facts on Stochastic Matrices. Let $\mathfrak{S}$ be the set of all $d \times d$-stochastic matrices (in $\mathfrak{M}(d, \mathbb{R})$) and $\mathfrak{S}_0$ the subset of doubly stochastic matrices in $\mathfrak{S}$. Clearly, $\mathfrak{S}$ and $\mathfrak{S}_0$ are compact convex sets.

The set $\mathfrak{S}_e$ of extreme points of $\mathfrak{S}$ can be identified with the set of stochastic matrices with 0 and 1 entries only and the set $\mathfrak{S}_{0e}$ of extreme points of $\mathfrak{S}_0$ is exactly the set of permutation matrices. It is easily shown that $\mathfrak{S}_e$ is in fact a semigroup and that $\mathfrak{S}_{0e}$ is the group of invertible elements of $\mathfrak{S}_e$.

Let G be a finite semigroup $\{x_1, \dots, x_d\}$ of order $d \geq 1$ with unit element. We consider the mapping D from G into $\mathfrak{S}$ defined by

$$D(x)_{ij} := \begin{cases} 1, & \text{if } x_i = x\, x_j \\ 0 & \text{otherwise} \end{cases}$$

for all $x \in G$ and $i, j = 1, \dots, d$.

Obviously, D is a semigroup homomorphism from G into the set $\mathfrak{S}_e$ of extreme points of $\mathfrak{S}$ which can be uniquely extended to a homomorphism D from $\mathcal{M}^1(G)$ into $\mathfrak{S}$ by the convention

$$D(\mu) := \sum_{x \in G} D(x)\, \mu(\{x\}),$$

valid for all $\mu \in \mathcal{M}^1(G)$.

Obviously, the representation D is norm preserving in the sense that

$$\|D(\mu)\| = \|\mu\| \qquad \text{for all } \mu \in \mathcal{M}^1(G).$$

In what follows it will be helpful to know that $D(G)$ is a finite subsemigroup of $\mathfrak{S}_e$.

Furthermore, we shall apply the homomorphism $\Delta : \mathfrak{S} \to \overline{\mathbb{R}}_+$ defined by

$$\Delta(S) := -\log|\det S| \qquad \text{for all } S \in \mathfrak{S}.$$

One has $\Delta(S) = \infty$ for singular $S \in \mathfrak{S}$ and $\Delta(S) = 0$ for $S \in \mathfrak{S}_0$.

In view of the above representation D of G one also introduces the variance H as the homomorphism

$$\mu \to H(\mu) := -\log|\det D(\mu)| \qquad \text{from } \mathcal{M}^1(G) \text{ into } \overline{\mathbb{R}}_+.$$

If G is a group, one observes that a measure $\mu \in \mathcal{M}^1(G)$ is regular iff $H(\mu) < \infty$. Plainly, $H(\mu) = 0$ iff $\mu = \varepsilon_x$ for some $x \in G$.

As for measures on a locally compact group G, one defines triangular systems $(S_{nj})_{j=1,\dots,k_n;\, n \geq 1}$ in $\mathfrak{S}$, their infinitesimality by the condition

$$\lim_{n \to \infty} \sup_{1 \leq j \leq k_n} \|S_{nj} - E\| = 0$$

(where E is the unit matrix in $\mathfrak{S}$), their commutativity by

$$S_{nj} S_{nk} = S_{nk} S_{nj} \qquad \text{for all } n \geq 1,\ 1 \leq j, k \leq k_n$$

and their convergence by saying that for the sequence $(S_n)_{n \geq 1}$ of n-th partial products $S_n := \prod_{j=1}^{k_n} S_{nj}$ $(n \geq 1)$ the limit $\lim_{n \to \infty} S_n$ exists.

6.6.8 Lemma. *Let* $S := (s_{ij})_{i,j=1,\dots,d} \in \mathfrak{S}$, *let* λ *be an eigenvalue of* S *and let* $\alpha := \inf_{i=1,\dots,d} s_{ii} > 0$ *(such that* $\|S - E\| = 2(1 - \alpha)$ *holds).*

Then

(i) $|\lambda| \leq 1$ *and* $|\lambda - \alpha| \leq 1 - \alpha$;

(ii) *for* $\alpha > \frac{1}{2}$ *or equivalently* $\|S - E\| < 1$ *one has*

(a) $1 - \operatorname{Re}\lambda \leq \dfrac{1 - |\lambda|}{\alpha}$ *and*

(b) $\operatorname{tr}(E - S)(1 - \|S - E\|) \leq -\log|\det S| \leq \dfrac{\operatorname{tr}(E - S)}{1 - \|S - E\|}$.

Proof. Since (i) follows directly from Geršgorin's famous matrix inequalities, we shall content ourselves with the proof of (ii). Let λ be an eigenvalue of S which by (i) satisfies

$$|\lambda|\leq 1 \quad \text{and} \quad |\lambda-\alpha|\leq 1-\alpha.$$

If $\alpha>\frac{1}{2}$, then $\operatorname{Re}\lambda>0$. But for all $\lambda_0\in\mathbb{C}$ the equalities

$$|\lambda_0|=|\lambda| \quad \text{and} \quad |\lambda_0-\alpha|=1-\alpha$$

imply $\operatorname{Re}\lambda\geq\operatorname{Re}\lambda_0$ and

$$\operatorname{Re}\lambda_0\geq 1-\frac{1-|\lambda|^2}{2\alpha}\geq 1-\frac{1-|\lambda|}{\alpha}.$$

From this Assertion (ii) (a) follows. Summing the inequalities over all eigenvalues λ of S and using $1-|\lambda|\leq -\log|\lambda|$, one obtains

$$\operatorname{tr}(E-S)\leq\frac{-\log|\det S|}{\alpha}\leq\frac{-\log|\det S|}{1-\|S-E\|}$$

or the left inequality of (b). We now conclude from $\alpha>\frac{1}{2}$ that

$$|\lambda|\geq 2\alpha-1=1-\|S-E\|$$

holds, so that

$$-\log|\lambda|\leq\frac{1-|\lambda|}{|\lambda|}\leq\frac{1-\operatorname{Re}\lambda}{1-\|S-E\|}.$$

Again by summing over all eigenvalues λ of S we get the right inequality of (b). □

6.6.9 Remark. The inequalities (ii)(b) of the lemma imply that on the set $\{S\in\mathfrak{S}: \|S-E\|<1\}$ one has $\lim_{n\to\infty}S_n=E$ iff $\lim_{n\to\infty}(-\log|\det S_n|)=0$.

6.6.10 Theorem. *Let G be a finite group and $(\mu_{nj})_{j=1,\dots,k_n;n\geq 1}$ an infinitesimal triangular system in $\mathscr{M}^1(G)$. Then $(\mu_{nj})_{j=1,\dots,k_n;n\geq 1}$ converges to a regular limit iff the accompanying system $(\nu_{nj})_{j=1,\dots,k_n;n\geq 1}$ converges to a regular limit, and in this case the limits of the systems coincide.*

Proof. By the discussion in 6.6.7 it suffices to prove that an infinitesimal triangular system $(S_{nj})_{j=1,\dots,k_n;n\geq 1}$ in $\mathfrak{S}$ converges to a regular limit iff the accompanying system $(\exp(S_{nj}-E))_{j=1,\dots,k_n;n\geq 1}$ converges to a regular limit and that in this case the limits of the systems coincide. First of all we choose $n_0\geq 1$ large enough that $\sup_{1\leq j\leq k_n}\|S_{nj}-E\|<1$ holds for all $n\geq n_0$. For all $n\geq n_0$ we have

$$\begin{aligned}&\textstyle\prod_{j=1}^{k_n}S_{nj}-\prod_{j=1}^{k_n}\exp(S_{nj}-E)\\ &=\textstyle\sum_{j=1}^{k_n}S_{n1}\cdot\dots\cdot S_{n,j-1}(S_{nj}-\exp(S_{nj}-E))\\ &\quad\cdot\exp(S_{n,j+1}-E)\cdot\dots\cdot\exp(S_{nk_n}-E)\end{aligned}$$

and hence, since $\|\exp(S_{nj}-E)\| \le 1$ for $\|S_{nj}\| \le 1$ $(j=1, \ldots, k_n)$,

$$\begin{aligned}&\|\prod_{j=1}^{k_n} S_{nj}-\prod_{j=1}^{k_n} \exp(S_{nj}-E)\|\\ &\le \sum_{j=1}^{k_n} \|S_{nj}-\exp(S_{nj}-E)\| \le \tfrac{1}{2}\sum_{j=1}^{k_n} \|S_{nj}-E\|^2 \exp\|S_{nj}-E\|\\ &\le 2\sup_{1\le j\le k_n} \|S_{nj}-E\| \sum_{j=1}^{k_n} \|S_{nj}-E\|.\end{aligned}$$

The theorem will be proved once we have shown the boundedness of

$$\Big(\sum_{j=1}^{k_n} \|S_{nj}-E\|\Big)_{n\ge 1}.$$

In fact, if $(\exp(S_{nj}-E))_{j=1,\ldots,k_n;n\ge 1}$ converges to a regular $S\in\mathfrak{S}$, then

$$\begin{aligned}&\lim_{n\to\infty} \det \prod_{j=1}^{k_n} \exp(S_{nj}-E)\\ &=\lim_{n\to\infty} \exp\Big(\sum_{j=1}^{k_n} \operatorname{tr}(S_{nj}-E)\Big)=\det S>0\end{aligned}$$

and $\sum_{j=1}^{k_n} \operatorname{tr}(S_{nj}-E)$ is bounded so that $\sum_{j=1}^{k_n} \|S_{nj}-E\|$ is also bounded.

Now let $(S_{nj})_{j=1,\ldots,k_n;n\ge 1}$ converge to $S\in\mathfrak{S}$. Then

$$-\log|\det S| = \lim_{n\to\infty} \sum_{j=1}^{k_n}(-\log|\det S_{nj}|)$$

and

$$\sum_{j=1}^{k_n}(-\log|\det S_{nj}|)$$

is bounded. But by (ii)(b) of Lemma 6.6.8 we obtain the boundedness of

$$\sum_{j=1}^{k_n} \operatorname{tr}(S_{nj}-E), \text{ and hence that of } \sum_{j=1}^{k_n} \|S_{nj}-E\|.$$

The proof is complete. □

6.6.11 Theorem. *Let G be a compact group. Then the following statements are equivalent:*

(i) *G is totally disconnected;*

(ii) *for any infinitesimal system $(\mu_{nj})_{j=1,\ldots,k_n;n\ge 1}$ in $\mathscr{M}^1(G)$ converging to a regular limit $\mu\in\mathscr{M}^1(G)$ the accompanying system $(\nu_{nj})_{j=1,\ldots,k_n;n\ge 1}$ converges also to μ.*

Proof. 1. (i) ⇒ (ii). By assumption, $G=\varprojlim_{\alpha\in\mathbb{A}} G_\alpha$ with finite groups G_α forming a projective system $(G_\alpha, p_{\alpha\beta}, \mathbb{A})$.

Let $(\mu_{nj})_{j=1,\ldots,k_n;n\ge 1}$ be an infinitesimal system in $\mathscr{M}^1(G)$ such that the corresponding sequence $(\mu_n)_{n\ge 1}$ (of row products) converges to a regular limit $\mu\in\mathscr{M}^1(G)$. Clearly, $(p_\alpha(\mu_{nj}))_{j=1,\ldots,k_n;n\ge 1}$ is an infinitesimal system in $\mathscr{M}^1(G_\alpha)$ with corresponding sequence $(p_\alpha(\mu_n))_{n\ge 1}$ for every $\alpha\in\mathbb{A}$.

Furthermore, $\lim_{n\to\infty} p_\alpha(\mu_n)=p_\alpha(\mu)$, and $p_\alpha(\mu)$ is a regular measure in $\mathscr{M}^1(G_\alpha)$ for all $\alpha\in\mathbb{A}$. Let $(\nu_{nj})_{j=1,\ldots,k_n;n\ge 1}$ be the accompanying system of $(\mu_{nj})_{j=1,\ldots,k_n;n\ge 1}$ with sequence $(\nu_n)_{n\ge 1}$ of row products. But the accompanying system of $(p_\alpha(\mu_{nj}))_{j=1,\ldots,k_n;n\ge 1}$ turns out to be the system $(p_\alpha(\nu_{nj}))_{j=1,\ldots,k_n;n\ge 1}$ and the sequence of row products corresponding to $(p_\alpha(\nu_{nj}))_{j=1,\ldots,k_n;n\ge 1}$ equals the sequence $(p_\alpha(\nu_n))_{n\ge 1}$ for all $\alpha\in\mathbb{A}$. Thus, Theorem 6.6.10 yields $\lim_{n\to\infty} p_\alpha(\nu_n)=p_\alpha(\mu)$ for every $\alpha\in\mathbb{A}$ and thus, $\lim_{n\to\infty} \nu_n=\nu$.

2. For the converse (ii) ⇒ (i) we shall prove a more general statement. If G is a non-totally disconnected, locally compact group, then there exists a (commutative) infinitesimal system $(\mu_{nj})_{j=1,\ldots,k_n;n\geq 1}$ in $\mathcal{M}^1(G)$ which converges to a regular limit $\mu\in\mathcal{M}^1(G)$ such that μ is not an accumulation point of the sequence $(\nu_n)_{n\geq 1}$ of row products corresponding to the accompanying system $(\nu_{nj})_{j=1,\ldots,k_n;n\geq 1}$ of $(\mu_{nj})_{j=1,\ldots,k_n;n\geq 1}$.

By Theorem 1 of [312], p. 192 the Lie projective group G is totally disconnected iff there is only the trivial continuous homomorphism from $\mathbb{R}$ into G. If G is not totally disconnected, then G_0 contains a closed subgroup H which by Theorem A is isomorphic to one of the groups $\mathbb{R}, \mathbb{T}$ or a solenoidal group $\mathbb{S}$. It therefore suffices to show the assertion for these three types of groups.

3. Let $G:=\mathbb{R}$. We choose sequences $(l_n)_{n\geq 1}$ in $\mathbb{N}$ with $\lim_{n\to\infty} l_n=\infty$, $(k_n)_{n\geq 1}$ in $\mathbb{N}$ defined by $k_n:=2l_n$ for all $n\geq 1$ and $(c_n)_{n\geq 1}$ in $]0,1[$ with $\lim_{n\to\infty} c_n=0$ such that

$$\lim_{n\to\infty} k_n[\cos(2\pi a c_n)-1]=-\infty \qquad \text{for all } a\in\mathbb{R}\setminus\{0\}.$$

The system $(\mu_{nj})_{j=1,\ldots,k_n;n\geq 1}$ with $\mu_{nj}:=\varepsilon_{c_n}$ for $j=1,\ldots,l_n$ and $\mu_{nj}=\varepsilon_{-c_n}$ for $j=l_n+1,\ldots,k_n$ $(n\geq 1)$ converges to ε_0, whereas the sequence $(\nu_n)_{n\geq 1}$ of row products

$$\nu_n=\exp_{\{0\}}\{k_n[\tfrac{1}{2}(\varepsilon_{c_n}+\varepsilon_{-c_n})-\varepsilon_0]\} \qquad (n\geq 1)$$

of the accompanying system $(\nu_{nj})_{j=1,\ldots,k_n;n\geq 1}$ converges to the zero measure. This can be seen as follows. For any $a\in\mathbb{R}^\wedge\cong\mathbb{R}$, $a\neq 0$, we get

$$\lim_{n\to\infty}\hat{\nu}_n(a)=\lim_{n\to\infty}\exp\{k_n[\cos(2\pi a c_n)-1]\}=0.$$

On the other hand, we have for every $g\in L^1_{\mathbb{C}}(\mathbb{R},\lambda)$ the limit relation

$$\begin{aligned}\lim_{n\to\infty}\nu_n(\hat{g})&=\lim_{n\to\infty}\iint g(x)\,e^{2\pi i t x}\,\lambda(dx)\,\nu_n(dt)\\ &=\lim_{n\to\infty}\int g(x)\,\hat{\nu}_n(x)\,\lambda(dx)=0.\end{aligned}$$

Since the set $L^1_{\mathbb{C}}(\mathbb{R},\lambda)^\wedge$ is dense in $\mathscr{C}^0(\mathbb{R})$, one concludes $\mathscr{T}_v$-$\lim \nu_n=0$.

4. Let $G:=\mathbb{T}:=\{e^{it}: t\in[0,2\pi[\}$. We choose a fixed nontrivial continuous homomorphism $\xi:\mathbb{R}\to\mathbb{T}$ and define for the triangular system $(\mu_{nj})_{j=1,\ldots,k_n;n\geq 1}$ in $\mathcal{M}^1(\mathbb{R})$ constructed in Part 3 a new triangular system $(\bar{\mu}_{nj})_{j=1,\ldots,k_n;n\geq 1}$ in $\mathcal{M}^1(\mathbb{T})$ by $\bar{\mu}_{nj}:=\xi(\mu_{nj})$ for all $j=1,\ldots,k_n$; $n\geq 1$. Clearly, $(\bar{\mu}_{nj})_{j=1,\ldots,k_n;n\geq 1}$ is infinitesimal and converges to ε_1. For every nontrivial character $\chi\in\mathbb{T}^\wedge$ we obtain

$$\phi:=\chi\circ\xi\in\mathbb{R}^\wedge\cong\mathbb{R},\qquad \phi\neq 0,$$

hence

$$\lim_{n\to\infty}\hat{\bar{\nu}}_n(\chi)=\lim_{n\to\infty}\hat{\nu}_n(\phi)=0$$

as has been shown in Part 3. Here $(\bar{\nu}_n)_{n\geq 1}$ denotes the sequence of row products of the system accompanying $(\bar{\mu}_{nj})_{j=1,\ldots,k_n;n\geq 1}$. One concludes $\mathscr{T}_v$-$\lim_{n\to\infty}\bar{\nu}_n=\omega_{\mathbb{T}}$.

5. Let $G:=\mathbb{S}$ and ξ a continuous homomorphism from $\mathbb{R}$ into $\mathbb{S}$ such that $\overline{\xi(\mathbb{R})}=\mathbb{S}$. Again we define a triangular system

$$(\bar{\mu}_{nj})_{j=1,\ldots,k_n;n\geq 1}$$

in $\mathscr{M}^1(\mathbb{S})$ by $\bar{\mu}_{nj} := \xi(\mu_{nj})$ for all $j=1,\ldots,k_n$; $n\geq 1$ and show similarly to Part 4 that $(\bar{\mu}_{nj})_{j=1,\ldots,k_n;n\geq 1}$ is infinitesimal, convergent to ε_e and that for any nontrivial character $\chi\in\mathbb{S}^\wedge$ we have

$$\phi := \chi\circ\xi\in\mathbb{R}^\wedge\cong\mathbb{R}, \qquad \phi\neq 0,$$

and hence

$$\lim_{n\to\infty}\hat{\bar{\nu}}_n(\chi) = \lim_{n\to\infty}\hat{\nu}_n(\phi) = 0, \qquad \text{i.e., } \mathscr{T}_v\text{-}\lim_{n\to\infty}\bar{\nu}_n = \omega_{\mathbb{S}}.$$

6. In order to complete the proof of the theorem it remains to show that the sequences $(l_n)_{n\geq 1}$ and $(c_n)_{n\geq 1}$ appearing in Part 3 in fact exist. One simply chooses

$$l_n := n^3 \quad \text{and} \quad c_n := \frac{1}{n} \qquad \text{for all } n\geq 1.$$

Then

$$\lim_{n\to\infty} l_n = \infty, \qquad \lim_{n\to\infty} c_n = 0,$$

and for every $a\in\mathbb{R}\setminus\{0\}$ one obtains

$$2n^3\left[\cos\left(\frac{2\pi a}{n}\right)-1\right] \leq -2n\pi^2 a^2$$

for sufficiently large $n\geq n_0$. Thus $\lim_{n\to\infty} k_n[\cos(2\pi a c_n)-1] = -\infty$. The proof is terminated. □

We are now going to study the limits of not necessarily commutative infinitesimal systems of probability measures on a compact group. In general, these limits are not necessarily embeddable in continuous convolution semigroups but rather in continuous convolution hemigroups in the sense of Definition 4.6.1.

Let G be a locally compact group.

6.6.12 Definition. A measure $\mu\in\mathscr{M}^1(G)$ is said to be *immersible* if there exists a continuous convolution hemigroup $(\mu_{s,t})_{s,t\in\mathbb{R}_+}$ in $\mathscr{M}^1(G)$ satisfying $\mu_{0,1} = \mu$.

The set of all immersible measures in $\mathscr{M}^1(G)$ will be denoted by $\mathscr{H}(G)$. Obviously, $\mathscr{E}_{\{e\}}(G)\subset\mathscr{H}(G)$.

The following result is a generalization of Theorem 6.6.5.

6.6.13 Theorem. *Let G be a totally disconnected compact group and*

$$(\mu_{nj})_{j=1,\ldots,k_n;n\geq 1}$$

an infinitesimal system in $\mathscr{M}^1(G)$ which converges to a regular limit measure $\mu\in\mathscr{M}^1(G)$. Then $\mu\in\mathscr{H}(G)$.

Proof. First of all we note that $G = \varprojlim_{\alpha\in\mathbb{A}} G_\alpha$ for a projective system $(G_\alpha, p_{\alpha\beta}, \mathbb{A})$ of finite groups G_α $(\alpha\in\mathbb{A})$. Let $(\mu_{nj})_{j=1,\ldots,k_n;n\geq 1}$ be an infinitesimal system in $\mathscr{M}^1(G)$ with corresponding sequence $(\mu_n)_{n\geq 1}$ of row products μ_n $(n\geq 1)$

converging to a regular measure $\mu\in\mathscr{M}^1(G)$. By $(\nu_{nj})_{j=1,\ldots,k_n;n\geq 1}$ we denote as usual the accompanying system corresponding to $(\mu_{nj})_{j=1,\ldots,k_n;n\geq 1}$.

For every $\alpha\in\mathbb{A}$ the family $(p_\alpha(\mu_{nj}))_{j=1,\ldots,k_n;n\geq 1}$ is an infinitesimal system in $\mathscr{M}^1(G_\alpha)$, and its corresponding sequence $(p_\alpha(\mu_n))_{n\geq 1}$ of row products

$$p_\alpha(\mu_n)=\ast_{j=1}^{k_n} p_\alpha(\mu_{nj}) \qquad (n\geq 1)$$

converges to a regular limit $p_\alpha(\mu)$. For $\alpha\in\mathbb{A}$ and $n\geq 1$ we introduce the measure

$$\mu_{nj}^\alpha:=p_\alpha(\mu_{nj}) \quad\text{and}\quad \nu_{nj}^\alpha:=p_\alpha(\nu_{nj})=\exp_{\{e_\alpha\}}(\mu_{nj}^\alpha-\varepsilon_{e_\alpha})$$

in $\mathscr{M}^1(G_\alpha)\,(j=1,\ldots,k_n)$ and the mapping q_n^α from $[0,1]$ into $\mathscr{M}^b(G_\alpha)$ by

$$q_n^\alpha(\tau):=\sum_{j=1}^{k_n} k_n 1_{\left[\frac{j-1}{k_n},\frac{j}{k_n}\right[}(\tau)(\mu_{nj}^\alpha-\varepsilon_{e_\alpha})$$

for all $\tau\in[0,1]$, where e_α denotes the unit element of G_α. From the proof of Theorem 6.6.10 we obtain for every $\alpha\in\mathbb{A}$ the existence of a constant $M:=M(\alpha)\in\mathbb{R}_+$ satisfying

$$\int_0^1\|q_n^\alpha(\tau)\|\,d\tau\leq\sum_{j=1}^{k_n}\|\mu_{nj}^\alpha-\varepsilon_{e_\alpha}\|\leq M<\infty.$$

Let $\alpha\in\mathbb{A}$ be fixed. For every $n\geq 1$ we define the $\mathscr{M}^b(G_\alpha)$-valued measure

$$\Lambda_n^\alpha:=q_n^\alpha\cdot\lambda_{[0,1]}$$

on $[0,1]$ in the sense of [43], p. 39. By the above boundedness condition the sequence $(\Lambda_n^\alpha)_{n\geq 1}$ is weakly relatively compact, so that there exists a subsequence $(\Lambda_{n_l}^\alpha)_{l\geq 1}$ of $(\Lambda_n^\alpha)_{n\geq 1}$ with limit Λ^α such that for every $m\geq 1$

$$\begin{aligned}&\lim_{l\to\infty}\int_s^t\cdot\ldots\cdot\int_s^{\tau_m}\Lambda_{n_l}^\alpha(d\tau_1)\cdot\ldots\cdot\Lambda_{n_l}^\alpha(d\tau_m)\\ &=\lim_{l\to\infty}\int_s^t\cdot\ldots\cdot\int_s^{\tau_m} q_{n_l}^\alpha(\tau_1)\cdot\ldots\cdot q_{n_l}^\alpha(\tau_m)\,d\tau_1\cdot\ldots\cdot d\tau_m\\ &=\int_s^t\cdot\ldots\cdot\int_s^{\tau_m}\Lambda^\alpha(d\tau_1)\cdot\ldots\cdot\Lambda^\alpha(d\tau_m).\end{aligned}$$

Now for $n\geq 1$ and $s,t\in\mathbb{R}_+$ with $s<t$ the product integral (in the sense of 4.6.7)

$$\mu_{s,t}^{\alpha,n}:=\int_s^{t}\!\!{}^{\frown}\exp(d\Lambda_n^\alpha)=\int_s^{t}\!\!{}^{\frown}\exp(q_n^\alpha(\tau)\,d\tau)$$

of the function q_n^α exists and admits a Peano representation

$$\mu_{s,t}^{\alpha,n}=\sum_{m\geq 0}\pi_{s,t}^{(m,n),\alpha}$$

with

$$\pi_{s,t}^{(m,n),\alpha}:=\int_s^t\cdot\ldots\cdot\int_s^{\tau_m}\Lambda_n^\alpha(d\tau_1)\cdot\ldots\cdot\Lambda_n^\alpha(d\tau_m),$$

where the series converges uniformly on any compact interval of $\mathbb{R}_+$ containing s and t with $s<t$.

From

$$\lim_{l\to\infty}\pi_{s,t}^{(m,n_l),\alpha}=\pi_{s,t}^{m,\alpha}:=\int_s^t\cdot\ldots\cdot\int_s^{\tau_m}\Lambda^\alpha(d\tau_1)\cdot\ldots\cdot\Lambda^\alpha(d\tau_m),$$

we obtain

$$\lim_{l\to\infty} \mu_{s,t}^{\alpha,n_l} = \mu_{s,t}^{\alpha} := \sum_{m\geq 0} \pi_{s,t}^{m,\alpha} =: \int_s^t{}^{\cap} \exp(d\Lambda^{\alpha})$$

for all $s, t \in \mathbb{R}_+$ with $s<t$, and $(\mu_{s,t}^{\alpha})_{s,t\in\mathbb{R}_+}$ forms a continuous convolution hemigroup in $\mathcal{M}^1(G_\alpha)$. By construction, the families $(q_n^{\alpha})_{\alpha\in\mathbb{A}}$, $(\Lambda_n^{\alpha})_{\alpha\in\mathbb{A}}$ (for $n\geq 1$) and $(\Lambda^{\alpha})_{\alpha\in\mathbb{A}}$ are projective systems so that for all $s, t\in\mathbb{R}_+$ with $s\leq t$ we have $p_{\alpha\beta}(\mu_{s,t}^{\beta}) = \mu_{s,t}^{\alpha}$ whenever $\alpha, \beta\in\mathbb{A}$ with $\alpha<\beta$, and $(\mu_{s,t}^{\alpha})_{\alpha\in\mathbb{A}}$ is a projective system of measures, whose projective limit $\mu_{s,t} := \varprojlim_{\alpha\in\mathbb{A}} \mu_{s,t}^{\alpha}$ exists in $\mathcal{M}^1(G)$.

It is easily verified that $(\mu_{s,t})_{s,t\in\mathbb{R}_+}$ is a continuous convolution hemigroup in $\mathcal{M}^1(G)$. Moreover, by the very definition of the measures Λ_n^{α} we have

$$\int_0^1{}^{\cap} \exp(d\Lambda_n^{\alpha}) = \mathop{\ast}_{j=1}^{k_n} \nu_{nj}^{\alpha} \qquad \text{for } n\geq 1.$$

Theorem 6.6.10 applied to the infinitesimal system $(\mu_{nj})_{j=1,\ldots,k_n; n\geq 1}$ in $\mathcal{M}^1(G_\alpha)$ therefore yields

$$\lim_{n\to\infty} \int_0^1{}^{\cap} \exp(d\Lambda_n^{\alpha}) = \lim_{n\to\infty} \mathop{\ast}_{j=1}^{k_n} \nu_{nj}^{\alpha} = p_\alpha(\mu).$$

On the other hand we have already shown that

$$\lim_{l\to\infty} \int_0^1{}^{\cap} \exp(d\Lambda_{n_l}^{\alpha}) = \int_0^1{}^{\cap} \exp(d\Lambda^{\alpha}) = \mu_{0,1}^{\alpha} = p_\alpha(\mu_{0,1}).$$

Thus, $p_\alpha(\mu_{0,1}) = p_\alpha(\mu)$ for all $\alpha\in\mathbb{A}$ or $\mu_{0,1} = \mu$. This shows that $\mu\in\mathcal{H}(G)$. □

6.6.14 Remark. If G is a non-totally disconnected, locally compact group such that G_0 is not arcwise connected, then there exists an infinitesimal system $(\mu_{nj})_{j=1,\ldots,k_n; n\geq 1}$ in $\mathcal{M}^1(G)$ which converges to a regular limit $\mu\notin\mathcal{H}(G)$.

This follows in complete analogy to Remark 6.6.6. One need only verify that for $G := \mathbb{S}$ there exists an $x\in G$ such that $\varepsilon_x\notin\mathcal{H}(G)$. For, if $\varepsilon_x\in\mathcal{H}(G)$ for all $x\in G$, then any two elements of G are connected by an arc. But this is impossible since G is not arcwise connected.

References and Comments

R 6.1 Poisson Embedding and Approximation

Most of the references concerning Poisson measures have been listed in connection with §2 of Chapter III. As basic papers we consider the publications [483] of Urbanik, [31] of Böge and [64] of Carnal. Further studies are contained in [209] of Hazod and Schmetterer and in [195] of Hazod. See also [230] of Heyer and [206] of Hazod. For the present work those results are collected which are based on the embedding and canonical representation problem discussed in Chapters III and IV. The characterization of Poisson forms for almost periodic groups is due to Siebert [436], [437]. For general locally compact groups the corresponding result (Theorem 6.1.5) is now well-known, but is nowhere written down explicitly. The observation that on a discrete group continuous convolution semigroups are Poisson is due to Martin-Löf [346]. The characterization of all groups with this property is a result of Hazod in [202]. A more straightforward approach has been given by Woll [505], who treated the problem within the framework of homogeneous spaces. The exponential principle was formulated for the first time in Heyer [232]. The proof of Theorem 6.1.9 based on Theorem 6.1.6 uses an idea of Hazod and Schmetterer [209]. For details, see Theorem 3.3.17. The example in (1) of 6.1.12 suggested by the discussion in Vorobev [492] appears in Böge [30] and also in the paper of Schmetterer [429]. The approximation of divisible probability measures on a compact group by Poisson measures can already be found in an early paper of Hlawka [245]. More recent references for density properties of the sets $\mathscr{P}_H(G)$, $\mathscr{P}_H^e(G)$ and $\mathscr{F}\mathscr{P}_H^e(G)$ are [200], [202] of Hazod and also [514] of Yuan. Here approximation procedures for contraction semigroups as part of the Hille-Yosida theory [508] are applied. A useful additional reference is the paper [115] of Ditzian. The proof of the fundamental lemma 6.1.17 is based on Chernoff's paper [71]; it makes possible an extended operator-theoretic treatment of the approximation problem. There is a separate theory of the product formulae for operator semigroups (Chernoff [71], Trotter [476]) and for convolution semigroups of measures (Hazod [196]) which is not discussed here. In the language of measures on a Lie group G a typical result can be formulated as follows: Given an $\{e\}$-continuous convolution semigroup $(\mu_t)_{t\in\mathbb{R}_+^*}$ in $\mathscr{M}^1(G)$ with corresponding infinitesimal generator N of the form $(a_i, a_{ij}, \eta)_{i,j=1,\ldots,n}$, where the data in the brackets define operators N_1, N_2 and N_3 resp., there exist semigroups $(\varepsilon_{x_t})_{t\in\mathbb{R}_+^*}$, $(\nu_t)_{t\in\mathbb{R}_+^*}$ and $(\pi_t)_{t\in\mathbb{R}_+^*}$ in $\mathscr{M}^1(G)$ with infinitesimal generators N_1, N_2 and N_3 resp. such that

$$\mu_t = \mathscr{T}_v\text{-}\lim_{n\to\infty} (\varepsilon_{x_{\frac{t}{n}}} * \nu_{\frac{t}{n}} * \pi_{\frac{t}{n}})^n \qquad \text{for all } t\in\mathbb{R}_+^*.$$

The example in Remark 6.1.21 is due to Hazod [202]. Poisson distributions with parameter as introduced in connection with the characterization theorems 6.1.22 and 6.1.25 were studied in the framework of compact Abelian groups for the first time by Urbanik in [483]. The extensions to almost periodic and arbitrary locally compact groups are new; the proof of Theorem 6.1.25 was communicated to the author by Siebert. It depends on an embedding result which makes use of the spectral theory for positive-semidefinite operators. For details concerning the method see Hazod [195]. The question of under what conditions the product of finitely many elementary Poisson measures on a group G is again a Poisson measure appears reasonable only for noncommutative G (see Property (2) of 6.1.12). In this case conditions on the given measures $\mu, \nu\in\mathscr{P}_{\{e\}}(G)$ can be imposed in such a way that $\mu * \nu\in\mathscr{P}_{\{e\}}(G)$ is satisfied. The problem has been discussed by Hazod in [207], and it is related to the Bang-Bang problem for stochastic matrices (see Johansen [277] and Hazod [203]). Finally, it should be noted that Poisson semigroups on compact groups admit very interesting applications (for example to random evolutions). In this connection ergodic theorems for Poisson semigroups are valuable tools. See the preparatory studies of Grenander [182], 2.3, and more far-reaching results in Bagget and Stroock [4].

R 6.2 Gauss Measures and their Characterizations

The references and comments of this section supplement those of §§2 and 3 of Chapter V, where Gauss measures on locally compact Abelian groups are treated. We concentrate here on the theory developed for arbitrary locally compact groups. The definition of a Gauss semigroup presented in 6.2.1 is due to Courrège [94], 8.1. Courrège defines convolution semigroups of local type for the group $\mathbb{R}^n$ $(n \geq 1)$ and characterizes them in various ways. The intuition leading to this definition stems from the theory of stochastic processes, particularly from that of the Brownian motion process, where the locality condition formulated in terms of the Brownian transition semigroup yields the a.s. continuity of the paths of the process. See P.A. Meyer [354], Chapitre XIII, §5. For arbitrary locally compact groups the definition was applied for the first time by Siebert in [437]. For locally compact Abelian groups this has been done by Forst in [149]. An extension of the definition to locally compact semigroups and a few properties is the contents of a recent paper of Yuan [511]. In the first part of the section we present the main results on Gauss measures which appear in [437]. We start with a collection of properties of the class $\mathscr{G}(G)$ of Gauss measures on G, characterize Gauss measures by the data of their canonical representations, here first for a Lie group G, and aim at a constructive proof that on a connected locally compact group $G \neq \{e\}$ Gauss measures always exist. This is Theorem 6.2.8, which can be proved more directly, as is pointed out in the subsequent remark. The construction in (a) of 6.2.9 has been carried out in Heyer [233], generalizing the constructions given in Carnal [64] and Heyer [230]; the construction in (b) relies on the solution of Hilbert's fifth problem as it is solved in [171]. Both constructions are based on the existence of normal distributions on $\mathbb{R}^n$ $(n \geq 1)$. See also the discussion in §2 of Chapter V. The main goal of Theorem 6.2.10 is the equivalence (i) $\Leftrightarrow$ (v), which for compact groups has been given by Carnal in [64]. The originator of the Gauss condition (GC) is Urbanik [484].

Corollaries 6.2.11 to 6.2.13 and Theorem 6.2.15 are of special importance. The independence of the definition of a Gauss measure of its embedding semigroup is proved in Corollary 6.2.11 for Lie projective, almost periodic groups.

The possible extension of the result to more general locally compact groups is still an open problem. In general, $\mathscr{G}(G)$ is not closed in $\mathscr{M}^1(G)$, but, again for Lie projective, almost periodic groups G, it is closed in $\mathscr{E}_{\{e\}}(G) \setminus \mathscr{D}(G)$. This is Corollary 6.2.12. Corollary 6.2.13 together with the subsequent remark apparently answers the question of whether $\mathscr{G}(G)$ can be a subsemigroup of $\mathscr{M}^1(G)$. The counterexample of 6.2.14 due to Carnal destroys this illusion.

Theorem 6.2.15 generalizes the work of Urbanik [484] and Carnal [64] in this direction as is noted in Remark 6.2.19, where a comparison of Gauss measures in the various senses is given. For Abelian groups this comparison is also contained in the paper [9] of Berg, which can be understood as a supplement to § 18 in the book [12] of Berg and Forst. In order to incorporate into the general theory the P-Gauss measures introduced in § 2 of Chapter V as well, Theorems 6.2.16 and 6.2.17 are proved. Corollary 6.2.18 gives a complete answer to the problem (compare the relevant pages of [182] and [12]). Urbanik's idea in [484] of defining Gauss measures within $\mathscr{E}_{\{e\}}(G)$ rather than $\mathscr{I}(G)$ is supported by his example given here in Remark 6.2.19: A non-Gauss measure on $\mathbb{T}^2$ is constructed whose image under all characters is normal on $\mathbb{T}$. In the second part of the section, starting with Theorem 6.2.20, we treat Gauss measures on arbitrary locally compact groups in some detail. Theorem 6.2.20 is hidden in the literature, but its proof can be found implicitly in Hazod [197] and Siebert [438]. It follows the proof due to Hazod of Corollary 6.2.25 that the set $\mathscr{G}^s(G)$ of symmetric Gauss measures on G is closed in the set of probability measures without idempotent factors. The method of proof is operator-theoretic and it uses the Trotter-Kato theory. (See for example [508]). The crucial point of the proof is the statement of Theorem 6.2.22, which relies on the development of Chapter IV.

Further results related to Gauss measures on locally compact groups can only be mentioned briefly. The theory of the circular normal distribution (initiated by von Mises [355] and Lévy [320]) has been advanced by Mardia in [344]. The first non-Abelian treatment appeared with Perrin [387], in whose work Gauss distributions on the sphere (viewed as the orthogonal group) play an important rôle. In connection with the theory of Brownian motion, Gauss measures on Lie groups have been studied by S. Ito in [267] (for the covering group of a Lie group), K. Ito in [264], [265] (for a Lie group) and Gorman in [176] for the rotation group.

An attempt to follow an operator-theoretic approach to Gauss measures on an arbitrary locally compact group via noncommutative Fourier transform theory was made by Grenander [180], whose ideas have been reformulated in Heyer [223].

There is still the classical theorem of Ibragimov [262] which suggests a possible definition of Gauss measures also for groups. If μ is a measure in $\mathscr{I}(\mathbb{R})$ with the property that $\mu * \nu = \lambda$ for $\nu \in \mathscr{M}^1(\mathbb{R})$ and $\lambda \in \mathscr{I}(\mathbb{R})$ implies $\nu \in \mathscr{I}(\mathbb{R})$, then $\mu \in \mathscr{G}(\mathbb{R})$.

Another classical approach to the normal distribution aims at a characterization within the context of stable measures $\mu \in \mathscr{M}^1(\mathbb{R})$ defined by the property that the set of image measures $T(\mu)$ for all proper affine mappings T of $\mathbb{R}$ is closed under the formation of convolution products. This approach has been generalized to arbitrary locally compact groups within the framework of positive-definite functions on groups by Schmidt in [432], [433] and by Parthasarathy and Schmidt in [386]. Most of the more profound results, however, are as yet valid for Abelian groups only.

We also know that the theorems of Cramér and Bernstein discussed in Chapter V fail to be true for locally compact groups G containing a torus group.

Weak versions of these theorems, however, can be established. In [206] Hazod reformulates and proves the theorems in terms of generating functionals of semigroups in $\mathcal{M}^1(G)$.

Finally, it should be mentioned that the classical theorem concerning subordination of Gauss semigroups on $\mathbb{R}$ can be extended to general locally compact groups as follows: Let $(\mu_t)_{t\in\mathbb{R}_+^*}$ be an $\{e\}$-continuous convolution semigroup in $\mathcal{M}^1(G)$ and $(F_s)_{s\in\mathbb{R}_+^*}$ a continuous convolution semigroup in $\mathcal{M}^1(\mathbb{R})$ such that F_s is concentrated on $\mathbb{R}_+$ for all $s\in\mathbb{R}_+^*$. The semigroups $(\mu_t)_{t\in\mathbb{R}_+^*}$ and $(F_s)_{s\in\mathbb{R}_+^*}$ are uniquely determined by their generating functionals A and B resp. For B there exist a measure $N\in\mathcal{M}_+(\mathbb{R}_+^*)$ and a constant $c\geq 0$ such that

$$B(f)=cf'(0)+\int_{0+}^{\infty}[f(x)-f(0)]\,N(dx)$$

for all $f\in\mathscr{C}^1(\mathbb{R}_+)$, where $\int_{0+}^{\infty}\frac{t}{1+t}N(dt)<\infty$. The functional B, or by its one-to-one correspondence, the pair (c,N), is called a subordinator of $(\mu_t)_{t\in\mathbb{R}_+^*}$. We now define

$$\lambda_s:=\int_{0+}^{\infty}\mu_t F_s(dt)$$

for all $s\in\mathbb{R}_+^*$ and obtain an $\{e\}$-continuous convolution semigroup $(\lambda_s)_{s\in\mathbb{R}_+^*}$ in $\mathcal{M}^1(G)$ with generating functional C of the form

$$C:=cA+\int_{0+}^{\infty}(\mu_t-\varepsilon_e)\,N(dt).$$

$(\lambda_s)_{s\in\mathbb{R}_+^*}$ is called the subordinated semigroup (corresponding to $(\mu_t)_{t\in\mathbb{R}_+^*}$ and $(F_s)_{s\in\mathbb{R}_+^*}$). In [204] Hazod shows that if the subordinated semigroup is Gaussian, then N is trivial in the sense that $\lambda_s=\mu_{cs}$ for all $s\in\mathbb{R}_+^*$. This theorem generalizes corresponding results of Bochner [29] and Carnal [64] from Abelian and compact groups resp. to arbitrary locally compact groups. See also Woll [505] and Hazod [208]. Preliminary material on subordinated semigroups is to be found in Feller [140]; further auxiliary results can be taken from the book [241] of Hille and Phillips.

R 6.3 Absolute Continuity and Diffusion of Gauss Semigroups

The construction of Gauss measures on a connected locally compact group G can be refined to a construction of absolutely ω_G-continuous Gauss measures, at least for certain subclasses of groups G. Consequently, one requires additional properties of the ω_G-densities involved. For Abelian groups the problem has been treated in § 5 of Chapter V. The relevant reference was Siebert [440]. See also Part I of the long paper [10] of Berg.

In the case of a non-Abelian group one needs deeper results from the structure theory of general locally compact groups. An outline of the presentation of this subject has been given in Heyer [233], Section 5. The starting point for a discussion in this direction is given in Urbanik [484], where it is shown that Gauss measures μ (in the sense of Urbanik) are full in the sense that $\operatorname{supp}(\mu)=G$. Although absolute continuity and fullness coincide for Gauss measures on $\mathbb{R}^n$ ($n\geq 1$), both properties are essentially different in the case of an arbitrary connected, locally compact group. See Remarks 6.3.15 and 6.3.16. In his paper [153] Furstenberg claims that on any connected Lie group, absolutely continuous Gauss measures always exist and indicates a proof via facts from the theory of parabolic differential equations as given in Friedman [152]. It was Azencott in [2], p. 134, who made the idea of the proof more precise by incorporating the important reference [268] of S. Ito. The first aim of the section is a complete presentation of this proof of Theorem 6.3.1 following Siebert [441]. It turns out that under the condition of (strict) positive-definiteness of the matrix in the infinitesimal generator of the given Gauss semigroup $(\mu_t)_{t\in\mathbb{R}_+^*}$ in $\mathscr{M}^1(G)$ the corresponding densities are strictly positive and analytic. General properties of the Laplacian (differential) operator on a Lie group and the fundamental solution of the corresponding heat equation are due to Nelson [370] and Nelson, Stinespring [371]. These authors use heavily the theory of elliptic differential operators. In [451] Stein introduced (for a compact Lie group) a different technique depending on a lemma of Sobolev. This technique was carried over to general Lie groups by Hulanicki in [259] (see also [260]). Further results in this direction appearing in [259] are described as follows: A Borel measurable function ϕ on G which is bounded on compact subsets of G is said to be submultiplicative, if $\phi\geq 1$, $\phi=\phi^*$ and if $\phi(xy)\leq\phi(x)\,\phi(y)$ for all $x, y\in G$. Let $(\mu_t)_{t\in\mathbb{R}_+^*}$ be an absolutely continuous Gauss semigroup in $\mathscr{M}^1(G)$ (for a Lie group G) such that $\mu_t=f_t\cdot\omega_G$ for all $t\in\mathbb{R}_+^*$. Then the densities $f_t(t\in\mathbb{R}_+^*)$ can be chosen such that the mapping $(t,x)\to f_t(x)$ is a $\mathscr{C}^\infty$-function on $\mathbb{R}_+\times G$, and for every $t_0\in\mathbb{R}_+^*$ there exists a constant $c\in\mathbb{R}_+^*$ satisfying $\int f_t\,\phi\,d\omega\leq c$ for all $t\leq t_0$.

The original proof of the boundedness statement is due to Nelson [370], a simplification has been provided by Gårding [157]. It should be added that generalizations to certain homogeneous spaces of the statement concerning the behavior at infinity of the semigroup $(f_t)_{t\in\mathbb{R}_+^*}$ can be found in the papers by Johnson [279], by Helgason [214] and by Azencott [3]. In [3], Section 8, Azencott extends part of Hunt's work [261] in various ways. He studies left invariant second-order elliptic differential operators on a Riemannian homogeneous space $M:=G/H$, where G is a connected Lie group and H is a closed subgroup of G, which fulfill a positive maximum principle and admit (variable) coefficients satisfying local Hölder conditions. Given the corresponding C_0-diffusion semigroups $(P_t)_{t\in\mathbb{R}_+^*}$ (of operators) Azencott obtains estimates for $|P_t f(x)|$ $(t\to\infty)$ whenever f is a bounded Borel function on M with compact support and $x\in M$.

We then take up the problem of establishing necessary and sufficient conditions for a Gauss semigroup on a Lie group to be absolutely continuous. We restrict ourselves (the technique employed does not yield more) to symmetric Gauss semigroups. Theorem 6.3.8 due to Wehn [498] solves the problem completely. Its solution makes use of the general theory of linear partial differential operators [247], in particular, Hörmander's hypoellipticity criterion [248]. This criterion has been used significantly also in the work of Bony [32], [33] on the analytic structure of axiomatic potential theories.

Brownian convolution semigroups involving absolutely continuous measures only can be analyzed fairly explicitly for the group of all proper affine mappings of $\mathbb{R}$, as is indicated in Grenander [182], 4.5, and also for the 3-dimensional rotation group G, as has been done by McKean [352]. McKean constructs the Brownian motion process on G by starting from a Brownian motion process on the Lie algebra $\mathfrak{L}(G)$ of G. His construction has not yet been extended to general Lie groups.

Having settled the question of the existence of absolutely continuous Gauss measures on a connected Lie group, we now establish in Theorem 6.3.14 the existence of full Gauss measures on an arbitrary connected locally compact group, following the work of Siebert [441].

Finally, diffuse or atomless Gauss semigroups on a locally compact group are taken up. The results presented have been proved by Hazod in [199]. Their forerunners are the classical contributions to this subject published by Blum and Rosenblatt [24], Hartmann and Wintner [193] and Huff [258].

In Theorem 6.3.21 the normality of all measures in the given semigroup $(\mu_t)_{t\in\mathbb{R}_+^*}$ in $\mathscr{M}^1(G)$ is assumed. This assumption might be redundant, but this has not yet been shown. At any rate, one has the following statement proved in [199]: If there exists a nondiffuse Gauss semigroup in $\mathscr{M}^1(G)$, then there exists also a Gauss semigroup of discrete measures in $\mathscr{M}^1(G)$.

R 6.4 Central Gauss Semigroups

Central Gauss measures appear for the first time in Stein [451], where they are studied for compact Lie groups. More precisely, Stein shows the existence of bi-invariant second-order differential operators, which are elliptic, formally self-adjoint and map constant functions to zero ([451], p. 35).

The existence of central $\{e\}$-continuous convolution semigroups on σ-compact, locally compact groups with compact commutator subgroup has been demonstrated by Siebert. His construction is contained in Theorem 6.4.1. The approach chosen for the presentation of central Gauss semigroups is due to Siebert [441]. In fact, we introduce the notion of a G-invariant quadratic form on $\mathfrak{D}(G)$ for a connected locally compact group G and show in Theorem 6.4.6 that strict G-invariant quadratic forms on $\mathfrak{D}(G)$ exist iff G is a Z-group. Thus, on connected Z-groups G central Gauss semigroups $(\mu_t)_{t\in\mathbb{R}_+^*}$ with $\operatorname{supp}(\mu_t)=G$ for all $t\in\mathbb{R}_+^*$ do in fact exist. The uniqueness of an equivalence class of strict G-invariant quadratic forms on $\mathfrak{D}(G)$ is equivalent to the fact that $\mathfrak{L}(G)$ does not admit a proper ideal. For the statement of Theorem 6.4.9 see [451], p. 36 and for its proof [441]. Instead of central Gauss semigroups on connected locally compact groups G one can study Gauss semigroups that are invariant under more general automorphisms of G. It can be shown that for every compact subgroup A of the group $A(G)$ of topological automorphisms of G there exists a Gauss semigroup $(\mu_t)_{t\in\mathbb{R}_+^*}$ in $\mathscr{M}^1(G)$ satisfying $\operatorname{supp}(\mu_t)=G$ and $\alpha(\mu_t)=\mu_t$ for all $t\in\mathbb{R}_+^*$ and all $\alpha\in A$.

In the special case of a compact connected group, absolutely continuous Gauss measures and semigroups admit properties of particular interest. Special densities with respect to Haar measure can be described in terms of Fourier expansions (Properties 6.4.10). The existence of central Gauss semigroups $(\mu_t)_{t\in\mathbb{R}_+^*}$ on a connected compact group with continuous densities for all $t\in[2, \infty[$

can be extended to all connected, almost periodic groups whose topology has a countable basis. The corresponding theorem 6.4.11 is due to Siebert [441].

By using the Freudenthal-Weil theorem it is shown in Theorem 6.4.12 that for connected almost periodic groups G whose topology has a countable basis, the existence of an absolutely continuous Gauss measure on G is equivalent to the local connectedness of G. This result is the analogue for compact groups of Theorem 5.5.13 valid for locally compact Abelian groups having a countable basis of their topology.

R 6.5 Convergence of Triangular Systems of Probability Measures

As was pointed out in the references and comments to Chapter V the classical theory of triangular systems of probability measures is extensively treated in the pioneering work of Gnedenko and Kolmogorov [172]. For a more up-to-date presentation Loève [325] is recommended. The central limit theorem involving the Lindeberg-Lévy-Feller condition has been reproved several times. Demonstrations using an idea of Trotter are given in Bauer [6] and Krickeberg [310]. Most recently J.A.Goldstein [174] gave a proof based on Chernoff's product formula [71]. For particular compact Abelian groups G a general form of the central limit problem on infinitesimal triangular systems of probability measures on G has been stated (but not completely proved) by Kloss [305], [306]. The first results for non-Abelian groups are due to Parthasarathy [376] for the rotation group, Heyer [229], [230] for general connected compact Lie groups and to Hlawka [245] and Carnal [66], [67] for arbitrary compact groups. At the time the papers [245] and [305] appeared, Wehn finished his thesis [496], which in a shorter form was published as [497]. A detailed discussion of the results in [496] has been given by Grenander in his book [182].

Our approach in the main text follows the ideas of Wehn [496] and Carnal [67]. Although by using the Fourier operator instead of the Fourier transform the results can be established for general locally compact groups, we content ourselves with their restrictions to almost periodic groups in order to unify the presentation. The more general setup has been chosen by Siebert in [436], and the modified one has been outlined recently in Heyer [233].

We start with the study of commutative triangular systems and aim at a description of their limiting behavior under general conditions on the given triangular system. Typical conditions of interest are (A) of Theorem 6.5.1 and (B) of Theorem 6.5.5. Relations between Condition (B) and the conditions introduced (for a Lie group) by Wehn in [496] are discussed in Remark 6.5.7. Under Condition (B) a general form of the accompanying laws theorem 6.5.5 can be proved. Its most elementary proof is due to Johansen [276]. In the following corollaries 6.5.6 to 6.5.9 conditions are given under which limits of infinitesimal triangular systems of probability measures are embeddable in an $\{e\}$-continuous convolution semigroup. Subsequently, convergence of such systems to the Dirac measure ε_e is taken up. Theorems 6.5.11 and 6.5.12 contain sufficient conditions for an infinitesimal triangular system to converge to an elementary Poisson measure with parameter in the sense of §1 of Chapter VI. Finally, we present necessary and sufficient conditions for the limit of an infinitesimal triangular system to converge to a Gauss measure. The results obtain for root compact Moore groups. Theorem 6.5.14 is the central limit theorem, Theorem 6.5.15 contains the necessity of the generalized Lindeberg condition. The proof of Theorem 6.5.15 is based on the corresponding results for Abelian locally compact groups due to Parthasarathy [377] and arbitrary compact groups due to Carnal [67], resp. In the special case of a compact group Carnal in [67] obtains, at least for triangular systems of symmetric probability measures, conditions for the convergence to an arbitrary given limit.

Concerning noncommutative infinitesimal triangular systems on a locally compact group there are only a few contributions. Wehn [496] has an approach to the solution of the evolution equation via such systems of probability measures and their corresponding systems of probability operators. Some of his results have been improved by Sobko [447], [448]. More recent attempts in this direction are indicated in Guth [188]. Results of the described kind due to Johansen for finite groups will be classified in the references and comments to § 6 of Chapter VI.

There is an extensive literature on limit theorems for probability measures on special Lie groups, which we want to refer to briefly. For the group of motions of $\mathbb{R}^n$ defined as the semidirect product $\mathbb{R}^n \times_\eta \mathfrak{SO}(n)$ with $\eta(a)(x) := ax$ for all $a \in \mathfrak{SO}(n)$, $x \in \mathbb{R}^n$ results are available by Tutubalin [480], Gorostiza [177], Crépel [97], [98] and Roynette [414]. More detailed information on this branch of the theory can be taken from the book [415] of Roynette. There are also central limit theorems available for random variables taking values in nilpotent and semisimple Lie groups due to Tutubalin [479], Virtser [491] and Virtser [490] resp.

In Roynette [415] one can also find central limit theorems for the Heisenberg group and for the diamant group. The classical central limit problem has been extended in a most complete way for Lobachewski spaces and more generally for homogeneous spaces arising from semisimple Lie groups with a finite center by Tutubalin [478], Getoor [162] and Gangolli [155]. In these papers moment conditions were established which assured the convergence of triangular systems of bi-invariant probability measures to the Gauss measure. See also Sazonov and Tutubalin [423], Chapter 6. It should be added that the papers [458] and [459] by Stroock and Varadhan and [138] by Feinsilver on additive processes taking values in a Lie group G also contain limit theorems for infinitesimal triangular systems of G-valued random variables.

Finally, limit theorems for infinitesimal triangular systems of positive-definite functions on a locally compact group have been studied in some detail by Parthasarathy in [381]. The results obtained are valid for large classes of locally compact groups and they refine previous studies by the author concerning the limiting behavior of infinitesimal triangular systems of probability measures on locally compact Abelian groups. More precisely, the following setup is chosen in [381]: Let G be a locally compact group having a countable basis of its topology and $(f_{nj})_{j=1,\dots,k_n;\, n\geq 1}$ a triangular system of normalized positive-definite functions on G.

The system $(f_{nj})_{j=1,\dots,k_n;\, n\geq 1}$ is said to be infinitesimal if

$$\lim_{n\to\infty} \sup_{1\leq j\leq k_n} \sup_{x\in K} |f_{nj}(x)-1| = 0$$

holds for every compact subset K of G. The problems arising are the possible convergence in the compact open topology of the sequence $(F_n)_{n\geq 1}$ defined by

$$F_n := \prod_{j=1}^{k_n} f_{nj} \quad (n \geq 1)$$

and the characterization of its limit.

Parthasarathy gives complete solutions to the problem in the cases of compact groups, connected, simply connected Lie groups and connected, semi-simple Lie groups. Furthermore, he obtains an analogue of the accompanying laws theorem.

R 6.6 Central Limit Theorems for Totally Disconnected Groups

There has been a strong demand for specifications of the general theory developed in § 5 of Chapter VI. Thus commutative infinitesimal triangular systems of probability measures must be studied in more detail for at least all totally disconnected compact groups. We start with a generalization of a result of Carnal [66] to root compact projective groups G. Theorem 6.6.1 characterizes the infinite divisibility of measures in $\mathscr{M}^1(G)$ by means of the infinite divisibility of their projections. Theorem 6.6.4 is proved in Carnal [66] on the basis of the preceding result. One concludes in Theorem 6.6.5 that on certain compact groups G every commutative infinitesimal system in $\mathscr{M}^1(G)$ has in fact a limit in $\mathscr{E}(G)$. Here the embedding theorem 3.5.8 is applied. The example in Remark 6.6.6 is due to Hazod [205]. It shows that for non-totally disconnected, locally compact groups G with non-arcwise connected G_0 there exist infinitesimal systems in $\mathscr{M}^1(G)$ which converge to a regular limit but not to a limit in $\mathscr{E}(G)$. Thus, the motivation has been given for a discussion of the central limit theorem for totally disconnected compact groups. Such groups allow a complete characterization of the domain of validity of the strong (unshifted) accompanying laws theorem. The first non-Abelian version of such a theorem is due to Hlawka [245].

We start by presenting auxiliary facts from Johansen [276] and prove Theorem 6.6.11 due to Hazod [205]. Although the first part of the demonstration of this theorem is based on the fact that totally disconnected compact groups are projective limits of finite groups and hence can be reduced to the case of a finite group, the converse requires a subtle construction of counterexamples assuring that the validity of the accompanying laws statement implies total disconnectedness. Using a result of Parthasarathy [381] (Theorem 5.12) one can prove the equivalence of the statements in Theorem 6.6.11 also for Abelian (not necessarily compact) locally compact groups.

Next we study the convergence of arbitrary, not necessarily commutative infinitesimal triangular systems in $\mathscr{M}^1(G)$ for a totally disconnected compact group G. In this situation the possible limit of the system is shown to be immersible in a continuous convolution hemigroup in $\mathscr{M}^1(G)$. The proof of Theorem 6.6.13 uses the product integral technique for measures on a finite group, as it has been presented in § 6 of Chapter IV. We observe in Remark 6.6.14 that for non-totally disconnected, locally compact groups the convergence of the given system to a regular limit does not assure its immersibility.

This has been only a choice of the results available on the central limit problem for finite or totally disconnected compact groups. The studies of Johansen [276] lead to much more in the case of a finite semigroup. Immersibility can be characterized by infinite factorizability, or by the approximation by finite convolution products of infinitely divisible measures, or by a concrete product integral representation, or by the solution of corresponding evolution equations. (For the terminology in a more general framework and further references see § 6 of Chapter IV or Heyer [235]).

The study of the behavior at infinity of continuous convolution hemigroups $(\mu_{s,t})_{s,t \in \mathbb{R}_+}$ in $\mathcal{M}^1(G)$ for a compact group G yields another characterization of total disconnectedness of G, as has been shown in Hazod [205]. In the case of a totally disconnected compact group G, this part of the theory enriches the discussion of convergence of n-th partial products

$$\nu_{k,n} = \mu_{k+1} * \cdots * \mu_n \qquad (0 \le k < n, n \ge 1)$$

for a sequence $(\mu_j)_{j \ge 1}$ in $\mathcal{M}^1(G)$ that was opened in § 2 of Chapter II. More precisely, G is totally disconnected iff for any sequence $(\mu_j)_{j \ge 1}$ in $\mathcal{M}^1(G)$ satisfying $\sup_{j \ge 1} \mu_j(\complement U) \le \text{const}(U) < 1$ for an open $U \in \mathfrak{B}(e)$ the corresponding sequence $(\nu_{k,n})_{0 \le k < n}$ converges as $n \to \infty$. The proof of this result relies on Theorem 2.3.5, which is due to Maksimov.

Finally, it should be mentioned that the paper [205] of Hazod also contains characterizations of discrete and of totally disconnected (locally compact) groups G in terms of the analyticity properties of all continuous convolution semigroups in $\mathcal{M}^1(G)$. Results in this direction support the general intuition that a large number of probabilistic properties can be used to characterize classes of locally compact groups.

Bibliography

1. Akemann, C.A., Walter, M.E.: Non-Abelian Pontriagin duality. Duke Math. J. **39**, 451 – 463 (1972)
2. Azencott, R.: Espaces de Poisson des groupes localement compacts. Lecture Notes in Math. Vol. 148. Berlin-Heidelberg-New York: Springer 1970
3. Azencott, R.: Behavior of diffusion semi-groups at infinity. Bull. Soc. Math. France **102**, 193 – 240 (1974)
4. Baggett, L., Stroock, D.: An ergodic theorem for Poisson processes on a compact group, with applications to random evolutions. J. Functional Analysis **16**, 405 – 414 (1974)
5. Bartfai, P.: Grenzverteilungssätze auf der Kreisperipherie und auf kompakten Abelschen Gruppen. Studia Sci. Math. Hungar. **1**, 71 – 85 (1966)
6. Bauer, H.: Wahrscheinlichkeitstheorie und Grundzüge der Maßtheorie. 2. Aufl. Berlin-New York: Walter de Gruyter 1974
7. Bellman, R.: Limit theorems for non-commutative operations I. Duke Math. J. **21**, 491 – 500 (1954)
8. Berg, Ch.: Sur les semi-groupes de convolution. In: Théorie du Potentiel et Analyse Harmonique, pp. 1 – 26. Lecture Notes in Math. Vol. 404. Berlin-Göttingen-Heidelberg: Springer 1974
9. Berg, Ch.: On the relation between Gaussian measures and convolution semigroups of local type. Math. Scand. **37**, 183 – 192 (1975)
10. Berg, Ch.: Potential theory on the infinite dimensional torus. Invent. Math. **32**, 49 – 100 (1976)
11. Berg, Ch.: On the support of the measures in a symmetric convolution semigroup. Math. Z. **148**, 141 – 146 (1976)
12. Berg, Ch., Forst, G.: Potential theory on locally compact Abelian groups. Berlin-Heidelberg-New York: Springer 1975
13. Berglund, J.F., Hofmann, K.H.: Compact semitopological semigroups and weakly almost periodic functions. Lecture Notes in Math. Vol. 42. Berlin-Heidelberg-New York: Springer 1967
14. Bergström, H.: Limit theorems for convolutions. Stockholm-Göteborg-Uppsala: Almquist & Wiksell 1963
15. Bernat, P., Conze, N., Duflo, M., Lévy-Nahas, M., Rais, M., Renouard, P., Vergue, M.: Représentations des groupes de Lie résolubles. Paris: Dunod 1972
16. Bernstein, S.N.: On a property characterizing the Gaussian law. In: Collected Works Vol. **IV**: Theory of Probability, pp. 394 – 395. Moscow: Nauka 1964 [Russian]
17. Bhattacharya, R.N.: Speed of convergence of the n-fold convolution of a probability measure on a compact group. Z. Wahrscheinlichkeitstheorie und verw. Gebiete **25**, 1 – 10 (1972/73)
18. Billingsley, P.: Convergence of probability measures. New York: John Wiley & Sons (1968)
19. Bingham, M.S.: Stochastic processes with independent increments taking values in an abelian group. Thesis, University of Sheffield (1969)
20. Bingham, M.S.: A note on stochastic processes with independent increments taking values in an abelian group. J. Appl. Prob. **6**, 449 – 452 (1969)
21. Bingham, M.S.: Stochastic processes with independent increments taking values in an abelian group. Proc. London Math. Soc. **22**, 507 – 530 (1971)
22. Bingham, M.S., Parthasarathy, K.R.: A probabilistic proof of Bochner's theorem on positive definite functions. J. London Math. Soc. **43**, 626 – 632 (1968)
23. Bingham, N.H.: Factorization theory and domains of attraction for generalized convolution algebras. Proc. London Math. Soc. **23**, 16 – 30 (1971)

24. Blum, J.R., Rosenblatt, M.: On the structure of infinitely divisible distributions. Pacific J. Math. **9**, 1 – 7 (1959)
25. Bochner, S.: Vorlesungen über Fouriersche Integrale. Leipzig: Akad. Verlagsgesellschaft m.b.H. 1932
26. Bochner, S.: On a theorem of Tannaka and Krein. Ann. of Math. **43**, 56 – 58 (1942)
27. Bochner, S.: Positive zonal functions on spheres. Proc. Nat. Acad. Sci. USA **40**, 1141 – 1147 (1954)
28. Bochner, S.: General analytic setting for the central limit theory of probability. In: Bulletin Calcutta Mathematical Society Golden Jubilee Commemoration Volume (1958), Part I, pp. 111 – 128
29. Bochner, S.: Harmonic analysis and the theory of probability. Berkeley: University of California Press 1960
30. Böge, W.: Über die Charakterisierung unendlich teilbarer Wahrscheinlichkeitsverteilungen. J. Reine Angew. Math. **201**, 150 – 156 (1959)
31. Böge, W.: Zur Charakterisierung sukzessiv unendlich teilbarer Wahrscheinlichkeitsverteilungen auf lokalkompakten Gruppen. Z. Wahrscheinlichkeitstheorie und verw. Gebiete **2**, 380 – 394 (1964)
32. Bony, J.M.: Principe du maximum, inégalité de Harnack et unicité du problème de Cauchy pour les opérateurs elliptiques dégénérés. Ann. Inst. Fourier (Grenoble) **19**, 272 – 304 (1969)
33. Bony, J.M.: Opérateurs elliptiques dégénérés associés aux axiomatiques de la théorie du potential. In: Potential Theory, C.I.M.E., 1° Ciclo, Stresa (1969) pp. 71 – 119. Rome: Edizioni Cremonese 1970
34. Bony, J.M., Courrège, Ph., Priouret, P.: Semi-groupes de Feller sur une variété à bord compacte et problèmes aux limites intégro-différentiels du second ordre donnant lieu au principe du maximum. Ann. Inst. Fourier (Grenoble) **18**, 369 – 521 (1968)
35. Boseck, R.: The 5th Hilbert Problem. Math. Nachr. **67**, 59 – 61 (1975)
36. Bossard, D.C.: Probability on locally compact abelian groups: Sums of independent random variables. Thesis, Dartmouth College, Hanover, New Hampshire (1967)
37. Bossard, D.C.: Limit theorems for measures on nonmetrizable locally compact abelian groups. Trans. Amer. Math. Soc. **159**, 185 – 205 (1971)
38. Bourbaki, N.: Eléments de Mathématique XXVI: Groupes et algèbres de Lie, Chapitre 1. Actual. Scient. Ind. 1285. Paris: Hermann 1960
39. Bourbaki, N.: Eléments de Mathématique XXXVII: Groupes et algèbres de Lie, Chapitres 2,3. Actual. Scient. Ind. 1349. Paris: Hermann 1972
40. Bourbaki, N.: Eléments de Mathématique XVII: Livre V: Espaces vectoriels topologiques. Chapitres 3 – 5. Actual. Scient. Ind. 1229. Paris: Hermann 1964
41. Bourbaki, N.: Eléments de Mathématique XIII: Livre VI: Intégration. Chapitres 1 – 4. 2e éd. Actual. Scient. Ind. 1175. Paris: Hermann 1965
42. Bourbaki, N.: Eléments de Mathématique XXI: Livre VI: Intégration. Chapitre 5. 2e éd. Actual. Scient. Ind. 1244. Paris: Hermann 1967
43. Bourbaki, N.: Eléments de Mathématique XXV: Livre VI: Intégration. Chapitre 6. Actual. Scient. Ind. 1281. Paris: Hermann 1959
44. Bourbaki, N.: Eléments de Mathèmatique XXIX: Livre VI: Intégration. Chapitre 7 – 8. Actual. Scient. Ind. 1306. Paris: Hermann (1963)
45. Bourbaki, N.: Eléments de Mathématique XXXV: Livre VI: Intégration. Chapitre 9. Actual. Scient. Ind. 1343. Paris: Hermann 1969
46. Bourbaki, N.: Eléments de Mathématique: Théorie spectrales. Chapitres 1 – 2. Actual. Scient. Ind. 1332. Paris: Hermann 1967
47. Brown, G., Moran, W.: Translation and power independence for Bernoulli convolutions. Colloq. Math. **27**, 301 – 313 (1973)
48. Brown, G., Moran, W.: In general, Bernoulli convolutions have independent powers. Studia Math. **47**, 141 – 152 (1973)
49. Brown, G., Moran, W.: A dichotomy for infinite convolutions of discrete measures. Proc. Cambridge Philos. Soc. **73**, 307 – 316 (1973)
50. Brown, G., Moran, W.: Sums of random variables in groups and the purity law. Z. Wahrscheinlichkeitstheorie und verw. Gebiete **30**, 227 – 234 (1974)

51. Bruhat, F.: Distributions sur un groupe localement compact et applications à l'étude des représentations des groupes p-adiques. Bull. Soc. math. France **89**, 43–75 (1961)
52. Burrell, Q.L., McCrudden, M.: Infinitely divisible distributions on connected nilpotent Lie groups I. J. Lond. Math. Soc. II. Ser. **7**, 584–588 (1974)
53. Burrell, Q.L., McCrudden, M.: Infinitely divisible distributions on connected nilpotent Lie groups II. J. Lond. Math. Soc. II. Ser. **9**, 193–196 (1974)
54. Byczkowska, H.: On the continuity of a Gaussian process with independent increments with values in an LCA group. Bull. Acad. Polon. Sci. Sér. Sci. Math. Astronom. Phys. **23**, 177–181 (1975)
55. Byczkowska, H., Byczkowski, T., Timoszyk, W.: Some properties of variance on compact groups. Bull. Acad. Polon. Sci. Sér. Sci. Math. Astronom. Phys. **20**, 945–947 (1972)
56. Byczkowska, H., Byczkowski, T., Timoszyk, W.: Certain properties of variance on compact topological groups. Raporty nr. 15 Instytut Matematyki i fizyki teoretycznej, Politechniki Wrocławskiej, Wrocław (1973)
57. Byczkowski, T.: On infinite products of random elements on compact topological semigroups. Bull. Acad. Polon. Sci. Sér. Sci. Math. Astronom. Phys. **21**, 735–740 (1973)
58. Byczkowski, T.: The notion of variance on compact topological semigroups. Bull. Acad. Polon. Sci. Sér. Sci. Math. Astronom. Phys. **21**, 945–950 (1973)
59. Byczkowski, T.: The invariance principle for group-valued random variables. Studia Math. **56**, 187–198 (1976)
60. Byczkowski, T., Woś, J.: An equivalence theorem for infinite products of independent random elements on locally compact semigroups. Bull. Acad. Polon. Sci. Sér. Sci. Math. Astronom. Phys. **22**, 583–585 (1974)
61. Byczkowski, T., Woś, J.: On infinite products of independent random elements on metric semigroups. Colloq. Math. **37** (1977)
62. Byczkowski, T., Woś, J.: Equivalence theorems for infinite products of independent random elements on metric semigroups. Preprint 1976
63. Carnal, H.: Distributions aléatoires indéfiniment divisibles sur les groupes compacts. C.R. Acad. Sci. Paris **255**, 1179–1180 (1962)
64. Carnal, H.: Unendlich oft teilbare Wahrscheinlichkeitsverteilungen auf kompakten Gruppen. Math. Ann. **153**, 351–383 (1964)
65. Carnal, H.: Non-validité du théorème de Lévy-Cramér sur le cercle. Publ. Inst. Statist. Univ. Paris **13**, 55–56 (1964)
66. Carnal, H.: Deux théorèmes sur les groupes stochastiques compacts. Comment. Math. Helv. **40**, 237–246 (1966)
67. Carnal, H.: Systèmes infinitésimaux sur les groupes topologiques compacts. In: Les Probalitités sur les Structures Algébriques, pp. 43–49. Paris: Editions du Centre National de la Recherche Scientifique 1970
68. Carruth, J.H., Lawson, J.D.: On the existence of one-parameter semigroups. Semigroup Forum **1**, 85–90 (1970)
69. Carruth, J.H., Hofmann, K.H., Mislove, M.W.: Errors in 'Elements of compact semigroups'. Semigroup Forum **5**, 285–322 (1972)
70. Cassels, J.W.S.: An introduction to Diophantine approximation. Cambridge Tracts in Math. and Math. Phys. Vol. 52. Cambridge at the University Press 1957
71. Chernoff, P.R.: Note on a product formula for operator semigroups. J. Functional Analysis **2**, 238–242 (1968)
72. Chevalley, C.: On the topological structure of solvable groups. Ann. of Math. **42**, 668–675 (1941)
73. Choquet, G.: Lectures on Analysis I. New York-Amsterdam: W.A. Benjamin, Inc. 1969
74. Choquet, G., Deny, J.: Sur l'équation de convolution $\mu = \mu * \sigma$. C.R. Acad. Sci. Paris **250**, 799–801 (1960)
75. Chow, H.L.: Probability measures on compact semitopological semigroups. Compositio Math. **28**, 209–212 (1974)
76. Chow, H.L.: Probability measures of finite order. J. London Math. Soc. II. Ser. **8**, 176–178 (1974)
77. Chow, H.L., Choy, S.T.L.: Probability measures on compact semitopological semigroups. Lecture Notes, University of Singapore 1976

78. Choy, S.T.L.: On a limit theorem of measures. Math. Scand. **29**, 256–258 (1971)
79. Choy, S.T.L.: Primitive idempotent measures on locally compact semigroups. Semigroup Forum **4**, 267–273 (1972)
80. Choy, S.T.L.: Primitive idempotent measures on compact semitopological semigroups. J. Austral. Math. Soc. **13**, 451–455 (1972)
81. Choy, S.T.L.: On central primitive idempotent measures. J. Austral. Math. Soc. **15**, 415–416 (1973)
82. Cigler, J.: Folgen normierter Maße auf kompakten Gruppen. Z. Wahrscheinlichkeitstheorie und verw. Gebiete **1**, 3–13 (1962)
83. Cigler, J.: Über die Grenzverteilung von Summen Markowscher Ketten auf endlichen Gruppen I. Z. Wahrscheinlichkeitstheorie und verw. Gebiete **1**, 415–420 (1963)
84. Cigler, J.: Über die Grenzverteilung von Summen Markowscher Ketten auf endlichen Gruppen II. Z. Wahrscheinlichkeitstheorie und verw. Gebiete **1**, 421–432 (1963)
85. Cohen, P.J.: On a conjecture of Littlewood and idempotent measures. Amer. J. Math. **82**, 191–212 (1960)
86. Collins, H.S.: Primitive idempotents in the semigroup of measures. Duke Math. J. **27**, 397–400 (1960)
87. Collins, H.S.: Idempotent measures on compact semigroups. Proc. Amer. Math. Soc. **13**, 442–446 (1962)
88. Collins, H.S.: Convergence of convolution iterates of measures. Duke Math. J., **29**, 259–264 (1962)
89. Corwin, L.: A "functional equation" for measures and a generalization of Gaussian measures. Bull. Amer. Math. Soc. **75**, 829–832 (1969)
90. Corwin, L.: Generalized Gaussian measure and a "functional equation" I. J. of Functional Analysis **5**, 412–427 (1970)
91. Corwin, L.: Generalized Gaussian measure and a "functional equation" II. J. of Functional Analysis **6**, 481–505 (1970)
92. Corwin, L.: Generalized Gaussian measures and a "functional equation" III. Measures on $\mathbb{R}^n$. Advances in Math. **6**, 239–251 (1971)
93. Courrège, Ph.: Théorie de la mesure. Paris: Centre de Documentation Universitaire 1962
94. Courrège, Ph.: Générateur infinitésimal d'un semi-groupe de convolution sur $\mathbb{R}^n$ et formule de Lévy-Khinchine. Bull. Sci. Math. 2ᵉ Sér. **88**, 3–30 (1964)
95. Cramér, H.: Über eine Eigenschaft der normalen Verteilungsfunktion. Math. Z. **41**, 405–414 (1936)
96. Cramér, H.: Random variables and probability distributions. London: Cambridge University Press 1937
97. Crépel, P.: Marches aléatoires sur le groupe des déplacements du plan. C.R. Acad. Sc. Paris Sér. A–B **278**, A961–A963 (1974)
98. Crépel, P.: Récurrence des marches aléatoires sur les groupes de Lie. In: Théorie ergodique, pp. 50–69. Lecture Notes in Math. Vol. 532. Berlin-Heidelberg-New York: Springer 1976
99. Csiszár, I.: A note on limiting distributions on topological groups. Publ. Math. Inst. Hung. Acad. Sc. **9**, 595–598 (1964)
100. Csiszár, I.: On infinite products of random elements and infinite convolutions of probability distributions on locally compact groups. Z. Wahrscheinlichkeitstheorie und verw. Gebiete **5**, 279–295 (1966)
101. Csiszár, I.: Some problems concerning measures on topological spaces and convolutions of measures on topological groups. In: Les Probabilités sur les Structures Algébriques, pp. 75–96. Paris: Edition du Centre National de la Recherche Scientifique 1970
102. Csiszár, I.: On the weak* continuity of convolution in a convolution algebra over an arbitrary topological group. Studia Sci. Math. Hung. **6**, 27–40 (1971)
103. Cuculescu, I.: Processes with independent increments with values in a locally compact group. Rev. Roumaine Math. Pures Appl. **12**, 817–827 (1967)
104. Cuppens, R.: Decomposition of multivariate probability. New York-San Francisco-London: Academic Press 1975
105. Da Prato, G.: Somma di generatori infinitesimali di semigruppi di contrazione e equazioni di evoluzione in spazi di Banach. Annali di Mat. pura ed appl. Sér. **4**, 78, 131-157 (1968)

106. Da Prato, G.: Equations opérationnelles dans les espaces de Banach (cas analytique). C.R. Acad. Sci. Paris Sér. A – B **266**, A277 – A279 (1968)
107. Davidson, R.: Arithmetic and other properties of certain Delphic semigroups. Z. Wahrscheinlichkeitstheorie und verw. Gebiete **10**, 120 – 145, 146 – 172 (1968)
108. Davidson, R.: More Delphic theory and practice. Z. Wahrscheinlichkeitstheorie und verw. Gebiete **13**, 191 – 203 (1969)
109. Deny, J.: Sur l'équation $\mu = \mu * \sigma$ de convolution. In: Séminaire Brelot-Choquet-Deny: Théorie du Potentiel **4**, 5.01 – 5.11 (1959/60)
110. Deny, J.: Notions sur les semi-groupes d'opérateurs linéaires. Cours fait à la Faculté d'Orsay rédigé par J. Faraut. Faculté des Sciences d'Orsay, Département de Mathématiques 1963
111. Deny, J.: Méthodes hilbertiennes en théorie du potentiel. In: Potential Theory, C.I.M.E., 1° Ciclo, Stresa (1969), pp. 121 – 201. Rome: Edizioni Cremonese 1970
112. Derriennic, Y.: Lois «zéro ou deux» pour les processus de Markov, applications aux marches aléatoires. Ann. Inst. H. Poincaré Sect. B **12**, 111 – 129 (1976)
113. Dieudonné, J.: Sur une propriété des groupes libres. J. Reine Angew. Math. **204**, 30 – 34 (1960)
114. Dieudonné, J.: Eléments d'analyse 2. Paris: Gauthier-Villars 1968
115. Ditzian, Z.: Exponential formulae for semigroups of operators in terms of the resolvent. Israel J. Math. **9**, 541 – 553 (1971)
116. Dixmier, J.: L'application exponentielle dans les groupes de Lie résolubles. Bull. Soc. Math. France **85**, 113 – 121 (1957)
117. Dixmier, J.: Quelques propriétés des groupes abéliens localement compacts. Bull. Sci. Math. II. Ser. **81**, 38 – 48 (1957)
118. Dixmier, J.: Les C*-algèbres et leurs représentations. Paris: Gauthier-Villars 1964
119. Dixmier, J.: Algèbres enveloppantes. Paris: Gauthier-Villars 1974
120. Dobrushin, R.L.: On Poisson laws for distributions of particles in space. Ukrain. Math. Z. **8**, 127 – 134 (1956) [Russian]
121. Doob, J.L.: Renewal theory from the point of view of the theory of probability. Trans. Amer. Math. Soc. **63**, 422 – 438 (1948)
122. Doss, S.: Sur la convergence stochastique dans les espaces uniformes. Ann. Sci. Ecole Norm. Sup. **71**, 87 – 100 (1954)
123. Drumm, N.: Die Lévy-Chintschin-Formel als Choquetsche Integraldarstellung. Dissertation, Saarbrücken (1976)
124. Duflo, M.: Semigroups of complex measures on a locally compact group. In: Non-Commutative Harmonic Analysis, pp. 56 – 64. Lecture Notes in Math. Vol. 466. Berlin-Heidelberg-New York: Springer 1975
125. Dugué, D.: Arithmétique des lois de probabilités. Mémorial des Sciences Mathématiques, Fasc. 137. Paris: Gauthier-Villars 1957
126. Dvoretzky, A., Wolfowitz, J.: Sums of random integers reduced modulo m. Duke Math. J. **18**, 501 – 507 (1951)
127. Egorov, B.A., Maksimov, V.M.: On a sequence of random variables with values in a compact commutative group. Theor. Probability Appl. **13**, 584 – 593 (1968)
128. Elfving, G.: Zur Theorie der Markoffschen Ketten. Acta Soc. Sci. Fennicae N. Ser. A **2**, 1 – 17 (1937)
129. Elfving, G.: Über die Interpolation von Markoffschen Ketten. Acta Soc. Sci. Fennicae Comment. Phys.-Math. **10**, 1 – 8 (1939)
130. Emerson, W.R.: Ratio properties in locally compact amenable groups. Trans. Amer. Math. Soc. **133**, 179 – 204 (1968)
131. Eymard, E.: L'algèbre de Fourier d'un groupe localement compact. Bull. Soc. Math. France **92**, 181 – 236 (1964)
132. Faraut, J.: Semi-groupes de mesures complexes et calcul symbolique sur les générateurs infinitésimaux de semi-groupes d'opérateurs. Ann. Inst. Fourier (Grenoble) **20**, 235 – 301 (1970)
133. Faraut, J.: Semigroups of invariant operators. In: Proceedings of the International Congress of Mathematicians, Vancouver 1974, Vol II, pp. 147 – 151
134. Faraut, J.: Représentation des Laplaciens généralisés, formule de Lévy-Khinchine. Preprint (1975)

135. Faraut, J., Harzallah, K.: Semi-groupes d'opérateurs invariants et opérateurs dissipatifs invariants. Ann. Inst. Fourier (Grenoble) **22**, 147–164 (1972)
136. Faraut, J., Harzallah, K.: Distances hilbertiennes invariantes sur un espace homogène. Ann. Inst. Fourier (Grenoble) **24**, 171–217 (1974)
137. Faucett, W.M.: Compact semigroups irreducibly connected between two idempotents. Proc. Amer. Math. Soc. **6**, 741–747 (1955)
138. Feinsilver, Ph.: Processes with independent increments on a Lie group. To appear in Trans. Amer. Math. Soc. (1978)
139. Fel'dman, G.M.: On generalized Poisson distribution on groups. Theor. Probability Appl. **20**, 641–644 (1975)
140. Feller, W.: An introduction to probability theory. Volume II. New York-London-Sydney: John Wiley & Sons 1966
141. Fichtner, K.H.: Gleichverteilungseigenschaften substochastischer Kerne und zufällige Punktfolgen. Math. Nachr. **62**, 251–260 (1974)
142. de Finetti, B.: Sulla funzione a incremento aleatorio I. Atti Accad. Naz. Lincei Rend. Cl. Sci. Fis. Mat. Natur. **10**, 163–168 (1929)
143. de Finetti, B.: Sulla funzione a incremento aleatorio II. Atti Accad. Naz. Lincei Rend. Cl. Sci. Fis. Mat. Natur. **10**, 325–329 (1929)
144. de Finetti, B.: Sulla funzione a incremento aleatorio III. Atti Accad. Naz. Lincei Rend. Cl. Sci. Fis. Mat. Natur. **10**, 548–553 (1929)
145. Fisher, M.J.: The embeddability of an invertible measure. Semigroup Forum **5**, 340–353 (1973)
146. Flusser, P.: A characterization of the distribution of three independent random variables with values in a locally compact abelian group. Sankhyā Ser. A **34**, 99–110 (1972)
147. Foguel, S.R.: Iterates of a convolution on a non abelian group. Ann. Inst. H. Poincaré Sect. B **11**, 199–202 (1975)
148. Forst, G.: Symmetric harmonic groups and translation invariant Dirichlet spaces. Invent. Math. **18**, 143–182 (1972)
149. Forst, G.: Convolution semigroups of local type. Math. Scand. **34**, 211–218 (1974)
150. Forst, G.: The Lévy-Hinčin representation of negative definite functions. Z. Wahrscheinlichkeitstheorie und verw. Gebiete **34**, 313–318 (1976)
151. Fourt, G.: Existence de mesures à puissances singulières à toutes leurs translatées. C.R. Acad. Sci. Paris Sér. A–B **274**, A648–A650 (1972)
152. Friedman, A.: Classes of solutions of linear systems of partial differential equations of parabolic type. Duke Math. J. **24**, 433–442 (1957)
153. Furstenberg, H.: A Poisson formula for semisimple Lie groups. Ann. of Math. **77**, 335–386 (1963)
154. Galmarino, A.R.: The equivalence theorem for compositions of independent random elements on locally compact groups and homogeneous spaces. Z. Wahrscheinlichkeitstheorie und verw. Gebiete **7**, 29–42 (1967)
155. Gangolli, R.: Isotropic infinitely divisible measures on symmetric spaces. Acta Math. **111**, 213–246 (1964)
156. Gangolli, R.: Positive-definite kernels on homogeneous spaces and certain stochastic processes related to Lévy's Brownian motion of several parameters. Ann. Inst. H. Poincaré Sect. B **3**, 121–225 (1967)
157. Gårding, L.: Vecteurs analytiques dans les représentations des groupes de Lie. Bull. Soc. Math. France **88**, 73–93 (1960)
158. Gerl, P.: Diskrete, mittelbare Gruppen. Monatsh. für Math. **77**, 307–318 (1973)
159. Gerl, P.: Probability measures on semigroups. Proc. Amer. Math. Soc. **40**, 527–532 (1973)
160. Gerl, P.: Gleichverteilung auf lokalkompakten Gruppen. Math. Nachr. **71**, 249–260 (1976)
161. Gerl, P.: Wahrscheinlichkeitsmaße auf diskreten Gruppen. Preprint 1976
162. Getoor, R.K.: Infinitely divisible probabilities on the hyperbolic plane. Pacific J. Math. **11**, 1287–1309 (1961)
163. Gihman, I.I., Skorohod, A.V.: The theory of stochastic processes II. Berlin-Heidelberg-New York: Springer 1975
164. Gilewski, J.: Generalized convolutions and delphic semigroups. Colloq. Math. **25**, 281–289 (1972)

165. Gilewski, J., Urbanik, K.: Generalized convolutions and generating functions. Bull. Acad. Polon. Sci. Ser. Sci. Math. Astronom. Phys. **16**, 481–487 (1968)
166. Gleason, A.M.: Arcs in locally compact groups. Proc. Nat. Acad. Sci. USA **36**, 663–667 (1950)
167. Glicksberg, I.: Convolution semigroups of measures. Pacific J. Math. **9**, 503–509 (1959)
168. Glicksberg, I.: Weak compactness and separate continuity. Pacific J. Math. **11**, 205–214 (1961)
169. Glicksberg, I.: Uniform boundedness for groups. Canad. J. Math. **14**, 269–276 (1962)
170. Glivenko, V.: Sul teorema limite della teoria delle funzioni caratteristiche. Giorn. Ist. Ital. Attuari **7**, 160–167 (1936)
171. Gluškov, V.M.: The structure of locally compact groups and Hilbert's fifth problem. American Mathematical Society Translations II. Ser. **15**, 55–93 (1957)
172. Gnedenko, B.V., Kolmogorov, A.N.: Grenzverteilungen von Summen unabhängiger Zufallsgrößen. Berlin: Akademie-Verlag 1960
173. Godement, R.: Les fonctions de type positif et la théorie des groupes. Trans. Amer. Math. Soc. **63**, 1–84 (1948)
174. Goldstein, J.A.: Semigroup-theoretic proofs of the central limit theorem and other theorems of analysis. Semigroup Forum **12**, 189–206 (1976). Corrigendum in Semigroup Forum **12**, 388 (1976)
175. Goodman, G.S.: An intrinsic time for non-stationary finite Markov chains. Z. Wahrscheinlichkeitstheorie und verw. Gebiete **16**, 165–180 (1970)
176. Gorman, C.D.: Brownian motion of rotation. Trans. Amer. Math. Soc. **94**, 103–117 (1960)
177. Gorostiza, G.: The central limit theorem for random motions of d-dimensional Euclidean space. Ann. of Prob. **1**, 603–612 (1973)
178. Greenleaf, F.P.: Invariant measures on topological groups and their applications. New York-Toronto-London: Van Nostrand-Reinhold Company 1969
179. Grenander, U.: Stochastic groups. Background. General discussion. Remarks on limit theorems. Ark. Mat. **4**, 163–183 (1960)
180. Grenander, U.: Stochastic groups. Fourier analysis of probability distributions on locally compact groups. Ark. Mat., **4**, 333–345 (1961)
181. Grenander, U.: Stochastic groups and related structures. In: Proceedings of the Fourth Berkeley Symposium on Mathematical Statistics and Probability Vol. II, pp. 171–184. Berkeley: University of California Press 1961
182. Grenander, U.: Probabilities on algebraic structures. Stockholm-Göteborg-Uppsala: Almquist & Wiksell 1963
183. Grosser, S., Moskowitz, M.: On central topological groups. Trans. Amer. Math. Soc. **127**, 317–340 (1967)
184. Grosser, S., Moskowitz, M.: Representation theory of central topological groups. Trans. Amer. Math. Soc. **129**, 361–390 (1967)
185. Grosser, S., Moskowitz, M.: Compactness conditions in topological groups. J. Reine Angew. Math. **246**, 1–40 (1971)
186. Guichardet, A.: Symmetric Hilbert spaces and related topics. Lecture Notes in Math. Vol. 261. Berlin-Heidelberg-New York: Springer 1972
187. Guivarc'h, Y.: Croissance polynomiale et périodes des fonctions harmoniques. Bull. Soc. Math. France **101**, 333–379 (1973)
188. Guth, W.: Sätze über zweiparametrige Familien von Wahrscheinlichkeitsmaßen auf lokalkompakten topologischen Gruppen. Dissertation, Wien (1974)
189. Gyires, B.: On the limit of distributions of sums of independent random variables. In: Transactions of the 1 st Congress of Hungarian Mathematicians, pp. 741–758 (1952) [Hungarian]
190. Halmos, P.: Measure theory. Princeton, N.J. -Toronto-London-New York, D. van Nostrand Company, Inc. 1950
191. Hannan, E.J.: Group representations and applied probability. Methuen's Review Series in Applied Probability Vol. 3. London: Methuen & Co. Ltd. 1965
192. Hanš, O.: Generalized random variables. In: Transactions of the 1st Prague Conference on Information Theory, Statistical Decision Functions, Random Processes, pp. 61–104. Prague: Publishing House of the Czechoslovakian Academy of Sciences 1957

193. Hartmann, P., Wintner, A.: On the infinitesimal generators of integral convolutions. Amer. Math. J. **64**, 273–298 (1942)
194. Harzallah, K.: Sur une démonstration de la formule de Lévy-Khinchine. Ann. Inst. Fourier (Grenoble) **19**, 527–532 (1962)
195. Hazod, W.: Über Wurzeln und Logarithmen beschränkter Maße. Z. Wahrscheinlichkeitstheorie und verw. Gebiete **20**, 259–270 (1971)
196. Hazod, W.: Eine Produktformel für Halbgruppen von Wahrscheinlichkeitsmaßen auf Lie-Gruppen. Monatsh. für Math. **76**, 295–299 (1972)
197. Hazod, W.: Über die Lévy-Hinčin-Formel auf lokalkompakten topologischen Gruppen. Z. Wahrscheinlichkeitstheorie und verw. Gebiete **25**, 301–322 (1973)
198. Hazod, W.: Über Faltungshalbgruppen von Wahrscheinlichkeitsmaßen. Österreich. Akad. Wiss. Math.-Natur.Kl. S.-B. II, **181**, 29–47 (1973)
199. Hazod, W.: Symmetrische Gauß-Verteilungen sind diffus. Manuscripta Math. **14**, 283–295 (1974)
200. Hazod, W.: Poissonmaße auf lokalkompakten Halbgruppen. Monatsh. für Math. **78**, 25–41 (1974)
201. Hazod, W.: Über Faltungshalbgruppen von Wahrscheinlichkeitsmaßen II: Einige Klassen topologischer Gruppen. Monatsh. für Math. **79**, 25–45 (1975)
202. Hazod, W.: Einige Sätze über unendlich teilbare Wahrscheinlichkeitsmaße auf lokalkompakten Gruppen. Arch. Math. (Basel) **26**, 297–312 (1975)
203. Hazod, W.: Bemerkungen zum Bang-Bang Problem für stochastische Matrizen. Z. Wahrscheinlichkeitstheorie und verw. Gebiete **35**, 39–43 (1976)
204. Hazod, W.: Subordination von Gauß- und Poissonmaßen. Z. Wahrscheinlichkeitstheorie und verw. Gebiete **35**, 45–55 (1976)
205. Hazod, W.: Probabilities on totally disconnected locally compact groups. In: Symposia Mathematica Vol. 21. London and New York: Academic Press 1977
206. Hazod, W.: Stetige Halbgruppen von Wahrscheinlichkeitsmaßen und erzeugende Distributionen. Lecture Notes in Math. Vol. 595. Berlin-Heidelberg-New York: Springer 1977
207. Hazod, W.: Über die Campbell-Hausdorff-Formel und über die Produkte von Poissonmaßen. Preprint 1975
208. Hazod, W.: Subordination von Faltungshalbgruppen. Preprint 1976
209. Hazod, W., Schmetterer, L.: Über Poissongesetze auf lokalkompakten Gruppen und verwandte Fragen. Studia Sci. Math. Hungar. **5**, 63–74 (1970)
210. Hazod, W., Schmetterer, L.: Über einige mit der Wahrscheinlichkeitstheorie zusammenhängende Probleme der Gruppentheorie. J. reine angew. Math. **262/263**, 144–152 (1973)
211. Heble, M.: Probability measures on certain compact semigroups. J. Math. Anal. Appl. **8**, 258–277 (1964)
212. Heble, M., Rosenblatt, M.: Idempotent measures on a compact topological semigroup. Proc. Amer. Math. Soc. **14**, 177–184 (1963)
213. Helgason, S.: Differential geometry and symmetric spaces. New York-London: Academic Press 1962
214. Helgason, S.: Fundamental solutions of invariant differential operators on symmetric spaces. Amer. J. Math. **86**, 565–601 (1974)
215. Helson, H.: Note on harmonic functions. Proc. Amer. Math. Soc. **4**, 686–691 (1953)
216. Herrmann, H.: Glättungseigenschaften der Faltung. Wiss. Z. Friedrich-Schiller-Univ. Jena, Math. Naturw. Reihe **14**, 221–234 (1965)
217. Herz, C.: The spectral theory of bounded functions. Trans. Amer. Math. Soc. **94**, 181–232 (1960)
218. Hewitt, E., Ross, K.A.: Abstract harmonic analysis I, II. Berlin-Göttingen-Heidelberg-New York: Springer 1963/70
219. Hewitt, E., Zuckerman, H.S.: Arithmetic and limit theorems for a class of random variables. Duke Math. J. **22**, 595–615 (1955)
220. Heyer, H.: Über Haarsche Maße auf lokalkompakten Gruppen. Arch. Math. (Basel) **17**, 347–351 (1966)
221. Heyer, H.: On the convergence principle of B.M. Kloss. Math. Scand. **19**, 211–216 (1966)
222. Heyer, H.: Allgemeine Grenzwertsätze der Wahrscheinlichkeitsrechnung. In: Studies in Mathematical Statistics: Theory and Applications, pp. 57–73. Budapest: Akadémiai Kiadó 1968

223. Heyer, H.: Fourier transforms and probabilities on locally compact groups. Jber. Dtsch. Math. Verein. **70**, 109–147 (1968)
224. Heyer, H.: L'analyse de Fourier non-commutative et applications à la théorie des probabilités. Ann. Inst. H. Poincaré Sect. B **4**, 143–164 (1968)
225. Heyer, H.: Das Äquivalenzprinzip der Wahrscheinlichkeitstheorie für lokalkompakte Gruppen und homogene Räume. Bayer. Akad. Wiss. Math.-Natur. Kl. S.-B. 1–11 (1968)
226. Heyer, H.: Remarques sur une axiomatique de la variance. In: Les Probabilités sur les Structures Algébriques, pp. 177–189. Paris: Editions du Centre National de la Recherche Scientifique 1970
227. Heyer, H.: Dualität lokalkompakter Gruppen. Lecture Notes in Math. Vol. 150. Berlin-Heidelberg-New York: Springer 1970
228. Heyer, H.: The duality principle and applications to probability theory on locally compact groups. In: Papers from the "Open House for Probabilists", pp. 118–128, Various Publ. Series No 21. Matematisk Institut, Aarhus Universitet (1972)
229. Heyer, H.: A central limit theorem for compact Lie groups. In: Papers from the "Open House for Probabilists", pp. 101–117, Various Publ. Series No 21. Matematisk Institut, Aarhus Universitet (1972)
230. Heyer, H.: Infinitely divisible probability measures on compact groups. In: Lectures on Operator Algebras, pp. 55–249. Lecture Notes in Math. Vol. 247. Berlin-Heidelberg-New York: Springer 1972
231. Heyer, H.: Groups with Chu duality. In: Probability and Information Theory II. Lecture Notes in Math. Vol. 296, pp. 181–215. Berlin-Heidelberg-New York: Springer 1973
232. Heyer, H.: Probabilistic characterization of certain classes of locally compact groups. In: Symposia Mathematica Vol. 16, pp. 315–355. London-New York: Academic Press 1975
233. Heyer, H.: Gauss distributions and central limit theorem for locally compact groups. In: Colloquia Mathematica Societatis János Bolyai, 11: Limit Theorems of Probability Theory. Amsterdam-London: North Holland Publishing Company 1975
234. Heyer, H.: The embedding of probability measures on a locally compact group. In: Atas do Nono Colóquio Brasileiro de Matemática, pp. 377–395. (Poços de Caldas, 9 a 28 di jullio de 1973)
235. Heyer, H.: Probability measures on almost periodic groups. In: Symposia Mathematica Vol. 21, pp. 281–311. London and New York: Academic Press 1977
236. Heyer, H., Rall, Ch.: Gaußsche Wahrscheinlichkeitsmaße auf Corwinschen Gruppen. Math. Z. **128**, 343–361 (1972)
237. Heyer, H., Tortrat, A.: Sur la divisibilité des probabilités dans un groupe topologique. Z. Wahrscheinlichkeitstheorie und verw. Gebiete **16**, 307–320 (1970)
238. Heyn, E.: Die Differentialgleichung dT/dt=P(t)T für Operatorfunktionen. Math. Nachr. **24**, 281–330 (1962)
239. Hida, T.: Stationary stochastic processes. Princeton, N.J.: Princeton University Press 1970
240. Hille, E.: On roots and logarithms of elements of a complex Banach algebra. Math. Ann. **136**, 46–57 (1958)
241. Hille, E., Phillips, R.S.: Functional analysis and semigroups. Amer. Math. Soc. Colloquium Publications, Vol. 31. Revised edition. Providence,R.I, Amer. Math. Soc. 1957
242. Hirsch, F.: Sur les semi-groupes d'opérateurs invariants par translations. C.R. Acad. Sci. Paris Sér. A–B **274**, A43–A46 (1971/72)
243. Hirsch, F., Roth, J.P.: Opérateurs dissipatifs et codissipatifs invariants sur un espace homogène. C.R. Acad. Sci. Paris Sér. A–B **274**, A1791–A1793 (1971/72)
244. Hirsch, F., Roth, J.P.: Opérateurs dissipatifs et codissipatifs invariants sur un espace homogène. In: Théorie du Potentiel et Analyse Harmonique, pp. 229–245. Lecture Notes in Math. Vol. 404. Berlin-Heidelberg-New York: Springer 1974
245. Hlawka, E.: Statistik auf kompakten Gruppen I. Österreich. Akad. Wiss. Math.-Natur. Kl. S.-B. II, **4**, 64–76 (1959)
246. Hochschild, G.: The structure of Lie groups. San Francisco: Holden-Day 1965
247. Hörmander, L.: Linear partial differential operators. Berlin-Göttingen-Heidelberg: Springer 1963
248. Hörmander, L.: Hypoelliptic second order differential equations. Acta Math. **119**, 147–171 (1967)

249. Hofmann, K.H.: Topologische Halbgruppen mit dichter submonogener Halbgruppe. Math. Z. **74**, 232–277 (1960)
250. Hofmann, K.H.: Einführung in die Theorie der Lie-Gruppen I, II. Vorlesungsausarbeitung, Universität Tübingen 1962/63
251. Hofmann, K.H.: Introduction to the theory of compact groups, Part I. 2nd printing. New Orleans: Tulane University 1966/67
252. Hofmann, K.H., Mostert, P.S.: Splitting in topological groups. Memoirs Amer. Math. Soc. Vol. 43 (1963)
253. Hofmann, K.H., Mostert, P.S.: Elements of compact semigroups. Columbus, Ohio: Charles E. Merrill Books, Inc. 1966
254. Horton, H.B.: A method for obtaining random numbers. Ann. Math. Stat. **19**, 81–85 (1948)
255. Horton, H.B., Smith III, R.T.: A direct method for producing random digits in any number system. Ann. Math. Stat. **20**, 82–90 (1949)
256. Hudson, W.N., Mason, J.D.: More on equivalence of infinitely divisible distributions. Ann. Probability **3**, 563–568 (1975)
257. Hudson, W.N., Tucker, G.: On admissible translates of infinitely divisible distributions. Z. Wahrscheinlichkeitstheorie und verw. Gebiete **32**, 65–72 (1975)
258. Huff, B.W.: On the continuity of infinitely divisible distributions. Sankhyā Ser. A **34**, 443–446 (1972)
259. Hulanicki, A.: Subalgebra of $L_1(G)$ associated with Laplacian on a Lie group. Colloq. Math. **31**, 259–287 (1974)
260. Hulanicki, A.: Commutative subalgebra of $L^1(G)$ associated with a subelliptic operator on a Lie group G. Bull. Amer. Math. Soc. **81**, 121–124 (1975)
261. Hunt, G.A.: Semi-groups of measures on Lie groups. Trans. Amer. Math. Soc. **81**, 264–293 (1956)
262. Ibragimov, I.A.: A theorem in the theory of infinitely divisible laws. Theor. Probability Appl. **1**, 440–444 (1956)
263. Ito, K.: On stochastic processes I. Japan. J. Math. **18**, 261–301 (1942)
264. Ito, K.: Stochastic differential equations in a differentiable manifold. Nagoya Math. J. **1**, 35–47 (1950)
265. Ito, K.: Brownian motions in a Lie group. Proc. Japan. Acad. **26**, 4–10 (1950)
266. Ito, K.: Stochastic processes. Lecture Notes Series No. 16. Matematisk Institut, Aarhus Universitet (1969)
267. Ito, S.: Brownian motions in a topological group and its covering group. Rend. Circ. Mat. Palermo **1**, 40–48 (1952)
268. Ito, S.: The fundamental solution of the parabolic equation in a differentiable manifold. Osaka Math. J. **5**, 75–92 (1953)
269. Ito, S.: Fundamental solutions of parabolic differential equations and boundary value problems. Japanese J. Math. **27**, 55–102 (1957)
270. Iwasawa, K.: On some types of topological groups. Ann. of Math. **50**, 507–558 (1949)
271. Johansen, S.: An application of extreme point methods to the representation of infinitely divisible distributions. Z. Wahrscheinlichkeitstheorie und verw. Gebiete **5**, 304–316 (1966)
272. Johansen, S.: The imbedding problem for finite Markov Chains II. Preprint 4, University of Copenhagen: Institute of Mathematical Statistics 1972
273. Johansen, S.: The imbedding problem for finite Markov Chains III. Preprint 8, University of Copenhagen: Institute of Mathematical Statistics 1972
274. Johansen, S.: The imbedding problem for finite Markov Chains IV. Preprint 10, University of Copenhagen: Institute of Mathematical Statistics 1972
275. Johansen, S.: The imbedding problem for Markov Chains. Preprint 10, University of Copenhagen: Institute of Mathematical Statistics 1973
276. Johansen, S.: A central limit theorem for finite semigroups and its applications to the imbedding problem for finite state Markov Chains. Z. Wahrscheinlichkeitstheorie und verw. Gebiete **26**, 171–190 (1973)
277. Johansen, S.: The Bang-Bang problem for stochastic matrices. Z. Wahrscheinlichkeitstheorie und verw. Gebiete **26**, 191–195 (1973)
278. Johansen, S.: Some results on the imbedding problem for finite Markov Chains. J. Lond. Math. Soc. II. Ser. **8**, 345–351 (1974)

279. Johnson, K.D.: Spectrum and semigroups of elliptic operators on homogeneous spaces. J. Differential Geometry **5**, 517–522 (1971).
280. Kac, M.: On a characterization of the normal distribution. Amer. J. Math. **61**, 726–728 (1939)
281. Kakehashi, T.: Stationary periodic distributions. J. Osaka Inst. Sci. Technology **1**, 21–25 (1949)
282. Kallianpur, G.: The topology of weak convergence of probability measures. J. Math. Mech. **10**, 947–969 (1961)
283. Kaplansky, I.: Lie algebras and locally compact groups. Chicago-London: The University of Chicago Press 1971
284. Karpelevich, F.I., Tutubalin, V.N., Shur, M.B.: Limit theorems for the compositions of distributions in the Lobachevsky plane and space. Theor. Probability Appl. **4**, 399–402 (1959)
285. Kato, T.: Linear evolution equations of "hyperbolic" type. J. Fac. Sci. Univ. Tokyo Sect. IA Math. **17**, 241–258 (1970)
286. Katznelson, Y.: An introduction to harmonic analysis. New York-London-Toronto: John Wiley & Sons 1968
287. Kawada, Y., Ito, K.: On the probability distribution on a compact group I. Proc. Phys.-Math. Soc. Japan **22**, 977–998 (1940)
288. Kawata, T.: Fourier analysis in probability theory. New York-London: Academic Press 1972
289. Kelley, J.L.: Averaging operators on $C_\infty(X)$. Illinois J. Math. **1**, 214–223 (1958)
290. Kendall, D.G.: Extreme-point methods in stochastic analysis. Z. Wahrscheinlichkeitstheorie und verw. Gebiete **1**, 295–300 (1963)
291. Kendall, D.G.: Delphic semigroups, infinitely divisible regenerative phenomena, and the arithmetic of p-functions. Z. Wahrscheinlichkeitstheorie und verw. Gebiete **9**, 163–195 (1968)
292. Kendall, D.G., Harding, E.F. (editors): Stochastic analysis. A tribute to the memory of Rollo Davidson. London-New York-Sydney-Toronto: John Wiley & Sons 1973
293. Kendall, D.G., Lamperti, J.: A remark on topologies for characteristic functions. Proc. Cambridge Philos. Soc. **68**, 703–705 (1970)
294. Kerstan, J., Matthes, K.: Gleichverteilungseigenschaften von Faltungspotenzen auf lokalkompakten abelschen Gruppen. Wiss. Z. Friedrich-Schiller-Univ. Jena, Math. Naturwiss. Reihe **14**, 457–462 (1966)
295. Kerstan, J., Matthes, K.: Gleichverteilungseigenschaften von Faltungen von Verteilungsgesetzen auf lokalkompakten abelschen Gruppen I. Math. Nachr. **37**, 267–312 (1968)
296. Kerstan, J., Matthes, K.: Gleichverteilungseigenschaften von Faltungen von Verteilungsgesetzen auf lokalkompakten abelschen Gruppen II. Math. Nachr. **41**, 121–132 (1969)
297. Kerstan, J., Matthes, K., Mecke, J.: Unbegrenzt teilbare Punktprozesse. Berlin: Akademie-Verlag 1974
298. Kesten, H.: Full Banach mean values on countable groups. Math. Scand. **7**, 146–156 (1959)
299. Khintchine, A.I.: Contribution à l'arithmétique des lois de distribution. Bull. Univ. d'Etat Moscou Sér. Int., Sect. A, Math. et Mécan. **1**, 6–17 (1937)
300. Kingman, J.F.C.: The imbedding problem for finite Markov chains. Z. Wahrscheinlichkeitstheorie und verw. Gebiete **1**, 14–24 (1962)
301. Kinzl, F.: Ein Null-Zwei-Gesetz auf abelschen Halbgruppen. To appear in Österreich. Akad. Wiss. Math.-Natur. Kl. S.-B. II (1977/78)
302. Klasa, J.: Limit problems on topological stochastic groups and Bohr compactification. In: Colloquia Mathematica Societatis János Bolyai, 11: Limit Theorems of Probability Theory, pp. 105–134. Amsterdam-London: North Holland Publishing Company 1975
303. Kloss, B.M.: Limit distributions for sums of independent variables with values in a bicompact group. Dokl. Akad. Nauk SSSR **109**, 453–455 (1956) [Russian]
304. Kloss, B.M.: Probability distributions on bicompact topological groups. Theor. Probability Appl. **4**, 237–270 (1959)
305. Kloss, B.M.: Limiting distributions on bicompact abelian groups. Theor. Probability Appl. **6**, 361–389 (1961)
306. Kloss, B.M.: Stable distributions on a class of locally compact groups. Theor. Probability Appl. **7**, 237–257 (1962)
307. Kloss, B.M.: Topology in a group and convergence of distributions. Theor. Probability Appl. **9**, 111–114 (1964)

308. Koutsky, Z.: Einige Eigenschaften der modulo k addierten Markowschen Ketten. In: Transactions of the 2nd Prague Conference on Information Theory, Random Functions and Statistical Decision Theory, pp. 263–278. Prague: Publishing House of the Czechoslovak Academy of Sciences 1959
309. Krein, M.G.: On positive functionals on almost periodic functions. Dokl. Acad. Nauk SSSR, N.S. **30**, 9–12 (1941)
310. Krickeberg, K.: Wahrscheinlichkeitstheorie. Stuttgart: B.G. Teubner 1963
311. Kuipers, L., Niederreiter, H.: Uniform distribution of sequences. New York: John Wiley & Sons 1974
312. Lashof, R.: Lie algebras of locally compact groups. Pacific J. Math. **7**, 1145–1162 (1957)
313. Le Cam, L.: Convergence in distribution of stochastic processes. In: University of California Publications in Statistics Vol. 2, pp. 207–236. Berkeley and Los Angeles: University of California Press 1957
314. Lesca, J.: Suites équiréparties dans un espace localement compact. Enseignement Math. 2e Sér. **17**, 311–328 (1971)
315. Letac, G.: Groupe de Stam d'une probabilité. Ann. Inst. H. Poincaré Sect. B **8**, 175–181 (1972)
316. Lévy, P.: Calcul des probabilités. Paris: Gauthier-Villars 1925.
317. Lévy, P.: Sur les integrales dont les éléments sont des variables aléatoires indépendantes. Ann. Scuola Norm. Sup. Pisa II. Ser. **3**, 337–366 (1934)
318. Lévy, P.: Sur l'arithmétique des lois de probabilités. J. Math. Pures et Appl. **103**, 17–40 (1938)
319. Lévy, P.: Sur l'arithmétique des lois de probabilité enroulées. C.R. Soc. Math. France 32–34 (1938)
320. Lévy, P.: L'addition des variables aléatoires définies sur une circonférence. Bull. Soc. Math. France **67**, 1–41 (1939)
321. Lévy, P.: Théorie de l'addition des variables aléatoires. 2e éd. Paris: Gauthier-Villars 1954
322. Linnik, Yu.V.: Décomposition des lois de probabilités. Paris: Gauthier-Villars 1962
323. Liukkonen, J.: Dual spaces of groups with precompact conjugacy classes. Trans. Amer. Math. Soc. **180**, 85–108 (1973)
324. Livšic, L.Z., Ostrovskiĭ, I.V., Čistjakov, G.P.: The arithmetic of probability laws. Probability Theory. Mathematical Statistics. Theoretical Cybernetics Vol. 12, pp. 5–42 Moskow: Akad. Nauk SSSR Vsesojuz. Inst. Naučn. i. Techn. Informacii 1975 [Russian]
325. Loève, M.: Probability theory. 3rd ed. Princeton, N.J.-Toronto-New York-London: D. van Nostrand 1963
326. Loomis, L.H.: An introduction to abstract harmonic analysis. Toronto-New York-London: D. van Nostrand 1953
327. Loynes, R.M.: Products of independent random elements in a topological group. Z. Wahrscheinlichkeitstheorie und verw. Gebiete, **1**, 446–455 (1963)
328. Loynes, R.M.: Fourier transforms and probability theory on a non-commutative locally compact topological group. Ark. Math. **5**, 37–42 (1963)
329. Lukacs, E.: Characteristic functions. 2nd ed. London: Griffin 1970
330. Maksimov, V.M.: On the convergence of products of independent random variables taking on values from an arbitrary finite group. Theor. Probability Appl. **12**, 619–637 (1967)
331. Maksimov, V.M.: Extension of Itô's theorem "on the independence of process jumps" to processes with independent increments defined on topological groups with a countable basis. Soviet Math. Dokl. **9**, 1101–1104 (1968)
332. Maksimov, V.M.: Necessary and sufficient convergence conditions for the convolution of non-identical distributions given on a finite group. Theor. Probability Appl. **13**, 287–298 (1968)
333. Maksimov, V.M.: A convergence property of products of independent random variables on compact Lie groups. In: Soviet-Japanese Symposium on Probability Theory, Khabarovsk 1969 [Russian]
334. Maksimov, V.M.: A convergence property of products of independent random variables on compact Lie groups. Math. USSR-Sb. **11**, 423–440 (1970)

335. Maksimov, V.M.: A contribution to the theory of variance for probability distributions on compact groups. Soviet Math. Dokl. **11**, 717–721 (1970)

336. Maksimov, V.M.: Probabilistic characterization of zero-dimensional compact groups. Soviet Math. Dokl. **11**, 1667–1671 (1970)

337. Maksimov, V.M.: Random processes with independent increments with values in an arbitrary finite group. Theor. Probability Appl. **15**, 215–228 (1970)

338. Maksimov, V.M.: Composition convergent sequences of measures on compact groups. Theor. Probability Appl. **16**, 55–73 (1971)

339. Maksimov, V.M.: Divisible distributions on finite groups. Theor. Probability Appl. **16**, 308–322 (1971)

340. Maksimov, V.M.: Mathematical expectation for probability distributions on compact groups and their application. Soviet Math. Dokl. **13**, 416–419 (1972)

341. Maksimov, V.M.: Nonhomogeneous semigroups of measures on compact Lie groups. Theor. Probability Appl. **17**, 601–619 (1972)

342. Maksimov, V.M.: The principle of convergence "almost everywhere" in Lie groups. Math. USSR-Sb. **20**, 543–555 (1973)

343. Marcinkiewicz, J.: Sur les variables aléatoires enroulées. C.R. Soc. Math. France 34–37 (1938)

344. Mardia, K.V.: Statistics of directional data. London and New York: Academic Press 1972

345. Martin-Löf, Per: The continuity theorem on a locally compact group. Theor. Probability Appl. **10**, 338–341 (1965)

346. Martin-Löf, Per: Probability theory on discrete semigroups. Z. Wahrscheinlichkeitstheorie und verw. Gebiete **4**, 78–102 (1965)

347. Masani, P.R.: Multiplicative Riemann integration in normed rings. Trans. Amer. Math. Soc. **61**, 147–192 (1947)

348. Masani, P.R.: On infinitely decomposable probability distributions and helical varieties in Hilbert space. In: Multivariate Analysis III, pp. 209–223. London and New York: Academic Press 1973

349. Maurer, J.: Bemerkung über Faltungspotenzen eines Wahrscheinlichkeitsmaßes auf kompakten Gruppen. Arch. Math. (Basel) **27**, 549–550 (1976)

350. Mayer, C.: Semi-groupes non homogènes et leurs générateurs. In: Séminaire Choquet (Initiation à l'Analyse) 7e année, n° B, pp. 8.01–8.10 (1967/68) Paris: Secrétariat mathématique, Inst. H. Poincaré 1968

351. Maxones, W., Rindler, H.: Asymptotisch gleichverteilte Wahrscheinlichkeitsmaße auf lokalkompakten Gruppen. To appear in Colloq. Math. (1977/78)

352. McKean, H.P.: Brownian motions on the 3-dimensional rotation group. Mem. Coll. Sci. Univ. Kyoto **33**, 25–38 (1960)

353. McKennon, K.: Multipliers, positive functionals, positive-definite functions, and Fourier-Stieltjes transforms. Memoirs of the Amer. Math. Soc. Vol. 111 (1971)

354. Meyer, P.A.: Processus de Markov. Lecture Notes in Math. Vol. 26. Berlin-Heidelberg-New York: Springer 1967

355. von Mises, R.: Über die "Ganzzahligkeit" der Atomgewichte und verwandte Fragen. Physikal. Z. **19**, 490–500 (1918)

356. Montgomery, D., Zippin, L.: Topological transformation groups. 4th printing. New York: Interscience 1966

357. Moore, C.C.: Groups with finite-dimensional irreducible representations. Trans. Amer. Math. Soc. **166**, 401–410 (1972)

358. Mostert, P.S., Shields, A.L.: On the structure of semigroups on a compact manifold with boundary. Ann. of Math. **65**, 117–143 (1957)

359. Mostert, P.S., Shields, A.L.: One-parameter semigroups in a semigroup. Trans. Amer. Math. Soc. **96**, 510–517 (1960)

360. Mukherjea, A.: Idempotent probabilities on semigroups. Z. Wahrscheinlichkeitstheorie und verw. Gebiete **11**, 142–146 (1969)

361. Mukherjea, A.: The convolution equation $P = P * Q$ of Choquet and Deny and relatively invariant measures on semigroups. Ann. Inst. Fourier (Grenoble) **21**, 87–97 (1971)

362. Mukherjea, A.: On the convolution equation P=P*Q of Choquet and Deny for probability measures on semigroups. Proc. Amer. Math. Soc. **32**, 457–463 (1972)
363. Mukherjea, A.: Limit theorems for probability measures on non-compact groups and semigroups. Z. Wahrscheinlichkeitstheorie verw. Gebiete **33**, 273–284 (1976)
364. Mukherjea, A.: Limit theorems for convolution iterates of a probability measure on completely simple or compact semigroups. Trans. Amer. Math. Soc. **225**, 355–370 (1977)
365. Mukherjea, A., Tserpes, N.A.: Idempotent measures on locally compact semigroups. Proc. Amer. Math. Soc. **29**, 143–150 (1971)
366. Mukherjea, A., Tserpes, N.A.: Measures on topological semigroups. Lecture Notes in Math. Vol. 547. Berlin-Heidelberg-New York: Springer 1976
367. Muthsam, H.: Über die Summe Markoffscher Ketten auf Halbgruppen. Monatsh. für Math. **76**, 43–54 (1972)
368. Nachbin, L.: The Haar integral. Princeton, N.J.-Toronto-New York-London: D. van Nostrand 1965
369. Nedoma, J.: Note on generalized random variables. In: Trans. 1st Prague Conference on Information Theory, Statistical Decision Functions, Random processes, pp. 139–143. Prague: Publishing House of the Czechoslovakian Academy of Sciences 1957
370. Nelson, E.: Analytic vectors. Ann. of Math. **70**, 572–615 (1959)
371. Nelson, E., Stinespring, W.F.: Representations of elliptic operators in an envelopping algebra. Amer. J. of Math. **81**, 547–560 (1959)
372. Neveu, J.: Théorie des semi-groupes de Markov. In: University of California Publications in Statistics Vol. 1, pp. 319–394. Berkeley and Los Angeles: University of California Press 1958
373. Numakura, K.: On bicompact semigroups. Math. J. Okayama Univ. **1**, 99–108 (1952)
374. Pakshirajan, R.P.: Regular measures and stochastic processes in topological groups. Technical Report No 2, Department of Mathematics, Eugene, Oregon (1959)
375. Pakshirajan, R.P.: An analogue of Kolmogorov's three-series theorem for abstract random variables. Pacific J. Math. **13**, 639–646 (1963)
376. Parthasarathy, K.R.: The central limit theorem for the rotation group. Theor. Probability Appl. **9**, 248–257 (1964)
377. Parthasarathy, K.R.: Probability measures on metric spaces. New York-London: Academic Press 1967
378. Parthasarathy, K.R.: A note on idempotent measures in topological groups. J. London Math. Soc. **42**, 534–536 (1967)
379. Parthasarathy, K.R.: On the imbedding of an infinitely divisible distribution in a one-parameter convolution semigroup. Theor. Probability Appl. **12**, 373–380 (1967)
380. Parthasarathy, K.R.: On the imbedding of an infinitely divisible distribution in a one parameter convolution semigroup. Sankhyā Ser. A **35**, 123–132 (1973)
381. Parthasarathy, K.R.: The central limit theorem for positive definite functions on locally compact groups. Multivariate Anal. **4**, 123–149 (1974)
382. Parthasarathy, K.R., Ranga Rao, R., Varadhan, S.R.S.: On the category of indecomposable distributions on topological groups. Trans. Amer. Math. Soc. **102**, 200–217 (1962)
383. Parthasarathy, K.R., Ranga Rao, R., Varadhan, S.R.S.: Probability distributions on locally compact abelian groups. Illinois J. Math. **7**, 337–369 (1963)
384. Parthasarathy, K.R., Sazonov, V.V.: On the representation of infinitely divisible distributions on locally compact abelian groups. Theor. Probability Appl. **9**, 108–111 (1964)
385. Parthasarathy, K.R., Schmidt, K.: Positive definite kernels, continuous tensor products, and central limit theorems of probability theory. Lecture Notes in Math. Vol. 272. Berlin-Heidelberg-New York: Springer 1972
386. Parthasarathy, K.R., Schmidt, K.: Stable positive definite functions. Trans. Amer. Math. Soc. **203**, 161–174 (1975)
387. Perrin, J.B.: Etude mathématique du mouvement Brownien de rotation. Ann. Sci. Ecole Norm. Sup. **45**, 1–51 (1928)
388. Pichon, G.: Groupes de Lie. Paris: Hermann 1973
389. Pontrjagin, L.S.: Topologische Gruppen I, II. Leipzig: B.G. Teubner 1957/1958.

390. Prakasa Rao, B.L.S.: On a characterization of probability distributions on locally compact abelian groups. Z. Wahrscheinlichkeitstheorie und verw. Gebiete **9**, 98–100 (1968)
391. Prékopa, A., Rényi, A., Urbanik, K.: On the limiting distribution of sums of independent random variables in commutative bicompact topological groups. Acta Math. Acad. Sci. Hungar. **7**, 255–261 (1959) [Russian]
392. Prohorov, Yu.V.: Convergence of stochastic processes and limit theorems in probability theory. Theor. Probability Appl. **1**, 157–214 (1956)
393. Pym, J.S.: Idempotent measures on semigroups. Pacific J. Math. **12**, 685–698 (1962)
394. Pym, J.S.: Idempotent probability measures on compact semitopological semigroups. Proc. Amer. Math. Soc. **21**, 499–501 (1969)
395. Raikov, D.A.: On the decomposition of Poisson laws. C.R. Acad. Sci. URSS N.S. **14**, 9–11 (1937)
396. Ramaswamy, S.: Semigroups of measures on Lie groups. J. Indian Math. Soc. **38**, 175–189 (1974)
397. Ranga Rao, R.: Relations between weak and uniform convergence of measures with applications. Ann. Math. Stat. **33**, 659–680 (1962)
398. Rao, C.R.: Characterization of the distribution of a random variable in linear structural relations. Sankhyā Ser. A **28**, 251–260 (1966)
399. Revuz, D.: Markov chains. Amsterdam-Oxford: North-Holland Publishing Company 1975
400. Richter, H.: Wahrscheinlichkeitstheorie. 2. neubearbeitete Auflage. Berlin-Heidelberg-New York: Springer 1966
401. Rickert, N.W.: Some properties of locally compact groups. J. Austral. Math. Soc. **7**, 433–454 (1967)
402. Rickert, N.W.: Arcs in locally compact groups. Math. Ann. **172**, 222–228 (1967)
403. Riesz, F., Nagy, B.Sz.: Leçons d'analyse fonctionnelle. 5e éd. Paris: Gauthier-Villars 1968
404. Rihl, P.: Poisson-Verteilungen auf lokalkompakten Gruppen. Dissertation, Wien 1964
405. Rindler, H.: Gleichverteilte Folgen in lokalkompakten Gruppen. Monatsh. für Math. **82**, 207–235 (1976)
406. Rivkind, Ya.I.: A limit theorem in probability theory on compact topological groups. Selected Translations in Mathematical Statistics and Probability. II. Ser. **2**, 201–217 (1962)
407. Robertson, L.C.: A note on the structure of Moore groups. Bull. Amer. Math. Soc. **75**, 594–599 (1969)
408. Rogalski, M.: Le théorème de Lévy-Khinčin. In: Séminaire Choquet (Initiation à l'Analyse) 3e année, pp. 2.01–2.18 (1963/64) Paris: Secrétariat mathématique, Inst. H. Poincaré 1964
409. Rosenblatt, M.: A central limit theorem and a strong mixing condition. Proc. Nat. Acad. Sci. USA **42**, 43–47 (1956)
410. Rosenblatt, M.: Limits of convolution sequences of measures on a compact topological semigroup. J. Math. Mech. **9**, 293–306 (1960)
411. Rosenblatt, M.: A strong mixing condition and a central limit theorem on compact groups. J. Math. Mech. **17**, 189–198 (1967). Corrigendum J. Math. Mech. **17**, 919 (1967)
412. Rosenblatt, M.: Markov processes. Structure and asymptotic behavior. Berlin-Heidelberg-New York: Springer 1971
413. Roth, J.P.: Sur les semi-groupes à contraction invariants sur un espace homogène. C.R. Acad. Sci. Paris Sér. A–B **277**, A1091–A1094 (1973)
414. Roynette, B.: Théorème central-limite pour le groupe de déplacements de $\mathbb{R}^d$. Ann. Inst. H. Poincaré Sect. B **10**, 391–398 (1974)
415. Roynette, B.: Marches aléatoires sur les groupes de Lie. Book manuscript (1976). To appear under the authors Guivarc'h, Y., Keane, M., Roynette, B. as Lecture Notes in Math. Berlin-Heidelberg-New York: Springer (1978)
416. Rudin, W.: Idempotent measures on abelian groups. Pacific J. Math. **9**, 195–209 (1959)
417. Rudin, W.: Fourier analysis on groups. New York-London: Interscience Publishers 1962
418. Rudin, W.: Idempotents in group algebras. Bull. Amer. Math. Soc. **69**, 224–227 (1963)
419. Ruhin, A.L.: Some statistical and probabilistic problems on groups. Proc. Steklov Inst. Math. **111**, 59–129 (1970)
420. Ruhin, A.L.: The Poisson law on groups. Litovsk Mat. Sb. **10**, 537–543 (1970)
421. Runnenburg, J.Th.: On Elfving's problem of imbedding a time-discrete Markov chain in a

time-continuous one for finitely many states I. Nederl. Akad. Wetensch. Proc. Ser. A **65**, 536–541 (1962)
422. Sagle, A.A., Walde, R.A.: Introduction to Lie groups and algebras. New York-London: Academic Press 1973
423. Sazonov, V.V., Tutubalin, V.N.: Probability distributions on topological groups. Theor. Probability Appl. **11**, 1–45 (1966)
424. Scheffer, C.L.: On Elfving's problem of imbedding a time-discrete Markov chain in a time-continuous one for finitely many states II. Nederl. Akad. Wetensch. Proc. Ser. A **65**, 542–548 (1962)
425. Schlesinger, L.: Neue Grundlagen für einen Infinitesimalkalkül der Matrizen. Math. Z. **33**, 33–61 (1931)
426. Schmetterer, L.: Über Markovsche Ketten auf einfachen endlichen Halbgruppen. Rev. Roum. Math. pures et appl. **12**, 1377–1386 (1967)
427. Schmetterer, L.: Über die Summe Markov'scher Ketten auf Halbgruppen. Monatsh. für Math. **71**, 223–230 (1967)
428. Schmetterer, L.: Sur les lois de Poisson. In: Les Probabilités sur les Structures Algébriques, pp. 311–318. Paris: Editions du Centre National de la Recherche Scientifique 1970
429. Schmetterer, L.: On Poisson laws and related questions. In: Proceedings of the 6th Berkeley Symposium on Mathematical Statistics and Probability Vol. II, pp. 169–185. Berkeley and Los Angeles: University of California Press 1970
430. Schmetterer, L.: On the embedding of infinitely divisible laws. In: Proceedings of the 4th Conference on Probability Theory, pp. 83–86, Braşov, Romania, Sept. 12–18, 1971. Bucarest: Editura Academiei Republicii Socialiste Romania
431. Schmidt, E.J.P. Georg: On multiplicative Lebesgue integration and families of evolution operators. Math. Scand. **29**, 113–133 (1971)
432. Schmidt, K.: On a characterisation of certain infinitely divisible positive definite functions and measures. J. London Math. Soc. II. Ser. **4**, 401–407 (1972)
433. Schmidt, K.: A class of infinitely divisible positive definite functions on topological groups. Z. Wahrscheinlichkeitstheorie und verw. Gebiete **25**, 97–102 (1972/73)
434. Schoenberg, I.J.: Metric spaces and positive definite functions. Trans. Amer. Math. Soc. **44**, 522–536 (1938)
435. Seidmann, T.I.: Approximation of operator semi-groups. J. Functional Analysis **5**, 160–167 (1970)
436. Siebert, E.: Wahrscheinlichkeitsmaße auf lokalkompakten maximal fastperiodischen Gruppen. Dissertation, Tübingen 1972
437. Siebert, E.: Stetige Halbgruppen von Wahrscheinlichkeitsmaßen auf lokalkompakten maximal fastperiodischen Gruppen. Z. Wahrscheinlichkeitstheorie und verw. Gebiete **25**, 269–300 (1973)
438. Siebert, E.: Über die Erzeugung von Faltungshalbgruppen auf beliebigen lokalkompakten Gruppen. Math. Z. **131**, 313–333 (1973)
439. Siebert, E.: Einbettung unendlich teilbarer Wahrscheinlichkeitsmaße auf topologischen Gruppen. Z. Wahrscheinlichkeitstheorie und verw. Gebiete **28**, 227–247 (1974)
440. Siebert, E.: Einige Bemerkungen zu den Gauß-Verteilungen auf lokalkompakten abelschen Gruppen. Manuscripta Math. **14**, 41–55 (1974)
441. Siebert, E.: Absolut-Stetigkeit und Träger von Gauß-Verteilungen auf lokalkompakten Gruppen. Math. Ann. **210**, 129–147 (1974)
442. Siebert, E.: Convergence and convolutions of probability measures on a topological group. Ann. Probability **4**, 433–443 (1976)
443. Siebert, E.: On the generation of convolution semigroups on arbitrary locally compact groups II. Arch. Math. (Basel) **28**, 139–148 (1977)
444. Siebert, E.: On the Lévy-Chintschin formula on locally compact maximally almost periodic groups. To appear in Math. Scand. (1977/78)
445. Šlosman, S.B.: Limit theorems of probability theory on compact Lie groups. Soviet Math. Dokl. **16**, 656–659 (1975)
446. Smith, W.L.: On some general renewal theorems for nonidentically distributed variables. In: Proceedings 4th Berkeley Symposium Mathematical Statistics Probability Vol. II, pp. 467–514. Berkeley and Los Angeles: University of California Press 1961

447. Sobko, G.M.: The first diffusion problem on differential manifolds. Theor. Probability Appl. **17**, 521–528 (1972)
448. Sobko, G.M.: A diffusion approximation of non-Markov random walks on differentiable manifolds. Theor. Probability Appl. **18**, 41–53 (1973)
449. Stam, A.J.: On shifting iterated convolutions I. Compositio math. **17**, 268–280 (1967)
450. Stam, A.J.: On shifting iterated convolutions II. Compositio math. **18**, 201–228 (1967)
451. Stein, E.M.: Topics in harmonic analysis related to the Littlewood-Paley theory. Ann. of Math. Studies No. 63. Princeton, N.J.: Princeton University Press 1970
452. Štěpán, J.: On the convex sets of probability measures. Časopis Pěst. Mat. **93**, 73–79 (1968)
453. Štěpán, J.: On the family of translations of a tight probability measure on a topological group. Z. Wahrscheinlichkeitstheorie und verw. Gebiete **15**, 131–138 (1970)
454. Štern, A.I.: Locally bicompact groups with finite dimensional irreducible representations. Math. USSR-Sb. **19**, 85–94 (1973)
455. Stone, C.: On a theorem by Dobrushin. Ann. Math. Stat. **39**, 1391–1401 (1968)
456. Stromberg, K.: A note on the convolution of semigroups. Math. Scand. **7**, 347–352 (1959)
457. Stromberg, K.: Probabilities on a compact group. Trans. Amer. Math. Soc. **94**, 295–309 (1960)
458. Stroock, D.W., Varadhan, S.R.S.: Diffusion processes with continuous coefficients I. Comm. Pure Appl. Math. **12**, 345–400 (1969)
459. Stroock, D.W., Varadhan, S.R.S.: Diffusion processes with continuous coefficients II. Comm. Pure Appl. Math. **12**, 479–530 (1969)
460. Stroock, D.W., Varadhan, S.R.S.: Limit theorems for random walks on Lie groups. Sankhyā Ser. A **35**, 277–294 (1973)
461. Takano, K.: Central convergence criterion in the multidimensional case. Ann. Inst. Statist. Math. **7**, 95–102 (1956)
462. Taylor, J.L.: Measure algebras. Regional Conference Series in Mathematics Nr. 16. Providence, R.I.: Amer. Math. Soc. 1973
463. Topsøe, F.: Topology and measure. Lecture Notes in Math. Vol. 133. Berlin-Heidelberg-New York: Springer 1970
464. Tortrat, A.: Lois tendues, convergence en probabilité et équation $P*P'=P$. C.R. Acad. Sci. Paris **258**, 3813–3816 (1964)
465. Tortrat, A.: Lois de probabilité sur un espace topologique complètement régulier et produits infinis à termes indépendants dans un groupe topologique. Ann. Inst. H. Poincaré Sect. B **1**, 217–237 (1965)
466. Tortrat, A.: Lois de probabilité dans les demi-groupes topologiques complètement simples de la forme $E\times G\times F$. C.R. Acad. Sci. Paris **260**, 4408–4411 (1965)
467. Tortrat, A.: Lois indéfiniment divisibles ($\mu\in\mathscr{I}$) dans un groupe topologique abélien métrisable X. Cas des espaces vectoriels. C.R. Acad. Sci. Paris **261**, 4973–4975 (1965)
468. Tortrat, A.: Lois tendues et convolutions dénombrables dans un groupe topologique X. Ann. Inst. H. Poincaré Sect. B **2**, 279–298 (1966)
469. Tortrat, A.: Lois tendues μ sur un demi-groupe topologique complètement simple X. Z. Wahrscheinlichkeitstheorie und verw. Gebiete **6**, 145–160 (1966)
470. Tortrat, A.: Structure des lois indéfiniment divisibles dans un espace vectoriel topologique. In: Symposium on Probability Methods in Analysis, pp. 299–328. Lecture Notes in Math. Vol. 31, Berlin-Heidelberg-New York 1967
471. Tortrat, A.: Sur la structure des lois indéfiniment divisibles (classe $\mathscr{I}(X)$) dans les espaces vectoriels X (sur le corps réel). Z. Wahrscheinlichkeitstheorie und verw. Gebiete **11**, 311–326 (1969)
472. Tortrat, A.: Convolutions dénombrables équitendues dans un groupe topologique X. In: Les Probabilités sur les Structures Algébriques, pp. 327–343. Paris: Editions du Centre National de la Recherche Scientifique 1970
473. Tortrat, A.: Sur la continuité de l'opération convolution dans un demi-groupe topologique X. C.R. Acad. Sci. Paris Sér. A–B **272**, A588–A591 (1971)
474. Tortrat, A.: Calcul des probabilités et introduction aux processus aléatoires. Paris: Masson et Cie 1971
475. Tortrat, A.: Sur les lois τ-régulières, leurs produits et leur convolution. In: Transactions

of the 6th Prague Conference on Information Theory, Statistical Decision Functions, Random Processes, pp. 839–849. Prague: Publishing House of the Czechoslovakian Academy of Sciences 1973
476. Trotter, H.: On the product of semigroups of operators. Proc. Amer. Math. Soc. **10**, 545–551 (1959)
477. Tserpes, N.A., Mukherjea, A.: Some problems on idempotent measures on semigroups. Bull. Austral. Math. Soc. **2**, 299–315 (1970)
478. Tutubalin, V.N.: On the limit behavior of compositions of measures in the plane and space of Lobachevsky. Theor. Probability Appl. **7**, 189–196 (1962)
479. Tutubalin, V.N.: Compositions of measures on the simplest nilpotent group. Theor. Probability Appl. **9**, 479–487 (1964)
480. Tutubalin, V.N.: The central limit theorem for random motions of a Euclidean space. Selected Translations in Mathematical Statistics and Probability **12**, 47–57 (1973)
481. Ullrich, M., Urbanik, K.: A limit theorem for random variables in compact topological groups. Colloq. Math. **7**, 191–198 (1960)
482. Urbanik, K.: On the limiting probability distribution on a compact topological group. Fund. Math. **44**, 253–261 (1957)
483. Urbanik, K.: Poisson distributions on compact Abelian topological groups. Colloq. Math. **6**, 13–24 (1958)
484. Urbanik, K.: Gaussian measures on locally compact Abelian topological groups. Studia Math. **19**, 77–88 (1960)
485. Urbanik, K.: Generalized convolutions. Studia Math. **23**, 217–245 (1964)
486. Urbanik, K.: Generalized convolutions II. Studia Math. **45**, 57–70 (1973)
487. Varadarajan, V.S.: Weak convergence of measures on separable metric spaces. Sankhyā **19**, 15–22 (1958)
488. Varadarajan, V.S.: Measures on topological spaces. Transl. Amer. Math. Soc. II. Ser. **48**, 161–228 (1965)
489. Varadarajan, V.S.: Lie groups, Lie algebras and their representations. Englewood Cliffs, N.J.: Prentice-Hall 1974
490. Virtser, A.D.: Central limit theorem for semisimple Lie groups. Theor. Probability Appl. **15**, 667–687 (1970)
491. Virtser, A.D.: Limit theorems for compositions of distributions on certain nilpotent Lie groups. Theor. Probability Appl. **19**, 86–105 (1974)
492. Vorobev, N.N.: The addition of independent random variables on finite groups. Mat. Sb. **34**, 89–126 (1954) [Russian]
493. von Waldenfels, W.: Fast positive Operatoren. Z. Wahrscheinlichkeitstheorie und verw. Gebiete **4**, 159–174 (1965)
494. Wang, S.P.: Compactness properties of topological groups II. Duke Math. J. **39**, 243–251 (1972)
495. Webb, G.F.: Product integral representation of time dependent nonlinear evolution equations in Banach spaces. Pacific J. Math. **32**, 269–281 (1970)
496. Wehn, D.F.: Limit distributions on Lie groups. Thesis, Yale (1959)
497. Wehn, D.F.: Probabilities on Lie groups. Proc. Nat. Acad. Sci. USA **48**, 791–795 (1962)
498. Wehn, D.F.: Some remarks on Gaussian distributions on a Lie group. Z. Wahrscheinlichkeitstheorie und verw. Gebiete **30**, 255–263 (1974)
499. Weil, A.: L'intégration dans les groupes topologiques et ses applications, 2e éd. Paris: Hermann et Cie 1951
500. Wendel, J.G.: Haar measure and the semigroup of measures on a compact group. Proc. Amer. Math. Soc. **5**, 923–929 (1954)
501. Whyburn, G.T.: Analytic topology. Amer. Math. Soc. Colloq. Publ. Vol. 28 (1942)
502. Wiener, N., Young, R.C.: The total variation of $g(x+h)-g(x)$. Trans. Amer. Math. Soc. **35**, 327–340 (1933)
503. Williamson, J.H.: Harmonic analysis on semigroups. J. London Math. Soc. **42**, 1–41 (1967)
504. Wolff, M.: Über Produkte abhängiger zufälliger Veränderlicher mit Werten in einer kompakten Halbgruppe. Z. Wahrscheinlichkeitstheorie und verw. Gebiete **35**, 253–264 (1976)
505. Woll jr., J.W.: Homogeneous stochastic processes. Pacific J. Math. **9**, 293–325 (1959)

506. Woll jr., J.W.: A property of homogeneous processes. Proc. Amer. Math. Soc. **13**, 131–133 (1962)
507. Yamabe, H.: On an extension of Helly's theorem. Osaka J. Math. **2**, 15–17 (1950)
508. Yosida, K.: Functional analysis. 3rd ed. Berlin-Heidelberg-New York: Springer 1971
509. Yuan, J.: Embedding of an infinitely divisible probability measure on a locally compact semigroup. Thesis, Tulane University, New Orleans (1974)
510. Yuan, J.: On the continuity of convolution. Semigroup Forum **10**, 367–372 (1975)
511. Yuan, J.: Gauss measures on semigroups. Semigroup Forum **11**, 165–169 (1975)
512. Yuan, J.: Embedding of an infinitely divisible probability measure over a connected solvable Lie group. Math. Z. **146**, 167–172 (1976)
513. Yuan, J.: On the convolution algebra of H-invariant measures. Pacific J. Math. **62**, 595–600 (1976)
514. Yuan, J.: On the construction of one-parameter semigroups in topological semigroups. Pacific J. Math. **65**, 285–292 (1976)
515. Yuan, J., Liang, Ta Chen: On the supports and absolute continuity of infinitely divisible probability measures. Semigroup Forum **12**, 34–44 (1976)

List of Symbols

Subject Index

Ergebnisse der Mathematik und ihrer Grenzgebiete

1. Bachmann: Transfinite Zahlen
2. Miranda: Partial Differential Equations of Elliptic Type
4. Samuel: Méthodes d'algèbre abstraite en géométrie algébrique
5. Dieudonné: La géométrie des groupes classiques
7. Ostmann: Additive Zahlentheorie. 1. Teil: Allgemeine Untersuchungen
8. Wittich: Neuere Untersuchungen über eindeutige analytische Funktionen
10. Suzuki: Structure of a Group and the Structure of its Lattice of Subgroups. Second edition in preparation
11. Ostmann: Additive Zahlentheorie. 2. Teil: Spezielle Zahlenmengen
13. Segre: Some Properties of Differentiable Varieties and Transformations
14. Coxeter/Moser: Generators and Relations for Discrete Groups
15. Zeller/Beckmann: Theorie der Limitierungsverfahren
16. Cesari: Asymptotic Behavior and Stability Problems in Ordinary Differential Equations
17. Severi: Il teorema di Riemann-Roch per curve, superficie e varietà questioni collegate
18. Jenkins: Univalent Functions and Conformal Mapping
19. Boas–Buck: Polynomial Expansions of Analytic Functions
20. Bruck: A Survey of Binary Systems
21. Day: Normed Linear Spaces
23. Bergmann: Integral Operators in the Theory of Linear Partial Differential Equations
25. Sikorski: Boolean Algebras
26. Künzi: Quasikonforme Abbildungen
27. Schatten: Norm Ideals of Completely Continuous Operators
30. Beckenbach/Bellman: Inequalities
31. Wolfowitz: Coding of Theorems of Information Theory
32. Constantinescu/Cornea: Ideale Ränder Riemannscher Flächen
33. Conner/Floyd: Differentiable Periodic Maps
34. Mumford: Geometric Invariant Theory
35. Gabriel/Zisman: Calculus of Fractions and Homotopy Theory
36. Putnam: Commutation y Properties of Hilbert Space Operators and Related Topics
37. Neumann: Varieties of Groups
38. Boas: Integrability Theorems for Trigonometric Transforms
39. Sz.-Nagy: Spektraldarstellung linearer Transformationen des Hilbertschen Raumes
40. Seligman: Modular Lie Algebras
41. Deuring: Algebren
42. Schütte: Vollständige Systeme modaler und intuitionistischer Logik
43. Smullyan: First-Order Logic
44. Dembowski: Finite Geometries
45. Linnik: Ergodic Properties of Algebraic Fields
46. Krull: Idealtheorie
47. Nachbin: Topology on Spaces of Holomorphic Mappings
48. A. Ionescu Tulcea/C. Ionescu Tulcea: Topics in the Theory of Lifting
49. Hayes/Pauc: Derivation and Martingales
50. Kahane: Séries de Fourier absolument convergentes
51. Behnke/Thullen: Theorie der Funktionen mehrerer komplexer Veränderlichen
52. Wilf: Finite Sections of Some Classical Inequalities
53. Ramis: Sous-ensembles analytiques d'une variété banachique complexe
54. Busemann: Recent Synthetic Differential Geometry
55. Walter: Differential and Integral Inequalities
56. Monna: Analyse non-archimédienne
57. Alfsen: Compact Convex Sets and Boundary Integrals
58. Greco/Salmon: Topics in $\mathfrak{m}$-Adic Topologies
59. López de Medrano: Involutions on Manifolds

60. Sakai: C*-Algebras and W*-Algebras
61. Zariski: Algebtaic Surfaces
62. Robinson: Finiteness Conditions and Generalized Soluble Groups, Part 1
63. Robinson: Finiteness Conditions and Generalized Soluble Groups, Part 2
64. Hakim: Topos annelés et schémas relatifs
65. Browder: Surgery on Simply-Connected Manifolds
66. Pietsch: Nuclear Locally Convex Spaces
67. Dellacherie: Capacités et processus stochastiques
68. Raghunathan: Discrete Subgroups of Lie Groups
69. Rourke/Sanderson: Introduction of Piecewise-Linear Topology
70. Kobayashi: Transformation Groups in Differential Geometry
71. Tougeron: Idéaux de fonctions différentiables
72. Gihman/Skorohod: Stochastic Differential Equations
73. Milnor/Husemoller: Symmetric Bilinear Forms
74. Fossum: The Divisor Class Group of a Krull Domain
75. Springer: Jordan Algebras and Algebraic Groups
76. Wehrfritz: Infinite Linear Groups
77. Radjavi/Rosenthal: Invariant Subspaces
78. Bognár: Indefinite Inner Product Spaces
79. Skorohod: Integration in Hilbert Space
80. Bonsall/Duncan: Complete Normed Algebras
81. Crossley/Nerode: Combinatorial Functors
82. Petrov: Sums of Independent Random Variables
83. Walker: The Stone-Čech Compactification
84. Wells/Williams: Embeddings and Extensions in Analysis
85. Hsiang: Cohomology Theory of Topological Transformation Groups
86. Olevskii: Fourier Series with Respect to General Orthogonal Systems
87. Berg/Forst: Potential Theory on Locally Compact Abelian Groups
88. Weil: Elliptic Functions according to Eisenstein and Kronecker
89. Lyndon/Schupp: Combinatorial Group Theory
90. Edwards/Gaudry: Littlewood-Paley and Multiplier Theory
91. Gunning: Riemann Surfaces and Generalized Theta Functions
92. Lindenstrauss/Tzafriri: Classical Banch Spaces I